Handbook of Neurochemistry and Molecular Neurobiology

Brain and Spinal Cord Trauma

Abel Lajtha (Ed.)

Handbook of Neurochemistry and Molecular Neurobiology
Brain and Spinal Cord Trauma

Volume Editors: Naren Banik and Swapan K. Ray

With 120 Figures and 28 Tables

Editor
Abel Lajtha
Director
Center for Neurochemistry
Nathan S. Kline Institute for Psychiatric Research
140 Old Orangeburg Road
Orangeburg
New York, 10962
USA

Volume Editors

Naren Banik
Department of Neurosciences
Division of Neurology
Medical University of South Carolina
96 Jonathan Lucas Street, Suite 309
Charleston, SC 29425
USA

Swapan K. Ray
Department of Pathology, Microbiology and Immunology
University of South Carolina School of Medicine
6439 Garners Ferry Road
Building 2, Room C11
Columbia, SC 29209
USA

Library of Congress Control Number: 2006922553

ISBN: 978-0-387-30343-7

Additionally, the whole set will be available upon completion under ISBN: 978-0-387-35443-9
The electronic version of the whole set will be available under ISBN: 978-0-387-30426-7
The print and electronic bundle of the whole set will be available under ISBN: 978-0-387-35478-1

© 2009 Springer Science+Business Media, LLC.

All rights reserved. This work may not be translated or copied in whole or in part without the written permission of the publisher (Springer Science+Business Media, LLC., 233 Spring Street, New York, NY 10013, USA), except for brief excerpts in connection with reviews or scholarly analysis. Use in connection with any form of information storage and retrieval, electronic adaptation, computer software, or by similar or dissimilar methodology now known or hereafter developed is forbidden.
The use in this publication of trade names, trademarks, service marks, and similar terms, even if they are not identified as such, is not to be taken as an expression of opinion as to whether or not they are subject to proprietary rights.

springer.com

Printed on acid-free paper

SPIN: 11417828 2109 - 5 4 3 2 1 0

Preface

This volume deals with nervous system injuries, repair and therapeutic approaches, and some neurodegenerative diseases not covered in other volumes of the Handbook. How current research changed our understanding of the epidemiology, pathophysiological mechanisms, cell demise leading to loss of function, whether in traumatic brain injury (TBI), spinal cord injury (SCI), or neurodegenerative diseases, and therapeutic approaches to ameliorate dysfunction of these devastating conditions are emphasized. Central nervous system (CNS) trauma (TBI and SCI combined) is one of the major causes of death in the United States and is one of the main killers (perhaps number one) of the young people below the mid-thirties. While TBI kills the majority of these victims, SCI, depending on the severity, leads to lifelong disability and despair at the time of their highest productivity. One of the most important areas in medicine now is the repair of the damaged tissue of injured brain and spinal cord so the function may be improved or restored. The major problem facing clinicians and researchers is the failure of brain and spinal cord to repair or regenerate the damaged areas, perhaps because these organs are so overly complex, functionally and structurally, containing different cell types, fiber pathways, messengers (neurochemicals), and other elements. In addition, damage to tissue in trauma is caused by not one, but many multi-destructive pathways.

In spite of this complexity, neuroscience has made significant advances in understanding of the delayed injury process (i.e. secondary damage), providing opportunities for target-based therapeutic intervention, improved imaging, blood vessel growth, regeneration, and tissue transplantation, including stem cell application. During the last two decades of vigorous research, scientists have identified several secondary injury factors, although many still remain unknown, that damage or destroy injured brain and spinal cord. They are trying to understand how these factors, including increased calcium and calcium-mediated events (namely lipases and proteases), free radicals, excitotoxicity, alterations in structural integrity of cell membrane, axonal damage in white matter, and reduction in blood flow, lead to apoptosis and necrosis of neurons and myelinating oligodendrocytes and ultimately destruction of tissue following injury. Although better imaging and pharmacological therapy using methylprednisolone and surgical manipulation, particularly in SCI, have slightly improved the clinical management of patients, we are still nowhere near acknowledging any real recovery of function. Yes, advances have been made over the years. Yet, we know little and much needs to be learned about how to reduce disability and restore function in TBI and SCI. Therefore, vigorous research is needed in other areas, including regeneration, protection of cells and preservation of axons, imaging by distension tensor imaging (DTI) and restoring blood supply that may help further improve function following TBI and SCI. Some of these detrimental pathways may be common to neurodegenerative disease such as Alzheimer's disease (AD), Parkinson's disease (PD), multiple sclerosis (MS), and amyotrophic lateral sclerosis (ALS), which will also benefit from this research attenuating dysfunction.

The chapters included in this volume clearly demonstrate the progress made in understanding the destructive mechanisms involved not only in TBI and SCI, but also in other neurodegenerative and neoplastic diseases. The work on enzymes (proteases and lipases), lipids, proteins, free radicals, and others have established roles as significant mediators of cell damage and disintegration of membrane architecture in diseases and injuries. We can learn more about how the blood-brain barrier (BBB) or spinal cord-blood barrier are maintained so that foreign elements can not enter and provide protection, how elements enter the CNS through the barrier that has been compromised in injury and inflammatory diseases (e.g. MS, AD, PD), and how to kill brain tumor cells by increasing the permeability. The contributors in this volume discussed the importance of microcirculation in terms of blood flow through microvessel growth following injury or in ischemic conditions, controlling angiogenesis in CNS tumor, and maintaining the BBB function. Other chapters have described cell death and degenerative processes in macular degeneration and diabetic retinopathy as well as in autoimmune diseases like optic neuritis and experimental allergic encephalomyelitis (EAE), which is the animal model of MS. Several chapters have elaborated on diseases, including adrenoleukodystorphy (a secondary demyelinating disease), peripheral neuropathy, and hyperammonemia. At the cellular level, white matter degeneration, including axon and cell death in CNS injuries as well as in neurodegenerative disease, has been emphasized. Although there is little or no CNS repair or regeneration in the brain, a chapter has been devoted to elucidate the potential therapeutic aspects of utilizing stem cell transplantation. Finally, chapters have been included on the rehabilitation of SCI patients and current management and ongoing clinical trials for treatment. The studies described in these chapters have also emphasized the importance of therapeutic strategies to combat dysfunction in injuries and diseases.

Our knowledge of the pathophysiology of CNS injuries and diseases has increased over the years only due to the tremendous advances that have been made in many areas of basic neuroscience research. The exciting findings of all contributors to this volume emphasize one most important aspect – applying basic neuroscience to the understanding of clinical problems of patients with diseases and injuries. Needless to say, although we have made substantial progress in some areas, further research is vital in areas such as regeneration/repair, axonal guidance, preservation and protection of cells and their processes, genetic regulation, growth factors, transplantation, and stem cell biology for attenuation of dysfunction through development of new therapeutic strategies. The advances made thus far on scientific research give us, the researchers, hope that further progress will help restore function and cure CNS diseases in the future.

We thank our authors for their excellent contributions to this volume. We also thank Ms. Denise Matzelle and Ms. Kristine Immediato for their patience and efforts in making this volume into its final form.

Naren L. Banik and Swapan K. Ray

Table of Contents

    Preface v

    Contributors xi

1   **Brain Tumor Angiogenesis** 1
    S. Lakka · J. S. Rao

2   **Adrenoleukodystrophy: Molecular, Metabolic, Pathologic, and Therapeutic Aspects** 13
    M. A. Contreras · I. Singh

3   **Hyperammonemia** 43
    V. Felipo

4   **Glutamate and Cytokine-Mediated Alterations of Phospholipids in Head Injury and Spinal Cord Trauma** 71
    A. A. Farooqui · L. A. Horrocks

5   **Kynurenines in the Brain: Preclinical and Clinical Studies, Therapeutic Considerations** 91
    C. Kiss · L. Vécsei

6   **Neurofibromatosis Type I: From Genetic Mutation to Tumor Formation** 107
    S. L. Thomas · G. H. De Vries

7   **Amyloid and Neurodegeneration: Alzheimer's Disease and Retinal Degeneration** 131
    A. Prakasam · C. Venugopal · A. Suram · J. Pacheco-Quinto · Y. Zhou · M. A. Pappolla · K. A. Sharpe · D. K. Lahiri · N. H. Greig · B. Rohrer · K. Sambamurti

8   **Diabetic Retinopathy** 165
    E. Bowie · C. E. Crosson

9   **Neurotransmitters and Electrophysiology in Traumatic Brain Injury** 179
    C. E. Dixon · A. E. Kline

© 2009 Springer Science+Business Media, LLC.

10 Free Radicals and Neuroprotection in Traumatic Brain and
    Spinal Cord Injury 203
    E. D. Hall

11 Neuroimmunology of Paraproteinemic Neuropathies 229
    A. A. Ilyas

12 Proteolytic Mechanisms of Cell Death in the Central Nervous System 249
    S. F. Larner · R. L. Hayes · K. K. W. Wang

13 Nitric Oxide in Experimental Allergic Encephalomyelitis 281
    S. Brahmachari · K. Pahan

14 The Blood–Brain Barrier—Biology, Development, and Brain Injury 303
    C. L. Keogh · K. R. Francis · V. R. Whitaker · L. Wei

15 Phospholipase A_2 in CNS Disorders: Implication on Traumatic
    Spinal Cord and Brain Injuries 321
    N.-K. Liu · W. Titsworth · X.-M. Xu

16 Axonal Damage due to Traumatic Brain Injury 343
    K. E. Saatman · G. Serbest · M. F. Burkhardt

17 Blood–Central Nervous System Barriers: The Gateway to
    Neurodegeneration, Neuroprotection and Neuroregeneration 363
    H. S. Sharma

18 Engineered Antibody Fragments as Potential Therapeutics against
    Misfolded Proteins in Neurodegenerative Diseases 459
    E. Kvam · A. Messer

19 Pathogenesis and Treatment of HIV-associated Dementia: Recent Studies
    in a SCID Mouse Model 471
    W. R. Tyor

20 Stem Cell Transplantation Therapy for Neurological Diseases 491
    X.-Y. Hu · J.-A. Wang · K. Francis · M. E. Ogle · L. Wei · S. P. Yu

21 Ubiquitin/Proteasome and Autophagy/Lysosome Pathways: Comparison
    and Role in Neurodegeneration 513
    N. Myeku · M. E. Figueiredo-Pereira

22 Experimental Autoimmune Encephalomyelitis in the Pathogenesis of
    Optic Neuritis: Is Calpain Involved? 525
    M. K. Guyton · A. W. Smith · S. K. Ray · N. L. Banik

23 Protein Carbonylation in Neurodegenerative and Demyelinating
    CNS Diseases 543
    O. A. Bizzozero

| 24 | **Clinical Considerations in Translational Research with Chronic Spinal Cord Injury: Intervention Readiness and Intervention Impact** 563
J. S. Krause · S. D. Newman · S. S. Brotherton |
|---|---|
| 25 | **Estrogen as a Promising Multi-Active Agent for the Treatment of Spinal Cord Injury** 581
E. A. Sribnick · D. D. Matzelle · S. K. Ray · N. L. Banik |
| 26 | **Immunotherapy Strategies for Lewy Body and Parkinson's Diseases** 599
L. Crews · B. Spencer · E. Masliah |
| 27 | **Clinical Outcomes After Spinal Cord Injury** 615
J. S. Krause · S. D. Newman |
| 28 | **Spinal Cord Injury – A Clinical Perspective** 633
A. Varma |
| 29 | **Cubing the Brain: Mapping Expression Patterns Genome-Wide** 649
M. H. Chin · D. J. Smith |
| | **Index** .... 657 |

Contributors

N. L. Banik
Department of Neurosciences
Division of Neurology
Medical University of South Carolina,
96 Jonathan Lucas Street, Suite 309
Charleston, SC 29425, USA
E-mail: baniknl@musc.edu

O. A. Bizzozero
Department of Cell Biology and Physiology,
University of New Mexico School of Medicine,
1 University of New Mexico, MSC08 4750,
Albuquerque, NM 87131, USA
E-mail: obizzozero@salud.unm.edu

E. Bowie
Department of Ophthalmology,
Storm Eye Institute,
Medical University of South Carolina,
Charleston, SC, USA

S. Brahmachari
Department of Neurological Sciences,
Rush University Medical Center, Chicago,
Illinois 60612, USA

S. S. Brotherton
Spinal Cord Injury Outcomes Research Group,
Medical University of South Carolina,
Charleston, SC, USA

M. F. Burkhardt
Spinal Cord and Brain Injury Research Center,
Department of Physiology, University of Kentucky,
Lexington, KY 40536, USA

M. H. Chin
Department of Human Genetics,
David Geffen School of Medicine at UCLA,
Los Angeles, CA 90095, USA

M. A. Contreras
Charles Darby Children's Research Institute,
Department of Pediatrics,
Medical University of South Carolina,
Charleston, SC 29425, USA

L. Crews
Department of Pathology
University of California San Diego, La Jolla,
CA 92093-0624, USA

C. E. Crosson
MUSC – Storm Eye Institute, 167 Ashley Ave. Room 511,
Charleston, SC 29425, USA
E-mail: crossonc@musc.edu

G. H. De Vries
Research Service, Edward Hines Jr. V.A. Hospital,
5th Avenue and Roosevelt Road,
Hines, IL 60141, USA
E-mail: George.Devries@med.va.gov

C. E. Dixon
Department of Neurosurgery,
University of Pittsburgh, Safar Center,
201 Hill Building, 3434 Fifth Avenue,
Pittsburgh, PA 15260, USA
E-mail: dixonec@upmc.edu

A. A. Farooqui
Department of Molecular and Cellular
Biochemistry, 1645 Neil Avenue,
The Ohio State University,
Columbus, Ohio 43210-1218, USA
E-mail: farooqui.1@osu.edu

V. Felipo
Laboratory of Neurobiology,
Centro de Investigación Príncipe,
Avda del Saler, 16, 46013 Valencia, Spain
E-mail: vfelipo@ochoa.fib.es

© 2009 Springer Science+Business Media, LLC.

M. E. Figueiredo-Pereira
Department of Biological Sciences,
Hunter College, City University of New York,
695 Park Avenue, New York,
N.Y. 10021, USA
E-mail: pereira@genectr.hunter.cuny.edu

K. R. Francis
Department of Pathology and Laboratory Medicine,
Medical University of South Carolina,
Charleston, SC 29425, USA

N. H. Greig
Cellular Neurobiology Branch, Intramural Research
Program, National Institute on Drug Abuse,
Baltimore, Maryland 21224, USA

M. K. Guyton
Departments of Microbiology & Immunology,
Medical University of South Carolina,
96 Jonathan Lucas Street, Suite 309,
Charleston, SC 29425, USA

E. D. Hall
Director, Spinal Cord & Brain Injury Research Center
(SCoBIRC), and SCoBIRC Endowed Professor of Anatomy
& Neurobiology, Neurology & Neurosurgery,
University of Kentucky Medical Center,
BBSRB 383, 741 S. Limestone Street, Lexington,
KY 40536-0509, USA
E-mail: edhall@uky.edu

R. L. Hayes
Department of Psychiatry,
Center for Traumatic Brain Injury Studies,
McKnight Brain Institute of the University of Florida,
Gainesville, FL, USA

L. A. Horrocks
Department of Molecular and Cellular Biochemistry,
1645 Neil Avenue, The Ohio State University,
Columbus, Ohio 43210-1218, USA

X.-Y. Hu
Department of Pathology College of Medicine,
Zhejiang University, Hangzhou, 310016,
China

A. A. Ilyas
Associate Professor, Department of Neurology &
Neuroscience, UMDNJ-New Jersey Medical School,
185 South Orange Avenue, Newark,
NJ 07103, USA
E-mail: ilyasaa@umdnj.edu

C. L. Keogh
Department of Pathology and Laboratory Medicine,
Medical University of South Carolina,
Charleston, SC 29425, USA

C. Kiss
Department of Neurology,
Albert Szent-Györgyi Medical and Pharmaceutical
Center, University of Szeged, POB 427, H-6701,
Szeged, Hungary

A. E. Kline
Departments of Neurosurgery,
Physical Medicine & Rehabilitation, Neurobiology,
Psychology, Center for Neuroscience,
Brain Trauma Research Center,
and Safar Center for Resuscitation Research,
University of Pittsburgh, Pittsburgh,
PA 15213, USA

J. S. Krause
Spinal Cord Injury Outcomes Research Group,
Medical University of South Carolina,
77 President Street, Suite 117, Charleston,
SC 29425, USA
E-mail: krause@musc.edu

E. Kvam
Division of Genetic Disorders,
Wadsworth Center,
New York State Dept. of Health,
Albany, NY 12208, USA

D. K. Lahiri
Department of Psychiatry,
Institute of Psychiatric Research,
Indiana University School of Medicine,
Indianapolis, IN-46202, USA

S. Lakka
Program of Cancer Biology,
Department of Cancer Biology and
Pharmacology, University of Illinois College
of Medicine at Peoria, Peoria,
IL, USA

S. F. Larner
Department of Neuroscience,
Center for Traumatic Brain Injury Studies,
McKnight Brain Institute of the University of Florida,
100 S. Newell Dr., PO Box 100244, Gainesville,
FL 32610, USA
E-mail: sflarner@ufl.edu

N.-K. Liu
Kentucky Spinal Cord Injury Research Center,
Departments of Neurological Surgery and Anatomical
Sciences and Neurobiology,
University of Louisville School of Medicine,
Louisville, KY 40292, USA

E. Masliah
Department of Neurosciences,
University of California, San Diego, La Jolla,
CA 92093-0624, USA
E-mail: emasliah@UCSD.edu

D. D. Matzelle
Division of Neurology,
Department of Neurosciences,
Medical University of South Carolina,
Charleston, SC 29425, USA

A. Messer
Wadsworth Center/ David Axelrod Inst.,
NY State Dept. Health, New Scotland Avenue,
P.O. Box 22002, Albany,
NY 12201-2002, USA
E-mail: messer@wadsworth.org

N. Myeku
Department of Biological Sciences,
Hunter College of the City University of New York,
New York, New York 10021, USA

S. D. Newman
Spinal Cord Injury Outcomes Research Group,
Medical University of South Carolina,
Charleston, SC, USA

M. E. Ogle
Department of Pharmaceutical and Biological Sciences,
Medical University of South Carolina, Charleston,
SC 29464, USA

J. Pacheco-Quinto
MUSC, Neurosciences,
173 Ashley Avenue, BSB 403, Charleston,
SC 29425, USA

K. Pahan
Department of Neurological Sciences,
Rush University Medical Center,
Cohn Research Building, Suite 320,
1735 West Harrison St., Chicago, IL 60612, USA
E-mail: Kalipada_Pahan@rush.edu

M. A. Pappolla
MUSC, Neurosciences, 173 Ashley Avenue,
BSB 403, Charleston, SC 29425, USA

A. Prakasam
MUSC, Neurosciences, 173 Ashley Avenue,
BSB 403, Charleston, SC 29425, USA

J. S. Rao
Program of Cancer Biology,
Department of Cancer Biology and Pharmacology,
University of Illinois College of Medicine,
One Illini Drive, Box 1649,
Peoria, IL 61605, USA
E-mail: jsr@uic.edu

S. K. Ray
Department of Pathology, Microbiology and
Immunology
University of South Carolina School of Medicine
6439 Garners Ferry Road
Building 2, Room C11
Columbia, SC 29209, USA
E-mail: raysk8@gw.med.sc.edu

B. Rohrer
MUSC, Neurosciences, 173 Ashley Avenue, BSB 403,
Charleston, SC 29425 and MUSC,
Ophthalmology, 167 Ashley Avenue, SEI,
Charleston, SC 29425, USA

K. E. Saatman
Spinal Cord and Brain Injury Research Center,
B367 BBSRB, 741 S. Limestone St., Lexington,
KY 40536-0509, USA
E-mail: k.saatman@uky.edu

K. Sambamurti
MUSC, Neurosciences, 173 Ashley Avenue,
BSB 403, Charleston, SC 29425, USA
E-mail: sambak@musc.edu

G. Serbest
Department of Neurosurgery,
University of Pennsylvania,
Philadelphia, PA 19104, USA

H. S. Sharma
Laboratory of Cerebrovascular Research,
Department of Surgical Sciences, Anesthesiology &
Intensive Care Medicine, University Hospital,
Uppsala University, SE-75185 Uppsala, Sweden
E-mail: Sharma@surgsci.uu.se

K. A. Sharpe
MUSC, Ophthalmology, 167 Ashley Avenue, SEI,
Charleston, SC 29425, USA

I. Singh
Charles Darby Children's Research Institute,
Department of Pediatrics,
Medical University of South Carolina,
Charleston, SC 29425, USA
E-mail: singhi@musc.edu

A. W. Smith
Departments of Microbiology & Immunology,
Medical University of South Carolina,
96 Jonathan Lucas Street, Suite 309,
Charleston, SC 29425, USA

D. J. Smith
Department of Molecular and Medical Pharmacology,
David Geffen School of Medicine at UCLA,
23-151A CHS, Los Angeles,
CA 90095, USA
E-mail: DSmith@mednet.ucla.edu

B. Spencer
Departments of Neurosciences,
University of California San Diego, La Jolla,
CA 92093-0624, USA

E. A. Sribnick
Neurological Surgery,
Emory University School of Medicine,
Atlanta, GA 30322, USA

A. Suram
MUSC, Neurosciences, 173 Ashley Avenue,
BSB 403, Charleston,
SC 29425, USA

S. L. Thomas
Research Service, Edward Hines Jr. V.A. Hospital,
5th Avenue and Roosevelt Road, Hines, IL 60141, USA

W. L. Titsworth
Kentucky Spinal Cord Injury Research Center,
Departments of Neurological Surgery and
Anatomical Sciences and Neurobiology,
University of Louisville School of Medicine,
Louisville, KY 40292, USA

W. R. Tyor
Professor, Department of Neurosciences and
Department of Microbiology and Immunology,
Medical University of South Carolina,
and Chief, Neurology Service,
Ralph H. Johnson VAMC, 109 Bee Street,
Charleston, SC 29401, USA
E-mail: william.tyor@va.gov

A. Varma
Assistant Professor of Neurosurgery,
Medical University of South Carolina,
Suite 428 CSB, 96 Jonathan Lucas Street,
Charleston, SC 29425, USA
E-mail: varma@musc.edu

L. Vécsei
Department of Neurology, Albert Szent-Györgyi
Medical and Pharmaceutical Center,
University of Szeged, POB 427, H-6701,
Szeged, Hungar and Neurology Research Group of the
Hungarian Academy of Sciences and University of
Szeged, POB 427, H-6701, Szeged, Hungary
E-mail: vecsei@nepsy.szote.u-szeged.hu

C. Venugopal
MUSC, Neurosciences, 173 Ashley Avenue,
BSB 403, Charleston, SC 29425, USA

J.-A. Wang
College of Medicine, Zhejiang University,
Hangzhou, 310016, China

K. K. W. Wang
Department of Psychiatry, Center for Traumatic
Brain Injury Studies, McKnight Brain Institute of the
University of Florida, Gainesville,
FL, USA

L. Wei
Medical University of South Carolina,
Department of Pathology and Laboratory Medicine,
165 Ashley Avenue, Charleston,
SC 29425, USA
E-mail: weil@musc.edu

V. R. Whitaker
Department of Pathology and Laboratory Medicine,
Medical University of South Carolina,
Charleston, SC 29425, USA

X.-M. Xu
James R Petersdorf Professor,
Department of Neurological Surgery,
University of Louisville School of Medicine,
511 S. Floyd Street, MDR 616, Louisville,
KY 40292, USA
E-mail: xmxu0001@louisville.edu

S. P. Yu
Department of Pharmaceutical and Biological Sciences,
280 Calhoun Street,
Medical University of South Carolina,
Charleston, SC 29464, USA
E-mail: yusp@musc.edu

Y. Zhou
MUSC, Neurosciences,
173 Ashley Avenue, BSB 403, Charleston,
SC 29425, USA

1 Brain Tumor Angiogenesis

S. Lakka · J. S. Rao

1	Introduction 2
2	Molecular Regulation of Glioma Angiogenesis 3
3	Genetic Alterations in Glioma Angiogenesis 5
4	Angiogenesis Inhibitors 6
5	Conclusion 8

© 2009 Springer Science+Business Media, LLC.

Abstract: The growth of malignant tumors of the central nervous system requires adequate supplies of oxygen and nutrients obtained by the formation of new capillaries from existing blood vessels. This process is called angiogenesis. Accordingly, anti-angiogenic therapy represents a promising modality for the treatment of brain tumors. This review describes the main biological events, molecular factors and critical signaling cascades implicated in brain tumor angiogenesis. We also summarized the key attributes of the emerging synthetic small molecular inhibitors and molecular targets with anti-angiogenic activities that make them potent anti-tumor agents. The studies reviewed here suggest that there is an urgent need for a new comprehensive treatment strategy combining anti-angiogenic agents with conventional cytoreductive treatments in the control of brain cancer.

List of Abbreviations: bFGF, basic fibroblast growth factor; EGF, epidermal growth factor; EGFR, Epidermal growth factor receptor; FGF, Fibroblast growth factors; *FKBP, FK506-binding protein*; FTS, farnesylthiosalicylic acid; HGA, High-grade astrocytoma; HIF, hypoxia-inducible transcription factor; HIF-1, Hypoxia-inducible factor 1; HIF-1alpha, Hypoxia-inducible factor 1alpha; IM, intramuscular; MMPs, matrix metalloproteinases; NABTC, North American Brain Tumor Consortium; PDGF, platelet derived growth factor; PDGF-R, Platelet derived growth factor receptor; PDGFR-beta, platelet-derived growth factor receptor-beta; PI3K, phosphatidyl inositol-3 kinase; PlGF, placental growth factor; PIP2, Phosphatidyl inositol bi phosphate; PIP3, phosphatidyl inositol tri phosphate; TNF-alpha, tumor necrosis factor alpha; TSP, Thrombospondin; PTEN, Phosphatese and tensin hemalog; VEGF, vascular endothelial growth factor; vHL, von Hippel–Lindau

1 Introduction

A tumor's ability to grow is linked to its ability to recruit a supply of new blood vessels. A tumor remains in a dormant state, the cellular proliferation rate balanced by the apoptotic rate, unable to grow in size beyond a few millimeters in the absence of the acquired angiogenic phenotype. In order to do this it sends out "signals" to already existing blood vessels to sprout new branches to feed it. This is the process of angiogenesis. In response to these chemical signals, endothelial cells divide and grow, breakdown their surrounding tissue barriers with enzymes called matrix metalloproteinases (MMPs), and migrate toward the tumor to form a connection to the body's blood supply. The mechanisms that lead to the tumors' switch to the angiogenic phenotype are unknown. The development of new blood vessels follows a well-defined pattern beginning with an initial increased vascular permeability leading to extravasation of plasma, plasma proteins, and deposition of pro angiogenic matrix proteins (Dvorak et al., 1995; Ferrara and Davis-Smyth, 1997; Dvorak et al., 1999; Dvorak, 2000). The process is complex and involves multiple steps and pathways with positive and negative signals (Fidler and Ellis, 1994; Folkman, 1995; Hanahan and Folkman, 1996). Endothelial cells then assemble themselves to form a central lumen, elaborate a new basement membrane, and eventually recruit pericytes and smooth muscle cells to surround the mature vessel. Thus angiogenesis is a complicated process involving a multitude of tightly controlled sequential events that lead to the development of new vasculature. Loss of the tight control over this highly regulated process can have disastrous effects in developmental, pathological, and physiological states. In neoplasia, the angiogenic balance is tipped in favor of new vessel growth (Hanahan and Folkman, 1996).

Glioblastomas are among the most dramatically neovascularized neoplasms; progression of a glioma to its more malignant form of glioblastoma is usually associated with striking neovascularization, evidenced by vasoproliferation, altered endothelial cell cytology, and endothelial cell hyperplasia. In the case of astrocytoma, the low-grade (WHO grade II) tumors show slight increases in vessel density, whereas high microvessel density and evidence of neo-angiogenesis is part of the histological definition of grade III and IV astrocytomas (Kleihues et al., 2000). In addition, glioblastomas (WHO grade IV astrocytomas) are characterized by two histological characteristics relevant to angiogenesis: glomeruloid vascular proliferation, a characteristic architecture of new vessels, and necrosis with pseudopalisading, a focus of hypoxia that stimulates angiogenesis (Brat and Van Meir, 2001). Endothelial proliferation is so frequently found within and adjacent to high-grade gliomas that it is one of the pathologic criteria for grading these tumors.

The obvious conclusion is that high-grade gliomas must stimulate the proliferation of endothelial cells and generate new blood vessels to support their growth.

Direct evidence of the angiogenic potential of gliomas was supplied by Brem (1976), who demonstrated that human gliomas transplanted into rabbit cornea elicit intense neovascularization, and by Klagsbrun et al. (1976) who demonstrated that glioblastoma cells are the most potent of all stimulators of angiogenesis in a chorioallantoic membrane assay. Low-grade gliomas are moderately vascularized tumors, whereas high-grade gliomas show prominent microvascular proliferations and areas of high vascular density (Plate and Risau, 1995). This view is further supported by the identification of tumor vessel density as an independent prognostic parameter for human astroglial tumors (Leon et al., 1996). Bovine retinal endothelial cells reorganize into thin capillary-like structures when cocultured with human glioblastoma cells (Chandrasekar et al., 2000; Nirmala et al., 2000). Telomerase-immortalized human microvascular endothelial cells undergo tubule formation when cocultured in the presence of human glioma cells, but not primary human astrocytes (Venetsanakos et al., 2002). Thus, tumor cells appear to control the differentiation of endothelial cells during tumor-induced angiogenesis. These data suggest that specific cellular interactions between endothelial cells and glioma cells may influence the release of autocrine/paracrine endothelial growth and differentiation factors to promote capillary-like structure formation. Other investigators have also examined endothelial cell differentiation into capillary-like structures in response to the conditioned media of different tumors. A number of factors mediate this neovascularization which includes angiogenic factors like angiopoietins, hypoxia inducible factor-1 (HIF-1) [which upregulates vascular endothelial growth factor (VEGF)], basic fibroblast growth factor (FGF), epidermal growth factor (EGF), placental growth factor (PlGF), platelet-derived growth factor (PDGF), tumor necrosis factor alpha (TNF-alpha), hepatocyte growth factor, and Ang-1, to name a few and inhibitors of angiogenesis such as thrombospondin-1 (TSP-1), platelet factor 4, angiostatin, and endostatin (Bello et al., 2004).

2 Molecular Regulation of Glioma Angiogenesis

Expression of hypoxia-inducible transcription factor (HIF) and the target genes that promote angiogenesis is critical for the growth and invasiveness of astrocytomas. In fact, dense collections of neoplastic astrocytoma cells seen palisading around necrotic foci in glioblastoma show some of the highest expression of hypoxia-inducible regulators of angiogenesis (Plate, 1999). Hypoxia is one of the most potent stimulators of VEGF expression, and it acts through a hypoxia-responsive element within the VEGF promoter to increase VEGF transcription (Shweiki et al., 1992). Expression of VEGF-A is mediated in part by binding of a HIF-1 complex to a HIF-1-binding site in the VEGF gene promoter. Regulation is finely tuned to the availability of oxygen because hypoxia-inducible factor-1alpha (HIF-1α), the oxygen-sensitive subunit of the HIF-1 complex, is stable in hypoxia, but is targeted for proteosomal degradation by the von Hippel–Lindau (vHL) complex during normoxia (Hon et al., 2002). HIF-2α encodes a basic helix–loop–helix PAS domain protein transcription factor that regulates vascular maturation, remodeling, and stabilization of angiogenesis in response to hypoxia (Wiesener et al., 1998). HIF-2α mRNA has been shown to be expressed in GBMs (Damert et al., 1997). Furthermore, HIF-1α expression in GBMs is localized in areas adjacent to necrotic zones and within infiltrating tumor cells. Expression of HIF-1α has been shown in adult high-grade astrocytoma (HGA) and oligodendroglioma (Zagzag et al., 2000). HIF-2α in HGA was differentially overexpressed by at the mRNA level ($p = 0.003$) and at the protein level as well ($p = 0.02$) (Khatua et al., 2003). Expression profiling of 13 childhood astrocytomas to determine the expression pattern angiogenesis-related genes revealed overexpression of HIF-2α as well as high-level expression of *FK506-binding protein (FKBP) 12* by HGA. Hypoxia enhances angiogenesis through FRAP/FKBP12 signaling. The VEGF gene in vitro is enhanced approximately ten times by hypoxia demonstrated by intensive expression of VEGF in palisading neoplastic astrocytic cells residing along necrotic foci (Plate et al., 1992).

Two central endothelium-specific growth factor families coordinate vascular development: the VEGFs and the angiopoietins (Ferrara, 1999; Jones et al., 2001). These were found to be highly overexpressed in astrocytomas. VEGF is a major angiogenic factor in gliomas, and shows increased expression with higher grades of astrocytic tumors, suggesting that it is a significant driving force in the increased microvessel density of these tumors. In addition to the ligands, the receptors for the VEGF signaling pathway also show upregulation in tumors. For example, both the high-affinity VEGF receptors VEGFR-1/Flt-1 and VEGFR-2/Flk-1 are not expressed in normal brain endothelium, but are progressively upregulated in grade II, III, and IV astrocytoma (Plate and Risau, 1995). The VEGFs are required for vasculogenesis and angiogenic sprouting and act through two high-affinity receptors, the class III protein tyrosine kinases PTKs VEGFR-1/Flt-1 8 (de Vries et al., 1992) and VEGFR-2 Flk-1/KDR 9 (Millauer et al., 1993) receptor tyrosine kinases (VEGFR-1, -2, and -3); VEGFR-2 mediates endothelial cell proliferation, differentiation, and microvascular permeability, which are almost exclusively expressed on microvascular endothelial cells. A dramatic upregulation of VEGF mRNA has been found in human glioblastoma tissue but not in normal brain tissue (Plate et al., 1992). Both temporal and spatial expression patterns of VEGF, Flk-1/KDR, and Flt-1 correlate significantly with the angiogenic activity in human gliomas (Plate et al., 1994; Samoto et al., 1995). There was also a significant upregulation of VEGF mRNA in the tissue of highly vascularized glioblastoma multiforme, capillary hemangioblastomas, and cerebral metastases (Berkman et al., 1993). More VEGF mRNA expression is found in neoplastic astrocytes of glioblastomas (Pietsch et al., 1997), which suggests a correlation between gliomas producing VEGF protein and malignant behavior (Oehring et al., 1999). Neuropilin-1 is a $VEGF_{165}$ and semaphorin receptor expressed by endothelial cells, and tumor cells Neuropilin-1 seems upregulated in endothelial cells and in neoplastic astrocytes of glioblastomas, with no difference in expression pattern between de novo glioblastomas and secondary glioblastomas. However Neuropilin-1 expression is found to a lesser degree in low-grade astrocytomas, suggesting a positive correlation between neuropilin-1 and angiogenesis in human astrocytic tumors (Broholm and Laursen, 2004). Genetic and pharmacological strategies demonstrated that high levels of activated Ras GTP in these cells regulate the increased VEGF expression by human astrocytoma cell lines (Feldkamp et al., 1999).

A second ligand–receptor system of importance in glioma angiogenesis is the angiopoietins, particularly Ang-1 and -2, and exerts their biological actions through the Tie2 receptor (Yancopoulos et al., 2000). The angiopoietins (Ang-1 and Ang-2) have been implicated in further remodeling of the initial microvasculature (Jones et al., 2001). Thus, functions of Ang-1 and VEGF may be complementary to each other, with VEGF providing most mitogenic activity and Ang-1 responsible for ensuring vascular stability (Suri et al., 1998; Thurston et al., 1999). Examination of the expression patterns of angiopoietics and their receptors suggest a role in human glioblastoma vasculature (Zagzag et al., 1999; Tse et al., 2003). Ang-2 and Tie-2 expression were absent in the normal brain vasculature but were induced in tumor endothelium of coopted tumor vessels prior to their regression (Stratmann et al., 1998). Increased Tie-2 expression has been observed with increasing grade of human astrocytoma, and treatment of glioma cell-derived mouse xenografts with ExTek (a dominant negative form of Tie2) resulted in a significant decrease in tumor growth (Zadeh et al., 2004).

The Fibroblast Growth Factor family includes 23 structurally related proteins that have broad mitogenic functions (Eckenstein 1994; Naski and Ornitz, 1998). FGF2 (basic FGF) is expressed in high amounts in both neurons and astrocytes (Eckenstein, 1994). Pro survival signaling by VEGF and FGF2 cooperate to decrease the endothelial cell susceptibility to intrinsic and extrinsic inducers of endothelial cell apoptosis in gliomas (Schmidt et al., 1999; Alavi et al., 2003). The presence of FGF2 in cerebrospinal fluids from children and adults with brain tumors has been shown to correlate with tumor microvessel formation (Li et al., 1994). There was a significant upregulation of mRNA for the FGF receptor in human glioblastoma tumor cells that was absent in normal brain cells and in endothelial cells from capillaries and large vessels within the tumor (Morrison et al., 1994). Both acidic FGF and transforming growth factor-alpha (TGF-α), together with epidermal growth factor receptor (EGFR), are found to be greatly overexpressed in 30 primary astrocytomas (Maxwell et al., 1991). Transfection of dominant negative FGFR2 or FGFR1 in glioma C6 cells demonstrated inhibition of tumor growth by both angiogenesis-dependent and -independent mechanisms (Auguste et al., 2001).

PDGF-B and its receptor (PDGF-R) are overexpressed in human gliomas and responsible for recruiting peri-endothelial cells to vessels. PDGF-B was demonstrated as a paracrine factor in U87MG gliomas, and that PDGF-B enhances glioma angiogenesis, at least in part, by stimulating VEGF expression in tumor endothelia and by recruiting pericytes to neovessels (Guo et al., 2003). Platelet-derived growth factor receptor-beta (PDGFR-beta) mRNA was not detectable in the vessels of normal human brain, but was expressed in the vasculature of low- and high-grade gliomas, particularly in endothelial cell proliferations in glioblastomas.

3 Genetic Alterations in Glioma Angiogenesis

Several endogenous negative regulators of angiogenesis, including thrombospondin (TSP), angiostatin, endostatin, and glioma-derived angiogenesis inhibitory factor (GD ± AIF) are considered to play significant roles in maintaining the quiescence of vessels. Genetic alterations in GBM are thought to contribute directly or indirectly to angiogenic dysregulation by tilting the balance toward an angiogenic phenotype through upregulation of proangiogenic factors and/or downregulation of angiogenesis inhibitors. Chromosomal abnormalities commonly found in gliomas include those that disrupt cell cycle arrest and death pathways and those that promote survival under hypoxic conditions (Maher et al., 2001; Korkolopoulou et al., 2004). Molecular alterations associated with progression to GBM and those that define genetic subsets include EGFR amplifications, p53 mutations, retinoblastoma pathway alterations [most commonly, p16 (CDKN2A) losses], and chromosome 10 alterations, including PTEN mutations. Although a wide range of genetic events ultimately lead to GBM, the vascular changes that evolve are remarkably similar.

The earliest genetic alteration in human astrocytoma progression is mutation of the p53 tumor suppressor gene, whereas one of the earliest phenotypic changes is the stimulation of neovascularization. A number of investigators have reported correlation between p53 alterations and progression of brain tumors. The P53 gene product regulates the development of tumor angiogenesis by affecting the expression of pro- and antiangiogenic factors. The expression of basic fibroblast growth factor (bFGF), a potent stimulator of angiogenesis, can be repressed by transfection with wild-type P53 (Ueba et al., 1994). Glioblastoma cells upon induction of wild-type, but not mutant, p53 expression, secreted a factor designated glioma-derived angiogenesis inhibitory factor (GD ± AIF), which was able to neutralize the angiogenicity of the factors produced by the parental cells as well as of bFGF (Van Meir et al., 1994). This is possibly due to loss of transactivation of one or more genes regulated by p53 (Vogelstein and Kinzler, 1992). p53-inducible gene, BAI1, is specifically expressed and was shown to be a strong inhibitor of the angiogenesis that accompanies progression of glia-derived tumors. The expression of this gene, BAI1 (brain-specific angiogenesis inhibitor 1) was absent or significantly reduced in eight of nine glioblastoma cell lines, suggesting BAI1 plays a significant role in angiogenesis inhibition, as a mediator of p53 (Nishimori et al., 1997). Recent studies suggest that BAI1 mRNA and protein expression are absent in a majority of human glioblastoma cell lines as well as GBM resection specimens (Kaur et al., 2003). Modulation of bFGF gene expression by wild-type and mutant forms of p53 depended on direct transcriptional regulation of the bFGF promoter. This study suggests that mutations of p53, which occur early in astrocytoma progression, cause direct upregulation of bFGF and could in part contribute to an angiogenic phenotype (Takahashi et al., 1990).

Tumor suppressor gene PTEN, located at chromosome 10q23.3, is mutated in 20–40% of GBMs and a variety of other cancers (Li et al., 1997; Steck et al., 1997). Reconstitution of PTEN expression in glioma cells induced TSP-1 and led to decreased vessel formation and tumorigenicity (Wen et al., 2001). PTEN has also been shown to directly inhibit angiogenesis in U87 glioma cells in an orthotopic brain tumor model (Wen et al., 2001). PTEN antagonizes the functions of the proto-oncogene phosphatidyl inositol-3 kinase (PI3K) by dephosphorylating its lipid substrates PIP2 (phosphatidyl inositol bi phosphate) and PIP3 (phosphatidyl inositol tri phosphate) (Maehama and Dixon, 1998). Activation of AKT, a downstream effector of PI3K, has been reported to regulate VEGF induction under hypoxic conditions in gliomas (Aoki et al., 1998). PTEN/MMAC gene transfer into human glioma cells that possess inactivating mutations of the PTEN/MMAC gene but also express either constitutively active EGFR (U87DeltaEGFR cells) or possess an inactivating mutation of p53 (U251 cells) still display inhibited angiogenesis in orthotopic and ectopic models of gliomas (Abe et al., 2003). Normal brain and lower-grade astrocytomas known to retain chromosome 10 stained strongly for TSP. There was very faint staining for TSP in 12 of 13 glioblastomas, the majority of which lose chromosome 10, indicating that the loss of tumor suppressors on chromosome 10 contributes to the aggressive malignancy of glioblastomas in part by releasing constraints on angiogenesis that are maintained by TSP-1 in lower-grade tumors (Hsu et al., 1996). TSP-1 is a multimodular secreted protein that associates with the extracellular matrix and possesses a variety of biological functions, including a potent antiangiogenic activity (Bornstein, 1995). TSP-1 expression was downregulated in a number of experimental tumor models, and this regulation has been correlated with the decreased expression of tumor suppressor genes (Hsu et al., 1996; Grossfeld et al., 1997). Expression of TSP-1 was shown to decrease the tumorigenicity of human glioma cells (Tenan et al., 2000).

Another frequent alteration in GBMs and glioma cell lines is amplification of the EGFR and/or expression of constitutive forms of EGFR (Schober et al., 1995). A high percentage (40–60%) of de novo GBMs are characterized by amplification of the EGFR gene (Ekstrand et al., 1991). EGFR stimulation can significantly increase the activity of c-Src, a signaling intermediate involved in cell cycle progression, motility, angiogenesis, and survival (Dehm and Bonham, 2004). EGFR-mediated pathways are intimately involved in tumor angiogenesis through upregulation of VEGF and other mediators of angiogenesis. Activation of EGFR by EGF or TGF-α has been shown to upregulate VEGF expression in human gliomas (Goldman et al., 1993). The PI3K pathway is activated by the EGFR and attenuated by the PTEN phosphatase, which are often amplified and deleted in gliomas, respectively, regulates angiogenic molecules. EGFR has also been recently shown to transcriptionally upregulate VEGF expression in human U87 glioma cells via a pathway involving ras/PI3K and Akt that was distinct from the pathway induced by hypoxia (Maity et al., 2000). The synthesis of HIF-2α is induced via a signal transduction pathway initiated by receptor tyrosine kinases, such as the EGFR. EGFR/FKBP12/HIF-2α pathway was shown to be an important promoter of angiogenesis in childhood HGA (Khatua et al., 2003).

4 Angiogenesis Inhibitors

Vascular network is one of the relatively new and promising approaches in the treatment of solid tumors. Angiogenesis, although essential for embryogenesis, is restricted in adults to ovulation, cyclical endometrial proliferation, and wound repair (Folkman and Shing, 1992). Therefore, inhibitors of angiogenesis are expected to be better tolerated than conventional cytotoxic cancer therapies that affect all rapidly growing cells. Owing to the invading normal host vasculature being the target, another potential attribute of an antiangiogenesis approach may be the avoidance of drug resistance, commonly associated with conventional anticancer modalities that target genetically unstable tumor cells (Boehm et al., 1997). Therapeutic strategies aimed at inhibiting various steps in the process of angiogenesis have significant clinical potential. Inhibitors of angiogenesis are separated into endogenous inhibitors such as angiostatin, TSP or alpha interferon, and natural or synthetic inhibitors such as thalidomide, antibodies against angiogenic growth factors, or inhibitors of tyrosine kinase receptors. Several agents have been developed to circumvent this vascular recruitment by malignant gliomas, and a few of these are being tested in humans.

In general, investigators design antiangiogenesis agents are using four strategies:
1. Inhibit normal endothelial cells directly.
2. Block the chemical signals that stimulate angiogenesis.
3. Block the ability of endothelial cells to breakdown their surrounding tissue barriers.
4. Block the action of proteins called integrins that are on the surface of endothelial cells and are responsible for the continued survival of endothelial cells undergoing angiogenesis.

Since gliomas are characterized by the production of (pro-) angiogenic factors, and reduction of their expression/availability or inhibition of their action are indirect approaches to reduce or block tumor angiogenesis. Preferential targets are VEGF and VEGFRs, but also bFGF, PDFR, and others. Early animal experimental studies have proved the efficacy of this concept, in particular for VEGF/VEGF action.

Inhibitors of VEGF receptors are promising agents in malignant gliomas, with the potential to inhibit angiogenesis and tumor growth, as well as to reduce peritumoral edema. ZK222584, an inhibitor of VEGF receptor tyrosine kinases, decreases glioma growth and vascularization in vivo (Goldbrunner et al., 2004). Other VEGF receptor inhibitors such as ZD6474 (Sandstrom et al., 2004) and CEP-7055, a pan-inhibitor of VEGF receptor (Ruggeri et al., 2003), have produced significant growth inhibition of glioblastoma xenografts in nude mice. The in vivo transfection of endothelial cells using a retrovirus encoding a dominant-negative mutant of the Flk-1/VEGF receptor was shown to inhibit VEGF-mediated signal transduction (Millauer et al., 1994). The aggressive, angiogenic, and VEGF secreting rat C6 glioblastoma tumor formation could be prevented in nude mice by inhibiting VEGF-mediated signaling (Saleh et al., 1996; Sasaki et al., 1999).

Chronically active Ras and PI3-K are common in most GBMs promote angiogenesis (Woods et al., 2002). Ras inhibitor trans-farnesylthiosalicylic acid (FTS) attenuated Ras signaling to extracellular signal-regulated kinase, PI3K, and Akt which in turn resulted in HIF-1alpha with a concomitant decrease in VEGF protein (Blum et al., 2005). The small-molecule tyrosine kinase inhibitor SU6668 was shown as more potent polyvalent inhibitor of angiogenic signaling pathways because it targets VEGFR, FGFR, and PDGFR. SU6668 treatment of orthotopic malignant gliomas potently inhibits tumor angiogenesis, impairs tumor microcirculation, induces regression of existing tumor blood vessels, and, thereby, suppresses tumor growth and prolongs survival (Farhadi et al., 2005). Targeting the Flk-1/KDR tyrosine kinase with a small molecule inhibitor, SU5416 represents an effective means to suppress tumor growth and progression by direct inhibition of tumor-induced angiogenesis and vascularization. SU5416 inhibits Flk-1 autophosphorylation, and thus, VEGF-induced endothelial proliferation (Vajkoczy et al., 1999). SU5416 is being evaluated in patients with recurrent malignant gliomas as part of a North American Brain Tumor Consortium (NABTC) trial. CP-673,451 a potent pharmacologic inhibitor of PDGFR-beta kinase- and PDGF-BB-stimulated autophosphorylation of PDGFR-beta in cells (IC(50) = 1 nmol/L) selectively inhibited PDGF-BB-stimulated angiogenesis in vivo (Roberts et al., 2005).

Endogenous inhibitors of angiogenesis such as endostatin and angiostatin, either delivered by gene transfer or by injection of the peptides, have been shown to be potent antiglioma agents in mice (Kirsch et al., 1998). Administration of a plasmid encoding an angiostatic peptide, endostatin, resulted in an 80% tumor volume reduction 1 week after treatment and enhanced survival time by up to 47%. Treated tumors exhibited a 40% decrease in tumor vessel density accompanied by alterations in tumor vessel ultrastructure (narrowed or collapsed lumens) (Barnett et al., 2004). Dr. Cheng's group showed that a single intramuscular (IM) injection of an AAV vector expressing angiostatin, a potent angiogenic inhibitor, effectively suppresses human glioma growth in the brain of nude mice. This study thus demonstrated that AAV-mediated antiangiogenesis gene therapy offers efficient and sustained systemic delivery of the therapeutic product, which in turn effectively suppresses glioma growth in the brain (Ma et al., 2002). In preclinical studies, cannabinoids can induce apoptosis, inhibit angiogenesis, and prevent tumor growth in glioma models. Administration of cannabinoids significantly lowered VEGF activity required for brain tumor blood vessels in laboratory mice and two patients with late-stage glioblastoma (Blazquez et al., 2004).

There are several factors that either inhibit endothelial cell activity or matrix breakdown, required for invasion of endothelial cells (as well as for tumor cells). Thalidomide is an example for inhibitors of endothelial cell activity, possibly through modulation of integrins (Eleutherakis-Papaiakovou et al., 2004). It was shown to prolong disease stabilization after conventional therapy in patients with recurrent glioblastoma (Morabito et al., 2004). SN38, known as an active metabolite of CPT-11 and a DNA topoisomerase I (Topo-1) inhibitor (O'Leary and Muggia, 1998), inhibited endothelial cell proliferation and tube formation in vitro at nontoxic concentrations. It was also shown to have an indirect antiangiogenic through HIF-1 and VEGF on malignant glioma cells (Kamiyama et al., 2005). A systemic antiangiogenic drug-based therapy targeted toward growth factor receptors within the tumor's vasculature demonstrated survival of 20–40% of the animals, whereas all nontreated controls and rats died within 40 days. In vitro, the drugs (Z)-3-[4-(dimethylamino) benzylidenyl] indolin-2-one (a PDGFR-beta and a fibroblast growth factor receptor 1 kinase inhibitor) and oxindole (a VEGF receptor 2 kinase inhibitor) inhibited endothelial cells from proliferating in response to the angiogenic factors produced by T9/9L glioma cells and prevented endothelial cell tubulogenesis (Jeffes et al., 2005).

Proteinases play an important role in the invasiveness of gliomas, mediating the degradation of the extracellular matrix and angiogenesis. Proteolysis of ECM components allows endothelial cells to migrate and releases store angiogenic molecules from the ECM (Bergers et al., 1998). Earlier studies of MMP inhibitors such as prinomastat have not shown significant activity. Newer metalloproteinase inhibitors such as SI-27 inhibit glioma invasion and angiogenesis in vivo and may have therapeutic potential (Yoshida et al., 2004). Our own extensive studies in this field have shown that proteases for extracellular matrix breakdown like cathepsin B, urokinase activator pathway and its receptor uPA(R), and MMPs like MMP-9 are required for glioma angiogenesis. Plasmid/adenoviral-mediated delivery of antisense or siRNA for these molecules for these proteases inhibited capillary network formation in a coculture assay in vitro and tumor cell-induced angiogenesis in vivo using the dorsal airsac assay (Gondi et al., 2003; Lakka et al., 2003, 2004;

Yanamandra et al., 2004; Gondi et al., 2004a, b; Lakka and Rao, 2005) These studies have also established that targeting two proteases at a time had more significant effects on tumor cell-induced angiogenesis in vitro and inhibited tumor growth in vivo.

Proliferating vascular cells require adhesion molecules in order to establish contact with the ECM, and they utilize this interaction to generate the traction required for changes in morphology and physical cell movement through the ECM in response to humoral stimuli (Ingber et al., 1995). Of these molecules, increased expression of $\alpha V \beta 3$ integrin has been reported in the regions of vascular proliferation in glioblastomas (Gladson, 1996). IS20I, a specific alphavbeta3 integrin inhibitor, strongly inhibited angiogenesis and suppressed intracranial tumor growth in vivo. The suppression of tumor growth is associated with a decrease in tumor vascularity, an increase in apoptosis, and a decrease in tumor cell proliferation (Bello et al., 2003). The alpha v-antagonist EMD 121974 suppresses brain tumor growth through induction of apoptosis in both brain capillary and brain tumor cells by preventing their interaction with the matrix proteins vitronectin and tenascin (Taga et al., 2002).

5 Conclusion

The inhibition of tumor angiogenesis could be an efficient therapeutic strategy for the treatment of malignant gliomas. However, glioma cells can also invade the brain diffusely over long distances without necessarily requiring angiogenesis. It was shown that glioma cells implanted in an orthotopic model could coopt the preexistent host vasculature to recruit their blood supply in the absence of neovascularization. It is likely that they will be more effective when used in combination with chemotherapy, radiotherapy, targeted molecular agents, or with other antiangiogenic therapies. For example, the combination of a direct angiogenesis inhibitor, such as endostatin, with an indirect antiangiogenic compound such as SU5416 (VEGF receptor inhibitor) produced greater antitumor effects than individual agents alone.

References

Abe T, Terada K, Wakimoto H, Inoue R, Tyminski E, et al. 2003. PTEN decreases in vivo vascularization of experimental gliomas in spite of proangiogenic stimuli. Cancer Res 63: 2300-2305.

Alavi A, Hood JD, Frausto R, Stupack DG, Cheresh DA. 2003. Role of Raf in vascular protection from distinct apoptotic stimuli. Science 301: 94-96.

Aoki M, Batista O, Bellacosa A, Tsichlis P, Vogt PK. 1998. The akt kinase: Molecular determinants of oncogenicity. Proc Natl Acad Sci USA 95: 14950-14955.

Auguste P, Gursel DB, Lemiere S, Reimers D, Cuevas P, et al. 2001. Inhibition of fibroblast growth factor/fibroblast growth factor receptor activity in glioma cells impedes tumor growth by both angiogenesis-dependent and -independent mechanisms. Cancer Res 61: 1717-1726.

Barnett FH, Scharer-Schuksz M, Wood M, Yu X, Wagner TE, et al. 2004. Intra-arterial delivery of endostatin gene to brain tumors prolongs survival and alters tumor vessel ultrastructure. Gene Ther 11: 1283-1289.

Bello L, Giussani C, Carrabba G, Pluderi M, Costa F, et al. 2004. Angiogenesis and invasion in gliomas. Cancer Treat Res 117: 263-284.

Bello L, Lucini V, Giussani C, Carrabba G, Pluderi M, et al. 2003. IS20I, a specific alphavbeta3 integrin inhibitor, reduces glioma growth in vivo. Neurosurgery 52: 177-185.

Bergers G, Hanahan D, Coussens LM. 1998. Angiogenesis and apoptosis are cellular parameters of neoplastic progression in transgenic mouse models of tumorigenesis. Int J Dev Biol 42: 995-1002.

Berkman RA, Merrill MJ, Reinhold WC, Monacci WT, Saxena A, et al. 1993. Expression of the vascular permeability factor/vascular endothelial growth factor gene in central nervous system neoplasms. J Clin Invest 91: 153-159.

Blazquez C, Gonzalez-Feria L, Alvarez L, Haro A, Casanova ML, et al. 2004. Cannabinoids inhibit the vascular endothelial growth factor pathway in gliomas. Cancer Res 64: 5617-5623.

Blum R, Jacob-Hirsch J, Amariglio N, Rechavi G, Kloog Y. 2005. Ras inhibition in glioblastoma down-regulates hypoxia-inducible factor-1alpha, causing glycolysis shutdown and cell death. Cancer Res 65: 999-1006.

Boehm T, Folkman J, Browder T, O'Reilly MS. 1997. Antiangiogenic therapy of experimental cancer does not induce acquired drug resistance. Nature 390: 404-407.

Bornstein P. 1995. Diversity of function is inherent in matricellular proteins: An appraisal of thrombospondin 1. J Cell Biol 130: 503-506.

Brat DJ, Van Meir EG. 2001. Glomeruloid microvascular proliferation orchestrated by VPF/VEGF: A new world of angiogenesis research. Am J Pathol 158: 789-796.

Brem S. 1976. The role of vascular proliferation in the growth of brain tumors. Clin Neurosurg 23: 440-453.

Broholm H, Laursen H. 2004. Vascular endothelial growth factor (VEGF) receptor neuropilin-1's distribution in astrocytic tumors. APMIS 112: 257-263.

Chandrasekar N, Jasti S, Alfred-Yung WK, Ali-Osman F, Dinh DH, et al. 2000. Modulation of endothelial cell morphogenesis in vitro by MMP-9 during glial-endothelial cell interactions. Clin Exp Metastasis 18: 337-342.

Damert A, Machein M, Breier G, Fujita MQ, Hanahan D, et al. 1997. Up-regulation of vascular endothelial growth factor expression in a rat glioma is conferred by two distinct hypoxia-driven mechanisms. Cancer Res 57: 3860-3864.

de Vries C, Escobedo JA, Ueno H, Houck K, Ferrara N, et al. 1992. The fms-like tyrosine kinase, a receptor for vascular endothelial growth factor. Science 255: 989-991.

Dehm SM, Bonham K. 2004. SRC gene expression in human cancer: The role of transcriptional activation. Biochem Cell Biol 82: 263-274.

Dvorak HF. 2000. VPF/VEGF and the angiogenic response. Semin Perinatol 24: 75-78.

Dvorak HF, Brown LF, Detmar M, Dvorak AM. 1995. Vascular permeability factor/vascular endothelial growth factor, microvascular hyperpermeability, and angiogenesis. Am J Pathol 146: 1029-1039.

Dvorak HF, Nagy JA, Feng D, Brown LF, Dvorak AM. 1999. Vascular permeability factor/vascular endothelial growth factor and the significance of microvascular hyperpermeability in angiogenesis. Curr Top Microbiol Immunol 237: 97-132.

Eckenstein FP. 1994. Fibroblast growth factors in the nervous system. J Neurobiol 25: 1467-1480.

Ekstrand AJ, James CD, Cavenee WK, Seliger B, Pettersson RF, et al. 1991. Genes for epidermal growth factor receptor, transforming growth factor alpha, and epidermal growth factor and their expression in human gliomas in vivo. Cancer Res 51: 2164-2172.

Eleutherakis-Papaiakovou V, Bamias A, Dimopoulos MA. 2004. Thalidomide in cancer medicine. Ann Oncol 15: 1151-1160.

Farhadi MR, Capelle HH, Erber R, Ullrich A, Vajkoczy P. 2005. Combined inhibition of vascular endothelial growth factor and platelet-derived growth factor signaling: Effects on the angiogenesis, microcirculation, and growth of orthotopic malignant gliomas. J Neurosurg 102: 363-370.

Feldkamp MM, Lau N, Rak J, Kerbel RS, Guha A. 1999. Normoxic and hypoxic regulation of vascular endothelial growth factor (VEGF) by astrocytoma cells is mediated by Ras. Int J Cancer 81: 118-124.

Ferrara N. 1999. Molecular and biological properties of vascular endothelial growth factor. J Mol Med 77: 527-543.

Ferrara N, Davis-Smyth T. 1997. The biology of vascular endothelial growth factor. Endocr Rev 18: 4-25.

Fidler IJ, Ellis LM. 1994. The implications of angiogenesis for the biology and therapy of cancer metastasis. Cell 79: 185-188.

Folkman J. 1995. Angiogenesis in cancer, vascular, rheumatoid and other disease. Nat Med 1: 27-31.

Folkman J, Shing Y. 1992. Angiogenesis. J Biol Chem 267: 10931-10934.

Gladson CL. 1996. Expression of integrin alpha v beta 3 in small blood vessels of glioblastoma tumors. J Neuropathol Exp Neurol 55: 1143-1149.

Goldbrunner RH, Bendszus M, Wood J, Kiderlen M, Sasaki M, et al. 2004. PTK787/ZK222584, an inhibitor of vascular endothelial growth factor receptor tyrosine kinases, decreases glioma growth and vascularization. Neurosurgery 55: 426-432.

Goldman CK, Kim J, Wong WL, King V, Brock T, et al. 1993. Epidermal growth factor stimulates vascular endothelial growth factor production by human malignant glioma cells: A model of glioblastoma multiforme pathophysiology. Mol Biol Cell 4: 121-133.

Gondi CS, Lakka SS, Dinh D, Olivero W, Gujrati M, et al. 2004a. Downregulation of uPA, uPAR and MMP-9 using small, interfering, hairpin RNA (siRNA) inhibits glioma cell invasion, angiogenesis and tumor growth. Neuron Glia Biol 1: 165-176.

Gondi CS, Lakka SS, Dinh DH, Olivero WC, Gujrati M, et al. 2004b. RNAi-mediated inhibition of cathepsin B and uPAR leads to decreased cell invasion, angiogenesis and tumor growth in gliomas. Oncogene 23: 8486-8496.

Gondi CS, Lakka SS, Yanamandra N, Siddique K, Dinh DH, et al. 2003. Expression of antisense uPAR and antisense uPA from a bicistronic adenoviral construct inhibits glioma cell invasion, tumor growth, and angiogenesis. Oncogene 22: 5967-5975.

Grossfeld GD, Ginsberg DA, Stein JP, Bochner BH, Esrig D, et al. 1997. Thrombospondin-1 expression in bladder cancer: Association with p53 alterations, tumor angiogenesis, and tumor progression. J Natl Cancer Inst 89: 219-227.

Guo P, Hu B, Gu W, Xu L, Wang D, et al. 2003. Platelet-derived growth factor-B enhances glioma angiogenesis by stimulating vascular endothelial growth factor expression in tumor endothelia and by promoting pericyte recruitment. Am J Pathol 162: 1083-1093.

Hanahan D, Folkman J. 1996. Patterns and emerging mechanisms of the angiogenic switch during tumorigenesis. Cell 86: 353-364.

Hon WC, Wilson MI, Harlos K, Claridge TD, Schofield CJ, et al. 2002. Structural basis for the recognition of hydroxyproline in HIF-1 alpha by pVHL. Nature 417: 975-978.

Hsu SC, Volpert OV, Steck PA, Mikkelsen T, Polverini PJ, et al. 1996. Inhibition of angiogenesis in human glioblastomas by chromosome 10 induction of thrombospondin-1. Cancer Res 56: 5684-5691.

Ingber DE, Prusty D, Sun Z, Betensky H, Wang N. 1995. Cell shape, cytoskeletal mechanics, and cell cycle control in angiogenesis. J Biomech 28: 1471-1484.

Jeffes EW, Zhang JG, Hoa N, Petkar A, Delgado C, et al. 2005. Antiangiogenic drugs synergize with a membrane macrophage colony-stimulating factor-based tumor vaccine to therapeutically treat rats with an established malignant intracranial glioma. J Immunol 174: 2533-2543.

Jones N, Iljin K, Dumont DJ, Alitalo K. 2001. Tie receptors: New modulators of angiogenic and lymphangiogenic responses. Nat Rev Mol Cell Biol 2: 257-267.

Kamiyama H, Takano S, Tsuboi K, Matsumura A. 2005. Anti-angiogenic effects of SN38 (active metabolite of irinotecan): Inhibition of hypoxia-inducible factor 1 alpha (HIF-1alpha)/vascular endothelial growth factor (VEGF) expression of glioma and growth of endothelial cells. J Cancer Res Clin Oncol 131: 205-213.

Kaur B, Brat DJ, Calkins CC, Van Meir EG. 2003. Brain angiogenesis inhibitor 1 is differentially expressed in normal brain and glioblastoma independently of p53 expression. Am J Pathol 162: 19-27.

Khatua S, Peterson KM, Brown KM, Lawlor C, Santi MR, et al. 2003. Overexpression of the EGFR/FKBP12/HIF-2alpha pathway identified in childhood astrocytomas by angiogenesis gene profiling. Cancer Res 63: 1865-1870.

Kirsch M, Strasser J, Allende R, Bello L, Zhang J, et al. 1998. Angiostatin suppresses malignant glioma growth in vivo. Cancer Res 58: 4654-4659.

Klagsbrun M, Knighton D, Folkman J. 1976. Tumor angiogenesis activity in cells grown in tissue culture. Cancer Res 36: 110-114.

Kleihues P, Burger PC, Collins V, Newcomb EW, Ohgaki H, et al. 2000. Glioblastoma. Pathology and Genetics of Tumours of the Nervous System. Kleihues P, Cavenee WK, editors. Lyon: IARC Press; pp. 29-39.

Korkolopoulou P, Patsouris E, Konstantinidou AE, Pavlopoulos PM, Kavantzas N, et al. 2004. Hypoxia-inducible factor 1alpha/vascular endothelial growth factor axis in astrocytomas. Associations with microvessel morphometry, proliferation and prognosis. Neuropathol Appl Neurobiol 30: 267-278.

Lakka SS, Rao JS. 2005. Role and Regulation of Proteases in Human Glioma. Proteases in the Brain. Lendeckel U, Hooper NM, editors. New York: Springer; pp. 151-177.

Lakka SS, Gondi CS, Yanamandra N, Dinh DH, Olivero WC, et al. 2003. Synergistic down-regulation of urokinase plasminogen activator receptor and matrix metalloproteinase-9 in SNB19 glioblastoma cells efficiently inhibits glioma cell invasion, angiogenesis, and tumor growth. Cancer Res 63: 2454-2461.

Lakka SS, Gondi CS, Yanamandra N, Olivero WC, Dinh DH, et al. 2004. Inhibition of cathepsin B and MMP-9 gene expression in glioblastoma cell line via RNA interference reduces tumor cell invasion, tumor growth and angiogenesis. Oncogene 23: 4681-4689.

Leon SP, Folkerth RD, Black PM. 1996. Microvessel density is a prognostic indicator for patients with astroglial brain tumors. Cancer 77: 362-372.

Li VW, Folkerth RD, Watanabe H, Yu C, Rupnick M, et al. 1994. Microvessel count and cerebrospinal fluid basic fibroblast growth factor in children with brain tumours. Lancet 344: 82-86.

Li J, Yen C, Liaw D, Podsypanina K, Bose S, et al. 1997. PTEN, a putative protein tyrosine phosphatase gene mutated in human brain, breast, and prostate cancer. Science 275: 1943-1947.

Ma HI, Lin SZ, Chiang YH, Li J, Chen SL, et al. 2002. Intratumoral gene therapy of malignant brain tumor in a rat model with angiostatin delivered by adeno-associated viral (AAV) vector. Gene Ther 9: 2-11.

Maehama T, Dixon JE. 1998. The tumor suppressor, PTEN/MMAC1, dephosphorylates the lipid second messenger, phosphatidylinositol 3,4,5-trisphosphate. J Biol Chem 273: 13375-13378.

Maher EA, Furnari FB, Bachoo RM, Rowitch DH, Louis DN, et al. 2001. Malignant glioma: Genetics and biology of a grave matter. Genes Dev 15: 1311-1333.

Maity A, Pore N, Lee J, Solomon D, O'Rourke DM. 2000. Epidermal growth factor receptor transcriptionally up-regulates vascular endothelial growth factor expression in human glioblastoma cells via a pathway involving phosphatidylinositol 3′-kinase and distinct from that induced by hypoxia. Cancer Res 60: 5879-5886.

Maxwell M, Naber SP, Wolfe HJ, Hedley-Whyte ET, Galanopoulos T, et al. 1991. Expression of angiogenic growth factor genes in primary human astrocytomas may contribute to their growth and progression. Cancer Res 51: 1345-1351.

Millauer B, Shawver LK, Plate KH, Risau W, Ullrich A. 1994. Glioblastoma growth inhibited in vivo by a dominant-negative Flk-1 mutant. Nature 367: 576-579.

Millauer B, Wizigmann-Voos S, Schnurch H, Martinez R, Moller NP, et al. 1993. High affinity VEGF binding and developmental expression suggest Flk-1 as a major regulator of vasculogenesis and angiogenesis. Cell 72: 835-846.

Morabito A, Fanelli M, Carillio G, Gattuso D, Sarmiento R, et al. 2004. Thalidomide prolongs disease stabilization after conventional therapy in patients with recurrent glioblastoma. Oncol Rep 11: 93-95.

Morrison RS, Yamaguchi F, Bruner JM, Tang M, McKeehan W, et al. 1994. Fibroblast growth factor receptor gene expression and immunoreactivity are elevated in human glioblastoma multiforme. Cancer Res 54: 2794-2799.

Naski MC, Ornitz DM. 1998. FGF signaling in skeletal development. Front Biosci 3: d781-d794.

Nirmala C, Jasti SL, Sawaya R, Kyritsis AP, Konduri SD, et al. 2000. Effects of radiation on the levels of MMP-2, MMP-9 and TIMP-1 during morphogenic glial-endothelial cell interactions. Int J Cancer 88: 766-771.

Nishimori H, Shiratsuchi T, Urano T, Kimura Y, Kiyono K, et al. 1997. A novel brain-specific p53-target gene, BAI1, containing thrombospondin type 1 repeats inhibits experimental angiogenesis. Oncogene 15: 2145-2150.

O'Leary J, Muggia FM. 1998. Camptothecins: A review of their development and schedules of administration. Eur J Cancer 34: 1500-1508.

Oehring RD, Miletic M, Valter MM, Pietsch T, Neumann J, et al. 1999. Vascular endothelial growth factor (VEGF) in astrocytic gliomas–a prognostic factor? J Neurooncol 45: 117-125.

Pietsch T, Valter MM, Wolf HK, von Deimling A, Huang HJ, et al. 1997. Expression and distribution of vascular endothelial growth factor protein in human brain tumors. Acta Neuropathol (Berl) 93: 109-117.

Plate KH. 1999. Mechanisms of angiogenesis in the brain. J Neuropathol Exp Neurol 58: 313-320.

Plate KH, Risau W. 1995. Angiogenesis in malignant gliomas. Glia 15: 339-347.

Plate KH, Breier G, Weich HA, Risau W. 1992. Vascular endothelial growth factor is a potential tumour angiogenesis factor in human gliomas in vivo. Nature 359: 845-848.

Plate KH, Breier G, Weich HA, Mennel HD, Risau W. 1994. Vascular endothelial growth factor and glioma angiogenesis: Coordinate induction of VEGF receptors, distribution of VEGF protein and possible in vivo regulatory mechanisms. Int J Cancer 59: 520-529.

Roberts WG, Whalen PM, Soderstrom E, Moraski G, Lyssikatos JP, et al. 2005. Antiangiogenic and antitumor activity of a selective PDGFR tyrosine kinase inhibitor, CP-673,451. Cancer Res 65: 957-966.

Ruggeri B, Singh J, Gingrich D, Angeles T, Albom M, et al. 2003. CEP-7055: A novel, orally active pan inhibitor of vascular endothelial growth factor receptor tyrosine kinases with potent antiangiogenic activity and antitumor efficacy in preclinical models. Cancer Res 63: 5978-5991.

Saleh M, Stacker SA, Wilks AF. 1996. Inhibition of growth of C6 glioma cells in vivo by expression of antisense vascular endothelial growth factor sequence. Cancer Res 56: 393-401.

Samoto K, Ikezaki K, Ono M, Shono T, Kohno K, et al. 1995. Expression of vascular endothelial growth factor and its possible relation with neovascularization in human brain tumors. Cancer Res 55: 1189-1193.

Sandstrom M, Johansson M, Andersson U, Bergh A, Bergenheim AT, et al. 2004. The tyrosine kinase inhibitor ZD6474 inhibits tumour growth in an intracerebral rat glioma model. Br J Cancer 91: 1174-1180.

Sasaki M, Wizigmann-Voos S, Risau W, Plate KH. 1999. Retrovirus producer cells encoding antisense VEGF prolong survival of rats with intracranial GS9L gliomas. Int J Dev Neurosci 17: 579-591.

Schmidt NO, Westphal M, Hagel C, Ergun S, Stavrou D, et al. 1999. Levels of vascular endothelial growth factor, hepatocyte growth factor/scatter factor and basic fibroblast growth factor in human gliomas and their relation to angiogenesis. Int J Cancer 84: 10-18.

Schober R, Bilzer T, Waha A, Reifenberger G, Wechsler W, et al. 1995. The epidermal growth factor receptor in glioblastoma: Genomic amplification, protein expression, and patient survival data in a therapeutic trial. Clin Neuropathol 14: 169-174.

Shweiki D, Itin A, Soffer D, Keshet E. 1992. Vascular endothelial growth factor induced by hypoxia may mediate hypoxia-initiated angiogenesis. Nature 359: 843-845.

Steck PA, Pershouse MA, Jasser SA, Yung WK, Lin H, et al. 1997. Identification of a candidate tumour suppressor gene, MMAC1, at chromosome 10q23.3 that is mutated in multiple advanced cancers. Nat Genet 15: 356-362.

Stratmann A, Risau W, Plate KH. 1998. Cell type-specific expression of angiopoietin-1 and angiopoietin-2 suggests a role in glioblastoma angiogenesis. Am J Pathol 153: 1459-1466.

Suri C, McClain J, Thurston G, McDonald DM, Zhou H, et al. 1998. Increased vascularization in mice overexpressing angiopoietin-1. Science 282: 468-471.

Taga T, Suzuki A, Gonzalez-Gomez I, Gilles FH, Stins M, et al. 2002. alpha v-Integrin antagonist EMD 121974 induces apoptosis in brain tumor cells growing on vitronectin and tenascin. Int J Cancer 98: 690-697.

Takahashi JA, Mori H, Fukumoto M, Igarashi K, Jaye M, et al. 1990. Gene expression of fibroblast growth factors in human gliomas and meningiomas: Demonstration of cellular source of basic fibroblast growth factor mRNA and peptide in tumor tissues. Proc Natl Acad Sci USA 87: 5710-5714.

Tenan M, Fulci G, Albertoni M, Diserens AC, Hamou MF, et al. 2000. Thrombospondin-1 is downregulated by anoxia and suppresses tumorigenicity of human glioblastoma cells. J Exp Med 191: 1789-1798.

Thurston G, Suri C, Smith K, McClain J, Sato TN, et al. 1999. Leakage-resistant blood vessels in mice transgenically overexpressing angiopoietin-1. Science 286: 2511-2514.

Tse V, Xu L, Yung YC, Santarelli JG, Juan D, et al. 2003. The temporal-spatial expression of VEGF, angiopoietins-1 and 2, and Tie-2 during tumor angiogenesis and their functional correlation with tumor neovascular architecture. Neurol Res 25: 729-738.

Ueba T, Nosaka T, Takahashi JA, Shibata F, Florkiewicz RZ, et al. 1994. Transcriptional regulation of basic fibroblast growth factor gene by p53 in human glioblastoma and hepatocellular carcinoma cells. Proc Natl Acad Sci USA 91: 9009-9013.

Vajkoczy P, Menger MD, Vollmar B, Schilling L, Schmiedek P, et al. 1999. Inhibition of tumor growth, angiogenesis, and microcirculation by the novel Flk-1 inhibitor SU5416 as assessed by intravital multi-fluorescence videomicroscopy. Neoplasia 1: 31-41.

Van Meir EG, Polverini PJ, Chazin VR, Su Huang HJ, de Tribolet N, et al. 1994. Release of an inhibitor of angiogenesis upon induction of wild type p53 expression in glioblastoma cells. Nat Genet 8: 171-176.

Veikkola T, Karkkainen M, Claesson-Welsh L, Alitalo K. 2000. Regulation of angiogenesis via vascular endothelial growth factor receptors. Cancer Res 60: 203-212.

Venetsanakos E, Mirza A, Fanton C, Romanov SR, Tlsty T, et al. 2002. Induction of tubulogenesis in telomerase-immortalized human microvascular endothelial cells by glioblastoma cells. Exp Cell Res 273: 21-33.

Vogelstein B, Kinzler KW. 1992. p53 function and dysfunction. Cell 70: 523-526.

Wen S, Stolarov J, Myers MP, Su JD, Wigler MH, et al. 2001. PTEN controls tumor-induced angiogenesis. Proc Natl Acad Sci USA 98: 4622-4627.

Wiesener MS, Turley H, Allen WE, Willam C, Eckardt KU, et al. 1998. Induction of endothelial PAS domain protein-1 by hypoxia: Characterization and comparison with hypoxia-inducible factor-1alpha. Blood 92: 2260-2268.

Woods SA, McGlade CJ, Guha A. 2002. Phosphatidylinositol 3′-kinase and MAPK/ERK kinase 1/2 differentially regulate expression of vascular endothelial growth factor in human malignant astrocytoma cells. Neuro-oncol 4: 242-252.

Yanamandra N, Gumidyala KV, Waldron KG, Gujrati M, Olivero WC, et al. 2004. Blockade of cathepsin B expression in human glioblastoma cells is associated with suppression of angiogenesis. Oncogene 23: 2224-2230.

Yancopoulos GD, Davis S, Gale NW, Rudge JS, Wiegand SJ, et al. 2000. Vascular-specific growth factors and blood vessel formation. Nature 407: 242-248.

Yoshida D, Takahashi H, Teramoto A. 2004. Inhibition of glioma angiogenesis and invasion by SI-27, an anti-matrix metalloproteinase agent in a rat brain tumor model. Neurosurgery 54: 1213-1220.

Zadeh G, Qian B, Okhowat A, Sabha N, Kontos CD, et al. 2004. Targeting the Tie2/Tek receptor in astrocytomas. Am J Pathol 164: 467-476.

Zagzag D, Hooper A, Friedlander DR, Chan W, Holash J, et al. 1999. In situ expression of angiopoietins in astrocytomas identifies angiopoietin-2 as an early marker of tumor angiogenesis. Exp Neurol 159: 391-400.

Zagzag D, Zhong H, Scalzitti JM, Laughner E, Simons JW, et al. 2000. Expression of hypoxia-inducible factor 1 alpha in brain tumors: Association with angiogenesis, invasion, and progression. Cancer 88: 2606-2618.

2 Adrenoleukodystrophy: Molecular, Metabolic, Pathologic, and Therapeutic Aspects

M. A. Contreras · I. Singh

1	Introduction	14
1.1	X-ALD as a Peroxisomal Disorder	14
1.2	Pathobiology of X-ALD	16
2	*Progression of Disease in X-ALD*	17
2.1	Molecular Basis of the Metabolic Disease	17
2.2	Molecular Basis of the Inflammatory Disease in X-ALD	20
3	*Therapeutics for X-ALD*	31
4	*Conclusions and Perspectives*	36

© 2009 Springer Science+Business Media, LLC.

Abstract: Adrenoleukodystrophy (ALD), an inherited metabolic disorder linked to chromosome X (X-ALD), is the most common peroxisomal disorder that affects boys in early childhood as a result of mutation/deletion of adrenoleukodystrophy protein (ALDP), a peroxisomal membrane protein, leading to deficient activity of peroxisomal very long-chain (VLC) acyl-CoA synthetase (VLCS). The pathognomonic accumulation of straight chain, saturated VLC fatty acids (i.e., made up of more than 22 carbon atoms) is the hallmark of this disease. The accumulation of VLC fatty acids is the result of an abnormality in their degradation. Reduced activity of the enzyme VLCS that activates VLC fatty acids to their CoA derivatives, the substrate for the β-oxidation cycle in peroxisomes, results in pathognomonic accumulation of VLC fatty acids in tissues such as the adrenals, testes, and brain in X-ALD patients. Consequently, they are the cause of the metabolic disease in X-ALD.

The characteristic metabolic abnormality of X-ALD can be detected during the prenatal stage, but clinical progression of X-ALD is slow until early childhood (6–8 years) where metabolic disease manifests as neurological disease leading to childhood X-ALD (cALD) characterized by neurodegeneration (loss of oligodendrocytes and myelin), loss of physical function, and death. Alternatively, X-ALD can develop as a late-onset disease as adrenomyeloneuropathy (AMN), characterized by a slower course of neurodegeneration that affects the peripheral nervous system, i.e., spinal cord nerve tracts. cALD and AMN accumulates VLC fatty acids; however, in cALD the metabolic disease quickly progresses into neuroinflammatory disease, leading to physical impairment and death due to severe neurological dysfunction. Both forms of X-ALD develop an inflammatory process that affects the white matter of brain in cALD and spinal cord nerve (axons) tracts in AMN. Since both variants of ALD can be derived from the same mutation in the *ALD* gene, these differing manifestations of the disease suggest the involvement of a secondary event/modifier gene.

The onset of clinical disease in cALD correlates with that of the inflammatory demyelinating disease process. The detection of cytokines/chemokines and accumulation of VLC fatty acids containing lipids in histologically normal areas suggest a role for these molecules in the induction of neuroinflammatory disease in X-ALD. Moreover, mounting experimental evidence indicates that inflammatory disease mediators secreted by activated glial cells (astrocytes, microglia, and macrophages) subsequently alter peroxisome functions, thus further compromising peroxisomal metabolism such as metabolism of H_2O_2, D-amino acids, and arachidonic acid metabolites (eicosanoids) and synthesis of plasmalogens and docosahexaenoic acid which in turn may compromise oligodendrocytes and myelin. The release of proinflammatory soluble products (cytokines, nitric oxide, eicosanoids, and free radicals) during the progression of inflammation is likely to play a role in the pathogenesis of inflammatory disease. These mediators have been identified in X-ALD and may be considered key players in the transition of metabolic to neuroinflammatory disease, and represent a target for pharmacologic intervention to ameliorate and/or halt the progression of the disease. Therefore, the ideal therapeutic approach for X-ALD should exert a dual role: correction of the metabolic defect and attenuation/halting of the inflammatory process.

In this chapter, we present an overview of the current knowledge of the molecular aspects related to the metabolic and inflammatory disease processes of X-ALD as well as some of the latest pharmacologic approaches that may become available, in the near future, for the treatment of this disease.

List of Abbreviations: ABC, ATP-binding cassette; ALDP, adrenoleukodystrophy protein; AMN, Adrenomyeloneuropathy; cALD, Childhood X-ALD; iNOS, inducible nitric oxide synthase; LCS, LC acyl-CoA synthetase; MS, multiple sclerosis; NF-κB, nuclear factor-κB; OPCs, oligodendrocyte precursor cells; P70R, PMP-70-related protein; PMP-70, peroxisomal membrane protein of 70 kDa; VCAM, vascular cell adhesion molecule; VLC, very long-chain; VLCS, VLC acyl-CoA synthetase; X-ALD, X-linked adrenoleukodystrophy

1 Introduction

1.1 X-ALD as a Peroxisomal Disorder

The peroxisome was the latest subcellular organelle to be discovered and its name derived from the fact that it has enzymatic activities that both produce (D-amino acid oxidases) and degrade (catalase) H_2O_2

Table 2-1
Metabolic functions of peroxisomes

A. Peroxisomes catabolic functions

β-Oxidation
  Long-chain fatty acids (>C16)
  Very long-chain fatty acids (>C22)
  Mono- and polyunsaturated fatty acids (erucic [C22:1 n-9], adrenic [C22:4 n-6], and tetracosanoic [C24:4 n-6] acids)
  Branched-chain fatty acids (pristanic acid)
  Arachidonic acid metabolites (eicosanoids, i.e., prostaglandins, thromboxanes, leukotrienes, HETE)
  Long-chain dicarboxylic acids
  Xenobiotics (aliphatic side chains)

α-Oxidation
  Phytanic acid

Others catabolic functions
  D-Amino acids oxidation (D-serine, D-aspartate, N-methyl-D-aspartate)
  L-α-Hydroxyacid oxidation (glycolate, DL-α-hydroxybutyrate)
  Hydrogen peroxide
  Polyamines
  Free radicals

B. Peroxisomes anabolic functions

Synthesis
  Bile acids (chenodeoxycholoyl-CoA, choloyl-CoA)
  Long-chain alcohol (hexadecanol)
  Plasmalogens (ether-phospholipids)
  Hydrogen peroxide
  Polyunsaturated fatty acids (docosahexaenoic [C22:6 n-3] and docosapentaenoic [C22:5 n-6] acids)
  Cholesterol
  NADH + H

(Baudhuin et al., 1964). Over the last 45 years peroxisomes have been recognized to perform important metabolic functions for lipids, hydrogen peroxide, D-amino acids, and uric acid metabolism (⯁ Table 2-1). The significance of peroxisomes to human health is further underscored by the identification of neurological disorders (known as peroxisomal disorders) (⯁ Table 2-2) caused by defects in the assembly or the deficiency of individual peroxisomal enzymes (Brown et al., 1993; Gould, et al., 2001; Wanders et al., 2001). Peroxisomal disorders are classified into two groups: A group of diseases caused by defects in the assembly/biogenesis of peroxisomes showing multiple abnormalities and a second group of diseases caused by deficiency of a single enzyme activity. The most common peroxisomal disorder caused by an individual enzyme deficiency is X-adrenoleukodystrophy (X-ALD), a childhood disorder characterized by abnormal very long-chain (VLC) fatty acids catabolism (Singh et al., 1984a; Lazo et al., 1988; Lazo et al., 1989), and concomitant accumulation of these fatty acids in tissues such as brain, adrenals, testes, and plasma (Singh et al., 1984b; Singh, 1997; Wanders, 1999; Moser et al., 2001). X-ALD is caused by mutations/deletions in the ALD gene that encodes a 74-kDa peroxisomal membrane protein, adrenoleukodystrophy protein (ALDP) (Mosser et al., 1993; Smith et al., 1999; Kemp et al., 2001). ALDP has significant homology to the members of the ATP-binding cassette (ABC) family of transporters (Valle and Gartner, 1993), but its precise function in the fatty acid β-oxidation pathway is not well understood at present. However, the absence of ALDP leads to accumulation of VLC fatty acids and supplementation of the *ALD* gene to ALD cells corrects the metabolic defect of VLC fatty acid accumulation (Braiterman et al., 1998).

X-ALD is caused by mutation in the *ABCD1* gene that inactivates the peroxisomal ALD protein (Mosser et al., 1993), hence altering its peroxisomal transport function (Gartner et al., 2002). The described

Table 2-2
Peroxisomal disorders

A. Peroxisome biogenesis disorders

Zellweger spectrum
1. Severe: Zellweger syndrome
2. Mild: Neonatal adrenoleukodystrophy Infantile Refsum disease
3. Classical rhizomelic chondrodysplasia punctata (RCDP type 1)

B. Single enzyme deficiencies

Disorders of fatty acid β-oxidation
1. X-Linked adrenoleukodystrophy
2. Acyl-CoA oxidase deficiency
3. D-Bifunctional protein deficiency
4. Thiolase deficiency
5. 2-methylacyl-CoA racemase deficiency

Disorders of ether-phospholipids (plasmalogens) biosynthesis
6. Dihydroxyacetonephosphate acyltransferase (DHAPAT) deficiency (RCDP type 2)
7. Alkyldihydroxyacetonephosphate synthase (alkyl DHAP synthase) deficiency (RCDP type 3)

Disorder of fatty acid α-oxidation
8. Phytanoyl-CoA hydroxylase deficiency

Disorder of isoprenoid biosynthesis
9. Mevalonate kinase deficiency

Disorder of L-lysine metabolism
10. L-Pipecolate oxidase deficiency

Disorder of glutaryl-CoA metabolism
11. Glutaryl-CoA oxidase deficiency

Disorder of hydrogen peroxide metabolism
12. Catalase deficiency

Disorder of glyoxylate detoxification
13. Alanine/glyoxylate aminotransferase deficiency

---

mutations are distributed all over the gene and include deletions, missense, nonsense, frameshift, and splice defects (Berger et al., 1994; Kemp et al., 1994; Mosser et al., 1994; Kok et al., 1995; Moser, 1997; Moser et al., 1999; Dvorakova et al., 2001; Kemp et al., 2001; Matsumoto et al., 2003; Coll et al., 2005; Montagna et al., 2005). No correlation has been observed between the nature of the mutation and the phenotype (Takano et al., 2000). A number of studies have reported that identical mutations in the *ALD* gene can be associated with each of the major clinical phenotypes, i.e., childhood X-ALD (cALD) and adrenomyeloneuropathy (AMN) (Berger et al., 1994; Kemp et al., 1994; Kok et al., 1995).

1.2 Pathobiology of X-ALD

Genetic-based accumulation of VLC fatty acids in amniocytes is the basis of prenatal diagnosis of X-ALD, and X-ALD children are born with ALD metabolic disease. Clinical disease develops during childhood (cALD), adolescence, or adulthood (AMN), indicating the involvement of a secondary event/modifier gene(s) for the disease. The fact that VLC fatty acids start to accumulate during fetal life (Moser and Moser, 1999) and that this process precedes the development of brain pathology in X-ALD, together with the fact that no correlation appears to exist between the nature of the mutations in ALD gene and the phenotypes (❯ *Figure 2-1*), suggest that a modifier gene might account for some of the phenotypic variability (Smith et al., 1999). Furthermore, genetic segregation analyses support the hypothesis of a modifier gene in X-ALD

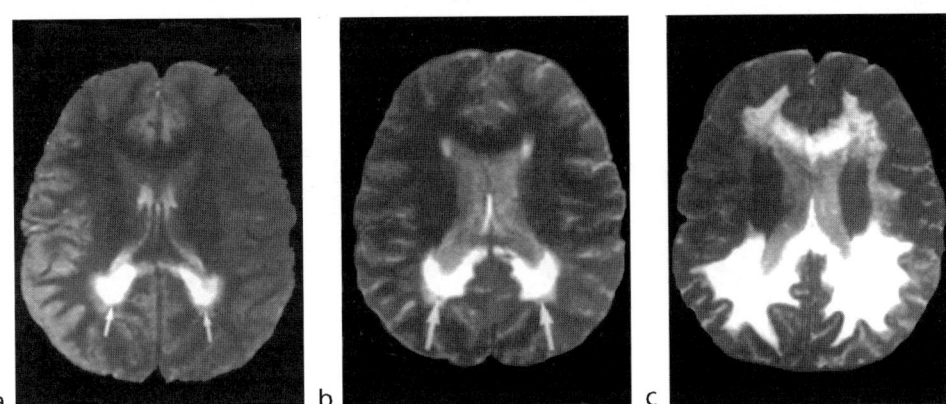

◘ Figure 2-1
Magnetic resonance imaging showing the course of progression of demyelination in the occipital cerebral form of adrenoleukodystrophy (cALD) over a period of 3 years (white arrows). (a) age seven; (b) age eight; (c) age nine. There is a marked neurological deterioration at (c) nine years of age. The progression of demyelinating lesions toward frontal lobes is seen from (a) to (c). Reprinted, with permission, from Elsevier (Aubourg, 1996)

(Maestri and Beaty, 1992; Moser et al., 1992). Therefore, a two hit hypothesis seems plausible for the development of clinical ALD: transition of the metabolic to inflammatory disease requires a trigger component/event (Wilkinson et al., 1987).

2 Progression of Disease in X-ALD

2.1 Molecular Basis of the Metabolic Disease

X-ALD belongs to the group of peroxisomal disorders characterized by impairment of a single enzyme function (Singh, 1997; Moser et al., 2001; Wanders et al., 2001). As mentioned above, the biochemical hallmark in X-ALD is the accumulation of VLC fatty acids in certain tissues and plasma (Moser et al., 2001), as a consequence of impaired metabolism in peroxisomes (Singh et al., 1984a; Lazo et al., 1988; Wanders et al., 1988). Initial studies indicated a defect in the peroxisomal β-oxidation system, a result of defective activation of VLC fatty acids to their CoA derivatives, the first and obligatory reaction in the pathway for its degradation (Lazo et al., 1988; Wanders et al., 1988). This reaction is catalyzed by a VLC acyl-CoA synthetase (VLCS), an enzyme localized in peroxisomes and endoplasmic reticulum (Lazo et al., 1990b; Smith et al., 2000). The gene for the rat liver VLCS has been identified (Uchiyama et al., 1996) and its sequence has surprisingly high homology to mouse fatty acid transport protein (Hirsch et al., 1998) than to others fatty acid synthetases (Suzuki et al., 1990). The high level of homology between VLCS and the fatty acid transport protein present in adipocyte (Schaffer and Lodish, 1994; Hirsch et al., 1998) adds another level of complexity in establishing the relationship between ALDP mutation and the abnormality of VLCS activity. The discovery of the VLCS gene (Uchiyama et al., 1996) and the availability of antibodies against this enzyme demonstrated that the enzyme was present in the peroxisomes of ALD patients and confirmed that its active site and its C-terminal end are oriented on the luminal side of the peroxisomal membrane (❯ *Figure 2-2*), facing the peroxisomal matrix (Lazo et al., 1990a; Steinberg et al., 1999; Smith et al., 2000). The other enzyme that has specificity for long-chain fatty acids is called LC acyl-CoA synthetase (LCS). The active site of this enzyme is localized on the cytosolic surface of the peroxisomal membrane (Mannaerts et al., 1982; Lazo et al., 1990a). Therefore, LC fatty acids are first activated to CoA, derived outside peroxisomes, before they reach the peroxisomal matrix for β-oxidation. On the other hand, VLC fatty acids reach the luminal side of the peroxisomal membrane as such (❯ *Figure 2-2*) and are activated by the

Figure 2-2
Molecular participants for the transport/activation of very long-chain fatty acids (VLCFA) prior to metabolism by the β-oxidation cycle in peroxisomes. The X represent an absent or nonfunctional ALDP present in X-ALD patients

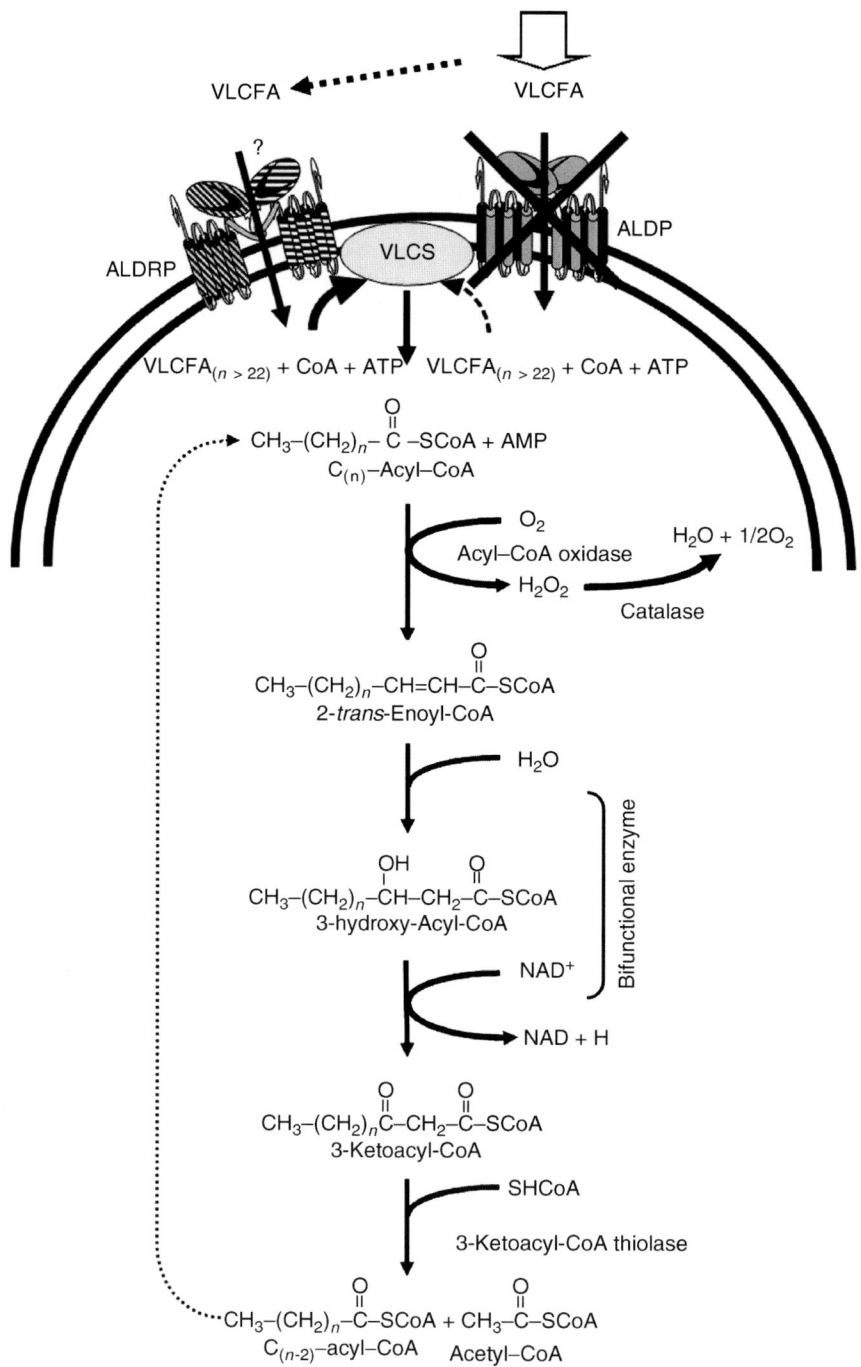

VLCS for their availability to the β-oxidation enzyme system in the peroxisomal matrix (Singh et al., 1992). These observations are consistent with the observed higher levels of transport of free lignoceric acid and palmitoyl-CoA as compared with lignoceroyl-CoA and palmitic acid in peroxisomes in in vitro studies (Singh et al., 1992).

The membrane of peroxisomes also contains integral membrane proteins with sequence homology to members of the ABC family of transporters (Kamijo et al., 1990; Mosser et al., 1993; Valle and Gartner, 1993; Berger et al., 1999). The individual function of these half transporter proteins, either as a homodimer or as a heterodimer, is not well understood at this time. Four genes encoding for peroxisomal ABC half transporters have been identified. Peroxisomal membrane protein of 70 kDa (PMP-70) is one of the most abundant in the peroxisomal membrane (Kamijo et al., 1990). As described above, ALDP is another gene product that based on protein sequence homology belongs to the ABC family of transporters (Mosser et al., 1993). Mutations at this gene are the cause of X-linked adrenoleukodystrophy. Although the function of this transporter is not well understood, the accumulation of VLC fatty acids and the impaired activity of the enzyme that activates them (i.e., VLCS) suggests that ALDP is essential for proper activation and metabolism of VLC fatty acids. The remaining peroxisomal transporters are two proteins related to PMP-70 and ALDP. The PMP-70-related protein (P70R) has homology to PMP-70 (Holzinger et al., 1997; Shani et al., 1997) and ALD-related protein (ALDRP) has homology to ALDP (Lombard-Platet et al., 1996). No disorder has been associated with deficiency of either P70R or ALDRP. However, overexpression of PMP-70, ALDRP, or P70R or induction of ALDRP leads to restoration of VLC fatty acid β-oxidation in human fibroblasts, indicating a promiscuous property of these transporters (Braiterman et al., 1998; Netik et al., 1999). The differential tissue expression of ALDRP and ALDP and of ALDP and PMP-70 suggests that the transporters may function as heterodimers as well as homodimers in the peroxisomal membrane (Holzinger et al., 1997; Troffer-Charlier et al., 1998; Berger et al., 1999). The pharmacological upregulation of ALDRP has added a new a dimension to the clinical treatment of this disease (Kemp et al., 1998; Singh et al., 1998a; Singh et al., 1998b; Pai et al., 2000). Formation of homodimer and heterodimer complexes between ALDP, ALDRP, and PMP-70 transporters has been found using different approaches (Liu et al., 1999; Guimaraes et al., 2004). Therefore, it is reasonable to suggest that heterodimer transporters might bring broad selectivity to the transport process in the peroxisome membrane (Valle and Gartner, 1993). It has been demonstrated that overexpression of PMP70 in CHO cells increased the β-oxidation of palmitic (C16:0) over lignoceric acid (C24:0), suggesting that PMP70 is involved in metabolic transport of long-chain acyl-CoA across peroxisomal membranes (Imanaka et al., 1999). On the other hand, transfection or transfer of ALDP or ALDRP genes corrects the defect in VLC fatty acid metabolism (❯ *Figure 2-2*) (Cartier et al., 1995; Braiterman et al., 1998; Flavigny et al., 1999; Netik et al., 1999). Since ALDP can bind (Contreras et al., 1996) and hydrolyze (Roerig et al., 2001; Tanaka et al., 2002) ATP, this suggests that this protein is a functional peroxisomal transporter. However, the association of any particular function in peroxisomes with hydrolysis of ATP by ALDP is not known at the present time.

The ALD gene encodes a 745-amino acid protein (ALDP) containing sequence homology to the ABC transporter proteins (Higgins, 1992; Mosser et al., 1993). ALDP normally localized in the peroxisomal membrane was found to be absent in the peroxisomal membrane of fibroblasts in ALD patients carrying mutation and deletions in the ALD gene (Contreras et al., 1994; Mosser et al., 1994). Further studies confirmed the transmembrane topology of ALDP in the peroxisomal membrane. ALDP consists of six membrane-spanning domains and an ATP-binding domain located on the outside of the peroxisomal membrane (Contreras et al., 1996). The last domain is functionally active in binding and hydrolysis of ATP (Contreras et al., 1996; Roerig et al., 2001; Tanaka et al., 2002). The fact that ALDP is a half transporter (Mosser et al., 1993) and that it may form heterodimers or homodimers (Valle and Gartner, 1993; Liu et al., 1999) suggests that its activity may be related to the assembly of or the transport of a cofactor(s) required for functional activity of the VLCS (Makkar et al., 2006; Yamada et al., 1999). However, how the affected ALDP correlates with reduced VLCS activity in the peroxisomal membrane is still unclear.

The importance of the saturated VLC fatty acids in human biology and the central nervous system (CNS) is recognized by genetic disorders that affect VLC fatty acids metabolism (Singh, 1997; Gould et al., 2001; Moser et al., 2001), as well as in the changes in the VLC fatty acid profile during brain maturation and aging (Svennerholm, 1968; Svennerholm and Stallberg-Stenhagen, 1968; Giusto et al., 1992). The abundance of saturated and monounsaturated VLC fatty acids in myelin sheaths prompted investigators to suggest a structural role for VLC fatty acids in myelin membranes (O'Brien et al., 1964; O'Brien and Sampson, 1965; Svennerholm and Stallberg-Stenhagen, 1968; Bourre et al., 1977a). Additional support for such a role comes from the studies that suggest a correlation between the maturational changes in enzyme activities of the brain microsomal fatty acid elongation system and the deposition of myelin (Bourre et al., 1977; Murad and Kishimoto, 1978) and that the brain of jumpy and quaking mouse mutants, myelin-deficient mutants, have marked low levels of VLC fatty acids (Bourre et al., 1977b; Suneja et al., 1991). In CNS, VLC fatty acids are esterified to various complex lipids (Svennerholm and Stallberg-Stenhagen, 1968), and under normal circumstances, a substantial amount is present in cerebrosides and sulfatides and sphingomyelin, both in mammalian and nonmammalian cells (Wertz and Downing, 1987; Wertz and Downing, 1989; Oh et al., 1997; Tvrdik et al., 2000). These complex lipids are essential for membrane and cellular functions, such as cell arrest (Hanada et al., 1992; Nickels and Broach, 1996), cell recognition and adhesion (Hakomori and Igarashi, 1995), cell signaling (Spiegel and Merrill, 1996; Testi, 1996), and plasma membrane functionality (Simons and Ikonen, 1997). Under pathological conditions in which VLC fatty acids accumulate, they are found esterified to lipid species such as cholesterol esters, gangliosides, sphingomyelin, phosphatidylcholine (Igarashi et al., 1976; Ramsey et al., 1977; Tsuji et al., 1981b; Antoku et al., 1984; Sharp et al., 1991), and in the myelin proteolipid protein (Bizzozero et al., 1991). Excessive accumulation of VLC fatty acids has been proposed as a possible mechanism that destabilizes membranes and myelin in the CNS of X-ALD patients (Knazek et al., 1983; Whitcomb et al., 1988; Ho et al., 1995; Di Biase et al., 1997). Interestingly, some young patients have increased plasma levels of VLC fatty acid for years and are asymptomatic until adulthood (Aubourg, 1999). This phenomena has also been observed in an animal model of X-ALD: the clinical phenotype does not develop although VLC fatty acid accumulates in brain (Forss-Petter et al., 1997; Kobayashi et al., 1997; Lu et al., 1997; Pujol et al., 2002). These two examples suggest that more than a moderate increase in VLC fatty acid levels is required to induce demyelination.

In addition to the decrease in degradation of VLC fatty acids, increase in elongation (synthesis) of VLC fatty acids has also been reported in X-ALD cells (Tsuji et al., 1981; Tsuji et al., 1984; Street et al., 1990). Enhanced endogenous biosynthesis of VLC fatty acids, owing to increased activity of a microsomal fatty acid elongation system, has been demonstrated in cultured fibroblasts from X-ALD patients (Tsuji et al., 1981a; Tsuji et al., 1984; Street et al., 1990). Since ALDP is a component of peroxisomes but not of microsomes, these findings suggest that changes observed in microsomal fatty acid elongation system are secondary to the primary defect in X-ALD (Kemp et al., 2005).

2.2 Molecular Basis of the Inflammatory Disease in X-ALD

Inflammatory disease of X-ALD is quite complex. Two clinical variants (cALD and AMN), caused by the same mutation, have inflammatory disease affecting different parts of the CNS (Braun et al., 1995; Krasemann et al., 1996; Wichers et al., 1999; O'Neill et al., 2001). In cALD, the inflammation affects the white matter and related cells in the brain, whereas AMN affects the axonal tracts of the spinal cord (Powers et al., 1992; Powers, 1995; Powers et al., 2000; Powers et al., 2001; van Geel et al., 2001). In addition the neuropathology of cALD is associated with the infiltration of vascular inflammatory immune cells (Griffin et al., 1985; Powers et al., 1992) and NO-mediated toxicity as evident from the detection of nitrotyrosine in the brain of X-ALD patients (Gilg et al., 2000). The molecular events associated with the transition from a relatively benign metabolic disease to a fatal neuroinflammatory disease in cALD are not well understood at present. Myelin breakdown products, as a result of myelin instability, containing complex lipids and/or acylated proteins enriched in VLC fatty acids are believed to play a role in this process possibly by inducing CNS resident immune cells (microglia) to produce cytokines/chemokines that further upregulate the cellular events to induce myelin breakdown (Dubois-Dalcq et al., 1999; Aubourg and Dubois-Dalcq, 2000). The infiltrating vascular immune cells may further amplify inflammatory disease leading to loss of oligodendrocytes, myelin, and axons in cALD. Accordingly, monocytes from ALD patients were reported to have altered intrinsic regulatory pathway of TNF-α production when subjected to stimulation with LPS

(Lannuzel et al., 1998; Di Biase et al., 2001). Furthermore, fibroblasts of ALD patients also show altered eicosanoid synthesis (Tiffany et al., 1990). Although the overall inflammatory disease of cALD has many similarities with multiple sclerosis (MS) in terms of immune response and associated inflammatory neurodegeneration, there are significant differences between the two (McGuinness et al., 1997). Lymphocyte infiltration in cALD, mostly T cells and macrophages, tend to be behind the demyelination edge and involved in the subsequent secondary tract degeneration (❯ *Figure 2-3*) (Powers et al., 1992). In contrast, in MS, the inflammatory infiltrates accumulate at the active demyelinating edge, suggesting a different mechanism of disease pathogenesis (Bar-Or et al., 1999). Moreover, the observed expression of proinflammatory cytokines such as TNF-α, IL-1β, and IFN-γ indicates a role for these cytokines in inflammatory demyelination (Powers et al., 1992; McGuinness et al., 1995). Studies from our laboratory, using a combination of super gene array, RT-PCR, and immunohistochemistry techniques, have demonstrated a broad presence of cytokines (IL-1α, TNF-α, IL-6, IL-2, IL-3, and GM-CF) (❯ *Figure 2-4*), chemokines (MCP-1, MCP-3, MIP-1b, SDF-2, fractalkine, eotaxin, MIP-2, MDC, HCC-4) (❯ *Figure 2-5*), chemokine receptors (CCR-1, CCR-2, CCR-4, CCR-5) (❯ *Figure 2-6*), and inducible nitric oxide synthase (iNOS) in regions of cALD brains such as the demyelinating plaque and the active edge of the demyelinating plaque compared with histological apparently unaffected brain areas, with appropriate age-matched control brain samples (Paintlia et al., 2003). Brain tissue in the inflammatory area had higher levels of proinflammatory cytokines (e.g., TNF-α, IL-1β, IL-6, IL-2) and chemokines (e.g., MCP-1, MCP-3, SDF-2). However, histologically normal-looking areas of cALD brain when compared with control brain had increased levels of proinflammatory cytokines (TNF-α, IL-1β, IL-6, IL-2), and chemokines (CCL-2, -4, -7, -11, -16, -21, and -22 and CC-1, -2, -4, -5), indicating a relationship between metabolic abnormality and inflammatory disease in the absence of vascular infiltrates (Paintlia et al., 2003). Furthermore, histopathological examination of cALD sections demonstrated cellular infiltration by macrophages and T cells in the demyelinating plaque and in the active edge of the demyelinating plaque, i.e., area around the plaque (Paintlia et al., 2003). Chemokines act as chemoattractants for vascular immune cells in neuroinflammatory disease (Owens et al., 2005). Studies with transgenic animals for chemokines document that chemokines are sufficient to recruit vascular immune cells into the CNS (Boztug et al., 2002), indicating that the observed higher levels of chemokines in histologically normal-looking areas of ALD brain may be responsible for recruitment of vascular immune cells to ALD brain. The detection of mRNA and the expressed iNOS protein as well as the presence of nitrotyrosine adducts in brain of cALD, as the result of the inflammatory events, support a role for iNOS in the progression of X-ALD brain pathology (Gilg et al., 2000).

The inflammatory mediator-induced downregulation of function of peroxisomes in hepatocytes (Beier et al., 1992), liver (Contreras et al., 2000), C6 glial cells (Khan et al., 1998), and CNS of experimental autoimmune encephalomyelitis (EAE) (an animal model of MS) (Singh et al., 2004) indicate a role for decreased peroxisomes in the pathobiology of neuroinflammatory disease. Cytokine-mediated compromise of peroxisomal integrity and impaired peroxisomal functions (Beier et al., 1992; Contreras et al., 2000; Khan et al., 2000) indicate overall negative consequences for myelin synthesis/repair processes. Higher levels of VLC fatty acids in demyelinating plaque and demyelinating inflammatory region as compared with histologically normal-looking brain region of X-ALD brain indicate that peroxisomal functions were altered by the inflammatory mediators (Paintlia et al., 2003). Accordingly, cytokines compromised the VLC fatty acid β-oxidation and other peroxisomal functions (Khan et al., 1998; Paintlia et al., 2003). Therefore, in cALD, possibly the decrease in peroxisomes and peroxisomal functions (e.g., VLC fatty acids β-oxidation, plasmalogen and docosahexaenoic acid biosyntheses) will further increase the accumulation of VLC fatty acids (Khan et al., 1998), thus amplifying the inflammatory response (production of cytokines) in brain tissue (❯ *Figure 2-7*). As expected, the plaque and inflammatory areas of cALD brain has 2.5- and 3.8-folds VLC fatty acids as compared with histologically normal-looking areas in X-ALD brain (❯ *Figure 2-4*) (Khan et al., 1998; Paintlia et al., 2003). Indeed, studies from our laboratory using cultured C6 glial cells expressing oligodendrocyte-like properties demonstrated that cytokine (TNF-α, IL-1β, INF-γ) treatment of these cells negatively affect the metabolism of VLC fatty acids, inhibiting their oxidation, and thereby inducing the accumulation of these fatty acids (❯ *Figures 2-8–11*) (Khan et al., 1998), suggesting that under activation microglial dysfunction might be enhanced. By using inhibitors of nitric oxide synthase, NO was identified

◘ Figure 2-3

(A) Childhood X-ALD (cALD) brain depicting samples collected from plaque (PA), plaque shadow (IA), and histologically normal-looking area (NLA) in white matter. (B) Hematoxylin staining (HE) demonstrated the accumulation of infiltrates (macrophages and lymphocytes) in vessel (V) and peri-vascular cuffs (PC) in (d) PA and (c) IA clearly visible compared with (b) NLA and (a) control. LFB staining of (h) PA, (g) IA, (f) NLA, and (e) control human brain demonstrated acute demyelination (D) in PA and to a lesser extent in IA. Reprinted, with permission, from Elsevier (Paintlia et al., 2003)

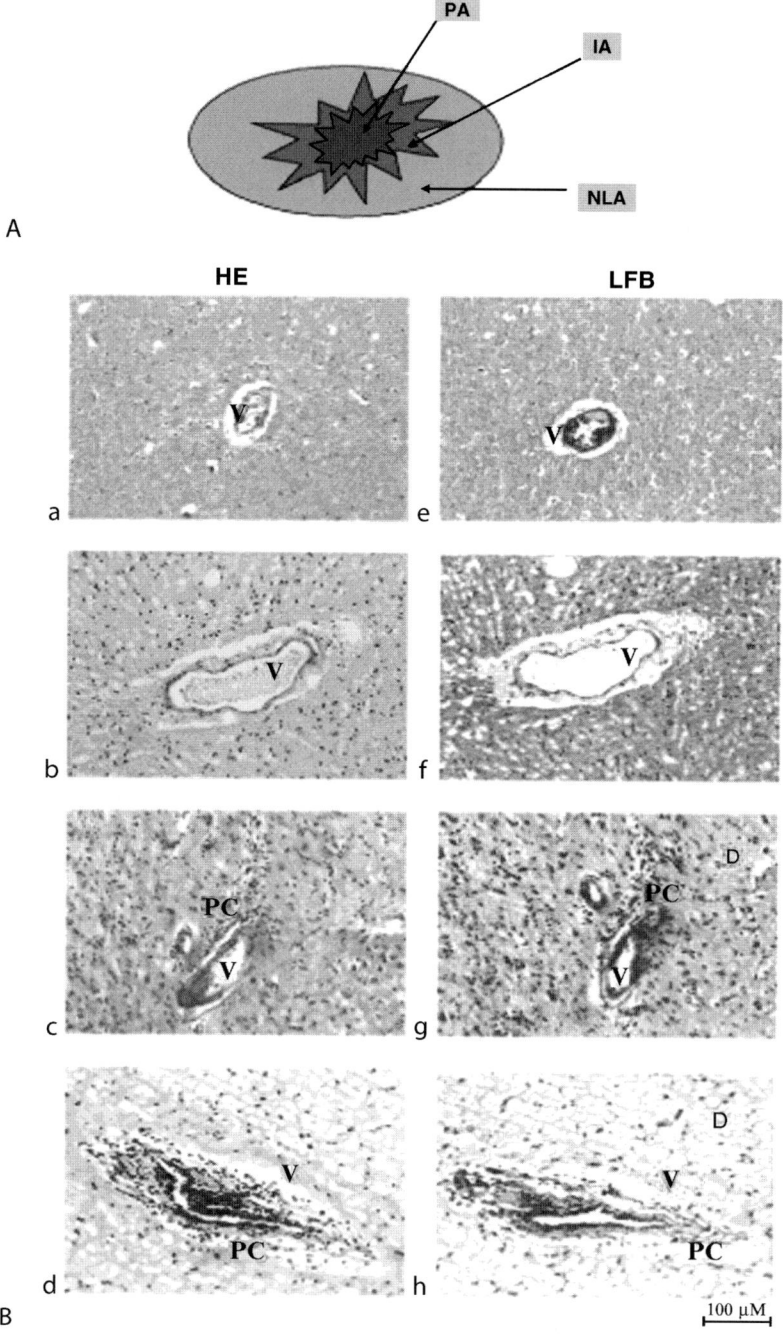

Figure 2-4
(a–f) Inflammatory cytokine expression in plaque (PA), inflammatory (IA), and normal-looking areas (NLA) of childhood X-ALD (cALD) and control brains determined by super gene array. (a) Interleukin-1α, (b) tumor necrosis factor-α, (c) interleukin-6, (d) interleukin-2, (e) interleukin-3, and (f) granulocyte–macrophage colony-stimulating factor. Data are expressed as ratio of cytokine/GAPDH mRNA. *, $P < 0.05$ versus control; **, $P < 0.01$ versus control; ***, $P < 0.001$ versus control; a, $P < 0.05$ versus cALD (PA); b, $P < 0.01$ versus cALD (PA). Reprinted, with permission, from Elsevier (Paintlia et al., 2003)

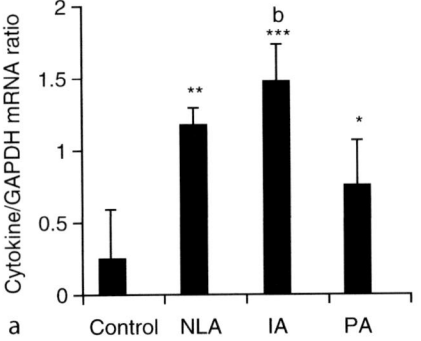
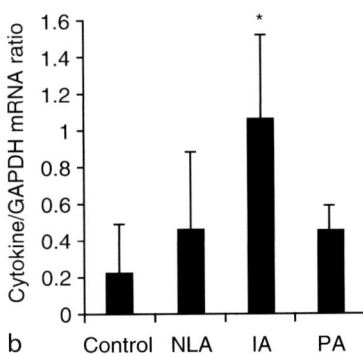
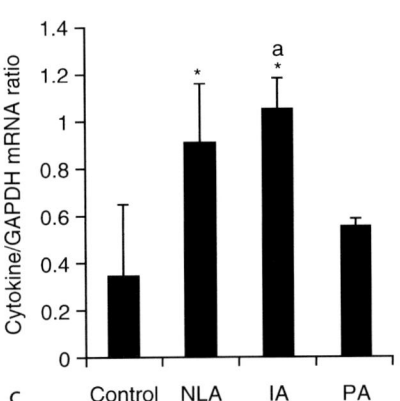
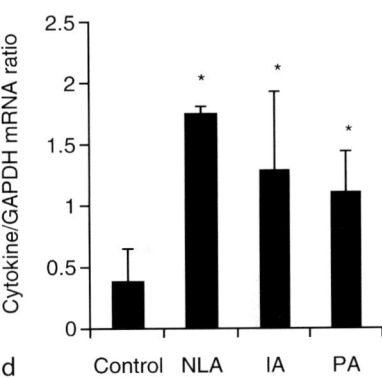
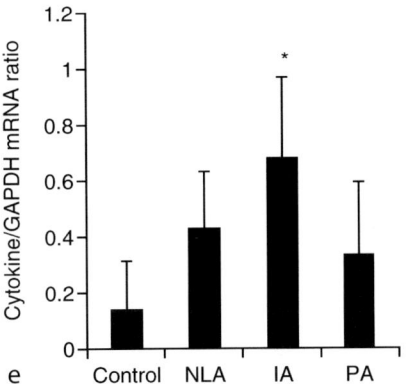
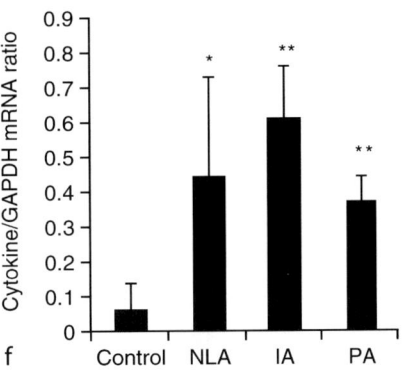

☐ Figure 2-5

(a–f) Chemokine expression in plaque (PA), inflammatory (IA), and normal-looking areas (NLA) of childhood X-ALD (cALD) and control brains determined by super gene array. (a) Monocyte chemotactic protein-1/CCL2, (b) monocyte chemotactic protein-3/CCL7, (c) macrophage inflammatory protein-1β/CCL4, (d) stromal-derived factor-2/SDF-2, (e) eotaxin/CCL11, and (f) fractalkine/CX3CL1. Data are expressed as ratio of chemokine/GAPDH mRNA. *, $P < 0.05$ versus control; **, $P < 0.01$ versus control; ***, $P < 0.001$ versus control; a, $P < 0.01$ versus cALD (NLA); b, $P < 0.001$ versus cALD (NLA); c, $P < 0.01$ versus cALD (PA). Reprinted, with permission, from Elsevier (Paintlia et al., 2003)

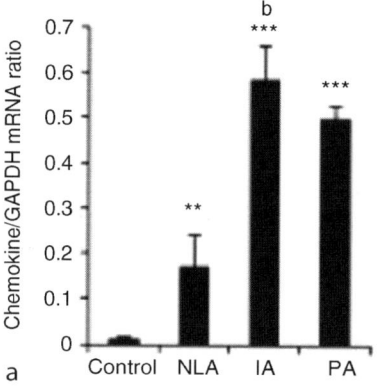
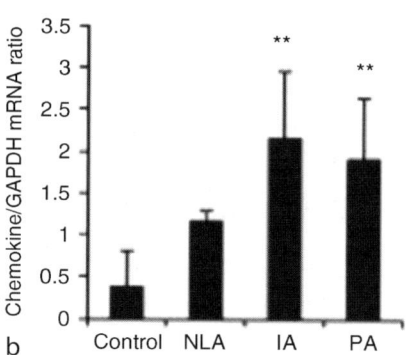
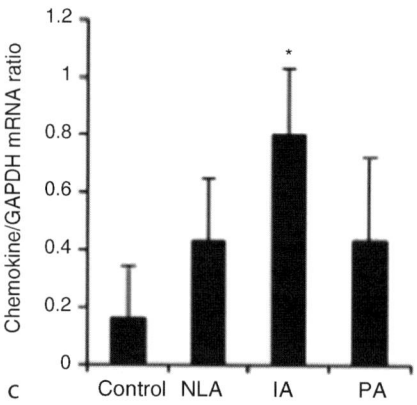
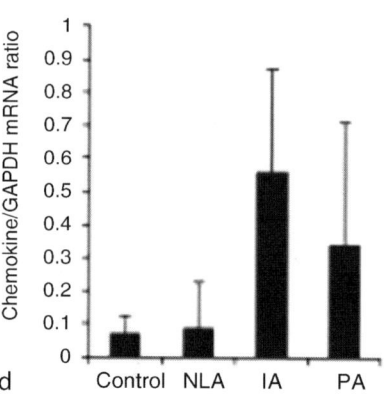
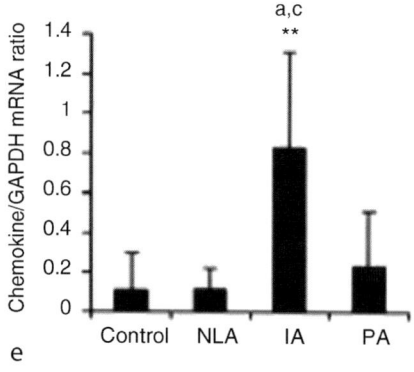
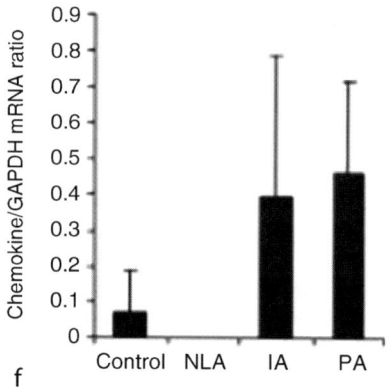

◻ Figure 2-6

Immunohistochemistry for chemokines (CCL2, CCL7) and chemokine receptor (CCR-2) in childhood X-ALD (cALD) and control brains. (a–d) CCL2, (e–h) CCL7, and (i–l) CCR-2. CCL2 was detected in (c) IA, (d) PA, and to a lesser extent in (b) NLA, but not in (a) control brain. CCL7 was detected in (g) IA and (f) NLA, and to a lesser extent in (h) PA, and rarely in (e) control brain. CCR-2 (receptor for CCL2 and CCL7) was detected in (k) IA, (l) PA, and to a lesser extent in (j) NLA, but not in (i) control brain. Reprinted, with permission, from Elsevier (Paintlia et al., 2003)

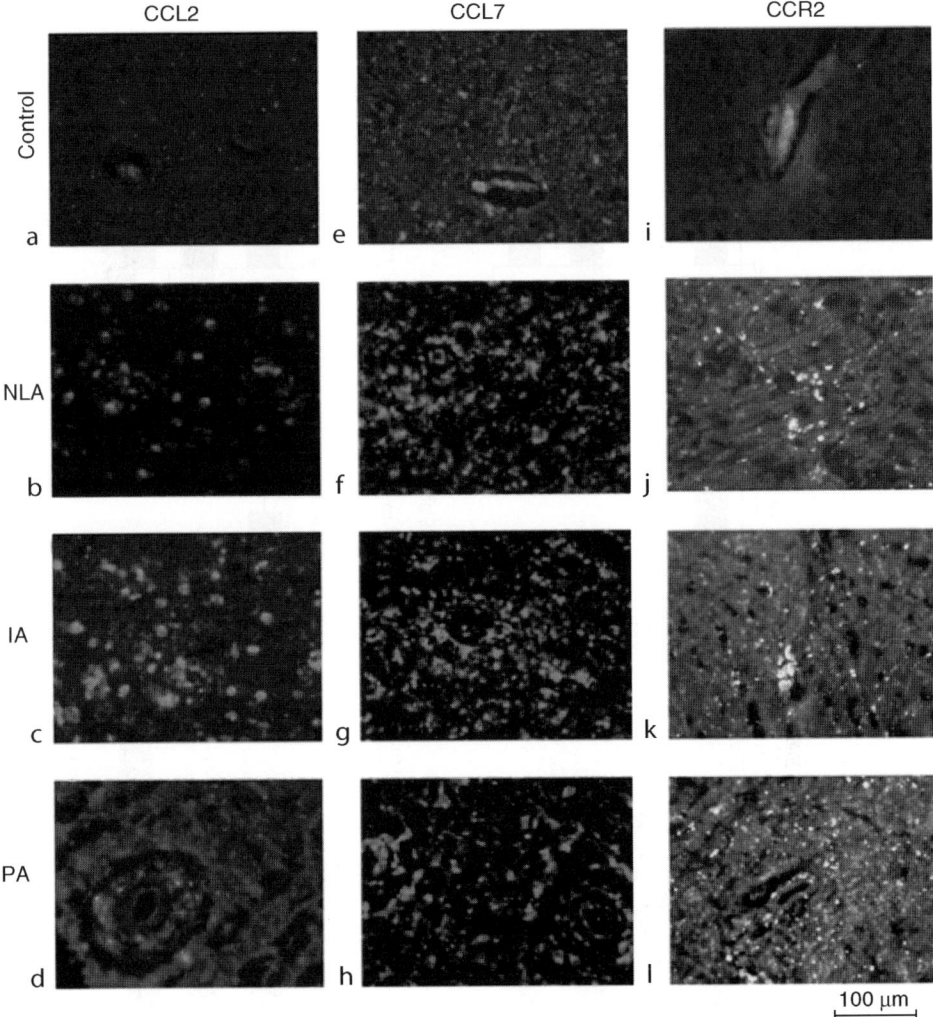

as one of the intermediates in this inhibition (Dobashi et al., 1997). These observations indicate that in addition to downregulation of peroxisomes/proteins, the inflammatory mediators (cytokines) induce inactivation of peroxisomal activities via NO-mediated mechanisms.

There is evidence to suggest that the mechanism of inhibition of VLC fatty acids β-oxidation in peroxisomes may be mediated by the inhibition of gene expression of peroxisomal proteins (Beier et al., 1997), and thus the activities of peroxisomal enzymes (Canonico et al., 1975; Yasmineh et al., 1991; Kim et al., 1995; DeLamatre et al., 1996; Dobashi et al., 1997; Keng et al., 2000; Stolz et al., 2002). The downregulation of

Figure 2-7

(a–f) Composition of very long-chain (VLC) fatty acids and lipids in childhood X-ALD (cALD) brain plaque area (PA), inflammatory area (IA), normal-looking area (NLA), and control brain (CTL). (a) VLC fatty acid (C26:0/C22:0), (b) cholesterol ester, (c) cholesterol and (d) sphingomyelin, (e, f) VLC fatty acids in sphingomyelin fraction (C26:0/C22:0 and C24:1/C22:0). *, $P < 0.001$ versus control; **, $P < 0.01$ versus control; a, $P < 0.001$ versus cALD (NLA); b, $P < 0.01$ versus cALD (NLA). Reprinted, with permission, from Elsevier (Paintlia et al., 2003)

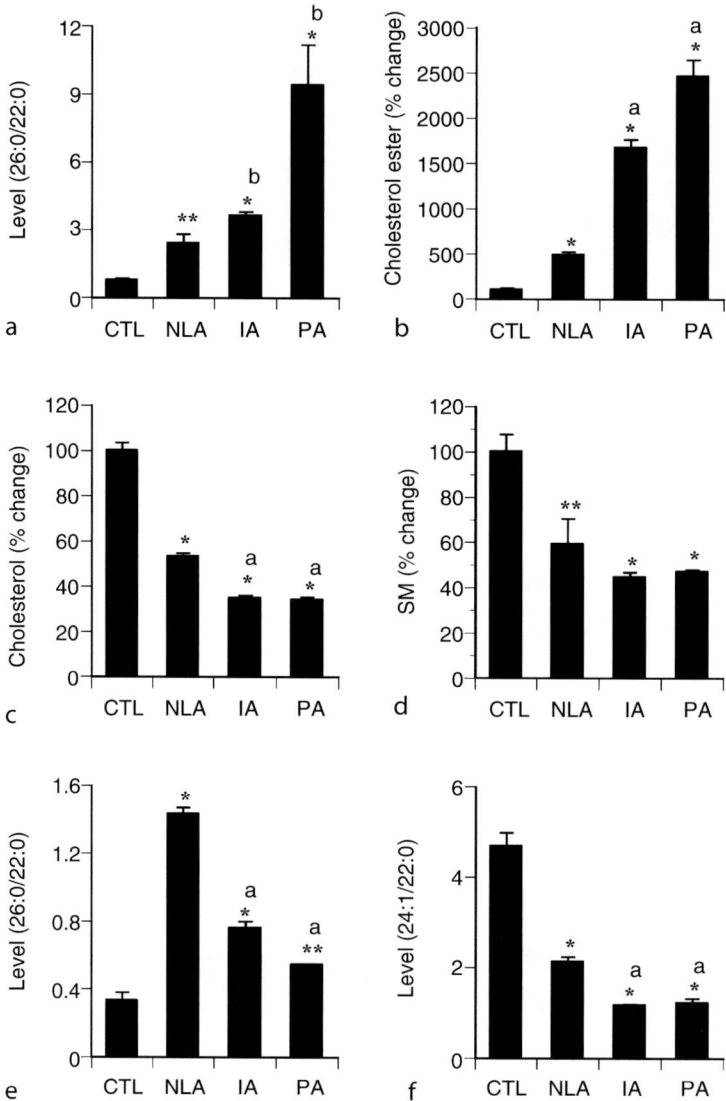

peroxisomal proteins/biogenesis under inflammatory disease conditions is supported by the observed decrease in peroxisomes/proteins in liver of rats treated with sublethal doses of endotoxin (Contreras et al., 2000). A sublethal injection of bacterial endotoxin produced drastic changes in peroxisomal structure and function. Structural changes were characterized by alteration in membrane lipid composition, and

Figure 2-8

Dose-dependent effect of TNF-α in the presence or absence of IFN-γ on β-oxidation of (a) lignoceric acid, (b) palmitic acid, and (c) induction of NO production. C6 glial cells received different concentrations of TNF-α (ng/mL) in the presence (*black square*) or absence (*white square*) of IFN-γ (10 U/mL) and β-oxidations of (a) lignoceric acid and (b) palmitic acid were measured at 72-h posttreatment. Activity of β-oxidation of lignoceric acid and palmitic acid in control C6 cells were 605 ± 52 and 4,920 ± 405 pmol/h/mg of protein, respectively. (c) NO was measured 24-h posttreatment. Reprinted, with permission, from Blackwell (Khan et al., 1998)

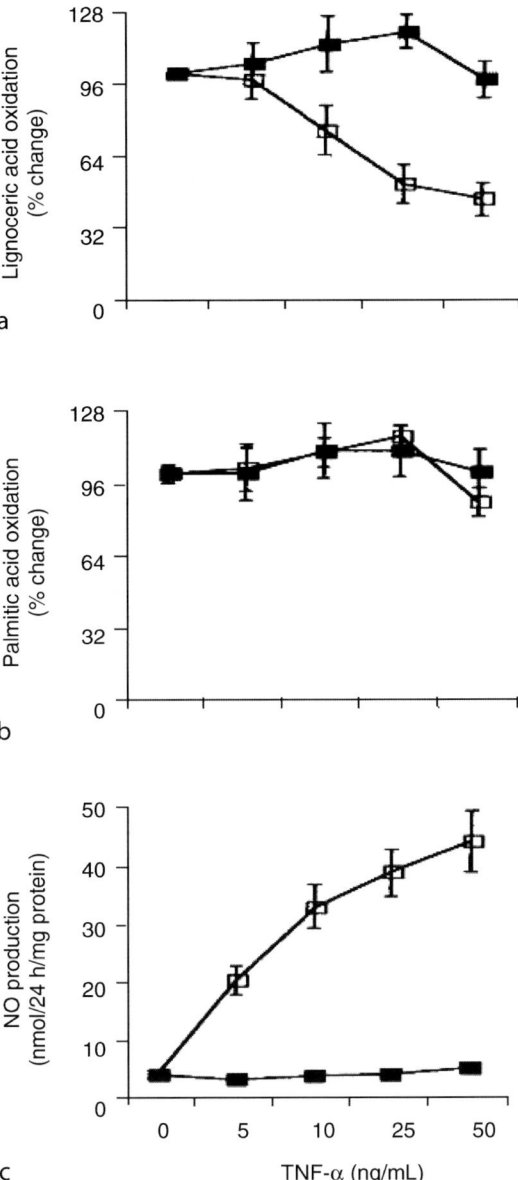

Figure 2-9

Inhibition of lignoceric acid β-oxidation by cytokine-induced NO production in C6 glial cells. C6 glial cells received different combinations of cytokines and LPS with (*black bars*) or without (*white bars*) 0.1 mM L-N-methylarginine (L-NMA). (a) β-oxidation of lignoceric acid (C24:0) was determined 72-h postincubation. Lignoceric acid β-oxidation in control C6 cells was 605 ± 52 pmol/h/mg of protein. (b) NO 24-h posttreatment. Treatments were: LPS, 0.5 μg/mL; TNF-α, 20 ng/mL; IL-1β, 20 ng/mL; IFN-γ, 20 U/mL. *, $P < 0.01$ versus control. Reprinted, with permission, from Blackwell (Khan et al., 1998)

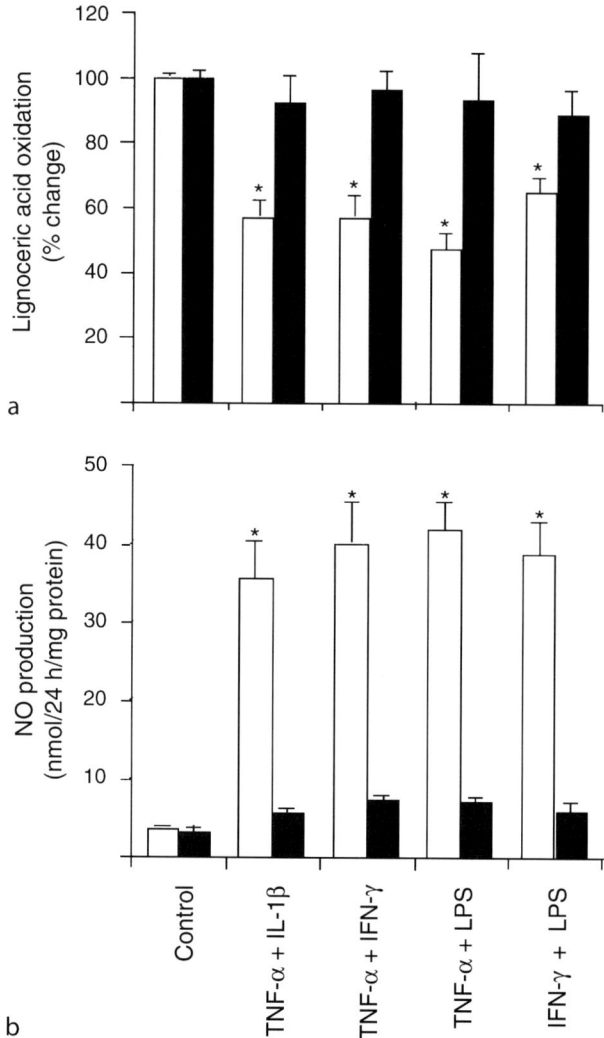

hence fluidity properties (Contreras et al., 2000; Khan et al., 2000). Changes in function were associated with a decrease in β-oxidation of VLC fatty acids. Overall, there was a marked decrease in the population of peroxisomes of normal density (Contreras et al., 2000).

Inflammatory features of EAE, an animal model for the human demyelinating disease MS, resemble the inflammatory disease of X-ALD (Moser, 1997). The induction of EAE following injection of myelin basic protein leads to loss of peroxisomal functions (fatty acids β-oxidation, DHAP-AT, and catalase), resulting in excessive accumulation of VLC fatty acids and decrease in plasmalogens (Singh et al., 2004). This decrease

◨ Figure 2-10
Cytokines induce accumulation of very long-chain (VLC) fatty acids in C6 glial cells. C6 glial cells received different combinations of cytokines and LPS with (*black bars*) or absence (*white bars*) of 0.1 mM L-NMA. Fatty acids levels were analyzed 72-h posttreatments. Levels of C26:0, C24:0, C22:0, C26:1, and C24:1 are expressed as ratios of (a) C26:0/C22:0; (b) C24:0/C22:0; (c) C26:1/C22:0, and (d) C24:1/C22:0. Treatments were: LPS, 0.5 μg/mL; TNF-α, 20 ng/mL; IL-1β, 20 ng/mL; and IFN-γ, 20 U/mL. *, $P < 0.01$; **, $P < 0.05$ versus control. Reprinted, with permission, from Blackwell (Khan et al., 1998)

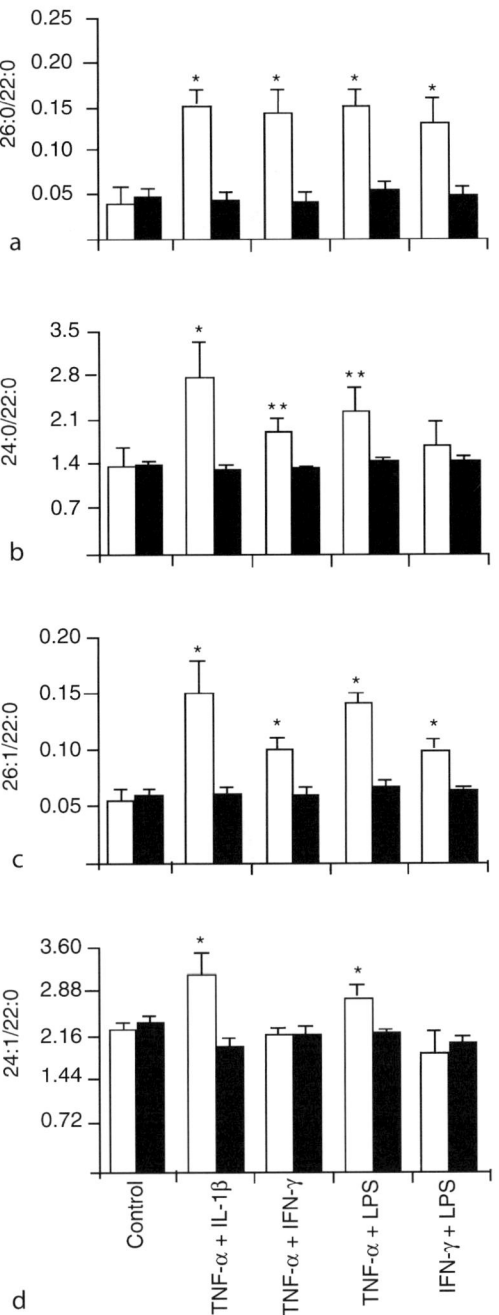

◘ Figure 2-11

Effect of cytokines or LPS on the accumulation of very long-chain (VLC) fatty acids and the induction of NO production in C6 glial cells. C6 glial cells received different cytokines and LPS treatments. Fatty acid levels were analyzed 72-h posttreatment. Levels of C26:0, C24:0, C22:0 are expressed as ratios of (a) C26:0/C22:0; and (b) C24:0/C22:0. (c) NO was measured 24-h posttreatment. Treatments were: LPS, 1.0 μg/mL; TNF-α 50 ng/mL; IL-1β, 50 ng/mL; and IFN-γ, 50 U/mL. Reprinted, with permission, from Blackwell (Khan et al., 1998)

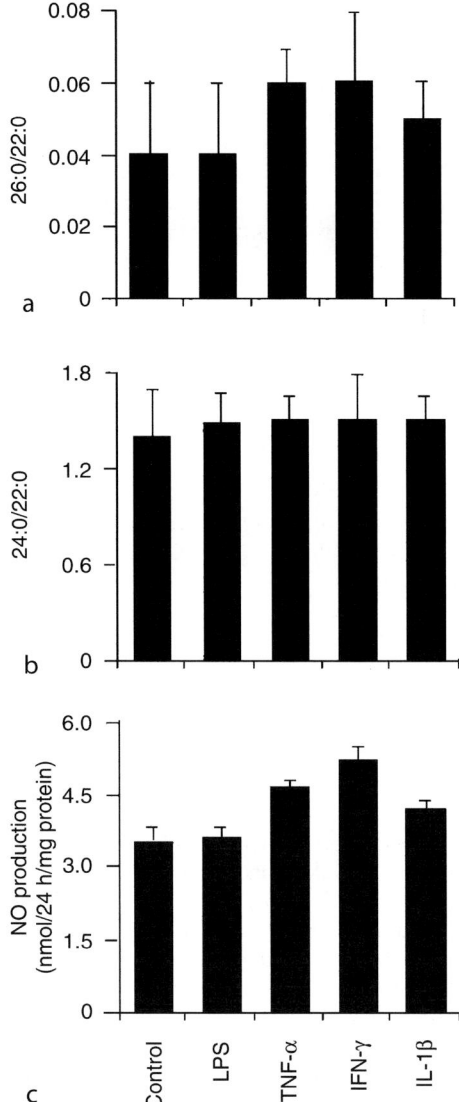

in peroxisomal functions was paralleled with a decrease in catalase and PMP-70 positive organelles in the brain and spinal cord of EAE animals. These observations indicate that the metabolic disease of cALD induces neuroinflammatory disease and that the inflammatory mediators further downregulate peroxisomal functions, thus creating a vicious circle resulting in progressive neuroinflammatory disease. Consequently, impairment of peroxisomal function compromises survival of cells, at least in part, as a result of derangement of peroxisomal metabolism such as arachidonic acid metabolites and synthesis of plasmalogen, thus limiting the availability of lipids for membrane turnover/repair. Therefore, therapy for inflammatory disease should be considered while developing treatment strategies for the metabolic disease aspects of X-ALD.

3 Therapeutics for X-ALD

X-ALD is associated with adrenal insufficiency and neurological disability. Adrenal insufficiency associated with X-ALD responds readily to steroid replacement therapy; however, there is no proven therapy for the neurological disability. The studies described above clearly indicate that ideal therapy should include both correction of the metabolic defect as well as attenuation/limiting/blocking of the neuroinflammatory disease including restriction of the transmigration of activated vascular immune cells into the CNS.

The dietary therapy with "Lorenzo's oil" does normalize the plasma levels of VLC fatty acids; however, its effects on the clinical status of X-ALD patients are not encouraging (Rasmussen et al., 1994; van Geel et al., 1999; Moser et al., 2005; Siva, 2005). The unsaturated fatty acids in Lorenzo's oil compete for chain elongation with saturated fatty acids resulting in reduced endogenous synthesis of VLC fatty acids (Rizzo et al., 1986; Wilson et al., 1992). The inability of Lorenzo's oil to correct the clinical disease may be because of its inability to attenuate the neuroinflammatory disease.

Bone marrow transplantation therapy was reported to be beneficial in a small number of patients; however, its value is limited because of the complexity of the protocol and complication-related fatality. Similar to other genetic diseases, gene therapy does not appear to be realistic in the near future for X-ALD. However, antiinflammatory drugs used as an adjunct to bone marrow transplantation and gene therapy may improve the outcome of these therapies.

Pharmacological induction of ALDRP, the ALDP-related protein that displays functional redundancy, with the prospect that its induction be able to correct the X-ALD biochemical phenotypes have been tested under experimental conditions (Kemp et al., 1998). Thyroid hormone, cAMP inhibitors, PPARα agonist, and plant flavonoids have been some of the compounds considered (Pahan et al., 1998; Fourcade et al., 2001; Fourcade et al., 2003; Gueugnon et al., 2003; Morita et al., 2005). Recent studies from Moser and colleagues and from our laboratory reported correction of VLC fatty acids levels in cultured skin fibroblasts from X-ALD patients using inhibitors of the mevalonate pathway (lovastatin, an inhibitor of HMG-CoA reductase, and phenylbutyrate and phenylacetate, inhibitors of mevalonate pyrophosphate decarboxylase) (Pahan et al., 1997; Kemp et al., 1998; Singh et al., 1998a; Singh et al., 1998b; Pai et al., 2000; Bugaut et al., 2003; Gondcaille et al., 2005). Phenylbutyrate also corrected the metabolic defect in X-ALD fibroblasts by upregulating the expression of *ALDR* gene product (ALDRP), which increases the VLC fatty acid β-oxidation, thus correcting the metabolic defect in X-ALD cultured skin fibroblasts (Gondcaille et al., 2005). The discovery of sterol regulatory elements in the promoter region of *ALDR* gene suggests that expression of this gene is regulated by cholesterol and/or cholesterol derivatives, and hence by lovastatin (Weinhofer et al., 2002). Furthermore, correlation of increased expression of ALDRP and increased β-oxidation activity suggests that this increased expression of ALDRP complements the function of ALDP in the β-oxidation of VLC fatty acids in peroxisomes (Kemp et al., 1998; Flavigny et al., 1999; Netik et al., 1999; Pujol et al., 2004). Subsequently, we tested the ability of lovastatin to decrease VLC fatty acid accumulation in X-ALD in a small group of X-ALD patients, most of them having the AMN form of the disease. The plasma levels of hexacosanoic acid (C26:0) showed a decline from pretreatment value within 1–3 months after starting a therapy with a dose of 40 mg of lovastatin per day (Singh et al., 1998a; Pai et al., 2000). In six patients, in whom the red blood cell membrane fatty acid composition was studied, a mean correction of 50% of the excess C26:0 was observed after six months of therapy, suggesting a sustained benefit. In patients who discontinued lovastatin therapy, the plasma C26:0 levels reverted to pretreatment values, suggesting a cause and effect relationship between these events in patients treated with lovastatin (Pai et al., 2000). Interestingly, the plasma levels of VLC fatty acids correlated with the levels of cholesterol. The patients who showed reversal of plasma cholesterol back to the original high levels also showed reversal of VLC fatty acids back to higher levels (Pai et al., 2000).

Inflammatory disease of X-ALD is quite complex since proinflammatory mediators (e.g., TNF-α) not only compromise the viability of oligodendrocytes and induce demyelination but also inhibit the biosynthesis of peroxisomes, thus further exacerbating the metabolic disease of X-ALD (Contreras et al., 2000; Cimini et al., 2003; Singh et al., 2004). As the inflammatory disease progresses in X-ALD brain so does the loss of peroxisomes. Studies from our laboratory documented that lovastatin also inhibits the activation of macrophages, astrocytes, and microglia for the induction of proinflammatory mediators (e.g., TNF-α, IL-1β, IFN-γ and iNOS), and this effect was reversed by farnesyl pyrophosphate but not by cholesterol, indicating that the observed effect of lovastatin is not due to depletion of the end product of the mevalonate pathway (cholesterol) (Pahan et al., 1997). These effects of statins were mediated by inhibition of isoprenylation of Ras/Rho small G proteins and in turn the inhibition of activation of nuclear factor-κB (NF-κB) signal transduction pathway. These studies identified for the first time a novel role of the metabolites of the mevalonate pathway in controlling/regulating induction of inflammatory mediators, which are known to play a role in the pathogenesis of neuroinflammatory diseases such as X-ALD and MS. Because of the unavailability of ALD mice with neurological symptoms/defects, we tested the effect of lovastatin on neuroinflammatory disease of animals with EAE, an animal model of MS. Treatment with lovastatin attenuated EAE disease by inhibiting induction of proinflammatory mediators (TNF-α, IL-1β, IFN-γ, and iNOS) in the brain and spinal cord of EAE animals (Stanislaus et al., 1999). Interestingly, lovastatin also attenuated/blocked the infiltration of activated mononuclear cells into CNS of EAE animals by inhibiting the expression of vascular cell adhesion molecule (VCAM) and E-selectin (Stanislaus et al., 2001; Prasad et al., 2005), a seminal event in neuroinflammatory diseases such as cALD, EAE, and MS. The inhibitory effects of lovastatin on binding of monocytes to endothelial cells was mediated by downregulation of vascular adhesion molecules via inhibition of phosphoinositide-3-kinase (PI3-kinase)/protein kinase B (Akt)/NF-κB pathway in the endothelial cells (Prasad et al., 2005). Therefore, statins protect the blood brain barrier, thus blocking the infiltration of activated vascular immune cells. We (Stanislaus et al., 2002) and others (Youssef et al., 2002) have also reported that statins (lovastatin) also inhibit the activation of T cells and, in fact, shift the immune response toward T_H2 bias instead of the T_H1 response observed in EAE (Stanislaus et al., 2002). On the basis of these encouraging results, we initiated a clinical trial for 30 patients with remitting/relapsing form of MS with a daily dose of 80 mg of simvastatin. Of the 28 patients who completed the study, three patients did not show any effect, two patients continued to worsen and 23 patients showed 43% decrease in number and 41% decrease in volume of Gd-positive MRI lesion following 6 months treatment. These results clearly document that statins not only attenuate the neuroinflammatory disease in EAE but also in MS (Vollmer et al., 2004).

On the basis of the pathobiology of X-ALD with T cells, perivascular cuffing of infiltrating cells, induction of proinflammatory mediators (i.e., cytokines, chemokines, and iNOS) in the brain, and the documented attenuation of these disease processes by statins in EAE and MS patients, statins may prove to be of therapeutic value for X-ALD patients. However, the efficacy of statins may be different with presymptomatic and symptomatic X-ALD patients because of the possible role of inflammatory disease process on the metabolism of VLC fatty acids (Khan et al., 1998) and that of the observed lower efficacy of statins when given to animals at an active stage of EAE disease (Nath et al., 2004). As described above, the development of effective therapy will require the use of drug(s) that correct both the metabolic as well the neuroinflammatory disease of X-ALD.

Recently, our laboratory reported that lovastatin may help in the restoration of impaired remyelination in the spinal cord of treated EAE animals by enhancing the survival and differentiation of oligodendrocyte precursor cells (OPCs) (Paintlia et al., 2005). Indeed, lovastatin treatment attenuates myelin breakdown, measured by the determination of myelin lipids (cholesterol esters, cerebroside, sulfatides, sphingomyelin, and cholesterol), which presented limited variation under lovastatin treatment, without producing significant changes in levels of cholesterol (❯ *Figure 2-12*). Furthermore, lovastatin enhanced the survival and differentiation of OPCs in the spinal cord of EAE animals, thus increasing the expression of myelin proteins (MBP, PLP, MOG, and MAG) and mRNA associated with differentiating oligodendrocytes (MyT1-L, GTX, PPAR-δ) during remission (❯ *Figure 2-13*). The lovastatin-induced proliferation and recruitment of OPCs in the spinal cord of treated EAE animals was further supported by immunohistochemistry for BrdU and NG2 (❯ *Figure 2-14*). These observations indicate that lovastatin treatment attenuates inflammatory

Figure 2-12
Lovastatin attenuates myelin breakdown and facilitates its restoration. Spinal cord homogenates were analyzed for myelin lipids associated with the demyelination/remyelination process. The levels of (a) cholesterol ester, (b) cerebrosides (nonhydroxy and hydroxy), (c) sulfatides (nonhydroxy and hydroxy), (d) sphingomyelin, and (e) cholesterol are shown in EAE, LOV-treated EAE (E + LOV), LOV-treated control (C + LOV), and control (CON) groups. **, $P < 0.01$; ***, $P < 0.001$ versus control; and #, $P < 0.05$; ###, $P < 0.001$ versus EAE. NS: nonsignificant. Reprinted, with permission, from the FASEB J (Paintlia et al., 2005)

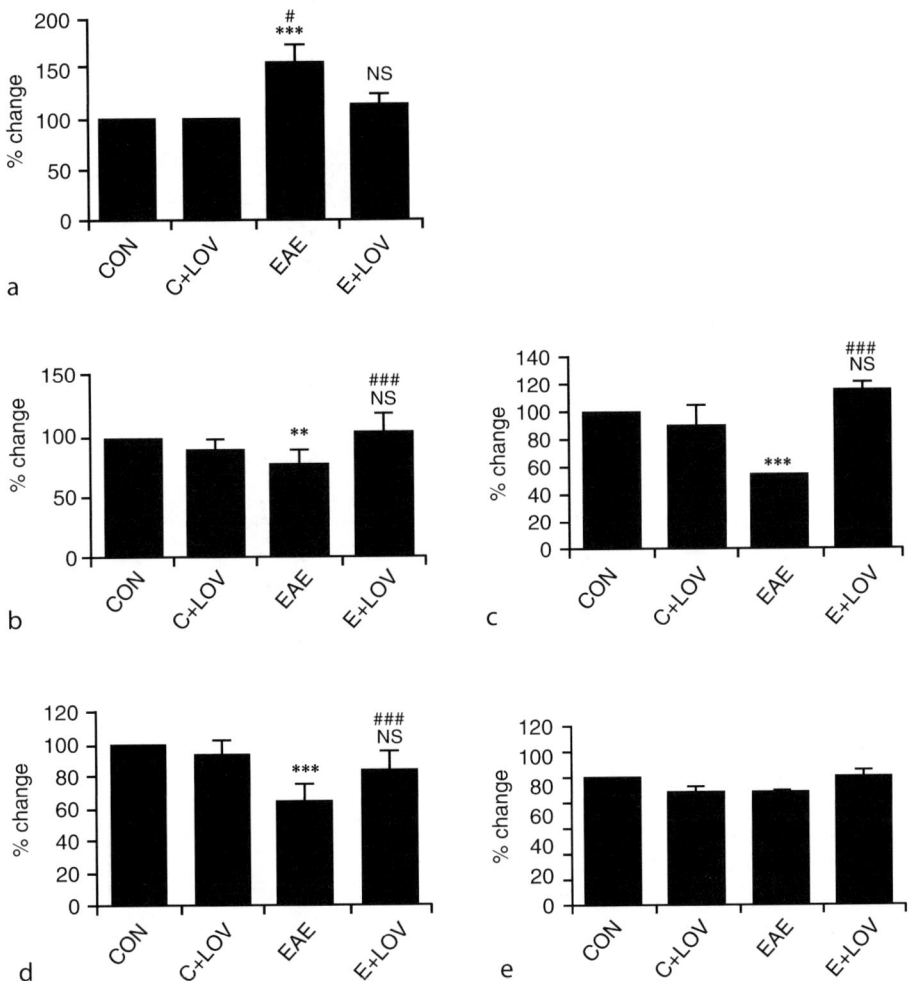

demyelination and restores remyelination by protecting the OPCs and promoting their proliferation and differentiation for the biosynthesis and expression of myelin constituent in differentiated oligodendrocytes. By attenuation of inflammatory disease and induction of a pro-remyelinating environment, lovastatin may promote both the proliferation and recruitment of OPCs in the CNS. These observed properties of lovastatin support its potential pharmacological application for the treatment of demyelinating diseases such as MS, Alzheimer disease, stroke, X-ALD, and HIV dementia. Although these studies suggest that while statin treatment can correct the metabolic defect in cultured fibroblasts (Singh et al., 1998b) and plasma of

◘ Figure 2-13
Lovastatin enhances the survival and differentiation of oligodendrocytes in the spinal cord of EAE animals. The expression of myelin proteins and mRNA associated with differentiating oligodendrocytes was observed in the spinal cord of animals. (a) Immunoblot of the protein levels of MBP, CNPase, and β-actin. (b) Spinal cord sections immunostained with anti-MBP and anti-CNPase antibodies. Weak immunofluorescence in demyelinated regions in the white matter is indicated (*arrowheads*) in EAE and E+LOV (400X magnification). (c) Quantification of immuno fluorescence for MBP and CNPase in immunostained sections. No demyelination was observed in the spinal cord of control animals (CON) LOV-treated control (C+LOV), or recovered E+LOV/R groups. Quantitative PCR analysis of mRNA for myelin proteins (d) MBP, (e) PLP, (f) MOG, and (g) MAG. *, $P < 0.05$; **, $P < 0.01$; ***, $P < 0.001$ versus control; #, $P < 0.05$, ##, $P < 0.01$, and ###, $P < 0.001$ versus EAE. NS; nonsignificant. Reprinted, with permission, from the FASEB J (Paintlia et al., 2005)

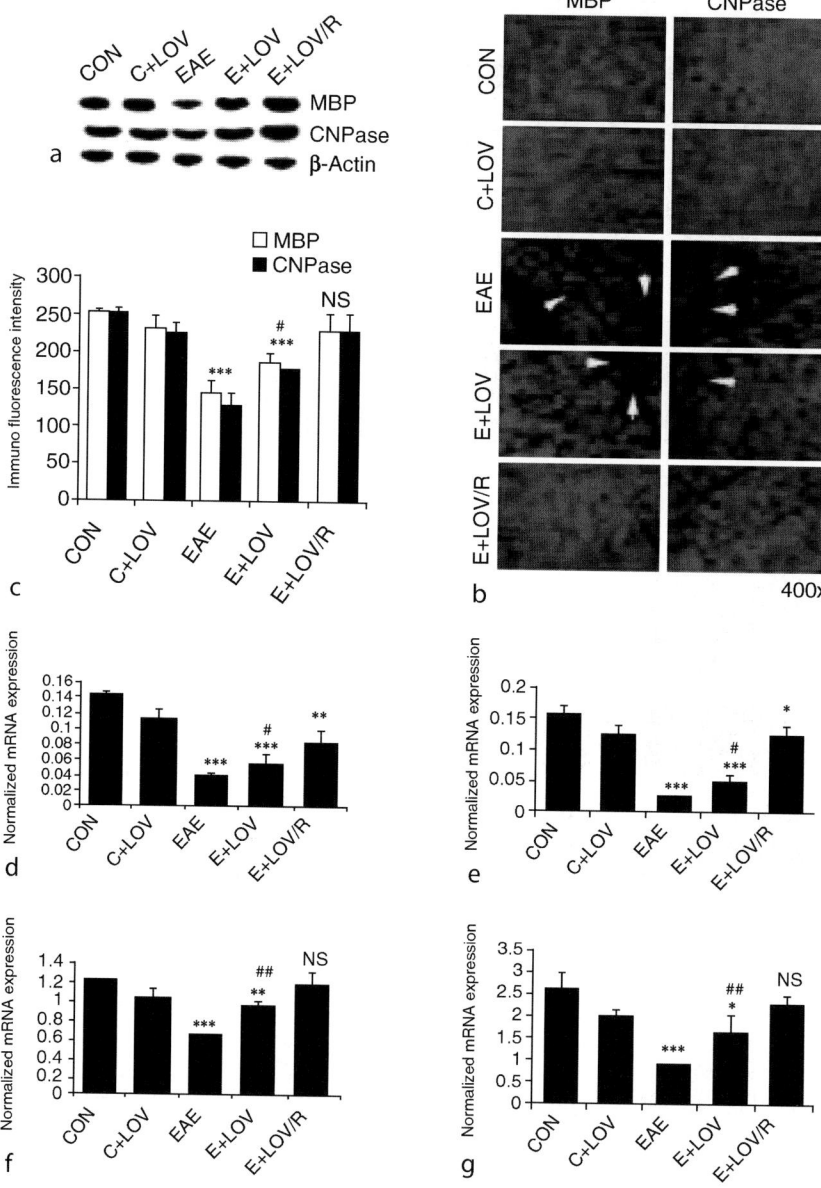

Figure 2-14

Lovastatin enhances proliferation and recruitment of oligodendrocyte precursor cells (OPCs) in the spinal cord of treated EAE animals. (a) Impaired recruitment and proliferation of OPCs in the spinal cord was detected by immunostaining with anti-BrdU and anti-NG2 antibodies: $NG2^+/BrDU^-$ cells (green; red arrowheads) in the white matter (WM) of spinal cord (400X magnification). (b) Average number of $NG2^+/BrDU^-$ cell counts in 10 fields/section. (c) Colocalization of NG2 and BrdU immunostaining in the spinal cord of recovered E+LOV/R animals (400X magnification). Red arrowheads indicate the colocalization of $NG2^+$ (green), $BrDU^-$ (red), and $NG2^+/BrDU^-$ (yellow) in proliferating OPCs, whereas a yellow arrowhead represents the migrated or resident OPC. (d) Number of $NG2^+/BrDU^-$ cell counts in 10 fields/section. (e) Immunoblotting of OPC-expressed proteins in spinal cord homogenates: A2B5 (70 kDa) and PDGF-αR (150 kDa) including β-actin. Quantitative PCR analysis of mRNA for (f) PDGF-αR, (g) SOX10, and (h) Shh. *, $P < 0.05$, **, $P < 0.01$, ***, $P < 0.001$ versus CON, and #, $P < 0.05$, ##, $P < 0.01$, and ###, $P < 0.001$ versus EAE. NS; nonsignificant. Reprinted, with permission, from the FASEB J (Paintlia et al., 2005)

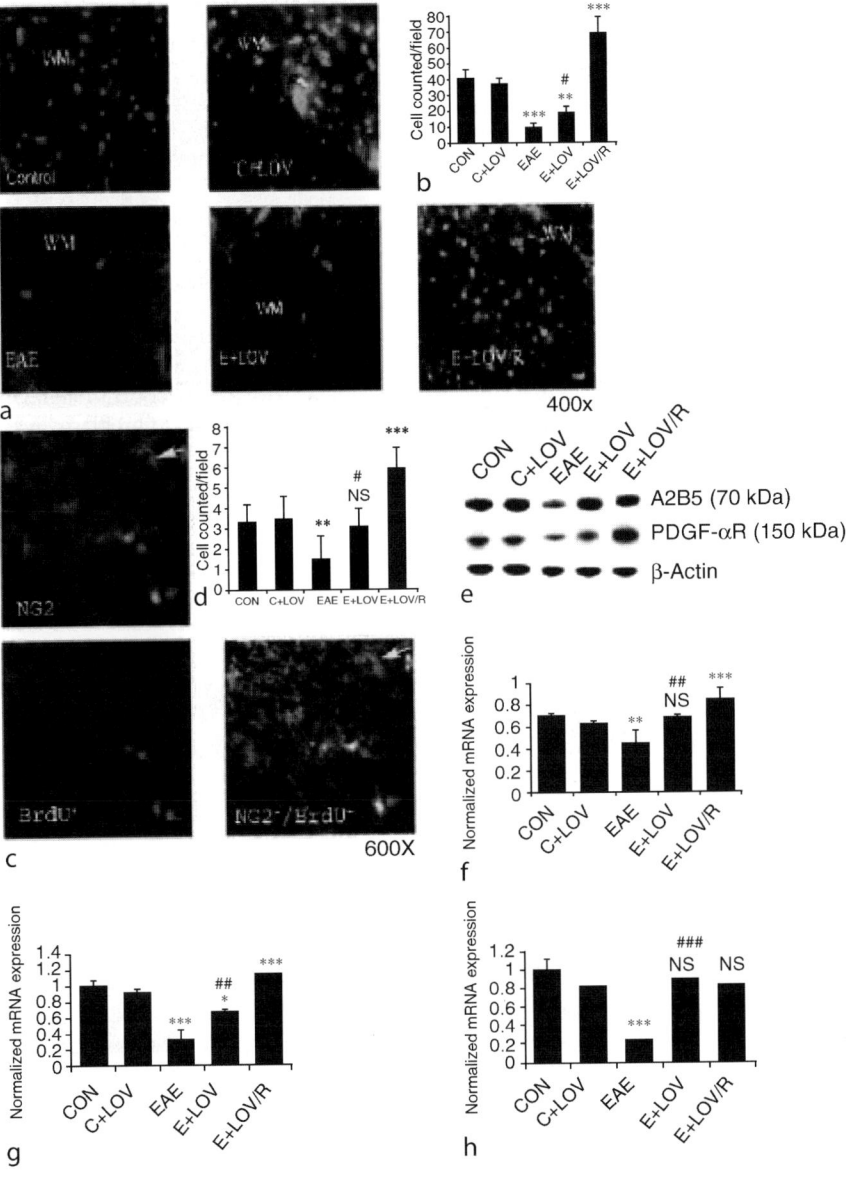

ALD patients (Pai et al., 2000; Singh et al., 1998a), down regulate the neuroinflammatory process (Stanislaus et al., 1999) and enhance remyelination (Paintlia et al., 2005) in EAE animals, it remains to be seen if the treatment will attenuate the neuroinflammatory and clinical disease of X-ALD patients.

4 Conclusions and Perspectives

Many years after of the discovery of the gene responsible for X-ALD, no correlation has been found between the mutation in the ALD gene and the clinical phenotype. This lack of correlation suggests that one or more modifier genes or a proinflammatory secondary event may play a role in the transition of metabolic disease starting in utero to neuroinflammatory disease manifesting in early childhood of X-ALD patients. Furthermore, endogenous factors of inflammatory events may also play a role in the severity of the disease. Emerging new data indicate that the inflammatory mediators induced by the inflammatory disease downregulate and inhibit peroxisomal activities indicating that these conditions may further exacerbate the accumulation of VLC fatty acids in the neural tissue of X-ALD patients as well as the molecular and cellular events that contribute to the progression of transition from metabolic to neuroinflammatory disease. Metabolic derangements in X-ALD brain induce activation of microglia/astrocytes which in turn release cytokines and chemokines resulting in initiation of inflammatory disease, followed by recruitment of vascular immune cells leading to upregulation of inflammatory mediators (e.g., iNOS) and the resulting neuropathology.

The tasks ahead include elucidation of the function and mechanism of action of ALDP peroxisomal membrane transporter and how this transporter is related to VLCS activity; elucidation of metabolites that are responsible for induction of neuroinflammatory disease and how they affect peroxisomal functions; and study of the relationship between VLC fatty acid accumulation and the genetic/environmental factors that play a role in the regulation and triggering of this process. Finally, understanding the mechanism and identifying the cell type, especially of CNS, responsible for upregulation of ALDP by pharmacological agents will establish the basis to identify drugs that correct the metabolic defect in X-ALD. Although Lorenzo's oil and statin treatments were reported to correct the metabolic defect in X-ALD fibroblasts in culture and plasma of X-ALD patients, it remains to be seen if such treatments will attenuate the neuroinflammatory and clinical disease of X-ALD patients. Ideally, any pharmacological intervention should include treatment for both the metabolic as well as the inflammatory disease.

Acknowledgements

We would like to thank Dr. Avtar K. Singh for helpful suggestions and for the revision of this manuscript. This work was supported by grant NS-22576 from the NIH.

References

Antoku Y, Sakai T, et al. 1984. Adrenoleukodystrophy: Abnormality of very long-chain fatty acids in erythrocyte membrane phospholipids. Neurology 34(11): 1499-1501.

Aubourg P. 1996. X-linked adrenoleukodystrophy. Handbook of Clinical Neurology: Neurodystrophies and Neurolipidoses. Vinken PJ, Bruyn GW, Moser HW, editors. Amsterdam: Elsevier; pp. 447-483.

Aubourg P. 1999. On the front of X-linked adrenoleukodystrophy. Neurochem Res 24(4): 515-520.

Aubourg P, Dubois-Dalcq M. 2000. X-linked adrenoleukodystrophy enigma: How does the ALD peroxisomal transporter mutation affect CNS glia? Glia 29(2): 186-190.

Bar-Or A, Oliveira EM, et al. 1999. Molecular pathogenesis of multiple sclerosis. J Neuroimmunol 100(1-2): 252-259.

Baudhuin P, Beaufay H, et al. 1964. Tissue fractionation studies. 17. Intracellular distribution of monoamine oxidase, aspartate aminotransferase, alanine aminotransferase,

D-amino acid oxidase and catalase in rat-liver tissue. Biochem J 92(1): 179-184.

Beier K, Volkl A, et al. 1992. Suppression of peroxisomal lipid β-oxidation enzymes of TNF-α. FEBS Lett 310(3): 273-276.

Beier K, Volkl A, et al. 1997. TNF-α downregulates the peroxisome proliferator activated receptor-α and the mRNAs encoding peroxisomal proteins in rat liver. FEBS Lett 412 (2): 385-387.

Berger J, Albet S, et al. 1999. The four murine peroxisomal ABC-transporter genes differ in constitutive, inducible and developmental expression. Eur J Biochem 265(2): 719-727.

Berger J, Molzer B, et al. 1994. X-linked adrenoleukodystrophy (ALD): A novel mutation of the ALD gene in 6 members of a family presenting with 5 different phenotypes. Biochem Biophys Res Commun 205(3): 1638-1643.

Bizzozero OA, Zuniga G, et al. 1991. Fatty acid composition of human myelin proteolipid protein in peroxisomal disorders. J Neurochem 56(3): 872-878.

Bourre JM, Paturneau-Jouas MY, et al. 1977b. Lignoceric acid biosynthesis in the developing brain. Activities of mitochondrial acetyl-CoA-dependent synthesis and microsomal malonyl-CoA chain-elongating system in relation to myelination. Comparison between normal mouse and dysmyelinating mutants (quaking and jimpy). Eur J Biochem 72(1): 41-47.

Bourre JM, Pollet S, et al. 1977b. Saturated and mono-unsaturated fatty acid biosynthesis in brain: Relation to development in normal and dysmyelinating mutant mice. Adv Exp Med Biol 83: 103-109.

Boztug K, Carson MJ, et al. 2002. Leukocyte infiltration, but not neurodegeneration, in the CNS of transgenic mice with astrocyte production of the CXC chemokine ligand 10. J Immunol 169(3): 1505-1515.

Braiterman LT, Zheng S, et al. 1998. Suppression of peroxisomal membrane protein defects by peroxisomal ATP binding cassette (ABC) proteins. Hum Mol Genet 7(2): 239-247.

Braun A, Ambach H, et al. 1995. Mutations in the gene for X-linked adrenoleukodystrophy in patients with different clinical phenotypes. Am J Hum Genet 56(4): 854-861.

Brown FR, Voigt R, et al. 1993. Peroxisomal disorders. Neurodevelopmental and biochemical aspects. Am J Dis Child 3rd, 147(6): 617-626.

Bugaut M, Fourcade S, et al. 2003. Pharmacological induction of redundant genes for a therapy of X-ALD: Phenylbutyrate and other compounds. Adv Exp Med Biol 544: 281-291.

Canonico PG, White JD, et al. 1975. Peroxisome depletion in rat liver during pneumococcal sepsis. Lab Invest 33(2): 147-150.

Cartier N, Lopez J, et al. 1995. Retroviral-mediated gene transfer corrects very long-chain fatty acid metabolism in adrenoleukodystrophy fibroblasts. Proc Natl Acad Sci USA 92(5): 1674-1678.

Cimini A, Bernardo A, et al. 2003. TNF-α downregulates PPARδ expression in oligodendrocyte progenitor cells: Implications for demyelinating diseases. Glia 41(1): 3-14.

Coll M, Palau N, et al. 2005. X-linked adrenoleukodystrophy in Spain. Identification of 26 novel mutations in the ABCD1 gene in 80 patients. Improvement of genetic counseling in 162 relative females. Clin Genet 67(5): 418-424.

Contreras M, Mosser J, et al. 1994. The protein coded by the X-adrenoleukodystrophy gene is a peroxisomal integral membrane protein. FEBS Lett 344(2-3): 211-215.

Contreras M, Sengupta TK, et al. 1996. Topology of ATP-binding domain of adrenoleukodystrophy gene product in peroxisomes. Arch Biochem Biophys 334(2): 369-379.

Contreras MA, Khan M, et al. 2000. Endotoxin induces structure-function alterations of rat liver peroxisomes: Kupffer cells released factors as possible modulators. Hepatology 31 (2): 446-455.

DeLamatre JG, Schillecj JT, et al. 1996. Influence of dietary fat on the effect of endotoxin on murine hepatic peroxisomes. Hepatology 24(3): 592-595.

Di Biase A, Avellino C, et al. 1997. Effects of exogenous hexacosanoic acid on biochemical myelin composition in weaning and post-weaning rats. Neurochem Res 22(3): 327-331.

Di Biase A, Merendino N, et al. 2001. Th 1 cytokine production by peripheral blood mononuclear cells in X-linked adrenoleukodystrophy. J Neurol Sci 182(2): 161-165.

Dobashi K, Pahan K, et al. 1997. Modulation of endogenous antioxidant enzymes by nitric oxide in rat C6 glial cells. J Neurochem 68(5): 1896-1903.

Dubois-Dalcq M, Feigenbaum V, et al. 1999. The neurobiology of X-linked adrenoleukodystrophy, a demyelinating peroxisomal disorder. Trends Neurosci 22(1): 4-12.

Dvorakova L, Storkanova G, et al. 2001. Eight novel ABCD1 gene mutations and three polymorphisms in patients with X-linked adrenoleukodystrophy: The first polymorphism causing an amino acid exchange. Hum Mutat 18(1): 52-60.

Flavigny E, Sanhaj A, et al. 1999. Retroviral-mediated adrenoleukodystrophy-related gene transfer corrects very long chain fatty acid metabolism in adrenoleukodystrophy fibroblasts: Implications for therapy. FEBS Lett 448(2-3): 261-264.

Forss-Petter S, Werner H, et al. 1997. Targeted inactivation of the X-linked adrenoleukodystrophy gene in mice. J Neurosci Res 50(5): 829-843.

Fourcade S, Savary S, et al. 2001. Fibrate induction of the adrenoleukodystrophy-related gene (ABCD2): Promoter analysis and role of the peroxisome proliferator-activated receptor PPARα. Eur J Biochem 268(12): 3490-3500.

Fourcade S, Savary S, et al. 2003. Thyroid hormone induction of the adrenoleukodystrophy-related gene (ABCD2). Mol Pharmacol 63(6): 1296-1303.

Gartner J, Dehmel T, et al. 2002. Functional characterization of the adrenoleukodystrophy protein (ALDP) and disease pathogenesis. Endocr Res 28(4): 741-748.

Gilg AG, Singh AK, et al. 2000. Inducible nitric oxide synthase in the central nervous system of patients with X-adrenoleukodystrophy. J Neuropathol Exp Neurol 59(12): 1063-1069.

Giusto NM, Roque ME, et al. 1992. Effects of aging on the content, composition and synthesis of sphingomyelin in the central nervous system. Lipids 27(11): 835-839.

Gondcaille C, Depreter M, et al. 2005. Phenylbutyrate upregulates the adrenoleukodystrophy-related gene as a nonclassical peroxisome proliferator. J Cell Biol 169(1): 93-104.

Gould SJ, Raymond GV, Valle D. 2001. The peroxisome biogenesis disorders. The Metabolic & Molecular Bases of Inherited Disease. Scriver CR, Beaudet AL, Sly WS, Valle D. New York: McGraw-Hill; II: 3181–3217.

Griffin DE, Moser HW, et al. 1985. Identification of the inflammatory cells in the central nervous system of patients with adrenoleukodystrophy. Ann Neurol 18(6): 660-664.

Gueugnon F, Lambert F, et al. 2003. Dehydroepiandrosterone induction of the Abcd2 and Abcd3 genes encoding peroxisomal ABC transporters: Implications for X-linked adrenoleukodystrophy. Adv Exp Med Biol 544: 245.

Guimaraes CP, Domingues P, et al. 2004. Mouse liver PMP70 and ALDP: Homomeric interactions prevail in vivo. Biochim Biophys Acta 1689(3): 235-243.

Hakomori S, Igarashi Y. 1995. Functional role of glycosphingolipids in cell recognition and signaling. J Biochem (Tokyo) 118(6): 1091-1103.

Hanada K, Nishijima M, et al. 1992. Sphingolipids are essential for the growth of Chinese hamster ovary cells. Restoration of the growth of a mutant defective in sphingoid base biosynthesis by exogenous sphingolipids. J Biol Chem 267(33): 23527-23533.

Higgins CF. 1992. ABC transporters: From microorganisms to man. Annu Rev Cell Biol 8: 67-113.

Hirsch D, Stahl A, et al. 1998. A family of fatty acid transporters conserved from mycobacterium to man. Proc Natl Acad Sci USA 95(15): 8625-8629.

Ho JK, Moser H, et al. 1995. Interactions of a very long chain fatty acid with model membranes and serum albumin. Implications for the pathogenesis of adrenoleukodystrophy. J Clin Invest 96(3): 1455-1463.

Holzinger A, Kammerer S, et al. 1997. cDNA cloning and mRNA expression of the human adrenoleukodystrophy related protein (ALDRP), a peroxisomal ABC transporter. Biochem Biophys Res Commun 239(1): 261-264.

Igarashi M, Belchis D, et al. 1976. Brain gangliosides in adrenoleukodystrophy. J Neurochem 27(1): 327-328.

Imanaka T, Aihara K, et al. 1999. Characterization of the 70-kDa peroxisomal membrane protein, an ATP binding cassette transporter. J Biol Chem 274(17): 11968-11976.

Kamijo K, Taketani S, et al. 1990. The 70-kDa peroxisomal membrane protein is a member of the Mdr (P-glycoprotein)-related ATP-binding protein superfamily. J Biol Chem 265(8): 4534-4540.

Kemp S, Ligtenberg MJ, et al. 1994. Identification of a two base pair deletion in five unrelated families with adrenoleukodystrophy: A possible hot spot for mutations. Biochem Biophys Res Commun 202(2): 647-653.

Kemp S, Pujol A, et al. 2001. ABCD1 mutations and the X-linked adrenoleukodystrophy mutation database: Role in diagnosis and clinical correlations. Hum Mutat 18(6): 499-515.

Kemp S, Valianpour F, et al. 2005. Elongation of very long-chain fatty acids is enhanced in X-linked adrenoleukodystrophy. Mol Genet Metab 84(2): 144-151.

Kemp S, Wei HM, et al. 1998. Gene redundancy and pharmacological gene therapy: Implications for X-linked adrenoleukodystrophy. Nat Med 4(11): 1261-1268.

Keng T, Privalle CT, et al. 2000. Peroxynitrite formation and decreased catalase activity in autoimmune MRL-lpr/lpr mice. Mol Med 6(9): 779-792.

Khan M., Contreras M, et al. 2000. Endotoxin-induced alterations of lipid and fatty acid compositions in rat liver peroxisomes. J Endotoxin Res 6(1): 41-50.

Khan M, Pahan K, et al. 1998. Cytokine-induced accumulation of very long-chain fatty acids in rat C6 glial cells: Implication for X-adrenoleukodystrophy. J Neurochem 71(1): 78-87.

Kim YM, Bergonia HA, et al. 1995. Loss and degradation of enzyme-bound heme induced by cellular nitric oxide synthesis. J Biol Chem 270(11): 5710-5713.

Knazek RA, Rizzo WB, et al. 1983. Membrane microviscosity is increased in the erythrocytes of patients with adrenoleukodystrophy and adrenomyeloneuropathy. J Clin Invest 72(1): 245-248.

Kobayashi T, Shinnoh N, et al. 1997. Adrenoleukodystrophy protein-deficient mice represent abnormality of very long chain fatty acid metabolism. Biochem Biophys Res Commun 232(3): 631-636.

Kok F, Neumann S, et al. 1995. Mutational analysis of patients with X-linked adrenoleukodystrophy. Hum Mutat 6(2): 104-115.

Krasemann EW, Meier V, et al. 1996. Identification of mutations in the ALD-gene of 20 families with adrenoleukodystrophy/adrenomyeloneuropathy. Hum Genet 97(2): 194-197.

Lannuzel A, Aubourg P, et al. 1998. Excessive production of tumour necrosis factor α by peripheral blood mononuclear cells in X-linked adrenoleukodystrophy. Eur J Paediatr Neurol 2(1): 27-32.

Lazo O, Contreras M, et al. 1988. Peroxisomal lignoceroyl-CoA ligase deficiency in childhood adrenoleukodystrophy and adrenomyeloneuropathy. Proc Natl Acad Sci USA 85(20): 7647-7651.

Lazo O, Contreras M, et al. 1989. Adrenoleukodystrophy: Impaired oxidation of fatty acids due to peroxisomal lignoceroyl-CoA ligase deficiency. Arch Biochem Biophys 270(2): 722-728.

Lazo O, Contreras M, et al. 1990a. Topographical localization of peroxisomal acyl-CoA ligases: Differential localization of palmitoyl-CoA and lignoceroyl-CoA ligases. Biochemistry 29(16): 3981-3986.

Lazo O, Contreras M, et al. 1990b. Cellular oxidation of lignoceric acid is regulated by the subcellular localization of lignoceroyl-CoA ligases. J Lipid Res 31(4): 583-595.

Liu LX, Janvier K, et al. 1999. Homo- and heterodimerization of peroxisomal ATP-binding cassette half-transporters. J Biol Chem 274(46): 32738-32743.

Lombard-Platet G, Savary S, et al. 1996. A close relative of the adrenoleukodystrophy (ALD) gene codes for a peroxisomal protein with a specific expression pattern. Proc Natl Acad Sci USA 93(3): 1265-1269.

Lu JF, Lawler AM, et al. 1997. A mouse model for X-linked adrenoleukodystrophy. Proc Natl Acad Sci USA 94(17): 9366-9371.

Maestri NE, Beaty TH. 1992. Predictions of a 2-locus model for disease heterogeneity: application to adrenoleukodystrophy. Am J Med Genet 44(5): 576-582.

Makkar RS, Contreras MA, et al. 2006. Molecular organization of peroxisomal enzymes: Protein-protein interactions in the membrane and in the matrix. Arch Biochem Biophys 451(2):128-40.

Mannaerts GP, Van Veldhoven P, et al. 1982. Evidence that peroxisomal acyl-CoA synthetase is located at the cytoplasmic side of the peroxisomal membrane. Biochem J 204(1): 17-23.

Matsumoto T, Tsuru A, et al. 2003. Mutation analysis of the ALD gene in seven Japanese families with X-linked adrenoleukodystrophy. J Hum Genet 48(3): 125-129.

McGuinness MC, Griffin DE, et al. 1995. Tumor necrosis factor-α and X-linked adrenoleukodystrophy. J Neuroimmunol 61(2): 161-169.

McGuinness MC, Powers JM, et al. 1997. Human leukocyte antigens and cytokine expression in cerebral inflammatory demyelinative lesions of X-linked adrenoleukodystrophy and multiple sclerosis. J Neuroimmunol 75(1-2): 174-182.

Montagna G, Di Biase A, et al. 2005. Identification of seven novel mutations in ABCD1 by a DHPLC-based assay in Italian patients with X-linked adrenoleukodystrophy. Hum Mutat 25(2): 222.

Morita M, Takahashi I, et al. 2005. Baicalein 5,6,7-trimethyl ether, a flavonoid derivative, stimulates fatty acid β-oxidation in skin fibroblasts of X-linked adrenoleukodystrophy. FEBS Lett 579(2): 409-414.

Moser AB, Moser HW. 1999. The prenatal diagnosis of X-linked adrenoleukodystrophy. Prenat Diagn 19(1): 46-48.

Moser HW. 1997. Adrenoleukodystrophy: Phenotype, genetics, pathogenesis and therapy. Brain 120(Pt 8): 1485-1508.

Moser HW, Kemp S, et al. 1999. Mutational analysis and the pathogenesis of variant X-linked adrenoleukodystrophy phenotypes. Arch Neurol 56(3): 273-275.

Moser HW, Moser AB, et al. 1992. Adrenoleukodystrophy: Phenotypic variability and implications for therapy. J Inherit Metab Dis 15(4): 645-664.

Moser HW, Raymond GV, et al. 2005. Follow-up of 89 asymptomatic patients with adrenoleukodystrophy treated with Lorenzo's oil. Arch Neurol 62(7): 1073-1080.

Moser HW, Smith KD, et al. 2001. X-Linked Adrenoleukodystrophy. The Metabolic & Molecular Bases of Inherited Disease. Scriver CR, Beaudet AL, Sly WS, Valle D, editors. New York: McGraw-Hill; II: 3257–3301.

Mosser J, Douar AM, et al. 1993. Putative X-linked adrenoleukodystrophy gene shares unexpected homology with ABC transporters. Nature 361(6414): 726-730.

Mosser J, Lutz Y, et al. 1994. The gene responsible for adrenoleukodystrophy encodes a peroxisomal membrane protein. Hum Mol Genet 3(2): 265-271.

Murad S, Kishimoto Y. 1978. Chain elongation of fatty acid in brain: A comparison of mitochondrial and microsomal enzyme activities. Arch Biochem Biophys 185(2): 300-306.

Nath N, Giri S, et al. 2004. Potential targets of 3-hydroxy-3-methylglutaryl coenzyme A reductase inhibitor for multiple sclerosis therapy. J Immunol 172(2): 1273-1286.

Netik A, Forss-Petter S, et al. 1999. Adrenoleukodystrophy-related protein can compensate functionally for adrenoleukodystrophy protein deficiency (X-ALD): Implications for therapy. Hum Mol Genet 8(5): 907-913.

Nickels JT, Broach JR. 1996. A ceramide-activated protein phosphatase mediates ceramide-induced G1 arrest of Saccharomyces cerevisiae. Genes Dev 10(4): 382-394.

O'Brien JS, Fillerup DL, et al. 1964. Brain lipids: I. Quantification and fatty acid composition of cerebroside sulfate in human cerebral gray and white matter. J Lipid Res 15: 109-116.

O'Brien JS, Sampson EL. 1965. Fatty acid and fatty aldehyde composition of the major brain lipids in normal human gray matter, white matter, and myelin. J Lipid Res 6(4): 545-551.

Oh CS, Toke DA, et al. 1997. ELO2 and ELO3, homologues of the Saccharomyces cerevisiae ELO1 gene, function in fatty acid elongation and are required for sphingolipid formation. J Biol Chem 272(28): 17376-17384.

O'Neill GN, Aoki M, et al. 2001. ABCD1 translation-initiator mutation demonstrates genotype-phenotype correlation for AMN. Neurology 57(11): 1956-1962.

Owens T, Babcock AA, et al. 2005. Cytokine and chemokine inter-regulation in the inflamed or injured CNS. Brain Res Brain Res Rev 48(2): 178-184.

Pahan K, Khan M, et al. 1998. Therapy for X-adrenoleukodystrophy: Normalization of very long chain fatty acids and inhibition of induction of cytokines by cAMP. J Lipid Res 39(5): 1091-1100.

Pahan K, Sheikh FG, et al. 1997. Lovastatin and phenylacetate inhibit the induction of nitric oxide synthase and cytokines in rat primary astrocytes, microglia, and macrophages. J Clin Invest 100(11): 2671-2679.

Pai GS, Khan M, et al. 2000. Lovastatin therapy for X-linked adrenoleukodystrophy: Clinical and biochemical observations on 12 patients. Mol Genet Metab 69(4): 312-322.

Paintlia AS, Gilg AG, et al. 2003. Correlation of very long chain fatty acid accumulation and inflammatory disease progression in childhood X-ALD: Implications for potential therapies. Neurobiol Dis 14(3): 425-439.

Paintlia AS, Paintlia MK, et al. 2005. HMG-CoA reductase inhibitor augments survival and differentiation of oligodendrocyte progenitors in animal model of multiple sclerosis. FASEB J 19(11): 1407-1421.

Powers JM. 1995. The pathology of peroxisomal disorders with pathogenetic considerations. J Neuropathol Exp Neurol 54(5): 710-719.

Powers JM, De Ciero DP, et al. 2000. Adrenomyeloneuropathy: A neuropathologic review featuring its noninflammatory myelopathy. J Neuropathol Exp Neurol 59(2): 89-102.

Powers JM, De Ciero DP, et al. 2001. The dorsal root ganglia in adrenomyeloneuropathy: Neuronal atrophy and abnormal mitochondria. J Neuropathol Exp Neurol 60(5): 493-501.

Powers JM, Liu Y, et al. 1992. The inflammatory myelinopathy of adrenoleukodystrophy: Cells, effector molecules, and pathogenetic implications. J Neuropathol Exp Neurol 51(6): 630-643.

Prasad R, Giri S, et al. 2005. Inhibition of phosphoinositide-3-kinase-Akt (protein kinase B)-nuclear factor-κB pathway by lovastatin limits endothelial-monocyte cell interaction. J Neurochem 94(1): 204-214.

Pujol A, Ferrer I, et al. 2004. Functional overlap between ABCD1 (ALD) and ABCD2 (ALDR) transporters: a therapeutic target for X-adrenoleukodystrophy. Hum Mol Genet 13(23): 2997-3006.

Pujol A, Hindelang C, et al. 2002. Late onset neurological phenotype of the X-ALD gene inactivation in mice: A mouse model for adrenomyeloneuropathy. Hum Mol Genet 11(5): 499-505.

Ramsey RB, Banik NL, et al. 1977. Galactolipid fatty acid composition in adrenoleukodystrophy. J Neurol Sci 32(1): 69-77.

Rasmussen M, Moser AB, et al. 1994. Brain, liver, and adipose tissue erucic and very long chain fatty acid levels in adrenoleukodystrophy patients treated with glyceryl trierucate and trioleate oils (Lorenzo's oil). Neurochem Res 19(8): 1073-1082.

Rizzo WB, Watkins PA, et al. 1986. Adrenoleukodystrophy: Oleic acid lowers fibroblast saturated C22–26 fatty acids. Neurology 36(3): 357-361.

Roerig P, Mayerhofer P, et al. 2001. Characterization and functional analysis of the nucleotide binding fold in human peroxisomal ATP binding cassette transporters. FEBS Lett 492(1-2): 66-72.

Schaffer JE, Lodish HF. 1994. Expression cloning and characterization of a novel adipocyte long chain fatty acid transport protein. Cell 79(3): 427-436.

Shani N, Jimenez-Sanchez G, et al. 1997. Identification of a fourth half ABC transporter in the human peroxisomal membrane. Hum Mol Genet 6(11): 1925-1931.

Sharp P, Johnson D, et al. 1991. Molecular species of phosphatidylcholine containing very long chain fatty acids in human brain: Enrichment in X-linked adrenoleukodystrophy brain and diseases of peroxisome biogenesis brain. J Neurochem 56(1): 30-37.

Simons K, Ikonen E. 1997. Functional rafts in cell membranes. Nature 387(6633): 569-572.

Singh I. 1997. Biochemistry of peroxisomes in health and disease. Mol Cell Biochem 167(1-2): 1-29.

Singh I, Khan M, et al. 1998a. Lovastatin for X-linked adrenoleukodystrophy. N Engl J Med 339(10): 702-703.

Singh I, Lazo O, et al. 1992. Transport of fatty acids into human and rat peroxisomes. Differential transport of palmitic and lignoceric acids and its implication to X-adrenoleukodystrophy. J Biol Chem 267(19): 13306-13013.

Singh I, Moser AE, et al. 1984a. Lignoceric acid is oxidized in the peroxisome: Implications for the Zellweger cerebrohepato-renal syndrome and adrenoleukodystrophy. Proc Natl Acad Sci USA 81(13): 4203-4207.

Singh I, Moser AE, et al. 1984b. Adrenoleukodystrophy: Impaired oxidation of very long chain fatty acids in white blood cells, cultured skin fibroblasts, and amniocytes. Pediatr Res 18(3): 286-290.

Singh I, Pahan K, et al. 1998b. Lovastatin and sodium phenylacetate normalize the levels of very long chain fatty acids in skin fibroblasts of X-adrenoleukodystrophy. FEBS Lett 426(3): 342-346.

Singh I, Paintlia AS, et al. 2004. Impaired peroxisomal function in the central nervous system with inflammatory disease of experimental autoimmune encephalomyelitis

animals and protection by lovastatin treatment. Brain Res 1022(1-2): 1-11.
Siva N. 2005. Positive effects with Lorenzo's oil. Lancet Neurol 4(9): 529.
Smith BT, Sengupta TK, et al. 2000. Intraperoxisomal localization of very-long-chain fatty acyl-CoA synthetase: Implication in X-adrenoleukodystrophy. Exp Cell Res 254(2): 309-320.
Smith KD, Kemp S, et al. 1999. X-linked adrenoleukodystrophy: Genes, mutations, and phenotypes. Neurochem Res 24 (4): 521-535.
Spiegel S, Merrill AH, Jr. 1996. Sphingolipid metabolism and cell growth regulation. FASEB J 10(12): 1388-1397.
Stanislaus R, Gilg AG, et al. 2002. Immunomodulation of experimental autoimmune encephalomyelitis in the Lewis rats by lovastatin. Neurosci Lett 333(3): 167-170.
Stanislaus R, Pahan K, et al. 1999. Amelioration of experimental allergic encephalomyelitis in Lewis rats by lovastatin. Neurosci Lett 269(2): 71-74.
Stanislaus R, Singh AK, et al. 2001. Lovastatin treatment decreases mononuclear cell infiltration into the CNS of Lewis rats with experimental allergic encephalomyelitis. J Neurosci Res 66(2): 155-162.
Steinberg SJ, Kemp S, et al. 1999. Role of very-long-chain acyl-coenzyme A synthetase in X-linked adrenoleukodystrophy. Ann Neurol 46(3): 409-412.
Stolz DB, Zamora R, et al. 2002. Peroxisomal localization of inducible nitric oxide synthase in hepatocytes. Hepatology 36(1): 81-93.
Street JM, Singh H, et al. 1990. Metabolism of saturated and polyunsaturated very-long-chain fatty acids in fibroblasts from patients with defects in peroxisomal β-oxidation. Biochem J 269(3): 671-677.
Suneja SK, Nagi MN, et al. 1991. Decreased long-chain fatty acyl CoA elongation activity in quaking and jimpy mouse brain: Deficiency in one enzyme or multiple enzyme activities? J Neurochem 57(1): 140-146.
Suzuki H, Kawarabayasi Y, et al. 1990. Structure and regulation of rat long-chain acyl-CoA synthetase. J Biol Chem 265 (15): 8681-8685.
Svennerholm L. 1968. Distribution and fatty acid composition of phosphoglycerides in normal human brain. J Lipid Res 9(5): 570-579.
Svennerholm L, Stallberg-Stenhagen S. 1968. Changes in the fatty acid composition of cerebrosides and sulfatides of human nervous tissue with age. J Lipid Res 9 (2): 215-225.
Takano H, Koike R, et al. 2000. Mutational analysis of X-linked adrenoleukodystrophy gene. Cell Biochem Biophys 32 Spring: 177-185.
Tanaka AR, Tanabe K, et al. 2002. ATP binding/hydrolysis by and phosphorylation of peroxisomal ATP-binding cassette proteins PMP70 (ABCD3) and adrenoleukodystrophy protein (ABCD1). J Biol Chem 277(42): 40142-40147.
Testi R. 1996. Sphingomyelin breakdown and cell fate. Trends Biochem Sci 21(12): 468-471.
Tiffany CW, Hoefler S, et al. 1990. Arachidonic acid metabolism in fibroblasts from patients with peroxisomal diseases: Response to interleukin 1. Biochim Biophys Acta 1096(1): 41-46.
Troffer- Charlier N, Doerflinger N, et al. 1998. Mirror expression of adrenoleukodystrophy and adrenoleukodystrophy related genes in mouse tissues and human cell lines. Eur J Cell Biol 75(3): 254-264.
Tsuji S, Ohno T, et al. 1984. Fatty acid elongation activity in fibroblasts from patients with adrenoleukodystrophy (ALD). J Biochem (Tokyo) 96(4): 1241-1247.
Tsuji S, Sano T, et al. 1981a. Increased synthesis of hexacosanoic acid (C23:0) by cultured skin fibroblasts from patients with adrenoleukodystrophy (ALD) and adrenomyeloneuropathy (AMN). J Biochem (Tokyo) 90(4): 1233-1236.
Tsuji S, Suzuki M, et al. 1981b. Abnormality of long-chain fatty acids in erythrocyte membrane sphingomyelin from patients with adrenoleukodystrophy. J Neurochem 36(3): 1046-1049.
Tvrdik P, Westerberg R, et al. 2000. Role of a new mammalian gene family in the biosynthesis of very long chain fatty acids and sphingolipids. J Cell Biol 149(3): 707-718.
Uchiyama A, Aoyama T, et al. 1996. Molecular cloning of cDNA encoding rat very long-chain acyl-CoA synthetase. J Biol Chem 271(48): 30360-30365.
Valle D, Gartner J. 1993. Human genetics. Penetrating the peroxisome. Nature 361(6414): 682-683.
van Geel BM, Assies J, et al. 1999. Progression of abnormalities in adrenomyeloneuropathy and neurologically asymptomatic X-linked adrenoleukodystrophy despite treatment with Lorenzo's oil. J Neurol Neurosurg Psychiatry 67(3): 290-299.
van Geel BM, Bezman L, et al. 2001. Evolution of phenotypes in adult male patients with X-linked adrenoleukodystrophy. Ann Neurol 49(2): 186-194.
Vollmer T, Key L, et al. 2004. Oral simvastatin treatment in relapsing-remitting multiple sclerosis. Lancet 363(9421): 1607-1608.
Wanders RJ. 1999. Peroxisomal disorders: Clinical, biochemical, and molecular aspects. Neurochem Res 24(4): 565-580.
Wanders RJ, Barth PG, 2001. Single peroxisomal enzyme deficiencies. The Metabolic & Molecular Bases of Inherited Disease. Scriver CR, Beaudet AL, Sly WS, Valle D, editors. New York: McGraw-Hill; II: 3219–3256.
Wanders RJ, van Roermund CW, et al. 1988. Direct demonstration that the deficient oxidation of very long chain fatty acids in X-linked adrenoleukodystrophy is due to an

impaired ability of peroxisomes to activate very long chain fatty acids. Biochem Biophys Res Commun 153(2): 618-624.

Weinhofer I, Forss-Petter S, et al. 2002. Cholesterol regulates ABCD2 expression: Implications for the therapy of X-linked adrenoleukodystrophy. Hum Mol Genet 11(22): 2701-2708.

Wertz PW, Downing DT. 1987. Covalently bound omega-hydroxyacylsphingosine in the stratum corneum. Biochim Biophys Acta 917(1): 108-111.

Wertz PW, Downing DT. 1989. ω-Hydroxyacid derivatives in the epidermis of several mammalian species. Comp Biochem Physiol B 93(2): 265-269.

Whitcomb RW, Linehan WM, et al. 1988. Effects of long-chain, saturated fatty acids on membrane microviscosity and adrenocorticotropin responsiveness of human adrenocortical cells in vitro. J Clin Invest 81(1): 185-188.

Wichers M, Kohler W, et al. 1999. X-linked adrenomyeloneuropathy associated with 14 novel ALD-gene mutations: No correlation between type of mutation and age of onset. Hum Genet 105(1-2): 116-119.

Wilkinson IA, Hopkins IJ, et al. 1987. Can head injury influence the site of demyelination in adrenoleukodystrophy? Dev Med Child Neurol 29(6): 797-800.

Wilson R, Tocher DR, et al. 1992. Effects of exogenous monounsaturated fatty acids on fatty acid metabolism in cultured skin fibroblasts from adrenoleukodystrophy patients. J Neurol Sci 109(2): 207-214.

Yamada T, Taniwaki T, et al. 1999. Adrenoleukodystrophy protein enhances association of very long-chain acyl-coenzyme A synthetase with the peroxisome. Neurology 52(3): 614-616.

Yasmineh WG, Parkin JL, et al. 1991. Tumor necrosis factor/cachectin decreases catalase activity of rat liver. Cancer Res 51(15): 3990-3995.

Youssef S, Stuve O, et al. 2002. The HMG-CoA reductase inhibitor, atorvastatin, promotes a Th2 bias and reverses paralysis in central nervous system autoimmune disease. Nature 420(6911): 78-84.

3 Hyperammonemia

V. Felipo

1	***Introduction***	**45**
1.1	General Introduction	45
1.2	Some Physiopathologically Relevant Properties of the Ammonia Molecule	45
1.2.1	Cell Membrane Permeability	45
1.2.2	Similarity to K^+	46
1.2.3	pH and Toxic Effects of Ammonia	46
1.3	Ammonia Is a Component of the Glutamate–Glutamine Cycle	47
1.4	Pathological Situations Associated with Hyperammonemia in Humans	48
1.5	Hyperammonemia Is a Main Contributor to the Origin and the Neurological Alterations in Hepatic Encephalopathy	48
1.6	Differential Alterations in Acute and Chronic Hyperammonemia	49
1.7	Animal and In Vitro Models of Hyperammonemia	49
2	***Molecular Mechanism of Acute Hyperammonemia Toxicity***	**50**
2.1	Acute Ammonia Intoxication Leads to Activation of NMDA Receptors in Brain In Vivo	50
2.2	Excessive Activation of NMDA Receptors Is Responsible for Ammonia-Induced Death of Animals	51
2.3	Acute Hyperammonemia Induces Depletion of ATP in Brain That Is Mediated by Activation of NMDA Receptors	52
2.4	Ammonia-Induced Activation of NMDA Receptors Leads to Activation of Calcineurin and Dephosphorylation and Activation of Na^+/K^+-ATPase in Brain, Thus Increasing ATP Consumption	52
2.5	Ammonia-Induced Activation of NMDA Receptors Impairs Mitochondrial Function, Thus Decreasing ATP Synthesis	54
2.6	Ammonia Induces Formation of Free Radicals	54
2.7	Ammonia-Induced Activation of NMDA Receptors Activates Calpain Which Degrades the Microtubule-Associated Protein MAP-2, Thus Altering the Microtubular Network	55
2.8	Hyperammonemia Increases Nitric Oxide Formation by Different Mechanisms and Increases Tyrosine Nitration of Brain Proteins	55
2.9	Ammonia-Induced Activation of NMDA Receptors Increases Nitric Oxide Formation, Which, in Turn, Reduces the Activity of Glutamine Synthetase, Thus Reducing Elimination of Ammonia in Brain	56
2.10	Effects of Acute Hyperammonemia on Non-NMDA Ionotropic Receptors	57
2.11	Acute Exposure to Ammonia Results in a Loss of Expression of Astrocytic Glutamate Transporters GLT-1 and GLAST	57
2.12	Exposure to Ammonia Induces Swelling of Astrocytes in Culture	58
3	***Effects of Chronic Hyperammonemia on Cerebral Function***	**58**
3.1	Acute and Chronic Moderate Hyperammonemia May Affect the Same Processes in Different, Even Opposite, Ways	58
3.2	Effects of Chronic Hyperammonemia on Glutamatergic Neurotransmission	59

3.2.1	Effects on NMDA Receptors	59
3.2.2	Alterations in Metabotropic Glutamate Receptors	60
3.3	Hyperammonemia and GABAergic Neurotransmission	61
3.4	Hyperammonemia and Serotoninergic Neurotransmission	61
3.5	Chronic Hyperammonemia Alters Phosphorylation of Neuronal Proteins	62
4	***Neurological Alterations in Hyperammonemia***	**63**
4.1	Neurological Alterations Associated to Perinatal Hyperammonemia	63
4.2	Alterations in Motor Function in Rats with Chronic Moderate Hyperammonemia	63
4.3	Impairment of Cognitive Function (Learning Ability) in Hyperammonemic Rats	64
4.3.1	Learning Ability Is Reduced in Rats with Chronic Hyperammonemia	64
4.3.2	Pharmacological Manipulation of Extracellular cGMP Concentration in Brain Restores Learning Ability in Rats with Chronic Liver Failure or Chronic Hyperammonemia	64

List of Abbreviations: CM, calmodulin; CsA, cyclosporin A; GC, guanylate cyclase; GS, glutamine synthetase; 5HIAA, 5-hydroxyindoleacetic acid; 5HT, Serotonin; iNOS, inducible nitric oxide synthase; LTP, long-term potentiation; mGluRs, metabotropic glutamate receptors; MPT, mitochondrial permeability transition; NAcc, nucleus accumbens; nNOS, neuronal nitric oxide synthase; NO, nitric oxide; NOS, nitric oxide synthase; PKC, protein kinase C

1 Introduction

1.1 General Introduction

Ammonia is produced in the body by a number (around 20) of enzymatic reactions and is a normal product of degradation of proteins, amino acids, and other physiological compounds. However, at high concentrations ammonia is toxic and leads to functional disturbances of the central nervous system and may lead to coma and death. To avoid these toxic effects, ammonia is maintained at low levels (50–100 µM in arterial blood). In ureotelic animals, including humans, the main mechanisms to keep low the concentration of ammonia is its detoxification by incorporation in urea, which is eliminated in urine. This process is carried out in the liver by the enzymes of the urea cycle. In other tissues, including the brain, which do not posses a functional urea cycle, ammonia is mainly detoxified by glutamine synthetase (GS), which incorporates ammonia in glutamate to form glutamine in a reaction that consumes ATP. Glutamine is then released to the bloodstream and may serve as a nontoxic "carrier" of ammonia from different tissues to the liver, where glutamine is broken down by glutaminase, releasing ammonia that is incorporated by the urea cycle in urea, which is eliminated in urine.

This chapter focuses mainly on the situations in which ammonia increases above normal levels (hyperammonemia) and especially on the current knowledge about the mechanisms by which hyperammonemia leads to altered cerebral function. The most common situation in which hyperammonemia may lead to altered cerebral function is hepatic disease (mostly liver cirrhosis). As the alteration in cerebral function and the associated neurological alterations in this situation are consequences of a previous hepatic failure this situation is called hepatic encephalopathy. This is a complex syndrome that covers a wide range of neuropsychiatric disturbances ranging from minimal changes in personality or altered circadian rhythms (sleep-waking cycle) to alterations in intellectual function, personality, consciousness, and motor coordination. Hepatic encephalopathy is usually reversible, but in the worse cases can lead to coma and death.

1.2 Some Physiopathologically Relevant Properties of the Ammonia Molecule

Ammonia (NH_3) is a small molecule that in aqueous solution is in equilibrium with the ammonium ion (NH_4^+).

$$NH_3 + H_2O \leftrightarrows NH_4^+ + OH^-$$

The equilibrium depends on the pH. The pK_a of the reaction at 37°C is 9.15. Under physiological conditions ammonia is therefore present mainly (approximately 98%) as NH_4^+.

1.2.1 Cell Membrane Permeability

NH_3 is a gas and can cross freely cellular membranes while NH_4^+ cannot. This property is the basis for one of the usual treatments of some hyperammonemic states: acidification of the intestinal lumen, which shifts the equilibrium of the above reaction to the right, increasing the formation of NH_4^+ and reducing the concentration of NH_3, the form that can more easily cross the membranes of intestinal cells and reach the bloodstream.

The existence of transporters that can transport NH_4^+ has been described in different tissues, including kidney and brain (Amlal and Soleimani, 1997; Ramirez et al., 1999; Marcaggi and Coles, 2001; Weiner, 2004). The possible role or alterations of these transporters in hyperammonemic situations has not been studied.

1.2.2 Similarity to K^+

The ionic size and charge of NH_4^+ in the hydrated state in aqueous solution are very similar to those of K^+, and many enzymes or channels using K^+ may also use NH_4^+. This property may contribute to the deleterious effects of high ammonia concentrations by interfering with the normal function of some of these enzymes or channels. One of these enzymes is phosphofructokinase. Abrahams and Younathan (1971) showed that ammonium ions are more efficient than potassium ions in activating rabbit skeletal muscle phosphofructokinase. Ammonia also substitutes K^+ in the function of another important enzyme, Na^+/K^+-ATPase, that can use either K^+ or NH_4^+ as counterion in ATP-dependent Na^+ transport (Towle and Holleland, 1987). NH_4^+ substitution for K^+ is a likely mechanism for NH_4^+ uptake by cells (Wall and Koger, 1994). Na^+/K^+-ATPase has an essential role in the control of membrane potential in neurons, and therefore of depolarization–repolarization of the cell membrane and of synaptic transmission. Moser (1987) showed that ammonium ions can also substitute for K^+ in the function of Na^+/K^+-ATPase in neurons. The alterations in Na^+/K^+-ATPase function may play a role in the pathological effects of hyperammonemia.

1.2.3 pH and Toxic Effects of Ammonia

Ammonia can act as an acid and as a base. NH_3 can cross the cellular membranes and may be protonated inside the cells to form NH_4^+. This may result in an increase in the intracellular pH. It was suggested that ammonia-induced alterations in intracellular pH could play a relevant role in the mechanisms of ammonia toxicity. In fact, studies in cultured cells have shown altered intracellular pH following treatment with ammonia; however, these studies were carried out using very high concentrations of ammonia (10–20 mM) that are never present in humans. However, in experimental acute ammonia intoxication with very large doses of ammonia in animals, the levels of ammonia in the brain did not reach more than 3–5 mM. The more common form of hyperammonemia is chronic hyperammonemia in which blood ammonia levels reach 0.1–0.3 mM (Patel et al., 2000).

Some studies have been carried out using nuclear magnetic resonance to assess whether hyperammonemia affects intracellular pH in the brain in vivo. Fitzpatrick and coworkers (1989) showed that acute intravenous infusion of ammonium acetate to rats to maintain a concentration of arterial blood ammonia of 0.5 mM for 50 min did not cause significant alterations in intracellular pH in the brain. Bates and coworkers (1989) used a rat model of acute liver failure in which plasma ammonia levels increased from 123 μM in controls to 564 μM in rats with liver failure. This change in ammonia concentration did not affect intracellular pH in the brain. These studies indicate that the brain is capable of maintaining normal intracellular pH when blood ammonia levels are increased 4–5 times over normal levels (at least up to 0.5–0.6 mM ammonia).

Patel and coworkers (2000) measured intracellular pH in the brain of patients with chronic liver disease with subclinical or overt hepatic encephalopathy. In patients with overt encephalopathy (with blood ammonia levels up to 286 μM), no alteration in intracellular pH in the brain was observed.

These studies indicate that altered intracellular pH does not play a role in the alterations in cerebral function in patients with chronic liver disease or with hyperammonemia at least up to increases in blood ammonia of 4–5 times over normal values, which includes most pathological situations in humans.

1.3 Ammonia Is a Component of the Glutamate–Glutamine Cycle

Glutamate is the main excitatory neurotransmitter in mammals. To avoid excessive activation of its receptors, glutamate must be removed from the extracellular fluid by specific transporters that transport it into both neurons and astrocytes. The uptake from the synaptic cleft is mainly carried out by astrocytic transporters. In astrocytes glutamate is converted to glutamine by glutamine synthetase (GS), which incorporates a molecule of ammonia in a molecule of glutamate and consumes one molecule of ATP

$$\text{Glutamate} + \text{NH}_4 + \text{ATP} \rightarrow \text{glutamine} + \text{ADP} + \text{P}_i$$

Glutamine formed in this reaction is released to the extracellular fluid, taken up by neurons and hydrolyzed there by glutaminase to glutamate, thus replenishing the neurotransmitter pool (❯ *Figure 3-1*). This

◘ Figure 3-1
The glutamate–glutamine cycle in brain. Glutamate is released from neurons into the synaptic cleft. To avoid excessive activation of its receptors, glutamate is rapidly removed from synaptic cleft by glutamate transporters (e.g., EAAT1 and EAAT2), which are mainly located on surrounding astrocytes. Within the astrocytes, glutamate and ammonia are combined to form glutamine by glutamine synthetase (GS), an astrocyte-specific enzyme. To replenish the neurotransmitter pool of glutamate, glutamine is released from astrocytes and taken up by glutamatergic neurons. Once glutamine is taken up into the neuron, phosphate-activated glutaminase (GLNase) splits it into glutamate and ammonia. Glutamate is then incorporated in synaptic vesicles that will release it to the synaptic cleft, starting a new cycle

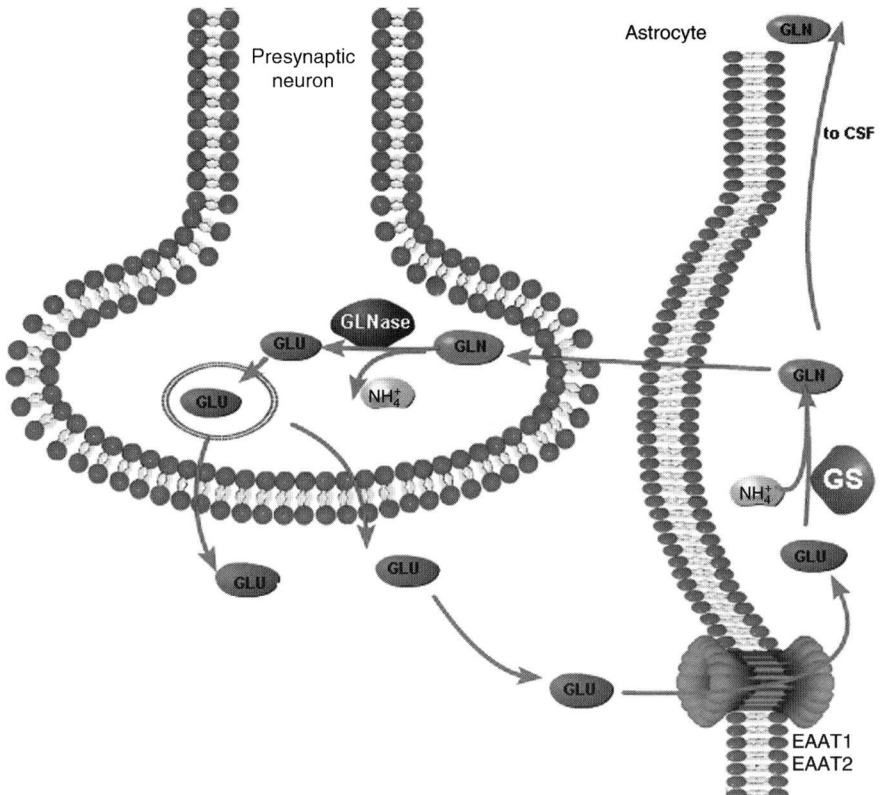

trafficking of glutamate and glutamine between astrocytes and neurons, usually called "the glutamate–glutamine cycle", is the major pathway by which the neurotransmitter pool of glutamate is recycled, and therefore this cycle is essential to maintain proper function of glutamatergic neurotransmission. Ammonia is a main player in this cycle and the increase in ammonia in hyperammonemia could affect its function, and therefore glutamatergic neurotransmission.

1.4 Pathological Situations Associated with Hyperammonemia in Humans

There are a number of pathological mechanisms and situations that may lead to hyperammonemia by different ways. The most common situation leading to hyperammonemia in humans is chronic liver disease (mainly liver cirrhosis). Acute liver failure leads to higher levels of hyperammonemia but is less frequent. Chronic liver failure may have different origins: alcohol ingestion, hepatitis virus, autoimmune origin, etc. Hyperammonemia in liver disease is mainly due to the reduced elimination of ammonia by the urea cycle in the liver. In acute liver failure this is due to the death of hepatic cells. In chronic liver disease a main contributor to the appearance of hyperammonemia is the formation of portal–systemic shunts that allow portal blood to circulate (and to reach, for example, the brain) without being properly detoxified by the liver. In addition to the reduced ammonia detoxification in the liver, patients with liver cirrhosis also present increased production of ammonia due to increased intestinal glutaminase activity (Romero-Gomez et al., 2004).

Congenital defects in enzymes of the urea cycles also result in hyperammonemia. Neonatal hyperammonemia may be due to congenital deficiencies in any of the urea cycle enzymes: ornithine transcarbamoylase (the most frequent), carbamoyl phosphate synthetase, argininosuccinate synthetase, argininosuccinase and arginase, or acetylglutamate synthetase, which synthesizes acetylglutamate, an essential activator of carbamoyl phosphate synthetase. Children with congenital defects in the urea cycle appear normal at birth, but rapidly develop hyperammonemia and, if not treated, may die soon after birth. If treated, they can survive, but hyperammonemia in the newborn period is commonly associated with central nervous system damage and can lead to mental retardation (see ❯ Section 4.1). Neonatal hyperammonemia may also occur in children with perinatal asphyxia or preterm delivery (Ballard et al., 1978; Yoshino et al., 1991).

Other situations resulting in hyperammonemia are Reye's syndrome, organic acidemias, and carnitine deficiencies. Hyperammonemia may also arise as complications of valproate therapy, hemodialysis, or leukemia treatment.

1.5 Hyperammonemia Is a Main Contributor to the Origin and the Neurological Alterations in Hepatic Encephalopathy

Clinical experience and basic research indicate that ammonia is a main factor responsible for hepatic encephalopathy in patients with liver disease. There are several evidences indicating that hyperammonemia is a main factor contributing to the neurological alterations found in hepatic encephalopathy.
- The clinical treatments that are effective in reversing the neurological alterations in patients with hepatic encephalopathy are those directed at reduction of ammonia levels such as reduction of protein intake, control of intestinal bacterial flora with antibiotics to reduce the formation of ammonia, and acidification of the pH in intestine and colon (with lactulose, etc) to maintain ammonia in the protonated form and reduce its transport to the blood stream.
- Precipitating factors of hepatic encephalopathy are associated with increased ammonia levels (high protein intake, gastrointestinal bleeding, etc).
- In congenital deficiencies of the urea cycle enzymes, hepatic function is normal except for detoxification of ammonia by incorporation into urea. Other metabolic alterations associated with liver failure are absent in these patients. However, these patients show encephalopathy similar to hepatic encephalopathy and its intensity increases in parallel with ammonia levels.

- In animal models, chronic hyperammonemia without liver failure reproduces most of the metabolic, neurochemical, and neurological alterations found in hepatic failure such as altered circadian rhythms and decreased learning ability.
- There is a good correlation between arterial blood ammonia concentration and the symptoms of hepatic encephalopathy (Clemmensen et al., 1999).

1.6 Differential Alterations in Acute and Chronic Hyperammonemia

Independently of the origin of hyperammonemia, all the above situations may lead to altered cerebral function. The effects of different hyperammonemic situations on cerebral function and the mechanisms responsible for these effects are different depending on the concentrations of ammonia reached, the duration of the exposure to hyperammonemia, and the age of the person. As mentioned above, neonatal hyperammonemia may induce mental retardation. In adults, acute and chronic hyperammonemia induce different alterations in cerebral function, and the mechanisms responsible for these alterations are also different.

Acute intoxication with large doses of ammonia may lead to rapid death. In acute liver failure, leading to high levels of ammonia in blood and brain, patients may die in a few days. In experimental animals, acute intoxication with high levels of ammonia leads to the rapid death of the animals.

In contrast, chronic hyperammonemia, as occurs for example in chronic liver disease, does not lead to death but impairs cerebral function leading to neurological alterations that include altered sleep-waking patterns, motor coordination, and impairment of intellectual capacity and learning ability. These alterations are observed both in patients with chronic liver disease with different grades of hepatic encephalopathy and in animal models of chronic liver failure or of chronic hyperammonemia without liver failure (see below).

In the following sections, we summarize the current state of knowledge about the molecular mechanisms by which acute or chronic hyperammonemia leads to these differential alterations in cerebral function.

1.7 Animal and In Vitro Models of Hyperammonemia

Most studies on the effects of hyperammonemia on cerebral function are directed to clarify the molecular mechanisms responsible for the cerebral and neurological alterations present in patients with liver disease and hepatic encephalopathy. For these studies it is usual to use models of chronic or acute liver failure that show, in addition to hyperammonemia, other alterations. The most frequently used model of chronic liver failure is the portacaval anastomosis in rats (Lee and Fisher, 1961). Other models of liver failure use hepatotoxic compounds such as thioacetamide in rats (Faff-Michalak and Albrecht, 1991) or galactosamine in rabbits (Pappas et al., 1984). However, although in these models the animals present hyperammonemia, other alterations are also present, making difficult to discern which effects are due to hyperammonemia and which due to other factors.

In this chapter we summarize mainly the studies carried out using models of "pure" hyperammonemia, without liver failure. There are several models of "pure" chronic hyperammonemia in animals in vivo. Some of them are described below.

One model induces sustained hyperammonemia by repeated injection of urease, which degrades urea and increases ammonia levels (Prior and Visek, 1973). Some authors use continuous infusion of ammonium salts (Tanigami et al., 2005). There is also a model of hyperammonemia in mice (sparse-fur) with a congenital defect that results in decreased activity of ornithine transcarbamoylase (one of the enzymes of the urea cycle) and in sustained hyperammonemia (Qureshi and Rao, 1997). Another model induces chronic moderate hyperammonemia by feeding rats a diet containing 20% ammonium acetate (Azorín et al., 1989).

To study the effects of acute hyperammonemia in animals, ammonium salts are injected intraperitoneally or, less frequently, intravenously.

In addition to these animal models of "pure" hyperammonemia, there is a large number of studies on the effects of ammonia on different aspects of cerebral function carried out in a large variety of in vitro systems. In some studies primary cultures of neurons or astrocytes are treated with ammonia. Other studies use slices from different areas of the brain, synaptosomes, etc. Some studies are performed ex vivo using slices, synaptosomes, etc, from animals with hyperammonemia induced in vivo.

It is important to point out that many studies on the effects of ammonia in vitro on different parameters in cultured cells or in vitro systems have been performed with very high concentrations (10, 20 mM) of ammonia that are never present in brain in vivo in patients with chronic or either acute liver failure. Care should be taken therefore when using this information to try to understand the alterations in cerebral function in patients with hyperammonemia due to liver disease, who show much lower levels of ammonia.

2 Molecular Mechanism of Acute Hyperammonemia Toxicity

Acute hyperammonemia causes severe neurological dysfunctions including seizures, stupor, and coma and may lead to death. Brain edema is a characteristic feature of acute liver failure and hyperammonemia and it is associated with high mortality due to brain herniation as a result of increased intracranial pressure. Brain edema is mainly attributed to ammonia-induced astrocyte swelling (see ❯ Section 2.12).

There is increasing evidence to support that the mechanisms responsible for cerebral dysfunction in acute hyperammonemia involve alterations in glutamatergic neurotransmission. A main contributor to the toxic effects of acute hyperammonemia in brain is the overactivation of NMDA type of glutamate receptors.

2.1 Acute Ammonia Intoxication Leads to Activation of NMDA Receptors in Brain In Vivo

Hermenegildo and coworkers (2000) showed by in vivo brain microdialysis in freely moving rats that acute ammonia intoxication (intraperitoneal injection of large doses of ammonia) leads to activation of NMDA receptors in rat brain in vivo. Activation of NMDA receptors in brain in vivo can be followed, under appropriate conditions, by measuring the activation of the glutamate–nitric oxide–cGMP pathway (❯ Figure 3-2).

Activation of NMDA receptors leads to increased intracellular Ca^{2+}, which binds to calmodulin (CM) and activates different enzymes, including neuronal nitric oxide synthase (nNOS), thus leading to increased formation of nitric oxide (NO) which, in turn, activates soluble guanylate cyclase (GC) and increases cGMP. Part of the cGMP formed is released to the extracellular space and, under appropriate conditions, the increase in extracellular cGMP is a good measure of the extent of activation of NMDA receptors.

Hermenegildo and coworkers (2000) showed by in vivo brain microdialysis that intraperitoneal injection of large doses of ammonia induces an increase in extracellular cGMP in the cerebellum, which is completely prevented if NMDA receptors are previously blocked by injecting MK-801. This indicates that acute ammonia intoxication leads to activation of NMDA receptors in brain, leading to activation of the glutamate–NO–cGMP pathway and to increased extracellular cGMP.

Only large doses of ammonia induced activation of NMDA receptors. A good correlation was found between the extent of activation of NMDA receptors in the brain (as measured by the increase in extracellular cGMP) and the severity of the neurological symptoms evoked by different doses of ammonia.

Hermenegildo and coworkers (2000) also tested using in vivo brain microdialysis whether the ammonia-induced activation of NMDA receptors would be mediated by an increase in extracellular glutamate in the brain. Acute ammonia intoxication leads to increased extracellular glutamate in the cerebellum. However, glutamate increase occurs later than activation of NMDA receptors (increase in cGMP) and is prevented by blocking NMDA receptors with MK-801, indicating that ammonia-induced increase of extracellular glutamate is not the cause for, but a consequence of, activation of NMDA receptors. Excessive activation of NMDA receptors leads to depletion of ATP (see ❯ Section 2.3). Under these conditions, the

Figure 3-2

The glutamate–nitric oxide–cyclic GMP pathway. Activation of ionotropic (mainly NMDA) glutamate receptors leads to increased intracellular calcium (Ca^{2+}), which, after binding to calmodulin (CM), activates nitric oxide synthase (NOS), leading to increased production of nitric oxide (NO), which in turn activates soluble guanylate cyclase (GC), resulting in increased formation of cGMP. Part of the cGMP formed is released to the extracellular space. cGMP is degraded by phosphodiesterases (PDE), which may be inhibited with zaprinast or sildenafil. GC may be also activated by NO formed by the NO-generating agent SNAP

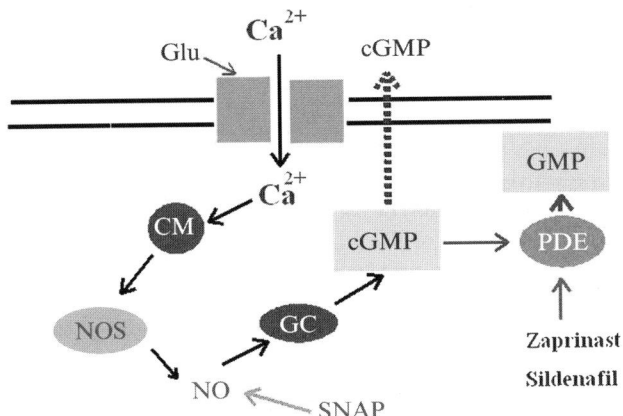

function of glutamate transporters is reversed (Sanchez-Prieto and Gonzalez, 1988) and, instead of taking up glutamate, cytosolic glutamate is released to the extracellular space. This mechanism may mediate the increase in glutamate induced by acute hyperammonemia.

If increased concentration of extracellular glutamate is not the reason for ammonia-induced activation of NMDA receptors, then how ammonia activates these receptors? The more likely explanation seems to be that the effect is mediated by ammonia-induced depolarization of neurons. The channel of the NMDA receptor is modulated by Mg^{2+} in a voltage-dependent manner (Mayer et al., 1984). Under normal conditions, Ca^{2+} can not enter the postsynaptic neuron because Mg^{2+} is blocking the NMDA receptor channel. However, when the neurons depolarize, the Mg^{2+} block is released and NMDA receptor may be activated by normal concentrations of extracellular glutamate without requiring an increase of glutamate. It has been shown that high ammonia concentrations lead to neuronal depolarization (Raabe, 1990); this would release the Mg^{2+} block of NMDA receptors, allowing increased activation without increasing extracellular glutamate.

2.2 Excessive Activation of NMDA Receptors Is Responsible for Ammonia-Induced Death of Animals

To assess whether activation of NMDA receptors is responsible for ammonia-induced death of animals it has been tested whether antagonists of this receptor prevent ammonia-induced death of mice and rats.

The NMDA receptor is a complex molecule with a channel and several sites for binding of agonists or modulators of the receptor. The protective effect of different compounds acting on different sites of the receptor has been tested. Ammonia-induced death of animals is nearly completely prevented by antagonists of the NMDA receptor acting as channel blockers (MK-801, ketamine, etc.), by competitive antagonists of the receptor (AP-5), and by compounds which inhibit the function of the receptor by acting at the glycine-binding site (ethanol, butanol) (Marcaida et al., 1992; Hermenegildo et al., 1996).

These results show that ten different antagonists of the NMDA receptor, acting on three different sites of the receptor prevent ammonia-induced death of animals. This indicates that ammonia-induced death of animals is mediated by excessive activation of NMDA receptors.

2.3 Acute Hyperammonemia Induces Depletion of ATP in Brain That Is Mediated by Activation of NMDA Receptors

Ammonia toxicity was first described in the 1890s in the laboratory of Pavlov (Hahn et al., 1893). A lot of work has been done during the last decades to clarify the molecular mechanism of acute ammonia toxicity. It has been repeatedly shown that acute intoxication with large doses of ammonia produces marked alterations in brain energy metabolism, including increased lactate, pyruvate, and mitochondrial [NAD^+]/[NADH] and decreased cytosolic [NAD^+]/[NADH] (Hindfelt and Siesjö, 1971; Hawkins et al., 1973; Hindfelt et al., 1977; Kosenko et al., 1993) and, at a later step, decreased ATP content (Schenker et al., 1967; Hindfelt et al., 1977; Kosenko et al., 1993, 1994).

The depletion of ATP and energy metabolites could be involved in the origin of ammonia-induced coma and death. As discussed above, acute ammonia toxicity is mediated by excessive activation of NMDA receptors. To assess whether this excessive activation is also responsible for ammonia-induced depletion of brain ATP, Kosenko and coworkers (1994) tested whether blocking NMDA receptors by injecting MK-801 prevents the depletion of brain ATP induced by i.p. injection of 7 mmol/kg of ammonium acetate to rats. In rats injected with ammonia the content of ATP in brain decreased from 1.95 to 0.86 μmol/g 15 min after ammonia injection. In contrast, in rats injected previously with MK-801 to block NMDA receptors, injection of ammonia did not induce any decrease of ATP (Kosenko et al., 1994).

These results indicate that ammonia-induced depletion of ATP in brain is mediated by excessive activation of NMDA receptors.

One of the consequences of this depletion of ATP is the reversal of the function of the glutamate transporter, which, instead of taking up glutamate, releases it from the cytosol to the extracellular space, leading to increased extracellular glutamate. Blocking NMDA receptors with MK-801 prevents ammonia-induced depletion of ATP and, therefore, also prevents ammonia-induced increase in extracellular glutamate (see above).

Ammonia-induced depletion of brain ATP is therefore mediated by excessive activation of NMDA receptors, which activates processes that may lead to increased consumption of ATP, decreased synthesis, or both. Some of the processes induced by ammonia and mediated by activation of NMDA receptors that contribute to ATP depletion in the brain are summarized below.

2.4 Ammonia-Induced Activation of NMDA Receptors Leads to Activation of Calcineurin and Dephosphorylation and Activation of Na^+/K^+-ATPase in Brain, Thus Increasing ATP Consumption

Activation of NMDA receptors leads to the opening of the ion channel allowing the entry of Ca^{2+} and Na^+ into the postsynaptic neuron. To maintain Na^+ homeostasis, Na^+ entering through the channel must be extruded from the neuron. This is carried out mainly by Na^+/K^+-ATPase, which consumes ATP. Activation of NMDA receptors would result therefore in increased activity of Na^+/K^+-ATPase to remove the excess of Na^+ and, therefore, lead to increased consumption of ATP (❯ Figure 3-3).

Kosenko and coworkers (1994) showed that in fact acute intoxication with ammonia leads to increased activity of Na^+/K^+-ATPase in the brain. The activity of Na^+/K^+-ATPase was increased by 76% in brains of rats 15 min after injection of 7 mmol/kg of ammonia. This increase was completely prevented by previous injection of MK-801, indicating that it is mediated by activation of NMDA receptors.

The fact that increased activity can be observed in vitro indicates that acute ammonia intoxication induces a covalent change in the Na^+/K^+-ATPase, which is maintained after homogenization of brain tissue. The activity of Na^+/K^+-ATPase is modulated with phosphorylation by protein kinase C (PKC); increased phosphorylation inhibits the enzyme while decreased phosphorylation increases the activity of Na^+/K^+-ATPase. This suggested that acute ammonia intoxication may be decreasing phosphorylation of Na^+/K^+-ATPase by PKC, thus resulting in increased activity. Kosenko and coworkers (1994) also showed that when PMA (an activator of PKC) is added to the samples from rats injected with ammonia, the activity of Na^+/K^+-ATPase is reduced, returning to the same values found in the controls.

◻ **Figure 3-3**
Some of the effects induced by acute ammonia toxicity and mediated by activation of NMDA receptors. Acute ammonia intoxication leads to activation of NMDA receptors in the brain allowing the entry of Ca^{2+} and Na^+ in the postsynaptic neuron. Calcium is taken up by mitochondria, resulting in mitochondrial calcium accumulation, impairment of mitochondrial respiration, decreased ATP synthesis, and increased formation of free radicals leading to oxidative stress. Cytosolic Ca^{2+} binds to calmodulin (CM) and activates both nitric oxide synthase (NOS) and calcineurin (CN). Activation of NOS increases the formation of nitric oxide (NO), which affects the activity or function of different enzymes and proteins both in neuronal mitochondria and cytosol and likely in neighboring astrocytes (e.g., glutamine synthetase, see ❯ *Figure 3-4*). The activation of CN leads to dephosphorylation of a residue of Na^+/K^+-ATPase phosphorylated by protein kinase (PKC). This results in activation of the ATPase, which facilitates the removal of the excess of Na^+ entering through the NMDA receptor, but increases the consumption of ATP. This, together with the reduced synthesis, leads to depletion of ATP

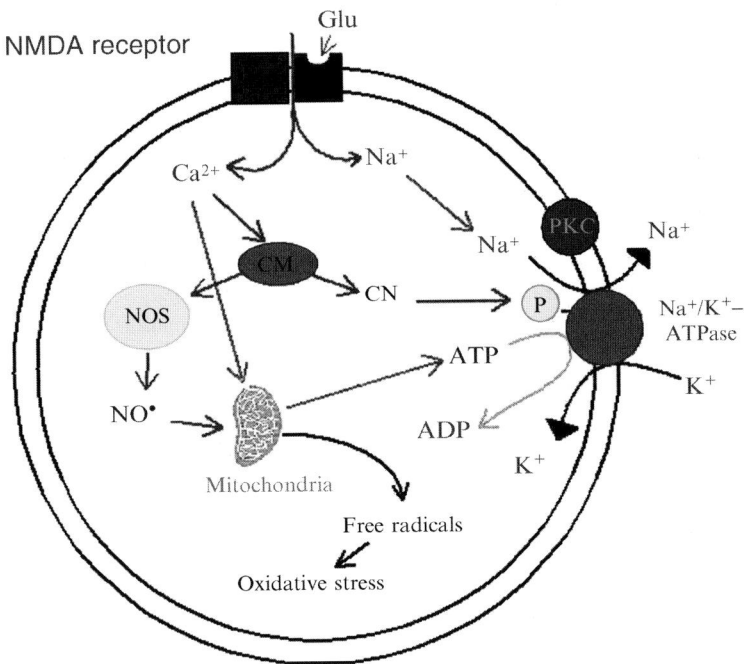

This indicates that acute ammonia intoxication leads to an NMDA receptor-mediated dephosphorylation of Na^+/K^+-ATPase and concomitant activation of the enzyme.

To analyze the mechanism by which activation of NMDA receptors leads to dephosphorylation of Na^+/K^+-ATPase, Marcaida and coworkers (1996) used primary cultures of neurons. They showed that addition of glutamate to the neurons in culture induces a rapid activation of Na^+/K^+-ATPase, which is prevented by blocking NMDA receptors with MK-801 and is reversed by activating PKC with PMA. This demonstrates that this system reproduces exactly the effects induced by acute ammonia intoxication on Na^+/K^+-ATPase in the rat brain. Using this model, Marcaida and coworkers (1996) showed that inhibitors of calcineurin, a Ca^{2+}/calmodulin-dependent protein phosphatase which is activated following activation of NMDA receptors, prevent activation of Na^+/K^+-ATPase by glutamate (❯ *Figure 3-3*).

The above studies indicate that acute ammonia intoxication leads to activation of NMDA receptors and increased intracellular Ca^{2+}, which binds to CM and activates calcineurin, which in turn dephosphorylates Na^+/K^+-ATPase in a phosphorylation site for PKC. Dephosphorylation of

Na$^+$/K$^+$-ATPase increases the activity of the enzyme and increases the consumption of ATP, thus contributing to depletion of ATP in the brain.

2.5 Ammonia-Induced Activation of NMDA Receptors Impairs Mitochondrial Function, Thus Decreasing ATP Synthesis

Excessive activation of NMDA receptors leads to the entry of Ca^{2+} from the extracellular space into the cytoplasm of the postsynaptic neuron. Much of this Ca^{2+} entering the cell is sequestered by mitochondria (❯ *Figure 3-3*). The uptake of calcium by mitochondria is driven by the mitochondrial membrane potential and will compete with the mitochondrial ATP synthase for protons, and it can be predicted that mitochondrial ATP synthesis may decrease as Ca^{2+} floods from the cytoplasm to the mitochondrial matrix.

Kosenko and coworkers (2000) showed that acute ammonia intoxication increased calcium content in brain mitochondria by 62% and this increase was prevented by MK-801, indicating that it is mediated by activation of NMDA receptors. Ammonia injection also reduced the calcium capacity of brain mitochondria and the calcium uptake rate and increased the spontaneous calcium efflux. These effects were also prevented by MK-801. The results indicate that ammonia-induced excessive activation of NMDA receptors in the rat brain in vivo leads to a rapid increase in the uptake of calcium by mitochondria, removing from the cytosol the calcium entering to the neuron through the NMDA receptor channel (❯ *Figure 3-3*). The initial uptake of calcium by mitochondria is followed by a strong alteration of calcium transport and this may be involved in the mechanism by which ammonia leads to the death of the animal.

The ammonia-induced increase in intramitochondrial Ca^{2+} described above may affect the function of different enzymes and of the respiratory chain and lead to increased formation of free radicals.

Kosenko and coworkers (1997a) also showed that state 3 respiration was inhibited in mitochondria from rats injected with ammonia suggesting that ammonia intoxication also inhibits the respiratory chain. Impairment of the respiratory chain may lead to decreased synthesis of ATP and to increased formation of free radicals.

2.6 Ammonia Induces Formation of Free Radicals

It has been shown that i.p. injection of large doses of ammonia leads to increased formation (more than twice of controls) of superoxide radical in brain mitochondria. Moreover, the activities of glutathione peroxidase, superoxide dismutase, and catalase were decreased in brain of rats injected with ammonia (Kosenko et al., 1996, 1997b). The content of glutathione was also reduced by approximately 50%, and lipid peroxidation was increased (Kosenko et al., 1999). This indicates that ammonia induces oxidative stress in the brain.

Blocking NMDA receptors with MK-801 prevented ammonia-induced effects on the activities of antioxidant enzymes, on glutathione content, on lipid peroxidation, and on superoxide formation, indicating that all these effects are a consequence of previous activation of NMDA receptors.

The above studies show that acute ammonia intoxication leads to serious disturbances in mitochondrial function at various levels: calcium content, transport and homeostasis, oxidative metabolism, respiratory chain, antioxidant enzymes, glutathione content, and superoxide formation. All these alterations are mediated by excessive activation of NMDA receptors and contribute to reduce ATP synthesis and to induce oxidative stress, which may be involved in the mechanism of acute ammonia toxicity.

Exposure to ammonia also increases the free-radical production in cultured astrocytes in a time-dependent manner (Murthy et al., 2001). Glutathione levels in cultured astrocytes were significantly decreased at the time of free-radical production (Murthy et al., 2000a), suggesting a decreased antioxidant capacity of astrocytes in hyperammonemia.

As summarized below (in ❯ *Section 2.8*), acute ammonia intoxication also increases the formation of NO, another free radical that contributes to ammonia toxicity.

The role of oxidative/nitrosative stress in hyperammonemia and hepatic encephalopathy has been recently reviewed by Norenberg et al. (2004).

2.7 Ammonia-Induced Activation of NMDA Receptors Activates Calpain Which Degrades the Microtubule-Associated Protein MAP-2, Thus Altering the Microtubular Network

Acute ammonia intoxication leads to degradation of the microtubule-associated protein MAP-2 in the rat brain, which is prevented by MK-801, indicating that it is mediated by activation of NMDA receptors (Felipo et al., 1993a). Calcium-dependent proteolysis of MAP-2 in brain extracts is inhibited by inhibitors of calpain, a calcium-dependent protease. This indicates that ammonia-induced activation of NMDA receptors leads to activation of calpain, which degrades MAP-2.

MAP-2 is also degraded in the brain during ischemia (Blomgren et al., 1995), a process in which NMDA receptors are also excessively activated.

MAP-2 binds to tubulin and promotes and maintain its polymerization, thus providing stability to the microtubules. It has been shown in primary cultures of neurons that activation of NMDA receptors leads to proteolysis of MAP-2 and derangement of the neuronal microtubular network (Miñana et al., 1998). It is therefore possible that ammonia-induced proteolysis of MAP-2 may also contribute to the toxic effects of ammonia.

2.8 Hyperammonemia Increases Nitric Oxide Formation by Different Mechanisms and Increases Tyrosine Nitration of Brain Proteins

Hyperammonemia results in stimulation of nNOS in the brain both in vitro and in vivo. Two different mechanisms may contribute to this effect: activation of NMDA receptors and subsequent increase in intracellular calcium and increased uptake of arginine, one of the substrates of nNOS. As summarized above, acute injection of ammonia leads to activation of NMDA receptors in brain in vivo. This leads to increased calcium and activation of nNOS in postsynaptic neurons and increased formation of NO (❯ *Figures 3-2* and ❯ *3-3*).

Exposure of rat cerebellar synaptosomal preparations to ammonia in vitro results in increased high affinity uptake of the nNOS substrate L-arginine. Increased capacity for L-arginine transport and a concomitant increase in nNOS mRNA is also observed in cerebellum of rats acutely injected with ammonia (Rao et al., 1997).

Moreover, ammonia can also increase the expression of the inducible form of nitric oxide synthase (iNOS). Schliess and coworkers (2002) showed that addition of ammonia to primary cultures of astrocytes increases the expression of iNOS resulting in increased formation of NO.

The increase in NO formation mediates some of the deleterious effects of hyperammonemia. This is supported by the fact that previous injection of nitroarginine, an inhibitor of NOS, prevents many metabolic effects of acute hyperammonemia as well as superoxide radical increase, and reduces ammonia toxicity (Kosenko et al., 1995, 1998).

Schliess and coworkers (2002) also showed that the increase in iNOS elicited by ammonia in astrocytes induces nitration of tyrosines in several proteins, including GS (see ❯ Section 2.9). Ammonia-induced nitration of most proteins was prevented by previous blocking of NMDA receptors with MK-801, and addition of NMDA induces a pattern of protein nitration similar to that induced by ammonia. This suggests that ammonia would activate NMDA receptors in astrocytes leading to iNOS-mediated increase in the formation of NO, which, in turn, leads to tyrosine nitration (and likely alteration of function) of a number of proteins.

Schliess and coworkers (2002) also showed that protein tyrosine nitration is increased in brain of rats acutely injected with ammonia as well as in rats with chronic liver failure (portacaval anastomosis). Increased tyrosine nitration of key proteins could therefore play a role in the mechanism of ammonia toxicity.

2.9 Ammonia-Induced Activation of NMDA Receptors Increases Nitric Oxide Formation, Which, in Turn, Reduces the Activity of Glutamine Synthetase, Thus Reducing Elimination of Ammonia in Brain

The main mechanism for ammonia detoxification in the brain is the reaction with glutamate to form glutamine. This reaction is catalyzed by GS and consumes ATP. Acute intoxication with ammonia leads to a rapid increase in glutamine content in the brain, thus contributing to reduce the concentration of ammonia and also to increase the consumption of ATP.

During the experiments of Kosenko and coworkers (1994), mentioned above, the authors found that in the group of rats injected only with MK-801 a significant increase of both the concentration of glutamine and the activity of GS in the brain, indicating that blocking NMDA receptors increases the activity of GS. Intraperitoneal injection of nitroarginine, an inhibitor of NOS, also results in a significant increase in glutamine content and in the activity of GS in brain (Kosenko et al., 1995).

GS from different organisms is modulated by oxidation. This fact and the above results suggested that GS in the brain is modulated by NMDA receptors and NO. Activation of NMDA receptors leads to increased formation of NO. This radical is a gas and can diffuse to neighboring astrocytes and react with GS in these cells, reducing the activity of the enzyme (❯ Figure 3-4).

To assess whether GS in astrocytes is actually modulated by NO, Miñana and coworkers (1997) used primary cultures of astrocytes and showed that the NO-generating agent SNAP inhibits the activity of GS in intact astrocytes in culture by approximately 50%.

◘ Figure 3-4

Proposed mechanism for the modulation of glutamine synthetase activity by NMDA receptors and nitric oxide. Acute ammonia toxicity leads to activation of NMDA receptors. This leads to activation of neuronal nitric oxide synthase nNOS and increases nitric oxide (NO) production. NO (a gas) can diffuse to neighboring astrocytes and modify covalently (likely by tyrosine nitration) glutamine synthetase (GS). This modification reduces GS activity. Other sources of NO, independent of NMDA receptors, also contribute to modulation of GS activity in rat brain in vivo. The covalent modification of GS induced by NO would be reversible, likely by an yet identified enzyme. Reducing the activation of NMDA receptors or of NOS reduces the tonic inhibition of GS by NO, resulting in enhanced activity of the enzyme. In addition to hyperammonemia, other agents or situations that alter NMDA receptor function or NO concentration also would alter GS activity and glutamine formation. (Modified from Kosenko et al., 2003)

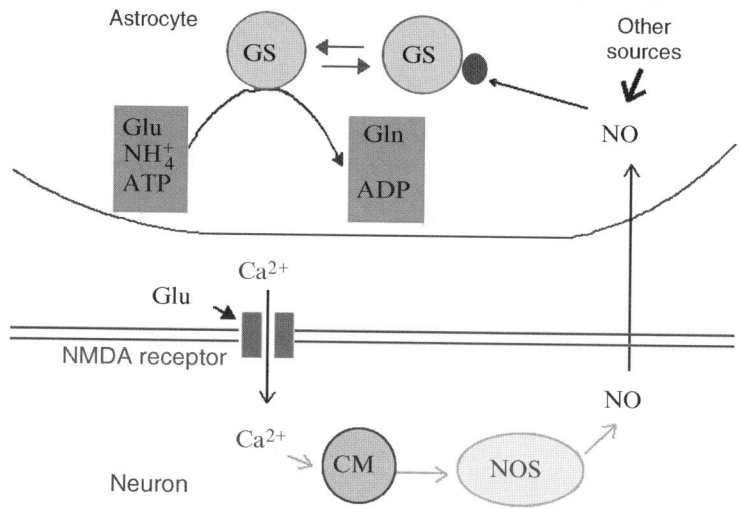

These results support the idea that in the brain, tonic activation of NMDA receptors maintains a tonic activation of nNOS, which produces a basal level of NO that maintains a tonic, basal, slight inhibition of GS in astrocytes (❯ *Figure 3-4*). Blocking NMDA receptors by injection of MK-801 (Kosenko et al., 1994) or inhibition of NOS with nitroarginine (Kosenko et al., 1995) eliminates this tonic inhibition and increases the activity of GS and the content of glutamine in brain. This tonic inhibition of GS by NMDA receptors activity and NO is acting under normal conditions. Agents or situations that alter the activation of NMDA receptors or NO concentration, including hyperammonemia, would therefore result in altered modulation of GS and formation of glutamine.

These results imply that GS in astrocytes is modulated by neuronal activity (see Kosenko et al., 2003) and that, in contrast with some reports, GS is not working at maximum rate under normal conditions and may be activated under certain conditions.

Moreover, these results also indicate that acute ammonia intoxication, which activates NMDA receptors and increases the formation of NO, leads to a reduction in the specific activity of GS. The increased content of glutamine following acute ammonia intoxication will be due to the increase in the concentration of the substrate (ammonia) of GS. This is supported by reports showing that when ammonia is injected after blocking of NMDA receptors with MK-801 (Kosenko et al., 1994) or after inhibition of NOS with nitroarginine (Kosenko et al., 1995), the increase in glutamine content in the brain is significantly higher than in rats injected only with ammonia.

On the basis of the above discussion, it may be proposed that one possible procedure to increase the removal of ammonia in brain in hyperammonemia (due to liver failure or to other reasons) may be the use of inhibitors of NOS, which will prevent inhibition of GS. In fact, in the above studies, the content of ammonia in the brain in rats injected with MK-801 and ammonia (Kosenko et al., 1994) or with nitroarginine and ammonia (Kosenko et al., 1995) was significantly lower than in the brain of rats injected only with ammonia.

As described above, Schliess and coworkers (2002) showed that addition of ammonia to astrocytes in culture elicits an increase in iNOS, which induces nitration of tyrosines in several proteins including GS. The increase in GS nitration was associated with a significant decrease of its activity by approximately 30% in cultured astrocytes. Schliess and coworkers (2002) also showed that tyrosine nitration of GS is increased in the brain of rats acutely injected with ammonia as well as in rats with chronic liver failure (portacaval anastomosis).

Schliess and coworkers (2002) propose that activation of NMDA receptors in astrocytes would mediate the effects of ammonia on GS nitration. It is considered that NMDA receptors are located essentially in neurons. The presence of NMDA receptors in astrocytes in vivo remains controversial. Some authors believe that maybe NMDA receptors could be expressed in astrocytes in culture but not in brain in vivo. Other authors suggest that NMDA receptors are expressed only in reactive astrocytes or microglia but not in normal astrocytes (Sugimoto et al., 1994; Gottlieb and Matute, 1997). It is likely that in brain in vivo modulation of GS by NMDA receptors should be attributed to neuronal receptors although the contribution of NMDA receptors in astrocytes should not be disregarded until specific studies are performed to clarify this possibility.

2.10 Effects of Acute Hyperammonemia on Non-NMDA Ionotropic Receptors

Acute exposure of rat cortical wedges to millimolar concentrations of ammonia reduces the grade of depolarization caused by AMPA (Lombardi et al., 1994). Also, the number of AMPA–kainate binding sites is reduced in the brains of rats with acute hyperammonemia induced by hepatic devascularization (Michalak et al., 1996). Although very few studies have been carried out on the alterations of AMPA or kainate receptors in hyperammonemia, it seems that these receptors are also altered.

2.11 Acute Exposure to Ammonia Results in a Loss of Expression of Astrocytic Glutamate Transporters GLT-1 and GLAST

Exposure of different types of rat brain preparations including hippocampal slices (Schmidt et al., 1990), cortical synaptosomes (Mena and Cotman, 1985), and primary cortical astrocyte cultures

(Norenberg et al., 1997; Chan and Butterworth, 1999) to ammonia results in a significant loss of high-affinity glutamate uptake capacity. In the case of cortical astrocytes, this loss seems to be due to decreased expression of GLAST (Chan et al., 2000), the glutamate transporter normally expressed by these cells. On the other hand, acute hyperammonemia associated with experimental acute liver failure results in a loss of expression of the major cortical glutamate transporter GLT-1 (Knecht et al., 1997), which is associated with increased extracellular glutamate concentration in brain. Increased extracellular brain glutamate is a consistent feature of acute hyperammonemia in a variety of experimental paradigms (see Monfort et al., 2002a for a review). The decrease in glutamate transporters, as well as the reversal of the function of some of them due to ATP depletion (see above), would be the main contributors to increased extracellular glutamate, which would maintain the toxic effects of acute ammonia intoxication after the initial activation of NMDA receptors.

2.12 Exposure to Ammonia Induces Swelling of Astrocytes in Culture

A serious complication in acute liver failure is brain edema. A major contributor to the edema seems to be astrocyte swelling, in which hyperammonemia seems to play a main role. Ganz and coworkers (1989) treated cortical brain slices with different concentrations of ammonia. Swelling was detected at ammonia concentrations of 5 and 10 mM, indicating that at high ammonia concentrations ammonia leads to water accumulation in cortical brain slices.

Norenberg and coworkers (1991) showed that treatment of cultured astrocytes with 10 mM ammonia for 4 days induced swelling. This effect was time- and dose dependent. The ammonia-induced swelling was reversible. These findings suggest that hyperammonemia leads to astrocyte swelling and may contribute to the brain edema in acute liver failure.

Exposure of cultured astrocytes to ammonia results in the mitochondrial permeability transition (MPT), a phenomenon associated with mitochondrial failure and subsequent cellular dysfunction. Treatment of cultured astrocytes with ammonia (5 mM) caused a time-dependent increase in astrocyte cell volume (swelling), which was completely inhibited by the MPT inhibitor cyclosporin A (CsA) (Rama Rao et al., 2003a).

Exposure of astrocytes to ammonia also induces aquaporin 4 (AQP4, a water channel) upregulation (Rama Rao et al., 2003b), which could increase water uptake and is also prevented by CsA (Rama Rao et al., 2003a).

These findings suggest that the MPT plays a significant role in the ammonia-induced astrocyte swelling and may contribute to the brain edema in acute liver failure.

Other possible mediators in the astrocyte swelling induced by ammonia are free radicals (Murthy et al., 2000b) and glutamine. Tanigami and coworkers (2005) showed that inhibition of glutamine synthesis with methionine sulfoximine reduces astrocyte swelling during ammonia infusion in anesthetized rats. Rats received a continuous i.v. infusion of ammonium acetate for 24 h to increase plasma ammonia from about 30–400 µmol/l. Hyperammonemia increased the number of swollen astrocytes in the cortex and methionine sulfoximine reduced this increase to control values.

Recent findings suggest a close interrelation between astrocyte swelling, NMDA receptor signaling, glutamate, oxidative stress, NO, and protein tyrosine nitration, which may result in mutual amplification of swelling and oxidative stress, thus contributing to the toxic effects of hyperammonemia (see Haussinger and Schliess, 2005; Norenberg et al., 2005 for recent reviews).

3 Effects of Chronic Hyperammonemia on Cerebral Function

3.1 Acute and Chronic Moderate Hyperammonemia May Affect the Same Processes in Different, Even Opposite, Ways

As mentioned before, the effects of different hyperammonemic situations on cerebral function and the mechanisms responsible for these effects are different depending on the concentrations of ammonia reached, the duration of the exposure to hyperammonemia, and the age of the person. In the above sections, we have summarized some reports on the molecular mechanisms involved in acute ammonia toxicity. Many of these reports indicate that excessive activation of the NMDA type of glutamate receptors and of associated signal transduction pathways play a main role in acute ammonia toxicity and ammonia-induced death. In contrast, in chronic hyperammonemia, the levels of ammonia are not high enough to induce death and this seems to allow the induction of an adaptive response which results in a reduced function of some signal transduction pathways associated with NMDA receptors. The impairment of these pathways may prevent some of the toxic effects due to ammonia-induced overactivation of NMDA receptors, but also reduces the normal physiological function of these pathways and this contributes (see ● Section 4) to at least some of the neurological alterations present in hyperammonemia and hepatic encephalopathy, including the impairment in cognitive function.

It is therefore clear that acute (large doses) and chronic moderate hyperammonemia may affect the same processes in different ways, by different mechanisms and can result in different (either opposite) effects on the same processes.

3.2 Effects of Chronic Hyperammonemia on Glutamatergic Neurotransmission

3.2.1 Effects on NMDA Receptors

Chronic Hyperammonemia Reduces the Binding of Ligands to NMDA Receptors The binding of [^{3}H] glutamate to NMDA-sensitive sites is reduced in synaptic membrane preparations from rats made chronically hyperammonemic by administration of ammonium salts (Rao et al., 1991) as well as in the brains of rats with chronic liver failure and hyperammonemia induced by portacaval anastomosis (Peterson et al., 1990). Chronic hyperammonemia also results in decreased binding of [^{3}H]MK-801 (a selective antagonist of NMDA receptors) to synaptosomal membranes from rat hippocampus (Marcaida et al., 1995) and in the brain of sparse-fur mice that present chronic hyperammonemia due to a congenital defect of the urea cycle enzyme ornithine transcarbamylase (Ratnakumari et al., 1995). A reduced binding of ligands to NMDA receptors is therefore a common finding in several different animal models of chronic hyperammonemia induced by very different procedures. This indicates that chronic hyperammonemia affects NMDA receptors. Immunoblotting studies failed to reveal any significant loss of expression of the NMDA-R1 subunit of the receptor due to chronic exposure to ammonia (Marcaida et al., 1995), suggesting a loss of receptor function or an alteration in the cell-surface expression of the receptor.

Chronic exposure of cultured rat cerebellar neurons to 1 mM ammonia likewise resulted in significant loss of [^{3}H]MK-801 binding of a similar magnitude to that observed in rat brain preparations (Marcaida et al., 1995). The authors suggest that, for this high ammonia concentration, the functional deficit was due to decreased PKC-mediated phosphorylation (Marcaida et al., 1995; Grau et al., 1996).

Chronic Hyperammonemia Impairs the Function of the Glutamate–Nitric oxide–cGMP Pathway, Associated with NMDA Receptors in Brain In Vivo Chronic moderate hyperammonemia in rats, similar to that present in patients with liver cirrhosis, impairs the glutamate–NO–cGMP pathway in cerebellum in vivo, as shown by brain microdialysis in freely moving rats by Hermenegildo and coworkers (1998). Microdialysis probes were inserted in the cerebellum of control and hyperammonemic rats without liver failure. Administration of NMDA through the microdialysis probe activates the glutamate–NO–cGMP pathway and increases cGMP formation. Part of the cGMP formed is released to the extracellular fluid (● Figure 3-2), and the increase in extracellular cGMP is a good measure of the function of the glutamate–NO–cGMP pathway in cerebellum in vivo. Hermenegildo and coworkers (1998) showed that the NMDA-induced increase in extracellular cGMP in cerebellum was significantly lower in hyperammonemic rats than in control rats, indicating that chronic hyperammonemia impairs the glutamate–NO–cGMP pathway in rat cerebellum in vivo.

The same authors showed that chronic exposure of cultured neurons to 0.1 mM ammonia (the increase found in brain of rat models of chronic hyperammonemia) led to a significant impairment the glutamate–NO–cGMP pathway (Hermenegildo et al., 1998). This treatment did not alter the increases in intracellular Ca^{2+} induced by 250 µM glutamate or NMDA, but results in significant reduction of the glutamate (NMDA) receptor-mediated formation of cGMP in a dose- and time-dependent manner. Exposure of the neurons to ammonia did not affect the activation of NOS but did cause a reduction in the synthesis of cGMP induced by the NO-generating agent SNAP indicating that chronic ammonia exposure impairs the activation of GC by NO.

To assess whether in rats in vivo the impairment also occurs at the level of activation of soluble GC by NO, an NO-generating agent, SNAP, was administered through the microdialysis probe to activate directly GC. The increase in extracellular cGMP induced by SNAP was also significantly reduced in hyperammonemic rats indicating that chronic moderate hyperammonemia impairs activation of soluble GC by NO in cerebellum in vivo, resulting in impairment of the glutamate–NO–cGMP pathway.

The glutamate–NO–cGMP pathway is also impaired in cerebellum in vivo in rats with chronic liver failure (and hyperammonemia) induced by portacaval anastomosis, as shown by brain microdialysis in freely moving rats by Monfort and coworkers (2001).

These results indicate that the function of the glutamate–NO–cGMP is altered in brain in vivo in animal models of chronic liver failure and chronic hyperammonemia. Moreover, the step of the pathway mainly affected is the activation of soluble GC by NO.

The modulation of soluble GC by NO is also altered in the brain of patients who died with hepatic encephalopathy. The activation of GC by the NO-generating agent SNAP was significantly lower in the cerebellum from cirrhotic patients than in controls (Corbalán et al., 2002).

The above results indicate that hyperammonemia is responsible for the alterations found in the modulation of GC in patients with liver cirrhosis. As will be described later (● Section 4.3), this alteration may contribute to the impairment of cognitive function in these patients.

Chronic Hyperammonemia also Impairs NMDA Receptor-Dependent Long-Term Potentiation in Hippocampus
NMDA receptor-dependent long-term potentiation (LTP) in the hippocampus is considered the molecular basis for some forms of learning and memory. LTP is impaired both in hippocampal slices from control rats exposed to ammonia in vitro (Muñoz et al., 2000; Monfort et al., 2004) and in hippocampal slices from rats with chronic moderate hyperammonemia (Monfort et al., 2005). Hyperammonemia reduces the magnitude of the potentiation and impairs its maintenance. Monfort and coworkers (2002b) showed that proper induction of LTP requires activation of the glutamate–NO–cGMP pathway, followed by activation of cGMP-dependent protein kinase (PKG), which phosphorylates and activates cGMP-degrading phosphodiesterase which, in turn, hydrolyzes cGMP to GMP, reducing cGMP concentration. This decrease in cGMP does not occur in hyperammonemia. The step impaired is the activation of the phosphodiesterase by PKG (Monfort et al., 2004, 2005). Moreover, the maintenance of the potentiation can be restored by addition of 8Br-cGMP (Monfort et al., 2004). These results again point out at alterations in cGMP in the brain as an important mediator in the deleterious effects of hyperammonemia.

NMDA receptor-dependent LTP in hippocampus is considered the basis of some forms of learning and memory. The impairment of this process (and of cGMP metabolism) may therefore be involved in the intellectual deficits present in patients with liver disease and hepatic encephalopathy.

3.2.2 Alterations in Metabotropic Glutamate Receptors

Although the effects of hyperammonemia and liver failure on ionotropic glutamate receptors have been studied by different groups, only very few studies have addressed the effects on metabotropic glutamate receptors (mGluRs). Lombardi and coworkers (1994) reported that ammonium acetate (2–4 mM) reduces the formation of inositol phosphates induced by tACPD, an agonist of mGluRs, in slices from cerebral cortex. However, ammonia potentiated the effects of mGluRs coupled to G proteins that inhibit adenylate cyclase, indicating that high ammonia concentrations (2–4 mM) differentially affect the effects induced by activation of different types of mGluRs in cerebral cortex.

Sáez and coworkers (1999) showed that in primary cultures of cerebellar neurons activation of mGluRs with tACPD leads to a significant increase in the phosphorylation of the microtubule-associated protein MAP-2 in control neurons. However, in neurons exposed to ammonia, tACPD induced a dephosphorylation of MAP-2 indicating that hyperammonemia alters the function of signal transduction pathways associated to mGluRs that modulate phosphorylation of MAP-2.

These in vitro studies suggest that the processes modulated by mGluRs may be altered in brains of hyperammonemic rats in vivo. Canales and coworkers (2003) showed, by using both immunohistochemistry and immunoblotting, that chronic hyperammonemia increases the content of mGluR1α but not of mGluR5 in the nucleus accumbens (NAcc) of rats.

Activation of mGluRs in the NAcc modulates locomotion. It is therefore possible that alterations in signal transduction associated to mGluRs contributes to the motor alterations found in liver disease and hepatic encephalopathy. One study of Canales and coworkers (2003) addressing this possibility is summarized in ❯ Section 4.2.

3.3 Hyperammonemia and GABAergic Neurotransmission

As mentioned above, hyperammonemia is considered the main contributor to the neurological alterations in hepatic encephalopathy. Schafer and Jones (1982) proposed that hepatic encephalopathy is caused by increased GABAergic tone, resulting in increased neuroinhibition. Originally this hypothesis was not related with ammonia but was proposed as an alternative mechanism to explain the alterations found in hepatic encephalopathy. As the present review deals on hyperammonemia, the possible alterations of GABAergic neurotransmission in liver disease due to reasons other than hyperammonemia are not discussed here. However, the same authors have recently modified the hypothesis to introduce a role for hyperammonemia in the increase in GABAergic tone. This part and other reports on the possible effects of hyperammonemia on GABAergic neurotransmission are reviewed below.

Norenberg and coworkers (1985) showed that ammonia concentrations in the range between 0.2 and 10 mM inhibits the uptake of GABA by astrocyte cultures. This could lead to increased extracellular GABA in hyperammonemia in vivo.

Takahashi and coworkers (1993) showed that ammonia potentiates the GABA-induced currents in dissociated rat cortical neurons with an EC_{50} of 0.2 mM, likely by increasing the affinity of GABA for the $GABA_A$ receptor. Subsequent studies on the effects of ammonia on the binding of ligands to $GABA_A$ receptors confirmed that ammonia in the 0.5–2.5 mM range may directly increase the binding of [^{3}H] muscimol to the $GABA_A$ receptor (Ha and Basile, 1996). Ammonia acts synergistically with benzodiazepines to enhance the binding of [^{3}H]muscimol to its receptor, suggesting an effect of ammonia on the coupling between these sites on the $GABA_A$ receptor complex (Basile, 2002). These reports suggest that hyperammonemia may increase GABAergic tone without affecting extracellular concentrations of GABA.

In contrast to the above results, Ahboucha and coworkers (2005) have shown that ammonia (at millimolar concentrations, IC_{50} = 4.8 mM) decreases the binding of ligands to the benzodiazepine site of GABA receptors in well-washed membranes from human cerebral cortex. This effect was not seen in cerebellum or hippocampus. Moreover, ammonia did not affect the binding of muscimol in any human brain area studied.

It remains therefore to be clearly established whether the effects of ammonia on GABA receptors occur in vivo in hyperammonemic syndromes and its possible contribution to the neurological alterations in hyperammonemia and hepatic encephalopathy.

3.4 Hyperammonemia and Serotoninergic Neurotransmission

Ammonia administration increases the transport of tryptophan across the blood–brain barrier of rats (Grippon et al., 1986). Tryptophan uptake into the brain is also increased in rats with chronic hyperammonemia (by injection of urease) that also shows increased production of the serotonin (5HT) metabolite 5-hydroxyindoleacetic acid (5HIAA) (Batshaw et al., 1986). This indicates that the metabolism of 5HT is increased in the brain in hyperammonemia. 5HIAA concentration is also increased in CSF of children with hyperammonemia due to congenital deficits in the urea cycle, and there is a good correlation between the plasma ammonia levels and 5HIAA in CSF of these children (Batshaw et al., 1990).

Some mechanisms proposed to explain this effect are: stimulation of the uptake of the precursor amino acid tryptophan, increased MAO-A expression in brain, and decreased storage and increased intracellular oxidation of 5HT leading to decreased availability for release from the presynaptic nerve terminals (Erecinska et al., 1987).

The binding of ligands to $5HT_2$ receptors is increased in brains from patients with chronic liver disease who died with hepatic coma (Rao et al., 1993). However, in sparse-fur mice that are hyperammonemic due to a congenital defect in ornithine transcarbamoylase, the binding of ligands to $5HT_2$ is decreased (Robinson et al., 1992). These data suggest that the effects of hyperammonemia on serotoninergic neurotransmission could be different in developing than in adult brain.

Brain concentrations of quinolinic acid, an excitotoxic metabolite of tryptophan, are increased in hyperammonemia (Moroni et al., 1986), and tryptamine, another metabolite of tryptophan, is increased in the brain in chronic liver failure (Young and Lal, 1980). These reports indicate that hyperammonemia also alters significantly serotoninergic neurotransmission.

3.5 Chronic Hyperammonemia Alters Phosphorylation of Neuronal Proteins

In addition to the alterations in neurotransmission, hyperammonemia also affects phosphorylation of several neuronal proteins. Altered phosphorylation would lead to altered function or activity of these proteins, resulting in altered neuronal function that may contribute to the neurological alterations in hyperammonemia.

Chronic exposure of cerebellar neurons in culture to 0.1 mM ammonia increases the basal phosphorylation of the microtubule-associated protein MAP-2. This effect is mimicked by treatment of neurons with MK-801 or with inhibitors of the protein phosphatase calcineurin indicating that tonic basal activation of NMDA receptors maintains a tonic activation of calcineurin which dephosphorylates MAP-2. Chronic exposure to ammonia would reduce tonic activation of NMDA receptors and of calcineurin and increases basal phosphorylation of MAP-2 (Sáez et al., 1999).

Chronic exposure to ammonia also alters the modulation of phosphorylation of MAP-2 by signal transduction pathways associated to different types of glutamate receptors. In primary cultures of cerebellar neurons, activation of NMDA receptors leads to a rapid increase in phosphorylation of MAP-2 followed by dephosphorylation by calcineurin (Sáez et al., 1999). NMDA-induced phosphorylation of MAP-2 is mediated by activation of NOS and MAP-kinase (Llansola et al., 2001). Chronic exposure to 0.1 mM ammonia inhibits the initial increase of MAP-2 phosphorylation induced by NMDA (Sáez et al., 1999).

Activation of mGluRs in the same neurons by t-ACPD increases phosphorylation of MAP-2. When these neurons are chronically exposed to 0.1 mM ammonia, activation of mGluRs leads to a decrease in phosphorylation of MAP-2 indicating that ammonia alters signal transduction pathways associated to mGluRs that modulate MAP-2 phosphorylation (Sáez et al., 1999). Hyperammonemia affects differentially the modulation of MAP-2 phosphorylation mediated by different mGluRs (Llansola et al., 2005).

An alteration in MAP-2 phosphorylation has also been shown in the brain of hyperammonemic rats (Felipo et al., 1993b). MAP-2 binds to tubulin and stimulates its polymerization, thus modulating microtubule stability. Ammonia-induced changes in MAP-2 phosphorylation may lead to alterations in the microtubular network and to neuronal dysfunction.

Chronic exposure to ammonia also alters phosphorylation of other neuronal proteins such as Na^+/K^+-ATPase, and NMDA receptors, or of a protein modulating this receptor (Kosenko et al., 1994; Marcaida et al., 1995; Grau et al., 1996). Ammonia-induced alterations in the phosphorylation of neuronal proteins may alter their activities, leading to neuronal dysfunction and contributing to the neurological alterations found in hyperammonemia.

4 Neurological Alterations in Hyperammonemia

The neurological alterations in different hyperammonemic situations and the mechanisms responsible for these alterations are different depending on the concentrations of ammonia reached, the duration of the exposure to hyperammonemia, and the age of the person.

Hyperammonemia is considered mainly responsible for many of the neurological alterations present in patients with liver disease and hepatic encephalopathy (● Section 1.5). Hepatic encephalopathy is a complex syndrome that covers a wide range of neuropsychiatric disturbances ranging from minimal changes in personality or altered circadian rhythms (sleep-waking cycle) to alterations in the intellectual function, personality, consciousness, and motor coordination.

The contribution of hyperammonemia to the alterations found in hepatic encephalopathy can be studied only using adequate animal models that can discern which effects are due to hyperammonemia and which are due to other factors associated with liver failure. However, until now only a few studies have been carried out to study the neurological alterations in these models. These studies and others on the effects of "pure" hyperammonemia are reviewed below. The studies performed using animal models of liver failure (portacaval anastomosis, thioacetamide injection, etc) that do not allow to discern which effects are due to hyperammonemia are not reviewed here.

4.1 Neurological Alterations Associated to Perinatal Hyperammonemia

Neonatal hyperammonemia may be due to congenital deficiencies in any of the urea cycle enzymes: ornithine transcarbamoylase (the most frequent), carbamoyl phosphate synthetase, argininosuccinate synthetase, argininosuccinase and arginase, or acetylglutamate synthetase, which synthesizes acetylglutamate, an essential activator of carbamoyl phosphate synthetase. Neonatal hyperammonemia may also occur in children with perinatal asphyxia or preterm delivery (Ballard et al., 1978; Yoshino et al., 1991). Children with congenital defects in the urea cycle appear normal at birth, but rapidly develop hyperammonemia. Blood ammonia levels may reach concentrations as high as 1 mM. If hyperammonemia is not treated promptly, severe neurological symptoms including seizures and coma occur. Surviving children have a high incidence of mental retardation and cerebral palsy. An inverse correlation of IQ with the duration and intensity of neonatal hyperammonemia has been reported (Msall et al., 1984; Colombo, 1994). In a study of 26 children with inherited urea cycle disorders who survived neonatal hyperammonemic coma and were maintained on a regimen of nitrogen restriction and stimulation of alternative nitrogen-excreting pathways, a significant negative correlation was observed between duration of hyperammonemic coma and IQ scores at 1 year of age (Msall et al., 1984). The mechanisms by which neonatal hyperammonemia induces these effects have not been studied.

4.2 Alterations in Motor Function in Rats with Chronic Moderate Hyperammonemia

One of the more characteristic neurological complications in hepatic encephalopathy is the impairment of motor coordination and function. Clinical signs of basal ganglia, corticospinal, and cerebellar dysfunctions that may contribute to motor function impairment are commonly detected in these patients.

Activation of mGluRs in the NAcc induces locomotion in rats. Signal transduction associated to mGluRs is altered in hyperammonemia (● Section 3.2.2). The locomotor activity induced by injection into the NAcc of DHPG, an agonist of group I mGluRs, was significantly increased in rats with chronic hyperammonemia without liver failure (Canales et al., 2003). Rats were made hyperammonemic by feeding them an ammonium-containing diet (Azorín et al., 1989). Both dopamine and glutamate have been reported to modulate motor function. Canales and coworkers (2003) also showed that the changes in extracellular dopamine and glutamate induced by activation of NAcc mGluRs are also altered in rats with chronic hyperammonemia. Injection of DHPG increased extracellular dopamine but not glutamate in the NAcc of control rats. In hyperammonemic rats, DHPG-induced increase in dopamine was significantly reduced, and extracellular glutamate increased sixfold. Blockade of mGluR1 completely prevented motor and neurochemical effects induced by DHPG. Moreover, the content of mGluR1, but not mGluR5, is increased in the NAcc of hyperammonemic rats, which may contribute to the altered response to DHPG. These results show that modulation of both motor function and extracellular concentration of neurotransmitters by mGluRs in the NAcc is altered in hyperammonemia. This may contribute to the alterations in motor function in hepatic encephalopathy.

4.3 Impairment of Cognitive Function (Learning Ability) in Hyperammonemic Rats

4.3.1 Learning Ability Is Reduced in Rats with Chronic Hyperammonemia

Another neurological alteration present in patients with hepatic encephalopathy is cognitive impairment. Cognitive function is also altered in animal models of chronic liver disease. When normal rats are moved from the cage where they usually live and placed in a novel cage, they explore it, resulting in increased motor activity. If rats are placed in the same cage during consecutive days, the exploratory behavior decreases since the animal remembers the novel cage and the interest for it lowers. In rats with chronic hyperammonemia with or without liver failure (portacaval anastomosis), this long-term habituation is impaired. This has been interpreted as consequence of a possibly impaired learning/memory capacity in these rats (Apelqvist et al., 1999).

Learning of a conditional discrimination task in a Y maze is impaired in rats with chronic hyperammonemia without liver failure (Aguilar et al., 2000) as well as in rats with chronic liver failure due to portacaval anastomosis (Erceg et al., 2005a), indicating that hyperammonemia is responsible for learning impairment.

4.3.2 Pharmacological Manipulation of Extracellular cGMP Concentration in Brain Restores Learning Ability in Rats with Chronic Liver Failure or Chronic Hyperammonemia

As mentioned above, activation of NMDA receptors leads to activation of the glutamate–NO–cGMP pathway and increases extracellular cGMP (❯ *Figure 3-2*). Several reports suggest that activation of this glutamate–NO–cGMP pathway is involved in some forms of learning (Danysz et al., 1995; Chen et al., 1997; Meyer et al., 1998).

As also summarized above, the function of the glutamate–NO–cGMP pathway is impaired in brains in vivo in animal models of chronic hyperammonemia and of chronic liver failure and also in autopsied brains from patients who died with hepatic encephalopathy.

Based on the above, Erceg and coworkers (2005a, b) hypothesized that (1) the alterations in the function of the glutamate–NO–cGMP pathway and the decrease in extracellular cGMP in the brain in hyperammonemia and liver disease may be responsible for the impairment in learning ability and intellectual function and that (2) pharmacological modulation of extracellular cGMP concentration may restore learning ability in hyperammonemia and hepatic encephalopathy. To assess this possibility they tried to reverse the impairment in learning ability of hyperammonemic rats by increasing extracellular cGMP by using three different treatments: (1) continuous intracerebral administration of zaprinast, an inhibitor of the phosphodiesterase that degrades cGMP; (2) chronic oral administration of sildenafil, an inhibitor of the phosphodiesterase that crosses the blood–brain barrier and (3) continuous intracerebral administration of cGMP.

Erceg and coworkers (2005a) administered intracerebrally zaprinast, an inhibitor of the phosphodiesterase that degrades cGMP, to control and hyperammonemic rats, continuously for 28 days, by using osmotic minipumps. Extracellular cGMP was significantly reduced in hyperammonemic rats and treatment with zaprinast increased extracellular cGMP to the same level present in control rats. The ability to learn a Y maze conditional discrimination task was significantly reduced in hyperammonemic rats. Continuous intracerebral administration of zaprinast completely restored the learning ability of hyperammonemic rats (Erceg et al., 2005a).

As zaprinast does not cross the blood–brain barrier it is not suitable for clinical treatment of patients. Erceg and coworkers (2005b) also showed that oral administration of sildenafil, an inhibitor of the phosphodiesterase that degrades cGMP and crosses the blood–brain barrier normalizes the function of the glutamate–NO–cGMP pathway and extracellular cGMP and also restored the ability to learn the Y maze conditional discrimination task both in rats with chronic hyperammonemia without liver failure and in rats with chronic liver failure (portacaval anastomosis) (Erceg et al., 2005b).

To further confirm that changes in extracellular cGMP are responsible for the changes in learning ability, Erceg and coworkers (2005a) tested whether increasing only extracellular cGMP without affecting intracellular cGMP is also able to restore learning ability in hyperammonemic rats. cGMP was administered intracerebrally to control and hyperammonemic rats, continuously for 28 days, by using osmotic minipumps. cGMP is not able to cross the cellular membrane, and therefore increases cGMP only in the extracellular fluid. Continuous intracerebral administration of cGMP to hyperammonemic rats completely restored the learning ability of hyperammonemic rats.

The above data indicate that the impairment of learning ability in hyperammonemic rats with or without chronic liver failure is due (at least for the Y maze test) to impairment of the glutamate–NO–cGMP pathway. As the function of this pathway is also altered in the brain of patients with liver cirrhosis (Corbalán et al., 2002), this alteration should also contribute to the cognitive impairment in these patients. Increasing extracellular cGMP by pharmacological means may be a new therapeutic approach to improve learning and memory in patients with hepatic encephalopathy.

Acknowledgments

This work was supported by grants from the Ministerio de Ciencia y Tecnología (SAF2002-00851 and SAF2005-06089) and from Ministerio de Sanidad (Red G03-155) of Spain and by grants from Consellería de Empresa, Universidad y Ciencia, Generalitat Valenciana (Grupos03/001) and GV04B-055 of Generalitat Valenciana.

References

Abrahams SL, Younathan ES. 1971. Modulation of the kinetic properties of phosphofructokinase by ammonium ions. J Biol Chem 246: 2464-2467.

Aguilar MA, Minarro J, Felipo V. 2000. Chronic moderate hyperammonemia impairs active and passive avoidance behavior and conditional discrimination learning in rats. Exp Neurol 161: 704-713.

Ahboucha S, Araqi F, Layrargues GP, Butterworth RF. 2005. Differential effects of ammonia on the benzodiazepine modulatory site on the GABA$_A$ receptor complex of human brain. Neurochem Int 47: 58-63.

Amlal H, Soleimani M. 1997. K^+/NH_4^+ antiporter: A unique ammonium carrying transporter in the kidney inner medulla. Biochim Biophys Acta 1323: 319-333.

Apelqvist G, Hindfelt B, Andersson G, Bengtsson F. 1999. Altered adaptive behaviour expressed in an open-field paradigm in experimental hepatic encephalopathy. Behav Brain Res 106: 165-173.

Azorín I, Miñana MD, Felipo V, Grisolía S. 1989. A simple animal model for hyperammonemia. Hepatology 10: 311-314.

Ballard RA, Vinocur B, Reynolds JW, Wennberg RP, Merritt A, et al. 1978. Transient hyperammonemia of the preterm infant. N Engl J Med 299: 920-925.

Basile AS. 2002. Direct and indirect enhancement of GABAergic neurotransmission by ammonia: Implications for the pathogenesis of hyperammonemic syndromes. Neurochem Int 41: 115-122.

Bates TE, Williams SR, Kauppinen RA, Gadian DG. 1989. Observation of cerebral metabolites in an animal model of acute liver failure in vivo: A ^{1}H and ^{31}P nuclear magnetic resonance study. J Neurochem 53: 102-110.

Batshaw ML, Heyes M, Djali S, Roke L, Robinson MB. 1990. Tryptophan (Trp), quinolinate (QUIN) and serotonin (5-HT). 1 Alterations in children with hyperammonemia (HA). Soc Neurosci Abstr 323: 1.

Batshaw ML, Hyman SL, Mellits ED, Thomas GH, De Muro R, et al. 1986. Behavioral and neurotransmitter changes in the urease-infused rat: A model of congenital hyperammonemia. Pediatr Res 20: 1310-1315.

Blomgren K, McRae A, Bona E, Saido TC, Karlsson JO, et al. 1995. Degradation of fodrin and MAP 2 after neonatal cerebral hypoxic-ischemia. Brain Res 684: 136-142.

Canales JJ, Elayadi A, Errami M, Llansola M, Cauli O, et al. 2003. Chronic hyperammonemia alters motor and neurochemical responses to activation of group I metabotropic glutamate receptors in the nucleus accumbens in rats in vivo. Neurobiol Dis 14: 380-390.

Chan H, Butterworth RF. 1999. Evidence for an astrocytic glutamate transporter deficit in hepatic encephalopathy. Neurochem Res 24: 1397-1401.

Chan H, Hazell AS, Desjardins P, Butterworth RF. 2000. Effects of ammonia on glutamate transporter (GLAST) protein and mRNA in cultured rat cortical astrocytes. Neurochem Int 37: 243-248.

Chen J, Zhang S, Zuo P, Tang L. 1997. Memory-related changes of nitric oxide synthase activity and nitrite level in rat brain. Neuroreport 8: 1771-1774.

Clemmensen JO, Larsen FS, Kondrup J, Hansen BA, Ott P. 1999. Cerebral herniation in patients with acute liver is correlated with arterial ammonia concentration. Hepatology 29: 648-653.

Colombo JP. 1994. N-acetylglutamate synthetase deficiency. Adv Exp Med Biol 364: 135-143.

Corbalán R, Chatauret N, Behrends S, Butterworth RF, Felipo V. 2002. Region selective alterations of soluble guanylate cyclase content and modulation in brain of cirrhotic patients. Hepatology 36: 1155-1162.

Danysz W, Zajaczkowski W, Parsons CG. 1995. Modulation of learning processes by ionotropic glutamate receptor ligands. Behav Pharmacol 6: 455-474.

Erceg S, Monfort P, Hernández-Viadel M, Rodrigo R, Montoliu C, et al. 2005a. Oral administration of sildenafil restores learning ability in rats with hyperammonemia and with portacaval shunts. Hepatology 41: 299-306.

Erceg S, Monfort P, Hernandez-Viadel M, Llansola M, Montoliu C, et al. 2005b. Restoration of learning ability in hyperammonemic rats by increasing extracellular cGMP in brain. Brain Res 1036: 115-121.

Erecinska M, Pastuszko A, Wilson DF, Nelson D. 1987. Ammonia-induced release of neurotransmitters from rat brain synaptosomes: Differences between the effects on amines and amino acids. J Neurochem 49: 1258-1265.

Faff-Michalak L, Albrecht J. 1991. Aspartate aminotransferase, malate dehydrogenase, and pyruvate carboxylase activities in rat cerebral synaptic and nonsynaptic mitochondria: Effects of in vitro treatment with ammonia, hyperammonemia and hepatic encephalopathy. Metab Brain Dis 6: 187-197.

Felipo V, Grau E, Miñana MD, Grisolía S. 1993a. Ammonium injection induces an N-methyl-D-aspartate receptor-mediated proteolysis of the microtubule-associated protein MAP-2. J Neurochem 60: 1626-1630.

Felipo V, Grau E, Miñana MD, Grisolia S. 1993b. Hyperammonemia decreases protein kinase C-dependent phosphorylation of the microtubule-associated protein 2 and increases its binding to tubulin. Eur J Biochem 214: 243-249.

Fitzpatrick SM, Hetherington HP, Behar KL, Shulman RG. 1989. Effects of acute hyperammonemia on cerebral amino acid metabolism and pH$_i$ in vivo, measured by ^{1}H and ^{31}P nuclear magnetic resonance. J Neurochem 52: 741-749.

Ganz R, Swain M, Traber P, Dal Canto M, Butterworth RF, et al. 1989. Ammonia-induced swelling of rat cerebral cortical slices: Implications for the pathogenesis of brain edema in acute hepatic failure. Metab Brain Dis 4: 213-223.

Gottlieb M, Matute C. 1997. Expression of ionotropic glutamate receptor subunits in glial cells of the hippocampal CA1 area following transient forebrain ischemia. J Cereb Blood Flow Metab 17: 290-300.

Grau E, Marcaida G, Montoliu C, Miñana MD, Grisolía S, et al. 1996. Effects of hyperammonemia on brain protein kinase C substrates. Metab Brain Dis 11: 205-216.

Grippon P, Le Poncin Lafitte M, Boschat M, Wang S, Faure G, et al. 1986. Evidence for the role of ammonia in the intracerebral transfer and metabolism of tryptophan. Hepatology 6: 682-686.

Ha JH, Basile AS. 1996. Modulation of ligand binding to components of the GABA$_A$ receptor complex by ammonia: Implications for the pathogenesis of hyperammonemic syndromes. Brain Res 720: 35-44.

Hahn M, Massen O, Nencki M, Pavlov L. 1893. Die Eck'sche Fistel zwischen der unteren Hohlvene und der Pfortader and ihre Folgen für den Organisms. Arch Exp Pothol Pharmakol 32: 161-210.

Haussinger D, Schliess F. 2005. Astrocyte swelling and protein tyrosine nitration in hepatic encephalopathy. Neurochem Int 47: 64-70.

Hawkins RA, Miller AL, Nielsen RC, Veech RL. 1973. The acute action of ammonia on rat brain metabolism in vivo. Biochem J 134: 1001-1008.

Hermenegildo C, Marcaida G, Montoliu C, Grisolía S, Miñana MD, et al. 1996. NMDA receptor antagonists prevent acute ammonia toxicity in mice. Neurochem Res 21: 1237-1244.

Hermenegildo C, Monfort P, Felipo V. 2000. Activation of NMDA receptors in rat brain in vivo following acute ammonia intoxication. Characterization by in vivo brain microdialysis Hepatology 31: 709-715.

Hermenegildo C, Montoliu C, Llansola M, Muñoz MD, Gaztelu JM, et al. 1998. Chronic hyperammonemia impairs glutamate-nitric oxide-cyclic GMP pathway in cerebellar neurons in culture and in the rat in vivo. Eur J Neurosci 10: 3201-3209.

Hindfelt B, Siesjö BK. 1971. Cerebral effects of acute ammonia intoxication. Scand J Clin Lab Invest 28: 365-374.

Hindfelt B, Plum F, Duffy TE. 1977. Effect of acute ammonia intoxication on cerebral metabolism in rats with portacaval shunts. J Clin Invest 59: 386-396.

Knecht K, Michalak A, Rose C, Rothstein JD, Butterworth RF. 1997. Decreased glutamate transporter (GLT-1) expression in frontal cortex of rats with acute liver failure. Neurosci Lett 229: 201-203.

Kosenko E, Felipo V, Montoliu C, Grisolia S, Kaminsky Y. 1996. Effects of acute hyperammonemia in vivo on oxidative metabolism in nonsynaptic rat brain mitochondria. Metab Brain Dis 12: 69-82.

Kosenko E, Felipo V, Montoliu C, Grisolía S, Kaminsky Y. 1997a. Effects of acute hyperammonemia in vivo on oxidative metabolism in nonsynaptic rat brain mitochondria. Metab Brain Dis 12: 69-82.

Kosenko E, Kaminsky Y, Kaminsky A, Valencia M, Lee L, et al. 1997b. Superoxide production and antioxidant enzymes in ammonia intoxication in rats. Free Radic Res 27: 637-644.

Kosenko E, Kaminsky Y, Grau E, Miñana MD, Grisolía S, et al. 1995. Nitroarginine, an inhibitor of nitric oxide synthetase, attenuates ammonia toxicity and ammonia-induced alterations in brain metabolism. Neurochem Res 20: 451-456.

Kosenko E, Kaminsky Y, Grau E, Miñana MD, Marcaida G, et al. 1994. Brain ATP depletion induced by acute ammonia intoxication in rats is mediated by activation of the NMDA receptor and of Na/K-ATPase. J Neurochem 63: 2172-2178.

Kosenko E, Kaminski Y, Lopata O, Muravyov N, Felipo V. 1999. Blocking NMDA receptors prevents the oxidative stress induced by acute ammonia intoxication. Free Radic Biol Med 26: 1369-1374.

Kosenko E, Kaminsky Y, Lopata O, Muravyov N, Kaminsky A, et al. 1998. Nitroarginine, an inhibitor of nitric oxide synthase, prevents changes in superoxide radical and antioxidant enzymes induced by ammonia intoxication. Metab Brain Dis 13: 29-41.

Kosenko E, Kaminsky Y, Stavroskaya IG, Felipo V. 2000. Alteration of mitochondrial calcium homeostasis by ammonia-induced activation of NMDA receptors in rat brain in vivo. Brain Res 880: 139-146.

Kosenko E, Kaminsky YG, Felipo V, Minana MD, Grisolia S. 1993. Chronic hyperammonemia prevents changes in brain energy and ammonia metabolites induced by acute ammonia intoxication. Biochim Biophys Acta 1180: 321-326.

Kosenko E, Llansola M, Montoliu C, Monfort P, Rodrigo R, et al. 2003. Glutamine synthetase activity and glutamine content in brain: Modulation by NMDA receptors and nitric oxide. Neurochem Int 43: 493-499.

Lee SH, Fisher B. 1961. Portacaval shunt in the rat. Surgery 50: 668-672.

Llansola M, Sáez R, Felipo V. 2001. NMDA-induced phosphorylation of the microtubule-associated protein MAP-2 is mediated by activation of nitric oxide synthase and MAP-kinase. Eur J Neurosci 13: 1283-91.

Llansola M, Erceg S, Felipo V. 2005. Chronic exposure to ammonia alters the modulation of phosphorylation of microtubule-associated protein MAP-2 by metabotropic glutamate receptors 1 and 5 in cerebellar neurons in culture. Neuroscience 133: 185-191.

Lombardi G, Mannaioni G, Leonadi P, Cherici G, Carla V, et al. 1994. Ammonium acetate inhibits ionotropic receptors and differentially affects metabotropic receptors for glutamate. J Neural Transm 97: 187-196.

Marcaggi P, Coles JA. 2001. Ammonium in nervous tissue: Transport across cell membranes, fluxes from neurons to glial cells, and role in signalling. Prog Neurobiol 64: 157-183.

Marcaida G, Felipo V, Hermenegildo C, Miñana MD, Grisolía S. 1992. Acute ammonia toxicity is mediated by the NMDA type of glutamate receptors. FEBS Lett 296: 67-68.

Marcaida G, Kosenko E, Miñana MD, Grisolía S, Felipo V. 1996. Glutamate induces a calcineurin-mediated dephosphorylation of Na^+/K^+-ATPase which results in its activation in cerebellar neurons in culture. J Neurochem 66: 99-104.

Marcaida G, Miñana MD, Burgal M, Grisolía S, Felipo V. 1995. Ammonia prevents activation of NMDA receptors in rat cerebellar neuronal cultures. Eur J Neurosci 7: 2389-2396.

Mayer ML, Westbrook GL, Guthrie PB. 1984. Voltage-dependent block by Mg^{2+} of NMDA responses in spinal cord neurons. Nature 309: 261-263.

Mena EE, Cotman CW. 1985. Pathologic concentrations of ammonium ions block L-glutamate uptake. Exp Neurol 89: 259-263.

Meyer RC, Knox J, Purwin DA, Spangler EL, Ingram DK. 1998. Combined stimulation of the glycine and polyamine sites of the NMDA receptor attenuates NMDA blockade-induced learning deficits of rats in a 14-unit-T-maze. Psychopharmacology 135: 290-295.

Michalak A, Rose C, Butterworth J, Butterworth RF. 1996. Neuroactive amino acids and glutamate (NMDA) receptors in frontal cortex of rats with experimental acute liver failure. Hepatology 24: 908-913.

Miñana MD, Kosenko E, Marcaida G, Hermenegildo C, Montoliu C, et al. 1997. Modulation of glutamine synthesis in cultured astrocytes by nitric oxide. Cell Mol Neurobiol 17: 433-445.

Miñana MD, Montoliu C, Llansola M, Grisolía S, Felipo V. 1998. Nicotine prevents glutamate-induced proteolysis of the microtubule-associated protein MAP-2 and glutamate neurotoxicity in primary cultures of cerebellar neurons. Neuropharmacology 37: 847-857.

Monfort P, Corbalán R, Martinez L, López-Talavera JC, Córdoba J, et al. 2001. Altered content and modulation of soluble guanylate cyclase in the cerebellum of rats with portacaval anastomosis. Neuroscience 104: 1127-1139.

Monfort P, Muñoz MD, El Ayadi A, Kosenko E, Felipo V. 2002a. Effects of hyperammonemia and liver disease on glutamatergic neurotransmission. Metab Brain Dis 17: 237-250.

Monfort P, Muñoz MD, Kosenko E, Felipo V. 2002b. Long-term potentiation in hippocampus involves sequential activation of soluble guanylate cyclase, cGMP-dependent protein kinase and cGMP-degrading phosphodiesterase. J Neurosci 22: 10116-10122.

Monfort P, Muñoz MD, Felipo V. 2004. Hyperammonemia impairs long-term potentiation in hippocampus by altering the modulation of cGMP-degrading phosphodiesterase by protein kinase G. Neurobiol Dis 15: 1-10.

Monfort P, Muñoz MD, Felipo V. 2005. Chronic hyperammonemia in vivo impairs long-term potentiation in hippocampus by altering activation of cGMP dependent-protein kinase and of phosphodiesterase 5. J Neurochem 94: 934-942.

Moroni F, Lombardi G, Carla V, Lal S, Etienne P, et al. 1986. Increase in the content of quinolinic acid in cerebrospinal fluid and frontal cortex of patients with hepatic failure. J Neurochem 47: 1667-1671.

Moser H. 1987. Electrophysiological evidence for ammonium as a substitute for potassium in activating the sodium pump in a crayfish sensory neuron. Can J Physiol Pharmacol 65: 141-145.

Msall M, Batshaw ML, Suss R, Brusilow SW, Mellits ED. 1984. Neurological outcome in children with inborn errors of urea synthesis. Outcome of urea cycle enzymopathies N Engl J Med 310: 1500-1505.

Muñoz MD, Monfort P, Gaztelu JM, Felipo V. 2000. Hyperammonemia impairs NMDA receptor-dependent long-term potentiation in the CA1 of rat hippocampus in vitro. Neurochem Res 25: 437-441.

Murthy RK, Bender AS, Dombro RS, Bai G, Norenberg MD. 2000a. Elevation of glutathione levels by ammonium ions in primary cultures of rat astrocytes. Neurochem Int 37: 255-268.

Murthy RK, Bai G, Dombro RS, Norenberg MD. 2000b. Ammonia-induced swelling in primary cultures of rat astrocytes: Role of free radicals. Soc Neurosci Abstr 26: 1893.

Murthy RK, Rama Rao KV, Bai G, Norenberg MD. 2001. Ammonia induced production of free radicals in primary cultures of rat astrocytes. J Neurosci Res 66: 282-288.

Norenberg MD, Baker L, Norenberg LO, Blicharska J, Bruce-Gregorios JH, et al. 1991. Ammonia-induced astrocyte swelling in primary culture. Neurochem Res 16: 833-836.

Norenberg MD, Huo Z, Neary JT, Roig-Cantisano A. 1997. The glial glutamate transporter in hyperammonemia and hepatic encephalopathy: Relation to energy metabolism and glutamatergic neurotransmission. Glia 21: 124-133.

Norenberg MD, Jayakumar AR, Rama Rao KV. 2004. Oxidative stress in the pathogenesis of hepatic encephalopathy. Metab Brain Dis 19: 313-329.

Norenberg MD, Mozes LW, Papendick RE, Norenberg LOB. 1985. Effects of ammonia on glutamate, GABA and rubidium uptake by astrocytes. Ann Neurol 18: 149.

Norenberg MD, Rama Rao KV, Jayakumar AR. 2005. Mechanisms of ammonia-induced astrocyte swelling. Metab Brain Dis 20: 303-318.

Pappas SC, Ferenci P, Schafer DF, Jones EA. 1984. Visual evoked potentials in a rabbit model of hepatic encephalopathy. II Comparison of hyperammonemic encephalopathy, postictal coma, and coma induced by synergistic neurotoxins Gastroenterology 86: 546-551.

Patel N, Forton DM, Coutts GA, Thomas HC, Taylor-Robinson SD. 2000. Intracellular pH measurements of the whole head and the basal ganglia in chronic liver disease: A phosphorous-31 MR spectroscopy study. Metab Brain Dis 15: 223-240.

Peterson C, Giguère JF, Cotman CW, Butterworth RF. 1990. Selective loss of NMDA ^{3}H-sensitive-glutamate binding sites in rat brain following portacaval anastomosis. J Neurochem 55: 386-390.

Prior RL, Visek WJ. 1973. Effects of modifying urea hydrolysis in acute nephrectomy on survival, ammonia in cecal contents and blood metabolites. Proc Soc Exp Biol Med 144: 184-188.

Qureshi IA, Rao KV. 1997. Sparse-fur (spf) mouse as a model of hyperammonemia: Alterations in the neurotransmitter systems. Adv Exp Med Biol 420: 143-158.

Raabe W. 1990. Effects of NH_4^+ on the function of the CNS. Adv Exp Med Biol 272: 99-120.

Rama Rao KV, Chen M, Simard JM, Norenberg MD. 2003a. Suppression of ammonia-induced astrocyte swelling by cyclosporin A. J Neurosci Res 74: 891-897.

Rama Rao KV, Chen M, Simard JM, Norenberg MD. 2003b. Increased aquaporin-4 expression in ammonia-treated cultured astrocytes. Neuroreport 14: 2379-2382.

Ramirez M, Fernandez R, Malnic G. 1999. Permeation of NH_3/NH_4^+ and cell pH in colonic crypts of the rat. Pflugers Arch 438: 508-515.

Rao VLR, Agrawal AK, Murthy CR. 1991. Ammonia-induced alterations in glutamate and muscimol binding to cerebellar synaptic membranes. Neurosci Lett 130: 251-254.

Rao VLR, Audet RM, Butterworth RF. 1997. Portacaval shunting and hyperammonemia stimulate the uptake of L-[^{3}H]arginine but not of L-[^{3}H]nitroarginine into rat brain synaptosomes. J Neurochem 68: 337-343.

Rao VLR, Giguère JF, Pomier Layrargues G, Butterworth RF. 1993. Increased activities of MAOA and MAOB in autopsied brain tissue from cirrhotic patients with hepatic encephalopathy. Brain Res 621: 349-352.

Ratnakumari L, Qureshi IA, Butterworth RF. 1995. Loss of [^{3}H]MK801 binding sites in brain in congenital ornithine transcarbamylase deficiency. Metab Brain Dis 10: 249-255.

Robinson MB, Anegawa NJ, Corry E, Qureshi IA, Coyle JT, et al. 1992. Brain serotonin2 and serotonin1A receptors are altered in the congenitally hyperammonemic sparse fur mouse. J Neurochem 58: 1016-1022.

Romero-Gomez M, Ramos-Guerrero R, Grande L, Collantes L, Corpas R, et al. 2004. Intestinal glutaminase activity is increased in liver cirrhosis and correlates with minimal hepatic encephalopathy. J Hepatol 41: 49-54.

Sáez R, Llansola M, Felipo V. 1999. Chronic exposure to ammonia alters pathways modulating phosphorylation of the microtubule-associated protein MAP-2 in cerebellar neurons in culture. J Neurochem 73: 2555-2562.

Sanchez-Prieto J, Gonzalez P. 1988. Occurrence of a large Ca^{2+}-independent release of glutamate during anoxia in isolated nerve terminals (synaptosomes). J Neurochem 50: 1322-1324.

Schafer DF, Jones EA. 1982. Hepatic encephalopathy and the γ-aminobutyric acid neurotransmitter system. Lancet 1: 18-29.

Schenker S, McCandless DW, Brophy E, Lewis MS. 1967. Studies on the intracerebral toxicity of ammonia. J Clin Invest 46: 838-848.

Schliess F, Gorg B, Fischer R, Desjardins P, Bidmon HJ, et al. 2002. Ammonia induces MK-801-sensitive nitration and phosphorylation of protein tyrosine residues in rat astrocytes. FASEB J 16: 739-741.

Schmidt W, Wolf G, Grungreiff K, Meier M, Reum T. 1990. Hepatic encephalopathy influences high-affinity uptake of transmitter glutamate and aspartate into the hippocampal formation. Metab Brain Dis 5: 19-31.

Sugimoto A, Takeda A, Kogure K, Onodera H. 1994. NMDA receptor (NMDAR1) expression in the rat hippocampus after forebrain ischemia. Neurosci Lett 170: 39-42.

Takahashi K, Kameda H, Kataoka M, Sanjou K, Harata N, et al. 1993. Ammonia potentiates $GABA_A$ response in dissociated rat cortical neurons. Neurosci Lett 151: 51-54.

Tanigami H, Rebel A, Martin LJ, Chen TY, Brusilow SW, et al. 2005. Effect of glutamine synthetase inhibition on astrocyte swelling and altered astroglial protein expression during hyperammonemia in rats. Neuroscience 131: 437-449.

Towle DW, Holleland T. 1987. Ammonium ion substitutes for K^+ in ATP-dependent Na^+ transport by basolateral membrane vesicles. Am J Physiol 252: R479-R489.

Wall SM, Koger LM. 1994. NH_4^+ transport mediated by Na^+/K^+-ATPase in rat inner medullary collecting duct. Am J Physiol 267: F660-F670.

Weiner ID. 2004. The Rh gene family and renal ammonium transport. Curr Opin Nephrol Hypertens 13: 533-540.

Yoshino M, Sakaguchi Y, Kuriya N, et al. 1991. A nationwide survey on transient hyperammonemia in newborn infants in Japan: Prognosis of life and neurological outcome. Neuropediatrics 22: 198-2002.

Young SN, Lal S. 1980. CNS tryptamine metabolism in hepatic coma. J Neural Transm 47: 153-161.

4 Glutamate and Cytokine-Mediated Alterations of Phospholipids in Head Injury and Spinal Cord Trauma

A. A. Farooqui · L. A. Horrocks

1	Introduction	72
2	Phospholipid Composition of Neural Membranes in Brain and Spinal Cord	73
3	Traumatic Injury to Head and Spinal Cord Releases Glutamate and Causes Upregulation of Cytokines	73
4	Glutamate-Mediated Alterations in the Phospholipid Metabolism of Neural Cells	75
5	Generation and Release of Cytokines in Head Injury and Spinal Cord Trauma	77
6	Cytokine-Mediated Alterations in Phospholipid Metabolism and Cross Talk Between Phospholipids and Sphingolipid-Generated Lipid Mediators	79
6.1	Phospholipid Alterations in Head Injury	80
6.2	Phospholipid Alterations in Spinal Cord Trauma	81
7	Interplay among Glutamate, Cytokines, and PLA_2-Derived Lipid Mediators	81
8	Prevention of Neurodegeneration by $cPLA_2$ Inhibitors	82
9	Conclusion	83

© 2009 Springer Science+Business Media, LLC.

Abstract: Glutamate and cytokines play a major role in neurodegeneration and demyelination that occur following head injury and spinal cord trauma. Glutamate and cytokines produce their effect by stimulating glutamate and cytokine receptors and enhancing neural membrane phospholipid degradation through caspase-mediated activation of phospholipase A_2. The stimulation of phospholipase A_2 not only results in alteration in neural membrane composition and changes in neural membrane permeability and fluidity but also generates proinflammatory mediators, eicosanoids, and platelet activating factor, and oxidative stress mediators, isoprostanes, and 4-hydroxynonenal (4-HNE). The generation of these mediators along with the decrease in ATP and glutathione content, and generation of reactive oxygen species may contribute to neural cell death in head and spinal cord injury. Involvement of phospholipase A_2 (PLA_2) in neural cell injury suggests that phospholipase A_2 inhibitors can limit neurodegeneration in head injury and spinal cord trauma and may therefore be used as potential drugs for the treatment of head and spinal cord injuries in animal models.

List of Abbreviations: $AACOCF_3$, arachidonoyl trifluoromethyl ketone; AMPA, α-amino-3-hydroxy-5-methyl-4-isoxazole propionate; BDNF, brain-derived neurotrophic factor; BEL, bromoenol lactone; CNQX, 6-cyano-7-nitroquinoxaline-2,3-dione; COX-2, cyclooxygenase-2; 4-HHE, 4-hydroxyhexanal; 4-HNE, 4-hydroxynonenal; IFN, interferons; iNOS, inducible nitric oxide synthase; JNK, c-Jun amino-terminal kinase; KA, kainate; PLA_2, phospholipases A_2; NMDA, N-methyl-D-aspartate; PLC, phospholipase C; PlsEtn, plasmenylethanolamines; PtdCho, phosphatidylcholines; PtdEtn, phosphatidylethanolamines; PtdH, phosphatidic acids; PtdIns, phosphatidylinositols; PtdSer, phosphatidylserines; SPM, synaptosomal plasma membrane; TGF, tumor growth factors; TNF, tumor necrosis factors

1 Introduction

Neural membranes contain three major categories of lipids: cholesterol, sphingolipids, and phospholipids. Among these lipids, phospholipids constitute the backbone of neural membranes. In adult brain, they account for about 20%–25% of the dry weight. Phospholipids not only constitute the backbone of neural membranes but also provide the membrane with suitable environment, fluidity, and ion permeability. They are also required for the proper functioning of integral membrane proteins, receptors, and ion channels (Farooqui et al., 2000a, 2009). Brain tissue contains five major classes of phospholipids. The first four (1,2-diacyl glycerophospholipids, 1-alk-1-enyl-2-acyl glycerophospholipids or plasmalogens, 1-alkyl-2-acyl glycerophospholipids, and phosphatidic acids) have a glycerol backbone with a fatty acid, usually polyunsaturated at carbon-2, and except for phosphatidic acid (PtdH) a phospho base (choline, ethanolamine, serine or inositol) at carbon-3 of the glycerol moiety. The fifth class, of which the only representative is sphingomyelin, contains ceramide linked to phosphocholine through its primary hydroxyl group (Farooqui et al., 2000a).

In neural membranes, phosphatidylethanolamines (PtdEtn), plasmenylethanolamines (PlsEtn), and phosphatidylserines (PtdSer) contain high levels of $22:6n-3$ acyl groups (docosahexaenoyl groups); whereas phosphatidylcholines (PtdCho), phosphatidylinositols (PtdIns), and phosphatidic acids (PtdH) contain high levels of $20:4n-6$ acyl groups (arachidonoyl groups). In brain tissue, each class of phospholipid exists as a heterogeneous mixture of molecular species, and the synthesis of different pools within a phospholipid subclass appears to be compartmentalized according to the fatty acid composition and the source of the head group (de novo synthesis versus modification and interconversion reactions) (Farooqui et al., 2000a, 2009). Thus, different pools of phospholipid molecular species may have different metabolic and physical properties depending upon their localization in different types of neural cells and membranes (Farooqui et al., 2000a, 2002a).

Receptors, enzymes, and ion channels that differentially protrude through the membrane, or localize predominantly on the intracellular or extracellular membrane surface penetrate the phospholipid bilayer to varying degrees. These membranes are highly interactive and dynamic. The interaction of an agonist with a receptor results in enhancement of phospholipid metabolism, which in turn results in the regulation of activities of membrane-bound enzymes, receptors, and ion channels (Farooqui et al., 2000a, 2009).

2 Phospholipid Composition of Neural Membranes in Brain and Spinal Cord

PtdCho and EtnGpl are two major phospholipid components of neural membranes in all regions. This is followed by Sphingomyelin (CerPCho), which is most enriched in white matter (Horrocks et al., 1981; Söderberg et al., 1990). Neural membranes of all regions also contain small quantities of PtdSer and PtdIns. Their EtnGpl/PtdCho ratio characterizes individual regions of human brain. This ratio is 1.7 for the white matter and pons, 1.4 for cerebellum and medulla oblongata, 0.8 for grey matter and caudate nucleus, and close to 1.0 in the hippocampus (Söderberg et al., 1990). The fatty acid composition of EtnGpl and PtdCho is specific for the different regions. Thus, PtdCho contains mostly saturated and 18:1 fatty acids, whereas EtnGpl are rich in polyunsaturated fatty acids (Söderberg et al., 1991). Comparative studies on molecular species in the glycerophospholipids containing choline, ethanolamine, serine, and inositol in the synaptosomal plasma membrane (SPM) and myelin membrane from the adult cerebral cortex of various animal species including human indicate that SPM contains more polyunsaturated fatty acids than myelin membranes, but comparisons of various phospholipid classes show differences that are more interesting. Thus, phospholipids containing ethanolamine and serine contain docosahexaenoic acid (22:6n−3) in SPM, but not in myelin; and glycerophospholipids containing choline contain more saturated fatty acids. The 18:1 content of all three phospholipid classes is much higher in myelin than in SPM, whereas the 20:4n−6 content is higher in ethanolamine-containing phospholipids and lower in serine-containing phospholipids (Porcellati, 1983). Like brain tissue, the major phospholipids of spinal cord include PtdCho (42.5%), PtdEtn (43.8%), PlsEtn (13%), PtdIns (1.5%), and sphingomyelin (3%). Among these phospholipids, PtdCho contains more saturated fatty acids, whereas PtdEtn and PlsEtn have the highest amount of polyunsaturated fatty acids.

Research on the phospholipid composition of brain and spinal cord neural membranes is still nascent (Farooqui et al., 2009). In order to determine the membrane composition in vivo under physiological conditions in neurons, astrocytes, oligodendrocytes, and microglial cells, one has to determine the phospholipid compositions expressed in terms of mol/surface area for a specific membrane in a particular subcellular organelle of a specific region of brain tissue. Most of the available information on neural membrane composition comes from membrane fractions that are often contaminated with membranes from other organelles of various neural cells (Farooqui et al., 2000a). Thus, knowledge of the phospholipid composition of neural membranes requires the identification of phospholipid molecular species.

3 Traumatic Injury to Head and Spinal Cord Releases Glutamate and Causes Upregulation of Cytokines

Traumatic injury to the brain and spinal cord consists of two broadly defined components: a primary component, attributable to the mechanical insult itself, and a secondary component, attributable to the series of systemic and local neurochemical and pathophysiological changes that occur in brain and spinal cord after the initial traumatic insult (Klussmann and Martin-Villalba, 2005). The primary injury causes a rapid deformation of brain and spinal cord tissues, leading to rupture of neural cell membranes, release of intracellular contents, and disruption of blood flow and breakdown of the blood–brain barrier. In contrast, secondary injury to the brain and spinal cord includes many neurochemical alterations, glial cell reactions involving both activated microglia and astroglia, and demyelination involving oligodendroglia (Beattie et al., 2000). Inflammatory reactions and oxidative stress are major components of secondary injury. Both these processes play a central role in regulating the pathogenesis of acute and chronic head injury and spinal cord trauma.

Among the most common mechanisms for excessive free radical generation, nonenzymatic iron-catalyzed Fenton reactions as well as enzymatic mechanisms involving lipoxygenases and cyclooxygenases may be the most important mechanisms associated with head injury and spinal cord trauma (Bazan et al., 1995; Hoffman et al., 2000). Neurochemically, secondary injury is characterized by the release of glutamate from intracellular stores (Demediuk et al., 1988; Panter et al., 1990; Sundström and Mo, 2002) and the overexpression of cytokines (Hayes et al., 2002; Ahn et al., 2004). In brain tissue, glutamate produces neural cell death through several mechanisms. Glutamate overstimulates both NMDA (N-methyl-D-aspartate) and AMPA (α-amino-3-hydroxy-5-methyl-4-isoxazole propionate) types of glutamate receptors resulting in the influx of Na^+, efflux of K^+, and a large Ca^{2+} influx into neurons (Stokes et al., 1983; Katayama et al., 1990). This process leads to an uncontrolled and sustained increase in cytosolic calcium levels, which is different from the transient increase in calcium that occurs during receptor stimulation. This increase in intracellular calcium may be responsible not only for the uncoupling of mitochondrial electron transport but also for stimulation of many calcium-dependent enzymes including lipases, phospholipases, calpains, nitric oxide synthase, protein phosphatases, and various protein kinases (Bazan et al., 1995; Pavel et al., 2001; Ray et al., 2003; Ellis et al., 2004; Arundine and Tymianski, 2004). Stimulation of these enzymes along with a rapid decrease in ATP level, changes in ion homeostasis, and alterations in cellular redox results in neural cell death in head injury and spinal cord trauma.

Another glutamate toxicity pathway is a transporter-mediated cell death that requires cellular expression of the cystine–glutamate antiporter system (Murphy et al., 1989). The competition between glutamate and cystine for the cystine–glutamate antiporter induces an imbalance in the homeostasis of cystine, an amino acid that is the precursor of glutathione. Thus, exposure of neural cells to high levels of glutamate results in an inability of neural cells to maintain intracellular levels of glutathione. This leads to a reduced ability of neural cells to protect against free-radical-mediated oxidative stress causing cell death (Murphy et al., 1989; Pereira and Resende de Oliveira, 2000).

Glutamate also produces its toxicity through delayed posttraumatic white matter degeneration (Park et al., 2004). This process involves glutamate release by the reversal of Na^+-dependent glutamate transport with subsequent activation of AMPA receptors and oligodendrocyte death (Li and Stys, 2000). Thus, the general response of brain and spinal cord tissue to traumatic injury is blood–brain barrier disruption, infiltration and activation of inflammatory cells, death of local neurons and glial cells caused by ischemia and hypoxia along with release of excitatory amino acids, gliosis, recruitment of endothelial cells, angiogenesis, compensatory responses in the surroundings of the most severely affected area, and formation of a glial scar.

Cytokines represent a broad, heterogeneous group of proteins and polypeptides involved in regulation of cell–cell interactions both in normal and pathological situations. They include interleukins (IL-1, IL-2, IL-6, and IL-12), interferons (IFN-γ), tumor necrosis factors (TNF-α and -β), tumor growth factors (TGF-α and -β), and colony-stimulating factors (Kim et al., 2001; Sun et al., 2004). In brain tissue, cytokines mediate cellular intercommunication through autocrine, paracrine, or endocrine mechanisms (Wilson et al., 2002). Their actions involve a complex network linked to feedback loops and cascades. Their overall response depends on the synergistic or antagonistic actions of various components. Thus, cytokines play an important role in neuronal development, maturation, survival, and regeneration. For instance, IL-1β, which is an astroglial growth factor at low concentrations, increases neuronal survival in dissociated spinal cord cultures and IL-2 enhances the viability of primary cultured neurons.

The overexpression of cytokines in head injury and spinal cord trauma (Bartholdi and Schwab, 1997) plays an important role not only in modulating the activities of protein kinases and phospholipase A_2 (PLA_2) but also in controlling the release of vasoactive factors such as nitric oxide through the increased expression of nitric oxide synthase. The short-term cytokine-mediated releases of low levels of nitric oxide along with low levels of PLA_2-generated lipid mediators are beneficial for neural cells. However, long-term release and high levels of nitric oxide and PLA_2-generated metabolites (eicosanoids and platelet activating factor) produce toxic substances like isoprostanes, 4-hydroxynonenals, and peroxynitrite that are harmful for the survival of neural cells (Zhang et al., 1999b; Hoffman et al., 2000; Chen et al., 2004; Dietrich et al., 2004; Hall et al., 2004).

Eicosanoids and platelet activating factor have a wide spectrum of biological actions including modulation of blood vessel contraction, capillary permeability, platelet aggregation, activation of inflammatory and oxidative stress cascades, and direct cytotoxicity (Maeda et al., 1997; Phillis et al., 2006). Such actions play a major role in the initiation and propagation of secondary injury in head and spinal cord trauma. Neural cells near the injury core can be affected directly by some of these metabolites, or indirectly by the resulting reduction in nervous tissue blood supply (Farooqui et al., 2004a). These events can affect all cell types in the CNS including neurons, astrocytes, oligodendrocytes, and endothelial cells. Glutamate release, upregulation of cytokines, and generation of PLA_2-derived metabolites are interrelated processes that depend not only on the release of the above signaling molecules but also on interactions between traumatized neural cells and infiltrating macrophages.

4 Glutamate-Mediated Alterations in the Phospholipid Metabolism of Neural Cells

As stated above, neural membrane glycerophospholipids contain arachidonic or docosahexaenoic acids at the *sn*-2 position of the glycerol moiety. Two enzymatic mechanisms release them. A direct mechanism for the release of arachidonic and docosahexaenoic acids involves cPLA$_2$ and PlsEtn-PLA$_2$ respectively (Farooqui et al., 2000b; Farooqui and Horrocks, 2001), whereas the indirect mechanism utilizes the phospholipase C (PLC)/diacylglycerol lipase pathway (Farooqui et al., 1989). This release of arachidonic and docosahexaenoic acids within the brain and spinal cord is of considerable physiological importance. These fatty acids are not only potent modulators of glutamatergic neurotransmission but also potential mediators of long-term potentiation, long-term depression, and gene expression (Okada et al., 1996; Gamoh et al., 1999).

Multiple forms of PLA$_2$ (secretory PLA$_2$, mol. mass 18 kDa; cPLA$_2$, mol. mass 85 kDa; calcium independent iPLA$_2$, mol mass 80 kDa; and PlsEtn-PLA$_2$, mol mass 39 kDa) are known to occur in brain and spinal cord tissues (Farooqui et al., 1997b; Sandhya et al., 1998; Ong et al., 1999; Farooqui and Horrocks, 2004; Ong et al., 2006). These multiple forms normally exist in a latent form in the cytosol. The treatment of cortical, striatal, hippocampal, and hypothalamic neurons, and cerebellar granule cells with glutamate, NMDA, kainate (KA) and other glutamate agonists results in a dose- and time-dependent increase in cPLA$_2$ activity and arachidonic acid release (Lazarewicz et al., 1988, 1990; Farooqui et al., 2003b) (❯ *Table 4-1*). An NMDA antagonist (2-amino-5-phosphonovalerate) and PLA$_2$ inhibitors (quinacrine and arachidonyltrifluoromethyl ketone) block this arachidonic acid release in a dose-dependent manner (Lazarewicz et al., 1988; Kim et al., 1995a; Farooqui et al., 2003b). Treatment of neuron-enriched cultures

◘ Table 4-1

Effect of glutamate and kainate on cPLA$_2$ and PlsEtn-PLA$_2$ activities in neuron-enriched cultures from rat cerebral cortex

Glutamate (μM)	cPLA$_2$ (pmol/min/mg protein)	Kainate (μM)	PlsEtn-PLA$_2$ (pmol/min/mg protein)
Control	5.0 ± 1.3	Control	3.2 ± 0.4
5.0	5.1 ± 1.5	–	–
10.0	9.2 ± 1.7	10.0	5.3 ± 0.4
20.0	16.8 ± 1.9	20.0	7.5 ± 0.5
30.0	22.9 ± 2.3	30.0	12.5 ± 0.8
40.0	30.5 ± 4.3	40.0	15.4 ± 1.3
50.0	35.7 ± 5.8	50.0	18.5 ± 1.5
75.0	27.3 ± 5.2	75.0	20.4 ± 1.8
Control	5.2 ± 1.2	Control	3.5 ± 0.5
Glu (50.0 μM)	33.0 ± 3.7	KA (50.0 μM)	17.6 ± 1.4
Glu (50.0 μM) + AACOCF$_3$ (25.0 μM)	5.8 ± 1.5	KA (50.0 μM) + BEL (5.0 μM)	4.5 ± 0.4
AACOCF$_3$ (25.0 μM)	6.7 ± 1.3	BEL (5.0 μM)	4.4 ± 0.4
Control	5.0 ± 1.0	Control	3.5 ± 0.9
Glu (50.0 μM)	35.5 ± 3.9	KA (50.0 μM)	17.3 ± 1.3
Glu (50.0 μM) + quinacrine (50.0 μM)	5.6 ± 1.2	KA (50.0 μM) + quinacrine (50.0 μM)	3.7 ± 0.7
Quinacrine (50.0 μM)	5.7 ± 1.4	Quinacrine (50.0 μM)	3.5 ± 0.4

Results are the means ± SEM for three different cultures (Farooqui et al., 2003a, b)

with glutamate and its analogs also results in increased PlsEtn-PLA$_2$ activity in a dose and time-dependent manner. 6-Cyano-7-nitroquinoxaline-2,3-dione (CNQX) and bromoenol lactone (BEL) block this increase, indicating that the stimulation of PlsEtn-PLA$_2$ is a receptor-mediated process (Farooqui et al., 2003a). Thus, the direct release of both arachidonic and docosahexaenoic acids from neural membrane phospholipids is receptor-mediated (Farooqui et al., 2003a, b).

The indirect mechanism of arachidonic acid release was found in primary neuronal cell cultures from fetal spinal cord (Farooqui et al., 1993). The treatment of neuron-enriched cultures with glutamate or NMDA results in a dose- and time-dependent stimulation of diacylglycerol and monoacylglycerol lipase activities. The NMDA antagonists, dextrorphan and MK-801 (◘ Table 4-2) as well as the diacylglycerol

◘ Table 4-2
Effects of NMDA and NMDA receptor antagonists on diacylglycerol and monoacylglycerol lipase activities of neuron-enriched cultures from rat spinal cord

Treatment	Diacylglycerol lipase (nmol/min/mg protein)	Monoacylglycerol lipase (nmol/min/mg protein)
Control	5.05 ± 0.6	11.90 ± 1.8
NMDA (50 μM)	20.80 ± 1.6	29.80 ± 2.5
NMDA (50 μM) + dextrorphan (50 μM)	5.27 ± 0.8	13.70 ± 1.8
Dextrorphan (50 μM)	6.34 ± 0.8	12.67 ± 1.8
Control	5.23 ± 0.6	12.56 ± 1.6
NMDA (50 μM)	18.76 ± 1.8	37.85 ± 3.3
NMDA (50 μM) + MK-801 (50 μM)	6.55 ± 0.8	13.65 ± 2.0
MK-801	5.50 ± 0.6	12.78 ± 1.8

Results are the means ± SEM for three different cultures (Farooqui et al., 1993, 2003b)

lipase inhibitor, RHC80267, block this increase (Farooqui et al., 2003b). Thus, both PLA$_2$-mediated and PLC/diacylglycerol lipase-mediated pathways participate in the release of arachidonic acid in neural cells. In brain tissue under normal conditions, arachidonic and docosahexaenoic acids modulate activities of various enzymes, ion channels and receptors, and gene expression (Farooqui and Horrocks, 2006).

The exposure of spinal cord neurons to increasing concentrations of arachidonic acid in vitro not only enhances oxidative stress and elevates intracellular calcium but also induces the production of nitric oxide and stimulation of caspase activities (Toborek et al., 1999, 2000). Furthermore, under traumatic situations, increased levels of arachidonic acid also set in motion an uncontrolled "arachidonic acid cascade." The latter includes the synthesis and accumulation of prostaglandins, leukotrienes, and thromboxanes (Liu et al., 2001), isoprostanes, and 4-hydroxynonenal (4-HNE), a peroxidized product of arachidonic acid (Farooqui and Horrocks, 2006; Phillis et al., 2006) (◘ Table 4-3).

4-HNE impairs the activities of key metabolic enzymes, including Na$^+$/K$^+$-ATPase, glucose-6-phosphate dehydrogenase, and several kinases. 4-HNE stimulates stress-activated protein kinases (Camandola et al., 2000; Tamagno et al., 2003) such as c-Jun amino-terminal kinase (JNK) and p38 mitogen-activated protein kinase. 4-HNE also disrupts transmembrane signaling and glucose and glutamate transporters in astrocytes (Mark et al., 1997).

The arachidonic acid cascade also potentiates the formation of free radicals and lipid hydroperoxides produced by the action of 12-lipoxygenase. The latter inhibit reacylation of phospholipids in neuronal membranes (Zaleska and Wilson, 1989). This inhibition may constitute another mechanism whereby oxidative processes contribute to necrotic cell death in neural cells (Farooqui et al., 2004a). In C6 glioma cells, arachidonic acid and its metabolites may promote cell death by changing apoptosis to necrosis through lipid peroxidation initiated by lipid hydroperoxides (Higuchi and Yoshimoto, 2002).

In contrast, docosahexaenoic acid exerts a neuroprotective effect in a rat model of compression spinal cord injury (Wu et al., 2005). It not only normalizes brain-derived neurotrophic factor (BDNF), reduces

◘ Table 4-3
Status of glutamate, cytokines, PLA$_2$ and its reaction products, and caspase activity in head and spinal cord injuries

Neurochemical parameter	Head injury	Spinal cord injury	Reference
Glutamate	Increased	Increased	Arundine and Tymianski (2004)
Cytokines	Increased	Increased	Wang et al. (1996); Hayes et al. (2002)
Phospholipid metabolism	Enhanced	Enhanced	Demediuk et al. (1985c); Farooqui et al. (2004a)
PLA$_2$ activity	Increased	Increased	Taylor (1988); Shohami et al. (1989)
Free fatty acids	Increased	Increased	Wei et al. (1982); Demediuk et al. (1985b); Demediuk et al. (1987); Shohami et al. (1989)
Eicosanoids	Increased	Increased	Ellis et al. (1981); Saunders and Horrocks (1987); Shohani et al. (1987); Xu et al. (1990); Liu et al. (2001)
Lipid peroxides	Increased	Increased	Anderson and Means (1985); Demediuk et al. (1985a)
Calcium ion influx	Increased	Increased	Stokes et al. (1983); Stokes and Somerson (1987); Hayes et al. (1992); Beer et al. (2000)
Caspase activity	Increased	Increased	McIntosh et al. (1998); Springer et al. (1999)
4-HNE	Increased	Increased	Springer et al. (1997); Zhang et al. (1999)
Oxidative stress	Increased	Increased	Hall (1996); Farooqui et al. (2004a)
Neuroinflammation	Increased	Increased	Farooqui et al. (2004a)
Apoptotic cell death	Increased	Increased	McIntosh et al. (1998); Springer et al. (1999); Beattie et al. (2000)

oxidative damage, and counteracts learning disability after traumatic brain injury (Wu et al., 2004) but also decreases TNF-α and IL-1 levels. Thus, rats fed docosahexaenoic acid have better locomotor recovery than those of the saline-treated group. The molecular mechanism associated with beneficial effects of docosahexaenoic acid in head and spinal cord injuries may include the action of a 15-lipoxygenase like enzyme on docosahexaenoic acid to produce docosatrienes, resolvins, and neuroprotectins (Hong et al., 2003; Marcheselli et al., 2003; Serhan et al., 2004; Serhan, 2005). Collectively, these lipid mediators are docosanoids (❯ *Figure 4-1*). They not only antagonize the effects of eicosanoids but also modulate leukocyte trafficking as well as downregulating the expression of cytokines in glial cells (Hong et al., 2003; Marcheselli et al., 2003; Serhan et al., 2004). Thus docosanoids slow down the inflammatory cycle induced and maintained by the action of cytokines on astrocytes (Farooqui and Horrocks, 2006). On the basis of these studies, the generation of docosanoids may be an endogenous protection mechanism induced in brain tissue under oxidative stress.

Under traumatic situations, peroxidation of docosahexaenoic acid in brain produces 4-hydroxyhexanal (4-HHE). In spite of the structural similarity between 4-HNE and 4-HHE, the neurochemical actions and efficacies of these aldehydes differ considerably. Thus, 4-HHE more effectively inhibits the mitochondrial ATP translocator than does 4-HNE (Picklo et al., 1999). Furthermore, in brain tissue, 4-HHE modulates the endothelial nitric oxide synthase through NF-κB activation. In contrast, 4-HNE inhibits NF-κB activation (Camandola et al., 2000; Lee et al., 2004).

5 Generation and Release of Cytokines in Head Injury and Spinal Cord Trauma

Traumatic injury to brain and spinal cord produces significant increases in the synthesis of multiple cytokines including TNF-α and IL-1β (Wang et al., 1996, 1997; Klusman and Schwab, 1997). Peripheral

◘ Figure 4-1
Interaction among phospholipid and sphingomyelin metabolism metabolites. Plasma membrane (PM), agonist (A), receptor (R), cytosolic phospholipase A_2 (cPLA$_2$), sphingomyelinase (SMase), arachidonic acid (AA), 4-hydroxynonenal (4-HNE), cyclooxygenase (COX), and prostaglandins (PG)

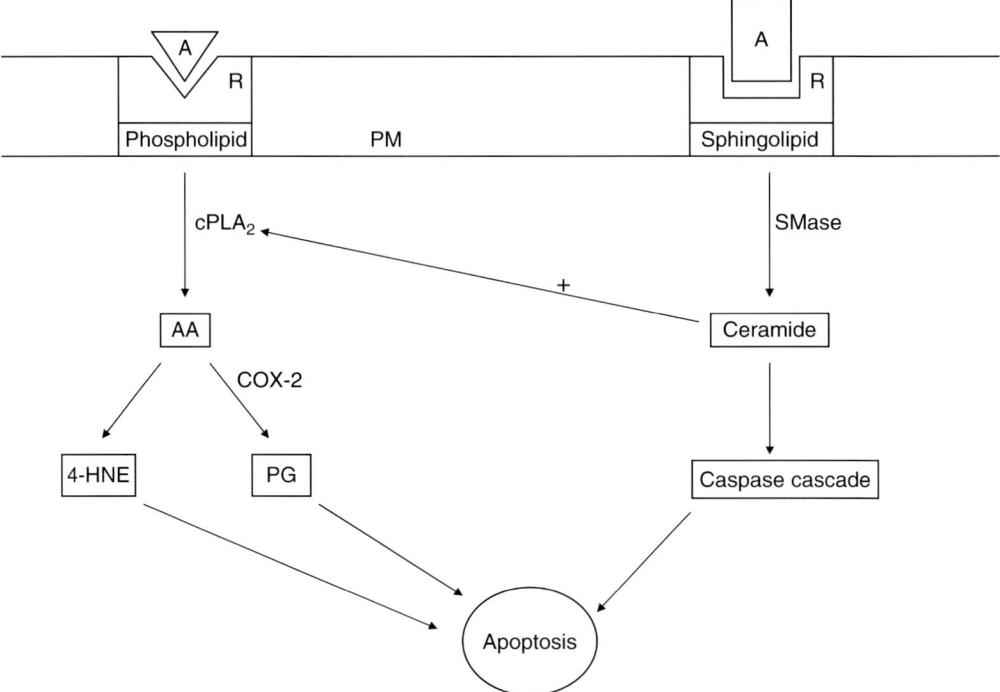

immune cells (macrophages and neutrophils), astrocytes, and microglial cells are the major sources of these cytokines. Increased cytokine activity in head and spinal cord injury plays an important role not only in inflammation and neurodegeneration but also in metabolic responses during compensatory regenerative processes (Vitkovic et al., 2000). Like glutamate, cytokines produce their effects by binding to specific membrane-associated receptors that are composed of an extracellular ligand-binding region, a membrane-spanning region, and an intracellular region that is activated by binding of cytokines, and hence delivering a signal to the nucleus (Rothwell and Relton, 1993). The magnitude and persistence of the elevations in cytokine levels may be related to the severity of trauma (Rothwell and Relton, 1993). Although TNF-α and IL-1β appear to trigger biologically indistinguishable effects by activating the same set of transcription factors, the two cytokines are structurally unrelated polypeptides that exert their effect through distinct and structurally unrelated cell-surface receptors.

The role of cytokine-induced inflammatory reactions in head injury and spinal cord trauma is quite complex. An early administration (at 1 day) of TNF-α has detrimental effects on the spinal cord, whereas delayed administration (at 4 days) reduces the extent of the lesions (Klusman and Schwab, 1997). Stereotaxic injections of recombinant TNF-α and IL-1β in brain parenchyma and spinal cord produce much stronger inflammatory responses in the spinal cord than in the cerebral cortex (Schnell et al., 1999). This may be due to differences in the vasculature of spinal cord and brain tissues. TNF-α is a master proinflammatory cytokine that modulates the induction of a large number of other inflammatory cytokines, chemokines, and adhesion molecules under traumatic conditions. The rapid increase of TNF-α levels in the injured brain may be due to its production by infiltrating macrophages and activated microglia as well as by the presence of an efficient transport system for TNF-α at the blood–brain barrier (Patrizio, 2004).

The mechanism of TNF-α and IL-1β actions is quite complex because they activate a number of signaling pathways, including phosphatases, kinases, phospholipases, oxygen radicals, and transcription factors (Jupp et al., 2003; Gomes-Leal et al., 2004). All these targets may participate in neurotoxicity induced by TNF-α and IL-1β in traumatized brain and spinal cord.

Another mechanism of TNF-α-mediated cytotoxicity involves the generation of oxygen radicals during the arachidonic acid cascade. This process results in oxidative stress, mitochondrial dysfunction, and calcium ion overload along with the release of cytochrome c and the activation of downstream caspases-9 and -3. Minocycline, a second-generation tetracycline that decreases activation and proliferation of microglia and macrophages, blocks the release of cytochrome c. Minocycline has been recently used for the treatment of spinal cord injury (Teng et al., 2004). An important factor in the signaling process induced by TNF-α is the involvement of the gene expression mediated by NF-κB. Traumatic injury to the spinal cord induces the expression of TNF-α and activation of NF-κB (Bethea et al., 1998; Xu et al., 1998). NF-κB in turn modulates the transcription of proinflammatory cytokines, chemokines, and proinflammatory enzymes such as cyclooxygenase-2 (COX-2) and inducible nitric oxide synthase (iNOS). These factors further intensify traumatic head and spinal cord injuries.

6 Cytokine-Mediated Alterations in Phospholipid Metabolism and Cross Talk Between Phospholipids and Sphingolipid-Generated Lipid Mediators

The proinflammatory effects of TNF-α and IL-1β manifest themselves by their ability to induce the enzymatic release of arachidonic acid from neural membrane phospholipids through the activation of PLA$_2$ isoforms. Several mechanisms of cPLA$_2$ stimulation by cytokines are possible. One molecular mechanism of cPLA$_2$ stimulation by TNF-α and IL-1β involves the phosphorylation of cPLA$_2$ by mitogen-activated protein kinase in the presence of agents that mobilize intracellular Ca^{2+} (Clark et al., 1995). Another mechanism involves TNF-α-mediated activation of caspase-3 and the proteolytic cleavage of cPLA$_2$ by caspase-3 (Wissing et al., 1997; Beer et al., 2000). A specific tetrapeptide inhibitor of caspase-3 (acetyl-Asp-Glu-Val-Asp-aldehyde) prevents the proteolytic cleavage and activation of cPLA$_2$, indicating that caspase-mediated cPLA$_2$ proteolysis retards cell injury and death. Also, arachidonoyl trifluoromethyl ketone (AACOCF$_3$), a potent inhibitor of cPLA$_2$ activity, inhibits neural cell injury and death (Wissing et al., 1997). These observations suggest that both caspase-3 and cPLA$_2$ are involved in neural injury (Farooqui et al., 2004a). In fact, nonneural cells that are resistant to TNF-α cytotoxicity have lower levels of cPLA$_2$ activity, whereas introduction of this enzyme into enzyme-deficient cells restores their TNF-α cytotoxicity (Hayakawa et al., 1993, 1996; Singh et al., 1998), suggesting that cPLA$_2$ plays an important role in cell injury and death.

A close interaction between cPLA$_2$-generated second messengers (arachidonic acid) and sphingomyelinase-generated second messenger (ceramide) occurs at several sites in cytokine-induced signal transduction process (❯ Figure 4-2). In nonneural cells, cytokines stimulate both sphingomyelinase and PLA$_2$ activities in a dose- and concentration-dependent manner. cPLA$_2$ and sphingomyelinase inhibitors (Vanags et al., 1997) can block this stimulation. Arachidonic acid stimulates sphingomyelinase (Robinson et al., 1997) and ceramide stimulates PLA$_2$ activity (Sato et al., 1999), suggesting a cross talk between receptor-mediated phospholipid and sphingomyelin signaling pathways. Phosphorylated sphingolipid metabolites, sphingosine 1-phosphate, and ceramide 1-phosphate are required for the activation and translocation of cyclooxygenase-2 and cPLA$_2$ (Chalfant and Spiegel, 2005), indicating that these metabolites of sphingolipid metabolism may act in concert to regulate generation of inflammatory mediators from neural membrane phospholipids. PlsEtn-PLA$_2$ is another target for ceramide-mediated apoptotic cell death (Latorre et al., 2003). Quinacrine (a PLA$_2$ inhibitor) and 1-O-octadecyl-2-methyl-rac-glycero-3-phosphocholine (a CoA-independent transacylase inhibitor) inhibit ceramide-mediated stimulation of PlsEtn-PLA$_2$. This suggests cross talk between PlsEtn degradation and sphingomyelin catabolism through the generation of arachidonic acid and ceramide (Farooqui and Horrocks, 2001). Fas-mediated apoptotic cell death requires the proteolytic degradation of calcium-independent PLA$_2$ (iPLA$_2$) by caspase-3. BEL, a specific inhibitor of

Figure 4-2
Interactions among excitotoxicity, glycerophospholipid metabolism, and cytokines. N-Methyl-D-aspartate receptor (NMDA-R), phosphatidylcholines (PtdCho), kainate receptor (KA-R), ethanolamine plasmalogens (PlsEtn), ethanolamine lysoplasmalogens (lysoPlsEtn), cytosolic phospholipase A_2 (cPLA$_2$), plasmalogen-selective phospholipase A_2 (PlsEtn-PLA$_2$), arachidonic acid (AA), docosahexaenoic acid (DHA), reactive oxygen species (ROS), nuclear factor κB (NF-κB), tumor necrosis factor-α (TNF-α), interleukin-I-1β (IL-1β), stimulation (+), and antagonism (−).

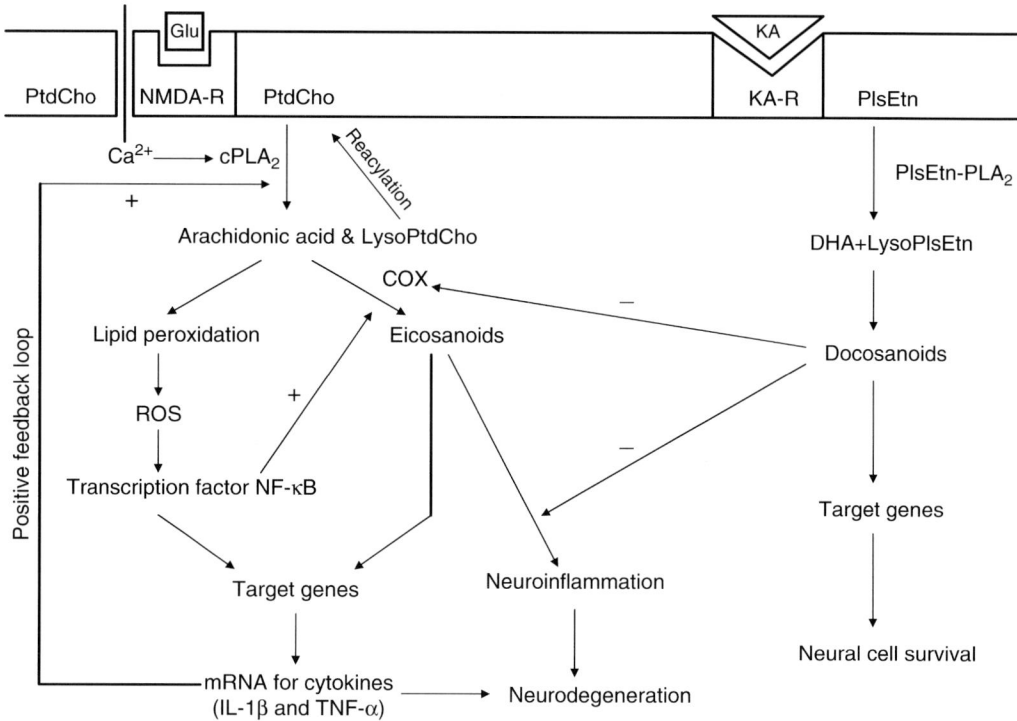

iPLA$_2$, blocks apoptotic cell death, indicating that iPLA$_2$ also plays an important role in cytokine-mediated cell death (Atsumi et al., 1998, 2000).

In glial cells, TNF-α mediated transactivation of NF-κB is sensitive to inhibitors of PLA$_2$ activity, indicating that PLA$_2$ activation is closely associated with the injury process (Won et al., 2005) and that PLA$_2$ inhibitors prevent neurodegeneration by blocking DNA fragmentation. (Vanags et al., 1997). Collective evidence thus suggests that levels, interactions, and communications between cytokines present in the CNS and infiltrating macrophages, along with the PLA$_2$ activity, mitochondrial abnormalities, and energy status of degenerating neurons and glial cells, determine the outcome of pathologic or regenerative sequelae following traumatic injury to brain and spinal cord (Farooqui and Horrocks, 1991, 1994).

6.1 Phospholipid Alterations in Head Injury

The neurochemical changes in head injury are accompanied by increased degradation of neural membrane phospholipids and accumulation of arachidonic acid and eicosanoids as well as leukotrienes (McIntosh et al., 1998; Schuhmann et al., 2003). The activations of cPLA$_2$ and the PLC/DAG-lipase pathway release arachidonic acid (Wei et al., 1982; Shohami et al., 1989 Dhillon et al., 1999). Changes in phospholipid composition occur in traumatic as well as fluid percussion models of head injury (Dhillon et al., 1994;

Homayoun et al., 1997, 2000; Schuhmann et al., 2003). In humans, free fatty acid levels in cerebrospinal fluid can be used as a predictive marker of outcome following traumatic brain injury (Pilitsis et al., 2003). Impact traumatic injury produces marked increases in free fatty acids (threefold) and diacylglycerols (twofold) in the injured cortex and to a lesser extent in the contralateral cortex compared with sham-operated rats. The stimulation of PLA_2 and PLC suggests that increased calcium influx, increased production of free radicals and eicosanoids, and impaired mitochondrial function and generation of ROS may produce neurodegeneration by apoptotic as well as necrotic cell death (Wingrave et al., 2003).

6.2 Phospholipid Alterations in Spinal Cord Trauma

In spinal cord injury, neurochemical changes include the release of glutamate in the extracellular compartment, increase in intracellular calcium, upregulation of cytokines, and degradation of membrane phospholipids by PLA_2 and PLC with generation of free fatty acids, diacylglycerols, eicosanoids, and lipid peroxides (Anderson et al., 1985; Demediuk et al., 1985b; Taylor, 1988) (> Table 4-3). About 10% of PlsEtn is lost during the first minute of compression trauma to spinal cord with an overall loss of 18% found at 30 min (Horrocks et al., 1985). Similar results have been reported in another model of spinal cord injury in rabbits (Lukácová et al., 1996). The loss of PlsEtn after compression and ischemic injuries is due to the stimulation of $PlsEtn-PLA_2$.

Stimulation of $PlsEtn-PLA_2$ may result in changes in membrane fluidity and permeability resulting in increased calcium ion influx, impaired mitochondrial function, and the subsequent generation of ROS. Low levels of ROS are second messengers and produce neurodegeneration by apoptosis, whereas high levels of ROS produce irreversible damage to cellular components and cause cell death by necrosis (Denecker et al., 2001). Necrosis normally occurs at the core of injury site whereas neural cells undergo apoptosis several hours or days after injury in the surrounding area.

7 Interplay among Glutamate, Cytokines, and PLA_2-Derived Lipid Mediators

At the molecular level, glutamate, cytokines, and PLA_2-generated lipid metabolites interact with each other to modulate neural cell injury. Thus, neurotoxic concentrations of glutamate and its analogs induce the expression of cytokines in brain tissue (Minami et al., 1991; Strijbos and Rothwell, 1995) and glutamate-mediated cell death can be attenuated by IL-1ra (an endogenous IL-1 competitive antagonist). It is interesting to note that TNF-α not only reduces excitotoxic cellular damage in vitro (Carlson et al., 1999) but also contributes to posttraumatic functional and structural damage (Knoblach et al., 1999). As stated above, cytokines often interact synergistically, producing an effect that is significantly greater than the sum of the effects of the individual cytokines. Thus, interferon-γ or interleukin-1β alone do not induce the generation of TNF-α but may generate a marked release of TNF-α when administered together. Similarly, the LPS-mediated generation of TNF-α is markedly increased by interferon-γ in cultured astrocytes (Jeohn et al., 1998; Turrin and Plata-Salaman, 2000). Furthermore, glutamate and cytokines intensify neurodegeneration by acting synergistically. Thus, TNF-α potentiates glutamate-induced neurotoxicity in an in vitro model of rat spinal cord injury. In this model, cytokines may act on AMPA receptors (a subtype of glutamate receptor) in microglia. This process may induce glial cells to release more glutamate and cytokines, thus exacerbating the outcome of traumatic injury (Hermann et al., 2001; Gomes-Leal et al., 2004). Both glutamate and cytokines stimulate PLA_2 activity not only in neuronal cultures but also in cell culture and animal models of head injury (Shohami et al., 1989) and spinal cord trauma (Taylor, 1988). As stated above, quinacrine, a PLA_2 inhibitor in cell culture, can block this increase in PLA_2 activity (Sanfeliu et al., 1990; Kim et al., 1995b; Farooqui et al., 2003b) (> Table 4-1). TNF-α and IL-1β are known to cause enhanced arachidonic acid metabolism. In turn, arachidonic acid metabolites increase TNF-α and IL-1β production by blood macrophages and microglial cells, implying that this positive feed-forward loop may amplify arachidonic acid metabolite-mediated events in the brain and spinal cord tissues

(Genis et al., 1992). Furthermore, in nonneural cells, PLA$_2$ inhibitors suppress the expression of cytokines (IL-2) suggesting a relationship between cytokines and PLA$_2$ activity (Ouyang and Kaminski, 1999).

Unlike glutamate, which becomes excitotoxic only after release, arachidonic acid and its metabolites elicit their effects in both the extracellular compartment and the intracellular spaces. Arachidonic acid and its metabolites produce a variety of detrimental effects on membrane structures, activities of membrane-bound enzymes, and neurotransmitter uptake systems (Farooqui et al., 1997a; Phillis et al., 2006). Thus arachidonic acid and its metabolites, which are generated by glutamate and cytokine receptor stimulation at the postsynaptic level, may cross the synaptic cleft to act at the presynaptic level and may thus act as retrograde messengers for the activation of protein kinase C-γ following head injury and spinal cord trauma. This process may be associated with regeneration and repair in head and spinal cord trauma (Farooqui and Horrocks, 2005).

8 Prevention of Neurodegeneration by cPLA$_2$ Inhibitors

PLA$_2$ inhibitors are good neuroprotective agents (Farooqui et al., 2002b, 2004b, 2005), acting at an early step in the biosynthesis of inflammatory mediators. The specificity of an inhibitor is the most important factor in choosing it as a therapeutic agent for the treatment of inflammation and oxidative stress during neurodegeneration (Farooqui et al., 2006). An ideal PLA$_2$ inhibitor should have regional specificity and should be able to reach the sites where cells are under oxidative stress and neurodegenerative processes are taking place. For the successful treatment of inflammatory and oxidative stress in neurological disorders, timely delivery is required of a well tolerated, chronically active, and specific inhibitor of cPLA$_2$ that can bypass or cross the blood–brain barrier.

During our studies on KA-mediated neurotoxicity, AACOCF$_3$ blocked the stimulation of cPLA$_2$ and reduced neurodegeneration in hippocampal slices (Lu et al., 2001). In nonneural cells, AACOCF$_3$ inhibits the expression of interleukin-2 (IL-2) at both the mRNA and protein levels, indicating that cPLA$_2$ may have marked effects on T-cell function (Amandi-Burgermeister et al., 1997). Similarly, LPS-mediated neurotoxicity in glial cells induces the upregulation of cPLA$_2$ and iPLA$_2$ activities as well as expression of iNOS. The inhibition of cPLA$_2$ by methyl arachidonoyl fluorophosphates or an antisense oligomer against cPLA$_2$ and inhibition of iPLA$_2$ by BEL prevents LPS-mediated iNOS gene expression and NF-κB activation, suggesting that LPS-induced iNOS gene expression may be mediated by both calcium-dependent and calcium-independent PLA$_2$ activities, and PLA$_2$ inhibitors may be useful in blocking LPS-mediated toxicity (Won et al., 2005). It remains to be seen whether the above mechanisms are involved in the beneficial effect of PLA$_2$ inhibitors in head injury and spinal cord trauma.

GM1 and GM3 gangliosides also inhibit cPLA$_2$ and PlsEtn-PLA$_2$ activities in brain (Yang et al., 1994). The GM3 type has the strongest effect. The mechanism of inhibition by gangliosides remains unknown. However, the orientation of N-acetylneuraminic acid residues in glycoconjugates is important for the inhibitory activity (Yang et al., 1994). Gangliosides not only stabilize neural membranes but also regulate calcium ion influx and enzymatic activities associated with signal transduction. Gangliosides may also inhibit apoptosis (Saito et al., 1999). Gangliosides may exert this antiapoptotic effect through augmentation of nerve growth factor receptor phosphorylation (Ferrari et al., 1995). Gangliosides have been used for the treatment of spinal cord injury (Geisler et al., 1991).

CDP-choline (citicoline) inhibits cPLA$_2$ activity and lowers the concentration of free fatty acids in a dose- and time-dependent manner (Adibhatla et al., 2002). This compound is an intermediate in PtdCho biosynthesis. It is an effective treatment for ischemic and head injuries in rats (Andersen et al., 1999; Krupinski et al., 2002; Dempsey and Rao, 2003; Cakir et al., 2005) and humans (Adibhatla et al., 2002, 2003; Adibhatla and Hatcher, 2005). It not only restores the concentration of PtdCho following ischemic injury by increasing PtdCho synthesis from diacylglycerols but also blocks the activation of cPLA$_2$ activity (Adibhatla et al., 2002). The inhibition of cPLA$_2$ may lead to reduced levels of arachidonic acid and reactive oxygen species. This promotes the stabilization of neural membranes. CDP-choline also protects cerebellar granule neurons from glutamate-mediated neurotoxicity (Mir et al., 2003), suggesting that CDP-choline may protect neurons from excitotoxicity. CDP-choline also improves verbal memory of aged human subjects (Alvarez et al., 1997). This cPLA$_2$ inhibitor could be used for treating acute neural trauma as well as neurodegenerative diseases.

The cPLA$_2$ inhibitors mentioned above are nonspecific. No compound with real clinical potential has emerged. Progress in this important area of research requires the design and synthesis of specific cPLA$_2$ inhibitors. The reaction catalyzed by cPLA$_2$ is a rate-limiting step for the generation of eicosanoids and platelet activating factor. High levels of these metabolites are responsible for oxidative stress and inflammatory reactions at the injury site. A novel cPLA$_2$ inhibitor should be harmless, cell permeable, and able to cross the blood–brain barrier without harm. It should inhibit oxidative and inflammatory reactions that are taking place at the injury site. Such studies are beginning to emerge in recent years (Miele, 2003; Farooqui et al., 2006).

9 Conclusion

Glutamate and cytokines mediate the abnormal phospholipid metabolism that is increasingly recognized for its central importance to acute neural trauma. The neurochemical consequences of disturbed phospholipid metabolism in head injury and spinal cord trauma include altered calcium ion homeostasis and the breakdown of membrane phospholipids owing to stimulation of cPLA$_2$ and PlsEtn-PLA$_2$ activities mediated by the caspase cascade. This results in changes in membrane fluidity and permeability, and the massive release of free arachidonic acid leading to the accumulation of eicosanoids, lipid peroxides, and 4-HNE. Reactive oxygen species generated during phospholipid degradation produce oxidative stress and are closely associated with the cell death, both apoptotic and necrotic, that occurs following acute neural trauma. The stimulation of PLA$_2$ isoforms mediated by glutamate and cytokines suggests that PLA$_2$ inhibitors can be developed as therapeutic agents for preventing neurodegeneration following head injury and spinal cord trauma.

Acknowledgments

We thank Siraj A Farooqui for providing figures and his help during preparation of this review.

References

Adibhatla RM, Hatcher JF. 2005. Cytidine 5′-diphosphocholine (CDP-choline) in stroke and other CNS disorders. Neurochem Res 30: 15-23.

Adibhatla RM, Hatcher JF, Dempsey RJ. 2002. Citicoline: Neuroprotective mechanisms in cerebral ischemia. J Neurochem 80: 12-23.

Adibhatla RM, Hatcher JF, Dempsey RJ. 2003. Phospholipase A$_2$, hydroxyl radicals, and lipid peroxidation in transient cerebral ischemia. Antioxid Redox Signal 5: 647-654.

Ahn MJ, Sherwood ER, Prough DS, Lin CY, De Witt DS. 2004. The effects of traumatic brain injury on cerebral blood flow and brain tissue nitric oxide levels and cytokine expression. J Neurotrauma 21: 1431-1442.

Alvarez XA, Laredo M, Corzo D, Fernández-Novoa L, Mouzo R, et al. 1997. Citicoline improves memory performance in elderly subjects. Methods Find Exp Clin Pharmacol 19: 201-210.

Amandi-Burgermeister E, Tibes U, Kaiser BM, Friebe WG, Scheuer WV. 1997. Suppression of cytokine synthesis, integrin expression and chronic inflammation by inhibitors of cytosolic phospholipase A$_2$. Eur J Pharmacol 326: 237-250.

Andersen M, Overgaard K, Meden P, Boysen G. 1999. Effects of citicoline combined with thrombolytic therapy in a rat embolic stroke model. Stroke 30: 1464-1470.

Anderson DK, Means ED. 1985. Iron-induced lipid peroxidation in spinal cord: Protection with mannitol and methylprednisolne. J Free Radic Biol Med 1: 59-64.

Anderson DK, Saunders RD, Demediuk P, Dugan LL, Braughler JM, et al. 1985. Lipid hydrolysis and peroxidation in injured spinal cord: Partial protection with methylprednisolone or vitamin E and selenium. Cent Nerv Syst Trauma 2: 257-267.

Arundine M, Tymianski M. 2004. Molecular mechanisms of glutamate-dependent neurodegeneration in ischemia and traumatic brain injury. Cell Mol Life Sci 61: 657-668.

Atsumi G, Murakami M, Kojima K, Hadano A, Tajima M, et al. 2000. Distinct roles of two intracellular phospholipase A_2s in fatty acid release in the cell death pathway. Proteolytic fragment of type IVA cytosolic phospholipase $A_{2\alpha}$ inhibits stimulus-induced arachidonate release, whereas that of type VI Ca^{2+}-independent phospholipase A_2 augments spontaneous fatty acid release. J Biol Chem 275: 18248-18258.

Atsumi G, Tajima M, Hadano A, Nakatani Y, Murakami M, et al. 1998. Fas-induced arachidonic acid release is mediated by Ca^{2+}-independent phospholipase A_2 but not cytosolic phospholipase A_2 which undergoes proteolytic inactivation. J Biol Chem 273: 13870-13877.

Bartholdi D, Schwab ME. 1997. Expression of pro-inflammatory cytokine and chemokine mRNA upon experimental spinal cord injury in mouse: An in situ hybridization study. Eur J Neurosci 9: 1422-1438.

Bazan NG, Rodriguez de Turco EB, Allan G. 1995. Mediators of injury in neurotrauma: Intracellular signal transduction and gene expression. J Neurotrauma 12: 791-814.

Beattie MS, Farooqui AA, Bresnahan JC. 2000. Review of current evidence for apoptosis after spinal cord injury. J Neurotrauma 17: 915-925.

Beer R, Franz G, Srinivasan A, Hayes RL, Pike BR, et al. 2000. Temporal profile and cell subtype distribution of activated caspase-3 following experimental traumatic brain injury. J Neurochem 75: 1264-1273.

Bethea JR, Castro M, Keane RW, Lee TT, Dietrich WD, et al. 1998. Traumatic spinal cord injury induces nuclear factor-κB activation. J Neurosci 18: 3251-3260.

Cakir E, Usul H, Peksoylu B, Sayin OC, Alver A, et al. 2005. Effects of citicoline on experimental spinal cord injury. J Clin Neurosci 12: 923-926.

Camandola S, Poli G, Mattson MP. 2000. The lipid peroxidation product 4-hydroxy-2,3-nonenal increases AP-1-binding activity through caspase activation in neurons. J Neurochem 74: 159-168.

Carlson NG, Wieggel WA, Chen J, Bacchi A, Rogers SW, et al. 1999. Inflammatory cytokines IL-1α, IL-1β, IL-6, and TNF-α impart neuroprotection to an excitotoxin through distinct pathways. J Immunol 163: 3963-3968.

Chalfant CE, Spiegel S. 2005. Sphingosine 1-phosphate and ceramide 1-phosphate: Expanding roles in cell signaling. J Cell Sci 118: 4605-4612.

Chen SF, Richards HK, Smielewski T, Johnstrom P, Salvador R, et al. 2004. Relationship between flow-metabolism uncoupling and evolving axonal injury after experimental traumatic brain injury. J Cereb Blood Flow Metab 24: 1025-1036.

Clark JD, Schievella AR, Nalefski EA, L-L. Lin 1995. Cytosolic phospholipase A_2. J Lipid Mediat Cell Signal 12: 83-117.

Demediuk P, Anderson DK, Horrocks LA, Means ED. 1985a. Mechanical damage to murine neuronal-enriched cultures during harvesting: Effects on free fatty acids, diglycerides, Na^+K^+-ATPase, and lipid peroxidation. In Vitro Cell Dev Biol 21: 569-574.

Demediuk P, Saunders RD, Anderson DK, Means ED, Horrocks LA. 1985b. Membrane lipid changes in laminectomized and traumatized cat spinal cord. Proc Natl Acad Sci USA 82: 7071-7075.

Demediuk P, Saunders RD, Clendenon NR, Means ED, Anderson DK, et al. 1985c. Changes in lipid metabolism in traumatized spinal cord. Prog Brain Res 63: 211-226.

Demediuk P, Daly MP, Faden AI. 1988. Free amino acid levels in laminectomized and traumatized rat spinal cord. Trans Am Soc Neurochem 19: 176.

Demediuk P, Saunders RD, Anderson DK, Means ED, Horrocks LA. 1987. Early membrane lipid changes in laminectomized and traumatized cat spinal cord. Neurochem Pathol 7: 79-89.

Dempsey RJ, Rao VLR. 2003. Cytidinediphosphocholine treatment to decrease traumatic brain injury-induced hippocampal neuronal death, cortical contusion volume, and neurological dysfunction in rats. J Neurosurg 98: 867-873.

Denecker G, Vercammen D, Declercq W, Vandenabeele P. 2001. Apoptotic and necrotic cell death induced by death domain receptors. Cell Mol Life Sci 58: 356-370.

Dhillon HS, Carman HM, Prasad RM. 1999. Regional activities of phospholipase C after experimental brain injury in the rat. Neurochem Res 24: 751-755.

Dhillon HS, Donaldson D, Dempsey RJ, Prasad MR. 1994. Regional levels of free fatty acids and Evans blue extravasation after experimental brain injury. J Neurotrauma 11: 405-415.

Dietrich WD, Chatzipanteli K, Vitarbo E, Wada K, Kinoshita K. 2004. The role of inflammatory processes in the pathophysiology and treatment of brain and spinal cord trauma. Mechanisms of Secondary Brain Damage from Trauma and Ischemia, Recent Advances of Our Understanding. Baethmann A, Eriskat J, Lehmberg J, Plesnila N, editors. Vienna: Springer-Verlag Wien; pp. 69-74.

Ellis EF, Wright KF, Wei EP, Kontos HA. 1981. Cyclooxygenase products of arachidonic acid metabolism in cat cerebral cortex after experimental concussive brain injury. J Neurochem 37: 892-896.

Ellis RC, Earnhardt JN, Hayes RL, Wang KK, Anderson DK. 2004. Cathepsin B mRNA and protein expression following contusion spinal cord injury in rats. J Neurochem 88: 689-697.

Farooqui AA, Horrocks LA. 1991. Excitatory amino acid receptors, neural membrane phospholipid metabolism and neurological disorders. Brain Res Brain Res Rev 16: 171-191.

Farooqui AA, Horrocks LA. 1994. Excitotoxicity and neurological disorders: Involvement of membrane phospholipids. Int Rev Neurobiol 36: 267-323.

Farooqui AA, Horrocks LA. 2001. Plasmalogens: Workhorse lipids of membranes in normal and injured neurons and glia. Neuroscientist 7: 232-245.

Farooqui AA, Horrocks LA. 2004. Brain phospholipases A_2: A perspective on the history. Prostaglandins Leukot Essent Fatty Acids 71: 161-169.

Farooqui AA, Horrocks LA. 2005. Signaling and interplay mediated by phospholipases A_2, C, and D in LA-N-1 cell nuclei. Reprod Nutr Dev 45: 613-631.

Farooqui AA, Horrocks LA. 2006. Phospholipase A_2-generated lipid mediators in brain: The good, the bad, and the ugly. Neuroscientist 12: 245-260.

Farooqui AA, Anderson DK, Horrocks LA. 1993. Effect of glutamate and its analogs on diacylglycerol and monoacylglycerol lipase activities of neuron-enriched cultures. Brain Res 604: 180-184.

Farooqui AA, Farooqui T, Horrocks LA. 2002a. Molecular species of phospholipids during brain development. Their occurrence, separation and roles. Brain Lipids and Disorders in Biological Psychiatry. Skinner ER, editor. Amsterdam: Elsevier Science B.V; pp. 147-158.

Farooqui AA, Ong WY, Lu XR, Horrocks LA. 2002b. Cytosolic phospholipase A_2 inhibitors as therapeutic agents for neural cell injury. Curr Med Chem—Anti-Inflammatory & Anti-Allergy Agents 1: 193-204.

Farooqui AA, Horrocks LA, Farooqui T. 2000a. Glycerophospholipids in brain: Their metabolism, incorporation into membranes, functions, and involvement in neurological disorders. Chem Phys Lipids 106: 1-29.

Farooqui AA, Ong WY, Horrocks LA, Farooqui T. 2000b. Brain cytosolic phospholipase A_2: Localization, role, and involvement in neurological diseases. Neuroscientist 6: 169-180.

Farooqui AA, Horrocks LA, Farooqui T. 2009. Choline and ethanolamine glycerophospholipids. Handbook of Neurochemistry. Tettamanti G, Goracci G, editors. New York: Springer; in press.

Farooqui AA, Ong WY, Horrocks LA. 2006. Inhibitors of brain phospholipase A_2 activity: Their neuropharmacologic effects and therapeutic importance for the treatment of neurologic disorders. Pharmacol Rev 58: 591-620.

Farooqui AA, Ong WY, Go ML, Horrocks LA. 2005. Inhibition of brain phospholipase A_2 by antimalarial drugs: Implications for neuroprotection in neurological disorders. Med Chem Rev Online 2: 379-392.

Farooqui AA, Ong WY, Horrocks LA. 2003a. Plasmalogens, docosahexaenoic acid, and neurological disorders. Peroxisomal Disorders and Regulation of Genes. Roels F, Baes M, de Bies S, editors. London: Kluwer Academic/Plenum Publishers; pp. 335-354.

Farooqui AA, Ong WY, Horrocks LA. 2003b. Stimulation of lipases and phospholipases in Alzheimer disease. Nutrition and Biochemistry of Phospholipids. Szuhaj B, van Nieuwenhuyzen W, editors. Champaign: AOCS Press; pp. 14-29.

Farooqui AA, Ong WY, Horrocks LA. 2004a. Biochemical aspects of neurodegeneration in human brain: Involvement of neural membrane phospholipids and phospholipases A_2. Neurochem Res 29: 1961-1977.

Farooqui AA, Ong WY, Horrocks LA. 2004b. Neuroprotection abilities of cytosolic phospholipase A_2 inhibitors in kainic acid-induced neurodegeneration. Curr Drug Targets Cardiovasc Haematol Disord 4: 85-96.

Farooqui AA, Rammohan KW, Horrocks LA. 1989. Isolation, characterization and regulation of diacylglycerol lipases from bovine brain. Ann N Y Acad Sci 559: 25-36.

Farooqui AA, Rosenberger TA, Horrocks LA. 1997a. Arachidonic acid, neurotrauma, and neurodegenerative diseases. Handbook of Essential Fatty Acid Biology. Yehuda S, Mostofsky DI, editors. Totowa, NJ: Humana Press; pp. 277-295.

Farooqui AA, Yang HC, Rosenberger TA, Horrocks LA. 1997b. Phospholipase A_2 and its role in brain tissue. J Neurochem 69: 889-901.

Ferrari G, Anderson BL, Stephens RM, Kaplan DR, Greene LA. 1995. Prevention of apoptotic neuronal death by GM1 ganglioside. Involvement of Trk neurotrophin receptors. J Biol Chem 270: 3074-3080.

Gamoh S, Hashimoto M, Sugioka K, Hossain MS, Hata N, et al. 1999. Chronic administration of docosahexaenoic acid improves reference memory-related learning ability in young rats. Neuroscience 93: 237-241.

Geisler FH, Dorsey FC, Coleman WP. 1991. Recovery of motor function after spinal-cord injury—a randomized, placebo-controlled trial with GM-1 ganglioside. N Engl J Med 324: 1829-1887.

Genis P, Jett M, Bernton EW, Boyle T, Gelbard HA, et al. 1992. Cytokines and arachidonic metabolites produced during human immunodeficiency virus (HIV)-infected macrophage-astroglia interactions: Implications for the neuropathogenesis of HIV disease. J Exp Med 176: 1703-1718.

Gomes-Leal W, Corkill DJ, Freire MA, Picanço-Diniz CW, Perry VH. 2004. Astrocytosis, microglia activation, oligodendrocyte degeneration, and pyknosis following acute spinal cord injury. Exp Neurol 190: 456-467.

Hall ED. 1996. Free radicals and lipid peroxidation in neurotrauma. Neurotrauma. Narayan RK, Wilberger JE, Povlishock JT, editors. New York: McGraw Hill; pp. 1405-1419.

Hall ED, Detloff MR, Johnson K, Kupina NC. 2004. Peroxynitrite-mediated protein nitration and lipid peroxidation in a mouse model of traumatic brain injury. J Neurotrauma 21: 9-20.

Hayakawa M, Ishida N, Takeuchi K, Shibamoto S, Hori T, et al. 1993. Arachidonic acid-selective cytosolic phospholipase A_2 is crucial in the cytotoxic action of tumor necrosis factor. J Biochem 268: 11290-11295.

Hayakawa M, Jayadev S, Tsujimoto M, Hannun YA, Ito F. 1996. Role of ceramide in stimulation of the transcription of cytosolic phospholipase A_2 and cyclooxygenase 2. Biochem Biophys Res Commun 220: 681-686.

Hayes KC, Hull TC, Delaney GA, Potter PJ, Sequeira KA, et al. 2002. Elevated serum titers of proinflammatory cytokines and CNS autoantibodies in patients with chronic spinal cord injury. J Neurotrauma 19: 753-761.

Hayes RL, Jenkins LW, Lyeth BG. 1992. Neurotransmitter-mediated mechanisms of traumatic brain injury: Acetylcholine and excitatory amino acids. J Neurotrauma 9: S173-S187.

Hermann GE, Rogers RC, Bresnahan JC, Beattie MS. 2001. Tumor necrosis factor-α induces cFOS and strongly potentiates glutamate-mediated cell death in the rat spinal cord. Neurobiol Dis 8: 590-599.

Higuchi Y, Yoshimoto T. 2002. Arachidonic acid converts the glutathione depletion-induced apoptosis to necrosis by promoting lipid peroxidation and reducing caspase-3 activity in rat glioma cells. Arch Biochem Biophys 400: 133-140.

Hoffman SW, Rzigalinski BA, Willoughby KA, Ellis EF. 2000. Astrocytes generate isoprostanes in response to trauma or oxygen radicals. J Neurotrauma 17: 415-420.

Homayoun P, Parkins NE, Soblosky J, Carey ME, Rodriguez de Turco EB, et al. 2000. Cortical impact injury in rats promotes a rapid and sustained increase in polyunsaturated free fatty acids and diacylglycerols. Neurochem Res 25: 269-276.

Homayoun P, Rodriguez de Turco EB, Parkins NE, Lane DC, Soblosky J, et al. 1997. Delayed phospholipid degradation in rat brain after traumatic brain injury. J Neurochem 69: 199-205.

Hong S, Gronert K, Devchand PR, Moussignac RL, Serhan CN. 2003. Novel docosatrienes and 17S-resolvins generated from docosahexaenoic acid in murine brain, human blood, and glial cells—autacoids in anti-inflammation. J Biol Chem 278: 14677-14687.

Horrocks LA, Demediuk P, Saunders RD, Dugan L, Clendenon NR, et al. 1985. The degradation of phospholipids, formation of metabolites of arachidonic acid, and demyelination following experimental spinal cord injury. Cent Nerv Syst Trauma 2: 115-120.

Horrocks LA, Van Rollins M, Yates AJ. 1981. Lipid changes in the ageing brain. The Molecular Basis of Neuropathology. Davison AN, Thompson RHS, editors. London: Edward Arnold Ltd.; pp. 601-630.

Jeohn GH, Kong LY, Wilson B, Hudson P, Hong JS. 1998. Synergistic neurotoxic effects of combined treatments with cytokines in murine primary mixed neuron/glia cultures. J Neuroimmunol 85: 1-10.

Jupp OJ, Vandenabeele P, MacEwan DJ. 2003. Distinct regulation of cytosolic phospholipase A_2 phosphorylation, translocation, proteolysis and activation by tumour necrosis factor-receptor subtypes. Biochem J 374: 453-461.

Katayama Y, Shimizu J, Suzuki S, Memezawa H, Kashiwagi F, et al. 1990. Role of arachidonic acid metabolism on ischemic brain edema and metabolism. Adv Neurol 52: 105-108.

Kim D, Sladek CD, Aguado-Velasco C, Mathiasen JR. 1995a. Arachidonic acid activation of a new family of K^+ channels in cultured rat neuronal cells. J Physiol 484: 643-660.

Kim DK, Rordorf G, Nemenoff RA, Koroshetz WJ, Bonventre JV. 1995b. Glutamate stably enhances the activity of two cytosolic forms of phospholipase A_2 in brain cortical cultures. Biochem J 310: 83-90.

Kim GM, Xu J, Xu JM, Song SK, Yan P, et al. 2001. Tumor necrosis factor receptor deletion reduces nuclear factor-κB activation, cellular inhibitor of apoptosis protein 2 expression, and functional recovery after traumatic spinal cord injury. J Neurosci 21: 6617-6625.

Klusman I, Schwab ME. 1997. Effects of pro-inflammatory cytokines in experimental spinal cord injury. Brain Res 762: 173-184.

Klussmann S, Martin-Villalba A. 2005. Molecular targets in spinal cord injury. J Mol Med 83: 657-671.

Knoblach SM, Fan L, Faden AI. 1999. Early neuronal expression of tumor necrosis factor-α after experimental brain injury contributes to neurological impairment. J Neuroimmunol 95: 115-125.

Krupinski J, Ferrer I, Barrachina M, Secades JJ, Mercadal J, et al. 2002. CDP-choline reduces pro-caspase and cleaved caspase-3 expression, nuclear DNA fragmentation, and specific PARP-cleaved products of caspase activation following middle cerebral artery occlusion in the rat. Neuropharmacology 42: 846-854.

Latorre E, Collado MP, Fernández I, Aragonés MD, Catalán RE. 2003. Signaling events mediating activation of brain ethanolamine plasmalogen hydrolysis by ceramide. Eur J Biochem 270: 36-46.

Lazarewicz JW, Wroblewski JT, Costa E. 1990. N-methyl-D-aspartate-sensitive glutamate receptors induce calcium-mediated arachidonic acid release in primary cultures of cerebellar granule cells. J Neurochem 55: 1875-1881.

Lazarewicz JW, Wroblewski JT, Palmer ME, Costa E. 1988. Activation of N-methyl-D-aspartate-sensitive glutamate receptors stimulates arachidonic acid release in primary cultures of cerebellar granule cells. Neuropharmacology 27: 765-769.

Lee JY, Je JH, Jung KJ, Yu BP, Chung HY. 2004. Induction of endothelial iNOS by 4-hydroxyhexenal through NF-κB activation. Free Radic Biol Med 37: 539-548.

Li S, Stys PK. 2000. Mechanisms of ionotropic glutamate receptor-mediated excitotoxicity in isolated spinal cord white matter. J Neurosci 20: 1190-1198.

Liu DX, Li LP, Augustus L. 2001. Prostaglandin release by spinal cord injury mediates production of hydroxyl radical, malondialdehyde and cell death: A site of the neuroprotective action of methylprednisolone. J Neurochem 77: 1036-1047.

Lu XR, Ong WY, Halliwell B, Horrocks LA, Farooqui AA. 2001. Differential effects of calcium-dependent and calcium-independent phospholipase A_2 inhibitors on kainate-induced neuronal injury in rat hippocampal slices. Free Radic Biol Med 30: 1263-1273.

Lukácová N, Halát G, Chavko M, Maršala J. 1996. Ischemia-reperfusion injury in the spinal cord of rabbits strongly enhances lipid peroxidation and modifies phospholipid profiles. Neurochem Res 21: 869-873.

Maeda T, Katayama Y, Kawamata T, Aoyama N, Mori T. 1997. Hemodynamic depression and microthrombosis in the peripheral areas of cortical contusion in the rat: Role of platelet activating factor. Acta Neurochir Suppl 70: 102-105.

Marcheselli VL, Hong S, Lukiw WJ, Tian XH, Gronert K, et al. 2003. Novel docosanoids inhibit brain ischemia-reperfusion-mediated leukocyte infiltration and pro-inflammatory gene expression. J Biol Chem 278: 43807-43817.

Mark RJ, Lovell MA, Markesbery WR, Uchida K, Mattson MP. 1997. A role for 4-hydroxynonenal, an aldehydic product of lipid peroxidation, in disruption of ion homeostasis and neuronal death induced by amyloid β-peptide. J Neurochem 68: 255-264.

McIntosh TK, Saatman KE, Raghupathi R, Graham DI, Smith DH, et al. 1998. The molecular and cellular sequelae of experimental traumatic brain injury: Pathogenetic mechanisms. Neuropathol Appl Neurobiol 24: 251-267.

Miele L. 2003. New weapons against inflammation: Dual inhibitors of phospholipase A_2 and transglutaminase. J Clin Invest 111: 19-21.

Minami M, Kuraishi Y, Satoh M. 1991. Effects of kainic acid on messenger RNA levels of IL-1β, IL-6, TNF-α and LIF in the rat brain. Biochem Biophys Res Commun 176: 593-598.

Mir C, Clotet J, Aledo R, Durany N, Argemi J, et al. 2003. CDP-choline prevents glutamate-mediated cell death in cerebellar granule neurons. J Mol Neurosci 20: 53-59.

Murphy TH, Miyamato M, Sastre A, Schaar RL, Coyle JT. 1989. Glutamate toxicity in a neuronal cell line involves inhibition of cystine transport leading to oxidative stress. Neuron 2: 1547-1558.

Okada M, Amamoto T, Tomonaga M, Kawachi A, Yazawa K, et al. 1996. The chronic administration of docosahexaenoic acid reduces the spatial cognitive deficit following transient forebrain ischemia in rats. Neuroscience 71: 17-25.

Ong WY, Horrocks LA, Farooqui AA. 1999. Immunocytochemical localization of cPLA$_2$ in rat and monkey spinal cord. J Mol Neurosci 12: 123-130.

Ong WY, Yeo JF, Ling SF, Farooqui AA. 2006. Immunocytochemical localization of calcium-independent phospholipase A_2 (iPLA$_2$) in rat and monkey spinal cord. J Neurocytol 34: 447-458.

Ouyang Y, Kaminski NE. 1999. Phospholipase A_2 inhibitors p-bromophenacyl bromide and arachidonyl trifluoromethyl ketone suppressed interleukin-2 (IL-2) expression in murine primary splenocytes. Arch Toxicol 73: 1-6.

Panter SS, Yum SW, Faden AI. 1990. Alteration in extracellular amino acids after traumatic spinal cord injury. Ann Neurol 27: 96-99.

Park E, Velumian AA, Fehlings MG. 2004. The role of excitotoxicity in secondary mechanisms of spinal cord injury: A review with an emphasis on the implications for white matter degeneration. J Neurotrauma 21: 754-774.

Patrizio M. 2004. Tumor necrosis factor reduces cAMP production in rat microglia. Glia 48: 241-249.

Pavel J, Lukácová N, Maršala J, Maršala M. 2001. The regional changes of the catalytic NOS activity in the spinal cord of the rabbit after repeated sublethal ischemia. Neurochem Res 26: 833-839.

Pereira CF, Resende de Oliveira C. 2000. Oxidative glutamate toxicity involves mitochondrial dysfunction and perturbation of intracellular Ca^{2+} homeostasis. Neurosci Res 37: 227-236.

Phillis JW, Horrocks LA, Farooqui AA. 2006. Cyclooxygenases, lipoxygenases, and epoxygenases in CNS: Their role and involvement in neurological disorders. Brain Res Brain Res Rev 52: 201-243.

Picklo MJ, Amarnath V, McIntyre JO, Graham DG, Montine TJ. 1999. 4-Hydroxy-2(E)-nonenal inhibits CNS mitochondrial respiration at multiple sites. J Neurochem 72: 1617-1624.

Pilitsis JG, Coplin WM, O'Regan MH, Wellwood JM, Diaz FG, et al. 2003. Free fatty acids in cerebrospinal fluids from patients with traumatic brain injury. Neurosci Lett 349: 136-138.

Porcellati G. 1983. Phospholipid metabolism in neural membranes. Sun GY, Bazan N, Wu JY, Porcellati G, Sun AY, editors. Neural Membranes. New York: Humana Press; pp. 3-35.

Ray SK, Hogan EL, Banik NL. 2003. Calpain in the pathophysiology of spinal cord injury: Neuroprotection with calpain inhibitors. Brain Res Rev 42: 169-185.

Robinson BS, Hii CST, Poulos A, Ferrante A. 1997. Activation of neutral sphingomyelinase in human neutrophils by polyunsaturated fatty acids. Immunology 91: 274-280.

Rothwell NJ, Relton JK. 1993. Involvement of cytokines in acute neurodegeneration in the CNS. Neurosci Biobehav Rev 17: 217-227.

Saito M, Saito M, Berg MJ, Guidotti A, Marks N. 1999. Gangliosides attenuate ethanol-induced apoptosis in rat cerebellar granule neurons. Neurochem Res 24: 1107-1115.

Sandhya TL, Ong WY, Horrocks LA, Farooqui AA. 1998. A light and electron microscopic study of cytoplasmic phospholipase A_2 and cyclooxygenase-2 in the hippocampus after kainate lesions. Brain Res 788: 223-231.

Sanfeliu C, Hunt A, Patel AJ. 1990. Exposure to N-methyl-D-aspartate increases release of arachidonic acid in primary cultures of rat hippocampal neurons and not in astrocytes. Brain Res 526: 241-248.

Sato T, Kageura T, Hashizume T, Hayama M, Kitatani K, et al. 1999. Stimulation by ceramide of phospholipase A_2 activation through a mechanism related to the phospholipase C-initiated signaling pathway in rabbit platelets. J Biochem (Tokyo) 125: 96-102.

Saunders R, Horrocks LA. 1987. Eicosanoids, plasma membranes, and molecular mechanisms of spinal cord injury. Neurochem Pathol 7: 1-22.

Schnell L, Fearn S, Schwab ME, Perry VH, Anthony DC. 1999. Cytokine-induced acute inflammation in the brain and spinal cord. J Neuropathol Exp Neurol 58: 245-254.

Schuhmann MU, Mokhtarzadeh M, Stichtenoth DO, Skardelly M, Klinge PA, et al. 2003. Temporal profiles of cerebrospinal fluid leukotrienes, brain edema and inflammatory response following experimental brain injury. Neurol Res 25: 481-491.

Serhan CN. 2005. Novel eicosanoid and docosanoid mediators: Resolvins, docosatrienes, and neuroprotectins. Curr Opin Clin Nutr Metab Care 8: 115-121.

Serhan CN, Arita M, Hong S, Gotlinger K. 2004. Resolvins, docosatrienes, and neuroprotectins, novel omega-3-derived mediators, and their endogenous aspirin-triggered epimers. Lipids 39: 1125-1132.

Shohami E, Shapira Y, Sidi A, Cotev S. 1987. Head injury induces increased prostaglandin synthesis in rat brain. J Cereb Blood Flow Metab 7: 58-63.

Shohami E, Shapira Y, Yadid G, Reisfeld N, Yedgar S. 1989. Brain phospholipase A_2 is activated after experimental closed head injury in the rat. J Neurochem 53: 1541-1546.

Singh I, Pahan K, Khan M, Singh AK. 1998. Cytokine-mediated induction of ceramide production is redox-sensitive. Implications to proinflammatory cytokine-mediated apoptosis in demyelinating diseases. J Biol Chem 273: 20354-20362.

Söderberg M, Edlund C, Kristensson K, Dallner G. 1990. Lipid compositions of different regions of the human brain during aging. J Neurochem 54: 415-423.

Söderberg M, Edlund C, Kristensson K, Dallner G. 1991. Fatty acid composition of brain phospholipids in aging and in Alzheimer's disease. Lipids 26: 421-425.

Springer JE, Azbill RD, Knapp PE. 1999. Activation of the caspase-3 apoptotic cascade in traumatic spinal cord injury. Nat Med 5: 943-946.

Springer JE, Azbill RD, Mark RJ, Begley JG, Wäg G, et al. 1997. 4-hydroxynonenal, a lipid peroxidation product, rapidly accumulates following traumatic spinal cord injury and inhibits glutamate uptake. J Neurochem 68: 2469-2476.

Stokes BT, Somerson SK. 1987. Spinal cord extracellular microenvironment. Can the changes resulting from trauma be graded? Neurochem Pathol 7: 47-55.

Stokes BT, Fox P, Hollinden G. 1983. Extracellular calcium activity in the injured spinal cord. Exp Neurol 80: 561-572.

Strijbos PJ, Rothwell NJ. 1995. Interleukin-1β attenuates excitatory amino acid-induced neurodegeneration in vitro: Involvement of nerve growth factor. J Neurosci 15: 3468-3474.

Sun D, Newman TA, Perry VH, Weller RO. 2004. Cytokine-induced enhancement of autoimmune inflammation in the brain and spinal cord: Implications for multiple sclerosis. Neuropathol Appl Neurobiol 30: 374-384.

Sundström E, Mo LL. 2002. Mechanisms of glutamate release in the rat spinal cord slices during metabolic inhibition. J Neurotrauma 19: 257-266.

Tamagno E, Robino G, Obbili A, Bardini P, Aragno M, et al. 2003. H_2O_2 and 4-hydroxynonenal mediate amyloid β-induced neuronal apoptosis by activating JNKs and $p38^{MAPK}$. Exp Neurol 180: 144-155.

Taylor WA. 1988. Effects of impact injury of rat spinal cord on activities of some enzymes of lipid hydrolysis. Dissertation. The Ohio State University, Columbus, Ohio.

Teng YD, Choi H, Onario RC, Zhu S, Desilets FC, et al. 2004. Minocycline inhibits contusion-triggered mitochondrial cytochrome c release and mitigates functional deficits after spinal cord injury. Proc Natl Acad Sci USA 101: 3071-3076.

Toborek M, Garrido R, Malecki A, Kaiser S, Mattson MP, et al. 2000. Nicotine attenuates arachidonic acid-induced overexpression of nitric oxide synthase in cultured spinal cord neurons. Exp Neurol 161: 609-620.

Toborek M, Malecki A, Garrido R, Mattson MP, Hennig B, et al. 1999. Arachidonic acid-induced oxidative injury to cultured spinal cord neurons. J Neurochem 73: 684-692.

Turrin NP, Plata-Salaman CR. 2000. Cytokine–cytokine interactions and the brain. Brain Res Bull 51: 3-9.

Vanags DM, Larsson P, Feltenmark S, Jakobsson PJ, Orrenius S, et al. 1997. Inhibitors of arachidonic acid metabolism reduce DNA and nuclear fragmentation induced by TNF plus cycloheximide in U937 cells. Cell Death Differ 4: 479-486.

Vitkovic L, Bockaert J, Jacque C. 2000. "Inflammatory" cytokines: Neuromodulators in normal brain?? J Neurochem 74: 457-471.

Wang CX, Nuttin B, Heremans H, Dom R, Gybels J. 1996. Production of tumor necrosis factor in spinal cord following traumatic injury in rats. J Neuroimmunol 69: 151-156.

Wang CX, Olschowka JA, Wrathall JR. 1997. Increase of interleukin-1β mRNA and protein in the spinal cord following experimental traumatic injury in the rat. Brain Res 759: 190-196.

Wei EP, Lamb RG, Kontos HA. 1982. Increased phospholipase C activity after experimental brain injury. J Neurosurg 56: 695-698.

Wilson CJ, Finch CE, Cohen HJ. 2002. Cytokines and cognition—the case for a head-to-toe inflammatory paradigm. J Am Geriatr Soc 50: 2041-2056.

Wingrave JM, Schaecher KE, Sribnick EA, Wilford GG, Ray SK, et al. 2003. Early induction of secondary injury factors causing activation of calpain and mitochondria-mediated neuronal apoptosis following spinal cord injury in rats. J Neurosci Res 73: 95-104.

Wissing D, Mouritzen H, Egeblad M, Poirier GG, Jäättelä M. 1997. Involvement of caspase-dependent activation of cytosolic phospholipase A_2 in tumor necrosis factor-induced apoptosis. Proc Natl Acad Sci USA 94: 5073-5077.

Won JS, Im YB, Khan M, Singh AK, Singh I. 2005. Involvement of phospholipase A_2 and lipoxygenase in lipopolysaccharide-induced inducible nitric oxide synthase expression in glial cells. Glia 51: 13-21.

Wu A, Ying Z, Gomez-Pinilla F. 2004. Dietary omega-3 fatty acids normalize BDNF levels, reduce oxidative damage, and counteract learning disability after traumatic brain injury in rats. J Neurotrauma 21: 1457-1467.

Wu A, Ying Z, Gomez-Pinilla F. 2005. Omega-3 fatty acids supplementation restores homeostatic mechanisms disrupted by traumatic brain injury. J Neurotrauma 22: 1212.

Xu J, Fan GS, Chen SW, Wu YJ, Xu XM, et al. 1998. Methylprednisolone inhibition of TNF-α expression and NF-κB activation after spinal cord injury in rats. Mol Brain Res 59: 135-142.

Xu J, Hsu CY, Liu TH, Hogan EL, Perot PL Jr, et al. 1990. Leukotriene B4 release and polymorphonuclear cell infiltration in spinal cord injury. J Neurochem 55: 907-912.

Yang H-C, Farooqui AA, Horrocks LA. 1994. Effects of sialic acid and sialoglycoconjugates on cytosolic phospholipases A_2 from bovine brain. Biochem Biophys Res Commun 199: 1158-1166.

Zaleska MM, Wilson DF. 1989. Lipid hydroperoxides inhibit reacylation of phospholipids in neuronal membranes. J Neurochem 52: 255-260.

Zhang DQ, Dhillon HS, Mattson MP, Yurek DM, Prasad RM. 1999. Immunohistochemical detection of the lipid peroxidation product 4-hydroxynonenal after experimental brain injury in the rat. Neurosci Lett 272: 57-61.

5 Kynurenines in the Brain: Preclinical and Clinical Studies, Therapeutic Considerations

C. Kiss · L. Vécsei

1	Introduction	92
2	L-Kynurenine and Neuroactive Metabolites of the Kynurenine Pathway	92
2.1	L-Kynurenine	92
2.2	Kynurenic Acid	92
2.3	Quinolinic Acid	93
2.4	3-Hydroxykynurenine	94
3	Enzymes of the Kynurenine Pathway	94
4	Alterations of the KP under Pathological Conditions	96
4.1	Diseases with Increased Cerebral QUIN Levels	96
4.2	Diseases with Increased Cerebral KYNA Levels	97
5	Therapeutic Approaches based on Kynurenines	97
5.1	KYNA Analogs	98
5.2	Prodrugs of KYNA and of KYNA Analogs	98
5.3	KP Enzyme Inhibitors	99
5.3.1	Manipulation of the KP with the Aim to Elevate the Cerebral Level of KYNA	99
5.3.2	KP Enzyme Inhibition with the Aim to Decrease Cerebral KYNA Levels	100
6	Conclusion	100

© 2009 Springer Science+Business Media, LLC.

5 Kynurenines in the brain: Preclinical and clinical studies, therapeutic considerations

Abstract: In mammalian cells, the kynurenine pathway (KP) is a major biochemical route for the conversion of tryptophan, yielding L-kynurenine (L-KYN) and ultimately several other metabolites, called kynurenines, which derive directly or indirectly from L-KYN. Kynurenines have been shown to be involved in various physiological and pathological processes. Alteration of KP metabolism is functionally significant and occurs in a variety of diseases of the central nervous system. The discovery of the importance of kynurenines in brain function under physiological and pathologic conditions has led to the identification of potential new drug targets exploiting the therapeutic potential of the pathway. Some of these compounds proved to provide neuroprotection in animal models of various human diseases which holds promise that their effectiveness will translate to the clinic in the future.

List of Abbreviations: EAA, excitatory amino acid; HD, Huntington's disease; IDO, indoleamine 2,3-dioxygenase ; KAT II KO, kynurenine-aminotransferase knockout; KP, The kynurenine pathway; KYNA, kynurenic acid; L-KYN, L-kynurenine; mNBA, meta-nitrobenzoylalanine; MS, Multiple sclerosis; NAD, nicotinamide adenine dinucleotide; NMDA, N-methyl-D-aspartate; oMBA, Ortho-methoxybenzoylalanine; PD, Parkinson's disease; QUIN, quinolinic acid; TDO, tryptophan dioxygenase; 3-HAO, 3-hydroxyanthranilate-oxygenase; 3-HK, 3-hydroxykynurenine

1 Introduction

The kynurenine pathway (KP) is a major route for the conversion of tryptophan, yielding L-kynurenine (L-KYN) and ultimately several other metabolites, called kynurenines, which are derived directly or indirectly from L-KYN (❯ *Figure 5-1*). The metabolic cascade was originally known to be a source of nicotinamide adenine dinucleotide (NAD) and nicotinamide adenine dinucleotide phosphate (NADP), two coenzymes of basic cellular processes, and only later it was discovered that some of its metabolites could exhibit neuromodulatory actions. Interest in the research of KP emerged as it turned out that two metabolites of the pathway, quinolinic acid (QUIN) and kynurenic acid (KYNA) exhibit activity at glutamate receptors (Stone and Perkins, 1981; Perkins and Stone, 1982). QUIN was shown to inhibit N-methyl-D-aspartate (NMDA) receptors, whereas KYNA proved to be a broad-spectrum antagonist of excitatory amino acid (EAA) receptors with particular high affinity to the glycine-coagonist site of the NMDA receptor (Kessler et al., 1989). Subsequently, studies from several laboratories have clarified the role of kynurenines in the brain function under physiological and pathological conditions. This has led to the identification of potential new drug targets for various neurological disorders.

2 L-Kynurenine and Neuroactive Metabolites of the Kynurenine Pathway

2.1 L-Kynurenine

L-KYN (❯ *Figure 5-1*) is a major compound of the KP, serving as a source for the synthesis of all the other metabolites of the pathway. L-KYN is present in the blood, brain, and peripheral organs in low micromolar concentrations, and gets transported through the blood–brain barrier by the neutral amino acid carrier (Fukui et al., 1991). Although L-KYN does not directly influence neuronal function, systemic or intracerebral administration of it decreases blood pressure (Lapin, 1976) and evokes convulsions (Lapin, 1978, 1981, 1982; Pinelli et al., 1984), probably by getting converted to its neuroactive metabolites.

2.2 Kynurenic Acid

KYNA (❯ *Figure 5-1*) was long known to be a side product of tryptophan metabolism but no particular biological function was assigned to it until more recently when in neurophysiological experiments it was shown to inhibit neurons (Perkins and Stone, 1982). Subsequently, KYNA was recognized as a broad-spectrum antagonist of ionotropic EAA receptors. Since EAA receptor activation takes place in a variety of

Figure 5-1
Chemical structure of L-kynurenine and neuroactive kynurenines

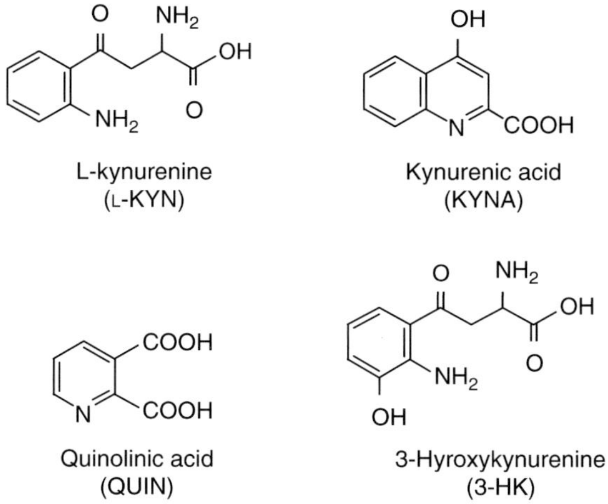

L-kynurenine
(L-KYN)

Kynurenic acid
(KYNA)

Quinolinic acid
(QUIN)

3-Hyroxykynurenine
(3-HK)

pathological states, KYNA has been tested as a neuroprotective agent. Indeed, at high concentrations, KYNA was shown to block excitotoxic damage and seizures induced by QUIN in rats (Foster et al., 1984) and to protect against various conditions like ischemia, traumatic brain injury (Germano et al., 1987; Andiné et al., 1988; Hicks et al., 1994; Salvati et al., 1999). Intracerebroventricular administration of KYNA was shown to dose-dependently evoke characteristic behavior in the rat: increased stereotypy and ataxia (Vécsei and Beal, 1990a, 1991). At lower concentrations, KYNA was recognized to act as a competitive antagonist at the glycine site of the NMDA receptor ($IC_{50} \cong 8$ μM) (Kessler et al., 1989) and as a noncompetive blocker at the α7-nicotonic receptor ($IC_{50} \cong 7$ μM) (Hilmas et al., 2001). These receptors could be the major sites of action of KYNA in the brain and thus endogenous KYNA could modify glutamatergic and cholinergic neurotransmission. Indeed, the level of KYNA seems to determine the vulnerability of the brain against excitotoxins, because intraperitoneal (i.p.) injection of amphetamine – which leads to reduction in the brain concentration of KYNA-potentiated quinolinate but not kainate excitotoxicity, and readjusting the level of KYNA to control levels by pharmacological manipulation of the KP, restored the vulnerability (Poeggeler et al., 1998; Rassoulpour et al., 2002). Furthermore, endogenous KYNA might influence glutamatergic neurotransmission presynaptically because modest elevations of KYNA in the rat striatum in vivo and in synaptosomes in vitro are able to inhibit glutamate release (Carpenedo et al., 2001).

Generation of kynurenine aminotransferase knockout (KAT II KO) mice – which have reduced brain KYNA levels early in development (<28 days) – has provided a useful tool for studying the role of KYNA in brain function and underlined the importance of endogenous KYNA in modulating glutamatergic and cholinergic neurotransmission. Intrastriatal injection of EAA receptor agonist QUIN induced significantly larger lesion in KAT II KO mice compared with wild-type mice (Sapko et al., 2003). Furthermore, the observation that KAT II KO mice have increased activity of α7-nicotinic receptor has proved the previous hypothesis that endogenous KYNA is an important regulator of α7-nicotinic receptor activity (Alkondon et al., 2003). The changes observed in the KAT II KO mice are evident early in development (<28 days), when cerebral KYNA levels are significanly decreased but tend to be reverted as the KYNA levels return to normal levels. These observations suggest that changes in endogenous KYNA levels have profound effects on the function of glutamatergic and cholineric receptors.

2.3 Quinolinic Acid

QUIN (▶ *Figure 5-1*) is present in the brain in concentrations similar to that of KYNA (50–100 nmol). It has pronounced effects on neuronal activity being an agonist at the NMDA receptor (Stone and Perkins,

1981), preferentially activating the NR2A and NR2B NMDA receptor subtypes (de Carvalho et al., 1996). Besides acting at the NMDA receptor, QUIN produces toxic free radicals (Rios and Santamaria, 1991). Due to the compound's excitotoxic and free-radical-generating property, injection of QUIN into the rat striatum leads to excitotoxic damage (Schwarcz et al., 1983) duplicating the neurochemical features of Huntington's disease (HD) (Beal et al., 1986). In addition, long-term exposure to submicromolar concentrations of QUIN results in neuronal death in vitro (Whetsell and Schwarcz, 1989). In pathological states, like neuroinflammatory diseases, dramatic increases of its level in the brain could occur which could result in neuronal damage.

2.4 3-Hydroxykynurenine

No important physiological function in the brain has so far been assigned to 3-hydroxykynurenine (3-HK) (❷ *Figure 5-1*), but in the primate lenses it could act as major UV filter together with its glucoside derivative, L-KYN and 4-(2-amino-3-hydroxyphenyl)-4-oxobutanoic acid O-β-D-glucoside and may be useful in protecting the retina from UV radiation (Vazquez et al., 2002). However, the autoxidation of these metabolites has been proposed in the processes leading to opacification of the lens and cataract formation (Chiarugi et al., 1999). In the brain, 3-HK can cause neuronal death in cell cultures by generating toxic free radicals. 3-HK is present in nanomolar concentrations in the mammalian brain, but under pathological conditions similarly to QUIN its level could increase dramatically reaching the micromolar range (0.8–1.2 μM) (Eastman and Guilarte, 1989). Chronic exposure of neuronal cultures to these levels of 3-HK could cause neuronal death (Okuda et al., 1996). 3-HK should get into neurons to induce toxicity, because the inhibition of its uptake by large neutral amino acid prevents from neuronal death (Okuda et al., 1998). 3-HK potentiates QUIN toxicity in the rat striatum, which suggests that these metabolites may act in concert to induce neuronal damage (Guidetti and Schwarcz, 1999).

3 Enzymes of the Kynurenine Pathway

❷ *Figure 5-2* summarizes the enzymes and metabolites of the KP. The first and rate-limiting step in the formation of L-KYN is the conversion of L-tryptophan to formyl kynurenine, which is catalyzed by two distinct heme-containing enzymes. In peripheral tissues and particularly in the liver, *tryptophan dioxygenase* (TDO; tryptophan pyrrolase; EC 1.13.1.2.) is mainly responsible for yielding L-KYN but in most mammalian organs including intestine, lung, epididymis, placenta, central nervous system, reticuloendothelial system, *indoleamine 2,3-dioxygenase* (IDO; EC 1.13.11.42) is able to convert tryptophan to formyl kynurenine. In the second step, formyl kynurenine is rapidly metabolized into L-KYN by *formamidase*, an enzyme abundant in most mammalian organs (Mehler and Knox, 1950).

TDO is able to metabolize L- but not D-tryptophan (Schutz et al., 1972), whereas IDO not only acts on L-tryptophan, but it can also cleave the indole ring of the D-tryptophan, L- or D-5-hydroxytryptophan, indoleamins like 5-HT, tryptamine, and melatonine (Hirata and Hayaishi, 1975; Yoshida and Hayaishi, 1987). The gene encoding TDO contains glucocorticoid-responsive elements (Danesh et al., 1983, 1987) which explains that glucocorticoid administration increases TDO formation in the liver, which causes a decrease in blood tryptophan levels with L-KYN accumulation. The transcription of the gene of IDO is under tight immunological control, and it contains two interferon-stimulated response elements (ISRE) and at least one gamma-interferon-activated sequence (GAS) (Dai and Gupta, 1990; Tone et al., 1990). Interferons and proinflammatory cytokines stimulate (Carlin et al., 1989; Taylor and Feng, 1991), whereas other cytokines and growth factors, like interleukin-4 or TGF-beta, inhibit IDO expression and activity (Musso et al., 1994; Yuan et al., 1998).

L-KYN is the key player of the pathway serving as a substrate of several enzymes: *kynurenine 3-hydroxylase* (EC 1.14.13.9; yielding 3-HK), *kynureninase* (EC 3.7.1.3; yielding anthranilic acid), and *kynurenine aminotransferases* (KATs; yielding KYNA). Kynurenine 3-hydroxylase is present in the liver, placenta, spleen, kidney, and brain (Erickson et al., 1992) and requires NADPH and molecular oxygen

Figure 5-2
The kynurenine pathway

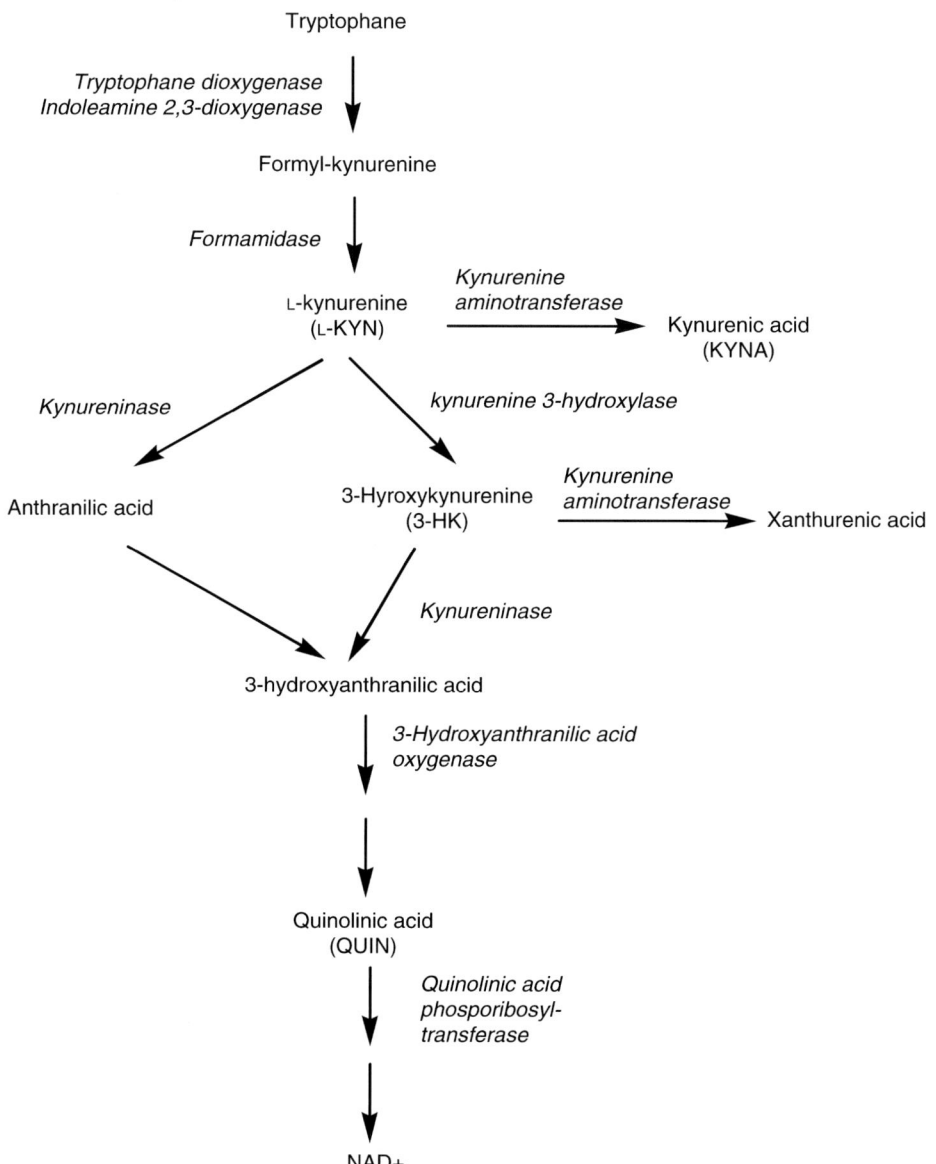

for the conversion of L-KYN to 3-HK. The enzyme has a high affinity for its substrate ($K_m \cong 1$ μM), implicating that it converts most of the available L-KYN under physiological conditions.

Kynureninase has been characterized from rodent liver, kidney, and spleen (Kawai et al., 1988). It is a pyridoxal phosphate-dependent enzyme that converts L-KYN and 3-HK into L-alanine and anthranilic or 3-hydroxyanthranilic acid, respectively. It has been shown that the affinity of the enzyme is tenfold higher for 3-HK than for L-KYN (K_m for L-KYN: 250 μM, K_m for 3-HK: 25 μM; Alberati-Giani et al., 1996; Toma et al., 1997).

3-Hydroxyanthranilate oxygenase (3-HAO, EC 1.13.11.6) has been purified from mammalian liver or kidney and it requires ferrous ions (Fe^{2+}) and sulfhydryl groups for its activity (Long et al., 1954; Decker et al., 1961; Koontz and Shiman, 1976). 3-HAO cleaves the benzene ring of 3-hydroxyanthranilate into α-amino-β-carboxymuconate-ε-semialdehyde, an unstable compound, which nonenzymatically gets transformed into QUIN. A portion of α-amino-β-carboxymuconate-ε-semialdehyde is metabolized by α-amino-β-carboxymuconate-ε-semialdehyde decarboxylase (EC 4.1.1.45) to produce picolinic acid or aminomuconic acids.

3-HAO has been shown to be present in the brain (Foster et al., 1986), and increased activity of the enzyme has been detected in various pathological states, like in the striatum of HD patients (Schwarcz et al., 1988a), in gerbil hippocampus after global ischemia (Saito et al., 1993), and in epileptic rats (Du et al., 1993). QUIN is metabolized by *quinolinic acid phosphoribosyltransferase* (QPRT; EC 2.4.2.19) yielding nicotinic acid mononucleotide and subsequent degradation products, including NAD^+. Since QPRT has a very low activity, the levels of QUIN in the extracellular space are essentially determined by its rate of synthesis.

KYNA is converted from L-KYN in the brain by two distinct KATs (Okuno et al., 1991; Guidetti et al., 1997). Arbitrarily termed KAT I and KAT II, these enzymes have substantially different pH optimum and substrate specificity. KAT I (EC 2.6.1.14; also called glutamine transaminase K) has an optimal pH of 9.5–10, whereas KAT II has a maximal activity in the neutral pH. KAT I prefers pyruvate as a cosubstrate and it is potently inhibited by glutamine. KAT II shows no preference for pyruvate, and it is not sensitive to inhibition by glutamine. These differences suggest that KAT II might be responsible for most part of the KYNA synthesis in the brain. Lesion and pharmacological studies have also confirmed that in most brain regions KYNA derives primarily from KAT II activity (Guidetti et al., 1997). Under pathological conditions, however, such as after exposure to mitochondrial toxins, a massive increase in KAT I immunostaining has been observed in the neurons in the affected brain areas (Csillik et al., 2002; Knyihár-Csillik et al., 1999). Both KATs have high affinity for their substrate (K_m in the millimolar range), suggesting that L-KYN bioavailability determines the rate for KYNA biosynthesis. Indeed, in experiments, when rat brain slices or human astrocyte cultures are exposed to L-KYN, the levels of KYNA increase linearly with L-KYN availability (Turski et al., 1989; Kiss et al., 2003).

No catabolic enzyme or cellular re-uptake system of KYNA exists in mammals (Turski et al., 1989); so KYNA is extruded from the brain by a probenecid-sensitive carrier system (Moroni et al., 1988). More recently, QUIN has been also shown to exit the brain by the same carrier system (Morrison et al., 1999).

All of the enzymes of the KP are detectable in the brain, but their activities are much lower than in the peripheral organs (Stone, 1993). KP enzymes are predominantly localized in glial cells according to immunocytochemical, lesion, and molecular biological studies (Guidetti et al., 1995; Schwarcz et al., 1996; Guillamin et al., 2001).

Astrocytes seem to produce most part of KYNA in the brain (Guillamin et al., 2001), whereas microglial cells are the primary source of metabolites of the QUIN branch of the pathway (Lehrmann et al., 2001).

4   Alterations of the KP under Pathological Conditions

Elucidation of the importance of the KP in brain function has facilitated extensive research, investigating the alterations of the pathway in various neurological disorders. In the literature, a recent extensive review is available on this topic (Stone, 2001). In a great number of diseases – especially in neuroimmunological disorders – dramatic increases in the level of QUIN have been detected, suggesting its role in the pathological processes. Whereas, in diseases with cognitive alterations, elevated levels of cerebral KYNA have been shown which could contribute to cognitive defects by inhibiting NMDA receptor function.

4.1   Diseases with Increased Cerebral QUIN Levels

Infection of the CNS with viruses, bacteria, parazites: Significant increases in QUIN levels in the CNS have been described in patients with AIDS–dementia complex (Heyes et al., 1991, 1998), Lyme borelliosis (Halperin and Heyes, 1992), in mice infected with Herpes simplex virus type 1, cerebral malaria, and

Toxoplasma gondii (Reinhard, 1998; Sanni et al., 1998; Fujigaki et al., 2002), in macaques with poliovirus infection and septicaemia (Heyes and Lackner, 1990; Heyes et al., 1992). Reinhard et al. (1994) listed many other infectious diseases in which alterations of the KP have been described.

Multiple sclerosis (MS): Experimental allergic encephalomyelitis is an autoimmune inflammatory disorder and serves as an animal model of human MS. The levels of neurotoxic kynurenine metabolites, 3-HK and QUIN, are elevated in the spinal cord of affected animals (Flanagan et al., 1995; Chiarugi et al., 2001). Since QUIN has been shown to be toxic to oligodendrocytes (Cammer, 2001), elevated levels of QUIN might contribute to the pathology of the disease.

Huntington's disease: QUIN has long been hypothesized to play an important role in HD, because intrastriatal injection of QUIN duplicates many of the distinct neuropathological features of the striatum of patients with HD (Schwarcz et al., 1983; Beal et al., 1996). However, no changes in QUIN levels in tissue samples have been detected in HD patients dying after a prolonged illness (Reynolds et al., 1998; Schwarcz et al., 1988b). Of potential interest is the finding that elevated cortical and striatal content of 3-HK and QUIN has been shown in early grade HD, suggesting a causal role of QUIN in nerve cell loss in HD (Guidetti et al., 2000, 2003).

Parkinson's disease (PD): 3-HK has been shown to be significantly increased in the putamen and substantia nigra in patients with PD, which could play a causal role in the neuronal loss in PD (Ogawa et al., 1992).

Ischaemia: Dramatic increases in cerebral QUIN levels have been observed several days after global ischemia in a gerbil model of the disease (Heyes and Nowak, 1990; Saito et al., 1993). The changes have been observed in brain areas exposed to ishemia, but not in areas with an uninterrupted blood supply (Saito et al., 1992).

Traumatic brain/spinal cord injury: Both in humans and in the animal model of the disease, massive increases (50-fold) in the level of QUIN in the CSF (Sinz et al., 1998) and in affected areas have been shown (Blight et al., 1995, 1997).

4.2 Diseases with Increased Cerebral KYNA Levels

Alzheimer's disease (AD): Increased level of KYNA has been measured in the putamen and caudate nucleus of AD patients, which could be responsible for the cognitive deficits seen in AD patients (Baran et al., 1999).

Down's syndrome: The cortical levels of KYNA have been shown to be elevated in the postmortem specimens of Down's syndrome patients that could play a causal role in the impaired memory and learning defects seen in these patients (Baran et al., 1996).

Schizophrenia: Elevated levels of KYNA have been detected in the prefrontal cortex (Brodmann area 9), but not in other cortical areas of schizophrenic patients (Schwarcz and Pellicciari, 2002). Increased KYNA levels may contribute to the hypofunction of glutamatergic neurotransmission that plays a critical role in schizophrenia (Tamminga, 1998).

5 Therapeutic Approaches based on Kynurenines

After KYNA was recognized as a broad-spectrum antagonist of EAA receptors, its neuroprotective potential has been investigated in a variety of disorders. KYNA itself poorly enters the brain due to its polar structure and lack of efficient transport across the blood–brain barrier, but if systematically given in high doses, it protects against anoxia (Simon et al., 1986), ischemia (Germano et al., 1987; Andiné et al., 1988; Salvati et al., 1999), traumatic brain injury (Hicks et al., 1994), and antagonizes the toxic effects of QUIN (Foster et al., 1984). There are several strategies to generate neuroprotective drugs that have superior therapeutic potency to KYNA.

One strategy to exploit the therapeutic potential of KYNA is to develop chemically related drugs with better bioavailability and higher potency on the glycine site of the NMDA receptor.

The second approach uses prodrugs of KYNA or its analogs, which readily penetrate the blood–brain barrier, and are hydrolyzed in the CNS to form active compounds.

Third way is manipulating the KP by administering compounds that block the activity of the KP enzymes. With this strategy, one can shift L-KYN metabolism to the KYNA or QUIN branch with the aim to inhibit or excite EAA receptors, respectively.

Recently, several reviews have been published on the therapeutic alternatives based on kynureninergic manipulations (Stone, 2000, 2001; Schwarcz and Pellicciari, 2002; Stone and Darlington, 2002).

5.1 KYNA Analogs

Several attempts have been made to use the KYNA structure to develop NMDA glycine site antagonists with better bioavailability and higher potency on the glycine site of the NMDA receptor. Substitution of KYNA with halogen atoms (7-chlorokynurenic acid and 5,7-dichlorokynurenic acid; Baron et al., 1990), replacement of the 4-hydroxy group of KYNA with amido substituents (MDL 100,748, L-689,560; Baron et al., 1992; Leeson et al., 1992), substitution with a phenyl or more complex lipophil groups at position 3 of the KYNA nucleus (MDL 104,653, L-701,324; Kulagowski et al., 1994; Bristow et al., 1996), or replacment of the six-membered nitrogen containing ring by a five-carbon ring (GV150526A or gavestinel; Glaxo, 1993) provided compounds that have superior neuroprotective abilities compared with KYNA. Gavestinel has already entered clinical trials but the Glycine Antagonist in Neuroprotection (GAIN) Americas trial, a randomized, double-blind placebo-controlled phase III trial has failed to show any benefit of gavestinel treatment in acute ischemic stroke patients (Sacco et al., 2001).

5.2 Prodrugs of KYNA and of KYNA Analogs

L-KYN readily penetrates the blood–brain barrier by the neutral amino acid transporter (Fukui et al., 1991) and serves as a precursor for the neuroprotectant KYNA in the brain. This explains that systemic administration of L-KYN has been shown to protect against cerebral ischemia or local injection of NMDA in neonatal rats (Nozaki and Beal, 1992) and to counteract pentylenetetrazol- and NMDA-induced seizures in mice (Vécsei et al., 1992), but only modest effects of L-KYN treatment havebeen observed on kainate-induced seizures in rats (Vécsei et al., 1990b). However, it should be noted that L-KYN gets converted not only to KYNA but also to 3-HK and QUIN which poses considerable limitations to L-KYN therapy and explains the poor efficacy of L-KYN treatment since systemical administration of L-KYN is not capable of selectively increasing the levels of the neuroprotectant KYNA. For this reason, more potent KYNA precursors have been searched that are not getting metabolized in the QUIN branch of the pathway.

L-4-Chlorokynurenine or 4,6-dichlorokynurenines are potent precursors that meet the previously mentioned criteria of not getting metabolized to QUIN. These compounds are converted to two NMDA glycine site antagonists, namely 7-chlorokynurenic acid and 5,7-dichlorokynurenic acid, respectively (Hokari et al., 1996). 4-Chlorokynurenine is also metabolized to 4-chloro-3-hydroxyanthranilic acid, which is a potent and selective inhibitor of 3-hydroxyanthranilic acid oxygenase, and thus the administration of 4-chlorokynurenine causes a reduction of QUIN synthesis besides inhibiting NMDA receptors. It penetrates into the brain easier than its metabolite7-chlorokynurenic acid, and provides several additional benefits compared with other glycine site NMDA receptor antagonists: it preferentially gets metabolized in brain areas where neurodegeneration takes place, allowing lower dosage of the drug (Lee and Schwarcz, 2001). 4-Chlorokynurenine diminishes brain damage induced by focal application of QUIN and malonate into the rat striatum and hippocampus (Guidetti et al., 2000; Wu et al., 1997) and inhibits convulsions and neurotoxicity after systemic application of kainate (Wu et al., 2002). The conversion of 4-chlorokynurenine to 7-chlorokynurenic acid has also been observed in human astrocytes, implicating that 4-chlorokynurenine therapy might be used in humans in diseases of excitotoxic origin (Kiss et al., 2003).

Another prodrug strategy uses D-glucose and D-galactose conjugates of KYNA and its analogs for allowing better penetration of these drugs into the brain, assuming that the conjugates are recognized by the glucose transporter facilitating their entry and get hydrolyzed in the brain to release the active compound (Battaglia et al., 2000a; Bonina et al., 2000). Systemic administration of 7-chlorokynurenic acid-D-glucopyranos-6′-yl ester or 7-chlorokynurenic acid-D-glucopyranos-3′-yl ester has provided protection against seizures induced by NMDA in mice.

A glucosamine–kynurenic acid conjugate has also been synthesized which induced stereotypy and ataxia in freely moving rats and reduced the evoked excitatory postsynaptic potentials in rat motor cortical slices after intracerebroventricular administration (Füvesi et al., 2004). Further studies are under way to characterize the bioavailability and therapeutic efficacy of the glucosamine–kynurenic acid conjugate after systemic administration.

5.3 KP Enzyme Inhibitors

5.3.1 Manipulation of the KP with the Aim to Elevate the Cerebral Level of KYNA

Modulation of the KP by inhibiting enzymes of QUIN synthesis is a rational approach to divert the kynurenine metabolism toward the neuroprotective KYNA. This therapy would be particularly useful in clinical situations when excessive EAA receptor activation takes place. The administration of *kynurenine 3-hydroxylase* inhibitors is the most rational approach to evoke marked elevation in cerebral KYNA levels with concomitant attenuation of 3-HK and QUIN formation. While application of compounds inhibiting *kynureninase* and 3-hydroxyanthranilic acid oxygenase – enzymes that act downstream from the 3-HK – has limited therapeutic potency, these drugs cause an increase in 3-HK levels with modest KYNA elevations.

Kynurenine 3-hydroxylase inhibitors: The first drug reported to exhibit *kynurenine 3-hydroxylase* inhibiting activity was nicotinylalanine (❯ *Figure 5-3*, Moroni et al., 1991), which is an analog of L-KYN. Subsequently, analogs with more potency and higher selectivity have been synthesized: *meta*-nitrobenzoylalanine (mNBA) is 1,000 times more active and by far more selective than nicotinylalanine (❯ *Figure 5-3*, Pellicciari et al., 1994). Modification of the aromatic ring region and the side chain yielded (R,S)-3,4-dichlorobenzoylalanine (PNU 156561, formerly known as FCE 288833A), which showed further improvements in its potency and selectivity, compared with mNBA (❯ *Figure 5-3*, Speciale et al., 1996). The

❑ Figure 5-3
Chemical structure of selected kynurenine 3-hydroxylase inhibitors

screening of a library of sulfonamides led to the identification of 3,4-dimethoxy-N-[4-(3-nitrophenyl) thiazol-2-yl]benzenesulfonamide (IC_{50} = 37 nM, Ro-61-8048), which is currently the most potent noncompetitive inhibitor of the *kynurenine 3-hydroxylase* (❿ *Figure 5 3*, Röver et al., 1997).

Systemic administration of *kynurenine 3-hydroxylase* inhibitors results in inhibition of maximal electroshock-induced seizures in rats and audiogenic seizures in DBA/2 mice (Russi et al., 1992; Carpenedo et al., 1994) and provides protection in a rat focal ischemia and a gerbil global ischemia model (Cozzi et al., 1999). Furthermore, these drugs significantly reduce blood and brain accumulation of QUIN in immunostimulated mice, suggesting that the application of *kynurenine 3-hydroxylase* inhibitors may be of benefit to the patient with neuroimmunological disorders (Chiarugi and Moroni, 1999b).

Kynureninase inhibitors: These compounds inhibit the QUIN branch of the pathway downstream of 3-HK, which explains the limited therapeutic potency of these drugs compared with the *kynurenine 3-hydroxylase* inhibitors. Application of these drugs also results in accumulation of 3-HK in the brain besides elevating cerebral KYNA levels. *ortho*-Methoxybenzoylalanine (oMBA) is a prototype inhibitor of kynurenase, which is chemically related to mNBA but preferentially inhibits *kynureninase* (Pellicciari et al., 1994). Administration of oMBA results in an increase in the amount of KYNA in the brain in vivo and antagonizes audiogenic convulsion in DBA2 mice (Carpenedo et al., 1994; Chiarugi et al., 1995). Although oMBA is a selective inhibitor of *kynureninase* in vitro, it also inhibits 3-hydroxyanthranilic acid oxygenase in vivo, making it difficult to study the metabolic effects of *kynureninase* inhibitors in vivo (Chiarugi and Moroni, 1999a).

3-Hydroxyanthranilic acid oxygenase inhibitors: Halogenated substrate analogs, such as 4-chloro-3-hydroxyanthranilic acid, were the first potent, selective, and competitive blockers reported to inhibit 3-hydroxyanthranilic acid oxygenase (Todd et al., 1989; Walsh et al., 1991); but more recently, dihalogenated analogs of 3-hydroxyanthranilic acid have also been described with even higher potency and selectivity (Linderberg et al., 1999). These drugs reduce functional deficits in animals exposed to experimental spinal cord injury and reduce QUIN accumulation in the brain of traumatized animals (Blight et al., 1995) and in the blood and brain of immunoactivated mice (Saito et al., 1994).

5.3.2 KP Enzyme Inhibition with the Aim to Decrease Cerebral KYNA Levels

Since certain cognition enhancers have been shown to decrease KYNA levels in the brain (Poeggeler et al., 1998; Rassoulpour et al., 1998) and others to prevent the KYNA antagonism of the NMDA receptor (Pittaluga et al., 1995), one might hypothesize that KYNA plays an important role in cognitive processes and pharmacological inhibition of cerebral KYNA levels might be a useful strategy to produce cognition enhancer drugs. Furthermore, the cognitive deficits might be attenuated with the use of these drugs in diseases with elevated cerebral KYNA levels such as schizophrenia, Alzheimer's disease, or Down's syndrome.

KAT inhibitors: KAT II is responsible for producing large part of KYNA present in the brain that makes it a prime target for influencing cerebral KYNA levels. α-Aminoadipate, quisqualate, DL-5-bromocriptine, and certain metabotropic glutamate receptor agonists, such as L-(+)-2-amino-4-phosphonobutyric acid (L-AP4), 4-carboxy-3-hydroxyphenylglycine (4C3HPG), and L-serine-o-phosphate (L-SOP) selectively block KAT II in vitro (Battaglia et al., 2000b). However, all KAT II inhibitors known so far influence other cellular processes besides acting on the KP, rendering it difficult to characterize the behavioral consequences of decreased KYNA content in the brain. Synthesis of more selective KAT II inhibitors is essential to study the importance of KYNA in the brain function.

6 Conclusion

The discovery of the importance of kynurenines in brain function under physiological and pathologic conditions has led to the identification of potential new drug targets exploiting the therapeutic potential of the pathway. Some of these compounds proved to provide neuroprotection in animal models of various human diseases, which holds promise that their effectiveness will translate to the clinic in the future.

References

Alberati-Giani D, Buchli R, Malherbe P, Broger C, Lang G, et al. 1996. Isolation and expression of a cDNA clone encoding human kynureninase. Eur J Biochem 239: 460-468.

Alkondon M, Pereira, EFR, Fawcett WP, Randall WR, Guidetti P, et al. 2003. Endogenous kynurenic acid regulates alpha7 nicotinic receptor activity in the hippocampus. Program No. 248.9. Abstract Viewer/Itinerary Planner. Washington, D.C.: Society for Neuroscience.

Andiné P, Lehmann A, Ellren K, Wennberg E, Kjellmer I, et al. 1988. The excitatory amino acid antagonist kynurenic acid administered after hypoxic-ischemia in neonatal rats offers neuroprotection. Neurosci Lett 90: 208-212.

Baran H, Cairns N, Lubec B, Lubec G. 1996. Increased kynurenic acid levels and decreased brain kynurenine aminotransferase I in patients with Down syndrome. Life Sci 58: 1891-1899.

Baran H, Jellinger K, Derecke L. 1999. Kynurenine metabolism in Alzheimer's disease. J Neural Transm 106: 165-181.

Baron BM, Harrison BL, Miller FP, McDonald IA, Salituro FG, et al. 1990. Activity of 5,7-dichlorokynurenic acid, a potent antagonist at the NMDA receptor-associated glycine binding site. Mol Pharmacol 38: 554-561.

Baron BM, Harrison BL, McDonald IA, Meldrum BS, Palfreyman MG, et al. 1992. Potent indole- and quinoline-containing NMDA antagonists at the strychnine-insensitive glycine binding site. J Pharmacol Exp Ther 262: 947-956.

Battaglia G, La Russa M, Bruno V, Arenare L, Ippolito R, et al. 2000a. Systemically administered D-glucose conjugates of 7-chlorokynurenic acid are centrally available and exert anticonvulsant activity in rodents. Brain Res 860: 149-156.

Battaglia G, Rassoulpour A, Wu HQ, Hodgkins PS, Kiss C, et al. 2000b. Some metabotropic glutamate receptor ligands reduce kynurenate synthesis in rats by intracellular inhibition of kynurenine aminotransferase II. J Neurochem 75: 2051-2060.

Beal MF, Kowall NW, Ellison DW, Mazurek MF, Swartz KJ, Martin JB, et al. 1986. Replication of the neurochemical characteristics of Huntington's disease by quinolinic acid. Nature 321: 168-171.

Blight AR, Cohen TI, Saito K, Heyes MP. 1995. Quinolinic acid accumulation and functional deficits following experimental spinal cord injury. Brain 118: 735-752.

Blight AR, Leroy Jr EC, Heyes MP. 1997. Quinolinic acid accumulation in injured spinal cord: Time course, distribution, and species differences between rat and guinea pig. J Neurotrauma 14: 89-98.

Bonina FP, Arenare L, Ippolito R, Boatto G, Battaglia G, et al. 2000. Synthesis, pharmacokinetics and anticonvulsant activity of 7-chlorokynurenic acid prodrugs. Int J Pharm 202: 79-88.

Bristow LJ, Flatman KL, Hutson PH, Kulagowski JJ, Leeson PD, et al. 1996. The atypical neuroleptic profile of the glycine/N-methyl-D-aspartate receptor antagonist, L-701324, in rodents. J Pharmacol Exp Ther 277: 578-585.

Cammer W. 2001. Oligodendrocyte killing by quinolinic acid in vitro. Brain Res 896: 157-160.

Carlin JM, Ozaki Y, Byrne GI, Brown RR, Borden EC. 1989. Interferons and indoleamine 2,3-dioxygenase: Role in antimicrobial and antitumor effects. Experientia 45: 535-541.

Carpenedo R, Chiarugi A, Russi P, Lombardi G, Carla V, et al. 1994. Inhibitors of kynurenine hydroxylase and kynureninase increase cerebral formation of kynurenate and have sedative and anticonvulsant activities. Neuroscience 61: 237-244.

Carpenedo R, Pittaluga A, Cozzi A, Attucci S, Galli A, et al. 2001. Presynaptic kynurenate-sensitive receptors inhibit glutamate release. Eur J Neurosci 13: 2141-2147.

Chiarugi A, Moroni F. 1999a. Effects of mitochondria and o-methoxybenzoylalanine on 3-hydroxyanthranilic acid dioxygenase activity and quinolinic acid synthesis. J Neurochem 72: 1125-1132.

Chiarugi A, Moroni F. 1999b. Quinolinic acid formation in immune-activated mice: Studies with (m-nitrobenzoyl)-Alanine (mNBA) and 3,4-dimethoxy-[N-4-(3-nitrophenyl) thiazol-2-yl]benzenesulfonamide (Ro 61-8048), two potent and selective inhibitors of kynurenine hydroxylase. Neuropharmacology 38: 1225-1233.

Chiarugi A, Carpenedo R, Molina MT, Mattoli L, Pellicciari R, et al. 1995. Comparison of the neurochemical and behavioural effects resulting from the inhibition of kynurenine hydroxylase and/or kynureninase. J Neurochem 65: 1176-1183.

Chiarugi A, Rapizzi E, Moroni F, Moroni F. 1999. The kynurenine metabolic pathway in the eye: Studies on 3-hydroxykynurenine, a putative cataractogenic compound. FEBS Lett 453: 197-200.

Chiarugi A, Cozzi A, Ballerini C, Massacesi L, Moroni F. 2001. Kynurenine 3-mono-oxygenase activity and neurotoxic kynurenine metabolites increase in the spinal cord of rats with experimental allergic encephalomyelitis. Neuroscience 102: 687-695.

Cozzi A, Carpenedo R, Moroni F. 1999. Kynurenine hydroxylase inhibitors reduce ischaemic brain damage. Studies with (m-nitrobenzoyl)-alanine (mNBA) and 3,4-dimethoxy-[N-4-(nitrophenyl)thiazol-2-yl]benzenesulfonamide (Ro61-8048) in models of focal or global brain ischaemia. J Cereb Blood Flow Metab 19: 771-777.

Csillik A, Knyihár E, Okuno E, Krisztin-Peva B, Csillik B, et al. 2002. Effect of 3-nitropropionic acid on kynurenine aminotransferase in the rat brain. Exp Neurol 177: 233-241.

Dai W, Gupta SL. 1990. Molecular cloning, sequencing and expression of human interferon-gamma-inducible indoleamine 2,3-dioxygenase cDNA. Biochem Biophys Res Commun 169: 1-8.

Danesh U, Hashimoto S, Renkawitz R, Schutz G. 1983. Transcriptional regulation of the tryptophan oxygenase gene in rat liver by glucocorticoids. J Biol Chem 258: 4750-4753.

Danesh U, Gloss B, Schmid W, Schutz G, Schule R, et al. 1987. Glucocorticoid induction of the rat tryptophan oxygenase gene is mediated by two widely separated glucocorticoid-responsive elements. EMBO J 6: 625-630.

de Carvalho LP, Bochet P, Rossier J. 1996. The endogenous agonist quinolinic acid and the non endogenous homoquinolinic acid discriminate between NMDAR2 receptor subunits. Neurochem Int 28: 445-452.

Decker RH, Kang HH, Leach FR, Henderson LM, 1961. Purification and properties of 3-hydroxyanthranilic acid oxidase. J Biol Chem 236: 3076-3082.

Du F, Williamson J, Bertram E, Lothman EW, Okuno E, et al. 1993. Kynurenine pathway enzymes in a rat model of chronic epilepsy: Immunohistochemical study of activated glial cells. Neuroscience 55: 975-989.

Eastman CL, Guilarte TR. 1989. Cytotoxicity of 3-hydroxykynurenine in a neuronal hybrid cell line. Brain Res 495: 225-231.

Erickson JB, Reinhard JFJ, Flanagan EM, Russo S. 1992. A radiometric assay for kynurenine 3-hydroxylase based on the release of ^3H_2O during hydroxylation of L-[3.5-^{3}H]-kynurenine. Anal Biochem 205: 257-262.

Flanagan EM, Erickson JB, Viveros OH, Chang SY, Reinhard Jr JF, 1995. Neurotoxin quinolinic acid is selectively elevated in spinal cords of rats with experimental allergic encephalomyelitis. J Neurochem 64: 1192-1196.

Foster AC, Vezzani A, French ED, Schwarcz R. 1984. Kynurenic acid blocks neurotoxicity and seizures induced in rats by the related brain metabolite quinolinic acid. Neurosci Lett 48: 273-278.

Foster AC, White RJ, Schwarcz R. 1986. Synthesis of quinolinic acid by 3-hydroxyanthranilic acid oxygenase in rat brain tissue in vitro. J Neurochem 47: 23-30.

Fujigaki S, Saito K, Takemura M, Maekawa N, Yamada Y, et al. 2002. L-tryptophan-L-kynurenine pathway metabolism accelerated by Toxoplasma gondii infection is abolished in gamma interferon-gene-deficient mice: Cross-regulation between inducible nitric oxide synthase and indoleamine-2,3-dioxygenase. Infect Immun 70: 779-786.

Fukui S, Schwarcz R, Rapoport SI, Takada Y, Smith QR. 1991. Blood-barrier transport of kynurenines: Implications for brain synthesis and metabolism. J Neurochem 56: 2007-2017.

Füvesi J, Somlai C, Németh H, Varga H, Kis Z, et al. 2004. Comparative study on the effects of kynurenic acid and glucosamine–kynurenic acid. Pharmacol Biochem Behav 77: 95-102.

Germano IM, Pitts LH, Meldrum BS, Bartkowski HM, Simon RP. 1987. Kynurenate inhibition of cell excitation decreases stroke size and deficits. Ann Neurol 22: 730-734.

Glaxo SPA. 1993. Patent 2266091.

Guidetti P, Eastman CL, Schwarcz R. 1995. Metabolism of [5-3H]kynurenine in the rat brain in vivo: Evidence for the existence of a functional kynurenine pathway. J Neurochem 65: 2621-2632.

Guidetti P, Okuno E, Schwarcz R. 1997. Characterization of rat brain kynurenine aminotransferases I and II. J Neurosci Res 50: 457-465.

Guidetti P, Schwarcz R. 1999. 3-Hydroxykynurenine potentiates quinolinate but not NMDA toxicity in the rat striatum. Eur J Neurosci 11: 3857-3863.

Guidetti P, Wu HQ, Schwarcz R. 2000. In situ produced 7-chlorokynurenate provides protection against quinolinate- and malonate-induced neurotoxicity in the rat striatum. Exp Neurol 163: 123-130.

Guidetti P, Luthi-Carter RE, Augood SJ, Schwarcz R. 2003. Quinolinic acid levels are increased in early grade Huntington's disease. Program No. 208.7. Abstract Viewer/Itinerary Planner. Washington, D.C.: Society for Neuroscience.

Guillemin GJ, Kerr SJ, Smythe GA, Smith DG, Kapoor V, et al. 2001. Kynurenine pathway metabolism in human astroyctes: A paradox for neuronal protection. J Neurochem 78: 842-853.

Halperin JJ, Heyes MP. 1992. Neuroactive kynurenines in Lyme borreliosis. Neurology 42: 43-50.

Heyes MP, Lackner A. 1990. Increased cerebrospinal fluid quinolinic acid, kynurenic acid and L-kynurenine in acute septicemia. J Neurochem 55: 338-341.

Heyes MP, Nowak Jr TS. 1990. Delayed increases in regional brain quinolinic acid follow transient ischemia in the gerbil. J Cereb Blood Flow Metab 10: 660-667.

Heyes MP, Brew BJ, Martin A, Price RW, Salazar AM, et al. 1991. Quinolinic acid in cerebrospinal fluid and serum in HIV-1 infection: Relationhip to clinical and neurologic status. Ann Neurol 29: 202-209.

Heyes MP, Saito K, Jocobowitz D, Markey SP, Takikawa O, et al. 1992. Poliovirus induces indoleamine-2,3-dioxygenase and quinolinic acid synthesis in macaque brain. FASEB J 6: 2977-2989.

Heyes MP, Saito K, Lackner A, Wiley CA, Achim CL, et al. 1998. Sources of the neurotoxin quinolinic acid in the brain of HIV-1 infected patients and retrovirus-infected macaques. FASEB J 12: 881-896.

Hicks RR, Smith DH, Gennarelli TA, McIntosh T. 1994. Kynurenate is neuroprotective following experimental brain injury in the rat. Brain Res 655: 91-96.

Hilmas C, Pereira EFR, Alkondon M, Rassoulpour A, Schwarcz R, et al. 2001. The brain metabolite kynurenic acid inhibits α7 nicotinic receptor activity and increases non-7 nicotinic receptor expression: Physiopathological implications. J Neurosci 21: 7463-7473.

Hirata H, Hayaishi O. 1975. Studies on Indoleamine 2,3-dioxygenase. Superoxide anion as substrate. J Biol Chem 250: 5960-5966.

Hokari M, Wu H-Q, Schwarcz R, Smith QR. 1996. Facilitated brain uptake of 4-chlorokynurenine and conversion to 7-chlorokynurenic acid. Neuroreport 8: 15-18.

Kawai J, Okuno E, Kido R. 1988. Organ distribution of rat kynureninase and changes of its activity during development. Enzyme 39: 181-189.

Kessler M, Terramani T, Lynch G, Baudry M. 1989. A glycine site associated with N-methyl-D-aspartic acid receptors: Characterization and identification of a new class of antagonists. J Neurochem 52: 1319-1328.

Kiss C, Ceresoli-Borroni G, Guidetti P, Zielke CL, Schwarcz R. 2003. Kynurenic acid production by human astrocytes. J Neural Transm 110: 1-14.

Knyihár-Csillik E, Okuno E, Vécsei L. 1999. Effects of in vivo sodium azide administration on the immunohistochemical localization of kynurenine aminotransferase in the rat brain. Neuroscience 94: 269-277.

Koontz WA, Shiman R. 1976. Beef kidney 3-hydroxyanthranilic acid oxygenase. J Biol Chem 251: 368-377.

Kulagowski JJ, Baker R, Curtis NR, Leeson PD, Mawer IM, et al. 1994. 3′-(Arylmethyl)- and 3′-(aryloxy)-3-phenyl-4-hydroxyquinolin-2(1H)-ones: Orally active antagonists of the glycine site on the NMDA receptor. J Med Chem 37: 1402-1405.

Lapin IP. 1976. Depressor effect of kynurenine and its metabolites in rats. Life Sci 19: 1479-1484.

Lapin IP. 1978. Stimulant and convulsive effects of kynurenines injected into brain ventricles in mice. J Neural Transm 42: 37-43.

Lapin IP. 1981. Kynurenines and seizures. Epilepsia 22: 257-265.

Lapin IP. 1982. Convulsant action of intracerebroventricularly administered kynurenine sulphate, quinolinic acid and other derivatives of succinic acid and effects of amino acids: Structure–activity relationships. Neuropharmacology 21: 1227-1233.

Lee SC, Schwarcz R. 2001. Excitotoxic injury stimulates prodrug-induced 7-chlorokynurenate formation in the rat striatum in vivo. Neurosci Lett 304: 185-188.

Leeson PD, Carling RW, Moore KW, Moseley AM, Smith JD, et al. 1992. 4-Amido-2-carboxytetrahydroquinolines. Structure–activity relationships for antagonism at the glycine site of the NMDA receptor. J Med Chem 35: 1954-1968.

Lehrmann E, Molinari A, Speciale C, Schwarcz R. 2001. Immunohistochemical visualization of newly formed quinolinate in the normal and excitotoxically lesioned rat striatum. Exp Brain Res 141: 389-397.

Linderberg M, Hellberg S, Björk S, Gotthammar B, Högberg T, et al. 1999. Synthesis and QSAR of substituted 3-hydroxyanthranilic acid derivatives as inhibitors of 3-hydroxyanthranilic acid dioxygenase (3-HAO). Eur J Med Chem 34: 729-744.

Long CL, Hill HN, Weinstock IM, Henderson LM. 1954. Studies on the enzymatic transformation of 3-hydroxyanthranilic acid to quinolinic acid. J Biol Chem 205: 411-413.

Mehler AH, Knox WE. 1950. Conversion of tryptophan to kynurenine in liver. II. The enzymatic hydrolysis of formylkynurenine. J Biol Chem 187: 431-433.

Moroni F, Russi P, Carla V, Lombardi G. 1988. Kynurenic acid is present in the rat brain and its content increases during development and aging processes. Neurosci Lett 94: 145-150.

Moroni F, Alesiani N, Galli A, Mori F, Pecorari R, et al. 1991. Thiokynurenates – a new group of antagonists of the glycine modulatory site of the NMDA receptor. Eur J Pharmacol 199: 227-232.

Morrison PF, Morishige GM, Beagles KE, Heyes MP. 1999. Quinolinic acid is extruded from the brain by a probenecid-sensitive carrier system: A quantitative analysis. J Neurochem 72: 2135-2144.

Musso T, Gusella GL, Brooks A, Varesio L. 1994. Interleukin-4 inhibits indoleamine 2,3-dioxygenase expression in human monocytes. Blood 83: 1408-1411.

Nozaki K, Beal MF. 1992. Neuroprotective effects of L-kynurenine on hypoxia-ischemia and NMDA lesions in neonatal rats. J Cereb Blood Flow Metab 12: 400-407.

Ogawa T, Matson WR, Beal MF, Myers RH, Bird ED, et al. 1992. Kynurenine pathway abnormalities in Parkinson's disease. Neurology 42: 1702-1706.

Okuda S, Nishiyama N, Saito H, Katsuki H. 1996. Hydrogen peroxide-mediated neuronal cell death induced by an endogenous neurotoxin, 3-hydroxykynurenine. Proc Natl Acad Sci USA 93: 12553-12558.

Okuda S, Nishiyama N, Saito H, Katsuki H. 1998. 3-Hydroxykynurenine, an endogenous oxidative stress generator, causes neuronal cell death with apoptotic features and region selectivity. J Neurochem 70: 299-307.

Okuno E, Nakamura M, Schwarcz R. 1991. Two kynurenine aminotransferases in human brain. Brain Res 542: 307-312.

Pellicciari R, Natalini B, Costantino G, Mahmoud MR, Mattoli L, et al. 1994. Modulation of the kynurenine pathway in search for new neuroprotective agents. Synthesis and preliminary evaluation of (m-nitrobenzoyl)alanine, a potent inhibitor of kynurenine 3-hydroxylase. J Med Chem 37: 647-655.

Perkins MN, Stone TW. 1982. An iontophoretic investigation of the actions of convulsant kynurenines and their interaction with the endogenous excitant quinolinic acid. Brain Res 247: 184-187.

Pinelli A, Ossi C, Colombo R, Tofanetti O, Spazzi L. 1984. Experimental convulsions in rats induced by intraventricular administration of kynurenine and structurally related compounds. Neuropharmacology 23: 333-337.

Pittaluga A, Pattarini R, Raiteri M. 1995. Putative cognition enhancers reverse kynurenic acid antagonism at hippocampal NMDA receptors. Eur J Pharmacol 272: 203-209.

Poeggeler B, Rassoulpour A, Guidetti P, Wu HQ, Schwarcz R. 1998. Dopaminergic control of kynurenate levels and NMDA toxicity in the developing rat striatum. Dev Neurosci 20: 146-153.

Rassoulpour A, Wu HQ, Poeggeler B, Schwarcz R. 1998. Systemic D-amphetamine administration causes a reduction of kynurenic acid levels in rat brain. Brain Res 802: 111-118.

Rassoulpour A., Guidetti P., Kiss C., Schwarcz R. 2002. Amphetamine potentiates quinolinate but not kainate excitotoxicity in the rat striatum. Society for Neuroscience Abstract Viewer/Itinerary Planner. Program No. 92.6.

Reinhard Jr JF, Erickson JB, Flanagan EM. 1994. Quinolinic acid in neurological disease: Opportunities for novel drug discovery. Adv Pharmacol 30: 85-126.

Reinhard Jr JF. 1998. Altered tryptophan metabolism in mice with herpes simplex virus encephalitis: Increases in spinal cord quinolinic acid. Neurochem Res 23: 661-665.

Reynolds GP, Pearson SJ, Halket J, Sandler M. 1988. Brain quinolinic acid in Huntington's disease. J Neurochem 50: 1959-1960.

Rios C, Santamaria A. 1991. Quinolinic acid is a potent lipid peroxidant in rat brain homogenates. Neurochem Res 16: 1139-1143.

Röver S, Cesura AM, Hugenin P, Kettler R, Szente A. 1997. Synthesis and biochemical evaluation of N-(4-phenylthiazol-2-yl)benzenesulfonamides as high-affinity inhibitors of kynurenine 3-hydroxylase. J Med Chem 40: 4378-4385.

Russi P, Alesiani M, Lombardi G, Davolio P, Pellicciari R, et al. 1992. Nicotinylalanine increases the formation of kynurenic acid in the brain and antagonizes convulsions. J Neurochem 59: 2076-2080.

Sacco RL, De Rosa JT, Haley Jr EC, Levin B, Ordronneau P, et al. 2001. The glycine antagonist in neuroprotection Americas investigators. Glycine antagonist in neuroprotection for patients with acute stroke: GAIN Americas: A randomized controlled trial. JAMA 285: 1719-1728.

Saito K, Nowak Jr TS, Markey SP, Heyes MP. 1992. Delayed increases in kynurenine pathway metabolism in damaged brain regions following transient cerebral ischemia. J Neurochem 60: 180-192.

Saito K, Nowak TS Jr, Markey SP, Heyes MP. 1993. Mechanism of delayed increases in kynurenine pathway metabolism in damaged brain regions following transient cerebral ischemia. J Neurochem 60: 180-192.

Saito K, Markey SP, Heyes MP. 1994. 6-Chloro-D,L-tryptophan, 4-chloro-3-hydroxyanthranilate and dexamethasone attenuate quinolinic acid accumulation in brain and blood following systemic immune activation. Neurosci Lett 178: 211-215.

Salvati P, Ukmar G, Dho L, Rosa B, Cini M, et al. 1999. Brain concentrations of kynurenic acid after a systemic neuroprotective dose in the gerbil model of global ischaemia. Prog Neuropsychopharmacol Biol Psychiatry 23: 741-752.

Sanni LA, Thomas SR, Tattam BN, Moore DE, Chaudhri G, et al. 1998. Dramatic changes in oxidative tryptophan metabolism along the kynurenine pathway in experimental cerebral and noncerebral malari. Am J Pathol 152: 611-619.

Sapko MT, Yu P, Guidetti P, Pellicciari R, Tagle DA, et al. 2003. Endogenous brain kynurenic acid modulates susceptibility to striatal quinolinic acid excitotoxicity. Program No. 805.20. 2003. Abstract Viewer/Itinerary Planner. Washington, D.C.: Society for Neuroscience.

Schutz G, Chow E, Feigelson P. 1972. Regulatory properties of hepatic tryptophan oxygenase. J Biol Chem 247: 5333-5337.

Schwarcz R, Whetsell Jr WO, Mangano RM. 1983. Quinolinic acid: An endogenous metabolite that produces axon-sparing lesions in rat brain. Science 219: 316-318.

Schwarcz R, Okuno E, White RJ, Bird ED, Whetsell WO. 1988a. 3OH-anthranilic acid oxygenase activity is increased in the brains of Huntington disease victims. Proc Natl Acad Sci USA 85: 4079-4081.

Schwarcz R, Tamminga CA, Kurlan R, Shoulson I. 1988b. CSF levels of quinolinic acid in Huntington's disease and schizophrenia. Ann Neurol 24: 580-582.

Schwarcz R, Pellicciari R. 2002. Manipulation of brain kynurenines: Glial targets, neuronal effects, and clinical opportunities. J Pharmacol Exp Ther 303: 1-10.

Simon RP, Young RS, Stout S, Cheng J. 1986. Inhibition of excitatory neurotransmission with kynurenate reduces brain edema in neonatal anoxia. Neurosci Lett 71: 361-364.

Sinz EH, Kochanek PM, Heyes MP, Wisniewski SR, Bell MJ, et al. 1998. Quinolinic acid is increased in CSF and associated with mortality after traumatic brain injury in humans. J Cereb Blood Flow Metab 18: 610-615.

Speciale C, Wu HQ, Cini M, Marconi M, Varasi M, et al. 1996. (R,S)-3,4-dichlorobenzoylalanine (FCE 28833A) causes a large and persistent increase in brain kynurenic acid levels in rats. Eur J Pharmacol 315: 263-267.

Stone TW. 1993. Neuropharmacology of quinolinic and kynurenic acids. Pharmacol Rev 45: 309-379.

Stone TW. 2000. Development and therapeutic potential of kynurenic acid and kynurenine derivatives for neuroprotection. Trends Pharmacol Sci 21: 149-154.

Stone TW. 2001. Kynurenines in the CNS: From endogenous obscurity to therapeutic importance. Prog Neurobiol 64: 185-218.

Stone TW, Perkins MN. 1981. Quinolinic acid: A potent endogenous excitant at amino acid receptors in CNS. Eur J Pharmacol 72: 411-412.

Stone TW, Darlington LG. 2002. Endogenous kynurenines as targets for drug discovery and development. Nat Rev Drug Discov 1: 609-620.

Schwarcz R, Ceresoli G, Guidetti P. 1996. Kynurenine metabolism in the rat brain in vivo. Effect of acute excitotoxic insults. Adv Exp Med Biol 398: 211-219.

Tamminga CA. 1998. Schizophrenia and glutamatergic transmission. Crit Rev Neurobiol 12: 21-36.

Taylor MW, Feng G. 1991. Relationship between interferon gamma, indoleamine 2,3-dioxygenase, and tryptophan catabolism. FASEB J 5: 2516-2522.

Todd WP, Carpenter BK, Schwarcz R. 1989. Preparation of 4-halo-3-hydroxyanthranilates and demonstration of their inhibition of 3-hydroxyanthranilate oxygenase in rat and human brain tissue. Prep Biochem 19: 155-165.

Toma S, Nakamura M, Tone S, Okuno E, Kido R, et al. 1997. Cloning and recombinant expression of rat and human kynureninase. FEBS Lett 408: 5-10.

Tone S, Takikawa O, Habara-Ohkubo A, Kadoya A, Yoshida R, et al. 1990. Primary structure of human indoleamine 2,3-dioxygenase deduced from the nucleotide sequence of its cDNA. Nucleic Acid Res 18: 367.

Turski WA, Gramsbergen JB, Traitler H, Schwarcz R. 1989. Rat brain slices produce and liberate kynurenic acid upon exposure to L-kynurenine. J Neurochem 52: 1629-1636.

Urenjak J, Obrenovitch TP. 2000. Kynurenine 3-hydroxylase inhibition in rats: Effects on extracellular kynurenic acid concentration and N-methyl-D-aspartate-induced depolarisation in the striatum. J Neurochem 75: 2427-2433.

Vazquez S, Aquilina JA, Jamie JF, Sheil MM, Truscott RJ. 2002. Novel protein modification by kynurenine in human lenses. J Biol Chem 277: 4867-4873.

Vécsei L, Beal MF. 1990a. Intracerebroventricular injection of kynurenic acid, but not kynurenine, induces ataxia and stereotyped behavior in rats. Brain Res Bull 25: 623-627.

Vécsei L, Beal MF. 1990b. Influence of kynurenine treatment on open-field activity, elevated plus-maze, avoidance behaviors and seizures in rats. Pharmacol Biochem Behav 37: 71-76.

Vécsei L, Beal MF. 1991. Comparative behavioral and pharmacological studies with centrally administered kynurenine and kynurenic acid in rats. Eur J Pharmacol 196: 239-246.

Vécsei L, Miller J, MacGarvey U, Beal MF. 1992. Kynurenine and probenecid inhibit pentylenetetrazol- and NMDLA-induced seizures and increase kynurenic acid concentrations in the brain. Brain Res Bull 28: 233-238.

Walsh JL, Todd WP, Carpenter BK, Schwarcz R. 1991. 4-Halo-3-hydroxyanthanilic acids: Potent competitive inhibitors of 3-hydroxyanthranilic acid oxygenase in vitro. Biochem Pharmacol 42: 985-990.

Watanabe Y, Yoshida R, Sono M, Hayaishi O. 1981. Immunohistochemical localization of indoleamine 2,3-dioxygenase in the argyrophilic cells of rabbit duodenum and thyroid gland. J Histochem Cytochem 29: 623-632.

Whetsell Jr WO, Schwarcz R. 1989. Prolonged exposure to submicromolar concentrations of quinolinic acid causes excitotoxic damage in organotypic cultures of rat corticostriatal system. Neurosci Lett 97: 271-275.

Wu HQ, Salituro FG, Schwarcz R. 1997. Enzyme-catalyzed production of the neuroprotective NMDA receptor antagonist 7-chlorokynurenic acid in the rat brain in vivo. Eur J Pharmacol 319: 13-20.

Wu HQ, Lee SC, Scharfman HE, Schwarcz R. 2002. L-4-Chlorokynurenine attenuates kainate-induced seizures and lesions in the rat. Exp Neurol 177: 222-232.

Yoshida R, Hayaishi O. 1987. Indoleamine 2,3-dioxygenase. Methods Enzymol 142: 188-195.

Yuan W, Collado-Hidalgo A, Yufit T, Taylor MW, Varga J. 1998. Modulation of cellular tryptophan metabolism in human fibroblasts by transforming growth factor beta: Selective inhibition of indoleamine 2,3-dioxygenase and tryptophanyl-tRNA synthetase gene expression. J Cell Physiol 177: 174-186.

6 Neurofibromatosis Type I: From Genetic Mutation to Tumor Formation

S. L. Thomas · G. H. De Vries

1	Introduction	108
2	NF1 Clinical Characteristics	108
2.1	Diagnostic Criteria	108
2.2	Peripheral Nerve Sheath Tumors	109
2.2.1	Dermal Neurofibroma	109
2.2.2	Plexiform Neurofibroma	109
2.2.3	Malignant Peripheral Nerve Sheath Tumor	110
2.3	Other Malignancies Associated with NF1	110
2.4	Other Complications Associated with NF1	110
3	Molecular and Cellular Characteristics	110
3.1	NF1 Gene and Neurofibromin	110
3.2	Expression of Neurofibromin	111
3.3	Loss of Heterozygosity for *NF1*	112
3.3.1	Animal Models of NF1 LOH	113
3.4	Mutations of the *NF1* Gene Predict Nonfunctional Neurofibromin	113
3.4.1	Genotype–Phenotype Correlations	114
3.5	Increased Ras Activation in Neurofibromin-Deficient Cells	114
3.6	Hypersensitivity to Growth Factors in Neurofibromin-Deficient Cells	115
3.7	Neurofibromin GAP-Related Domain Activity and Cell Growth	116
3.8	Other Possible Functions of Neurofibromin	116
4	Paradigm for Peripheral Nerve Sheath Tumor Formation in NF1	117
4.1	Aberrant Receptor Expression	117
4.2	Aberrant Angiogenesis	119
4.2.1	Proangiogenic and Antiangiogenic Factor Expression	119
4.2.2	In Vivo Studies Targeting Angiogenesis	120
4.3	Role of Mast Cells in NF1 Tumor Formation	120
4.4	Secondary Genetic Events	120
4.4.1	Animal Models of Secondary Genetic Events	121
5	Therapeutic Approaches for the Treatment of NF1 Tumors	121

© 2009 Springer Science+Business Media, LLC.

Abstract: Neurofibromatosis type I (NF1) is one of the most common genetic disorders of the nervous system. Inheritance occurs through a germline deletion or a loss-of-function mutation of the neurofibromatosis 1 (*NF1*) gene. The *NF1* gene codes for the protein neurofibromin, which contains a GTPase-activating protein (GAP)-related domain (NF1-GRD) and functions as a negative regulator of small G proteins, including Ras. The hyperactivation of Ras that results from the loss of neurofibromin is thought to contribute to the various types of lesions that are prevalent in NF1. Individuals with NF1 are predisposed to a number of pathologies including peripheral nerve sheath tumors, hyperpigmentation of the skin, lesions of the iris, skeletal lesions, vascular abnormalities, learning disorders, and an increased incidence of several malignancies. Malignant peripheral nerve sheath tumors (MPNST) are one of the most common features of NF1. These tumors constitute a major cause of morbidity and mortality for individuals with NF1 since the only current treatment for these tumors (surgical removal) is not effective. This chapter discusses the clinical characteristics of NF1, mutations of the *NF1* gene, animal models of NF1, expression and function of neurofibromin, and current therapeutic approaches to the disease. The current understanding of the etiology of MPNST formation is also summarized.

List of Abbreviations: ANGPT1, angiopoietin-1; bFGF, basic fibroblast growth factor; cAMP, cyclic adenosine monophosphate; CNS, central nervous system; DNA, deoxyribonucleic acid; EGF, epidermal growth factor; EGFR, epidermal growth factor receptor; FGF2, fibroblast growth factor 2; FGF7, fibroblast growth factor 7; FGFR3, fibroblast growth factor receptor 3; FTI, farnesyltransferase inhibitor; GAP, GTPase-activating protein; GDP, guanosine diphosphate; GM-CSF, granulocyte–macrophage colony-stimulating factor; GRD, GAP-related domain; GGF2, glial growth factor 2; GROa, growth-related oncoprotein α; GTP, guanosine triphosphate; GTPase, guanosine triphosphatase; HGF, hepatocyte growth factor; IGF-1, insulin-like growth factor 1; LOH, loss of heterozygosity; MAPK, mitogen-activated protein kinase; MDK, midkine; MPNST, malignant peripheral nerve sheath tumor; mRNA, messenger ribonucleic acid; NF1, neurofibromatosis type I; *NF1*, neurofibromatosis 1 gene; NF1-GRD, NF1 GAP-related domain; NRP-1, neuropilin 1; PDGF, platelet-derived growth factor; PEDF, pigment epithelium-derived factor; PGE, prostaglandin; PI3K, phosphatidylinositol-3-kinase; PKA, protein kinase A; PlGF, placental growth factor; *Rb*, retinoblastoma gene; RNA, ribonucleic acid; SCF, stem cell factor; SPARC, secreted protein acidic and rich in cysteine; TGF-β1, transforming growth factor β1; TGF-βRII, TGFβ receptor 2; TIMP-1, tissue inhibitor of metalloproteinase 1; TIMP-2, tissue inhibitor of metalloproteinase 2; TSP-1, thrombospondin-1; uPA, urokinase plasminogen activator; uPAR, urokinase plasminogen activator receptor; VEGF, vascular endothelial growth factor; VEGFC, vascular endothelial growth factor C

1 Introduction

Neurofibromatosis type I (NF1) is a common genetic disorder of the nervous system that causes tumor predisposition, predominantly affecting cells of neural crest origin. The incidence of disease is approximately 1 in 3,000 to 1 in 5,000 individuals of which approximately 50% of affected individuals have inherited the disorder whereas the other 50% of individuals have sporadic cases of NF1 (Riccardi and Friedman, 1999). Inheritance occurs through a germline deletion or a loss-of-function mutation in one allele of the neurofibromatosis 1 (*NF1*) gene (Friedman, 1999a). Individuals with NF1 are predisposed to a number of pathologies including peripheral nerve sheath tumors, hyperpigmentation of the skin, lesions of the iris, skeletal lesions, vascular abnormalities, learning disorders, and an increased incidence of several malignancies (Riccardi and Friedman, 1999).

2 NF1 Clinical Characteristics

2.1 Diagnostic Criteria

Diagnosis of NF1 is based on the presence of various types of lesions. NF1 diagnosis is made when two or more of the following criteria are met: first-degree relative with NF1, six or more café-au-lait macules (hyperpigmentation of the skin), two or more neurofibromas of any type or one plexiform neurofibroma (benign peripheral nerve sheath tumors), freckling in the axillary or inguinal regions, optic glioma (astrocyte tumor), two or more Lisch nodules (lesion of the iris), or an osseous lesion (National Institutes of Health Consensus Development Conference, 1988).

2.2 Peripheral Nerve Sheath Tumors

The most common type of tumor found in patients with NF1 is the benign peripheral nerve sheath tumor, known as a neurofibroma. Two types of benign neurofibromas that commonly develop in individuals with NF1 are dermal and plexiform (see ❂ *Table 6-1*). Plexiform neurofibromas can undergo malignant

◘ Table 6-1
Characteristics of NF1-associated peripheral nerve sheath tumors

Tumor type	Location	Age of onset
Benign peripheral nerve sheath tumor		
Dermal/cutaneous neurofibroma	Dermis	Puberty
Plexiform neurofibroma	Subcutaneous/visceral	Any age
Malignant peripheral nerve sheath tumor (neurofibrosarcoma, neurogenic sarcoma)	Subcutaneous/visceral	Any age

progression to become malignant peripheral nerve sheath tumors (MPNST) (Wiestler and Radner, 1994; Korf, 1999a). Both types of neurofibromas are heterogeneous tumors comprised of a variety of cell types including Schwann cells, fibroblasts, perineurial cells, mast cells, endothelial cells, and smooth muscle cells (Stefansson et al., 1982; Peltonen et al., 1988; Carr and Warren, 1993; Nurnberger and Moll, 1994). However, Schwann cells are the major cellular component of neurofibromas with estimates of 40%–80% of the total cell population in tumor tissue (Stefansson et al., 1982; Peltonen et al., 1988).

2.2.1 Dermal Neurofibroma

Dermal neurofibromas are one of the most common features of NF1. These tumors grow within the dermis and are associated with the terminal nerve endings in the skin. These tumors often occur in large numbers with an age of onset around puberty. The number of dermal neurofibromas tends to increase with age so that most adults with NF1 present with these lesions. Dermal neurofibromas do not undergo malignant progression (Wiestler and Radner, 1994; Korf, 1999a; Woodruff, 1999).

2.2.2 Plexiform Neurofibroma

Plexiform neurofibromas are a major cause of morbidity in NF1. These tumors exhibit growth along a visceral or subcutaneous peripheral nerve. These tumors may also form on cranial nerves or spinal nerve roots. Plexiform neurofibromas can grow as discrete nodules along a nerve or more commonly as a diffuse growth associated with hypertrophy of the surrounding soft tissue. These tumors have a sporadic growth rate with periods of rapid growth followed by periods with no growth. They can become very large in size resulting in compression of the surrounding tissues. Plexiform tumors have a prepubertal onset and are often present in the first 5 years of life. It has been estimated that up to 44% of individuals with NF1 develop plexiform tumors. A further complication of plexiform neurofibroma is the potential to undergo malignant transformation into an MPNST (Wiestler and Radner, 1994; Korf, 1999a, b; Woodruff, 1999; Waggoner et al., 2000).

2.2.3 Malignant Peripheral Nerve Sheath Tumor

Approximately 2%–5% of individuals with NF1 develop MPNST. MPNST often develops in the second and third decades of life but have been detected in NF1 patients between the ages of 7 and 63 years. The aggressive and metastatic growth of these tumors can be life threatening and constitutes a major cause of mortality in individuals with NF1 (Wiestler and Radner, 1994; Korf, 1999a; Woodruff, 1999; Ferner and Gutmann, 2002).

2.3 Other Malignancies Associated with NF1

Patients with NF1 have an increased incidence of developing several malignancies of both neural crest and nonneural crest tissues. Some of the most common malignancies seen in individuals with NF1 involve neural crest-derived chromaffin cells of the adrenal medulla (pheochromocytoma), embryonic muscle cells (rhabdomyosarcoma), astrocytes (malignant glioma, often involving the optic nerve), and myeloid cells (juvenile chronic myeloid leukemia) (Gutmann and Gurney, 1999; Korf, 2000). Malignancy is the most common cause of death in individuals with NF1 (Friedman, 1999b).

2.4 Other Complications Associated with NF1

Other complications associated with NF1 include vascular abnormalities, skeletal lesions, and central nervous system (CNS) disorders. NF1 is associated with a high incidence of vasculopathy that most often affects arterial vessels. Vascular anomalies include stenosis, occlusion, and aneurysm. There is a high incidence of hypertension associated with NF1, which often results from renovascular stenosis or pheochromocytoma. Other problems that result from vasculopathy include hemorrhaging from aneurysms and infarction from occlusions (Ferner, 1994; Friedman, 1999c). These vascular lesions exhibit overgrowth of cells in the intimal layer of the vessel, which results in narrowing of the vessel lumen. The cell type responsible for the vascular lesions has been found to be smooth muscle cells rather than Schwann cells (Hamilton and Friedman, 2000). Vascular disorders contribute to the increased mortality rate in young individuals with NF1 (Rasmussen et al., 2001).

In addition, mutation of the *NF1* gene produces effects on the development of the skeletal system. Common skeletal defects include macrocephaly, short stature, scoliosis, and dysplasia of the flat and long bones. Scoliosis is estimated to occur in at least 10% of individuals with NF1 and may be severe or progressive in nature and require bracing or surgery. Another NF1-associated skeletal defect that can be severe is bowing or pseudoarthrosis of the long bones resulting in deformity and the development of fractures, which may compromise movement of the affected limb. Additional skeletal defects include the erosion or overgrowth of bone because of the presence of a local neurofibroma (Fairbank, 1994; Riccardi, 1999).

NF1 is also associated with various effects on the CNS. Some of the common neurologic complications include learning disability/mental retardation, epilepsy, glioma (optic pathway/cerebral/brain stem), and deformity of the skull (sphenoid wing dysplasia) (Hughes, 1994; Gutmann, 1999).

3 Molecular and Cellular Characteristics

3.1 NF1 Gene and Neurofibromin

In 1990, the identity of the gene mutated in individuals with NF1 was reported by three laboratories (Cawthon et al., 1990; Viskochil et al., 1990; Wallace et al., 1990). The *NF1* gene contains 60 exons and spans more than 300 kb of genomic DNA on chromosome 17q11.2 (Li et al., 1995) (❷ *Figure 6-1*). The *NF1* gene codes for the 2,818-amino acid protein neurofibromin (Xu et al., 1990b; Marchuk et al., 1991), which contains a domain with sequence homology to the catalytic domains of mammalian and yeast

◨ **Figure 6-1**
Diagram of the *NF1* gene with exons represented as *rectangular boxes*. The *scale bar* in the left corner indicates the size of the exons. Introns are not drawn to scale. The *arrow* represents the transcription start site. The exons of the *NF1* GAP-related domain (NF1-GRD) are striped. The alternatively spliced exons are filled in

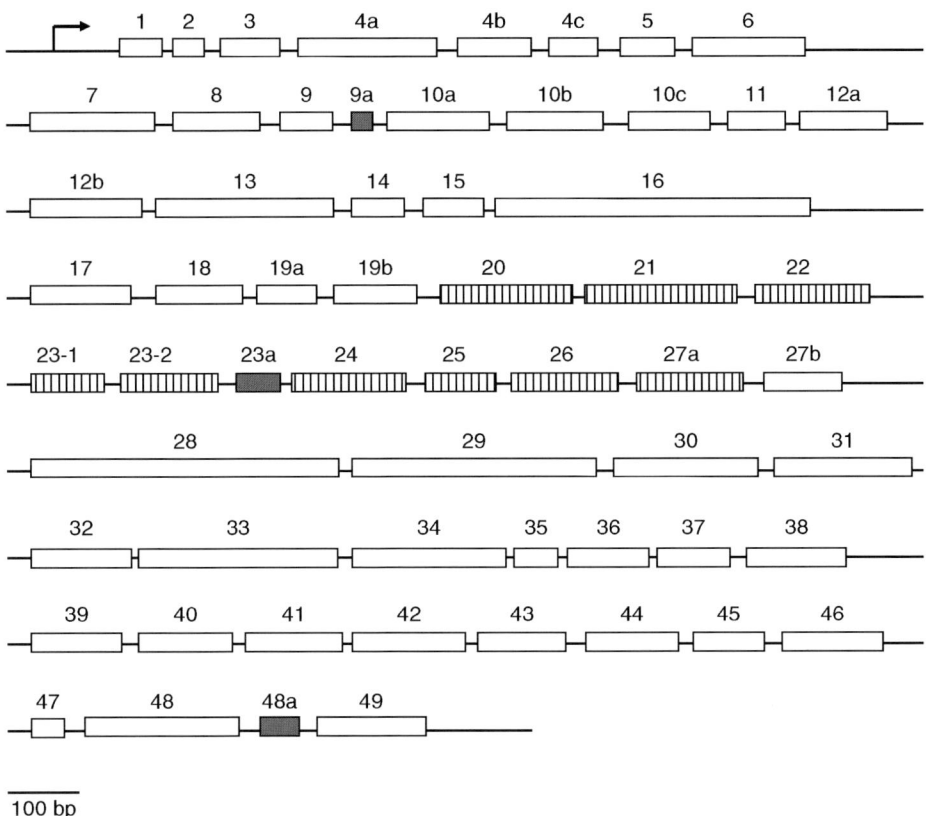

GTPase-activating proteins (GAPs) (Ballester et al., 1990; Xu et al., 1990a). GAP proteins negatively regulate small G proteins, including Ras, by accelerating the intrinsic GTPase activity of these proteins. Neurofibromin increases guanosine triphosphate (GTP) hydrolysis by Ras up to 100,000-fold by stabilizing the transition state between Ras-GTP and Ras-GDP. Thus, loss of functional neurofibromin results in the hyperactivation of Ras (Basu et al., 1992; DeClue et al., 1992).

3.2 Expression of Neurofibromin

Neurofibromin is a cytosolic protein (Daston et al., 1992; Golubic et al., 1992) that has been found to colocalize in some cell types with cytoskeletal components, including tubulin, actin, and keratin (Gregory et al., 1993; Xu and Gutmann, 1997; Koivunen et al., 2000; Li et al., 2001; Malminen et al., 2002). Studies in rat, mouse, and chicken have demonstrated that neurofibromin is ubiquitously expressed during early embryonic development (Daston and Ratner, 1992; Baizer et al., 1993; Huynh et al., 1994; Gutmann et al., 1995c) with expression in the adult animal the highest in the nervous system and adrenal gland (Daston and Ratner, 1992; Daston et al., 1992; Golubic et al., 1992; Gutmann et al., 1995c).

There are at least four different isoforms of neurofibromin that result from insertions of alternatively spliced exons 9a, 23a, and 48a. Type I neurofibromin, the original isoform identified (Cawthon et al., 1990;

Viskochil et al., 1990; Wallace et al., 1990), does not contain any of the alternatively spliced exons. Type II neurofibromin contains a 63-nucleotide/21-amino acid insertion (exon 23a) in the *NF1* GAP-related domain (NF1-GRD) between exons 23 and 24 (Nishi et al., 1991; Andersen et al., 1993). The insertion in type II neurofibromin alters the structure of the NF1-GRD (Nishi et al., 1991), resulting in a decrease in GAP activity (Andersen et al., 1993) and loss of association with microtubules (Gutmann et al., 1995a). The type III isoform contains a 54-nucleotide/18-amino acid insertion (exon 48a) near the carboxyl terminus between exons 48 and 49 (Gutman et al., 1993). A fourth isoform of neurofibromin (type IV) contains a 30-nucleotide/10-amino acid insertion (exon 9a) near the amino terminus between exons 9 and 10 (Danglot et al., 1995; Geist and Gutmann, 1996). Type III and type IV neurofibromin are expected to have the same GAP activity as that of type I because the NF1-GRD sequence is the same in these three isoforms.

The alternatively spliced isoforms of neurofibromin have tissue-specific and/or developmental stage-specific expression. Both the type I and type II isoforms have been found to be expressed in every tissue analyzed in the adult mouse with each isoform having a predominant expression in certain tissues. Predominant expression of type I was found in brain, spinal cord, and testis; while type II was the predominant isoform in the adrenal gland, kidney, ovary, and lung (Mantani et al., 1994). Studies of neurofibromin expression in the nervous system have found that the type I isoform is the major form expressed by CNS neurons, whereas type II is the major form expressed by CNS glial cells and neural crest-derived tissues including Schwann cells, neurons of the dorsal root ganglia, and adrenal medulla (Teinturier et al., 1992; Gutmann et al., 1995c). However, the type I/type II ratio can switch as cells differentiate. The type III isoform is expressed in human and rat skeletal, cardiac, and smooth muscle but not in brain, spleen, liver, or kidney (Gutman et al., 1993; Gutmann et al., 1995b). Expression of type III neurofibromin in muscle is high during embryogenesis and begins to decline one week after birth. Therefore, Gutmann and coworkers (1995b) suggest that type III neurofibromin may play a role in muscle differentiation. Type IV neurofibromin is expressed in CNS neurons, but not in CNS glial cells or neural crest-derived cells (Danglot et al., 1995; Geist and Gutmann, 1996). Expression of type IV neurofibromin is increased during late embryogenesis and early postnatal development in a region-specific pattern that correlates with CNS neuronal differentiation (Geist and Gutmann, 1996). These studies suggest that neurofibromin isoforms have tissue-type specific expression and may be involved in cellular differentiation.

3.3 Loss of Heterozygosity for *NF1*

Inheritance of NF1 occurs through a germline loss of one *NF1* allele. Loss of the second *NF1* allele (loss of heterozygosity) is thought to be required for the formation of both benign and malignant peripheral nerve sheath tumors, as well as other malignancies in individuals with NF1. Loss of heterozygosity (LOH) for *NF1* has been shown to be common in NF1-associated malignancies. Several studies have reported *NF1* LOH in MPNST (Skuse et al., 1989; Menon et al., 1990; Glover et al., 1991; Xu et al., 1992; Legius et al., 1993). One study demonstrated homozygous inactivation of the *NF1* gene in an MPNST in which only one chromosome 17 was present in the tumor and contained a large deletion in the *NF1* gene (Legius et al., 1993). Furthermore, a study of *NF1* gene loss in NF1-associated pheochromocytoma found that LOH was present in all seven of the tumors examined and five of the tumors had LOH for loci on both 17p and 17q suggesting loss of the entire chromosome 17 (Xu et al., 1992). Additional studies in children with NF1 and malignant myeloid disorders found that *NF1* LOH is present in bone marrow cells in some individuals. One study reported that the normal *NF1* allele from the unaffected parent was absent in bone marrow samples from 5 of 11 individuals (Shannon et al., 1994). Another study found that bone marrow from 8 of 18 children contained truncating mutations of the NF1 gene and 5 of the 8 individuals with truncating mutations had lost the normal *NF1* allele from the unaffected parent (Side et al., 1997). Importantly, these studies have demonstrated that homozygous inactivation of *NF1* is involved in at least some cases of NF1-associated MPNST, pheochromocytoma, and leukemia.

The requirement of homozygous inactivation of *NF1* for the development of benign dermal and plexiform neurofibromas is less clear. Several studies have found that *NF1* LOH was not present in dermal or plexiform neurofibromas (Skuse et al., 1989; Menon et al., 1990; Lothe et al., 1993; Shimizu et al., 1993).

On the other hand, two studies found the presence of *NF1* LOH in a subset of neurofibromas studied (Colman et al., 1995; Serra et al., 1997). Similarly, a dermal neurofibroma from an individual with NF1 was found to contain a point mutation in the remaining *NF1* allele (Sawada et al., 1996). A recent study of 38 dermal neurofibromas from nine individuals with NF1 found that somatic mutations of the *NF1* gene were present in 29 of the tumors studied (Maertens et al., 2006). Another study looked at particular cell types in a neurofibroma and found evidence of *NF1* LOH in a subset of Schwann cells but not in fibroblasts (Serra et al., 2000). Thus, LOH for the *NF1* locus appears to be involved in the formation of at least some benign neurofibromas. The actual frequency of LOH in neurofibromas may be underestimated due to the heterogeneous composition of the tumors, the presence of normal tissue in the sample analyzed, the limited number of intragenic markers for *NF1* in early studies, and the use of techniques that cannot detect small genetic alterations like point mutations.

3.3.1 Animal Models of NF1 LOH

Further evidence that LOH is required for peripheral nerve sheath tumor formation comes from genetic mouse models of NF1. Initial attempts at homozygous deletion of the *NF1* gene (*NF1*−/−) resulted in embryonic lethality due to cardiac defects (Brannan et al., 1994; Jacks et al., 1994; Lakkis and Epstein, 1998). Heterozygous deletion of the *NF1* gene (*NF1*+/−) resulted in viable mice that develop late-onset malignancies including pheochromocytoma and myeloid leukemia, but did not develop peripheral nerve sheath tumors. In this model, loss of the wild-type *NF1* allele occurred in all cases of pheochromocytoma and myeloid leukemia (Jacks et al., 1994). To analyze the *NF1*−/− phenotype in viable mice, Cichowski and coworkers developed chimeric mice with both *NF1*+/+ and *NF1*−/− cells. The *NF1* chimeric mice developed multiple plexiform neurofibromas that were often associated with dorsal root ganglia and peripheral nerves in the limbs. The plexiform neurofibromas developed from *NF1*−/− cells but did not show evidence of malignant transformation (Cichowski et al., 1999). Other animal models were developed to examine the targeted loss of *NF1* in Schwann cells. Mice were engineered with *NF1*−/− Schwann cells in an *NF1*+/+ genetic background and *NF1*−/− Schwann cells in an *NF1* +/− background (Zhu et al., 2002). The complete loss of *NF1* in Schwann cells in the presence of an *NF1* +/+ genetic background is sufficient to induce plexiform neurofibroma formation in vivo. However, an *NF1*+/− genetic background contributed to the potential of *NF1*−/− Schwann cells to form neurofibromas resulting in an increased frequency and size of lesions compared to mice with *NF1*−/− Schwann cells in an *NF1*+/+ background. Thus, the presence of various *NF1* heterozygous cell types in the tumor microenvironment contributes to neurofibroma formation (Zhu et al., 2002). These animal models of NF1 demonstrate that the loss of both *NF1* alleles is required for plexiform neurofibroma formation. On the other hand, the animals in these models do not develop the common cutaneous (café-au-lait spots and dermal neurofibromas) and skeletal lesions that are diagnostic of NF1. Therefore, either loss of the *NF1* gene is not sufficient to produce these features of NF1 or there are significant species differences in the response to the loss of *NF1*.

3.4 Mutations of the *NF1* Gene Predict Nonfunctional Neurofibromin

NF1 mutations have been shown to occur throughout the length of the *NF1* gene and can be deletions, insertions, or point mutations. Several hundred mutations have been identified, most of which are novel. Recurrent mutations have been reported, but no true mutational hotspots in the *NF1* gene have been found. Most *NF1* mutations, independent of the region in which they occur, are predicted to truncate neurofibromin and produce an unstable or nonfunctional protein.

Although the NF1-GRD (exons 20–27a) of the *NF1* gene is highly conserved among GAP proteins and is the only region of the gene with a known function, the majority of reported mutations do not occur in this region (reviewed in Upadhyaya et al., 1994; Shen et al., 1996; Thomson et al., 2002). For example, exon 31 is one of the most frequently mutated exons in the *NF1* gene. There have been reports of nonsense, missense, insertion, and deletion mutations occurring in exon 31. One recurrent mutation in exon 31 is a cytosine to thymidine transition (C5839T) that results in a stop codon (reviewed in Upadhyaya et al., 1994; Shen et al., 1996; Thomson et al., 2002). This mutation occurs in an estimated 1% of cases of NF1. Several other recurrent mutations occur throughout the length of the NF1 gene, none of which have a high frequency of occurrence (Thomson et al., 2002).

Within the NF1-GRD various types of mutations have been found (Li et al., 1992; Upadhyaya et al., 1997). One large screening of NF1 patients found that 6% of individuals had a mutation in the NF1-GRD region. The mutations consisted of missense, nonsense, splice site mutations, and small deletions, most of which were predicted to result in premature termination of neurofibromin synthesis (Upadhyaya et al., 1997). Furthermore, two studies tested the effect of NF1-GRD missense substitutions on GAP function and found that activity was reduced 200–400-fold compared to wild-type NF1-GRD (Li et al., 1992; Upadhyaya et al., 1997). A study of one family with a missense substitution in an arginine residue that is critical for GAP activity (R1276P) found that the mutation did not affect neurofibromin folded structure, expression level, or association with Ras; however, the mutation resulted in a complete loss of neurofibromin GAP activity (8,000-fold reduction) (Klose et al., 1998). The affected individuals in the family presented with a classical NF1 phenotype including café-au-lait spots, freckling, scoliosis, learning disability, and minor mental retardation. The mother developed multiple cutaneous neurofibromas, an optic glioma, and multiple peripheral nerve sheath tumors (one of which became malignant) (Klose et al., 1998).

3.4.1 Genotype–Phenotype Correlations

Given the complex nature of the clinical features and progression of NF1 and the high frequency of novel mutations of the *NF1* gene, only one correlation has been made between genotype and phenotype. It was recently reported that a 3-bp inframe deletion in exon 17 is associated with the absence of cutaneous and plexiform neurofibromas in 21 unrelated individuals (Upadhyaya et al., 2007). Another suggested genotype–phenotype correlation involves the gross deletion of the *NF1* gene including surrounding sequences (>700 kb), which is estimated to occur in up to 10% of cases of NF1 (Carey and Viskochil, 1999; Thomson et al., 2002). Individuals with large deletions encompassing the entire *NF1* gene often have a severe phenotype including craniofacial dysmorphism, severe learning disability or mental retardation, and a large number of dermal neurofibromas with early onset (Kayes et al., 1992, 1994; Wu et al., 1995, 1997; Cnossen et al., 1997; Leppig et al., 1997; Upadhyaya et al., 1998; Riva et al., 2000). However, not all individuals with gross deletions of the *NF1* gene have these characteristics and not all individuals with these severe features have a gross deletion of the *NF1* gene (Tonsgard et al., 1997). In addition, there are several examples in which individuals with the same *NF1* mutation do not exhibit the same clinical features (Shen et al., 1996; Upadhyaya et al., 1994, 1997). Thus, the developmental timing and location of the *NF1* gene mutation may have a major impact on the phenotype that is expressed.

3.5 Increased Ras Activation in Neurofibromin-Deficient Cells

Increased Ras activation has been well established in neurofibromin-deficient cells. MPNST Schwann cell lines have decreased GAP activity along with elevated Ras-GTP (Basu et al., 1992; DeClue et al., 1992). In addition, Ras-GTP levels were increased in both benign and malignant NF1 peripheral nerve sheath tumors compared to non-NF1 schwannomas (Guha et al., 1996; Feldkamp et al., 1999). Another study found that Ras-GTP was elevated in 12%–62% of Schwann cells but not fibroblasts from five dissociated neurofibroma cultures (Sherman et al., 2000). Further evidence that loss of neurofibromin results in increased Ras activation comes from a study of *NF1*-null mouse Schwann cells that demonstrated elevated Ras-GTP in *NF1*−/− compared to *NF1*+/+ and *NF1*+/− Schwann cells (Kim et al., 1995). Furthermore, neurofibromin-deficient cell types other than Schwann cells have dysregulated Ras activation. An NF1-associated astrocytoma expressed elevated Ras-GTP compared to normal white matter (Lau et al., 2000). Also, *NF1*-deficient mouse oligodendrocyte progenitors have increased basal and growth factor stimulated Ras-GTP (Bennett et al., 2003).

Some evidence suggests that neurofibromin may not be essential for the regulation of Ras in certain cells types. It has been found that melanoma and neuroblastoma cell lines in which neurofibromin is deficient express normal or only moderately elevated Ras-GTP (Johnson et al., 1993; The et al., 1993). These studies suggest the possibility that Schwann cells may be more sensitive to the loss of neurofibromin than other cell types, which could explain the major involvement of Schwann cells in NF1 pathology.

Interestingly, increased Ras activation does not always result in increased cell growth. There have been reports that both human and mouse neurofibromin-deficient Schwann cells do not show increased proliferation when cultured under standard conditions (Kim et al., 1995; Rosenbaum et al., 2000). In addition, expression of oncogenic *ras* induces growth arrest in primary Schwann cells (Ridley et al., 1988) and neural crest-derived pheochromocytoma cells (Bar-Sagi and Feramisco, 1985; Noda et al., 1985) and induces senescence in primary fibroblasts (Serrano et al., 1997). Therefore, it is likely that an additional genetic alteration in a gene involved in the regulation of cell cycle is required for uncontrolled cell growth to occur.

Ras activation has differential effects on cell proliferation and survival depending on the degree and duration of activation (Marshall, 1995; Sewing et al., 1997). In neurofibromin-deficient cells, p120GAP and possibly other unidentified GAPs are present and functional. Recent evidence suggests that neurofibromin-deficient cells still have some control over Ras activation as Ras activity does decrease back to basal levels after growth factor stimulation (Farrer, R.G., Farrer, J.R., and De Vries, G.H., unpublished observation). In a recent study, platelet-derived growth factor (PDGF) stimulation produced an increase in Ras activation in both normal human Schwann cells and an MPNST-derived Schwann cell line; however, the duration of Ras activation was longer in MPNST Schwann cells taking 50 min to return to basal levels compared to 20 min in normal Schwann cells (Farrer, R.G., Farrer, J.R., and De Vries, G.H., submitted for publication). Thus, the activation of Ras may be qualitatively different when neurofibromin is lost than when Ras is oncogenic and constitutively active. In NF1, the increased duration of Ras activation due to loss of neurofibromin may sensitize the cells to growth factor stimulation; however, the Ras activation may not be long enough in duration to induce growth arrest or senescence.

3.6 Hypersensitivity to Growth Factors in Neurofibromin-Deficient Cells

Although neurofibromin-deficient cells do not always possess a growth advantage under standard culture conditions, several different neurofibromin-deficient cell types have been reported to possess an increased growth potential in response to specific growth factors. Various *NF1*-deficient cells isolated from knockout mice are hypersensitive to growth factors, for example, fibroblasts to TGF-β (Yang et al., 2006), oligodendrocyte progenitors to fibroblast growth factor 2 (FGF2) (Bennett et al., 2003), endothelial cells to basic fibroblast growth factor (bFGF) (Wu et al., 2006), mast cells to Kit ligand (Ingram et al., 2000), hematopoietic progenitor cells to GM-CSF (granulocyte–macrophage colony-stimulating factor) (Bollag et al., 1996; Largaespada et al., 1996; Zhang et al., 1998b) and astrocytes to coculture with neurons (Gutmann et al., 1999). NF1−/− mouse Schwann cells hyperproliferate in response to forskolin and serum-free culture conditions, but exhibit decreased proliferation in response to glial growth factor 2 (GGF2) or contact with axons (Kim et al., 1995, 1997). Likewise, human cells that are neurofibromin-deficient are hypersensitive to growth factors, for example, Schwann cells to PDGF-BB (Badache and De Vries, 1998) and Kit ligand/SCF (Badache et al., 1998), fibroblasts to PDGF-BB and transforming growth factor β1 (TGF-β1) (Kadono et al., 1994), and leukemia cells to GM-CSF (Emanuel et al., 1996). Human neurofibromin-deficient Schwann cells respond to PDGF-BB with an increase in intracellular Ca^{2+} and phosphorylation of calmodulin kinase II, whereas these responses to PDGF-BB were not observed in normal human Schwann cells (Dang and DeVries, 2005). Moreover, neurofibromin-deficient Schwann cells display an increased growth in vivo. Peripheral nerve sheath tumor-derived Schwann cells exhibit growth when xenografted in sciatic nerves of immunodeficient mice, whereas normal Schwann cells do not exhibit growth under these same conditions (Muir et al., 2001; Perrin et al., 2007). Furthermore, *NF1* knockout in Schwann cells is sufficient to induce tumor formation in mice (Zhu et al., 2002). These studies provide evidence that neurofibromin-deficient cells possess an increased growth potential.

3.7 Neurofibromin GAP-Related Domain Activity and Cell Growth

Studies have demonstrated that expression of the neurofibromin GAP-related domain (NF1-GRD) regulates cell growth. Transduction of the NF1-GRD in human neurofibromin-deficient Schwann cells resulted in a reduction in Ras activation and a greater than 50% decrease in cell proliferation (Thomas and De Vries, 2007). Similarly, expression of the NF1-GRD in mouse *NF1−/−* hematopoietic cells and fibroblasts restored normal growth to the cells (Hiatt et al., 2001). In addition, it was found that a human colon cancer cell line with an oncogenic K-*ras* mutation could be suppressed from forming a tumor in vivo by a construct expressing the full-length *NF1* gene or the GRD of the *NF1* gene (Li and White, 1996). Since oncogenic Ras cannot be regulated by GAP activity, the authors suggest that neurofibromin and Raf-1 compete for binding to Ras. Thus, overexpression of the *NF1* gene resulted in a suppression of Raf-1 activation and inhibition of cell growth (Li and White, 1996). Similarly, overexpression of a 91-amino acid fragment of the NF1-GRD in v-Ha-Ras-transformed NIH/3T3 cells reduced the ability of the cells to form colonies in soft agar (Nur-E-Kamal et al., 1993).

3.8 Other Possible Functions of Neurofibromin

There is growing evidence that neurofibromin may have other functions besides the regulation of Ras activation. Neurofibromin was found to regulate the adenylyl cyclase/cAMP/PKA signaling pathway (Guo et al., 1997; The et al., 1997). In *Drosophila*, loss of *NF1* results in flies that have a small body size and defective potassium current in response to pituitary adenylyl cyclase-activating polypeptide. These defects are not diminished by manipulating Ras signaling, but instead are rescued by increasing signaling through the cAMP/PKA pathway (Guo et al., 1997; The et al., 1997). The authors concluded that *NF1* negatively regulates the *rutabaga*-encoded adenylyl cyclase instead of the Ras pathway in fruit flies (Guo et al., 1997). Later studies demonstrated defects in adenylyl cyclase activity due to neurofibromin-deficiency in mammalian cells. One study found that adenylyl cyclase activity was decreased in the brains of *NF1−/−* mice compared to *NF1+/−* mice (Tong et al., 2002). Another study examined cyclic adenosine monophosphate (cAMP) activation in astrocytes and found that the loss of *NF1* in mouse astrocytes resulted in reduced cAMP production in response to pituitary adenylyl cyclase-activating polypeptide (Dasgupta et al., 2003). On the basis of the effects of stimulating the cAMP pathway at different points along the pathway, it was determined that neurofibromin most likely regulates adenylyl cyclase activation. Furthermore, reconstitution of the NF1-GRD could only partially restore the defect in cAMP suggesting that a region of neurofibromin outside of the NF1-GRD is involved in the regulation of adenylyl cyclase (Dasgupta et al., 2003). On the other hand, studies of the loss of *NF1* in Schwann cells have reported the opposite effect in which the loss of *NF1* results in increased levels of cAMP (Kim et al., 2001; Dang, I., and De Vries G.H., unpublished observation). One report found that *NF1−/−* mouse Schwann cells have a threefold increase in intracellular cAMP relative to *NF1+/+* Schwann cells (Kim et al., 2001). Similarly, unpublished data from our laboratory demonstrate a twofold increase in cAMP and expression of additional adenylyl cyclase subtypes in MPNST-derived Schwann cell lines compared to normal human Schwann cells (Dang and De Vries G.H., unpublished observation). These studies provide evidence that the loss of *NF1* is involved in the abnormal regulation of the adenylyl cyclase/cAMP/PKA signaling pathway, although the effects on this pathway appear to be different in Schwann cells compared to other cell types.

Neurofibromin has been suggested to play a role in the structure or signaling of the cytoskeleton. Neurofibromin has been shown to associate with cytoplasmic microtubules (Gregory et al., 1993) through an interaction with the NF1-GRD region of neurofibromin (Gregory et al., 1993; Xu and Gutmann, 1997). In addition, tubulin inhibits neurofibromin GAP activity suggesting that the GRD binds to both Ras and tubulin (Bollag et al., 1993). Another study found that neurofibromin associates with different cytoskeletal components depending on the phase of differentiation in neurons. During early differentiation, neurofibromin colocalized with F-actin; whereas in a later phase of differentiation, neurofibromin colocalized with microtubules (Li et al., 2001). Further evidence of the involvement of neurofibromin in the cytoskeleton was provided by reports that neurofibromin colocalizes with keratin intermediate filaments in human keratinocytes during differentiation in culture and in human epidermis during early fetal development (Koivunen et al., 2000; Malminen et al., 2002). In developing epidermis, neurofibromin was not found to associate with actin or tubulin (Malminen et al., 2002). These studies suggest a role for neurofibromin in the cytoskeleton during cellular differentiation; however, the biological significance of this function for neurofibromin remains to be resolved.

4 Paradigm for Peripheral Nerve Sheath Tumor Formation in NF1

In light of the literature, the current model for peripheral nerve sheath tumor formation in NF1 involves the loss of heterozygosity of the *NF1* gene in Schwann cells, paracrine signaling with the *NF1* haploinsufficient microenvironment, and the accumulation of secondary genetic events leading to malignant transformation (❯ *Figure 6-2*). NF1 inheritance occurs through germline *NF1* haploinsufficiency in all cells of the body. Somatic LOH for the *NF1* gene in Schwann cells is thought to initiate peripheral nerve sheath tumor formation. Multiple changes result from the loss of *NF1* in Schwann cells including hyperactivation of Ras, increased growth factor receptor expression, and changes in angiogenic factor expression that promote the induction of angiogenesis. In addition, *NF1*-null Schwann cells secrete chemotactic factors that recruit mast cells. *NF1*-haploinsufficient mast cells are hyperresponsive to paracrine signals and secrete proangiogenic factors and mitogens for other cells in the microenvironment, including Schwann cells. As a result, Schwann cells that have lost both copies of *NF1* are stimulated to hyperproliferate. Endothelial cells proliferate and migrate to form new microvessels. Fibroblasts are stimulated to overproduce extracellular matrix components. Perineurial cells lose contact with one another and proliferate, resulting in breakdown of the perineurium. Thus, benign neurofibroma formation is initiated. Furthermore, secondary genetic events affecting the p53 and pRb pathways may occur within a plexiform neurofibroma leading to malignant transformation and MPNST formation. Evidence for each of the events involved in this paradigm for NF1 peripheral nerve sheath tumor formation is discussed in detail in the following sections.

4.1 Aberrant Receptor Expression

Peripheral nerve sheath tumors have changes in receptor expression that may be important for tumor growth and progression. For instance, the stem cell factor (SCF) receptor, c-Kit, was found to be highly expressed in neurofibromin-deficient human Schwann cell lines and tissue but not in normal primary Schwann cells or non-NF1 schwannoma-derived cells (Ryan et al., 1994; Badache et al., 1998; Dang et al., 2005). Furthermore, proliferation of neurofibromin-deficient Schwann cells can be stimulated by treatment with SCF suggesting that an aberrant SCF/c-Kit autocrine and/or paracrine loop may be involved in Schwann cell hyperplasia (Badache et al., 1998). In addition, mast cells in neurofibroma tissue were found to express c-Kit RNA, which may be associated with the increased number of mast cells in neurofibroma tissue (Hirota et al., 1993).

It has been suggested that the epidermal growth factor receptor (EGFR) is important for NF1 tumor growth. EGFR is expressed by MPNST-derived Schwann cell lines, neurofibroma tumor tissue, and MPNST tumor tissue (DeClue et al., 2000). In addition, treatment of MPNST cell lines with epidermal growth factor (EGF) resulted in MAP kinase activation and induced cell growth under low serum conditions (DeClue et al., 2000). Transcriptional profiling studies demonstrated that EGFR is expressed by MPNST cell lines but not by normal human Schwann cells (Lee et al., 2004; Miller et al., 2006; Thomas and De Vries, 2007). EGFR was also found to be expressed by cell lines derived from $NF1+/-:p53+/-$ mice. These cell lines responded to EGF with increased proliferation and activation of both the MAPK (mitogen-activated protein kinase) and PI3K (phosphatidylinositol-3-kinase) pathways (Li et al., 2002).

The PDGF receptors may also be important to the growth of NF1 tumors. Both PDGF receptors α and β were found to have increased expression in MPNST-derived Schwann cell lines compared to non-NF1 Schwann cell lines on Western blot analysis (Badache and De Vries, 1998). It was also found that PGDF-BB is a mitogen for MPNST-derived Schwann cells (Badache and De Vries, 1998). In another study, immunohistochemistry

Figure 6-2

Diagram of peripheral nerve sheath tumor formation in NF1. The early events in NF1 tumor formation involve the loss of heterozygosity (LOH) of the *NF1* gene in Schwann cells and paracrine signaling between the *NF1*-null Schwann cells and the *NF1*-haploinsufficient microenvironment. Tumor formation requires multiple interactions among various *NF1*-deficient cell types resulting in hyperproliferation of *NF1*-deficient cells, breakdown of the perineurium, mast cell infiltration, and the induction of angiogenesis

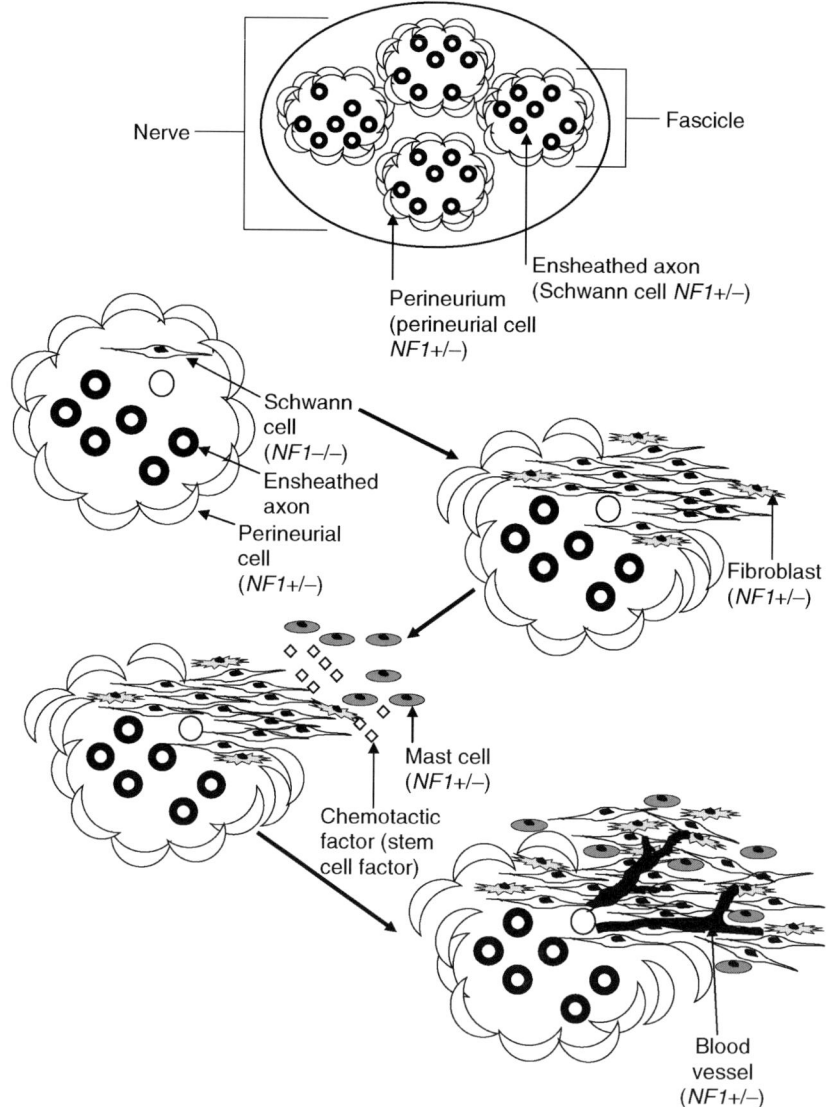

was used to show that PDGF receptor α has increased expression in MPNST tissue compared to dermal and plexiform neurofibroma tissue (Holtkamp et al., 2004). In addition, transcriptional profiling studies found that PDGF receptor α has increased expression in an MPNST cell line compared to normal human Schwann cells (Lee et al., 2004; Thomas and De Vries, 2007). Similarly, PDGF receptor β mRNA was detected in NF1 neurofibroma tissue but not in NF1 nontumor tissue or normal control skin (Kadono et al., 2000).

Several other growth factor receptors were found to have increased expression in NF1 tumor tissue or cell lines. Two studies implicate HGF/c-MET signaling in NF1 tumor progression by showing increased c-MET and hepatocyte growth factor (HGF) expression in MPNST tissue compared to benign neurofibroma tissue (Rao et al., 1997; Watanabe et al., 2001). In addition, TGF-βRII expression was found to be increased in MPNST tissue compared to benign neurofibroma tissue in five of eight samples analyzed (Watanabe et al., 2001). Furthermore, gene expression profiling found that the bFGF receptor FGFR3 (fibroblast growth factor receptor 3) is increased in neurofibroma-derived Schwann cells, whereas the VEGF receptor NRP-1 is increased in both neurofibroma- and MPNST-derived Schwann cells (Thomas and De Vries, 2007).

Abnormal prostaglandin signaling has also been implicated in the growth of MPNST Schwann cell lines. The prostaglandin E_2 (PGE_2) receptor EP2 is expressed by MPNST cell lines but not by normal human Schwann cells (Deadwyler et al., 2004). MPNST Schwann cell lines also secrete more PGE_2 than normal human Schwann cells, and the PGE-inducible synthetic enzyme COX-2 is expressed by MPNST cell lines but not by normal Schwann cells. Furthermore, treatment of the cell lines with PGE_2 receptor antagonists resulted in a decrease in cell proliferation (Deadwyler et al., 2004).

It has been observed that neurofibromas increase in growth during puberty and pregnancy and dermal neurofibromas first appear during puberty (Dugoff and Sujansky, 1996; Huson et al., 1988). In support of these observations, hormone receptors have been detected in neurofibroma tissue (Cunha et al., 2003; McLaughlin and Jacks, 2003). Progesterone receptor expression was detected in 75% of neurofibromas analyzed but was not detected in normal nerve. Dermal neurofibromas had a higher frequency of being positive for progesterone receptor than plexiform neurofibromas with 86% and 50% respective positive tumors (McLaughlin and Jacks, 2003). In addition, growth hormone receptor was detected in 65% of dermal neurofibromas analyzed (Cunha et al., 2003). These studies suggest that hormones may be important in neurofibroma development.

4.2 Aberrant Angiogenesis

NF1-associated neurofibroma and MPNST have been found to be highly vascular tumors (Arbiser et al., 1998; Angelov et al., 1999). In vivo studies have shown that MPNST xenografts stimulate angiogenesis in immunodeficient mice (Lee et al., 1990; Angelov et al., 1999; Perrin et al., 2007), and Schwann cells isolated from neurofibromas stimulate angiogenesis in the chick chorioallantoic membrane assay (Sheela et al., 1990). *NF1−/−* mouse Schwann cells have also been shown to stimulate angiogenesis in vivo (Kim et al., 1997; Wu et al., 2005).

4.2.1 Proangiogenic and Antiangiogenic Factor Expression

Several proangiogenic factors were shown to be expressed in peripheral nerve sheath tumor tissue including bFGF (Ratner et al., 1990; Kawachi et al., 2003), HGF (Krasnoselsky et al., 1994; Rao et al., 1997; Watanabe et al., 2001), IGF-1 (insulin-like growth factor 1) (Hansson et al., 1988), MDK (midkine) (Mashour et al., 2001; Miller et al., 2006), and VEGF (vascular endothelial growth factor) (Arbiser et al., 1998; Angelov et al., 1999; Kawachi et al., 2003), although the cellular source of these factors was not determined. Neurofibromin-deficient human Schwann cells were shown to express increased proangiogenic factors including bFGF (Lee et al., 2004; Thomas and De Vries, 2007), MDK (Miller et al., 2006), ANGPT1 (angiopoietin-1), FGF7 (fibroblast growth factor 7), GROa (growth-related oncoprotein α), HGF, PlGF (placental growth factor), VEGFC (vascular endothelial growth factor C), and uPA (urokinase plasminogen activator) (Thomas and De Vries, 2007). Furthermore, Schwann cells isolated from *NF1*-knockout mouse embryos express increased mRNA for the proangiogenic factors MDK, bFGF, and PDGF-B (Mashour et al., 2001). These studies provide evidence of aberrant proangiogenic factor expression in NF1.

There have only been a handful of studies examining antiangiogenic factor expression in NF1. It was reported that normal human Schwann cells and neurofibromin-deficient Schwann cells secrete pigment epithelium-derived factor (PEDF) (Lertsburapa and De Vries, 2004), which is responsible for the antiangiogenic effect of normal Schwann cell-conditioned media (Crawford et al., 2001). In addition,

Mashour and coworkers have reported increased transcript for thrombospondin 1 (TSP-1) in *NF1−/−* mouse Schwann cells (Mashour et al., 2001). Only two studies have provided evidence of decreased antiangiogenic factor expression in neurofibromin-deficient cells. Gene profiling demonstrated a decrease in SPARC (secreted protein acidic and rich in cysteine) transcript in both neurofibroma- and MPNST-derived Schwann cells (Lee et al., 2004; Thomas and De Vries, 2007). Other proteins that have antiangiogenic properties and were found to be decreased in neurofibromin-deficient Schwann cells include TIMP-1 (tissue inhibitor of metalloproteinase 1), TIMP-2 (tissue inhibitor of metalloproteinase 2), and uPAR (urokinase plasminogen activator receptor) (Thomas and De Vries, 2007). These studies demonstrate that aberrant antiangiogenic factor expression may be important for the growth and progression of NF1 tumors.

4.2.2 In Vivo Studies Targeting Angiogenesis

Two in vivo studies have examined the ability of antiangiogenic treatments to inhibit the vascularization and growth of xenografted MPNST tissue. Lee and coworkers grew MPNST tissue in the subrenal capsule of nude mice and found that heparin alone stimulated tumor growth and vascularity, hydrocortisone alone reduced tumor growth and vascularity, and a combination treatment of heparin and hydrocortisone resulted in a greater reduction in tumor growth and vascularity than hydrocortisone alone (Lee et al., 1990). Another study used subcutaneous implantation of MPNST tissue in immunodeficient mice and found that a small molecule inhibitor of VEGF receptor 2 reduced tumor vascularity resulting in a decrease in tumor volume (Angelov et al., 1999). The results of these studies suggest that inhibition of angiogenesis may be useful for the treatment of malignant tumors associated with NF1. However, neither of the studies grew the tumors in their native environment, the peripheral nerve, so it is unknown whether these treatments would be effective within the nerve environment.

4.3 Role of Mast Cells in NF1 Tumor Formation

There are several types of evidence suggesting the importance of mast cells to the development of NF1 tumors. First, there are reports that mast cells have a higher concentration in neurofibroma tissue than in normal nerve tissue (Riccardi, 1981; Johnson et al., 1989; Hirota et al., 1993). Second, treatment of NF1 patients with the mast cell-stabilizing drug ketotifen resulted in a decrease in the rate of neurofibroma growth and a decrease in neurofibroma-related symptoms (Riccardi, 1987). Further evidence comes from a study by Zhu and coworkers in which an *NF1+/−* genetic background contributed to the ability of *NF1−/−* Schwann cells to form neurofibromas. Tumors that formed under these conditions had a substantial number of infiltrating mast cells and an increased frequency and size compared to the tumors that formed in mice with *NF1−/−* Schwann cells with an *NF1+/+* background (Zhu et al., 2002). The authors suggest that the infiltrating *NF1+/−* mast cells may contribute to neurofibroma formation.

Advances have been made in determining the signaling events between Schwann cells and mast cells that lead to neurofibroma formation. Neurofibromin-deficient human Schwann cells and *NF1−/−* mouse Schwann cells have been shown to produce SCF/Kit ligand (Ryan et al., 1994; Yang et al., 2003). Mast cells in neurofibroma tissue were found to express a high level of SCF receptor, c-Kit, mRNA (Hirota et al., 1993). Furthermore, *NF1* haploinsufficiency in mouse mast cells resulted in increased proliferation and survival when the cells were stimulated with SCF/Kit ligand (Ingram et al., 2000). A study by Yang and coworkers(2003) demonstrated that *NF1−/−* Schwann cells stimulate the migration of *NF1+/−* mast cells by the secretion of SCF/Kit ligand (Yang et al., 2003). Moreover, the mechanism by which mast cells respond to Kit ligand involves the PI3K/Rac2 pathway (Ingram et al., 2001; Yang et al., 2003).

4.4 Secondary Genetic Events

Secondary genetic events have been postulated to occur in NF1-associated neurofibromas and are clearly involved in the progression of plexiform neurofibroma to MPNST. Mutations or deletions of genes controlling cell-cycle progression and apoptosis are common in MPNST. There are reports of mutations in the *p53* tumor suppressor gene and deletions in the *p53* chromosomal region (chromosome 17p) in MPNST from individuals with NF1. One study found that five out of six MPNST samples exhibited a deletion on the short arm of chromosome 17, whereas no loss or deletions of 17p were detected in the 30 NF1-associated benign tumors analyzed. Two of the MPNST with deletions in 17p were found to have a point mutation in the remaining *p53* allele suggesting that *p53* function was completely lost in these tumors (Menon et al., 1990). Another study reported a similar finding that four malignant NF1-associated tumors had LOH for markers in the region of *TP53* along with a mutation in the remaining *p53* allele in three of the four tumors analyzed (Legius et al., 1994). Other studies have shown nuclear accumulation of p53 in MPNST tissue that is thought to occur when p53 is nonfunctional due to a gene mutation (Kindblom et al., 1995; Kourea et al., 1999a; Birindelli et al., 2001).

Another chromosomal region that is frequently lost in MPNST is 9p21. The three tumor suppressor genes that map to the 9p21 chromosomal band ($p16^{INK4a}$, $p14^{ARF}$, and $p15^{INK4b}$) are involved in the regulation of the pRb and p53 pathways (Serrano et al., 1993; Hannon and Beach, 1994; Kamijo et al., 1998; Pomerantz et al., 1998; Zhang et al., 1998a). Several reports have provided evidence that loss of 9p21 may be important for the malignant transformation of plexiform neurofibroma to MPNST (Berner et al., 1999; Kourea et al., 1999b; Nielsen et al., 1999; Perrone et al., 2003). One study found that half of the NF1-associated MPNSTs examined had homozygous deletion of the $p16^{INK4a}$ gene which was not found to be deleted in benign neurofibroma tissue. No abnormalities in promoter methylation or coding region mutations were found in tumors without gene deletion (Nielsen et al., 1999). Two other studies reported similar findings that deletions on chromosome band 9p21 occur with a high frequency in both NF1-associated and sporadic MPNST without detection of coding region mutation or promoter methylation (Berner et al., 1999; Kourea et al., 1999b). Another study looked for alterations in the $p16^{INK4a}$, $p14^{ARF}$, and $p15^{INK4b}$ genes including homozygous deletion, mutation, and promoter methylation using MPNST samples from 14 individuals with NF1 and 12 sporadic cases. The authors found that 43% of the NF1 samples contained homozygous deletion of $p14^{ARF}/p16^{INK4a}$ and 43% contained homozygous deletion of $p15^{INK4b}$, while 71% of the samples with deletions had lost the entire 9p21 region. The results were very similar for sporadic MPNST. Only one case of promoter methylation was found in NF1 MPNST tissue and no gene mutations where found in the cases that did not have deletions (Perrone et al., 2003). Further evidence for the involvement of the pRb pathway in transformation comes from a study that demonstrated LOH for the *Rb* (retinoblastoma) gene in NF1-associated MPNST samples (Mawrin et al., 2002). Thus, there is good evidence that loss of the p53 and/or pRb pathways is involved in the formation or progression of MPNST in NF1.

4.4.1 Animal Models of Secondary Genetic Events

Animal models provide further evidence that loss of *p53*, in addition to the loss of *NF1*, contributes to the formation of MPNST. Two groups created mouse models of NF1 tumor development by generating recombinant mice with inactivated *NF1* and *p53* alleles linked on the same chromosome (Cichowski et al., 1999; Vogel et al., 1999). The *NF1*+/−:*p53*+/− mutant mice had a high incidence of soft tissue sarcomas, the majority of which exhibited histologic features of MPNST and expressed the Schwann cell marker S100. In addition, the tumors exhibited LOH at both the *NF1* and *p53* loci. MPNST did not develop when only *NF1* or *p53* was inactivated or when *NF1* and *p53* mutations occurred together on opposite chromosomes suggesting that the complete inactivation of both genes cooperate in the development of MPNST (Cichowski et al., 1999; Vogel et al., 1999).

5 Therapeutic Approaches for the Treatment of NF1 Tumors

Benign peripheral nerve sheath tumors are disfiguring and can be painful and cause loss of nerve or organ function. There is also the risk that benign plexiform neurofibromas will progress to malignancy and become life threatening. Currently, the only effective treatment for these tumors is surgical resection;

however, the tumors can grow back after resection or may be inoperable due to the location or nerve involvement of the tumor. Therefore, a therapy to inhibit the growth of peripheral nerve sheath tumors is desperately needed for individuals with NF1. Current approaches that are being evaluated include inhibition of Ras activation, angiogenesis, and fibroblast function.

Since the hyperactivation of Ras is associated with the loss of *NF1*, there is interest in the use of farnesyltransferase inhibitors (FTIs) for NF1. FTIs block the activation of proteins, including Ras, which require posttranslational farnesylation for activity. A phase I trial of the FTI tipifarnib was conducted in 17 children with NF1 and plexiform neurofibromas (Widemann et al., 2006). The maximum tolerated dose was determined and this dose was found to reduce farnesyltransferase activity to 30% of baseline in peripheral-blood mononuclear cells. Tipifarnib is currently being evaluated in a phase II trial in individuals with NF1 (Widemann et al., 2006).

Another approach to the treatment of NF1 involves the use of an antifibrotic to inhibit the aberrant function of neurofibromin-deficient fibroblasts. A preclinical study demonstrated a reduction in survival of human neurofibroma xenografts in immunodeficient mice when treated with the antifibrotic agent pirfenidone (Babovic-Vuksanovic et al., 2004). Pirfenidone was then studied in a phase II clinical trial in 24 patients with NF1. Of the 17 individuals who completed 2 years of treatment, the tumor volume remained stable in 11, increased in 3 ($\geq$15% volume increase), and decreased in 3 (15%–29% volume decrease) (Babovic-Vuksanovic et al., 2006).

A third approach to control NF1 tumors involves inhibiting aberrant angiogenesis. A phase I clinical trial of the antiangiogenic drug thalidomide was conducted in 20 patients with NF1. The study assessed the safety and efficacy of thalidomide for the treatment of plexiform neurofibroma. Of the 12 individuals who completed the 1-year study, four had minor reductions in tumor size and seven had improvement of symptoms including pain and paresthesias. Thalidomide was determined to be well tolerated with the most frequent adverse side effect being transient drowsiness (Gupta et al., 2003).

The results of these trials suggest that inhibiting Ras activation, angiogenesis, and aberrant fibroblast function should be further evaluated for the treatment of NF1 tumors, perhaps in combination with other therapeutic approaches. In support of this view, Hirokawa and coworkers (2005) reported that the drug FK229 which suppresses Ras transformation, completely inhibits the growth of an MPNST cell line in vitro and in vivo. The pivotal role of Ras is also supported by a recent study which showed that Ras activation inhibits the transcription factor "forkhead box 0," which is responsible for suppressing the abnormal growth of tumorigenic cells (Lavery et al., 2007). Other targets for therapeutic intervention include mast cells and the hypersensitivity of neurofibromin-deficient cells to growth factors. Given the exciting new investigative tools and the current intense research efforts, it is anticipated that new and more effective treatments for NF1 will be forthcoming.

Acknowledgments

The NF1 research in the author's laboratory was supported by the U.S. Army Medical Research and Material Command (DAMD17-98-8607) and by grants from the Massachusetts. Bay Area Neurofibromatosis, Inc., Illinois Neurofibromatosis Inc., and Minnesota Neurofibromatosis Inc.

References

Andersen LB, Ballester R, Marchuk DA, Chang E, Gutmann DH, et al. 1993. A conserved alternative splice in the von Recklinghausen neurofibromatosis (NF1) gene produces two neurofibromin isoforms, both of which have GTPase-activating protein activity. Mol Cell Biol 13: 487-495.

Angelov L, Salhia B, Roncari L, McMahon G, Guha A. 1999. Inhibition of angiogenesis by blocking activation of the vascular endothelial growth factor receptor 2 leads to decreased growth of neurogenic sarcomas. Cancer Res 59: 5536-5541.

Arbiser JL, Flynn E, Barnhill RL. 1998. Analysis of vascularity of human neurofibromas. J Am Acad Dermatol 38: 950-954.

Babovic-Vuksanovic D, Ballman K, Michels V, McGrann P, Lindor N, et al. 2006. Phase II trial of pirfenidone in adults with neurofibromatosis type 1. Neurology 67: 1860-1862.

Babovic-Vuksanovic D, Petrovic L, Knudsen BE, Plummer TB, Parisi JE, et al. 2004. Survival of human neurofibroma in immunodeficient mice and initial results of therapy with pirfenidone. J Biomed Biotechnol 2004: 79-85.

Badache A, De Vries GH. 1998. Neurofibrosarcoma-derived Schwann cells overexpress platelet-derived growth factor (PDGF) receptors and are induced to proliferate by PDGF BB. J Cell Physiol 177: 334-342.

Badache A, Muja N, De Vries GH. 1998. Expression of Kit in neurofibromin-deficient human Schwann cells: Role in Schwann cell hyperplasia associated with type 1 neurofibromatosis. Oncogene 17: 795-800.

Baizer L, Ciment G, Hendrickson SK, Schafer GL. 1993. Regulated expression of the neurofibromin type I transcript in the developing chicken brain. J Neurochem 61: 2054-2060.

Ballester R, Marchuk D, Boguski M, Saulino A, Letcher R, et al. 1990. The NF1 locus encodes a protein functionally related to mammalian GAP and yeast IRA proteins. Cell 63: 851-859.

Bar-Sagi D, Feramisco JR. 1985. Microinjection of the ras oncogene protein into PC12 cells induces morphological differentiation. Cell 42: 841-848.

Basu TN, Gutmann DH, Fletcher JA, Glover TW, Collins FS, et al. 1992. Aberrant regulation of ras proteins in malignant tumour cells from type 1 neurofibromatosis patients. Nature 356: 713-715.

Bennett MR, Rizvi TA, Karyala S, McKinnon RD, Ratner N. 2003. Aberrant growth and differentiation of oligodendrocyte progenitors in neurofibromatosis type 1 mutants. J Neurosci 23: 7207-7217.

Berner JM, Sorlie T, Mertens F, Henriksen J, Saeter G, et al. 1999. Chromosome band 9p21 is frequently altered in malignant peripheral nerve sheath tumors: Studies of CDKN2A and other genes of the pRB pathway. Genes Chromosomes Cancer 26: 151-160.

Birindelli S, Perrone F, Oggionni M, Lavarino C, Pasini B, et al. 2001. Rb and TP53 pathway alterations in sporadic and NF1-related malignant peripheral nerve sheath tumors. Lab Invest 81: 833-844.

Bollag G, Clapp DW, Shih S, Adler F, Zhang YY, et al. 1996. Loss of NF1 results in activation of the Ras signaling pathway and leads to aberrant growth in haematopoietic cells. Nat Genet 12: 144-148.

Bollag G, McCormick F, Clark R. 1993. Characterization of full-length neurofibromin: Tubulin inhibits Ras GAP activity. EMBO J 12: 1923-1927.

Brannan CI, Perkins AS, Vogel KS, Ratner N, Nordlund ML, et al. 1994. Targeted disruption of the neurofibromatosis type-1 gene leads to developmental abnormalities in heart and various neural crest-derived tissues. Genes Dev 8: 1019-1029.

Carey JC, Viskochil DH. 1999. Neurofibromatosis type 1: A model condition for the study of the molecular basis of variable expressivity in human disorders. Am J Med Genet 89: 7-13.

Carr NJ, Warren AY. 1993. Mast cell numbers in melanocytic naevi and cutaneous neurofibromas. J Clin Pathol 46: 86-87.

Cawthon RM, Weiss R, Xu GF, Viskochil D, Culver M, et al. 1990. A major segment of the neurofibromatosis type 1 gene: cDNA sequence, genomic structure, and point mutations. Cell 62: 193-201.

Cichowski K, Shih TS, Schmitt E, Santiago S, Reilly K, et al. 1999. Mouse models of tumor development in neurofibromatosis type 1. Science 286: 2172-2176.

Cnossen MH, van der Est MN, Breuning MH, Asperen van CJ, Breslau-Siderius EJ, et al. 1997. Deletions spanning the neurofibromatosis type 1 gene: Implications for genotype-phenotype correlations in neurofibromatosis type 1? Hum Mutat 9: 458-464.

Colman SD, Williams CA, Wallace MR. 1995. Benign neurofibromas in type 1 neurofibromatosis (NF1) show somatic deletions of the NF1 gene. Nat Genet 11: 90-92.

Crawford SE, Stellmach V, Ranalli M, Huang X, Huang L, et al. 2001. Pigment epithelium-derived factor (PEDF) in neuroblastoma: A multifunctional mediator of Schwann cell antitumor activity. J Cell Sci 114: 4421-4428.

Cunha KS, Barboza EP, Da Fonseca EC. 2003. Identification of growth hormone receptor in localised neurofibromas of patients with neurofibromatosis type 1. J Clin Pathol 56: 758-763.

Dang I, De Vries GH. 2005. Schwann cell lines derived from malignant peripheral nerve sheath tumors respond abnormally to platelet-derived growth factor-BB. J Neurosci Res 79: 318-328.

Dang I, Nelson JK, Devries GH. 2005. c-Kit receptor expression in normal human Schwann cells and Schwann cell lines derived from neurofibromatosis type 1 tumors. J Neurosci Res 82: 465-471.

Danglot G, Regnier V, Fauvet D, Vassal G, Kujas M, et al. 1995. Neurofibromatosis 1 (NF1) mRNAs expressed in the central nervous system are differentially spliced in the 5′ part of the gene. Hum Mol Genet 4: 915-920.

Dasgupta B, Dugan LL, Gutmann DH. 2003. The neurofibromatosis 1 gene product neurofibromin regulates pituitary adenylate cyclase-activating polypeptide-mediated signaling in astrocytes. J Neurosci 23: 8949-8954.

Daston MM, Ratner N. 1992. Neurofibromin, a predominantly neuronal GTPase activating protein in the adult, is ubiquitously expressed during development. Dev Dyn 195: 216-226.

Daston MM, Scrable H, Nordlund M, Sturbaum AK, Nissen LM, et al. 1992. The protein product of the neurofibromatosis type 1 gene is expressed at highest abundance in neurons, Schwann cells, and oligodendrocytes. Neuron 8: 415-428.

Deadwyler GD, Dang I, Nelson J, Srikanth M, De Vries GH. 2004. Prostaglandin E2 metabolism is activated in Schwann cell lines derived from human NF1 malignant peripheral nerve sheath tumors. Neuron Glia Biology 1: 149-155.

DeClue JE, Heffelfinger S, Benvenuto G, Ling B, Li S, et al. 2000. Epidermal growth factor receptor expression in neurofibromatosis type 1-related tumors and NF1 animal models. J Clin Invest 105: 1233-1241.

DeClue JE, Papageorge AG, Fletcher JA, Diehl SR, Ratner N, et al. 1992. Abnormal regulation of mammalian p21ras contributes to malignant tumor growth in von Recklinghausen (type 1) neurofibromatosis. Cell 69: 265-273.

Dugoff L, Sujansky E. 1996. Neurofibromatosis type 1 and pregnancy. Am J Med Genet 66: 7-10.

Emanuel PD, Shannon KM, Castleberry RP. 1996. Juvenile myelomonocytic leukemia: Molecular understanding and prospects for therapy. Mol Med Today 2: 468-475.

Fairbank J. 1994. Orthopaedic manifestations of neurofibromatosis. Huson SM, Hughes RAC, editors. The Neurofibromatoses: A Pathogenetic and Clinical Overview. Chapman & Hall; London: pp. 275-304.

Feldkamp MM, Angelov L, Guha A. 1999. Neurofibromatosis type 1 peripheral nerve tumors: Aberrant activation of the Ras pathway. Surg Neurol 51: 211-218.

Ferner RE. 1994. Medical complications of neurofibromatosis 1. Huson SM, Hughes RAC, editors. The Neurofibromatoses: A Pathogenetic and Clinical Overview. Chapman & Hall; London: pp. 316-330.

Ferner RE, Gutmann DH. 2002. International consensus statement on malignant peripheral nerve sheath tumors in neurofibromatosis. Cancer Res 62: 1573-1577.

Friedman JM. 1999a. Clinical genetics. Friedman JM, Gutmann DH, MacCollin M, Riccardi VM, editors. Neurofibromatosis: Phenotype, Natural History, and Pathogenesis. The Johns Hopkins University Press; Baltimore: pp. 110-118.

Friedman JM. 1999b. Epidemiology of neurofibromatosis type 1. Am J Med Genet 89: 1-6.

Friedman JM. 1999c. Vascular and endocrine abnormalities. Friedman JM, Gutmann DH, MacCollin M, Riccardi VM, editors. Neurofibromatosis: Phenotype, Natural History, and Pathogenesis. The Johns Hopkins University Press; Baltimore: pp. 274-296.

Geist RT, Gutmann DH. 1996. Expression of a developmentally-regulated neuron-specific isoform of the neurofibromatosis 1 (NF1) gene. Neurosci Lett 211: 85-88.

Glover TW, Stein CK, Legius E, Andersen LB, Brereton A, et al. 1991. Molecular and cytogenetic analysis of tumors in von Recklinghausen neurofibromatosis. Genes Chromosomes Cancer 3: 62-70.

Golubic M, Roudebush M, Dobrowolski S, Wolfman A, Stacey DW. 1992. Catalytic properties, tissue and intracellular distribution of neurofibromin. Oncogene 7: 2151-2159.

Gregory PE, Gutmann DH, Mitchell A, Park S, Boguski M, et al. 1993. Neurofibromatosis type 1 gene product (neurofibromin) associates with microtubules. Somat Cell Mol Genet 19: 265-274.

Guha A, Lau N, Huvar I, Gutmann D, Provias J, et al. 1996. Ras-GTP levels are elevated in human NF1 peripheral nerve tumors. Oncogene 12: 507-513.

Guo HF, The I, Hannan F, Bernards A, Zhong Y. 1997. Requirement of Drosophila NF1 for activation of adenylyl cyclase by PACAP38-like neuropeptides. Science 276: 795-798.

Gupta A, Cohen BH, Ruggieri P, Packer RJ, Phillips PC. 2003. Phase I study of thalidomide for the treatment of plexiform neurofibroma in neurofibromatosis 1. Neurology 60: 130-132.

Gutman DH, Andersen LB, Cole JL, Swaroop M, Collins FS. 1993. An alternatively-spliced mRNA in the carboxy terminus of the neurofibromatosis type 1 (NF1) gene is expressed in muscle. Hum Mol Genet 2: 989-992.

Gutmann DH. 1999. Abnormalities of the nervous system. Friedman JM, Gutmann DH, MacCollin M, Riccardi VM, editors. Neurofibromatosis: Phenotype, Natural History, and Pathogenesis. The Johns Hopkins University Press; Baltimore: pp. 190-202.

Gutmann DH, Cole JL, Collins FS. 1995a. Expression of the neurofibromatosis type 1 (NF1) gene during mouse embryonic development. Prog Brain Res 105: 327-335.

Gutmann DH, Geist RT, Rose K, Wright DE. 1995b. Expression of two new protein isoforms of the neurofibromatosis type 1 gene product, neurofibromin, in muscle tissues. Dev Dyn 202: 302-311.

Gutmann DH, Geist RT, Wright DE, Snider WD. 1995c. Expression of the neurofibromatosis 1 (NF1) isoforms in developing and adult rat tissues. Cell Growth Differ 6: 315-323.

Gutmann DH, Gurney JG. 1999. Other malignancies. Neurofibromatosis: Phenotype, Natural History, and Pathogenesis. Friedman JM, Gutmann DH, MacCollin M, Riccardi VM, editors. Baltimore: The Johns Hopkins University Press; pp. 231-249.

Gutmann DH, Loehr A, Zhang Y, Kim J, Henkemeyer M, et al. 1999. Haploinsufficiency for the neurofibromatosis 1 (NF1) tumor suppressor results in increased astrocyte proliferation. Oncogene 18: 4450-4459.

Hamilton SJ, Friedman JM. 2000. Insights into the pathogenesis of neurofibromatosis 1 vasculopathy. Clin Genet 58: 341-344.

Hannon GJ, Beach D. 1994. p15INK4B is a potential effector of TGF-β-induced cell cycle arrest. Nature 371: 257-261.

Hansson HA, Lauritzen C, Lossing C, Petruson K. 1988. Somatomedin C as tentative pathogenic factor in neurofibromatosis. Scand J Plast Reconstr Surg Hand Surg 22: 7-13.

Hiatt KK, Ingram DA, Zhang Y, Bollag G, Clapp DW. 2001. Neurofibromin GTPase-activating protein-related domains restore normal growth in Nf1−/− cells. J Biol Chem 276: 7240-7245.

Hirokawa Y, Nakajima H, Hanemann CO, Kurtz A, Fram S, et al. 2005 Signal therapy of NF1-deficient tumor xenograph in mice by the anti-PAK drug FK228. Cancer Bio Ther 4: 379-381.

Hirota S, Nomura S, Asada H, Ito A, Morii E, et al. 1993. Possible involvement of c-kit receptor and its ligand in increase of mast cells in neurofibroma tissues. Arch Pathol Lab Med 117: 996-999.

Holtkamp N, Mautner VF, Friedrich RE, Harder A, Hartmann C, et al. 2004. Differentially expressed genes in neurofibromatosis 1-associated neurofibromas and malignant peripheral nerve sheath tumors. Acta Neuropathol (Berl) 107: 159-168.

Hughes RAC. 1994. Neurological complications of neurofibromatosis 1. The Neurofibromatoses: A athogenetic and Clinical Overview. Huson SM, Hughes RAC, editors. London: Chapman & Hall; pp. 204-232.

Huson SM, Harper PS, Compston DA. 1988. Von Recklinghausen neurofibromatosis. A clinical and population study in south-east Wales. Brain 111 (Pt 6): 1355-1381.

Huynh DP, Nechiporuk T, Pulst SM. 1994. Differential expression and tissue distribution of type I and type II neurofibromins during mouse fetal development. Dev Biol 161: 538-551.

Ingram DA, Hiatt K, King AJ, Fisher L, Shivakumar R, et al. 2001. Hyperactivation of p21(ras) and the hematopoietic-specific Rho GTPase, Rac2, cooperate to alter the proliferation of neurofibromin-deficient mast cells in vivo and in vitro. J Exp Med 194: 57-69.

Ingram DA, Yang FC, Travers JB, Wenning MJ, Hiatt K, et al. 2000. Genetic and biochemical evidence that haploinsufficiency of the Nf1 tumor suppressor gene modulates melanocyte and mast cell fates in vivo. J Exp Med 191: 181-188.

Jacks T, Shih TS, Schmitt EM, Bronson RT, Bernards A, et al. 1994. Tumour predisposition in mice heterozygous for a targeted mutation in Nf1. Nat Genet 7: 353-361.

Johnson MD, Kamso-Pratt J, Federspiel CF, Whetsell WO, Jr. 1989. Mast cell and lymphoreticular infiltrates in neurofibromas. Comparison with nerve sheath tumors. Arch Pathol Lab Med 113: 1263-1270.

Johnson MR, Look AT, DeClue J E, Valentine MB, Lowy DR. 1993. Inactivation of the NF1 gene in human melanoma and neuroblastoma cell lines without impaired regulation of GTP.Ras. Proc Natl Acad Sci USA 90: 5539-5543.

Kadono T, Kikuchi K, Nakagawa H, Tamaki K. 2000. Expressions of various growth factors and their receptors in tissues from neurofibroma. Dermatology 201: 10-14.

Kadono T, Soma Y, Takehara K, Nakagawa H, Ishibashi Y, et al. 1994. The growth regulation of neurofibroma cells in neurofibromatosis type-1: Increased responses to PDGF-BB and TGF-β1. Biochem Biophys Res Commun 198: 827-834.

Kamijo T, Weber JD, Zambetti G, Zindy F, Roussel MF, et al. 1998. Functional and physical interactions of the ARF tumor suppressor with p53 and Mdm2. Proc Natl Acad Sci USA 95: 8292-8297.

Kawachi Y, Xu X, Ichikawa E, Imakado S, Otsuka F. 2003. Expression of angiogenic factors in neurofibromas. Exp Dermatol 12: 412-417.

Kayes LM, Burke W, Riccardi VM, Bennett R, Ehrlich P, et al. 1994. Deletions spanning the neurofibromatosis 1 gene: Identification and phenotype of five patients. Am J Hum Genet 54: 424-436.

Kayes LM, Riccardi VM, Burke W, Bennett RL, Stephens K. 1992. Large de novo DNA deletion in a patient with sporadic neurofibromatosis 1, mental retardation, and dysmorphism. J Med Genet 29: 686-690.

Kim HA, Ling B, Ratner N. 1997. Nf1-deficient mouse Schwann cells are angiogenic and invasive and can be induced to hyperproliferate: Reversion of some phenotypes by an inhibitor of farnesyl protein transferase. Mol Cell Biol 17: 862-872.

Kim HA, Ratner N, Roberts TM, Stiles CD. 2001. Schwann cell proliferative responses to cAMP and Nf1 are mediated by cyclin D1. J Neurosci 21: 1110-1116.

Kim HA, Rosenbaum T, Marchionni MA, Ratner N, DeClue JE. 1995. Schwann cells from neurofibromin deficient mice exhibit activation of p21ras, inhibition of cell proliferation and morphological changes. Oncogene 11: 325-335.

Kindblom LG, Ahlden M, Meis-Kindblom JM, Stenman G. 1995. Immunohistochemical and molecular analysis of p53, MDM2, proliferating cell nuclear antigen and Ki67 in benign and malignant peripheral nerve sheath tumours. Virchows Arch 427: 19-26.

Klose A, Ahmadian MR, Schuelke M, Scheffzek K, Hoffmeyer S, et al. 1998. Selective disactivation of neurofibromin GAP activity in neurofibromatosis type 1. Hum Mol Genet 7: 1261-1268.

Koivunen J, Yla-Outinen H, Korkiamaki T, Karvonen SL, Poyhonen M, et al. 2000. New function for NF1 tumor suppressor. J Invest Dermatol 114: 473-479.

Korf BR. 1999a. Neurofibromas and malignant tumors of the peripheral nerve sheath. Neurofibromatosis: Phenotype, Natural History, and Pathogenesis. Friedman JM,

Gutmann DH, MacCollin M, Riccardi VM, editors. Baltimore: The Johns Hopkins University Press; pp. 142-161.

Korf BR. 1999b. Plexiform neurofibromas. Am J Med Genet 89: 31-37.

Korf BR. 2000. Malignancy in neurofibromatosis type 1. Oncologist 5: 477-485.

Kourea HP, Cordon-Cardo C, Dudas M, Leung D, Woodruff JM. 1999a. Expression of p27(kip) and other cell cycle regulators in malignant peripheral nerve sheath tumors and neurofibromas: The emerging role of p27(kip) in malignant transformation of neurofibromas. Am J Pathol 155: 1885-1891.

Kourea HP, Orlow I, Scheithauer BW, Cordon-Cardo C, Woodruff JM. 1999b. Deletions of the INK4A gene occur in malignant peripheral nerve sheath tumors but not in neurofibromas. Am J Pathol 155: 1855-1860.

Krasnoselsky A, Massay MJ, De Frances MC, Michalopoulos G, Zarnegar R, et al. 1994. Hepatocyte growth factor is a mitogen for Schwann cells and is present in neurofibromas. J Neurosci 14: 7284-7290.

Lakkis MM, Epstein JA. 1998. Neurofibromin modulation of ras activity is required for normal endocardial-mesenchymal transformation in the developing heart. Development 125: 4359-4367.

Largaespada DA, Brannan CI, Jenkins NA, Copeland NG. 1996. Nf1 deficiency causes Ras-mediated granulocyte/macrophage colony stimulating factor hypersensitivity and chronic myeloid leukaemia. Nat Genet 12: 137-143.

Lau N, Feldkamp MM, Roncari L, Loehr AH, Shannon P, et al. 2000. Loss of neurofibromin is associated with activation of RAS/MAPK and PI3-K/AKT signaling in a neurofibromatosis 1 astrocytoma. J Neuropathol Exp Neurol 59: 759-767.

Lavery W, Hall V, Yager J, Rottgers A, Wells M, et al. 2007 Phosphatidylinositol-3-kinase and Akt nonautonomously promote perineurial glial growth in Drosophila peripheral nerves. J Neurosci 27: 279-288.

Lee JK, Choi B, Sobel RA, Chiocca EA, Martuza RL. 1990. Inhibition of growth and angiogenesis of human neurofibrosarcoma by heparin and hydrocortisone. J Neurosurg 73: 429-435.

Lee PR, Cohen JE, Tendi EA, Farrer R, De Vries GH, et al. 2004. Transcriptional profiling in an MPNST-derived cell line and normal human Schwann cells. Neuron Glia Biology 1: 135-147.

Legius E, Dierick H, Wu R, Hall BK, Marynen P, et al. 1994. TP53 mutations are frequent in malignant NF1 tumors. Genes Chromosomes Cancer 10: 250-255.

Legius E, Marchuk DA, Collins FS, Glover TW. 1993. Somatic deletion of the neurofibromatosis type 1 gene in a neurofibrosarcoma supports a tumour suppressor gene hypothesis. Nat Genet 3: 122-126.

Leppig KA, Kaplan P, Viskochil D, Weaver M, Ortenberg J, et al. 1997. Familial neurofibromatosis 1 microdeletions: Cosegregation with distinct facial phenotype and early onset of cutaneous neurofibromata. Am J Med Genet 73: 197-204.

Lertsburapa T, De Vries GH. 2004. In vitro studies of pigment epithelium-derived factor in human Schwann cells after treatment with axolemma-enriched fraction. J Neurosci Res 75: 624-631.

Li C, Cheng Y, Gutmann DA, Mangoura D. 2001. Differential localization of the neurofibromatosis 1 (NF1) gene product, neurofibromin, with the F-actin or microtubule cytoskeleton during differentiation of telencephalic neurons. Brain Res Dev Brain Res 130: 231-248.

Li H, Velasco-Miguel S, Vass WC, Parada LF, DeClue JE. 2002. Epidermal growth factor receptor signaling pathways are associated with tumorigenesis in the Nf1:p53 mouse tumor model. Cancer Res 62: 4507-4513.

Li Y, Bollag G, Clark R, Stevens J, Conroy L, et al. 1992. Somatic mutations in the neurofibromatosis 1 gene in human tumors. Cell 69: 275-281.

Li Y, O'Connell P, Breidenbach HH, Cawthon R, Stevens J, et al. 1995. Genomic organization of the neurofibromatosis 1 gene (NF1). Genomics 25: 9-18.

Li Y, White R. 1996. Suppression of a human colon cancer cell line by introduction of an exogenous NF1 gene. Cancer Res 56: 2872-2876.

Lothe RA, Saeter G, Danielsen HE, Stenwig AE, Hoyheim B, et al. 1993. Genetic alterations in a malignant schwannoma from a patient with neurofibromatosis (NF1). Pathol Res Pract 189: 465-471; discussion 471-464.

Maertens O, Brems H, Vandesompele J, De Raedt T, Heyns I, et al. 2006. Comprehensive NF1 screening on cultured Schwann cells from neurofibromas. Hum Mutat 27: 1030-1040.

Malminen M, Peltonen S, Koivunen J, Peltonen J. 2002. Functional expression of NF1 tumor suppressor protein: Association with keratin intermediate filaments during the early development of human epidermis. BMC Dermatol 2: 10.

Mantani A, Wakasugi S, Yokota Y, Abe K, Ushio Y, et al. 1994. A novel isoform of the neurofibromatosis type-1 mRNA and a switch of isoforms during murine cell differentiation and proliferation. Gene 148: 245-251.

Marchuk DA, Saulino AM, Tavakkol R, Swaroop M, Wallace MR, et al. 1991. cDNA cloning of the type 1 neurofibromatosis gene: Complete sequence of the NF1 gene product. Genomics 11: 931-940.

Marshall CJ. 1995. Specificity of receptor tyrosine kinase signaling: Transient versus sustained extracellular signal-regulated kinase activation. Cell 80: 179-185.

Mashour GA, Ratner N, Khan GA, Wang HL, Martuza RL, et al. 2001. The angiogenic factor midkine is aberrantly expressed in NF1-deficient Schwann cells and is a mitogen for neurofibroma-derived cells. Oncogene 20: 97-105.

Mawrin C, Kirches E, Boltze C, Dietzmann K, Roessner A, et al. 2002. Immunohistochemical and molecular analysis of p53, RB, and PTEN in malignant peripheral nerve sheath tumors. Virchows Arch 440: 610-615.

McLaughlin ME, Jacks T. 2003. Progesterone receptor expression in neurofibromas. Cancer Res 63: 752-755.

Menon AG, Anderson KM, Riccardi VM, Chung RY, Whaley JM, et al. 1990. Chromosome 17p deletions and p53 gene mutations associated with the formation of malignant neurofibrosarcomas in von Recklinghausen neurofibromatosis. Proc Natl Acad Sci USA 87: 5435-5439.

Miller SJ, Rangwala F, Williams J, Ackerman P, Kong S, et al. 2006. Large-scale molecular comparison of human Schwann cells to malignant peripheral nerve sheath tumor cell lines and tissues. Cancer Res 66: 2584-2591.

Muir D, Neubauer D, Lim IT, Yachnis AT, Wallace MR. 2001. Tumorigenic properties of neurofibromin-deficient neurofibroma Schwann cells. Am J Pathol 158: 501-513.

National Institutes of Health Consensus Development Conference 1988. Neurofibromatosis. Conference statement. Arch Neurol 45: 575–578.

Nielsen GP, Stemmer-Rachamimov AO, Ino Y, Moller MB, Rosenberg AE, et al. 1999. Malignant transformation of neurofibromas in neurofibromatosis 1 is associated with CDKN2A/p16 inactivation. Am J Pathol 155: 1879-1884.

Nishi T, Lee PS, Oka K, Levin VA, Tanase S, et al. 1991. Differential expression of two types of the neurofibromatosis type 1 (NF1) gene transcripts related to neuronal differentiation. Oncogene 6: 1555-1559.

Noda M, Ko M, Ogura A, Liu DG, Amano T, et al. 1985. Sarcoma viruses carrying ras oncogenes induce differentiation-associated properties in a neuronal cell line. Nature 318: 73-75.

Nur-E-Kamal MS, Varga M, Maruta H. 1993. The GTPase-activating NF1 fragment of 91 amino acids reverses v-Ha-Ras-induced malignant phenotype. J Biol Chem 268: 22331-22337.

Nurnberger M, Moll I. 1994. Semiquantitative aspects of mast cells in normal skin and in neurofibromas of neurofibromatosis types 1 and 5. Dermatology 188: 296-299.

Peltonen J, Jaakkola S, Lebwohl M, Renvall S, Risteli L, et al. 1988. Cellular differentiation and expression of matrix genes in type 1 neurofibromatosis. Lab Invest 59: 760-771.

Perrin GQ, Fishbein L, Thomson SA, Thomas SL, Stephens K, et al. 2007. Plexiform-like neurofibromas develop in the mouse by intraneural xenograft of an NF1 tumor-derived Schwann cell line. J Neurosci Res (March 4, online in advance of print).

Perrone F, Tabano S, Colombo F, Dagrada G, Birindelli S, et al. 2003. p15INK4b, p14ARF, and p16INK4a inactivation in sporadic and neurofibromatosis type 1-related malignant peripheral nerve sheath tumors. Clin Cancer Res 9: 4132-4138.

Pomerantz J, Schreiber-Agus N, Liegeois NJ, Silverman A, Alland L, et al. 1998. The Ink4a tumor suppressor gene product, p19Arf, interacts with MDM2 and neutralizes MDM2's inhibition of p53. Cell 92: 713-723.

Rao UN, Sonmez-Alpan E, Michalopoulos GK. 1997. Hepatocyte growth factor and c-MET in benign and malignant peripheral nerve sheath tumors. Hum Pathol 28: 1066-1070.

Rasmussen SA, Yang Q, Friedman JM. 2001. Mortality in neurofibromatosis 1: An analysis using U.S. death certificates. Am J Hum Genet 68: 1110-1118.

Ratner N, Lieberman MA, Riccardi VM, Hong DM. 1990. Mitogen accumulation in von Recklinghausen neurofibromatosis. Ann Neurol 27: 298-303.

Riccardi VM. 1981. Cutaneous manifestation of neurofibromatosis: Cellular interaction, pigmentation, and mast cells. Birth Defects Orig Artic Ser 17: 129-145.

Riccardi VM. 1987. Mast-cell stabilization to decrease neurofibroma growth. Preliminary experience with ketotifen. Arch Dermatol 123: 1011-1016.

Riccardi VM. 1999. Skeletal system. Neurofibromatosis: Phenotype, Natural History, and Pathogenesis. Friedman JM, Gutmann DH, MacCollin M, Riccardi VM, editors. Baltimore: The Johns Hopkins University Press; pp. 250-273.

Riccardi VM, Friedman JM. 1999. Clinical and epidemiological features. Neurofibromatosis: Phenotype, Natural History, and Pathogenesis. Friedman JM, Gutmann DH, MacCollin M, Riccardi VM, editors. Baltimore: The Johns Hopkins University Press; pp. 29-86.

Ridley AJ, Paterson HF, Noble M, Land H. 1988. Ras-mediated cell cycle arrest is altered by nuclear oncogenes to induce Schwann cell transformation. EMBO J 7: 1635-1645.

Riva P, Corrado L, Natacci F, Castorina P, Wu BL, et al. 2000. NF1 microdeletion syndrome: Refined FISH characterization of sporadic and familial deletions with locus-specific probes. Am J Hum Genet 66: 100-109.

Rosenbaum T, Rosenbaum C, Winner U, Muller HW, Lenard HG, et al. 2000. Long-term culture and characterization of human neurofibroma-derived Schwann cells. J Neurosci Res 61: 524-532.

Ryan JJ, Klein KA, Neuberger TJ, Leftwich JA, Westin EH, et al. 1994. Role for the stem cell factor/KIT complex in Schwann cell neoplasia and mast cell proliferation associated with neurofibromatosis. J Neurosci Res 37: 415-432.

Sawada S, Florell S, Purandare SM, Ota M, Stephens K, et al. 1996. Identification of NF1 mutations in both alleles of a dermal neurofibroma. Nat Genet 14: 110-112.

Serra E, Puig S, Otero D, Gaona A, Kruyer H, et al. 1997. Confirmation of a double-hit model for the NF1 gene in benign neurofibromas. Am J Hum Genet 61: 512-519.

Serra E, Rosenbaum T, Winner U, Aledo R, Ars E, et al. 2000. Schwann cells harbor the somatic NF1 mutation in neurofibromas: Evidence of two different Schwann cell subpopulations. Hum Mol Genet 9: 3055-3064.

Serrano M, Hannon GJ, Beach D. 1993. A new regulatory motif in cell-cycle control causing specific inhibition of cyclin D/CDK4. Nature 366: 704-707.

Serrano M, Lin AW, McCurrach ME, Beach D, Lowe SW. 1997. Oncogenic ras provokes premature cell senescence associated with accumulation of p53 and p16INK4a. Cell 88: 593-602.

Sewing A, Wiseman B, Lloyd AC, Land H. 1997. High-intensity Raf signal causes cell cycle arrest mediated by p21Cip1. Mol Cell Biol 17: 5588-5597.

Shannon KM, O'Connell P, Martin GA, Paderanga D, Olson K, et al. 1994. Loss of the normal NF1 allele from the bone marrow of children with type 1 neurofibromatosis and malignant myeloid disorders. N Engl J Med 330: 597-601.

Sheela S, Riccardi VM, Ratner N. 1990. Angiogenic and invasive properties of neurofibroma Schwann cells. J Cell Biol 111: 645-653.

Shen MH, Harper PS, Upadhyaya M. 1996. Molecular genetics of neurofibromatosis type 1 (NF1). J Med Genet 33: 2-17.

Sherman LS, Atit R, Rosenbaum T, Cox AD, Ratner N. 2000. Single cell Ras-GTP analysis reveals altered Ras activity in a subpopulation of neurofibroma Schwann cells but not fibroblasts. J Biol Chem 275: 30740-30745.

Shimizu E, Shinohara T, Mori N, Yokota J, Tani K, et al. 1993. Loss of heterozygosity on chromosome arm 17p in small cell lung carcinomas, but not in neurofibromas, in a patient with von Recklinghausen neurofibromatosis. Cancer 71: 725-728.

Side L, Taylor B, Cayouette M, Conner E, Thompson P, et al. 1997. Homozygous inactivation of the NF1 gene in bone marrow cells from children with neurofibromatosis type 1 and malignant myeloid disorders. N Engl J Med 336: 1713-1720.

Skuse GR, Kosciolek BA, Rowley PT. 1989. Molecular genetic analysis of tumors in von Recklinghausen neurofibromatosis: Loss of heterozygosity for chromosome 17. Genes Chromosomes Cancer 1: 36-41.

Stefansson K, Wollmann R, Jerkovic M. 1982. S-100 protein in soft-tissue tumors derived from Schwann cells and melanocytes. Am J Pathol 106: 261-268.

Teinturier C, Danglot G, Slim R, Pruliere D, Launay JM, et al. 1992. The neurofibromatosis 1 gene transcripts expressed in peripheral nerve and neurofibromas bear the additional exon located in the GAP domain. Biochem Biophys Res Commun 188: 851-857.

The I, Hannigan GE, Cowley GS, Reginald S, Zhong Y, et al. 1997. Rescue of a Drosophila NF1 mutant phenotype by protein kinase A. Science 276: 791-794.

The I, Murthy AE, Hannigan GE, Jacoby LB, Menon AG, et al. 1993. Neurofibromatosis type 1 gene mutations in neuroblastoma. Nat Genet 3: 62-66.

Thomas SL, De Vries GH. 2007. Angiogenic expression profile of normal and neurofibromin-deficient human Schwann cells. Neurochem Res (in press).

Thomson SA, Fishbein L, Wallace MR. 2002. NF1 mutations and molecular testing. J Child Neurol 17: 555-561.

Tong J, Hannan F, Zhu Y, Bernards A, Zhong Y. 2002. Neurofibromin regulates G protein-stimulated adenylyl cyclase activity. Nat Neurosci 5: 95-96.

Tonsgard JH, Yelavarthi KK, Cushner S, Short MP, Lindgren V. 1997. Do NF1 gene deletions result in a characteristic phenotype? Am J Med Genet 73: 80-86.

Upadhyaya M, Huson SM, Davies M, Thomas N, Chuzhanova N, et al. 2007. An absence of cutaneous neurofibromas associated with a 3-bp inframe deletion in exon 17 of the NF1 gene (c.2970-2972 delAAT): Evidence of a clinically significant NF1 genotype-phenotype correlation. Am J Hum Genet 80: 140-151.

Upadhyaya M, Osborn MJ, Maynard J, Kim MR, Tamanoi F, et al. 1997. Mutational and functional analysis of the neurofibromatosis type 1 (NF1) gene. Hum Genet 99: 88-92.

Upadhyaya M, Ruggieri M, Maynard J, Osborn M, Hartog C, et al. 1998. Gross deletions of the neurofibromatosis type 1 (NF1) gene are predominantly of maternal origin and commonly associated with a learning disability, dysmorphic features and developmental delay. Hum Genet 102: 591-597.

Upadhyaya M, Shaw DJ, Harper PS. 1994. Molecular basis of neurofibromatosis type 1 (NF1): Mutation analysis and polymorphisms in the NF1 gene. Hum Mutat 4: 83-101.

Viskochil D, Buchberg AM, Xu G, Cawthon RM, Stevens J, et al. 1990. Deletions and a translocation interrupt a cloned gene at the neurofibromatosis type 1 locus. Cell 62: 187-192.

Vogel KS, Klesse LJ, Velasco-Miguel S, Meyers K, Rushing EJ, et al. 1999. Mouse tumor model for neurofibromatosis type 1. Science 286: 2176-2179.

Waggoner DJ, Towbin J, Gottesman G, Gutmann DH. 2000. Clinic-based study of plexiform neurofibromas in neurofibromatosis 1. Am J Med Genet 92: 132-135.

Wallace MR, Marchuk DA, Andersen LB, Letcher R, Odeh HM, et al. 1990. Type 1 neurofibromatosis gene: Identification of a large transcript disrupted in three NF1 patients. Science 249: 181-186.

Watanabe T, Oda Y, Tamiya S, Masuda K, Tsuneyoshi M. 2001. Malignant peripheral nerve sheath tumour arising within neurofibroma. An immunohistochemical analysis in the comparison between benign and malignant components. J Clin Pathol 54: 631-636.

Widemann BC, Salzer WL, Arceci RJ, Blaney SM, Fox E, et al. 2006. Phase I trial and pharmacokinetic study of the farnesyltransferase inhibitor tipifarnib in children with refractory solid tumors or neurofibromatosis type I and plexiform neurofibromas. J Clin Oncol 24: 507-516.

Wiestler OD, Radner H. 1994. Pathology of neurofibromatosis 1 and 2. The Neurofibromatoses: A Pathogenetic and Clinical Overview. Huson SM, Hughes RAC, editors. London: Chapman & Hall; pp. 135-159.

Woodruff JM. 1999. Pathology of tumors of the peripheral nerve sheath in type 1 neurofibromatosis. Am J Med Genet 89: 23-30.

Wu BL, Austin MA, Schneider GH, Boles RG, Korf BR. 1995. Deletion of the entire NF1 gene detected by the FISH: Four deletion patients associated with severe manifestations. Am J Med Genet 59: 528-535.

Wu BL, Schneider GH, Korf BR. 1997. Deletion of the entire NF1 gene causing distinct manifestations in a family. Am J Med Genet 69: 98-101.

Wu M, Wallace MR, Muir D. 2005. Tumorigenic properties of neurofibromin-deficient Schwann cells in culture and as syngrafts in Nf1 knockout mice. J Neurosci Res 82: 357-367.

Wu M, Wallace MR, Muir D. 2006. Nf1 haploinsufficiency augments angiogenesis. Oncogene 25: 2297-2303.

Xu GF, Lin B, Tanaka K, Dunn D, Wood D, et al. 1990a. The catalytic domain of the neurofibromatosis type 1 gene product stimulates ras GTPase and complements ira mutants of Scerevisiae. Cell 63: 835-841.

Xu GF, O'Connell P, Viskochil D, Cawthon R, Robertson M, et al. 1990b. The neurofibromatosis type 1 gene encodes a protein related to GAP. Cell 62: 599-608.

Xu H, Gutmann DH. 1997. Mutations in the GAP-related domain impair the ability of neurofibromin to associate with microtubules. Brain Res 759: 149-152.

Xu W, Mulligan LM, Ponder MA, Liu L, Smith BA, et al. 1992. Loss of NF1 alleles in phaeochromocytomas from patients with type I neurofibromatosis. Genes Chromosomes Cancer 4: 337-342.

Yang FC, Chen S, Clegg T, Li X, Morgan T, et al. 2006. Nf1+/− mast cells induce neurofibroma like phenotypes through secreted TGF-β signaling. Hum Mol Genet 15: 2421-2437.

Yang FC, Ingram DA, Chen S, Hingtgen CM, Ratner N, et al. 2003. Neurofibromin-deficient Schwann cells secrete a potent migratory stimulus for Nf1+/− mast cells. J Clin Invest 112: 1851-1861.

Zhang Y, Xiong Y, Yarbrough WG. 1998a. ARF promotes MDM2 degradation and stabilizes p53: ARF-INK4a locus deletion impairs both the Rb and p53 tumor suppression pathways. Cell 92: 725-734.

Zhang YY, Vik TA, Ryder JW, Srour EF, Jacks T, et al. 1998b. Nf1 regulates hematopoietic progenitor cell growth and ras signaling in response to multiple cytokines. J Exp Med 187: 1893-1902.

Zhu Y, Ghosh P, Charnay P, Burns DK, Parada LF. 2002. Neurofibromas in NF1: Schwann cell origin and role of tumor environment. Science 296: 920-922.

7 Amyloid and Neurodegeneration: Alzheimer's Disease and Retinal Degeneration

A. Prakasam · C. Venugopal · A. Suram · J. Pacheco-Quinto · Y. Zhou · M. A. Pappolla · K. A. Sharpe · D. K. Lahiri · N. H. Greig · B. Rohrer · K. Sambamurti

1	Introduction	133
2	Alzheimer's Disease	134
2.1	Amyloid β Protein	134
2.2	Risk Factors for AD	134
3	Retina as a Model for Neurodegeneration in AD	135
4	Eye Structure	135
5	Retina	137
5.1	Retinal Maintenance and Turnover	138
5.2	Oxygen Requirement	139
5.2.1	Specialized Proteins in Retinal Oxygen Delivery	139
6	Normal Aging of the Eye	139
7	Retinal Degeneration (Overview)	140
8	Age-Related Macular Degeneration	141
8.1	Dry Macular Degeneration	141
8.2	Wet Macular Degeneration	141
9	Risk Factors for AMD	142
9.1	Age	142
9.2	Sex	142
9.3	Education	142
9.4	Sunlight	142
9.5	Smoking	143
9.6	Estrogen	143
9.7	Antioxidant	143
9.8	Cardiovascular Risk Factors	143
9.9	Amyloid β	144
9.10	Apolipoprotein E	144
9.11	Cholesterol Metabolism	144
9.12	Inflammation	146
10	Biochemistry of Drusen	146

© 2009 Springer Science+Business Media, LLC.

11	***Genetics of Macular Degeneration***	*147*
11.1	Macular Hole	147
11.2	Cone-Rod Dystrophy (CRD)	148
11.3	Sorsby's Macular Dystrophy	148
11.4	Best Disease	148
11.5	Stargardt's Disease	148
11.6	Juvenile Retinoschisis (X-Linked)	149
11.7	Genetic Risk Factors for AMD	149
12	***Animal Models for Retinal Degeneration***	*149*
12.1	Light Damage Model of Photoreceptor Degeneration	150
13	***Mechanisms of Photoreceptor Degeneration***	*152*
14	***Shared Features of the AD and AMD***	*152*

Abstract: Deposits of specific proteins are believed to cause a large group of degenerative diseases. Given that the neurodegenerative disease, prion, is an infectious protein deposit disorder, a common theme of degeneration mediated by toxic aggregates has been developed. However, the mechanisms of cellular toxicity induced by protein deposits remain a mystery and although new diseases are being added to the list and treatments for these diseases remain elusive. Alzheimer's disease (AD) and age-related macular degeneration (AMD) are two prototypic neurodegenerative diseases that are characterized by extracellular protein deposits in the brain and the eye, respectively. Indeed, the amyloid β peptide (Aβ) deposited in the core of senile plaques and cerebrovascular amyloid of AD are also found in drusen—extracellular deposits found between the retinal pigmented epithelium (RPE) and Bruch's membrane—of patients suffering from AMD. Since amyloid deposits in the brain are fostered by multiple genetic and environmental factors that cause AD, and certain forms of Aβ are neurotoxic, it has been attributed a central place in the pathogenesis of AD. Finding Aβ in the drusen suggests that the two diseases may share common pathogenic mechanisms and that lessons learnt from one disease can be applied to the other even though clinically AD does not appear to increase the risk of AMD and vice versa. However, AD patients suffer an increase in the risk of glaucoma, a retinal disease affecting the ganglion cells that collect the integrated information from the eye and communicate with the brain via the optic nerve. Since the cause and mechanisms of neurodegeneration are poorly understood in both diseases, this is a very important fundamental and unanswered question. In this chapter, we discuss the salient features of AD and retinal degeneration (including AMD and glaucoma) and attempt to integrate the major features to identify common pathogenic pathways that cause both diseases.

List of Abbreviations: (Aβ40), 40-residue species; (Aβ), amyloid β peptide; (ABCR), ATP-binding cassette transporter; (ACAT), acyltransferase; (AD), Alzheimer's disease; (AMD), age-related macular degeneration; (APP), Aβ precursor; (ApoE), apolipoprotein E; (BBS), Bardet-Biedl syndrome; (CFB), complement factor B; (CFB), complement factor B; (CFH), complement factor H; *(CFH)*, complement factor H; (CHM), choroideremia gene; (CNS), central nervous system; (CRD), Cone-Rod Dystrophy; (CRP), C-reactive protein; (CRP), C-reactive protein; (cygb), cytoglobin; (DHRD), Doyne honeycomb retinal dystrophy; (ML), Malattia Leventinese; (FAD), familial AD; (GA), Geographic Atrophy; (HDL), high-density lipoprotein; (JMD), juvenile forms of macular degeneration; (MAC), membrane attack complex; (MAPT), microtubule-associated protein tau; (MCI), mild cognitive impairment; (NFTs), neurofibrillary tangles; (Ngb), neuroglobin; (NSAIDs), nonsteroidal anti-inflammatory drugs; (OCA), oculocutaneous albinism; (PUFA), polyunsaturated fatty acid-rich; (ROS), reactive oxygen species; (ROS), rod outer segment(RP), retinitis pigmentosa; (RPE), retinal-pigmented epithelium

1 Introduction

Alzheimer's disease (AD) and age-related macular degeneration (AMD) are major medical problems that are caused by neurodegeneration in selected areas of the brain and retina, respectively. Both diseases have a complex and progressive course with age as the major risk factor and share some pathological characteristics, such as the deposition of the amyloid β protein (Aβ) in the senile plaques of AD and drusen of AMD. Activation of inflammatory mechanisms and alteration of cholesterol metabolism are believed to be the contributing factors for both diseases, but the underlying causes remain poorly understood, hindering prevention and treatment efforts. An important consideration is that statistics do not bring the two diseases together in spite of the shared risk factors, but the two diseases may nevertheless share common pathogenic mechanisms. AD is associated with swelling of the retinal ganglion cells and there the clinical risk of glaucoma is increased in AD, suggesting that the two diseases may share some aspects of pathogenesis, even though glaucoma is not associated with protein aggregation and deposition (Bayer et al., 2002). Glaucoma may therefore be a good model for the mechanisms of degeneration in AD that is independent of Aβ deposition and may provide a valuable clue to its role in neuronal dysfunction and degeneration. In this article, we review the major aspects of retinal degeneration, particularly AMD, and discuss the pathogenic pathways potentially shared with AD.

2 Alzheimer's Disease

We have recently published a number of reviews that discuss various aspects of AD (Sambamurti et al., 2002, 2004) and have also recently proposed novel hypotheses for the mechanisms of neurodegeneration for this disease (Sambamurti et al., 2006). Briefly, AD is a leading cause of dementia, which currently affects 12 million people worldwide (4.5 million in America), and this number is likely to triple with the aging of the baby-boom generation by 2050 (Hodes et al., 1996; Bennett et al., 2005, 2006). The prevalence rate for AD is about 7% for individuals aged 65 or more, and the risk doubles every 5 years after age 65 (McCullagh et al., 2001). Research on AD has been extensive and for the first time brought the processing and turnover of proteins to center stage in the context of degenerative diseases.

2.1 Amyloid β Protein

The major focus of AD has been the small 4 kDa amyloid β protein (Aβ) found deposited in the amyloid plaque, which is a major hallmark of the disease (Hardy and Selkoe, 2002). The major reason for this focus is that together with the neurofibrillary tangles (NFTs) amyloid plaques not only define the disease, but all mutations that cause familial AD (FAD) alter the metabolism of the Aβ precursor (APP) to increase the levels of a longer 42-residue version (Aβ42) relative to the normally major 40-residue species (Aβ40). Moreover, the Aβ42 readily self-aggregates to form neurotoxic oligomeric intermediates, making it the center of focus in AD pathogenesis (Hardy, 2006; Sambamurti et al., 2006). Cumulative literature has demonstrated that Aβ is generated by the proteolytic processing of APP by β-secretase (a.k.a. BACE-1 or memapsin-2) to the soluble fragment, sAPPβ, and membrane-bound fragment, CTFβ, followed by the intramembrane cleavage of the latter by γ-secretase to the secreted fragment Aβ and intracellular fragment CTFγ (a.k.a. AICD, AID, CTFε) (Sambamurti et al., 2002). A large body of literature shows that the vast majority of FAD mutations are on presenilins 1 (PS1) and 2 (PS2), which are believed to constitute the catalytic subunits of γ-secretase (Fraering et al., 2004; Lazarov et al., 2006). Since several mutations reduce CTFγ, we have developed a novel working hypothesis that inhibition of γ-secretase, rather than accumulation of Aβ42, is the key trigger in AD (Sambamurti et al., 2006). It is widely hypothesized that toxic Aβ forms trigger the hyperphosphorylation of the microtubule-associated protein tau (MAPT), a key component of NFT (Lee and Trojanowski, 2006). Several academic and pharmaceutical efforts are being targeted toward treatment of AD by reducing the levels of amyloid, including inhibitors of Aβ aggregation such as Alzhemed (Neurochem, Inc.), immunotherapy (Elan corporation), production such as β-secretase (a.k.a. BACE), γ-secretase, and activators of α-secretase with agents such as M1 agonists or 5-HT agonists. Although the outcome of some of these trials appears to be promising, none of these compounds appear to have completely stopped the progression of the disease. Since AD follows a long and torturous course, it is likely that the disease process involves multiple levels of damage and a cascade of triggers, which may require a combination of drugs for treatment. This makes the study of risk factors for AD extremely important. Some of these risk factors affect the course of other diseases and may therefore provide an insight into the multiple targets involved in this multifactorial disease.

2.2 Risk Factors for AD

The amyloid–tau hypothesis and its variants discussed earlier are the best studied and supported. However, other factors are known to affect the pathogenesis of AD such as inheritance of the E variant of apolipoprotein E (ApoE), dietary cholesterol intake, homocysteine, glucocorticoids, estrogen, diabetes, hypertension, head injury, and stroke (Atwood et al., 2002; Hardy, 2006; Sambamurti et al., 2006). Several of these factors foster either the production or deposition of amyloid, but also have other amyloid-independent effects on the brain as well as other peripheral tissues.

The largest genetic risk factor is ApoE, the sole cholesterol transporter of the brain. Interestingly, the lack of ApoE results in the failure of amyloid plaque formation in a transgenic mouse. Moreover, amyloid plaque formation in human ApoE+APP double transgenic mice follows a pattern of E4 > E3 > E2, which parallels the pattern of AD risk in humans (Holtzman et al., 2000). ApoE is proteolytically processed to smaller fragments, which accumulate and may be independently involved in AD pathogenesis (Huang et al., 2001).

Recent studies have identified an important new risk factor—SORL1 (a.k.a. LR11, SORLA). SORL1 is an ApoE receptor, which appears to bind APP and play a role in its trafficking away from endosomal compartments that favor β-secretase processing and Aβ production (Rogaeva et al., 2007). Reduced expression (noted in some AD subjects) of the SORL1 protein results in increased production of Aβ40 as well as Aβ42, without changes in APP levels or secretion.

A high-cholesterol diet, particularly in combination with copper in the diet, results in an increase in Aβ production and deposition (Refolo et al., 2000; Sparks et al., 2000). Conversely, treatment with drugs that reduce cholesterol synthesis results in a reduced Aβ yield (Refolo et al., 2001). However, evidence indicates that changes in Aβ can be mediated by isoprenoids, which are also inhibited by the commonly used cholesterol synthesis inhibitors, the statins (Vassar et al., 1999; Zhou et al., 2008).

Estrogen replacement therapy was reported to be protective against AD pathogenesis. However, this finding remains controversial, given the failure of hormone replacement trials. Nevertheless, given that the failed trials use a combination of estrogen and progesterone, the effects of progesterone need to be studied in greater detail. The latter was shown to antagonize the beneficial effects of estrogen in animal models and may therefore explain the failure of these trials (Simpkins and Singh, 2004).

3 Retina as a Model for Neurodegeneration in AD

One of the major problems in studying AD is the lack of animal models that present the progressive neurodegeneration in AD and the difficulties in routine noninvasive monitoring of neurodegeneration in the brain of patients. The retina is a neuron-rich extension of the central nervous system (CNS) susceptible to age-associated degeneration and the progression in patients can be closely monitored. It is the only nervous tissue that lends itself to noninvasive visual examination. Thus, in addition to its vital role in the independent life of man, it also serves as an excellent system to understand the degeneration of the CNS, particularly with age. Three major degenerative diseases of the retina—Diabetic retinopathy, Glaucoma, and AMD—are associated with aging and are responsible for most age-associated blindness. Recent studies have suggested that amyloid deposition, the signature of AD, is also found in eye of AMD (Johnson et al., 2002; Yoshida et al., 2005), suggesting that molecular mechanisms of neurodegeneration may be conserved in both diseases as discussed in greater detail later. Correlations between AD and diabetes have been reported in the literature, with AD being described as type-III diabetes of the brain (Lester-Coll et al., 2006; Pilcher, 2006). However, there is poor literature on diabetic retinopathy and AD, probably due to the high level of morbidity associated with diabetes. In order to understand the mechanisms of neurodegeneration, it is important to get a clear understanding of the structure of the eye, not only in relation to its functional regions but also as it relates to the various intricate mechanisms for the long-term maintenance of this tissue.

4 Eye Structure

The eye is a highly specialized organ for detecting and processing the images of the external world. It is a ball-shaped structure contained in the cavity of the orbit, where it is protected from injury and securely placed in a position to maximize the visual range. The eyeball is a fluid-filled organ of about one inch in diameter in humans, surrounded by three outer layers of tissue (❷ *Figure 7-1.*) The outer layer of the eye is formed by the hard white fibrous tissue, sclerotic or sclera, in the posterior part and the transparent cornea in the anterior end. The sclera is extremely hard and forms the protective shell that gives shape to the eye. The cornea covers both the iris and the pupil and is part of the eye's refractive system, focusing the images onto the photoreceptors of the retina. The middle layer consists of the iris and the ciliary body in the

Figure 7-1

Structure of the eye. The eye is a globe-shaped structure, which is connected to the rest of the body at the optic disc via the optic nerve. It is a high-energy organ richly supplied by blood vessels of the choroids to feed the retina and to maintain the other key structural elements. The eye contains two major fluid-filled chambers that cushion and protect the lens and retina. In the anterior chamber, this fluid is dynamically replenished by the blood and drained via the Schlemm's canal. The optic disk marks a blind spot where the optic nerve exits the eye and is thus devoid of photoreceptors. In contrast, the macular region of the retina has a very high density of photoreceptors and constitutes the key area of detailed visual processing. A central depression in the macula called the fovea contains primarily the cone photoreceptors whereas the rods are enriched in the peripheral retina

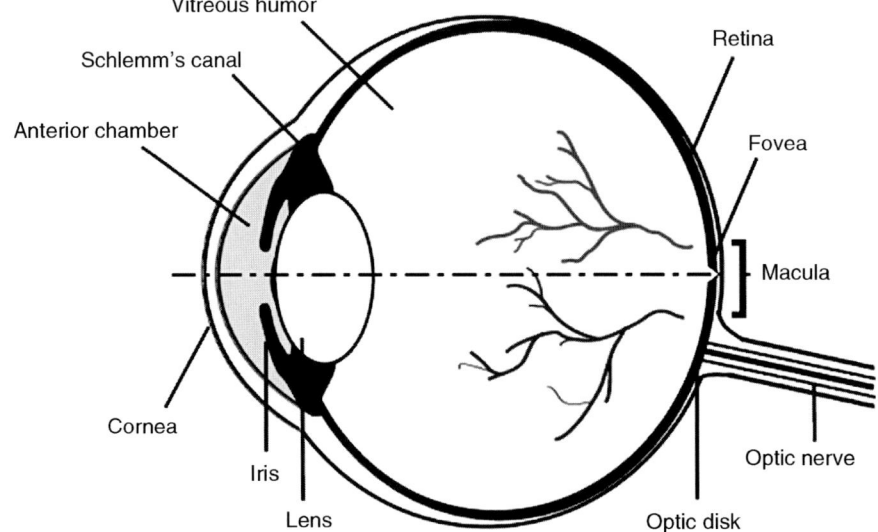

anterior part of the eye, and the choroids in the posterior part. The colored iris is a circular muscle that controls the amount of light that enters the eye. The ciliary body consists of a muscular component that connects to the lens via zonule fibers and adjusts the power of the lens for focusing. It also contains the ciliary vasculature that continuously produces aqueous humor, which is secreted into the anterior chamber to feed and maintain the cornea, iris, and lens. The choroid consists of vasculature that merges with the iris. It provides about 65% of the blood flow that feeds and oxygenates the retina, the remaining 35% is provided by retinal blood vessels. Finally, the retina, or the sensory part of the eye, forms the internal layer. It is responsible for the absorption of photons by the photoreceptors, the conversion of this signal into an electrochemical signal, and the transmission of the signal to the brain.

The eye consists of two main chambers, the anterior chamber and the vitreous chamber, with the former being filled by aqueous humor, the latter with a more viscous fluid, the vitreous humor. Aqueous humor is drained through Schlemm's canal at the sclero-corneal junction (Brubaker, 1991). Failure of drainage results in an increase in intraocular pressure leading to glaucoma. The vitreous humor on the other hand occupies the largest space in the eyeball between the lens and the retina and provides a cushioned support for the rest of the eye, as well as a clear unobstructed path for light to travel to the retina. Suspended in the eye right behind the iris is the lens, which is composed of tightly packed transparent fibrous cells enclosed with an elastic epithelial layer that focuses the image on the retina with help from the muscles of the ciliary body and the zonule fibers. Thus, light rays are refracted onto the retina through the transparent cornea, the aqueous humor, the lens, and the vitreous humor to form an image on the retina, which ultimately processes all the visual information and sends it to the brain for further processing.

5 Retina

This neuron-rich tissue is an extension of the brain and thus develops from the neural tub (Gray, 1901; Mann, 1964; Hilfer and Yang, 1980; Hendrickson and Yuodelis, 1984). It is a paper thin transparent tissue of ~250 μm and in man decreases by about 0.53 μm/year with age (Alamouti and Funk, 2003). The retina consists of six types of neurons—rods, cones, bipolar cells, ganglion cells, horizontal cells, and amacrine cells—distributed in a laminar fashion (Purves et al., 2001; Wassle, 2004). The retinal system consists of three nuclear layers and two synaptic layers (❯ *Figure 7-2*) as described in order from inside out.

The ganglion cell layer consists of cell bodies of the retinal ganglion cells as well as displaced amacrine cells. The retinal ganglion cell axons form the optic nerve, which connects the retina with the brain.

◘ Figure 7-2
Section of the retina showing retinal layers. **Diagram showing circuitry of the retina with a layered arrangement with the outer (*top*) layer representing the choroid followed by the retinal pigmented epithelium (RPE), photoreceptors (rods: R and cones: C), which are the light detectors of the retina. The photoreceptor, bipolar cell (B), and ganglion cell (G) are connected in series, representing a direct pathway through the retina. Horizontal cells (H) and amacrine cells (A) mediate lateral interactions in the outer and inner plexiform layers, respectively. Retinal ganglion cell axons funnel the visual information through the optic nerve to the brain for further processing**

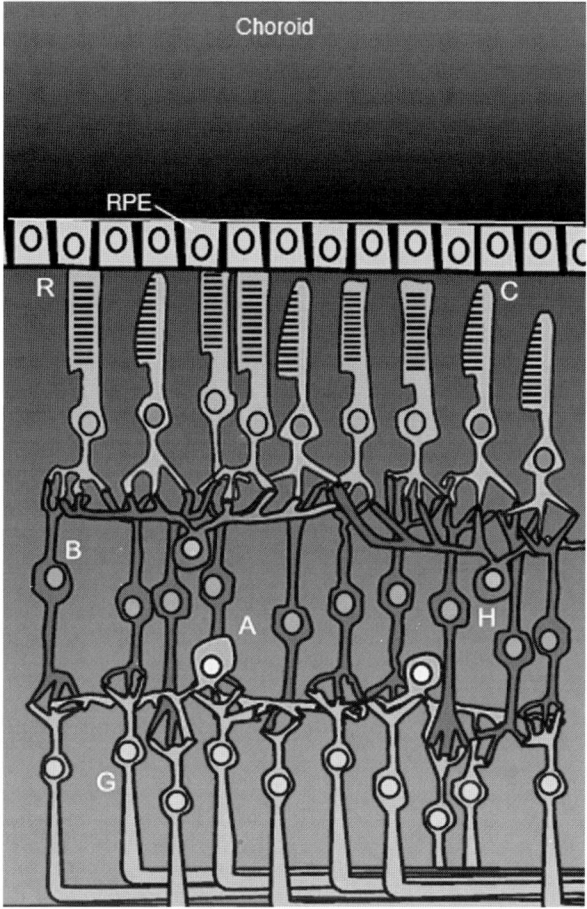

The inner nuclear layer contains the cell bodies of the bipolar, amacrine, and horizontal cells as well as the only glial type in the retina, the Mueller glial cells. The outer nuclear layer consists of the cell bodies of the rods and cone photoreceptors. The three layers of nuclei are separated by two layers of neuropil. The inner plexiform layer constitutes the synaptic connections between the ganglion cells and the bipolar cells interconnected by amacrine cells; whereas the outer plexiform layer includes the synaptic contacts between the photoreceptor cells and the bipolar cells interconnected by horizontal cells. Finally, on the distal side of the photoreceptors lies the retinal pigmented epithelium (RPE), a single cell layer of epithelial cells that are densely packed with pigment granules.

The retina contains three basic types of glial cells, Mueller glial cells, astroglia, and microglia. The Muller glial cells constitute the radial glia of the retina, spanning across the entire thickness of the retina and forming the inner and outer limiting membrane (Reichenbach et al., 1995). These cells appear to play an important role in the distribution of potassium, but their functions are otherwise poorly understood. The retina also has astrocytes and microglia. Astrocytes appear to have their origins in the brain rather than in the developing retina, entering the retina through the optic nerve (Schnitzer, 1988), and are located primarily in the nerve fiber layer. Microglia on the other hand appear to be distributed throughout the entire retina. They also enter the retina during development, and due to their mesodermal origin appear in the retina together with the mesenchymal precursors of retinal blood vessels.

The detection of light begins with the photoreceptor cells and follows an elaborate pattern of integration in the retina before being presented to the brain. This part of vision will not be discussed in detail as it has been reviewed in detail in several text books as, for example, Purves et al. (2001). Although most vision in human beings in daytime is in color and therefore mediated by cones, the number of cones is relatively small (5% of all photoreceptors) in comparison to rods (95% of all photoreceptors). However, the distribution of rods and cones is not uniform in the human eye. The center of the human eye consists of a distinct depression of approximately 1.2 mm not seen, for example, in the mouse eye, called the fovea centralis described by Polyak (1941) and is part of a darkened yellowish circular central area called the macula lutea (Kolb et al., 2002). This region is very rich in cones and the very central region termed the foveola of ~300 μm is completely devoid of rods. The depression is rich in photoreceptors but other layers are moved to the side, allowing light to reach its target without any dispersion. In addition, this layer lacks capillaries to avoid light scattering and thus completely depends on the choroid for its oxygen, nutrients, and waste disposal, making it more susceptible to accumulated metabolic waste. Defects in this part of the eye results in impairment of central acute vision and interferes with driving and other daily activities as in AMD.

5.1 Retinal Maintenance and Turnover

Photoreceptors, like other neuronal tissues, constitutively use high levels of oxygen due to their continuous need to transport Na^+ ions against the concentration gradient and maintain their resting potential (Besharse et al., 1986; Ames, 1992). The very high neuronal density in the eye exacerbates the high-energy requirement, which is higher than even the brain per unit mass. In addition, the photoreceptors constantly shed and renew their outer segment membranes to maintain a functional signal transduction cascade in the presence of an environment rich in reactive oxygen species and light, which is damaging to proteins. In particular, the rod outer segments (ROS) are composed of stacks of double-membrane disks surrounded by a plasma membrane, which is renewed throughout life (Williams, 1991). The older outer stacks are replaced continuously by younger freshly generated inner stacks, with the entire stack being replaced in approximately 10 days in the primate retina (Young, 1976). The RPE cell layer efficiently degrades the shed disks after phagocytosis (Fraser-Bell et al., 2006). RPE is composed of a densely packed single layer of hexagonal cells that lies in close association with the rod and cone photoreceptors (Bok, 1993). Each cell consists of an outer nonpigmented part containing a large oval nucleus and an inner pigmented portion, partially ensheathing the photoreceptor outer segments (Boyer et al., 2000). A pivotal function of the RPE is the regeneration of the visual chromophore, 11-*cis* retinal. The visual pigment, rhodopsin, consists of an apoprotein opsin and the vitamin A-derived chromophore 11-*cis* retinal. Vision is initiated when rhodopsin absorbs a photon, leading to the photoisomerization of 11-*cis* retinal to all-*trans*-retinal. In order to regenerate functional photopigment, all-*trans*-retinal has to exit the binding pocket in rhodopsin and get reisomerized to 11-*cis* retinal. This reisomerization process occurs within the RPE through many complex intermediate steps. A failure of any one of the enzymatic reactions involved in 11-*cis* retinal production and turnover can lead to loss of visual function, degeneration of the retina, and ultimately complete blindness (Strauss, 2005). Taken together, although the RPE has no photoreceptive or neural function, it is essential for the support, viability, and function of photoreceptor cells.

5.2 Oxygen Requirement

Due to its limited storage capacity and high and continuous energy utilization, the eye primarily uses glucose as its energy source and generates energy by aerobic respiration. The photoreceptor and RPE layers of the retina are primarily provided oxygen by diffusion from the choroidal circulation across Bruch's membrane. This process is facilitated by the high permeability of the choriocapillaris ensuring that the outer retinal layers are provided with near-arterial levels of oxygen (Bill et al., 1980). The inner cell layers are provided oxygen and nutrients by blood circulating through capillaries that cover most of the eye. However, the central fovea with a particularly high cone density is devoid of capillaries, making it more susceptible to limitations in oxygen and nutrients. The specific adaptations of the foveal region to the limited oxygen supply are not yet fully understood, although studies in primates and rabbits confirm that these zones contain low oxygen tension. Oxygen is known to be the most supply-limited metabolite in the retina (Anderson, 1963) and is acutely important as demonstrated by extensive retinal damage upon even brief ischemia. An additional example of the importance of oxygen in retinal disease is retinitis pigmentosa (RP), a disease in which the retina exhibits an increase in oxygen metabolism (Anderson, 1968). On the flip side, significant improvements have been reported in the electrophysiological responses in the treatment of RP patients with hyperbaric oxygen therapy (Vingolo et al., 1999).

5.2.1 Specialized Proteins in Retinal Oxygen Delivery

In addition to an efficient circulatory system, the eye expresses an interesting group of oxygen-binding proteins—neuroglobin (Ngb) and cytoglobin (cygb). Ngb and Cygb are members of the globin family related to hemoglobin and myoglobin and are believed to function as either oxygen traps to facilitate the transport of oxygen to the mitochondria of neurons from the blood or as scavengers of reactive oxygen species (Schmidt et al., 2003; Sun et al., 2003; Bentmann et al., 2005; Ostojic et al., 2006). Ngb and Cygb are localized in the brain and endocrine tissues, which also need a high level of oxygen and are conserved from pufferfish to man (Awenius et al., 2001; Zhang et al., 2002a). The vascularized retina shows expression of Ngb in the photoreceptor, ganglion cells, as well as outer and inner plexiform layers, which are also rich in mitochondria (Bentmann et al., 2005). In contrast, the avascular guinea pig retinas show expression of Ngb primarily in the photoreceptor inner segments (Bentmann et al., 2005). This finding is consistent with the notion that the tissues that efficiently use mitochondrial energy metabolism to generate energy need use Ngb to supply the oxygen. Alternatively, the same tissues that may be producing the highest levels of reactive oxygen species, may be using Ngb to scavenge reactive oxidizing species, providing a cellular defense against oxidative stress. Thus, it is very likely that Ngb is used to both feed and preserve the neuron during oxidative metabolism.

6 Normal Aging of the Eye

The normal human eye undergoes several structural changes with aging. Some common age-related changes are crystalline lens opacification, progressive optical aberrations, and loss of pupillary reactivity (Guirao et al., 1999). Indeed, cataract formation in the lens is almost an inevitable consequence of aging and most elderly individuals undergo surgery to correct this problem. Several studies suggest that the lens of the cataract patient accumulates Aβ, which may contribute to its opacity (Frederikse et al., 1996; Frederikse and Zigler, 1998; Goldstein et al., 2003). It has been proposed that Aβ facilitates αB-crystallin aggregation and amyloid formation in the lens, which ultimately results in cataract formation (Sandilands et al., 2002). Trisomy 21, which leads to Down's syndrome, is also associated with an increase in the incidence of cataracts at a young age and a mouse model of Alzheimer's disease develops cataracts (Melov et al., 2005). However, since increase in opacity is an obvious cause of visual impairment and this does not constitute a clear model for neurodegeneration, we will not be discussing cataracts in detail.

Normal aging is also associated with visual impairment caused by dysfunction and loss of the retinal neurons. The sensitivity of the peripheral field of vision declines more rapidly than that of the central field and scotopic or dim light vision using rods is more greatly affected than photopic bright light vision using cones (Gao and Hollyfield, 1992; Curcio et al., 2000), resulting in a decrease in rod density in the central retina between the ages of 34 and 90 years by 30%; on the other hand, the number of cones in the macular area remains stable. However, loss of a significant number of rod cells is followed by a loss of cones, suggesting that the loss of rods associated with aging can secondarily lead to a loss of cones (Cideciyan et al., 1998; Hicks and Sahel, 1999; Curcio et al., 2000).

7 Retinal Degeneration (Overview)

The human retina is a delicate and sophisticated neuron-rich organ, which detects and processes visual images and delivers them in a digitized form to the brain. This extreme sophistication involves the intimate and continuous cooperation of a number of different cell types, which are constantly being remodeled and require the continuous support of cells such as the RPE for turnover of the shed domains as discussed earlier. These maintenance systems are very elaborate and require a number of genes and developmental cues for their formation and proper function. Failure or reduced efficiency in any of the maintenance systems leads to the degeneration of one or more of the retinal neuronal networks to impair vision. As expected, this complex system is vulnerable to environmental as well as genetic changes resulting in a number of clinical conditions that lead to impaired vision and ultimately blindness. Three of the most frequent retinal diseases are AMD, diabetic retinopathy, and glaucoma. In AMD, the macula and fovea become compromised, leading to loss of central vision. One of the hallmarks of AMD is the deposition of drusen between the RPE and choroid of the macula. There two forms/stages of AMD, so-called early or dry AMD and late or wet AMD. In patients with dry AMD, the light-sensitive neurons in the macula slowly degenerate, leading to gradual central vision loss in the affected eye. On the other hand, wet AMD occurs due to angiogenesis of the choroidal blood vessels. These new vessels, which tend to be very fragile and often leak blood and fluid, start to grow under the macula, and thus separate the neural retina from its supporting cells of the retinal pigment epithelium (Haddad et al., 2006). This causes rapid photoreceptor cell death. An interesting connection between AD and AMD is the recent finding that drusen contain deposits of Aβ surrounded by ApoE, as in the case of the senile plaque (Johnson et al., 2002). Neovascularization leading to vision loss is also a hallmark of diabetic retinopathy caused by uncontrolled hyperglycemia. The third major degenerative disease of the retina is glaucoma, which is characterized by degeneration of the ganglion cells, rather than photoreceptors (Klein et al., 1995). One of the causes of glaucoma is the failure of control of intraocular pressure due to poor drainage of the aqueous humor through Schlemm's canal (Jackson and Owsley, 2003). However, glaucoma is a syndrome that does not require the increase in intraocular pressure and may be caused by multiple different mechanisms (McKinnon, 2003; Tatton et al., 2003). There is a major clinical connection between glaucoma and AD, with an increase in the frequency of open-angle glaucoma in AD patients, suggesting that some of the disease mechanisms may be conserved even though amyloid or other extracellular deposits are not observed in glaucoma (Tamura et al., 2006). Other forms of retinal degeneration such as retinitis pigmentosa (RP) provide a deep insight into the important structural and supportive components of the eye, which may play a role in major forms of retinal degeneration. RP comes in many forms and is due to a large number of genetic mutations. Most of the faulty genes that have been discovered concern the rod photoreceptors. The rods of the peripheral retina begin to degenerate in early stages of the disease. Patients become night blind gradually as more and more of the rod-rich peripheral retina becomes damaged. Eventually, patients are reduced to tunnel vision with only the fovea spared the disease process.

8 Age-Related Macular Degeneration

AMD is the most common cause of irreversible vision loss in the elderly. It is estimated that the prevalence of AMD in the USA is as high as 9 million, with 1.8 million individuals already suffering severe loss of vision from advanced AMD. The incidence increases with age; 1.2% of individuals aged 43–86 years develop AMD per year, whereas 3.7% of individuals greater than 75 develop AMD each year (Tomany et al., 2004b; Brown et al., 2005). As the population ages, the prevalence of AMD will continue to grow. The loss of visual acuity seen in even milder cases of AMD has a major impact on the quality of life as well as causing a significant economic burden to society (Brown et al., 2005). As mentioned earlier, there two forms/stages of AMD, dry or nonexudative AMD and wet or exudative AMD.

8.1 Dry Macular Degeneration

Dry macular degeneration is a milder form of the disease and accounts for the vast majority of affected individuals. It is characterized by slow rod photoreceptor degeneration in the macula, followed by loss of cones, leading to a loss of central vision (Curcio et al., 1996). The disease is characterized by the deposition of a yellowish hyaline material, known as drusen, between the basal lamina of the RPE and the inner collagenous layer of Bruch's membrane (Sarks, 1976; Green and Enger, 1993). Drusen size, number, composition, and degree of confluence are significant risk factors for the development of AMD. Dry AMD may occur in three stages in one or both eyes:

Early AMD is identified by several small drusen or a few medium-sized drusen. However, no obvious symptoms or vision loss are seen at this stage. Intermediate AMD is characterized by the presence of many medium-sized drusen or one or more large, irregular-shaped drusen (soft drusen). Symptoms at that stage may include a blurred or blind spot (scotoma) or distortion of images in the central field of vision. Also, more light and higher contrast may be needed for clear vision. Finally, Geographic Atrophy (GA) represents the most advanced stage, characterized by extensive visual impairment. GA constitutes the ultimate stage of dry macular degeneration. It is characterized by the presence of extensive soft drusen accompanied by photoreceptor cells degeneration as well as breakdown of RPE cells and surrounding choroid tissue in the macula. The depigmented areas of the retina grow to form large areas resembling the continents in a map, hence the name geographic atrophy. Blind spots may become larger and distortion more severe and may eventually encompass the entire center field, causing the patient to rely exclusively upon the peripheral field for sight. Detailed activities including driving are severely impaired at this stage.

There are two types of drusen; hard and soft drusen. Hard drusen affects young people and is not considered a sign AMD. Soft drusen appears in people over 55 years and represents the hallmark of AMD (Gass, 2003). The degree of loss of central vision associated with retinal atrophy is correlated with the level of drusen found in the AMD patients. It has been proposed that drusen may lead to dysfunction and/or degeneration of the RPE and retina by inducing ischemia and/or restricting the exchange of nutrients and waste products between the neural retina and choroid. In addition, drusen themselves may have a detrimental effect on vision, particularly with respect to contrast sensitivity. Mild loss of visual acuity is observed with dry AMD, which may progress to the more damaging wet form of the disease.

8.2 Wet Macular Degeneration

It is now accepted that wet AMD arises from or is linked to the dry form based on the discovery of common genetic risk factors (Donoso et al., 2006) and the co-occurrence of drusen. There are approximately 200,000 new cases of wet AMD diagnosed each year accounting for about 10% of the AMD-affected patients, but represent 90% of the patients with loss of vision. It is characterized by the abnormal outgrow of vasculature from the choroid (choroidal neovascularization) that breaches Bruch's membrane and enters the retina. Peripheral vision is usually intact, so AMD does not result in total blindness, but the visual dysfunction is quite severe, as the macula is responsible for the detailed vision involved in most activities of daily living. Treatment of neovascularization by a new generation of drugs has become a promising method of reducing or preventing the ravages of wet AMD (Bylsma and Guymer, 2005).

9 Risk Factors for AMD

AMD appears to be a complex multifactorial disease and may actually represent a syndrome of diseases that affect the macula (see Retnet.org for a list of genes associated with autosomal dominant, autosomal recessive, or X-linked macular degeneration). The disease process appears to be influenced by a number of environmental insults making it difficult to identify one central cause. Indeed, the genetics of macular degeneration and other forms of retinal degeneration point to multiple targets and multiple causes, each of which may be involved in a subset of patients, and ultimately lead to the common endpoint of failed central visual processing mediated by the macular region of the eye. Although there are some shared risk factors between AD and AMD, there are some surprisingly distinct risk factors that may influence our use of one of these diseases as a model for the other.

9.1 Age

Like AD, the largest and most consistent risk factor for AMD is aging. Approximately 10% of patients 66–74 years of age show signs of macular degeneration. The prevalence increases to 30% in patients 75–85 years of age.

9.2 Sex

Like AD, women suffer from a higher risk of developing AMD, and the disease lasts longer and is more severe in women.

9.3 Education

Intense reading implies that the subject's eyes are exposed to strain, particularly in the macula, due to the constant focus on small objects. However, unlike AD, literacy as well as socioeconomic status does not appear to influence the incidence of AMD. For example, the Beaver Dam Eye Study found little relationship between age-related maculopathy and measures of social class (Klein et al., 1994a, b).

9.4 Sunlight

The visible spectrum ranges from 370 to 730 nm. Shorter wavelengths of light, including ultraviolet (UV) light, which damage the DNA, are mostly filtered out by the lens, with only a small percentage of UV light being able to reach the retina. Long-term exposure to blue to ultraviolet light may increase your risk of developing macular degeneration (Rattner and Nathans, 2006). Recently, the Rotterdam study as well as a study on watermen who work on the Chesapeake Bay concluded that long-term chronic exposure to sunlight or its blue component, especially later in life, may be related to the development of AMD (Taylor et al., 1993; Tomany et al., 2004b). Although this hypothesis remains to be fully tested, a useful preventive measure is to wear glasses that filter blue light.

9.5 Smoking

The only environmental agent unequivocally linked to macular degeneration is tobacco smoking. Smokers are up to four times more likely than nonsmokers to develop AMD (Schmidt et al., 2006; Weale, 2006). History of smoking has been shown to be a stronger risk factor for exudative AMD in carriers of the ApoE E2 allele than carriers of ApoE4 and the most common ApoE E3/3 genotype (Schmidt-Erfurth et al., 2005). More recently, smoking was shown to enhance neovascularization in animal models. One possible mechanism for the smoking-mediated retinopathy may be an increase in the levels of oxidative stress (Husain et al., 2002; Hyman and Neborsky, 2002).

9.6 Estrogen

Like AD, the incidence of AMD is higher in women. Moreover, in the case of AD, the protective role of estrogen remains controversial (Defay et al., 2004; Evans et al., 2004; Fraser-Bell et al., 2006). The Eye Disease study group reported a reduced risk of exudative AMD among women on hormone replacement therapy compared with those who had never taken estrogen replacement, and the Rotterdam study reported a higher risk of advanced AMD among women with early menopause compared with those aged 45 years at menopause (Vingerling et al., 1995), lending support to the hypothesis that estrogen may play a protective role and be used as a therapeutic agent.

9.7 Antioxidant

The NEI-sponsored AREDS study demonstrated that subjects at high risk for AMD as well as those with early AMD symptoms benefited from supplements containing high levels of antioxidants and Zinc (Bartlett and Eperjesi, 2003). The unique phagocytotic function of RPE and the need to efficiently eliminate or recycle the polyunsaturated fatty acid-rich (PUFA) shed outer segments expose the RPE to high levels of oxidative stress (Snodderly, 1995). Oxidation of PUFA initiates a chain reaction producing an abundance of reactive oxygen species (ROS), including lipid aldehyde radicals (Esterbauer et al., 1991a, b; Srivastava et al., 1995). Furthermore, RPE cells contain an abundance of photosensitizers, and exposure to intense visible lights induces generation of ROS (Gaillard et al., 1995; Rozanowska et al., 1995). To cope with these toxic oxygen intermediates, the RPE has evolved effective defenses against oxidative damage. It is particularly rich in antioxidants such as vitamin E, superoxide dismutase, catalase, glutathione-S-transferases, glutathione, and ascorbate (Newsome et al., 1990). However, with increasing age, the RPE antioxidative capability appears to be reduced (Liles et al., 1991; Tate et al., 1993). Thus, it is likely that aging RPE cells are more susceptible to oxidative damage (Akeo et al., 1996; Ueda et al., 1996; Kayatz et al., 1999).

9.8 Cardiovascular Risk Factors

A number of cardiovascular risk factors also increase the risk of AD as well as AMD, possibly mediated by inflammatory cascades (Donoso et al., 2006; Seddon et al., 2006a). Diets rich in omega 3 fatty acids are protective against the pathogenesis of AD as well as AMD (Cole et al., 2005; Seddon et al., 2006b). A high-cholesterol diet results in high risk for both AD and AMD, although the clinical statistics and mechanisms remain highly debated topics (Cousins et al., 2002; Sambamurti et al., 2004). Interestingly, there is a large increase in the levels of ApoE and triglyceride-containing VLDL particles just before amyloid deposition in the blood (Burgess et al., 2006; Haines et al., 2006). Another important cardiovascular risk factor for AMD is hypertension (Hyman et al., 2000). Exudative or neovascular AMD is associated with both treated as well as untreated hypertension.

9.9 Amyloid β

Several authors have attempted to associate AD and AMD on the basis of genetic, environmental, and inflammatory characteristics (Anderson et al., 2002; Yoshida et al., 2005; Luibl et al., 2006), as both processes are considered to be chronic neurodegenerative disorders with common characteristics to aging, unknown etiology, and more specifically the presence of senile plaques (extracellular formations with an Aβ peptide fibers) both in the gray matter of the brain and the macula of the retina. In addition, at the histological level, AD exhibits neurofibrillar tangles (aggregates of hyperphosphorylated tau protein) together with other less specific alterations. Although the etiology of both pathologies is not clearly established, it is known that their origin comprises multiple causes, including genetic and acquired factors. New evidence indicates that, in AMD, substructural elements within drusen contain Aβ, which is a major component of senile plaques and cerebrovascular deposits in the brains of patients with AD. Aβ assemblies are most prevalent in eyes with moderate or high drusen loads and it is suggested that Aβ might be associated with the more advanced stages of AMD (Anderson et al., 2004).

9.10 Apolipoprotein E

ApoE has been identified as a major risk gene for AD, including typical late-onset (sporadic) AD. The ApoE gene codes for a lipid-binding protein, transports of lipids (fatty acids and cholesterol) between cells (Alaupovic, 1982). Additionally, ApoE is critical for the coordination of cholesterol in the repair, growth, and maintenance of myelin and neuronal membranes during development or after injury, as it is the only such transporter in the brain (Boyles et al., 1989).

The ApoE gene exists in three isoforms—E2, E3, and E4. In the normal human population, the E3 allele is the most common and is considered to be the ancestral allele followed by the E4 allele (Roses, 1995). The E4 allele, associated with elevated cholesterol, has also been shown to result in increased risk of coronary heart disease, decreased longevity, and increased incidence of AD (Anderson et al., 2002; Anderson et al., 2004). However, in AMD, the E4 allele appears to confer protection, whereas the E2 allele is apparently associated with a moderately increased incidence of the disease, completely reversing the trend observed in AD (Arnold and Sarks, 2002). A lower frequency of the E4 allele carriers was observed in the AMD group compared with control subjects (Souied et al., 1998).

The ApoE protein is associated with the RPE and Bruch's membrane and accumulates in the cytoplasm of a subpopulation of RPE cells, many of which flank or overlie drusen (Anderson et al., 2004). ApoE expression is upregulated by astrocytic glia in response to neuronal injury and neurodegenerative disease (Bennett et al., 1996; Bishop et al., 2004). Notably, a synergistic combination of environment and genetics is observed between smoking and the E2 allele of ApoE (Brubaker, 1991).

A growing body of evidence suggests that the ApoE allele may also be associated with other retinal diseases, such as glaucoma (Kuhrt et al., 1997). ApoE upregulation by Muller glia has been reported in the degenerating human retina, where increased ApoE immunoreactivity was found in the subretinal space of the detached retinas (Blanks et al., 1989) and in Muller cells of retinas affected by glaucoma or AMD (Bok, 1993). However, the study of genetic association conducted by different groups gave conflicting results. One group found a positive association between the E2 allele and intraocular pressure (Junemann et al., 2004). Another group noted an increase in the risk of normal tension glaucoma in carriers of the E2 allele (Vickers et al., 2002) whereas others failed to detect a meaningful correlation (Lake et al., 2004; Ressiniotis et al., 2004). Since genetic associations represent a loose clustering of populations of a certain genotype with a disease, these weak correlations need to be analyzed in the context of other risk factors before any firm conclusions can be derived.

9.11 Cholesterol Metabolism

Cholesterol is an essential component of cell membranes for maintaining their structure and functions. Like most essential metabolites in the human body, cholesterol levels are tightly regulated, and a failure of this regulation leads to an increase in the risk for atherosclerosis. There has been increasing bodies of evidence showing that cholesterol metabolism is linked closely to the pathophysiology of several other diseases, including AD (Sambamurti et al., 2004) and AMD (Tomany et al., 2004b; Malek et al., 2005). Recent epidemiological studies have revealed the possibility of a link between cholesterol metabolism based on an association of an elevated serum total cholesterol level with a risk for the development of AD and mild cognitive impairment (MCI) (Notkola et al., 1998; Kivipelto et al., 2001). In addition to Aβ, senile plaques also contain large quantities of cholesyerol (Mori et al., 2001). Studies in animal models fed high-cholesterol diets show an increase in the deposition of Aβ in the brain serving as a model for AD (Sparks et al., 1994; Refolo et al., 2000). Similarly, recent studies have identified drusen deposits in transgenic mice expressing the E4 allele of human ApoE when fed a high-cholesterol diet (Malek et al., 2005).

Cholesterol influences the activity of the enzymes involved in the metabolism of the APP and in the production of Aβ. The posttranslational cleavage of APP by βand γ-secretases results in amyloidogenic products that aggregate as extracellular plaques, whereas cleavage by α-secretase results in nonamyloidogenic or soluble APP. In animal studies, dietary cholesterol accelerates Aβ deposition in the brain whereas cholesterol-lowering drugs lower it (Ohm and Meske, 2006). Other in vitro studies have shown that a high-cholesterol environment results in reduced production of soluble APP (Michikawa, 2006). A change in membrane properties, including stiffness and fluidity, has been suggested to influence activities of membrane-bound proteins and enzymes, including secretases. The high-cholesterol content in lipid rafts, membrane regions where these secretases are located, facilitates the clustering of the β and γ-secretases with their substrates into an optimum configuration, thereby promoting the undesirable pathogenetic cleavage of APP. The significance of this in human beings is unknown (Shobab et al., 2005).

The enzymatic conversion of CNS cholesterol to 24S-hydroxycholesterol, which readily crosses the blood–brain barrier, is the major pathway for brain cholesterol transport, elimination, and maintenance of brain cholesterol homeostasis (Lutjohann and von Bergmann, 2003). Interestingly, polymorphisms in the CYP46 gene (which encodes cholesterol 24S-hydroxylase) influence both Aβ peptide load in the brain and the genetic risk for late-onset sporadic AD. Moreover, 24S-hydroxycholesterol induces neurotoxic effects and increased concentrations have been detected in the CNS of AD patients (Kolsch et al., 2003).

Increases in levels of cholesterol esters have been linked to high Aβ levels in vivo and in cultured cells (Puglielli et al., 2001; Hutter-Paier et al., 2004). Conversely, the lack or inhibition of acyl-CoA: cholesterol O-acyltransferase (ACAT), which leads to decreased levels of cholesterol esters and increased intracellular free cholesterol, inhibits Aβ production. It is therefore possible that cholesterol esters, rather than cholesterol itself, alter the production of Aβ, possibly by interacting with β-secretase activity.

Cholesterol is also an important constituent of drusen and although the link between AMD and serum cholesterol seems plausible, epidemiologic studies failed to demonstrate a clear relationship. However, given the variation in susceptibility of human populations and the large number of ways in which serum cholesterol levels are regulated, one cannot rule out an association. Many studies did find elevated levels of high-density lipoprotein (HDL) cholesterol in patients with AMD (Dasch et al., 2005), because the cholesterol gene transporter ApoE is associated with AMD, which influences cholesterol levels in the CNS (Tomany et al., 2004b). In Bruch's membrane of the macula, there is a progressive accumulation of lipids stainable by oil red O, and Bruch's membrane/choroid extracts contain phospholipids, triglycerides, and esterified and unesterified cholesterol. It is thought that lipid accumulation renders Bruch's membrane increasingly hydrophobic with age, impeding diffusion between the RPE and choroidal vessels (Curcio, 2001).

A strong correlation between the levels of cholesterol and a number of diseases has been revealed over time and a number of cholesterol-lowering drugs with a range of positive and negative effects have become routinely used in clinical practice. Thus, hypercholesterolemia has been identified as a risk factor for a number of new diseases other than atherosclerosis. However, the mechanisms of pathogenesis that underlie this correlation remain unclear and the ultimate targets remain elusive. More detailed studies are hampered by limitations in the model systems, the large number of pathways that are involved in cholesterol metabolism, and the biologically active intermediates of cholesterol synthesis as well as metabolites of cholesterol.

9.12 Inflammation

Inflammation has been proposed as a possible driving force of AD and AMD pathology (Johnson et al., 2002; Klein et al., 2005a; Umeda et al., 2005a). Generally, inflammation is considered as a double-edged sword: whereas low tissue levels promote healing, high levels seriously damage tissue. Many compounds have been identified in brains of AD patients, which are known to promote and sustain inflammatory responses. They include Aβ, the pentraxins, C-reactive protein (CRP), amyloid P, complement proteins; inflammatory cytokines such as interleukin-1 (IL-1), interleukin-6 (IL-6), and tumor necrosis factor-alpha (TNFα); protease inhibitors such as alpha-2-macroglobulin and alpha-1-antichymotrypsin; and the prostaglandin generating cyclooxygenases—COX-1 and COX-2 (Klein et al., 2005a). Specifically, Aβ accumulation leads to a site-specific activation of glia resulting in the secretion of proinflammatory cytokines to clear Aβ deposits; however, the progressive accumulation of Aβ and its aggregation into insoluble plaques may induce a chronic proinflammatory response leading to compromised neuronal function (Donoso et al., 2006). More than 20 epidemiological studies suggest that the chronic use of nonsteroidal anti-inflammatory drugs (NSAIDs) greatly reduces the risk of AD, and recent epidemiological studies indicate that such use of NSAIDs also reduces the risk of AMD (Chen et al., 2003).

The hypothesis that inflammation is as a key mediator in AMD got a boost by the recent observations that genetic variations in complement factor H (CFH) and B (CFB) (Edwards et al., 2005; Haines et al., 2005; Klein et al., 2005a; Gold et al., 2006) are present in ~60% of all AMD patients. CFH and CFB together with complement factor I and D are regulatory molecules of the alternative complement pathway, a key component of innate immunity. CFB and CFD are activators and CFH and CFI are inhibitors in the complement cascade. This indicates that AMD involves the activation of the alternative pathway by unknown mechanisms, whereas classical pathway plays a major role in AD (McGeer et al., 2005).

10 Biochemistry of Drusen

The clinical hallmark of AMD is the appearance of drusen, localized deposits lying between the basement membrane of the RPE and Bruch's membrane (Pauleikhoff, 1992; Pauleikhoff et al., 1992a, b; Smith et al., 2005). Similarly cholesterol, in the free as well as esterified form, accumulates in drusen from AMD eyes (Curcio et al., 2005a). ApoE and lipoprotein particles are also found in this structure (Li et al., 2005; Curcio et al., 2005b).

There are three lines of evidence suggesting the way in which faulty protein turnover occurs in AMD as in AD (Swanson et al., 1996). First, the mutation in rhodopsin results in defective signal cascade followed by faulty turnover and deposition. Second, the gene responsible for Sorsby's fundus dystrophy encodes an inhibitor of metalloproteinases and leads to deposition of unprocessed extracellular matrix proteins. Third, the deficiency of enzymes involved in posttranslational modifications of proteins has been shown to cause retinal disease. For example, defects in the choroideremia gene (CHM), which encodes a subunit of geranylgeranyl transferase, lead to impairment in the prenylation of the Rab proteins (Sandberg and Gaudio, 2006), and loss of cathepsin D leads to neuronal ceroid lipofuscinosis, a progressive form of retinal degeneration (Siintola et al., 2006).

Unlike AD, where the predominant proteins deposited in the brain are Aβ in plaques and tau in tangles, drusen in AMD has a complex mix of proteins with no single protein being identified as characteristically abundant. Proteomics has identified over 129 proteins in drusen, but a comparison of drusen from normal aged eyes and AD did not reveal any differences (Crabb et al., 2002). Several studies show that drusen contains a variety of immunomodulatory proteins, which indicates that the process of drusen formation involves local inflammatory events, including the complement cascade (Johnson et al., 2001; Anderson et al., 2002; Johnson et al., 2002; Leu et al., 2002; Dentchev et al., 2003; Umeda et al., 2005b). Inflammation and complement are also implicated in AD and recent studies have identified another important distinguishing feature of AD, amyloid deposits, in AMD as well (Mullins et al., 2000; Johnson et al., 2002; Dentchev et al., 2003; Anderson et al., 2004; Yoshida et al., 2005; Luibl et al., 2006). The findings suggest that the proposed mechanisms of impaired processing in AD may also be conserved in AMD but the proteins in drusen are different, probably due to the tissue-specific composition of the retina. Nevertheless, since Aβ oligomers are detected in the drusen, it is possible that the atrophy and degeneration associated with drusen in AMD is partly be mediated by these oligomers (Luibl et al., 2006).

11 Genetics of Macular Degeneration

AMD usually starts in one eye but there is a high risk (over 40%) that the disease will affect the other eye within 5 years (Klein et al., 2004). Thus, although the eye is an immunologically protected site with a strong blood–eye barrier, there is a co-occurrence of the disease in the two eyes suggesting systemic or genetic involvement. Not surprisingly, AMD appears to be heritable as an autosomal dominant trait in a significant proportion (~50%) of afflicted individuals and there is a very high level of concordance of the disease in monozygotic twins, suggesting a strong genetic component (Klein et al., 1994a; Grizzard et al., 2003; Seddon et al., 2005). Mutations that cause macular defects are associated with loss of visual acuity, but spare night vision and peripheral vision. In contrast, diseases such as retinitis pigmentosa, which affects rods, present with rod photoreceptor degeneration and work well in the rod-dominant mouse models (Lolley, 1994; Keen and Inglehearn, 1996; LaVail et al., 1998).

Genetic factors play a substantial role in the etiology of AMD and associated macular characteristics. Most of the genetic associations have been documented in an extensive recent review (Haddad et al., 2006). Recently, substantial progress has been made in determining the genetic basis of monogenic eye disorders (Allikmets, 1999; Allikmets et al., 1999). Some of the more relevant genes are listed in ❯ Table 7-1.

Table 7-1
Genes that influence AMD/AD

Name	AMD	AD	References
ApoE cholesterol and fatty acid transporter	E4 protective	E4 strong risk factor	Klaver et al. (1998); Baird et al. (2004)
ELOVL4 long-chain fatty acid synthesis	Dominant mutations Stargardt-like disease		Edwards et al. (2001); Zhang et al. (2001)
Complement factor H (CFH; Y402H). Attenuates complement alternative pathway	Variants increase AMD risk		Edwards et al. (2005); Klein et al. (2005b); Haines et al. (2006)
CFB and C2 similar to CFH	1 risk and 2 protective haplotypes		Gold et al. (2006)
EFEMP1 (Arg345Trp)	Malattia Leventinese (ML) and Doyne honeycomb retinal dystrophy (DHRD)		Stone et al. (1999); Li and Wong (2001); Zhang et al. (2001)
VMD2 (11q13) anion channel	Best disease		Lotery et al. (2001)

An important approach to understanding the genetics of AD, which suffers from several of the limitations of studying AMD, has been to examine the genetics of early onset forms of the disease. A number of juvenile forms of macular degeneration (JMD) have been described and some of the mutations have been identified as discussed later. One will expect that the numerous types of JMD with identifiable causes will provide insight into the possible causes of AMD.

11.1 Macular Hole

A small tear in the center of the macula is often associated with aging due to detachment of the vitreous humor. This sometimes affects younger individuals due to eye injury. The macular hole can often correct itself but surgical procedures may be required to treat this condition. Although the symptoms include the failure of vision at the macula, macular holes do not constitute the progressive degeneration pathway. The macular hole can occur clustered in families, suggesting that some forms of this disease may be inherited (Lalin et al., 2004).

11.2 Cone-Rod Dystrophy (CRD)

Unlike typical macular degeneration, this group of diseases affects the macula as well as the periphery and manifests a loss of visual acuity in light and darkness. Patients with CRD first experience central vision loss, followed by night blindness and peripheral vision loss. Central vision loss begins in the first decade of life with the onset of night blindness occurring sometime after age 20. Little visual function remains after the age of 50. One group of CRD termed Bardet–Biedl syndrome (BBS) has been linked to a family of genes, BBS1–12, which are linked to defects in the sensory cilia and affect several cell types (Nishimura et al., 2005; Ross et al., 2005; Chiang et al., 2006; Dollfus et al., 2006; Stoetzel et al., 2007).

11.3 Sorsby's Macular Dystrophy

Sorsby's disease is also called cystoid macular degeneration and is characterized by early drusen followed by neovascularization and a sudden loss of visual acuity (Hamilton et al., 1989; Felbor et al., 1997b; Weber et al., 2002). Autosomal dominant mutations in the tissue inhibitor of metalloproteinase 3 (TIMP3) are linked to Sorsby's disease (Felbor et al., 1997b). A mouse model expressing the pathogenic TIMP3 mutation, S156C, results in abnormalities in the Bruch's membrane and adjacent RPE cells, an early feature of the clinical disease. The mutation appears to represent a hyperactive form of TIMP3, suggesting that the mutation may inhibit metalloproteinase activity. In addition to confirming the pathogenic nature of the mutation, the mouse model suggests that the loss of proteolytic turnover may affect the maintenance of homeostasis and thus lead to accumulation of drusen and subsequent cellular damage (Weber et al., 2002).

11.4 Best Disease

Best disease (also known as vitelliform macular dystrophy) is another early onset form of macular degeneration. The disease manifests with a yellow cyst under the macula in early childhood, but the cyst eventually releases a yellow deposit. The macula and RPE cells atrophy leading to loss of visual acuity later in life. The genetic disease is inherited in the autosomal dominant pattern and has been linked to a chloride channel membrane protein named Bestrophin on chromosome 11 (Allikmets et al., 1999; Musarella, 2001; Stohr et al., 2005).

11.5 Stargardt's Disease

Stargardt's disease is the most common form of inherited juvenile retinal degeneration. Like other forms of macular degeneration, central vision mediated by cone-dominant macula is reduced, but peripheral vision is preserved. Stargardt's disease is linked to the ABCR4 gene responsible for the transport of the retinal pigment (Musarella, 2001). A fluorescent yellow deposit containing bis-retinoid accumulates is deposited and causes cells to degenerate even in a mouse model lacking ABCR4 (Radu et al., 2004). Other loci have also been identified for Stargardt's disease such as ELOV4, a gene involved in the biosynthesis of very long-chain fatty acids, and successful mouse models for the disease have already been developed (Stone et al., 1994; Edwards et al., 2001; Zhang et al., 2001).

Some heterozygous carriers of the Stargardt disease mutations in ABCR4 also suffer from AMD. In addition, ABCR variants Gly1961Glu or Asp2177Asn were found in ~3.5% of ~1,200 AMD patients versus

~1% of ~1,200 controls (Allikmets, 2000). When different members of the fibulin gene family were screened for sequence variants in a cohort of 402 AMD patients, fibulin 5 missense variants were identified in 7 of the AMD patients (~1.5%) but no variants were seen in 429 controls (Stone et al., 2004). Current evidence suggests that the genes coding for the retinal-specific ATP-binding cassette transporter (ABCR) and fibulin 5, a close relative of fibulin 3/EFEMP1 (the protein that is altered in a JMD—malattia leventinese), might play a causal role in some patients with AMD.

11.6 Juvenile Retinoschisis (X-Linked)

Juvenile retinoschisis is a sex-linked form of macular degeneration, which almost exclusively affects male children. The disease manifests as a progressive loss of central vision as well as peripheral vision due to retinal degeneration. The fovea develops streaks and appears blistered and gets neovascularized. The disease is also associated with loss of the vitreous humor, which may lead to retinal detachment. The gene responsible for the disease has been identified as XLRS-1, a discoidin domain-containing protein secreted by photoreceptors and bipolar cells (Musarella, 2001).

11.7 Genetic Risk Factors for AMD

Several genome-wide linkage studies for AMD have detected linkage to chromosome 1q32. Recent reports by several groups show that there is an association between a common missense variant (Y402H) in CFH and AMD in the USA (Edwards et al., 2005; Hageman et al., 2005; Sepp et al., 2006). The sequence changes in the gene region of CFH apparently impair its binding to receptors on the cells of the retina and the choroids, thereby prevented it from inhibiting the inflammatory pathways (Haines et al., 2005; Klein et al., 2005b). Base substitution in CFH may also lead to decreased binding to CRP and heparin. If unchecked, the alternative pathway can foster inflammation in the retina and the surrounding blood vessels (Charbel Issa et al., 2005). The single nucleotide polymorphism Y402H present in approximately every third copy of the gene and the resulting substitution of histidine for tyrosine at codon 402 of the CFH protein leads to more than a twofold increase in risk of AMD. CFH protein has been detected in choriocapillaris and within soft drusen of AMD patients (Hageman et al., 2005). Other complement regulatory proteins such as CFB and component 2 (C2) have also been implicated in AMD (Gold et al., 2006). CFB is a positive activator of the alternative complement pathway and competes with CFH for binding to C3b62 (Gold et al., 2006). On binding to C3b, CFB is cleaved by complement factor D, forming the active C3bBb protease. C3bBb cleaves additional molecules of C3 to C3b (hence its name, "C3 convertase"), generating a positive feedback loop of activated complement and releasing the proinflammatory peptide C3a. C3bBb also proteolytically activates complement factor 5 (C5), which leads ultimately to the assembly of the membrane attack complex (MAC) and to the production of additional proinflammatory peptides. C2 has a somewhat analogous role to BF in the context of the classical pathway of complement activation. Proteolytically activated C2 binds to activated C4 to form the complex C4bC2a, which then proteolytically activates C3 and C5 (Rattner and Nathans, 2006).

Other recent studies have identified changes in the HTRA1 promoter, which is associated with increased expression of this secreted serine protease in AMD (Dewan et al., 2006; Yang et al., 2006). HTRA1 is implicated in cartilage or extracellular matrix degradation and may be involved in stimulating an inflammatory cascade.

12 Animal Models for Retinal Degeneration

The major advantage of studying retinal degeneration as a model for neurodegeneration in the brain is the easy access to the eye for noninvasive assessment. This had led to the detailed analysis of a number of important genetic and environmental factors for retinal degeneration. Based on these models, retinal degeneration models have been developed in several animal systems. Genetic models in mice have become valuable in studying various aspects of retinal degeneration and are listed in ❷ *Tables 7-2* and ❷ *7-3*. Since environmental models can be readily created and studied, one of these is considered in detail later.

◻ Table 7-2
Transgenic mouse models for AMD

Name	Genetic modification	AMD features	References
APO B100	Transgenic: expressing human APO B100 gene	Basal laminar/linear deposits	Espinosa-Heidmann et al. (2004)
APOE genotype	ApoE TR mice fed a HF-C diet	Human AMD model with CNV	Malek et al. (2005)
ELOVL4	Transgenic: expressing human mutant ELOVL4 (elongation of very long-chain fatty acid) construct;	Lipofuscin, PR degeneration, PR and RPE atrophy, abnormal ERG	Karan et al. (2005)
APO*E3-Leiden	Transgenic: expressing human APO* E3-Leiden gene (dysfunctional apolipoprotein);	BLD, drusen	Kliffen et al. (2000)
TIMP3	22q12.1–q13.2 Knock-in TIMP3 allele;	Sorsby Fundus Dystrophy BLD	Felbor et al. (1997a); Weber et al. (2002)
mcd/mcd	Transgenic: expressing mutated human CatD (cathepsinD)	Lipofuscin, BLD, PR and RPE atrophy, GA, RPE proliferation, abnormal ERG	Rakoczy et al. (2002); Zhang et al. (2002b)
RPE65/VEGF	Transgenic: expressing murine VEGF	Intrachoroidal neovascularization	Schwesinger et al. (2001)
rho/rtTA-TRE/VEGF	Transgenic: expressing human VEGF	Retinal neovascularization	Okamoto et al. (1997)
Rho-hPK1	Transgenic: expressing CNV	Retinal neovascularization	Tanaka et al. (2005)

Abbreviations: A2E, N-retinylidene-N-retinylethanolamine; BLD, basal laminar/linear deposits; CNV, choroidal neovascularization; ERG, electroretinography; GA, geographic atrophy; RPE, retinal pigment epithelial; PR, photoreceptor

12.1 Light Damage Model of Photoreceptor Degeneration

A number of studies have shown that the development of AMD is significantly correlated with exposure to sunlight (Taylor et al., 1990; Tomany et al., 2004a), suggesting that light damage is a good model for the disease. Retinal light damage is under investigation in experimental animal models to identify the pathological cellular and molecular events (Wenzel et al., 2005). Light as an environmental factor has been shown to be toxic to rodent rod photoreceptors if the retina is exposed to intermediate to high light levels over a long period of time (Penn and Anderson, 1992). Oxidative stress has been implicated as the main trigger for photoreceptor cell death (Li et al., 1985; Noell et al., 1987; Tanito et al., 2002; Ohira et al., 2003). In addition, it has been suggested that photoreceptors (which are the main oxygen consuming cells) in the retina degenerate as they are hyperoxic (Penn and Anderson, 1992), and the choroidal vasculature (the main oxygen supplier to the photoreceptors) cannot autoregulate or adequately provide the oxygen requirement in the photoreceptors under these conditions of continuous use. Thus, the photoreceptors in these three models may be challenged in addition to the individual stressors, by an increase in oxygen. The resulting cascade that leads to degeneration is not fully understood, and is considered in the next section.

◘ Table 7-3
Knockout mouse models for AMD

Name	Genetic modification	AMD features	Reference
ABCR4−/−1p21−p22	Gene encodes a transmembrane rim protein located in the discs of rod and foveal cone outer segments	Lipofuscin fluorophore, A2E, atrophy	Weng et al. (1999); Mata et al. (2000); Radu et al. (2004)
ApoE(−)	Knockout: murine ApoE gene (apolipoprotein E)	BLD	Dithmar et al. (2000)
CCL2−/−	CCL2−/−: Knockout	Lipofuscin, A2E, BLD, drusen, PR atrophy, CNV	Kuziel et al. (1997); Ambati et al. (2003)
CCR2−/−	CCR2−/−: Knockout mouse	PR and RPE atrophy, CNV	Hahn et al. (2004)
cp−/−Heph−/Y (ceruloplasmin)	cp gene crossed with spontaneous sex-linked anemia (sla) mutation in C57BL/6 mouse	PR and RPE atrophy, CNV	Hahn et al. (2004)
TIMP3	22q12.1−q13.2 Knock-in TIMP3 allele	Sorsby Fundus dystrophy BLD	Felbor et al. (1997a); Weber et al. (2002)
LDLR	Kockout: LDL receptor	Degeneration of Bruch's membrane with accumulation of lipid particles	Rudolf et al. (2004)
VLDLR (Vldlrtm1 Her)	Knockout: disrupted mouse VLDLR gene	Retinal spots and subretinal neovascularization	Frykman et al. (1995); Heckenlively et al. (2003)

Albinism is a genetic condition resulting from lack of pigmentation in the eyes, skin, and hair. Patients suffering from oculocutaneous albinism (OCA) inherit mutations in tyrosinase (type-I) or a tyrosinase transport protein (type-II) and are unable to synthesize the melanin pigment. These patients lack the protective shield provided by melanin and are highly photosensitive and also generally suffer from partial-sightedness—either near-sighted or far-sighted. Melanin in RPE and choroids may protect the macular region from photooxidative effect by its antioxidant capability and there has always been an inverse correlation between the melanin content of the eye and the incidence of AMD.

A frequently used animal model for studying retinal degeneration is light-induced loss of photoreceptors in albino mice and rats (Anderson and Penn, 2004; Richards et al., 2006). Photochemical damage of retina has been the most extensively studied form of light damage owing to its ability to cause damage under ambient conditions and its potential role in retinal damage and the mechanism is extremely complicated and is still relatively poorly understood. Exposing animals to constant light first results in photostasis, a mechanisms whereby the outer segments are shortened in order to adapt to the changed light intensity. Continued exposure leads to disruption of outer segment structure, pyknosis of mitochondria, followed by photoreceptor cell loss (Wu et al., 2006). Complete loss of photoreceptors will subsequently trigger extensive retinal remodeling (Marc et al., 2003). Damage triggers neurodegenerative pathways including an increase in oxidative stress, inflammation, autophagy, as well as apoptosis and a misregulation in glucose metabolism (Lohr et al., 2006). In addition to direct photochemical damage to the photoreceptors, damage might be exacerbated indirectly by damaging effects to the RPE produced by the increased demand on the RPE to phagocytose the damaged photoreceptor outer segments. Damage to the rods and cones has been shown in other models to be associated with accumulation of incompletely digested phagosomes to form lipofuscin granules and A2E beneath the RPE layer (Finnemann and Silverstein, 2001; Finnemann et al., 2002). One hypothesis is that accumulation of these granules inhibits lysosomal function and glycosaminoglycan catabolism, resulting in a build-up of undigested material in the RPE. This accumulation of material may finally lead to the formation of drusen in AMD (Glickman, 2002; Yoshida et al., 2005).

13 Mechanisms of Photoreceptor Degeneration

Photoreceptors are very stable, but at the same time extremely fragile cells. What makes them extremely fragile is their specialization, the conversion of light to a neurochemical signal, and the energy requirements to carry out this task (Stone et al., 1999). Photoreceptor degeneration can result from any changes that alter the signal transduction cascade associated with the photocycle, influence the energy metabolism or the oxygen tension in the outer retina, or disturb the phagocytotic process by the retinal pigment epithelium (Pierce, 2001).

Cell death serves a number of different functions to maintain tissue homeostasis as well as to aid in tissue remodeling. Three different types of cell death have been distinguished—apoptosis, autophagy, and necrosis. Apoptosis (Type I cell death) is an ordered process whereby the entire cell is disassembled into membrane-enclosed vesicles that are then recognized and removed by phagocytes. This removal process prevents intracellular components from being released from the dying cells to avoid activation of the immune response (Danial and Korsmeyer, 2004). Autophagy (Type II) is a process by which a cell compartmentalizes and digests itself, thus providing a mechanism by which cell death can be controlled in damaged cells (Klionsky and Emr, 2000; Levine and Klionsky, 2004). Finally, Necrosis (Type III) refers to a process in which cells swell and rupture, releasing their intracellular components into the surrounding tissue, causing an inflammatory response. This inflammatory response can either trigger repair processes or further tissue damage (Proskuryakov et al., 2003).

Although there is clear evidence that apoptotic mechanisms are involved in photoreceptor cell death (Travis, 1998), they seem not to be sufficient to account for all the cell loss occurring in photoreceptor dystrophies. We have discussed this issue previously with the example of three mouse models of photoreceptor dystrophy, the rd1 (Farber, 1995) the rd2 mouse (Travis et al., 1989), and the light damage model in the albino mouse; and the reader is referred to this manuscript (Lohr et al., 2006). In short, in all three models, based on the current literature, additional mechanisms other then caspase-mediated apoptosis were required to account for the demise of the photoreceptors. These additional mechanisms were found to include inflammation as well as autophagy, as well as a misregulation in glucose metabolism possibly leading to necrosis (Lohr et al., 2006). However, it is unclear which events in degeneration trigger which pathways. It would be of great interest to further explore the possibility that amyloid released in the retina triggers lysosome activity (autophagy) and that increased lysosome activity is correlated with improved photoreceptor cell survival in the presence of noxic stimuli such as Aβ-overload or excessive light exposure.

14 Shared Features of the AD and AMD

AD and AMD clearly represent two important diseases of the central nervous system and share an important property of premature neurodegeneration due to as yet unidentified causes. Aging is the major risk factor for both diseases, although AD and AMD do not appear to be an inevitable consequence of aging with a number of individuals having a perfectly healthy mind and a healthy retina until their death. The most interesting correlation between AD and AMD is the presence of extracellular proteinaceous deposits (plaques or drusen) in the two diseases. In addition, the finding that ApoE and Aβ, hallmarks of AD, are deposited in the drusen associated with AMD raises the possibility that common mechanisms of degeneration are involved in both disorders (Anderson et al., 2001; Dentchev et al., 2003; Anderson et al., 2004). Both diseases show a large inflammation component as indicated by the role of complement factors above and multiple different pathways in AD as suggested in the selected examples (Emmerling et al., 2000; Cole et al., 2005; Craft, 2005; Rosenberg, 2005; Walker and Lue, 2005; Papadopoulos et al., 2006). Please note that we are excluding a number of seminal contributions in this area due to the large number of publications in this field. Both diseases also show a vascular component as shown by the importance of cerebrovascular amyloid (Khalil et al., 1996; Wyss-Coray et al., 2001; Chauhan and Siegel, 2003; Vagnucci and Li, 2003; Irizarry, 2004; Lott and Head, 2005). A third important correlation is that a high-cholesterol or high-fat diet is a risk factor for both diseases, although the exact mechanisms and properties of this relationship remain to be discovered (Sparks et al., 2000; Teunissen et al., 2002; Flirski and Sobow, 2005; Donaldson and Pulido, 2006; Wolozin et al., 2006). However, despite the fact that AD and AMD share so many risk factors, the correlation between the two diseases appears to be very poor and visual structures appear to be spared in AD (Kergoat et al., 2002). Indeed the defects associated with AD are in the ganglion cell and long nerve fibers, which support our hypothesis of faulty protein trafficking, but do not simultaneously generate the signature deposits of AD or AMD (Hedges et al., 1996; Blanks et al., 1996a, b). One explanation may lie in the strange and unexplained relationship between AD, AMD, and ApoE. As discussed earlier, there are three forms of ApoE—E2, E3, and E4—coded by three codominant alleles with different biochemical properties and pathologic consequences. The E3 allele is considered to be the ancestral allele, and E2 and E4 are considered as variants, on the basis of single point mutations. APOE's polymorphism is of particular interest within the framework of neurodegeneration, for it is strongly associated with the risk of Alzheimer's disease and may be associated with various other neurodegenerative disorders. Moreover, ApoE is expressed in lesions that characterize Alzheimer's disease, Down's syndrome, and prion diseases. The E4 allele, associated with elevated cholesterol concentrations, has also shown to increase the risk of coronary heart disease, decrease longevity, and to result in the occurrence of AD (Mahley and Huang, 1999). ApoE polymorphisms have also been correlated with the incidence of age-related macular degeneration, the leading cause of irreversible blindness in many countries. In age-related macular degeneration, however, the E4 allele appears to confer protection, whereas the E2 allele is apparently associated with a moderately increased incidence of the disease (Klaver et al., 1998; Souied et al., 1998). The different alleles associated with AD and AMD may therefore confound the statistics of correlation between sporadic forms of AD and AMD. To our knowledge, an important unexplored area is the study of AMD correlations in FAD families, as the diseases causing mutations are inherited in the eyes and the brain. Moreover, the effect of the E4 allele appears to be attenuated in FAD. Thus, the alleles of ApoE may play a role in disease pathogenesis, which is not really related to its capacity to chaperone Aβ and foster its aggregation.

There are a number of reasons to believe significant association between APP immunoreactivity and RPE cytoplasm, which flank or overlie drusen. First, studies detected the deposition of Aβ in the retina of AMD patients (Johnson et al., 2002). Second, anti-Aβ labeled structures resembling amyloid vesicles can be identified intracellularly in the RPE cell cytoplasm. Third, cultured human RPE cells label with APP and Aβ antibodies and contain transcripts for all three APP isoforms, as well as β-secretase (Loffler et al., 1995; Dentchev et al., 2003). Fourth, the Aβ-degrading enzyme, neprilysin, expressed in the RPE, appears to play an important role in retinal maintenance and protection (Yoshida et al., 2005). These results point strongly toward an RPE origin for Aβ peptides. All of these findings are consistent with the conclusion that the RPE has the capacity to synthesize significant amounts of APP and to generate Aβ through enzymatic processing. Thus, the available evidence suggests that the amyloid vesicles in drusen are derived from degenerate RPE cells, which contain Aβ and, perhaps, other molecules highly resistant to proteolytic degradation

The human lens is also vulnerable to age-dependent degenerative changes and shows progressive deposition of insoluble protein and extensive oxidative damage. Early onset cataracts and AD are typical comorbid disorders in adults with Down's syndrome and in those with familial Danish dementia, an AD variant with cerebral Aβ amyloidosis (Frederikse and Ren, 2002). H_2O_2 or UV treatment of monkey eyes (a model for cataract) was shown to induce APP, which brings together AD and lens degeneration (Frederikse et al., 1996). Conversely, αB-crystallin, an abundant cytosolic lens protein and small heat-shock protein with molecular chaperone properties, is expressed in brain of individuals with Alzheimer's disease. Moreover, Aβ interacts with αB-crystallin in vitro and in Aβ-expressing transgenic *Caenorhabditis elegans* (Goldstein et al., 2003). Although the lens fibers are not neurons, they do share the property of being long and thin and therefore vulnerable to agents that impair protein transport.

Despite the shared similarities mentioned earlier, AMD has not been classified as an amyloid disease. Among the principal differences is the fact that classical amyloid diseases typically exhibit large amounts of amyloid fibrils. For example, in the case of AD, the characteristic plaques consist primarily of fibrillar Aβ peptide whereas amyloid fibrils are elongated, 6–15 nm-wide, rod-like structures of indeterminate length, which are characterized by a common cross β structure (Sunde and Blake, 1997). In addition to their related structural features, amyloid fibrils display characteristic tinctorial properties, such as thioflavin T and Congo red staining (Krebs et al., 2005). Although amyloid proteins such as the Aβ peptide, transthyretin, immunoglobulin light chains, and amyloid A are found in drusen and sub-RPE deposits (Anderson et al., 2002, 2004), electron microscopy studies have yielded sparse evidence of the presence of bona fide amyloid fibrils. Nevertheless, recent studies have detected oligomeric forms of Aβ in the drusen of AMD patients and suggest that this may play a causative role in the atrophy.

Acknowledgments

We thank the American Health Assistance Foundation for providing grants to KS and BR and the NIH (NIA) for support to KS. We also thank Ms Meera Parasuraman and Ms Christina Demos for reading the manuscript and providing useful criticism.

References

Akeo K, Hiramitsu T, Kanda T, Karasawa Y, Okisaka S. 1996. Effects of superoxide dismutase and catalase on growth of retinal pigment epithelial cells in vitro following addition of linoleic acid or linoleic acid hydroperoxide. Ophthalmic Res 28: 8-18.

Alamouti B, Funk J. 2003. Retinal thickness decreases with age: An OCT study. Br J Ophthalmol 87: 899-901.

Alaupovic P. 1982. The role of apolipoproteins in lipid transport processes. Ric Clin Lab 12: 3-21.

Allikmets R. 1999. Molecular genetics of age-related macular degeneration: Current status. Eur J Ophthalmol 9: 255-265.

Allikmets R. 2000. Further evidence for an association of ABCR alleles with age-related macular degeneration. The International ABCR Screening Consortium. Am J Hum Genet 67: 487-491.

Allikmets R, Seddon JM, Bernstein PS, Hutchinson A, Atkinson A, et al. 1999. Evaluation of the Best disease gene in patients with age-related macular degeneration and other maculopathies. Hum Genet 104: 449-453.

Ambati J, Anand A, Fernandez S, Sakurai E, Lynn BC, et al. 2003. An animal model of age-related macular degeneration in senescent ccl-2- or ccr-2- deficient mice. Nat Med 9: 1390-1397.

Ames A 3rd. 1992. Energy requirements of CNS cells as related to their function and to their vulnerability to ischemia: A commentary based on studies on retina. Can J Physiol Pharmacol 70 Suppl: S158-S164.

Anderson Jr. B 1963. Photocoagulation in the treatment of ocular diseases. N C Med J 24: 24-27.

Anderson Jr. B 1968. Ocular effects of changes in oxygen and carbon dioxide tension. Trans Am Ophthalmol Soc 66: 423-474.

Anderson DH, Ozaki S, Nealon M, Neitz J, Mullins RF, et al. 2001. Local cellular sources of apolipoprotein E in the human retina and retinal pigmented epithelium: Implications for the process of drusen formation. Am J Ophthalmol 131: 767-781.

Anderson DH, Mullins RF, Hageman GS, Johnson LV. 2002. A role for local inflammation in the formation of drusen in the aging eye. Am J Ophthalmol 134: 411-431.

Anderson DH, Talaga KC, Rivest AJ, Barron E, Hageman GS, et al. 2004. Characterization of beta amyloid assemblies in drusen: The deposits associated with aging and age-related macular degeneration. Exp Eye Res 78: 243-256.

Anderson RE, Penn JS. 2004. Environmental light and heredity are associated with adaptive changes in retinal DHA levels that affect retinal function. Lipids 39: 1121-1124.

Arnold J, Sarks S. 2002. Age related macular degeneration. Clin Evid 8: 614-628.

Atwood CS, Robinson SR, Smith MA. 2002. Amyloid-beta: Redox-metal chelator and antioxidant. J Alzheimers Dis 4: 203-214.

Awenius C, Hankeln T, Burmester T. 2001. Neuroglobins from the zebrafish Danio rerio and the pufferfish Tetraodon nigroviridis. Biochem Biophys Res Commun 287: 418-421.

Baird PN, Guida E, Chu DT, Vu HT, Guymer RH. 2004. The epsilon2 and epsilon4 alleles of the apolipoprotein gene are associated with age-related macular degeneration. Invest Ophthalmol Vis Sci 45: 1311-1315.

Bartlett H, Eperjesi F. 2003. Age-related macular degeneration and nutritional supplementation: A review of randomised controlled trials. Ophthalmic Physiol Opt 23: 383-399.

Bayer AU, Ferrari F, Erb C. 2002. High occurrence rate of glaucoma among patients with Alzheimer's disease. Eur Neurol 47: 165-168.

Bennett DA, Schneider JA, Arvanitakis Z, Kelly JF, Aggarwal NT, et al. 2006. Neuropathology of older persons without cognitive impairment from two community-based studies. Neurology 66: 1837-1844.

Bennett DA, Schneider JA, Buchman AS, Mendes de Leon C, Bienias JL, et al. 2005. The Rush Memory and Aging Project: Study design and baseline characteristics of the study cohort. Neuroepidemiology 25: 163-175.

Bennett J, Tanabe T, Sun D, Zeng Y, Kjeldbye H, et al. 1996. Photoreceptor cell rescue in retinal degeneration (rd) mice by in vivo gene therapy. Nat Med 2: 649-654.

Bentmann A, Schmidt M, Reuss S, Wolfrum U, Hankeln T, et al. 2005. Divergent distribution in vascular and avascular mammalian retinae links neuroglobin to cellular respiration. J Biol Chem 280: 20660-20665.

Besharse JC, Spratt G, Forestner DM. 1986. Light-evoked and kainic-acid-induced disc shedding by rod photoreceptors: Differential sensitivity to extracellular calcium. J Comp Neurol 251: 185-197.

Bill A, Tornquist P, Alm A. 1980. Permeability of the intraocular blood vessels. Trans Ophthalmol Soc UK 100: 332-336.

Bishop PN, Holmes DF, Kadler KE, McLeod D, Bos KJ. 2004. Age-related changes on the surface of vitreous collagen fibrils. Invest Ophthalmol Vis Sci 45: 1041-1046.

Blanks JC, Hinton DR, Sadun AA, Miller CA. 1989. Retinal ganglion cell degeneration in Alzheimer's disease. Brain Res 501: 364-372.

Blanks JC, Schmidt SY, Torigoe Y, Porrello KV, Hinton DR, et al. 1996b. Retinal pathology in Alzheimer's disease. II. Regional neuron loss and glial changes in GCL. Neurobiol Aging 17: 385-395.

Blanks JC, Torigoe Y, Hinton DR, Blanks RH. 1996a. Retinal pathology in Alzheimer's disease. I. Ganglion cell loss in foveal/parafoveal retina. Neurobiol Aging 17: 377-384.

Bok D. 1993. The retinal pigment epithelium: A versatile partner in vision. J Cell Sci Suppl 17: 189-195.

Boyer MM, Poulsen GL, Nork TM. 2000. Relative contributions of the neurosensory retina and retinal pigment epithelium to macular hypofluorescence. Arch Ophthalmol 118: 27-31.

Boyles JK, Zoellner CD, Anderson LJ, Kosik LM, Pitas RE, et al. 1989 A role for apolipoprotein E, apolipoprotein A-I, and low density lipoprotein receptors in cholesterol transport during regeneration and remyelination of the rat sciatic nerve. J Clin Invest 83: 1015-1031.

Brown MM, Brown GC, Stein JD, Roth Z, Campanella J, et al., eds. 2005. Age-related macular degeneration: Economic burden and value-based medicine analysis.

Brubaker RF. 1991 Flow of aqueous humor in humans [The Friedenwald Lecture]. Invest Ophthalmol Vis Sci 32: 3145-3166.

Burgess BL, McIsaac SA, Naus KE, Chan JY, Tansley GH, et al. 2006. Elevated plasma triglyceride levels precede amyloid deposition in Alzheimer's disease mouse models with abundant A beta in plasma. Neurobiol Dis 24: 114-127.

Bylsma GW, Guymer RH. 2005. Treatment of age-related macular degeneration. Clin Exp Optom 88: 322-334.

Charbel Issa P, Scholl HP, Holz FG, Knolle P, Kurts C. 2005. [The complement system and its possible role in the pathogenesis of age-related macular degeneration (AMD)]. Ophthalmologe 102: 1036-1042.

Chauhan NB, Siegel GJ. 2003. Intracerebroventricular passive immunization with anti-Abeta antibody in Tg2576. J Neurosci Res 74: 142-147.

Chen C, Wu L, Jiang F, Liang J, Wu DZ. 2003. Scotopic sensitivity of central retina in early age-related macular degeneration. Yan Ke Xue Bao 19: 15-19.

Chiang AP, Beck JS, Yen HJ, Tayeh MK, Scheetz TE, et al. 2006. Homozygosity mapping with SNP arrays identifies TRIM32, an E3 ubiquitin ligase, as a Bardet–Biedl syndrome gene (BBS11). Proc Natl Acad Sci USA 103: 6287-6292.

Cideciyan AV, Hood DC, Huang Y, Banin E, Li ZY, et al. 1998. Disease sequence from mutant rhodopsin allele to rod and cone photoreceptor degeneration in man. Proc Natl Acad Sci USA 95: 7103-7108.

Cole GM, Lim GP, Yang F, Teter B, Begum A, et al. 2005. Prevention of Alzheimer's disease: Omega-3 fatty acid and phenolic anti-oxidant interventions. Neurobiol Aging 26 Suppl 1: 133-136.

Cole SL, Grudzien A, Manhart IO, Kelly BL, Oakley H, Vassar R. 2005. Statins cause intracellular accumulation of amyloid precursor protein, beta-secretase-cleaved fragments, and amyloid beta-peptide via an isoprenoid-dependent mechanism. J Biol Chem 280: 18755-70.

Cousins SW, Espinosa-Heidmann DG, Alexandridou A, Sall J, Dubovy S, et al. 2002. The role of aging, high fat diet and blue light exposure in an experimental mouse model for basal laminar deposit formation. Exp Eye Res 75: 543-553.

Crabb JW, Miyagi M, Gu X, Shadrach K, West KA, et al. 2002. Drusen proteome analysis: An approach to the etiology of age-related macular degeneration. Proc Natl Acad Sci USA 99: 14682-14687.

Craft S. 2005. Insulin resistance syndrome and Alzheimer's disease: Age- and obesity-related effects on memory, amyloid, and inflammation. Neurobiol Aging 26(Suppl 1): 65-69.

Curcio CA, Medeiros NE, Millican CL. 1996. Photoreceptor loss in age-related macular degeneration. Invest Ophthalmol Vis Sci 37: 1236-1249.

Curcio CA. 2001. Photoreceptor topography in ageing and age-related maculopathy. Eye 15: 376-383.

Curcio CA, Owsley C, Jackson GR. 2000. Spare the rods, save the cones in aging and age-related maculopathy. Invest Ophthalmol Vis Sci 41: 2015-2018.

Curcio CA, Presley JB, Malek G, Medeiros NE, Avery DV, et al. 2005a. Esterified and unesterified cholesterol in drusen and basal deposits of eyes with age-related maculopathy. Exp Eye Res 81: 731-741.

Curcio CA, Presley JB, Millican CL, Medeiros NE. 2005b. Basal deposits and drusen in eyes with age-related maculopathy: Evidence for solid lipid particles. Exp Eye Res 80: 761-775.

Danial NN, Korsmeyer SJ. 2004. Cell death: Critical control points. Cell 116: 205-219.

Dasch B, Fuhs A, Schmidt J, Behrens T, Meister A, et al. 2005. Serum levels of macular carotenoids in relation to age-related maculopathy: The Muenster Aging and Retina Study (MARS). Graefes Arch Clin Exp Ophthalmol 243: 1028-1035.

Defay R, Pinchinat S, Lumbroso S, Sutan C, Delcourt C. 2004. Sex steroids and age-related macular degeneration in older French women: The POLA study. Ann Epidemiol 14: 202-208.

Dentchev T, Milam AH, Lee VM, Trojanowski JQ, Dunaief JL. 2003. Amyloid-beta is found in drusen from some age-related macular degeneration retinas, but not in drusen from normal retinas. Mol Vis 9: 184-190.

Dewan A, Liu M, Hartman S, Zhang SS, Liu DT, et al. 2006. HTRA1 promoter polymorphism in wet age-related macular degeneration. Science 314: 989-992.

Dithmar S, Curcio CA, Le NA, Brown S, Grossniklaus HE. 2000. Ultrastructural changes in Bruch's membrane of apolipoprotein E-deficient mice. Invest Ophthalmol Vis Sci 41: 2035-2042.

Dollfus H, Muller J, Stoetzel C, Laurier V, Bonneau D, et al. 2006. [Bardet–Biedl syndrome: A unique family for a major gene (BBS10)]. Med Sci (Paris) 22: 901-904.

Donaldson MJ, Pulido JS. 2006. Treatment of nonexudative (dry) age-related macular degeneration. Curr Opin Ophthalmol 17: 267-274.

Donoso LA, Kim D, Frost A, Callahan A, Hageman G. 2006. The role of inflammation in the pathogenesis of age-related macular degeneration. Surv Ophthalmol 51: 137-152.

Edwards AO, Donoso LA, Ritter R 3rd. 2001. A novel gene for autosomal dominant Stargardt-like macular dystrophy with homology to the SUR4 protein family. Invest Ophthalmol Vis Sci 42: 2652-2663.

Edwards AO, Ritter R, Abel KJ, 3rd. Manning A, Panhuysen C, et al. 2005. Complement factor H polymorphism and age-related macular degeneration. Science 308: 421-424.

Emmerling MR, Watson MD, Raby CA, Spiegel K. 2000. The role of complement in Alzheimer's disease pathology. Biochim Biophys Acta 1502: 158-171.

Espinosa-Heidmann DG, Sall J, Hernandez EP, Cousins SW. 2004. Basal laminar deposit formation in APO B100 transgenic mice: Complex interactions between dietary fat, blue light, and vitamin E. Invest Ophthalmol Vis Sci 45: 260-266.

Esterbauer H, Dieber-Rotheneder M, Striegl G, Waeg G. 1991a. Role of vitamin E in preventing the oxidation of low-density lipoprotein. Am J Clin Nutr 53: 314S-321S.

Esterbauer H, Puhl H, Dieber-Rotheneder M, Waeg G, Rabl H. 1991b. Effect of antioxidants on oxidative modification of LDL. Ann Med 23: 573-581.

Evans JR, Fletcher AE, Wormald RP. 2004. Causes of visual impairment in people aged 75 years and older in Britain: An add-on study to the MRC Trial of Assessment and Management of Older People in the Community. Br J Ophthalmol 88: 365-370.

Farber DB. 1995. From mice to men: The cyclic GMP phosphodiesterase gene in vision and disease. The Proctor Lecture Invest Ophthalmol Vis Sci 36: 263-275.

Felbor U, Doepner D, Schneider U, Zrenner E, Weber BH. 1997a. Evaluation of the gene encoding the tissue inhibitor of metalloproteinases-3 in various maculopathies. Invest Ophthalmol Vis Sci 38: 1054-1059.

Felbor U, Suvanto EA, Forsius HR, Eriksson AW, Weber BH. 1997b. Autosomal recessive Sorsby fundus dystrophy revisited: Molecular evidence for dominant inheritance. Am J Hum Genet 60: 57-62.

Finnemann SC, Silverstein RL. 2001. Differential roles of CD36 and alphavbeta5 integrin in photoreceptor phagocytosis by the retinal pigment epithelium. J Exp Med 194: 1289-1298.

Finnemann SC, Leung LW, Rodriguez-Boulan E. 2002. The lipofuscin component A2E selectively inhibits phagolysosomal degradation of photoreceptor phospholipid by the retinal pigment epithelium. Proc Natl Acad Sci USA 99: 3842-3847.

Flirski M, Sobow T. 2005. Biochemical markers and risk factors of Alzheimer's disease. Curr Alzheimer Res 2: 47-64.

Fraering PC, Ye W, Strub JM, Dolios G, La Voie MJ, et al. 2004. Purification and characterization of the human gamma-secretase complex. Biochemistry 43: 9774-9789.

Fraser-Bell S, Wu J, Klein R, Azen SP, Varma R. 2006. Smoking, alcohol intake, estrogen use, and age-related macular degeneration in Latinos: The Los Angeles Latino Eye Study. Am J Ophthalmol 141: 79-87.

Frederikse PH, Garland D, Zigler Jr. JS, Piatigorsky J. 1996. Oxidative stress increases production of beta-amyloid precursor protein and beta-amyloid (Abeta) in mammalian lenses, and Abeta has toxic effects on lens epithelial cells. J Biol Chem 271: 10169-10174.

Frederikse PH, Ren XO. 2002. Lens defects and age-related fiber cell degeneration in a mouse model of increased AbetaPP gene dosage in Down syndrome. Am J Pathol 161: 1985-1990.

Frederikse PH, Zigler Jr. JS 1998. Presenilin expression in the ocular lens. Curr Eye Res 17: 947-952.

Frykman PK, Brown MS, Yamamoto T, Goldstein JL, Herz J. 1995. Normal plasma lipoproteins and fertility in gene-targeted mice homozygous for a disruption in the gene encoding very low density lipoprotein receptor. Proc Natl Acad Sci USA 92: 8453-8457.

Gaillard ER, Atherton SJ, Eldred G, Dillon J. 1995. Photophysical studies on human retinal lipofuscin. Photochem Photobiol 61: 448-453.

Gao H, Hollyfield JG. 1992. Aging of the human retina. Differential loss of neurons and retinal pigment epithelial cells. Invest Ophthalmol Vis Sci 33: 1-17.

Gass JD. 2003. Drusen and disciform macular detachment and degeneration. 1972. Retina 23: 409-436.

Glickman RD. 2002. Phototoxicity to the retina: Mechanisms of damage. Int J Toxicol 21: 473-490.

Gold B, Merriam JE, Zernant J, Hancox LS, Taiber AJ, et al. 2006. Variation in factor B (BF) and complement component 2 (C2) genes is associated with age-related macular degeneration. Nat Genet 38: 458-462.

Goldstein LE, Muffat JA, Cherny RA, Moir RD, Ericsson MH, et al. 2003. Cytosolic beta-amyloid deposition and supranuclear cataracts in lenses from people with Alzheimer's disease. Lancet 361: 1258-1265.

Gray H. 1901. Anatomy, descriptive and surgical. Philadelphia, PA: Running Press.

Green WR, Enger C. 1993. Age-related macular degeneration histopathologic studies. The 1992 Lorenz E. Zimmerman Lecture. Ophthalmology 100: 1519-1535.

Grizzard SW, Arnett D, Haag SL. 2003. Twin study of age-related macular degeneration. Ophthalmic Epidemiol 10: 315-322.

Guirao A, Gonzalez C, Redondo M, Geraghty E, Norrby S, et al. 1999. Average optical performance of the human eye as a function of age in a normal population. Invest Ophthalmol Vis Sci 40: 203-213.

Haddad S, Chen CA, Santangelo SL, Seddon JM. 2006. The genetics of age-related macular degeneration: A review of progress to date. Surv Ophthalmol 51: 316-363.

Hageman GS, Anderson DH, Johnson LV, Hancox LS, Taiber AJ, et al. 2005. A common haplotype in the complement regulatory gene factor H (HF1/CFH) predisposes individuals to age-related macular degeneration. Proc Natl Acad Sci USA 102: 7227-7232.

Hahn P, Qian Y, Dentchev T, Chen L, Beard J, et al. 2004. Disruption of ceruloplasmin and hephaestin in mice causes retinal iron overload and retinal degeneration with features of age-related macular degeneration. Proc Natl Acad Sci USA 101: 13850-13855.

Haines JL, Hauser MA, Schmidt S, Scott WK, Olson LM, et al. 2005. Complement factor H variant increases the risk of age-related macular degeneration. Science 308: 419-421.

Haines JL, Schnetz-Boutaud N, Schmidt S, Scott WK, Agarwal A, et al. 2006. Functional candidate genes in age-related macular degeneration: Significant association with VEGF, VLDLR, and LRP6. Invest Ophthalmol Vis Sci 47: 329-335.

Hamilton WK, Ewing CC, Ives EJ, Carruthers JD. 1989. Sorsby's fundus dystrophy. Ophthalmology 96: 1755-1762.

Hardy J. 2006. Has the amyloid cascade hypothesis for Alzheimer's disease been proved? Curr Alzheimer Res 3: 71-73.

Hardy J, Selkoe DJ. 2002. The amyloid hypothesis of Alzheimer's disease: Progress and problems on the road to therapeutics. Science 297: 353-356.

Heckenlively JR, Hawes NL, Friedlander M, Nusinowitz S, Hurd R, et al. 2003. Mouse model of subretinal neovascularization with choroidal anastomosis. Retina 23: 518-22.

Hedges TR, 3rd, Perez Galves R, Speigelman D, Barbas NR, Peli E, et al. 1996. Retinal nerve fiber layer abnormalities in Alzheimer's disease. Acta Ophthalmol Scand 74: 271-275.

Hendrickson AE, Yuodelis C. 1984. The morphological development of the human fovea. Ophthalmology 91: 603-612.

Hicks D, Sahel J. 1999. The implications of rod-dependent cone survival for basic and clinical research. Invest Ophthalmol Vis Sci 40: 3071-3074.

Hilfer SR, Yang JW. 1980 Accumulation of CPC-precipitable material at apical cell surfaces during formation of the optic cup. Anat Rec 197: 423-433.

Hodes RJ, Cahan V, Pruzan M. 1996. The National Institute on Aging at its twentieth anniversary: Achievements and promise of research on aging. J Am Geriatr Soc 44: 204-206.

Holtzman DM, Bales KR, Tenkova T, Fagan AM, Parsadanian M, et al. 2000. Apolipoprotein E isoform-dependent amyloid deposition and neuritic degeneration in a mouse model of Alzheimer's disease. Proc Natl Acad Sci USA 97: 2892-2897.

Huang Y, Liu XQ, Wyss-Coray T, Brecht WJ, Sanan DA, et al. 2001. Apolipoprotein E fragments present in Alzheimer's disease brains induce neurofibrillary tangle-like intracellular inclusions in neurons. Proc Natl Acad Sci USA 98: 8838-8843.

Husain D, Ambati B, Adamis AP, Miller JW. 2002. Mechanisms of age-related macular degeneration. Ophthalmol Clin North Am 15: 87-91.

Hutter-Paier B, Huttunen HJ, Puglielli L, Eckman CB, Kim DY, et al. 2004. The ACAT inhibitor CP-113,818 markedly reduces amyloid pathology in a mouse model of Alzheimer's disease. Neuron 44: 227-238.

Hyman L, Neborsky R. 2002. Risk factors for age-related macular degeneration: An update. Curr Opin Ophthalmol 13: 171-175.

Hyman L, Schachat AP, He Q, Leske MC. 2000. Hypertension, cardiovascular disease, and age-related macular degeneration. Age-Related Macular Degeneration Risk Factors Study Group. Arch Ophthalmol 118: 351-358.

Irizarry MC. 2004. Biomarkers of Alzheimer disease in plasma. NeuroRx 1: 226-234.

Jackson GR, Owsley C. 2003. Visual dysfunction, neurodegenerative diseases, and aging. Neurol Clin 21: 709-728.

Johnson LV, Leitner WP, Rivest AJ, Staples MK, Radeke MJ, et al. 2002. The Alzheimer's A beta-peptide is deposited at sites of complement activation in pathologic deposits associated with aging and age-related macular degeneration. Proc Natl Acad Sci USA 99: 11830-11835.

Johnson LV, Leitner WP, Staples MK, Anderson DH. 2001. Complement activation and inflammatory processes in Drusen formation and age related macular degeneration. Exp Eye Res 73: 887-896.

Junemann A, Bleich S, Reulbach U, Henkel K, Wakili N, et al. 2004. Prospective case control study on genetic association of apolipoprotein epsilon2 with intraocular pressure. Br J Ophthalmol 88: 581-582.

Karan G, Lillo C, Yang Z, Cameron DJ, Locke KG, et al. 2005. Lipofuscin accumulation, abnormal electrophysiology, and photoreceptor degeneration in mutant ELOVL4 transgenic mice: A model for macular degeneration. Proc Natl Acad Sci USA 102: 4164-4169.

Kayatz P, Heimann K, Schraermeyer U. 1999. Tracing of benzidine-reactive substances in ROS, RPE and choroid after light-induced peroxidation. Graefes Arch Clin Exp Ophthalmol 237: 763-774.

Keen TJ, Inglehearn CF. 1996. Mutations and polymorphisms in the human peripherin-RDS gene and their involvement in inherited retinal degeneration. Hum Mutat 8: 297-303.

Kergoat H, Kergoat MJ, Justino L, Chertkow H, Robillard A, et al. 2002. Visual retinocortical function in dementia of the Alzheimer type. Gerontology 48: 197-203.

Khalil Z, Chen H, Helme RD. 1996. Mechanisms underlying the vascular activity of beta-amyloid protein fragment (beta A(4)25-35) at the level of skin microvasculature. Brain Res 736: 206-216.

Kivipelto M, Helkala EL, Hanninen T, Laakso MP, Hallikainen M, et al. 2001. Midlife vascular risk factors and late-life mild cognitive impairment: A population-based study. Neurology 56: 1683-1689.

Klaver CC, Kliffen M, van Duijn CM, Hofman A, Cruts M, et al. 1998. Genetic association of apolipoprotein E with age-related macular degeneration. Am J Hum Genet 63: 200-206.

Klein ML, Mauldin WM, Stoumbos VD. 1994a. Heredity and age-related macular degeneration. Observations in monozygotic twins. Arch Ophthalmol 112: 932-937.

Klein R, Klein BE, Knudtson MD, Wong TY, Shankar A, et al. 2005a. Systemic markers of inflammation, endothelial dysfunction, and age-related maculopathy. Am J Ophthalmol 140: 35-44.

Klein R, Klein BE, Moss SE. 1995. Age-related eye disease and survival. The Beaver Dam Eye Study. Arch Ophthalmol 113: 333-339.

Klein R, Klein BE, Wang Q, Moss SE. 1994b. Is age-related maculopathy associated with cataracts? Arch Ophthalmol 112: 191-196.

Klein RJ, Zeiss C, Chew EY, Tsai JY, Sackler RS, et al. 2005b. Complement factor H polymorphism in age-related macular degeneration. Science 308: 385-389.

Klein WL, Stine WB, Jr. Teplow DB. 2004. Small assemblies of unmodified amyloid beta-protein are the proximate neurotoxin in Alzheimer's disease. Neurobiol Aging 25: 569-580.

Kliffen M, Lutgens E, Daemen MJ, de Muinck ED, Mooy CM, et al. 2000. The APO(*)E3-Leiden mouse as an animal model for basal laminar deposit. Br J Ophthalmol 84: 1415-1419.

Klionsky DJ, Emr SD. 2000. Autophagy as a regulated pathway of cellular degradation. Science 290: 1717-1721.

Kolb H, Fernandez E, Nelson R. 2002. Webvision: The organization of the retina and visual system. vol 2006. http://webvision.med.utah.edu/index.html.

Kolsch H, Lutjohann D, von Bergmann K, Heun R. 2003. The role of 24S-hydroxycholesterol in Alzheimer's disease. J Nutr Health Aging 7: 37-41.

Krebs MR, Bromley EH, Donald AM. 2005. The binding of thioflavin-T to amyloid fibrils: Localisation and implications. J Struct Biol 149: 30-37.

Kuhrt H, Hartig W, Grimm D, Faude F, Kasper M, et al. 1997. Changes in CD44 and ApoE immunoreactivities due to retinal pathology of man and rat. J Hirnforsch 38: 223-229.

Kuziel WA, Morgan SJ, Dawson TC, Griffin S, Smithies O, et al. 1997. Severe reduction in leukocyte adhesion and monocyte extravasation in mice deficient in CC chemokine receptor 2. Proc Natl Acad Sci USA 94: 12053-12058.

Lake S, Liverani E, Desai M, Casson R, James B, et al. 2004. Normal tension glaucoma is not associated with the common apolipoprotein E gene polymorphisms. Br J Ophthalmol 88: 491-493.

Lalin SC, Chang S, Flynn H, Von Fricken M, Del Priore LV. 2004. Familial idiopathic macular hole. Am J Ophthalmol 138: 608-611.

La Vail MM, Yasumura D, Matthes MT, Lau-Villacorta C, Unoki K, et al. 1998. Protection of mouse photoreceptors by survival factors in retinal degenerations. Invest Ophthalmol Vis Sci 39: 592-602.

Lazarov VK, Fraering PC, Ye W, Wolfe MS, Selkoe DJ, et al. 2006. Electron microscopic structure of purified, active gamma-secretase reveals an aqueous intramembrane chamber and two pores. Proc Natl Acad Sci USA 103: 6889-6894.

Lee VM, Trojanowski JQ. 2006. Progress from Alzheimer's tangles to pathological tau points towards more effective therapies now. J Alzheimers Dis 9: 257-262.

Lester-Coll N, Rivera EJ, Soscia SJ, Doiron K, Wands JR, et al. 2006. Intracerebral streptozotocin model of type 3 diabetes: Relevance to sporadic Alzheimer's disease. J Alzheimers Dis 9: 13-33.

Leu ST, Batni S, Radeke MJ, Johnson LV, Anderson DH, et al. 2002. Drusen are cold spots for proteolysis: Expression of matrix metalloproteinases and their tissue inhibitor proteins in age-related macular degeneration. Exp Eye Res 74: 141-154.

Levine B, Klionsky DJ. 2004. Development by self-digestion: Molecular mechanisms and biological functions of autophagy. Dev Cell 6: 463-477.

Li C, Wong WH. 2001. Model-based analysis of oligonucleotide arrays: Expression index computation and outlier detection. Proc Natl Acad Sci USA 98: 31-36.

Li CM, Chung BH, Presley JB, Malek G, Zhang X, et al. 2005. Lipoprotein-like particles and cholesteryl esters in human Bruch's membrane: Initial characterization. Invest Ophthalmol Vis Sci 46: 2576-2586.

Li ZY, Tso MO, Wang HM, Organisciak DT. 1985. Amelioration of photic injury in rat retina by ascorbic acid: A histopathologic study. Invest Ophthalmol Vis Sci 26: 1589-1598.

Liles MR, Newsome DA, Oliver PD. 1991. Antioxidant enzymes in the aging human retinal pigment epithelium. Arch Ophthalmol 109: 1285-1288.

Loffler KU, Edward DP, Tso MO. 1995. Immunoreactivity against tau, amyloid precursor protein, and beta-amyloid in the human retina. Invest Ophthalmol Vis Sci 36: 24-31.

Lohr HR, Kuntchithapautham K, Sharma AK, Rohrer B. 2006. Multiple, parallel cellular suicide mechanisms participate in photoreceptor cell death. Exp Eye Res 83: 380-389.

Lolley RN. 1994. The rd gene defect triggers programmed rod cell death. The Proctor Lecture. Invest Ophthalmol Vis Sci 35: 4182-4191.

Lotery AJ, Jacobson SG, Fishman GA, Weleber RG, Fulton AB, et al. 2001. Mutations in the CRB1 gene cause Leber congenital amaurosis. Arch Ophthalmol 119: 415-420.

Lott IT, Head E. 2005. Alzheimer disease and Down syndrome: Factors in pathogenesis. Neurobiol Aging 26: 383-389.

Luibl V, Isas JM, Kayed R, Glabe CG, Langen R, et al. 2006. Drusen deposits associated with aging and age-related macular degeneration contain nonfibrillar amyloid oligomers. J Clin Invest 116: 378-385.

Lutjohann D, von Bergmann K. 2003. 24S-hydroxycholesterol: A marker of brain cholesterol metabolism. Pharmacopsychiatry 36 Suppl 2: S102-S106.

Mahley RW, Huang Y. 1999. Apolipoprotein E: From atherosclerosis to Alzheimer's disease and beyond. Curr Opin Lipidol 10: 207-217.

Malek G, Johnson LV, Mace BE, Saloupis P, Schmechel DE, et al. 2005. Apolipoprotein E allele-dependent pathogenesis: A model for age-related retinal degeneration. Proc Natl Acad Sci USA 102: 11900-11905.

Mann I. 1964. Geographic ophthalmology. a review of the possibilities. Arch Ophthalmol 72: 632-636.

Marc RE, Jones BW, Watt CB, Strettoi E. Neural remodeling in retinal degeneration. Prog Reti Eye Res 2003 Sep; 22 (5):607-55.

Mata NL, Weng J, Travis GH. 2000. Biosynthesis of a major lipofuscin fluorophore in mice and humans with ABCR-mediated retinal and macular degeneration. Proc Natl Acad Sci USA 97: 7154-7159.

McCullagh S, Oucherlony D, Protzner A, Blair N, Feinstein A. 2001. Prediction of neuropsychiatric outcome following mild trauma brain injury: An examination of the Glasgow Coma Scale. Brain Inj 15: 489-497.

McGeer EG, Klegeris A, McGeer PL. 2005. Inflammation, the complement system and the diseases of aging. Neurobiol Aging 26 Suppl 1: 94-97.

McKinnon SJ. 2003. Glaucoma: Ocular Alzheimer's disease? Front Biosci 8: s1140-s1156.

Melov S, Wolf N, Strozyk D, Doctrow SR, Bush AI. 2005. Mice transgenic for Alzheimer disease beta-amyloid develop lens Cataracts that are rescued by antioxidant treatment. Free Radic Biol Med 38: 258-261.

Michikawa M. 2006. Role of cholesterol in amyloid cascade: Cholesterol-dependent modulation of tau phosphorylation and mitochondrial function. Acta Neurol Scand Suppl 185: 21-26.

Mori T, Paris D, Town T, Rojiani AM, Sparks DL, et al. 2001. Cholesterol accumulates in senile plaques of Alzheimer disease patients and in transgenic APP(SW) mice. J Neuropathol Exp Neurol 60: 778-785.

Mullins RF, Russell SR, Anderson DH, Hageman GS. 2000. Drusen associated with aging and age-related macular degeneration contain proteins common to extracellular deposits associated with atherosclerosis, elastosis, amyloidosis, and dense deposit disease. FASEB J 14: 835-346.

Musarella MA. 2001. Molecular genetics of macular degeneration. Doc Ophthalmol 102: 165-177.

Newsome DA, Dobard EP, Liles MR, Oliver PD. 1990. Human retinal pigment epithelium contains two distinct species of superoxide dismutase. Invest Ophthalmol Vis Sci 31: 2508-2513.

Nishimura DY, Swiderski RE, Searby CC, Berg EM, Ferguson AL, et al. 2005. Comparative genomics and gene expression analysis identifies BBS9, a new Bardet–Biedl syndrome gene. Am J Hum Genet 77: 1021-1033.

Notkola IL, Sulkava R, Pekkanen J, Erkinjuntti T, Ehnholm C, et al. 1998. Serum total cholesterol, apolipoprotein E epsilon 4 allele, and Alzheimer's disease. Neuroepidemiology 17: 14-20.

Ohira A, Tanito M, Kaidzu S, Kondo T. 2003. Glutathione peroxidase induced in rat retinas to counteract photic injury. Invest Ophthalmol Vis Sci 44: 1230-1236.

Ohm TG, Meske V. 2006. Cholesterol, statins and tau. Acta Neurol Scand Suppl 185: 93-101.

Okamoto N, Tobe T, Hackett SF, Ozaki H, Vinores MA, et al. 1997. Transgenic mice with increased expression of vascular endothelial growth factor in the retina: A new model of intraretinal and subretinal neovascularization. Am J Pathol 151: 281-291.

Ostojic J, Sakaguchi DS, de Lathouder Y, Hargrove MS, Trent 3rd. JT, et al. 2006. Neuroglobin and cytoglobin: Oxygen-binding proteins in retinal neurons. Invest Ophthalmol Vis Sci 47: 1016-1023.

Papadopoulos V, Lecanu L, Brown RC, Han Z, Yao ZX. 2006. Peripheral-type benzodiazepine receptor in neurosteroid biosynthesis, neuropathology and neurological disorders. Neuroscience 138: 749-756.

Pauleikhoff D. 1992. [Drusen in Bruch's membrane. Their significance for the pathogenesis and therapy of age-associated macular degeneration]. Ophthalmologe 89: 363-386.

Pauleikhoff D, Chen J, Bird AC, Wessing A. 1992a. [The Bruch membrane and choroid. Angiography and functional characteristics in age-related changes]. Ophthalmologe 89: 39-44.

Pauleikhoff D, Wormald RP, Wright L, Wessing A, Bird AC. 1992b. Macular disease in an elderly population. Ger J Ophthalmol 1: 12-15.

Penn JS, Anderson DH. 1992. Effects of light history on the rat retina. New York, NY: Pergamon Press.

Pierce EA. 2001. Pathways to photoreceptor cell death in inherited retinal degenerations. Bioessays 23: 605-618.

Pilcher H. 2006. Alzheimer's disease could be "type 3 diabetes". Lancet Neurol 5: 388-389.

Polyak SL. 1941. The Retina. Chicago, IL: University of Chicago Press.

Proskuryakov SY, Konoplyannikov AG, Gabai VL. 2003. Necrosis: A specific form of programmed cell death? Exp Cell Res 283: 1-16.

Puglielli L, Konopka G, Pack-Chung E, Ingano LA, Berezovska O, et al. 2001. Acyl-coenzyme A: Cholesterol acyltransferase modulates the generation of the amyloid beta-peptide. Nat Cell Biol 3: 905-912.

Purves D, Augustine GJ, Fitzpatrick D, Katz LC, La Mantia A-S, et al. 2001. Neuroscience. Second edn. Sunderland, MA: Sinauer Associates, Inc. pp. 681.

Radu RA, Mata NL, Bagla A, Travis GH. 2004. Light exposure stimulates formation of A2E oxiranes in a mouse model of Stargardt's macular degeneration. Proc Natl Acad Sci USA 101: 5928-5933.

Rakoczy PE, Zhang D, Robertson T, Barnett NL, Papadimitriou J, et al. 2002. Progressive age-related changes similar to age-related macular degeneration in a transgenic mouse model. Am J Pathol 161: 1515-1524.

Rattner A, Nathans J. 2006. An evolutionary perspective on the photoreceptor damage response. Am J Ophthalmol 141: 558-562.

Refolo LM, Malester B, La Francois J, Bryant-Thomas T, Wang R, et al. 2000. Hypercholesterolemia accelerates the Alzheimer's amyloid pathology in a transgenic mouse model. Neurobiol Dis 7: 321-331.

Refolo LM, Pappolla MA, La Francois J, Malester B, Schmidt SD, et al. 2001. A cholesterol-lowering drug reduces beta-amyloid pathology in a transgenic mouse model of Alzheimer's disease. Neurobiol Dis 8: 890-899.

Reichenbach A, Siegel A, Rickmann M, Wolff JR, Noone D, et al. 1995. Distribution of Bergmann glial somata and processes: Implications for function. J Hirnforsch 36: 509-517.

Ressiniotis T, Griffiths PG, Birch M, Keers S, Chinnery PF. 2004. The role of apolipoprotein E gene polymorphisms in primary open-angle glaucoma. Arch Ophthalmol 122: 258-261.

Richards A, Emondi AA, Rohrer B. 2006. Long-term ERG analysis in the partially light-damaged mouse retina reveals regressive and compensatory changes. Vis Neurosci 23: 91-97.

Rogaeva E, Meng Y, Lee JH, Gu Y, Kawarai T, et al. 2007. The neuronal sortilin-related receptor SORL1 is genetically associated with Alzheimer disease. Nat Genet.

Rosenberg PB. 2005. Clinical aspects of inflammation in Alzheimer's disease. Int Rev Psychiatry 17: 503-514.

Roses AD. 1995. On the metabolism of apolipoprotein E and the Alzheimer diseases. Exp Neurol 132: 149-156.

Ross AJ, May-Simera H, Eichers ER, Kai M, Hill J, et al. 2005. Disruption of Bardet–Biedl syndrome ciliary proteins perturbs planar cell polarity in vertebrates. Nat Genet 37: 1135-1140.

Rozanowska M, Jarvis-Evans J, Korytowski W, Boulton ME, Burke JM, et al. 1995. Blue light-induced reactivity of retinal age pigment. In vitro generation of oxygen-reactive species. J Biol Chem 270: 18825-18830.

Rudolf M, Ivandic B, Winkler J, Schmidt-Erfurth U. 2004. [Accumulation of lipid particles in Bruch's membrane of LDL receptor knockout mice as a model of age-related macular degeneration]. Ophthalmologe 101: 715-719.

Sambamurti K, Granholm AC, Kindy MS, Bhat NR, Greig NH, et al. 2004. Cholesterol and Alzheimer's disease: Clinical and experimental models suggest interactions of different genetic, dietary and environmental risk factors. Curr Drug Targets 5: 517-528.

Sambamurti K, Greig NH, Lahiri DK. 2002. Advances in the cellular and molecular biology of the beta-amyloid protein in Alzheimer's disease. Neuromolecular Med 1: 1-31.

Sambamurti K, Suram A, Venugopal C, Prakasam A, Zhou Y, et al. 2006. A partial failure of membrane protein turnover may cause Alzheimer's disease: A new hypothesis. Curr Alzheimer Res 3: 81-90.

Sandberg MA, Gaudio AR. 2006. Reading speed of patients with advanced retinitis pigmentosa or choroideremia. Retina 26: 80-88.

Sandilands A, Hutcheson AM, Long HA, Prescott AR, Vrensen G, et al. 2002. Altered aggregation properties of mutant gamma-crystallins cause inherited cataract. EMBO J 21: 6005-6014.

Sarks SH. 1976. Ageing and degeneration in the macular region: A clinico-pathological study. Br J Ophthalmol 60: 324-341.

Schmidt M, Giessl A, Laufs T, Hankeln T, Wolfrum U, et al. 2003. How does the eye breathe? Evidence for neuroglobin-mediated oxygen supply in the mammalian retina. J Biol Chem 278: 1932-1935.

Schmidt S, Hauser MA, Scott WK, Postel EA, Agarwal A, et al. 2006. Cigarette smoking strongly modifies the association of LOC387715 and age-related macular degeneration. Am J Hum Genet 78: 852-864.

Schmidt-Erfurth U, Michels S, Michels R, Aue A. 2005. Anecortave acetate for the treatment of subfoveal choroidal neovascularization secondary to age-related macular degeneration. Eur J Ophthalmol 15: 482-485.

Schnitzer J. 1988. Immunocytochemical studies on the development of astrocytes, Muller (glial) cells, and oligodendrocytes in the rabbit retina. Brain Res Dev Brain Res 44: 59-72.

Schwesinger C, Yee C, Rohan RM, Joussen AM, Fernandez A, et al. 2001. Intrachoroidal neovascularization in transgenic mice overexpressing vascular endothelial growth factor in the retinal pigment epithelium. Am J Pathol 158: 1161-1172.

Seddon JM, Cote J, Page WF, Aggen SH, Neale MC. 2005. The US twin study of age-related macular degeneration: Relative roles of genetic and environmental influences. Arch Ophthalmol 123: 321-327.

Seddon JM, Gensler G, Klein ML, Milton RC. 2006a. C-reactive protein and homocysteine are associated with dietary and behavioral risk factors for age-related macular degeneration. Nutrition 22: 441-443.

Seddon JM, George S, Rosner B. 2006b. Cigarette smoking, fish consumption, omega-3 fatty acid intake, and associations with age-related macular degeneration: The US Twin Study of Age-Related Macular Degeneration. Arch Ophthalmol 124: 995-1001.

Sepp T, Khan JC, Thurlby DA, Shahid H, Clayton DG, et al. 2006. Complement factor H variant Y402H is a major risk determinant for geographic atrophy and choroidal neovascularization in smokers and nonsmokers. Invest Ophthalmol Vis Sci 47: 536-540.

Shobab LA, Hsiung GY, Feldman HH. 2005. Cholesterol in Alzheimer's disease. Lancet Neurol 4: 841-852.

Siintola E, Partanen S, Stromme P, Haapanen A, Haltia M, et al. 2006. Cathepsin D deficiency underlies congenital human neuronal ceroid-lipofuscinosis. Brain 129: 1438-1445.

Simpkins JW, Singh M. 2004. Consortium for the Assessment of Research on Progestins and Estrogens (CARPE) Fort Worth, Texas August 1–3, 2003. J Womens Health (Larchmt) 13: 1165-1168.

Smith RT, Chan JK, Nagasaki T, Sparrow JR, Barbazetto I. 2005. A method of drusen measurement based on reconstruction of fundus background reflectance. Br J Ophthalmol 89: 87-91.

Snodderly DM. 1995. Evidence for protection against age-related macular degeneration by carotenoids and antioxidant vitamins. Am J Clin Nutr 62: 1448S-1461S.

Souied EH, Benlian P, Amouyel P, Feingold J, Lagarde JP, et al. 1998. The epsilon4 allele of the apolipoprotein E gene as a potential protective factor for exudative age-related macular degeneration. Am J Ophthalmol 125: 353-359.

Sparks DL, Kuo YM, Roher A, Martin T, Lukas RJ. 2000. Alterations of Alzheimer's disease in the cholesterol-fed rabbit, including vascular inflammation. Preliminary observations. Ann NY Acad Sci 903: 335-344.

Sparks DL, Scheff SW, Hunsaker 3rd. JC, Liu H, Landers T, et al. 1994. Induction of Alzheimer-like beta-amyloid immunoreactivity in the brains of rabbits with dietary cholesterol. Exp Neurol 126: 88-94.

Srivastava RA, Ito H, Hess M, Srivastava N, Schonfeld G. 1995. Regulation of low density lipoprotein receptor gene expression in HepG2 and Caco2 cells by palmitate, oleate, and 25-hydroxycholesterol. J Lipid Res 36: 1434-1446.

Stoetzel C, Muller J, Laurier V, Davis EE, Zaghloul NA, et al. 2007. Identification of a novel BBS gene (BBS12) highlights the major role of a vertebrate-specific branch of chaperonin-related proteins in Bardet–Biedl syndrome. Am J Hum Genet 80: 1-11.

Stohr H, Milenkowic V, Weber BH. 2005. [VMD2 and its role in Best's disease and other retinopathies]. Ophthalmologe 102: 116-121.

Stone EM, Braun TA, Russell SR, Kuehn MH, Lotery AJ, et al. 2004. Missense variations in the fibulin 5 gene and age-related macular degeneration. N Engl J Med 351: 346-353.

Stone EM, Lotery AJ, Munier FL, Heon E, Piguet B, et al. 1999. A single EFEMP1 mutation associated with both Malattia Leventinese and Doyne honeycomb retinal dystrophy. Nat Genet 22: 199-202.

Stone EM, Nichols BE, Kimura AE, Weingeist TA, Drack A, et al. 1994. Clinical features of a Stargardt-like dominant progressive macular dystrophy with genetic linkage to chromosome 6q. Arch Ophthalmol 112: 765-772.

Strauss O. 2005. The retinal pigment epithelium in visual function. Physiol Rev 85: 845-81.

Sun Y, Jin K, Peel A, Mao XO, Xie L, et al. 2003. Neuroglobin protects the brain from experimental stroke in vivo. Proc Natl Acad Sci USA 100: 3497-3500.

Sunde M, Blake C. 1997. The structure of amyloid fibrils by electron microscopy and X-ray diffraction. Adv Protein Chem 50: 123-159.

Swanson DA, Freund CL, Ploder L, McInnes RR, Valle D. 1996. A ubiquitin C-terminal hydrolase gene on the proximal short arm of the X chromosome: Implications for X-linked retinal disorders. Hum Mol Genet 5: 533-538.

Tamura H, Kawakami H, Kanamoto T, Kato T, Yokoyama T, et al. 2006. High frequency of open-angle glaucoma in Japanese patients with Alzheimer's disease. J Neurol Sci 246: 79-83.

Tanaka M, Machida S, Ohtaka K, Tazawa Y, Nitta J. 2005. Third-order neuronal responses contribute to shaping the negative electroretinogram in sodium iodate-treated rats. Curr Eye Res 30: 443-453.

Tanito M, Masutani H, Nakamura H, Oka S, Ohira A, et al. 2002. Attenuation of retinal photooxidative damage in thioredoxin transgenic mice. Neurosci Lett 326: 142-146.

Tate Jr. DJ, Newsome DA, Oliver PD. 1993. Metallothionein shows an age-related decrease in human macular retinal pigment epithelium. Invest Ophthalmol Vis Sci 34: 2348-2351.

Tatton W, Chen D, Chalmers-Redman R, Wheeler L, Nixon R, et al. 2003. Hypothesis for a common basis for neuroprotection in glaucoma and Alzheimer's disease: Anti-apoptosis by alpha-2-adrenergic receptor activation. Surv Ophthalmol 48 Suppl 1: S25-S37.

Taylor HR, Munoz B, West S, Bressler NM, Bressler SB, et al. 1990. Visible light and risk of age-related macular degeneration. Trans Am Ophthalmol Soc 88: 163-173; discussion 173–178.

Taylor A, Jacques PF, Dorey CK. 1993. Oxidation and aging: Impact on vision. Toxicol Ind Health 9: 349-371.

Teunissen CE, de Vente J, Steinbusch HW, De Bruijn C. 2002. Biochemical markers related to Alzheimer's dementia in serum and cerebrospinal fluid. Neurobiol Aging 23: 485-508.

Tomany SC, Cruickshanks KJ, Klein R, Klein BE, Knudtson MD. 2004a. Sunlight and the 10-year incidence of age-related maculopathy: The Beaver Dam Eye Study. Arch Ophthalmol 122: 750-757.

Tomany SC, Wang JJ, Van Leeuwen R, Klein R, Mitchell P, et al. 2004b. Risk factors for incident age-related macular degeneration: Pooled findings from 3 continents. Ophthalmology 111: 1280-1287.

Travis GH. 1998. Mechanisms of cell death in the inherited retinal degenerations. Am J Hum Genet 62: 503-508.

Travis GH, Brennan MB, Danielson PE, Kozak CA, Sutcliffe JG. 1989. Identification of a photoreceptor-specific mRNA encoded by the gene responsible for retinal degeneration slow (rds). Nature 338: 70-73.

Ueda T, Ueda T, Armstrong D. 1996. Preventive effect of natural and synthetic antioxidants on lipid peroxidation in the mammalian eye. Ophthalmic Res 28: 184-192.

Umeda S, Ayyagari R, Allikmets R, Suzuki MT, Karoukis AJ, et al. 2005a. Early-onset macular degeneration with drusen in a cynomolgus monkey (Macaca fascicularis) pedigree: Exclusion of 13 candidate genes and loci. Invest Ophthalmol Vis Sci 46: 683-691.

Umeda S, Suzuki MT, Okamoto H, Ono F, Mizota A, et al. 2005b. Molecular composition of drusen and possible involvement of anti-retinal autoimmunity in two different forms of macular degeneration in cynomolgus monkey (Macaca fascicularis). FASEB J 19: 1683-1685.

Vagnucci AH, Jr., Li WW. 2003. Alzheimer's disease and angiogenesis. Lancet 361: 605-608.

Vassar R, Bennett BD, Babu-Khan S, Kahn S, Mendiaz EA, et al. 1999. Beta-secretase cleavage of Alzheimer's amyloid precursor protein by the transmembrane aspartic protease BACE. Science 286: 735-741.

Vickers JC, Craig JE, Stankovich J, McCormack GH, West AK, et al. 2002. The apolipoprotein epsilon4 gene is associated with elevated risk of normal tension glaucoma. Mol Vis 8: 389-393.

Vingerling JR, Dielemans I, Hofman A, Grobbee DE, Hijmering M, et al. 1995. The prevalence of age-related maculopathy in the Rotterdam Study. Ophthalmology 102: 205-210.

Vingolo EM, De Mattia G, Giusti C, Forte R, Laurenti O, et al. 1999. Treatment of nonproliferative diabetic retinopathy with Defibrotide in noninsulin-dependent diabetes mellitus: A pilot study. Acta Ophthalmol Scand 77: 315-320.

Walker DG, Lue LF. 2005. Investigations with cultured human microglia on pathogenic mechanisms of Alzheimer's disease and other neurodegenerative diseases. J Neurosci Res 81: 412-425.

Wassle H. 2004. Parallel processing in the mammalian retina. Nat Rev Neurosci 5: 747-757.

Weale R. 2006 Smoking and age-related maculopathies. Lancet 368: 1235-1236.

Weber BH, Lin B, White K, Kohler K, Soboleva G, et al. 2002. A mouse model for Sorsby fundus dystrophy. Invest Ophthalmol Vis Sci 43: 2732-2740.

Weng J, Mata NL, Azarian SM, Tzekov RT, Birch DG, et al. 1999. Insights into the function of Rim protein in photoreceptors and etiology of Stargardt's disease from the phenotype in abcr knockout mice. Cell 98: 13-23.

Wenzel A, Grimm C, Samardzija M, Reme CE. 2005. Molecular mechanisms of light-induced photoreceptor apoptosis and neuroprotection for retinal degeneration. Prog Retin Eye Res 24: 275-306.

Williams DS. 1991. Actin filaments and photoreceptor membrane turnover. Bioessays 13: 171-178.

Wolozin B, Manger J, Bryant R, Cordy J, Green RC, et al. 2006. Re-assessing the relationship between cholesterol, statins and Alzheimer's disease. Acta Neurol Scand Suppl 185: 63-70.

Wu J, Marmorstein AD, Peachey NS. 2006 Functional abnormalities in the retinal pigment epithelium of CFTR mutant mice. Exp Eye Res 83: 424-428.

Wyss-Coray T, Lin C, Yan F, Yu GQ, Rohde M, et al. 2001. TGF-beta1 promotes microglial amyloid-beta clearance and reduces plaque burden in transgenic mice. Nat Med 7: 612-618.

Yang Z, Camp NJ, Sun H, Tong Z, Gibbs D, et al. 2006. A variant of the HTRA1 gene increases susceptibility to age-related macular degeneration. Science 314: 992-993.

Yoshida T, Ohno-Matsui K, Ichinose S, Sato T, Iwata N, et al. 2005. The potential role of amyloid beta in the pathogenesis of age-related macular degeneration. J Clin Invest 115: 2793-2800.

Young RW. 1976 Visual cells and the concept of renewal. Invest Ophthalmol Vis Sci 15: 700-725.

Zhang HB, Liu DP, Liang CC. 2002a The control of expression of the alpha-globin gene cluster. Int J Hematol 76: 420-426.

Zhang K, Kniazeva M, Han M, Li W, Yu Z, et al. 2001. A 5-bp deletion in ELOVL4 is associated with two related forms of autosomal dominant macular dystrophy. Nat Genet 27: 89-93.

Zhang X, Hargitai J, Tammur J, Hutchinson A, Allikmets R, et al. 2002b. Macular pigment and visual acuity in Stargardt macular dystrophy. Graefes Arch Clin Exp Ophthalmol 240: 802-809.

Zhou Y, Suram A, Venugopal C, Prakasam A, Lin S, Su Y, Li B, Paul SM, Sambamurti K, 2008. Geranylgeranyl pyrophosphate stimulates gamma-secretase to increase the generation of Abeta and APP-CTF gamma. FASEB J 22(1): 47-54.

8 Diabetic Retinopathy

E. Bowie · C. E. Crosson

1	Epidemiology	166
2	Risk Factors	166
3	**Clinical Presentation**	**166**
3.1	Nonproliferative (Background) Diabetic Retinopathy	166
3.2	Proliferative Diabetic Retinopathy	168
4	**Retinal Physiology and Pathophysiology of Diabetic Retinopathy**	**168**
4.1	Anatomy and Physiology of the Retina	168
4.2	Background Retinopathy: Early Development	169
4.3	Macular Edema	169
4.4	Proliferative Retinopathy: Angiogenesis	170
4.5	Advanced Glycation End-Products and Receptors	170
4.6	Vascular Endothelial Growth Factor	171
5	**Therapies**	**172**
5.1	Blood Glucose Control	172
5.2	Laser	172
5.3	Vitrectomy	172
5.4	Steroids	173
5.5	Protein Kinase C Inhibitors	173
5.6	Antiangiogenic Agents	174
6	Summary	174

© 2009 Springer Science+Business Media, LLC.

8 Diabetic retinopathy

Abstract: Diabetes is a metabolic disorder that results in microvascular changes within the retina. Although advances in the treatment of diabetic retinopathy have been made, this disease remains one of the leading causes of blindness. This chapter describes some of the current concepts in the diagnosis, risk factors, pathogenesis, and treatment of diabetic retinopathy.

List of Abbreviations: AGEs, advanced glycation end-products; ETDRS, early treatment diabetic retinopathy study; HIF-1, hypoxia-inducible factor 1; PlGF, placenta growth factor; PKC, protein kinase C; RPE, retinal pigment epithelium; VEGF, vascular endothelial growth factor

1 Epidemiology

Diabetic retinopathy is the leading ocular complication associated with diabetes mellitus and the leading cause of blindness in adults 20–74 years of age (Fong et al., 2003). Approximately 4.1 million adults 40 years and older in the USA have diabetic retinopathy and in diabetics 40 years of age and older 40.3% have diabetic retinopathy (Kempen et al., 2004). Vision-threatening retinopathy affects 8.2% of diabetics. The estimated prevalence rates for diabetic retinopathy and vision-threatening retinopathy in the general US population are 3.4% and 0.75%, respectively (Kempen et al., 2004). Among patients with type 1 diabetes, the prevalence of retinopathy in blacks and whites is 74.9% and 82.3%, respectively (Roy et al., 2004). Vision-threatening retinopathy affects 30% of black and 32.2% of white patients diagnosed with type 1 diabetes (Roy et al., 2004). Of the 50,000 new cases of blindness each year, 50% are due to diabetes, and diabetic retinopathy is the leading cause (Jawa et al., 2004).

2 Risk Factors

Duration of diabetes is the main risk factor for development and progression of diabetic retinopathy. In patients with type 1 diabetes, 99% will develop some retinopathy 20 years after diagnosis and 53% will progress on to the proliferative stage of diabetic retinopathy (Jawa et al., 2004). In patients with type 2 diabetes, 60% will develop some retinopathy 20 years after diagnosis and 5% will progress on to the proliferative stage of diabetic retinopathy (Jawa et al., 2004). Poor glycemic control increases the risk of diabetic retinopathy. The Diabetes Control and Complications Trial (DCCT) and the UK Prospective Diabetes Study (UKPDS) demonstrated that in type 1 and type 2 diabetic patients intensive glucose control significantly decreased the risk for the development of diabetic retinopathy (Fong et al., 2003). These studies showed that for every 1% decrease in HbA1c, there was a 35% decreased risk in microvascular disease (Fong et al., 2003). Race may also play a role in the prevalence of diabetic retinopathy. A higher frequency of diabetic retinopathy has been noted in African-American and Hispanic patients (Emanuele et al., 2005). Other risk factors for diabetic retinopathy include hypertension, hyperlipidemia, and the presence of renal disease (Berinstein et al., 1997; Roy, 2000; Chew, 2003; Lightman and Towler, 2003). Anemia and pregnancy have been shown to contribute to the development of diabetic retinopathy (Davis et al., 1998).

3 Clinical Presentation

3.1 Nonproliferative (Background) Diabetic Retinopathy

In the early stages of the disease, no retinal abnormalities can be seen on dilated fundus examination and patients may not have any symptoms. Mild nonproliferative retinopathy is characterized by microaneurysms, dot-blot hemorrhages, and exudates. Moderate nonproliferative diabetic retinopathy will have the features of mild retinopathy, but in addition there may be cotton wool spots, venous beading, and intraretinal microvascular abnormalities. Severe nonproliferative diabetic retinopathy is characterized by intraretinal hemorrhages in four quadrants of the retina, venous beading in two or more quadrants, and intraretinal microvascular abnormalities in one or more quadrants. Patients may complain of decreased vision if they have diabetic macula edema or macula ischemia (❯ *Figure 8-1*). Fifty percent of patients with severe nonproliferative diabetic retinopathy will develop proliferative diabetic retinopathy within 15 months (Ferris, 1996).

Diabetic macula edema is a significant cause of vision loss in patients with diabetes mellitus. The Early Treatment Diabetic Retinopathy Study (ETDRS) defined clinically significant diabetic macula edema as

◘ **Figure 8-1**
Fundus photos of eyes from normal and diabetic individuals. (a) Normal fundus photograph. (b) Nonproliferative diabetic retinopathy with macular edema and hard exudates. (c) Proliferative diabetic retinopathy with preretinal hemorrhage. (d) Proliferative diabetic retinopathy with neovascularization of the iris. (e) Fundus photograph of proliferative diabetic retinopathy with neovascularization and preretinal hemorrhage. (f) Fluorescein angiography photograph showing proliferative diabetic retinopathy with leakage from retinal neovascularization

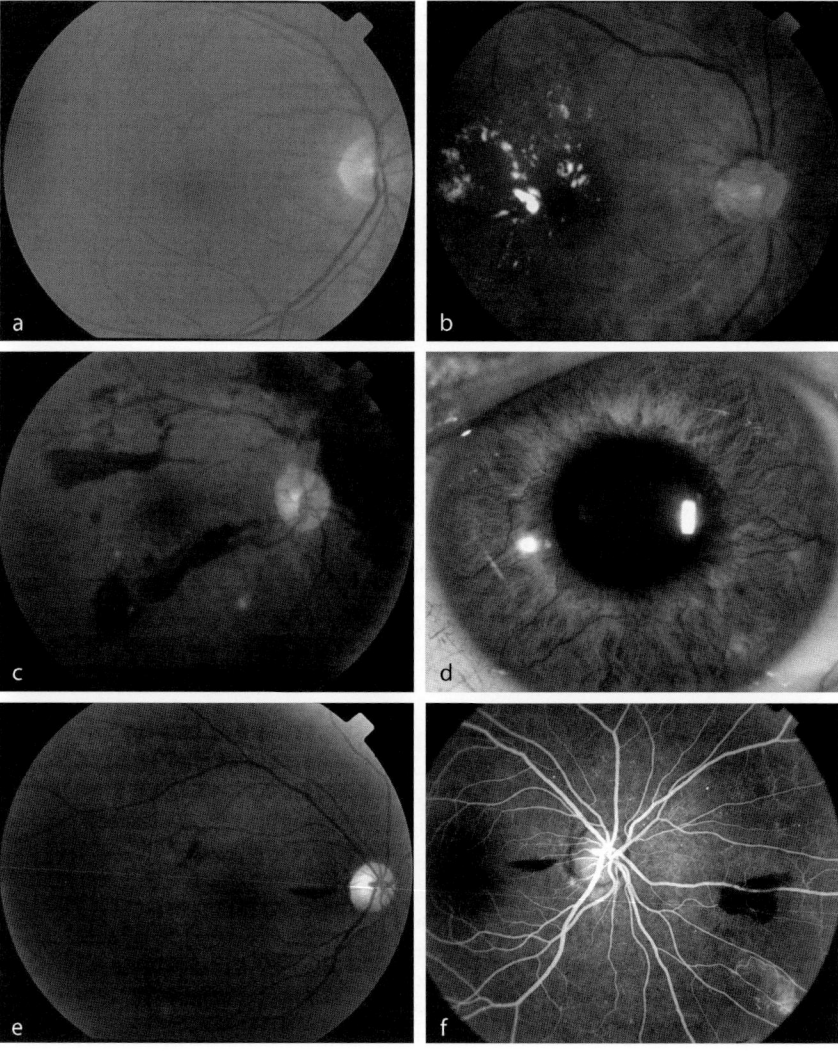

retina thickening at or within 500 μm of the center of the fovea, hard exudates associated with retina thickening at or within 500 μm of the center of the fovea, or thickening within one disc area of the center of the fovea (Early Treatment Diabetic Retinopathy Study Group, 1985). The Wisconsin Epidemiologic Study of Diabetic Retinopathy (WESDR) demonstrated a 10-year rate of developing macula edema as 20.1% of patients with type 1 diabetes, 25.4% in type 2 diabetics on insulin, and 13.9% for type 2 patients on oral therapy (Klein et al., 1995a). ◉ *Figure 8-2* shows optical coherence tomography (OCT) images of the macula from a normal and diabetic individual.

◘ Figure 8-2
Optical coherence tomography (OCT) images of the macular region from a normal (a) and diabetic individual (b). Note the dramatic increase in retinal thickness within the central foveal region (Images courtesy of E. Sharp, MD)

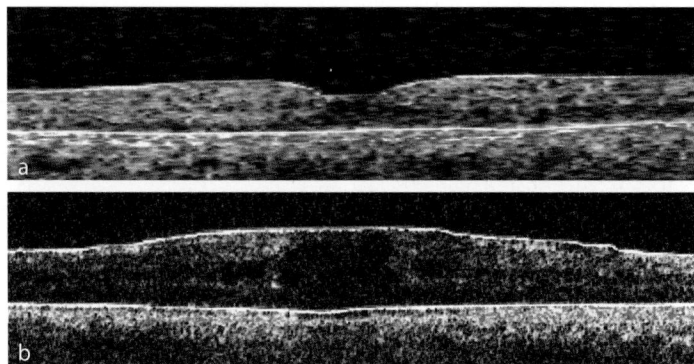

3.2 Proliferative Diabetic Retinopathy

Proliferative diabetic retinopathy is characterized by the growth of new retinal vessels and/or fibrous tissue on the optic disc, retina, or iris. These new vessels are fragile and often bleed resulting in preretinal or vitreous hemorrhage. As a result, diabetic patients will present with increased floaters or a sudden loss of vision depending on the severity of the hemorrhage. The neovascularization and fibrous growth can result in traction retinal detachment and severe loss of vision. In diabetic individuals, new blood vessel growth in the iris can also occur (◉ *Figure 8-1*). If left untreated, the growth of these vessels in the anterior segment can lead to the development of neovascular glaucoma. Neovascular glaucoma usually presents with pain and decreased vision owing to the elevated intraocular pressure and, if not treated, the individual will likely develop optic neuropathy and eventual blindness.

4 Retinal Physiology and Pathophysiology of Diabetic Retinopathy

4.1 Anatomy and Physiology of the Retina

The retina is composed of primary neurons (photoreceptors), secondary neurons (bipolar cells), and third order neurons (ganglion cells) that convert light into neuronal signals and covey these signals to the brain. In addition, the retina contains interneurons (horizontal and amacrine cells), glial cells (Müller cells, astrocytes, and microglia), and retinal pigment epithelium (RPE). Histological characterization of the retina classically divides the retina into seven distinct layers: the RPE, photoreceptor layer, outer nuclear layer (cell bodies of the photoreceptors and horizontal cells), outer plexiform layer (synaptic connections between photoreceptors and bipolar cells), inner nuclear layer (cell bodies of bipolar, amacrine, and Müller cells), inner plexiform layer (synaptic connection between bipolar and ganglion cells), and the ganglion cell layer (cell bodies of ganglion cells and astrocytes). In the retina, glutamate is the major excitatory neurotransmitter (Massey, 1990). However, unlike most neurons, photoreceptors and bipolar cells continuously release the neurotransmitter glutamate in the dark, and the absorbance of light by photoreceptors induces a hyperpolarization suppressing glutamate release (Dowling, 1987). While this neural network along with the high density of cells within the retina creates a sensory system that is exquisitely sensitive to light and has a high spatial resolution, it is also metabolically very active. To meet the metabolic demands of the retina, blood is supplied by two anatomically distinct systems: the choriocapillaris and the inner retinal vessels. The choriocapillaris supplies the metabolic needs of the photoreceptor and outer nuclear layers, whereas the inner retinal vessels supply the metabolic needs of the inner retina.

4.2 Background Retinopathy: Early Development

In diabetic retinopathy the earliest clinical signs of disease are microaneurysms and small intraretinal hemorrhages within the inner retinal vessels (Frank, 2004). These lesions lead to the progressive obliteration of retinal microvessels, focal ischemia (cotton wool spots), macular edema, and eventual unregulated angiogenesis. These vasculopathies are thought to result from cell death in isolated capillaries that then become nonperfused (Engerman and Kern, 1995).

In normal primate retinas, the primary cells contributing to vascular structure and function are the endothelial cells and pericytes (Provis, 2001). Inner retinal vascular endothelial cells form what is termed the inner blood–retinal barrier. Ultrastructurally, this paracellular barrier is formed by the expression of the zonula occludens and associated proteins between adjacent endothelial cells to form tight junctions (Felinski and Antonetti, 2005). These junctional complexes limit the diffusion of hydrophilic substances from the blood to the retinal environment. Pericytes lie between the endothelium and basement membrane of the retinal vessel. Owing to their location and strong expression of α-smooth muscle actin, pericytes are thought to have a contractile function and a role in regulating blood flow (Bandopadhyay et al., 2001). Studies have shown that in diabetes both retinal pericytes and endothelial cells undergo apoptotic cell death and that these apoptotic events are specific for vascular cells and precede histological and clinical evidence of retinopathy (Mizutani et al., 1996). In addition to endothelial cells and pericytes, retinal astrocytes also play an essential role in retinal vascular development and the maintenance of vascular integrity by controlling the extracellular environment and the secretion of angiogenic and antiangiogenic cytokines (Provis, 2001).

4.3 Macular Edema

The loss of retinal endothelial cells and pericytes is thought to result in focal hypoxic and ischemic regions within the retina. Support for this hypothesis comes from animal studies which have shown that in diabetic animals the inner retinal layers exhibit a significant reduction in PO_2 levels during the early stages of diabetes (Linsenmeier et al., 1998). In addition, human studies have shown that hyperoxia can reverse early contrast sensitivity deficits observed in diabetic individuals (Harris et al., 1996).

The retinal response to metabolic stress has been the focus of intense research for the past several years. Studies have shown that the retina reacts to hypoxia via two functionally distinct responses: vasodilatation of inner retinal vessels and alterations in gene expression. Both of these events can lead to the development of macular edema. While macular edema is normally not associated with retinal degeneration, its development significantly affects visual acuity in approximately 9% of diabetics (Klein et al., 1995b). In diabetic individuals, generalized retinal vasodilatation, and blunted vasoconstriction in response to increasing oxygen has been associated with the development of macular edema (Kristinsson et al., 1997; Bek, 1999). This vasodilatation likely contributes to macular edema by increasing the hydrostatic pressure within the retinal vessels moving more fluid into the retinal extracellular space according to Starling's law (for review see Stefansson, 2006).

In diabetic retinopathy, changes in expression of retinal protein have been primarily focused on the peptide vascular endothelial growth factor (VEGF). As detailed later, VEGF is the principle cytokine involved in angiogenesis during the proliferative phase of diabetic retinopathy. However, the original studies on VEGF demonstrated that this molecule also increases the permeability of endothelial cells (Senger et al., 1983). In the eye, elevated levels of VEGF have been shown to decrease the amount and location of tight junction proteins in vascular endothelial cells, and possibly to increase the permeability of the paracellular pathway between endothelial cells (Antonetti et al., 1999). More recent studies have shown that VEGF also increases permeability across the RPE which forms the outer blood–retina barrier (Hartnett et al., 2003). This increase in RPE permeability contributes to macular edema by an increase in fluid movement that is related to the perfusion pressure in the choriocapillaris, and disrupting the transport process across the RPE that actively removes fluid from the subretinal environment. Although VEGF is thought to be the primary cytokine involved in regulating the permeability of the blood–retina barriers, other factors, such as hepatocyte growth factor (HGF) and insulin-like growth factor (IGF-1) also have been shown to modulate the permeability of the retinal vessels or RPE (Brausewetter et al., 2001; Clermont et al., 2006).

4.4 Proliferative Retinopathy: Angiogenesis

The progression of diabetic retinopathy from the early background stage to the proliferative or angiogenic stage has been associated with baseline hyperglycemia, glycemic control, blood pressure, and elevated levels of various cytokines (Stratton et al., 2001). However, the primary factor controlling vascular development in infants, or angiogenesis in diabetic individuals, is low PO_2 levels within the retinal tissue. The cells responsible for sensing and initiating the response to declining PO_2 levels appear to be the glial cells within the retina. Both Müller cells and astrocytes have been shown to significantly increase VEGF expression and secretion in response to hypoxia (Stone et al., 1995; Behzadian et al., 1998; Eichler et al., 2000). This secreted VEGF activates specific VEGF receptors on the endothelial cells to stimulate proliferation and new vessel growth (see later). Direct evidence for the role of VEGF in angiogenesis has come from studies demonstrating that neutralizing agents can block ocular neovascularization (Adamis et al., 1996, 2006; Spaide and Fisher, 2006). However, vitreal administration of VEGF alone increased permeability but did not induce retinal neovascularization (Hofman et al., 2000). These studies provide evidence that VEGF is necessary for retinal neovascularization, and that the regulation of this process is complex, involving input from other factors that act as coagonists or functional antagonists with VEGF. Studies have identified platelet-derived growth factor (PDGF), IGF-1, and placenta growth factor (PlGF) as possible coagonists and pigment epithelial-derived factor (PEDF) as a functional antagonist (Khaliq et al., 1998; Gao et al., 2001; Mori et al., 2002; Poulaki et al., 2004).

Another potential factor involved in endothelial cell proliferation is stretch of the vascular cells. In diabetic retinopathy the loss of autoregulatory capabilities within retinal vessels exposes these vessels to greater mechanical stretch as blood moves through the retina during the cardiac cycle. In vitro studies have shown that cellular stretch is an effective stimulus for endothelial proliferation (Davies and Tripathi, 1993). In addition, cellular stretch has been shown to upregulate the expression of VEGF and its receptors, and induce apoptosis in retinal pericytes (Suzuma et al., 2002; Beltramo et al., 2006). These biomechanical studies provide a possible explanation for the association of systemic hypertension as an independent risk factor for diabetic retinopathy.

4.5 Advanced Glycation End-Products and Receptors

The duration and severity of hyperglycemia is the primary factor in the development of diabetic retinopathy (Diabetes Control and Complications Trial Research Group, 1995; UK Prospective Diabetes Study Group, 1998). One result of hyperglycemia is the accelerated formation of nonenzymatic, advanced glycation end-products (AGEs). These products form when proteins or lipids interact with aldose sugars for an extended period of time, resulting in the glycation of proteins or lipids. In the extracellular environment, accumulation of AGEs results in the formation of protein cross-links increasing the stiffness of the basement membranes, reducing the interactions between basement membrane components and proteoglycan, as well as limiting the adherence of endothelial cells to basement membrane structures (Goldin et al., 2006). AGEs can also form within the intracellular environment altering cellular signaling and metabolism that are critical for maintaining homeostasis (Kass et al., 2001).

Soluble AGEs can also influence cell function by binding to surface receptor, RAGE, on a variety of cells. Action of the AGE receptor has been shown to stimulate multiple signaling pathways, such as MAP kinases and NFκB (Schmidt et al., 1994; Schiekofer et al., 2003). Functionally, these receptors have been linked to the generation of reactive oxygen species, upregulation of VEGF, increased vascular permeability, and increased leukocyte adhesion to vascular endothelial cells. Soluble AGEs also reduce endothelium-derived nitric oxide. Nitric oxide has been shown to inhibit many of the events that contribute to the development of vascular lesions. A number of different compounds that inhibit AGE/RAGE axis have been investigated as therapies for the vascular dysfunction that develop in diabetes. These include aminoguanidine, 4,5-dimethyl-3-phenacylthiozolium chloride (ALT-711) and soluble forms of extracellular ligand-binding domain of RAGE (Corman et al., 1998; Park et al., 1998; Kass et al., 2001). These studies have provided evidence that the recognition and binding of AGEs to RAGE contribute to many of the microvascular and macrovascular complications of diabetes.

4.6 Vascular Endothelial Growth Factor

The family of VEGF proteins has been linked to abnormal vessel growth from preexisting vessels (i.e., angiogenesis) in cancer, myocardial infarction, macular degeneration, retinopathy of prematurity, and diabetic retinopathy. Proteins included in the VEGF family are: PlGF, VEGF-A, VEGF-B, VEGF-C, VEGF-D and the viral homolog VEGF-E. However, most studies have focused on VEGF-A and the angiogenic actions of this protein (Ferrara, 2004). The human VEGF-A gene is organized in eight exons, separated by seven introns and is localized in chromosome 6p21.3 (Houck et al., 1991; Tischer et al., 1991; Vincenti et al., 1996). Alternative exon splicing results in the secretion of four different VEGF-A isoforms ($VEGF_{121}$, $VEGF_{165}$, $VEGF_{189}$, $VEGF_{206}$). The predominant isoform is $VEGF_{165}$, which lacks the amino acids encoded by exon 6 (Tischer et al., 1991; Vincenti et al., 1996). The VEGF isoforms vary in their affinity for heprin and therefore extracellular matrix material. Larger isoforms, such as $VEGF_{189}$, exhibit high affinity for heprin and are found almost exclusively bound to extracellular matrix, whereas the smaller isoforms, such as $VEGF_{121}$, are freely diffusible. The 165 isoform shows intermediate binding to heparin (Houck et al., 1992; Park et al., 1993). These bound VEGFs form a reservoir of biologically active proteins that can be hydrolyzed by various proteases to release a soluble, biologically active 110 amino acid VEGF-fragment (Houck et al., 1992).

The cellular actions of VEGF are mediated by binding to receptor tyrosine kinases. Three VEGF receptors have been identified, VEGFR-1 (FLT-1), VEGFR-2 (KDR), and VEGFR-3 (FLT-4). VEGFR-1 and VEGFR-2 are expressed in the cell surface of most blood endothelial cells, whereas VEGFR-3 is primarily restricted to lymphatic cells. These receptors do exhibit some selectivity for the endogenous VEGF family of proteins: VEGF-A binds to VEGFR-1 and VEGFR-2; PlGF and VEGF-B only bind to VEGFR-1; VEGF-C and VEGF-D bind to VEGFR-2 and VEGFR-3 (Ferrara, 2004). In addition to the classic VEGF receptors, $VEGF_{165}$ has been shown to interact with neuropilin-1 to enhance VEGF interaction with VEGFR-2 (Fuh et al., 2000). Although our understanding of the signaling pathway used by VEGF receptors remains incomplete, it is generally agreed that the activation VEGFR-2 receptor is primarily responsible for the development of angiogenesis within adult tissues.

As stated earlier, the primary factor regulating VEGF expression is the level of PO_2 within a tissue. The primary mediator in hypoxic-induced transcription is hypoxia-inducible factor 1 (HIF-1). HIF-1 is a heterodimer composed of an inducibly expressed HIF-1α subunit and a constitutively expressed HIF-1β subunit. In response to hypoxia HIF-1α levels increase, and it binds to specific enhancer elements resulting in increased gene transcription (Semenza, 2002). Under normoxic conditions, HIF-1α is subjected to rapid ubiquitination and proteasomal degradation. Under hypoxic conditions this process is blocked resulting in a rapid increase in HIF-1α levels as the cellular O_2 concentration falls (Sutter et al., 2000). The basis for O_2-dependent regulation is the hydroxylation of proline residues on HIF-1α by HIF-1α prolyl hydroxylases (Bruick and McKnight, 2001). Prolyl hydroxylation of HIF-1α permits the binding of the von Hippel–Lindau tumor suppressor protein allowing the complex to be targeted for proteasomal degradation.

In addition to PO_2, growth factors are also important regulators of VEGF gene expression. Studies have shown that EGF, TGF-β, IGF-I, FGF, and PDGF, can upregulate VEGF mRNA expression (Frank et al., 1995; Warren et al., 1996). Also, inflammatory cytokines such as IL-1-α and IL-6 have been associated with the expression of VEGF (Borg et al., 2005). Hence, multiple factors within a tissue work in concert to regulate VEGF expression and secretion and ultimately angiogenesis.

5 Therapies

5.1 Blood Glucose Control

Early detection of retinopathy is the key to preventing vision loss. A routine dilated eye examination is effective in detecting and treating diabetic retinopathy. Type 1 diabetics should have a dilated eye examination within 3–5 years after diagnosis. Type 2 diabetics should have an eye examination at the time of diagnosis since 21% will have developed retinopathy at the time of the initial diagnosis. As found with most of the complications associated with diabetes, intensive control of blood sugar in diabetic individuals decreases retinal microvascular abnormalities, macular edema, and progression of diabetic retinopathy (Diabetes Control and Complications Trial Research Group, 1995; UK Prospective Diabetes Study Group, 1998; Stratton et al., 2000). In addition, the control of other systemic factors, such as hypertension and hyperlipidemia, has also been shown to decrease the rate of progression of diabetic retinopathy.

5.2 Laser

The ETDRS demonstrated that focal laser photocoagulation in affected retinal regions can decrease the rate of moderate vision loss by 50% in individuals with diabetic macula edema (Early Treatment Diabetic Retinopathy Study Group, 1985). In patients at a high risk for progressing to the proliferative stage of diabetic retinopathy, the Diabetic Retinopathy Study (DRS) demonstrated that panretinal laser photocoagulation (scatter laser photocoagulation in the midperipheral retinal regions) results in the regression of new blood vessels, decreasing the risk of severe vision loss by 50% (Campbell et al., 1965). Side effects from panretinal laser photocoagulation include scotomas, corresponding to the laser burns, decreased dark adaptation, choroidal neovascular membrane, mydriasis due to damage of uveal tract nerves, choroidal effusion, and problems with accommodation.

Although the precise mechanisms responsible for the improved visual outcome following laser therapy are not completely understood, studies have shown that PO_2 levels increase around the photocoagulation regions (Stefansson, 2006). This rise in PO_2 leads to vasoconstriction and lower perfusion pressure that ultimately reduces fluid movement into the retinal environment. Elevated retinal PO_2 levels will also reduce HIF-1-mediated changes in gene expression. Studies have shown that in individuals following panretinal laser photocoagulation, VEGF levels are reduced (Lip et al., 2000). More recent studies have shown that panretinal laser photocoagulation can increase levels of the antiangiogenic factor PEDF (Hattenbach et al., 2005).

5.3 Vitrectomy

Based on the results of the Diabetic Retinopathy Vitrectomy Study (DRVS), vitrectomy is indicated for nonclearing vitreous hemorrhage and tractional retinal detachments (Diabetic Retinopathy Vitrectomy Study Group, 1990). Studies have also shown that following vitrectomy, retinal neovascularization does not progress and macular edema can be reduced in diabetic individuals (Tachi and Ogino, 1996; Otani and Kishi, 2002). This improvement may be due to the removal of epiretinal membranes and the ability of these extracellular matrix structures to serve as reservoirs for VEGF and other cytokines, or the improved oxygen delivery to the retina. Additional studies have shown that vitrectomy can increase the overall retinal oxygen tension in eyes with vascular disorders (Stefansson et al., 1990).

5.4 Steroids

Intravitreal glucocorticoids have been used for the treatment of diabetic macula edema. Although large scale trials have yet to be conducted, case studies have provided evidence that intravitreal administration of triamcinolone acetonide can decrease retinal thickness and improve visual acuity (Martidis et al., 2002; Jonas et al., 2003). The anti-inflammatory actions of glucocorticoids likely contribute to the reduction in diabetic macular edema by suppressing the formation and activity of inflammatory cytokines in the retina. However, recent studies have provided evidence that glucocorticoids and these associated cellular responses reduce retinal edema by regulating the expression and phosphorylation of tight junction proteins in retinal vascular endothelial cells (Antonetti et al., 2002). Glucocorticoids may also influence blood–retinal barriers indirectly by reduced VEGF expression and increased PEDF expression (Tombran-Tink et al., 2004; Matsuda et al., 2005). The anti-inflammatory effects of glucocorticoids result from three mechanisms. First, the activated glucocorticoid receptor binds to specific DNA sequences, glucocorticoid receptor responsive elements that are located in the promoter regions of target genes to induce transcription (i.e., transactivation). Second, negative regulation of gene expression (i.e., transrepression) can be achieved by activated receptor binding to other proinflammatory transcription factors and the resulting complex inhibiting the transcription of relevant genes. This transrepression has been shown to inhibit the expression of various genes, such as IL-1and TNF-α. The third mechanism is glucocorticoid signaling through membrane-associated receptors and second messengers (so-called nongenomic pathways). Glucocorticoids may also alter cellular responses by changing histone acetylation and chromatin structure (Barnes, 1998).

5.5 Protein Kinase C Inhibitors

Studies have shown that hyperglycemia increases serum and cellular levels of diacylglycerol (Sheetz and King, 2002). In addition, AGEs have also been shown to increase diacylglycerol in retinal cells (Denis et al., 2002). Diacylglycerol is the endogenous activator for multiple forms of protein kinase C (PKC). However, the increase in activity of the PKC$_\beta$ isoform is thought to play a central role in the development of diabetic complications such as cell proliferation and permeability changes. Direct activation of PKC$_\beta$ has also been shown to alter retinal blood flow (Shiba et al., 1993). Ruboxistaurin mesylate, LY333531, is a relatively selective inhibitor of PKC$_\beta$ and has been investigated for inhibition of progression of diabetic retinopathy and for the inhibition of diabetic macular edema (The PKC-DRS Study Group, 2005). The results from these studies demonstrated that ruboxistaurin is well tolerated and reduced the risk of visual loss, but did not prevent the progression of diabetic retinopathy. The effects on visual loss were more pronounced in those individuals with the greatest macular edema at the time therapy was initiated. More recent studies have shown that PKC inhibition can also ameliorate the diabetes-induced changes that occur in retinal hemodynamic abnormalities (Aiello et al., 2006).

The inhibition of AGE-related events and PKC activation have both been shown to reduce abnormalities associated with diabetes. Although biochemical analysis of these pathways suggests no common link in the pathogenesis of diabetes, studies have provided evidence that initial activation of each pathway is related to hyperglycemia-induced alterations in mitochondrial metabolism and the generation of mitochondrial forms of superoxide dismutase. The over production of superoxide dismutase leads to reduced activity in the glycolytic pathway and results in an elevation in metabolic intermediates being used in the formation of AGEs and diacylglycerol (for review see Brownlee, 2001). Thus, long-term antioxidant therapy may be eventually useful in the treatment of diabetic retinopathy; however, not all studies have supported this conclusion.

5.6 Antiangiogenic Agents

Pegaptanib sodium (Macugen) and ranibizumab (Lucentis) are anti-VEGF blocking agents, which have been approved by the FDA for the treatment of neovascular age-related macular degeneration. Pegaptanib is an RNA aptamer that is designed to bind VEGF-165, whereas ranibizumab is a humanized monoclonal antibody fragment that recognizes all VEGF isoforms (Chakravarthy et al., 2006; Ng et al., 2006). As studies have concluded that VEGF plays a role in diabetic retinopathy, a number of investigators have hypothesized that anti-VEGF would also be effective in treating the ocular complications found in diabetics. Recent clinical trials have shown that pegaptanib treatment can produce a significant benefit in the treatment of diabetic macular edema and a regression of neovascularization when present in the eyes of diabetic individuals (Cunningham et al., 2005; Adamis et al., 2006). However, the need for multiple injections may limit the usage of these agents in their current formulations.

6 Summary

Diabetic retinopathy is one of the leading causes of blindness in developed countries. The use of large-scale clinical trials has shown that tight control of blood glucose and systemic hypertension can limit the development of vision-threatening retinopathy along with data demonstrating the efficacy of laser and vitrectomy in treating this disease. However, future treatments for diabetic retinopathy will rely on increasing our understanding of the pathogenic mechanisms responsible for the early vascular changes observed in the retina.

References

Adamis AP, Altaweel M, Bressler NM, Cunningham ET Jr, Davis MD, et al. 2006. Changes in retinal neovascularization after pegaptanib (Macugen) therapy in diabetic individuals. Ophthalmology 113: 23-28.

Adamis AP, Shima DT, Tolentino MJ, Gragoudas ES, Ferrara N, et al. 1996. Inhibition of vascular endothelial growth factor prevents retinal ischemia-associated iris neovascularization in a nonhuman primate. Arch Ophthalmol 114: 66-71.

Aiello LP, Clermont A, Arora V, Davis MD, Sheetz MJ, et al. 2006. Inhibition of PKC β by oral administration of ruboxistaurin is well tolerated and ameliorates diabetes-induced retinal hemodynamic abnormalities in patients. Invest Ophthalmol Vis Sci 47: 86-92.

Antonetti DA, Barber AJ, Hollinger LA, Wolpert EB, Gardner TW. 1999. Vascular endothelial growth factor induces rapid phosphorylation of tight junction proteins occludin and zonula occluden 1. A potential mechanism for vascularpermeability in diabetic retinopathy and tumors. J Biol Chem 274: 23463-23467.

Antonetti DA, Wolpert EB, De Maio L, Harhaj NS, Scaduto RC Jr. 2002. Hydrocortisone decreases retinal endothelial cell water and solute flux coincident with increased content and decreased phosphorylation of occludin. J Neurochem 80: 667-677.

Bandopadhyay R, Orte C, Lawrence JG, Reid AR, De Silva S, et al. 2001. Contractile proteins in pericytes at the blood–brain and blood–retinal barriers. J Neurocytol 30: 35-44.

Barnes PJ. 1998. Anti-inflammatory actions of glucocorticoids: Molecular mechanisms. Clin Sci (Lond) 94: 557-572.

Behzadian MA, Wang XL, Al-Shabrawey M, Caldwell RB. 1998. Effects of hypoxia on glial cell expression of angiogenesis-regulating factors VEGF and TGF-β. Glia 24: 216-225.

Bek T. 1999. Diabetic maculopathy caused by disturbances in retinal vasomotion. A new hypothesis. Acta Ophthalmol Scand 77: 376-380.

Beltramo E, Berrone E, Giunti S, Gruden G, Perin PC, et al. 2006. Effects of mechanical stress and high glucose on pericyte proliferation, apoptosis and contractile phenotype. Exp Eye Res 83: 989-994.

Berinstein DM, Stahn RM, Welty TK, Leonardson GR, Herlihy JJ. 1997. The prevalence of diabetic retinopathy and associated risk factors among Sioux Indians. Diabetes Care 20: 757-759.

Borg SA, Kerry KE, Royds JA, Battersby RD, Jones TH. 2005. Correlation of VEGF production with IL1α and IL6 secretion by human pituitary adenoma cells. Eur. J Endocrinol 152: 293-300.

Brausewetter F, Jehle PM, Jung MF, Boehm BO, Brueckel J, et al. 2001. Microvascular permeability is increased in both types of diabetes and correlates differentially with serum levels of insulin-like growth factor 1 (IGF-1) and vascular endothelial growth factor (VEGF). Horm Metab Res 33: 713-720.

Brownlee M. 2001. Biochemistry and molecular cell biology of diabetic complications. Nature 414: 813-820.

Bruick RK, McKnight SL. 2001. A conserved family of prolyl-4-hydroxylases that modify HIF. Science 294: 1337-1340.

Campbell CJ, Koester CJ, Curtice V, Noyori KS, Rittler MC. 1965. Clinical studies in laser photocoagulation. Arch Ophthalmol 74: 57-65.

Chakravarthy U, Soubrane G, Bandello F, Chong V, Creuzot-Garcher C, et al. 2006. Evolving European guidance on the medical management of neovascular age related macular degeneration. Br J Ophthalmol 90: 1188-1196.

Chew EY. 2003. Epidemiology of diabetic retinopathy. Hosp Med 64: 396-399.

Clermont AC, Cahill M, Salti H, Rook SL, Rask-Madsen C, et al. 2006. Hepatocyte growth factor induces retinal vascular permeability via MAP-kinase and PI-3 kinase without altering retinal hemodynamics. Invest Ophthalmol Vis Sci 47: 2701-2708.

Corman B, Duriez M, Poitevin P, Heudes D, Bruneval P, et al. 1998. Aminoguanidine prevents age-related arterial stiffening and cardiac hypertrophy. Proc Natl Acad Sci USA 95: 1301-1306.

Cunningham ET Jr, Adamis AP, Altaweel M, Aiello LP, Bressler NM, et al. 2005. A phase II randomized double-masked trial of pegaptanib, an anti-vascular endothelial growth factor aptamer, for diabetic macular edema. Ophthalmology 112: 1747-1757.

Davies PF, Tripathi SC. 1993. Mechanical stress mechanisms and the cell. An endothelial paradigm. Circ Res 72: 239-245.

Davis MD, Fisher MR, Gangnon RE, Barton F, Aiello LM, et al. 1998. Risk factors for high-risk proliferative diabetic retinopathy and severe visual loss: Early treatment diabetic retinopathy study report #18. Invest Ophthalmol Vis Sci 39: 233-252.

Denis U, Lecomte M, Paget C, Ruggiero D, Wiernsperger N, et al. 2002. Advanced glycation end-products induce apoptosis of bovine pericytes in culture: Involvement of diacylglycerol/ceramide production and oxidative stress induction. Free Radic Biol Med 33: 236-247.

Diabetes Control and Complications Trial Research Group. 1995. Progression of retinopathy with intensive versus conventional treatment in the Diabetes Control and Complications Trial. Ophthalmology 102: 647–661.

Diabetic Retinopathy Vitrectomy Study Group. 1990. Early vitrectomy for severe vitreous hemorrhage in diabetic retinopathy. Four-year results of a randomized trial: Diabetic Retinopathy Vitrectomy Study Report 5. Arch Ophthalmol 108: 958–964. [Erratum: 1990. 108: 1452]

Dowling JE. 1987. The Retina: An Approachable Part of the Brain. Cambridge, MA: Belknap Press.

Early Treatment Diabetic Retinopathy Study Group. 1985. Photocoagulation for diabetic macular edema: Early Treatment Diabetic Retinopathy Study Report Number 1. Arch Ophthalmol 103: 1796-1806.

Eichler W, Kuhrt H, Hoffmann S, Wiedemann P, Reichenbach A. 2000. VEGF release by retinal glia depends on both oxygen and glucose supply. Neuroreport 11: 3533-3537.

Emanuele N, Sacks J, Klein R, Reda D, Anderson R, et al. 2005. Ethnicity, race, and baseline retinopathy correlates in the veterans affairs diabetes trial. Diabetes Care 28: 1954-1958.

Engerman RL, Kern TS. 1995. Retinopathy in animal models of diabetes. Diabetes Metab Rev 11: 109-120.

Felinski EA, Antonetti DA. 2005. Glucocorticoid regulation of endothelial cell tight junction gene expression: Novel treatments for diabetic retinopathy. Curr Eye Res 30: 949-957.

Ferrara N. 2004. Vascular endothelial growth factor: Basic science and clinical progress. Endocr Rev 25: 581-611.

Ferris F. 1996. Early photocoagulation in patients with either type I or type II diabetes. Trans Am Ophthalmol Soc 94: 505-537.

Fong DS, Aiello L, Gardner TW, King GL, Blankenship G, et al. 2003. Diabetic retinopathy. Diabetes Care 26: 226-229.

Frank RN. 2004. Diabetic retinopathy. N Engl J Med 350: 48-58.

Frank S, Hubner G, Breier G, Longaker MT, Greenhalgh DG, et al. 1995. Regulation of vascular endothelial growth factor expression in cultured keratinocytes. Implications for normal and impaired wound healing. J Biol Chem 270: 12607-12613.

Fuh G, Garcia KC, de Vos AM. 2000. The interaction of neuropilin-1 with vascular endothelial growth factor and its receptor FLT-1. J Biol Chem 275: 26690-26695.

Gao G, Li Y, Zhang D, Gee S, Crosson C, et al. 2001. Unbalanced expression of VEGF and PEDF in ischemia-induced retinal neovascularization. FEBS Lett 489: 270-276.

Goldin A, Beckman JA, Schmidt AM, Creager MA. 2006. Advanced glycation end-products: Sparking the development of diabetic vascular injury. Circulation 114: 597-605.

Harris A, Arend O, Danis RP, Evans D, Wolf S, et al. 1996. Hyperoxia improves contrast sensitivity in early diabetic retinopathy. Br Med J 80: 209-213.

Hartnett ME, Lappas A, Darland D, McColm JR, Lovejoy S, et al. 2003. Retinal pigment epithelium and endothelial cell interaction causes retinal pigment epithelial barrier dysfunction via a soluble VEGF-dependent mechanism. Exp Eye Res 77: 593-599.

Hattenbach LO, Beck KF, Pfeilschifter J, Koch F, Ohrloff C, et al. 2005. Pigment-epithelium-derived factor is upregulated in photocoagulated human retinal pigment epithelial cells. Ophthalmic Res 37: 341-346.

Hofman P, Blaauwgeers HG, Tolentino MJ, Adamis AP, Nunes Cardozo BJ, et al. 2000. VEGF-A induced hyperpermeability of blood–retinal barrier endothelium in vivo is predominantly associated with pinocytotic vesicular transport and not with formation of fenestrations. Vascular endothelial growth factor-A. Curr Eye Res 21: 637-645.

Houck KA, Ferrara N, Winer J, Cachianes G, Li B, et al. 1991. The vascular endothelial growth factor family: Identification of a fourth molecular species and characterization of alternative splicing of RNA. Mol Endocrinol 5: 1806-1814.

Houck KA, Leung DW, Rowland AM, Winer J, Ferrara N. 1992. Dual regulation of vascular endothelial growth factor bioavailability by genetic and proteolytic mechanisms. J Biol Chem 267: 26031-26037.

Jawa A, Kcomt J, Fonseca VA. 2004. Diabetic nephropathy and retinopathy. Med Clin North Am 88: 1001-1036.

Jonas JB, Kreissig I, Sofker A, Degenring RF. 2003. Intravitreal injection of triamcinolone for diffuse diabetic macular edema. Arch Ophthalmol 121: 57-61.

Kass DA, Shapiro EP, Kawaguchi M, Capriotti AR, Scuteri A, et al. 2001. Improved arterial compliance by a novel advanced glycation end-product crosslink breaker. Circulation 104: 1464-1470.

Kempen JH, O'Colmain BJ, Leske MC, Haffner SM, Klein R, et al. 2004. The prevalence of diabetic retinopathy among adults in the United States. Arch Ophthalmol 122: 552-563.

Khaliq A, Foreman D, Ahmed A, Weich H, Gregor Z, et al. 1998. Increased expression of placenta growth factor in proliferative diabetic retinopathy. Lab Invest 78: 109-116.

Klein R, Klein BE, Moss SE, Cruickshanks KJ. 1995a. The Wisconsin Epidemiologic Study of Diabetic Retinopathy. XV. The long-term incidence of macular edema. Ophthalmology 102: 7-16.

Klein R, Klein BE, Moss SE, Cruickshanks KJ. 1995b. The Wisconsin Epidemiologic Study of Diabetic Retinopathy. XV. The long-term incidence of macular edema. Ophthalmology 102: 7-16.

Kristinsson JK, Gottfredsdottir MS, Stefansson E. 1997. Retinal vessel dilatation and elongation precedes diabetic macular edema. Br J Ophthalmol 81: 274-278.

Lightman S, Towler HM. 2003. Diabetic retinopathy. Clin Cornerstone 5: 12-21.

Linsenmeier RA, Braun RD, McRipley MA, Padnick LB, Ahmed J, et al. 1998. Retinal hypoxia in long-term diabetic cats. Invest Ophthalmol Vis Sci 39: 1647-1657.

Lip PL, Belgore F, Blann AD, Hope-Ross MW, Gibson JM, et al. 2000. Plasma VEGF and soluble VEGF receptor FLT-1 in proliferative retinopathy: Relationship to endothelial dysfunction and laser treatment. Invest Ophthalmol Vis Sci 41: 2115-2119.

Martidis A, Duker JS, Greenberg PB, Rogers AH, Puliafito CA, et al. 2002. Intravitreal triamcinolone for refractory diabetic macular edema. Ophthalmology 109: 920-927.

Massey SC. 1990. Cell types using glutamate as a neurotransmitter in the vertebrate retina. Prog Retin Res 9: 399-426.

Matsuda S, Gomi F, Oshima Y, Tohyama M, Tano Y. 2005. Vascular endothelial growth factor reduced and connective tissue growth factor induced by triamcinolone in ARPE19 cells under oxidative stress. Invest Ophthalmol Vis Sci 46: 1062-1068.

Mizutani M, Kern TS, Lorenzi M. 1996. Accelerated death of retinal microvascular cells in human and experimental diabetic retinopathy. J Clin Invest 97: 2883-2890.

Mori K, Gehlbach P, Ando A, Dyer G, Lipinsky E, et al. 2002. Retina-specific expression of PDGF-B versus PDGF-A: Vascular versus nonvascular proliferative retinopathy. Invest Ophthalmol Vis Sci 43: 2001-2006.

Ng EW, Shima DT, Calias P, Cunningham ET Jr, Guyer DR, et al. 2006. Pegaptanib, a targeted anti-VEGF aptamer for ocular vascular disease. Nat Rev Drug Discov 5: 123-132.

Otani T, Kishi S. 2002. A controlled study of vitrectomy for diabetic macular edema. Am J Ophthalmol 134: 214-219.

Park JE, Keller GA, Ferrara N. 1993. The vascular endothelial growth factor isoforms (VEGF): Differential deposition into the subepithelial extracellular matrix and bioactivity of extracellular matrix-bound VEGF. Mol Biol Cell 4: 1317-1326.

Park L, Raman KG, Lee KJ, Lu Y, Ferran LJ Jr, et al. 1998. Suppression of accelerated diabetic atherosclerosis by the soluble receptor for advanced glycation end-products. Nat Med 4: 1025-1031.

Poulaki V, Joussen AM, Mitsiades N, Mitsiades CS, Iliaki EF, et al. 2004. Insulin-like growth factor-I plays a pathogenetic role in diabetic retinopathy. Am J Pathol 165: 457-469.

Provis JM. 2001. Development of the primate retinal vasculature. Prog Retin Eye Res 20: 799-821.

Roy MS. 2000. Diabetic retinopathy in African Americans with type 1 diabetes: The New Jersey 725. II. Risk factors. Arch Ophthalmol 118: 105-115.

Roy MS, Klein R, O'Colmain BJ, Klein BE, Moss SE, et al. 2004. The prevalence of diabetic retinopathy among adult type 1 diabetic persons in the United States. Arch Ophthalmol 122: 546-551.

Schiekofer S, Andrassy M, Chen J, Rudofsky G, Schneider J, et al. 2003. Acute hyperglycemia causes intracellular formation of CML and activation of ras, p42/44 MAPK, and nuclear factor-κB in PBMCs. Diabetes 52: 621-633.

Schmidt AM, Hasu M, PopovD, Zhang JH, Chen J, et al. 1994. Receptor for advanced glycation end-products (AGEs) has a central role in vessel wall interactions and gene activation in response to circulating AGE proteins. Proc Natl Acad Sci USA 91: 8807-8811.

Semenza GL. 2002. Signal transduction to hypoxia-inducible factor 1. Biochem Pharmacol 64: 993-998.

Senger DR, Galli SJ, Dvorak AM, Perruzzi CA, Harvey VS, et al. 1983. Tumor cells secrete a vascular permeability factor that promotes accumulation of ascites fluid. Science 219: 983-985.

Sheetz MJ, King GL. 2002. Molecular understanding of hyperglycemia's adverse effects for diabetic complications. J Am Med Assoc 288: 2579-2588.

Shiba T, Inoguchi T, Sportsman JR, Heath WF, Bursell SE, et al. 1993. Correlation of diacylglycerol level and protein kinase C activity in rat retina to retinal circulation. Am J Physiol 265: E783-E793.

Spaide RF, Fisher YL. 2006. Intravitreal bevacizumab (Avastin) treatment of proliferative diabetic retinopathy complicated by vitreous hemorrhage. Retina 26: 275-278.

Stefansson E. 2006. Ocular oxygenation and the treatment of diabetic retinopathy. Surv Ophthalmol 51: 364-380.

Stefansson E, Novack RL, Hatchell DL. 1990. Vitrectomy prevents retinal hypoxia in branch retinal vein occlusion. Invest Ophthalmol Vis Sci 31: 284-289.

Stone J, Itin A, Alon T, Pe'er J, Gnessin H, et al. 1995. Development of retinal vasculature is mediated by hypoxia-induced vascular endothelial growth factor (VEGF) expression by neuroglia. J Neurosci 15(Pt 1): 4738-4747.

Stratton IM, Adler AI, Neil HA, Matthews DR, Manley SE, et al. 2000. Association of glycaemia with macrovascular and microvascular complications of type 2 diabetes (UKPDS 35): Prospective observational study. Br Med J 321: 405-412.

Stratton IM, Kohner EM, Aldington SJ, Turner RC, Holman RR, et al. 2001. UKPDS 50: Risk factors for incidence and progression of retinopathy in Type II diabetes over 6 years from diagnosis. Diabetologia 44: 156-163.

Sutter CH, Laughner E, Semenza GL. 2000. Hypoxia-inducible factor 1α protein expression is controlled by oxygen-regulated ubiquitination that is disrupted by deletions and missense mutations. Proc Natl Acad Sci USA 97: 4748-4753.

Suzuma I, Suzuma K, Ueki K, Hata T, Feener EP, et al. 2002. Stretch-induced retinal vascular endothelial growth factor expression is mediated by phosphatidylinositol 3-kinase and protein kinase C (PKC)-zeta but not by stretch-induced ERK1/2, Akt, Ras, or classical/novel PKC pathways. J Biol Chem 277: 1047-1057.

Tachi N, Ogino N. 1996. Vitrectomy for diffuse macular edema in cases of diabetic retinopathy. Am J Ophthalmol 122: 258-260.

The PKC-DRS Study Group. 2005. The effect of ruboxistaurin on visual loss in patients with moderately severe to very severe nonproliferative diabetic retinopathy: Initial results of the Protein Kinase C β Inhibitor Diabetic Retinopathy Study (PKC-DRS) multicenter randomized clinical trial. Diabetes 54: 2188–2197.

Tischer E, Mitchell R, Hartman T, Silva M, Gospodarowicz D, et al. 1991. The human gene for vascular endothelial growth factor. Multiple protein forms are encoded through alternative exon splicing. J Biol Chem 266: 11947-11954.

Tombran-Tink J, Lara N, Apricio SE, Potiuri P, Gee S, et al. 2004. Retinoic acid and dexamethasone regulate the expression of PEDF in retinal and endothelial cells. Exp Eye Res 78: 945-955.

UK Prospective Diabetes Study Group. 1998. Tight blood pressure control and risk of macrovascular and microvascular complications in type 2 diabetes: UKPDS 38. Br Med J 317: 703–713. [Erratum: 1999. 318: 329]

Vincenti V, Cassano C, Rocchi M, Persico G. 1996. Assignment of the vascular endothelial growth factor gene to human chromosome 6p21.3. Circulation 93: 1493-1495.

Warren RS, Yuan H, Matli MR, Ferrara N, Donner DB. 1996. Induction of vascular endothelial growth factor by insulin-like growth factor 1 in colorectal carcinoma. J Biol Chem 271: 29483-29488.

9 Neurotransmitters and Electrophysiology in Traumatic Brain Injury

C. E. Dixon · A. E. Kline

1	Introduction	180
2	*Acetylcholine*	180
2.1	Acute Cholinergic Responses to Traumatic Brain Injury	180
2.2	Chronic Cholinergic Responses to Traumatic Brain Injury	181
2.3	Acetylcholine Therapies for Traumatic Brain Injury	181
3	*Catecholamines*	182
3.1	Catecholaminergic Responses to Traumatic Brain Injury	182
3.2	Catecholaminergic Therapies for Traumatic Brain Injury	184
4	*Serotonin*	186
4.1	Serotonergic Responses to Traumatic Brain Injury	186
4.2	Serotonergic Therapies for Traumatic Brain Injury	186
5	*Glutamate*	188
5.1	Glutamate Mechanisms of Traumatic Brain Injury	188
5.2	Inhibiting Glutamate Release	188
5.3	Prevention of Glutamate Receptor Activation	188
6	*Gamma-Aminobutyric Acid*	189
7	*Adenosine*	190
7.1	Adenosine Responses to Traumatic Brain Injury	190
7.2	Adenosine Therapies for Traumatic Brain Injury	190
7.3	A_2AR/D_2R Interactions	190
8	*Electrophysiology in TBI Research*	191
8.1	TBI-Induced Alterations in Spontaneous Brain Electrical Activity	192
8.2	Evoked Potentials in Traumatic Brain Injury	192
8.3	Long-Term Potentiation in TBI Models	194

© 2009 Springer Science+Business Media, LLC.

Abstract: Normal brain function requires the careful orchestration of neurotransmission and neuroelectric activity. In the traumatically injured brain, disruption of these processes may underlie, in part, subsequent brain dysfunction. This chapter focuses on two areas of traumatic brain injury (TBI) research: neurotransmitters and electrophysiology. Reviewed here are the TBI-induced responses to specific neurotransmitter systems and interventions to modulate these responses. Also, the effects of TBI on spontaneous and evoked electrical activity are reviewed.

List of Abbreviations: 2-CA, 2-chloroadenosine; 5-HIAA, 5-hydroxyindoleacetic acid; 5-HT, 5-hydroxytryptamine; 8-OH-DPAT, 8-hydroxy-2-(di-n-propylamino)tetralin; A_1AR, A_1 adenosine receptor; A_{2A}, adenosine-2a receptor subtype; ACh, acetylcholine; AMPA, [alpha]-amino-3-hydroxy-5-methylisoazole-4-proprionic acid; CAP, compound action potential; ChAT, choline acetyltransferase; CCI, controlled cortical impact; CPP, 3-(2-carboxypiperazin-4-yl)propyl-l-phosphonic acid; DA, dopamine; DAI, diffuse axonal injury; DAT, dopamine transporter; EP, evoked potential; EPSP, excitatory postsynaptic potential; GABA, gamma-aminobutyric acid; LTP, long-term potentiation; MWM, Morris water maze; M_1, muscarinic receptor subtype 1; M_2, muscarinic receptor subtype 2; mGluR, metabotropic glutamate receptors; NMDA, N-methyl-D-aspartate; NR, NMDA receptor; CSD, cortical spreading depression; TBI, traumatic brain injury; TH, tyrosine hydroxylase; VAChT, vesicular ACh transporter

1 Introduction

It is well established that traumatic brain injury (TBI), regardless of the level of severity, results in significant disability due to its attendant neurobehavioral deficits. Neuropsychological studies of outcome following TBI have indicated that the most enduring deficits include impaired learning and memory. Impairment in memory specifically involves reduced ability to learn new information, which is greater following severe TBI, although it is present even after mild TBI. This chapter will cover recent developments in the roles of neurotransmitter and electrophysiological alterations following TBI. The neurotransmitter section will focus on the cholinergic, catecholaminergic, serotonergic, glutamatergic, GABAergic, and adenosinergic systems as prototypic neurotransmitter systems for hippocampal and memory function. The second part of this chapter will examine the three broad categories of electrophysiological studies in TBI research (1) studies that measure spontaneous brain electrical potentials as assessed by changes in brain activity, (2) studies of environmental evoked changes in electric potentials as a measure of neuronal structural pathway integrity, and (3) measurements of persistent changes of evoked field potentials as indices of synaptic plasticity.

2 Acetylcholine

2.1 Acute Cholinergic Responses to Traumatic Brain Injury

Numerous studies have shown alterations in acetylcholine (ACh) levels after both clinical and experimental TBI. For instance, increased ACh levels have been detected in CSF after TBI in humans (Tower and McEachern, 1949; Sachs, 1957; Haber and Grossman, 1980), dogs (Bornstein, 1946; Ruge, 1954; Sachs, 1957; Metz, 1971), cats (Bornstein, 1946), and rats (Sachs, 1957; Robinson, 1990). Gorman et al. (1989) reported an acute (within 5–10 min after fluid-percussion brain injury) and significant increase (74% above control) of hippocampal ACh levels detected via microdialysis. Saija et al. (1988a, b) found that ACh turnover rate, as measured by a phosphoryl [$^2H_9$]choline method, increased by 73% relative to controls in the brainstem at 12 min after moderate fluid-percussion TBI in the rat. Moreover, the levels were still elevated 50% 4 h later. These data indicate that moderate TBI produces an early release of ACh and increased cholinergic neuronal activity in some brain regions for several hours after injury.

Potential mechanisms mediating why acute increases in ACh can be pathologic after TBI include, but are not limited to, brain injury-induced muscarinic receptor activation contributing to excitotoxic processes by modulation of ionic fluxes via altered protein kinase C activity (Yang et al., 1993) or inositol 1,4,5-triphosphate-associated G-protein-coupled PI turnover (Delahunty, 1992). Acute increases in ACh may also contribute to cerebrovascular pathology. Numerous studies have shown that cerebral blood flow can be regulated by cholinergic mechanisms (Scremin et al., 1973; Molnar et al., 1991; Scremin, 1991). Furthermore, Scremin et al. (1997) reported that cholinergic agonists and antagonists could modulate CBF responses to trauma. Chen et al. (1998) demonstrated that posttraumatic edema in the contused hemisphere (24 h after injury) and the acetylcholinesterase inhibitor ENA713 can attenuate disruption of the blood–brain barrier (4 h after injury).

2.2 Chronic Cholinergic Responses to Traumatic Brain Injury

Several lines of evidence indicate that TBI in rats produces disturbances in cognitive performance, which may be related to chronic decreases in cholinergic neurotransmission. Microdialysis-based studies have shown a reduction in the release of ACh evoked by the muscarinic autoreceptor antagonist, scopolamine (Dixon et al., 1995, 1996) 2 weeks after TBI. TBI can also produce increased sensitivity to disruption of spatial memory function by scopolamine, which occurs concurrently with a reduction in scopolamine-evoked ACh release (Dixon et al., 1994). Time-dependent loss of choline acetyltransferase (ChAT) enzymatic activity (Gorman et al., 1996) and ChAT immunohistochemical staining (Leonard et al., 1994; Schmidt and Grady, 1995; Sinson et al., 1997) has been reported after TBI. Using a controlled cortical impact (CCI) model of brain injury, Ciallella and colleagues have reported chronic changes in vesicular ACh transporter (VAChT) (❶ *Figure 9-1*) and M_2 cholinergic muscarinic receptor, two proteins involved in cholinergic neurotransmission (Ciallella et al., 1998). To determine if transcriptional regulation may play a role in the protein-level alterations, Ciallella and colleagues examined chronic mRNA changes in VAChT and M_2 4 weeks after moderate brain trauma in rats. The data were compared with changes in protein levels that occurred at the same time point. Changes in VAChT and M_2 mRNA levels were evaluated by RT-PCR. At 4 weeks postinjury, both immunohistochemical and Western blot methods demonstrated an increase in hippocampal VAChT protein. An increase in VAChT mRNA was also observed. Immunohistochemistry demonstrated a loss of M_2; however, there was no significant change in M_2 mRNA levels compared with sham controls. Changes in VAChT and M_2 protein have been demonstrated to persist for up to 1 year after TBI (Dixon et al., 1999a). These changes may represent a compensatory response of cholinergic neurons to increase the efficiency of ACh neurotransmission chronically after TBI through differential transcriptional regulation.

TBI-induced alterations of cholinergic neurotransmission in humans have been determined from both postmortem brain analyses and treatments with cholinergic agonists in TBI survivors. Recent studies of pre- and postsynaptic markers of cholinergic transmission were examined postmortem from patients who died as a result of TBI (Dewar and Graham, 1996; Murdoch et al., 1998). ChAT activity, M_1, M_2, and nicotinic receptor binding sites were assayed in head-injured patients and age-matched controls. ChAT activity was reduced in the head-injured group compared with the control group. In contrast, there was no difference between the head-injured and control group in the levels of M_1, M_2, or nicotinic receptor binding. However, the correlation between ChAT activity and synaptophysin immunoreactivity suggests a deficit in cholinergic terminals in the postmortem human brain after TBI.

2.3 Acetylcholine Therapies for Traumatic Brain Injury

Pharmacologically increasing levels of ACh can attenuate posttraumatic spatial memory performance deficits. For example, increasing ACh synthesis by increasing the availability of choline using CDP-choline treatment has been reported to attenuate posttraumatic spatial memory performance deficits (Dixon et al., 1997). Chronic postinjury administration of BIBM 99, a selective muscarinic M_2 receptor antagonist that increases ACh release by blocking presynaptic autoreceptors, attenuates spatial memory deficits after TBI (Pike and Hamm, 1995). Similarly, chronic postinjury administration of MDL 26,479 (suritozole), a

◘ Figure 9-1

TBI causes an increase in VAChT protein levels 2 and 4 weeks after injury. Semiquantitative graph of ipsilateral hippocampal homogenates at various times after moderate TBI ($n = 4–6$). (*ANOVA single factor, $p < 0.05$)

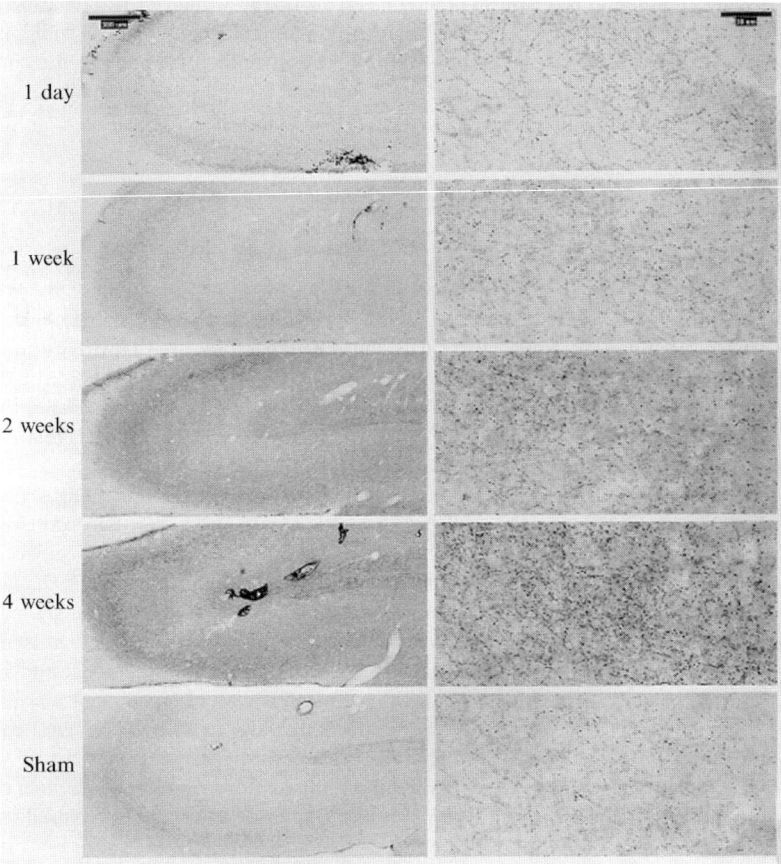

negative modulator at the gamma-aminobutyric acid (GABA) receptor, which enhances cholinergic function, attenuates spatial memory deficits after TBI (O'Dell and Hamm, 1995). Verbois et al. (2003) reported that twice-daily nicotine injections significantly attenuated TBI-induced cognitive dysfunction.

3 Catecholamines

3.1 Catecholaminergic Responses to Traumatic Brain Injury

Alterations in the catecholamine systems have been found in various brain regions after experimental TBI and have been shown to be time-dependent (Dunn-Meynell et al., 1994; McIntosh et al., 1994; Massucci et al., 2004). In an early study assessing the time course of brain catecholamine concentration after TBI in rats, McIntosh et al. (1994) reported a decrease in DA tissue concentrations in the injured cortex beginning at 1 h and persisting for 2 weeks after fluid-percussion brain injury. In a more recent study, Massucci and colleagues investigated the temporal changes in DA tissue levels and metabolism at 1 h or 1, 7, 14, and 28 days after CCI or sham injury in rats. DA and DOPAC levels were measured by high-performance liquid chromatography in the frontal cortex and striatum. DA levels were significantly increased at 1 h in the contralateral frontal cortex and at 1 day in the ipsilateral frontal cortex versus respective sham groups (◉ *Figure 9-2*). DA and DOPAC levels were significantly increased bilaterally at 1 h in the striatum of TBI versus sham control. These data indicate that TBI induces an early increase in DA and DOPAC, which returns to sham levels over time (Massucci et al., 2004). DA receptors have also been shown to change after TBI. For instance, concussive brain injury in rats has been demonstrated to produce an initial downregulation of D_1 receptors followed by an upregulation (Henry et al., 1997). Furthermore, both tyrosine hydroxylase (TH), the rate-limiting enzyme in the synthesis of catecholamines, and the DA

◘ **Figure 9-2**
Mean (±SEM) DA levels ((illimole per milligram tissue) at 1 h or 1, 7, 14, and 28 days after cortical impact injury in the rodent FC and striatum. (a) *$p < 0.05$ versus respective sham controls at 1 h in the contralateral and 1 day in the ipsilateral FC. (b) *$p < 0.05$ versus respective shams in both the ipsilateral and contralateral striatum at 1 h post-TBI. [Reprinted with permission from Massucci et al. (2004)]

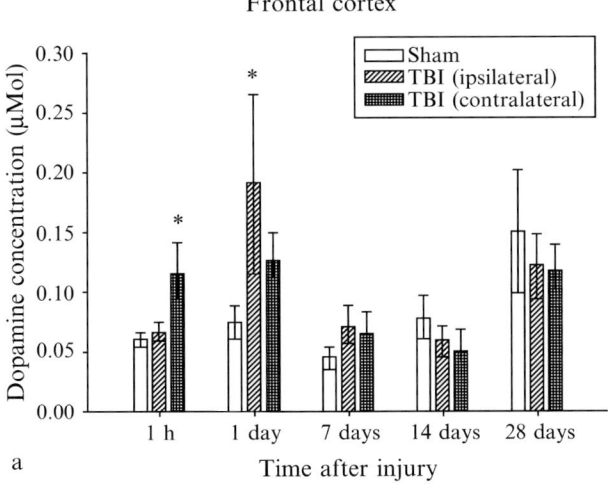

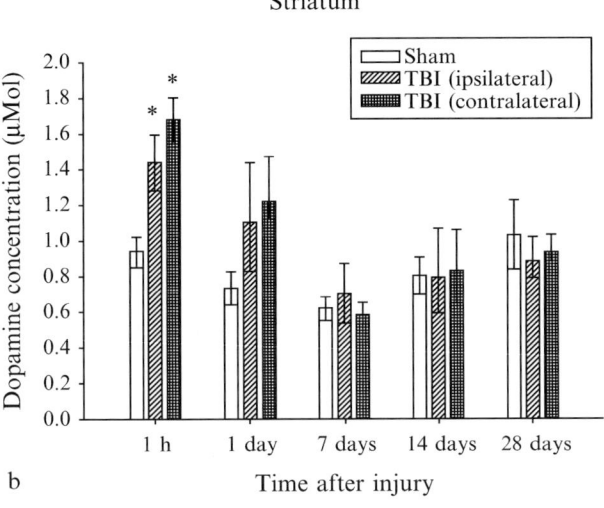

transporter (DAT), which plays a critical role in maintaining DA homeostasis in the CNS, are temporally altered after CCI injury (Yan et al., 2001, 2002; Wagner et al., 2005b; Wilson et al., 2005). Specifically, Yan and colleagues reported chronic regional changes in TH protein levels by immunohistochemistry and Western blots. An increased immunohistochemical expression of TH protein was observed bilaterally at 2 and 4 weeks after TBI, and is increased more in the contralateral than in the ipsilateral hemisphere. The upregulation of the TH protein was further confirmed by Western blot analysis. TH protein upregulation may reflect a compensatory response of dopaminergic neurons to upregulate the synthesizing capacity and increase the efficiency of DA neurotransmission chronically after TBI (Yan et al., 2001). Yan and colleagues have also reported that DAT protein expression is decreased in the ipsilateral (i.e., injured) cortex at 7 days and bilaterally at 28 days in the injured versus sham controls rats. The authors suggest that the decrease in DAT protein levels may reflect a TBI-induced downregulation of DA transporter and/or loss of dopaminergic fibers (Yan et al., 2002). In a follow-up study, Wilson and colleagues reported a decrease in DAT protein expression, but no differences in synaptosomal uptake (K_m, V_{max}) in the ipsilateral striatum of injured versus sham control rats at 2 and 4 weeks (Wilson et al., 2005). Using the same injury model, Wagner and colleagues have reported gender differences in the alteration of DAT protein expression (Wagner et al., 2005a). Using fast scan cyclic voltammetry (FSCV) and Western blot at 14 days after CCI injury in rats, Wagner and colleagues showed that striatal-evoked DA overflow was lower for injured rats in the ipsilateral hemisphere. TBI rats exhibited a decrease in V_{max} (52% of naïve) for DA clearance. DAT expression was proportionally decreased in the striatum ipsilateral to injury compared with controls (60% of naïves). No change in dopamine D_2 receptor expression was observed (Wagner et al., 2005b).

3.2 Catecholaminergic Therapies for Traumatic Brain Injury

Extensive evidence in animal studies suggest that enhancement of recovery from TBI can be achieved by pharmacological stimulation of the catecholaminergic system. Seminal studies by Feeney and colleagues have shown that a single dose of d-amphetamine given 24 h after a sensorimotor cortex injury produces an immediate and enduring acceleration in beam-walking recovery in rats (Feeney et al., 1982) and that multiple doses of amphetamine restore binocular depth perception in cats with bilateral visual cortex injury (Hovda et al., 1989). Beneficial effects of other catecholamine agonists on functional outcome in rat and/or cat following cortical injury have also been reported (Feeney and Sutton, 1987; Goldstein and Davis, 1990; Feeney, 1991; Sutton and Feeney, 1992; Feeney et al., 1993; Kline et al., 1994; Phillips et al., 2003; Feeney et al., 2004). The dopamine D_2 receptor agonist methylphenidate is reported to have pharmacological properties similar to amphetamine, but without the undesirable sympathomimetic effect of the latter. It has been shown to enhance recovery of motor function after sensorimotor cortex injury when administered singly followed by multiple-symptom relevant experience (Kline et al., 1994). Methylphenidate has also been shown to confer beneficial functional effects with delayed and multiple treatments after CCI injury (Kline et al., 2000). This positive finding with delayed treatment suggests that strategies that enhance catecholamine neurotransmission during the chronic postinjury phase may be a useful adjunct in ameliorating some of the neurobehavioral sequelae following TBI in humans. The partial dopamine D_2 agonist amantadine has also been shown to exert beneficial effects after experimental TBI. Specifically, rats treated with amantadine beginning 1 day after CCI injury had significantly less spatial memory performance deficits than saline vehicle controls (Dixon et al., 1999b). Biochemical studies have demonstrated that amantadine increases release of DA into extracellular pools by blocking reuptake and by facilitating the synthesis of DA (Gerlak et al., 1970; von Voigtlander and Moore, 1971; Bak et al., 1972; Gianutsos et al., 1985). In addition to acting presynaptically, amantadine has been demonstrated to act postsynaptically to increase the density of postsynaptic DA receptors (Gianutsos et al., 1985) or to alter their conformation (Allen, 1983). Evidence of a postsynaptic mechanism is clinically promising since the mechanisms of actions may not depend solely on the presence of surviving postsynaptic terminals. Because the mechanism of action of amantadine differs from other DA-releasing drugs (see Gualtieri et al., 1989, for review), it is likely that the dopaminergic effects of amantadine are a combination of pre- and postsynaptic effects. The dopamine receptor agonist bromocriptine has also been reported to attenuate working memory deficits after experimental TBI. Kline et al. (2002a) have shown that the administration of bromocriptine (5 mg/kg, i.p.) beginning 24 h after CCI, with continued daily injections until all behavioral assessments were completed, improved both reference (◉ *Figure 9-3*) and working memory versus vehicle-treated controls. Additionally,

◘ Figure 9-3
Mean (±SE) latency (sec) to locate either a submerged (hidden) or raised (visible) platform in a spatial learning acquisition paradigm in the Morris water maze. The injury/bromocriptine group (5 mg/kg i.p., 1–20 days after cortical impact) located the hidden platform significantly quicker than the injury/vehicle group on the first day of training and maintained that difference at several time points throughout the testing period (p's < 0.05). *Significantly different from injury/vehicle. No differences were observed between the two CCI injured groups or the sham/vehicle control group in the latency to locate the visible platform ($p > 0.05$). [Reprinted with permission from Kline et al. (2002a)]

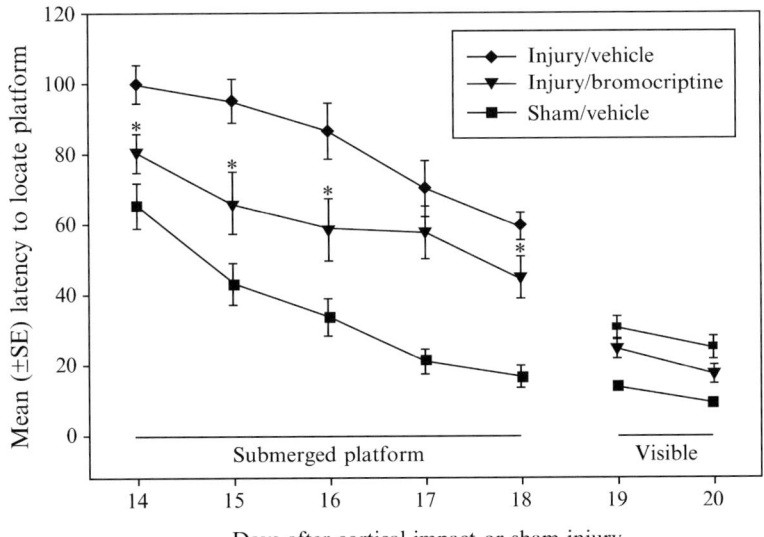

the injured bromocriptine-treated group exhibited significantly more morphologically intact CA3 neurons than the injured vehicle-treated group (55.60 ± 3.10% versus 38.34 ± 7.78% (Kline et al., 2002a). In a more recent study, Kline and colleagues reported significant decreases in malondialdehyde in both the striatum and substantia nigra in groups provided a single bromocriptine treatment versus saline. These data suggest that TBI-induced oxidative stress is attenuated by acute bromocriptine treatment, which may, in part, explain the benefits observed in cognitive and histological outcome (Kline et al., 2004b). Lastly, chronic treatment with the DA enhancer L-deprenyl has been reported to attenuate cognitive deficits following TBI (Zhu et al., 2000).

Clinically, amantadine, bromocriptine, and methylphenidate are three of the most commonly used treatments for TBI patients. Several uncontrolled clinical and case studies suggest that amantadine may improve the neurobehavioral deficits that often accompany TBI (Gualtieri, 1988; Gualtieri et al., 1989). Amantadine has been shown to be effective for both cognitive and behavioral symptoms in a series of chronic patients who were several years out from the original injury (Kraus and Maki, 1997). Recent studies have shown neurobehavioral improvements with amantadine in both adult (Kraus et al., 2005) and pediatric (Beers et al., 2005) TBI populations. Bromocriptine treatment may be beneficial in certain instances following TBI. Pulaski and Emmett (1994) and Karli et al. (1999) have reported functional improvement following TBI when bromocriptine was combined with therapy. Powell et al. (1996) reported significant improvements in cognitive function and motivation following chronic treatment with bromocriptine. In a double-blind, placebo-controlled crossover study, McDowell and colleagues (1998) found that bromocriptine treatment improved performance on clinical measures of executive function and in dual-task performance. More recently, Passler and Riggs (2001) have reported positive outcomes in TBI-vegetative state patients treated with bromocriptine. Ample evidence also exists showing that methylphenidate benefits recovery from TBI, especially in patients with predominant attention and memory deficits. Furthermore, the magnitude of the methylphenidate effect may depend on the severity of the initial TBI. In patients who demonstrated a positive behavioral change, methylphenidate also reduced brain injury-related anger and improved memory (Mooney and Haas, 1993). Speech et al. (1993) failed to find any significant cognitive improvement in a sample of predominantly severe TBI patients treated with methylphenidate approximately 3 years after injury. However, in a recent double-blind study, a positive effect for methylphenidate on overall outcome as well as on specific measures of vigilance and immediate verbal memory was found (Plenger et al., 1996). Several recent studies have shown benefits in cognitive performance in TBI patients treated with methylphenidate (Whyte et al., 1997, 2004; Lee et al., 2005; Kim et al., 2006).

4 Serotonin

4.1 Serotonergic Responses to Traumatic Brain Injury

In contrast to the extensive empirical study of the cholinergic and catecholaminergic systems following TBI, the serotonergic (5-hydroxytryptamine, 5-HT) neurotransmitter system has received notably less attention. The available studies suggest that serotonergic responses after cortical freeze lesions have revealed widespread decreases in 5-HT levels and increased cortical 5-HT metabolism, as evidenced by increased 5-hydroxyindoleacetic acid (5-HIAA) levels in the ipsilateral (lesioned) hemisphere (Pappius and Dadoun, 1987). The increase is temporally related to the depression of local cerebral glucose utilization (Pappius, 1981). Administration of the 5-HT synthesis inhibitor *p*-chlorophenylalanine attenuated the depression of local cerebral glucose utilization and prevented postinjury increases in 5-HT levels (Pappius et al., 1988). Based on their collective findings, the authors suggested that activation of the serotonergic system may play a role in postinjury cerebral metabolic dysfunction, and that blocking this system may be a beneficial strategy in the treatment of brain injury. Busto et al. (1997) reported that 5-HT levels increased from 18.85 ± 7.12 pm/mL (mean $\pm$ SD) to 65.78 ± 11.36 pm/mL in the first 10 min after fluid-percussion brain injury. The levels of 5-HT remained significantly higher than controls for the first 90-min sampling period. In parallel to the rise in 5-HT levels, a significant 71% decrease in extracellular 5-HIAA levels was noted in the first 10 min after fluid-percussion injury. These findings suggest that following TBI there is a rapid rise in extracellular 5-HT levels in cortical regions proximal to the injury site. Because 5-HT potentiation of excitatory amino acids has been reported in cat neocortex (Nedergaard et al., 1987), the trauma-induced release of 5-HT might negatively impact recovery by promoting excitotoxic processes.

4.2 Serotonergic Therapies for Traumatic Brain Injury

Several studies have shown that postinjury treatment with selective 5-HT_{1A} receptor agonists, such as repinotan HCL or 8-hydroxy-2-(di-*n*-propylamino)tetralin (8-OH-DPAT), confers neuroprotection and improves neurobehavioral performance in experimental models of TBI (Alessandri et al., 1999; Kline et al., 2001b, 2002b, 2004a). Kline and colleagues examined the effects of repinotan HCL in spatial memory and histopathological outcome using a CCI model of TBI, which has been shown to produce many of the characteristics of human TBI (Kline and Dixon, 2001a). The data revealed that a 4-h continuous infusion of repinotan HCL (10 µg/kg/h i.v.) commencing 5 min after cortical impact injury significantly attenuated cognitive deficits detected on a Morris water maze (MWM) test relative to the vehicle-treated group. Moreover, repinotan HCL attenuated hippocampal CA1 and CA3 cell loss and decreased cortical lesion volume compared with the vehicle-treated group (Kline et al., 2001b). Follow-up studies by Kline and colleagues have shown that a single and early administration of the classic 5-HT_{1A} receptor agonist,

8-OH-DPAT, provides similar beneficial effects on functional and histological outcome after cortical impact injury. Specifically, a dose of 0.5 mg/kg of 8-OH-DPAT administered intraperitoneally 15 min after cortical impact significantly attenuated water maze performance relative to the vehicle controls (◆ *Figure 9-4*).

◻ **Figure 9-4**
Mean (±SE) latency (sec) to locate either a submerged or visible platform in a spatial learning acquisition paradigm in the Morris water maze. *$p < 0.05$ versus TBI + vehicle (Scheffé). [Reprinted with permission from Kline et al. (2002b)]

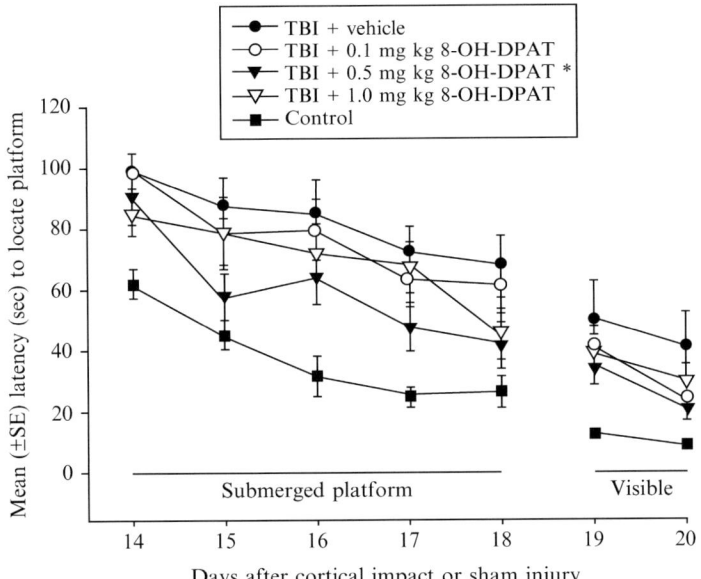

Additionally, 8-OH-DPAT attenuated hippocampal CA3 cell loss (Kline et al., 2002b). Because 8-OH-DPAT induces rapid, mild, and transient hypothermia, a follow-up study by Kline and colleagues sought to decipher the potential neurobehavioral and histological benefit of hypothermia by actively keeping 8-OH-DPAT-treated rats warm with the use of a heating lamp. The data showed that both groups of treated rats, regardless of whether allowed to cool spontaneously or actively kept at normothermic temperatures, recovered significantly quicker than the vehicle-treated group, but did not differ from one another (Kline et al., 2004a). These data suggest that the neuroprotective effect afforded by early treatment with the 5-HT$_{1A}$ receptor agonist 8-OH-DPAT was not mediated by concomitant hypothermia. Additional support for the efficacy of 5-HT$_{1A}$ receptor agonist treatment comes from the results of a rat model of acute subdural hematoma, in which Alessandri et al. (1999) treated rats with 0.01 or 0.003 mg/kg repinotan (BAY X 3702) or vehicle 15 min before (i.v.) and after (continuous infusion) the injection of 400 µl of autologous blood into the subdural space. The 4-h treatment with repinotan reduced ischemic brain damage as evidenced by cortical lesion volumes of 59.01 ± 39 mm^3 (low dose) and 60.8 ± 49 mm^3 (high dose), which were significantly smaller than the 106.2 ± 33 mm^3 of the vehicle-treated group (Alessandri et al., 1999). While the exact cause(s) attributed to the beneficial effects observed in our TBI studies following 5-HT$_{1A}$ receptor agonism is not yet known, the results from other studies investigating 5-HT$_{1A}$ receptor agonists may shed some light onto possible mechanisms. Electrophysiological studies have shown that 5-HT$_{1A}$ receptor agonism induces neuronal hyperpolarization via the activation of G-protein coupled inwardly rectifying K$^+$ channels (Prehn et al., 1991; Andrade, 1992) and/or the observed decrease in glutamate release following brain insult (Prehn et al., 1993; De Vry et al., 1997; Mauler et al., 2001). The latter effect is likely mediated via activation of presynaptic 5-HT$_{1A}$ receptors located on glutamatergic terminals (Raiteri et al., 1991; Matsuyama et al., 1996). Additionally, a recent study has reported that activation of 5-HT$_{1A}$ receptors may contribute to neuroprotection by directly interacting with voltage-gated Na$^+$ channels to reduce Na$^+$ influx (Melena et al., 2000).

5 Glutamate

5.1 Glutamate Mechanisms of Traumatic Brain Injury

Excitotoxicity is an important mechanism in secondary neuronal injury following TBI. Glutamate, aspartate, and glycine are some of the most abundant excitatory neurotransmitters and most commonly implicated in excitotoxic injury. Glutamate is the most prominent of these amino acids and activates receptors that are classified according to specific agonists. The receptors that modulate ionic channels are divided onto those stimulated by N-methyl-D-aspartate (NMDA receptors), or those stimulated by [alpha]-amino-3-hydroxy-5-methylisoazole-4-proprionic acid (AMPA) or kainic acid (together referred to as non-NMDA receptors). Glutamate also acts at metabotropic receptors that act via second messengers systems (Bittigau and Ikonomidou, 1997). Excitotoxicity resulting from the TBI-induced release of excitatory amino acids and the damaging effects of increased intracellular calcium associated with glutamate receptor activation has been the target of intense investigation for over a decade. [3H]Glutamate binding to the NMDA receptor has been reported to be significantly decreased at 3 h post-TBI in the hippocampus (Miller et al., 1990). Closed head injury in mice has been found to produce a short-lived (<1 h) hyperactivation of glutamate NMDA receptors followed by a profound and long-lasting (<7 days) loss of function (Biegon et al., 2004). Significant decreases in the NMDA receptor subunits NR1, NR2A, and NR2B protein have been reported at 6 and 12-h postinjury in rat hippocampus followed by complete recovery of NR1, NR2A, and NR2B subunit protein to the levels of sham controls at 24 h post-CCI injury (Kumar et al., 2002). More persistent NR1 and NR2 decreases have been reported following lateral fluid-percussion TBI (Osteen et al., 2004).

5.2 Inhibiting Glutamate Release

Inhibiting glutamate release has been demonstrated to attenuate functional and morphological deficits after experimental TBI. Sun and Faden (1995) reported that treatment with 619C89, a sodium channel blocker that inhibits glutamate release, 15 min postinjury can attenuate behavioral deficits and hippocampal neuronal loss. Riluzole (2-amino-6-trifluoromethoxy benzothiazole) has properties that include inhibition of glutamate release and has also been reported to attenuate functional (McIntosh et al., 1996) and morphological damage (Stover et al., 2000) after TBI. Treatment with galanin, a neuropeptide that inhibits neurotransmitter release by opening potassium channels and closing N-type calcium channels, attenuates motor, but not MWM deficits following TBI (Liu et al., 1994). However, blocking postinjury ion fluxes with tetrodotoxin was not found to be behaviorally neuroprotective (Di et al., 1996).

5.3 Prevention of Glutamate Receptor Activation

As summarized later, treatment with NMDA and non-NMDA receptor blockers has led to improved functional recovery and reduced hippocampal cell death and cortical damage following experimental TBI. NMDA-receptor blockade has been reported in several models to attenuate neurological deficits following TBI. Hayes et al. (1988) reported behavioral protection by the noncompetitive NMDA antagonist phencyclidine following TBI. Studies by Faden et al. (1989) found behavioral protection by dextromethorphan, a noncompetitive NMDA antagonist and 3-(2-carboxypiperazin-4-yl)propyl-l-phosphonic acid (CPP), a competitive NMDA antagonist. The NMDA antagonist, MK-801, has been repeatedly shown to be neuroprotective (McIntosh et al., 1988, 1989; Hamm et al., 1993; Lewen et al., 1999). Cortical and hippocampal damage has been attenuated by pretreatment with the NMDA antagonist CPP or the non-NMDA antagonist NBQX (Ikonomidou and Turski, 1996). While this study reported that posttreatment of these agents was generally not effective, a delayed treatment with NBQX, but not CPP, beginning 1 and 7 h after TBI prevented hippocampal damage. Ramacemide hydrochloride, an NMDA receptor-associated ionophore blocker, has been reported to reduce posttraumatic cortical lesion volume, but failed to improve MWM function (Smith et al., 1997). Gacycliddine, a noncompetitive NMDA receptor antagonist, has been shown to attenuate neuronal death and deficits in MWM performance following frontal cortex contusion (Smith et al., 2000). Treatment with kynurenate, an NMDA and non-NMDA antagonist, has been reported to attenuate hippocampal CA3 neuronal loss after TBI (Hicks et al., 1994). Animals treated with HU-211, a synthetic, nonpsychotropic cannabinoid that acts as a noncompetitive NMDA receptor antagonist, have been reported as possessing enhanced motor function recovery after TBI (Shohami et al., 1993). The NMDA antagonist ketamine has been demonstrated to attenuate postinjury neurological (Shapira et al., 1994) and cognitive deficits (Smith et al., 1993). Treatment with CP-98,113, an NMDA receptor blocker, 15 min postinjury has been shown to attenuate neurologic motor function and MWM performance (Okiyama et al., 1998). Increasing inhibitory function through stimulation of GABA(A) receptors by administering bicucullin has been reported to attenuate MWM deficits after TBI (O'Dell et al., 2000).

Magnesium, in addition to having an essential role in normal cell function, also has an antagonist effect on the NMDA receptor by blocking the NMDA receptor ion channel. Feldman et al. (1996) reported that administration of magnesium 1 h postinjury produced significant improvement in neurological function at 18 and 48 h after injury. Magnesium has been demonstrated to be neuroprotective to both functional and morphological deficits following TBI. Magnesium chloride used to treat animals 30 min after injury has been reported to protect against neurological deficits (McIntosh et al., 1989). One-hour posttreatment with magnesium chloride has been reported to reduce TBI-induced damage to the cortex, but did not alter posttraumatic cell loss in the CA3 region of the ipsilateral cortex (Bareyre et al., 2000). Administration of $MgSO_4$ has been reported to attenuate recovery of function when administered up to 12 h postinjury (Heath and Vink, 1999). In a recent study, the administration of magnesium (1.0 mmol/kg) at 15 min and 24 h after medial frontal cortex contusion injury significantly reduced the behavioral impairments observed on a bilateral tactile stimulation test and enhanced reference memory versus saline controls (Hoane, 2005).

Many noncompetitive NMDA receptor antagonists have undesirable side effects. NPS 1506 is a new noncompetitive NMDA receptor antagonist, which appears to be less toxic than earlier agents (Leoni et al., 2000). NPS 1506 has been reported to attenuate MWM performance and hippocampal CA3 cell death, but did not affect cortical lesion loss (Leoni et al., 2000).

TBI-induced glutamate release can also damage tissue by activating metabotropic glutamate receptors (mGluR), which are coupled to second messenger cascades through G-proteins. Modulation of mGluRs has been shown to confer neuroprotection. Gong et al. (1995) reported that intraventricular administration of alpha-methyl-4-carboxyphenylglycine (MCPG), a mGluR antagonist, before TBI attenuated motor and MWM deficits. Administration of 2-methyl-6-(phenylethynyl)-pyridine (MTEP), an mGluR5 antagonist, has been reported to reduce lesion volume and to attenuate motor and MWM deficits following TBI (Movsesyan et al., 2001). Administration of a group II mGluR agonist 30 min after TBI can improve behavioral recovery (Allen et al., 1999). Selective mGluR1 antagonists have also been reported to attenuate motor deficits and MRI-assessed lesion volume (Faden et al., 2001).

6 Gamma-Aminobutyric Acid

GABA is a major inhibitory neurotransmitter in the CNS. A reversible increase in the extracellular GABA concentration was shown by microdialysis in concussive brain injury in rats (Nilsson et al., 1990). TBI-induced loss of gamma-aminobutyric acid A (GABAA) receptors has been suggested by a significant loss of the [11C]flumazenil binding sites in the frontoparietal cortex and hippocampus ipsilaterally, 12 h from the fluid-percussion injury (Sihver et al., 2001). Improved effects on mortality and cognitive outcome were observed in rats subjected to moderate level FPI after pretreatment with diazepam (O'Dell et al., 2000). Postinjury treatment did not enhance the survival but induced significantly better recovery of cognitive function. Postinjury administration of MDL 26,479 (suritozole), a negative modulator at the GABAA receptor, which enhances cholinergic function, aided in attenuating spatial memory deficits after TBI in the rat (O'Dell and Hamm, 1995).

7 Adenosine

7.1 Adenosine Responses to Traumatic Brain Injury

Adenosine is a ubiquitous nucleoside with numerous physiologic roles, many of which are related to cellular energy utilization. It is produced primarily by adenosine triphosphate (ATP) breakdown and is interconverted to 5′-adenosine monophosphate by adenosine kinase; adenosine breakdown is initiated by adenosine deaminase and proceeds to sequentially yield the purine metabolites inosine, hypozanthine, zanthine, and finally uric acid (Geiger et al., 1997). Increases in these metabolic products were reported by Headrick et al. (1994) after fluid-percussion injury in rats and also by our group after CCI injury in rats (Bell et al., 1998). Recent studies support enhanced production of adenosine during secondary insults after severe TBI in humans (Bell et al., 2001). Adenosine can act at four subtypes of receptors (A_1, A_{2A}, A_{2B}, and A_3) found on the surface of many cell types. G-proteins couple adenosine receptor to effector binding to effector systems (Rudolphi et al., 1992). The A_1 adenosine receptor (A_1AR) has the highest affinity for adenosine and is suggested to mediate antiexcitotoxic effects, along with a number of other potentially beneficial actions. Notable among the effects of adenosine at the A_1AR are potent antiepileptic effects (Knutsen and Murray, 1997). Experimental TBI in A_1AR knockout mice produces profound seizure activity compared with either wild-type or heterozygote littermates (Kochanek et al., 2006).

7.2 Adenosine Therapies for Traumatic Brain Injury

Adenosine receptor agonists increase CBF in normal brain. This occurs across species with intracarotid (Joshi et al., 2002), intravenous (Soricelli et al., 1995), topical (Ibayashi et al., 1991), or parenchymal (Van Wylen et al., 1989) administration. However, some investigators have reported differing results depending on the dose administered (Sollevi et al., 1987; Stange et al., 1997). Vasodilation by adenosine occurs through activation of adenosine-2a (A_{2A}) receptors with adenyl cyclase activation in vascular smooth muscle (Kalaria and Harik, 1988). A_2B receptor activation linked to nitric oxide synthase activity may also participate (Shin et al., 2000; Ngai et al., 2001). Beneficial effects of the nonselective agonist 2-chloroadenosine (2-CA) on cerebral energy charge and functional outcome were shown after intraventricular or intraparenchymal administration in experimental TBI (Headrick et al., 1994; Varma et al., 2002). 2-CA can produce persistent increases in CBF in uninjured brain and can attenuate posttraumatic hypoperfusion (Kochanek et al., 2005) (❷ *Figure 9-5*). Varma et al. (2002) reported improved motor function after CCI in mice treated with the nonselective agonist 2-CA and the attenuation of CA3 hippocampal cell death by treatment with the selective A_1 receptor agonist CCPA. Consistent with this finding, detrimental effects of the A_1 receptor antagonist DPCPX were observed. Caffeine (1,3,7-trimethylxanthine) is a broad-spectrum adenosine receptor antagonist, and most of its pharmacological actions are attributed to its ability to block adenosine A_1 and A_2 receptors (Fredholm, 1995). Caffeine at doses of 50–150 mg/kg significantly exacerbated the neurological deficit and mortality of animals after head injury (Al-Moutaery et al., 2003). However, it has been reported that coadministration of caffeine (10 mg/kg) and a low amount of alcohol (0.65 g/kg; caffeinol) reduced cortical tissue loss and improved working memory (Dash et al., 2004).

7.3 A_2AR/D_2R Interactions

Binding of adenosine to A_{2A} receptors increases intracellular cAMP—which is the key signal transduction pathway for A_{2A}—although a few reports have suggested a role for protein kinase C (Dixon et al., 1997). Putative detrimental consequences of A_{2A} receptor activation include augmented presynaptic glutamate

Figure 9-5

Acute effects of injection of either saline vehicle (*left* three vertical panels) or the nonselective adenosine receptor agonist 2-CA (12 nmol, *right* three vertical panels) into the *left* dorsal hippocampus of rats on CBF (mL 100 g^{-1} min^{-1}, assessed by the continuous arterial spin-labeled MRI) after controlled cortical impact at a moderate (2.0 mm deformation) injury level. The *left* hemisphere appears as the *left* side of the image. The *top* horizontal panels show unlabeled control (Mc) images reflecting anatomy. Pseudocolor images are CBF maps. Numbers to the *left* of the images are times after injection, with time after CCI indicated in parentheses. Some local swelling in the Mc images at the injury site is seen after moderate injury. Cerebral blood flow was not affected at either 3 h 12 min or 5 h 3 min after injection of saline, but at both 2 h 58 min and 4 h 9 min after 2-CA administration CBF was increased in the ipsilateral hippocampus, overlying cortex, and brain regions in proximity to the contusion in these representative examples at the moderate injury level. [Reprinted with permission from Kochanek, et al. (2005)]

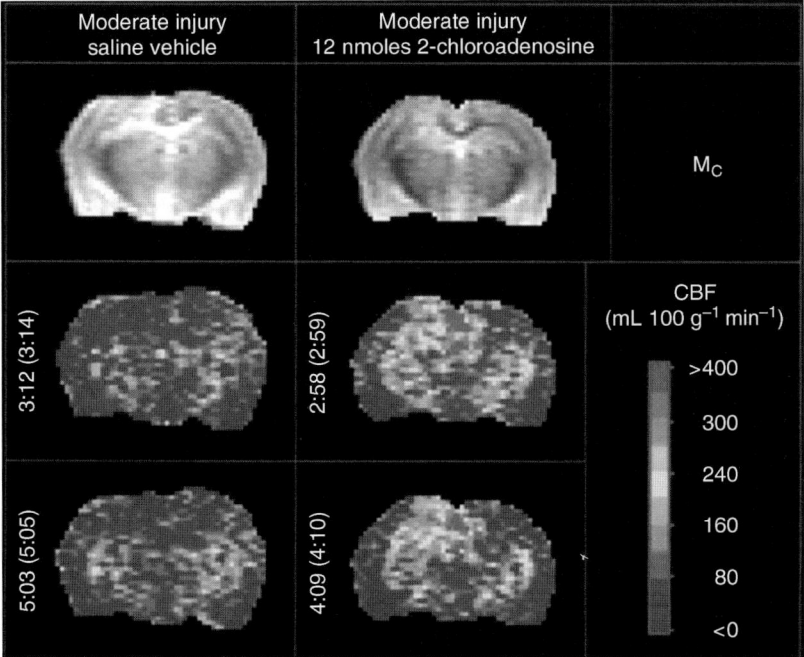

release (Popoli et al., 2002) and A_1 receptor desensitization (Dixon et al., 1997). It is well known that adenosine receptors are densely expressed in the striatum and exert a modulatory influence on dopamine neurotransmission (Moreau and Huber, 1999; Svenningsson et al., 1999). Recently, the understanding of the role of adenosine receptors in basal ganglia and their anatomical and functional relationships with the striatal dopamine receptor has increased, providing evidence of an antagonist interaction between A_2A/D_2 receptors in the striatum. Furthermore, an antagonistic interaction between A_{2A} and D_2 has been reported in human striatal tissue (Diaz-Cabiale et al., 2001). Interplay between A_{2A} and D_2 receptors has been observed via transmembrane protein–protein interactions and synergistic effects at the second messenger level. Chronically, A_2AR antagonists may be beneficial for TBI recovery by increasing the sensitivity of D_2 receptors.

8 Electrophysiology in TBI Research

The electrophysiological studies in TBI research fall into three broad categories (1) studies that measure spontaneous brain electrical potentials as assessed by changes in brain activity, (2) studies of environmental evoked changes in electric potentials as a measure of neuronal structural pathway integrity, and (3) measurements of persistent changes of evoked field potentials as indices of synaptic plasticity.

8.1 TBI-Induced Alterations in Spontaneous Brain Electrical Activity

The electroencephalogram (EEG) noninvasively measures the electrical activity of the brain, via electrodes applied to the scalp. There is a long history of relating changes in the brain electrical activity to severity of the injury and outcome (Dow et al., 1944; Dawson et al., 1951). Modernization of signal processing techniques has lead to more precise quantification of the EEG signals. Such quantification includes measurement of the spectral components of the recorded waveforms. The percent of alpha variability (PAV) has recently been found to be a better predictor of outcome than neurological status (Glasgow Como Scale) at hospital admission (Vespa et al., 2002). While the acute EEG response is not available in humans, a generalized decrease in EEG amplitude has been consistently reported in several models of brain injury (Denny-Brown and Russell, 1941; West et al., 1982; Dixon et al., 1987; McIntosh et al., 1987). Decreases in power spectra of delta, alpha, beta, and theta frequency band have been observed in both anesthetized (Dixon et al., 1988) and unanesthetized rats (West et al., 1982). In TBI patients, new technologies are becoming available to measure EEG. The bispectral index (BIS) of the electroencephalogram is a weighted sum of electroencephalographic subparameters containing time domain, frequency domain, and higher-order spectral information. Recent clinical studies suggest that BIS monitoring may a potentially useful predictor of outcome (Fábregas et al., 2004).

Cortical spreading depression (CSD) is a neural phenomenon present in several animal species. It was first described by Leão (1944) as a consequence of stimulation of one point on the cortical surface. CSD consists of a reversible "wave" of reduction (depression) of the spontaneous and evoked cortical electrical activity, with a simultaneous slow DC-potential change of the tissue. Once produced in a single cortical point, CSD concentrically propagates for the entire cortical surface, with propagation velocities of the order of a few mm/min (Leão, 1944; Gorji, 2001). CSD in normal brain does not lead to cell death (Nedergaard and Hansen, 1988). In fact, in models of ischemia, CSD has been found to have preconditioning properties associated with neuronal protection against a subsequent, lethal insult (Kawahara et al., 1995, Kobayashi et al., 1995; Matsushima et al., 1996; Yanamoto et al., 1998; Horiguchi et al., 2005a, b). However, the occurrence of CSD after an ischemic brain injury is associated with increases in final infarct volume (Mies et al., 1993; Back et al., 1996; Busch et al., 1996; Takano et al., 1996). This pathology can be attenuated by glutamate receptor blockers (Marrannes et al., 1988; Ohta et al., 2001). The occurrence and propagation of spontaneous depolarization-like events in the injured human brain has been recently documented (Mayevsky et al., 1996; Strong et al., 2002) and may be an important secondary injury mechanism.

8.2 Evoked Potentials in Traumatic Brain Injury

An evoked potential (EP) is an electrical potential recorded following a stimulus. EP amplitudes are typically lower than those recorded from spontaneous EEG and require computerized signal averaging techniques to resolve the potentials from background electrical activity. The stimuli used to produce EPs include auditory stimuli, electrical nerve stimuli, and visual stimuli. These multimodal stimuli have been traditionally used in brain trauma research to study brain function and as possible predictors of outcome (Greenberg et al., 1981; Carter and Butt, 2005). EPs have some advantages compared with EEG. EPs are less affected than EEG by sedative drugs frequently used in ICU (Newlon et al., 1983; Drummond et al., 1987; Garcia-Larrea et al., 1993). They are less global than EEG and test the integrity of specific multi-synaptic pathways. For example, brainstem auditory evoked potentials (BAEPs) evaluate the functional state of brainstem auditory pathways in the pons, the lower part of the mesencephalon including the lateral lemniscus up to the inferior colliculi and are used as a screening test for poor prognosis (Attia and Cook, 1998).

Short-latency somatosensory evoked potentials (SEPs) have been studied extensively in TBI patients and provide information about the brainstem somatosensory pathways, thalamo-cortical projections, and the primary somatosensory cortex itself (Mauguiere et al., 1983). The presence of a clearly defined long-latency SEP in patients in coma is associated with a favorable outcome (Pfurtscheller et al., 1985). SEPs are transiently suppressed following concussion in rats (◗ *Figure 9-6*). Since the first reports on BAEPs and short-latency

◘ Figure 9-6
Long-latency SEP recorded in rat before and after fluid percussion TBI

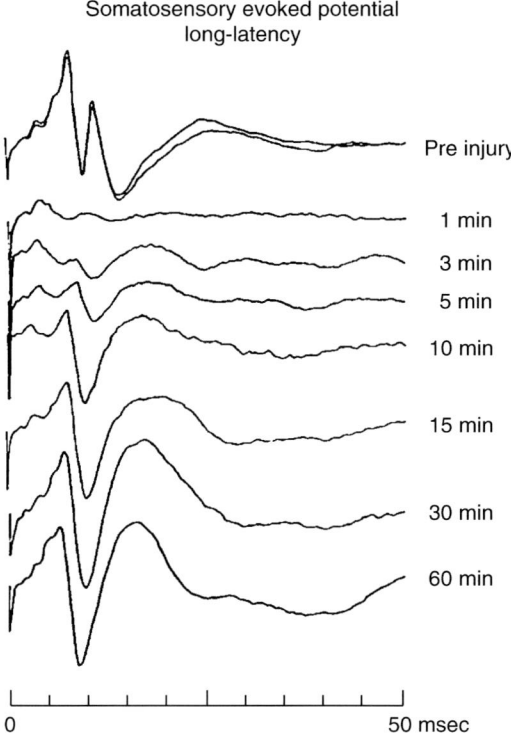

SEPs in the prognosis of coma by Rappaport et al. (1977) and Greenberg et al. (1981), the use of multi-modality EPs as a prognostic tool has progressively increased (Narayan et al., 1981; Walser et al., 1985; Cant and Shaw, 1986; Ganes and Lundar, 1988; Lindsay et al., 1990; Barelli et al., 1991; Cusumano et al., 1992; Zentner and Rohde, 1992; Facco et al., 1993, 1998; Madl et al., 1996, 2000; Pohlmann-Eden et al., 1997; Sleigh et al., 1999; Sherman et al., 2000; Rothstein, 2000).

Diffuse axonal injury (DAI) is a hallmark morphological feature of TBI (Povlishock and Katz, 2005) and is associated with loss of axonal function in white matter regions. DAI may contribute to neurotransmission deficits via diffuse deafferentation. Recently, stimulated compound action potential (CAP) recordings have used as an electrophysiological correlate of axonal integrity (Baker et al., 2002) in an experimental model of TBI. Baker et al. (2002) observed decreased CAP amplitudes that persisted for up to 7 days after TBI. CAPs evoked in the corpus callosum have also been used to distinguish between the TBI response to two axonal populations; small unmyelinated axons and large myelinated axons (Reeves et al., 2005). Their findings suggest the differential vulnerabilities of axons to brain injury and also that damage to unmyelinated fibers may play a significant role in morbidity associated with brain injury.

8.3 Long-Term Potentiation in TBI Models

One mechanism of chronic memory dysfunction after TBI is a decreased ability to modulate synaptic strength. Long-term potentiation (LTP) is a persistent, use-dependent increase in the efficiency of synaptic transmission (Bliss and Lomo, 1973). LTP is considered an electrophysiological correlate of synaptic plasticity and memory. In the hippocampus, LTP is typically induced by the delivery of high-frequency electrical stimulation to fibers that project from area CA3 to area CA1 and produce glutamate release into the synapse resulting in depolarization of the postsynaptic neuron. LTP is characterized by a persistent increase in the size of the synaptic response. A limited number of studies have examined hippocampal LTP in rats subjected to TBI induced by fluid percussion. One of the earliest studies found that TBI suppressed LTP, as evidenced by reductions in posttetanic increases in population spikes as well as EPSPs at 2–3 h after injury (Miyazaki et al., 1992). Reeves et al. (2005) observed increased changes in excitability as well as latency shifts in tetanus-induced population spike latency shifts and greater titanic stimulation elevated EPSP to spike ratios in injured compared with control animals. In this report, these parameters returned to control levels by 7 days after midline fluid-percussion TBI. Sick et al. (1998) observed that TBI in rats inhibited the expression of hippocampal LTP as early as 4 h and as late as 48 h after injury in both dorsal and ventral hippocampus ipsilateral to the site of injury (◉ *Figure 9-7*). Maintenance of LTP has been reported to be disrupted for up to 8-weeks postinjury (Sanders et al., 2000).

◘ Figure 9-7
Examples of fEPSPs recorded from stratum radiatum of hippocampal slices harvested from a sham-operated animal and from an animal 4 h following mild fluid-percussion brain injury. EPSPs were recorded before, and 60 min following, a 1 sec (100 Hz) tetanus delivered to the Schaffer collaterals. [Reprinted with permission from Sick et al. (1998)]

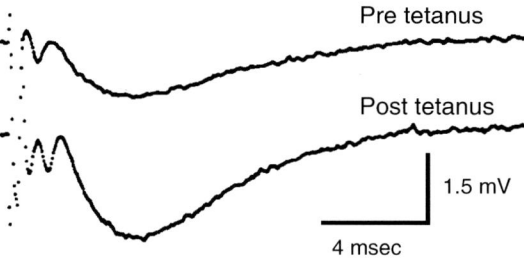

References

Al-Moutaery K, Al- Deeb S, Ahmad-Khan H, Tariq M. 2003. Caffeine impairs short-term neurological outcome after concussive head injury in rats. Neurosurgery 53(3): 704-711; discussion 711–712.

Alessandri B, Tsuchida E, Bullock RM. 1999. The neuroprotective effect of a new serotonin receptor agonist, BAY X3702, upon focal ischemic brain damage caused by acute subdural hematoma in the rat. Brain Res 845: 232-235.

Allen JW, Ivanova SA, Fan L, Espey MG, Basile AS, et al. 1999. Group II metabotropic glutamate receptor activation attenuates traumatic neuronal injury and improves neurological recovery after traumatic brain injury. J Pharmacol Exp Ther 290: 112-120.

Allen RM. 1983. Role of amantadine in the management of neuroleptic-induced extrapyramidal syndromes: Overview and pharmacology. Clin Neuropharmacol 6: S64-S73.

Andrade R. (1992). Electrophysiology of 5-HT$_{1A}$ receptors in the rat hippocampus and cortex. Drug Dev Res 26: 275-286.

Attia J, Cook DJ. 1998. Prognosis in anoxic and traumatic coma. Crit Care Clin 14: 497-511.

Back T, Ginsberg MD, Dietrich WD, Watson BD. 1996. Induction of spreading depression in the ischemic hemisphere following experimental middle cerebral artery occlusion: Effect on infarct morphology. J Cereb Blood Flow Metab 16: 202-213.

Bak IJ, Hassler R, Kim JS, Kataoka K. 1972. Amantadine actions on acetylcholine and GABA in striatum and substantia nigra of rat in relation to behavioral changes. J Neural Transm 33: 45-61.

Baker AJ, Phan N, Moulton RJ, Fehlings MG, Yucel Y, et al. 2002. Attenuation of the electrophysiological function of the corpus callosum after fluid percussion injury in the rat. J Neurotrauma 19: 587-599.

Barelli A, Valente MR, Clemente A, Bozza P, Proietti R, et al. 1991. Serial multimodality-evoked potentials in severely head-injured patients: Diagnostic and prognostic implications. Crit Care Med 19: 1374-1381.

Bareyre FM, Saatman KE, Raghupathi R, McIntosh TK. 2000. Postinjury treatment with magnesium chloride attenuates cortical damage after traumatic brain injury in rats. J Neurotrauma 17: 1029-1039.

Beers SR, Skold A, Dixon CE, Adelson PD. 2005. Neurobehavioral effects of amantadine after pediatric traumatic brain injury: A preliminary report. J Head Trauma Rehabil 20: 450-463.

Bell MJ, Kochanek PM, Carcillo JA, Mi Z, Schiding JK, et al. 1998. Interstitial adenosine, inosine, and hypoxanthine are increased after experimental traumatic brain injury in the rat. J Neurotrauma 15: 163-170.

Bell MJ, Robertson CS, Kochanek PM, Goodman JC, Gopinath SP, et al. 2001. Interstitial brain adenosine and xanthine increase during jugular venous oxygen desaturations in humans after traumatic brain injury. Crit Care Med 29: 399-404.

Biegon A, Fry PA, Paden CM, Alexandrovich A, Tsenter J, et al. 2004. Dynamic changes in N-methyl-D-aspartate receptors after closed head injury in mice: Implications for treatment of neurological and cognitive deficits. Proc Natl Acad Sci USA 101: 5117-5122.

Bittigau P, Ikonomidou C. 1997. Glutamate in neurologic diseases. J Child Neurol 12: 471-485.

Bliss TVP, Lomo T. 1973. Long-lasting potentiation of synaptic transmission in the dentate area of the anaesthetized rabbit following stimulation of the perforant path. J Physiol 232: 331-356.

Bornstein MB. 1946. Presence and action of acetylcholine in experimental brain trauma. J Neurophysiol 9: 349-366.

Busch E, Gyngell ML, Eis M, Hoehn Berlage M, Hossmann KA. 1996. Potassium-induced cortical spreading depressions during focal cerebral ischemia in rats: Contribution to lesion growth assessed by diffusion-weighted NMR and biochemical imaging. J Cereb Blood Flow Metab 16: 1090-1099.

Busto R, Dietrich WD, Globus MY-T, Alonso O, Ginsberg MD. 1997. Extracellular release of serotonin following fluid-percussion brain injury in rats. J Neurotrauma 14, 35-42.

Cant BR, Shaw NA. 1986. Central somatosensory conduction time: Method and clinical applications. Cracco RQ, Bodis-Wollner I, editors. Evoked Potentials. Alan R Liss; New York: pp. 58-67.

Carter BG, Butt W. 2005. Are somatosensory evoked potentials the best predictor of outcome after severe brain injury? A systematic review. Intensive Care Med 31: 765-775.

Chen Y, Shohami E, Bass R, Weinstock M. 1998. Cerebroprotective effects of ENA713, a novel acetylcholinesterase inhibitor, in closed head injury in the rat. Brain Res 784: 18-24.

Ciallella JR, Yan HQ, Ma X, Wolfson BM, Marion DW, et al. 1998. Chronic effects of traumatic brain injury on hippocampal vesicular acetylcholine transporter and M_2 muscarinic receptor protein in rats. Exp Neurol 142: 11-19.

Cusumano S, Paolin A, Di Paola F, Boccaletto F, Simini G, et al. 1992. Assessing brain function in post-traumatic coma by means of bit-mapped SEPs, BAEPs, CT, SPET and clinical sources. Prognostic implications. Electroencephalogr Clin Neurophysiol 84: 499-514.

Dash PK, Moore AN, Moody MR, Treadwell R, Felix JL, et al. 2004. Post-trauma administration of caffeine plus ethanol reduces contusion volume and improves working memory in rats. J Neurotrauma 21: 1573-1583.

Dawson RE, Webster JE, Gurdjian ES. 1951. Serial electroencephalography in acute head injuries. J Neurosurg 8: 613-630.

Delahunty TM. 1992. Traumatic brain injury enhances muscarnic receptor-linked inositol phosphate production in the rat. Brain Res 594: 307-310.

Denny-Brown D, Russell WR. 1941. Experimental cerebral concussion. Brain 64: 93-164.

De Vry J, Dietrich H, Glaser T, Heine H-G, Horváth E, et al. 1997. BAY x 3702. Drugs Future 22: 341-349.

Dewar D, Graham DI. 1996. Depletion of choline acetyltransferase activity but preservation M1 and M2 muscarinic receptor binding sites in temporal cortex following head injury: A preliminary human postmortem study. J Neurotrauma 13: 181-187.

Di X, Lyeth BG, Hamm RJ, Bullock MR. 1996. Voltage-dependent Na^+/K^+ ion channel blockade fails to ameliorate behavioral deficits after traumatic brain injury in the rat. J Neurotrauma 13: 497-504.

Diaz-Cabiale Z, Hurd Y, Guidolin D, Finnman UB, Zoli M, et al. 2001. Adenosine A2A agonist CGS 21680 decreases the affinity of dopamine D2 receptors for dopamine in human striatum. Neuroreport 12: 1831-1834.

Dixon AK, Widdowson L, Richardson PJ. 1997. Desensitisation of the adenosine A1 receptor by the A2A receptor in the rat striatum. J Neurochem 69: 315-321.

Dixon CE, Lighthall JW, Anderson TE. 1988. A physiologic, histopathologic, and cineradiographic characterization of a new fluid percussion model of experimental brain injury in the rat. J Neurotrauma 5: 91-104.

Dixon CE, Ma X, Marion DW. 1997. Effects of CDP-choline treatment on neurobehavioral deficits after TBI and on hippocampal and neocortical acetylcholine release. J Neurotrauma 14: 161-169.

Dixon CE, Bao J, Long DA, Hayes RL. 1996. Reduced evoked release of acetylcholine in the rodent hippocampus following traumatic brain injury. Pharmacol Biochem Behav 53: 579-686.

Dixon CE, Hamm RJ, Taft WC, Hayes RL. 1994. Increased anticholinergic sensitivity following closed skull impact and controlled cortical impact traumatic brain injury in the rat. J Neurotrauma 11: 275-287.

Dixon CE, Kochanek PM, Yan HQ, Schiding JK, Griffith R, et al. 1999a. One-year study of spatial memory performance, brain morphology and cholinergic markers after moderate controlled cortical impact in rats. J Neurotrauma 16: 109-122.

Dixon CE, Kraus MF, Kline AE, Ma X, Yan HQ, et al. 1999b. Amantadine improves water maze performance without affecting motor behavior following traumatic brain injury in rats. Restor Neurol Neurosci 14: 285-294.

Dixon CE, Liu SJ, Jenkins LW, Bhattachargee M, Whitson J, et al. 1995. Time course of increased vulnerability of cholinergic neurotransmission following traumatic brain injury in the rats. Behav Brain Res 70: 125-131.

Dixon CE, Lyeth BG, Povlishock JT, Findling RL, Hamm RJ, et al. 1987. A fluid percussion model of experimental brain injury in the rat. J Neurosurg 67: 110-119.

Dow RS, Ulett G, Raaf J. 1944. Electroencephalographic studies immediately following head injury. Am J Psychiatry 101: 174-183.

Drummond JC, Todd MM, Schubert A, Sang H. 1987. Effect of the acute administration of high dose pentobarbital on human brain stem auditory and median nerve somatosensory evoked responses. Neurosurgery 20: 830-835.

Dunn-Meynell A, Pan S, Levin BE. 1994. Focal traumatic brain injury causes widespread reductions in rat brain norepinephrine turnover from 6 to 24 h. Brain Res 660: 88-95.

Fábregas N, Gambus PL, Valero R, Carrero EJ, Salvador L, et al. 2004. Can bispectral index monitoring predict recovery of consciousness in patients with severe brain injury? Anesthesiology 101(1): 43-51.

Facco E, Munari M, Baratto F, Behr AU, Giron GP. 1993. Multimodality evoked potentials (auditory, somatosensory and motor) in coma. Neurophysiol Clin 23: 237-258.

Facco E, Behr AU, Munari M, Baratto F, Volpin SM, et al. 1998. Auditory and somatosensory evoked potentials in coma following spontaneous cerebral hemorrhage: Early prognosis and outcome. Electroencephalogr Clin Neurophysiol 107: 332-338.

Faden AI, Demediuk P, Panter SS, Vink R. 1989. The role of excitatory amino acids and NMDA receptors in traumatic brain injury. Science 244: 798-800.

Faden AI, O'Leary DM, Fan L, Bao W, Mullins PG, et al. 2001. Selective blockade of the mGluR1 receptor reduces traumatic neuronal injury in vitro and improves outcome after brain trauma. Exp Neurol 167: 435-444.

Feeney DM. 1991. Pharmacologic modulation of recovery after brain injury: A reconsideration of diaschisis. J Neurol Rehabil 5: 113-128.

Feeney DM, Sutton RL. 1987. Pharmacotherapy for recovery of function after brain injury. Crit Rev Neurobiol 3: 135-197.

Feeney DM, De Smet AM, Rai S. 2004. Noradrenergic modulation of hemiplegia: Facilitation and maintenance of recovery. Restor Neurol Neurosci 22: 175-190.

Feeney DM, Gonzalez A, Law WA. 1982. Amphetamine, haloperidol, and experience interact to affect rate of recovery after motor cortex injury. Science 217: 855-857.

Feeney DM, Weisend MP, Kline AE. 1993. Noradrenergic pharmacotherapy, intracerebral infusion and adrenal transplantation promote functional recovery after cortical damage. J Neural Transplant Plast 4: 199-213.

Feldman Z, Gurevitch B, Artru AA, Oppenheim A, Shohami E, et al. 1996. Effect of magnesium given 1 hour after head trauma on brain edema and neurological outcome. J Neurosurg 85: 131-137.

Fredholm BB. 1995. Adenosine, adenosine receptors and the action of caffeine. Pharmacol Toxicol 76: 93-101.

Ganes T, Lundar T. 1988. EEG and evoked potentials in comatose patients with severe brain damage. Electroencephalogr Clin Neurophysiol 69: 6-13.

Garcia-Larrea L, Fischer C, Artru F. 1993. Effect of anesthetics on sensory evoked potentials. Neurophysiol Clin 23: 141-162.

Geiger JD, Parkinson FE, Kowaluk EL. 1997. Regulators of endogenous adenosine levels as therapeutic agents. Purinergic Approaches in Experimental Therapeutics. Jacobson KA, Jarvis MF, editors. New York: Wiley-Liss; pp. 55-84.

Gerlak RP, Clark R, Stump JM, Vernier VG. 1970. Amantadine-dopamine interaction: Possible mode of action in Parkinsonism. Science 169: 203-204.

Gianutsos G, Chute S, Dunn JP. 1985. Pharmacological changes in dopaminergic systems induced by long-term administration of amantadine. Eur J Pharmacol 110: 357-361.

Goldstein LB, Davis JN. 1990. Post-lesion practice and amphetamine-facilitated recovery of beam-walking in the rat. Res Neurol Neurosci 1: 311-314.

Gong Q-Z, Delahunty TM, Hamm RJ, Lyeth BG. 1995. Metabotropic glutamate antagonist, MCPG, treatment of traumatic brain injury in rats. Brain Res 700: 299-302.

Gorji A. 2001. Spreading depression: A review of the clinical relevance. Brain Res Brain Res Rev 38: 33-60.

Gorman LK, Fu K, Hovda DA. 1989. Analysis of acetylcholine release following concussive brain injury in the rat. J Neurotrauma 6: 203

Gorman LK, Fu K, Hovda DA, Murray M, Traystman RJ. 1996. Effects of traumatic brain injury on the cholinergic system in the rat. J Neurotrauma 13: 457-463.

Greenberg RP, Newlon PG, Hyatt MS, Narayan RK, Becker DP. 1981. Prognostic implications of early multimodality evoked potentials in severely head-injured patients. A prospective study. J Neurosurg 55: 227-236.

Gualtieri CT. 1988. Pharmacotherapy and the neurobehavioural sequelae of traumatic brain injury. Brain Inj 2: 101-129.

Gualtieri T, Chandler M, Coons TB, Brown LT. 1989. Amantadine: A new clinical profile for traumatic brain injury. Clin Neuropharmacol 12: 258-270.

Haber B, Grossman RG. 1980. Acetylcholine metabolism in intracranial and lumbar cerebrospinal fluid and in blood. Neurobiology and Cerebrospinal Fluid. Wood JH, editor. New York: Plenum Press; pp. 345-350.

Hamm RJ, O'Dell DM, Pike BR, Lyeth BG. 1993. Cognitive impairment following traumatic brain injury: The effect of pre- and post-injury administration of scopolamine and MD-801. Brain Res Cogn Brain Res 1: 223-226.

Hayes RL, Jenkins LW, Lyeth BG, Balster RL, Robinson SE, et al. 1988. Pretreatment with phencyclidine, an N-methyl-D-aspartate antagonist, attenuates long-term behavioral deficits in the rat produced by traumatic brain injury. J Neurotrauma 5: 259-274.

Headrick JP, Bendall MR, Faden AI, Vink R. 1994. Dissociation of adenosine levels from bioenergetic state in experimental brain trauma: Potential role in secondary injury. J Cereb Blood Flow Metab 14: 853-861.

Heath DL, Vink R. 1999. Improved motor outcome in response to magnesium therapy received up to 24 hours after traumatic diffuse axonal brain injury in rats. J Neurosurg 90: 504-509.

Henry JM, Talukder NK, Lee AB, Walker ML. 1997. Cerebral trauma-induced changes in corpus striatal dopamine receptor subtypes. J Invest Surg 10: 218-286.

Hicks RR, Smith DH, Gennarelli TA, McIntosh T. 1994. Kynurenate is neuroprotective following experimental brain injury in the rat. Brain Res 655: 91-96.

Hoane MR. 2005. Treatment with magnesium improves reference memory but not working memory while reducing GFAP expression following traumatic brain injury. Restor Neurol Neurosci 23: 67-77.

Horiguchi T, Kis B, Rajapakse N, Shimizu K, Busija DW. 2005a. Cortical spreading depression (CSD)-induced tolerance to transient focal cerebral ischemia in halothane anesthetized rats is affected by anesthetic level but not ATP-sensitive potassium channels. Brain Res 1062: 127-133.

Horiguchi T, Snipes JA, Kis B, Shimizu K, Busija DW. 2005b. The role of nitric oxide in the development of cortical spreading depression-induced tolerance to transient focal cerebral ischemia in rats. Brain Res 1039: 84-89.

Hovda DA, Sutton RL, Feeney DM. 1989. Amphetamine-induced recovery of visual cliff performance after bilateral visual cortex ablation in cats: Measurements of depth perception thresholds. Behav Neurosci 103: 574-584.

Ibayashi S, Ngai AC, Meno JR, Winn HR. 1991. Effects of topical adenosine analogs and forskolin on rat pial arterioles in vivo. J Cereb Blood Flow Metab 11: 72-76.

Ikonomidou C, Turski L. 1996. Prevention of trauma-induced neurodegeneration in infant and adult rat brain: Glutamate antagonists. Metab Brain Dis 11: 125-141.

Joshi S, Duong H, Mangla S, Wang M, Libow AD, et al. 2002. In nonhuman primates intracarotid adenosine, but not sodium nitroprusside, increases cerebral blood flow. Anesth Analg 94: 393-399.

Kalaria RN, Harik SI. 1988. Adenosine receptors and the nucleoside transporter in human brain vasculature. J Cereb Blood Flow Metab 8: 32-39.

Karli DC, Burke DT, Kim HJ, Calvanio R, Fitzpatrick M, et al. 1999. Effects of dopaminergic combination therapy for frontal lobe dysfunction in traumatic brain injury rehabilitation. Brain Inj 13: 63-68.

Kawahara N, Ruetzler CA, Klatzo I. 1995. Protective effect of spreading depression against neuronal damage following cardiac arrest cerebral ischaemia. Neurol Res 17: 9-16.

Kim YH, Ko MH, Na SY, Park SH, Kim KW. 2006. Effects of single-dose methylphenidate on cognitive performance in patients with traumatic brain injury: A double-blind placebo-controlled study. Clin Rehabil 20: 24-30.

Kline AE, Dixon CE. 2001a. Contemporary in vivo models of brain trauma and a comparison of injury responses. Head Trauma: Basic, Preclinical, and Clinical Directions. Miller LP, Hayes RL, editors. New York: John Wiley & Sons; pp. 65-84.

Kline AE, Chen MJ, Tso-Olivas DY, Feeney DM. 1994. Methylphenidate treatment following ablation-induced hemiplegia in rat: Experience during drug action alters effects on recovery of function. Pharmacol Biochem Behav 48: 773-779.

Kline AE, Massucci JL, Marion DW, Dixon CE. 2002a. Attenuation of working memory and spatial acquisition deficits after a delayed and chronic bromocriptine treatment regimen in rats subjected to traumatic brain injury by controlled cortical impact. J Neurotrauma 19: 415-425.

Kline AE, Yu J, Massucci JL, Zafonte RD, Dixon CE. 2002b. Protective effects of the 5-HT$_{1A}$ receptor agonist 8-hydroxy-2-(di-n-propylamino)tetralin (8-OH-DPAT) against traumatic brain injury-induced cognitive deficits and neuropathology in adult male rats. Neurosci Lett 333: 179-182.

Kline AE, Massucci JL, Dixon CE, Zafonte RD, Bolinger BD. 2004a. The therapeutic efficacy conferred by the 5-HT$_{1A}$ receptor agonist 8-hydroxy-2-(di-n-propylamino)tetralin (8-OH-DPAT) after experimental traumatic brain injury is not mediated by concomitant hypothermia. J Neurotrauma 21: 175-185.

Kline AE, Massucci JL, Ma X, Zafonte RD, Dixon CE. 2004b. Bromocriptine reduces lipid peroxidation and enhances spatial learning and hippocampal neuron survival in a rodent model of focal brain trauma. J Neurotrauma 21: 1712-1722.

Kline AE, Yan HQ, Bao J, Marion DW, Dixon CE. 2000. Chronic methylphenidate treatment enhances water maze performance following traumatic brain injury in rats. Neurosci Lett 280: 163-166.

Kline AE, Yu J, Horváth E, Marion DW, Dixon CE. 2001b. The selective 5-HT$_{1A}$ receptor agonist repinotan HCL attenuates histopathology and spatial learning deficits following traumatic brain injury in rats. Neuroscience 106: 547-555.

Knutsen LJS, Murray TF. 1997. Adenosine and ATP in epilepsy. Purinergic Approaches in Experimental Therapeutics, Chapter 22. Jacobson KA, Jarvis MF, editors. New York: Wiley-Liss; pp. 423-447.

Kobayashi S, Harris VA, Welsh FA. 1995. Spreading depression induces tolerance of cortical neurons to ischemia in rat brain. J Cereb Blood Flow Metab 15: 721-727.

Kochanek PM, Hendrich KS, Jackson EK, Wisniewski SR, Melick JA, et al. 2005. Characterization of the effects of adenosine receptor agonists on cerebral blood flow in uninjured and traumatically injured rat brain using continuous arterial spin-labeled magnetic resonance imaging. J Cereb Blood Flow Metab 25: 1596-1612.

Kochanek PM, Vagni VA, Janesko KL, Washington CB, Crumrine PK, et al. 2006. Adenosine A1 receptor knockout mice develop lethal status epilepticus after experimental traumatic brain injury. J Cereb Blood Flow Metab 26: 565-575.

Kraus MF, Maki PM. 1997. Effect of amantadine hydrochloride on symptoms of frontal lobe dysfunction in brain injury: Case studies and review. J Neuropsychiatry Clin Neurosci 9: 222-230.

Kraus MF, Smith GS, Butters M, Donnell AJ, Dixon C, et al. 2005. Effects of the dopaminergic agent and NMDA receptor antagonist amantadine on cognitive function, cerebral glucose metabolism and D2 receptor availability in chronic traumatic brain injury: A study using positron emission tomography (PET). Brain Inj 19: 471-479.

Kumar A, Zou L, Yuan X, Long Y, Yang K. 2002. N-Methyl-D-aspartate receptors: Transient loss of NR1/NR2A/NR2B subunits after traumatic brain injury in a rodent model. J Neurosci Res 67: 781-786.

Leão AAP. 1944. Spreading depression of activity in cerebral cortex. J Neurophysiol 7: 359-390.

Lee H, Kim SW, Kim JM, Shin IS, Yang SJ, et al. 2005. Comparing effects of methylphenidate, sertraline and placebo on neuropsychiatric sequelae in patients with traumatic brain injury. Hum Psychopharmacol 20: 97-104.

Leonard JR, Maris DO, Grady MS. 1994. Fluid percussion injury causes loss of forebrain choline acetyltransferase and nerve growth factor receptor immunoreactive cells in the rat. J Neurotrauma 11: 379-392.

Leoni MJ, Chen XH, Mueller AL, Cheney J, McIntosh TK, et al. 2000. NPS 1506 attenuates cognitive dysfunction and hippocampal neuron death following brain trauma in the rat. Exp Neurol 166: 442-449.

Lewen A, Fredriksson A, Li GL, Olsson Y, Hillered L. 1999. Behavioural and morphological outcome of mild cortical contusion trauma of the rat brain: Influence of NMDA-receptor blockade. Acta Neurochir (Wien) 141: 193-202.

Lindsay K, Pasaoglu A, Hirst D, Allardyce G, Kennedy I, et al. 1990. Somatosensory and auditory brain-stem conduction after head injury: A comparison with clinical features in prediction of outcome. Neurosurgery 26: 278-285.

Liu S, Lyeth BG, Hamm RJ. 1994. Protective effect of galanin on behavioral deficits in experimental traumatic brain injury. J Neurotrauma 11: 73-82.

Madl C, Kramer L, Domanovits H, Woolard RH, Gervais H, et al. 2000. Improved outcome prediction in unconscious cardiac arrest survivors with sensory evoked potentials compared with clinical assessment. Crit Care Med 28: 721-726.

Madl C, Kramer L, Yeganehfar W, Eisenhuber E, Kranz A, et al. 1996. Detection of nontraumatic comatose patients with no benefit of intensive care treatment by recording of sensory evoked potentials. Arch Neurol 53: 512-516.

Marrannes R, Willems R, De Prins E, Wauquier A. 1988. Evidence for a role of the N-methyl-D-aspartate (NMDA) receptor in cortical spreading depression in the rat. Brain Res 457: 226-240.

Massucci JL, Kline AE, Ma X, Zafonte RD, Dixon CE. 2004. Time dependent alterations in dopamine tissue levels and metabolism after experimental traumatic brain injury in rats. Neurosci Lett 372: 127-131.

Matsushima K, Hogan MJ, Hakim AM. 1996. Cortical spreading depression protects against subsequent focal cerebral ischemia in rats. J Cereb Blood Flow Metab 16: 221-226.

Matsuyama S, Nei K, Tanaka C. 1996. Regulation of glutamate release via NMDA and 5-HT1A receptors in guinea pig dentate gyrus. Brain Res 728: 175-180.

Mauguiere F, Desmedt JE, Courjon J. 1983. Neural generators of N18 and P14 far-field somatosensory evoked potentials studied in patients with lesion of thalamus or thalamo-cortical radiations. Electroencephalogr Clin Neurophysiol 56: 283-292.

Mauler F, Fahrig T, Horváth E, Jork R. 2001. Inhibition of evoked glutamate release by the neuroprotective 5-HT$_{1A}$ receptor agonist BAY x 3702 in vitro and in vivo. Brain Res 888: 150-157.

Mayevsky A, Doron A, Manor T, Meilin S, Zarchin N, et al. 1996. Cortical spreading depression recorded from the human brain using a multiparametric monitoring system. Brain Res 740: 268-274.

McDowell S, Whyte J, D'Esposito M. 1998. Differential effect of a dopaminergic agonist on prefrontal function in traumatic brain injury patients. Brain 121: 1155-1164.

McIntosh TK, Yu T, Gennarelli TA. 1994. Alterations in regional brain catecholamine concentrations after experimental brain injury in the rat. J Neurochem 63: 1426-1433.

McIntosh TK, Faden AI, Yamakami I, Vink R. 1988. Magnesium deficiency exacerbates and pretreatment improves outcome following traumatic brain injury in rats: 31P magnetic resonance spectroscopy and behavioral studies. J Neurotrauma 5: 17-31.

McIntosh TK, Noble L, Andrews B, Faden AI. 1987. Traumatic brain injury in the rat: Characterization of a midline fluid-percussion model. Cent Nerv Syst Trauma 4: 119-134.

McIntosh TK, Vink R, Yamakami I, Faden AI. 1989. Magnesium protects against neurological deficit after brain injury. Brain Res 482: 252-260.

McIntosh TK, Smith DH, Voddi J, Perri BR, Stutzmann JM. 1996. Riluzole, a novel neuroprotective agent, attenuates both neurologic motor and cognitive dysfunction following experimental brain injury in the rat. J Neurotrauma 13: 767-780.

Melena J, Chidlow G, Osborne NN. 2000. Blockade of voltage-sensitive Na^+ channels by the 5-HT1A receptor agonist 8-OH-DPAT: Possible significance for neuroprotection. Eur J Pharmacol 406: 319-324.

Metz B. 1971. Acetylcholine and experimental brain injury. J Neurosurg 35: 523-528.

Mies G, Iijima T, Hossmann KA. 1993. Correlation between peri-infarct DC shifts and ischaemic neuronal damage in rat. Neuroreport 4: 709-711.

Miller LP, Lyeth BG, Jenkins LW, Oleniak L, Panchision D, et al. 1990. Excitatory amino acid receptor subtype binding following traumatic brain injury. Brain Res 526: 103-107.

Miyazaki S, Katayama Y, Lyeth BG, Jenkins L, De Wit DS, et al. 1992. Enduring suppression of hippocampal long-term potentiation following traumatic brain injury in rat. Brain Res 585: 335-339.

Molnar L, Hegedus K, Fekete I. 1991. Difference between the cerebrovascular effect of purinergic Co-ATP and that of the cholinesterase inhibitor, physostigmine, in vivo. Eur J Pharmacol 209: 81-86.

Mooney GF, Haas LJ. 1993. Effect of methylphenidate on brain injury-related anger. Arch Phys Med Rehabil 74: 153-160.

Moreau JL, Huber G. 1999. Central adenosine A(2A) receptors: An overview. Brain Res. Brain Res Rev 31: 65-82.

Movsesyan VA, O'Leary DM, Fan L, Bao W, Mullins PGM, et al. 2001. mGluR5 antagonists 2-methyl-6-(phenylethynyl)-pyridine and (E)-2-methyl-6(2-phenylethenyl)-pyridine reduce traumatic neuronal injury in vitro and in vivo by antagonizing N-methyl-D-aspartate receptors. J Pharmacol Exp Ther 296: 41-47.

Murdoch I, Perry EK, Court JA, Graham DI, Dewar D. 1998. Cortical cholinergic dysfunction after human head injury. J Neurotrauma 15: 295-305.

Ngai AC, Coyne EF, Meno JR, West GA, Winn HR. 2001. Receptor subtypes mediating adenosine-induced dilation of cerebral arterioles. Am J Physiol Heart Circ Physiol 280: H2329-H2335.

Narayan RK, Greenberg RP, Miller JD, Enas GG, Choi SC, et al. 1981. Improved confidence of outcome prediction in severe head injury. A comparative analysis of the clinical examination, multimodality evoked potentials, CT scanning and intracranial pressure. J Neurosurg 54: 751-762.

Nedergaard M, Hansen AJ. 1988. Spreading depression is not associated with neuronal injury in the normal brain. Brain Res 449: 395-398.

Nedergaard S, Engberg I, Flatman JA. 1987. The modulation of excitatory amino acids responses by serotonin in the cat neocortex in vitro. Cell Mol Neurobiol 7: 367-379.

Newlon PG, Greenberg RP, Enas GG, Becker DP. 1983. Effects of therapeutic pentobarbital coma on multimodality evoked potentials recorded from severely head-injured patients. Neurosurgery 12: 613-619.

Nilsson P, Hillered L, Ponten U, Ungerstedt U. 1990. Changes in cortical extracellular levels of energy-related metabolites and amino acids following concussive brain injury in rats. J Cereb Blood Flow Metab 10: 631-637.

O'Dell DM, Hamm RJ. 1995. Chronic post injury administration of MDL 26,479 (suritozol) a negative modulator at the GABAA receptor, and cognitive impairment in rats following traumatic brain injury. J Neurosurg 83: 878-883.

O'Dell DM, Gibson CJ, Wilson MS, De Ford SM, Hamm RJ. 2000. Positive and negative modulation of the GABA(A) receptor and outcome after traumatic brain injury in rats. Brain Res 861: 325-332.

Ohta K, Graf R, Rosner G, Heiss WD. 2001. Calcium ion transients in periinfarct depolarizations may deteriorate ion homeostasis and expand infarction in focal cerebral ischemia in cats. Stroke 32: 535-543.

Okiyama K, Smith DH, White WF, McIntosh TK. 1998. Effects of the NMDA antagonist CP-98,113 on regional cerebral edema and cardiovascular, cognitive, and neurobehavioral function following experimental brain injury in the rat. Brain Res 792: 291-298.

Osteen CL, Giza CC, Hovda DA. 2004. Injury-induced alterations in N-methyl-D-aspartate receptor subunit composition contribute to prolonged 45calcium accumulation following lateral fluid percussion. Neuroscience 128: 305-322.

Pappius HM. 1981. Local cerebral glucose utilization in thermally traumatized rat brain. Ann Neurol 9: 484-491.

Pappius HM, Dadoun R. 1987. Effects of injury on the indoleamines in cerebral cortex. J Neurochem 49: 321-325.

Pappius HM, Dadoun R, McHugh M. 1988. The effect of p-chlorophenylalanine on cerebral metabolism and biogenic amine content of traumatized brain. J Cereb Blood Flow Metab 8: 324-334.

Passler MA, Riggs RV. 2001. Positive outcomes in traumatic brain injury-vegetative state: Patients treated with bromocriptine. Arch Phys Med Rehabil 82: 311-315.

Pfurtscheller G, Schwarz G, Gravenstein N. 1985. Clinical relevance of long-latency SEPs and VEPs during coma and emergence from coma. Electroencephalogr Clin Neurophysiol 62: 88-98.

Phillips JP, Devier DJ, Feeney DM. 2003. Rehabilitation pharmacology: Bridging laboratory work to clinical application. J Head Trauma Rehabil 18: 342-356.

Pike BR, Hamm RJ. 1995. Post injury administration of BIBN 99, a selective muscarinic M_2 receptor antagonist, improves cognitive performance following traumatic brain injury in rats. Brain Res 686: 37-43.

Plenger PM, Dixon CE, Castillo RM, Frankowski RF, Yablon SA, et al. 1996. Subacute methylphenidate treatment for moderate to moderately severe traumatic brain injury: A preliminary double-blind placebo-controlled study. Arch Phys Med Rehabil 77: 536-540.

Pohlmann-Eden B, Dingethal K, Bender H-J, Koelfen W. 1997. How reliable is the predictive value of SEP (somatosensory evoked potentials) patterns in severe brain damage with special regard to the bilateral loss of cortical responses. Intensive Care Med 23: 301-308.

Popoli P, Pintor A, Domenici MR, Frank C, Tebano MT, et al. 2002. Blockade of striatal adenosine A2A receptor reduces, through a presynaptic mechanism, quinolinic acid-induced excitotoxicity: Possible relevance to neuroprotective interventions in neurodegenerative diseases of the striatum. J Neurosci 22: 1967-1975.

Povlishock JT, Katz DI. 2005. Update of neuropathology and neurological recovery after traumatic brain injury. J Head Trauma Rehabil 20: 76-94.

Powell JH, Al-Adawi S, Morgan J, Greenwood RJ. 1996. Motivational deficits after brain injury: Effects of bromocriptine in 11 patients. J Neurol Neurosurg Psychiatry 60: 416-421.

Prehn JH, Backhauss C, Karkoutly C, Nuglisch J, Peruche B, et al. 1991. Neuroprotective properties of 5-HT_{1A} receptor agonists in rodent models of focal and global cerebral ischemia. Eur J Pharmacol 203: 213-222.

Prehn JH, Welsch M, Backhauss C, Nuglisch J, Ausmeier F, et al. 1993. Effects of serotonergic drugs in experimental brain ischemia: Evidence for a protective role of serotonin in cerebral ischemia. Brain Res 630: 10-20.

Pulaski KH, Emmett L. 1994. The combined intervention of therapy and bromocriptine mesylate to improve functional performance after brain injury. Am J Occup Ther 48: 263-270.

Raiteri M, Maura G, Barzizza A. 1991. Activation of presynaptic 5-hydroxytryptamine1-like receptors on glutamatergic terminals inhibits *N*-methyl-D-aspartate-induced cyclic GMP production in rat cerebellar slices. J Pharmacol Exp Ther 257: 1184-1188.

Rappaport M, Hall K, Hopkins K, Belleza T, Berrol S, et al. 1977. Evoked brain potentials and disability in brain-damaged patients. Arch Phys Med Rehabil 58: 333-338.

Reeves TM, Phillips LL, Povlishock JT. 2005. Myelinated and unmyelinated axons of the corpus callosum differ in vulnerability and functional recovery following traumatic brain injury. Exp Neurol 196: 126-137.

Robinson SE. 1986. 6-Hydroxydopamine lesion of the ventral noradrenergic bundle blocks the effect of amphetamine on hippocampal acetylcholine. Brain Res 397: 181-184.

Robinson SE, Martin RM, Davis TR, Gyenes CA, Ryland JE, Enters EK. 1990. The effect of acetylcholine depletion on behavior following traumatic brain injury. Brain Res 509(1): 41-46.

Rothstein TL. 2000. The role of evoked potentials in anoxic-ischemic coma and severe brain trauma. J Clin Neurophysiol 17: 486-497.

Rudolphi KA, Schubert P, Parkinson FE, Fredholm BB. 1992. Adenosine and brain ischemia. Cerebrovasc Brain Metab Rev 4(4): 346-369, 1992.

Ruge D. 1954. The use of cholinergic blocking agents in the treatment of craniocerebral injuries. J Neurosurg 11: 77-83.

Sachs EJr. 1957. Acetylcholine and serotonin in the spinal fluid. J Neurosurg 14: 22-27.

Saija A, Hayes RL, Lyeth BG. 1988a. Effect of concussive head injury on central cholinergic neurons. Brain Res 452: 303-311.

Saija A, Robinson SE, Lyeth BG. 1988b. Effect of scopolamine and traumatic brain injury on central cholinergic neurons. J Neurotrauma 5: 161-169.

Sanders MJ, Sick TJ, Perez-Pinzon MA, Dietrich WD, Green EJ. 2000. Chronic failure in the maintenance long-term potentiation following fluid percussion injury in the rat. Brain Res 861: 69-76.

Schmidt RH, Grady MS. 1995. Loss of forebrain cholinergic neurons following fluid-percussion injury: Implications for cognitive impairment in closed head injury. J Neurosurg 3: 496-502.

Scremin OU. 1991. Pharmacological control of the cerebral circulation. Ann Rev Pharmacol Toxicol 31: 229-251.

Scremin OU, Li MG, Jenden DJ. 1997. Cholinergic modulation of cerebral cortical blood flow changes induced by trauma. J Neurotrauma 14: 573-586.

Scremin OU, Rovere AA, Raynald AC, Giardini A. 1973. Cholinergic control of blood flow in the cerebral cortex of the rat. Stroke 4: 232-239.

Shapira Y, Lam AM, Eng CC, Laohaprasit V, Michel M. 1994. Therapeutic time window and dose response of the beneficial effects of ketamine in experimental head injury. Stroke 25: 1637-1643.

Sherman AL, Tirschwell DL, Micklesen PJ, Longstreth WT, Robinson LR. 2000. Somatosensory potentials, CSF creatine kinase BB activity, and awakening after cardiac arrest. Neurology 54: 889-894.

Shin HK, Shin YW, Hong KW. 2000. Role of adenosine A(2B) receptors in vasodilation of rat pial artery and cerebral blood flow autoregulation. Am J Physiol Heart Circ Physiol 278(2): H339-H344.

Shohami E, Novikov M, Mechoulam R. 1993. A nonpsychotropic cannabinoid, HU-211, has cerebroprotective effects after closed head injury in the rat. J Neurotrauma 10: 109-119.

Sick TJ, Perez-Pinzon MA, Feng ZZ. 1998. Impaired expression of long-term potentiation in hippocampal slices 4 and 48 h following mild fluid-percussion brain injury in vivo. Brain Res 785: 287-292.

Sihver S, Marklund N, Hillered L, Langstrom B, Watanabe Y, et al. 2001. Changes in mACh, NMDA and GABA(A) receptor binding after lateral fluid-percussion injury: In vitro autoradiography of rat brain frozen sections. J Neurochem 78: 417-423.

Sinson G, Perri BR, Trojanowski JQ, Flamm ES, McIntosh TK. 1997. Improvement of cognitive deficits and decreased cholinergic neuronal cell loss and apoptotic cell death following neurotrophin infusion after experimental traumatic brain injury. J Neurosurg 86: 511-518.

Sleigh JW, Havill JH, Frith R, Kersei D, Marsh N, et al. 1999. Somatosensory evoked potentials in severe traumatic brain injury: A blinded study. J Neurosurg 91: 577-580.

Smith DH, Okiyama K, Gennarelli TA, McIntosh TK. 1993. Magnesium and ketamine attenuate cognitive dysfunction following experimental brain injury. Neurosci Lett 157: 211-214.

Smith DH, Perri BR, Raghupathi R, Saatman KE, McIntosh TK. 1997. Remacemide hydrochloride reduces cortical lesion volume following brain trauma in the rat. Neurosci Lett 231: 135-138.

Smith JS, Fulop ZL, Levinsohn SA, Darrell RS, Stein DG. 2000. Effects of the novel NMDA receptor antagonist gacyclidine on recovery from medial frontal cortex contusion injury in rats. Neural Plast 7: 73-91.

Sollevi A, Ericson K, Ericksson L, Lindqvist C, Langerkranser M, et al. 1987. Effect of adenosine on human cerebral blood flow as determined by positron emission tomography. J Cereb Blood Flow Metab 7: 673-678.

Soricelli A, Postiglione A, Cuocolo A, De Chiara S, Ruocco A, et al. 1995. Effect of adenosine on cerebral blood flow as evaluated by single-photon emission computed tomography in normal subjects and in patients with occlusive carotid disease. A comparison with acetazolamide. Stroke 26: 1572-1576.

Speech TJ, Rao SM, Osmon DC, Sperry LT. 1993. A double-blind control study of methylphenidate treatment in closed head injury. Brain Inj 7: 333-338.

Stange K, Greitz D, Ingvar M, Hindmarsh T, Sollevi A. 1997. Global cerebral blood flow during infusion of adenosine in humans: Assessment by magnetic resonance imaging and positron emission tomography. Acta Physiol Scand 160: 117-122.

Stover JF, Beyer TF, Unterberg AW. 2000. Riluzole reduces brain swelling and contusion volume in rats following controlled cortical impact injury. J Neurotrauma 17: 1171-1178.

Strong AJ, Fabricius M, Boutelle MG, Hibbins SJ, Hopwood SE, et al. 2002. Spreading and synchronous depressions of cortical activity in acutely injured human brain. Stroke 33: 2738-2743.

Sun FY, Faden AI. 1995. Neuroprotective effects of 619C89, a use-dependent sodium channel blocker, in rat traumatic brain injury. Brain Res 673: 133-140.

Sutton RL, Feeney DM. 1992. A-noradrenergic agonists and antagonists affect recovery and maintenance of beam-walking ability after sensorimotor cortex ablation in the rat. Restor Neurol Neurosci 4: 1-11.

Svenningsson P, Le Moine C, Fisone G, Fredholm BB. 1999. Distribution, biochemistry and function of striatal adenosine A2A receptors. Prog Neurobiol 59: 355-396.

Takano K, Latour LL, Formato JE, Carano RA, Helmer KG, et al. 1996. The role of spreading depression in focal ischemia evaluated by diffusion mapping. Ann Neurol 39: 308-318.

Tower DB, McEachern D. 1949. Cholinesterase patterns and acetylcholine in the cerebrospinal fluids of patients with craniocerebral trauma. Can J Res 27: 105-119.

Van Wylen DG, Park TS, Rubio R, Berne RM. 1989. The effect of local infusion of adenosine and adenosine analogues on local cerebral blood flow. J Cereb Blood Flow Metab 9: 556-562.

Varma MR, Dixon CE, Jackson EK, Peters GW, Melick JA, et al. 2002. Administration of adenosine receptor agonists or antagonists after controlled cortical impact in mice: Effects on function and histopathology. Brain Res 951: 191-201.

Verbois SL, Hopkins DM, Scheff SW, Pauly JR. 2003. Chronic intermittent nicotine administration attenuates traumatic brain injury-induced cognitive dysfunction. Neuroscience 119: 1199-1208.

Vespa PM, Boscardin WJ, Hovda DA, McArthur DL, Nuwer MR, et al. 2002. Early and persistent impaired percent alpha variability on continuous electroencephalography monitoring as a predictive of poor outcome after traumatic brain injury. J Neurosurg 97: 84-92.

von Voigtlander PF, Moore KE. 1971. Dopamine: Release from the brain in vivo by amantadine. Science 174: 408-410.

Wagner AK, Chen X, Kline AE, Li Y, Zafonte RD, et al. 2005a. Gender and environmental enrichment impact dopamine transporter expression after experimental traumatic brain injury. Exp Neurology 195: 475-483.

Wagner AK, Sokoloski JE, Ren D, Chen X, Khan AS, et al. 2005b. Controlled cortical impact injury affects dopaminergic transmission in the rat striatum. J Neurochem 95: 457-465.

Walser H, Mattle H, Keller HM, Janzer R. 1985. Early cortical median nerve somatosensory evoked potentials. Prognostic value in anoxic coma. Arch Neurol 42: 32-38.

West M, Parkinson D, Havlicek V. 1982. Spectral analysis of the electroencephalographic response to experimental concussion in the rat. Electroencephalogr Clin Neurophysiol 53: 192-200.

Whyte J, Hart T, Schuster K, Fleming M, Polansky M, et al. 1997. Effects of methylphenidate on attentional function after traumatic brain injury. A randomized, placebo-controlled trial. Am J Phys Med Rehabil 76: 440-450.

Whyte J, Hart T, Vaccaro M, Grieb-Neff P, Risser A, et al. 2004. Effects of methylphenidate on attention deficits after traumatic brain injury: A multidimensional, randomized, controlled study. Am J Phys Med Rehabil 83: 401-420.

Wilson MS, Chen X, Ma X, Ren D, Wagner AK, et al. 2005. Synaptosomal dopamine uptake in rat striatum following controlled cortical impact. J Neurosci Res 80: 85-91.

Yan HQ, Kline AE, Ma X, Li Y, Dixon CE. 2002. Traumatic brain injury reduces dopamine transporter protein expression in the rat frontal cortex. Neuroreport 13: 1899-1901.

Yan HQ, Kline AE, Ma X, Hooghe-Peters EL, Marion DW, et al. 2001. Tyrosine hydroxylase, but not dopamine beta-hydroxylase, is increased in rat frontal cortex after traumatic brain injury. Neuroreport 12: 2323-2327.

Yanamoto H, Hashimoto N, Nagata I, Kikuchi H. 1998. Infarct tolerance against temporary focal ischemia following spreading depression in rat brain. Brain Res 784: 239-249.

Yang K, Taft WC, Dixon CE, Todaro CA, Yu RK, et al. 1993. Alterations in protein kinase C in rat hippocampus following traumatic brain injury. J Neurotrauma 10: 287-295.

Zentner J, Rohde V. 1992. The prognostic value of somatosensory and motor evoked potentials in comatose patients. Neurosurgery 31: 429-434.

Zhu J, Hamm RJ, Reeves TM, Povlishock JT, Phillips LL. 2000. Postinjury administration of L-deprenyl improves cognitive function and enhances neuroplasticity after traumatic brain injury. Exp Neurol 166: 136-152.

10 Free Radicals and Neuroprotection in Traumatic Brain and Spinal Cord Injury

E. D. Hall

1	Introduction	204
2	Chemistry of Reactive Oxygen Species, Free Radical Production, and Oxidative Damage	205
2.1	Superoxide Radical	205
2.2	Iron and Formation of Hydroxyl Radical	205
2.3	Lipid Peroxidation	206
2.4	Measurement of Lipid Peroxidation	207
2.5	Other Forms of Oxidative Damage	208
2.6	Antioxidant Defense Mechanisms	208
2.6.1	Antioxidant Enzymes	208
2.6.2	Endogenous Small Molecule Antioxidants	208
2.7	Peroxynitrite-Mediated Oxygen Radical Formation and Oxidative Damage	209
3	Interaction of Oxidative Damage with Other Secondary Injury Mechanisms	210
3.1	Role in the Disruption of Ion Homeostasis	210
3.2	Role in Excitotoxicity	211
3.3	Role in Posttraumatic Mitochondrial Dysfunction	211
3.4	Role in Microvascular Dysfunction: Loss of Autoregulation and Blood–Brain (Spinal Cord) Compromise	212
4	Pharmacological Antioxidants That Have Been Explored in CNS Injury Models	214
4.1	Protective Effects of Antioxidants in Spinal Cord Injury	215
4.1.1	High-Dose Methylprednisolone Inhibition of Lipid Peroxidation	215
4.1.2	NASCIS II Clinical Trial of High-Dose Methylprednisolone	216
4.1.3	Non-Glucocorticoid Antioxidant 21-Aminosteroid Tirilazad	216
4.1.4	NASCIS III	217
4.2	Protective Effects of Antioxidants in Traumatic Brain Injury	218
4.2.1	Effects of SOD in TBI	218
4.2.2	Effects of Tirilazad in TBI	219
4.2.3	Effects of More Potent, Brain-Penetrable Antioxidants in TBI	219
5	Future Directions	220
5.1	Improved Antioxidant Drugs	220
5.1.1	Scavengers of Peroxynitrite or Peroxynitrite-Derived Radicals	220
5.1.2	Dual Mechanism Antioxidants	221
5.2	Oxidative Damage Biomarkers	222
6	Summary	222

© 2009 Springer Science+Business Media, LLC.

Abstract: This chapter reviews the considerable body of evidence that supports the role of reactive oxygen species (ROS), reactive nitrogen species (RNS) and their derived oxygen free radicals in the pathophysiology of acute traumatic brain injury (TBI) and spinal cord injury (SCI). Free radical-induced oxidative damage to membrane lipids and proteins occurs in the injured brain of spinal cord within the first minutes and hours and has been implicated in the disruption of neuronal ion homeostasis, exacerbation of glutamate-mediated excitotoxicity, mitochondrial respiratory dysfunction and microvascular structural and functional damage. Lipid peroxidation (LP) is a key mechanism of the radical-induced secondary injury. Several free radical scavengers and LP inhibitors have been shown to be neuroprotective in animals models of TBI and/or SCI strongly implicating free radical-induced LP as an important target for inhibition of secondary CNS injury. The glucocorticoid steroid methylprednisolone which has been shown to decrease post-traumatic LP in the injured spinal cord when administered in large doses has been shown to improve neurological recovery in SCI clinical trials. The non-glucocorticoid 21-aminosteroid tirilazad has also shown evidence of efficacy in SCI patients and in a subset of TBI patients who have post-traumatic subarachnoid hemorrhage. Improved antioxidants approaches are presented which either more potently inhibit LP or that scavenge the RNS peroxynitrite (PN) or its highly reactive free radical products. Finally, the feasibility of developing compounds with dual antioxidant mechanisms is discussed.

List of Abbreviations: LP, lipid peroxidation; RNS, reactive nitrogen species; ROS, reactive oxygen species; SCI, spinal cord injury; TBI, Traumatic brain injury

1 Introduction

Traumatic brain injury (TBI) and spinal cord injury (SCI) represent two of the most catastrophic consequences that human beings can suffer. In most cases of moderate or severe TBI, and in many mild cases, the neurological, economic, and social consequences are devastating to the patient, his or her family, and to society as a whole. In the USA, there are approximately 1.2 million TBIs that occur yearly of which about 58,000 are severe (Glasgow Coma Score = 3–8) and 64,000 moderate (Glasgow Coma Score = 9–12). In the case of SCI, there are about 11,000 each year in the USA. Although TBI and SCI can victimize active individuals at any age, most occur in young adults in the second and third decades of life. Those who survive their initial injuries can now expect to live long lives due to improvements in medical and surgical care. Nevertheless, the need for intensive rehabilitation and prolonged disability exacts a significant toll on the individual, his or her family, and society. Effective ways of maintaining or recovering function could markedly improve the outlook for those with TBI or SCI by enabling higher levels of independence and productivity.

The potential for pharmacological intervention to preserve neurological function after TBI and SCI exists due to the fact that most of the neurodegeneration that follows these injuries is not due to the primary mechanical injury (i.e., shearing of blood vessels, and nerve cells), but rather to secondary injury events. For instance, most SCIs do not involve actual physical transection of the cord, but rather the spinal cord is damaged as a result of a contusive, compressive, or stretch injury. Some residual white matter, containing portions of the ascending sensory and descending motor tracts, remains intact which allows for the possibility of neurological recovery. However, during the first minutes and hours following injury, a secondary degenerative process is initiated by the primary mechanical injury that is proportional to the magnitude of the initial insult. Nevertheless, the initial anatomical continuity of the injured spinal cord in the majority of cases, together with our present knowledge of many of the factors involved in the secondary injury process, has lead to the notion that pharmacological treatments which interrupt the secondary cascade, if applied early, could improve spinal cord tissue survival, and thus preserve the necessary anatomic substrates for functional recovery to take place. Similarly, the outcome after TBI is mainly determined by the extent of the potentially treatable secondary pathophysiology and neurodegeneration. One of the most completely validated mechanisms in secondary tissue damage following brain or spinal cord injury involves reactive oxygen species (ROS) and reactive nitrogen species (RNS) which are generated by multiple mechanisms and give rise to highly reactive oxygen radicals that damage neuronal, glial, and microvascular elements largely via a process known as lipid peroxidation (LP). The purpose of this chapter is to review the basics of posttraumatic free radical formation and lipid peroxidative cellular damage, their role in the acute pathophysiology of CNS injury, and the neuroprotective effects of free radical scavenging and antioxidant pharmacological compounds in experimental models of TBI and SCI as well as their partial success to date in clinical trials.

2 Chemistry of Reactive Oxygen Species, Free Radical Production, and Oxidative Damage

2.1 Superoxide Radical

The primary radical for consideration is superoxide anion ($O_2^{\cdot-}$). Within the injured nervous system, a number of possible sources of superoxide may be operative within the first minutes and hours after injury including: the arachidonic acid cascade (i.e., prostaglandin synthase and 5-lipoxygenase activity), enzymatic of autoxidation of biogenic amine neurotransmitters (e.g., dopamine, norepinephrine, 5-hydroxytryptamine), "mitochondrial leak," xanthine oxidase activity, and the oxidation of extravasated hemoglobin. Activated microglia and infiltrating neutrophils and macrophages provide additional sources of superoxide.

Superoxide, which is formed by the single electron reduction of oxygen, may act as either an oxidant or reductant. While, superoxide itself is reactive and can, for example, reduce ferric iron (Fe^{+++}) to ferrous iron (Fe^{++}) or cause the release of Fe^{++} from ferritin, its direct reactivity toward biological substrates in aqueous environments is questioned. Moreover, once formed, superoxide undergoes spontaneous dismutation to form H_2O_2 in a reaction that is markedly accelerated by the enzyme superoxide dismutase (SOD) (Halliwell and Gutteridge, 2007).

$$O_2^{\cdot-} + O_2^{\cdot-} + 2H^+ \xrightarrow{SOD} H_2O_2 + O_2$$

In solution, superoxide actually exists in equilibrium with the hydroperoxyl radical ($HO_2^{\cdot}$).

$$O_2^{\cdot-} + H^+ \rightarrow HO_2^{\cdot}$$

The pKa of this reaction is 4.8 and the relative concentrations of $O_2^{\cdot-}$ and $HO_2^{\cdot}$ depend upon the H^+ concentration. Therefore, at a pH around 6.8, the ratio of $O_2^{\cdot-}/HO_2^{\cdot}$ is 100/1 whereas at a pH of 5.8 the ratio is only 10/1. Thus, under conditions of tissue acidosis of a magnitude known to occur within the severely injured nervous system, a significant amount of the superoxide formed will exist as hydroperoxyl radical. Compared with superoxide, $HO_2^{\cdot}$ is considerably more lipid soluble and is a far more powerful oxidizing or reducing agent (9). Therefore, as the pH of a solution falls and the equilibrium between $O_2^{\cdot-}$ and $HO_2^{\cdot}$ shifts in favor of $HO_2^{\cdot}$, superoxide becomes more reactive, particularly toward lipids. In addition, while the spontaneous (non-SOD catalyzed) dismutation of $O_2^{\cdot-}$ to H_2O_2 is exceedingly slow at neutral pH, $HO_2^{\cdot}$ will dismutate to H_2O_2 far more readily at acidic pH values since the rate constant for $HO_2^{\cdot}$ dismutation is on the order of 10^8 times greater than for $O_2^{\cdot-}$. Thus, in an acidic environment, $O_2^{\cdot-}$ is (1) converted to the more-reactive, more lipid-soluble $HO_2^{\cdot}$ and, (2) its rate of dismutation to H_2O_2 is greatly increased. In the context of the injured brain or spinal cord, this is a pathophysiologically relevant issue since the tissue pH typically falls due to the typical occurrence of lactic acid accumulation.

2.2 Iron and Formation of Hydroxyl Radical

The CNS is an extremely rich source of iron and its regional distribution varies in parallel with the sensitivity of various regions to in vitro lipid peroxidation (Zaleska and Floyd, 1985). Under normal circumstances, low molecular weight forms of redox active iron in the brain are maintained, as in other tissues, at extremely low levels. Extracellularly in plasma, the iron transport protein transferrin tightly binds iron in the Fe^{+++} form. Intracellularly, Fe^{+++} is sequestered by the iron storage protein ferritin. While both ferritin and transferrin have very high affinity for iron at neutral pH and effectively maintain iron in a noncatalytic state (Halliwell and Gutteridge, 1999), both proteins readily give up their iron at pH values of 6.0 or less. In the case of ferritin, its iron can also be released by reductive mobilization by $O_2^{\cdot-}$. Once iron is released from ferritin or transferrin, it can actively catalyze oxygen radical reactions. Therefore, within the traumatized CNS environment, where pH is typically lowered, conditions are favorable for the potential release of iron from storage proteins (Halliwell and Gutteridge, 2007).

A second source of catalytically active iron is hemoglobin. Hemorrhage resulting from mechanical trauma is an obvious source of hemoglobin. While hemoglobin itself has been reported to stimulate oxygen radical reactions, it is more likely that iron released from hemoglobin is responsible for hemoglobin-mediated LP (Sadrzadeh et al., 1984; Sadrzadeh and Eaton, 1988). Iron is released from hemoglobin by H_2O_2 or by lipid hydroperoxides (LOOH; see below) and this release is further enhanced as the pH falls to 6.5 or below. Therefore, hemoglobin may catalyze oxygen radical formation and lipid peroxidation either directly or through the release of iron by H_2O_2, LOOH, and/or acidic pH.

Free iron or iron chelates participate in free radical reactions at at least two levels. The autoxidation of Fe^{II} results in the formation of $O_2^{\cdot-}$ (Halliwell and Gutteridge, 2007).

$$Fe^{++} + O_2 \rightarrow Fe^{+++} + O_2^{\cdot-}$$

The reverse reaction or reduction of Fe^{+++} to Fe^{++} by $O_2^{\cdot-}$ also occurs and competes with the dismutation of $O_2^{\cdot-}$. In theory, Fe^{++} autoxidation could result in the redox cycling of iron due to the reaction of $O_2^{\cdot-}$ produced with Fe^{+++}. Secondly, Fe^{++} is also oxidized in the presence of H_2O_2 to form hydroxyl radical (•OH) (Fenton reaction) or perhaps a ferryl ion (Fe^{+++} -OH).

$$Fe^{++} + H_2O_2 \rightarrow Fe^{+++} + \bullet OH + OH^-$$

$$Fe^{++} + H_2O_2 \rightarrow Fe^{+++}OH + OH^-$$

Either •OH or Fe^{+++} OH are extraordinarily potent initiators of lipid peroxidation (LP) which has been shown to be a key mechanism of posttraumatic oxidative damage in SCI or TBI models (Hall and Braughler, 1993).

2.3 Lipid Peroxidation

In addition to its role in the formation of •OH and/or Fe^{+++} OH, iron also profoundly promotes the process of LP. Indeed, the role of iron in the initiation of LP has been a subject of considerable interest (Halliwell and Gutteridge, 1999). There is no question that iron catalyzes the formation of various radicals as described above. Studies carried out using the salicylate trapping method have in fact demonstrated the early occurrence of increased •OH levels in the injured brain (Hall et al., 1993; Globus et al., 1995).

"Initiation" of lipid peroxidation occurs when a radical species (R•) attacks and removes an allylic hydrogen from a polyunsaturated fatty acid (LH) resulting in a radical chain reaction:

$$LH + R\bullet \rightarrow L\bullet + RH$$

In the process, the initiating radical is quenched by receipt of an electron (hydrogen) from the polyunsaturated fatty acid. This, however, converts the latter into a lipid or alkyl radical (L•). This sets the stage for a series of "propagation" reactions which begins when the alkyl radical takes on a mole of oxygen creating a lipid peroxyl radical (LOO•):

$$L\bullet + O_2 \rightarrow LOO\bullet$$

The peroxyl radical then reacts with a neighboring LH within the membrane and steals its electron forming a lipid hydroperoxide (LOOH) and a second alkyl radical (L•).

$$LOO\bullet + LH \rightarrow LOOH + L\bullet$$

Once lipid peroxidation begins, iron may participate in driving the process as lipid hydroperoxides (LOOH) formed through initiation are decomposed by reactions with either ferrous iron (Fe^{++}) or ferric iron (Fe^{+++}). In the case of Fe^{++}, the reaction results in the formation of a lipid alkoxyl radical (LO•):

$$LOOH + Fe^{++} \rightarrow LO\bullet + OH^- + Fe^{+++}$$

If, however, the reaction involves Fe^{+++}, the lipid hydroperoxide is converted back into a lipid peroxyl radical (LOO•):

$$LOOH + Fe^{+++} \rightarrow LOO\bullet + Fe^{++}$$

Both of the reactions of LOOH with iron have acidic pH optima. Either alkoxyl (LO•) or peroxyl (LOO•) radicals arising from LOOH decomposition by iron can initiate so called lipid hydroperoxide dependent lipid peroxidation resulting in "chain branching" reactions.

$$LOO\bullet + LH \rightarrow LOOH + L\bullet$$

or

$$LO\bullet + LH \rightarrow LOH + L\bullet$$

Thus, in a membrane, LP need not be initiated by a primordial inorganic oxygen radical provided that LOOH and a source of iron are available. In that regard, the presence of preexisting LOOH within normal healthy cell membranes is a subject for consideration. One would suspect that cells would not have evolved defense mechanisms (see ❯ Section 2.4) against radical damage and LP unless a need existed. During normal cellular energy metabolism, oxygen radicals and H_2O_2 are produced, and estimated steady state concentrations are on the order of 10^{-9} M or higher depending upon the tissue and radical in question (Halliwell and Gutteridge, 1999). Thus, unless cellular defenses against radical production and LP are 100% efficient, a certain fraction of the radicals would be expected to escape cellular scavenging processes and result in LP. Indeed, it has been proposed that oxidation of fatty acids within membranes and their hydrolysis by phospholipases is involved in the normal turnover of peroxidized membrane lipids (Sevanian and Kim, 1985). From the standpoint of LP during CNS trauma, it may be unnecessary to distinguish between true initiation by a primordial inorganic radical or the occurrence of lipid hydroperoxide-dependent LP triggered by the release of iron since sufficient LOOH may preexist in normal membranes to allow for the latter to occur (Zhang et al., 1993). In any event, CNS tissue appears to provide an especially avid environment for the occurrence of oxygen radical generation and uncontrolled LP reactions. Reasons for this include a high content of iron in many brain regions (Zaleska and Floyd, 1985) and a high proportion of membrane phospholipids containing polyunsaturated fatty acids such as linoleic acid (18:2), arachidonic acid (20:4), and docosahexaenoic acid (22:6) that are sensitive to LP (Hall and Braughler, 1993).

2.4 Measurement of Lipid Peroxidation

A plethora of methods have been developed to measure LP-mediated oxidative damage which are beyond the scope of this brief consideration. However, a brief consideration of the major methods and LP markers is needed in order to enhance the understanding of the following text. Although techniques exist for following the initiation of LP (e.g., alkyl radicals), the initial propagation of LP reactions (e.g., LOOH), the most widely used approaches target various end-products of LP. Ultimately, peroxidized polyunsaturated fatty acids will undergo a series of reactions and break down to form several aldehyde-containing species. Two of the most popular in CNS injury studies is either the 3 carbon malondialdehyde (MDA) or the 9 carbon 4-hydroxynonenal (4-HNE). The former is viewed as the less reliable endpoint since MDA is also formed during enzymatic LP reactions that occur during the arachidonic acid cascade (Halliwell and Gutteridge, 1999). This is not the case with 4-HNE which is purely a product of radical-catalyzed, nonenzymatic LP. Furthermore, 4-HNE has high affinity for covalent binding to cellular proteins and in the process can result in potent protein dysfunction and neuronal cell death (Kruman et al., 1997). The most common approaches in use today for directly measuring 4-HNE are immunoblotting or immunohistochemistry (Hall et al., 2004; Singh et al., 2006).

2.5 Other Forms of Oxidative Damage

As pointed out, the CNS is exquisitely sensitive to LP because of its high content of peroxidation-susceptible lipids and the high levels of iron. Moreover, end products of LP can bind to proteins, modifying their structure and compromising function. However, primary radical-mediated oxidative damage can also occur in proteins and nucleic acids (e.g., DNA, RNA). For instance, iron-catalyzed, •OH mechanisms can target certain basic amino acids (e.g., lysine, arginine, histidine) leading to the formation of "protein carbonyl" moieties that can be measured by spectrophotometric or immunoblot assays after reaction of the carbonyl groups with dinitrophenylhydrazine (DNPH). This approach has become increasingly popular in CNS injury studies. However, it should be noted that not all of the protein carbonyl that is measured is due to primary protein oxidation, but also includes carbonyl containing 4-HNE and MDA that is bound to proteins. Thus, the use of the protein carbonyl assay is actually a combined measurement of protein oxidation and LP.

Another form of protein oxidative damage involves the oxidation of cysteine sulfhydryl groups which can lead to the formation of abnormal disulfide bridges and changes in protein structure and function.

Nucleic acids, both DNA and RNA, are also susceptible to oxidative modification by inorganic and organic (i.e., lipid) radicals. The typical assay of this form of damage involves the measurement of guanine oxidation product 8-hydroxyguanine. In addition to potentially compromising DNA replication, transcription, and mRNA translation, DNA oxidative damage also triggers DNA repair mechanisms that can greatly stress cellular function and survival. One such mechanism concerns the activation of poly ADP ribose polymerase (PARP) whose action can lead to severe depletion of cellular stores of ATP. In addition, DNA-protein cross-linking can occur (e.g., thymine–tyrosine) (Halliwell and Gutteridge, 2007). However, compared to the numerous studies that have documented posttraumatic LP in TBI and SCI models, very little examination of nucleic acid oxidation has occurred.

2.6 Antioxidant Defense Mechanisms

2.6.1 Antioxidant Enzymes

As mentioned here, several endogenous antioxidant defense mechanisms have evolved in order to protect cells against the continuous onslaught of ROS and free radicals formed during cellular metabolism. These include multiple antioxidant enzymes that convert ROS or free radicals to less reactive products as well as radical scavenging small molecules. In regards to the antioxidant enzymes, SOD above acts to convert superoxide radical to H_2O_2. The enzyme exists in at least three isoforms: Cu/Zn SOD which is cytoplasmic, MnSOD which is mitochondrial, and EcSOD which is extracellular. Hydrogen peroxide, in turn can be decomposed by either catalase or glutathione peroxidase (GSHPx) to H_2O. In the case of lipid hydroperoxides (LOOH), the enzyme phospholipid glutathione peroxidase (PHGPx) converts them to lipid alcohols (LOH). Both GSHPx and PHGPx require reduced glutathione (GSH) for their activity which is converted to an oxidized form (GSSG). Working in concert with these two GSH-requiring peroxidases is the enzyme glutathione reductase (GSHRx) which reduces GSSG $\rightarrow$ GSH. Glutathione synthetase which produces GSH can also be listed among the antioxidant enzymes (Halliwell and Gutteridge, 2007).

2.6.2 Endogenous Small Molecule Antioxidants

Multiple small molecule antioxidants participate in the defense of the CNS and all tissues against oxidative damage. Perhaps the most important of these is vitamin E (α-tocopherol) which probably constitutes our principal protector against LP. It donates an electron to lipid peroxyl radicals and in the process is converted to vitamin E radical. However, this radical is fairly stable and is not believed to be capable of initiating oxidative damage. Another antioxidant small molecule is vitamin C (ascorbic acid; ascorbate) which has a large number of documented antioxidant mechanisms, among which is the donation of an electron to vitamin E radical resulting in a regeneration of the reduced (active) form of vitamin E (Halliwell and Gutteridge, 2007).

A simplified summary of the chemistry of these small molecule and enzymatic antioxidant reactions and interactions is as follows:

(1) Vitamin E + LOO• → Vitamin E• + LOOH
(2) H_2O_2 + GSH $\xrightarrow{GSHPx}$ H_2O + GSSG or LOOH + GSH $\xrightarrow{PHGPx}$ LOH + GSSG
(3) GSSG + NADPH $\xrightarrow{GSHRx}$ 2GSH + $NADP^+$
(4) Vitamin E• + Vitamin → C Vitamin E + Vitamin C•

Several CNS injury studies have made use of this chemistry to indirectly measure posttraumatic LP in SCI or TBI models in terms of the depletion of the reduced forms of vitamin E, vitamin C, or GSH. Such work has clearly indicated that after severe degrees of CNS injury and intense ROS generation, the endogenous antioxidant defenses are largely overwhelmed. On the contrary, pharmacological antioxidant efficacy has been measured in terms of a preservation of the tissue levels of the reduced forms of these antioxidants. For a more detailed discussion of the complexities of ROS and free radical formation, oxidative damage, and antioxidant defense mechanisms, the reader is directed to the very fine monograph by Halliwell and Gutteridge (2007).

2.7 Peroxynitrite-Mediated Oxygen Radical Formation and Oxidative Damage

Another ROS and mechanism of hydroxyl radical formation in addition to that produced by the iron-dependent Fenton reaction has been identified and is becoming increasingly appreciated as a critical player in posttraumatic pathophysiology and neurodegeneration, peroxynitrite (PN) (Beckman et al., 1990; Radi et al., 1991). It is known that many types of neurons, endothelial cells, neutrophils, macrophages, and microglia can produce two radicals, superoxide and nitric oxide (•NO), from nitric oxide synthase (NOS). The two species can combine to produce PN in its anionic form ($ONOO^-$). The rate constant for the reaction of $O_2^{•-}$ and •NO is faster than that for the reactivity of $O_2^{•-}$ with SOD, thus favoring the formation of $ONOO^-$. In the physiological pH range, $ONOO^-$ will largely undergo protonation (pKa = 6.8) to become peroxynitrous acid (ONOOH). However, ONOOH is an unstable acid that can readily decompose to give hydroxyl radical and nitrogen dioxide (•NO_2).

$$O_2^{•-} + •NO \rightarrow ONOO^- + H^+ \rightarrow ONOOH \rightarrow •NO_2 + •OH$$

Via this mechanism, another source of the highly reactive •OH may be operative within the injured CNS. Moreover, •NO_2 may similarly initiate lipid peroxidation or nitrate and inactivate cellular proteins. A particularly attractive aspect of this scenario is that peroxynitrite has a relatively long half-life and thus is potentially more diffusible compared to either superoxide or •OH. Therefore, it may offer a mechanism by which free radical damage may occur at a site remote from the actual location of oxygen radical formation (Beckman, 1991).

Probably more important physiologically, PN will react with carbon dioxide (CO_2) to form nitrosoperoxocarbonate ($ONOOCO_2^-$) which can decompose into •NO_2 and carbonate radical (•CO_3) (Squadrito and Pryor, 1998; Squadrito and Pryor, 2002).

$$ONOO^- + CO_2 \rightarrow ONOOCO_2^- \rightarrow •NO_2 + •CO_3$$

Each of the PN-derived radicals (•OH, •NO_2, and •CO_3) can initiate LP cellular damage by abstraction of an electron from a hydrogen atom bound to an allylic carbon in polyunsaturated fatty acids or cause protein carbonylation by reaction with susceptible amino acids (e.g., lysine, cysteine, arginine). Additionally, •NO$_2$ can nitrate the 3 position of tyrosine residues in proteins; 3-nitrotyrosine (3-NT) is a specific footprint of PN-induced cellular damage. Collectively, these oxidative mechanisms underlie the demonstrated neurotoxic effects of PN reported in neuronal cell culture models (Kruman et al., 1997; Neely et al., 1999). An increase in 3-NT and 4-HNE (LP) has been shown in the injured spinal cord (Xu et al., 2001a; Bao and Liu, 2002; Bao et al., 2003; Liu et al., 2005; Xiong and Rabchevsky, 2007) and brain (Mesenge et al., 1998; Hall et al., 2004). ❯ *Figure 10-1* summarizes the oxidative damage mechanisms that are generated by PN and other ROS.

◘ **Figure 10-1**
Overview of oxidative damage mechanisms caused by reactive oxygen species (ROS) and reactive nitrogen species (RNS)

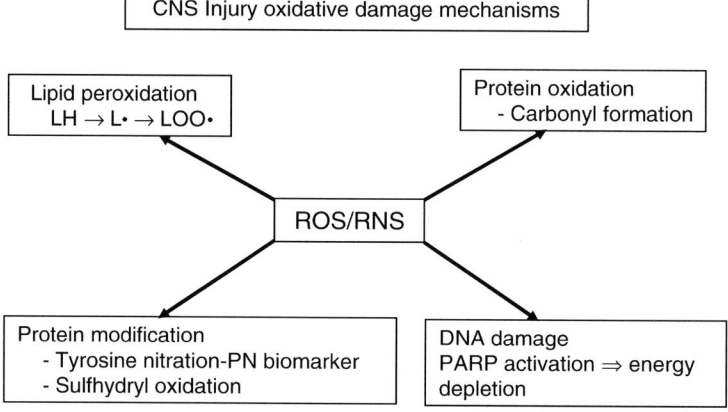

3 Interaction of Oxidative Damage with Other Secondary Injury Mechanisms

The secondary injury cascade following TBI or SCI is a complex process of interwoven molecular and pathophysiological processes that collectively lead to microvascular, glial, and neuronal degeneration. In that context, ROS and oxidative damage mechanisms have been linked to a number of these processes. It is obviously important to understand the relationship of oxidative damage with other secondary injury mechanisms.

3.1 Role in the Disruption of Ion Homeostasis

One of the best established aspects of posttraumatic pathophysiology concerns the occurrence of disruptions in neuronal ionic physiology related to Na$^+$, K$^+$, and Ca^{++}. Although the initial changes in ion distributions are triggered by mechanical depolarization, opening of voltage-dependent ion channels, and release of excitatory neurotransmitters (e.g., glutamate and aspartate), oxidative membrane damage has been linked to exacerbations of ion homeostatic dysfunction. One aspect of this dysfunction concerns the sensitivity of the plasma and endoplasmic reticular membrane Ca^{++} ATPase (i.e., Ca^{++} pump) to peroxidation-induced damage. Inhibition of this enzyme interferes with Ca^{++} extrusion. Similarly, oxidative inactivation of the membrane Na$^+$/K$^+$-ATPase can lead to intracellular Na$^+$ accumulation which will then reverse the direction of the Na$^+$/Ca^{++} exchanger (antiporter) and further exacerbate intracellular Ca^{++} accumulation (Rohn et al., 1993a; Rohn et al., 1993b; Rohn et al., 1995; Rohn et al., 1996).

3.2 Role in Excitotoxicity

In addition to the association of lipid peroxidation with the loss of Ca^{++} there appears to be an intimate reciprocal association between the excitotoxic and oxygen radical-lipid peroxidative mechanisms of neuronal degeneration. First of all, free radical mechanisms have been demonstrated to potentiate glutamate release. For instance, when brain slices are exposed to the O_2^- -generating system xanthine plus xanthine oxidase, there is an enhancement in glutamate and aspartate release which is antagonized by various free radical scavengers (Pellegrini-Giampietro et al., 1990). Moreover, peroxidation inhibitors have been shown to attenuate NMDA-induced damage in cortical cell cultures, implying the involvement of lipid peroxidation in glutamate excitotoxicity (Monyer et al., 1990). A role of LP in neuronal dysfunction has also been demonstrated in synaptosomes from the injured hemisphere where an increase in LP products occurs coincidently with an impairment of glutamate and glucose uptake (Sullivan et al., 1998). Thus, these and other examples have disclosed that there is a clear association between glutamate excitotoxicity, calcium overload, and free radical formation and membrane oxidative damage. Indeed, each of these processes can enhance the others as again shown in ❯ Figure 10-2.

◘ Figure 10-2
Interplay of excitotoxic, intracellular calcium overload, mitochondrial dysfunction, and oxidative damage mechanisms in secondary injury to the brain or spinal cord

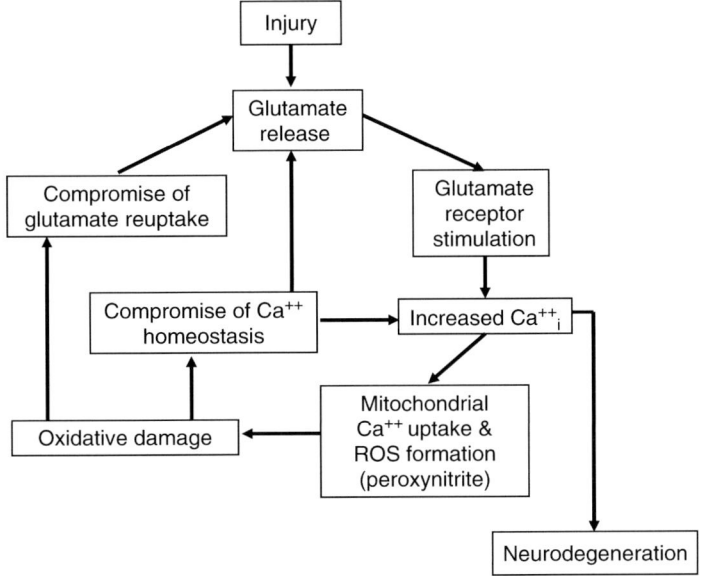

3.3 Role in Posttraumatic Mitochondrial Dysfunction

Mitochondrial dysfunction plays a key role in the posttraumatic death cascade (Finkel, 2001; Hunot and Flavell, 2001). It is clear that this is directly related to Ca^{++} ions that alter mitochondrial function and increase ROS production (Mattson et al., 1995; Wang and Thayer, 1996; Kristal and Dubinsky, 1997; White and Reynolds, 1997; Verweij et al., 1997; Stout et al., 1998; Nicholls and Budd, 2000; Ward et al., 2000; Rego et al., 2001). Following CNS injury, loss of mitochondrial homeostasis, increased mitochondrial ROS production, and disruption of synaptic homeostasis have been shown to occur (Azbill et al., 1997; Matsushita and Xiong, 1997; Sullivan et al., 1999a; Sullivan et al., 1999c), implicating a pivotal role for mitochondrial dysfunction in the neuropathological sequelae of SCI and TBI. Importantly, several reports have solidified this theory by demonstrating that therapeutic intervention with cyclosporin A following experimental TBI significantly reduces mitochondrial dysfunction (Sullivan et al., 1999b) and cortical damage (Scheff and Sullivan, 1999; Sullivan et al., 2000b; Sullivan et al., 2000c), as well as cytoskeletal changes and axonal dysfunction (Okonkwo et al., 1999; Okonkwo and Povlishock, 1999). Furthermore, maintaining mitochondrial bioenergetics by dietary supplementation with creatine has also proved effective in ameliorating neuronal cell death by reducing mitochondrial ROS production and maintaining ATP levels following TBI (Sullivan et al., 2000a).

Recently, evidence has begun to accumulate which shows that the particular ROS that is being formed by mitochondria is PN. Nitric oxide has been shown to be present in mitochondria (Lopez-Figueroa et al., 2000; Zanella et al., 2002), and a mitochondrial NOS isoform (mtNOS) has been isolated, Although probably playing a physiological role in mitochondria, dysregulation of mitochondrial •NO generation, and the aberrant production of its toxic metabolite PON, appear to play a role in many, if not all, of the major acute and chronic neurodegenerative conditions (Heales et al., 1999). Exposure of mitochondria to Ca^{++}, which is known to cause them to become dysfunctional, leads to PN generation which in turn triggers mitochondrial Ca^{++} release (i.e., limits their Ca^{++} uptake or buffering capacity) (Bringold et al., 2000). Both PN forms, $ONOO^-$ and $ONOOCO_2$, have been shown to deplete mitochondrial antioxidant stores and to cause protein nitration (Valdez et al., 2000). The relatively long half-life of PN in comparison to most other short-lived ROS also allows for mtNOS-derived PN to diffuse from one cell to another. Accordingly, coculture studies have shown that astrocyte-derived •NO (probably due to PN formation) can bring about damage to neuronal mitochondrial respiratory complexes II, III, and IV (Stewart et al., 2000; Stewart et al., 2002).

❯ *Figure 10-3* illustrates the critical aspects of posttraumatic PN-mediated mitochondrial oxidative damage and dysfunction after CNS injury. As shown, $O_2^{\cdot-}$ radical production is a byproduct of the mitochondrial electron transport chain during ATP generation. Electrons escape from the chain and reduce O_2 to $O_2^{\cdot-}$. Normally, cells convert $O_2^{\cdot-}$ to H_2O_2 utilizing manganese superoxide dismutase (MnSOD), which is also localized to the mitochondria. However, if pathophysiological insults such as mechanical trauma trigger an increase in intracellular Ca^{++}, causing an increase in mtNOS activity and •NO liberation, PN formation is a certainty since the rate constant for reaction of •NO with O_2^- greatly exceeds the rate constant for dismutation of O_2^- by MnSOD (Beckman, 1991). The PN can then damage mitochondria by tyrosine nitration and by causing LP and the production of 4-HNE that conjugates to mitochondrial membrane proteins, impairing their function (Keller et al., 1997a; Keller et al., 1997b; Mark et al., 1997; Sullivan et al., 1998). Such oxidative injury results in significant alterations in mitochondrial function.

Increased 3-NT has been found in injured brain (Mesenge et al., 1998; Hall et al., 2004) and spinal cord (Bao and Liu, 2003) indicative of a role of PON in posttraumatic pathophysiology and neurodegeneration. We have recently shown that posttraumatic mitochondrial failure in the injured brain (Singh et al., 2006) and spinal cord (Xiong et al., 2007) is paralleled by a coincident increase in mitochondrial oxidative damage markers including 3-NT, specific for PN the LP marker 4-HNE and protein carbonyl.

3.4 Role in Microvascular Dysfunction: Loss of Autoregulation and Blood–Brain (Spinal Cord) Compromise

There is also compelling experimental support for an important role of ROS in the microvascular pathophysiology of acute TBI and SCI. It was first demonstrated that there is an almost immediate postinjury increase in brain microvascular $O_2^{\cdot-}$ production together with a compromise of autoregulatory function in fluid percussion TBI models (Kontos and Povlishock, 1986; Kontos and Wei, 1986). Scavengers of $O_2^{\cdot-}$ reduce the posttraumatic $O_2^{\cdot-}$ levels and protect against the loss of autoregulatory responsiveness. Other investigators using the salicylate trapping method have observed a rapid rise in brain hydroxyl radical (•OH) levels in a mouse diffuse and a rat focal TBI model (Hall and Braughler, 1993; Hall et al., 1994); (Hall and Braughler, 1993; Smith et al., 1994; Globus et al., 1995). Here again, the CNS microvasculature appears to be the initial source of posttraumatic radical production (Hall and Braughler, 1993; Hall et al., 1994). Elegant microdialysis studies using salicylate trapping have also demonstrated an increase in hydroxyl radical levels in the injured rat spinal cord during the first minutes after contusion injury (Liu et al., 2001).

Figure 10-3

Schematic diagram of mitochondrial organization and the electron transport chain showing the site of formation of superoxide (O_2^-) and nitric oxide (•NO). These two radicals interact to form peroxynitrite anion (ONOO$^-$). The rate constant for this reaction is faster than the rate constant for the dismutation of superoxide by the mitochondrial manganese superoxide dismutase (MnSOD). Thus, the formation of peroxynitrite is favored which can lead to the formation of highly reactive free radicals (•OH, •NO$_2$, •CO$_3$) which can cause oxidative damage to the inner mitochondrial membrane (IMM) and the complexes involved in ATP production. OMM = outer mitochondrial membrane.

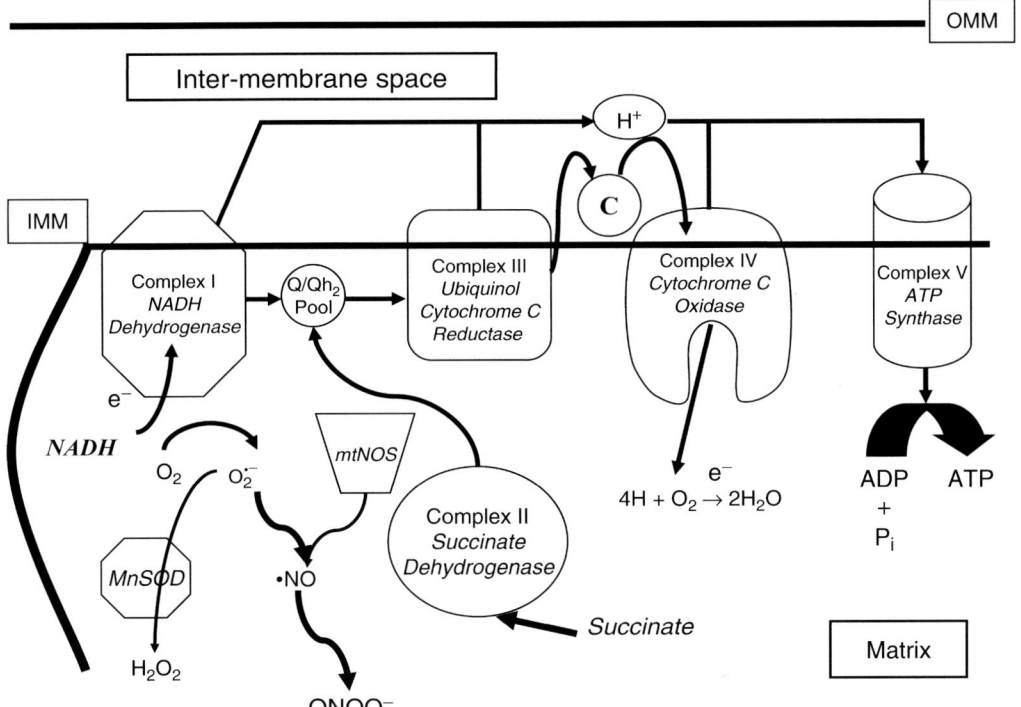

The most studied mechanism of oxidative damage in models of TBI concerns ROS-induced LP. Work using the rat focal contusion injury model, has shown that there is an increase in brain lipid hydroperoxide levels that is measurable within 30 min postinjury, following closely behind the increase in •OH (Smith et al., 1994). Moreover, on the heels of the increase in LP markers, there is an opening of the blood–brain barrier (BBB) suggesting that the initial site of ROS-induced LP is the microvascular endothelium. Others have confirmed the posttraumatic increase in LP products in rats after focal contusion injury and its association with brain edema mechanisms (Nishio et al., 1997). Related to the posttraumatic microvascular damage is the pathophysiological process of vasogenic brain edema that represents a disruption of blood–brain barrier integrity resulting in brain parenchymal sodium and protein accumulation and osmotic fluid expansion of the brain extracellular space. Clinically, this is reflected by an increase in intracranial pressure, which can cause secondary compressive injury to vital brain structures.

Biochemical indices of early oxygen radical reactions in the bluntly injured spinal cord (contusion or compression injuries) have also been repeatedly demonstrated using a variety of analytical techniques and appear to be temporally linked to the occurrence of pathophysiological changes in spinal cord blood flow (SCBF) (Young and Flamm, 1982; Hall et al., 1984; Hall and Wolf, 1986; Hall et al., 1992; Hall et al., 1995). Evidence that oxidative damage is causally related to the injury-induced hypoperfusion comes from the fact that pretreatment of animals prior to SCI with high doses of the antioxidants vitamin E or ascorbic acid prevents posttraumatic decreases in SCBF (Hall and Wolf, 1986; Hall et al., 1992). Similarly, acute post-injury treatment with various pharmacological antioxidant compounds (see below) has been shown to prevent decreases in SCBF in the injured spinal cord and to improve spinal tissue energy metabolism (Young and Flamm, 1982; Hall et al., 1984; Hall and Wolf, 1986; Hall et al., 1995).

4 Pharmacological Antioxidants That Have Been Explored in CNS Injury Models

The most convincing evidence that supports the role of ROS, free radicals, and oxidative damage in the secondary pathophysiology of SCI and TBI is derived from studies showing the protective effects of various pharmacological antioxidants in models of CNS injury. A large number of such compounds have been demonstrated to be neuroprotective and/or to improve neurological recovery in experimental models. Some of this literature will be discussed below. However, ❯ Figure 10-4 displays several examples of pharmacological lipid LP inhibitors, pharmacological antioxidants that have specifically been shown to have efficacy in SCI and TBI models and in clinical trials in the case of some.

◘ Figure 10-4

Chemical structures of lipid peroxidation inhibitors shown to be neuroprotective in CNS injury models. The glucocorticoid steroid methylprednisolone and its non-glucocorticoid analog U72099 are not true chemical antioxidants, but rather act via a membrane stabilizing (decreased phospholipid fluidity) action which limits the propagation of LP reactions (Hall, 1992; Hall et al., 1992). The 21-aminosteroid tirilazad works by a combination of membrane stabilizing and lipid peroxyl radical (LOO•) scavenging (Hall et al., 1994; Hall, 1997). Vitamin E scavenges LOO• by donation of an electron from the hydroxyl functional group on the phenolic chroman ring structure. The long aliphatic tail is involved in membrane localization. U78517F is a dual LOO• scavenger that couples the antioxidant ring structure of vitamin E and the antioxidant amino group of tirilazad. U101033 is a pyrrolopyrimidine compound that scavenges LOO• by formation of a radical cation mechanism (Hall et al., 1997)

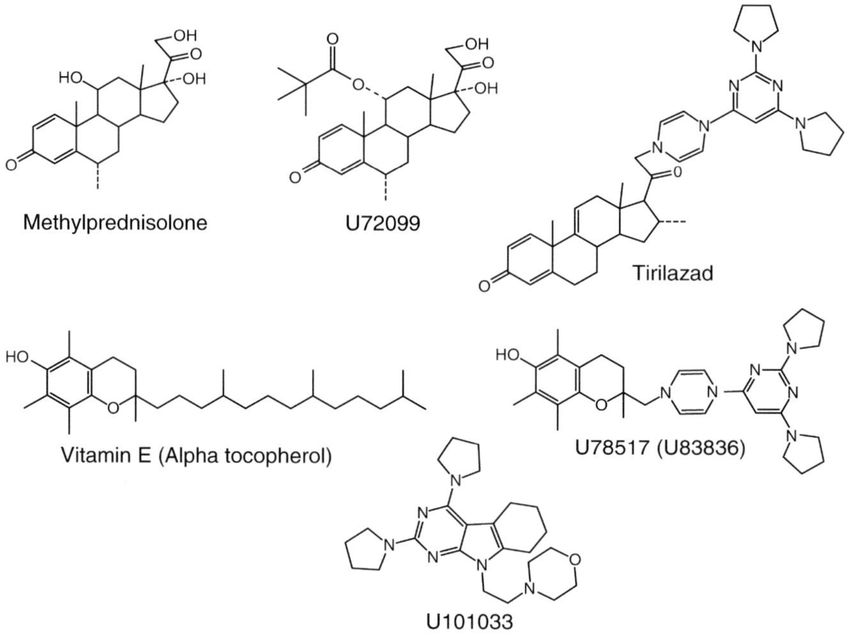

4.1 Protective Effects of Antioxidants in Spinal Cord Injury

4.1.1 High-Dose Methylprednisolone Inhibition of Lipid Peroxidation

Increasing knowledge of the posttraumatic LP mechanism in the 1970s and early 1980s prompted the search for a neuroprotective pharmacologic strategy aimed at antagonizing oxygen radical-induced LP in a safe and effective manner. Attention was focused on the hypothetical possibility that glucocorticoid steroids might be effective inhibitors of posttraumatic LP based upon their high lipid solubility and known ability to intercalate into artificial membranes between the hydrophobic polyunsaturated fatty acids of the membrane phospholipids and to thereby limit the propagation of LP chain reactions throughout the phospholipid bilayer (Demopoulos et al., 1980; Hall and Braughler, 1981; Hall and Braughler, 1982; Hall, 1992).

In experiments in spinal cord-injured cats, it was observed that the administration of an i.v. bolus of the glucocorticoid steroid methylprednisolone (MP) could indeed inhibit posttraumatic LP in spinal cord tissue (Hall and Braughler, 1981), but that the doses required for this effect were much higher (30 mg/kg) than those empirically employed in the clinical treatment of acute CNS injury up to that time. Further experimental studies, also conducted in cat SCI models, showed that the 30-mg/kg dose of MP not only prevented LP, but in parallel inhibited posttraumatic spinal cord ischemia (Young and Flamm, 1982; Hall et al., 1984), supported aerobic energy metabolism (i.e., reduced lactate and improved ATP and energy charge) (Anderson et al., 1982; Braughler and Hall, 1983a; Braughler and Hall, 1984), improved recovery of extracellular calcium (i.e., reduced intracellular overload) (Young and Flamm, 1982), and attenuated calpain-mediated neurofilament loss (Braughler and Hall, 1984). However, the central effect in this protective scenario is the inhibition of posttraumatic LP. With many of these therapeutic parameters (LP, secondary ischemia, aerobic energy metabolism), the dose-response for MP follows a sharp U-shaped pattern. The neuro- and vasoprotective effect is partial with a dose of 15 mg/kg, it is optimal at 30 mg/kg, and diminishes at higher doses (60 mg/kg) (Hall, 1992).

The antioxidant neuroprotective action of MP is closely linked to the drug's tissue pharmacokinetics (Braughler and Hall, 1982; Braughler and Hall, 1983a; Braughler and Hall, 1983b; Hall, 1992). For instance, when MP tissue levels are at their peak following administration of a 30 mg/kg IV dose, lactate levels in the injured cord are suppressed. When tissue MP levels decline, spinal tissue lactate rises. However, the administration of a second dose (15 mg/kg IV) at the point at which the levels after the first dose have declined by 50%, acts to maintain the suppression of lactate seen at the peak of the first dose and to more effectively maintain ATP generation and energy charge and protect spinal cord neurofilaments from degradation (Braughler and Hall, 1983a; Braughler and Hall, 1984). This prompted the hypothesis that prolonged MP therapy might better suppress the secondary injury process and lead to better outcomes compared to the effects of a single large IV dose. Indeed, subsequent experiments in a cat spinal injury model demonstrated that animals treated with MP using a 48-h antioxidant dosing regimen had improved recovery of motor function over a 4-week period (Anderson et al., 1985; Braughler et al., 1987).

It should be pointed out that all of the original preclinical studies defining the antioxidant neuroprotective pharmacology showing that high-dose MP could exert antioxidant and related neuroprotective effects were conducted in cat models of blunt (nontransecting) SCI that were the standard in the experimental SCI field prior to the 1990s. Since that time, rat contusion and compression models have become the standard, and several investigators have tested the ability of high-dose MP (usually 30 mg/kg i.v. as a starting dose) to lessen posttraumatic pathophysiology and neurodegeneration and/or to improve neurological recovery. Several of these studies have replicated in some fashion, or another, the neuroprotective properties of MP in the injured rat spinal cord. Specifically, high-dose MP has been reported in rat SCI models to attenuate posttraumatic LP (Taoka et al., 2001), decrease lactate accumulation (Farooque et al., 1996), prevent hypoperfusion (Holtz et al., 1990), attenuate vascular permeability (Xu et al., 1992), decrease inflammatory markers (Xu et al., 2001b), and improve neurological recovery (Holtz et al., 1990; Behrmann et al., 1994) reminiscent of similar effects shown earlier in cat models (Hall and Braughler, 1982; Hall, 1992). In contrast, failures of high-dose MP to improve neurological recovery in rat SCI models have also appeared in the literature (Koyanagi and Tator, 1997; Rabchevsky et al., 2002). However, of considerable concern in the extrapolation of the cat MP dosing parameters to the rat is the lack of any definition thus far of the relative pharmacokinetics of MP in rats. This is in striking contrast to the documentation of the uptake and elimination of MP from the cat spinal cord and correlation of plasma and spinal tissue levels with the neuroprotective actions (Braughler and Hall, 1982; Braughler and Hall, 1983a; Braughler and Hall, 1983b; Hall, 1992). The likelihood that the precise dose–response relationship and requirements for repeated dosing defined in cats is also optimal for the rat is exceedingly small. Thus, the interpretation of rat SCI studies with MP, whether positive or negative, is difficult without the necessary pharmacokinetic correlation.

4.1.2 NASCIS II Clinical Trial of High-Dose Methylprednisolone

The above-reviewed experimental studies with high-dose MP inspired the second National Acute Spinal Cord Injury Study (NASCIS II) (Bracken et al., 1990) even though the earlier NASCIS trial, which came to be known as NASCIS I, had failed to show any efficacy of lower MP doses even when administered over a 10-day period (Bracken et al., 1984; Bracken et al., 1985). The NASCIS II trial compared 24 h of dosing with MP versus placebo for the treatment of acute SCI. A priori trial hypotheses included the prediction that SCI patients treated within the first 8 postinjury h would respond better to pharmacotherapy than patients treated after 8 h. Indeed, the results demonstrated the effectiveness of 24 hours of intensive MP dosing (30 mg/kg IV bolus plus a 23-hour infusion at 5.4 mg/kg per hour) when treatment was initiated within 8 hours. Significant benefit was observed in individuals with both neurologically complete (i.e., "plegic") and incomplete (i.e., "paretic") injuries. Moreover, the functional benefits were sustained at 6-week, 6-month, and 1-year follow-up (Bracken et al., 1990; Bracken et al., 1992; Bracken, 1993; Bracken and Holford, 1993). The high-dose regimen actually improved function below the level of the injury and lowered the level of the functional injury (Bracken and Holford, 1993). Although predictable side effects of steroid therapy were noted, including GI bleeding, wound infections, and delayed healing, these were not significantly more frequent than those recorded in placebo-treated patients (Bracken et al., 1990). Another finding was the fact that delay in the initiation of MP treatment until after 8 h is actually associated with decreased neurological recovery (Bracken and Holford, 1993). Thus, treatment within the 8 h window is beneficial whereas dosing after 8 h can be detrimental.

4.1.3 Non-Glucocorticoid Antioxidant 21-Aminosteroid Tirilazad

Methylprednisolone is a potent glucocorticoid that possesses a number of glucocorticoid receptor-mediated antiinflammatory actions. Despite the above-discussed role of antiinflammatory effects of MP in the injured spinal cord, the principal neuroprotective mechanism appears to be the inhibition of posttraumatic LP that is not mediated via glucocorticoid receptor-mediated activity (Braughler et al., 1988; Hall et al., 1994; Hall, 1997). This prompted speculation that modifying the steroid molecule to enhance the anti-LP effect, while eliminating the steroid's glucocorticoid effects, would result in more targeted antioxidant therapy devoid of the typical side effects of steroid therapy. This rationale led to the discovery of U72099 which is a non-glucocorticoid analog of MP which duplicated the LP inhibitory effects of MP. However, further efforts lead to the development of the 21-aminosteroids or "lazaroids" which greatly surpassed the antioxidant efficacy of either MP or U72099. One of these, tirilazad, was selected for development. ◗ *Figure 10-5* compares the structures of the glucocorticoid, MP, U72099 and the non-glucocorticoid 21-aminosteroid tirilazad. Further studies suggest that tirilazad retains its efficacy in promoting posttraumatic recovery after experimental spinal cord injury even when initiation of treatment is delayed to 4 hours postinjury (Anderson et al., 1991). However, while there is still some residual efficacy apparent when administration is withheld until 8 h postinjury, it is much less than when treatment is begun within the first 4 h. Thus, the opportunity for limiting the impact of posttraumatic lipid peroxidation appears to be 8 h.

● Figure 10-5
Chemical structures of different nitroxide "spin traps" that have been shown to be neuroprotective in CNS injury models

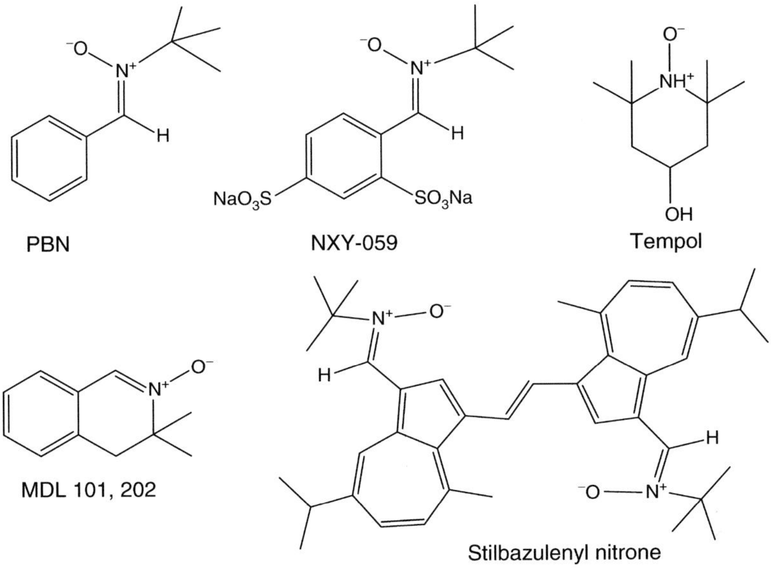

PBN

NXY-059

Tempol

MDL 101, 202

Stilbazulenyl nitrone

4.1.4 NASCIS III

The demonstrated efficacy of a 24-h dosing regimen of MP in human SCI in NASCIS II (Bracken et al., 1990), and the discovery of tirilazad (Braughler et al., 1988; Hall et al., 1994; Hall, 1997), led to the organization and conduct of NASCIS III (Bracken et al., 1997; Bracken et al., 1998). In the NASCIS III trial, three groups of patients were evaluated. The first (active control) group was treated with the 24-hour MP dosing regimen that had previously been shown to be effective in NASCIS II. The second group was also treated with MP, except that the duration of MP infusion was prolonged to 48 hours. The purpose was to determine whether extension of the MP infusion from 24 to 48 h resulted in greater improvement in neurological recovery in acute SCI patients. The third group of patients was treated with a single 30 mg/kg IV bolus of MP followed by the 48-hour administration of tirilazad. No placebo group was included because it was deemed ethically inappropriate to withhold at least the initial large bolus of MP. Another objective of the study was to ascertain whether treatment initiation within 3 hours following injury was more effective than when therapy was delayed until 3–8 h post-SCI.

Upon completion of the NASCIS III trial, it was found that all three treatment arms produced comparable degrees of recovery when treatment was begun within the shorter 3-h window. When the 24-h dosing of MP was begun more than 3 h post-SCI, recovery was poorer in comparison to the cohort treated within 3 h following SCI. However, in the 3-to-8 h post-SCI cohort, when MP dosing was extended to 48 h, significantly better recovery was observed than with the 24 h dosing. In the comparable tirilazad cohort (3–8 h post-SCI), recovery was slightly, but not significantly better than in the 24-hour MP group, and poorer than in the 48-hour MP group. These results showed that: (1) initiation of high-dose MP treatment within the first 3 h is optimal; (2) the non-glucocorticoid tirilazad is as effective as 24-hour MP therapy; and (3) if treatment is initiated more than 3 hours post-SCI, extension of the MP dosing regimen, from 24 hours to 48 hours is indicated. However, in comparison with the 24-hour dosing regimen, significantly more glucocorticoid-related immunosuppression-based side effects were seen with more prolonged dosing (i.e., the incidence of severe sepsis and pneumonia significantly increased). In contrast, tirilazad showed no evidence of steroid-related side effects suggesting that this non-glucocorticoid

21-aminosteroid would be safer for extension of dosing beyond the 48-hour limit used in NASCIS III (Bracken et al., 1997; Bracken et al., 1998). In any event, the beneficial and apparent neuroprotective effects of high-dose MP and tirilazad in human SCI arguably represents the best evidence of acute pharmacological neuroprotection to date and strongly validates the role of free radical-oxidative damage mechanisms in posttraumatic secondary CNS injury.

4.2 Protective Effects of Antioxidants in Traumatic Brain Injury

During the past 20 years, there has been an intense effort to discover and develop pharmacological agents for acute treatment of TBI. This has included multiple compounds that possess free radical scavenging/antioxidant properties including polyethylene glycol-conjugated superoxide dismutase (PEG-SOD), the LP inhibitor tirilazad (Marshall et al., 1998; Langham et al., 2000; Narayan et al., 2002), and more recently the mixed glutamate antagonist/antioxidant compound dexanabinol (Maas et al., 2006). However, these trials, as a whole, have been therapeutic failures in that no overall benefit has been documented in moderate and severe TBI patient populations. These failures can be attributed to several factors. Perhaps most importantly, the preclinical assessment of compounds destined for acute TBI trials has been woefully inadequate in regards to the definition of neuroprotective dose–response relationships, pharmacokinetic-pharmacodynamic correlations, therapeutic window, and optimum dosing regimen and treatment duration. However, a number of other issues related to design of the clinical trials is also believed to be involved (Narayan et al., 2002).

4.2.1 Effects of SOD in TBI

As mentioned earlier, the earliest studies of free radical scavenging compounds in TBI models were carried out with Cu/Zn SOD based upon the work of Kontos and colleagues who showed that posttraumatic microvascular dysfunction was initiated by $O_2^{\cdot-}$ generated as a by-product of the arachidonic acid cascade which is massively activated during the first minutes and hours after TBI (Kontos and Povlishock, 1986; Kontos and Wei, 1986; Kontos, 1989). Their work showed that administration of SOD prevented the posttraumatic microvascular dysfunction. This lead to clinical trials in which the more metabolically polyethylene glycol (PEG)-conjugated SOD was examined in moderate and severe TBI patients when administered within the first 8 hours after injury. Although an initial small phase II study showed a positive trend, subsequent multicenter phase III studies failed to show a significant benefit in terms of increased survival or improved neurological outcomes (Muizelaar et al., 1995). Although many explanations for these negative results may be postulated, one reason may be that a large protein like SOD is unlikely to have much brain penetrability and therefore its radical scavenging effects may be limited to the microvasculature. A second reason may be that attempting to scavenge the short-lived inorganic radical $O_2^{\cdot-}$ may be associated with a very short therapeutic window. Indeed, the time course of posttraumatic •OH formation in the injured rodent brain has been shown to largely run its course by the end of the first hour after TBI (Hall et al., 1993; Smith et al., 1994) A more rational strategy would be to inhibit the LP that is triggered by the initial burst of inorganic radicals. A comparison of the time course of LP with that of posttraumatic •OH shows that LP reactions continue to build beyond the first posttraumatic hour (Smith et al., 1994) and may continue for 3–4 days (Hall et al., 2004).

Despite the failure of PEG-SOD in human TBI, experimental studies have shown that transgenic mice that over-express Cu/Zn SOD are significantly protected against post-TBI pathophysiology and neurodegeneration (Chan et al., 1995; Mikawa et al., 1996; Lewen et al., 2000; Kim et al., 2002; Xiong et al., 2005). This fully supports the importance of posttraumatic $O_2^{\cdot-}$ in posttraumatic secondary injury, despite the fact that targeting this primordial radical is probably not the best antioxidant strategy for acute CNS injury compared to trying to stop the downstream LP process that is initiated by the early increase in $O_2^{\cdot-}$ and •OH.

4.2.2 Effects of Tirilazad in TBI

The protective efficacy of 21-aminosteroid LP inhibitor tirilazad has also been evaluated in multiple animal models of acute head injury. For instance, the compound significantly improves the early (1–4 h) neurological recovery and survival of mice subjected to severe concussive head injury (Hall et al., 1988). Tirilazad has similarly been shown to exert beneficial effects on motor recovery and survival in a rat model of moderately severe fluid percussion head injury (McIntosh et al., 1992). In a cat model of severe concussive head injury, tirilazad significantly reduced posttraumatic lactic acid accumulation in both the cerebral cortex and subcortical white matter indicative of a positive action on aerobic energy metabolism (Dimlich et al., 1990). The locus of this particular effect is unclear. It most likely involves a protection of neuronal mitochondrial function. While the compound is largely localized in the microvascular endothelium, the posttraumatic disruption of the BBB is known to allow the successful penetration of tirilazad into the brain parenchyma as noted earlier (Hall et al., 1992). Other mechanistic data derived from the rat controlled cortical impact and the mouse concussive head injury models has definitively shown that a major effect of tirilazad is to lessen posttraumatic BBB opening (Hall et al., 1992; Smith et al., 1994). In the mouse model, the BBB protection occurred together with an attenuation of the early posttraumatic rise in brain •OH levels measured via the salicylate trapping method (Hall et al., 1992; Smith et al., 1994). Thus, the effect of tirilazad to protect the BBB is paralleled by a reduced formation of •OH and/or a protection of the microvascular endothelium from •OH-induced LP.

Based upon the strength of this data, tirilazad was evaluated in two phase III multicenter clinical trials for its ability to improve neurological recovery in moderate and closed head injury patients, one in North America and the other in Europe. In both trials, TBI patients were treated within 4 hours after injury with vehicle or tirilazad (2.5 mg/kg i.v. q6h for 5 days). The North American trial was never published due to a major confounding imbalance in the randomization of the patients to placebo or tirilazad in regards to injury severity and pretreatment neurological status. In contrast, the European trial had much better randomization balance and has been published (Marshall et al., 1998). The results failed to show a significant beneficial effect of tirilazad in either moderate or severe patients. However, male TBI patients with traumatic subarachnoid hemorrhage (SAH) showed significantly less mortality after treatment with tirilazad (34%) compared to placebo (43%, $p < 0.026$) (Lyons et al., 1994; Marshall et al., 1998). This result is consistent with the fact that this compound is highly effective in animal models of SAH (Hall et al., 1994). However, additional trials would be required in order to establish the neuroprotective utility of tirilazad in the traumatic SAH subset of TBI patients.

4.2.3 Effects of More Potent, Brain-Penetrable Antioxidants in TBI

As noted in the preceding section, tirilazad largely localizes in the cerebral microvasculature and therefore much of its neuroprotective action is probably mediated by protection against secondary free radical-mediated vascular pathophysiology (e.g., loss of microvascular autoregulatory disturbances, BBB compromise, and delayed post-SAH cerebral vasospasm). Evidence for a direct neuroprotective action is weak. Similarly, the brain uptake of the highly lipophilic vitamin E is limited and several weeks of vitamin E supplementation have been shown to be required to elevated CNS tissue levels. Thus, efforts have been directed at the discovery of more brain-penetrable LP inhibitors that would be capable of protecting the neural parenchyma from posttraumatic oxidative damage mechanisms. Two classes of compounds have been discovered which are more potent inhibitors of LP in in vitro oxidative damage paradigms and in *in vivo* models of TBI. These are the 2-methylaminochromans which couple the hydrophilic chroman antioxidant ring structure of vitamin E with the antioxidant amino functionality of tirilazad. The prototype of this series is U78517 (❷ *Figure 10-4*). Since U78517 has an asymmetric carbon, it is actually a racemic mixture. The (−)enantiomer, U83836E, has been employed in in vivo studies including a mouse diffuse TBI model in which it more potently improved early neurological recovery compared to tirilazad. (Hall, 1997). Another series of LP inhibitors have been discovered, the pyrrolopyrimidines typified by

U101033 (▶ *Figure 10-4*), which has antioxidant potency similar to U78517, but even greater brain and cellular permeability (Hall et al., 1997).

5 Future Directions

5.1 Improved Antioxidant Drugs

The role of ROS generation and the resulting oxidative damage mechanisms in secondary CNS injury is arguably one of the more established aspects of secondary CNS injury process based upon the fact that a multitude of free radical scavengers and lipid peroxidation-inhibiting antioxidants have been shown to reduce posttraumatic damage in preclinical TBI and/or SCI models. Since the discovery and attempted neuroprotective development of PEG-SOD and tirilazad, discussed above, considerable interest shifted to a variety of very potent nitroxide-based "spin-trapping" agents (▶ *Figure 10-5*). The prototype of this antioxidant class is α-phenyl-N-tert-butylnitrone (PBN) which was shown to be protective in animal models of TBI (Marklund et al., 2001a; Marklund et al., 2001b; Marklund et al., 2001c; Marklund et al., 2002; Gahm et al., 2005). Other nitroxides shown in ▶ *Figure 10-5*, including tempol, the cyclic compound MDL 101,202, and the dual nitroxide compound stilbazulenyl nitrone (STAZN) and NXY-059 (disulfonyl analog of PBN) have also been shown to reduce postischemic brain damage in rodent stroke models (Ginsberg et al., 2003; Green et al., 2003). However, only tempol has been studied and shown to exert a beneficial effect in SCI (Hillard et al., 2004) and TBI models (Beit-Yannai et al., 1996; Zhang et al., 1998c).

5.1.1 Scavengers of Peroxynitrite or Peroxynitrite-Derived Radicals

Following the initiation of neuroprotective clinical trials with agents that were mainly discovered to deal with superoxide radical and iron-dependent LP, respectively, there has been increasing interest in targeting PN and/or the PN-derived radicals (•OH, •NO_2 and •CO_3) which can initiate lipid peroxidative (LP) and protein oxidative cellular damage. As noted earlier, increased 3-NT has been found in injured CNS tissue in models of TBI (Hall et al., 2004) and SCI (Bao and Liu, 2003), indicative of a role of PN in each of these acute insults. Compared to other short-lived ROS (e.g., O_2^-, H_2O_2, •OH), PN possesses a much longer half-life of $\sim$1 sec which enables it to diffuse across intracellular and cellular membranes. In addition, PN's relatively long lifespan makes it a more practical target for successful scavenging. ▶ *Figure 24-6* displays the chemical structures of several compounds reported to scavenge PN many of which have been shown to have neuroprotective properties in CNS injury animal models including uric acid (Scott et al., 2005), dimethylthiourea (DMTU) (Whiteman and Halliwell, 1997), the indoleamine melatonin (Blanchard et al., 2000), the pyrrolopyrimidine antioxidant U101033 (Rohn and Quinn, 1998), and penicillamine (Hall et al., 1999). Perhaps the two best characterized in regards to chemical scavenging mechanism are penicillamine and melatonin. Both of these compounds have been shown to be stoichiometric scavengers of peroxynitrite (Hall et al., 1999; Zhang et al., 1999). This means that one molecule of penicillamine or melatonin decomposes one molecule of PON, but in so doing each compound is destroyed in the process.

Although such an inefficient scavenging mechanism may be adequate to produce neuroprotective effects, a catalytic scavenging mechanism wherein one molecule of the antioxidant can decompose many molecules of the oxidant without being destroyed would constitute a more desirable PN scavenging pharmacophore. Such a compound is the nitroxide tempol, which is able to catalytically scavenge the PN-derived radical species •NO_2 and •CO_3 (Carroll et al., 2000) (▶ *Figure 10-7*). Other nitroxides (▶ *Figure 10-5*) may share this property although the structure–activity relationship for this potent and efficient radical scavenging property remains to be established. In the case of tempol, it has been shown to exert a beneficial effect in rodent TBI models (Beit-Yannai et al., 1996; Zhang et al., 1998b) that may be a manifestation of its ability to efficiently scavenge highly reactive PN-derived radicals.

Figure 10-6
Chemical structures of compound purported to scavenge PN. The chemical mechanisms for the reaction of the compound with PN have only been defined for penicillamine (Hall et al., 1999) and melatonin (Zhang et al., 1998a; Zhang et al., 1999; Blanchard et al., 2000)

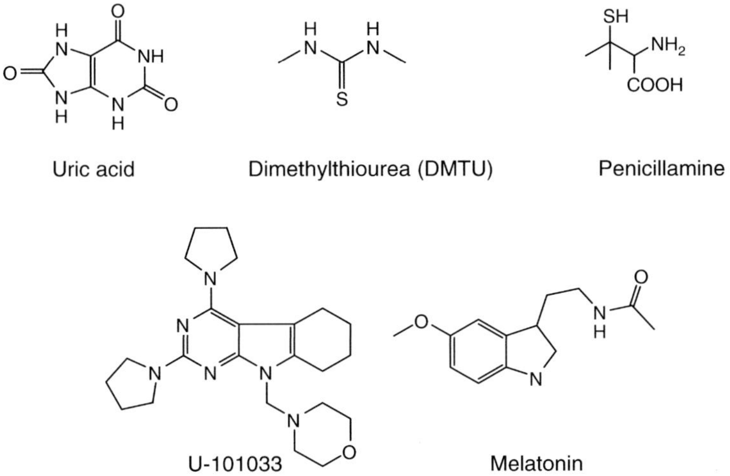

Uric acid   Dimethylthiourea (DMTU)   Penicillamine

U-101033   Melatonin

Figure 10-7
Chemical mechanisms for catalytic scavenging of the PN-derived radicals •NO_2, •CO_3, and •OH by the nitroxide compound tempol (Carroll et al., 2000)

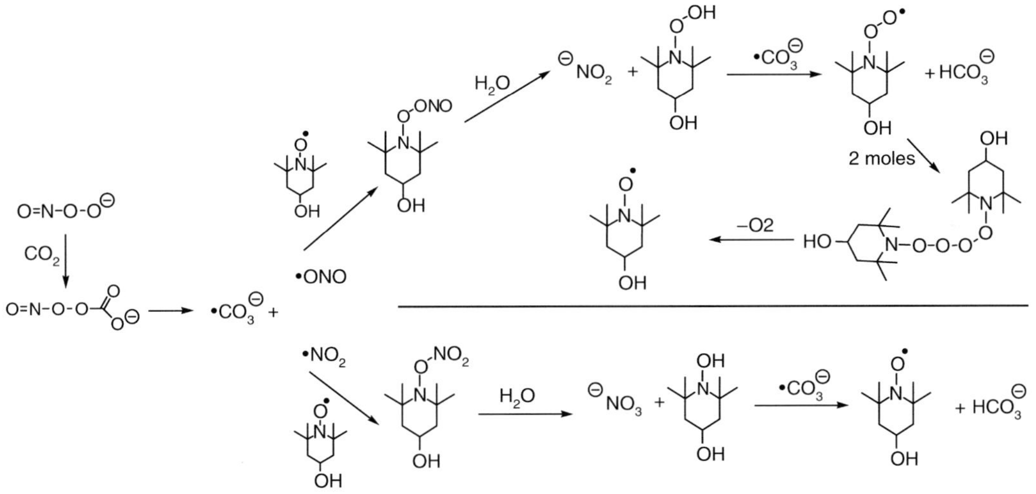

5.1.2 Dual Mechanism Antioxidants

Perhaps an even better antioxidant approach would be to combine two mechanistically additive or complementary antioxidant pharmacophores into a single compound such as a dual lipid peroxidation inhibitor or a lipid peroxidation inhibitor coupled to a moiety that will either inhibit •NO production or scavenge PON or PON-derived radicals. Two examples of dual antioxidant compounds are the

Figure 10-8
Chemical structures of dual mechanism antioxidants U78517 (Hall et al., 1991) and BN 80933 (Chabrier et al., 1999)

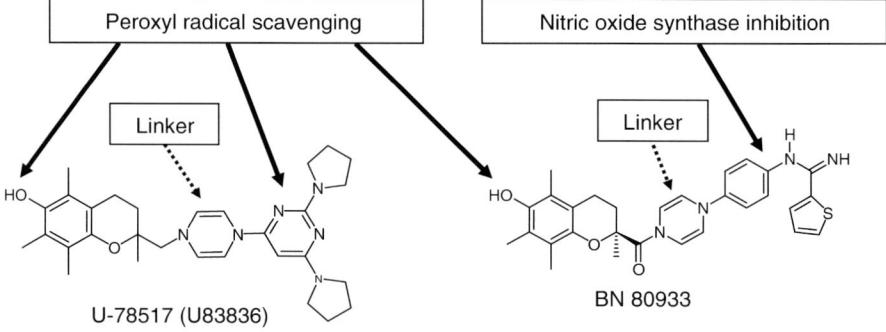

2-methylaminochroman U-78517 (Hall et al., 1991) and the compound BN-80933 (Chabrier et al., 1999) (❯ *Figure 10-8*). These two compounds link the peroxyl radical scavenging chroman ring structure of vitamin E (❯ *Figure 10-4*) via a piperazine bridge to the peroxyl radical scavenging amine moiety of the 21-aminosteroid tirilazad in the case of U-78517 or with an inhibitor nitric oxide synthase (NOS) in the case of BN-80933 which would decrease production of •NO and consequently reduce PON formation. U-78517F has been shown to be more effective than vitamin E or the vitamin E chroman in protecting culture spinal cord neurons from oxidative damage (Hall et al., 1991). Similarly, BN-80933 has been shown to be more neuroprotective than other single antioxidant mechanism compounds in rodent stroke and TBI models than either the trolox or NOS-inhibiting moieties alone (Chabrier et al., 1999).

5.2 Oxidative Damage Biomarkers

A key need to improve the translation of preclinical neuroprotective effects of antioxidant drugs into successful mechanism-based clinical trials is the identification of measurable biomarkers that allow the pathophysiology of ROS-mediated oxidative damage and drug effects thereon to be determined. The most promising of these are the LP-related isoprostanes (Pratico et al., 2002) and neuroprostanes (Morrow and Roberts, 2002). The validation of these for monitoring of posttraumatic injury is ongoing and should lead to a quantifiable means for the more efficient clinical evaluation of antioxidant pharmacotherapies in regards to neuroprotective efficacy, dose-response, therapeutic window, and optimal dosing regimen determinations.

6 Summary

This chapter has reviewed the current state of knowledge regarding the role of oxygen radical generation and lipid peroxidation in acute neurotrauma. Three criteria required to establish the pathophysiological importance of oxygen radical reactions have been met, at least in part. First of all, oxygen radical generation and lipid peroxidation are early biochemical events subsequent to either TBI or SCI. Secondly, they are linked to pathophysiological processes including loss of membrane ion homeostatic mechanisms, disturbances in microvascular autoregulation with resulting cerebral hypoperfusion, SAH-induced vasospasm, vasogenic edema, and failure of mitochondrial energy metabolism. Third, and most convincing, is the repeated observation that compounds that inhibit lipid peroxidation or scavenge oxygen radicals can block posttraumatic pathophysiology and promote functional recovery and survival in experimental studies.

The relative importance of oxygen radicals and oxidative damage in secondary CNS injury ultimately depends upon whether it can be convincingly demonstrated that early application of effective antioxidant can promote survival and neurological recovery after CNS injury in humans. The results of the NASCIS II clinical trial, which have shown that an antioxidant dosing regimen with methylprednisolone begun within 8 hours after spinal cord injury, can enhance chronic neurological recovery strongly supports the significance of lipid peroxidation as a posttraumatic degenerative mechanism at least in the context of the injured spinal cord. However, in the case of TBI, no compound with antioxidant properties tested in phase III trials has succeeded in producing a statistically significant improvement in survival or recovery including PEG-SOD (Muizelaar et al., 1995), tirilazad (Marshall et al., 1998), or dexanabinol (Maas et al., 2006). However, some optimism can be extracted from the finding that tirilazad has been shown to significantly improve survival in moderate or severe TBI patients whose pathophysiology includes traumatic SAH (Marshall et al., 1998). This suggests that oxygen radical mechanisms may be more important to inhibit in certain subsets of TBI patients who have hemorrhagic complications. Thus, further work is needed to better define the specific ROS and pathophysiological processes that involve oxidative damage mechanisms, to design improved antioxidant compounds that more effectively target those processes, and then design clinical trials that focus on subsets of TBI patients who are most likely to benefit from antioxidant neuroprotective pharmacotherapy.

References

Anderson DK, Means ED, Waters TR, Green ES. 1982. Microvascular perfusion and metabolism in injured spinal cord after methylprednisolone treatment. J Neurosurg 56(1): 106-113.

Anderson DK, Hall ED, Braughler JM, McCall JM, Means ED. 1991. Effect of delayed administration of U74006F (tirilazad mesylate) on recovery of locomotor function after experimental spinal cord injury. J Neurotrauma 8(3): 187-192.

Anderson DK, Saunders RD, Demediuk P, Dugan LL, Braughler JM, et al. 1985. Lipid hydrolysis and peroxidation in injured spinal cord: Partial protection with methylprednisolone or vitamin E and selenium. Cent Nerv Syst Trauma 2(4): 257-267.

Azbill RD, Mu X, Bruce-Keller AJ, Mattson MP, Springer JE. 1997. Impaired mitochondrial function, oxidative stress and altered antioxidant enzyme activities following traumatic spinal cord injury. Brain Res 765(2): 283-290.

Bao F, Liu D. 2002. Peroxynitrite generated in the rat spinal cord induces neuron death and neurological deficits. Neuroscience 115(3): 839-849.

Bao F, Liu D. 2003. Peroxynitrite generated in the rat spinal cord induces apoptotic cell death and activates caspase-3. Neuroscience 116(1): 59-70.

Bao F, De Witt DS, Prough DS, Liu D. 2003. Peroxynitrite generated in the rat spinal cord induces oxidation and nitration of proteins: Reduction by Mn (III) tetrakis (4-benzoic acid) porphyrin. J Neurosci Res 71(2): 220-227.

Beckman JS. 1991. The double-edged role of nitric oxide in brain function and superoxide-mediated injury. J Dev Physiol 15(1): 53-59.

Beckman JS, Beckman TW, Chen J, Marshall PA, Freeman BA. 1990. Apparent hydroxyl radical production by peroxynitrite: Implications for endothelial injury from nitric oxide and superoxide. Proc Natl Acad Sci USA 87(4): 1620-1624.

Behrmann DL, Bresnahan JC, Beattie MS. 1994. Modeling of acute spinal cord injury in the rat: Neuroprotection and enhanced recovery with methylprednisolone, U-74006F and YM-14673. Exp Neurol 126(1): 61-75.

Beit-Yannai E, Zhang R, Trembovler V, Samuni A, Shohami E. 1996. Cerebroprotective effect of stable nitroxide radicals in closed head injury in the rat. Brain Res 717(1–2): 22-28.

Blanchard B, Pompon D, Ducrocq C. 2000. Nitrosation of melatonin by nitric oxide and peroxynitrite. J Pineal Res 29(3): 184-192.

Bracken MB. 1993. Pharmacological treatment of acute spinal cord injury: Current status and future projects. J Emerg Med 11 Suppl 1: 43-48.

Bracken MB, Holford TR. 1993. Effects of timing of methylprednisolone or naloxone administration on recovery of segmental and long-tract neurological function in NASCIS 2. J Neurosurg 79(4): 500-507.

Bracken MB, Collins WF, Freeman DF, Shepard MJ, Wagner FW, et al. 1984. Efficacy of methylprednisolone in acute spinal cord injury. Jama 251(1): 45-52.

Bracken MB, Shepard MJ, Collins WF, Jr., Holford TR, Baskin DS, et al. 1992. Methylprednisolone or naloxone treatment after acute spinal cord injury: 1-year follow-up data. Results of the second national acute spinal cord injury study. J Neurosurg 76(1): 23-31.

Bracken MB, Shepard MJ, Collins WF, Holford TR, Young W, et al. 1990. A randomized, controlled trial of methylprednisolone or naloxone in the treatment of acute spinal-cord injury. Results of the second national acute spinal cord injury study. N Engl J Med 322(20): 1405-1411.

Bracken MB, Shepard MJ, Hellenbrand KG, Collins WF, Leo LS, et al. 1985. Methylprednisolone and neurological function 1 year after spinal cord injury. Results of the national acute spinal cord injury study. J Neurosurg 63(5): 704-713.

Bracken MB, Shepard MJ, Holford TR, Leo-Summers L, Aldrich EF, et al. 1997. Administration of methylprednisolone for 24 or 48 hours or tirilazad mesylate for 48 hours in the treatment of acute spinal cord injury. Results of the third national acute spinal cord injury randomized controlled trial. National acute spinal cord injury study. Jama 277(20): 1597-1604.

Bracken MB, Shepard MJ, Holford TR, Leo-Summers L, Aldrich EF, et al. 1998. Methylprednisolone or tirilazad mesylate administration after acute spinal cord injury: 1-year follow up. Results of the third national acute spinal cord injury randomized controlled trial. J Neurosurg 89(5): 699-706.

Braughler JM, Hall ED. 1982. Correlation of methylprednisolone levels in cat spinal cord with its effects on (Na++K+)-ATPase, lipid peroxidation, and alpha motor neuron function. J Neurosurg 56(6): 838-844.

Braughler JM, Hall ED. 1983a. Lactate and pyruvate metabolism in injured cat spinal cord before and after a single large intravenous dose of methylprednisolone. J Neurosurg 59(2): 256-261.

Braughler JM, Hall ED. 1983b. Uptake and elimination of methylprednisolone from contused cat spinal cord following intravenous injection of the sodium succinate ester. J Neurosurg 58(4): 538-542.

Braughler JM, Hall ED. 1984. Effects of multi-dose methylprednisolone sodium succinate administration on injured cat spinal cord neurofilament degradation and energy metabolism. J Neurosurg 61(2): 290-295.

Braughler JM, Hall ED, Means ED, Waters TR, Anderson DK. 1987. Evaluation of an intensive methylprednisolone sodium succinate dosing regimen in experimental spinal cord injury. J Neurosurg 67(1): 102-105.

Braughler JM, Chase RL, Neff GL, Yonkers PA, Day JS, et al. 1988. A new 21-aminosteroid antioxidant lacking glucocorticoid activity stimulates adrenocorticotropin secretion and blocks arachidonic acid release from mouse pituitary tumor (AtT-20) cells. J Pharmacol Exp Ther 244(2): 423-427.

Bringold U, Ghafourifar P, Richter C. 2000. Peroxynitrite formed by mitochondrial NO synthase promotes mitochondrial Ca^{2+} release. Free Radic Biol Med 29(3–4): 343-348.

Carroll RT, Galatsis P, Borosky S, Kopec KK, Kumar V, et al. 2000. 4-Hydroxy-2,2,6,6-tetramethylpiperidine-1-oxyl (Tempol) inhibits peroxynitrite-mediated phenol nitration. Chem Res Toxicol 13(4): 294-300.

Chabrier PE, Auguet M, Spinnewyn B, Auvin S, Cornet S, et al. 1999. BN 80933, a dual inhibitor of neuronal nitric oxide synthase and lipid peroxidation: A promising neuroprotective strategy. Proc Natl Acad Sci USA 96(19): 10824-10829.

Chan PH, Epstein CJ, Li Y, Huang TT, Carlson E, et al. 1995. Transgenic mice and knockout mutants in the study of oxidative stress in brain injury. J Neurotrauma 12(5): 815-824.

Demopoulos HB, Flamm ES, Pietronigro DD, Seligman ML. 1980. The free radical pathology and the microcirculation in the major central nervous system disorders. Acta Physiol Scand Suppl 492: 91-119.

Dimlich RV, Tornheim PA, Kindel RM, Hall ED, Braughler JM, et al. 1990. Effects of a 21-aminosteroid (U-74006F) on cerebral metabolites and edema after severe experimental head trauma. Adv Neurol 52: 365-375.

Farooque M, Hillered L, Holtz A, Olsson Y. 1996. Effects of methylprednisolone on extracellular lactic acidosis and amino acids after severe compression injury of rat spinal cord. J Neurochem 66(3): 1125-1130.

Finkel E. 2001. The Mitochondrion: Is it central to apoptosis? Science 292(5517): 624-626.

Gahm C, Danilov A, Holmin S, Wiklund PN, Brundin L, et al. 2005. Reduced neuronal injury after treatment with NG-nitro-L-arginine methyl ester (L-NAME) or 2-sulfophenyl-N-tert-butyl nitrone (S-PBN) following experimental brain contusion. Neurosurgery 57(6): 1272-1281; discussion 1272–1281.

Ginsberg MD, Becker DA, Busto R, Belayev A, Zhang Y, et al. 2003. Stilbazulenyl nitrone, a novel antioxidant, is highly neuroprotective in focal ischemia. Ann Neurol 54(3): 330-342.

Globus MY, Alonso O, Dietrich WD, Busto R, Ginsberg MD. 1995. Glutamate release and free radical production following brain injury: Effects of posttraumatic hypothermia. J Neurochem 65(4): 1704-1711.

Green AR, Ashwood T, Odergren T, Jackson DM. 2003. Nitrones as neuroprotective agents in cerebral ischemia, with particular reference to NXY-059. Pharmacol Ther 100(3): 195-214.

Hall ED. 1992. The neuroprotective pharmacology of methylprednisolone. J Neurosurg 76(1): 13-22.

Hall ED. 1997. Lazaroid: Mechanisms of action and implications for disorders of the CNS. Neuroscientist 3: 42-51.

Hall ED, Braughler JM. 1981. Acute effects of intravenous glucocorticoid pretreatment on the in vitro peroxidation of cat spinal cord tissue. Exp Neurol 73(1): 321-324.

Hall ED, Braughler JM. 1982. Glucocorticoid mechanisms in acute spinal cord injury: A review and therapeutic rationale. Surg Neurol 18(5): 320-327.

Hall ED, Braughler JM. 1993. Free radicals in CNS injury. Res Publ Assoc Res Nerv Ment Dis 71: 81–105.

Hall ED, Andrus PK, Yonkers PA. 1993. Brain hydroxyl radical generation in acute experimental head injury. J Neurochem 60(2): 588594.

Hall ED, Wolf DL. 1986. A pharmacological analysis of the pathophysiological mechanisms of posttraumatic spinal cord ischemia. J Neurosurg 64(6): 951-961.

Hall ED, Kupina NC, Althaus JS. 1999. Peroxynitrite scavengers for the acute treatment of traumatic brain injury. Ann N Y Acad Sci 890: 462-468.

Hall ED, McCall JM, Means ED. 1994. Therapeutic potential of the lazaroids (21-aminosteroids) in acute central nervous system trauma, ischemia and subarachnoid hemorrhage. Adv Pharmacol 28: 221-268.

Hall ED, Wolf DL, Braughler JM. 1984. Effects of a single large dose of methylprednisolone sodium succinate on experimental posttraumatic spinal cord ischemia. Dose-response and time-action analysis. J Neurosurg 61(1): 124-130.

Hall ED, Detloff MR, Johnson K, Kupina NC. 2004. Peroxynitrite-mediated protein nitration and lipid peroxidation in a mouse model of traumatic brain injury. J Neurotrauma 21(1): 9-20.

Hall ED, Yonkers PA, McCall JM, Braughler JM. 1988. Effects of the 21-aminosteroid U74006F on experimental head injury in mice. J Neurosurg 68(3): 456-461.

Hall ED, Yonkers PA, Taylor BM, Sun FF. 1995. Lack of effect of postinjury treatment with methylprednisolone or tirilazad mesylate on the increase in eicosanoid levels in the acutely injured cat spinal cord. J Neurotrauma 12(3): 245-256.

Hall ED, Yonkers PA, Andrus PK, Cox JW, Anderson DK. 1992. Biochemistry and pharmacology of lipid antioxidants in acute brain and spinal cord injury. J Neurotrauma 9 Suppl 2: S425-S442.

Hall ED, Andrus PK, Smith SL, Fleck TJ, Scherch HM, et al. 1997. Pyrrolopyrimidines: Novel brain-penetrating antioxidants with neuroprotective activity in brain injury and ischemia models. J Pharmacol Exp Ther 281(2): 895-904.

Hall ED, Braughler JM, Yonkers PA, Smith SL, Linseman KL, et al. 1991. U-78517F: A potent inhibitor of lipid peroxidation with activity in experimental brain injury and ischemia. J Pharmacol Exp Ther 258(2): 688-694.

Halliwell B, Gutteridge JMC. 2007. Free radicals in biology and medicine (4th Ed.): Oxford University Press. 1-851 pp.

Heales SJ, Bolanos JP, Stewart VC, Brookes PS, Land JM, et al. 1999. Nitric oxide, mitochondria and neurological disease. Biochim Biophys Acta 1410(2): 215-228.

Hillard VH, Peng H, Zhang Y, Das K, Murali R, et al. 2004. Tempol, a nitroxide antioxidant, improves locomotor and histological outcomes after spinal cord contusion in rats. J Neurotrauma 21(10): 1405-1414.

Holtz A, Nystrom B, Gerdin B. 1990. Effect of methylprednisolone on motor function and spinal cord blood flow after spinal cord compression in rats. Acta Neurol Scand 82(1): 68-73.

Hunot S, Flavell RA. 2001. APOPTOSIS: Death of a monopoly? Science 292(5518): 865-866.

Keller JN, Mark RJ, Bruce AJ, Blanc E, Rothstein JD, et al. 1997a. 4-Hydroxynonenal, an aldehydic product of membrane lipid peroxidation, impairs glutamate transport and mitochondrial function in synaptosomes. Neuroscience 80(3): 685-696.

Keller JN, Pang Z, Geddes JW, Begley JG, Germeyer A, et al. 1997b. Impairment of glucose and glutamate transport and induction of mitochondrial oxidative stress and dysfunction in synaptosomes by amyloid beta-peptide: Role of the lipid peroxidation product 4-hydroxynonenal. J Neurochem 69(1): 273-284.

Kim CD, Shin HK, Lee HS, Lee JH, Lee TH, et al. 2002. Gene transfer of Cu/Zn SOD to cerebral vessels prevents FPI-induced CBF autoregulatory dysfunction. Am J Physiol Heart Circ Physiol 282(5): H1836-H1842.

Kontos HA. 1989. Oxygen radicals in CNS damage. Chem Biol Interact 72(3): 229-255.

Kontos HA, Povlishock JT. 1986. Oxygen radicals in brain injury. Cent Nerv Syst Trauma 3(4): 257-263.

Kontos HA, Wei EP. 1986. Superoxide production in experimental brain injury. J Neurosurg 64(5): 803-807.

Koyanagi I, Tator CH. 1997. Effect of a single huge dose of methylprednisolone on blood flow, evoked potentials, and histology after acute spinal cord injury in the rat. Neurol Res 19(3): 289-299.

Kristal BS, Dubinsky JM. 1997. Mitochondrial permeability transition in the central nervous system: Induction by calcium cycling-dependent and -independent pathways. J Neurochem 69: 524-538.

Kruman I, Bruce-Keller AJ, Bredesen D, Waeg G, Mattson MP. 1997. Evidence that 4-hydroxynonenal mediates oxidative stress-induced neuronal apoptosis. J Neurosci 17(13): 5089-5100.

Langham J, Goldfrad C, Teasdale G, Shaw D, Rowan K. 2000. Calcium channel blockers for acute traumatic brain injury. Cochrane Database Syst Rev (2): CD000565.

Lewen A, Matz P, Chan PH. 2000. Free radical pathways in CNS injury. J Neurotrauma 17(10): 871-890.

Liu D, Li L, Augustus L. 2001. Prostaglandin release by spinal cord injury mediates production of hydroxyl radical, malondialdehyde and cell death: A site of the neuroprotective action of methylprednisolone. J Neurochem 77(4): 1036-1047.

Liu D, Bao F, Prough DS, Dewitt DS. 2005. Peroxynitrite generated at the level produced by spinal cord injury induces peroxidation of membrane phospholipids in normal rat cord: Reduction by a metalloporphyrin. J Neurotrauma 22(10): 1123-1133.

Lopez-Figueroa MO, Caamano C, Morano MI, Ronn LC, Akil H, et al. 2000. Direct evidence of nitric oxide presence within mitochondria. Biochem Biophys Res Commun 272(1): 129-133.

Lyons WE, George EB, Dawson TM, Steiner JP, Snyder SH. 1994. Immunosuppressant FK506 promotes neurite outgrowth in cultures of PC12 cells and sensory ganglia. Proc Natl Acad Sci USA 91(8): 3191-3195.

Maas AI, Murray G, Henney H III, Kassem N, Legrand V, et al. 2006. Efficacy and safety of dexanabinol in severe traumatic brain injury: Results of a phase III randomised, placebo-controlled, clinical trial. Lancet Neurol 5(1): 38-45.

Mark RJ, Lovell MA, Markesbery WR, Uchida K, Mattson MP. 1997. A role for 4-hydroxynonenal, an aldehydic product of lipid peroxidation, in disruption of ion homeostasis and neuronal death induced by amyloid beta-peptide. J Neurochem 68(1): 255-264.

Marklund N, Clausen F, McIntosh TK, Hillered L. 2001b. Free radical scavenger posttreatment improves functional and morphological outcome after fluid percussion injury in the rat. J Neurotrauma 18(8): 821-832.

Marklund N, Lewander T, Clausen F, Hillered L. 2001c. Effects of the nitrone radical scavengers PBN and S-PBN on in vivo trapping of reactive oxygen species after traumatic brain injury in rats. J Cereb Blood Flow Metab 21(11): 1259-1267.

Marklund N, Sihver S, Langstrom B, Bergstrom M, Hillered L. 2002. Effect of traumatic brain injury and nitrone radical scavengers on relative changes in regional cerebral blood flow and glucose uptake in rats. J Neurotrauma 19(10): 1139-1153.

Marklund N, Clausen F, Lewen A, Hovda DA, Olsson Y, et al. 2001a. Alpha-Phenyl-tert-N-butyl nitrone (PBN) improves functional and morphological outcome after cortical contusion injury in the rat. Acta Neurochir (Wien) 143(1): 73-81.

Marshall LF, Maas AI, Marshall SB, Bricolo A, Fearnside M, et al. 1998. A multicenter trial on the efficacy of using tirilazad mesylate in cases of head injury. J Neurosurg 89(4): 519-525.

Matsushita M, Xiong G. 1997. Projections from the cervical enlargement to the cerebellar nuclei in the rat, studied by anterograde axonal tracing. J Comp Neurol 377(2): 251-261.

Mattson MP, Barger SW, Begley JG, Mark RJ. 1995. Calcium, free radicals, and excitotoxic neuronal death in primary cell culture. Methods Cell Biol 46: 187-216.

McIntosh TK, Thomas M, Smith D, Banbury M. 1992. The novel 21-aminosteroid U74006F attenuates cerebral edema and improves survival after brain injury in the rat. J Neurotrauma 9(1): 33-46.

Mesenge C, Charriaut-Marlangue C, Verrecchia C, Allix M, Boulu RR, et al. 1998. Reduction of tyrosine nitration after N(omega)-nitro-L-arginine-methylester treatment of mice with traumatic brain injury. Eur J Pharmacol 353(1): 53-57.

Mikawa S, Kinouchi H, Kamii H, Gobbel GT, Chen SF, et al. 1996. Attenuation of acute and chronic damage following traumatic brain injury in copper, zinc-superoxide dismutase transgenic mice. J Neurosurg 85(5): 885-891.

Monyer H, Hartley DM, Choi DW. 1990. 21-Aminosteroids attenuate excitotoxic neuronal injury in cortical cell cultures. Neuron 5(2): 121-126.

Morrow JD, Roberts LJ. 2002. The isoprostanes: Their role as an index of oxidant stress status in human pulmonary disease. Am J Respir Crit Care Med 166(12 Pt 2): S25-S30.

Muizelaar JP, Kupiec JW, Rapp LA. 1995. PEG-SOD after head injury. J Neurosurg 83(5): 942.

Narayan RK, Michel ME, Ansell B, Baethmann A, Biegon A, et al. 2002. Clinical trials in head injury. J Neurotrauma 19(5): 503-557.

Neely MD, Sidell KR, Graham DG, Montine TJ. 1999. The lipid peroxidation product 4-hydroxynonenal inhibits neurite outgrowth, disrupts neuronal microtubules, and modifies cellular tubulin. J Neurochem 72(6): 2323-2333.

Nicholls DG, Budd SL. 2000. Mitochondria and neuronal survival. Physiol Rev 80(1): 315-360.

Nishio S, Yunoki M, Noguchi Y, Kawauchi M, Asari S, et al. 1997. Detection of lipid peroxidation and hydroxyl radicals in brain contusion of rats. Acta Neurochir Suppl (Wien) 70: 84-86.

Okonkwo DO, Povlishock JT. 1999. An intrathecal bolus of cyclosporin A before injury preserves mitochondrial integrity and attenuates axonal disruption in traumatic brain injury. J Cereb Blood Flow Metab 19(4): 443-451.

Okonkwo DO, Buki A, Siman R, Povlishock JT. 1999. Cyclosporin A limits calcium-induced axonal damage following traumatic brain injury. Neuroreport 10(2): 353-358.

Pellegrini-Giampietro DE, Cherici G, Alesiani M, Carla V, Moroni F. 1990. Excitatory amino acid release and free radical formation may cooperate in the genesis of ischemia-induced neuronal damage. J Neurosci 10(3): 1035-1041.

Pratico D, Reiss P, Tang LX, Sung S, Rokach J, et al. 2002. Local and systemic increase in lipid peroxidation after moderate experimental traumatic brain injury. J Neurochem 80(5): 894-898.

Rabchevsky AG, Fugaccia I, Sullivan PG, Blades DA, Scheff SW. 2002. Efficacy of methylprednisolone therapy for the injured rat spinal cord. J Neurosci Res 68(1): 7-18.

Radi R, Beckman JS, Bush KM, Freeman BA. 1991. Peroxynitrite-induced membrane lipid peroxidation: The cytotoxic potential of superoxide and nitric oxide. Arch Biochem Biophys 288(2): 481-487.

Rego AC, Ward MW, Nicholls DG. 2001. Mitochondria control ampa/kainate receptor-induced cytoplasmic calcium deregulation in rat cerebellar granule cells. J Neurosci 21(6): 1893-1901.

Rohn TT, Quinn MT. 1998. Inhibition of peroxynitrite-mediated tyrosine nitration by a novel pyrrolopyrimidine antioxidant. Eur J Pharmacol 353(2–3): 329-336.

Rohn TT, Hinds TR, Vincenzi FF. 1993a. Inhibition of the Ca pump of intact red blood cells by t-butyl hydroperoxide: Importance of glutathione peroxidase. Biochim Biophys Acta 1153(1): 67-76.

Rohn TT, Hinds TR, Vincenzi FF. 1993b. Ion transport ATPases as targets for free radical damage. Protection by an aminosteroid of the Ca2+ pump ATPase and Na+/K+ pump ATPase of human red blood cell membranes. Biochem Pharmacol 46(3): 525-534.

Rohn TT, Hinds TR, Vincenzi FF. 1995. Inhibition by activated neutrophils of the Ca2+ pump ATPase of intact red blood cells. Free Radic Biol Med 18(4): 655-667.

Rohn TT, Hinds TR, Vincenzi FF. 1996. Inhibition of Ca2+-pump ATPase and the Na+/K+-pump ATPase by iron-generated free radicals. Protection by 6,7-dimethyl-2,4-DI-1- pyrrolidinyl-7H-pyrrolo[2,3-d] pyrimidine sulfate (U-89843D), a potent, novel, antioxidant/free radical scavenger. Biochem Pharmacol 51(4): 471-476.

Sadrzadeh SM, Eaton JW. 1988. Hemoglobin-mediated oxidant damage to the central nervous system requires endogenous ascorbate. J Clin Invest 82(5): 1510-1515.

Sadrzadeh SM, Graf E, Panter SS, Hallaway PE, Eaton JW. 1984. Hemoglobin. A biologic fenton reagent. J Biol Chem 259(23): 14354-14356.

Scheff SW, Sullivan PG. 1999. Cyclosporin A significantly ameliorates cortical damage following experimental traumatic brain injury in rodents. J Neurotrauma 16(9): 783-792.

Scott GS, Cuzzocrea S, Genovese T, Koprowski H, Hooper DC. 2005. Uric acid protects against secondary damage after spinal cord injury. Proc Natl Acad Sci USA 102(9): 3483-3488.

Sevanian A, Kim E. 1985. Phospholipase A2 dependent release of fatty acids from peroxidized membranes. J Free Radic Biol Med 1(4): 263-271.

Singh IN, Sullivan PG, Deng Y, Mbye LH, Hall ED. 2006. Time course of posttraumatic mitochondrial oxidative damage and dysfunction in a mouse model of focal traumatic brain injury: Implications for neuroprotective therapy. J Cereb Blood Flow Metab 26: 1407-1418.

Smith SL, Andrus PK, Zhang JR, Hall ED. 1994. Direct measurement of hydroxyl radicals, lipid peroxidation, and blood-brain barrier disruption following unilateral cortical impact head injury in the rat. J Neurotrauma 11(4): 393-404.

Squadrito GL, Pryor WA. 1998. Oxidative chemistry of nitric oxide: The roles of superoxide, peroxynitrite, and carbon dioxide. Free Radic Biol Med 25(4–5): 392-403.

Squadrito GL, Pryor WA. 2002. Mapping the reaction of peroxynitrite with CO2: Energetics, reactive species, and biological implications. Chem Res Toxicol 15(7): 885-895.

Stewart VC, Sharpe MA, Clark JB, Heales SJ. 2000. Astrocyte-derived nitric oxide causes both reversible and irreversible damage to the neuronal mitochondrial respiratory chain. J Neurochem 75(2): 694-700.

Stewart VC, Heslegrave AJ, Brown GC, Clark JB, Heales SJ. 2002. Nitric oxide-dependent damage to neuronal mitochondria involves the NMDA receptor. Eur J Neurosci 15(3): 458-464.

Stout AK, Raphael HM, Kabterewucz BI, Klann E, Reynolds IJ. 1998. Glutamate-induced neuron death requires mitochondrial calcium uptake. Nat Neursci 1(5): 366-373.

Sullivan PG, Thompson MB, Scheff SW. 1999b. Cyclosporin A attenuates acute mitochondrial dysfunction following traumatic brain injury. Exp Neurol 160: 226-34.

Sullivan PG, Thompson MB, Scheff SW. 1999c. Cyclosporin A attenuates acute mitochondrial dysfunction following traumatic brain injury. Exp Neurol 160(1): 226-234.

Sullivan PG, Thompson M, Scheff SW. 2000c. Continuous infusion of cyclosporin A postinjury significantly ameliorates cortical damage following traumatic brain injury. Exp Neurol 161(2): 631-637.

Sullivan PG, Geiger JD, Mattson MP, Scheff SW. 2000a. Dietary supplement creatine protects against traumatic brain injury. Ann Neurol 48(5): 723-729.

Sullivan PG, Keller JN, Mattson MP, Scheff SW. 1998. Traumatic brain injury alters synaptic homeostasis: Implications for impaired mitochondrial and transport function. J Neurotrauma 15(10): 789-798.

Sullivan PG, Bruce-Keller AJ, Rabchevsky AG, Christakos S, Clair DK, et al. 1999a. Exacerbation of damage and altered NF-kappaB activation in mice lacking tumor necrosis factor receptors after traumatic brain injury. J Neurosci 19(15): 6248-6256.

Sullivan PG, Rabchevsky AG, Hicks RR, Gibson TR, Fletcher-Turner A, et al. 2000b. Dose-response curve and optimal dosing regimen of cyclosporin A after traumatic brain injury in rats. Neuroscience 101(2): 289-295.

Taoka Y, Okajima K, Uchiba M, Johno M. 2001. Methylprednisolone reduces spinal cord injury in rats without affecting tumor necrosis factor-alpha production. J Neurotrauma 18(5): 533-543.

Valdez LB, Alvarez S, Arnaiz SL, Schopfer F, Carreras MC, et al. 2000. Reactions of peroxynitrite in the mitochondrial matrix. Free Radic Biol Med 29(3–4): 349-356.

Verweij BH, Muizelaar JP, Federico CV, Peterson PL, Xiong Y, et al. 1997. Mitochondrial dysfunction after experimental and human brain injury and its possible reversal with a selective N-type calcium channel antagonist (SNX-111). Neurol Res 19(3): 334-339.

Wang GJ, Thayer SA. 1996. Sequestration of glutamate-induced Ca2+ loads by mitochondria in cultured rat hippocampal neurons. J Neurophysiol 76(3): 1611-1621.

Ward MW, Rego AC, Frenguelli BG, Nicholls DG. 2000. Mitochondrial membrane potential and glutamate excitotoxicity in cultured cerebellar granule cells. J Neurosci 20(19): 7208-7219.

White RJ, Reynolds IJ. 1997. Mitochondria accumulate Ca2+ following intense glutamate stimulation of cultured rat forebrain neurones. J Physiol (Lond) 498(Pt 1): 31-47.

Whiteman M, Halliwell B. 1997. Thiourea and dimethylthiourea inhibit peroxynitrite-dependent damage: Nonspecificity as hydroxyl radical scavengers. Free Radic Biol Med 22(7): 1309-1312.

Xiong Y, Rabchevsky AG, Hall ED. 2007. Role of peroxynitrite in secondary oxidative damage after spinal cord injury. J Neurochem 100: 639-649.

Xiong Y, Shie FS, Zhang J, Lee CP, Ho YS. 2005. Prevention of mitochondrial dysfunction in posttraumatic mouse brain by superoxide dismutase. J Neurochem 95(3): 732-744.

Xu J, Qu ZX, Hogan EL, Perot PL, Jr. 1992. Protective effect of methylprednisolone on vascular injury in rat spinal cord injury. J Neurotrauma 9(3): 245-253.

Xu J, Gyeong-Moon K, Chen S, Yan P, Hinan A, et al. 2001a. iNOS and nitrotyrosine expression after spinal cord injury. J Neurotrauma 18(5): 523-532.

Xu J, Kim GM, Ahmed SH, Yan P, Xu XM, et al. 2001b. Glucocorticoid receptor-mediated suppression of activator protein-1 activation and matrix metalloproteinase expression after spinal cord injury. J Neurosci 21(1): 92-97.

Young W, Flamm ES. 1982. Effect of high-dose corticosteroid therapy on blood flow, evoked potentials, and extracellular calcium in experimental spinal injury. J Neurosurg 57(5): 667-673.

Zaleska MM, Floyd RA. 1985. Regional lipid peroxidation in rat brain in vitro: Possible role of endogenous iron. Neurochem Res 10(3): 397-410.

Zanella B, Calonghi N, Pagnotta E, Masotti L, Guarnieri C. 2002. Mitochondrial nitric oxide localization in H9c2 cells revealed by confocal microscopy. Biochem Biophys Res Commun 290(3): 1010-1014.

Zhang JR, Andrus PK, Hall ED. 1993. Age-related regional changes in hydroxyl radical stress and antioxidants in gerbil brain. J Neurochem 61(5): 1640-1647.

Zhang H, Squadrito GL, Pryor WA. 1998a. The reaction of melatonin with peroxynitrite: Formation of melatonin radical cation and absence of stable nitrated products. Biochem Biophys Res Commun 251(1): 83-87.

Zhang H, Squadrito GL, Uppu R, Pryor WA. 1999. Reaction of peroxynitrite with melatonin: A mechanistic study. Chem Res Toxicol 12(6): 526-534.

Zhang R, Shohami E, Beit-Yannai E, Bass R, Trembovler V, et al. 1998b. Mechanism of brain protection by nitroxide radicals in experimental model of closed-head injury. Free Radic Biol Med 24(2): 332-340.

Zhang R, Shohami E, Beit-Yannai E, Bass R, Trembovler V, et al. 1998c. Mechanism of brain protection by nitroxide radicals in experimental model of closed-head injury. Free Radic Biol Med 24(2): 332-340.

11 Neuroimmunology of Paraproteinemic Neuropathies

A. A. Ilyas

1	Introduction	230
2	Paraproteinemias	230
3	Neuropathies Associated with Paraproteinemia	231
4	Neural Antigens for Paraproteins in Patients with Neuropathies	232
4.1	Glycoprotein Antigens	232
4.2	Glycosphingolipid Antigens	232
5	Methods for Detecting Antibodies to Neural Antigens	232
6	Reactivity of Neuropathy-Associated IgM Paraproteins with Neural Antigens	233
6.1	Demyelinating Neuropathy Associated with Anti-MAG/SGPG IgM Paraproteins	234
6.2	Neuropathy Associated with Anti-Ganglioside IgM Paraproteins	236
6.3	Neuropathy Associated with Anti-Sulfatide IgM Paraproteins	236
6.4	Neuropathy Associated with IgM Paraproteins to Other Antigens	237
7	Neuropathy Associated with Nonmalignant IgG and IgA Monoclonal Gammopathy	237
8	Neuropathies Associated with Malignant Monoclonal Gammopathy	238
8.1	Neuropathy Associated with Waldenstrom's Macroglobulinemia (WM)	239
8.2	Neuropathy Associated with Multiple Myeloma	239
8.3	Neuropathy Associated with Osteosclerotic Myeloma (POEMS Syndrome)	239
8.4	Neuropathy Associated with Cryoglobulinemia	239
9	Animal Models of Peripheral Neuropathies	239
9.1	Experimental Neuropathy Associated with Anti-MAG/SGPG Antibodies	239
9.2	Experimental Neuropathies Induced by Antibodies to Gangliosides	240
9.3	Experimental Neuropathy Induced by Anti-Sulfatide Antibodies	241
10	Therapy	241
11	Conclusions	242

© 2009 Springer Science+Business Media, LLC.

List of Abbreviations: ELISA, enzyme-linked immunosorbent assay; IVIg, intravenous immunoglobulin; MGUS, monoclonal gammopathy of unknown significance; MAG, myelin-associated glycoprotein; PMP-22, peripheral nerve myelin protein-22; SGGLs, sulfated glucuronyl glycosphingolipids; SGPG, sulfoglucuronyl paragloboside; SGLPG, sulfoglucuronyl lactosaminyl paragloboside; TLC, thin-layer chromatogram; WM, Waldenstrom's macroglobulinemia

1 Introduction

Paraproteinemic neuropathies are a diverse group of disorders closely connected with the presence of an excessive amount of monoclonal antibody termed a paraprotein or M-protein or monoclonal gammopathy. Paraproteins are produced by the uncontrolled clonal proliferation of plasma cells. A paraprotein or M-protein can be associated with several malignant disorders including multiple myeloma, Waldenstrom's macroglobulinemia, and other lymphoproliferative disorders. When no underlying disease is detected, the term monoclonal gammopathy of unknown significance (MGUS) is used. MGUS is the most common of the paraproteinemias. The prevalence of MGUS in the general population increases with age and is seen in about 1% of patients older than 50 years of age and reaches approximately 3% in those older than 70 years. The incidence of MGUS seems to be higher in African-American populations. MGUS is found in approximately 5% of all patients with polyneuropathy and 10% of patients with idiopathic peripheral neuropathies. This prevalence is six to ten times that in the general population. Several different syndromes can be recognized based on the types of paraproteinemia and peripheral neuropathy. More than 70% of patients with neuropathy associated with IgM paraproteinemia have antibodies reactive against oligosaccharide moieties attached to glycoproteins and glycolipids. In about 50–60% of the patients with neuropathy, the IgM paraproteins reacting with sulfated glucuronic acid moiety shared the myelin-associated glycoprotein (MAG), several other adhesion proteins, and sulfoglucuronyl glycosphingolipids. IgM paraproteins from other patients not reactive to MAG are directed against various gangliosides, sulfatide, and chondroitin sulfates. Unlike IgM paraproteins, the antigen targets of IgG and IgA paraproteins remain unknown. Patients with IgM gammopathy and antineural anibodies can be associated with particular clinical syndromes. These include a chronic, predominantly sensory, ataxic, demyelinating neuropathy with an IgM paraprotein reacting with oligosaccharide HNK-1-moiety shared by MAG and other glycoconjugates, sensory ataxic neuropathy with IgM paraproteins directed against disialosyl group bearing gangliosides such as GD1b, GD2, and GD3, and sensory neuropathy with paraproteins directed against sulfatide. Multifocal motor neuropathy is associated with IgM paraproteins directed against GM1 and other glycoconjugates containing a Gal(1β-3)GalNAc moiety (◉ *Table 11-1*). Because of the causal relationship between the IgM paraproteins and neuropathy, treatments in these patients have been directed at reducing the concentration of paraproteins by depleting the clones of plasma cells. Several recent open trials in patients with neuropathy associated with gammopathy and anti-MAG or anti-GM1 IgM antibodies have shown beneficial effects of rituximab, a monoclonal antibody directed against the B cell surface membrane marker CD20.

2 Paraproteinemias

Immunoglobulins are produced by B lymphocytes activated to become plasma cells. A paraprotein is a monoclonal immunoglobulin or immunoglobulin light chain in the blood or urine resulting from a clonal proliferation of plasma cells or B-lymphocytes (synonyms: monoclonal spike, M-protein, monoclonal protein). It may consist of whole immunoglobulin, or only the light chain: in the latter case, most or all of the paraprotein is in the urine, because of the low molecular weight. IgM paraproteins are detected by serum protein electrophoresis, immunoelectrophoresis, or immunofixation electrophoresis, the latter being the most sensitive technique. Paraproteinemias can be malignant or nonmalignant. Malignant disorders include multiple myeloma, Waldenstrom's macroglobulinemia, B-cell lymphoma, systemic amyloidosis,

◘ Table 11-1
IgM paraprotein reactivities with neural antigens in patients with neuropathies

Antigens	Nerve pathology	Clinical features
MAG, P0 PMP-22, SGPG	Demyelinating	Chronic, predominantly sensory or S>M
Gangliosides		
GM1	Predominantly demyelinating or mixed	Chronic, purely motor, asymmetric
Gangliosides with disialosyl moieties (GD1b, GD2, GD3, GT1b, GQ1b)	demyelinating	purely sensory ataxia, asymmetric or symmetric
GD1a	Demyelinating	Motor
Sulfatide	Predominantly Demyelinating	Sensorimotor
Chondroitin sulfate C	Axonal	Sensorimotor
Cytoskeletal Proteins	Axonal	Sensorimotor

MAG, myelin-associated glycoprotein; M, motor neuropathhy; S, sensory neuropathy; SGPG, sulfoglucuronyl paragloboside

and other lymphoproliferative disorders. The most common of the paraproteinemias is monoclonal gammopathy of unknown significance (MGUS). MGUS is characterized by the presence of a serum monoclonal protein value of 3 g/dL, fewer than 10% plasma cells in the bone marrow, no or a small amount of monoclonal protein in the urine, and absence of lytic bone lesions, anemia, hypercalcemia, or renal insufficiency related to the plasma-cell proliferative process (Kyle and Rajkumar, 2003). Sixty percent of patients with paraproteinemias have MGUS. Paraproteins occur in serum or urine of about 1% of patients aged 50 years or older and about 3% of those 70 or older (Yeung et al., 1991; Kyle and Rajkumar, 2003). IgG is the most common paraprotein in MGUS accounting for about 60% of MGUS. The frequency of paraproteins is higher among blacks than whites (Cohen et al., 1998; Landgren et al., 2006). The disease affects men more often than women, and the prognosis of the former appears to be worse in some studies. Although MGUS is considered nonmalignant, long-term follow-up studies have shown that the risk of progression from MGUS to a lymphoproliferative disorder (usually multiple myeloma, less often amyloidosis or Waldenström's macroglobulinemia) is about 1% per year (Kyle et al., 2002; Montoto et al., 2003). Periodic screening therefore is prudent to search for the early development of malignant gammopathy.

3 Neuropathies Associated with Paraproteinemia

Peripheral neuropathies can be associated with malignant disorders such as multiple myeloma, Waldenstrom's macroglobulinemia, and lymphoma. However, two thirds of patients with peripheral neuropathy and paraproteinemia do not have an underlying malignant disease, and thus are associated with MGUS. The association of neuropathy and benign monoclonal gammopathy is well established. Kelly et al. (1981) reported that a benign paraprotein was present in 10% of patients with idiopathic peripheral neuropathy compared with a frequency of 1–3% in the control population. Subsequent studies have reported higher prevalence of neuropathy associated with MGUS (Kahn et al., 1980; Nobile-Orazio et al., 1987; Vrethem et al., 1993). The neuropathic syndromes associated with paraprotenemias are heterogeneous. While IgG is the most common paraprotein in patients with MGUS (IgG 70%, IgM 15%, IgA 12%, biclonal 3%), IgM is more common in those with peripheral neuropathy (60%) followed by IgG (30%) and IgA (10%) and the light chain predominantly being kappa type (Kahn et al., 1980; Kyle and Rajkumar, 2003).

Paraproteins from neuropathy patients who react with neural antigens are almost always of the IgM class. In more than two thirds of patients with neuropathy and IgM paraproteinemia, the monoclonal antibodies react with oligosaccharide moieties attached to glycoproteins and glycolipids. The antigenic specificity of IgG and IgA paraproteins in patients with neuropathy remain largely uncharacterized. Neuropathies associated with paraproteinemia have been the subject of excellent reviews (Latov, 1995; Ropper and Gorson, 1998; Vital, 2001).

4 Neural Antigens for Paraproteins in Patients with Neuropathies

4.1 Glycoprotein Antigens

The myelin-associated glycoprotein (MAG) is a minor constituent of nerve myelin. It is concentrated in periaxonal Schwann cell membranes and other uncompacted regions of PNS myelin. MAG is a 100-kDa glycoprotein and contains about 30% by weight carbohydrate. MAG in many species contains the adhesion-related, HNK-1 carbohydrate epitope (O'Shannessy et al., 1985). The selective localization of MAG in the periaxonal membranes of myelin-forming Schwann cells suggests that it functions in glia–axon interactions (Quarles, 2002). MAG is a common antigen for IgM in patients with paraproteinemic neuropathy.

Po glycoprotein accounts for over half the protein of the PNS myelin and the peripheral nerve myelin protein-22 (PMP-22), which accounts for less than 5% are expressed throughout the compact myelin and like MAG, express HNK-1 oligosaccharide epitope in some species (O'shannessy et al., 1985; Quarles, 2002). These proteins are involved in myelin formation and compaction. The HNK-1 oligosaccharide epitope is also expressed on other adhesion glycoproteins of nerves such as NCAM and J1 (Kruse et al., 1984; Quarles and Weiss, 1999).

4.2 Glycosphingolipid Antigens

Glycosphingolipids are glycolipids containing the amino alcohol sphingosine. Glycosphingolipids can be further subdivided as neutral or acidic. Most acidic glycosphingolipids are gangliosides containing sialic acid; other acidic glycosphingolipids contain a sulfate or sulfated glucuronic acid groups. Gangliosides are a class of complex glycosphingolipids containing sialic acid that are particularly concentrated in neural plasma membranes (Ledeen, 1989). Glycosphingolipids are amphipathic molecules that occur primarily in cell-surface membranes with their oligosaccharides exposed to the extracellular environment. Glycosphingolipids function as antigens, mediate cell–cell and cell–substrate interactions, and mediate and modulate signal transduction (Hakomori, 2003). Recent studies indicate that glycosphingolipids assemble with signal transducers and other membrane proteins to form a functional unit termed "glycosynapse", through which glycosylation-dependent cell adhesion coupled with signal transduction takes place (Hakomori, 2002).

Structures of some of the major gangliosides and sulfated glycosphingolipids of neural tissue are shown in ❯ *Figure 11-1*. The glycosphingolipids exhibit species and tissue differences. PNS myelin contains large proportions of lactotetraose and lactohexaose gangliosides, LM1 and Hex-LM1, and sulfoglucuronylglycosphingolipids (SGGLs), which occur in negligible amounts in the CNS myelin (Svennerholm et al., 1994; Ogawa-Goto and Abe, 1998). LM1, Hex-LM1, and SGGLs also exhibit species differences (Ilyas et al., 1986, 1988a). SGGLs were first identified in human peripheral nerves because they react with anti-MAG IgM paraprotein (Ilyas et al., 1984b, 1986, 1990, 1992; Chou et al., 1986; Ariga et al., 1987; Jungalwala, 1994). Two SGGLs have been characterized as sulfoglucuronylparagloboside (SGPG) (❯ *Figure 11-2*) and sulfoglucuronyl lactosaminyl paragloboside (SGLPG) (❯ *Figure 11-1*).

5 Methods for Detecting Antibodies to Neural Antigens

There are three procedures commonly used for the detection and quantification of autoantibodies to neural antigens in autoimmune neuropathies. These are: Western blotting, thin-layer chromatogram (TLC)-overlay procedure, and the enzyme-linked immunosorbent assay (ELISA). Western blotting is employed

Figure 11-1

Glycolipid antigens for IgM paraprotein. LM1, sialosylparaglosboside SGPG = sulfoglucuronyl paragloboside; SGLPG = sulfoglucuronyl lactosaminyl paragloboside

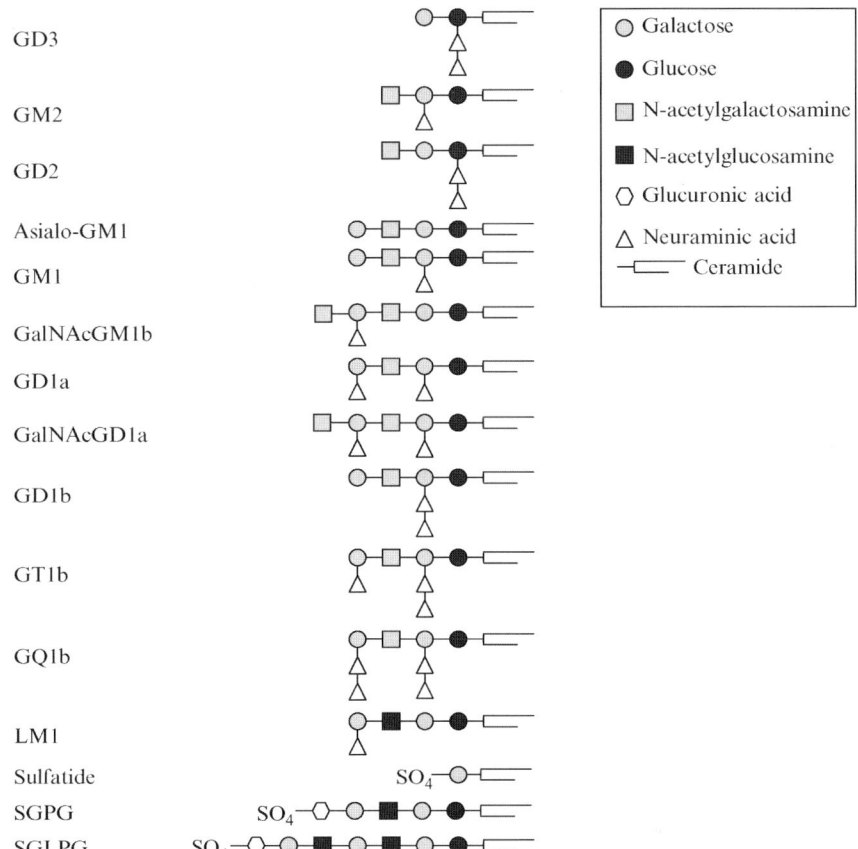

for the detection of protein/glycoprotein antigens while TLC-immunostaining is used for the detection of glycolipid antigens separated on aluminum-backed TLC plates (◗ Figure 11-3 and ◗ Figure 11-4). The ELISA is the most commonly used to quantify autoantibodies to proteins and glycolipids. Although ELISA is quantitative and convenient for assaying several samples, it requires purified antigens and can give false positives and negatives. Therefore, it is critical that appropriate controls are used and data are confirmed by the overlay procedures (Quarles, 1989).

6 Reactivity of Neuropathy-Associated IgM Paraproteins with Neural Antigens

More than 70% of the IgM paraprotein associated with neuropathy react with oligosaccharide moieties of glycoproteins and glycolipids including gangliosides. In about 50% of the patients, IgM paraprotein react with MAG and SGPG. IgM paraproteins not reactive with MAG/SGPG often react with various other neural antigens including gangliosides, sulfatide, chondroitin sulfates, and cytoskeletal proteins (Quarles and Weiss, 1999; Willison and Yuki, 2002).

Figure 11-2
Structure of SGPG

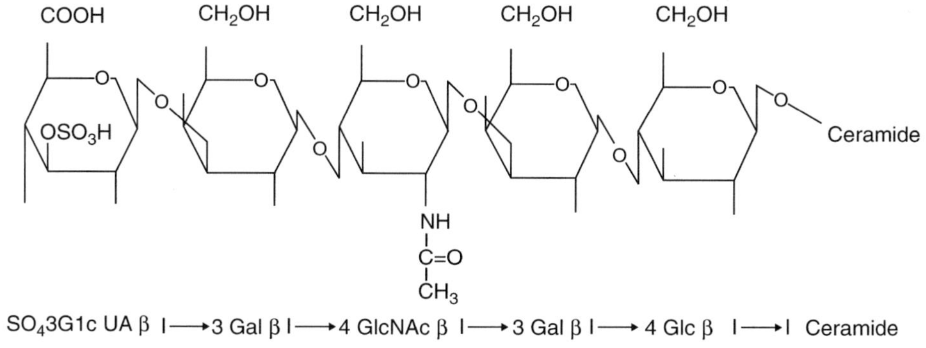

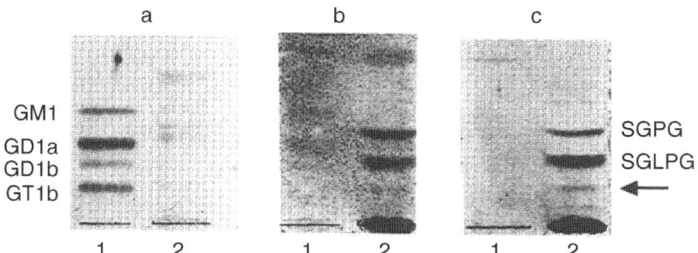

SO$_4$3Glc UA β 1⟶3 Gal β 1⟶4 GlcNAc β 1⟶3 Gal β 1⟶4 Glc β 1⟶1 Ceramide

Figure 11-3
Binding of serum IgM from two patients with neuropathy associated with monoclonal gammopathy to SGPG and SGLPG on TLC. Lanes 1 contained bovine brain gangliosides. Lanes 2 contained dog sciatic nerve acidic glycolipid fractions. Panel a was stained with orcinol. Panels b and c were immunostained with test serum and peroxidase-conjugated antihuman IgM (From Ilyas et al., J Neuroimmunol 1992; 37: 85–92. Elsevier Science Publishers B.V.)

6.1 Demyelinating Neuropathy Associated with Anti-MAG/SGPG IgM Paraproteins

Latov and associates first reported that an IgM paraprotein in a patient with sensorimotor neuropathy and IgM gammopathy reacted with an antigen in the peripheral nerve myelin (Latov et al., 1980). The myelin antigen was subsequently characterized to be the myelin-associated glycoprotein (MAG) (Braun et al., 1982). The epitope on MAG was later shown to be on the oligosaccharide moiety of MAG (Ilyas et al., 1984b). The oligosaccharide epitope of MAG also react with the monoclonal antibody HNK-1 (McGarry et al., 1983). The oligosaccharide HNK-1 epitope on MAG is shared with a number of other glycoproteins including Po glycoprotein, PMP-22, and J1 glycoprotein (Bollensen et al., 1988; Hammer et al., 1993; Snipes et al., 1993; Quarles and Weiss, 1999). The anti-MAG paraproteins were first shown by Ilyas et al. (1984b) to crossreact with acidic glycolipids in the ganglioside fraction from human peripheral nerve. The reactive glycolipids were identified as novel sulfoglucuronyl glycosphingolipids (SGGLs) (Chou et al., 1986; Ariga et al., 1987; Jungalwala, 1994). The major SGGL is sulfoglucuronylparagloboside (SGPG) (◉ Figure 11-2). The terminal sulfated glucuronic acid in SGPG is a critical part of the epitope for all anti-MAG/SGPG IgM paraproteins and for monoclonal antibody HNK-1 (Ilyas et al., 1984a, 1986, 1990, 1992).

In over 50% of patients with neuropathy and IgM gammopathy, the monoclonal antibody reacts with oligosaccharide epitopes shared by MAG, Po protein, PMP-22 protein, and SGPG (Quarles et al., 1986;

Figure 11-4

Species distribution of SGGLs. Binding of an anti-MAG IgM paraprotein from a polyneuropathy patient to SGPG and SGLPG from peripheral nerves of several species after separation on TLC. Lane HB contained human brain gangliosides. Each of the other samples is from sciatic nerve unless otherwise indicated. Lanes, H (human), B (bovine cauda equine), C (cat), Ch (chicken), D (dog), G (guinea pig), Rb (rabbit), R (rat). Panel A depicts resorcinol stained gangliosides. Panel B depicts an autoradiogram of the same chromatogram after overlaying with a 1:200 dilution of serum from a neuropathy patient followed by radioiodinated goat antihuman IgM (μ–chain specific). (From Ilyas et al., Brain Res 1986: 385; 1–9. Elsevier Science Publishers B.V.)

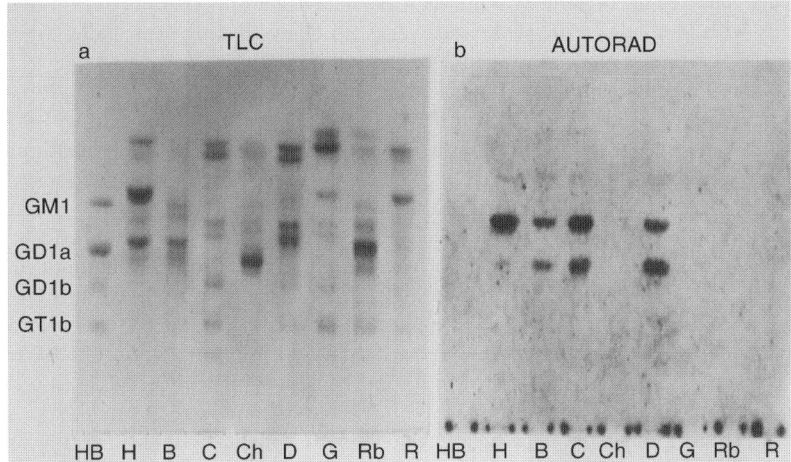

Latov, 1995; Quarles and Weiss, 1999; Nobile-Orazio, 2004). Almost 80–90% of the patients are men, in the age range of 40–75 years. The neuropathy in patients with anti-MAG/SGPG paraproteins is quite homogenous. Most patients have a chronic, slowly progressive, predominantly sensory, ataxic, demyelinating neuropathy (Latov, 1995; Van den Berg et al., 1996; Chassande et al., 1998; Quarles and Weiss, 1999; Nobile-Orazio, 2004). Neurophysiological examination typically shows a widespread slowing of sensorimotor nerve conduction velocity with marked delay of distal latencies, suggesting that nerve fiber endings are particularly affected (Kaku et al., 1994).

Pathologic studies on sural nerve biopsies in patients with anti-MAG/SGPG IgM paraproteins have revealed segmental demyelination without inflammatory infiltrates, amyloid deposits, or vasculitis. Teased fiber studies almost invariably show segmental demyelination. Deposits of IgM and complement on myelin sheaths have been demonstrated by direct immunofluorescence (Takatsu et al., 1985; Hays et al., 1988).

Myelin widening is the hallmark of neuropathy associated with anti-MAG IgM paraproteinemia. Ultrastructural examination shows the presence of a variable number of fibers with widely spaced myelin lamellae due to the increased distance between the major dense lines in most patients with neuropathy associated with anti-MAG IgM paraprotein (Mendell et al., 1985; Vital et al., 1989). Analysis of IgM deposits in myelin and myelin widening in adjacent sections from nerve biopsies of patients with anti-MAG IgM paraproteins revealed a good correlation between the myelin widening and the IgM penetration pattern into myelinated fibers (Ritz et al., 1999). These studies support the hypothesis that myelin widening is due to the intercalation of IgM paraproteins into myelin lamellae. Studies on sural nerve of patients with anti-MAG paraproteins showed altered neurafilament spacing in the axons of demyelinated fibers, suggesting a different pathogenic mechanism of anti-MAG IgM paraprotein in neuropathy (Lunn et al., 2002).

Heterogeneity in the fine specificities of anti-MAG/SGPG IgM paraproteins has been demonstrated using derivatives of SGPG (Ilyas, 1986, 1990). For example, some anti-MAG/SGPG paraproteins can react with GPG, a desulfated derivative of SGPG, while other paraproteins require a sulfate group for binding to SGPG. Some anti-MAG/SGPG paraproteins also crossreact with sulfatide (Ilyas et al., 1992).

Some anti-MAG IgM paraproteins also react with a faster migrating 3rd lipid antigen (Ilyas et al., 1986, 1992; Kusunoki et al., 1987) which has been characterized as lysophosphatidylinositol (LysoPI) (Suzuki et al., 2001). Subsequent studies also demonstrated that there is variability in binding of anti-MAG IgM paraprotein with various peripheral nerves glycoconjugates that share HNK-1 epitope (van den Berg et al., 1996; Weiss et al., 1999). While most anti-SGPG paraproteins crossreact with MAG, some do not. Lopate et al. (2001) reported that anti-MAG antibodies have different binding patterns to neural tissues, including axonal binding. These antigen and tissue binding variability of anti-MAG/SGPG paraproteins may relate to clinical heterogeneity in patients with this type of neuropathy. High antibody titers to MAG, but not to SGPG, have been reported to correlate with the degree of demyelination (Trojaborg et al., 1995). In another study, reactivity with MAG was associated exclusively with demyelinating neuropathy, whereas reactivity with SGPG was also found in some axonopathies (Chassande et al., 1998).

Although we have come a long way in elucidating the mechanism of anti-MAG IgM paraproteins in patients with neuropathy, we still know very little about the development of anti-MAG IgM paraproteins in these patients. A very recent study on immunoglobulin gene analysis in polyneuropathy associated with IgM monoclonal gammopathy revealed that V genes associated with bacterial responses appear overrepresented (Eurelings et al., 2006). A large majority of patients showed somatic mutations in their immunoglobulin variable region genes that are likely the result of antigenic stimulation. This report extends previously reported observations that anti-MAG IgM paraproteins are somatically mutated (Ayadi et al., 1992; Spatz et al., 1992; Lee et al., 1994). Interestingly, 25 of 28 patients analyzed thus far use V_H genes associated with polysaccharide responses (Eureling et al., 2006). These studies suggest that polyneuropathy associated with monoclonal gammopathy may be caused by an immune response to bacterial antigens. Crossreactions of anti-MAG IgM paraproteins with some bacterial polypeptides has been reported (Brouet et al., 1994), suggesting that infections may have triggered monoclonal gammopathy. Association of cytomegalovirus (CMV) infection in a high proportion of patients with neuropathy and anti-MAG antibodies was reported by Yuki et al. (1998) but it was not confirmed by other investigators (Lunn et al., 1999; Irie et al., 2000).

6.2 Neuropathy Associated with Anti-Ganglioside IgM Paraproteins

Ilyas et al. (1985b) reported the first case of neuropathy associated with IgM gammopathy in which the monoclonal antibody reacted with several gangliosides containing a disialosyl group, including GD2, GD3, GD1b, GT1b, and GQ1b (❯ *Figure 11-5*). Subsequently, several more patients were reported to have IgM paraproteins with similar reactivity. These patients have a chronic sensory ataxic neuropathy in which large sensory fibers are affected (Daune et al., 1992; Dalakas and Quarles, 1996; Willison et al., 2001).

Freddo et al. (1986) first reported that in a patient with lower motor neuron disease and IgM gammopathy, the paraprotein reacted with GM1 and GD1b. Since then several patients with predominantly motor neuropathy and anti-GM1 IgM gammopathy have been reported (Latov et al., 1988; Ilyas et al., 1988a). The IgM paraprotein in most cases also reacts with GD1b and asialo-GM1, glycolipids that share Galβ1-3GalNAc moiety with GM1 (Latov et al., 1988).

Two patients with neuropathy and gammopathy with IgM reactivity to GM2, GalNAc-GM1b, and GalNAc-GD1a were first reported by Ilyas et al. (1985a, 1988b). Subsequently, more patients with neuropathy and IgM gammopathy have been reported with IgM reactivity with GM2, GalNAc-GM1b, and GalNAc-GD1a (Willison and Yuki, 2002) or with GM1 and GM2 (Ilyas et al., 1988b; Willison and Yuki, 2002).

IgM paraprotein reactivity with GD1a has been occasionally reported in patients with predominantly motor neuropathy (Bollensen et al., 1989; Carpo et al., 1996). A patient with sensory motor neuropathy and lymphoma had an IgM paraprotein reativity with sialosyllactosaminylparagloboside (Miyatani et al., 1987).

6.3 Neuropathy Associated with Anti-Sulfatide IgM Paraproteins

Sulfatide is a major glycosphingolipid of myelin sheath. It is synthesized by Schwann cells in the PNS and by oligodendrocytes in the CNS. It has been demonstrated to be on the surface of dorsal root ganglion neurons

◘ Figure 11-5
Binding of a patient's IgM paraprotein to gangliosides containing a disialosyl group on TLC. (*Left panel*) Resorcinol stained gangliosides. Lane 1 contained GM1 and GD1a standards. Lanes 2 and 3 contained the ganglioside fraction from human brain. (*Right panel*) Autoradiogram of the chromatogram after overlaying with the patient's serum (1:300) followed by radioiodinated goat antihuman IgM (From Ilyas et al., Ann Neurol 1985; 18: 655–659; Little Brown and Company)

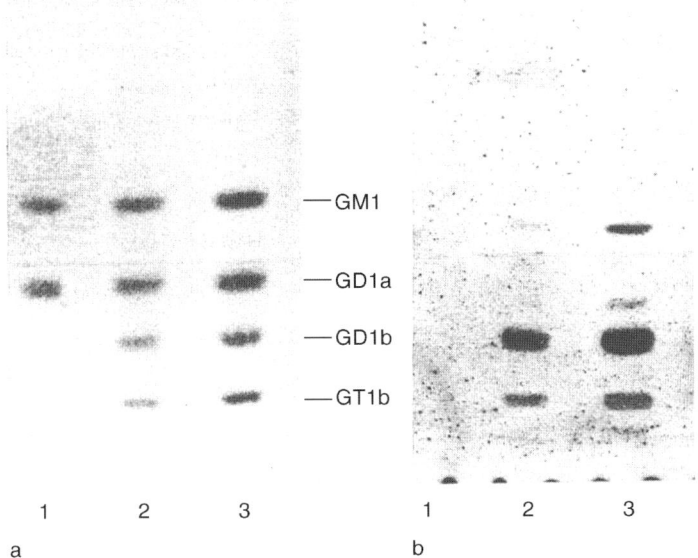

(Quattrini et al., 1992). Sulfatide has been shown to react with IgM from some patients with neuropathy associated with IgM gammopathy (❷ *Figure 11-6*, Pestronk et al., 1991; Ilyas et al., 1992; Quattrini et al., 1992). In a large study, 5% of patients with IgM gammopathy had high titer anti-sulfatide antibodies (Nobile-Orazio et al., 1994). These patients have widened myelin lamellae identical to those observed in patients with anti-MAG/SGPG gammopathy. Moreover, IgM and complement deposits in myelin were detected by direct immunofluorescence (Nobile-Orazi et al., 1994; Ferrari et al., 1998).

Lopate et al. (1997) reported that patients with IgM paraproteinemia and anti-sulfatide antibodies have a demyelinating neuropathy with significant motor involvement. However, in another study, most patients with neuropathy associated with IgM gammopathy and anti-sulfatide antibodies were reported to have an axonal, predominantly sensory neuropathy (Erb et al., 2000).

6.4 Neuropathy Associated with IgM Paraproteins to Other Antigens

Some patients with neuropathy and monoclonal gammopathy have IgM paraproteins that react with chondroitin sulfate C (Sherman et al., 1983; Yee et al., 1989). Reactivity with trisulfated heparin disaccharide has been described in patients with painful, predominantly sensory, axonal neuropathy associated with IgM gammopathy (Pestronk et al., 2003a). Antibodies to cytoskeletal proteins have also been occasionally described in patients with neuropathy and IgM gammopathy (Dellagi et al., 1982; Nobile-Orazio et al., 1994).

7 Neuropathy Associated with Nonmalignant IgG and IgA Monoclonal Gammopathy

The prevalence of neuropathy is much lower in patients with IgG gammopathy than with IgM gammopathy (Nobile-Orazio et al., 1992; Vretham et al., 1993). Neuropathy associated with IgG gammopathy is

◘ Figure 11-6
Binding of anti-MAG/SGPG IgM paraproteins from two patients with monoclonal gammopathy to sulfatide. Lane 1 contained bovine brain ganglioside standards. Lanes 2–4 contained sulfatide. Panel B and C were stained with two patients sera and peroxidase–conjugated antihuman IgM. (From Ilyas et al., J Neuroimmunol 1992; 37: 85–93) (Elsevier Science Publishers B.V.)

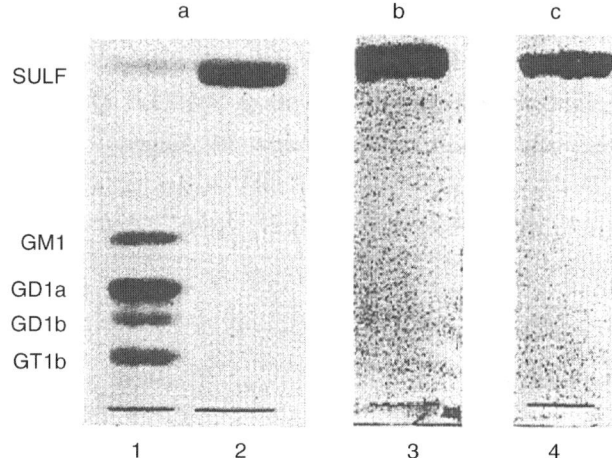

heterogenous. Di Troia and associates reported that in 17 patients with neuropathy and IgG gammopathy, 10 patients had a chronic demyelinating neuropathy clinically indistinguishable from chronic inflammatory demyelinating polyneuropathy (CIDP) while 7 had a predominantly sensory axonal or mixed neuropathy (Di Troia et al., 1999). Pathological studies of nerve biopsies from patients with neuropathy and IgG gammopathy show either demyelination or axonal degeneration.

Neuropathy associated with IgA MGUS is not common. Patients with IgA gammopathy also have demyelinating, axonal or mixed neuropathies (Nemni et al., 1991; Yeung et al., 1991; Simmons et al., 1993). Deposits of IgA paraproteins in the myelin sheath of patients with neuropathy associated with IgA gammopathy have been reported (Vallat et al., 2000; Mehndiratta et al., 2004). Moreover, widening of myelin lamellae identical to those commonly described in neuropathy associated with IgM gammopathy and anti-MAG antibodies has been reported in a patient with IgA gammopathy (Vallat et al., 2000). These studies suggest the possible pathogenic role of IgA paraproteins in neuropathy although the target antigen for IgA autoantibodies remains unknown.

Plasma exchange was effective on patients with IgG and IgA paraproteinemic neuropathies (Dyck et al., 1991). IVIg was beneficial in 8 (40%) of 20 patients with neuropathy associated with IgG MGUS (Gorson et al., 2002). Moreover, more than 60% of the patients with neuropathy and IgG gammopathy improved with immunotherapy (Di Troia et al., 1999).

8 Neuropathies Associated with Malignant Monoclonal Gammopathy

Although most of the patients with paraproteinemic neuropathy are associated with MGUS, some are associated with malignant paraproteinemia. These lymphoproliferative disorders include Waldenstrom's macroglobulinemia, myeloma, and cryoglobulinemia (Latov, 1995; Ropper and Gorson, 1998; Vital, 2001). Recent studies indicate that benign gammopathy progresses into malignant forms at 1% per year (Kyle and Rajkumar, 2003). Moreover, patients with neuropathy and benign gammopathy also progress into malignant gammopathy (Smith, 1994; Ponsford et al., 2000; Eurelings et al., 2005).

8.1 Neuropathy Associated with Waldenstrom's Macroglobulinemia (WM)

Waldenstrom's macroglobulinemia (WM) is a rare chronic B-cell lymphoproliferative disorder characterized by a monoclonal IgM and morphological evidence of lymphoplasmacytic lymphoma (Ghobrial et al., 2003). Peripheral neuropathy occurs in 5% to 10% of cases with WM (Dimopoulos and Alexanian, 1994). Neuropathy associated with WM and IgM paraproteins are thought to be of autoimmune origin. In some cases, the IgM paraprotein has been shown to react with myelin sheaths of peripheral nerve (Vital et al., 1985) and with MAG (Nobile-Orazi et al., 1994), suggesting that anti-MAG IgM paraproteins in these patients may have a causative role. Moreover, prominent widening of myelin lamellae in patients with neuropathy associated with WM and anti-MAG antibodies has been reported (Vital et al., 1997). Plasma exchange is beneficial in slowing the progress of the neuropathy associated with WM (Dalakas et al., 1983).

8.2 Neuropathy Associated with Multiple Myeloma

Multiple myeloma is a cancer of plasma cells. It is the most common hematological malignancy in patients with monoclonal gammopathy, with an annual incidence of three per 100,000 persons (Bataille and Harousseau, 1997). The paraprotein in multiple myeloma is usually IgG or IgA. Peripheral neuropathy affects less than 5% of patients with multiple myeloma and the neuropathy may be sensory, motor, or sensorimotor.

8.3 Neuropathy Associated with Osteosclerotic Myeloma (POEMS Syndrome)

Osteosclerotic myeloma constitutes approximately 3% of myelomas, but 50% of these have peripheral neuropathy (Miralles et al., 1992; Latov, 1995). POEMS (polyneuropathy, organomegaly, endocrinopathy, monoclonal gammopathy, skin changes) syndrome is a rare multisystemic disease that occurs in association with an IgG-lambda or IgA-lambda monoclonal gammopathy. POEMS syndrome is characterized by a chronic, progressive sensorimotor polyneuropathy with predominantly motor disability (Dispenzieri et al., 2003).

8.4 Neuropathy Associated with Cryoglobulinemia

In cryoglobulinemia, immunoglobulins (usually IgG and IgM) reversibly precipitate when cooled. The proteins may be monoclonal immunoglobulins (type I), or monoclonal and polyclonal immunoglobulins (type II), or polyclonal immunoglobulins (type III). Types II and III are called mixed cryoglobulinemia. Cryoglobulinemia is associated with many underlying systemic illnesses but when no underlying disease can be identified, it is called "essential cryoglobulinemia." The prevalence of peripheral neuropathies is up to 70% of patients with essential mixed cryoglobulinemia (Gemignani et al., 1992). Sensory neuropathy, often in the form of selective small fiber sensory neuropathy, is by far the commonest form of neuropathy associated with cryoglobulinemia (Gemignani et al., 2005). Many patients with mixed cryoglobulinemia have circulating antibodies against hepatitis C virus (Agnello et al., 1992). Nerve biopsies demonstrate vasculitis of the vasa nervorum (Vital et al., 1988; Gemignani et al., 1992).

9 Animal Models of Peripheral Neuropathies

9.1 Experimental Neuropathy Associated with Anti-MAG/SGPG Antibodies

Evidence in support of a causal role of anti-MAG/SGPG antibodies in peripheral neuropathy includes deposits of anti-MAG/SGPG IgM paraproteins and complement on affected myelin sheaths (Latov, 1995; Quarles and Weiss, 1999). Moreover, the IgM paraprotein deposits are found in myelin with wide spacing of lamellae that is the hallmark of this form of neuropathy (Mendell et al., 1985; Vital et al., 1989). Injection of serum from paraproteinemic patients with neuropathy and anti-MAG/SGPG antibodies and complement into feline and rabbit peripheral nerve produce myelin destruction (Hays et al., 1987; Willison et al., 1988; Monaco et al., 1995). Systemic transfusion of anti-MAG paraproteins was shown to cause segmental demyelination in the chicken (Tatum, 1993). There was binding of antibody to myelin sheaths, and widening of myelin lamellae, similar to the pathology of human disorder. However, systemic transfusion of anti-MAG paraproteins into guinea pigs, rabbits, and marmosets did not produce lesions in these animals (Steck et al., 1985). In another study, intraneural injections of rat anti-SGPG antibody induced demyelination in rat sciatic nerve, along with mild to moderate clinical symptoms (Maeda et al., 1991).

Active immunization experiments have also been performed with MAG and SGPG. Immunization of cats with human MAG did not induce peripheral neuropathy (Kahn et al., 1989) presumably due to the fact that the cats developed antibodies to the peptide part of MAG but not to the crucial oligosaccharide moiety that is recognized by anti-MAG/SGPG antibodies. Immunization of rabbits with purified SGPG produced mild neurological symptoms and electrophysiological dysfunction (Kohriyama et al., 1988), but demyelination of peripheral nerves was not observed. Maeda et al. (1991) reported that immunization of Lewis rats with SGPG induced high titer anti-SGPG antibodies but there was no evidence of sciatic nerve damage. Moreover, electrophysiological studies revealed no detectable conduction block or slowed conduction velocity in the sciatic nerves of sensitized rats. However, in a subsequent study by the same group, it was found that sensitization of Lewis rats with SGPG induced minor but clear clinical signs of neuropathy, consisting of mild tail muscle tone loss and walking disabilities. Electrophysiological examination of the sciatic nerves revealed nerve conduction abnormalities that consisted of conduction block and a mild decrease in conduction velocity (Yamawaki et al., 1996). In a more recent study, active sensitization of Lewis rats with purified SGPG produced high levels of anti-SGPG but did not produce any clinical and pathological changes in the sensitized animals (Ilyas et al., 2002). One reason for the failure to induce experimental neuropathy in the rats and rabbits is that these animals do not express HNK-1 epitopes on glycoproteins and express very low levels of sulfated glucuronyl glycosphingolipids (Quarles and Weiss, 1999). Thus, further study of immunization of animals such as the cats that express high levels of HNK-1 epitope on myelin proteins (O'Shannessy et al., 1985) and high levels of SGGLs in peripheral nerves (❯ *Figure 11-4*; Ilyas et al., 1986), with purified SGPG may produce an experimental model that closely resembles the human disease.

9.2 Experimental Neuropathies Induced by Antibodies to Gangliosides

Many attempts to induce peripheral neuropathy by systemic injection of anti-ganglioside antibodies have not been successful. However, a mild predominantly axonal neuropathy was induced in mice with intraperitoneal implantation of a hybridoma secreting antibody with GD1a ganglioside (Sheikh et al., 2004), presumably due to increased permeability of the blood–nerve barrier by the hybridoma.

IgM paraproteins that bind to disialosyl containing gangliosides including GD1b are strongly associated with a chronic sensory ataxic neuropathy (Dalakas and Quarles, 1996; Willison et al., 2001). Immunization of Japanese white rabbits with GD1b ganglioside induced acute sensory neuropathy that resembles human neuropathy associated with anti-GD1b autoantibodies (Kusunoki et al., 1996). The rabbits exhibited areflexia and ataxia with normal strength, and the pathological findings included axonal degeneration in the dorsal column of the spinal cord, dorsal roots, and sciatic nerve.

A reproducible experimental neuropathy by sensitization of rodents and rabbits with GM1 has not been observed. Thomas et al. (1991) immunized rabbits with GM1 that developed paralysis or subclinical neuropathy. However, immunization of Japanese white rabbits with ganglioside GM1 or bovine brain gangliosides induced high titer antibodies and an acute motor axonal neuropathy (Yuki et al., 2001). In two recent studies, however, immunization of New Zealand white rabbits with GM1 induced high titer anti-GM1 antibodies but not peripheral neuropathy (Lopez et al., 2002; Dasgupta et al., 2004). Sensitization of Lewis rats with KLH and *Campylobacter jejuni* LPS or with KLH and GM1 induced high titer anti-GM1 antibodies but the animals showed no overt signs of neuropathy (Wirguin et al., 1997; Ilyas and Chen, unpublished results). More recently, induction of high titer human IgM and IgG anti-GM1 antibodies in transgenic mice in response to lipopolysaccharides from *Campylobacter jejuni* did not induce peripheral neuropathy (Lee et al., 2004). These data suggest that high titer anti-GM1 ganglioside antibodies by themselves do not cause neuropathy.

9.3 Experimental Neuropathy Induced by Anti-Sulfatide Antibodies

Nardelli et al. (1995) reported that systemic injection of anti-sulfatide IgM paraproteins from a patient with demyelinating neuropathy into newborn rabbits reproduced demyelinating nerve lesion very similar to that in the donor's neuropathy. Moreover, IgM paraproteins were deposited at the nodes of Ranvier and Schmidt-Lanterman incisures. Also, sensitization of guinea pigs with sulfatide induced high titer anti-sulfatide antibodies and peripheral neuropathy in most animals. Demyelination of peripheral nerves and immunoglobulin deposits in myelin sheaths were found in symptomatic animals only (Qin and Guan, 1997).

10 Therapy

Treatments in paraproteinemic neuropathy patients have been directed at reducing the concentration of paraproteins by depleting the clones of plasma cells. Despite the obvious casual link between the paraprotein and neuropathy, therapies including steroids, cytotoxic agents such as cyclophosphamide, chlorambucil, fludarabine, cladribine, plasma exchange, and high dose intravenous immunoglobulin have offered only minimal and transient benefit in a small number of patients. Moreover, the response varies between patients, depending on the degree of reduction, the type of neuropathy, and the damage already present. Furthermore, these agents are toxic and have substantial side effects, which limit their protracted use (Latov et al., 1995; Nobile-Orazio, 2005).

Intravenous immunoglobulin (IVIg) has been reported to be effective in some patients with IgM paraproteinemic neuropathy including with anti-MAG IgM paraproteins (Dalakas et al., 1996; Mariette et al., 1997; Comi et al., 2002). Plasma exchange was found to be more effective in patients with IgG or IgA than in those with IgM paraproteins (Dyck et al., 1991).

Rituximab, a human mouse chimeric monoclonal antibody (IgG1 kappa isotype), is specific for the common B cell antigen CD20. It depletes pre-B and mature B lymphocytes without altering neutrophils or hematopoietic stem cells. Rituximab induces antibody dependent cell and complement mediated cytotoxicity, and reduces peripheral B lymphocyte counts by almost 90% within three days (Onrust et al., 1999; Maloney et al., 2002). In humans with indolent B cell lymphomas, rituximab can be safely administered, is well tolerated, promotes selective B cell depletion, and lowers the serum IgM levels. Rituximab treatment was initially approved for the treatment of non-Hodgkin's B-cell lymphoma. Recently, its use has been expanded to nonmalignant disorders, specifically those that are immune mediated. The use of rituximab in immune mediated polyneuropathy is based on the evidence that reduction in autoantibody levels by immunomodulating agents improves symptoms. Several open trials in patients with neuropathy associated with gammopathy and anti-MAG and anti-GM1 antibodies have shown beneficial effects of rituximab. Treatment of 21 patients with paraproteinemic neuropathy associated with anti-MAG and anti-GM1 IgM praproteins with rituximab was followed by improved strength, a reduction in serum IgM autoantibodies, and a reduction in total levels of IgM (Levine and Pestronk, 1999; Pestronk et al., 2003). The results of another study of rituximab, suggest a beneficial effect in nine patients with anti-MAG neuropathy that were resistant to other therapies (Renaud et al., 2003). A follow-up study by Renaud et al. has shown that a higher dose of rituximab was well tolerated and led to clinical improvement in four of eight patients, along with improvement of nerve conduction velocities and a reduction of anti-MAG antibody titers (Renaud et al., 2006). Recently, Goldfarb et al., reported a patient with an autonomic and painful sensory neuropathy associated with an IgM lambda monoclonal gammopathy, who was responsive to rituximab. Treatment resulted in a decline in total IgM and improvement in the patient's painful neuropathy and dysautonomia (Goldfarb et al., 2005). More recently, Kelly reported a patient with chronic PN associated with IgM gammopathy and anti-MAG antibodies, who improved after rituximab treatment (Kelly, 2006).

11 Conclusions

Great progress has been made in our understanding of the pathogenesis of paraproteinemic neuropathies since the original identification of anti-MAG IgM paraproteins in patients with demyelinating polyneuropathy almost 25 years ago. It is now well established that more than 70% of the patients with paraproteinemic neuropathies have IgM paraprotein reactivity with various glycoconjugates of peripheral nerves. The correlations of anti-glycoconjugate IgM paraprotein specificity and clinical symptoms support the notion that the paraproteins play an important role in the pathogenesis of neuropathies. Despite strong evidence that anti-MAG/SGPG IgM paraproteins may be pathogenic, several attempts to establish a clear causal relationship by active immunization of rabbits and rats with SGPG have not been very successful, in part due to the fact that the important HNK-1 oligosaccharide moiety that is the target of IgM paraproteins from patients with neuropathy is species restricted. Thus, there is a need to establish an experimental animal model of neuropathy associated with anti-MAG/SGPG antibodies. While IgM paraproteins from most neuropathy patients react with various glycoconjugate antigens, the antigen specificity of IgG and IgA paraproteins remains unknown. Response to therapy in IgM gammopathy and neuropathy is poor. Although several open trials in patients with neuropathy associated with gammopathy and IgM anti-glycoconjugate antibodies have shown beneficial effects of rituximab, a placebo randomized trial is needed to determine the efficacy of rituximab.

Acknowledgements

I am grateful to Dr. Yajuan Gu for her invaluable assistance with ❿ *Figure 11-1* and for review of the manuscript.

References

Agnello V, Chung RT, Kaplan LM. 1992. A role for hepatitis C virus infection in type II cryoglobulinemia. N Engl J Med 327: 1490-1495.

Ariga T, Kohriyama T, Freddo L, Latov N, Saito M, et al. 1987. Characterization of sulfated glucuronic acid containing glycolipids reacting with IgM M-proteins in patients with neuropathy. J Biol Chem 262: 848-853.

Ayadi H, Mihaesco, E, Congy N, Roy JP, Gendron MC, et al. 1992. H chain V region sequences of three human monoclonal IgM with anti-myelin-associated glycoprotein activity. J Immunol 148: 2812-2816.

Bataille R, Harousseau JL. 1997. Multiple myeloma. N Engl J Med 336(23): 1657-1664.

Bollensen E, Schipper HI, Steck AJ. 1989. Motor neuropathy with activity of monoclonal IgM antibody to GD1a ganglioside. J Neurol 236: 353-355.

Bollensen E, Steck AJ, Schachner M. 1988. Reactivity with the peripheral myelin glycoprotein P0 in serum from patients with monoclonal IgM gammopathy and polyneuropathy. Neurology 38: 1266-1270.

Braun PE, Frail DE, Latov N. 1982. Myelin-associated glycoprotein is the antigen for a monoclonal IgM in polyneuropathy. J Neurochem 39: 1261-1265.

Brouet JC, Mariette X, Gendron MC, Dubreuil ML. 1994. Monoclonal IgM from patients with peripheral demyelinating neuropathies crossreact with bacterial polypeptides. Clin Exp Immunol 96: 466-469.

Carpo M, Nobile-Orazi E, Meucci N, Gamba M, Barbieri S, et al. 1996. Anti-GD1a antibodies in peripheral motor syndromes. Ann Neurol 39: 539-543.

Chassande B, Leger JM, Younes-Chennoufi AB, Bengoufa D, Maisonobe T, et al. 1998. Peripheral neuropathy associated with IgM monoclonal gammopathy: Correlations between M-protein antibody activity and clinical/electrophysiological features in 40 cases. Muscle Nerve 21: 55-62.

Chou DK, Ilyas AA, Evans JE, Costello C, Quarles RH, et al. 1986. Structure of sulfated glucuronyl glycolipids in the nervous system reacting with HNK-1 antibody and some IgM paraproteins in neuropathy. J Biol Chem 261: 11717-11725.

Cohen HJ, Crawford J, Rao MK, Pieper CF, Currie MS. 1998. Racial differences in the prevalence of monoclonal gammopathy in a community-based sample of the elderly. Am J Med 104: 439-444. [Erratum, Am J Med 1998; 105: 362.]

Comi G, Roveri L, Swan A, Willison H, Bojar M, et al. 2002. A randomised controlled trial of intravenous immunoglobulin in IgM paraprotein associated demyelinating neuropathy. J Neurol 249: 1370-1377.

Connolly AM, Pestronk A, Mehta S, Yee WC, Green BJ, et al. 1997. Serum IgM monoclonal autoantibody binding to the 301 to 314 amino acid epitope of beta-tubulin: Clinical association with slowly progressive demyelinating polyneuropathy. Neurology 48: 243-248.

Dalakas MC, Quarles RH. 1996. Autoimmune ataxic neuropathies (sensory ganglionopathies): Are glycolipids the responsible autoantigens? [editorial]. Ann Neurol 39: 419-422.

Dalakas MC, Flaum MA, Rick M, Engel WK, Gralnick HR. 1983. Treatment of polyneuropathy in Waldenström's macroglobulinemia: Role of paraproteinemia and immunologic studies. Neurology 33: 1406-1410.

Dalakas MC, Quarles RH, Farrer RG, Damrosia J, Soueidan S, et al. 1996. A controlled study of intravenous immunoglobulin in demyelinating neuropathy with IgM gammapathy. Ann Neurol 40: 792-795.

Dasgupta S, Li D, Yu RK. 2004. Lack of apparent neurological abnormalities in rabbits sensitized by gangliosides. Neurochem Res 29(11): 2147-2152.

Daune GC, Farrer RG, Dalakas MC, Quarles RH. 1992. Sensory neuropathy associated with monoclonal immunoglobulin M to GD1b ganglioside. Ann Neurol 31: 683-685.

Dellagi K, Brouet JC, Perreau J, Paulin D. 1982. Human monoclonal IgM with autoantibody reactivity against intermediate filaments. Proc Natl Acad Sci 79: 446-450.

Dimopoulos MA, Alexanian R. 1994. Waldenström's macroglobulinemia. Blood 83: 1452-1459.

Dimopoulos MA, Panayiotidis P, Moulopoulos LA, Sfikakis P, Dalakas M. 2000. Waldenstrom's macroglobulinemia; clinical features, complications, and management. J Clin Oncol 18: 214-226.

Di Troia A, Carpo M, Meucci N, Pellegrino C, Allaria S, et al. 1999. Clinical features and anti-neuronal activity in neuropathy associated with IgG monoclonal gammopathy of undetermined significance. J Neurol Sci 164: 64-71.

Dispenzieri A, Kyle RA, Lacy MQ, Rajkumar SV, Therneau TM, et al. 2003. POEMS syndrome: Definitions and long-term outcome. Blood 101: 2496-2506.

Dyck PJ, Low PA, Windebank AJ, Jaradeh SS, Gosselin S, et al. 1991. Plasma exchange in polyneuropathy associated with monoclonal gammopathy of undetermined significance. N Engl J Med 325: 1482-1486.

Ellie E, Vital A, Steck A, Boiron JM, Vital C, et al. 1996. Neuropathy associated with "benign" anti-myelin-associated glycoprotein IgM gammopathy: Clinical, immunological, neurophysiological pathological findings and response to treatment in 33 cases. J Neurol 243: 34-43.

Erb S, Ferracin F, Fur P, Rosler KM, Hess CW, et al. 2000. Polyneuropathy attributes: A comparison between patients with anti-MAG and anti-sulfatide antibodies. J Neurol 247: 767-772.

Eurelings M, Notermans NC, de Donk Van NWCJ, Lokhorst HM. 2001. Risk factors for hematological malignancy in polyneuropathy associated with monoclonal gammopathy. Muscle Nerve 24: 1295-1302.

Eurelings M, Notermans NC, Lokhorst HM, Kessel van B, Jacobs BC, et al. 2006. Immunoglobulin gene analysis in polyneuropathy associated with IgM monoclonal gammopathy. J Neuroimmunol 175: 152-159.

Ferrari S, Morbin M, Nobile-Orazio E, Musso A, Tomelleri G, et al. 1998. Antisulfatide polyneuropathy: Antibody-mediated complement attack on peripheral myelin Acta Neuropathol 96: 569-574.

Freddo L, Yu RK, Latov N, Donofrio PD, Hays AP, et al. 1986. Gangliosides GM1 and GD1b are antigens for IgM M-protein in a patient with motor neuron disease. Neurology 36: 454-458.

Gemignani F, Brindani F, Alfiesi S, Giuberti T, Allegri I, et al. 2005. Clinical spectrum of cryoglobulinaemic neuropathy. J Neurol Neurosurg Psychiatry 76: 1410-1414.

Gemignani F, Pevesi G, Fiocchi A, Manganelli P, Ferraccioli G, et al. 1992. Peripheral neuropathy in essential mixed cryoglobulinemia. J. Neurol Neurosurg Psychiatry 55: 116-120.

Ghobrial IM, Gertz MA, Fonseca R. 2003. Waldenstrom macroglobulinemia. Lancet Oncol 4: 679-685.

Goldfarb AR, Weimer LH, Brannagan TH. 2005. Rituximab treatment of an IgM monoclonal autonomic and sensory neuropathy. Muscle Nerve 31: 510-515.

Gorson KC, Ropper AH, Weinberg DH, Weinstein R. 2002. Efficacy of intravenous immunoglobulin in patients with IgG monoclonal gammopathy and polyneuropathy. Arch Neurol 59: 766-772.

Hakomori S. 2002. The glycosynapse. Proc Natl Acad Sci 99(1): 225-232.

Hakomori S. 2003. Structure, organization, and function of glycosphingolipids in membrane. Curr Opin Hematol 10(1): 16-24.

Hammer JA, O'Shannessy DJ, De Leon M, Gould R, Zand D, et al. 1993. Immunoreactivity of PMP-22, P0, and other 19 to 28 kDa glycoproteins in peripheral nerve myelin of mammals and fish with HNK1 and related antibodies. J Neurosci Res 35: 546-558.

Hays AP, Lee SS, Latov N. 1988. Immune reactive C3d on the surface of myelin sheaths in neuropathy. J Neuroimmunol 18: 231-244.

Hays AP, Latov N, Takatsu M, Sherman WH. 1987. Experimental demyelination of nerve induced by serum of patients with neuropathy and an anti-MAG IgM M-protein. Neurology 37: 242-256.

Ilyas AA, Chen ZW, Prineas JW. 2002. Generation and characterization of antibodies to sulfated glucuronyl glycolipids in Lewis rats. J Neuroimmunol 127: 54-58.

Ilyas AA, Quarles RH, Brady RO. 1984a. The monoclonal antibody HNK-1 reacts with a peripheral nerve ganglioside. Biochem Biophys Res Commun 122: 1206-1211.

Ilyas AA, Cook SD, Dalakas MC, Mithen FA. 1992. Anti-MAG IgM paraproteins from some patients with polyneuropathy associated with IgM paraproteinemia also react with sulfatide. J Neuroimmunol 37: 85-92.

Ilyas AA, Dalakas MC, Brady RO, Quarles RH. 1986. Sulfated glucuronyl glycolipids reacting with anti-myelin-associated glycoprotein monoclonal antibodies including IgM paraproteins in neuropathy: Species distribution and partial characterization of epitopes. Brain Res 385: 1-9.

Ilyas AA, Quarles, RH, Dalakas MC, Brady RO. 1985a. Polyneuropathy with monoclonal gammopathy: Glycolipids are frequently antigens for IgM paraproteins. Proc Natl Acad Sci USA 82: 6697-6700.

Ilyas AA, Chou DH, Jungalwala FB, Costello C, Quarles RH. 1990. Variability in the structural requirements for binding of human monoclonal anti-MAG IgM antibodies and HNK-1 to sphingolipid antigens. J Neurochem 55: 594-601.

Ilyas AA, Quarles RH, Dalakas MC, Fishman PH, Brady RO. 1985b. Monoclonal IgM in a patient with paraproteinemic neuropathy binds to gangliosides containing disialosyl groups. Ann Neurol 18: 655-659.

Ilyas AA, Willison HJ, Dalakas MC, Whitaker JN, Quarles RH. 1988a. Identification and characterization of gangliosides reacting with IgM paraproteins in three patients with neuropathy associated with biclonal gammopathy. J Neurochem 51: 851-858.

Ilyas AA, Li S-C, Chou DKH, Li YT, Dalakas MC, et al. 1988b. Gangliosides GM2, IV^{4}GalNAcGM1b, and IV^{4}GalNAcGD1a as antigens for monoclonal IgM in neuropathy associated with gammopathy. J Biol Chem 263: 4369-4373.

Ilyas AA, Quarles RH, McIntosh TD, Dobersen MJ, Dalakas MC, et al. 1984b. IgM in a human neuropathy related to paraproteinemia binds to a carbohydrate determinant in the myelin-associated glycoprotein and to a ganglioside. Proc Natl Acad Sci USA 81: 1225-1229.

Irie S, Kanazawa N, Ogino M, Saito T, Funato T. 2000. No cytomegalovirus DNA in sera from patients with anti-MAG/SGPG antibody-associated neuropathy. Ann Neurol 47(2): 274-275.

Jungalwala FB. 1994. Expression and biological functions of sulfoglucuronylglycolipids (SGGLs) in the nervous system—a review. Neurochem Res 19: 945-957.

Kahn SN, Riches PG, Kohn J. 1980. Paraproteinaemia in neurological disease: Incidence, associations, and classification of monoclonal immunoglobulins. J Clin Pathol 33: 617-621.

Kahn SN, Stanton NL, Sumner AJ, Brown MJ, Spitalnik SL, et al. 1989. Analysis of feline immune response to human-myelin-associated glycoprotein. J Neurol Sci 89: 141-148.

Kaku DA, England JD, Sumner AJ. 1994. Distal accentuation of conduction slowing in polyneuropathy associated with antibodies to myelin-associated glycoprotein and sulphated glucuronyl paragloboside. Brain 117: 941-947.

Kelly JJ. 2006. Chronic peripheral neuropathy response to rituximab. Rev Neurol Dis 3(2): 78-81.

Kelly JJ, Adelman LS, Beman E, Bahn I. 1988. Polyneuropathies associated with IgM monoclonal gammopathies. Arch Neural 45: 1355-1359.

Kelly JJ, Kyle RA, O'Brien PC, Dyck PJ. 1981. Prevalence of monoclonal proteins in peripheral neuropathy. Neurology 31: 1480-1483.

Kohriyama T, Ariga T, Yu RK. 1988. Preparation and characterization of antibodies against a sulfated glucuronic acid-containing glycosphingolipid. J Neurochem 51: 869-877.

Kruse J, Mailhammer R, Wernecke H, Faissner A, Sommer I, et al. 1984. Neural cell adhesion molecules and myelin-associated glycoprotein share a common carbohydrate moiety recognized by monoclonal antibodies L2 and HNK-1. Nature 311: 153-155.

Kusunoki S, Kohriyama T, Pachner AR, Latov N, Yu RK. 1987. Neuropathy and IgM paraproteinemia: differential binding of IgM M-proteins to peripheral nerve glycolipids. Neurology 37: 1795-1797.

Kusunoki S, Shimizu J, Chiba A, Ugawa Y, Hitoshi S, et al. 1996. Experimental sensory neuropathy induced by sensitization with ganglioside GD1b. Ann Neurol 39: 424-431.

Kyle RA, Rajkumar SV. 2003. Monoclonal gammopathies of undetermined significance: A review. Immunol Rev 194: 112-139.

Kyle RA, Therneau TM, Rajkumar SV, Offord JR, et al. 2002. A long-term study of prognosis in monoclonal gammopathy of undetermined significance. N Engl J Med 346: 564-569.

Landgren O, Gridley G, Turesson I, Caporaso NE, Goldin LR, et al. 2006. Risk of monoclonal gammopathy of undetermined significance (MGUS) and subsequent multiple myeloma among African American and white veterans in the United States. Blood 107: 904-906.

Latov N. 1995. Pathogenesis and therapy of neuropathies associated with monoclonal gammopathies. Ann Neurol 37: S32-S42.

Latov N, Hays AP, Donofrio PD, Liao J, Ito H, et al. 1988. Monoclonal IgM with unique specificity to gangliosides GM1 and GD1b and to lacto-N-tetraose associated with human motor neuron disease. Neurology 38(5): 763-768.

Latov N, Sherman WH, Nemni R, Galassi G, Shyong JS, et al. 1980. Plasma cell dyscrasia and peripheral neuropathy with a monoclonal antibody to peripheral nerve myelin. N Engl J Med 303: 618-621.

Ledeen RW. 1989. Biosynthesis, metabolism, and biological effects of gangliosides. In: Margolis RU, Margolis RK (eds). Neurobiology of Glycoconjugates, Chapter 2. New York: Plenum; pp. 43-83.

Lee G, Ware RR, Latov N, 1994. Somatically mutated member of the human V lambda VIII gene family encodes anti-myelin-associated glycoprotein (MAG) activity. J Neuroimmunol 51: 45-52.

Lee G, Jeong Y, Wirguin I, Hays AP, Willison HJ, et al. 2004. Induction of human IgM and IgG anti-GM1 antibodies in transgenic mice in response to lipopolysaccharides from Campylobacter jejuni. J Neuroimmunol 146(2): 63-75.

Levine TD, Pestronk A. 1999. IgM antibody-related polyneuropathies: B-cell depletion chemotherapy using rituximab. Neurology 52: 1701-1704.

Li F, Pestronk A, Griffin J, Feldman EL, Cornblath D, et al. 1991. Polyneuropathy syndromes associated with serum antibodies to sulfatide and myelin-associated glycoprotein. Neurology 41: 357-362.

Lopate G, Kornberg AJ, Yue J, Choksi R, Pestronk A. 2001. Anti-myelin associated glycoprotein antibodies: Variability in patterns of IgM binding to peripheral nerve. J Neurol Sci 188(1–2): 67-72.

Lopate G, Parks BJ, Goldstein JM, Yee WC, Friesenhahn GM, et al. 1997. Polyneuropathies associated with high titre antisulphatide antibodies: Characteristics of patients with and without serum monoclonal proteins. J Neurol Neurosurg Psychiatry 62: 581-585.

Lopez PHH, Villa AM, Sica REP, Nores GA. 2002. High affinity as a disease determinant factor in anti-GM1 antibodies: Comparative characterization of experimentally induced vs. disease-associated antibodies. J Neuroimmunol 128: 69-76.

Lunn MP, Crawford TO, Hughes RAC, Griffin JW, Sheikh KA. 2002. Anti-myelin-associated glycoprotein antibodies alter neurofilament spacing. Brain 125: 904-911.

Lunn MP, Muir P, Brown LJ, Mac Mahon EM, Gregson NA, et al. 1999. Cytomegalovirus is not associated with IgM anti-myelin-associated glycoprotein/sulphate-3-glucuronyl paragloboside antibody-associated neuropathy. Ann Neurol 46(2): 267-270.

Maeda Y, Brosnan CF, Miyatani N, Yu RK. 1991a. Preliminary studies on sensitization of Lewis rats with sulfated glucuronyl paragloboside. Brain Res 541: 257-264.

Maeda Y, Bigbee JW, Maeda R, Miyatani N, Kalb RG, et al. 1991b. Induction of demyelination by intraneural injection of antibodies against sulfoglucuronyl paragloboside. Exp Neurol 113: 221-225.

Maloney DG, Smith B, Rose A. 2002. Rituximab: Mechanism of action and resistance. Semin Oncol 29 (Suppl 2): 2-9.

Mariette X, Chastang C, Clavelou P, Louboutin JP, Leger JM, et al. 1997. A randomised clinical trial comparing interferon-a and intravenous immunoglobulin in polyneuropathy associated with monoclonal IgM. J Neurol Neurosurg Psychiatry 63: 28-34.

McGarry RC, Helfand SL, Quarles RH, Roder JC. 1983. Recognition of myelin-associated glycoprotein by the monoclonal antibody HNK-1. Nature 306: 376-378.

Mehndiratta MM, Sen K, Tatke M, Bajaj BK. 2004. IgA monoclonal gammopathy of undetermined significance with peripheral neuropathy. J Neurol Sci 221: 99-104.

Mendell JR, Sahenk Z, Whitaker JN, Trapp BD, Yates AJ, et al. 1985. Polyneuropathy and IgM monoclonal gammopathy: Studies on the pathogenetic role of antimyelin-associated glycoprotein antibody. Ann Neurol 17: 243-254.

Miralles GD, O'Fallon JR, Talley NJ. 1992. Plasma-cell dyscrasia with polyneuropathy. The spectrum of POEMS syndrome. N Engl J Med 327: 1919-1923.

Miyatani N, Baba H, Sato S, Kanamura K, Yuasa T, et al. 1987. Antibody to sialosyllactosaminylparagloboside in a patient with IgM paraproteinemia and polyradiculoneuropathy. J Neuroimmunol 14: 189-196.

Monaco S, Ferrari S, Bonetti B, Moretto G, Kirshfink M, et al. 1995. Experimental induction of myelin changes by anti-MAG antibodies and terminal complement complex. J Neuropathol Exper Neurol 54: 96-104.

Montoto S, Rozman M, Rosinol L, Nadal E, Gine E, et al. 2003. Malignant transformation in IgM monoclonal gammopathy of undetermined significance. Semin Oncol 30: 178-181.

Nardelli E, Bassi A, Mazzi G, Anzini P, Rizzuto N. 1995. Systemic passive transfer studies using IgM monoclonal antibodies to sulfatide. J Neuroimmunol 63(1): 29-37.

Nemni R, Mamoli A, Fazio R, Camerlingo M, Quattrini A, et al. 1991. Polyneuropathy associated with IgA monoclonal gammopathy: A hypothesis of its pathogenesis. Acta Neuropathol (Berl) 81(4): 371-376.

Nobile-Orazio E. 2004. IgM paraproteinaemic neuropathies. Curr Opin Neurol 17: 599-605.

Nobile-Orazio E. 2005. Treatment of dysimmune neuropathies J Neurol 252: 385-395.

Nobile-Orazio E, Barbieri S, Baldini L, Marmiroli P, Carpo M, et al. 1992. Peripheral neuropathy in monoclonal gammopathy of undetermined significance: Prevalence and immunopathogenetic studies. Acta Neurol Scand 85(6): 383-390.

Nobile-Orazio E, Manfredini E, Carpo M, Meucci N, Monaco S, et al. 1994. Frequency and clinical correlates of antineural IgM antibodies in neuropathy associated with IgM monoclonal gammopathy. Ann Neurol 36: 416-424.

Ogawa-Goto K, Abe T. 1998. Gangliosides and glycosphingolipids of peripheral nervous system myelin—a minireview. Neurochem Res 23: 305-310.

Onrust SV, Lamb HM, Balfour JA. 1999. Rituximab. Drugs 58: 79-88.

O'Shannessy DJ, Willison HJ, Inuzuka T, Doberson MJ, Quarles RH. 1985. The species distribution of nervous system antigens that react with anti-myelin-associated glycoprotein antibodies. J Neuroimmunol 9: 255-268.

Pestronk A, Choksi R, Logigian E, Al-Lozi MT. 2003a. Sensory neuropathy with monoclonal IgM binding to a trisulfated heparin disaccharide. Muscle Nerve 27: 188-195.

Pestronk A, Florence J, Miller T, Choksi R, Al-Lozi MT, et al. 2003b. Treatment of IgM antibody associated polyneuropathies using rituximab. J Neurol Neurosurg psychiatry 74: 485-489.

Ponsford S, Willison H, Veitch J, Morris R, Thomas PK. 2000. Long-term clinical and neurophysiological follow-up of patients with peripheral neuropathy associated with benign monoclonal gammopathy. Muscle Nerve 23: 164-174.

Qin Z, Guan Y. 1997. Experimental polyneuropathy produced in guinea-pigs immunized against sulfatide. Neuroreport 8: 2867-2870.

Quarles RH. 1989. Human monoclonal antibodies associated with neuropathy. Meth Enzymol 179: 291-299.

Quarles RH. 2002. Myelin sheaths: Glycoproteins involved in their formation, maintenance and degeneration. Cell Mol Life Sci 59: 1851-1871.

Quarles RH, Weiss MD. 1999. Autoantibodies associated with peripheral neuropathy. Muscle Nerve 22: 800-822.

Quarles RH, Ilyas AA, Willison HJ. 1986. Antibodies to glycolipids in demyelinating diseases of the peripheral nervous system. Chem Phys Lipids 42: 235-248.

Quattrini A, Carbo M, Dhaliwal JL, Sadiq AS, Lugaresi A, et al. 1992. Anti-sulfatide antibodies in neurological disease: Binding to rat dorsal root ganglia neurons. J Neurol Sci 112: 152-159.

Renaud S, Fuhr P, Gregor M, Schweikert K, Lorenz D, et al. 2006. High-dose rituximab and anti-MAG neuropathy. Neurology 66(5): 742-744.

Renaud S, Gregor M, Fuhr P, Lorenz D, Deuschl G, et al. 2003. Rituximab in the treatment of polyneuropathy associated with anti-MAG antibodies. Muscle Nerve 27: 611-615.

Ritz MF, Erne B, Ferracin F, Vitale A, Vital C, et al. 1999. Anti-MAG IgM penetration myelin fibers correlates with the extent of myelin widening. Muscle Nerve 22: 1030-1037.

Ropper AH, Gorson KC. 1998. Neuropathies associated with paraproteinemia. N Engl J Med 1338: 1601-1607.

Sheikh KA, Zhang G, Gong Y, Schnaar RL, Griffin JW. 2004. An antiganglioside antibody-secreting hybridoma induces neuropathy in mice. Ann Neurol 56: 228-239.

Sherman WH, Latov N, Hays AP, Takatsu M, Nemni R, et al. 1983. Monoclonal IgMk antibody precipitating with chondroitin sulphate C from patients with axonal polyneuropathy and epidermolysis. Neurology 33: 192-201.

Simmons Z, Bromberg MB, Feldman EL, Blaivas M. 1993. Polyneuropathy associated with IgA Monoclonal gammaopathy of undetermined significance. Muscle Nerve 16: 77-93.

Smith IS. 1994. The natural history of chronic demyelinating neuropathy associated with benign IgM paraproteinemia. A clinical and neurophysiological study. Brain 117: 949-957.

Snipes GJ, Suter U, Shooter EM. 1993. Human peripheral myelin protein-22 carries the L2/HNK-1 carbohydrate adhesion epitope. J Neurochem 61: 1961-1964.

Spatz LA, Williams M, Brender B, Desai R, Latov N. 1992. DNA sequence analysis and comparison of the variable heavy and light chain regions of two IgM, monoclonal, anti-myelin associated glycoprotein antibodies. J Neuroimmunol 36: 29-39.

Steck AJ, Murray N, Justafre JC, Meier C, Toyka KV, et al. 1985. Passive transfer studies in demyelinating neuropathy with IgM monoclonal antibodies to myelin-associated glycoprotein. J Neurol Neurosurg Psychiatry 48: 927-929.

Suzuki M, Suetake K, Kasama T, Ariga T, Shiina M, et al. 2001. Characterization of a phospholipid antigen reacting with serum antibody in patients with peripheral neuropathies and paraproteinemia. J Neurochem 79: 970-975.

Svennerholm L, Bostrom K, Fredman P, Jungbjer B, Lekman A, et al. 1994. Gangliosides and allied glycosphingolipids in peripheral nerve and spinal cord. Biochim Biophys Acta 1214: 115-123.

Takatsu M, Hays AP, Latov N, Abrams GM, Nemni R, et al. 1985. Immunofluorescence study of patients with neuropathy and IgM M proteins. Ann Neural 18: 173-181.

Tatum AH. 1993. Experimental paraprotein neuropathy, demyelination by passive transfer of human IgM anti-myelin associated glycoprotein. Ann Neurol 33: 502-506.

Thomas FP, Adapon PH, Goldberg GP, Latov N, Hays AP. 1989. Localization of neural epitopes that bind to IgM monoclonal autoantibodies (M-proteins) from two patients with motor neuron disease. J Neuroimmunol 21: 31-39.

Thomas FP, Trojaborg W, Nagy C, Santoro M, Sadiq SA, et al. 1981. Experimental autoimmune neuropathy with anti-GM1 antibodies and immunoglobulin deposits at the nodes of Ranvier. Acta Neuropathol (Berl) 82: 378-383.

Trojaborg W, Hays AP, Berg van den L, Younger DS, Latov N. 1995. Motor conduction parameters in neuropathies associated with anti-MAG antibodies and other types of demyelinating and axonal neuropathies. Muscle Nerve 18: 730-735.

Vallat JM, Desproges-Gotteron MJ, Leboutet A, Loubet N, Gualde RT. 1980. Cryoglobulinemic neuropathy: A pathological study. Ann Neural 8: 179-185.

Vallat JM, Tabaraud F, Sindou P, Preux PM, Vandenberghe A, et al. 2000. Myelin widening and MGUS-IgA: An immunoelectron microscopic study. Ann Neurol 47: 808-811.

Van den Berg L, Hays AP, Nobile-Orazio E, Kinsella LJ, Manfredini E, et al. 1996. Anti-MAG and anti-SGPG antibodies in neuropathy. Muscle Nerve 19: 637-643.

Vital A. 2001. Paraproteinemic neuropathies. Brain Pathol 11: 399-407.

Vital A, Vital C, Julien J, Baquey A, Steck AJ. 1989. Polyneuropathy associated with IgM monoclonal gammopathy. Immunological and pathological study in 31 patients. Acta Neuropathol 79: 160-167.

Vital C, Deminiere C, Bourgouin B, Lagueny A, David B, et al. 1985. Waldenstrom's macroglobulinemia and peripheral neuropathy: Deposition of M-component and kappa light chain in the endoneurium. Neurology 35(4): 603-606.

Vital C, Deminiere C, Lagueny A, Bergouignan FX, Pellegrin JL, et al. 1988. Peripheral neuropathy with essential mixed cryoglobulinemia: Biopsies from 5 cases. Acta Neuropathol (Berl) 75(6): 605-610.

Vital C, Vital A, Deminiere C, Julien J, Lagueny A, et al. 1997. Myelin modifications in 8 cases of peripheral neuropathy with Waldenstrom's macroglobulinemia and anti-MAG activity. Ultrastruct Pathol 21: 509-516.

Vital C, Vital A, Ferrer X, Viallard JF, Pellegrin JL, et al. 2003. Crow–Fukase (POEMS) syndrome: A study of peripheral nerve biopsy in five new cases. J Peripher Nerv Syst 8: 136-144.

Vrethem M, Cruz M, Huang W, Malm C, Holmgren H, et al. 1993. Clinical, neurophysiological and immunological evidence of polyneuropathy in patients with monoclonal gammopathies. J Neurol Sci 114: 193-199.

Weiss MD, Dalakas MC, Lauter CJ, Willison HJ, Quarles RH. 1999. Variability in the binding of anti-MAG and anti-SGPG antibodies to target antigens in demyelinating neuropathy and IgM paraproteinemia. J Neuroimmunol 95: 174-184.

Willison HJ, Yuki N. 2002. Peripheral neuropathies and antiglycolipid antibodies. Brain 125: 2591-2625.

Willison HJ, Paterson G, Veitch J, Inglis G, Barnett SC. 1993. Peripheral neuropathy with monoclonal IgM anti-Pr2 cold agglutinins. J Neurol Neurosurg Psychiatry 56: 1178-1183.

Willison HJ, O'Leary CP, Veitch J, Blumhardt LD, Busby M, et al. 2001. The clinical and laboratory features of chronic sensory ataxic neuropathy with anti-disialosyl IgM antibodies. Brain 124: 1968-1977.

Willison HJ, Trapp BD, Bacher JD, Dalakas MC, Griffin JW, et al. 1988. Demyelination induced by intraneural injection of human anti-myelin-associated glycoprotein antibodies. Muscle Nerve 11: 1169-1176.

Wirguin I, Briani C, Suturkova-Milosevic L, Fisher T, Della-Latta P, et al. 1997. Induction of anti-GM1 ganglioside antibodies by *Campylobacter jejuni* lipopolysaccharides. J Neuroimmunol 78: 138-142.

Yamawaki M, Vasques A, Ben-Younas A, Yoshino H, Kanda T, et al. 1996. Sensitization of Lewis rats with sulfoglucuronyl paragloboside: Electrophysiological and immunological studies of an animal model of peripheral neuropathy. J Neurosci Res 44: 58-65.

Yee WC, Hahn AF, Hearn SA, Rupar AR. 1989. Neuropathy in IgM lambda paraproteinaemia. Immunoreactivity to neural proteins and chondroitin sulfate. Acta Neuropathol (Berl) 78: 57-64.

Yeung KB, Thomas PK, King RHM, Waddy H, Will RG, et al. 1991. The clinical spectrum of peripheral neuropathies associated with benign monoclonal IgM, IgG and IgA paraproteinemia. J Neurol 238: 383-391.

Yuki N, Yamamoto T, Hirata K. 1998. Correlation between cytomegalovirus infection and IgM anti-MAG/SGPG antibody associated neuropathy. Ann Neurol 44: 408-410.

Yuki N, Yamada M, Koga M, Odaka M, Susuki K, et al. 2001. Animal model of axonal Guillain-Barré syndrome induced by sensitization with GM1 ganglioside. Ann Neurol 49: 712-720.

12 Proteolytic Mechanisms of Cell Death in the Central Nervous System

S. F. Larner · R. L. Hayes · K. K. W. Wang

1	Introduction	250
2	Proteases	251
2.1	Calpains	251
2.1.1	Calpain Family	251
2.1.2	Calpain Inhibitors	253
2.2	Caspases	254
2.2.1	Caspase Family	254
2.2.2	Caspase Inhibitors	255
2.3	Cathepsins	257
2.3.1	Cathepsin Family	257
2.3.2	Cathepsin Inhibitors	259
2.4	Matrix Metalloproteinases	260
2.4.1	Matrix Metalloproteinase Family	260
2.4.2	Matrix Metalloproteinase Inhibitors	261
2.5	Serine Proteases	262
2.5.1	Serine Family	263
2.5.2	Serine Protease Inhibitors (Serpins)	265
3	Others	266
3.1	Proteasome	266
3.2	Necroptosis	267
4	Summary	267

© 2009 Springer Science+Business Media, LLC.

Abstract: The proteases implicated in programmed cell death herein discussed are involved in many normal and critical physiological functions within the central nervous system. However, when they are overexpressed following activation of the programmed cell death process they can be devastating to cells and the organism, even under controlled conditions. Because proteases (calpains, caspases, cathepsins, matrix metalloproteinases, and serine proteases) are widely expressed and extremely powerful, elaborate cellular safety precautions regulate their expression and activity. All proteases, for example, are translated as zymogens and are only activated under proscribed conditions. However, following injury, it is likely these regulatory controls are either disrupted and overwhelmed, thereby leading to the overexpression, overactivation, and possibly aberrant release of the proteases, or they are activated under energically and genetically controlled conditions leading to the demise of the cell. The relatively ubiquitous distribution of proteases and their documented role in the proteolysis of vital substrates make them strong candidates for involvement in both primary and secondary injury cascades. Active research continues not only into the proteolytic mechanisms of each member of the different families but also into their control mechanisms, both endogenous and synthetic, in the hope that therapeutic agents will be uncovered so that programmed cell death can be controlled for the benefit of the suffering patients.

List of Abbreviations: ALS, amyotrophic lateral sclerosis; BBB, blood–brain barrier; BIR, baculoviral IAP repeat; CARD, caspase recruitment domain; CNS, central nervous system; DED, death effector domain; DISC, death-inducing signaling complex; FADD, Fas-associated death domain protein; iNOS, inducible nitric oxide synthase; IAPs, inhibitors of apoptosis proteins; IBM, IAP-binding motif; MAP, mitogen-activated protein; OGD, Oxygen–glucose deprivation; PAI-1, plasminogen activator inhibitor-1; PARs, protease-activated receptors; PARP, poly-ADP-ribose-polymerase; PCD, Programmed cell death; PKA, protein kinase A; TBI, traumatic brain injury; SCI, spinal cord injury; TNFR, tumor necrosis factor receptor; TRADD, TNF-associated receptor with death domain; ZBGs, zinc binding groups

1 Introduction

Programmed cell death (PCD) is a naturally occurring conserved dedicated molecular program essential for the development and maintenance of multicellular organisms. It is designed to eradicate excess or potentially dangerous cells. It is genetically regulated, allowing multicellular organisms to tightly control cell number and tissue mass as well as to protect the organism against rogue cells that threaten homeostasis. As an active molecular process it often requires an upregulation of proteins for initiation (Zipfel et al., 2000). A term originally coined for this phenomenon, apoptosis, was adopted in 1972 by Currie and colleagues (Kerr et al., 1972). Interest in the process, however, was modest until the early 1990s when a number of observations noted that cell death was not incidental, but a critical factor in the pathobiology of many acute and chronic neurodegenerative disorders. It is now widely recognized as a normal part of the physiological processes that occur during development as well as during the pathological progress of a disease (Yuan et al., 2003). The terms apoptosis and PCD were commonly used synonymously; however, as research progressed a number of different forms of PCD were uncovered. For example, it was originally believed that necrosis or oncosis was accidental and that the cellular machinery had no active role. However, this manner of cell death has also been found to be regulated. It is now recognized that different signals and different PCD pathways can be evoked and that each is genetically controlled.

Apoptosis refers to a particular physiological change the cell undergoes as it is dying. This morphology derived from the activation of proteases, such as members of the caspase family, is common to many but not all cell deaths. It should be noted though that the results of a caspase-independent apoptosis often bear little morphological resemblance to its classical definition (Hengartner, 2000).

Various mechanisms of cell death with combined characteristics of necrosis and apoptosis have been proposed, such as oncosis and autophagic cell death. The hallmark of oncosis, often referred to as necrosis, is cellular swelling leading to rupture of the plasma membrane with extracellular leakage of cell components. When necrotic cells lyse, they provoke a substantial inflammatory response (Lockshin and Zakeri, 2001). Now, since it is generally understood that the genetic machinery can be evoked in necrosis, such as occurs with the activation of calpains, the necrosis versus apoptosis process is considered more of a continuum than sharply delineated patterns of cell death.

The finding in PCD research that inhibiting caspase-activated apoptosis did not necessarily protect a cell lead to the discovery of compensatory caspase-independent PCD pathways. These other pathways resemble both the apoptotic and necrotic PCD processes and even autophagic cell death. The evidence for alternative death pathways highlights the complexity of death signaling and that overlapping pathways initiated by a single stimulus is the rule rather than the exception. Examination of potential PCD pathway overlaps with different phenotypic outcomes allows for better understanding of the events that determine what commits a cell to PCD and which proteases are involved in upstream signaling or downstream execution within the cell (Leist and Jaattela, 2001).

From a clinical point of view the recognition that there are genetically regulated steps a cell follows to its demise and with the understanding that these pathways in human pathologies are frequently disabled offers hope that the PCD progression may be open to therapeutic intervention (Kass and Orrenius, 1999).

The focus of this chapter will be therefore limited to the better-studied proteolytic enzymes that govern PCD in the central nervous system (CNS). Proteases, also known as proteinases or proteolytic enzymes, refer to those proteins whose catalytic function is to hydrolyze (breakdown) proteins. The proteolytic mechanisms used in peptide bond cleavage involve making either an amino acid residue (serine or cysteine peptidases) or a water molecule (aspartic or metallo peptidases) nucleophilic, enabling it to split the peptide carbonyl group. Proteases constitute between 1% and 5% of the genetic code in all organisms. These enzymes are involved in a multitude of physiological reactions from digestion necessary for cellular maintenance to the more highly regulated cascades that can be destructive to the cell as part of a signaling pathway like apoptosis.

2 Proteases

2.1 Calpains

Calpains, a family of calcium-activated neutral cytosolic cysteine proteases, are involved in a wide variety of basic physiological and pathological processes (Sorimachi et al., 1997; Huang and Wang, 2001). Currently, two major isoenzymes of calpain are known to play a major role in cell death processes, μ-calpain (calpain-1) and m-calpain (calpain-2) (Suzuki et al., 1981). Calpain proteolysis is relatively selective for subsets of cellular proteins including cytoskeletal proteins (e.g., spectrin; Nath et al., 1996a), calmodulin binding proteins (Wang et al., 1989; McGinnis et al., 1998), signal transduction enzymes, membrane proteins, and transcription factors (Saido et al., 1994; Goll et al., 2003). The recent increase in interest in calpains can be attributed to the growing recognition that pathological cascades that follow CNS insult such as traumatic brain injury (TBI), cerebral ischemia, and spinal cord injury (SCI) are, at least in part, mediated by increased intracellular calcium and subsequently the activation of calpains. Although calpain activation has been historically assumed to result in oncotic necrosis, several recent investigations have suggested that calpains may also contribute to different models of apoptosis (Squier et al., 1994; Behrens et al., 1995; Nath et al., 1996b; Jordan et al., 1997; Waterhouse et al., 1998; Pike et al., 1998a; Tan et al., 2006).

Calpains' role in cell death is not as well characterized as that for caspases, but it is believed calpains are activated by any insult that stimulates a rise in cytosolic calcium whether from the endoplasmic reticulum (ER), the mitochondria, or via an influx of extracellular calcium. Furthermore, recent work suggests calpains can be activated via a mitogen-activated protein (MAP) kinase signaling pathway in the absence of calcium influxes. Due to their abundance, known cleaveage of important intracellular signaling and structural proteins as well as regulators of apoptosis, and due to their cross-talk with caspases, their activity has to be tightly regulated both temporally and spatially.

2.1.1 Calpain Family

Family members. Fourteen mammalian genes have been identified that encode for the isoform-specific large subunit (calpain 1–3, 5–15) and two genes for the small regulatory subunits. At least two calpains have more than one splice variant, eight for calpain-10 and two for calpain-3. These large subunit isoforms are often classified into two groups: ubiquitous and tissue specific. Two of the isoforms, calpain-1 (μ-calpain) and calpain-2 (m-calpain), ubiquitously distributed in all mammalian cells, have been fairly well characterized (Huang and Wang, 2001). Except for a difference in Ca^{2+} concentrations required for in vitro activation, they have similar biochemical features. The other members, many of which were originally considered to be tissue specific, are now believed to be ubiquitous and have been described as possessing a typical (i.e., a four-domain large subunit associated with a two-domain regulatory subunit) structure.

With the recent discovery of several new members including calpain-5, -6, -7 and -10 (Dear et al., 1997; Matena et al., 1998; Franz et al., 1999; Ma et al., 2001), more research has focused on the roles these newcomers play in tissue physiology. For example in recently completed studies, for human tissue-specific calpain-5 (htra-3) (Dear et al., 1997; Waghray et al., 2004), the highest mRNA levels in human tissues were found in the testis, lung, liver, kidney, colon, and brain (Waghray et al., 2004). Calpain-5 is not required for development, but some of the homozygous knockout offspring from heterozygous matings were small and sickly and did not survive to adulthood (Franz et al., 2004). To date the physiological role of calpain-5 is still unknown but in a recently completed study it was demonstrated that calpain-5, like its cousins calpain-1 and -2, may be involved in necrotic cell death in mammalian cells.

Structure. To protect cells from unintended proteolysis, calpains exist in the cytosol as inactive proenzymatic-heterodimers consisting of a large 80 kDa catalytic subunit and a common 28 kDa small regulatory subunit. The catalytic subunit is isoform specific and ranges in size from 73 kDa (for calpain-5) to 117 kDa (for calpain-15). A typical calpain contains six domains (D), four (I–IV) of which reside on the large subunit and two on the smaller subunit (V–VI). Calcium, which appears to be required for calpain activation, was originally thought to bind only at the "calmodulin-like" domains (penta-EF hand-rich domains IV and VI), but this is no longer believed to be accurate. In the presence of Ca^{2+}, the large subunit D-I is cleaved off, activating the enzyme. The catalytic active site resides in D-II and contains the catalytic triad of Cys, His, and Asn and two non-EF Ca^{2+}-binding sites, which reportedly serve as a calcium-switch capable of activating calpain-1 and -2 (Huang and Wang, 2001; Hosfield et al., 2004). D-III has been reported to bind calcium via C2-like folding (Khorchid and Ikura, 2002; Moldoveanu et al., 2002). D-IV contains four calmodulin-like calcium binding EF-hand motifs, which confers additional calcium proteolytic activity regulatory capabilities. The amino acid sequence of calpain-5 has been found to be similar to other members of the calpain family except in the calmodulin calcium-binding EF-hand structure of D-IV. This EF-hand structure has been replaced by a sequence termed domain T (Barnes and Hodgkin, 1996; Franz et al., 1999). It was determined, through multiple amino acid sequence alignments that there is a high degree of similarity in D-II and D-III across species, from nematode to arthropods to mammals (Barnes and Hodgkin, 1996; Waghray et al., 2004; Kim et al., 2005).

Function. Activated calpains cleave a wide variety of substrates including cytoskeletal proteins like αII-spectrin (α-fodrin), talin, microtubule-associated proteins, membrane proteins including growth factor receptors, adhesion molecules (e.g., integrin), and ion transporters (e.g., Ca^{2+}-ATPase) as well as other enzymes such as protein kinase C (Sorimachi et al., 1997; Huang and Wang, 2001; Goll et al., 2003). Interestingly, the unique breakdown products of spectrin (150- and 145-kDa fragments) are used as prognostic markers of calpain activation after TBI (Pike et al., 1998b; 2001).

Calpain involvement in a number of cellular functions such as cell migration and cytoskeletal remodeling (Wu et al., 2006) is required for normal embryonic development. In calpain knockout mice the consequences of a disruption of the calpain-4 (small subunit) homozygous gene was embryonic lethality due to defects in the cardiovascular system, hemorrhaging, and accumulation of erythroid progenitors (Arthur et al., 2000). Active calpains have been associated with the pathogenesis of a wide range of disorders including cataract formation, diabetes mellitus, muscular dystrophies, neurodegenerative disorders such as Alzheimer's disease (Nixon, 2000; Huang and Wang, 2001), Huntington's disease (Gafni and Ellerby, 2002; Gafni et al., 2004; Schilling et al., 2006), prion disorders (Yadavalli et al., 2004), and secondary degeneration resulting from acute insults including myocardial infarction, stroke, TBI, and SCI (Huang and Wang, 2001; Branca, 2004). Calpains have been implicated in both apoptotic and necrotic cell deaths (Solary et al., 1998; Wang, 2000; Tan et al., 2006). Calpains are regulated by a variety of factors including intracellular calcium fluxes, autoproteolytic cleavage, and phospholipid binding (Bozoky et al., 2005), via the endogenous inhibitor calpastatin (Sorimachi et al., 1997), by phosphorylation through protein kinase A (PKA) (Shiraha et al., 2002), and perhaps by its intracellular distribution.

2.1.2 Calpain Inhibitors

Endogenous. In vivo calpain activity is tightly regulated by the endogenous inhibitor calpastatin, the product of a single gene. However, at least eight different calpastatin isoforms have been uncovered ranging from 17.5 kDa to 84 kDa, the results of different promoters or alternative splicing mechanisms (Goll et al., 2003). Calpastatin inhibits calpain by interacting with several sites on the calpain protein. When breakdown in calpain–calpastatin binding occurs, several important pathological disorders follow.

Synthetic. To further understand calpains and to manage the pathophysiological diseases they are associated with, a large diversity of synthetic calpain inhibitors have been developed, many of them offering protection from ischemia and toxicant-induced cell death in a variety of models. The first inhibitors, peptide aldehydes (e.g., leupeptin) (Umezawa et al., 1973), provided the starting point to later synthesized compounds that have exhibited improved membrane permeability and calpain specificity. For example, calpeptin, found to be more potent than leupeptin (Tsujinaka et al., 1988), when intravenously infused into an injured rat's spinal cord inhibited degradation of neurofilament (NF) 68 and NF200 and reduced cell death (Banik et al., 1998; Ray and Banik, 2003). Other inhibitors have since been discovered, including α-dicarbonyls (originally developed as serine protease inhibitors), nonpeptide quinolinecarboxamides, and α-mercaptoacrylic acids (Wang et al., 1996); phosphorus derivatives (Liu et al., 2004b); epoxysuccinates (relatively nonspecific cysteine protease inhibitors), acyloxymethyl ketones, and halomethylketones (cell permeable and selective for calpain over cathepsins); sulfonium methyl ketones (calpain-2 specific), diazomethyl ketones, and Z-Leu-Abu-CONHEt (AK275); 27-mer calpastatin peptide; all offering protection from ischemic and toxicant-induced cell death in diverse models (Liu et al., 2004b).

The epoxysuccinate inhibitor E64, a commonly used, generally nonselective, irreversible calpain inhibitor, and its analogues (Liu et al., 2004b) have been widely used to inhibit a variety of proteases including calpains, cathepsins B, H, and L (Barrett et al., 1982), and matrix metalloproteinases (Tsubokawa et al., 2006).

Calpain inhibitors I and II, PD150606 (3-(4-iodophenyl)-2-mercapto-(Z)-2-propenoic acid (a Ca^{2+}-binding site inhibitor)), nonpeptidyl inhibitors including chloroacetic acid N0-(6,7-dichloro-4-phenyl-3-oxo-3,4-dihydroquinoxalin-2-yl) hydrazide (SJA-7029), and several peptidyl α-keto amide inhibitors were found to protect renal proximal tubules from oncosis and to promote the recovery of mitochondrial function and active ion transport (Liu et al., 2004b). Another calpain inhibitor, MDL28170 (Cbz-Val-Phe-H), protected hippocampal neurons and reduced focal cerebral ischemic injury and death in rats (Hong et al., 1994; Rami et al., 2000). Furthermore, CEP-4143 inhibited calpain-1 activation and cytoskeletal degradation, improved neurological function, and enhanced axonal survival after SCI in female rats (Schumacher et al., 2000). The reversible inhibitor, dipeptidyl aldehyde SJA6017 (Fukiage et al., 1997; Kupina et al., 2001; Inoue et al., 2003), has been shown to have neuroprotective efficacy in the rat retinal ischemia model when intravenously administered. However, it requires a relatively high dose when orally administered to be efficacious and shows poor bioavailability (Shirasaki et al., 2005). A more chemically stable modified version of SJA6017, ketoamide SNJ-2008, showed good oral bioavailability and a higher retinato-plasma ratio and appears to be a good candidate for further development (Shirasaki et al., 2006).

Even with all the preclinical successes, currently none of the calpain inhibitors have been found to be selective for the different calpains in cell cultures or in vivo (Pignol et al., 2006), nor sufficiently potent or stable. The current diverse list however provides researchers with a starting point for characterizing the activity and function of calpains, and due to their importance in disease they will continue to stimulate development.

Alternatives. An alternate approach to controlling calpain activation uses molecular biological techniques that selectively decrease specific calpains. This approach has shown potential in cloning and sequencing calpain family members and includes the development of knockout and transgenic animal models and the use of antisense and siRNA technologies. Zhang et al. (1996) successfully employed antisense oligonucleotides to decrease expression of the calpain small subunit. Similar approaches could be used to target specific calpain members prior to cell injury to determine their specific role in the process of oncosis.

2.2 Caspases

The family of *c*ysteine-dependent *asp*artate-specific prote*ases*, or caspases, has been found to be critical mediators of apoptosis. To date 14 members have been identified, 12 in humans. Caspase-11 is restricted to the mouse and caspase-13 represents the bovine homolog to human caspase-4. The caspase family plays a key role in the implementation of apoptosis (Jacobson et al., 1997) if one maintains the strict morphological criteria of apoptosis including chromatin condensation (Leist and Jaattela, 2001). The implementation of apoptosis in mammalian cells generally requires the coordinated action of caspases since these proteases are responsible for the cleavage of key substrates that result in the systematic and orderly disassembly of dying cells. Caspases, synthesized as inactive proenzymes, are activated during apoptosis by cleavage of the N-terminal prodomain to yield the mature form containing both large, approximately 20 kDa, and small, approximately 10 kDa, subunits complexed to form an active tetramer (Rathmell and Thompson, 1999).

2.2.1 Caspase Family

Family members. All caspase zymogens must be proteolytically processed to become active (Cohen, 1997) with perhaps the exception of caspase-9 (Rodriguez and Lazebnik, 1999; Stennicke et al., 1999). The caspase family is divided into 3 broad functional categories: initiator caspases (caspase-2, -4, -8, -9, -10, and -12), the downstream effector caspases (caspase-3, -6, and -7), and a third group that generally functions as both initiator and inflammatory caspases (caspase-1, -5, and -11) and is associated with immune responses to microbial pathogens. The initiator caspases are upstream in the apoptotic pathway and respond to apoptotic stimuli by undergoing autoproteolytic activation. The downstream effector caspases, processed by the active initiator caspases, are responsible for dismantling cellular structure via substrate cleavage (Degterev et al., 2001). The inflammatory caspases are activated after assembly of an intracellular complex, termed the inflammasome, which results in the cleavage and activation of the proinflammatory cytokines IL-1β and IL-18 (Martinon and Tschopp, 2004).

Structure. The constitutively expressed procaspases are believed to have little or no enzymatic activity thus preventing the triggering of PCD. There is a high degree of homology in the protein structure of the caspases with the main structural differences localized to the active sites. Zymogens require a minimum of two cleavages to be converted to a mature or active enzyme, one separating the N-terminal prodomain from the large subunit (p20) and small subunit (p10) and a second to separate the two subunits. The initiator caspase prodomain motif is conserved and contains either the caspase recruitment domain (CARD) or the death effector domain (DED). The caspases activated predominantly via the extrinsic or death receptor pathway (caspases-8 and -10) have DED as prodomains, whereas the caspases activated by the intrinsic or mitochondrial pathway (caspases-2 and -9) have CARDs as prodomains. Executioner (effector) caspases (caspases-3, -6, and -7) have short prodomains.

When activated the large subdomain of approximately 16 to 22 kDa and small subdomain of approximately 8–12 kDa (Cohen, 1997) form heterotetramers possessing enzymatic activity. The active sites are directionally opposite except for caspase-9, which has but one active site (Degterev et al., 2001). The differences in active sites among the caspases lead to variability of substrate specificity with three distinct classes being identified. These are group I caspases (caspase 1, 4, 5) that prefer the tetrapeptide sequence WEHD; group II (caspase 2, 3, and 7) prefer DEXD, while group III (caspase 6, 8, 9) favor (l/V)EXD (Degterev et al., 2001). This allows the caspases to act on different substrates, in different cell types, or different cellular compartments.

Function. Most studies on PCD have focused on two canonical pathways (Kaufmann and Hengartner, 2001). The extrinsic pathway initiates apoptosis with the ligation of ligands, FasL or TNFα, to death domain-containing members of the tumor necrosis factor receptor (TNFR) superfamily, such as Fas receptor and TNFR-1, respectively. The increased expression and interaction of the Fas receptor (Beer et al., 2000; Qiu et al., 2002) and TNFR-1 (Taupin et al., 1993; Beer et al., 2000) have been demonstrated in TBI. The Fas death-inducing signaling complex (DISC), containing the adaptor protein Fas-associated death domain protein (FADD), and the TNFR-1 DISC, containing the TNF-associated receptor with death domain (TRADD), and their respective procaspases-8 and -10, leads to the autoproteolytic processing of these zymogens. Depending on the context, caspases-8 and -10 will proteolytically activate the executioner caspases-3, -6, and -7, which are responsible for dismantling cellular proteins directly, or by initiating an apoptotic cascade by cleaving Bid (to tBid) that will activate the mitochondrial or intrinsic pathway. This latter pathway activates caspase-9, and then, eventually, the same executioner caspases (Hengartner, 2000).

Activation of the intrinsic pathway such as after TBI involves altering mitochondria homeostasis and release of cytochrome c as well as other mitochondrial polypeptides such as SMAC/Diablo and AIF into the cytosol (Xiong et al., 1997; Raghupathi et al., 2000). Once released cytochrome c promotes the assembly of the apoptosome macromolecule whose other members include Apaf-1, dATP, and the recruitment and activation of caspase-9. This assemblage then activates the executioner caspases (Hengartner, 2000).

A novel pathway to caspase activation emerged with the discovery that calpain-2 and caspase-7 activated the endoplasmic reticulum (ER) bound caspase-12, independent of the intrinsic and extrinsic pathways (Nakagawa and Yuan, 2000; Nakagawa et al., 2000; Rao et al., 2001). This third pathway initiated under ER stress, such as that might occur with the disruption of Ca^{2+} homeostasis, is mediated through ER-specific proteins. It has been reported that the ER chaperone GRP78 (BiP) constitutively associates with procaspase-7 (Reddy et al., 2003) forming a complex with caspases-7 and -12 (Rao et al., 2002). Under ER stress, BiP is released and caspase-7 cleaves caspase-12 (Rao et al., 2001), activating the caspase apoptotic cascade thus coupling ER stress to the cell death program. The activation of caspase-12 (an initiator) by caspase-7 (an effector) is unusual. Caspase-4, also an ER bound caspase and only found in humans, is also believed to be activated (caspase-12-like) by this pathway. As previously discussed, calpain-2 and caspase-7 are activated by TBI (Huang and Wang, 2001; Larner et al., 2005), suggesting that both play an active role in the activation of caspase-12 (Larner et al., 2004).

Activated executioner caspases cleave numerous targets including cytoskeleton proteins (nuclear lamins, actin, gelsolin), DNA repair regulator (PARP), and cell cycle proteins (Rb and p21-activated kinase [PAK]) (Enari et al., 1998). They also precipitate the degradation of nuclear DNA by activating DNase (CAD) by cleaving the inhibitor of caspase-activated deoxyribonuclease protein (ICAD/DFF45) (Enari et al., 1998; Sakahira et al., 1998; Liu et al., 2001). The cleavage and inactivation of ICAD allows CAD to enter the nucleus and fragment the DNA, causing the characteristic 50–200 kb DNA laddering fragments seen in apoptotic cells. Caspase-dependent cleavage of p21 (Fujita et al., 1998) and Rb (Zhang et al., 1999) during DNA-damage-induced apoptosis suggests that there is a link between apoptosis and cell cycle progression. Caspases are also involved in extracellular apoptotic events including the cleavage of apoptotic bodies and exposure of phosphatidylserine on the cell surface. Recent studies have also demonstrated that the antiapoptotic proteins Bcl-2, Bcl-X_L, and XIAP are also substrates for caspase cleavage (Cheng et al., 1997; Fattman et al., 1997; Grandgirard et al., 1998). Cleavage of these proteins releases a C-terminal product lacking the BH4 domain converting them into proapoptotic death effectors. The exact contribution of caspase substrate cleavage to the biochemistry and morphology of apoptosis is still under investigation.

2.2.2 Caspase Inhibitors

Endogenous. Understanding the mechanisms of direct caspase inhibition is largely based on the study of viruses and the methods they use to prevent cell death. Genetic analysis of the baculovirus genome led to the identification of a direct caspase inhibitor, p35 (Ekert et al., 1999), the viral protein used to suppress the host cell's death response strategy. Similar to viruses, cells have their own class of inhibitors of apoptosis proteins (IAPs), which are characterized by the presence of a homologous domain named the baculoviral IAP repeat (BIR) domain. At least eight IAP-like molecules have been described in humans: NAIP, c-IAP1, c-IAP2, XIAP/hILP-1, Survivin, and BRUCE/Apollon (Rothe et al., 1995; Duckett et al., 1996;

Liston et al., 1996; Ambrosini et al., 1997; Hauser et al., 1998), Ts-XIAP/hILP-2, and the most recent member, ML-IAP/Livin/KIAP (Lin et al., 2000; Vucic et al., 2000; Kasof and Gomes, 2001).

Several of the human IAPs reportedly bind directly and inhibit specific members of the caspase family. For example, XIAP, c-IAP1, c-IAP2, and survivin directly inhibit caspases-3, -7, and -9 (Sanna et al., 2002). In contrast, NAIP does not seem to bind caspases (Roy et al., 1997), even though inhibition of caspase-3 and -7 has been reported (Shin et al., 2001) but may act through caspase-dependent and -independent pathways (Mercer et al., 2000). NAIP along with ML-IAP activates JNK1. When JNK1 is catalytically inactivated, it blocks the ability of NAIP and ML-IAP to protect against ICE- and TNF-α-induced apoptosis, indicating that JNK1 activation is necessary for the antiapoptotic effect of these IAPs (Sanna et al., 2002).

Ts-XIAP (ILP-2), unlike XIAP (ILP-1), potently inhibits apoptosis induced by active caspase-9 due to interactions that suggest restricted specificity (Richter et al., 2001). BRUCE/Apollon associates with both the precursors as well as the mature forms of Smac/DIABLO (an inhibitor of IAPs), HtrA2 (serine protease), and caspase-9. Through these associations, BRUCE promotes the degradation of Smac thereby inhibiting caspase-9 activation. In response to apoptotic stimuli, BRUCE is degraded by proteasomes, or cleaved by caspases and HtrA2, depending on the specific stimulus and the cell type, thus preventing its inhibitory activity (Qiu and Goldberg, 2005).

Some IAPs such as ML-IAP also contain a RING domain that has E3 ubiquitin ligase activity thereby promoting, through ubiquitination, the degradation of Smac/DIABLO. Since Smac/DIABLO promotes apoptosis by inhibiting IAP-caspase interactions, degradation of Smac/DIABLO allows IAPs to more effectively block caspase activity thereby promoting cell survival (Crnkovic-Mertens et al., 2006).

The cell death regulators need not be proteins. Two microRNAs identified in *Drosophila* have been shown to be capable of controlling translation of apoptotic proteins quickly and reversibly. The first, bantam, suppresses *hid* translation independent of the Ras/MAPK pathway (Brennecke et al., 2003) while promoting cell growth and proliferation by some unknown mechanism (Xu et al., 2004). The other, mir-14, regulates the *Drosophila* caspase Drice (Xu et al., 2003). It will be interesting to discover whether mammalian apoptotic components are also regulated by microRNAs.

Synthetic. Designed caspase inhibitors demonstrated neuroprotective capabilities in many animal models of ischemia and TBI (Wang, 2000). For example, the potent pan-caspase inhibitor M826 is neuroprotective against neonatal hypoxic–ischemic brain injury (Han et al., 2002) while MX1013 reduced cortical damage in an ischemia/reperfusion brain injury model (Yang et al., 2003b). Recently, the chemically optimized and drug-like caspase inhibitor, IDN-6556, was reported to suppress apoptosis in a model of hepatic injury (Linton et al., 2005). Another caspase inhibitor, IDN-5370, was found to be protective of cortical and synaptic neurons by substantially reducing infarct size in rodent cardiac ischemia/reperfusion models (Kreuter et al., 2004). The discovery of a class of drugs that selectively inhibit inflammatory caspases (caspase-1, -4, and -5) may help to control autoimmune diseases like rheumatoid arthritis. Among the inhibitors of the inflammatory caspases, VX-740 (pralnacasan) is in Phase II clinical studies for rheumatoid arthritis and other diseases (Leung-Toung et al., 2002).

Minocycline, a broad-spectrum tetracycline antibiotic homolog, in use for more than 30 years and specifically designed to cross the blood–brain barrier (BBB), was found to inhibit cytochrome *c* release. Recently this antibiotic was reportedly protective in disease models of acute brain injury, multiple sclerosis (MS), amyotrophic lateral sclerosis (ALS), AD, stroke and more. Finally, dexamethasone was found to decrease caspase-3 activation in a rat model of bacterial meningitis, demonstrating that it could decrease brain injury as measured by neurobehavioral performance (Irazuzta et al., 2005).

There are also potent peptide-based inhibitors that use sequences corresponding to specific caspase-substrate cleavage sites. Each caspase specifically recognizes a 4–5 amino acid sequence on the target substrate including the required aspartic acid (D) in the P1 position (Thornberry et al., 1992). Z-YVAD-fmk is a synthetic peptide that irreversibly inhibits caspase-1 activity. Another version of the inhibitor YVAD-CHO inhibited motoneuron death in vitro and in vivo (Milligan et al., 1995). The other sequences used in the peptide-based inhibitors include Z-VDVAD-fmk (caspase-2), Z-DEVD (caspase-3), Z-LEVD-fmk (caspase-4), Z-WEHD-fmk (caspase-5), Z-VEID-fmk (caspase-6), Z-IETD-fmk (caspase-8), Z-LEHD-fmk (caspase-9), Z-AEVD-fmk (caspase-10), Z-ATAD-fmk (caspase-12), and Z-LEED-fmk (caspase-13).

Other drugs being tested include the pan-caspase inhibitors boc-Aspartyl(OMe)-FMK (BAF) and z-VAD-FMK (zVAD). They inhibit NGF deprivation-induced cell death of sympathetic neurons and prevent apoptosis in cerebellar granule neurons and cortical neurons. They were also found effective in reducing infarct volume in an occlusion model of reperfusion injury (Deshmukh, 1998). zVAD has, also, provided substantial protection in rodent models of myocardial infarction, hepatic injury, sepsis, ALS, and several other diseases (Endres et al., 1998; Wiessner et al., 2000; Rabuffetti et al., 2000).

2.3 Cathepsins

The cathepsin family of proteases has also been implicated in both necrosis and apoptosis. In healthy cells, cathepsins reside and become fully active in the acidic environment of lysosomes, helping to break down phagocytosed molecules so they can be recycled as building blocks for de novo biosynthesis. When the lysosomal membrane loses integrity under certain conditions, cathepsins leak out and cleave substrates normally not exposed to these proteases, eventually leading to cell death. Cathepsins come in three classes distinguished by the residue (cysteine, serine, and aspartate) in the active site. The cysteine proteases include cathepsins B, C, H, F, L, K, O, S, V, X, and W, the serine proteases (cathepsins A and G), and the aspartic proteases (cathepsins D and E). Of the known cathepsins B, L and D have, currently, been found to play the more important roles in PCD and will be the primary focus of this discussion.

2.3.1 Cathepsin Family

Family members. Of the cysteine cathepsins, the most prominent and ubiquitously expressed and involved in PCD are cathepsins B and L. These cathepsins have less restricted substrate specificities than caspases. Specificity, defined by the S2 binding pocket, is usually displayed by a preference for bulky hydrophobic side chains. The common feature of all cysteine proteases is the utilization of the catalytic mechanism of a Cys residue as a nucleophile and a His residue as a general base for proton shuttling. Among the cysteine cathepsins, Cath L is the most active. It has shown to be involved in several intra- and extracellular degradation processes in normal and pathological states.

The other prominent lysosomal protease actively involved in PCD has a catalytic center formed by two aspartate residues and belongs to the aspartic class. The most prominent and ubiquitously expressed is cathepsin D. Until recently the main accepted function of Cath D was to degrade proteins in lysosomes at an acidic pH. However, it has been shown that Cath D is also involved in the degradation of the extracellular matrix (ECM) and basement membrane (Liaudet-Coopman et al., 2006).

Structure. Although the processing and activation mechanism of cathepsins remains to be fully understood, their activity is regulated in several ways (Turk et al., 2000). All cathepsins are synthesized as inactive preproenzymes with N-terminal proregions that act as potent reversible inhibitors preventing inappropriate activation. Cath B, for example, is synthesized as an inactive preproenzyme (39 kDa) and upon reaching the lysosomal compartment is processed to the inactive proenzyme (37 kDa) (Mort and Buttle, 1997). Removal of the propeptide from its N-terminus converts the inactive proenzyme into the single-chain active form (30 kDa). Finally this single chain is cleaved into two components, a heavy (25 kDa) and a light (5 kDa) chain, held together by a disulfide bond (Mort and Buttle, 1997). This two-chain form takes on a two-lobe appearance with one lobe displaying a mostly alpha helical arrangement while the other is more of a small beta barrel. The active site sits in a cleft between these two lobes.

Similarly Cath D is, also, synthesized as a preproenzyme that undergoes several proteolytic cleavages during biosynthesis to produce the mature form and it likewise has a bilobed appearance (Liaudet-Coopman et al., 2006). This cleavage is consumated by Cath B and Cath L (Liaudet-Coopman et al., 2006). Using a catalytically inactive Cath D expressed in a Cath D-deficient cell line, it was recently shown that the Cath D maturation process was independent of its catalytic function (Liaudet-Coopman et al., 2006). Increased amounts of intermediate and mature Cath D were detected in the brains of Cath B/L double-knockout mice

(Felbor et al., 2002). Studies in estrogen receptor-positive breast cancer cell lines revealed that this protease is highly regulated by estrogens and certain growth factors (i.e., IGF1, EGF) (Rochefort et al., 1989).

The interaction between domains has a hydrophilic as well as a hydrophobic character and is specific for each cathepsin. Most of cysteine cathepsins exhibit predominantly endopeptidase activity (e.g., Cath L) binding substrates in the active site that extends along the whole length of the two-domain interface (Stoka et al., 2005). Without a single binding site, specificity is not involved but instead recognition is a result of contributions of all the interactions along the active site. The additional features of the exopeptidases (e.g., Cath B) reduce the number of substrate binding sites thereby preventing the binding of longer peptide substrates. For the exopeptidase cathepsins, specificity is more dependent on the exclusive interactions of the free-chain termini than side-chain recognition (Stoka et al., 2005).

Function. Cathepsins play important physiological roles in a number of processes including homeostatic protein degradation (Turk et al., 2000), antigen presentation (Chapman et al., 1997), bone remodeling (Kakegawa et al., 1993), and hormone processing (Dunn et al., 1991). Cathepsins have also been identified as contributors to several diseases and injury paradigms including several neuropathologies such as AD, MS, ALS, and myoclonus epilepsy, and ischemic cell death (Bever and Garver, 1995; Mackay et al., 1997; Kikuchi et al., 2003; Houseweart et al., 2003).

The primary function of Cath B is homeostatic protein turnover and digestion of cellular debris (Mort and Buttle, 1997; Turk et al., 2000). Because Cath B is capable of hydrolyzing complex carbohydrates, nucleic acids, lipids, and ECM components under a broad pH optimum, it needs to be sequestered in lysosomes, away from potential substrates (Linebaugh et al., 1999). How it functions proteolytically needs further resolution, but what is known, in neuropathologies, is that elevated levels of Cath B mRNA and protein are usually followed by increased activity.

There is strong evidence that Cath D's role extends beyond protein turnover and includes the degradation of the ECM and basement membrane. Due to this, Cath D is now viewed as an active participant in releasing ECM-bound growth factors, such as FGF-1, a potent mitogen and angiogenic factor. It may perform this function without the need of an active site as demonstrated by a proteolytic inactive, mutated Cath D that was still mitogenic. The results suggest that pro-Cath D may act by triggering, either directly or indirectly, a yet to be identified cell surface receptor (Liaudet-Coopman et al., 2006).

Various pathological stresses or responses to extralysosomal and intralysosomal stimuli, such as global cerebral ischemia (Nitatori et al., 1995; Uchiyama, 2001), induce the release of cathepsins into the cytoplasm, where their proteolytic functions promote cell degradation (Boya et al., 2003; Fusek and Vetvicka, 2005). The inhibition of cathepsins leads to significant neuroprotection in experimental stroke confirming their involvement in neurodegeneration (Yamashima et al., 1998; Seyfried et al., 2001). Although cathepsins have mainly been associated with autophagy (Uchiyama, 2001) and necrosis (Yamashima, 2000; Tsukada et al., 2001), recent evidence in nonneuronal cells points to a critical involvement in apoptosis through mitochondrial homeostatic alterations and activation of proapoptotic members of the Bcl-2 family (Guicciardi et al., 2000; Stoka et al., 2001; Boya et al., 2003; Dietrich et al., 2004).

Indirect stimuli may activate caspase-8 following tumor necrosis factor family ligand such as TNFα binding to the death receptors (Guicciardi et al., 2000; Foghsgaard et al., 2001; Stoka et al., 2005). Released cathepsins B and L cleaved Bid, generating tBid and the release of mitochondrial cytochrome *c* and eventually the formation of the apoptosome and activation of caspases-9 and -3. Cath B was deemed essential in TNF-induced apoptotic cell death of HT22 hippocampal cells and cerebellar granular cells following microglial activation (Kingham and Pocock, 2001). Benchoua et al. (2004) showed that Cath B and caspases-1 and -11 were activated by focal cerebral ischemia with similar spatiotemporal patterns colocalizing in neurons in the damaged cortex.

In transient cerebral ischemia and SCI, Cath B appears to play a functional role in cell death (Ellis et al., 2005). In monkeys (Yamashima et al., 1996; Kohda et al., 1996) the CA1 region of the hippocampus underwent neuronal death following cerebral ischemia by postinjury day 5, and Nitatori et al. (1995) established that the cell death in these models was apoptotic. Both calpains and Cath B were quickly activated and it appears that Cath B was redistributed from the lysosomes to the cytosol (Yamashima et al., 2003).

The function of Cath D in apoptosis is not yet fully understood and needs further investigation. Cath D can either prevent apoptosis, noted under certain physiological conditions with Cath D knock-out mice experiments, or it can promote apoptosis induced by cytotoxic agents via different mechanisms. Evidence suggests that Cath D is a key mediator of apoptosis induced by apoptotic agents such as IFN-gamma, FAS/APO, TNF-α, oxidative stress, adriamycin, and etoposide, cisplatin and 5-fluorouracil, as well as staurosporine (Liaudet-Coopman et al., 2006). Released Cath D activated the proapoptotic Bcl-2 homolog Bax, inducing both the tBid and Bax activation pathways. The critical step is the Bak and/or Bax-dependent mitochondrial membrane permeabilization. Once in the cytosol, the signaling pathway is cathepsin-specific.

2.3.2 Cathepsin Inhibitors

Endogenous. Lysosomal cysteine cathepsin activity is regulated by endogenous protein inhibitors, the cystatins, and the more recently discovered thyropins (Stoka et al., 2005). The cystatins superfamily of homologous proteins is subdivided into three inhibitory families: the stefins (family 1), the cystatins (family 2), and the kininogens (family 3) (Turk and Bode, 1991). Cystatins' major cell protection role is against proteases released from lysosomes and against invading microorganisms and parasites that use cysteine proteases to enter the body (Turk et al., 1997, 2002). The two superfamilies differ in their inhibitory specificity against various cathepsins.

Cystatin C has been implicated in late-onset AD and other neuropathologies (Nagai et al., 2005). In an odd twist one study showed that cystatin C injected into the rat hippocampus induced neuronal cell death in the granule cell layer of dentate gyrus in vivo. This was confirmed when the cystatin C neurotoxicity was inhibited by a simultaneous injection of Cath B. In vitro cytotoxicity was also studied and cystatin C was found to induce neuronal cell death in a dose-dependent manner, upregulating active caspase-3. The results suggest that cystatin C participates in apoptotic neuronal cell death by inhibiting the activity of cysteine proteases. On the other hand the overexpression of stefin A (a cystatin-type cathepsin inhibitor) decreased apoptosis by directly inhibiting Cath B (Jones et al., 1998) and possibly several other cathepsins (Stoka et al., 2005).

The other superfamily of cathepsin inhibitors, the thyropins (Lenarcic and Bevec, 1998), includes the MHC class II-associated p41 invariant chain (Ii) fragment, a selective inhibitor (Bevec et al., 1996). The thyropin p41 fragment specifically inhibits Cath L apparently by its novel structure. Presumably, the other thyropins also exhibit high specificity for their target enzymes, in contrast to the relatively nonselective cystatins.

Several other cellular proteins protect cells against cathepsin-mediated apoptosis. The chaperone heat shock protein-70 promotes cell survival by inhibiting lysosomal membrane permeabilization (Nylandsted et al., 2004), and the transcription factor NF-kB protects cells by specifically inhibiting Cath B through the serine protease inhibitor Spi2A (Liu et al., 2004c).

Synthetic and Natural. While the role of the different cathepsins in apoptotic cell death varies, they offer a target for potential therapeutic intervention. To date, the neuroprotective potential of cathepsin inhibitor treatment following traumatic or ischemic CNS injury has received sparse attention. However, a few published studies indicate that inhibition of these proteases reduces cell death in models of cerebral ischemia. For example, the administration of CA-074 and E-64c immediately following transient ischemia in monkeys reduced Cath B and Cath L activity levels sparing CA1 neurons from cell death (Tsuchiya et al., 1999; Yoshida et al., 2002). Other studies show that this neuroprotection extends to other neuronal populations as well (e.g., cortical, caudate-putamine, Purkinje) (Yoshida et al., 2002). In the rat MCAO model, cathepsin proteolytic activity increased (Seyfried et al., 1997) but an IV injection of CP-1, a member of the peptidyl diazomethane family specific for Cath B and L, not calpains or caspases, reduced infarct volume and improved functional scores (Seyfried et al., 2001). A study of CA-074 Me (cell permeable derivative of CA-074) found that it prevented all apoptotic related events in WEH1 fibrosarcoma cells (Foghsgaard et al., 2001). Benchoua et al. (2004) were able to show the neuroprotective effect of CA-074 in acute focal ischemia.

Reports demonstrating participation of Cath D in apoptosis execution have relied heavily on the pharmacologic inhibitor pepstatin A. Pepstatin A is an inhibitor that appears to bind to nonactive sites of aspartic proteases. Treatment with this inhibitor partially inhibited apoptosis caused by bile salts, naphthazarin, TNFα, IFN-γ, sphingosine, staurosporine, Fas stimulation, and withdrawal of essential neurotrophic factor. However, pepstatin is not entirely specific and is also known to inhibit Cath E, pepsin, and rennin and many microbial aspartic proteases (Emert-Sedlak et al., 2005).

2.4 Matrix Metalloproteinases

The family of matrix metalloproteinases (MMPs) (or metalloproteases, also called matrixins) is divided into two subgroups (a) metallocarboxypeptidases and (b) metalloendopeptidases. MMPs get their name from the one feature they share in common, a conserved motif of three histidine residues (HExxHxxGxxH), which bind a metal ion such as Zn^{2+} or Ca^{2+} in their active site followed by a methionine that initiates a turn in the molecule (Handsley et al., 2005). Presently the 28 known members of the MMP family are characterized by strong structural similarities and, for most of the MMPs, overlapping substrate specificities.

MMPs play important roles in tumor metastasis, embryonic development, and wound healing, generally as part of the matrix degradation process whereby they regulate the microenvironment of the cell. By breaking down the interactions between the cell and the ECM, these proteases are able to trigger anoikis (a form of apoptosis). Their importance in CNS cell death is still being investigated but what is known is that overexpression is often pathogenic. For instance, the degradation of myelin-basic protein in neuroinflammatory diseases (Chandler et al., 1995), the opening of the BBB following brain injury after ischemia, hemorrhagic brain injury leading to secondary vasogenic brain edema (Rosenberg, 1995), and stoke (Copin et al., 2005) are examples where MMPs have a role in cell death. Among the MMPs, MMP-2, -3, and -9 have been found to convert the inactive precursor form of IL-1β into a biological active form implicated in brain damage following cerebral ischemia (Copin et al., 2005). Key questions remain including which MMPs are involved in which disease states.

2.4.1 Matrix Metalloproteinase Family

Family members. The MMP family is part of a larger group of metalloproteinases or metzincins. They were initially classified into collagenases, gelatinases, and stromelysins on the basis of substrate specificity. As further members were discovered, distinction was made for the membrane-type MMPs, which are characterized by a C-terminal transmembrane domain that anchors them to the cell membrane (Sato et al., 1994). There are also a number of odd MMPs that do not readily fit into any of the earlier four classes. Other members of the extended metzincin family include the serralysins, the astacins, and the adamalysins (reprolysins). These related groups are also growing in number but since there is very limited data on their role in PCD they will not be covered in this chapter.

Structure. All MMPs are synthesized as preproenzymes and secreted, in most cases, as inactive pro-MMPs. They share a basic structural organization comprising a signal peptide that targets them for secretion, a propeptide domain, and an N-terminal catalytic domain. Most MMPs, except MMP-7, -23, and -26, have a hinge region and C-terminal hemopexin-like domain. The propeptide domain is about 80 amino acids long with a conserved unique sequence (PRCGxPD) containing a conserved cysteine residue or cysteine switch, which regulates the catalytic zinc in the inactive condition (Van Wart and Birkedal-Hansen, 1990). Some MMPs (MT-MMPs, MMP-11, -21, -23, -28) have a furin recognition site as part of this domain, which when cleaved activates the enzyme. Other MMP subgroups contain unique features such as a transmembrane domain, a cytoplasmic tail, and an MT-loop (MT1-, MT2-, MT3-, MT5-MMP), a GPI anchor (MT4-, MT6-MMP), fibronectin type II repeats (MMP-2, -9), and an N-terminal signal anchor, a cysteine array and an Ig-like domain (MMP-23) (Handsley et al., 2005).

To function properly MMPs depend on the metal ions, Ca^{2+} and Zn^{2+}, as cofactors to bind their targets. The two ions are utilized differently. The Zn^{2+} ion is bound by three histidine residues arranged in a conserved sequence in the active site of the enzyme. The Ca^{2+} ions are used to maintain the molecule's conformation. Both are indispensable parts of the catalytic domain and are required for stability and enzymatic expression. As part of its catalytic activity the Zn^{2+} ion normally coordinates with a water molecule. This water molecule enhances protease reactivity by stabilizing the negative charge of the tetrahedral transition state to promote hydrolysis.

Function. Although recent evidence suggests that MMPs are involved in processing nonmatrix proteins, they are best known for their involvement in remodeling the ECM. Evidence points to the involvement of MMP-9 in granule cell migration and apoptosis observed in the developing cerebellum through degradation of ECM proteins (Vaillant et al., 2003). MMPs activated beyond the normal remodeling process are known to promote neuronal apoptosis in conjunction with caspase activation via anoikis. This type of PCD is induced by inadequate or altered cell-to-matrix interactions. When the critical basement membrane survival signals are removed by cleavage, the adhesion molecules, cadherins and integrins, stimulate modifications causing specific signaling and structural components to interrupt the survival signals emanating from the ECM and neighboring cells (Cardone et al., 1997). This loss could lead to the rounding up of cells typical of apoptosis (Mannello et al., 2005). Reducing cell contact disruption has been noted to promote cell survival (Copin et al., 2005). As a case in point, accumulating data suggest that when the BBB integrity is disrupted due to free radicals (Lohmann et al., 2004; Maier et al., 2006) there is an amplification of infiltrates and demyelination (Yong et al., 2001).

The fact that MMPs are known to process a wide range of nonmatrix proteins, such as soluble molecules and cell surface receptors, suggests that MMPs are involved in other PCD pathways as well (Mannello et al., 2005). For example, neurotrophin activity is regulated by proteolytic cleavage. While the proforms of the neurotrophins (i.e., nerve growth factor and brain-derived neurotrophic factor) preferentially activate p75NTR to promote neuronal apoptosis, MMP-cleaved neurotrophins activate Trk receptors to promote cell survival (Lee et al., 2001; Teng et al., 2005). Other studies have shown that MMP-1, MMP-7, and MMP-9 trigger neuronal apoptosis when S-nitrosylation activates the MMPs, resulting in further oxidation of posttranslational modifications with pathological activity. (Yong et al., 2001; Gu et al., 2002). And in another study MMPs were found to induce caspase-mediated endothelial cell death after hypoxia-reperfusion (Lee and Lo, 2004). Another nonmatrix protein, known to be involved in PCD that is cleaved by MMPs is the proinflammatory cytokine interleukin-1β (Schonbeck et al., 1998).

Interestingly, the translocation and localization of MMP-2 and -9 in the nucleus suggest that MMPs have a role in the nucleus. The ability to cleave the nuclear protein poly-ADP-ribose-polymerase (PARP) indicates that MMPs can inactivate PARP similar to caspase-3 (Scovassi and Diederich, 2004). In this respect, MMPs and caspases may have analogous nuclear functions, a discovery that needs further exploration.

There has also been an unexpected finding, MMPs with proteolytic activities that appear to play opposing PCD roles. It has been demonstrated that MMP-7 is able to release membrane bound Fas ligand (FasL) a known inducer of apoptosis in Fas receptor expressing cells (Powell et al., 1999). On the other hand, MMP-7 inhibits apoptosis by cleaving proheparin-binding epidermal growth factor (pro-HB-EGF) into a biologically active protein that promotes among other things cell survival by stimulating the ErbB4 receptor (Yu et al., 2002). An explanation for this conundrum still needs to be ascertained.

2.4.2 Matrix Metalloproteinase Inhibitors

Endogenous. The MMPs are inhibited by a family of four specific endogenous "tissue inhibitor of metalloproteinases" (TIMP-1 through -4) and the membrane bound "reversion-inducing cysteine-rich protein with kazal motifs" (RECK) (Handsley et al., 2005). TIMP-1, -2, and -4 are diffusible, secreted proteins whereas TIMP-3 is matrix-associated (Leco et al., 1994; Brew et al., 2000). TIMPs share a similar motif in which the essential N-terminal domain cysteine binds noncovalently to the active site Zn^{2+} resulting in inhibition (Bode et al., 1999). The variability of the C-terminal domain sequence among the four TIMPs could be responsible for their distinct properties. TIMPs inhibit most MMPs without major selectivity, although MMP-specific association occasionally has been reported. They do differ in other ways such as tissue distribution and transcriptional regulation, suggesting they each have complex, separate, and specific physiological functions independent of MMP-inhibitory activities (Handsley et al., 2005).

RECK is a membrane-anchored inhibitor of MMP-9, MMP-2, and MT1-MMP (Takahashi et al., 1998; Oh et al., 2001). Since MMP processing to active form occurs at the plasma membrane, the location of RECK is key to its inhibitory activity. RECK does not bear any structural resemblance to the TIMPs and apparently acts quite differently. Its expression has been detected in a wide range of tissues (Handsley et al., 2005).

Synthetic and Natural. A number of rationally designed inhibitors have shown some promise in the treatment of pathologies where MMP involvement is suspected. However, to date the major disappointment has been the lack of clinical success for such compounds in spite of the encouraging effectiveness in animal models. Due to the number of these compounds, this section will just summarize some of the major findings.

Much of the early medicinal chemistry focused on a series of succinyl hydroxamates. Side chain modifications of early leads led to the discovery of three broad-spectrum inhibitors that displayed efficacy in animal models, BB-94, BB-1101, and later BB-2516. Later N-sulfonyl amino acid hydroxamates were, also, identified as MMP inhibitors and the orally available, broad-spectrum inhibitor CGS 27023A was the first such compound to enter development. The aryl hydroxamates and the corresponding heterocyclic analog compounds with alternative zinc binding groups (ZBGs) have drawn considerable interest. The most successful have been the carboxylic acid and thiol ZBGs. Although the carboxylate group is a less effective ZBG, many carboxylates were still effectual and appeared promising as clinical candidates. For example, a derivative of the antiinflammatory drug Fenbufen was identified and provided the starting point for a series of carboxylic acid inhibitors such as the biphenyl derivative BAY 12–9566. Other discoveries include thiadiazole, a selective inhibitor of MMP-3, the pyrimidine-2,4,6-trione, and the selective inhibitor of MMP-2, thiirane. The MMP inhibitors actually evaluated in clinical trials include BB-2516, BAY 12–9566, CGS-27023A, D2163, and Ro 32–3555; however, most of these performed poorly. The toxicity of the drugs is the primary reason for their failure to show expected results. The reasons behind the largely disappointing clinical results are unclear, especially in light of their success in animal models (summarized from Whittaker and Ayscough, 2001).

New drugs, nevertheless, continue to be evaluated. Recently the broad-spectrum MMP inhibitor CH6631, a member of a novel class of nonpeptidic hydroxamic acid inhibitors, was used in MCAO mice. The inhibitor reduced α-spectrin degradation and cytochrome c release, increased levels of PARP and the precursor form of IL-1β, and was found to protect the ischemic brain by decreasing nuclear DNA degradation and cerebral infarct. These results also suggested that MMPs may act directly or indirectly on calpain activity (Copin et al., 2005). Another drug that recently showed efficacy, also discussed under calpain inhibitors, is E64d. A single dose of E64d 30 min before the induction of focal ischemia not only prevented calpain and Cath B activation but also reduced MMP-9 activation in both neuronal cells and neurovascular endothelial cells after transient MCAO in rats (Tsubokawa et al., 2006). There is still the need to further investigate the inhibitor on a poststroke basis.

Timing of drug therapy is also critical as Zhao et al. (2006) noted. MMPs participated in delayed cortical responses after focal cerebral ischemia in rats. MMP-9, in particular, was found upregulated in the peri-infarct cortex 7–14 days after stroke, colocalizing with markers of neurovascular remodeling. MMP inhibitor treatment applied 7 days after stroke suppressed neurovascular remodeling, thereby increasing ischemic brain injury and impairing functional recovery when measured at 14 days postinjury.

Natural product MMP inhibitors include the tetracyclines, doxycycline, and minocycline. It was discovered that with chemical modifications it was possible to separate MMP inhibitory activity from antibiotic activity (Suomalainen et al., 1992). Doxycycline, for example, at subantibiotic doses, inhibits MMP activity and has been used in various experimental systems for this purpose and, clinically, for the treatment of periodontal disease. It is the only MMP inhibitor available for this purpose.

2.5 Serine Proteases

There are a number of distinct serine proteases and their inhibitors within the nervous system. However, the identities of all the serine proteases that play a significant role in the CNS have not been established nor have the identity of those who are active during PCD been confirmed. Consequently, this section will limit discussion to those clearly implicated: Htra2/Omi, thrombin, and tissue-type plasminogen activator (tPA).

2.5.1 Serine Family

Family members. The endopeptidases that make up serine proteases are grouped into 8 clans subdivided into 34 families with approximately 700 individual enzymes. They are distinguished by their ability to catalyze the hydrolysis of covalent peptidic bonds based on a nucleophilic attack of the targeted bond by a serine.

Structure. Serine proteases are translated as zymogens and are activated under very controlled conditions. When activated, in many cases, the nucleophilic property of the attacking serine is improved by the presence of a histidine, held in a proton acceptor state by an aspartate. Side chains of serine, histidine, and aspartate are aligned building a catalytic triad common to most serine proteases. Also, it was discovered that two other amino acids, a glycine and another serine, are involved in creating what is called an oxyanion hole contributing to a second stabilizing effect.

Function—Htra2/Omi. The three members of the Htra/Omi serine protease group in humans, Htra1 Htra2, and Htra3 (Li et al., 2002), share homology with the bacterial Htr (high-temperature requirement) A protein. The mammalian Htra2/Omi reportedly has been localized to three different cellular compartments, the nucleus (Gray et al., 2000), the endoplasmic reticulum (Faccio et al., 2000b), and the mitochondria (Suzuki et al., 2001). Htra2/Omi mRNA has been shown to be upregulated during kidney ischemia in humans (Faccio et al., 2000a) and in a variety of cellular stresses such as high temperature and tunicamycin treatments (Gray et al., 2000). Like many other serine proteases Htra2/Omi contains a C-terminal PDZ domain and a central serine protease domain. The serine protease domain is located in the central region of the molecule and the PDZ domain that regulates its protease activity is located in the C-terminal region.

The nuclear-encoded mitochondrial HtrA2/Omi is autocatalytically processed into a 36-kDa activated protein fragment and released into the cytosol as a homotrimer. The cleavage of the mitochondrial-targeting sequences prior to release generates a processed active Htra2/Omi with a new apoptogenic N-terminal with an amino acid sequence of AVPS, referred to as the IAP-binding motif (IBM) (Hegde et al., 2002). A recent report identified Htra2/Omi as a p53-targeted gene. A p53-mediated upregulation of Htra2/Omi correlates with c-IAP1 cleavage in etoposide-induced cell death (Jin et al., 2003). Omi/HtrA2 is efficient in inactivating IAPs, such as c-IAP1 and XIAP, thus promoting caspase activity (Vaux and Silke, 2003; Jin et al., 2003; Yang et al., 2003a). This reported ability to bind and cleave IAPs when released from the mitochondria appears crucial in cytochrome c-dependent caspase activation (Seong et al., 2004) making this protease indispensable in promoting apoptosis (Chai et al., 2001; Li et al., 2002).

Function—Thrombin. Thrombin like tPA (see below) is a trypsin-like serine protease of the tissue kallikrein family. Both have active sites with similar substrate specificities. Their targets vary due to the differences in the binding sites (Lijnen and Collen, 1998; Stone and le Bonniec, 1998). Accumulating evidence demonstrates that thrombin and its receptors trigger a variety of responses in neurons and glial cells, from being neuroprotective to involvement in several neuropathologies that lead to cell death.

The adverse effects of intracerebral hemorrhage due to hypertension, ischemic infarct, ruptured blood vessels, or trauma (Xue and Del Bigio, 2001) remain a significant clinical problem (Mendelow, 1993). The blood's negative effects have been attributed to proteolytic enzymes such as thrombin. When activated, thrombin converts fibrinogen into fibrin, a protein involved in blood coagulation (Wang and Reiser, 2003). Under normal conditions, large molecules like thrombin are excluded from the CNS by the BBB. Under various cerebrovascular injuries, like TBI and ischemia, the breakdown of the BBB enables thrombin to directly enter the brain and under these pathological conditions it can be neurotoxic (Gingrich and Traynelis, 2000). With the detection of prothrombin mRNA in brain tissue and in neuronal and glial cell lines, it is now believed that the thrombin effects may be locally produced as well (Deschepper et al., 1991; Dihanich et al., 1991).

In the CNS, thrombin has been implicated in synaptic plasticity as well as in several neurodegenerative diseases such as AD (Vaughan et al., 1994). The physiological effects of thrombin are mediated by a family of G-protein-coupled receptors known as the protease-activated receptors (PARs), which are widely distributed in brain tissue and are noted to exert many physiological and pathological functions in the CNS (McKinney et al., 1983; Niclou et al., 1994; Bartha et al., 2000; Cocks and Moffatt, 2000). PAR-1, -3, and -4 can be activated by low concentrations of thrombin (Smith-Swintosky et al., 1997; Hollenberg and Compton, 2002; Vergnolle et al., 2003) and are expressed by neurons at various densities in numerous brain regions and in glial cells in culture (Striggow et al., 2001; Wang et al., 2002a). Since these cells express all three PAR subtypes this coexpression suggests distinct functions and/or cross-talk (Coughlin, 2000; Macfarlane et al., 2001; Hollenberg and Compton, 2002).

The precise role of PARs in normal signaling is unclear. However, when thrombin was injected into the brain, edema was generated, which was preventable by several thrombin inhibitors (Lee et al., 1995; Colon et al., 1996; Lee et al., 1996; Lee et al., 1997). In addition, thrombin increases show a dose-dependent toxicity to neurons in brain slices (Striggow et al., 2000). A study by Xue and Del Bigio (2001) demonstrated in vivo that thrombin injections into the rat striatum was correlatively related with dose-dependent tissue necrosis, cell death, and inflammation within 48 h, similar in magnitude to that seen after autologous whole blood infusions (Xue and Del Bigio, 2000a, b). PAR-1, in particular, mediated this neurotoxicity in primary motor neurons in culture (Turgeon et al., 1999) and in focal cerebral ischemia in vivo (Junge et al., 2003). It was noted to activate caspase-3-linked apoptosis in a motor neuronal cell culture line (Smirnova et al., 1998). PAR-4 appears to be the primary receptor involved in the death of dopaminergic neurons in mesencephalic cultures (Choi et al., 2003). Another thrombin influence appears to occur during intracerebral hemorrhage or during BBB breakdown. By interacting with the NMDA receptor, it appears to exacerbate glutamate-mediated cell death and may possibly be a participant in posttraumatic seizures (Wang and Reiser, 2003).

The discovery that PARs are expressed in microglial cells suggests a role for thrombin in brain inflammation. Brain inflammation is a common feature in neurodegenerative diseases such as AD (McGeer and McGeer, 2002), stroke (Balduini et al., 2004), and PD (Teismann and Schulz, 2004). The essential feature of CNS inflammation is the activation of microglia, the major immune cells whenever brain injury occurs (Aloisi, 2001; Nakajima and Kohsaka, 2001). Thrombin-activated microglia produce and release proinflammatory cytokines like TNF-α, interleukin(IL)-6, IL-12, nitric oxide (NO), and inducible nitric oxide synthase (iNOS) and cyclooxygenase (COX)-2 (Möller et al., 2000; Suo et al., 2002; Lee et al., 2006). There are still questions as to which PARs are involved for which effect. It has been suggested that PAR-4 may be involved in thrombin-induced proinflammatory effects (Suo et al., 2003). The results of one study indicated that thrombin induced the degeneration of rat cortical neurons both directly in the absence of microglia and indirectly in their presence. This suggests that the neurotoxicity of thrombin may be mediated both by its proteolytic activity and, independent of this activity, by the activation of the microglia (Lee et al., 2006). Nevertheless, these studies confirm thrombin's important pathophysiological role in the inflammatory response in the brain. In addition, thrombin contributes to neuronal cell death in AD due to its role in cleaving apoE4 into fragments that are themselves neurotoxic (Marques et al., 1996; Tolar et al., 1999).

It is important to emphasize that numerous studies reveal a clear demonstrable difference between thrombin's protective and apoptotic effects in neurons and astrocytes. At low concentrations, thrombin promotes cell survival, while at high concentrations it induces apoptosis. In experiments that mimicked in vivo conditions following cerebrovascular injury, thrombin distinctly protected rat primary astrocytes and hippocampal neurons from cell death, which was induced by glucose withdrawal or oxidative stress (Vaughan et al., 1995). However, high concentrations of thrombin killed both cultured astrocytes and neurons (Turgeon et al., 1998; Kudo et al., 2000; Striggow et al., 2000; Ohyama et al., 2001; Friedmann et al., 2001; Reimann-Philipp et al., 2001) and can eventually lead to brain damage (Xue and Del Bigio, 2001).

Thrombin was shown to apoptotically kill primary murine spinal cord motoneurons through a process involving RhoA activation (Smirnova et al., 1998; Smirnova et al., 2001) but RhoA activation was also shown to be protective. The apoptotic pathway involved a more rapid and robust change in available RhoA than the protective pathway (Donovan and Cunningham, 1998). In the apoptotic case, caspase-3 and caspase-1 activation followed the stimulation of PAR-1 and the activation of connected pathways. It was the continual presence of thrombin that was the most important event in the induction of apoptosis, not its high concentration per se.

Function—Tissue-type plasminogen activator. Tissue-type plasminogen activator (tPA) is involved in extracellular proteolysis both during development and in response to physiological and pathological events. Normally it is expressed at low levels in specific areas of the CNS (Sappino et al., 1993). There are events, however, where synthesis is increased such as when synaptic plasticity is required for long-term potentiation and motor learning as well as in kindling and seizures (Qian et al., 1993; Carroll et al., 1994; Seeds et al., 1995; Yepes et al., 2002). The actual role tPA plays in PCD is still controversial and needs additional research. What is known is that brain endothelia produced tPA converts plasminogen into plasmin (Ware et al., 1995), which lyses blood clots by digesting fibrin. Thus, tPA's action via plasmin works opposite to that of thrombin and its fibrin created blood clots. It has also been reported that tPA increases the effectiveness of thrombin related brain damage and cellular loss following MCAO in rats (Figueroa et al., 1998; Wang et al., 1998).

In contrast to its exacerbating effect on excitotoxic necrosis (Lebeurrier et al., 2004; Benchenane et al., 2005), the actual effect of tPA on apoptotic neuronal death continues to be controversial. The data from a number of studies, both in vivo and in vitro, report that tPA acts proapoptotically (Lu et al., 2002; Liu et al., 2004a; Medina et al., 2005; Takahashi et al., 2005), whereas in others it acts antiapoptotically such as in cases when tPA is injected to aid in the evacuation of intracerebral hematomas (Schaller et al., 1995; Flavin and Zhao, 2001). Cases have been reported where there was no reaction at all (Ferrer et al., 2001). The reason for such discrepancies remains unknown, although for in vitro experiments this may reflect differences in experimental protocols or the different cell cultures types. Since tPA has been designated as a potential therapeutic for several brain disorders, it is important to clarify the apoptotic function of tPA. A recent work using serum deprivation demonstrated that tPA, like thrombin, can dose-dependently protect cultured cortical neurons against apoptosis, although it required the PI-3 kinase pathway and occurred independent of its proteolytic activity (Liot et al., 2006).

Following a modest oxygen–glucose deprivation (OGD) injury, tPA was found to have a direct neuroprotective effect (Flavin and Zhao, 2001). In this study the protection was not significantly impeded by the serine protease inhibitor, PAI-1, whereas the tPA antibody, at a concentration previously found to neutralize about 80% of tPA activity (Flavin et al., 2000), abolished the protective effect (Flavin and Zhao, 2001). This contrasts with a study by the same laboratory in which they reported that PAI-1 protected the hippocampal neurons from death triggered by factors present in the medium conditioned by activated microglia. When exogenous tPA was added it worsened the outcome. This suggests that the interaction of tPA with some soluble microglial factors and with neurons was damaging to the cells (Flavin et al., 2000). As far as it is known PAI-1 only inhibits the tPA proteolytic cleavage of plasminogen, yet it had no impact on OGD injury. The authors suggested that tPA neuroprotection in OGD was mediated through a nonproteolytic mechanism. So similar to thrombin, the proteolytic effect of tPA can be either protective or neurotoxic whereas the nonproteolytic effects of tPA appear to be neuroprotective (Flavin and Zhao, 2001).

2.5.2 Serine Protease Inhibitors (Serpins)

Serpins (*serine protease inhibitors*) are a diverse group of proteins that form a covalent bond with the serine protease, inhibiting its function. There are a number of endogenous serpins, the most noted being protease nexin-1 (PN-1), neuroserpin (PI-12), plasminogen activator inhibitor-1 (PAI-1), and antithrombin (AT). The synthetic and natural inhibitors of note include ucf-101 and hirudin.

Endogenous. Plasmin formation results in proteolysis of fibronectin and laminin followed by cell detachment and apoptosis (anoikis). PN-1 expressed by transfected cells significantly inhibits the activity of plasmin and tPA by forming inhibitory complexes that prevent anoikis. PN-1, a specific thrombin inhibitor, is present in the brain (Reinhard et al., 1994) and has been shown to inhibit tumor cell-mediated ECM destruction (Bergman et al., 1986), to modulate neurite outgrowth (Monard, 1988), and to promote neuronal cell survival (Houenou et al., 1995). PN-1 secretion by a variety of anchorage dependent cells including astrocytes (Monard et al., 1992) and neuroblastoma cells (Vaughan and Cunningham, 1993) suggests that it may play a major protective role against active thrombin and other serine proteases

(Choi et al., 1990). A study by Rossignol et al (2004) showed that in wild-type Chinese hamster ovary fibroblasts (CHO-K1) that constitutively expressed tPA, anoikis was inhibited with the gene transfer of PN-1 or PAI-1, implying that they may be important antiapoptotic factors for adherent cells.

PI-12, an axonally secreted serpin family member (Osterwalder et al., 1996; Hastings et al., 1997), is primarily expressed in the embryonic and adult nervous systems (Osterwalder et al., 1996; Hastings et al., 1997; Krueger et al., 1997). The tissue distribution is different from PAI-1 and PN-1. Whereas PAI-1, a tPA inhibitor secreted primarily by endothelium cells lining blood vessels, is barely found in the adult mouse brain (Sawdey and Loskutoff, 1991), PI-12 is identified in well-defined areas of the CNS. Likewise, in contrast to PN-1, which is found mainly in glial cells, (Mansuy et al., 1993) PI-12 expression is primarily detected in neurons.

Studies of PI-12 showed it reacting preferentially with tPA; however, unidentified targets may exist (Hastings et al., 1997). Studies in animal models of MCAO with reperfusion (Yepes et al., 2000) demonstrated that PI-12 levels increase in the ischemic penumbra within 6 h after cerebral ischemia and remained elevated for up to 7 days. These data suggest that an endogenous neuroprotectant PI-12 may, during the course of cerebral ischemia, protect apoptotically prone cells (Yepes et al., 2000). It has been observed that ischemia-induced increases of PI-12 (Yepes et al., 2003; Yepes and Lawrence, 2004) blocked microglial activation thereby reducing inflammation (Cinelli et al., 2001) and appeared to preserve the integrity of the vascular basement membrane (Yepes et al., 2000).

AT, a plasma-derived glycoprotein, is a potent, complex anticoagulant demonstrating beneficial properties in animal models and small cohorts of patients with coagulation disorders, AT has also exhibited antiinflammatory properties. Activated thrombin contributes to inflammation through microglia activation, and AT prevents its activity by controlling its binding to its target. By forming an AT-thrombin complex with an irreversible inhibition, it acts progressively thereby inhibiting coagulation slowly but effectively. It has been proposed that AT be regarded more as a physiological reaction modulator rather than as a fast acting, tight-binding inhibitor (Roemisch et al., 2002).

Synthetic and Natural. The compound, ucf-101, isolated in a high-throughput screen, is a specific inhibitor of Htra2/Omi with very little reactivity against other serine proteases. When tested in caspase-$9^{-/-}$ null fibroblasts, ucf-101 inhibited Htra2/Omi-induced cell death (Cilenti et al., 2003).

Another anticoagulant hirudin, the active ingredient in the salivary secretion of leeches, functions as a specific inhibitor of thrombin by reducing its neurotoxicity effect. In an in vivo ischemic insult study to CA1 hippocampal cells (Striggow et al., 2000) and one of thrombin intracerebroventricular infusion (Mhatre et al., 2004), thrombin was effectively neutralized by hirudin. However, hirudin was unable to block thrombin-induced neuronal cell death or NO production in cocultures of microglia and neurons. This suggests that microglial activation by thrombin is, at least in part, due to a nonproteolytic function and further supports the perception that thrombin needs to be proteolytically active to be neurotoxic (Lee et al., 2006).

3 Others

3.1 Proteasome

Proteasomes are abundant barrel-shaped multiprotein enzyme complexes present in the cytoplasm and nucleus of all eukaryotic cells. These nonlysosomal complexes are responsible for degradation of the majority of intracellular proteins into short polypeptides and amino acids in an ATP-driven reaction. In eukaryotes, the proteasome is composed of about 28 distinct subunits, which form a highly ordered ring-shaped structure (20S ring) of about 700 kDa. In eukaryotic organisms there are a number of different types all showing different specificities. The molecule's active sites are on the inner surfaces while the terminal apertures restrict substrate access to these active sites. Because it removes damaged or misfolded proteins from cells, as well as various short-lived proteins that regulate cell cycle, cell growth, and differentiation, it has been implicated in a diverse range of cellular processes. By regulating protein turnover through timely degradation and recycling, the proteasome plays a critical role in the maintenance of cellular homeostasis. Due to its critical role it has become a therapeutic target.

The therapeutic compound MLN519 (1R-[1S,4R,5S]-1-(1-hydroxy-2-methylpropyl-6-oxa-2-azabicyclo [3.2.1] heptane-3,7-dione) is a synthetic analog of clastolactacystin β-lactone and was developed as a cell-permeable small molecule and potent inhibitor. MLN519 targets the proteasome complex and has shown encouraging results in treating brain injury (Williams et al., 2006). Although a key role of the proteasome is to degrade ubiquinated proteins, cells also use this system to activate transcriptional regulatory proteins such as NF-κB. Brain ischemia, caused by cerebral blood flow loss, induces secondary injury processes including a neuroinflammatory response that can eventually lead to cell death (Williams et al., 2006). Studies conducted to assess the efficacy of MLN519 following transient MCAO and reperfusion injury in rats (Williams et al., 2003; Williams et al., 2004; Williams et al., 2005) found neuroprotective effects. The compound appeared to reduce neuronal and astrocytic degeneration, decrease cortical infarct volume, and improve neurological recovery out to 72 h postinjury (Williams et al., 2003). The MCAO injury recovery was then verified out to two weeks postinjury. The treatment was associated with an improved survivability, reduced brain tissue loss, improved weight gain, and better overall neurological outcome when compared with control animals (Williams et al., 2005). The favorable therapeutic window of MLN519 is contributed, in part, to its antiinflammatory mechanism of action. Specifically, the 6–10 h therapeutic window of MLN519 correlates with peak increases in cytokines (TNF-α, IL-1β, and IL-6) and cellular adhesion molecule (ICAM-1 and E-selectin) mRNA (Williams et al., 2004). Data support findings that MLN519 treatment inhibits proteasome activation by reducing NF-κB-induced cytokine and cell adhesion molecule expression, thereby blocking leukocyte infiltration into the injured brain. This suggests that the major action site is most likely nonneuronal endothelial cells (Williams et al., 2003).

3.2 Necroptosis

A new pathway to necrosis has recently been described although it comes without a proteolytical mechanism. Necrotic cell death, often considered to be a passive, nonregulated process, has consistently drawn less attention than apoptosis. Junying Yuan and colleagues (Degterev et al., 2005), who identified this new cellular pathway, which they termed necroptosis, demonstrated its involvement in an ischemic brain injury model. The extrinsic apoptotic pathway is generally triggered when the Fas/TNFR (death) receptor family is stimulated by their corresponding ligands. In this study the authors demonstrated that even when apoptotic signaling along this pathway was inhibited, by blocking caspase activation, it was still possible to activate necroptosis. They showed that although necroptosis displays some necrotic morphology, it is an active PCD process triggered by death receptor signaling and autophagy activation (Degterev et al., 2005).

A growing number of studies report evidence for nonapoptotic cell death with features of necrosis but with few tools to investigate the potential mechanisms behind these observations researchers have not been able to address this issue. Degterev et al. (2005) screened about 15,000 compounds and found inhibitors that prevented necroptosis. They identified necrostatin-1 (Nec-1), a tryptophan derivative, and a more potent derivative 7-Cl-Nec-1. Nec-1 blocked this form of cell death during an ischemic brain injury in a mouse stroke model, demonstrating that necroptosis is a regulated cell death pathway. Consistent with this, MCAO injured mice treated with a combination of the more potent 7-Cl-Nec-1 and the caspase inhibitor zVAD.fmk demonstrated even greater neuroprotection than those treated with either alone. Interestingly, necroptosis seemed to occur later than apoptosis and 7-Cl-Nec-1 could protect against neuronal death even if administered several hours after the artery occlusion (Degterev et al., 2005). This work needs to be interpreted with care because the presumed proteolytic target of Nec-1 is unknown and the existence of a distinct necrotic pathway is yet to be formally proven.

4 Summary

The proteases involved in PCD are as varied as they are numerous and in the CNS they are often difficult to investigate. As the physiology of cell death continues to be examinined, our understanding of the complexity of the mechanisms involved has continued to grow. As pointed out in the Necroptosis section even though a protease has not been uncovered our understanding of the cell death process is still being expanded. The proteolytical activity in PCD in this discussion highlighted only the most well studied (❯ *Table 12-1*). While the calpains and the caspases have garnered the most attention cathepsins, matrix

Table 12-1
Summary of proteases

Protease family	Number	Most prominent	Endogenous inhibitors	PCD action
Calpains	14	Calpain-1, -2	Calpastatin	Oncosis, apoptosis
Caspases	14	Caspase-3, -7, -8, -9, -10, -12	NAIP, XIAP, cIAP1, cIAP2 survivin, Ts-XIAP, BRUCE, ML-IAP	Apoptosis
Cathepsins	15	Cath B, Cath D, Cath L	Cystatin superfamily Thyropin superfamily	Oncosis, autophagy, apoptosis
Matrixins or matrix Metalloproteinases	28	MMP-1, -2, -3, -7, -9	TIMP-1, -2, -3, -4; RECK	Apoptosis
Serine proteases	700	Htra2/Omi, thrombin, tPA	PN-1, PI-12, PAI-1, AT	Apoptosis
Proteasome	1	Proteasome	Unknown	Apoptosis

metalloproteinases, and the serine proteases are beginning to get substantial consideration. The one important conclusion this literature review offers is that the proteolytic field has grown extremely complex and interrelated. It is no longer possible to focus on only one protease or one family during a neuropathological insult and understand the full extent of the injury. This not only applies to the complex interactions that occur after a neuropathological event but in the design of treatments. As a consequence, as a better understanding of the proteolytic events following PCD occurs it should be possible to recognize when cell death should be allowed to proceed, but also to allow for improved treatment and to better control this devastating condition when it is permitted to transpire unchecked.

Disclosure Statement: "Drs. Kevin K. W. Wang and Ronald L. Hayes own stock, receive royalties from, and are executive officers of Banyan Biomarkers Inc. as such may benefit financially as a result of the outcomes of this research or work reported in this publication."

References

Aloisi F. 2001. Immune function of microglia. Glia 36: 165-179.

Ambrosini G, Adida C, Altieri DC. 1997. A novel anti-apoptosis gene, survivin, expressed in cancer and lymphoma. Nat Med 3: 917-921.

Arthur JS, Elce JS, Hegadorn C, Williams K, Greer PA. 2000. Disruption of the murine calpain small subunit gene, Capn4: Calpain is essential for embryonic development but not for cell growth and division. Mol Cell Biol 20: 4474-4481.

Balduini W, Carloni S, Mazzoni E, Cimino M. 2004. New therapeutic strategies in perinatal stroke. Curr Drug Targets CNS Neurol Disord 3: 315-323.

Banik NL, Shields DC, Ray S, Davis B, Matzelle D et al. 1998. Role of calpain in spinal cord injury: Effects of calpain and free radical inhibitors. Ann NY Acad Sci 844: 131-137.

Barnes TM, Hodgkin J. 1996. The tra-3 sex determination gene of Caenorhabditis elegans encodes a member of the calpain regulatory protease family. EMBO J 15: 4477-4484.

Barrett AJ, Kembhavi AA, Brown MA, Kirschke H, Knight CG, et al. 1982. L-trans-Epoxysuccinyl-leucylamido(4-guanidino)butane (E-64) and its analogues as inhibitors of cysteine proteinases including cathepsins B, H and L. Biochem J 201: 189-198.

Bartha K, Domotor E, Lanza F, Adam-Vizi V, Machovich R. 2000. Identification of thrombin receptors in rat brain capillary endothelial cells. J Cereb Blood Flow Metab 20: 175-182.

Beer R, Franz G, Schopf M, Reindl M, Zelger B, et al. 2000. Expression of Fas and Fas ligand after experimental traumatic brain injury in the rat. J Cereb Blood Flow Metab 20: 669-677.

Behrens MM, Marinez JL, Moratilla C, Renart J. 1995. Apoptosis induced by protein kinase-C inhibition in a neuroblastoma cell line. Cell Growth Differ 6: 1375-1380.

Benchenane K, Berezowski V, Fernandez-Monreal M, Brillault J, Valable S, et al. 2005. Oxygen glucose deprivation switches the transport of tPA across the blood–brain barrier from an LRP dependent to an increased LRP-independent process. Stroke 36: 1065-1070.

Benchoua A, Braudeau J, Reis A, Couriaud C, Onteniente B. 2004. Activation of proinflammatory caspases by cathepsin B in focal cerebral ischemia. J Cereb Blood Flow Metab 24: 1272-1279.

Bergman BL, Scott RW, Bajpai A, Watts S, Baker JB. 1986. Inhibition of tumor-cell-mediated extracellular matrix destruction by a fibroblast proteinase inhibitor, protease nexin I. Proc Natl Acad Sci USA 83: 996-1000.

Bevec T, Stoka V, Pungercic G, Dolenc I, Turk V. 1996. Major histocompatibility complex class II-associated p41 invariant chain fragment is a strong inhibitor of lysosomal cathepsin L. J Exp Med 183: 1331-1338.

Bever CT Jr, Garver DW. 1995. Increased cathepsin B activity in multiple sclerosis brain. J Neurol Sci 131: 71-73.

Bode W, Fernandez-Catalan C, Grams F, Gomis-Ruth FX, Nagase H, et al. 1999. Insights into MMP–TIMP interactions. Ann NY Acad Sci 878: 73-91.

Boya P, Andreau K, Poncet D, Zamzani N, Perfettini JL, et al. 2003. Lysosomal membrane permeabilization induces cell death in a mitochondrion-dependent fashion. J Exp Med 197: 1323-1334.

Bozoky Z, Alexa A, Tompa P, Friedrich P. 2005. Multiple interactions of the 'transducer' govern its function in calpain activation by Ca2+. Biochem J 388: 741-744.

Branca D. 2004. Calpain-related diseases. Biochem Biophys Res Commun 322: 1098-1104.

Brennecke J, Hipfner DR, Stark A, Russell RB, Cohen SM. 2003. Bantam encodes a developmentally regulated microRNA that controls cell proliferation and regulates the pro-apoptotic gene hid in Drosophila. Cell 113: 25-36.

Brew K, Dinakarpandian D, Nagase H. 2000. Tissue inhibitors of metalloproteinases: Evolution, structure and function. Biochim Biophys Acta 1477: 267-283.

Cardone MH, Salvesen GS, Widmann C, Johnson G, Frisch SM. 1997. The regulation of anoikis: MEKK-1 activation requires cleavage by caspases. Cell 90: 315-323.

Carroll PM, Tsirka SE, Richards WG, Frohman MA, Strickland S. 1994. The mouse tissue plasminogen activator gene 5' flanking region directs appropriate expression in development and a seizure-enhanced response in the CNS. Development 120: 3173-3183.

Chai J, Shiozaki E, Srinivasula SM, Wu Q, Datta P, et al. 2001. Structural basis of caspase-7 inhibition by XIAP. Cell 104: 769-780.

Chandler S, Coates R, Gearing A, Lury J, Wells G, et al. 1995. Matrix metalloproteinases degrade myelin basic protein. Neurosci Lett 201: 223-226.

Chapman HA, Riese RJ, Shi GP. 1997. Emerging roles for cysteine proteases in human biology. Annu Rev Physiol 59: 63-88.

Cheng EH, Kirsch DG, Clem RJ, Ravi R, Kastan MB, et al. 1997. Conversion of bcl-2 to a Bax-like death effector by caspases. Science 278: 1966-1968.

Choi BH, Suzuki M, Kim T, Wagner SL, Cunningham DD. 1990. Protease nexin-1. Localization in the human brain suggests a protective role against extravasated serine proteases. Am J Pathol 137: 741-747.

Choi SH, Lee DY, Ryu JK, Kim J, Joe EH, et al. 2003. Thrombin induces nigral dopaminergic neurodegeneration in vivo by altering expression of death-related proteins. Neurobiol Dis 14: 181-193.

Cilenti L, Lee Y, Hess S, Srinivasula S, Park KM, et al. 2003. Characterization of a novel and specific inhibitor for the pro-apoptotic protease Omi/HtrA2. J Biol Chem 278: 11489-11494.

Cinelli P, Madani R, Tsuzuki N, Vallet P, Arras M, et al. 2001. Neuroserpin, a neuroprotective factor in focal ischemic stroke. Mol Cell Neurosci 18: 443-457.

Cocks TM, Moffatt JD. 2000. Protease-activated receptors: Sentries for inflammation? Trends Pharmacol Sci 21: 103-108.

Cohen GM. 1997. Caspases: The executioners of apoptosis. Biochem J 326: 1-16.

Colon GP, Lee KR, Keep RF, Chenevert TL, Betz AL, et al. 1996. Thrombin-soaked gelatin sponge and brain edema in rats. J Neurosurg 85: 335-339.

Copin JC, Goodyear MC, Gidday JM, Shah AR, Gascon E, et al. 2005. Role of matrix metalloproteinases in apoptosis after transient focal cerebral ischemia in rats and mice. Eur J Neurosci 22: 1597-1608.

Coughlin SR. 2000. Thrombin signalling and protease-activated receptors. Nature 407: 258-264.

Crnkovic-Mertens I, Semzow J, Hoppe-Seyler F, Butz K. 2006. Isoform-specific silencing of the Livin gene by RNA interference defines Livin B as key mediator of apoptosis inhibition in HeLa cells. J Mol Med 84: 232-240.

Dear N, Matena K, Vingron M, Boehm T. 1997. A new subfamily of vertebrate calpains lacking a calmodulin-like domain: Implications for calpain regulation and evolution. Genomics 45: 175-184.

Degterev A, Huang Z, Boyce M, Li Y, Jagtap P, et al. 2005. Chemical inhibitor of nonapoptotic cell death with therapeutic potential for ischemic brain injury. Nat Chem Biol 1: 112-119.

Degterev A, Lugovskoy A, Cardone M, Mulley B, Wagner G, et al. 2001. Identification of small-molecule inhibitors of interaction between the BH3 domain and Bcl-xL. Nat Cell Biol 3: 173-182.

Deschepper CF, Bigornia V, Berens ME, Lapointe MC. 1991. Production of thrombin and antithrombin III by brain and astroglial cell cultures. Mol Brain Res 11: 355-358.

Deshmukh M. 1998. Caspases in ischaemic brain injury and neurodegenerative disease. Apoptosis 3: 387-394.

Dietrich N, Thastrup J, Holmberg C, Gyrd-Hansen M, Fehrenbacher N, et al. 2004. JNK2 mediates TNF-induced cell death in mouse embryonic fibroblasts via regulation of both caspase and cathepsin protease pathways. Cell Death Differ 11: 301-313.

Dihanich M, Kaser M, Reinhard E, Cunningham D, Monard D. 1991. Prothrombin mRNA is expressed by cells of the nervous system. Neuron 6: 575-581.

Donovan FM, Cunningham DD. 1998. Signaling pathways involved in thrombin-induced cell protection. J Biol Chem 273: 12746-12752.

Duckett CS, Nava VE, Gedrich RW, Clem RJ, Van Dongen JL, et al. 1996. A conserved family of cellular genes related to the baculovirus iap gene and encoding apoptosis inhibitors. EMBO J 15: 2685-2689.

Dunn AD, Crutchfield HE, Dunn JT. 1991. Thyroglobulin processing by thyroidal proteases. Major sites of cleavage by cathepsins B, D, and L. J Biol Chem 266: 20198-20204.

Ekert PG, Silke J, Vaux DL. 1999. Inhibition of apoptosis and clonogenic survival of cells expressing crmA variants: Optimal caspase substrates are not necessarily optimal inhibitors. EMBO J 18: 330-338.

Ellis RC, O'Steen WA, Hayes RL, Nick HS, Wang KK, Anderson DK. 2005. Cellular localization and enzymatic activity of cathepsin B after spinal cord injury in the rat. Exp Neurol 193: 19-28.

Emert-Sedlak L, Shangary S, Rabinovitz A, Miranda MB, Delach SM, et al. 2005. Involvement of cathepsin D in chemotherapy-induced cytochrome c release, caspase activation, and cell death. Mol Cancer Ther 4: 733-742.

Enari M, Sakahira H, Yokoyama H, Okawa K, Iwamatsu A, et al. 1998. A caspase-activated DNase that degrades DNA during apoptosis, and its inhibitor ICAD. Nature 391: 43-50.

Endres M, Namura S, Shimizu-Sasamata M, Waeber C, Zhang L, et al. 1998. Attenuation of delayed neuronal death after mild focal ischemia in mice by inhibition of the caspase family. J Cereb Blood Flow Metab 18: 238-247.

Faccio L, Fusco C, Viel A, Zervos AS. 2000b. Tissue-specific splicing of Omi stress-regulated endoprotease leads to an inactive protease with a modified PDZ motif. Genomics 68: 343-347.

Faccio L, Fusco C, Chen A, Martinotti S, Bonventre JV, et al. 2000a. Characterization of a novel human serine protease that has extensive homology to bacterial heat shock endoprotease HtrA and is regulated by kidney ischemia. J Biol Chem 275: 2581-2588.

Fattman CL, An B, Dou QP. 1997. Characterization of interior cleavage of retinoblastoma protein in apoptosis. J Cell Biochem 67: 399-408.

Felbor U, Kessler B, Mothes W, Goebel HH, Ploegh HL, et al. 2002. Neuronal loss and brain atrophy in mice lacking cathepsins B and L. Proc Natl Acad Sci USA 99: 7883-7888.

Ferrer I, Puig B, Goutan E, Gombau L, Munoz-Canoves P. 2001. Methylazoximethanol acetate-induced cell death in the granule cell layer of the developing mouse cerebellum is associated with caspase-3 activation, but does not depend on the tissue-type plasminogen activator. Neurosci Lett 299: 77-80.

Figueroa BE, Keep RF, Betz AL, Hoff JT. 1998. Plasminogen activators potentiate thrombin-induced brain injury. Stroke 29: 1202-1207.

Flavin MP, Zhao G. 2001. Tissue plasminogen activator protects hippocampal neurons from oxygen-glucose deprivation injury. J Neurosci Res 63: 388-394.

Flavin MP, Zhao G, Ho LT. 2000. Microglial tissue plasminogen activator (tPA) triggers neuronal apoptosis in vitro. Glia 29: 347-354.

Foghsgaard L, Wissing D, Mauch D, Lademann U, Bastholm L, et al. 2001. Cathepsin B acts as a dominant execution protease in tumor cell apoptosis induced by tumor necrosis factor. J Cell Biol 153: 999-1010.

Franz T, Vingron M, Boehm T, Dear TN. 1999. Calpain-7: A highly divergent vertebrate calpain with a novel C-terminal domain. Mamm Genome 10: 318-321.

Franz T, Winckler L, Boehm T, Dear TN. 2004. Capn5 is expressed in a subset of T cells and is dispensable for development. Mol Cell Biol 24: 1649-1654.

Friedmann I, Yoles E, Schwartz M. 2001. Thrombin attenuation is neuroprotective in the injured rat optic nerve. J Neurochem 76: 641-649.

Fujita N, Nagahashi A, Nagashima K, Rokudai S, Tsuruo T. 1998. Acceleration of apoptotic cell death after the cleavage of Bcl-XL protein by caspase-3-like proteases. Oncogene 17: 1295-1304.

Fukiage C, Azuma M, Nakamura Y, Tamada Y, Nakamura M, et al. 1997. SJA6017, a newly synthesized peptide aldehyde inhibitor of calpain: Amelioration of cataract in cultured rat lenses. Biochim Biophys Acta 1361: 304-312.

Fusek M, Vetvicka V. 2005. Dual role of cathepsin D: Ligand and protease. Biomed Pap Med Fac Univ Palacky Olomouc Czech Repub 149: 43-50.

Gafni J, Ellerby LM. 2002. Calpain activation in Huntington's disease. J Neurosci 22: 4842-4849.

Gafni J, Hermel E, Young JE, Wellington CL, Hayden MR, et al. 2004. Inhibition of calpain cleavage of huntingtin reduces toxicity: Accumulation of calpain/caspase fragments in the nucleus. J Biol Chem 279: 20211-20220.

Gingrich MB, Traynelis SF. 2000. Serine proteases and brain damage: Is there a link? Trends Neurosci 23: 399-407.

Goll DE, Thompson VF, Li H, Wei W, Cong J. 2003. The calpain system. Physiol Rev 83: 731-801.

Grandgirard D, Studer E, Monney L, Belser T, Fellay I, et al. 1998. Alphaviruses induce apoptosis in Bcl-2-overexpressing cells: Evidence for a caspase-mediated, proteolytic inactivation of Bcl-2. EMBO J 17: 1268-1278.

Gray CW, Ward RV, Karran E, Turconi S, Rowles A, et al. 2000. Characterization of human HtrA2, a novel serine protease involved in the mammalian cellular stress response. Eur J Biochem 267: 5699-5710.

Gu Z, Kaul M, Yan B, Kridel SJ, Cui J, et al. 2002. S-nitrosylation of matrix metalloproteinases: Signaling pathway to neuronal cell death. Science 297: 1186-1190.

Guicciardi ME, Deussing J, Miyoshi H, Bronk SF, Svingen PA, et al. 2000. Cathepsin B contributes to TNF-mediated hepatocyte apoptosis by promoting mitochondrial release of cytochrome c. J Clin Invest 106: 1127-1137.

Han BH, Xu D, Choi J, Han Y, Xanthoudakis S, et al. 2002. Selective, reversible caspase-3 inhibitor is neuroprotective and reveals distinct pathways of cell death after neonatal hypoxic-ischemic brain injury. J Biol Chem 277: 30128-30136.

Handsley MM, Cross J, Gavrilovic J, Edwards DR. 2005. The matrix metalloproteinases and their inhibitors. Matrix Metalloproteinases in the Central Nervous System. Conant K, Gottschall PE, editors. London, UK: Imperial College Press; pp. 3-16.

Hastings GA, Coleman TA, Haudenschild CC, Stefansson S, Smith EP, et al. 1997. Neuroserpin, a brain-associated inhibitor of tissue plasminogen activator is localized primarily in neurons. Implications for the regulation of motor learning and neuronal survival. J Biol Chem 272: 33062-33067.

Hauser HP, Bardroff M, Pyrowolakis G, Jentsch S. 1998. A giant ubiquitin-conjugating enzyme related to IAP apoptosis inhibitors. J Cell Biol 141: 1415-1422.

Hegde R, Srinivasula SM, Zhang Z, Wassell R, Mukattash R, et al. 2002. Identification of Omi/HtrA2 as a mitochondrial apoptotic serine protease that disrupts inhibitor of apoptosis protein-caspase interaction. J Biol Chem 277: 432-438.

Hengartner MO. 2000. The biochemistry of apoptosis. Nature 407: 770-776.

Hollenberg MD, Compton SJ. 2002. International Union of Pharmacology. XXVIII. Proteinase-activated receptors. Pharmacol Rev 54: 203-217.

Hong SC, Goto Y, Lanzino G, Soleau S, Kassell NF, et al. 1994. Neuroprotection with a calpain inhibitor in a model of focal cerebral ischemia. Stroke 25: 663-669.

Hosfield CM, Elce JS, Jia Z. 2004. Activation of calpain by Ca^{2+}: Roles of the large subunit N-terminal and domain III–IV linker peptides. J Mol Biol 343: 1049-1053.

Houenou LJ, Turner PL, Li L, Oppenheim RW, Festoff BW. 1995. A serine protease inhibitor, protease nexin I, rescues motoneurons from naturally occurring and axotomy-induced cell death. Proc Natl Acad Sci USA 92: 895-899.

Houseweart MK, Vilaythong A, Yin XM, Turk B, Noebels JL, et al. 2003. Apoptosis caused by cathepsins does not require Bid signaling in an in vivo model of progressive myoclonus epilepsy (EPM1). Cell Death Differ 10: 1329-1335.

Huang Y, Wang KK. 2001. The calpain family and human disease. Trends Mol Med 7: 355-362.

Inoue J, Nakamura M, Cui Y, Sakai Y, Sakai O, et al. 2003. Structure-activity relationship study and drug profile of N-(4-fluorophenylsulfonyl)-L-valyl-L-leucinal (SJA6017) as a potent calpain inhibitor. J Med Chem 46: 868-871.

Irazuzta J, Pretzlaff RK, DeCourten-Myers G, Zemlan F, Zingarelli B. 2005. Dexamethasone decreases neurological sequelae and caspase activity. Intensive Care Med 31: 146-150.

Jacobson MD, Weil M, Raff MC. 1997. Programmed cell death in animal development. Cell 88: 347-354.

Jin S, Kalkum M, Overholtzer M, Stoffel A, Chait BT, et al. 2003. CIAP1 and the serine protease HTRA2 are involved in a novel p53-dependent apoptosis pathway in mammals. Genes Dev 17: 359-367.

Jones B, Roberts PJ, Faubion WA, Kominami E, Gores GJ. 1998. Cystatin A expression reduces bile salt-induced apoptosis in a rat hepatoma cell line. Am J Physiol 275: G723-G730.

Jordan J, Galindo MF, Miller RJ. 1997. Role of calpain and interleukin-B converting enzyme-like proteases in the B-amyloid-induced death of rat hippocampal neurons in culture. J Neurochem 68: 1612-1621.

Junge CE, Sugawara T, Mannaioni G, Alagarsamy S, Conn PJ, et al. 2003. The contribution of protease activated receptor 1 to neuronal damage caused by transient focal cerebral ischemia. Proc Natl Acad Sci USA 100: 13019-13024.

Kakegawa H, Nikawa T, Tagami K, Kamioka H, Sumitani K, et al. 1993. Participation of cathepsin L on bone resorption. FEBS Lett 321: 247-250.

Kasof GM, Gomes BC. 2001. Livin, a novel inhibitor of apoptosis protein family member. J Biol Chem 276: 3238-3246.

Kass GEN, Orrenius S. 1999. Calcium signaling and cytotoxicity. Environ Health Perspect 107: 25-35.

Kaufmann SH, Hengartner MO. 2001. Programmed cell death: Alive and well in the new millennium. Trends Cell Biol 11: 526-534.

Kerr JF, Wyllie AH, Currie AR. 1972. Apoptosis: A basic biological phenomenon with wide-ranging implications in tissue kinetics. Br J Cancer 26: 239-257.

Khorchid A, Ikura M. 2002. How calpain is activated by calcium. Nat Struct Biol 9: 239-241.

Kikuchi H, Yamada T, Furuya H, Doh-ura K, Ohyagi Y, et al. 2003. Involvement of cathepsin B in the motor neuron degeneration of amyotrophic lateral sclerosis. Acta Neuropathol 105: 462-468.

Kim HW, Chang ES, Mykles DL. 2005. Three calpains and ecdysone receptor in the land crab Gecar cinus lateralis: Sequences, expression and effects of elevated ecdysteriod induced by eyestalk ablation. J Exp Biol 208: 3177-3197.

Kingham PJ, Pocock JM. 2001. Microglial secreted cathepsin B induces neuronal apoptosis. J Neurochem 76: 1475-1784.

Kohda Y, Yamashima T, Sakuda K, Yamashita J, Ueno T, et al. 1996. Dynamic changes of cathepsins B and L expression in monkey hippocampus after transient ischemia. Biochem Biophys Res Commun 228: 616-622.

Kreuter M, Langer C, Kerkhoff C, Reddanna P, Kania AL, et al. 2004. Stroke, myocardial infarction, acute and chronic inflammatory diseases: Caspases and other apoptotic molecules as targets for drug development. Arch Immunol Ther Exp (Warsz) 52: 141-155.

Krueger SR, Ghisu GP, Cinelli P, Gschwend TP, Osterwalder T, et al. 1997. Expression of neuroserpin, an inhibitor of tissue plasminogen activator, in the developing and adult nervous system of the mouse. J Neurosci 17: 8984-8996.

Kudo A, Suzuki M, Kubo Y, Watanabe M, Yoshida K, et al. 2000. Intrathecal administrationof thrombin inhibitor ameliorates cerebral vasospasm. Use of a drug delivery system releasing hirudin. Cerebrovasc Dis 10: 424-430.

Kupina NC, Nath R, Bernath EE, Inoue J, Mitsuyoshi A, et al. 2001. The novel calpain inhibitor SJA6017 improves functional outcome after delayed administration in a mouse model of diffuse brain injury. J Neurotrauma 18: 1229-1240.

Larner SF, McKinsey DM, Hayes RL, Wang KKW. 2005. Caspase 7: Increased expression and activation after traumatic brain injury in rats. J Neurochem 94: 97-108.

Larner SF, Hayes RL, McKinsey DM, Pike BR, Wang KK. 2004. Increased expression and processing of caspase-12 after traumatic brain injury in rats. J Neurochem 88: 78-90.

Lebeurrier N, Vivien D, Ali C. 2004. The complexity of tissue-type plasminogen activator: Can serine protease inhibitors help in stroke management? Expert Opin Ther Targets 8: 309-320.

Leco KJ, Khokha R, Pavloff N, Hawkes SP, Edwards DR. 1994. Tissue inhibitor of metalloproteinases-3 (TIMP-3) is an extracellular matrix-associated protein with a distinctive pattern of expression in mouse cells and tissues. J Biol Chem 269: 9352-9360.

Lee DY, Park KW, Jin PK. 2006. Thrombin induces neurodegeneration and microglial activation in the cortex in vivo and in vitro: Proteolytic and non-proteolytic actions. Biochem Biophys Res Commun 346: 727-738.

Lee KR, Kawai N, Kim S, Sagher O, Hoff JT. 1997. Mechanisms of edema formation after intracerebral hemorrhage: Effects of thrombin on cerebral blood flow, blood-brain barrier permeability, and cell survival in a rat model. J Neurosurg 86: 272-278.

Lee KR, Betz AL, Keep RF, Chenevert TL, Kim S, et al. 1995. Intracerebral infusion of thrombin as a cause of brain edema. J Neurosurg 83: 1045-1050.

Lee KR, Colon GP, Betz AL, Keep RF, Kim S, et al. 1996. Edema from intracerebral hemorrhage: The role of thrombin. J Neurosurg 84: 91-96.

Lee R, Kermani P, Teng KK, Hempstead BL. 2001. Regulation of cell survival by secreted proneurotrophins. Science 294: 1945-1948.

Lee SR, Lo EH. 2004. Induction of caspase-mediated cell death by matrix metalloproteinases in cerebral endothelial cells after hypoxia-reoxygenation. J Cereb Blood Flow Metab 24: 720-727.

Leist M, Jaattela M. 2001. Four deaths and a funeral: From caspases to alternative mechanisms. Nat Rev Mol Cell Biol 2: 589-598.

Lenarcic B, Bevec T. 1998. Thyropins—new structurally related proteinase inhibitors. Biol Chem 379: 105-111.

Leung-Toung R, Li W, Tam TF, Karimian K. 2002. Thiol-dependent enzymes and their inhibitors: A review. Curr Med Chem 9: 979-1002.

Li W, Srinivasula SM, Chai J, Li P, Wu JW, et al. 2002. Structural insights into the pro-apoptotic function of mitochondrial serine protease HtrA2/Omi. Nat Struct Biol 9: 436-441.

Liaudet-Coopman E, Beaujouin M, Derocq D, Garcia M, Glondu-Lassis M, et al. 2006. Cathepsin D: Newly discovered functions of a long-standing aspartic protease in cancer and apoptosis. Cancer Lett 237: 167-179.

Lijnen HR, Collen D. 1998. t-plasminogen activator. Handbook of Proteolytic Enzymes. Barrett AJ, Rawlings ND, Woessner JF, editors. San Diego, CA: Academic Press; pp. 184-190.

Lin JH, Deng G, Huang Q, Morser J. 2000. KIAP, a novel member of the inhibitor of apoptosis protein family. Biochem Biophys Res Commun 279: 820-831.

Linebaugh BE, Sameni M, Day NA, Sloane BF, Keppler D. 1999. Exocytosis of active cathepsin B enzyme activity at pH 7.0, inhibition and molecular mass. Eur J Biochem 264: 100-109.

Linton SD, Aja T, Armstrong RA, Bai X, Chen LS, et al. 2005. First-in-class pan caspase inhibitor developed for the treatment of liver disease. J Med Chem 48: 6779-6782.

Liot G, Roussel BD, Lebeurrier N, Benchenane K, Lopez-Atalaya JP, et al. 2006. Tissue-type plasminogen activator rescues neurones from serum deprivation-induced apoptosis through a mechanism independent of its proteolytic activity. J Neurochem 98: 1458-1464.

Liston P, Roy N, Tamai K, Lefebvre C, Baird S, et al. 1996. Suppression of apoptosis in mammalian cells by NAIP and a related family of IAP genes. Nature 379: 349-353.

Liu D, Cheng T, Guo H, Fernandez JA, Griffin JH, et al. 2004a. Tissue plasminogen activator neurovascular toxicity is controlled by activated protein C. Nat Med 10: 1379-1383.

Liu N, Wang Y, Ashton-Rickardt PG. 2004c. Serine protease inhibitor 2A inhibits caspase-independent cell death. FEBS Lett 569: 49-53.

Liu PK, Grossman RG, Hsu CY, Robertson CS. 2001. Ischemic injury and faulty gene transcripts in the brain. Trends Neurosci 24: 581-588.

Liu X, Van Vleet T, Schnellmann RG. 2004b. The role of calpain in oncotic cell death. Annu Rev Pharmacol Toxicol 44: 349-370.

Lockshin RA, Zakeri Z. 2001. Programmed cell death and apoptosis: Origins of the theory. Nat Rev Mol Cell Biol 2: 545-550.

Lohmann C, Krischke M, Wegener J, Galla HJ. 2004. Tyrosine phosphatase inhibition induces loss of blood-brain barrier integrity by matrix metalloproteinase-dependent and -independent pathways. Brain Res 995: 184-196.

Lu W, Bhasin M, Tsirka SE. 2002. Involvement of tissue plasminogen activator in onset and effector phases of experimental allergic encephalomyelitis. J Neurosci 22: 10781-10789.

Ma H, Fukiage C, Kim YH, Duncan MK, Reed NA, et al. 2001. Characterization and expression of calpain 10. A novel ubiquitous calpain with nuclear localization. J Biol Chem 276: 28525-28531.

Macfarlane SR, Seatter MJ, Kanke T, Hunter GD, Plevin R. 2001. Proteinase-activated receptors. Pharmacol Rev 53: 245-282.

Mackay EA, Ehrhard A, Moniatte M, Guenet C, Tardif C, et al. 1997. A possible role for cathepsins D, E and B in the processing of beta-amyloid precursor protein in Alzheimer's disease. Eur J Biochem 244: 414-425.

Maier CM, Hsieh L, Crandall T, Narasimhan P, Chan PH. 2006. Evaluating therapeutic targets for reperfusion-related brain hemorrhage. Ann Neurol 59: 929-938.

Mannello F, Luchetti F, Falcieri E, Papa S. 2005. Multiple roles of matrix metalloproteinases during apoptosis. Apoptosis 10: 19-24.

Mansuy IM, Putten van der H, Schmid P, Meins M, Botteri FM, et al. 1993. Variable and multiple expression of Protease Nexin-1 during mouse organogenesis and nervous system development. Development 119: 1119-1134.

Marques MA, Tolar M, Harmony JA, Crutcher KA. 1996. A thrombin cleavage fragment of apolipoprotein E exhibits isoform-specific neurotoxicity. Neuroreport 7: 2529-2532.

Martinon F, Tschopp J. 2004. Inflammatory caspases: Linking an intracellular innate immune system to autoinflammatory diseases. Cell 117: 561-574.

Matena K, Boehm T, Dear N. 1998. Genomic organization of mouse Capn5 and Capn6 genes confirms that they are a distinct calpain subfamily. Genomics 48: 117-120.

McGeer PL, McGeer EG. 2002. Local neuroinflammation and the progression of Alzheimer's disease. J Neurovirol 8: 529-538.

McGinnis KM, Whitton MM, Gnegy ME, Wang KK. 1998. Calcium/calmodulin-dependent protein kinase IV is cleaved by caspase-3 and calpain in SH-SY5Y human neuroblastoma cells undergoing apoptosis. J Biol Chem 273: 19993-20000.

McKinney M, Snider RM, Richelson E. 1983. Thrombin binding to human brain and spinal cord. Mayo Clin Proc 58: 829-881.

Medina MG, Ledesma MD, Dominguez JE, Medina M, Zafra D, et al. 2005. Tissue plasminogen activator mediates amyloid-induced neurotoxicity via Erk1/2 activation. EMBO J 24: 1706-1716.

Mendelow AD. 1993. Mechanisms of ischemic brain damage with intracerebral hemorrhage. Stroke 24: I-115-I-117.

Mercer EA, Korhonen L, Skoglosa Y, Olsson PA, Kukkonen JP, et al. 2000. NAIP interacts with hippocalcin and protects neurons against calcium- induced cell death through caspase-3-dependent and -independent pathways. EMBO J 19: 3597-3607.

Mhatre M, Nguyen A, Kashani S, Pham T, Adesina A, et al. 2004. Thrombin, a mediator of neurotoxicity and memory impairment. Neurobiol Aging 25: 783-793.

Milligan CE, Prevette D, Yaginuma H, Homma S, Cardwell C, et al. 1995. Peptide inhibitors of the ICE protease family arrest programmed cell death of motoneurons in vivo and in vitro. Neuron 15: 385-393.

Moldoveanu T, Hosfield CM, Lim D, Elce JS, Jia Z, et al. 2002. A Ca(2+) switch aligns the active site of calpain. Cell 108: 649-660.

Möller T, Hanisch UK, Ransom BR. 2000. Thrombin-induced activation of cultured rodent microglia. J Neurochem 75: 1539-1547.

Monard D. 1988. Cell-derived proteases and protease inhibitors as regulators of neurite outgrowth. Trends Neurosci 11: 541-544.

Monard D, Suidan HS, Nitsch C. 1992. Relevance of the balance between glia-derived nexin and thrombin following lesion in the nervous system. Ann N Y Acad Sci 674: 237-242.

Mort JS, Buttle DJ. 1997. Cathepsin B. Int J Biochem Cell Biol 29: 715-720.

Nagai A, Ryu JK, Terashima M, Tanigawa Y, Wakabayashi K, et al. 2005. Neuronal cell death induced by cystatin C in vivo and in cultured human CNS neurons is inhibited with cathepsin B. Brain Res 1066: 120-128.

Nakagawa T, Yuan J. 2000. Cross-talk between two cysteine protease families. Activation of caspase-12 by calpain in apoptosis. J Cell Biol 150: 887-894.

Nakagawa T, Zhu H, Morishima N, Li E, Xu J, et al. 2000. Caspase-12 mediates endoplasmic-reticulum-specific apoptosis and cytotoxicity by amyloid-beta. Nature 403: 98-103.

Nakajima K, Kohsaka S. 2001. Microglia: Activation and their significance in the central nervous system. J Biochem (Tokyo) 130: 169-175.

Nath R, McGinnis KJ, Nadimpalli R, Stafford D, Wang KKW. 1996a. Effects of ICE-like proteases and calpain inhibitors on neuronal apoptosis. Neuroreport 8: 249-255.

Nath R, Raser KJ, Stafford D, Hajimohammadreza I, Posner A, et al. 1996b. Non-erythroid alpha-spectrin breakdown by calpain and interleukin 1 beta-converting-enzyme-like protease(s) in apoptotic cells: Contributory roles of both protease families in neuronal apoptosis. Biochem J 319: 683-690.

Niclou S, Suidan HS, Brown-Luedi M, Monard D. 1994. Expression of the thrombin receptor mRNA in rat brain. Cell Mol Biol 40: 421-428.

Nitatori T, Sato N, Waguri S, Karasawa Y, Araki H, et al. 1995. Delayed neuronal death in the CA-1 pyramidal cell layer of the gerbil hippocampus following transient ischemia is apoptosis. J Neurosci 15: 1001-1011.

Nixon RA. 2000. A "protease activation cascade" in the pathogenesis of Alzheimer's disease. Ann N Y Acad Sci 924: 117-131.

Nylandsted J, Gyrd-Hansen M, Danielewicz A, Fehrenbacher N, Lademann U, et al. 2004. Heat shock protein 70 promotes cell survival by inhibiting lysosomal membrane permeabilization. J Exp Med 200: 425-435.

Oh J, Takahashi R, Kondo S, Mizoguchi A, Adachi E, et al. 2001. The membrane-anchored MMP inhibitor RECK is a key regulator of extracellular matrix integrity and angiogenesis. Cell 107: 789-800.

Ohyama H, Hosomi N, Takahashi T, Mizushige K, Kohno M. 2001. Thrombin inhibition attenuates neurodegeneration and cerebral edema formation following transient forebrain ischemia. Brain Res 902: 264-271.

Osterwalder T, Cinelli P, Baici A, Pennella A, Krueger SR, et al. 1996. Neuroserpin, an axonally secreted serine protease inhibitor. EMBO J 15: 2944-2953.

Pignol B, Auvin S, Carré D, Marin J-G, Chabrier P-E. 2006. Calpain inhibitors and antioxidants act synergistically to prevent cell necrosis: Effects of the novel dual inhibitors (cysteine protease inhibitor and antioxidant) BN 82204 and its pro-drug BN 82270. J Neurochem 98: 1217-1228.

Pike BR, Flint J, Dutta S, Johnson E, Wang KK, et al. 2001. Accumulation of non-erythroid alpha II-spectrin and calpain-cleaved alpha II-spectrin breakdown products in cerebrospinal fluid after traumatic brain injury in rats. J Neurochem 78: 1297-1306.

Pike BR, Zhao X, Newcomb JK, Posmantur RM, Wang KK, et al. 1998b. Regional calpain and caspase-3 proteolysis of alpha-spectrin after traumatic brain injury. Neuroreport 9: 2437-2442.

Pike BR, Zhao X, Newcomb JK, Wang KKW, Posmantur RM, et al. 1998a. Temporal relationships between De Novo protein synthesis, calpain and caspase-3 (CPP32) protease activation, and DNA fragmentation during apoptosis in septo-hippocampal cultures. J Neurosci Res 52: 505-520.

Powell WC, Fingleton B, Wilson CL, Boothby M, Matrisian LM. 1999. The metalloproteinase Matrilysin (MMP-7) proteolytically generates active soluble Fas ligand and potentiates epithelial cell apoptosis. Curr Biol 9: 1441-1447.

Qian Z, Gilbert ME, Colicos MA, Kandel ER, Kuhl D. 1993. Tissue plasminogen activator is induced as an immediate-early gene during seizure, kindling and longterm potentiation. Nature 361: 453-457.

Qiu J, Whalen MJ, Lowenstein P, Fiskum G, Fahy B, et al. 2002. Upregulation of the Fas receptor death inducing signaling complex after traumatic brain injury in mice and humans. J Neurosci 22: 3504-3511.

Qiu XB, Goldberg AL. 2005. The membrane-associated inhibitor of apoptosis protein, BRUCE/Apollon, antagonizes both the precursor and mature forms of Smac and caspase-9. J Biol Chem 280: 174-182.

Rabuffetti M, Sciorati C, Tarozzo G, Clementi E, Manfredi AA, et al. 2000. Inhibition of caspase-1-like activity by Ac-Tyr-Val-Ala-Asp-chloromethyl ketone induces long-lasting neuroprotection in cerebral ischemia through apoptosis reduction and decrease of proinflammatory cytokines. J Neurosci 20: 4398-4404.

Raghupathi R, Graham DI, McIntosh TK. 2000. Apoptosis after traumatic brain injury. J Neurotrauma 17: 927-938.

Rami A, Agarwal R, Botez G, Winckler. 2000. mu-Calpain activation, DNA fragmentation, and synergistic effects of caspase and calpain inhibitors in protecting hippocampal neurons from ischemic damage. Brain Res 866: 299-312.

Rao RV, Hermel E, Castro-Obregon S, del Rio G, Ellerby LM, et al. 2001. Coupling endoplasmic reticulum stress to the cell death program. Mechanism of caspase activation. J Biol Chem 276: 33869-33874.

Rao RV, Peel A, Logvinova A, del Rio G, Hermel E, et al. 2002. Coupling endoplasmic reticulum stress to the cell death program: Role of the ER chaperone GRP78. FEBS Lett 514: 122-128.

Rathmell JC, Thompson CB. 1999. The central effectors of cell death in the immune system. Annu Rev Immunol 17: 781-828.

Ray SK, Banik NL. 2003. Calpain and its involvement in the pathophysiology of CNS injuries and diseases: Therapeutic potential of calpain inhibitors for prevention of neurodegeneration. Curr Drug Targets CNS Neurol Disord 2: 173-189.

Reddy RK, Mao C, Baumeister P, Austin RC, Kaufman RJ, et al. 2003. Endoplasmic reticulum chaperone protein GRP78 protects cells from apoptosis induced by topoisomerase inhibitors: Role of ATP binding site in suppression of caspase-7 activation. J Biol Chem 278: 20915-20924.

Reimann-Philipp U, Ovase R, Weigel PH, Grammas P. 2001. Mechanisms of cell death in primary cortical neurons and PC12 cells. J Neurosci Res 64: 654-660.

Reinhard E, Suidan HS, Pavlik A, Monard D. 1994. Glia-derived nexin/protease nexin-1 is expressed by a subset of neurons in the rat brain. J Neurosci Res 37: 256-270.

Richter BW, Mir SS, Eiben LJ, Lewis J, Reffey SB, et al. 2001. Molecular cloning of ILP-2, a novel member of the inhibitor of apoptosis protein family. Mol Cell Biol 21: 4292-4301.

Rochefort H, Cavailles V, Augereau P, Capony F, Maudelonde T, et al. 1989. Overexpression and hormonal regulation of pro-cathepsin D in mammary and endometrial cancer. J Steroid Biochem 34: 177-182.

Rodriguez J, Lazebnik Y. 1999. Caspase-9 and APAF-1 form an active holoenzyme. Genes Dev 13: 3179-3184.

Roemisch J, Gray E, Hoffmann JN, Wiedermann CJ. 2002. Antithrombin: A new look at the actions of a serine protease inhibitor. Blood Coagul Fibrinolysis 13: 657-670.

Rosenberg GA. 1995. Matrix metalloproteinases in brain injury. J Neurotrauma 12: 833-842.

Rossignol P, Ho-Tin-Noe B, Vranckx R, Bouton MC, Meilhac O, et al. 2004. Protease nexin-1 inhibits plasminogen activation-induced apoptosis of adherent cells. J Biol Chem 279: 10346-10356.

Rothe M, Pan M-G, Henzel WJ, Ayres TM, Goeddel DV. 1995. The TNFR2-TRAF signaling complex contains two novel proteins related to baculoviral inhibitor of apoptosis proteins. Cell 83: 1243-1252.

Roy N, Deveraux QL, Takahashi R, Salvesen GS, Reed JC. 1997. The c-IAP-1 and c-IAP-2 proteins are direct inhibitors of specific caspases. EMBO J 16: 6914-6925.

Saido TC, Sorimachi H, Suzuki K. 1994. Calpain: New perspectives in molecular diversity and physiological-pathological involvement. FASEB J 8: 814-822.

Sakahira H, Enari M, Nagata S. 1998. Cleavage of CAD inhibitor in CAD activation and DNA degradation during apoptosis. Nature 391: 96-99.

Sanna MG, da Silva Correia J, Ducrey O, Lee J, Nomoto K, et al. 2002. IAP suppression of apoptosis involves distinct mechanisms: The TAK1/JNK1 signaling cascade and caspase inhibition. Mol Cell Biol 22: 1754-1766.

Sappino AP, Madani R, Huarte J, Belin D, Kiss JZ, et al. 1993. Extracellular proteolysis in the adult murine brain. J Clin Invest 92: 679-685.

Sato H, Takino T, Okada Y, Cao J, Shinagawa A, et al. 1994. A matrix metalloproteinase expressed on the surface of invasive tumour cells. Nature 370: 61-65.

Sawdey MS, Loskutoff DJ. 1991. Regulation of murine type 1 plasminogen activator inhibitor gene expression in vivo. Tissue specificity and induction by lipopolysaccharide, tumor necrosis factor-alpha, and transforming growth factor-beta. J Clin Invest 88: 1346-1353.

Schaller C, Rohde V, Meyer B, Hassler W. 1995. Stereotactic puncture and lysis of spontaneous intracerebral hemorrhage using recombinant tissue-plasminogen activator. Neurosurgery 36: 328-335.

Schilling B, Gafni J, Torcassi C, Cong X, Row RH, et al. 2006. Huntingtin phosphorylation sites mapped by mass spectrometry: Modulation of cleavage and toxicity. J Biol Chem 281: 23686-23697.

Schonbeck U, Mach F, Libby P. 1998. Generation of biologically active IL-1β by matrix metalloproteinases: A novel caspase-1-independent pathway of IL-1β processing. J Immunol 161: 3340-3346.

Schumacher PA, Siman RG, Fehlings MG. 2000. Pretreatment with calpain inhibitor CEP-4143 inhibits calpain I activation and cytoskeletal degradation, improves neurological function, and enhances axonal survival after traumatic spinal cord injury. J Neurochem 74: 1646-1655.

Scovassi AI, Diederich M. 2004. Modulation of ploy(ADP-ribosylation) in apoptotic cells. Biochem Pharmacol 68: 1041-1047.

Seeds NW, Williams BL, Bickford PC. 1995. Tissue plasminogen activator induction in Purkinje neurons after cerebellar motor learning. Science 270: 1992-1994.

Seong Y, Choi J, Park H, Kim K, Ahn S, et al. 2004. Autocatalytic processing of HtrA2/Omi is essential for induction of caspase-dependent cell death through antagonizing XIAP. J Biol Chem 279: 37588-37596.

Seyfried D, Han Y, Zheng Z, Day N, Moin K, et al. 1997. Cathepsin B and middle cerebral artery occlusion in the rat. J Neurosurg 87: 716-723.

Seyfried DM, Veyna R, Han Y, Li K, Tang N, et al. 2001. A selective cysteine protease inhibitor is non-toxic and cerebroprotective in rats undergoing transient middle cerebral artery ischemia. Brain Res 901: 94-101.

Shin S, Sung BJ, Cho YS, Kim HJ, Ha NC, et al. 2001. An anti-apoptotic protein human survivin is a direct inhibitor of caspase-3 and -7. Biochemistry 40: 1117-1123.

Shiraha H, Glading A, Chou J, Jia Z, Wells A. 2002. Activation of m-calpain (calpain II) by epidermal growth factor is limited by protein kinase A phosphorylation of m-calpain. Mol Cell Biol 22: 2716-2727.

Shirasaki Y, Miyashita H, Yamaguchi M. 2006. Exploration of orally available calpain inhibitors. Part 3: Dipeptidyl alpha-ketoamide derivatives containing pyridine moiety. Bioorg Med Chem 14: 5691-5698.

Shirasaki Y, Miyashita H, Yamaguchi M, Inoue J, Nakamura M. 2005. Exploration of orally available calpain inhibitors: Peptidyl alpha-ketoamides containing an amphiphile at P3 site. Bioorg Med Chem 13: 4473-4484.

Smirnova IV, Citron BA, Arnold PM, Festoff BW. 2001. Neuroprotective signal transduction in model motor neurons exposed to thrombin: G-protein modulation effects on neurite outgrowth, Ca^{2+} mobilization, and apoptosis. J Neurobiol 48: 87-100.

Smirnova IV, Zhang SX, Citron BA, Arnold PM, Festoff BW. 1998. Thrombin is an extracellular signal that activates intracellular death protease pathways inducing apoptosis in model motor neurons. J Neurobiol 36: 64-80.

Smith-Swintosky VL, Cheo-Isaacs CT, D'Andrea MR, Santulli RJ, Darrow AL, et al. 1997. Protease-activated receptor-2 (PAR-2) is present in the rat hippocampus and is associated with neurodegeneration. J Neurochem 69: 1890-1896.

Solary E, Eymin B, Droin N, Haugg M. 1998. Proteases, proteolysis, and apoptosis. Cell Biol Toxicol 14: 121-132.

Sorimachi H, Ishiura S, Suzuki K. 1997. Structure and physiological function of calpains. Biochem J 328: 721-732.

Squier MKT, Miller ACK, Malkinson AM, Cohen JJ. 1994. Calpain activation in apoptosis. J Cell Physiol 159: 229-237.

Stennicke HR, Deveraux QL, Humke EW, Reed JC, Dixit VM, et al. 1999. Caspase-9 can be activated without proteolytic processing. J Biol Chem 274: 8359-8362.

Stoka V, Turk B, Turk V. 2005. Lysosomal cysteine proteases: Structural features and their role in apoptosis. IUBMB Life 57: 347-353.

Stoka V, Turk B, Schendel SL, Kim TH, Cirman T, et al. 2001. Lysosomal protease pathways to apoptosis. Cleavage of bid, not pro-caspases, is the most likely route. J Biol Chem 276: 3149-3157.

Stone SR, le Bonniec BF. 1998. Thrombin. Handbook of Proteolytic Enzymes. Barrett AJ, Rawlings ND, Woessner JF, editors. San Diego, CA: Academic Press; pp. 168-174.

Striggow F, Riek M, Breder J, Henrich-Noack P, Reymann KG, et al. 2000. The protease thrombin is an endogenous mediator of hippocampal neuroprotection against ischemia at low concentrations but causes degeneration at high concentrations. Proc Natl Acad Sci USA 97: 2264-2269.

Striggow F, Riek-Burchardt M, Kiesel A, Schmidt W, Henrich-Noack P, et al. 2001. Four different types of protease-activated receptors are widely expressed in the brain and are up-regulated in hippocampus by severe ischemia. Eur J Neurosci 14: 595-608.

Suo Z, Wu M, Citron BA, Gao C, Festoff BW. 2003. Persistent protease-activated receptor 4 signaling mediates thrombin-induced microglial activation. J Biol Chem 278: 31177-31183.

Suo Z, Wu M, Ameenuddin S, Anderson HE, Zoloty JE, et al. 2002. Participation of protease-activated receptor-1 in thrombin-induced microglial activation. J Neurochem 80: 655-666.

Suomalainen K, Sorsa T, Golub LM, Ramamurthy N, Lee HM, et al. 1992. Specificity of the anticollagenase action of tetracyclines: Relevance to their anti-inflammatory potential. Antimicrob Agents Chemother 36: 227-229.

Suzuki K, Tsuji S, Kubota S, Kimura Y, Imahori K. 1981. Limited autolysis of Ca^{2+}-activated neutral protease (CANP) changes its sensitivity to Ca^{2+} ions. J Biochem (Tokyo) 90: 275-278.

Suzuki Y, Imai Y, Nakayama H, Takahashi K, Takio K, et al. 2001. A serine protease, HtrA2, is released from the mitochondria and interacts with XIAP, inducing cell death. Mol Cell 8: 613-621.

Takahashi C, Sheng Z, Horan TP, Kitayama H, Maki M, et al. 1998. Regulation of matrix metalloproteinase-9 and inhibition of tumor invasion by the membrane-anchored glycoprotein RECK. Proc Natl Acad Sci 95: 13221-13226.

Takahashi H, Nagai N, Urano T. 2005. Role of tissue plasminogen activator/plasmin cascade in delayed neuronal death after transient forebrain ischemia. Neurosci Lett 381: 189-193.

Tan Y, Wu C, De Veyra T, Greer PA. 2006. Ubiquitous calpains promote both apoptosis and survival signals in responses to different cell death stimuli. J Biol Chem 281: 17689-17698.

Taupin V, Toulmond S, Serrano A, Benavides J, Zavala F. 1993. Increase in IL-6, IL-1 and TNF levels in rat brain following traumatic lesion. Influence of pre and post-traumatic treatment with Ro5 4864, a peripheral-type (p site) benzodiazepine ligand. J Neuroimmunol 42: 177-185.

Teismann P, Schulz JB. 2004. Cellular pathology of Parkinson's disease: Astrocytes, microglia and inflammation. Cell Tissue Res 318: 149-161.

Teng HK, Teng KK, Lee R, Wright S, Tevar S, et al. 2005. ProBDNF induces neuronal apoptosis via activation of a receptor complex of p75NTR and sortilin. J Neurosci 25: 5455-5463.

Thornberry NA, Bull HG, Calaycay JR, Chapman KT, Howard AD, et al. 1992. A novel heterodimeric cysteine protease is required for interleukin-1 beta processing in monocytes. Nature 356: 768-674.

Tolar M, Keller JN, Chan S, Mattson MP, Marques MA, et al. 1999. Truncated apolipoprotein E (ApoE) causes increased intracellular calcium and may mediate ApoE neurotoxicity. J Neurosci 19: 7100-7110.

Tsubokawa T, Yamaguchi-Okada M, Calvert JW, Solaroglu I, Shimamura N, et al. 2006. Neurovascular and neuronal protection by E64d after focal cerebral ischemia in rats. J Neurosci Res 84: 832-840.

Tsuchiya K, Kohda Y, Yoshida M, Zhao L, Ueno T, et al. 1999. Postictal blockade of ischemic hippocampal neuronal death in primates using selective cathepsin inhibitors. Exper Neurol 155: 187-194.

Tsujinaka T, Kajiwara Y, Kambayashi J, Sakon M, Higuchi N, et al. 1988. Synthesis of a new cell penetrating calpain inhibitor (calpeptin). Biochem Biophys Res Commun 153: 1201-1208.

Tsukada T, Watanabe M, Yamashima T. 2001. Implication of CAD and DNase II in ischemic neuronal necrosis specific for the primate hippocampus. J Neurochem 79: 1196-1206.

Turgeon VL, Milligan CE, Houenou LJ. 1999. Activation of the protease activated thrombin receptor (PAR)-1 induces motoneuron degeneration in the developing avian embryo. J Neuropathol Exp Neurol 58: 499-504.

Turgeon VL, Lloyd ED, Wang S, Festoff BW, Houenou LJ. 1998. Thrombin perturbs neurite outgrowth and induces apoptotic cell death in enriched chick spinal motoneuron cultures through caspase activation. J Neurosci 18: 6882-6891.

Turk B, Turk D, Salvesen GS. 2002. Regulating cysteine protease activity: Essential role of protease inhibitors as guardians and regulators. Curr Pharm Des 8: 1623-1637.

Turk B, Turk D, Turk V. 2000. Lysosomal cysteine proteases: More than scavengers. Biochim Biophys Acta 1477: 98-111.

Turk B, Turk V, Turk D. 1997. Structural and functional aspects of papain-like cysteine proteinases and their protein inhibitors. Biol Chem 378: 141-150.

Turk V, Bode W. 1991. The cystatins: Protein inhibitors of cysteine proteinases. FEBS Lett 285: 213-219.

Uchiyama Y. 2001. Autophagic cell death and its execution by lysosomal cathepsins. Arch Histol Cytol 64: 233-246.

Umezawa H, Miyano T, Murakami T, Takita T, Aoyagi T. 1973. Chemistry of enzyme inhibitors of microbial origins. Pure Appl Chem 33: 129-144.

Vaillant C, Meissirel C, Mutin M, Belin F, Lund LR, et al. 2003. MMP-9 deficiency affects axonal outgrowth, migration, and apoptosis in the developing cerebellum. Mol Cell Neurosci 24: 395-408.

Van Wart HE, Birkedal-Hansen H. 1990. The cysteine switch: A principle of regulation of metalloproteinase activity with potential applicability to the entire matrix metalloproteinase gene family. Proc Natl Acad Sci USA 87: 5578-5582.

Vaughan PJ, Cunningham DD. 1993. Regulation of protease nexin-1 synthesis and secretion in cultured brain cells by injury-related factors. J Biol Chem 268: 3720-3727.

Vaughan PJ, Pike CJ, Cotman CW, Cunningham DD. 1995. Thrombin receptor activation protects neurons and astrocytes from cell death produced by environmental insults. J Neurosci 15: 5389-5401.

Vaughan PJ, Su J, Cotman CW, Cunningham DD. 1994. Protease nexin-1, a potent thrombin inhibitor, is reduced around cerebral blood vessels in Alzheimer's disease. Brain Res 668: 160-170.

Vaux DL, Silke J. 2003. HtrA2/Omi, a sheep in wolf's clothing. Cell 115: 251-253.

Vergnolle N, Ferazzini M, D'Andrea MR, Buddenkotte J, Steinhoff M. 2003. Proteinase-activated receptors: Novel signals for peripheral nerves. Trends Neurosci 26: 496-500.

Vucic D, Stennicke HR, Pisabarro MT, Salvesen GS, Dixit VM. 2000. ML-IAP, a novel inhibitor of apoptosis that is preferentially expressed in human melanomas. Curr Biol 10: 1359-1366.

Waghray A, Wang DS, McKinsey D, Hayes RL, Wang KK. 2004. Molecular cloning and characterization of rat and human calpain-5. Biochem Biophys Res Commun 324: 46-51.

Wang H, Reiser G. 2003. Thrombin signaling in the brain: The role of protease-activated receptors. Biol Chem 384: 193-202.

Wang H, Ubl JJ, Reiser G. 2002a. The four subtypes of protease-activated receptors, co-expressed in rat astrocytes, evoke different physiological signaling. Glia 37: 53-63.

Wang KK. 2000. Calpain and caspase: Can you tell the difference? Trends Neurosci 23: 20-26.

Wang KK, Nath R, Posner A, Raser KJ, Buroker-Kilgore M, et al. 1996. An alpha mercaptoacrylic acid derivative is a selective nonpeptide cell-permeable calpain inhibitor and is neuroprotective. Proc Natl Acad Sci USA 93: 6687-6692.

Wang KKW, Villalobo A, Roufogalis BD. 1989. Calmodulin-binding proteins as calpain substrates. Biochem J 262: 693-706.

Wang YMF, Tsirka SE, Strickland S, Stieg PE, Soriano SG, et al. 1998. Tissue plasminogen activator (tPA) increases neuronal damage after focal cerebral ischemia in wild-type and tPA-deficient mice. Nature Med 4: 228-231.

Ware JH, Dibenedetto AJ, Pittman RN. 1995. Localization of tissue plasminogen activator mRNA in adult rat brain. Brain Res Bull 37: 275-281.

Waterhouse NJ, Finucane DM, Green DR, Elce JS, Kumar S, et al. 1998. Calpain activation is upstream of caspases in radiation-induced apoptosis. Cell Death Differ 5: 1051-1061.

Whittaker M, Ayscough A. 2001. Matrix metalloproteinases and their inhibitors – current status and future challenges. Celltransmissions (Sigma-Aldrich Inc.) 17: 3-14.

Wiessner C, Sauer D, Alaimo D, Allegrini PR. 2000. Protective effect of a caspase inhibitor in models for cerebral ischemia in vitro and in vivo. Cell Mol Biol (Noisy-le-grand) 46: 53-62.

Williams AJ, Dave JR, Tortella FC. 2006. Neuroprotection with the proteasome inhibitor MLN519 in focal ischemic brain injury: Relation to nuclear factor kappaB (NF-kappaB), inflammatory gene expression, and leukocyte infiltration. Neurochem Int 49: 106-112.

Williams AJ, Berti R, Dave JR, Elliot PJ, Adams J, et al. 2004. Delayed treatment of ischemia/reperfusion brain injury: Extended therapeutic window with the proteosome inhibitor MLN519. Stroke 35: 1186-1191.

Williams AJ, Hale SL, Moffett JR, Dave JR, Elliott PJ, et al. 2003. Delayed treatment with MLN519 reduces infarction and associated neurologic deficit caused by focal ischemic brain injury in rats via antiinflammatory mechanisms involving nuclear factor-kappaB activation, gliosis, and leukocyte infiltration. J Cereb Blood Flow Metab 23: 75-87.

Williams AJ, Myers TM, Cohn SI, Sharrow KM, Lu XC, et al. 2005. Recovery from ischemic brain injury in the rat following a 10 h delayed injection with MLN519. Pharmacol Biochem Behav 81: 182-189.

Wu M, Yu Z, Fan J, Caron A, Whiteway M, et al. 2006. Functional dissection of human protease μ-calpain cell migration using RNAi. FEBS Lett 580: 3246-3256.

Xiong Y, Gu Q, Peterson PL, Muizelaar JP, Lee CP. 1997. Mitochondrial dysfunction and calcium perturbation induced by traumatic brain injury. J Neurotrauma 14: 23-34.

Xu P, Guo M, Hay BA. 2004. MicroRNAs and the regulation of cell death. Trends Genet 20: 617-624.

Xu P, Vernooy SY, Guo M, Hay BA. 2003. The Drosophila microRNA Mir-14 suppresses cell death and is required for normal fat metabolism. Curr Biol 13: 790-795.

Xue M, Del Bigio MR. 2000a. Intracerebral injection of autologous whole blood in rats: Time course of inflammation and cell death. Neurosci Lett 283: 230-232.

Xue M, Del Bigio MR. 2000b. Intracortical hemorrhage injury in rats: Relationship between blood fractions and cell death. Stroke 31: 1721-1727.

Xue M, Del Bigio MR. 2001. Acute tissue damage after injections of thrombin and plasmin into rat striatum. Stroke 32: 2164-2169.

Yadavalli R, Guttmann RP, Seward T, Centers AP, Williamson RA, et al. 2004. Calpain-dependent endoproteolytic cleavage of PrPSc modulates scrapie prion propagation. J Biol Chem 279: 21948-21956.

Yamashima T. 2000. Implication of cysteine proteases calpain, cathepsins and caspase in ischemic neuronal death of primates. Prog Neurobiol 62: 273-295.

Yamashima T, Kohda Y, Tsuchiya K, Ueno T, Yamashita J, et al. 1998. Inhibition of ischaemic hippocampal neuronal death in primates with cathepsin B inhibitor CA-074: A novel strategy for neuroprotection based on 'calpain-cathepsin hypothesis'. Eur J Neurosci 10: 1723-1733.

Yamashima T, Saido TC, Takita M, Miyazawa A, Yamano J, et al. 1996. Transient brain ischaemia provokes Ca^{+2}, PIP2 and calpain responses prior to delayed neuronal death in monkeys. Eur J Neurosci 9: 1932-1944.

Yamashima T, Tonchev AB, Tsukada T, Saido TC, Imajoh-Ohmi S, et al. 2003. Sustained calpain activation associated with lysosomal rupture executes necrosis of the postischemic CA1 neurons in primates. Hippocampus 13: 791-800.

Yang Q, Church-Hajduk R, Ren J, Newton ML, Du C. 2003a. Omi/HtrA2 catalytic cleavage of inhibitor of apoptosis (IAP) irreversibly inactivates IAPs and facilitates caspase activity in apoptosis. Genes Dev 17: 1487-1496.

Yang W, Guastella J, Huang JC, Wang Y, Zhang L, et al. 2003b. MX1013, a dipeptide caspase inhibitor with potent in vivo antiapoptotic activity. Br J Pharmacol 140: 402-412.

Yepes M, Lawrence DA. 2004. Neuroserpin: A selective inhibitor of tissue-type plasminogen activator in the central nervous system. Thromb Haemost 91: 457-464.

Yepes M, Sandkvist M, Coleman TA, Moore E, Wu JY, et al. 2002. Regulation of seizure spreading by neuroserpin and tissue-type plasminogen activator is plasminogen-independent. J Clin Invest 109: 1571-1578.

Yepes M, Sandkvist M, Moore EG, Bugge TH, Strickland DK, et al. 2003. Tissue-type plasminogen activator induces opening of the blood-brain barrier via the LDL receptor-related protein. J Clin Invest 112: 1533-1540.

Yepes M, Sandkvist M, Wong MK, Coleman TA, Smith E, et al. 2000. Neuroserpin reduces cerebral infarct volume and protects neurons from ischemia-induced apoptosis. Blood 96: 569-576.

Yong VW, Power C, Forsyth P, Edwards DR. 2001. Metalloproteinases in biology and pathology of the nervous system. Nat Rev Neurosci 2: 502-511.

Yoshida M, Yamashima T, Zhao L, Tsuchiya K, Kohda Y, et al. 2002. Primate neurons show different vulnerability to transient ischemia and response to cathepsin inhibition. Acta Neuropathol 104: 267-272.

Yu WH, Woessner JF Jr, McNeish JD, Stamenkovic I. 2002. CD44 anchors the assembly of matrilysin/MMP-7 with heparin binding epidermal growth factor precursor and ErbB4 and regulates female reproductive organ remodelling. Genes Dev 16: 307-323.

Yuan J, Lipinski M, Degterev A. 2003. Diversity in the mechanisms of neuronal cell death. Neuron 40: 401-413.

Zhang W, Lane RD, Mellgren RL. 1996. The major calpain isozymes are long livedproteins. Design of an antisense strategy for calpain depletion in cultured cells. J Biol Chem 271: 18825-18830.

Zhang Y, Fujita N, Tsuruo T. 1999. Caspase-mediated cleavage of p21Waf1/Cip1 converts cancer cells from growth arrest to undergoing apoptosis. Oncogene 18: 1131-1138.

Zhao BQ, Wang S, Kim HY, Storrie H, Rosen BR, et al. 2006. Role of matrix metalloproteinases in delayed cortical responses after stroke. Nat Med 12: 441-445.

Zipfel GJ, Babcock DJ, Lee JM, Choi DW. 2000. Neuronal apoptosis after CNS injury: The roles of glutamate and calcium. J Neurotrauma 17: 857-869.

13 Nitric Oxide in Experimental Allergic Encephalomyelitis

S. Brahmachari · K. Pahan

1	Introduction	282
2	**Biochemistry of NO**	283
2.1	Nitric Oxide Synthases	283
2.2	Molecular Targets of NO	283
2.2.1	Soluble Guanylyl Cyclase: A Major Physiological Target	283
2.2.2	Activation of MAP Kinases	283
2.2.3	Peroxynitrite Formation	284
2.2.4	Protein Tyrosination	284
2.2.5	Protein Nitrosylation	284
2.2.6	Alteration of Oxidative Stress	284
3	**Role of NO in the Disease Process of EAE**	285
3.1	Involvement of NO in the Pathogenesis of EAE	285
3.2	Role of NO in the Switching of Th Cells	285
3.3	Role of NO in CNS Infiltration	286
3.3.1	Involvement of NO in Blood-Brain-Barrier Permeability	286
3.3.2	Effect of NO on Adhesion Molecules	287
3.4	NO in Gliosis	287
3.4.1	Induction of Inducible Nitric Oxide Synthases in Microglia	288
3.4.2	Involvement of NO in Microgliosis	289
3.4.3	Induction of Inducible Nitric Oxide Synthases in Astrocytes	290
3.4.4	Involvement of NO in Astrogliosis	291
3.5	Role of NO in Demyelination	291
3.5.1	Oligodendrocyte Damage by Peroxynitrite and Oxidative Stress	291
3.5.2	Cysteine modifications by NO: Damage to Myelin Sheath	292
3.5.3	NO-Mediated Glutamate Excitotoxicity and Oligodendrocyte Damage	293
3.6	Protective Role of NO in EAE	294
3.6.1	Inhibition of Leucocyte Infiltration by NO	294
3.6.2	Role of NO as an Inhibitor of NF-κB in Glial Cells	295
3.6.3	Immunosuppressive Role of NO	295
4	*Clinical Perspective of NO in EAE*	296
5	*Conclusion*	296

© 2009 Springer Science+Business Media, LLC.

Abstract: Nitric oxide (NO) is a biologically precious molecule responsible for diverse functions in physiology and pathophysiology of many organs including the brain. Accordingly, NO contributes significantly to both neurodegeneration and neuroprotection in experimental allergic encephalomyelitis (EAE), the animal model of multiple sclerosis (MS). The neurodegenerative aspects of NO in EAE are marked by enhanced CNS infiltration, gliosis, and demyelination whereas the neuroprotection is mainly attributed by diminished adhesion and CNS infiltration, inhibition of proinflammatory factors, and immunosuppression. In this chapter, we have made an attempt to illuminate this bidirectional role of NO in EAE and discuss the therapeutic importance of this molecule in EAE and MS.

List of Abbreviations: AOE, antioxidant enzymes; AP-1, activator protein-1; APC, antigen-presenting cell; BBB, blood-brain barrier; cGMP, cyclic guanosine-3′, 5′-monophosphate; CNS, central nervous system; LPS, lipopolysaccharides; EAE, experimental allergic encephalomyelitis; ERK, extracellular signal-regulated kinase; GAPDH, glyceraldehydes-3-phosphate dehydrogenase; GC, guanylate cyclase; GPX, glutathione peroxidase; GSNO, S-nitroso glutathione; ICAM, intercellular cell adhesion molecule; iNOS, nitric oxide synthase; IFN-γ, interferon-gamma; IL-12R, interleukin-12 receptor; IL-1β, interleukin-1beta; LFA1, lymphocyte function-associated antigen 1; MAPK, mitogen-activated protein kinase; MBP, myelin basic protein; MEK, MAP kinase kinase; MS, multiple sclerosis; MMP-9, matrix metalloproteinase-9; NO, nitric oxide; NF-κB, nuclear factor-kappaB; NMDA, N-methyl-D-aspartate NMDA; ONOO$^-$, peroxynitrite; PKG, protein kinase G; PARP, poly (ADP-ribose) polymerase; Raf, MEK kinase; ROS, reactive oxygen species; SAPK, stress-activated protein kinase; SCI, spinal cord injury; SNAP, S-nitroso-N-acetylpenicillamine; SNP, sodium nitroprusside; SOD, superoxide dismutase; TNF-α, tumor necrosis factor-alpha; VCAM, vascular cell adhesion molecule; VLA4, very-late antigen 4

1 Introduction

Multiple sclerosis (MS) is the most common T cell-mediated autoimmune demyelinating disease of the central nervous system (CNS) that specifically devastates younger population of the northern European ancestry (Martin et al., 1992). Experimental allergic encephalomyelitis (EAE) is an animal model for MS that can be induced in mice or rodents by immunization with myelin proteins or by adoptive transfer of myelin-reactive T cells (Tuohy et al., 1988; Banik, 1992; Benveniste, 1997). In particular strain of mice, the relapsing-remitting type of EAE and its pathophysiology closely resemble with the human disease MS (Benveniste, 1997). Induction of the disease is primarily characterized by generation of autoreactive T cells recognizing myelin proteins as self-antigens, infiltration of these T cells and associated mononuclear cells into the CNS, followed by inflammation, gliosis and oligodendrocyte damage. Similar to MS, various proinflammatory molecules are also implicated in the disease process of EAE. This model is widely used to identify new therapeutic approach against MS.

Among several molecules implicated in the disease pathology of EAE, nitric oxide (NO) has been discussed most widely. Since its discovery, the physiological importance of this molecule is gradually increasing. Several literatures indicate that this molecule plays a vital role in the pathogenesis of MS and other neurological diseases including Alzheimer's disease, Parkinson's disease, and HIV-associated dementia. Although the involvement of NO in the pathophysiology of EAE is evident from many previous studies, the accurate role of this molecule in this disease is still ambiguous because of its complex biochemistry and multiple actions. NO plays critical roles at almost all the pathophysiological stages of EAE, beginning with CNS infiltration of lymphocytes to demyelination. However, the role of NO in developing inflammation in the CNS appears to be the most significant. Although, NO has been identified as a primary proinflammatory signal in the pathogenesis of EAE, it does not necessarily commit NO to be only neurodegenerative. Neuroprotective role of NO has also been suggested in many literatures. Diversity of NO in its physiological manifestations may underlie its contradictory role in the pathogenesis of EAE. In this chapter, we will illustrate cellular, molecular, and biochemical perspectives of the involvement of NO in both physiology and pathophysiology of EAE. Our ultimate goal is to unravel the functional complicacy of NO in EAE pathogenesis in order to consider NO as a potential target for designing new drugs.

2 Biochemistry of NO

NO, a free radical with an unshared electron, was first identified as the endothelium-derived-relaxing factor (Furchgott and Zawadzki, 1980). Since then, the molecule has been known as a key signaling molecule in the regulation of vasodilation and immune response. However, emergence of NO as a key player in the pathophysiology of neurodegenerative diseases brings about a revolutionary change in the understanding of CNS diseases. Depending on the environment in which it is generated, NO can mediate regulatory physiological functions such as vasodilation, immunosuppression, and inhibition of platelet activation as well as can generate reactive nitrogen intermediates like peroxynitrites and others, which play crucial roles in nitrosative and oxidative stress, DNA damage, cytotoxicity, and neurodegeneration. In order to interpret the role of NO in pathophysiology of EAE, it is very necessary to understand the complex biochemistry of NO in the mammalian system.

2.1 Nitric Oxide Synthases

NO is synthesized enzymatically as a coproduct, by the action of nitric oxide synthase (NOS), a heme protein, which catalyzes sequential five-electron oxidation of L-arginine to L-citrulline (Michel and Feron, 1997). There are three isoforms of NOS protein family, which are endothelial NOS (eNOS or NOS III), neuronal NOS (nNOS or NOS I), and inducible NOS (iNOS or NOS II). These three isoforms are the products of three distinct genes. Among these three isoforms, eNOS and nNOS are constitutively and predominantly expressed in vascular endothelial cells and neuronal tissues, respectively, and their expression is calcium (Ca^{2+}) calmodulin (CAM) dependent, while the expression of iNOS is inducible and independent of Ca^{2+} (Popp et al., 1998; Ignarro, 1999) Interestingly, in case of nNOS and eNOS, CAM is regulated by Ca^{2+}; therefore, CAM binding to these two NOS isoforms is reversible. However, CAM binding to iNOS is not regulated by Ca^{2+}, therefore CAM binds irreversibly to iNOS and activates it (Cho et al., 1992). Once iNOS is activated, it catalyzes huge production of NO in the range of micromolar level which is 100- to 1,000-fold higher than what produced by other two NOS isoforms (Alderton et al., 2001). This explains why iNOS-generated NO has become an important area of study in neuroinflammatory diseases like EAE and MS. In this chapter, we will mainly, but not exclusively, focus our discussion on iNOS-generated NO.

2.2 Molecular Targets of NO

NO is a major physiologically active free radical. Because of its unpaired electron, this molecule is highly reactive. Therefore, it can react with vast ranges of biomolecules. It is such a strong free radical that less than 1 μM of local concentration of NO permits indirect reactions. Here we will particularly focus on those cellular effects and molecular modifications which are relevant to pathophysiology of EAE.

2.2.1 Soluble Guanylyl Cyclase: A Major Physiological Target

The primary signaling pathway of NO is the activation of the soluble guanylyl cyclase (sGC). NO, at nanomolar concentration, reacts with the iron of the heme group of sGC, thereby changes its conformation and thus activates this enzyme. The sGC then catalyzes the production of cyclic guanosine-3′, 5′-monophosphate (cGMP) from guanosine-5′-triphosphate (GTP). The cGMP in turn activates protein kinase G (PKG), a cGMP-dependent protein kinase, which then phosphorylates downstream factors resulting in various cellular responses such as gliosis, vasodilation by lowering cytosolic Ca^{+2}, and many others.

2.2.2 Activation of MAP Kinases

NO activates $p21^{ras}$, a monomeric G protein that responds to many extracellular signals via the classical Raf-MEK-ERK pathway. This cascade plays a key role in proliferation, differentiation, and apoptosis by modulating cyclins, cyclin-dependent kinases, and their inhibitors (Lander et al., 1996; Luth et al., 2001). NO can directly interact with JNKs, stress-activated protein kinases (SAPK), and p38 MAPK (Lander et al., 1996). There are evidences showing that NO also inhibits JNK and SAPK pathways by S-nitrosylation (Park et al., 2000). In contrast, in the vascular system, endogenous NO can enhance JNK activity (Go et al., 1999). Because NO-induced activation of JNK, SAPK, and p38 MAPK are involved in ROS-mediated damage and the ROS-mediated damage is detrimental for EAE, NO-mediated activation of MAP kinase signaling might play a pivotal role in the pathophysiology of EAE.

2.2.3 Peroxynitrite Formation

NO reacts very fast with superoxide anion (O_2^-) to form peroxynitrite ($ONOO^-$) and thereby O_2^- escapes from being scavenged by superoxide dismutase (SOD). This happens because the affinity of O_2^- is higher for NO than for SOD (Cudd and Fridovich, 1982; Huie and Padmaja, 1993). The half-life of $ONOO^-$ is approximately 1–2 seconds, thus it quickly decomposes into hydroxyl radicals and other toxic products (Beckman et al., 1990). Peroxynitrite-mediated damages have been found in MS lesions in brain and spinal cord (Whiteman et al., 2002). Peroxynitrite may directly or indirectly damage myelin-producing oligodendrocytes in demyelinating disorders such as MS and EAE (Merril et al., 1993).

2.2.4 Protein Tyrosination

Protein tyrosination means the addition of nitro group (NO_2) to the tyrosine residue of protein to form 3-nitro-tyrosine. Local environment of tyrosine residue affects tyrosination, such as proximity to negatively charged residues increases susceptibility to tyrosination (Souza et al., 1999). It is considered as one of the posttranslational modifications that may lead to conformational change, prevention of phosphorylation, and loss of function (Cassina et al., 2000; Newman et al., 2002; Amici et al., 2003). Peroxynitrite, although controversial, is suspected to be a main player for tyrosination of protein. Peroxynitrite may mediate protein tyrosination under inflammatory conditions. Although this process is not severe, only 1–5 out of 10,000 tyrosine residues may be affected under inflammatory conditions (Koprowski et al., 2001), the event that is very relevant to EAE and MS.

2.2.5 Protein Nitrosylation

S-nitrosylation is considered to be an important posttranslational modification that regulates activity and stability of proteins (Chung et al., 2004; Hess et al., 2005). S-nitrosylation is defined by reaction between NO and thiol groups of cysteine residues in a favorable environment induced by surrounding amino acid residues. It mostly leads to the impairment of protein function. Activity of NF-κB is inhibited by S-nitrosylation (Reynaert et al., 2004). Because transcriptional induction of iNOS is dependent on NF-κB, S-nitrosylation of this transcription factor may be considered as an autoregulatory mechanism for iNOS-induced NO production. S-nitrosylation is also known to prevent dimerization of e-NOS, thereby modulating NO production from eNOS (Ravi et al., 2004). S-nitrosylation might play critical role in oligodendrocyte death, which is the most fatal outcome of EAE or MS (Boullerne et al., 1999).

2.2.6 Alteration of Oxidative Stress

Reactive oxygen species (ROS) are involved in the pathogenesis of many human disorders. Therefore, the regulation of ROS is very important. Fortunately, we have been blessed with four different antioxidant enzymes (AOEs) to combat with excessive ROS. It has been found that NO can alter the generation of ROS by modulating the activity of AOEs. Both exogenous and endogenous NO decrease the activities of catalase, glutathione peroxidase (GPX), and Mn-SOD and increase the activity of Cu-Zn-SOD in C6 glial cells behaving closely to oligodendroglia (Dobashi et al., 1997). Down-regulation of mRNA of catalase and GPX and upregulation of mRNA of Cu-Zn-SOD indicate that theses enzymes are regulated by NO at the transcriptional level (Dobashi et al., 1997). In contrast, decreased activity of Mn-SOD by NO is probably because of nitration of tyrosine residues (Dobashi et al., 1997). Because ROS are always cytotoxic to CNS cells and EAE is a neuroinflammatory disease, NO-mediated modulation of AOEs may play a role in the disease process of EAE.

3   Role of NO in the Disease Process of EAE

EAE is a neuroinflammatory and neurodegenerative disease characterized by inflammation, gliosis, and oligodendrocyte damage resulting in demyelination and axonal degeneration. T lymphocytes, macrophages, microglia, and astrocytes play key roles in the pathogenesis of EAE. Because inflammation is a key feature of EAE and MS, and NO is well characterized for its proinflammatory role, this molecule has been found to be critically involved in EAE.

3.1   Involvement of NO in the Pathogenesis of EAE

NO is considered as a primary proinflammatory molecule, mediating inflammatory reactions in EAE and MS. Elevated mRNA expression of iNOS has been reported in the brain of MS patients (Bo et al., 1994). On the other hand, iNOS expression decreases when inflammation is reduced. Demyelinated plaques of MS patients show decreased expression of iNOS during reduced inflammation (Liu et al., 2001). Although, iNOS plays vital role in stimulating many inflammatory cascades, its implication is still controversial in EAE.

Apart from neuroinflammation, NO also contributes critically to BBB disruption and lymphocyte infiltration and also to oligodendrocyte damage and demyelination. Adhesion of autoimmune T lymphocytes on BBB endothelium and its disruption facilitates the entry of these cells into the CNS. Nitrite and nitrate levels have been found significantly higher in the CSF of MS patients compared to serum (Mayhan, 1996; Yuceyar et al., 2001). Although, the mechanism behind NO-induced regulation of BBB integrity and adhesion of cells on BBB endothelium is poorly known, we will discuss the matter critically in following sections of this chapter.

Oxidative and nitrosative stress conditions increase the severity of EAE by causing oligodendrocyte damage leading to demyelination and axonal degeneration. For example, S-nitrosylation affects stability of myelin proteins, and protein tyrosination has been found in plaques of MS lesions (Koprowski et al., 2001; Bizzozero et al., 2005). Therefore, it is likely that, NO could play as a master regulatory molecule in pathogenesis of EAE, as this molecule seems to be intriguingly involved in almost each and every pathophysiological steps of EAE.

3.2   Role of NO in the Switching of Th Cells

Th1/Th2 homeostasis is very critical with respect to immune regulation and a misbalance of which may lead to various pathological conditions. Excessive induction of Th1 is primarily associated with inflammatory diseases and EAE, in particular, is a Th1-mediated autoimmune disease. It has been observed that NO at low concentration specifically favors Th1 differentiation (Niebdala et al., 2002). Both exogenously provided or endogenous NO produced by antigen presenting cells (APCs), increases cGMP production via activation of sGC. This then leads to the upregulation of IL-12Rβ2 which strongly facilitates the differentiation of Th1 type by interacting with its ligand IL-12 produced by surrounding APCs (Niebdala et al., 2002). How cGMP upregulates IL-12Rβ2 is not yet known, but it is suspected that it might utilize the Raf/MEK/ERK pathway (Niebdala et al., 2002). Interestingly, NO at low concentration is usually neuroprotective, but in this situation, probably because of the physiological environment or the cell-type, Th1 bias by low concentration of NO may lead to inflammatory disease like EAE.

3.3 Role of NO in CNS Infiltration

3.3.1 Involvement of NO in Blood-Brain-Barrier Permeability

Dysfunction or disruption of BBB is a key process that allows infiltration of neuroantigen specific T lymphocytes and associated mononuclear cells into the CNS. Therefore, maintenance of BBB integrity is essential to control extravasation or infiltration. BBB consists of monolayer of endothelial cells that regulate passage of several molecules or cells across it. A major cause behind the disruption of BBB integrity is NO. However, the precise mechanism by which NO mediates leakage of BBB is not fully clear. There are several hypotheses and evidences, which explain the role of NO in the disruption of BBB integrity. Among them, the most fascinating story is NO-mediated energy deprivation in endothelial cells. Metabolic activities in BBB endothelial cells are very high and the maintenance of BBB tight junction is energy dependent (Oldendorf et al., 1977; Staddon et al., 1995). Therefore, high ATP content in endothelial cells seems to be essential for the maintenance of the tight junction. NO has been shown to play potential roles in lowering of ATP content of BBB endothelium. A most convincing mechanism is inhibition of classical glycolytic enzyme glyceraldehydes-3-phosphate dehydrogenase (GAPDH) resulting in inhibition of endothelial glycolysis and thus ATP generation (Hurst et al., 2001). The thiol group of active cysteine residue in GAPDH is S-nitrosylated by NO and thereby inhibited (❯ Figure 13-1). Under normal physiological condition, glutathione (GSH) redox system serves to prevent S-nitrosylation of GAPDH, but during inflammatory condition, the protection by GSH is superseded by oxidative stress (Padgett and Whorton, 1997).

◘ Figure 13-1

NO in BBB damage: a schematic representation of various pathophysiological processes involved in NO-mediated BBB damage leading to infiltration of T cells and associated mononuclear cells into the CNS

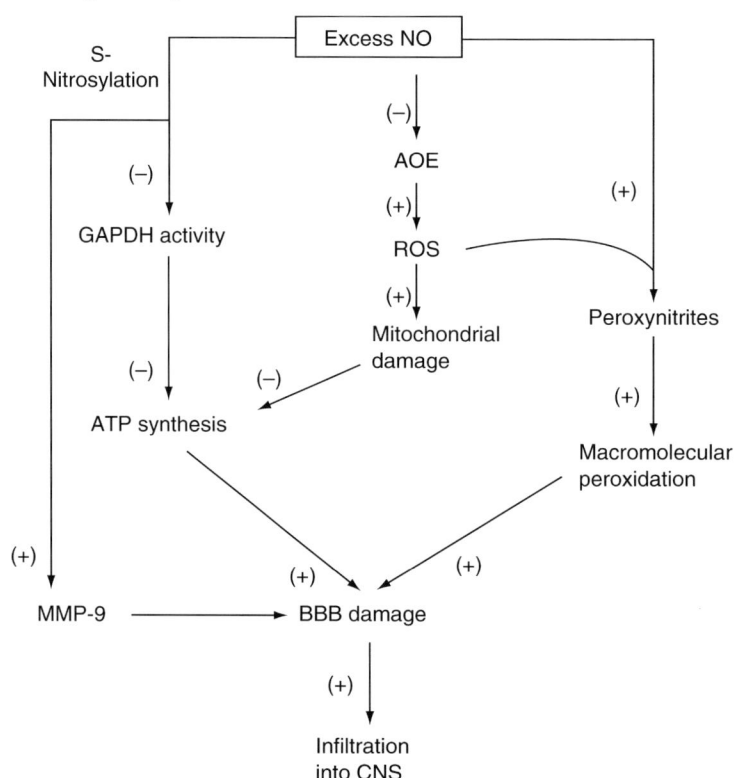

Other mechanisms have also been proposed by which NO can limit endothelial ATP synthesis. NO might affect oxidative phosphorylation by inhibiting components of mitochondrial respiratory chain, particularly cytochrome oxidase in BBB endothelium, thereby reducing ATP synthesis (❯ *Figure 13-1*) (Hurst and Clark, 1997).

Although NO-mediated energy deprivation in endothelial cells is considered as a potential mechanism for BBB damage during inflammatory diseases like EAE, this cannot convincingly explain rapid detrimental effect on BBB by NO. Lipid peroxidation by NO may cause prompt damage in BBB because of membrane disruption. Inhibition of mitochondrial respiration by NO gives rise to O_2^- that reacts with NO to generate $ONOO^-$, the molecule responsible for BBB damage by lipid peroxidation (❯ *Figure 13-1*) (Radi et al., 1991). Protein nitrotyrosination may also contribute to disruption of BBB (Beckman et al., 1993). Activation of matrix metalloproteinases-9 (MMP-9) by S-nitrosylation might play a role in blood-brain-barrier damage (Gu et al., 2002; Shiqemori et al., 2006).

3.3.2 Effect of NO on Adhesion Molecules

Extravasation of T lymphocytes into the CNS primarily needs three steps, namely, rolling, arrest and adhesion, and diapedesis. During neuroinflammation in EAE and MS, several integrins are upregulated on neuroantigen-specific T cells, and adhesion molecules are upregulated on BBB endothelium, to provide a conducive environment for extravasation to occur. Interaction between E & P-selectin and their glycoprotein ligand counterpart facilitates rolling of neuroantigen-specific T lymphocytes on BBB endothelium (Piccio et al., 2002). On the other hand, activation of integrins such as VLA-4 ($\alpha 4\beta 1$), LFA-1 ($\alpha L\beta 2$) is very crucial for lymphocyte arrest during circulation (Engelhardt and Ransohoff, 2005). Interactions of VLA-4 and LFA-1 integrins with the adhesion molecules VCAM-1 and ICAM-1 respectively are strongly implicated in adhesion of T lymphocytes on BBB endothelium (Engelhardt and Ransohoff, 2005). Therefore, blocking of VLA-4 integrin by functional blocking antibodies has been found to inhibit EAE (Yednock et al., 1992; Baron et al., 1993; Engelhardt et al., 1998). Although in EAE or MS, numerous chemokines are produced by BBB endothelium, and neuroantigen-primed T lymphocytes also express several chemokine receptors, involvement of these molecules in migration T lymphocytes are controversial.

Whether NO regulates the expression of various adhesion molecules required for extravasation of neuroantigen-specific T lymphocytes across the BBB during EAE is an area that needs to be investigated. There is no direct evidence whether NO facilitates T cell extravasation by upregulating the adhesion molecules during induction of EAE. However, it has been shown that, NOS inhibitors inhibit polyinosinic-polycytidylic acid (poly IC) induced mRNA expression of VCAM-1 and E-selectin in human vascular endothelial cells and thereby prevent poly IC-stimulated leukocyte adhesion to endothelial cells (Faruqi et al., 1997). Poly IC is a synthetic double-stranded RNA that simulates viral-infected state of a cell. Because viral infection may result in autoimmune diseases, it is suspected that, NO might facilitate the infiltration of T cells into the CNS by upregulating those adhesion molecules. However, this interpretation is a mere hypothesis. In contrary, there are evidences showing reduction of T lymphocyte adhesion to human brain endothelial cells via NO, which will be discussed in details in the subsequent sections of this chapter. Therefore, implication of NO in T cell adhesion to BBB endothelium during EAE is contradictory and it needs further experimental evidences to support any hypothesis.

3.4 NO in Gliosis

An important pathophysiological hallmark of EAE is gliosis. Gliosis is a result of over-activity of glial cells characterized by release of various neurotoxic substances that augment or initiate inflammatory response leading to oligodendrocyte damage and axonal degeneration. Any kind of neurodegenerative insult may cause gliosis. In EAE, neuroantigen-specific T lymphocytes and associated mononuclear cells interact with astrocytes and microglia in the CNS and subsequently cause gliosis. Recent studies have implicated NO in gliosis.

Both astrocytes and microglia have been shown to express iNOS (Jana et al., 2001; Jana et al., 2005). Because of its inducible nature, upregulation of iNOS in glia leads to huge production of NO, which is mostly cytotoxic. Although glial iNOS can be regulated at transcriptional, posttranscriptional, translational, and posttranslational level, like other inducible genes, the most important is transcriptional regulation of iNOS. There are various signaling mechanisms and transcription factors involved in the regulation iNOS in glial cells (◐ Table 13-1). However, we will restrict our discussions particularly to those mechanisms that are relevant to EAE.

◘ Table 13-1
Involvement of different signaling pathways for the induction of iNOS in glial cells

Cell type	Signaling pathways
Human astrocytes	(1) p38—NF-κB/AP-1/C/EBP—iNOS (Hua et al., 2002; Auch et al., 2004; Jana et al., 2005)
	(2) JNK—AP-1—iNOS (Jana et al., 2005)
	(3) PKR—NF-κB—iNOS (Auch et al., 2004)
	(4) p38—C/EBPβ—iNOS (Auch et al., 2004
	(5) p21Ras fernesylation—NF-κB—iNOS (Pahan et al., 2000)
Rat astrocytes	(1) JAK2—STAT1α—iNOS (Dell'Albani et al., 2001)
	(2) p38—NF-κB/ATF-2/C/EBPβ—iNOS (Bhat et al., 1998; Chen et al., 1998; Pahan et al., 1998; Cvetkovic et al., 2004)
	(3) ROS—NF-κB (?)—iNOS (Pahan et al., 1998a; Pannu et al., 2004)
	(4) Ceramide—Ras/ERK—NF-κB—iNOS (Pahan et al., 1998)
	(5) Isoprenylation of Ras/Rac—NF-κB—iNOS —(Pahan et al., 1997)
	(6) cAMP↓ — NF-κB—iNOS (Pahan et al., 1997a)
	(7) cAMP↓ - p38— NF-κB/ATF-2/AP-1/C/EBP (?)—iNOS (Bhat et al., 2002; Won et al., 2004)
	(8) PP1/PP2A↓ — NF-κB—iNOS (Pahan et al., 1998b)
	(9) PI-3K↓ — TF (?)—iNOS (Pahan et al., 1999, 2000)
	(10) G6PD—C/EBP-delta—iNOS (Won et al., 2003)
C6 glial cells	(1) ROS—NF-κB (?)—iNOS (Pahan et al., 1998a; Pannu et al., 2004)
	(2) Ceramide—Ras/ERK—NF-κB—iNOS (Pahan et al., 1998)
	(3) cAMP ↓— p38— NF-κB/ATF-2/AP-1/C/EBP (?)—iNOS (Bhat et al., 2002; Won et al., 2004) PP1/PP2A↓ — NF-κB—iNOS (Pahan et al., 1998b)
	(4) PI-3K↓ — TF (?)—iNOS (Pahan et al., 1999, 2000)
	(5) G6PD—C/EBP-delta—iNOS (Won et al., 2003)
BV-2 microglia	(1) VLA-4-VCAM-1—C/EBPβ—iNOS (Dasgupta et al., 2002; Dasgupta et al., 2005)
	(2) CD40-CD40L—NF-κB—iNOS (Jana et al., 2001)
Mouse primary microglia	(1) CD40-CD40L—NF-κB—iNOS (Jana et al., 2001)
	(2) IL-12Rβ1/ IL-12Rβ2—NF-κB—iNOS (Pahan et al., 2001)
	(3) p38/ERK—NF-κB(?)—iNOS (Bhat et al., 1998)
Rat microglia	(1) p38/ERK/JNK—NF-κB(?)—iNOS (Bhat et al., 2002; Bhat et al., 2003)

3.4.1 Induction of Inducible Nitric Oxide Synthases in Microglia

Although not exclusive, microglia is an important source of glial NO in EAE and other neurodegenerative diseases. There are several inducers of microglial iNOS. Bacterial endotoxin LPS has been known as a potent inducer of iNOS in microglia (Simmons and Murphy, 1992). However, LPS-induced induction of iNOS in microglia does not have any pathophysiological relevance to EAE. Among the proinflammatory cytokines which are involved in the pathophysiology of EAE, IFNγ alone or in combination with TNF-α or IL-1β can induce iNOS in microglia (Jana et al., 2001). However, microglia isolated from adult human brain has been found as a poor inducer of iNOS in response to LPS and proinflammatory cytokines.

Because, neuroantigen-specific T cells play key roles in the pathogenesis of EAE and MS, studies were performed to investigate whether these T cells interact with microglia and have any effect on the induction of microglial iNOS. It has been demonstrated that, myelin basic protein (MBP)-primed T cells, but not the normal T cells, stimulate the induction of iNOS in mouse microglial cells by cell–cell contact (Dasgupta et al., 2002). Importantly, when neuroantigen-primed T cells were placed in inserts to prevent direct contact between T cells and microglia, it did not stimulate induction of iNOS in microglia and moreover, supernatants of T cells were found to be very weak inducer of iNOS in microglial cells. These observations strongly suggest that direct contact between T cells and microglia is essential for the induction of iNOS in mouse microglia.

3.4.2 Involvement of NO in Microgliosis

Microglia, the CNS-resident professional macrophages, comprise only 2–5% of the total brain cells (Lawson et al., 1990; Gonzalez-Scarano and Baltuch, 1999). However, in response to neurodegenerative insults, population of microglia may increase up to 12% (Gonzalez-Scarano and Baltuch, 1999). Activation of microglia has been implicated in almost all neurodegenerative diseases including EAE (Pahan et al., 1997; Gonzalez-Scarano and Martin-Garcia, 2005). Neurodegenerative plaques in CNS have been found to contain activated microglia (Gonzalez-Scarano and Baltuch, 1999; Dauer and Przedborski, 2003; Rock et al., 2004) Although activated microglia remove dead cells of CNS by phagocytosis and secrete some neurotrophic factors necessary for neuronal survival (Gonzalez-Scarano and Baltuch, 1999; Carson, 2002), if overactivated, they also produce several proinflammatory molecules including NO. Microgliosis is the result of overactivation of microglia associated with excessive production of NO that ultimately supersedes the beneficial effects of microglial activation. One of the hallmarks of microgliosis is the upregulation of a surface molecule CD11b. In various neurodegenerative diseases including EAE or MS, increased expression of CD11b corresponds to severe activation of microglia (Ling and Wong, 1993; Gonzalez-Scarano and Baltuch, 1999; Rock et al., 2004). Morphologically, microgliosis is characterized by intense ramification and cytoskeletal rearrangement resulting in changes in shape and motility which is directly related to the upregulation of CD11b (Ling and Wong, 1993; Gonzalez-Scarano and Baltuch, 1999; Rock et al., 2004). During microgliosis, the cytoplasmic domain of CD11b is predicted to interact with cytoskeleton protein (Gonzalez-Scarano and Baltuch, 1999). Immunohistochemical evidences of upregulation of CD11b in the spinal cord of EAE mice (Dasgupta et al., 2003) suggest that microgliosis may be involved in the pathophysiology of MS.

In recent studies, inhibition of LPS-mediated upregulation of CD11b expression in-vivo in mouse striatum by carboxy-PTIO, a NO scavenger indicates that NO is probably involved in upregulation of CD11b in microglia (Roy et al., 2006). Further investigations demonstrate a key role of NO in the expression of CD11b in mouse microglia (❶ Figure 13-2). NO employs soluble guanylyl cyclase (sGC) as its immediate target to upregulate the expression of CD11b (❶ Figure 13-2). The sGC catalyzes the formation of cGMP which in turn activates PKG and subsequently stimulates CD11b expression (Roy et al., 2006). 8-Br-cGMP, a cell permeable analog of cGMP has been also shown to upregulate CD11b expression in microglia (Roy et al., 2006). Stabilization of cGMP by phosphodiesterase inhibitor, MY5445 also increases CD11b expression. Similarly, if PKG is inhibited, NO-induced upregulation of CD11b is also inhibited (Roy et al., 2006). These evidences suggest the essential role of downstream components of sGC in NO-mediated upregulation of CD11b in microglia. Further downstream components of CD11b expression involves phosphorylation of transcription factor CREB, followed by its recruitment to corresponding promoter element of CD11b and its subsequent transcriptional activation. This was evidenced by antisense knockdown of CREB and subsequent abrogation of NO-induced CD11b expression in microglia (Roy et al., 2006). Analysis of mouse CD11b promoter shows the presence of six cAMP response elements. Phosphorylation of CREB by PKG has also been reported. Because microgliosis plays an important role in the pathogenesis of EAE, understanding of the NO-sGC-cGMP-PKG-CREB-CD11b pathway in microglia may be helpful in designing therapeutic compounds to mitigate microgliosis and thereby suppress EAE or MS.

◘ Figure 13-2
Induction of iNOS in glia in response to various stimuli followed by gliosis via activation of NO-GC-cGMP-PKG pathway

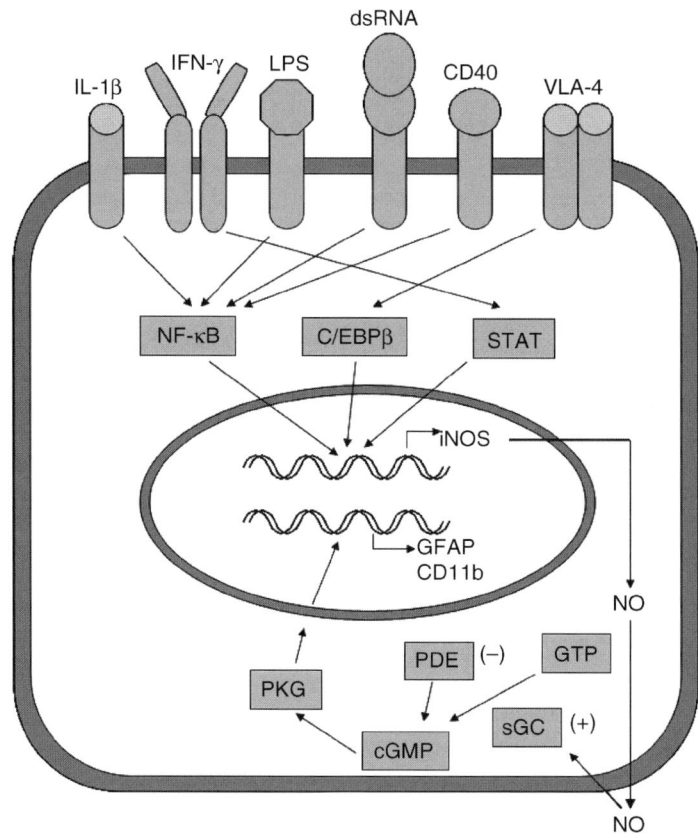

3.4.3 Induction of Inducible Nitric Oxide Synthases in Astrocytes

Astrocytes in the healthy brain do not express iNOS but following ischemic, traumatic, neurotoxic, or inflammatory damage the reactive astrocytes express iNOS in mouse, rat, and human (Stojkovic et al., 1998; Dell'Albani et al., 2001; Dasgupta et al., 2002; Auch et al., 2004). Because astrocytes are the major cells in the CNS, once iNOS is induced in these cells, volumetrically activated astrocytes become a major source of NO for neurotoxic or neurodegenerative insults. Therefore, understanding the regulation of iNOS expression in astrocytes is an important area of study as it may help in identifying targets for therapeutic intervention in NO-mediated neurological disorders. We have summarized various signaling cascades that converge to activate several transcription factors required for the transcription of iNOS in rat, mouse, and human astrocytes (❯ Table 13-1). For details, please consult recent reviews (Saha and Pahan, 2006; Saha and Pahan, 2006a). Although rat and mouse astrocytes express iNOS and produce excessive amount of NO in response to LPS and other proinflammatory cytokines, only IL-1β, alone or in combination with other cytokines, is capable of inducing iNOS in human astrocytes (Hua et al., 2002; Jana et al., 2005). It was found that among three cytokines (TNF-α, IL-1β and IFN-γ) tested, only IL-1β was capable of inducing the activation of CCAAT/enhancer-binding proteinβ (C/EBPβ) in human astrocytes suggesting an essential role of C/EBPβ in the expression of iNOS in human astrocytes.

3.4.4 Involvement of NO in Astrogliosis

Astrocytes, the major glial cell population within CNS, are present ten times more in numbers than neuron in adult brain. Although astrocytes are important in the maintenance of CNS homeostasis, in response to any neurodegenerative insults, astrocytes react vigorously leading to astrogliosis (Eng and Ghirnikar, 1994). The hallmark of astrogliosis is upregulation of glial fibrillary acidic protein (GFAP) which is considered as a marker protein of astrogliosis (Eng and Ghirnikar, 1994). GFAP is the primary intermediate filament (IF) of mature astrocytes. It plays vital role in controlling shape and motility of astrocytes. GFAP is also important for astrocyte-neuron cross-talk and the GFAP-mediated astrocytic processes modulate efficiency of synaptic transmission (Liedtke et al., 1996). In addition to many physiological roles as mentioned above, GFAP also possesses CNS-damaging activity. In neurodegenerative diseases, increased GFAP expression correlates with the severity of astrogliosis (Eng et al., 1992; Eng and Ghirnikar, 1994). Although activation of astrocytes has many beneficial effects in the CNS, including recovery of injured CNS by actively monitoring and controlling extracellular water, pH and ion homeostasis, once astrocytes are activated in neurodegenerative microenvironments, it goes beyond the control, and the detrimental effects outclass the beneficial effects (Eng and Ghirnikar, 1994). Astrogliosis has been implicated in many neurodegenerative diseases including demyelinating diseases like EAE. Immunohistochemical studies have revealed upregulation of both iNOS and GFAP in the spinal cord of EAE mice which suggest that NO and astrogliosis are intriguingly involved in the pathophysiology of EAE (Dasgupta et al., 2003).

Similar to microgliosis, the role of NO in astrogliosis is also evident from inhibition of LPS-mediated upregulation of GFAP expression in vitro in cultured astrocytes and in vivo in mouse striatum by PTIO (Brahmachari et al., 2006). Here as well, NO upregulates GFAP expression via sGC-cGMP-PKG pathway (● Figure 13-2) (Brahmachari et al., 2006). However, the downstream molecules of PKG involved in the upregulation of GFAP are not known. Because astrocytes are the major glial cells and astrogliosis plays a role in the pathogenesis of EAE and MS, understanding of the signaling pathway leading to astrogliosis is significant in the light of clinical perspective of MS.

3.5 Role of NO in Demyelination

EAE or MS is a demyelinating disease in which myelin and myelin producing cells (oligodendrocytes) are destroyed. Myelin, the multilayered membrane surrounding the axon, is actively involved in the transmission of nerve impulse and therefore disruption of myelin effectively interrupts the nerve conduction. In MS plaques, oligodendrocytes appear in dying and necrotic morphology with swollen nuclei and disrupted plasma and mitochondrial membranes (Parkinson et al., 1997). The presence of actively phagocytosing cells in MS plaques confirms the destruction of myelin and also suggests that reactive nitrogenous species may be involved in the pathophysiology of oligodendrocyte damage. In subsequent sections, we will discuss about different cellular and biochemical mechanisms by which NO either directly or indirectly mediates damage to oligodendrocytes and myelin (● Figure 13-3).

3.5.1 Oligodendrocyte Damage by Peroxynitrite and Oxidative Stress

Among three glial cells, oligodendrocytes are the most susceptible ones to NO-mediated damage. The destructive action of NO is primarily caused by peroxynitrite ($ONOO^-$). Once this is formed, it causes nitration of tyrosine residues, known as protein tyrosination, and ultimately leading to conformational change and loss of function of target proteins. This type of protein modification by $ONOO^-$ can be easily detected by using antibodies against nitrotyrosine. Therefore it is often considered as a footprint of pathologic activity of NO during inflammation. Increased level of nitrotyrosine has been detected in the spinal cord of EAE mice (Koprowski et al., 2001). Uric acid, a natural scavenger of $ONOO^-$ reduces demyelination and axonal damage in EAE, suggesting the detrimental role of $ONOO^-$ in oligodendrocyte damage (Scott et al., 2002). Although $ONOO^-$ significantly increases DNA strand break and the activity of poly (ADP-ribose) polymerase (PARP) in oligodendrocyte culture, the cytotoxic effect of $ONOO^-$ on

Figure 13-3
NO in demyelination; an outline of different mechanisms for NO-mediated oligodendrocyte death and axonal damage

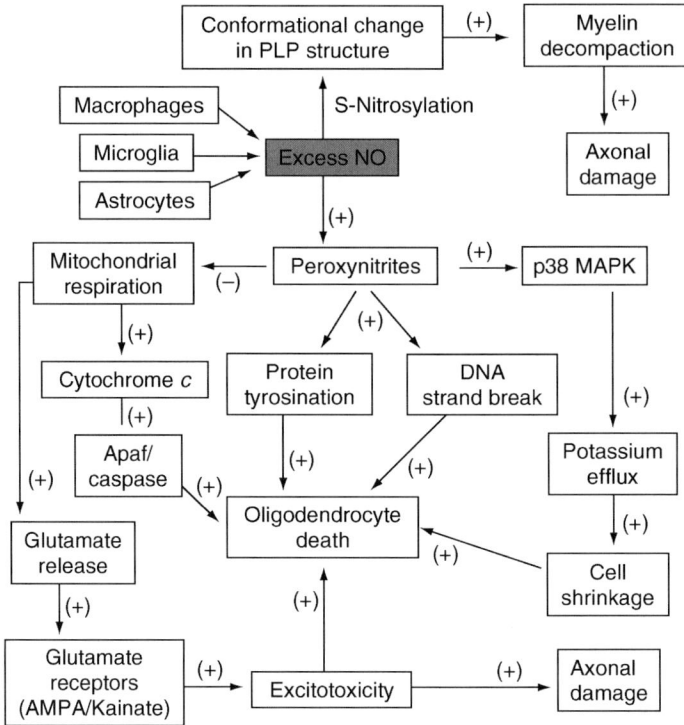

oligodendrocytes is probably independent of PARP activity. The detrimental effect of ONOO$^-$ can be reduced by pretreatment of oligodendrocytes with 17β-estradiol, but the mechanism of this protection is not clearly understood (Takao et al., 2004). The damage of myelin sheath by ONOO$^-$ may also involve lipid peroxidation where both the membrane lipids and the protein components of the membrane are affected.

Possibility of another pathway in oligodendrocyte damage by ONOO$^-$ is implicated, but not yet studied in EAE. Production of ONOO$^-$ in neurodegenerative microenvironment augments release of zinc from intracellular stores which in turn blocks mitochondrial respiration, enhances release of cytochrome c, and activates Apaf/caspase pathway leading to apoptotic death of oligodendrocytes and increased ROS production (Bossy-Wetzel et al., 2004). Excessive ROS further elevate the level of ONOO$^-$ via O_2^- and subsequently, the elevated level of ONOO$^-$ activates p38 MAP kinase pathway leading to potassium efflux and cell shrinkage, thereby further aggravating the death process in oligodendrocytes (Bossy-Wetzel et al., 2004).

NO may also lead to mitochondrial DNA damage in oligodendrocytes by modulating the Mn-SOD, a mitochondrial antioxidant enzyme. NO deactivates Mn-SOD probably by tyrosination and thereby increases oxidative stress, which subsequently inhibits mitochondrial respiratory chain and ultimately causes DNA damage (Dobashi et al., 1997).

3.5.2 Cysteine modifications by NO: Damage to Myelin Sheath

NO can even directly damage myelin sheath, independent of ONOO$^-$. Excessive production of NO causes myelin decompaction associated with S-nitrosylation of cysteine-rich proteolipid protein (PLP) (Bizzozero et al., 2004). PLP is the most abundant protein of myelin sheath surrounding the axon and is important for structural integrity of the sheath. Because PLP stabilizes intraperiod line in the myelin, it is suspected that nitrosylation of thiol groups of PLP may result in alteration of structural conformation and thereby triggers myelin decompaction (Bizzozero et al., 2004). Interestingly, myelin of peripheral nervous system, the stability of which depends on proteins other than PLP, does not undergo decompaction when exposed to NO. This suggests why CNS myelin is more vulnerable to attack by NO than the peripheral myelin. In agreement with these findings, elevated levels of S-nitrosothiols were found in CSF of MS patients and increased levels of anti-S-nitrosocysteine antibodies were also observed in MS patients as well as in animals with EAE (Boullerne et al., 1995; Boullerne et al., 2002). In EAE animals, serum levels of anti-S-nitrosocysteine antibodies reached the peak 7 days before the onset of clinical symptoms and the antibody titer correlated with the extent of demyelination (Boullerne et al., 1995). In patients with relapsing-remitting MS, the serum level of anti-S-nitrosocysteine antibody was maximum during relapse, while during remission, the level of the antibody was found to be normal. The level of the antibody was also elevated during acute phase of MS as well as in progressive phases (Boullerne et al., 2002). Together, these findings suggest that cysteine modification by NO plays a crucial role in the damage of myelin sheath in neurodegenerative environment of MS or EAE. Serum level of anti-S-nitrosocysteine antibody may also be used as a marker for clinical activity. Structural studies of PLP and other myelin proteins following exposure to NO will be helpful in further elucidating the mechanism.

3.5.3 NO-Mediated Glutamate Excitotoxicity and Oligodendrocyte Damage

NO may also impart its deleterious effect on functional integrity and survival of oligodendrocytes by affecting glutamate release and uptake. NO, at pathophysiological concentration, inhibits mitochondrial respiration at the level of cytochrome oxidase, which subsequently stimulates Ca^{+2}–independent release of glutamate by impairing Na^+, K^+-ATP-ase resulting in decrease of Na^+ gradient and reversal of the Na^+-glutamate transporter (Bredt, 1999; Ignarro, 1999). Release of excessive glutamate potentiates the activation of glutamate receptors on oligodendrocytes and neurons, thereby causing excitotoxic cell damage. Oligodendrocytes, particularly, are more susceptible to glutamate-mediated excitotoxic cell damage than neurons (McDonald et al., 1998; Pitt et al., 2000; Rosin et al., 2004). The main glutamate transporters in oligodendrocytes are suppressed in EAE and MS (Pitt et al., 2003). This explains the possibility that NO can prolong the excitotoxic damage to oligodendrocytes in a neurodegenerative milieu by increasing the release of glutamate as well as by suppressing its reuptake.

Glutamate excitotoxicity to oligodendrocytes is mediated by AMPA (α-amino-3-hydroxy-5-methyl-4-isoxazolepropionic acid)/kainate type of excitatory glutamate receptor, the only glutamate receptor expressed by oligodendrocytes (Pitt et al., 2000). Inhibition of AMPA/kainate receptors by an antagonist NBQX significantly increases the survival of oligodendrocytes in EAE mice (Pitt et al., 2000), suggesting the potential role of theses receptors in the death of oligodendrocytes in EAE.

Axonal damage, another important pathological lesion of EAE, also involves glutamate excitotoxicity. Axonal damage is assessed by the presence of abnormal dephosphorylation of heavy chain neurofilament H (NF-H), a well studied immunohistochemical marker of demyelinated axons in MS (Trapp et al., 1998). The role of glutamate excitotoxicity in axonal damage is evident from substantial reduction of abnormal dephosphorylation of NF-H and simultaneously amelioration of the clinical symptoms of EAE by treatment with the antagonist NBQX (Pitt et al., 2000).

Hypoxic condition may aggravate NO-glutamate-mediated excitotoxic damage to oligodendrocytes and neurons. Potentiation of excitotoxic damage by synergistic action of NO and hypoxia can be prevented by blocking N-methyl-D-aspartate (NMDA) receptor (Mander et al., 2005). Thus, co-existence of inflammatory and hypoxic environment sensitizes neurons and oligodendrocytes to NO and glutamate mediated excitotoxicity. It is important to mention that, pattern of certain lesions in MS patients strikingly resemble the injury found after white matter stroke suggesting a fact that hypoxic injury is an essential component of inflammatory demyelinating lesions of MS (Lassmann, 2003). Taken together, it can be inferred that NO and associated hypoxia-mediated glutamate excitotoxicity play a vital role in oligodendrocyte death and axonal damage found in EAE and MS.

3.6 Protective Role of NO in EAE

Whatever we have discussed till now, is all about neurotoxic role of NO in EAE. However, the story will remain incomplete unless we look at the other side of the coin – the protective role of NO in EAE. Although most of the literatures describe NO as a potent neurotoxic substance, there are many reports showing the evidence of protective role of NO in EAE. Whether NO is proinflammatory or anti-inflammatory is decided by many factors, such as the environment in which it is generated, concentration, and many others. In the subsequent sections, we will delineate the neuroprotective role of NO in EAE (❯ Figure 13-4).

◻ Figure 13-4
A mechanistic outline delineating protective effects of NO in EAE

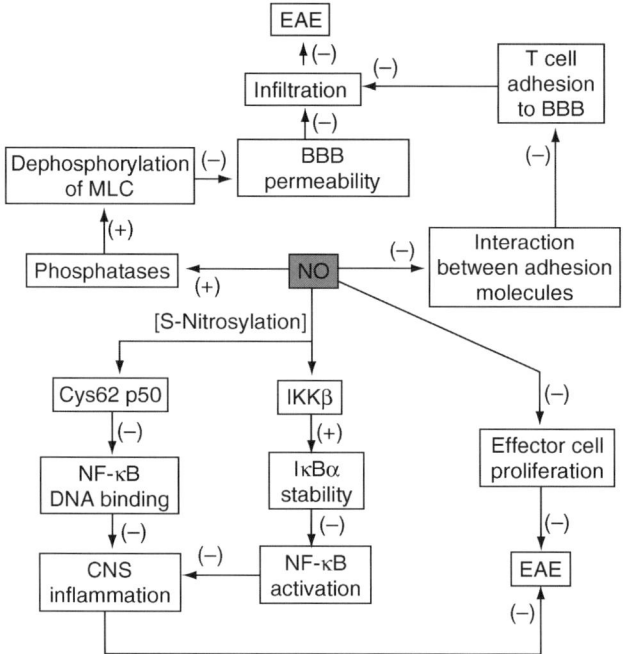

3.6.1 Inhibition of Leucocyte Infiltration by NO

Adhesion of T cells to BBB endothelium is one of the factors that regulate infiltration into the CNS. Some groups of researchers have shown that NO negatively regulates T cell adhesion and thereby reduces the infiltration. In an in-vitro model of blood-brain barrier using primary cultures of human brain microvessel endothelial cells, NO decreases T cell adhesion via sGC-cGMP-dependent pathway (Wong et al., 2005). This finding was further supported by increased adhesion of T cells to endothelium by using NO donors and 8-Br-cGMP, a cell permeable analogue of cGMP. On the other hand, as expected, blocking sGC activity by its inhibitor increased the adhesion of T cells (Wong et al., 2005). NO has also been shown to modulate the interaction between polymorphonuclear leucocytes and the brain microvessel endothelial cells by decreasing the binding of these leucocytes to the adhesion molecules E-selectin and ICAM-1 (Wong et al., 2004).

Interestingly, one group of researchers has recently shown the involvement of mevalonate pathway in stabilizing blood-brain-barrier integrity (Kuhlmann et al., 2006). Activation of mevalonate pathway leads to the formation of isoprenoids, like geranylgeranyl pyrophosphate (GGPP) and farnesyl pyrophosphate (FPP). Geranylgeranylation of small signaling molecules like Rho has been demonstrated to downregulate mRNA of eNOS, thereby decreasing the production of NO (Kuhlmann et al., 2006). Inhibition of mevalonate pathway by fluvastatin, an HMG-CoA reductase inhibitor, increased NO production from eNOS, but the effect of fluvastatin was reversed by the addition of GGPP, but not FPP, thereby suggesting a possible role of geranylgeranylation in the downregulation of eNOS (Kuhlmann et al., 2006). Geranylgeranylation of small molecules like Rho GTPase family interferes with the endothelial contractile machinery and thereby disrupts the BBB (Kuhlmann et al., 2006). Upregulation of eNOS by fluvastatin and subsequent production of NO may stabilize the tight junction of BBB by activating phosphatase and thereby increasing the dephosphorylation of myosin light chain in the BBB endothelium (Kuhlmann et al., 2006).

Although the mechanisms described above strongly suggest a protective role of NO on BBB integrity, the effect of NO on BBB is still controversial. It is probably the amplitude of NO levels that may determine whether NO enhances or decreases BBB integrity.

3.6.2 Role of NO as an Inhibitor of NF-κB in Glial Cells

Till now, we have seen that NF-κB positively regulates the expression of iNOS in glia. On the other hand, there are evidences demonstrating that NO can also inhibit NF-κB activity in glial cells therefore suggesting an important auto-regulatory circuit in the maintenance of NO homeostasis in the brain (Colasanti et al., 2000). Two mechanisms have been implicated by which NO may inhibit the action of NF-κB. Firstly, NO is capable of suppressing NF-κB activity by stabilizing IκBα, the inhibitor of NF-κB (Peng et al., 1995). For example, NO donors, sodium nitroprusside (SNP) and S-nitrosoglutathione (GSNO) inhibit TNF-α-stimulated activation of NF-κB in human vascular endothelial cells (Peng et al., 1995). NO stabilizes IκBα by preventing its degradation from NF-κB (❷ *Figure 13-4*) probably by nitrosylating and thus inactivating IKKβ and also by upregulating the mRNA expression of IκBα without affecting the mRNA levels of p50 and p65, the two subunits of NF-κB (Peng et al., 1995; Kapahi et al., 2000). These effects are independent of activation of guanylyl cyclase, as the cell permeable analogue 8-Br-cGMP has no effects on NF-κB activation (Peng et al., 1995).

Secondly, NO has been reported to inhibit NF-κB DNA binding activity by S-nitrosylation of cysteine residues both in vitro and in vivo (DelaTorre et al., 1998; DelaTorre et al., 1999). Specifically, the Cys62 residue of p50 subunit located within a polypeptide loop directly interacts with NF-κB binding DNA motif via contact with phosphates in the DNA backbone. NO-donors such as S-nitrose-N-acetyl-penicillamine (SNAP) and sodium nitroprusside (SNP) inhibit NF-κB DNA binding activity by nitrosylating the Cys62 residue of p50 subunit and this inhibition is prevented by Cys62Ser mutant therefore suggesting the importance of Cys62 in NO-mediated NF-κB inactivation (DelaTorre et al., 1998; DelaTorre et al., 1999). These two mechanisms distinctly characterize NO as an anti-inflammatory molecule. However, in the earlier sections, we have already discussed proinflammatory role of NO in the context of EAE. Possibly, the physiological environment and the cell-type in which NO is generated, strongly regulate its biological role. Therefore, it is probably the balance between pro- and anti-inflammatory activities of NO that decides the resultant outcome of NO in EAE.

3.6.3 Immunosuppressive Role of NO

Depending on the dose, NO could be immunosuppressive. In a recent study, glucocorticoid receptor (GR) deficiency has been shown to increase resistance to EAE through the production of NO (Marchetti et al., 2002). Binding of glucocorticoids (GC) to its receptor causes the translocation of GR-complex to the nucleus where it binds to glucocorticoid-responsive elements containing promoters of many molecules including the iNOS and thereby downregulates its transcription. Deficiency in GR leads to threefold to sixfold increase in level of nitrites and simultaneous reduction in proliferation of splenic and lymph node cells which correspond to increased resistance to EAE (Marchetti et al., 2002).

However, inhibition of NO by a specific inhibitor of inducible nitric oxide synthase results in upregulation of effector cell proliferation and subsequent reversal of resistance to EAE (Marchetti et al., 2002), suggesting that, NO may protect EAE by suppressing the proliferation of effector cells. Neuroantigen-specific T cells and macrophages are the primary effector cells in EAE. NO has been shown to suppress proliferation of both cell types. Downregulation of proliferation involves suppression of cytokine production, scavenging superoxides or apoptosis of macrophages and T cells, and it may utilize a complex mechanism involving multiple signaling pathways such as activation or inactivation of ion channels, G-proteins, protein-tyrosine kinase, janus kinases (Jak1, Jak2, Jak3), p38 MAP kinases, c-Jun-NH_2 –terminal kinases, caspase-1 and caspase-3, metalloproteases, and phosphoproteases (Bogdan, 2001). Furthermore, excessive NO may trigger inactivation of GR-signaling via a feedback loop, thereby exacerbating the level of NO at the critical time of antigen-priming, resulting in the suppression of effector cells (Marchetti et al., 2002). Therefore, critical regulation of iNOS-NO may protect EAE or MS by suppressing the proliferation of neuroantigen-primed T cells and macrophages.

4 Clinical Perspective of NO in EAE

Generally, it is thought that aberrant induction of iNOS in the CNS microenvironment complicates the pathophysiology of neurodegenerative diseases with severe detrimental outcomes. Because once NO is generated, it reacts with superoxides to form peroxynitrites ($ONOO^-$), the most dangerous nitrogenous species. Accordingly, oligodendrocytes are very much vulnerable to $ONOO^-$. According to Bagasra et al (1995), the mRNA for iNOS was detectable in all brains examined from MS patients and EAE mice but not from control brains. Similar to EAE, increased levels of iNOS and nitrotyrosine have been observed after spinal cord injury (SCI) (Isaksson et al., 2005). The absence of significant damage after SCI in $iNOS^{-/-}$ demonstrates the functional importance of aberrant NO production from iNOS. Therefore, it appears that inhibition of iNOS could be a promising therapeutic target in EAE. Consistently, EAE in both rat and mouse model was prevented by inhibiting the iNOS by aminoguanidine (Zhao et al., 1996). Similar trends of recovery from EAE were also observed in other studies with either iNOS inhibitors or NO scavengers (Hooper et al., 1994). Ding et al. (1998) have also shown that selective inhibition of iNOS in the CNS by intraventricular administration of antisense oligonucleotides blocks the disease process of EAE in mice.

However, contradictory findings about the role of iNOS and NO have also been reported in studies with EAE. The increase in clinical symptoms of EAE and mortality rates in $iNOS^{-/-}$ mice compared with that of wild-type mice suggest that expression of iNOS may also be beneficial in EAE (Fenyk-Melody et al., 1998). Therefore, presently, there are two distinct scenarios regarding the involvement of iNOS/NO in the disease process of EAE. It seems that chemical inhibitors or antisense oligonucleotides at the doses used do not inhibit iNOS completely; therefore, partial inhibition of iNOS by its chemical inhibitors or antisense oligonucleotides targeting iNOS is able to protect animals from EAE. On the other hand, complete inhibition of iNOS by gene knock-out worsens clinical symptoms of EAE. These observations suggest that, NO produced up to a certain level from the activation of iNOS is protective in EAE.

It is worth mentioning that MS and EAE, in particular, are T cell-mediated neuroinflammatory disorders and that NO inhibits the activation of T cells and thereby may act as an immunosuppressant. However, once iNOS is activated, it remains active for long time; thereby producing excessive amount of NO for prolonged time period, and NO, produced in excess is potentially cytotoxic. Therefore, partial inhibition of iNOS by pharmacological compounds to inhibit the excessive production of NO, but not the complete knockdown is a plausible therapeutic approach for EAE and MS.

5 Conclusion

It is evident from our discussions that NO exerts both pro- and antiinflammatory activities in EAE. NO, on one side, facilitates the entry of neuroantigen-primed T cells into the CNS by disrupting the BBB integrity and most importantly induces gliosis, oligodendrocyte damage, and axonal degeneration, whereas on the other hand, the same molecule inhibits proliferation of T cells and their infiltration into the CNS and suppresses its own cytotoxic action by an autoregulatory pathway involving the inhibition of NF-κB. There are many factors like physiological environment, cell type, and concentration which influence the exact role of NO in this disease. Because of its dual nature, it is extremely difficult to pinpoint the action of NO in a distinctive manner in a particular process. Although the partial inhibition of iNOS seems to be beneficial in EAE, the functional duality of this molecule poses a real challenge for scientists to design NO-based therapeutic compounds for the treatment of MS. Further studies are needed to extricate the functional complicacy of NO in the physiology and pathophysiology of EAE, and thereby brightening the prospect of NO as a potential therapeutic target in MS.

Acknowledgments

This study was supported by grants from NIH (NS39940 and NS48923), National Multiple Sclerosis Society (RG3422A1/1), and Michael J. Fox Foundation for Parkinson Research.

References

Alderton WK, Cooper CE, Knowles RG. 2001. Nitric oxide synthases: Structure, function and inhibition. Biochem J 357: 597.

Amici M, Lupidi G, Angeletti M, Fioretti E, Eleuteri AM. 2003. Peroxynitrite-induced oxidation and its effects on isolated proteasomal systems. Free Radic Biol Med 34: 987.

Auch CJ, Saha RN, Sheikh FG, Liu X, Jacobs BL, et al. 2004. Role of protein kinase R in double-stranded RNA-induced expression of nitric oxide synthase in human astroglia. FEBS Lett 563: 223.

Bagasra O, Michaels FH, Zheng YM, Bobroski LE, Spitsin SV, et al. 1995. Activation of the inducible form of nitric oxide synthase in the brains of patients with multiple sclerosis. Proc Natl Acad Sci USA 92: 12041.

Banik NL. 1992. Pathogenesis of myelin breakdown in demyelinating diseases: Role of proteolytic enzymes. Crit Rev Neurobiol 6: 257.

Baron JL, Madri JA, Ruddle NH, Hashim G, Janeway CA Jr. 1993. Surface expression of alpha 4 integrin by CD4 T cells is required for their entry into brain parenchyma. J Exp Med 177: 57.

Beckman JS, Beckman TW, Chen J, Marshall PA, Freeman BA. 1990. Apparent hydroxyl radical production by peroxynitrite: Implications for endothelial injury from nitric oxide and superoxide. Proc Natl Acad Sci USA 87: 1620.

Beckman JS, Carson M, Smith CD, Koppenol WH. 1993. ALS, SOD and peroxynitrite. Nature 364: 584.

Benveniste, EN. 1997. Role of macrophages/microglia in multiple sclerosis and experimental allergic encephalomyelitis. J Mol Med 75: 165.

Bhat NR, Shen Q, Fan F. 2003. TAK1-mediated induction of nitric oxide synthase gene expression in glial cells. J Neurochem 87: 238.

Bhat NR, Feinstein DL, Shen Q, Bhat AN. 2002. p38 MAPK-mediated transcriptional activation of inducible nitric-oxide synthase in glial cells. Roles of nuclear factors, nuclear factor kappa B, cAMP response element-binding protein, CCAAT/enhancer- binding protein-beta, and activating transcription factor-2. J Biol Chem 277: 29584.

Bhat NR, Zhang P, Lee JC, Hogan EL. 1998. Extracellular signal-regulated kinase and p38 subgroups of mitogen-activated protein kinases regulate inducible nitric oxide synthase and tumor necrosis factor-alpha gene expression in endotoxin-stimulated primary glial cultures. J Neurosci 18: 1633.

Bizzozero OA, De Jesus G, Howard TA. 2004. Exposure of rat optic nerves to nitric oxide causes protein S-nitrosation and myelin decompaction. Neurochem Res 29: 1675.

Bizzozero OA, De Jesus G, Bixler HA, Pastuszyn A. 2005. Evidence of nitrosative damage in the brain white matter of patients with multiple sclerosis. Neurochem Res 30: 139.

Bo L, Dawson TM, Wesselingh S, Mork S, Choi S, et al. 1994. Induction of nitric oxide synthase in demyelinating regions of multiple sclerosis brains. Ann Neurol 36: 778.

Bogdan C. 2001. Nitric oxide and the regulation of gene expression. Trends Cell Biol 11: 66.

Bossy-Wetzel E, Talantova MV, Lee WD, Scholzke MN, Harrop A, et al. 2004. Crosstalk between nitric oxide and zinc pathways to neuronal cell death involving mitochondrial dysfunction and p38- activated K+ channels. Neuron 41: 351.

Boullerne AI, Nedelkoska L, Benjamins JA. 1999. Synergism of nitric oxide and iron in killing the transformed murine oligodendrocyte cell line N20.1. J Neurochem 72: 1050.

Boullerne AI, Petry KG, Meynard M, Geffard M. 1995. Indirect evidence for nitric oxide involvement in multiple sclerosis by characterization of circulating antibodies directed against conjugated S-nitrosocysteine. J Neuroimmunol 60: 117.

Boullerne AI, Rodriguez JJ, Touil T, Brochet B, Schmidt S, et al. 2002. Anti-S-nitrosocysteine antibodies are a predictive marker for demyelination in experimental autoimmune encephalomyelitis: Implications for multiple sclerosis. J Neurosci 22: 123.

Brahmachari S, Fung YK, Pahan K. 2006. Induction of glial fibrillary acidic protein expression in astrocytes by nitric oxide. J Neurosci 26: 4930.

Bredt DS. 1999. Endogenous nitric oxide synthesis: Biological functions and pathophysiology. Free Radic Res 31: 577.

Carson MJ. 2002. Microglia as liaisons between the immune and central nervous systems: Functional implications for multiple sclerosis. Glia 40: 218.

Cassina AM, Hodara R, Souza JM, Thomson L, Castro L, et al. 2000. Cytochrome c nitration by peroxynitrite. J Biol Chem 275: 21409.

Chen CC, Wang JK, Chen WC, Lin SB. 1998. Protein kinase C eta mediates lipopolysaccharide-induced nitric-oxide synthase expression in primary astrocytes. J Biol Chem 273: 19424.

Cho HJ, Xie QW, Calaycay J, Mumford RA, Swiderek KM, et al. 1992. Calmodulin is a subunit of nitric oxide synthase from macrophages. J Exp Med 176: 599.

Chung KK, Thomas B, Li X, Pletnikova O, Troncoso JC, et al. 2004. S-nitrosylation of parkin regulates ubiquitination and compromises parkin's protective function. Science 304: 1328.

Cudd A, Fridovich I. 1982. Electrostatic interactions in the reaction mechanism of bovine erythrocyte superoxide dismutase. J Biol Chem 257: 11443.

Cvetkovic I, Miljkovic D, Vuckovic O, Harhaji L, Nikolic Z, et al. 2004. Taxol activates inducible nitric oxide synthase in rat astrocytes: The role of MAP kinases and NF-kappaB. Cell Mol Life Sci 61: 1167.

Dasgupta S, Jana M, Liu X, Pahan K. 2002. Myelin basic protein-primed T cells induce nitric oxide synthase in microglial cells. Implications for multiple sclerosis. J Biol Chem 277: 39327.

Dasgupta S, Zhou Y, Jana M, Banik NL, Pahan K. 2003. Sodium phenylacetate inhibits adoptive transfer of experimental allergic encephalomyelitis in SJL/J mice at multiple steps. J Immunol 170: 3874.

Dauer W, Przedborski S. 2003. Parkinson's disease: Mechanisms and models. Neuron 39: 889.

Dell'Albani P, Santangelo R, Torrisi L, Nicoletti VG, de Vellis J, et al. 2001. JAK/STAT signaling pathway mediates cytokine-induced iNOS expression in primary astroglial cell cultures. J Neurosci Res 65: 417.

Dela Torre A, Schroeder RA, Bartlett ST, Kuo PC. 1998. Differential effects of nitric oxide-mediated S-nitrosylation on p50 and c-jun DNA binding. Surgery 124: 137.

Dela Torre A, Schroeder RA, Punzalan C, Kuo PC. 1999. Endotoxin-mediated S-nitrosylation of p50 alters NF-kappa B-dependent gene transcription in ANA-1 murine macrophages. J Immunol 162: 4101.

Ding M, Zhang M, Wong JL, Rogers NE, Ignarro LJ, et al. 1998. Antisense knockdown of inducible nitric oxide synthase inhibits induction of experimental autoimmune encephalomyelitis in SJL/J mice. J Immunol 160: 2560.

Dobashi K, Pahan K, Chahal A, Singh I. 1997. Modulation of endogenous antioxidant enzymes by nitric oxide in rat C6 glial cells. J Neurochem 68: 1896.

Eng LF, Ghirnikar RS. 1994. GFAP and astrogliosis. Brain Pathol 4: 229.

Eng LF, Yu AC, Lee YL. 1992. Astrocytic response to injury. Prog Brain Res 94: 353.

Engelhardt B, Ransohoff RM. 2005. The ins and outs of T-lymphocyte trafficking to the CNS: Anatomical sites and molecular mechanisms. Trends Immunol 26: 485.

Engelhardt B, Laschinger M, Schulz M, Samulowitz U, Vestweber D, et al. 1998. The development of experimental autoimmune encephalomyelitis in the mouse requires alpha4-integrin but not alpha4beta7-integrin. J Clin Invest 102: 296.

Faruqi TR, Erzurum SC, Kaneko FT, Di Corleto PE. 1997. Role of nitric oxide in poly(I- C)-induced endothelial cell expression of leukocyte adhesion molecules. Am J Physiol 273: H2490.

Fenyk-Melody JE, Garrison AE, Brunnert SR, Weidner JR, Shen F, et al. 1998. Experimental autoimmune encephalomyelitis is exacerbated in mice lacking the NOS2 gene. J Immunol 160: 2940.

Furchgott RF, Zawadzki JV. 1980. The obligatory role of endothelial cells in the relaxation of arterial smooth muscle by acetylcholine. Nature 288: 373.

Go YM, Patel RP, Maland MC, Park H, Beckman JS, et al. 1999. Evidence for peroxynitrite as a signaling molecule in flow-dependent activation of c- Jun NH(2)-terminal kinase. Am J Physiol 277: H1647.

Gonzalez-Scarano F, Baltuch G. 1999. Microglia as mediators of inflammatory and degenerative diseases. Annu Rev Neurosci 22: 219.

Gonzalez-Scarano F, Martin-Garcia J. 2005. The neuropathogenesis of AIDS. Nat Rev Immunol 5: 69.

Gu Z, Kaul M, Yan B, Kridel SJ, Cui J, et al. 2002. S-nitrosylation of matrix metalloproteinases: Signaling pathway to neuronal cell death. Science 297: 1186.

Hess DT, Matsumoto A, Kim SO, Marshall HE, Stamler JS. 2005. Protein S-nitrosylation: Purview and parameters. Nat Rev Mol Cell Biol 6: 150.

Hooper DC, Bagasra O, Marini JC, Zborek A, Ohnishi ST, et al. 1994. Proc Natl Acad Sci USA 94: 2528.

Hua LL, Zhao ML, Cosenza M, Kim MO, Huang H, et al. 2002. Role of mitogen-activated protein kinases in inducible nitric oxide synthase and TNFalpha expression in human fetal astrocytes. J Neuroimmunol 126: 180.

Huie RE, Padmaja S. 1993. The reaction of NO with superoxide. Free Radic Res Commun 18: 195.

Hurst RD, Clark JB. 1997. Nitric oxide-induced blood-brain barrier dysfunction is not mediated by inhibition of mitochondrial respiratory chain activity and/or energy depletion. Nitric Oxide 1: 121.

Hurst RD, Azam S, Hurst A, Clark JB. 2001. Nitric-oxide-induced inhibition of glyceraldehyde-3-phosphate dehydrogenase may mediate reduced endothelial cell monolayer integrity in an in vitro model blood-brain barrier. Brain Res 894: 181.

Ignarro LJ. 1999. Nitric oxide: A unique endogenous signaling molecule in vascular biology. Biosci Rep 19: 51.

Isaksson J, Farooque M, Olsson Y. 2005. Improved functional outcome after spinal cord injury in iNOS-deficient mice. Spinal Cord 43: 167.

Jana M, Anderson JA, Saha RN, Liu X, Pahan K. 2005. Regulation of inducible nitric oxide synthase in proinflammatory cytokine-stimulated human primary astrocytes. Free Radic Biol Med 38: 655.

Jana M, Liu X, Koka S, Ghosh S, Petro TM, et al. 2001. Ligation of CD40 stimulates the induction of nitric-oxide synthase in microglial cells. J Biol Chem 276: 44527.

Kapahi P, Takahashi T, Natoli G, Adams SR, Chen Y, et al. 2000. Inhibition of NF-kappa B activation by arsenite through reaction with a critical cysteine in the activation loop of Ikappa B kinase. J Biol Chem 275: 36062.

Koprowski H, Spitsin SV, Hooper DC. 2001. Prospects for the treatment of multiple sclerosis by raising serum levels of uric acid, a scavenger of peroxynitrite. Ann Neurol 49: 139.

Kuhlmann CR, Lessmann V, Luhmann HJ. 2006. Fluvastatin stabilizes the blood-brain barrier in vitro by nitric oxide-dependent dephosphorylation of myosin light chains. Neuropharmacology 51: 907.

Lander HM, Jacovina AT, Davis RJ, Tauras JM. 1996. Differential activation of mitogen- activated protein kinases by nitric oxide-related species. J Biol Chem 271: 19705.

Lassmann H. 2003. Hypoxia-like tissue injury as a component of multiple sclerosis lesions. J Neurol Sci 206: 187.

Lawson LJ, Perry VH, Dri P, Gordon S. 1990. Heterogeneity in the distribution and morphology of microglia in the normal adult mouse brain. Neuroscience 39: 151.

Liedtke W, Edelmann W, Bieri PL, Chiu FC, Cowan NJ, et al. 1996. GFAP is necessary for the integrity of CNS white matter architecture and long-term maintenance of myelination. Neuron 17: 607.

Ling EA, Wong WC. 1993. The origin and nature of ramified and amoeboid microglia: A historical review and current concepts. Glia 7: 9.

Liu JS, Zhao ML, Brosnan CF, Lee SC. 2001. Expression of inducible nitric oxide synthase and nitrotyrosine in multiple sclerosis lesions. Am J Pathol 158: 2057.

Luth HJ, Holzer M, Gartner U, Staufenbiel M, Arendt T. 2001. Expression of endothelial and inducible NOS-isoforms is increased in Alzheimer's disease, in APP23 transgenic mice and after experimental brain lesion in rat: Evidence for an induction by amyloid pathology. Brain Res 13: 57.

Mander P, Borutaite V, Moncada S, Brown GC. 2005. Nitric oxide from inflammatory-activated glia synergizes with hypoxia to induce neuronal death. J Neurosci Res 79: 208.

Marchetti B, Morale MC, Brouwer J, Tirolo C, Testa N, et al. 2002. Exposure to a dysfunctional glucocorticoid receptor from early embryonic life programs the resistance to experimental autoimmune encephalomyelitis via nitric oxide-induced immunosuppression. J Immunol 168: 5848.

Martin R, McFarland HF, McFarlin DE. 1992. Immunological aspects of demyelinating diseases. Annu Rev Immunol 10: 153.

Mayhan WG. 1996. Role of nitric oxide in histamine-induced increases in permeability of the blood-brain barrier. Brain Res 743: 70.

McDonald JW, Althomsons SP, Hyrc KL, Choi DW, Goldberg MP. 1998. Oligodendrocytes from forebrain are highly vulnerable to AMPA/kainate receptor-mediated excitotoxicity. Nat Med 4: 291.

Merrill JE, Ignarro LJ, Sherman MP, Melinek J, Lane TE. 1993. Microglial cell cytotoxicity of oligodendrocytes is mediated through nitric oxide. J Immunol 151: 2132.

Michel T, Feron O. 1997. Nitric oxide synthases: Which, where, how, and why? J Clin Invest 100: 2146.

Newman DK, Hoffman S, Kotamraju S, Zhao T, Wakim B, et al. 2002. Nitration of PECAM-1 ITIM tyrosines abrogates phosphorylation and SHP-2 binding. Biochem Biophys Res Commun 296: 1171.

Niedbala W, Wei XQ, Campbell C, Thomson D, Komai-Koma M, et al. 2002. Nitric oxide preferentially induces type 1 T cell differentiation by selectively up-regulating IL-12 receptor beta 2 expression via cGMP. Proc Natl Acad Sci USA 99: 16186.

Oldendorf WH, Cornford ME, Brown WJ. 1977. The large apparent work capability of the blood-brain barrier: A study of the mitochondrial content of capillary endothelial cells in brain and other tissues of the rat. Ann Neurol 1: 409.

Padgett CM, Whorton AR. 1997. Glutathione redox cycle regulates nitric oxide-mediated glyceraldehyde-3-phosphate dehydrogenase inhibition. Am J Physiol 272: C99.

Pahan K, Raymond JR, Singh I. 1999. Inhibition of phosphatidylinositol 3-kinase induces nitric-oxide synthase in lipopolysaccharide- or cytokine-stimulated C6 glial cells. J Biol Chem 274: 7528.

Pahan K, Liu X, Wood C, Raymond JR. 2000. Expression of a constitutively active form of phosphatidylinositol 3-kinase inhibits the induction of nitric oxide synthase in human astrocytes. FEBS Lett 472: 203.

Pahan K, Sheikh FG, Namboodiri AM, Singh I. 1997. Lovastatin and phenylacetate inhibit the induction of nitric oxide synthase and cytokines in rat primary astrocytes, microglia, and macrophages. J Clin Invest 100: 2671.

Pahan K, Sheikh FG, Namboodiri AM, Singh I. 1998b. Inhibitors of protein phosphatase 1 and 2A differentially regulate the expression of inducible nitric-oxide synthase in rat astrocytes and macrophages. J Biol Chem 273: 12219.

Pahan K, Namboodiri AM, Sheikh FG, Smith BT, Singh I. 1997a. Increasing cAMP attenuates induction of inducible nitric-oxide synthase in rat primary astrocytes. J Biol Chem 272: 7786.

Pahan K, Sheikh FG, Khan M, Namboodiri AM, Singh I. 1998. Sphingomyelinase and ceramide stimulate the expression of inducible nitric-oxide synthase in rat primary astrocytes. J Biol Chem 273: 2591.

Pahan K, Liu X, McKinney MJ, Wood C, Sheikh FG, et al. 2000. Expression of a dominant-negative mutant of p21 (ras) inhibits induction of nitric oxide synthase and activation of nuclear factor-kappaB in primary astrocytes. J Neurochem 74: 2288.

Pahan K, Sheikh FG, Liu X, Hilger S, McKinney M, et al. 2001. Induction of nitric-oxide synthase and activation of NF-kappaB by interleukin-12 p40 in microglial cells. J Biol Chem 276: 7899.

Pannu R, Won JS, Khan M, Singh AK, Singh I. 2004. A novel role of lactosylceramide in the regulation of lipopolysaccharide/interferon-gamma-mediated inducible nitric oxide synthase gene expression: Implications for neuroinflammatory diseases. J Neurosci 24: 5942.

Park HS, Huh SH, Kim MS, Lee SH, Choi EJ. 2000. Nitric oxide negatively regulates c- Jun N-terminal kinase/stress-activated protein kinase by means of S-nitrosylation. Proc Natl Acad Sci USA 97: 14382.

Parkinson JF, Mitrovic B, Merrill JE. 1997. The role of nitric oxide in multiple sclerosis. J Mol Med 75: 174.

Peng HB, Libby P, Liao JK. 1995. Induction and stabilization of I kappa B alpha by nitric oxide mediates inhibition of NF-kappa B. J Biol Chem 270: 14214.

Piccio L, Rossi B, Scarpini E, Laudanna C, Giagulli C, et al. 2002. Molecular mechanisms involved in lymphocyte recruitment in inflamed brain microvessels: Critical roles for P-selectin glycoprotein ligand-1 and heterotrimeric G(i)-linked receptors. J Immunol 168: 1940.

Pitt D, Werner P, Raine CS. 2000. Glutamate excitotoxicity in a model of multiple sclerosis. Nat Med 6: 67.

Pitt D, Nagelmeier IE, Wilson HC, Raine CS. 2003. Glutamate uptake by oligodendrocytes: Implications for excitotoxicity in multiple sclerosis. Neurology 61: 1113.

Popp R, Fleming I, Busse R. 1998. Pulsatile stretch in coronary arteries elicits release of endothelium-derived hyperpolarizing factor: A modulator of arterial compliance. Circ Res 82: 696.

Radi R, Beckman JS, Bush KM, Freeman BA. 1991. Peroxynitrite-induced membrane lipid peroxidation: The cytotoxic potential of superoxide and nitric oxide. Arch Biochem Biophys 288: 481.

Ravi K, Brennan LA, Levic S, Ross PA, Black SM. 2004. S-nitrosylation of endothelial nitric oxide synthase is associated with monomerization and decreased enzyme activity. Proc Natl Acad Sci USA 101: 2619.

Reynaert NL, Ckless K, Korn SH, Vos N, Guala AS, et al. 2004. Nitric oxide represses inhibitory kappaB kinase through S-nitrosylation. Proc Natl Acad Sci USA 101: 8945.

Rock RB, Gekker G, Hu S, Sheng WS, Cheeran M, et al. 2004. Role of microglia in central nervous system infections. Clin Microbiol Rev 17: 942.

Rosin C, Bates TE, Skaper SD. 2004. Excitatory amino acid induced oligodendrocyte cell death in vitro: Receptor-dependent and -independent mechanisms. J Neurochem 90: 1173.

Roy A, Fung YK, Liu X, Pahan K. 2006. Up-regulation of microglial CD11b expression by nitric oxide. J Biol Chem 281: 14971.

Saha RN, Pahan K. 2006. Regulation of inducible nitric oxide synthase gene in glial cells. Antioxid Redox Signal 8: 929.

Saha RN, Pahan K. 2006a. Signals for the induction of nitric oxide synthase in astrocytes. Neurochem Int 49: 154.

Scott GS, Virag L, Szabo C, Hooper DC. 2003. Peroxynitrite-induced oligodendrocyte toxicity is not dependent on poly (ADP-ribose) polymerase activation. Glia 41: 105.

Scott GS, Spitsin SV, Kean RB, Mikheeva T, Koprowski H, et al. 2002. Therapeutic intervention in experimental allergic encephalomyelitis by administration of uric acid precursors. Proc Natl Acad Sci USA 99: 16303.

Shigemori Y, Katayama Y, Mori T, Maeda T, Kawamata T. 2006. Matrix metalloproteinase-9 is associated with blood-brain barrier opening and brain edema formation after cortical contusion in rats. Acta Neurochir Suppl 96: 130.

Simmons ML, Murphy S. 1992. Induction of nitric oxide synthase in glial cells. J Neurochem 59: 897.

Staddon JM, Herrenknecht K, Smales C, Rubin LL. 1995. Evidence that tyrosine phosphorylation may increase tight junction permeability. J Cell Sci 108: 609.

Stojkovic T, Colin C, Le Saux F, Jacque C. 1998. Specific pattern of nitric oxide synthase expression in glial cells after hippocampal injury. Glia 22: 329.

Takao T, Flint N, Lee L, Ying X, Merrill J, et al. 2004. 17beta-estradiol protects oligodendrocytes from cytotoxicity induced cell death. J Neurochem 89: 660.

Touil T, Deloire-Grassin MS, Vital C, Petry KG, Brochet B. 2001. In vivo damage of CNS myelin and axons induced by peroxynitrite. Neuroreport 12: 3637.

Trapp BD, Peterson J, Ransohoff RM, Rudick R, Mork S, et al. 1998. Axonal transection in the lesions of multiple sclerosis. N Engl J Med 338: 278.

Tuohy, VK, Sobel, RA, Less MB. 1988. Myelin proteolipid protein-induced experimental allergic encephalomyelitis. J Immunol 140: 1868.

Whiteman M, Ketsawatsakul U, Halliwell B. 2002. A reassessment of the peroxynitrite scavenging activity of uric acid. Ann N Y Acad Sci 962: 242.

Won JS, Im YB, Singh AK, Singh I. 2004. Dual role of cAMP in iNOS expression in glial cells and macrophages is mediated by differential regulation of p38-MAPK/ATF-2 activation and iNOS stability. Free Radic Biol Med 37: 1834.

Won JS, Im YB, Key L, Singh I, Singh AK. 2003. The involvement of glucose metabolism in the regulation of inducible nitric oxide synthase gene expression in glial cells: Possible role of glucose-6-phosphate dehydrogenase and CCAAT/enhancing binding protein. J Neurosci 23: 7470.

Wong D, Prameya R, Dorovini-Zis K, Vincent SR. 2004. Nitric oxide regulates interactions of PMN with human brain microvessel endothelial cells. Biochem Biophys Res Commun 233: 142.

Wong D, Prameya R, Wu V, Dorovini-Zis K, Vincent SR. 2005. Nitric oxide reduces T lymphocyte adhesion to human brain microvessel endothelial cells via a cGMP-dependent pathway. Eur J Pharmacol 514: 91.

Yednock TA, Cannon C, Fritz LC, Sanchez-Madrid F, Steinman L, et al. 1992. Prevention of experimental autoimmune encephalomyelitis by antibodies against alpha 4 beta 1 integrin. Nature 356: 63.

Yuceyar N, Taskiran D, Sagduyu A. 2001. Serum and cerebrospinal fluid nitrite and nitrate levels in relapsing-remitting and secondary progressive multiple sclerosis patients. Clin Neurol Neurosurg 103: 206.

Zhao W, Tilton RG, Corbett JA, McDaniel ML, Misko TP, et al. 1996. Experimental allergic encephalomyelitis in the rat is inhibited by aminoguanidine, an inhibitor of nitric oxide synthase. J Neuroimmunol 64: 123.

14 The Blood–Brain Barrier— Biology, Development, and Brain Injury

C. L. Keogh · K. R. Francis · V. R. Whitaker · L. Wei

1	Introduction	304
2	Function of the BBB	304
3	The Junctional Complex	305
4	Development of the BBB	305
4.1	Tight Junctions and Permeability	306
4.2	Vascularity	306
4.3	Transporter Expression	306
5	BBB Permeability and Measurement	307
6	Factors Influencing the BBB	307
6.1	Vascular Endothelial Growth Factor	308
6.2	Fibroblast Growth Factor	309
6.3	Tumor Necrosis Factor-alpha	310
7	Ion Channels and BBB Functional Regulation	310
7.1	Analyzing BBB Ion Channel Components and Functionality	311
7.2	Potassium Transport and Regulation by the BBB	311
7.3	Na+ Regulation	311
7.4	Ca2+, Cl−, and Regulation of Other Ionic Concentrations	312
7.5	Amino Acids and Neurotransmitter Regulation	312
7.6	Astrocyte Regulation of BBB Through Ion Channel Activity	313
7.7	Water Transport and the BBB	313
7.8	P-Glycoprotein	313
8	Brain Disorders and the BBB	314
9	Conclusion	315

© 2009 Springer Science+Business Media, LLC.

Abstract: The blood–brain barrier is a complex structure that serves to regulate the transport of molecules to and from the brain. This chapter will describe the unique properties of the blood–brain barrier, including the endothelial cells and tight junctional complexes, the regulatory set of transporters, junctional complex components, and receptors, all of which play a major role in regulating cerebral homeostasis. The development of the blood–brain barrier will be discussed, with particular attention paid to vascularity and other influences on growth and regulation. Pathological processes such as cerebral ischemia, which affects the integrity and permeability of the blood–brain barrier, will be described. Discussion of the challenges that these pathological processes present helps to further elucidate the mechanism of blood–brain barrier vascularity and transporter regulation. Finally, therapeutic mechanisms to repair the blood–brain barrier and the role of astrocytes and particular growth factors, such as VEGF and bFGF, which play a role in facilitating regeneration, are offered.

List of Abbreviations: AA, amino acids; AIB, a-aminoisobutyric acid; AQP4, Aquaporin-4; BBB, blood–brain barrier; bFGF, basic fibroblast growth factor; BMSCs, bone marrow stromal cells; ECs, endothelial cells; eNOS, endothelial nitric oxide synthase; FAK, focal adhesion kinase; GFAP, glial fibrillary acidic protein; GLUT-1, glucose transporter-1; Hif-1, hypoxia inducible factor-1; MCA, middle cerebral artery; MMPs, matrix-metalloproteinases; PKC, protein kinase C; PS, permeability-surface area; RAFTK, related adhesion focal tyrosine kinase; NO, nitric oxide; 3MG, 3-0-methylglucose; TNFα, tumor necrosis factor-alpha; VEGF, vascular endothelial growth factor

1 Introduction

The unique properties of the cerebral microvascular endothelium comprise what is known as the blood–brain barrier (BBB). This structure is composed of highly specialized brain capillary endothelial cells (ECs), which form tight junctional complexes that are surrounded by pericytes, perivascular macrophages, and astrocytic endfeet (Rieckmann and Engelhardt, 2003; Ubogu et al., 2006b). The result is a highly regulated and complex barrier that blocks the passage of molecules between the blood and the brain, thus preserving ionic homeostasis in the microenvironment (Petty and Lo, 2002). The structural basis of the low permeability of the BBB to water-soluble materials lies in the extreme tightness of the intercellular junctions, the near absence of pinocytotic vesicles, and the lack of fenestrae. Under pathological conditions, a change in BBB permeability almost certainly involves alterations in some or all of these structures.

The majority of blood vessels in the body contain leaky interendothelial cell junctions that allow the passage of many circulating factors (McCarty, 2005). The endothelial layer of the BBB differs from other ECs in that it is regulated and selective. There are fewer endocytotic vescicles, resulting in a low transcellular flux, and the ECs of the BBB are coupled by specialized tight junctions, which allows for a more restricted paracellular flux with a high electrical resistance (Rubin and Staddon, 1999; McCarty, 2005). The complexity of the BBB presents a problem when studying BBB mechanisms in an in vivo setting. Brain injury such as cerebral ischemia results in a disruption of the BBB. This chapter serves to briefly introduce BBB biology and development, factors associated with BBB permeability, ionic transport mechanisms of the BBB, and finally pathological changes of the BBB following brain injury, particularly cerebral ischemia.

2 Function of the BBB

The BBB exerts control over the microenvironment of brain cells, providing the brain with a supply of essential nutrients while mediating the removal of waste (Abbott et al., 2006). The BBB restricts not only diffusional movement but also bulk or convective flow of water and entrained solutes, flows driven by hydrostatic and osmotic pressure gradients. This is accomplished in two ways. First, the filtration rate, which indicates the speed of flow, is exceedingly low, resulting in half-times of convection on the order of many minutes. Second, the reflection coefficients (indicators of osmotic effectiveness) of polar solutes—including Na+ and Cl−—at the BBB are close to 1.0, indicating that fluid passing across the intact BBB is virtually devoid of solutes. Such fluid dilutes the endothelial cell and interstitial fluids and immediately sets up a countering osmotic gradient. This regulatory system has almost certainly gone awry when edema fluid accumulates in brain. In addition to its barrier properties, the BBB has carrier systems that facilitate the transendothelial movement of solutes such as glucose, lactate, and phenylalanine and involve transport proteins (transporters) on both the luminal and abluminal membranes. Transport proteins have been found in the BBB for glucose (GLUT-1) and monocarboxylic acids such as lactate (MCT-2) (Rubin and Staddon, 1999). P-glycoprotein is a functional transporter of the BBB encoded by the human multidrug-resistance gene MDR1 (Chen et al., 2003; Cutler et al., 2006). Glucose, the main substrate for nervous tissue, is transported across the BBB by glucose transporter-1 (GLUT-1), part of a family of facilitative transmembrane transport proteins (Maurer et al., 2006). The Na+-dependent X− transport system for acidic amino acids (AA) such as glutamate has been localized to the BBB, possibly operating mostly to clear excitatory AA from brain to blood. Another main function of the BBB is to regulate the entrance of toxins into the brain. This is achieved by a complex set of transporters expressed on ECs (Rieckmann and Engelhardt, 2003) that transport similarly to a drug efflux pump back to the blood lipophilic molecules that have entered the brain while providing an important protective mechanism (Rubin and Staddon, 1999; Cutler et al., 2006). Its expression in the BBB limits the distribution of a variety of substances (Jin et al., 2006).

3 The Junctional Complex

Tight and adherens junctions together form the junctional complex that exists between ECs (Petty and Lo, 2002). This intricate association is the molecular basis for the extremely impermeable properties of the BBB (McCarty, 2005). Tight junctions form a seal that prevents molecules from paracellular (between cells) diffusion down the concentration gradient, thus mediating the gate function of the BBB and eliminating intracellular space (Petty and Lo, 2002; McCarty, 2005). The important components of tight junctions are the transmembrane proteins occludins and claudins, junctional adhesion molecules, and cytoplasmic proteins such as zona occludens and cingulin, which are linked to an actin-based cytoskeleton (Petty and Lo, 2002). Together, claudins and occludins form the tight junction strands that are observed by electron microscopy and also may modulate permeability (McCarty, 2005). The family of junctional adhesion molecules are counter receptors for members of the integrin family of cell adhesion proteins and may be integral in the migration of leukocytes across ECs (McCarty, 2005). The cytoplasmic proteins mentioned allow for intracellular signaling during and after formation of the tight junctions; specifically the zona occludens proteins act as scaffolding proteins and connect different components at the junctions (McCarty, 2005). Tyrosine kinases, protein kinase C (PKC), Ca2+, heterotrimeric G proteins, calmodulin, CAMP, lipid second messengers, and phospholipase C, all of which are molecules involved in most intracellular signaling pathways, have been implicated in BBB paracellular permeability (Petty and Lo, 2002). Both phosphorylation of transmembrane and accessory proteins and changes in cAMP influence the regulation of tight junctions (Petty and Lo, 2002).

The adherens junctions hold the ECs together, forming a continuous "belt" that maintains the entire junctional complex (Petty and Lo, 2002). This adhesive cell–cell association is found in an abundance of tissues, with a cadherin dimer as the main mediator of the adhesion (Brown and Davis, 2002). Proper functioning of the adherens junction is crucial for successful formation of the tight junction (Farkas et al., 2005). E-cadherin is present in epithelial and ECs and is associated in a complex at the basolateral end of the cell with α-catenin, β-catenin, and actin (Brown and Davis, 2002). The intimacy of these two junctions is exhibited in the finding that ZO-1 and α-catenin are able to interact (Farkas et al., 2005). The adherens junctions stabilize the cell–cell interactions in the tight junction zone (Abbott et al., 2006).

4 Development of the BBB

The final step during the complex development of cerebral vascular growth, which involves BBB differentiation, termed barriergenesis, is a long and complex process that involves a series of early and late stages and is induced by the embryonic neural tube (Engelhardt, 2003; Rieckmann and Engelhardt, 2003). Due to limitations in experimental techniques, it is difficult to study the development of the BBB. It has been shown that ECs first appear in the brain in a discontinuous and unstructured manner; the first marker for brain ECs can be detected on embryonic day 10.5 (E10.5) in mice (Rubin and Staddon, 1999). Studies thus far have been able to conclude that brain ECs adopt a different phenotype for the BBB when compared to peripheral ECs, which seems to be influenced or dictated by the microenvironment of the brain (Rubin and Staddon, 1999).

4.1 Tight Junctions and Permeability

Tight junctions have been shown to appear from E10 in mice and are arguably fully functional at E16 (Bauer et al., 1993; Finnie et al., 2006). Studies done with freeze-fracture analysis have shown that the density of proteins in the junctional strands increases at day E18 in the rat, along with an increase in the association of tight junction proteins, indicating a conformational change to the adult form of junctions (Kniesel et al., 1996; Engelhardt, 2003). The proportion of the junction that has unfused parts was shown to be higher prenatally and decreased with aging (Stewart and Hayakawa, 1987). Immunohistochemical studies on the developing human brain have investigated the localization of tight junction proteins occludin and claudin-5 (Virgintino et al., 2004). The study concluded that a shift of localization occurs during fetal development, demonstrating that expression of these proteins is an early phenomenon in brain development, which is followed by assembly of the proteins at the appropriate locations (Virgintino et al., 2004). Tightening of the BBB occurs independently of vascular proliferation (Engelhardt, 2003).

During the embryonic period, the BBB is assumed to be deficient, partly due to the fact that the capillaries differ from mature brain capillaries (Himwich, 1970). The tightness of the BBB occurs regionally and increases gradually (Engelhardt, 2003). It has traditionally been thought that the BBB in young animals may be less effective than that of mature animals (Himwich, 1970). The BBB matures stepwise and it is surmised that in a newborn with an immature and unformed BBB, substances may penetrate the brain more freely than in the adult brain (Himwich, 1970). While groups disagree on the permeability of the early BBB, it seems more likely that fetal BBB permeability to macromolecules is similar to that of the adult (Engelhardt, 2003). Studies as early as 1988 concluded that the immature brain had a well-formed BBB even at an embryonic stage and ultrastructural studies showed the presence of tight junctions as early as the day of birth (Dziegielewska et al., 1988).

4.2 Vascularity

The vascular system within the CNS develops by angiogenesis, peaking in the early postnatal period and decreasing in adulthood (Engelhardt, 2003). While the factors that have been identified as mediators of angiogenesis in the brain such as vascular endothelial growth factor (VEGF) and basic fibroblast growth factor (bFGF) are involved in the growth of ECs, an integral part of the BBB, it is likely that they are not involved in the creation of highly specialized ECs because their mechanisms of action also apply outside the CNS (Engelhardt, 2003). It has been observed that VEGF is downregulated during the 18th week of gestation in the developing human brain, which may be related to preparation for a functional BBB (Virgintino et al., 2003). Determining the factors that play a role in BBB development and differentiation is the biggest remaining obstacle (Engelhardt, 2003).

4.3 Transporter Expression

GLUT-1 is one of the earliest BBB markers (Engelhardt, 2003). It is expressed in brain ECs that are part of the very first vascular sprouts in the neural tube (Engelhardt, 2003). Increasing demand for glucose by the developing brain results in an increased expression of GLUT-1 through development and then a subsequent decrease into adulthood (Engelhardt, 2003). This change in development seems to be partly associated with the change and development of BBB tightness (Dermietzel et al., 1992). Expression of gene products encoding P-glycoprotein has been found in ECs during early brain angiogenesis, which may be important in protecting the brain from maternally ingested lipophilic molecules, which the placental barrier cannot halt (Engelhardt, 2003). Overall, the changes that accompany BBB differentiation are associated with the induction of specific metabolic pathways that are involved in the strict control of substances to the brain (Engelhardt, 2003).

5 BBB Permeability and Measurement

The distribution of material from blood to brain involves many steps besides passage across the BBB. The material is first brought to the capillaries by the blood, and its concentration in plasma water within the capillary is the driving force for influx. As the material flows down the capillary network and passes into the surrounding tissue, its concentration in plasma drops. If the material is bound to plasma proteins and/or is carried by blood cells, then some of the material passing out of plasma water may be replaced from these "reservoirs," thereby maintaining a relatively constant concentration and driving force within the capillary. Once the material has entered the endothelium from the plasma, it can be metabolized and pass back into the blood or move further into the tissue. Within the interstitium the substance can diffuse, be carried by convection if interstitial fluid is flowing, bind to extracellular and membrane proteins, pass into the cells, and efflux back into the ECs and blood. The blood–tissue distribution of a molecule in extracellular fluid is, thus, partly a function of the size of the extracellular space, its effective diffusion coefficient in that space, the rate and direction of interstitial fluid flow, and time. A compound avidly taken up by brain cells [e.g., 3-0-methylglucose (3MG) and a-aminoisobutyric acid (AIB)] is, however, trapped close to its site of passage across the BBB and further distribution through the interstitium or back into blood is limited. As for time-dependency, spread into surrounding tissue can be greatly reduced by using short experimental periods and accentuated by using longer periods of experimentation.

Many studies of "BBB permeability" with models of ischemia measure a parameter such as a distribution space or produce a view of tissue staining or ultrastructural distribution of an electron dense marker or its reaction product (as is the case with HRP) and not the permeability-surface area (PS) product, which is the physiological indicator of capillary permeability. Properly designed experiments determine an influx or transfer constant (K), which is a function of both flow and PS. Then, a capillary model and its equation are assumed and used to calculate PS from K and suitable value of flow. The Krogh single capillary model and the Renkin–Crone equation are the only tools currently available for this task and are assumed in this application.

For several reasons, BBB permeability needs to be measured in an exacting manner, one that yields the PS product. First, it allows the quantitative comparison of data among experimental groups and research reports. Second, if BBB permeability is thought to be part of the pathology of a particular lesion, then the PS product is what should be measured to confirm that suspicion. Third, if a particular treatment is postulated to modify BBB permeability, then the PS product needs to be measured for the testing. Fourth, to deliver a drug or other therapeutic agent to neurons or glia at an effective concentration, the permeability of the BBB—be it normal or altered—to that drug must be relatively high in that condition, something that needs to be ascertained properly. Fifth, it could be that a distributional step or process other than BBB permeation limits drug delivery to the target site and effective drug treatment; this can only be determined by accurately measuring BBB permeability.

6 Factors Influencing the BBB

Experimentation has focused on studying factors that stimulate the growth and completion of the BBB. The existence and proximity of astrocytes seems to at least in part influence ECs to adapt a brain phenotype, whereas the accumulation of P-glycoprotein, which is important for the brain EC phenotype, occurs before the appearance of astrocytes in vivo (Rubin and Staddon, 1999). This observation implies a multifactorial developmental and regulatory process for the formation and completion of the BBB. A glial-expressed protein, the αvβ8 integrin, is a novel neural cell-expressed factor, which has been shown to regulate function of the BBB (McCarty, 2005). Physiochemical properties such as lipophilicity and molecular weight are the major influencing factors that limit the extent to which molecules can cross the BBB (Cutler et al., 2006). Trafficking of certain cells across the BBB is a complex process that requires the coordination of selectins, integrins, cell adhesion molecules, chemokines, and matrix metalloproteinases (MMPs) (Ubogu et al., 2006a, b). Interactions with astrocytes seem to play a crucial role in the development and maintenance of the BBB (Rieckmann and Engelhardt, 2003). Astrocytic endfeet reside close to the EC plasma membrane (Hamm et al., 2004). Studies using an in vitro BBB model of a coculture with brain ECs and astrocytes showed that the removal of astrocytes from the culture resulted in an increased permeability of small tracers across the EC monolayer (Hamm et al., 2004). The removal also resulted in an opening of tight junctions, visualized by electron microscopy. Through contact with their endfeet, astrocytes have also been shown to induce the transdifferentiation of nonneural ECs into cells with BBB properties in vitro (Hayashi et al., 1997).

6.1 Vascular Endothelial Growth Factor

VEGF is a secreted protein, which is important in development, differentiation, and adaptation of the vascular system. VEGF plays a key role in new vessel formation by inducing proliferation and migration of ECs. It is ubiquitously expressed and binds to the endothelial tyrosine receptors VEGFR-1 (Flt-1) and VEGFR-2 (Flk-1). Following brain injury, increased VEGF expression is seen in neurons, astroglia, and inflammatory cells. In addition, VEGF receptor expression is increased on ECs. It has been well documented that tissue hypoxia, often the result of brain injury, leads to an upregulation in the expression of VEGF at the site of injury (Mayhan, 1999; Zhang et al., 2000, 2002; Wei et al., 2001a; Schoch et al., 2002; Hayashi et al., 2003; Krum and Khaibullina, 2003; Fischer et al., 2004). The activation of specific genes as a result of tissue hypoxia is implicated in the pathogenesis of vascular leakage leading to brain edema (Fischer et al., 2002; Schoch et al., 2002).

VEGF is also involved in angiogenesis after cerebral injury (Wei et al., 2001a; Zhang et al., 2002; Hayashi et al., 2003). Angiogenesis, or the sprouting of new blood vessels from the preexisting vascular network, occurs to restore blood flow and the delivery of nutrients to hypoxic tissue (Wei et al., 2003). While the mechanism for angiogenesis remains unclear, expression of growth factors such as VEGF, bFGF, angiopoietin-1 and -2, and their receptors is increased in angiogenesis (Beck et al., 2000; Wei et al., 2001a; Hayashi et al., 2003). These factors are also known as angiogenic factors and are important in the formation of new vessels after injury. Studies have shown that the blocking of VEGF activity leads to less EC proliferation and increased cell death after cerebral insult (Mayhan, 1999; Krum and Khaibullina, 2003). Although VEGF is a permeability factor, it is also important for new vessel formation in angiogenesis.

After an injury to the brain, tissue hypoxia leads to damage of the BBB. Hypoxia activates the transcription factor hypoxia inducible factor-1 (Hif-1), and it has been shown that Hif-1 activation enhances VEGF transcription (Hippenstiel et al., 1998; Fischer et al., 1999; Schoch et al., 2002). VEGF expression is activated at the level of transcription, the increase in stability of mRNA, and preferential translation (Schoch et al., 2002). It is thought that injured tissue releases VEGF to increase nutrient delivery by increasing blood flow and permeability and promoting angiogenesis (Breslin et al., 2003). Due to the expression of VEGF after hypoxia and its ability to stimulate proliferation and migration of ECs, it is the growth factor that is most likely involved in ischemia-induced vascular leakage. Schoch et al. have observed that exposing mice to a hypoxic environment leads to an increase in the levels of VEGF mRNA and protein and an increase in vascular leakage. They then correlated this effect to the severity of the hypoxic stimulus (Schoch et al., 2002). In hypoxic tissue, it has also been seen that VEGF induces rearrangement and changes in expression of proteins that make up endothelial tight junctions (Fischer et al., 2002, 2004; Avraham et al., 2003).

To evaluate the affect of VEGF on the break down of the BBB after injury, many groups have observed the effects of administration of VEGF to an animal brain or the effects of blocking VEGF activity via the administration of a neutralizing antibody. Proescholdt et al. examined the effects of chronic VEGF exposure in the normal rat brain. After 6 days of continuous exposure to VEGF, focal breakdown of the BBB to both small and large molecules was observed (Proescholdt et al., 1999). The latter results indicate that VEGF indeed plays a role in inducing vascular permeability. Mayhan et al. also saw that administration of VEGF to rat brains after cerebral trauma leads to an increase in BBB permeability and dilation of cerebral arterioles (Mayhan, 1999). Krum et al. sought to observe the importance of endogenous VEGF by administering an antibody to neutralize VEGF activation in the rat brain. In their studies, they saw reduced vascularization after cerebral injury. With VEGF blocked, the resulting injury was larger, there was less proliferation of ECs and astrocytes, and increased cell death (Krum and Khaibullina, 2003). These findings indicate that not only is VEGF involved in vascular leakage, but it also plays a role in the process of wound repair.

It has been observed that BBB disruption occurs rapidly after cerebral ischemia, whereas regeneration of microvessels occurs later (Zhang et al., 2000; Avraham et al., 2003). Zhang et al. compared the seemingly contradictory roles of VEGF by administering VEGF in rat brain at 1 or 48 h after cerebral ischemia. Early administration of VEGF (1 h) leads to exacerbation of BBB leakage, increases in hemorrhagic transformation, and increases in ischemic cell damage. In contrast, late administration of VEGF (48 h) showed enhanced microvascular plasma perfusion and improved recovery of neurological function (Zhang et al., 2000). These findings indicate that VEGF has a pleiotropic effect in the injured brain.

To better understand the complicated interactions between VEGF and the cerebral vascular network, it is important to understand the cellular mechanisms that are involved in VEGF signaling. To study the BBB as a single unit, in vitro models are often used. This allows for direct study of the ECs that comprise the BBB. Binding of VEGF to its receptors causes receptor dimerization, which results in stimulation of several intracellular kinases shown to be associated with increases in EC proliferation and differentiation (Lal et al., 2001; Fischer et al., 2004). VEGF binding leads to tyrosine phosphorylation of focal adhesion kinase (FAK) and related adhesion focal tyrosine kinase (RAFTK) and activation of their subsequent signaling cascades (Avraham et al., 2003). Involvement of FAK and RAFTK suggests VEGF plays a role in regulation of focal adhesion complexes, which are a component of the BBB. VEGF-induced permeability of the BBB involves the synthesis and release of nitric oxide (NO), which is associated with vasodilation and increased vascular permeability (Mayhan, 1999; Fischer et al., 2002). Many studies implicate the activation of endothelial nitric oxide synthase (eNOS) in the increase of NO release seen as a result of vascular hyperpermeability (Mayhan, 1999; Lal et al., 2001; Fischer et al., 2002; Breslin et al., 2003; Fischer et al., 2004). Several intracellular signaling pathways have been shown to be involved in eNOS activation by hypoxia and VEGF. They include activation of the PI-3/Akt pathway, activation of PLCγ, subsequent release of intracellular calcium, and activation of PKG (Lal et al., 2001; Avraham et al., 2003; Breslin et al., 2003; Fischer et al., 2004). Future studies with help to clarify the mechanisms of VEGF activation and determine if reactions similar to the ones mentioned here take place in vivo.

The interactions of VEGF and the BBB are complex. Hypoxia-induced VEGF signaling activates a number of proteins by a number of pathways. The effects of VEGF on the cerebral vascular network are numerous and relate to both BBB hyperpermeability and EC proliferation and migration in vascular repair. This is important when considering manipulating VEGF expression as a means of therapy after cerebral injury or in pathological cerebral diseases. More recently, VEGF has been investigated for therapeutic purposes for the treatment of neurodegenerative diseases such as lysosomal storage disease in newborns (Young et al., 2004). The administration of VEGF opened the BBB in newborn mice, allowing for successful bone marrow transplantation and resulting ultimately in increased life span compared with saline control mice.

6.2 Fibroblast Growth Factor

bFGF is another well-known endothelial growth factor. Its expression is also increased after hypoxia (Hayashi et al., 2003; Wei et al., 2003). In addition to evaluating VEGF's effects on vascularization, Krum et al. observed the expression of bFGF as well. After blocking VEGF activity, an increase in bFGF-expressing cells was seen. Despite this increase, vascularization was still reduced as determined by decreased angiogenesis and astrocyte proliferation (Krum and Khaibullina, 2003). Astrocytes are closely associated with brain microvessels and are known to influence function of the BBB. It is thought that alterations of astrocytes may be involved in pathological changes of the BBB, and an increase in bFGF has been shown to correspond to increased glial fibrillary acidic protein (GFAP), which is a marker for reactive astrocytes (Reuss et al., 2003). Reuss et al. tested the importance of bFGF in BBB leakage by observing the changes after cerebral insult in a mouse strain deficient for bFGF. The experiments showed that when bFGF is knocked out, there is a decrease in GFAP expression and levels of tight junction proteins resulting in a leaky BBB (Reuss et al., 2003). The results suggest that bFGF is important in regulating astrocyte proliferation, which in turn is involved in BBB integrity.

6.3 Tumor Necrosis Factor-alpha

The inflammatory cytokine tumor necrosis factor-alpha (TNFα) is known to modulate inflammatory, degenerative, vascular, and traumatic responses in the injured brain (Pan and Kastin, 2002; Brown et al., 2003). TNFα has also been shown to be an activator of EC inflammatory responses, indicating involvement in influencing changes in BBB permeability (Yang et al., 1999; Franzen et al., 2003). Sibson et al. evaluated changes in the brain after TNFα injection in a rat model. They observed that vasoconstriction, disrupted tissue homeostasis, and damage to the BBB occur as adverse effects of TNFα in the brain (Sibson et al., 2002). The mechanism by which TNFα induces changes in BBB permeability is not known well, but it has been suggested that activation of soluble guanylate cyclase and protein tyrosine kinase is involved (Mayhan, 2002). In addition, activation of the urokinase plasminogen activator system and cytoskeletal rearrangements have been shown to occur as a result of activation of cerebral ECs by TNFα (Franzen et al., 2003). More recently, it has also been shown that the effects of TNFα on the regulation of the CBF and the breakdown of the BBB may be mediated by NO, which may contribute to a swelling and damage of the astrocytic endfeet associated with the BBB (Farkas et al., 2006). To date, results indicate that TNFα plays a role in altering BBB permeability after cerebral trauma. It appears that the involvement of TNFα in inducing BBB permeability is complex and may interact with pathways involved in VEGF signaling after cerebral insult.

7 Ion Channels and BBB Functional Regulation

Ion flux across the BBB is strictly controlled through ion-specific channels and transporters. While many channels and transporters have been identified and at least partially characterized, the full complement of BBB transport mechanisms has not been determined. The most extensively studied include the Na+ Cl cotransporter, Na+/H+ exchanger, Cl−/HCO3−exchanger, Na/K-ATPase pump, K+ channels, and Cl− channels (Somjen, 2002). These transporters or channels can be found on both the capillary luminal and brain interstitial abluminal surfaces of BBB ECs. Importantly, ion transporter expression can differ between surfaces, creating ion transport polarity. For example, the Na/K-ATPase pump is found only on the endothelial surface directly opposite the brain parenchyma (Somjen, 2002). This polarity is vital to maintain differing ion, AA, and neurotransmitter concentrations between the blood and brain, allowing for normal brain function.

Fluctuations of ionic concentrations across the BBB will occur during periods of CNS activity. However, during nonpathological conditions, brain pH is maintained principally through Na+ extrusion to the abluminal surface via the Na/K-ATPase pump and luminal uptake of Na+, K+, and Cl− through luminal transporters (Ennis et al., 1996). Other mechanisms at maintaining pH and homeostasis also occur including HCO3− secretion into the brain interstitium (Taylor et al., 2006). Disruption of ion-dependent transport and ionic concentrations at the BBB has been shown to be an important factor for the initiation and progression of pathologic events within the brain. It is unknown if there are changes in the concentration of specific transporters at the BBB under pathologic conditions. While not meant to be a comprehensive listing, this section will present some of the known and most important mechanisms of ion channel regulation and BBB transport.

7.1 Analyzing BBB Ion Channel Components and Functionality

Determination of ion channel function and expression within the BBB has principally been achieved through in vitro analysis. Most commonly, a primary culture of isolated brain microvessels is used. However, in vitro culture is a less than ideal method for analysis of endothelial ion channels. Over time, cultured endothelium are likely to exhibit ion channel features different from in vivo due to simulated conditions. In vitro culture of microvessels also limits the degree of ion channel experimentation, which can be performed by patch clamp electrophysiology, an important technique for determining ion channel function and expression. Patch clamp analysis has been routinely used for measurement of abluminal ion channels (Vigne et al., 1989; Hoyer et al., 1991). However, there is no method currently available to effectively measure luminal ion channel activity in vitro.

7.2 Potassium Transport and Regulation by the BBB

While the exact transport mechanisms underlying K+ flux at the BBB are not fully understood, a number of mechanisms have been elucidated. Potassium homeostasis is vital for maintenance of cell membrane potential and regulation of neuronal action potential. Potassium is principally transported to the capillary lumen from the brain (Hansen et al., 1977). Potassium efflux across the BBB luminal surface to the bloodstream is known to occur during ischemic or hypoxic events as BBB permeability is increased. This is believed to act as a buffering mechanism for removal of excessive K+ from brain parenchyma. Abluminal K+ efflux is achieved primarily through activation of the Na/K-ATPase transporter expressed on brain endothelium (Vorbrodt et al., 1982). However, this process is ATP dependent and will eventually fail during energy stressed events such as ischemia, preventing K+ efflux.

Influx of K+ across the abluminal surface also occurs primarily through the Na/K-ATPase. Regulation of K+ influx into the brain across the BBB is at least partly regulated through Na/K-ATPases on the abluminal surface, particularly through expression of the α3 subunit (Keep et al., 1999). This influx has also been shown to occur during pathologic conditions (Stummer et al., 1994). However, this K+ influx has been shown to be negated by elevated efflux through the Na/K-ATPase (Stummer et al., 1995). BBB regulation of brain parenchymal K+ levels can also be effected by the Na,K,2Cl-cotransporter and 3Na-2K exchange pump. Studies suggest an integral role for the 3Na-2K exchange pump during K+ efflux from the brain interstitium (Schielke et al., 1991).

Glial regulation of extracellular K+ also functions to maintain K+ homeostasis. Potassium channels have been shown to be highly expressed in rodent astroctye perivascular endfeet near blood vessels (Higashi et al., 2001). This is principally achieved through glial uptake via the Na+/K+ or Cl−/K+ cotransporters or ATP-dependent Na+/K+-ATPase activity. Glia are also able to regulate K+ levels through K+ diffusion between adjacent glial cells. This process is referred to as buffering and is unique to glial cells (Gee and Keller, 2005). Glial function in BBB regulation will be discussed further later.

7.3 Na+ Regulation

Many different transporters and channels within brain capillary ECs can regulate Na+ concentrations within the brain. While no voltage-gated sodium channels have been identified from brain capillary ECs, sodium channels sensitive to amiloride, a potassium-sparing diuretic, have been studied. Amiloride-sensitive Na+ channels were first identified in kidney tubules and subsequently in brain capillaries (Vigne et al., 1989). These channels were found to be selective for Na+, also able to transport K+, and localized to the luminal surface. The ATP-dependent Na/K-ATPase transporter was discussed earlier and shown to be vital to the maintenance of Na+/K+ homeostasis within the brain. This enzyme has been shown to be ouabain sensitive within brain capillaries (Harik et al., 1985).

The Na+/H+ exchanger, also ATP-dependent, is a primary means of brain interstitial sodium regulation and has been shown to malfunction during brain ischemia due to ATP depletion. This transport mechanism has previously been shown to be sensitive to amiloride in isolated microvessels (Betz, 1983). This study also determined that Na+/H+ exchange principally occurs abluminally. A study using rate kinetics to determine sodium transport to the brain from the BBB found that sodium transport to brain parenchyma principally occurs through the luminal Na+/H+ exchanger and abluminal Na+/K+-ATPase and Na+/H+ exchanger (Ennis et al., 1996).

The Na,K,2Cl-cotransporter is ubiquitously expressed on many cell types, including brain capillary ECs. This transporter functions primarily to maintain intracellular pH and has been implicated indirectly to Na+ influx during cell death events (Cala and Maldonado, 1994). PKC was identified as a regulator of the Na,K,2Cl-cotransporter on the abluminal surface of BBB endothelium (Yang et al., 2006). Research suggests that the Na+/H+ exchanger, Na,K,2Cl-cotransporter, and the Na/K-ATPase can act in concert to regulate Cl− and K+ accumulation during cell injury events (Somjen, 2002).

A number of sodium-dependent transporters can be found within the BBB. These transporters have been shown to function in AA transport. This primarily occurs across the abluminal BBB surface, as luminal carriers are sodium independent (Oldendorf, 1971; Christensen, 1979). It was further shown that luminal transport of AAs was less specific than abluminal, where elevated sodium flux increased phenylalanine transport (Sanchez del Pino et al., 1992). These studies exhibit the need for the regulation of intracellular Na+ levels for normal AA transport.

7.4 Ca2+, Cl−, and Regulation of Other Ionic Concentrations

Ca2+ has long been known to function in cell signaling and cell death mechanisms. Activation of Ca2+ transport to the brain interstitium by inflammatory mediators has long been known. Some K+ channels have been shown to be regulated by elevated Ca2+ transport (Ningaraj et al., 2002; Csanady and Adam-Vizi, 2003). Ca2+ release to the brain interstitium has also been shown to cause astrocytic activation (Paemeleire, 2002), likely affecting astrocytic regulation of other ion concentrations. Abluminal Ca-dependent ATPase expression has also been determined in the BBB (Vorbrodt, 1988). Cl− transport across the BBB mediates maintenance of homeostatic pH. This transport has been shown to occur through the Na+-dependent HCO-3/Cl- transporter (Sipos et al., 2005). Regulation of Cl− concentration within the brain interstitium by the Na,K,2Cl-cotransporter is particularly important during brain development, with less importance later in development as an increasing proportion of neurons exhibit inhibitory, GABAergic properties (Delpire, 2000).

7.5 Amino Acids and Neurotransmitter Regulation

The BBB is essential for maintaining appropriate levels of AA and neurotransmitters within the brain. One essential neurotransmitter regulated by the BBB is l-glutamate, essential for excitatory neurotransmission. l-Glutamate is produced via the Kreb's cycle and synaptic recycling. This AA is then stored within presynaptic vesicles and when triggered to release binds postsynaptic glutamatergic receptors (Smith, 2000). While normal levels of l-glutamate are essential for normal neural activity, overproduction or release of l-glutamate can produce excitotoxic responses in neural tissues and trigger cell death. For this reason, maintenance of brain glutamate levels must be tightly regulated by the BBB.

As stated earlier, AA transport across the BBB membrane can be regulated by a Na+ gradient. The concentration of AA is much lower within the brain interstitium than blood plasma (Hawkins et al., 2006). Due to this concentration difference, multiple mechanisms are present for AA transport other than sodium-dependent mechanisms. These include facilitative carriers specific for chemical characteristics of different AA.

Large essential AA is principally transported via the L1 system. This carrier system is found on both the luminal and abluminal membrane. The L1 transporter has high affinity for essential AA, the principal carrier system within the BBB for these vital compounds (Boado et al., 1999). The cationic carrier y+ system exhibits high affinity for AA with positively charged side chains. This carrier system is primarily abluminal, but it is also found on the luminal membrane. Also, the system is present on the luminal membrane and mediates only glutamine transport (Lee et al., 1998). Glutamate transport from the brain to plasma is mediated by the xG system. Facilitative transport of glutamate was found only at the luminal surface, increasing blood glutamate levels (Lee et al., 1998).

7.6 Astrocyte Regulation of BBB Through Ion Channel Activity

Astrocytes, a glial cell subtype, also form an integral component of the BBB. Ninety-nine percent of the ECs forming the BBB are found within close proximity to astrocytes. These neural cells extend processes with specialized functions capable of affecting BBB permeability to water and ion flux through aquaporin water channel and inward rectifying K+ channel expression (Wolburg and Lippoldt, 2002). In vitro coculture of primary isolated brain capillary ECs and astrocytes produced a BBB-like structure expressing ZO proteins and high endothelial electrical resistance (Cecchelli et al., 1999). Addition of astrocyte-derived medium has also been used in vitro to achieve a BBB-like formation morphologically, though high endothelial electrical resistance was not achieved (Rist et al., 1997). Increased vascular permeability has also been shown to be induced at least partially through astrocyte activation (Abbott et al., 2006), likely through ionic regulation of BBB permeability. These cells have been shown to induce K+ uptake intracellularly to alleviate cellular stress (Walz and Hertz, 1983). The precise mechanisms of astrocytic regulation of endothelial properties and BBB permeability are still poorly understood.

7.7 Water Transport and the BBB

Recent research suggests that an intimate connection between the BBB and water transport could be extremely important during both homeostatic and pathologic conditions. Water transport across biological membranes is achieved through transport by the aquaporin family of transporters. Water functions within the brain to allow entry of glucose and similar compounds, maintain homeostatic pH, and to remove waste. Aquaporins are ubiquitously expressed by mammalian cells, but specific transporters can be found in cell types within the brain. Aquaporin-4 (AQP4) currently appears to be the most important member of the family concerning BBB permeability to water during pathologic conditions (Manley et al., 2000). This transporter is expressed within different regions of the brain, including the hippocampus, cerebellum, and hypothalamic nuclei (Jung et al., 1994).

AQP4 has also been shown to be expressed within astrocytic endfeet processes known to regulate BBB permeability (Nielsen et al., 1997). AQP4 expression at the junction between the BBB and astrocytes suggests a potential regulation of water influx to the brain cavity. A close tie between AQP4 and GLUT-1 expression was identified in a hypertensive rat model, with increased AQP4 and decreased GLUT-1 expression as hypertension progressed (Ishida et al., 2006). Another study found that expression of the AQP1 transporter is very low in primary cultured brain microvessels but increases with cell passage, indicating loss of brain microvessel phenotype (Dolman et al., 2005). This study further showed a lack of AQP4 expression in brain microvessels, although this disagrees with previous research suggesting a slight AQP4 expression in brain microvessels (Amiry-Moghaddam and Ottersen, 2003). Further evidence suggests that AQP4 levels increase during brain edema (Taniguchi et al., 2000; Saadoun et al., 2002). Current research suggests an interaction between aquaporin expression and BBB permeability during a pathologic condition, though the mechanisms remain elusive.

7.8 P-Glycoprotein

As stated previously, P-glycoprotein is an ATP-dependent efflux transporter. It is a member of the ABC (ATP-binding cassette) transporter superfamily, which acts as an energy-dependent efflux pump. The precise function of P-glycoprotein in normal tissues is not fully known (Virgintino et al., 2002). Immunohistochemical analysis of human brain indicated that P-glycoprotein was primarily expressed in the ECs lining the small vessels of the cortex and in the microvascular pericytes that envelop the ECs (Virgintino et al., 2002). The study confirmed the role of P-glycoprotein as an important transport system of the BBB. A loss of glutathione during a pathologic insult can affect the expression levels of P-glycoprotein, which may affect the normal transport of molecules during certain conditions (Hong et al., 2006). P-glycoprotein activity has been shown to decline with aging (Toornvliet et al., 2006).

8 Brain Disorders and the BBB

The disruption of the BBB and changes in BBB permeability are a major part of the pathology of stroke and other CNS disorders such as multiple sclerosis (Rubin and Staddon, 1999). Breakdown of BBB integrity is associated with the transmigration of numerous immune system cells such as monocytes and lymphocytes (Perry et al., 1997; Petty and Lo, 2002). It is the result of vascular hyperpermeability, which is mediated by enhanced transcytosis, gap formation between ECs, and the induction of fenestrations in the endothelium (Fischer et al., 1999). This section serves to focus on the pathology of the BBB following cerebral ischemia.

In acute stroke, brain insult is attributed to abolishment or interruption of cerebral blood flow and the breakdown of the BBB (Borlongan et al., 2004). It has been shown that even 1 h of ischemia can result in marked increases in BBB permeability (Albayrak et al., 1997). The work of Dietrich et al. (1988), in which the middle cerebral artery (MCA) was photochemically thrombosed in rats, is often cited as showing early ischemic BBB damage because significant leakage of HRP was observed after 15 min. Leakage of HRP was, however, observed on both the ipsilateral and contralateral sides of the brain as well as in other organs, raising questions about this model. In contrast, Garcia et al. (1994) saw HRP extravasation only from 3–24 h after unilateral MCA occlusion in rats. Betz and coworkers (Ennis et al., 1990; Menzies et al., 1993; Betz et al., 1994) found in rats that the PS products of AIB and urea were normal or somewhat reduced up to 4 h after MCA occlusion but were significantly higher than normal after 24 h. Opening of the BBB to circulating albumin appeared to begin 4–6 h after occlusion. Suggestive of Na+-linked edema formation, a small increase in Na+ permeability, a small decrease in K+ permeability, and unchanged Cl− permeability at 3 h after MCA occlusion were detected. In a set of experiments with the reversible two-vessel occlusion model in rats, Preston and Foster (1997) reported that 10 min of occlusion plus hypotension produced two- to fivefold increases in influx of coadministered 3H-inulin and 14C-sucrose in several brain regions following 6 h of reperfusion. They suggested that this increase in BBB permeability was most consistent with the opening of large water-filled pores or channels in the capillary wall. Brain/blood ratios of 14C-sucrose were found to be 2–5 times higher than normal after 15 or 30 min of global ischemia plus 3 or 6 h of reperfusion in three of seven brain regions (Yoshizumi et al., 1993). Also measuring brain/blood ratios in a rat 2-h MCA-occlusion model, Rosenberg et al. (1998) found that the 14C-sucrose ratios were several times higher in the forebrain on the ischemic side than the contralateral side after 3 and 48 h of reperfusion but were similar on both sides at 6–24 h. Studies of neuroinflammatory mediators have indicated that signals for microvascular activation and angiogenesis are generated soon after the onset of stroke (Petty and Lo, 2002). Treatment strategies aimed at blocking BBB disruption or restoring BBB function are investigated to ameliorate stroke deficit (Borlongan et al., 2004), yet it is also important to fundamentally understand the biology of the intact BBB so that therapeutics in the future will be more appropriately targeted and more efficient.

Transplantation of mouse bone marrow stromal cells (BMSCs) intrastriatally following MCA occlusion resulted in alterations in BBB permeability (Borlongan et al., 2004). This study concluded that the robust effect of the BMSCs on the restoration of the BBB resulted from a secretion of trophic factors by the grafts, which in turn rescued viable, injured tissue in the stroke area (Borlongan et al., 2004). Reperfusion after stroke may lead to a sudden generation of reactive oxygen species and a hemodynamic surge, which can cause pronounced BBB leakage (Aoki et al., 2002). Recently, MMPs have been implicated in vascular matrix degradation and vascular integrity. MMP activity has been associated with free radical production at vascular sites following mouse focal cerebral ischemia, and MMP-9 knockout mice were protected against BBB disruption and edema after a transient cerebral ischemic event (Gasche et al., 2001; Asahi et al., 2001;

Aoki et al. 2002). Yang et al. recently demonstrated that MMPs (MMP-2 and -9) affected BBB permeability following reperfusion by degrading tight junction proteins occludin and claudin-5 (Yang et al., 2007). This effect was partially reversed following the administration of a general MMP inhibitor, which reduced the destruction of the tight junction proteins. Interestingly, they also found that in the later stage of BBB disruption, the tight junction proteins appeared in the astrocytes near the microvessels. As mentioned previously, it is thought that astrocytes influence the formation and regulation of brain ECs in the BBB. In this light, it is not surprising that astroglial cells also play critical regulatory roles following an ischemic event (Petty and Lo, 2002).

9 Conclusion

The intricate and complicated nature of the BBB renders it an invaluable characteristic of the brain and an extremely difficult structure to study. Under traumatic conditions, an increase in BBB permeability can be harmful, yet the "window" that exists due to the leakage allows for the passage of therapeutics. Other pathologies such as neurodegenerative diseases call for the impossible transport of drugs into the brain. An increasing amount of research is being dedicated to methods of drug delivery to breach the BBB (McCarty, 2005).

References

Abbott NJ, Ronnback L, Hansson E. 2006. Astrocyte-endothelial interactions at the blood-brain barrier. Nat Rev Neurosci 7(1): 41-53.

Albayrak S, Zhao Q, Siesjo BK, Smith ML. 1997. Effect of transient focal ischemia on blood-brain barrier permeability in the rat: Correlation to cell injury. Acta Neuropathol (Berl) 94(2): 158-163.

Amiry-Moghaddam M, Ottersen OP. 2003. The molecular basis of water transport in the brain. Nat Rev Neurosci 4(12): 991-1001.

Aoki T, Sumii T, Mori T, Wang X, Lo EH. 2002. Blood-brain barrier disruption and matrix metalloproteinase-9 expression during reperfusion injury: Mechanical versus embolic focal ischemia in spontaneously hypertensive rats. Stroke 33(11): 2711-2717.

Asahi M, Wang X, Mori T, Sumii T, Jung JC, et al. 2001. Effects of matrix metalloproteinase-9 gene knock-out on the proteolysis of blood-brain barrier and white matter components after cerebral ischemia. J Neurosci 21(19): 7724-7732.

Avraham HK, Lee TH, Koh Y, Kim TA, Jiang S, et al. 2003. Vascular endothelial growth factor regulates focal adhesion assembly in human brain microvascular endothelial cells through activation of the focal adhesion kinase and related adhesion focal tyrosine kinase. J Biol Chem 278(38): 36661-36668.

Bauer HC, Bauer H, Lametschwandtner A, Amberger A, Ruiz P, et al. 1993. Neovascularization and the appearance of morphological characteristics of the blood-brain barrier in the embryonic mouse central nervous system. Brain Res Dev Brain Res 75(2): 269-278.

Beck H, Acker T, Wiessner C, Allegrini PR, Plate KH. 2000. Expression of angiopoietin-1, angiopoietin-2, and tie receptors after middle cerebral artery occlusion in the rat. Am J Pathol 157(5): 1473-1483.

Betz AL. 1983. Sodium transport in capillaries isolated from rat brain. J Neurochem 41(4): 1150-1157.

Betz AL, Keep RF, Beer ME, Ren XD. 1994. Blood-brain barrier permeability and brain concentration of sodium, potassium, and chloride during focal ischemia. J Cereb Blood Flow Metab 14(1): 29-37.

Boado RJ, Li JY, Nagaya M, Zhang C, Pardridge WM. 1999. Selective expression of the large neutral amino acid transporter at the blood-brain barrier. Proc Natl Acad Sci USA 96(21): 12079-12084.

Borlongan CV, Lind JG, Dillon-Carter O, Yu G, Hadman M, et al. 2004. Bone marrow grafts restore cerebral blood flow and blood brain barrier in stroke rats. Brain Res 1010(1–2): 108-116.

Breslin JW, Pappas PJ, Cerveira JJ, Hobson RW, 2nd, Duran WN. 2003. VEGF increases endothelial permeability by separate signaling pathways involving ERK-1/2 and nitric oxide. Am J Physiol Heart Circ Physiol 284(1): H92-H100.

Brown RC, Davis TP. 2002. Calcium modulation of adherens and tight junction function: A potential mechanism for blood-brain barrier disruption after stroke. Stroke 33(6): 1706-1711.

Brown RC, Mark KS, Egleton RD, Huber JD, Burroughs AR, et al. 2003. Protection against hypoxia-induced increase in blood-brain barrier permeability: Role of tight junction proteins and NFkappaB. J Cell Sci 116(Pt 4): 693-700.

Cala PM, Maldonado HM. 1994. pH regulatory Na/H exchange by Amphiuma red blood cells. J Gen Physiol 103(6): 1035-1053.

Cecchelli R, Dehouck B, Descamps L, Fenart L, Buee-Scherrer VV, et al. 1999. In vitro model for evaluating drug transport across the blood-brain barrier. Adv Drug Deliv Rev 36(2–3): 165-178.

Chen C, Liu X, Smith BJ. 2003. Utility of Mdr1-gene deficient mice in assessing the impact of P-glycoprotein on pharmacokinetics and pharmacodynamics in drug discovery and development. Curr Drug Metab 4(4): 272-291.

Christensen HN. 1979. Developments in amino acid transport, illustrated for the blood-brain barrier. Biochem Pharmacol 28(13): 1989-1992.

Csanady L, Adam-Vizi V. 2003. Ca(2+)- and voltage-dependent gating of Ca(2+)- and ATP-sensitive cationic channels in brain capillary endothelium. Biophys J 85(1): 313-327.

Cutler L, Howes C, Deeks NJ, Buck TL, Jeffrey P. 2006. Development of a P-glycoprotein knockout model in rodents to define species differences in its functional effect at the blood-brain barrier. J Pharm Sci 95(9): 1944-1953.

Delpire E. 2000. Cation-Chloride Cotransporters in Neuronal Communication. News Physiol Sci 15:309–312.

Dermietzel R, Krause D, Kremer M, Wang C, Stevenson B. 1992. Pattern of glucose transporter (Glut 1) expression in embryonic brains is related to maturation of blood-brain barrier tightness. Dev Dyn 193(2): 152-163.

Dietrich WD, Prado R, Watson BD, Nakayama H. 1988. Middle cerebral artery thrombosis: Acute blood-brain barrier consequences. J Neuropathol Exp Neurol 47(4): 443-451.

Dolman D, Drndarski S, Abbott NJ, Rattray M. 2005. Induction of aquaporin 1 but not aquaporin 4 messenger RNA in rat primary brain microvessel endothelial cells in culture. J Neurochem 93(4): 825-833.

Dziegielewska KM, Hinds LA, Mollgard K, Reynolds ML, Saunders NR. 1988. Blood-brain, blood-cerebrospinal fluid and cerebrospinal fluid-brain barriers in a marsupial (Macropus eugenii) during development. J Physiol 403: 367-388.

Engelhardt B. 2003. Development of the blood-brain barrier. Cell Tissue Res 314(1): 119-129.

Ennis SR, Keep RF, Schielke GP, Betz AL. 1990. Decrease in perfusion of cerebral capillaries during incomplete ischemia and reperfusion. J Cereb Blood Flow Metab 10(2): 213-220.

Ennis SR, Ren XD, Betz AL. 1996. Mechanisms of sodium transport at the blood-brain barrier studied with in situ perfusion of rat brain. J Neurochem 66(2): 756-763.

Farkas A, Szatmari E, Orbok A, Wilhelm I, Wejksza K, et al. 2005. Hyperosmotic mannitol induces Src kinase-dependent phosphorylation of beta-catenin in cerebral endothelial cells. J Neurosci Res 80(6): 855-861.

Farkas E, Sule Z, Toth-Szuki V, Matyas A, Antal P, et al. 2006. Tumor necrosis factor-alpha increases cerebral blood flow and ultrastructural capillary damage through the release of nitric oxide in the rat brain. Microvasc Res 72(3): 113-119.

Finnie JW, Blumbergs PC, Cai Z, Manavis J, Kuchel TR. 2006. Neonatal mouse brain exposure to mobile telephony and effect on blood-brain barrier permeability. Pathology 38(3): 262-263.

Fischer S, Clauss M, Wiesnet M, Renz D, Schaper W, et al. 1999. Hypoxia induces permeability in brain microvessel endothelial cells via VEGF and NO. Am J Physiol 276(4 Pt 1): C812-820.

Fischer S, Wiesnet M, Marti HH, Renz D, Schaper W. 2004. Simultaneous activation of several second messengers in hypoxia-induced hyperpermeability of brain derived endothelial cells. J Cell Physiol 198(3): 359-369.

Fischer S, Wobben M, Marti HH, Renz D, Schaper W. 2002. Hypoxia-induced hyperpermeability in brain microvessel endothelial cells involves VEGF-mediated changes in the expression of zonula occludens-1. Microvasc Res 63(1): 70-80.

Franzen B, Duvefelt K, Jonsson C, Engelhardt B, Ottervald J, et al. 2003. Gene and protein expression profiling of human cerebral endothelial cells activated with tumor necrosis factor-alpha. Brain Res Mol Brain Res 115(2): 130-146.

Garcia JH, Liu KF, Yoshida Y, Lian J, Chen S, et al. 1994. Influx of leukocytes and platelets in an evolving brain infarct (Wistar rat). Am J Pathol 144(1): 188-199.

Gasche Y, Copin JC, Sugawara T, Fujimura M, Chan PH. 2001. Matrix metalloproteinase inhibition prevents oxidative stress-associated blood-brain barrier disruption after transient focal cerebral ischemia. J Cereb Blood Flow Metab 21(12): 1393-1400.

Gee JR, Keller JN. 2005. Astrocytes: regulation of brain homeostasis via apolipoprotein E. Int J Biochem Cell Biol 37(6): 1145-1150.

Hamm S, Dehouck B, Kraus J, Wolburg-Buchholz K, Wolburg H, et al. 2004. Astrocyte mediated modulation of blood-brain barrier permeability does not correlate with a loss of tight junction proteins from the cellular contacts. Cell Tissue Res 315(2): 157-166.

Hansen AJ, Lund-Andersen H, Crone C. 1977. K+-permeability of the blood-brain barrier, investigated by aid of a K+-sensitive microelectrode. Acta Physiol Scand 101(4): 438-445.

Harik SI, Doull GH, Dick AP. 1985. Specific ouabain binding to brain microvessels and choroid plexus. J Cereb Blood Flow Metab 5(1): 156-160.

Hawkins RA, O'Kane RL, Simpson IA, Vina JR. 2006. Structure of the blood-brain barrier and its role in the transport of amino acids. J Nutr 136(1 Suppl): 218S-226S.

Hayashi T, Noshita N, Sugawara T, Chan PH. 2003. Temporal profile of angiogenesis and expression of related genes in the brain after ischemia. J Cereb Blood Flow Metab 23(2): 166-180.

Hayashi Y, Nomura M, Yamagishi S, Harada S, Yamashita J, et al. 1997. Induction of various blood-brain barrier properties in non-neural endothelial cells by close apposition to co-cultured astrocytes. Glia 19(1): 13-26.

Higashi K, Fujita A, Inanobe A, Tanemoto M, Doi K, et al. 2001. An inwardly rectifying K(+) channel, Kir4.1, expressed in astrocytes surrounds synapses and blood vessels in brain. Am J Physiol Cell Physiol 281(3): C922-931.

Himwich WA. 1970. Developmental Neurobiology. Journal title 770.

Hippenstiel S, Krull M, Ikemann A, Risau W, Clauss M, et al. 1998. VEGF induces hyperpermeability by a direct action on endothelial cells. Am J Physiol 274(5 Pt 1): L678-684.

Hong H, Lu Y, Ji ZN, Liu GQ. 2006. Up-regulation of P-glycoprotein expression by glutathione depletion-induced oxidative stress in rat brain microvessel endothelial cells. J Neurochem 98(5): 1465-1473.

Hoyer J, Popp R, Meyer J, Galla HJ, Gogelein H. 1991. Angiotensin II, vasopressin and GTP[gamma-S] inhibit inward-rectifying K+ channels in porcine cerebral capillary endothelial cells. J Membr Biol 123(1): 55-62.

Ishida H, Takemori K, Dote K, Ito H. 2006. Expression of glucose transporter-1 and aquaporin-4 in the cerebral cortex of stroke-prone spontaneously hypertensive rats in relation to the blood-brain barrier function. Am J Hypertens 19(1): 33-39.

Jin JS, Sakaeda T, Kakumoto M, Nishiguchi K, Nakamura T, et al. 2006. Effect of Therapeutic Moderate Hypothermia on Multi-drug Resistance Protein 1-Mediated Transepithelial Transport of Drugs. Neurol Med Chir (Tokyo) 46(7): 321-327.

Jung JS, Bhat RV, Preston GM, Guggino WB, Baraban JM, et al. 1994. Molecular characterization of an aquaporin cDNA from brain: Candidate osmoreceptor and regulator of water balance. Proc Natl Acad Sci USA 91(26): 13052-13056.

Keep RF, Ulanski LJ, 2nd, Xiang J, Ennis SR, Lorris Betz A. 1999. Blood-brain barrier mechanisms involved in brain calcium and potassium homeostasis. Brain Res 815(2): 200-205.

Kniesel U, Risau W, Wolburg H. 1996. Development of blood-brain barrier tight junctions in the rat cortex. Brain Res Dev Brain Res 96(1–2): 229-240.

Krum JM, Khaibullina A. 2003. Inhibition of endogenous VEGF impedes revascularization and astroglial proliferation: Roles for VEGF in brain repair. Exp Neurol 181(2): 241-257.

Lal BK, Varma S, Pappas PJ, Hobson RW 2nd, Duran WN. 2001. VEGF increases permeability of the endothelial cell monolayer by activation of PKB/akt, endothelial nitric-oxide synthase, and MAP kinase pathways. Microvasc Res 62(3): 252-262.

Lee WJ, Hawkins RA, Vina JR, Peterson DR. 1998. Glutamine transport by the blood-brain barrier: A possible mechanism for nitrogen removal. Am J Physiol 274(4 Pt 1): C1101-1107.

Manley GT, Fujimura M, Ma T, Noshita N, Filiz F, et al. 2000. Aquaporin-4 deletion in mice reduces brain edema after acute water intoxication and ischemic stroke. Nat Med 6(2): 159-163.

Maurer MH, Geomor HK, Burgers HF, Schelshorn DW, Kuschinsky W. 2006. Adult neural stem cells express glucose transporters GLUT1 and GLUT3 and regulate GLUT3 expression. FEBS Lett 580(18): 4430-4434.

Mayhan WG. 1999. VEGF increases permeability of the blood-brain barrier via a nitric oxide synthase/cGMP-dependent pathway. Am J Physiol 276(5 Pt 1): C1148-1153.

Mayhan WG. 2002. Cellular mechanisms by which tumor necrosis factor-alpha produces disruption of the blood-brain barrier. Brain Res 927(2): 144-152.

McCarty JH. 2005. Cell biology of the neurovascular unit: Implications for drug delivery across the blood-brain barrier. Assay Drug Dev Technol 3(1): 89-95.

Menzies SA, Betz AL, Hoff JT. 1993. Contributions of ions and albumin to the formation and resolution of ischemic brain edema. J Neurosurg 78(2): 257-266.

Nielsen S, Nagelhus EA, Amiry-Moghaddam M, Bourque C, Agre P, et al. 1997. Specialized membrane domains for water transport in glial cells: High-resolution immunogold cytochemistry of aquaporin-4 in rat brain. J Neurosci 17(1): 171-180.

Ningaraj NS, Rao M, Hashizume K, Asotra K, Black KL. 2002. Regulation of blood-brain tumor barrier permeability by calcium-activated potassium channels. J Pharmacol Exp Ther 301(3): 838-851.

Oldendorf WH. 1971. Uptake of radiolabeled essential amino acids by brain following arterial injection. Proc Soc Exp Biol Med 136(2): 385-386.

Paemeleire K. 2002. Calcium signaling in and between brain astrocytes and endothelial cells. Acta Neurol Belg 102(3): 137-140.

Pan W, Kastin AJ. 2002. TNFalpha transport across the blood-brain barrier is abolished in receptor knockout mice. Exp Neurol 174(2): 193-200.

Perry VH, Anthony DC, Bolton SJ, Brown HC. 1997. The blood-brain barrier and the inflammatory response. Mol Med Today 3(8): 335-341.

Petty MA, Lo EH. 2002. Junctional complexes of the blood-brain barrier: Permeability changes in neuroinflammation. Prog Neurobiol 68(5): 311-323.

Preston E, Foster DO. 1997. Evidence for pore-like opening of the blood-brain barrier following forebrain ischemia in rats. Brain Res 761(1): 4-10.

Proescholdt MA, Heiss JD, Walbridge S, Muhlhauser J, Capogrossi MC, et al. 1999. Vascular endothelial growth factor (VEGF) modulates vascular permeability and inflammation in rat brain. J Neuropathol Exp Neurol 58(6): 613-627.

Reuss B, Dono R, Unsicker K. 2003. Functions of fibroblast growth factor (FGF)-2 and FGF-5 in astroglial differentiation and blood-brain barrier permeability: Evidence from mouse mutants. J Neurosci 23(16): 6404-6412.

Rieckmann P, Engelhardt B. 2003. Building up the blood-brain barrier. Nat Med 9(7): 828-829.

Rist RJ, Romero IA, Chan MW, Couraud PO, Roux F, et al. 1997. F-actin cytoskeleton and sucrose permeability of immortalised rat brain microvascular endothelial cell monolayers: Effects of cyclic AMP and astrocytic factors. Brain Res 768(1–2): 10-18.

Rosenberg GA, Estrada EY, Dencoff JE. 1998. Matrix metalloproteinases and TIMPs are associated with blood-brain barrier opening after reperfusion in rat brain. Stroke 29(10): 2189-2195.

Rubin LL, Staddon JM. 1999. The cell biology of the blood-brain barrier. Annu Rev Neurosci 22(11–28.

Saadoun S, Papadopoulos MC, Davies DC, Krishna S, Bell BA. 2002. Aquaporin-4 expression is increased in oedematous human brain tumours. J Neurol Neurosurg Psychiatry 72(2): 262-265.

Sanchez del Pino MM, Hawkins RA, Peterson DR. 1992. Neutral amino acid transport by the blood-brain barrier. Membrane vesicle studies. J Biol Chem 267(36): 25951-25957.

Schielke GP, Moises HC, Betz AL. 1991. Blood to brain sodium transport and interstitial fluid potassium concentration during early focal ischemia in the rat. J Cereb Blood Flow Metab 11(3): 466-471.

Schoch HJ, Fischer S, Marti HH. 2002. Hypoxia-induced vascular endothelial growth factor expression causes vascular leakage in the brain. Brain 125(Pt 11): 2549-2557.

Sibson NR, Blamire AM, Perry VH, Gauldie J, Styles P, et al. 2002. TNF-alpha reduces cerebral blood volume and disrupts tissue homeostasis via an endothelin- and TNFR2-dependent pathway. Brain 125(Pt 11): 2446-2459.

Sipos H, Torocsik B, Tretter L, Adam-Vizi V. 2005. Impaired regulation of pH homeostasis by oxidative stress in rat brain capillary endothelial cells. Cell Mol Neurobiol 25(1): 141-151.

Smith QR. 2000. Transport of glutamate and other amino acids at the blood-brain barrier. J Nutr 130(4S Suppl): 1016S-1022S.

Somjen GG. 2002. Ion regulation in the brain: Implications for pathophysiology. Neuroscientist 8(3): 254-267.

Stewart PA, Hayakawa EM. 1987. Interendothelial junctional changes underlie the developmental 'tightening' of the blood-brain barrier. Brain Res 429(2): 271-281.

Stummer W, Keep RF, Betz AL. 1994. Rubidium entry into brain and cerebrospinal fluid during acute and chronic alterations in plasma potassium. Am J Physiol 266(6 Pt 2): H2239-2246.

Stummer W, Betz AL, Keep RF. 1995. Mechanisms of brain ion homeostasis during acute and chronic variations of plasma potassium. J Cereb Blood Flow Metab 15(2): 336-344.

Taniguchi M, Yamashita T, Kumura E, Tamatani M, Kobayashi A, et al. 2000. Induction of aquaporin-4 water channel mRNA after focal cerebral ischemia in rat. Brain Res Mol Brain Res 78(1–2): 131-137.

Taylor CJ, Nicola PA, Wang S, Barrand MA, Hladky SB. 2006. Transporters involved in regulation of intracellular pH in primary cultured rat brain endothelial cells. J Physiol 576(Pt 3): 769-785.

Toornvliet R, van Berckel BN, Luurtsema G, Lubberink M, Geldof AA, et al. 2006. Effect of age on functional P-glycoprotein in the blood-brain barrier measured by use of (R)-[(11)C]verapamil and positron emission tomography. Clin Pharmacol Ther 79(6): 540-548.

Ubogu EE, Callahan MK, Tucky BH, Ransohoff RM. 2006. Determinants of CCL5-driven mononuclear cell migration across the blood-brain barrier. Implications for therapeutically modulating neuroinflammation. J Neuroimmunol 179(1–2): 132-144.

Ubogu EE, Cossoy MB, Ransohoff RM. 2006. The expression and function of chemokines involved in CNS inflammation. Trends Pharmacol Sci 27(1): 48-55.

Vigne P, Champigny G, Marsault R, Barbry P, Frelin C, et al. 1989. A new type of amiloride-sensitive cationic channel in endothelial cells of brain microvessels. J Biol Chem 264(13): 7663-7668.

Virgintino D, Errede M, Robertson D, Capobianco C, Girolamo F, et al. 2004. Immunolocalization of tight junction proteins in the adult and developing human brain. Histochem Cell Biol 122(1): 51-59.

Virgintino D, Errede M, Robertson D, Girolamo F, Masciandaro A, et al. 2003. VEGF expression is developmentally regulated during human brain angiogenesis. Histochem Cell Biol 119(3): 227-232.

Virgintino D, Robertson D, Errede M, Benagiano V, Girolamo F, et al. 2002. Expression of P-glycoprotein in human cerebral cortex microvessels. J Histochem Cytochem 50(12): 1671-1676.

Vorbrodt AW. 1988. Ultrastructural cytochemistry of blood-brain barrier endothelia. Prog Histochem Cytochem 18(3): 1-99.

Vorbrodt AW, Lossinsky AS, Wisniewski HM. 1982. Cytochemical localization of ouabain-sensitive, K+-dependent p-nitro-phenylphosphatase (transport ATPase) in the mouse central and peripheral nervous systems. Brain Res 243(2): 225-234.

Walz W, Hertz L. 1983. Functional interactions between neurons and astrocytes. II. Potassium homeostasis at the cellular level. Prog Neurobiol 20(1–2): 133-183.

Wei L, Boyle MP, Yu SP 2001. Roles of VEGF in angiogenesis and functional recovery after focal ischemia in mice. J Cereb Blood Flow Metab 21(2001) S315

Wei L, Erinjeri JP, Rovainen CM, Woolsey TA. 2001. Collateral growth and angiogenesis around cortical stroke. Stroke 32(9): 2179-2184.

Wei L, Yin K, Lee JM, Chao J, Yu SP, et al. 2003. Restorative potential of angiogenesis after ischemic stroke. Maiese, editor. Kenneth Neuronal and Vascular Plasticity, London: Kluwer Academic/Plenum Publishers; pp. 75–94.

Wolburg H, Lippoldt A. 2002. Tight junctions of the blood-brain barrier: Development, composition and regulation. Vascul Pharmacol 38(6): 323-337.

Yang GY, Gong C, Qin Z, Liu XH, Lorris Betz A. 1999. Tumor necrosis factor alpha expression produces increased blood-brain barrier permeability following temporary focal cerebral ischemia in mice. Brain Res Mol Brain Res 69(1): 135-143.

Yang T, Roder KE, Bhat GJ, Thekkumkara TJ, Abbruscato TJ. 2006. Protein kinase C family members as a target for regulation of blood-brain barrier Na,K,2Cl-cotransporter during in vitro stroke conditions and nicotine exposure. Pharm Res 23(2): 291-302.

Yang Y, Estrada EY, Thompson JF, Liu W, Rosenberg GA. 2007. Matrix metalloproteinase-mediated disruption of tight junction proteins in cerebral vessels is reversed by synthetic matrix metalloproteinase inhibitor in focal ischemia in rat. J Cereb Blood Flow Metab.

Yoshizumi H, Fujibayashi Y, Kikuchi H. 1993. A new approach to the integrity of dual blood-brain barrier functions of global ischemic rats. Barrier and carrier functions. Stroke 24(2): 279-284; discussion 284–275.

Young PP, Fantz CR, Sands MS. 2004. VEGF disrupts the neonatal blood-brain barrier and increases life span after non-ablative BMT in a murine model of congenital neurodegeneration caused by a lysosomal enzyme deficiency. Exp Neurol 188(1): 104-114.

Zhang ZG, Zhang L, Jiang Q, Zhang R, Davies K, et al. 2000. VEGF enhances angiogenesis and promotes blood-brain barrier leakage in the ischemic brain. J Clin Invest 106(7): 829-838.

Zhang ZG, Zhang L, Tsang W, Soltanian-Zadeh H, Morris D, et al. 2002. Correlation of VEGF and angiopoietin expression with disruption of blood-brain barrier and angiogenesis after focal cerebral ischemia. J Cereb Blood Flow Metab 22(4): 379-392.

15 Phospholipase A₂ in CNS Disorders: Implication on Traumatic Spinal Cord and Brain Injuries

N.-K. Liu · W. Titsworth · X.-M. Xu

1	Introduction	322
2	Classification, Structure, and Properties of PLA$_2$	323
2.1	sPLA$_2$	323
2.2	cPLA$_2$	325
2.3	iPLA$_2$	326
2.4	PAF-AH	326
3	PLA$_2$ Distribution in the Mammalian CNS	326
3.1	sPLA$_2$	326
3.2	cPLA$_2$	327
3.3	iPLA$_2$	327
3.4	PAF-AH	328
4	PLA$_2$ in Neurotrauma	328
4.1	Traumatic Spinal Cord Injury	328
4.2	Traumatic Brain Injury	330
4.3	Ischemia	332
5	Mechanisms Underlying PLA$_2$-Mediated CNS Injuries	332
5.1	Membrane Breakdown	333
5.2	Inflammation	334
5.3	Oxidation	334
5.4	Excitatory Neurotoxicity	334
5.5	Apoptosis	334
6	Conclusion and Future Directions	335

© 2009 Springer Science+Business Media, LLC.

Abstract: Phospholipases A_2 (PLA$_2$s) are a diverse family of lipolytic enzymes which hydrolyze the acyl bond at the *sn*-2 position of glycerophospholipids to produce free fatty acids and lysophospholipids. These products are precursors of bioactive eicosanoids and platelet-activating factor (PAF). The hydrolysis of membrane phospholipids by PLA$_2$ is a rate-limiting step for generation of eicosanoids and PAF. To date, more than 27 isoforms of PLA$_2$ have been found in the mammalian system which can be classified into four major categories: secretory PLA$_2$ (sPLA$_2$), cytosolic PLA$_2$ (cPLA$_2$), Ca^{2+}-independent PLA$_2$ (iPLA$_2$), and PAF-acetylhydrolases (PAF-AH). Multiple isoforms of PLA$_2$ are found in the mammalian central nervous system (CNS) including the brain and spinal cord. Under physiological conditions, PLA$_2$s are involved in diverse cellular responses, including phospholipid digestion and metabolism, host defense, and signal transduction. However, under pathological situations, increased PLA$_2$ activity and excessive production of free fatty acids and their metabolites may lead to the loss of membrane integrity, inflammation, oxidative stress, and subsequent neuronal injury. There is emerging evidence that PLA$_2$ plays a key role in the secondary injury process after traumatic or ischemic injuries in the brain and spinal cord. Importantly, PLA$_2$ may act as a convergence molecule that mediates multiple key mechanisms of the secondary injury since it can be induced by multiple toxic factors such as inflammatory cytokines, free radicals, and excitatory amino acid, and its activation and metabolites can exacerbate the secondary injury. Blocking PLA$_2$ action may represent a novel and efficient strategy to block multiple injury pathways associated with the CNS secondary injury. This review outlines the current knowledge of the PLA$_2$ in the CNS with an emphasis placed on the possible roles of PLA$_2$ in mediating CNS injuries particularly the traumatic and ischemic injuries in the brain and spinal cord.

List of Abbreviations: AACOCF$_3$, arachidonyltrifluoromethyl ketone; AA, arachidonic acid; AMPA, α-amino-3-hydroxy-5-methylisoxazole-4-propionate; CNQX, 6-cyano-7-nitroquinoxaline-2,3-dione; CNS, central nervous system; cPLA$_2$, cytosolic PLA$_2$; CSF, cerebrospinal fluid; DHA, docosahexaenoic acid; EAA, excitatory amino acid; EGF, epidermal growth factor; EPA, eicosapentaenoic acid; 4-HNE, 4-hydroxynonenal; IFN-γ, interferon-γ; IL-1, interleukin-1; iPLA$_2$, Ca^{2+}-independent PLA$_2$; KA, kainic acid; LPS, lipopolysaccharide; lyso-PC, lysophosphatidylcholine; MAPK, mitogen-activated protein kinase; M-CSF, macrophage-colony stimulating factor; NMDA, *N*-methy-D-aspartate; OGD, oxygen and glucose deprivation; PAF-AH, platelet-activating factor acetylhydrolases; PAF, platelet-activating factor; PDGF, platelet-derived growth factor; PGE$_2$, prostaglandin E$_2$; PKC, protein kinase C; PLA$_2$, Phospholipase A$_2$; PlsEtn, plasmenylethanolamine; ROS, reactive oxygen species; SCI, spinal cord injury; sPLA$_2$, secretory PLA$_2$; TBI, traumatic brain injury; TGF-β, transforming growth factor-β; TNF-α, tumor necrosis factor-α; TXA$_2$, thromboxane A$_2$

1 Introduction

Phospholipases A_2 (PLA$_2$s) are a diverse family of lipolytic enzymes which hydrolyze the acyl bond at the *sn*-2 position of glycerophospholipids to produce free fatty acids and lysophospholipids (Farooqui et al., 1997; Murakami et al., 1997; Kudo and Murakami, 2002) (❱ *Figure 15-1*). These products are precursors of bioactive eicosanoids and platelet-activating factor (PAF), which are well-known mediators of inflammation and tissue damage implicated in pathological states of numerous acute and chronic neurological disorders including spinal cord injury (SCI) and traumatic brain injury (TBI) (Bonventre, 1996; Farooqui et al., 1997, 2006; Yedgar et al., 2006). The hydrolysis of membrane phospholipids by PLA$_2$ is a rate-limiting step for generation of eicosanoids and PAF (Farooqui et al., 1997, 1999). Stimulation of PLA$_2$ is thought to be an important event in production of lipid inflammatory mediators. Under physiological conditions, PLA$_2$s are involved in diverse cellular responses, including phospholipid digestion and metabolism, host defense, and signal transduction. However, in pathological situations, increased PLA$_2$ activity and excessive production of free fatty acids such as arachidonic acid (AA), and proinflammatory mediators such as eicosanoids and PAF, may lead to the loss of membrane integrity, inflammation, oxidative stress, and subsequent neuronal injury (Bazan et al., 1995; Farooqui et al., 1997, 2006; Murakami et al., 1997; Phillis and O'Regan, 2004). This review outlines the current knowledge of the PLA$_2$ in the central nervous system (CNS) with an emphasis placed on the possible role of PLA$_2$ in mediating CNS injuries following ischemia and trauma.

Figure 15-1
Effect of PLA$_2$ on the metabolic pathway of lipid mediator production. AA, arachidonic acid; EPA, eicosapentaenoic acid; DHA, docosahexaenoic acid; lyso-PA, lysophosphatidic acid; lyso-PC, lysophosphatidylcholine; PAF, platelet-activating factor

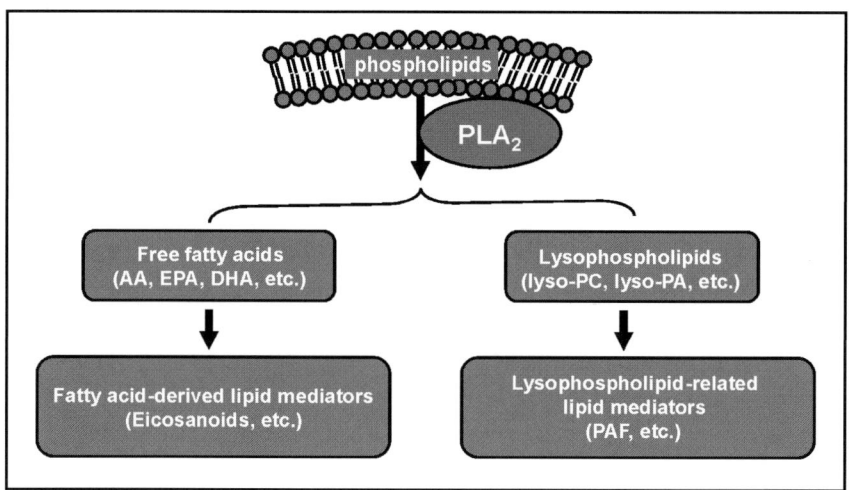

2 Classification, Structure, and Properties of PLA$_2$

To date, more than 27 isoforms of PLA$_2$s have been found in the mammalian system which can be classified into four major categories: secretory PLA$_2$ (sPLA$_2$), cytosolic PLA$_2$ (cPLA$_2$), Ca^{2+}-independent PLA$_2$ (iPLA$_2$), and PAF acetylhydrolases (PAF-AH) (Murakami et al., 1997; Kudo and Murakami, 2002; Murakami and Kudo, 2002; Schaloske and Dennis, 2006) (❯ *Table 15-1*). sPLA$_2$s, in which ten isozymes have been identified, have a low molecular mass of about 14-18 kDa and require the presence of submillimolar to millimolar concentrations of Ca^{2+} for effective hydrolysis of a substrate phospholipid without any fatty acid selectivity (Mayer and Marshall, 1993; Dennis, 1994; Murakami et al., 1995; Schaloske and Dennis, 2006). Members of the cPLA$_2$ isoform have a higher molecular mass (85-110 kDa) and selectively hydrolyze phospholipids containing AA and require a submicromolar concentration of Ca^{2+} for optimal activity (Dennis, 1994; Clark et al., 1995; Murakami et al., 1997). cPLA$_2$s consist of six isoforms, among which cPLA$_2\alpha$ plays an essential role in the initiation of AA metabolism. Intracellular activation of cPLA$_2\alpha$ is tightly regulated by Ca^{2+} and phosphorylation (Clark et al., 1995; Schaloske and Dennis, 2006). iPLA$_2$, containing seven isozymes, are intracellular enzymes with higher molecular mass ranging from 28 to 91 kDa that show no fatty acid selectivity and do not require Ca^{2+} for its activity. iPLA$_2$ is generally considered as a housekeeping enzyme for the maintenance of membrane phospholipids (Murakami et al., 1997; Murakami and Kudo, 2002; Schaloske and Dennis, 2006). The PAF-AH family represents a unique group of PLA$_2$ that contains four enzymes exhibiting unusual substrate specificity toward PAF and/or oxidized phospholipids (Murakami et al., 1997; Murakami and Kudo, 2002; Schaloske and Dennis, 2006).

2.1 sPLA$_2$

The mammalian sPLA$_2$s can be divided into at least six groups containing ten isoforms (IB, IIA, IIC, IID, IIE, IIF, III, V, X, and XIII). They are low-molecular-weight (14-18 kDa), structurally related isozymes that contain an N-terminal secretion signal peptide (Mayer and Marshall, 1993; Dennis, 1994; Murakami et al., 1995; Schaloske and Dennis, 2006). These isozymes are synthesized intracellularly and then secreted into the

Table 15-1
A summary of mammalian PLA$_2$ enzymes

Families	Group	Other name	Size (kD)	Ca^{2+} requirement	Catalytic site	sn-2 FA Preference	CNS location	Human chromosome
sPLA$_2$	IB	Pancreatic PLA$_2$	14	mM	His/Asp dyad	No	Brain, SC	12q23-24
	IIA	Synovial PLA$_2$	14	mM	His/Asp dyad	No	Brain, SC	1p34-36
	IIC		15	mM	His/Asp dyad	No	Brain, SC	1p34-36
	IID		14	mM	His/Asp dyad	No	NRA	1p34-36
	IIE		14	mM	His/Asp dyad	No	Brain, NRA	1p34-36
	IIF		16	mM	His/Asp dyad	No	NRA	1p34-36
	III		55	mM	His/Asp dyad	No	NRA	22q
	V		14	mM	His/Asp dyad	No	Brain, SC	1p34-36
	X		14	mM	His/Asp dyad	No	Brain, NRA	16p12-13
	XII		19	mM	His/Asp dyad	No	NRA	4q25
cPLA$_2$	IVA	cPLA$_2\alpha$	85	µM	Ser/Asp dyad	AA	Brain, SC	1q25
	IVB	cPLA$_2\beta$	110	µM	Ser/Asp dyad	To be confirmed	Brain, NRA	15
	IVC	cPLA$_2\gamma$	60	None	Ser/Asp dyad	To be confirmed	Brain, NRA	19
	IVD	cPLA$_2\delta$	92-93	µM	Ser/Asp dyad	To be confirmed	Brain, NRA	15
	IVE	cPLA$_2\varepsilon$	100	µM	Ser/Asp dyad	To be confirmed	NRA	15
	IVF	cPLA$_2\zeta$	96	µM	Ser/Asp dyad	To be confirmed	NRA	15
iPLA$_2$	VIA1	iPLA$_2$	84-85	None	Ser/His/Asp triad	No	Brain, SC	22q13.1
	VIA2	iPLA$_2\beta$	88-90	None	Ser/His/Asp triad	No	Brain, SC	22q13.1
	VIB	iPLA$_2\gamma$	88-91	None	Ser/His/Asp triad	No	NRA	7q31
	VIC	iPLA$_2\delta$	146	None	Ser/His/Asp triad	No	NRA	NRA
	VID	iPLA$_2\varepsilon$	53	None	Ser/His/Asp triad	No	NRA	NRA
	VIE	iPLA$_2\zeta$	57	None	Ser/His/Asp triad	No	NRA	NRA
	VIF	iPLA$_2\eta$	28	None	Ser/His/Asp triad	No	NRA	NRA
PAF-AH	VIIA	Plasma PAF-AH	45	None	Ser/His/Asp triad	Acetate	NRA	ND
	VIIB	PAF-AH II	40	None	Ser/His/Asp triad	Acetate	Brain, NRA	N.D.
	VIIIA	PAF-AH-Iα_1 (Ib)	26	None	Ser/His/Asp triad	Acetate	Brain, NRA	ND
	VIIIB	PAF-AH-Iα_2 (Ib)	26	None	Ser/His/Asp triad	No	Brain, NRA	11q23

Source: Adopted mainly from Kudo and Murakami (2002), Schaloske and Dennis (2006), and Ghosh et al. (2006) with modifications and additions

Note: ND, not determined; NRA, no report available; SC, spinal cord

extracellular space and can act extracellularly. However, binding of sPLA$_2$s to heparin-sulfate proteoglycan allows for both internalization of the enzyme and membrane binding to varying extents (Murakami et al., 1999, 2001). They have a highly conserved catalytic domain and a Ca^{2+}-binding loop and require the presence of submillimolar to millimolar concentrations of Ca^{2+} for effective hydrolysis of a substrate phospholipid without any fatty acid selectivity (Murakami et al., 1995; Kudo and Murakami, 2002; Murakami and Kudo, 2002; Schaloske and Dennis, 2006). sPLA$_2$ binds to two types of cell surface receptors, namely the N type, identified in neurons, and the M type, identified in skeletal muscles, of sPLA$_2$ receptors although this nomenclature is merely academic since neither receptor is limited to these tissues and the expression has been shown widely for both types (Hanasaki and Arita, 2002). sPLA$_2$ IIA is a well-known sPLA$_2$ isoform which can be detected in large amount at various inflamed sites and in the plasma of patients with rheumatoid arthritis or septic shock, as well as in experimental animal models of inflammation (Murakami et al., 1995). The levels of sPLA$_2$ IIA in sera or exudate are well correlated with the severity of inflammatory diseases (Murakami and Kudo, 2002). Accordingly, sPLA$_2$ IIA is referred to as the "inflammatory PLA$_2$." The expression of sPLA$_2$ IIA is markedly induced by proinflammatory cytokines such as interleukin-1 (IL-1), tumor necrosis factor-α (TNF-α), and lipopolysaccharide (LPS) (Murakami et al., 1997; Kudo and Murakami, 2002). Anti-inflammatory glucocorticoids are some of the most potent suppressants of the induction of sPLA$_2$ IIA expression (Murakami et al., 1997; Kudo and Murakami, 2002). Transforming growth factor (TGF-β) and platelet-derived growth factor (PDGF) also inhibit sPLA$_2$ IIA expression (Murakami et al., 1997). sPLA$_2$s have been implicated in a number of biological processes, such as modification of eicosanoid generation, inflammation, host defense, and neurotoxicity (Murakami et al., 1995; Sun et al., 2004; Farooqui et al., 2006; Schaloske and Dennis, 2006).

2.2 cPLA$_2$

The mammalian cPLA$_2$s consist of six isoforms, cPLA$_2\alpha$, cPLA$_2\beta$, cPLA$_2\gamma$, cPLA$_2\delta$, cPLA$_2\varepsilon$, and cPLA$_2\xi$, which are classified into group IVA, IVB, IVC, IVD, IVE, and IVF, respectively (Ghosh et al., 2006; Kita et al., 2006; Schaloske and Dennis, 2006). These enzymes are high-molecular-weight, intracellular proteins (Murakami et al., 1997; Murakami and Kudo, 2002; Schaloske and Dennis, 2006). The genes for cPLA$_2\beta$, cPLA$_2\delta$, cPLA$_2\varepsilon$, and cPLA$_2\xi$ form a gene cluster on human chromosome 15 (chromosome 2 in mouse), whereas cPLA$_2\alpha$ and cPLA$_2\gamma$ map to human chromosomes 1 (mouse chromosome 1) and 19 (mouse chromosome 7), respectively. cPLA$_2\alpha$ is the most extensively studied enzyme of the group IV and prefers AA over other fatty acids. cPLA$_2\alpha$ has a Ca^{2+}-binding domain (CaLB) (also known as C2 domain) in the N-terminal portion, a catalytic domain at the C-terminal portion of the protein and multiple phosphorylation sites including two consensus sites (S505 and S727) for phosphorylation by mitogen-activated protein kinase (MAPK) family and a S515 site for Ca^{2+}/calmodulin. The catalytic action of cPLA$_2\alpha$ is Ca^{2+}-independent, but a submicromolar concentration of free intracellular Ca^{2+} ([Ca^{2+}]$_i$) is required for translocation of the enzyme from the cytosol to the nuclear or other cellular membranes. cPLA$_2\alpha$ activity is regulated by Ca2$^+$-dependent translocation, phosphorylation and an additional component involving G protein. Because of the presence of a Ca^{2+}-dependent phospholipid-binding domain at the N-terminal region, increased intracellular Ca^{2+} can cause translocation of cPLA$_2\alpha$ in a Ca^{2+}-dependent manner from cytosol to the membrane where its substrate, phospholipids, is localized (Clark et al., 1995; Murakami and Kudo, 2002). This step is essential for the activation of cPLA$_2\alpha$. The C-terminal region of cPLA$_2$ contains the phosphorylation site and catalytic site. The activation of cPLA$_2$ requires sustained dual phosphorylation of Ser-505 and Ser-727 by MAP kinase and protein kinase C (PKC) (Clark et al., 1995; Hirabayashi and Shimizu, 2000; Hirabayashi et al., 2004). Ca^{2+}/calmodulin kinase II has also been found to bind cPLA$_2\alpha$ and phosphorylate it on Ser-515, resulting in its activation independent of MAPK pathway (Murakami and Kudo, 2002). For long-time regulation, the expression of cPLA$_2\alpha$ is induced by a number of the proinflammatory cytokines such as IL-1, TNF-α, and interferon-γ (IFN-γ), and growth factors such as macrophage-colony stimulating factor (M-CSF), epidermal growth factor (EGF), and stem cell factor (Clark et al., 1995; Murakami et al., 1997). Conversely, expression of cPLA$_2\alpha$ can be suppressed by glucocorticoids (Murakami et al., 1997).

cPLA$_2\beta$ contains C2 domain and is Ca^{2+}-dependent (Pickard et al., 1999; Song et al., 1999) whereas cPLA$_2\gamma$ lacks the C2 domain and acts in a Ca^{2+}-independent manner (Underwood et al., 1998; Murakami and Kudo, 2002; Schaloske and Dennis, 2006). cPLA$_2\gamma$ appears to be constitutively associated with membranes (Tucker et al., 2005). cPLA$_2\beta$ appears to have only low intrinsic activity (Lucas and Dennis, 2004). Both cPLA$_2\beta$ and cPLA$_2\gamma$ show little specificity for the *sn*-2 fatty acid (Song et al., 1999; Stewart et al., 2002).

cPLA$_2\delta$, cPLA$_2\varepsilon$, and cPLA$_2\xi$ have been recently discovered and cloned from mammalian tissues (Murakami and Kudo, 2002; Ohto et al., 2005; Schaloske and Dennis, 2006). All of them have a similar primary structure: C2 domain on N-terminal side followed by a catalytic domain with the catalytic dyad S and D conserved (Ghosh et al., 2006; Kita et al., 2006). When transiently expressed in mammalian cells, they exhibited Ca^{2+}-dependent phospholipase activity (Ohto et al., 2005; Ghosh et al., 2006; Kita et al., 2006).

2.3 iPLA$_2$

iPLA$_2$s belong to group VI and are also high-molecular-weight, intracellular enzymes not requiring Ca^{2+} for their activities. These enzymes consist of six isoforms, VIA, VIB, VIC, VID, VIE, and VIF (Schaloske and Dennis, 2006). The most extensively studied enzyme of this group is iPLA$_2$ VIA (Larsson et al., 1998; Kudo and Murakami, 2002; Schaloske and Dennis, 2006). The human group VIA gene resides on chromosome 22q13.1. Group VIA isoform has several splice variants, all with ankyrin repeats (Tang et al., 1997; Larsson et al., 1998; Kudo and Murakami, 2002; Schaloske and Dennis, 2006). Group VIB is less well studied and lacks ankyrin repeats but consists of a signal motif for peroxisome localization (Mancuso et al., 2000; Kudo and Murakami, 2002; Schaloske and Dennis, 2006). iPLA$_2$s exhibit no specificity for AA-containing phospholipids (Kudo and Murakami, 2002; Murakami and Kudo, 2002). Although iPLA$_2$s are generally considered as a housekeeping enzyme for the maintenance of membrane phospholipids, recent studies suggest that iPLA$_2$s are involved in stimulus-coupled AA release (Murakami et al., 1997; Murakami and Kudo, 2002; Schaloske and Dennis, 2006). Group VIC, VID, VIE, and VIF are recently discovered novel iPLA$_2$ enzymes (van Tienhoven et al., 2002; Jenkins et al., 2004; Schaloske and Dennis, 2006).

2.4 PAF-AH

Two groups of serine PLA$_2$s (group VII and VIII) can hydrolyze the acetyl group from the *sn*-2 position of PAF and were therefore originally named PAF acetylhydrolases (PAF-AH) (Murakami et al., 1997; Schaloske and Dennis, 2006). They represent a family of C^{2+}-independent PLA$_2$s. One of these enzymes is secreted, namely the group VIIA PLA$_2$. This enzyme is also known as plasma PAF-AH or lipoprotein-associated PLA$_2$ (Lp-PLA$_2$). The two other enzymes, group VIIB and group VIII PLA$_2$ are intracellular enzymes (Schaloske and Dennis, 2006).

3 PLA$_2$ Distribution in the Mammalian CNS

Multiple isoforms of PLA$_2$s are found in the mammalian CNS including the brain (Lautens et al., 1998; Molloy et al., 1998; Kishimoto et al., 1999; Ong et al., 1999b) and spinal cord (Ong et al., 1999a; Samad et al., 2001). Additionally, they have been localized to neurons (Kishimoto et al., 1999; Ong et al., 1999a) and glial cells (Oka and Arita, 1991; Stephenson et al., 1994; Lautens et al., 1998; Kishimoto et al., 1999).

3.1 sPLA$_2$

sPLA$_2$s are present in all regions of mammalian brain. The highest activities of sPLA$_2$ are found in medulla oblongata, pons, and hippocampus; moderate activities in the hypothalamus, thalamus, and cerebral cortex; and lowest activities in the cerebellum and olfactory bulb (Thwin et al., 2003). Molloy et al. (1998) found that mRNAs for sPLA$_2$ IIA and IIC were expressed in all regions of normal rat brain using RT-PCR. sPLA$_2$ V mRNA was found at low levels in most areas of the brain, but at very high levels in the hippocampus (Molloy et al., 1998). sPLA$_2$ IB mRNA was not detected in the rat brain (Molloy et al., 1998). However, Kolko et al. (2005) reported that sPLA$_2$ IB mRNA was detected in the rat and human brain at very high levels as well as in neurons in primary cultures using various detection methods. The distribution of sPLA$_2$ IB seems to be mainly neuronal, with the highest abundance occurring in the cerebral cortex and hippocampus (Kolko et al., 2005). sPLA$_2$ IIA and V were also detected in the rat cerebellum using immunostaining and in situ hybridization methods (Shirai and Ito, 2004). sPLA$_2$ IIA is associated with the endoplasmic reticulum in perinuclear regions of Purkinje cell somata and sPLA$_2$ V was localized in Bergmann glia cells (Shirai and Ito, 2004). Recently, Kolko et al. (2006) found the presence of sPLA$_2$ IIE, V, and X in the rat brain as well as in neurons in primary cultures using RT-PCR, in situ hybridization and immunohistochemistry. The distribution of sPLA$_2$ IIE, V, and X seems to be mainly neuronal, with the highest abundance occurring in the cerebral cortex and hippocampus (Kolko et al., 2006).

In the spinal cord, sPLA$_2$ activity was detected in the normal rat spinal cord homogenate (Svensson et al., 2005). Western blot revealed that sPLA$_2$ IIA and V are expressed in the normal rat spinal cord (Svensson et al., 2005). mRNAs of sPLA$_2$s (IB, IIA, IIC, and V) are also detected in the normal rat spinal cord with RT-PCR (Lucas et al., 2005).

3.2 cPLA$_2$

cPLA$_2$s are also present in the brain and spinal cord. cPLA$_2$ activity is present throughout the rat brain (Yang et al., 1999). RT-PCR analysis revealed that cPLA$_2$ mRNA is present in the normal rat brain, with the highest levels in the brainstem, hippocampus, striatum, and midbrain as well as relatively low levels in the cerebellum (Molloy et al., 1998). Recently, Western blot analysis showed that cPLA$_2$ protein is expressed in the monkey brain (Weerasinghe et al., 2006). Immunohistochemical localization studies have indicated that cPLA$_2$ is mainly present in astrocytes of the gray matter of human brain but not in those of the whiter matter (Stephenson et al., 1994). However, some of immunohistochemical localization studies (Farooqui et al., 2000) have indicated that cPLA$_2$ is present not only in neurons in the normal rat brain (Ong et al., 1999b), but also in neurons and astrocytes in the kainate-lesioned brain (Sandhya et al., 1998). In vitro experiments demonstrated that cPLA$_2$ is present in cultured neurons and astrocytes (Luo et al., 1998). In addition, Northern blot and *in situ* hybridization analysis further demonstrated that the cPLA$_2$ mRNA is much more abundant in neurons than glial cells with higher levels in distinct regions of the rat brain such as the cerebral cortex, hippocampus, amygdala, several thalamic and hypothalamic nuclei, and cerebellum (Kishimoto et al., 1999).

cPLA$_2$α activity is uniformly distributed in various regions of the rat brain (Farooqui et al., 2000, 2006). From immunostaining and *in situ* hybridization studies, cPLA$_2$α is localized in somata and dendrites of Purkinje cells, whereas cPLA$_2$β is present in the granule cells of rat cerebellum (Shirai and Ito, 2004). More recently, the mRNAs for cPLA$_2$β and cPLA$_2$δ have been identified by RT-PCR in human brain tissues (Pickard et al., 1999; Song et al., 1999; Farooqui et al., 2000). Northern blot analysis showed that cPLA$_2$β mRNA is expressed ubiquitously and strongly in brain whereas cPLA2γ mRNA is expressed modestly in brain (Underwood et al., 1998; Pickard et al., 1999; Song et al., 1999).

cPLA$_2$ activity is present in the cytosolic fraction of the rat spinal cord (Farooqui et al., 2000). cPLA$_2$ immunoreactivity is found in the dorsal horn and motor neurons of the rat and monkey (Ong et al., 1999a), and also present in the rat spinal cord as shown by Western blot (Samad et al., 2001). Lucas et al. (2005) also found that cPLA$_2$ activity was present in the normal rat spinal cord homogenates and demonstrated that both the mRNA and protein of cPLA$_2$ are expressed in the normal rat spinal cord. We have recently shown that cPLA$_2$ protein is expressed in the normal rat spinal cord and is mainly localized in neurons and oligodendrocytes but not astrocytes in the normal rat spinal cord (Liu et al., 2006).

3.3 iPLA$_2$

The brain cytosolic fraction contains an 80-kDa iPLA$_2$ activity. iPLA$_2$ activity is present in all brain regions with the highest activity in striatum, hypothalamus, and hippocampus and is the dominant PLA$_2$ activity in the cytosolic fraction of rat brain throughout all postnatal ages (Yang et al., 1999). The gene encoding iPLA$_2$

has been identified in the rat brain and is strongly expressed in all brain regions (Molloy et al., 1998). In situ hybridization studies in the rat brain have indicated that iPLA$_2$ mRNA was ubiquitously expressed in the brain with strong signals in the cerebellum, olfactory bulb, hippocampal CA1-3, dentate gyrus, and brain stem, with moderate signals in the cerebral cortex, thalamus, and midbrain, and with weak signals in the amygdala and basal forebrain (Shirai and Ito, 2004). In the rat cerebellum, iPLA$_2$ mRNA was strongly expressed in the Purkinje cells and granule cells, and moderately to weakly in the molecular layer (Shirai and Ito, 2004). Rat cerebellum contains iPLA$_2$-1 and iPLA$_2$-2 or iPLA$_2$-3, but not iPLA$_2$-ankyrin-1 or iPLA$_2$-ankyrin-2 (Shirai and Ito, 2004). Both mRNA and protein of iPLA$_2\gamma$ (VIB) were detected in the brain (Kinsey et al., 2005).

Recently, Ong et al. (2005) reported in the normal monkey brain that iPLA$_2$ immunoreactivity was observed in structures derived from the telencephalon, including the cerebral neocortex, amygdala, hippocampus, caudate nucleus, putamen, and nucleus accumbens, whereas structures derived from the diencephalon, including the thalamus, hypothalamus, and globus pallidus were lightly labeled. The midbrain, vestibular, trigeminal and inferior olivary nuclei, and the cerebellar cortex were densely labeled. Immunoreactivity was observed on the nuclear envelope of neurons, and dendrites and axon terminals at the electron microscopic level. Western blot analysis showed higher levels of iPLA$_2$ protein in the cytosolic, than the nuclear fraction, but little or no protein was found in the membrane fraction. Similarly, subcellular fractionation studies of iPLA$_2$ activity in the rat brain cortical cell cultures showed greater enzymatic activity in the cytosolic, than the nuclear fraction, and the least activity in nonnuclear membranes.

iPLA$_2$ mRNA is constitutively expressed in the human spinal cord (Larsson Forsell et al., 1999) and is also present in the normal rat spinal cord (Svensson and Yaksh, 2002). Recently, Lucas et al. (2005) demonstrated that the activity and protein and mRNA expression of iPLA$_2$ are present in the normal rat spinal cord using iPLA$_2$ activity assay, RT-PCR, and Western blot methods.

3.4 PAF-AH

Bovine soluble fraction contains at least two types of PAF-AH, namely PAF-AH (I) and PAF-AH (II) (Hattori et al., 1993). Brain intracellular PAF-AH (I6, GVIII) is a tertiary G-protein-complex-like heterotrimeric enzyme which is composed of $\alpha 1$, $\alpha 2$, and β subunits and is implicated in stages of brain development such as the formation of the brain cortex. Northern blot confirm that all of $\alpha 1$, $\alpha 2$, and β mRNAs of PAF-AH (IB) were expression in the adult rat brain and in cultured neurons; only $\alpha 2$, and β mRNAs were expressed in cultured astroglia and microglia (Watanabe et al., 1998). All three subunits of PAF-AH (I, IB, GVIII) are expressed in the embryonic brain, whereas only the $\alpha 2$ and β subunits are detected in adult brain (Manya et al., 1998). PAF-AH activity was also found in the rat brain (Shmueli et al., 1999).

4 PLA$_2$ in Neurotrauma

Acute CNS traumatic injuries such as SCI and TBI trigger a secondary injury by multiple injury processes including ischemia, inflammation, free radical generation, and excitatory amino acid (EAA) release (Hall and Braughler, 1986; Young, 1993; Buki et al., 2000; Park et al., 2004), which have been shown to induce PLA$_2$ activation (McIntosh et al., 1998; Phillis and O'Regan, 2004; Farooqui et al., 2006). Overstimulation of PLA$_2$ may result in membrane phospholipid degradation, generation of proinflammatory mediators such as eicosanoids and PAF, formation of free radicals, and subsequent lipid peroxidation and neuronal cell death (Bonventre, 1996; Farooqui et al., 1997, 1999, 2006). Therefore, PLA$_2$ may act as a converging molecule in mediating the secondary injury process after the initial trauma.

4.1 Traumatic Spinal Cord Injury

Acute SCI triggers a secondary injury by multiple injury processes including inflammation, free radical-induced cell death and glutamate excitotoxicity (Hall and Braughler, 1986; Young, 1993; Buki et al., 2000; Park et al., 2004). Following SCI, inflammatory cells such as polymorphonuclear neutrophils (PMN), macrophages and lymphocytes quickly infiltrate into the traumatized cord, and proinflammatory cytokines such as TNF-α and IL-1β, and neurotoxic factors from leukocytes such as nitric oxide, hydrogen peroxide, and myeloperxidase are accumulated (Popovich et al., 1997; Wang et al., 1997; Carlson et al., 1998; Streit et al., 1998; Xu et al., 1998; Lee et al., 2000; Mautes et al., 2000; Heinecke, 2002). Free radical generation and lipid peroxidation were also found to be early events subsequent to SCI (Hall and Braughler, 1986; Hall, 1989; Braughler and Hall, 1992; Farooque et al., 1996; Springer et al., 1997). The EAAs such as glutamate and aspartate are released rapidly following SCI and their extracellular concentrations increased to neurotoxic levels within minutes post SCI (Panter et al., 1990; Liu et al., 1991; Farooque et al., 1996; Xu et al., 1998; Liu et al., 1999; McAdoo et al., 1999; Heinecke, 2002). The interplay between multiple harmful substances likely perpetuates a progressive course of secondary injury, resulting in cell death, axonal destruction, and functional loss.

A number of evidence suggests that the previously mentioned harmful substances generated in the injured cord might induce PLA_2 activation and/or expression. *In vitro* and *in vivo* experiments also indicate that increased PLA_2 activity and its metabolic products can in turn exacerbate inflammation (Bonventre, 1996; Murakami et al., 1997), oxidation (Bonventre, 1996; Murakami et al., 1997), and neurotoxicity (Clapp et al., 1995; Bonventre, 1996) suggesting that PLA_2 may serve as a common mediator in the progression of secondary SCI.

It has been demonstrated in several experimental SCI models that the degradation of membrane phospholipids, along with the generation of free fatty acids, eicosanoids, and lipid peroxides, increased following SCI (Demediuk et al., 1985; Demediuk et al., 1989; Murphy et al., 1994; Cherian et al., 1996; Lukacova et al., 1996), suggesting that PLA_2 activity increased following SCI. For example, within first few minutes of SCI, the presence of free fatty acids has increased in the grey matter as a result of PLA_2 activity and in the white matter to a lesser extent at a later time period (Demediuk et al., 1985; Faden et al., 1987). During the first minute of compression trauma to the spinal cord, 10% of the plasmenylethanolamine (PlsEtn) is lost with an overall loss of 18% found at 30 min after the compression injury (Horrocks et al., 1985). The hydrolysis of membrane phospholipids by PLA_2 is a rate-limiting step for generation of proinflammatory mediators such as eicosanoids and PAF (Farooqui et al., 1997; Farooqui et al., 1999). Eicosanoids such as thromboxane A_2 (TXA_2) and prostaglandin E_2 (PGE_2) also increased in the injured cord tissue within a few minutes after SCI (Tonai et al., 1999; Resnick et al., 2001). The metabolism of free fatty acid represents a source of reactive oxygen species (ROS). The generation of free fatty acids in SCI is closely associated with increases in free radical formation observed in the lesioned region of injured spinal cord (Hamada et al., 1996; Azbill et al., 1997). Application of pathophysiological concentrations of free fatty acids has been demonstrated in vitro to induce oxidative injury to spinal cord cell cultures (Toborek et al., 1999). Low concentrations of free fatty acid, AA, support cultured neurons to survive whereas higher concentrations are neurotoxic (Okuda et al., 1994). Neurotoxic effects of AA have also been observed in hippocampal neurons and cortical neurons (Okuda et al., 1994; Li et al., 1997) as well as oligodendrocytes (Wang et al., 2004). The bioactive eicosanoids from AA induced by PLA_2 have been implicated as mediators of secondary injury via a host of mechanisms.

Lysophosphatidylcholine (lyso-PC) and PAF are also metabolic products mediated by PLA_2. Injection of lyso-PC into the spinal cord causes demyelination as well as expression of a number of chemokines and cytokines (Ousman and David, 2000, 2001), which occurred in the injured cord after SCI. PAF levels have been shown to increase 20-fold after SCI induced by stroke (Lindsberg et al., 1990). Intrathecal administration of PAF leads to reduced spinal cord blood flow and motor deficits, an effect which can be blocked by the PAF receptor antagonist, WEB 2170 (Faden and Halt, 1992). Treatment with WEB 2170 after acute spinal cord contusion resulted in significant increases in white matter sparing as well as decreases in proinflammatory cytokine mRNA levels within the lesion epicenter (Hostettler and Carlson, 2002; Hostettler et al., 2002). Treatment of PAF receptor antagonist, BN52021 also improves behavioral function after SCI (Xiao et al., 1998). In vitro experiments showed that low concentrations of PAF resulted in neuronal differentiation and sprouting, while higher concentrations were neurotoxic (Kornecki and Ehrlich, 1988). PAF induced not only cultured neuronal death in a dose-dependent manner (Xu and Tao, 2004; Bate et al., 2007) but also death of oligodendrocytes and astrocytes (Hostettler et al., 2002).

It has been shown that the release of high levels of EAAs such as glutamate and aspartate in experimental SCI is an important mechanism inducing the secondary injury (Park et al., 2004). Growing evidence indicates that PLA$_2$ mediates EAA-induced neuronal death and tissue damage. Marked increases in PLA$_2$ activity and AA release have been reported after treatments of neuronal cultures with glutamate, N-methy-D-aspartate (NMDA) and kainic acid (KA) (Dumuis et al., 1988; Kim et al., 1995; Farooqui et al., 2001). This increased PLA$_2$ activity can be inhibited by a PLA$_2$ inhibitor, mepacrine as well as a KA/α-amino-3-hydroxy-5-methylisoxazole-4-propionate (AMPA) receptor antagonist, 6-cyano-7-nitroquinoxaline-2,3-dione (CNQX) (Farooqui et al., 2001). It has been hypothesized that the glutamate release induces NMDA receptor activation which, in turn, results in intracellular Ca^{2+} increase, stimulates PLA$_2$ activation, and eventually neuronal death (Klein, 2000). In addition, glutamate release in the spinal cord can be suppressed by PLA$_2$ inhibitors indomethacin by 40%, arachidonyltrifluoromethyl ketone (AACOCF$_3$) by 45%, and 4-bromophenacyl bromide by 36% suggesting that increased PLA$_2$-mediated EAA release is in a positive feedback manner (Sundstrom and Mo, 2002). Thus, the excessive stimulation of NMDA receptors as occurs in the spinal cord trauma may result in stimulation of PLA$_2$ activity leading to alterations in membrane composition, permeability, and fluidity leading to neuronal and glial cell death. Indeed, *in vivo* and *in vitro* experiments showed that exogenous administration of PLA$_2$ induced neuronal death and tissue damage (Clapp et al., 1995; Kolko et al., 1996, 1999). PLA$_2$ has been reported to mediate myelin breakdown and axonal degeneration (De et al., 2003).

Recently, we found that PLA$_2$ activity increased following SCI which peaked at 4 h postinjury and remained significantly increased at 7 days (Liu et al., 2006). The expression of cPLA$_2$, an important isotype of PLA$_2$, was also increased and peaked at 3 and 7 days postinjury. Immunohistochemical studies revealed that cPLA$_2$ immunoreactivity was also markedly increased in both the injured gray and white matter at 7 days after injury (❯ *Figure 15-2*). Immunofluorescence double labeling further confirmed that the increased cPLA$_2$ was localized mainly in neurons, swollen axons, oligodendrocytes, and a subpopulation of microglia. *In vitro* experiments showed that both sPLA$_2$ and melittin, an activator of endogenous PLA$_2$, induced spinal neuronal death in a dose-dependent manner, an effect that could be substantially reversed by mepacrine, a PLA$_2$ inhibitor. When sPLA$_2$ was directly microinjected into the normal rat spinal cord, it induced tissue damage, demyelination, and sustained impairment in motor function. Such sPLA$_2$-induced demyelination, however, could be effectively attenuated with mepacrine, a PLA$_2$ inhibitor. Injections of sPLA$_2$ also induced the expression of inflammatory cytokines TNF-α and IL-1β, as well as 4-hydroxynonenal (4-HNE), a product of lipid peroxidation and a marker of oxygen free radical-mediated membrane injury.

In summary, PLA$_2$ can be activated by several key injury mediators such as inflammatory cytokines, free radicals, and EAAs that have been shown to increase following traumatic SCI. Furthermore, the increases in PLA$_2$ activity can further increase inflammation, oxidation, and EAA release. This indicates that PLA$_2$ activation may play a central role in this positive feedback loop triggered by traumatic SCI resulting in neuronal and glial cell death, tissue damage, and corresponding electrophysiological and behavioral impairments. Thus, PLA$_2$ may acts as a convergence molecule that mediates multiple key mechanisms of the secondary injury. Blocking PLA$_2$ action may represent a novel and efficient strategy to block multiple injury pathways and the positive feedback loop.

4.2 Traumatic Brain Injury

TBI also triggers secondary or delayed cell death by multiple injury processes including ischemia, inflammation, elevated intracellular Ca^{2+}, generation of free radicals, and glutamate release, which have been shown to induce PLA$_2$ activation (McIntosh et al., 1998; Phillis and O'Regan, 2004; Farooqui et al., 2006). Like SCI, TBI-induced PLA$_2$ activation results in membrane phospholipid degradation, generation of proinflammatory mediators such as eicosanoids and PAF, formation of free radicals, and subsequent lipid peroxidation and cell death. Following a closed head injury in rats, PLA$_2$ was activated and PGE$_2$ was subsequently formed (Shohami et al., 1989). The release of free fatty acids and membrane phospholipid degradation due to PLA$_2$ activation were found in the brain following traumatic injury (Dhillon et al., 1994; Homayoun et al., 1997). An increase in free fatty acids in human cerebrospinal fluid (CSF) following brain

Figure 15-2

Distribution of cPLA$_2$ immunoreactivity (IR) in the spinal cords of sham-operated and spinal cord injured rats. *Left column*: In the spinal cord of sham-operated rats, a basal level of cPLA$_2$-IR in ventral horn neurons (arrows) and white matter glial cells (arrowheads) was found. *Right column*: In the spinal cord at 7 days post-injury, a marked increase in cPLA$_2$-IR was found in neurons (arrows) and glial cells (arrowheads) at the injury epicenter (0 mm) as well as 5 mm rostral (−5 mm; b) and caudal (+5 mm; f). Note that the gray matter (outlined with a dashed line) was completely degenerated at the injury epicenter. Scale bar=50 μm (a-f)

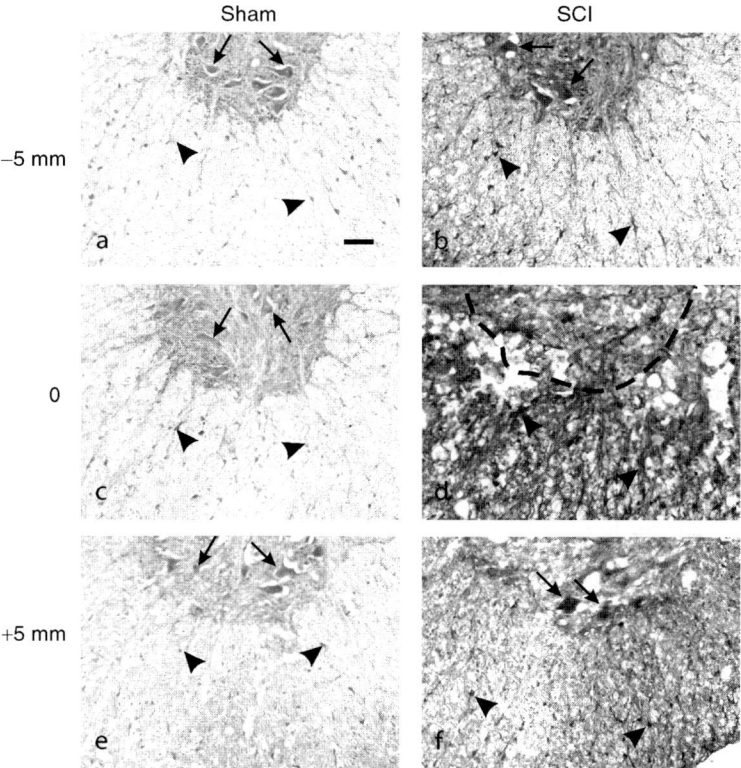

injury has been reported (Pilitsis et al., 2003). Upregulation of cPLA$_2$ and 4-HNE was found in the brain after transection which can be reversed by mepacrine, a PLA$_2$ inhibitor (Lu et al., 2001a).

Under both experimental and clinical settings, the level of extracellular EAAs such as glutamate and aspartate increased following TBI (Faden et al., 1989; Palmer et al., 1993; Globus et al., 1995; Bullock et al., 1998). It has been shown that both competitive and noncompetitive NMDA and non-NMDA receptor antagonists are efficacious in the treatment of experimental brain injury (Bullock and Fujisawa, 1992). Increased level of extracellular glutamate following TBI causes overstimulation of glutamate receptors that may result in secondary events such as PLA$_2$ activation, degradation of membrane phospholipids, and accumulation of free fatty acids, leading to neuronal cell death (McIntosh et al., 1998; Park et al., 2004).

It is possible that EAA-mediated PLA$_2$ activation and abnormal phospholipid metabolism may represent a common mechanism involved in traumatic spinal cord and brain injuries. Pronounced increases in prostaglandin F$_{2\alpha}$, prostaglandin D$_2$, leukotrienes, and thromboxane B$_2$ have been reported to occur in brain tissues after KA injection (Baran et al., 1987). Several studies showed that glutamate, NMDA and KA result in a dose-dependent increase in PLA$_2$ activity and AA release in brain neuronal cultures (Dumuis et al., 1988; Kim et al., 1995; Farooqui et al., 2001). An increase in cPLA$_2$ immunoreactivity in the rat hippocampus occurred following an intravenous administration of kainite and the increased cPLA$_2$ was mainly localized in neurons and astrocytes (Sandhya et al., 1998). Such a cPLA$_2$ increase was associated with an increase in lipid peroxidation as evidenced by accumulation of 4-HNE, a product of lipid peroxidation and marker for oxygen free radical-induced membrane injury (Lu et al., 2001a). Mepacrine, a PLA$_2$ inhibitor, blocked not only KA-induced cPLA$_2$ expression at both mRNA and protein levels, but also 4-HNE expression and neural damage in the hippocampus (Lu et al., 2001b; Ong et al., 2003). In addition, AACOCF3, a cPLA$_2$ specific inhibitor, also blocked KA-induced increase of cPLA$_2$ and 4-HNE (Lu et al., 2001b).

4.3 Ischemia

Ischemia is a key mechanism of the secondary injury after acute traumatic injuries (Tator, 1991; Tator and Fehlings, 1991; Gennarelli, 1993). Posttraumatic ischemia may result in energy failure that initiates a complex series of metabolic events, ultimately cause neuronal death. One such critical metabolic event is the activation of PLA$_2$ which can result in hydrolysis membrane phospholipids, release of free fatty acid, generation of oxygen free radicals, and formation of eicosanoids (Nakano et al., 1990; Phillis and O'Regan, 2004; Muralikrishna Adibhatla and Hatcher, 2006).

In both experimental models of brain (Yoshida et al., 1983; Abe et al., 1988; Nakano et al., 1990; Narita et al., 2000) and spinal cord ischemia (Halat et al., 1987), significant increases in the level of free fatty acids, indirectly reflecting PLA$_2$ activation, were found. Significant increases in PLA$_2$ activities were also reported *in vivo* following brain ischemia (Rordorf et al., 1991; Yagami et al., 2002a; Adibhatla et al., 2006) and *in vitro* in brain neurons cultured under ischemic conditions such as oxygen and glucose deprivation (OGD) (Arai et al., 2001). Biphasic increased expression of sPLA$_2$ IIA is observed in ischemic rat forebrain (Lauritzen et al., 1994). An early increase in sPLA$_2$ IIA mRNA occurred at 1-6 h post-ischemia and a late phase of greater induction of sPLA$_2$ IIA appeared between 7 and 20 days post-ischemia. Recently, increased expression of PLA$_2$ IIA has been confirmed at both mRNA and protein levels after brain ischemia (Lin et al., 2004; Adibhatla et al., 2006). Cytokines such as TNF-α and IL-1β have been shown to mediate the ischemia-induced PLA$_2$ activation and sPLA$_2$ IIA expression in transient focal rat cerebral ischemic model (Adibhatla et al., 2006). Indoxam, a specific sPLA$_2$ inhibitor, was shown to offer protection against the ischemia-induced damage (Yagami et al., 2002a). In vitro experiments showed that increased sPLA$_2$ activity was associated with ischemia-induced apoptosis (Yagami et al., 2002a).

It has been shown that inhibition of cPLA$_2$ attenuated OGD-induced neuronal death in an *in vitro* brain ischemic model, indicating the involvement of cPLA$_2$ in ischemic injury (Arai et al., 2001). *In vivo* studies have demonstrated that cPLA$_2$ activity and expression at both the mRNA and protein levels were significantly increased after brain ischemia (Clemens et al., 1996; Saluja et al., 1997; Stephenson et al., 1999). Several immunohistochemical studies showed that increased cPLA$_2$ was mainly localized in astrocytes and activated microglia in hippocampus (Clemens et al., 1996; Stephenson et al., 1999). The contribution of cPLA$_2$ to ischemic injury has also been demonstrated in cPLA$_2$ gene knockout mice (Bonventre et al., 1997; Sapirstein and Bonventre, 2000; Tabuchi et al., 2003). After a transient middle cerebral artery occlusion, cPLA$_2$ knockout mice develop smaller infarcts, less brain edema, and less neurological deficits than control mice, indicating a reduced susceptibility of cPLA$_2$ knockout mice to ischemic brain injuries.

In summary, ischemia induces PLA$_2$ activation which could result in deleterious effects such as the loss of membrane integrity through excessive phospholipids hydrolysis, formation of eicosanoids, cytotoxic products, ROS, and induction of apoptosis of affected cells (Sapirstein and Bonventre, 2000; Farooqui et al., 2006).

5 Mechanisms Underlying PLA$_2$-Mediated CNS Injuries

Multiple injury insults such as ischemia, proinflammatory cytokines, free radicals, and EAAs are suggested to play roles in PLA$_2$ activation following CNS injuries. Overactivation of PLA$_2$ in turn can exacerbate CNS injuries by attacking cellular membranes, releasing proinflammatory mediators, generating free radicals, increasing release of excitotoxic neurotransmitters and enhancing apoptosis (O'Regan et al., 1995; Farooqui et al., 1997; Klein, 2000) (❯ *Figure 15-3*). The following are possible mechanisms that may explain, at least in part, the actions of PLA_2 in mediating CNS injuries.

◘ Figure 15-3
Possible mechanisms underlying PLA_2-mediated secondary injury after the initial CNS trauma

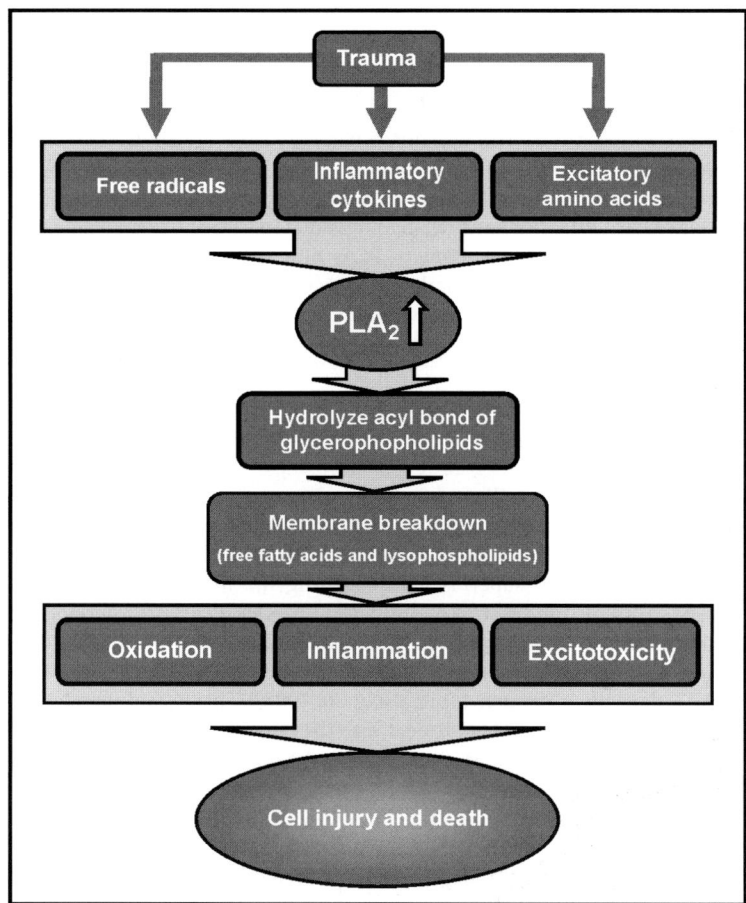

5.1 Membrane Breakdown

Phospholipids are the main components of the neural cell bilayer membrane. They not only constitute the backbone of neural membrane, but also provide the membrane with suitable environment, fluidity, and ion permeability, which are required for the proper function of integral membrane proteins, receptors, and ion channels. PLA_2 activation induces directly phospholipid degradation and membrane breakdown through hydrolysis of neural membrane phospholipids, resulting in alternation of membrane function such as fluidity and permeability, behavior of transporters and receptors, and ion homeostasis, and eventually leading to functional failure of excitable membranes (Farooqui et al., 1997, 2004; Klein, 2000).

5.2 Inflammation

Inflammation has been implicated in many CNS neurological disorders including SCI and TBI. PLA_2 may serve as a key molecule that controls the biosynthesis of several well-known bioactive mediators of inflammation such as eicosanoids (prostaglandins, thromboxanes, leukotrienes, and lipoxins) and PAF (Farooqui et al., 1997, 1999) in a rate-limiting manner. Proinflammatory cytokines such as TNF-α and IL-1β may also induce PLA_2 expression and the activated PLA_2 may further induce the expression of cytokines in a positive feedback manner (Murakami et al., 1997; Liu et al., 2006).

5.3 Oxidation

Oxidative injury is a common mechanism of damage in CNS neurological disorders. Free radicals induce not only lipid peroxidation of neural membrane, but also the oxidation of proteins, RNAs, and DNAs. The metabolism of free fatty acids as PLA_2 metabolites represents a source of ROS. Application of pathophysiological concentrations of free fatty acids has been demonstrated to induce oxidative injury to spinal cord cell cultures (Toborek et al., 1999). Microinjections of PLA_2 into normal spinal cord induced expression of 4-HNE, a product of lipid peroxidation and marker for oxygen free radical-mediated membrane injury (Liu et al., 2006). Mepacrine, a PLA_2 inhibitor, reduced brain injury-induced expression of $cPLA_2$ and 4-HNE (Lu et al., 2001a).

5.4 Excitatory Neurotoxicity

Elevated levels of EAAs have been implicated in the pathogenesis of neural injury and death in many neurological disorders. For example, application of PLA_2 to the rat ischemic cerebral cortex resulted in a significant increase in EAA levels and a PLA_2 inhibitor mepacrine significantly decreased the ischemia-evoked efflux of EAA into cortical superfusates, suggesting the involvement of PLA_2 in EAA release (O'Regan et al., 1995). Administration of PLA_2 inhibitors, $AACOCF_3$ or 4-bromophenacyl bromide also inhibits glutamate release in the spinal cord (Sundstrom and Mo, 2002). However, the exact mechanism of PLA_2 action on EAA release remains unknown. It has been suggested that PLA_2 disrupts an artificial planar lipid bilayer in a Ca^{2+}-dependent manner (O'Regan et al., 1996). This loss of membrane integrity allows EAAs to diffuse from the intracellular compartment into the synaptic cleft. For example, glutamate, which has an intracellular concentration of 10 mM, diffuses from the intracellular compartment into the synaptic cleft where the extracellular concentration is 1 μM (Farooqui et al., 1997). Inhibitors of PLA_2 activity such as 4-bromophenacyl bromide, a nonselective PLA_2 inhibitor, 7,7-dimethyleicosadienoic, a $sPLA_2$ inhibitor, $AACOCF_3$, a $cPLA_2$ inhibitor, and HELSS, an $iPLA_2$ inhibitor, all reduced the efflux of both glutamate and aspartate (Phillis and O'Regan, 1996), suggesting the involvement of multiple isoforms of PLA_2 in EAA release.

5.5 Apoptosis

In recent years, apoptosis has been identified as an important mechanism of cell death in many neurological disorders including SCI (Crowe et al., 1997; Liu et al., 1997; Beattie et al., 2000) and TBI (McIntosh et al., 1998; Waldmeier, 2003). Several excellent reviews focusing on PLA_2s and apoptosis (Cummings et al., 2000; Taketo and Sonoshita, 2002; Balsinde et al., 2006) show the involvement of multiple isoforms of PLA_2 in apoptosis. Apoptosis or programmed cell death is associated with changes in glycerophospholipid metabolism. Cells undergoing apoptosis generally release free fatty acids including AA, which parallels the reduction in cell viability (Taketo and Sonoshita, 2002; Balsinde et al., 2006), suggesting the involvement of PLA_2 in apoptosis.

cPLA$_2$ mediates apoptosis induced by multiple harmful agents such as oxidative stress, TNF-α, or ischemia, but has no role in Fas-induced apoptosis (Cummings et al., 2000; Taketo and Sonoshita, 2002). TNF-α stimulates cPLA$_2$ via caspase-3, in turn, induces AA release and caspase-3 downstream, resulting in apoptosis (Cummings et al., 2000). Recently, a study with cPLA$_2$ antisense oligonucleotides and a selective inhibitor of cPLA$_2$ activity, MAFP, also showed that cPLA$_2$ mediated Aβ-induced neural apoptosis (Kriem et al., 2005).

Growing evidence suggests that iPLA$_2$ plays a more important role in apoptosis. Several lines of evidence indicate that iPLA$_2$ mediates some signal transduction processes associated with apoptosis such as Fas-induced AA release and membrane remodeling (Taketo and Sonoshita, 2002). Inhibition of iPLA$_2$ decreased Fas-induced cell death (Taketo and Sonoshita, 2002). Other functions for iPLA$_2$ in apoptosis may include various cellular regulation as ion channel activity and membrane phospholipid remodeling (Taketo and Sonoshita, 2002). iPLA$_2$ may induce apoptosis by its bioactive lipid metabolites, lyso-PC and AA when their concentrations inside cells rise above certain levels (Balsinde et al., 2006).

sPLA$_2$ IB, III, and IIA have been recently shown to induced neuronal cell death via apoptosis (Yagami et al., 2002a, 2002b; DeCoster, 2003). In contrast, recombinant sPLA$_2$ appears to prevent apoptosis of mast cells, although AA had no effect on the survival and catalytically inactive sPLA$_2$ (Taketo and Sonoshita, 2002). These results suggest that the inhibition of apoptosis is caused by binding of sPLA$_2$s to a cell surface receptor and subsequent intracellular signaling (Taketo and Sonoshita, 2002). Therefore, it may be helpful to characterize the role of each sPLA$_2$ in apoptosis regarding the binding to the sPLA$_2$ receptor.

6 Conclusion and Future Directions

The studies presented in this review indicate that PLA$_2$ plays a central role in mediating CNS traumatic injuries as well as ischemic neurological disorders. PLA$_2$ activation is induced after these injuries and ischemia, and overstimulation of PLA$_2$ attacks cellular membrane, releases proinflammatory mediators, generates free radicals, releases EAAs, and enhances apoptosis. It is conceivable that PLA$_2$ may act as a common or convergence molecule that mediates multiple processes of the secondary injury because it can be induced by multiple toxic factors that are generated following the initial injury, and its activation and metabolites can exacerbate the secondary injury. Increasing evidence suggests that multiple isoforms of PLA$_2$s are involved in mediating the secondary injury process. They clearly represent important targets for developing new protective strategies for treatments of these injuries. However, the specific role and precise mechanism of individual PLA$_2$s as well as the interaction between these PLA$_2$ isoforms within a complex pathophysiological environment of the secondary injury remain to be elucidated. Further studies are needed to identify the type, action, and signaling mechanism of the PLA$_2$s in different cells of the mammalian CNS as well as therapeutic interventions following traumatic and ischemic CNS injuries.

Acknowledgments

This work was supported by NIH NS36350, NS52290, the Kentucky Spinal Cord and Head Injury Research Trust #4-16, the Daniel Heumann Fund for Spinal Cord Research (XMX), F31 NS5657401 (WLT), and the Paralysis Project of America (NKL). We are also thankful for the Norton Healthcare, Kentucky Spinal Cord and Head Injury Research Trust Board, The Commonwealth of Kentucky Bucks for Brains Program and University of Louisville through the James R. Petersdorf Endowment. We also appreciate the use of the Center's Core facility supported by NIH COBRE RR15576.

References

Abe K, Yuki S, Kogure K. 1988. Strong attenuation of ischemic and postischemic brain edema in rats by a novel free radical scavenger. Stroke 19: 480-485.

Adibhatla RM, Hatcher JF, Larsen EC, Chen X, Sun D, et al. 2006. CDP-choline significantly restores phosphatidylcholine levels by differentially affecting phospholipase A2 and

CTP: Phosphocholine cytidylyltransferase after stroke. J Biol Chem 281: 6718-6725.

Arai K, Ikegaya Y, Nakatani Y, Kudo I, Nishiyama N, et al. 2001. Phospholipase A2 mediates ischemic injury in the hippocampus: A regional difference of neuronal vulnerability. Eur J Neurosci 13: 2319-2323.

Azbill RD, Mu X, Bruce-Keller AJ, Mattson MP, Springer JE. 1997. Impaired mitochondrial function, oxidative stress and altered antioxidant enzyme activities following traumatic spinal cord injury. Brain Res 765: 283-290.

Balsinde J, Perez R, Balboa MA. 2006. Calcium-independent phospholipase A2 and apoptosis. Biochim Biophys Acta 1761: 1344-1350.

Baran H, Heldt R, Hertting G. 1987. Increased prostaglandin formation in rat brain following systemic application of kainic acid. Brain Res 404: 107-112.

Bate C, Rumbold L, Williams A. 2007. Cholesterol synthesis inhibitors protect against platelet-activating factor-induced neuronal damage. J Neuroinflammation 4: 5.

Bazan NG, Rodriguez de Turco EB, Allan G. 1995. Mediators of injury in neurotrauma: Intracellular signal transduction and gene expression. J Neurotrauma 12: 791-814.

Beattie MS, Farooqui AA, Bresnahan JC. 2000. Review of current evidence for apoptosis after spinal cord injury. J Neurotrauma 17: 915-925.

Bonventre JV. 1996. Roles of phospholipases A2 in brain cell and tissue injury associated with ischemia and excitotoxicity. J Lipid Mediat Cell Signal 14: 15-23.

Bonventre JV, Huang Z, Taheri MR, O'Leary E, Li E, et al. 1997. Reduced fertility and postischaemic brain injury in mice deficient in cytosolic phospholipase A2. Nature 390: 622-625.

Braughler JM, Hall ED. 1992. Involvement of lipid peroxidation in CNS injury. J Neurotrauma 9(Suppl 1): S1-S7.

Buki A, Okonkwo DO, Wang KK, Povlishock JT. 2000. Cytochrome c release and caspase activation in traumatic axonal injury. J Neurosci 20: 2825-2834.

Bullock R, Fujisawa H. 1992. The role of glutamate antagonists for the treatment of CNS injury. J Neurotrauma 9 (Suppl 2): S443-S462.

Bullock R, Zauner A, Woodward JJ, Myseros J, Choi SC, et al. 1998. Factors affecting excitatory amino acid release following severe human head injury. J Neurosurg 89: 507-518.

Carlson SL, Parrish ME, Springer JE, Doty K, Dossett L. 1998. Acute inflammatory response in spinal cord following impact injury. Exp Neurol 151: 77-88.

Cherian L, Kuruvilla A, Chandy MJ, Abraham J. 1996. Changes in phospholipids and acetylcholinesterase during early phase of injury to spinal cord - an experimental study in rats. Indian J Physiol Pharmacol 40: 134-138.

Clapp LE, Klette KL, De Coster MA, Bernton E, Petras JM, et al. 1995. Phospholipase A2-induced neurotoxicity in vitro and in vivo in rats. Brain Res 693: 101-111.

Clark JD, Schievella AR, Nalefski EA, Lin LL. 1995. Cytosolic phospholipase A2. J Lipid Mediat Cell Signal 12: 83-117.

Clemens JA, Stephenson DT, Smalstig EB, Roberts EF, Johnstone EM, et al. 1996. Reactive glia express cytosolic phospholipase A2 after transient global forebrain ischemia in the rat. Stroke 27: 527-535.

Crowe MJ, Bresnahan JC, Shuman SL, Masters JN, Beattie MS. 1997. Apoptosis and delayed degeneration after spinal cord injury in rats and monkeys. Nat Med 3: 73-76.

Cummings BS, McHowat J, Schnellmann RG. 2000. Phospholipase A(2)s in cell injury and death. J Pharmacol Exp Ther 294: 793-799.

De S, Trigueros MA, Kalyvas A, David S. 2003. Phospholipase A2 plays an important role in myelin breakdown and phagocytosis during Wallerian degeneration. Mol Cell Neurosci 24: 753-765.

DeCoster MA. 2003. Group III secreted phospholipase A2 causes apoptosis in rat primary cortical neuronal cultures. Brain Res 988: 20-28.

Demediuk P, Daly MP, Faden AI. 1989. Changes in free fatty acids, phospholipids, and cholesterol following impact injury to the rat spinal cord. J Neurosci Res 23: 95-106.

Demediuk P, Saunders RD, Anderson DK, Means ED, Horrocks LA. 1985. Membrane lipid changes in laminectomized and traumatized cat spinal cord. Proc Natl Acad Sci USA 82: 7071-7075.

Dennis EA. 1994. Diversity of group types, regulation, and function of phospholipase A2. J Biol Chem 269: 13057-13060.

Dhillon HS, Donaldson D, Dempsey RJ, Prasad MR. 1994. Regional levels of free fatty acids and Evans blue extravasation after experimental brain injury. J Neurotrauma 11: 405-415.

Dumuis A, Sebben M, Haynes L, Pin JP, Bockaert J. 1988. NMDA receptors activate the arachidonic acid cascade system in striatal neurons. Nature 336: 68-70.

Faden AI, Chan PH, Longar S. 1987. Alterations in lipid metabolism, Na+, K+-ATPase activity, and tissue water content of spinal cord following experimental traumatic injury. J Neurochem 48: 1809-1816.

Faden AI, Demediuk P, Panter SS, Vink R. 1989. The role of excitatory amino acids and NMDA receptors in traumatic brain injury. Science 244: 798-800.

Faden AI, Halt P. 1992. Platelet-activating factor reduces spinal cord blood flow and causes behavioral deficits after intrathecal administration in rats through a specific receptor mechanism. J Pharmacol Exp Ther 261: 1064-1070.

Farooque M, Hillered L, Holtz A, Olsson Y. 1996. Changes of extracellular levels of amino acids after graded compression trauma to the spinal cord: An experimental study in the rat using microdialysis. J Neurotrauma 13: 537-548.

Farooqui AA, Litsky ML, Farooqui T, Horrocks LA. 1999. Inhibitors of intracellular phospholipase A2 activity: Their neurochemical effects and therapeutical importance for neurological disorders. Brain Res Bull 49: 139-153.

Farooqui AA, Ong WY, Horrocks LA. 2004. Biochemical aspects of neurodegeneration in human brain: Involvement of neural membrane phospholipids and phospholipases A2. Neurochem Res 29: 1961-1977.

Farooqui AA, Ong WY, Horrocks LA. 2006. Inhibitors of brain phospholipase A2 activity: Their neuropharmacological effects and therapeutic importance for the treatment of neurologic disorders. Pharmacol Rev 58: 591-620.

Farooqui AA, Ong WY, Horrocks LA, Farooqui T. 2000. Brain cystosolic phospholipase A2: Localization, role, and involvement in neurological diseases. Neuroscientist 6: 169-180.

Farooqui AA, Yang HC, Rosenberger TA, Horrocks LA. 1997. Phospholipase A2 and its role in brain tissue. J Neurochem 69: 889-901.

Farooqui AA, Yi Ong W, Lu XR, Halliwell B, Horrocks LA. 2001. Neurochemical consequences of kainate-induced toxicity in brain: Involvement of arachidonic acid release and prevention of toxicity by phospholipase A(2) inhibitors. Brain Res Brain Res Rev 38: 61-78.

Gennarelli TA. 1993. Mechanisms of brain injury. J Emerg Med 11(Suppl 1): 5-11.

Ghosh M, Tucker DE, Burchett SA, Leslie CC. 2006. Properties of the Group IV phospholipase A2 family. Prog Lipid Res 45: 487-510.

Globus MY, Alonso O, Dietrich WD, Busto R, Ginsberg MD. 1995. Glutamate release and free radical production following brain injury: Effects of posttraumatic hypothermia. J Neurochem 65: 1704-1711.

Halat G, Lukacova N, Chavko M, Marsala J. 1987. Effects of incomplete ischemia and subsequent recirculation on free palmitate, stearate, oleate and arachidonate levels in lumbar and cervical spinal cord of rabbit. Gen Physiol Biophys 6: 387-399.

Hall ED. 1989. Free radicals and CNS injury. Crit Care Clin 5: 793-805.

Hall ED, Braughler JM. 1986. Role of lipid peroxidation in post-traumatic spinal cord degeneration: A review. Cent Nerv Syst Trauma 3: 281-294.

Hamada Y, Ikata T, Katoh S, Tsuchiya K, Niwa M, et al. 1996. Roles of nitric oxide in compression injury of rat spinal cord. Free Radic Biol Med 20: 1-9.

Hanasaki K, Arita H. 2002. Phospholipase A2 receptor: A regulator of biological functions of secretory phospholipase A2. Prostaglandins Other Lipid Mediat 68-69: 71-82.

Hattori M, Arai H, Inoue K. 1993. Purification and characterization of bovine brain platelet-activating factor acetylhydrolase. J Biol Chem 268: 18748-18753.

Heinecke JW. 2002. Tyrosyl radical production by myeloperoxidase: A phagocyte pathway for lipid peroxidation and dityrosine cross-linking of proteins. Toxicology 177: 11-22.

Hirabayashi T, Murayama T, Shimizu T. 2004. Regulatory mechanism and physiological role of cytosolic phospholipase A2. Biol Pharm Bull 27: 1168-1173.

Hirabayashi T, Shimizu T. 2000. Localization and regulation of cytosolic phospholipase A(2). Biochim Biophys Acta 1488: 124-138.

Homayoun P, Rodriguez de Turco EB, Parkins NE, Lane DC, Soblosky J, et al. 1997. Delayed phospholipid degradation in rat brain after traumatic brain injury. J Neurochem 69: 199-205.

Horrocks LA, Demediuk P, Saunders RD, Dugan L, Clendenon NR, et al. 1985. The degradation of phospholipids, formation of metabolites of arachidonic acid, and demyelination following experimental spinal cord injury. Cent Nerv Syst Trauma 2: 115-120.

Hostettler ME, Carlson SL. 2002. PAF antagonist treatment reduces pro-inflammatory cytokine mRNA after spinal cord injury. Neuroreport 13: 21-24.

Hostettler ME, Knapp PE, Carlson SL. 2002. Platelet-activating factor induces cell death in cultured astrocytes and oligodendrocytes: Involvement of caspase-3. Glia 38: 228-239.

Jenkins CM, Mancuso DJ, Yan W, Sims HF, Gibson B, et al. 2004. Identification, cloning, expression, and purification of three novel human calcium-independent phospholipase A2 family members possessing triacylglycerol lipase and acylglycerol transacylase activities. J Biol Chem 279: 48968-48975.

Kim DK, Rordorf G, Nemenoff RA, Koroshetz WJ, Bonventre JV. 1995. Glutamate stably enhances the activity of two cytosolic forms of phospholipase A2 in brain cortical cultures. Biochem J 310(Pt 1): 83-90.

Kinsey GR, Cummings BS, Beckett CS, Saavedra G, Zhang W, et al. 2005. Identification and distribution of endoplasmic reticulum iPLA2. Biochem Biophys Res Commun 327: 287-293.

Kishimoto K, Matsumura K, Kataoka Y, Morii H, Watanabe Y. 1999. Localization of cytosolic phospholipase A2 messenger RNA mainly in neurons in the rat brain. Neuroscience 92: 1061-1077.

Kita Y, Ohto T, Uozumi N, Shimizu T. 2006. Biochemical properties and pathophysiological roles of cytosolic phospholipase A2s. Biochim Biophys Acta 1761: 1317-1322.

Klein J. 2000. Membrane breakdown in acute and chronic neurodegeneration: Focus on choline-containing phospholipids. J Neural Transm 107: 1027-1063.

Kolko M, Bruhn T, Christensen T, Lazdunski M, Lambeau G, et al. 1999. Secretory phospholipase A2 potentiates glutamate-induced rat striatal neuronal cell death in vivo. Neurosci Lett 274: 167-170.

Kolko M, Christoffersen NR, Barreiro SG, Miller ML, Pizza AJ, et al. 2006. Characterization and location of secretory phospholipase A2 groups IIE, V, and X in the rat brain. J Neurosci Res 83: 874-882.

Kolko M, Christoffersen NR, Varoqui H, Bazan NG. 2005. Expression and induction of secretory phospholipase A2 group IB in brain. Cell Mol Neurobiol 25: 1107-1122.

Kolko M, DeCoster MA, de Turco EB, Bazan NG. 1996. Synergy by secretory phospholipase A2 and glutamate on inducing cell death and sustained arachidonic acid metabolic changes in primary cortical neuronal cultures. J Biol Chem 271: 32722-32728.

Kornecki E, Ehrlich YH. 1988. Neuroregulatory and neuropathological actions of the ether-phospholipid platelet-activating factor. Science 240: 1792-1794.

Kriem B, Sponne I, Fifre A, Malaplate-Armand C, Lozac'h-Pillot K, et al. 2005. Cytosolic phospholipase A2 mediates neuronal apoptosis induced by soluble oligomers of the amyloid-beta peptide. FASEB J 19: 85-87.

Kudo I, Murakami M. 2002. Phospholipase A2 enzymes. Prostaglandins Other Lipid Mediat 68-69: 3-58.

Larsson Forsell PK, Kennedy BP, Claesson HE. 1999. The human calcium-independent phospholipase A2 gene multiple enzymes with distinct properties from a single gene. Eur J Biochem 262: 575-585.

Larsson PK, Claesson HE, Kennedy BP. 1998. Multiple splice variants of the human calcium-independent phospholipase A2 and their effect on enzyme activity. J Biol Chem 273: 207-214.

Lauritzen I, Heurteaux C, Lazdunski M. 1994. Expression of group II phospholipase A2 in rat brain after severe forebrain ischemia and in endotoxic shock. Brain Res 651: 353-356.

Lautens LL, Chiou XG, Sharp JD, Young WS III, Sprague DL, et al. 1998. Cytosolic phospholipase A2 (cPLA2) distribution in murine brain and functional studies indicate that cPLA2 does not participate in muscarinic receptor-mediated signaling in neurons. Brain Res 809: 18-30.

Lee YB, Yune TY, Baik SY, Shin YH, Du S, et al. 2000. Role of tumor necrosis factor-alpha in neuronal and glial apoptosis after spinal cord injury. Exp Neurol 166: 190-195.

Li Y, Maher P, Schubert D. 1997. A role for 12-lipoxygenase in nerve cell death caused by glutathione depletion. Neuron 19: 453-463.

Lin TN, Wang Q, Simonyi A, Chen JJ, Cheung WM, et al. 2004. Induction of secretory phospholipase A2 in reactive astrocytes in response to transient focal cerebral ischemia in the rat brain. J Neurochem 90: 637-645.

Lindsberg PJ, Yue TL, Frerichs KU, Hallenbeck JM, Feuerstein G. 1990. Evidence for platelet-activating factor as a novel mediator in experimental stroke in rabbits. Stroke 21: 1452-1457.

Liu D, Thangnipon W, McAdoo DJ. 1991. Excitatory amino acids rise to toxic levels upon impact injury to the rat spinal cord. Brain Res 547: 344-348.

Liu D, Xu GY, Pan E, McAdoo DJ. 1999. Neurotoxicity of glutamate at the concentration released upon spinal cord injury. Neuroscience 93: 1383-1389.

Liu NK, Zhang YP, Titsworth WL, Jiang X, Han S, et al. 2006. A novel role of phospholipase A2 in mediating spinal cord secondary injury. Ann Neurol 59: 606-619.

Liu XZ, Xu XM, Hu R, Du C, Zhang SX, et al. 1997. Neuronal and glial apoptosis after traumatic spinal cord injury. J Neurosci 17: 5395-5406.

Lu XR, Ong WY, Halliwell B. 2001a. The phospholipase A2 inhibitor quinacrine prevents increased immunoreactivity to cytoplasmic phospholipase A2 (cPLA2) and hydroxynonenal (HNE) in neurons of the lateral septum following fimbria-fornix transection. Exp Brain Res 138: 500-508.

Lu XR, Ong WY, Halliwell B, Horrocks LA, Farooqui AA. 2001b. Differential effects of calcium-dependent and calcium-independent phospholipase A(2) inhibitors on kainate-induced neuronal injury in rat hippocampal slices. Free Radic Biol Med 30: 1263-1273.

Lucas KK, Dennis EA. 2004. The ABC's of Group IV cytosolic phospholipase A2. Biochim Biophys Acta 1636: 213-218.

Lucas KK, Svensson CI, Hua XY, Yaksh TL, Dennis EA. 2005. Spinal phospholipase A2 in inflammatory hyperalgesia: Role of group IVA cPLA2. Br J Pharmacol 144: 940-952.

Lukacova N, Halat G, Chavko M, Marsala J. 1996. Ischemia-reperfusion injury in the spinal cord of rabbits strongly enhances lipid peroxidation and modifies phospholipid profiles. Neurochem Res 21: 869-873.

Luo J, Lang JA, Miller MW. 1998. Transforming growth factor beta1 regulates the expression of cyclooxygenase in cultured cortical astrocytes and neurons. J Neurochem 71: 526-534.

Mancuso DJ, Jenkins CM, Gross RW. 2000. The genomic organization, complete mRNA sequence, cloning, and expression of a novel human intracellular membrane-associated calcium-independent phospholipase A(2). J Biol Chem 275: 9937-9945.

Manya H, Aoki J, Watanabe M, Adachi T, Asou H, et al. 1998. Switching of platelet-activating factor acetylhydrolase catalytic subunits in developing rat brain. J Biol Chem 273: 18567-18572.

Mautes AE, Weinzierl MR, Donovan F, Noble LJ. 2000. Vascular events after spinal cord injury: Contribution to secondary pathogenesis. Phys Ther 80: 673-687.

Mayer RJ, Marshall LA. 1993. New insights on mammalian phospholipase A2(s); comparison of arachidonoyl-selective and -nonselective enzymes. FASEB J 7: 339-348.

McAdoo DJ, Xu GY, Robak G, Hughes MG. 1999. Changes in amino acid concentrations over time and space around an impact injury and their diffusion through the rat spinal cord. Exp Neurol 159: 538-544.

McIntosh TK, Saatman KE, Raghupathi R, Graham DI, Smith DH, et al. 1998. The Dorothy Russell Memorial Lecture. The molecular and cellular sequelae of experimental traumatic brain injury: Pathogenetic mechanisms. Neuropathol Appl Neurobiol 24: 251-267.

Molloy GY, Rattray M, Williams RJ. 1998. Genes encoding multiple forms of phospholipase A2 are expressed in rat brain. Neurosci Lett 258: 139-142.

Murakami M, Kambe T, Shimbara S, Yamamoto S, Kuwata H, et al. 1999. Functional association of type IIA secretory phospholipase A(2) with the glycosylphosphatidylinositol-anchored heparan sulfate proteoglycan in the cyclooxygenase-2-mediated delayed prostanoid-biosynthetic pathway. J Biol Chem 274: 29927-29936.

Murakami M, Koduri RS, Enomoto A, Shimbara S, Seki M, et al. 2001. Distinct arachidonate-releasing functions of mammalian secreted phospholipase A2s in human embryonic kidney 293 and rat mastocytoma RBL-2H3 cells through heparan sulfate shuttling and external plasma membrane mechanisms. J Biol Chem 276: 10083-10096.

Murakami M, Kudo I. 2002. Phospholipase A2. J Biochem (Tokyo) 131: 285-292.

Murakami M, Kudo I, Inoue K. 1995. Secretory phospholipases A2. J Lipid Mediat Cell Signal 12: 119-130.

Murakami M, Nakatani Y, Atsumi G, Inoue K, Kudo I. 1997. Regulatory functions of phospholipase A2. Crit Rev Immunol 17: 225-283.

Muralikrishna Adibhatla R, Hatcher JF. 2006. Phospholipase A2, reactive oxygen species, and lipid peroxidation in cerebral ischemia. Free Radic Biol Med 40: 376-387.

Murphy EJ, Behrmann D, Bates CM, Horrocks LA. 1994. Lipid alterations following impact spinal cord injury in the rat. Mol Chem Neuropathol 23: 13-26.

Nakano S, Kogure K, Abe K, Yae T. 1990. Ischemia-induced alterations in lipid metabolism of the gerbil cerebral cortex. I. Changes in free fatty acid liberation. J Neurochem 54: 1911-1916.

Narita K, Kubota M, Nakane M, Kitahara S, Nakagomi T, et al. 2000. Therapeutic time window in the penumbra during permanent focal ischemia in rats: Changes of free fatty acids and glycerophospholipids. Neurol Res 22: 393-400.

Ohto T, Uozumi N, Hirabayashi T, Shimizu T. 2005. Identification of novel cytosolic phospholipase A(2)s, murine cPLA(2){delta}, {epsilon}, and {zeta}, which form a gene cluster with cPLA(2){beta}. J Biol Chem 280: 24576-24583.

Oka S, Arita H. 1991. Inflammatory factors stimulate expression of group II phospholipase A2 in rat cultured astrocytes. Two distinct pathways of the gene expression. J Biol Chem 266: 9956-9960.

Okuda S, Saito H, Katsuki H. 1994. Arachidonic acid: Toxic and trophic effects on cultured hippocampal neurons. Neuroscience 63: 691-699.

Ong WY, Horrocks LA, Farooqui AA. 1999a. Immunocytochemical localization of cPLA2 in rat and monkey spinal cord. J Mol Neurosci 12: 123-130.

Ong WY, Lu XR, Ong BK, Horrocks LA, Farooqui AA, et al. 2003. Quinacrine abolishes increases in cytoplasmic phospholipase A2 mRNA levels in the rat hippocampus after kainate-induced neuronal injury. Exp Brain Res 148: 521-524.

Ong WY, Sandhya TL, Horrocks LA, Farooqui AA. 1999b. Distribution of cytoplasmic phospholipase A2 in the normal rat brain. J Hirnforsch 39: 391-400.

Ong WY, Yeo JF, Ling SF, Farooqui AA. 2005. Distribution of calcium-independent phospholipase A2 (iPLA 2) in monkey brain. J Neurocytol 34: 447-458.

O'Regan MH, Alix S, Woodbury DJ. 1996. Phospholipase A2-evoked destabilization of planar lipid membranes. Neurosci Lett 202: 201-203.

O'Regan MH, Smith-Barbour M, Perkins LM, Phillis JW. 1995. A possible role for phospholipases in the release of neurotransmitter amino acids from ischemic rat cerebral cortex. Neurosci Lett 185: 191-194.

Ousman SS, David S. 2000. Lysophosphatidylcholine induces rapid recruitment and activation of macrophages in the adult mouse spinal cord. Glia 30: 92-104.

Ousman SS, David S. 2001. MIP-1alpha, MCP-1, GM-CSF, and TNF-alpha control the immune cell response that mediates rapid phagocytosis of myelin from the adult mouse spinal cord. J Neurosci 21: 4649-4656.

Palmer AM, Marion DW, Botscheller ML, Swedlow PE, Styren SD, et al. 1993. Traumatic brain injury-induced excitotoxicity assessed in a controlled cortical impact model. J Neurochem 61: 2015-2024.

Panter SS, Yum SW, Faden AI. 1990. Alteration in extracellular amino acids after traumatic spinal cord injury. Ann Neurol 27: 96-99.

Park E, Velumian AA, Fehlings MG. 2004. The role of excitotoxicity in secondary mechanisms of spinal cord injury: A review with an emphasis on the implications for white matter degeneration. J Neurotrauma 21: 754-774.

Phillis JW, O'Regan MH. 1996. Mechanisms of glutamate and aspartate release in the ischemic rat cerebral cortex. Brain Res 730: 150-164.

Phillis JW, O'Regan MH. 2004. A potentially critical role of phospholipases in central nervous system ischemic, traumatic, and neurodegenerative disorders. Brain Res Brain Res Rev 44: 13-47.

Pickard RT, Strifler BA, Kramer RM, Sharp JD. 1999. Molecular cloning of two new human paralogs of 85-kDa cytosolic phospholipase A2. J Biol Chem 274: 8823-8831.

Pilitsis JG, Coplin WM, O'Regan MH, Wellwood JM, Diaz FG, et al. 2003. Measurement of free fatty acids in cerebrospinal fluid from patients with hemorrhagic and ischemic stroke. Brain Res 985: 198-201.

Popovich PG, Wei P, Stokes BT. 1997. Cellular inflammatory response after spinal cord injury in Sprague-Dawley and Lewis rats. J Comp Neurol 377: 443-464.

Resnick DK, Nguyen P, Cechvala CF. 2001. Regional and temporal changes in prostaglandin E2 and thromboxane B2 concentrations after spinal cord injury. Spine J 1: 432-436.

Rordorf G, Uemura Y, Bonventre JV. 1991. Characterization of phospholipase A2 (PLA2) activity in gerbil brain: Enhanced activities of cytosolic, mitochondrial, and microsomal forms after ischemia and reperfusion. J Neurosci 11: 1829-1836.

Saluja I, Song D, O'Regan MH, Phillis JW. 1997. Role of phospholipase A2 in the release of free fatty acids during ischemia-reperfusion in the rat cerebral cortex. Neurosci Lett 233: 97-100.

Samad TA, Moore KA, Sapirstein A, Billet S, Allchorne A, et al. 2001. Interleukin-1beta-mediated induction of Cox-2 in the CNS contributes to inflammatory pain hypersensitivity. Nature 410: 471-475.

Sandhya TL, Ong WY, Horrocks LA, Farooqui AA. 1998. A light and electron microscopic study of cytoplasmic phospholipase A2 and cyclooxygenase-2 in the hippocampus after kainate lesions. Brain Res 788: 223-231.

Sapirstein A, Bonventre JV. 2000. Phospholipases A2 in ischemic and toxic brain injury. Neurochem Res 25: 745-753.

Schaloske RH, Dennis EA. 2006. The phospholipase A2 superfamily and its group numbering system. Biochim Biophys Acta 1761: 1246-1259.

Shirai Y, Ito M. 2004. Specific differential expression of phospholipase A2 subtypes in rat cerebellum. J Neurocytol 33: 297-307.

Shmueli O, Cahana A, Reiner O. 1999. Platelet-activating factor (PAF) acetylhydrolase activity, LIS1 expression, and seizures. J Neurosci Res 57: 176-184.

Shohami E, Shapira Y, Yadid G, Reisfeld N, Yedgar S. 1989. Brain phospholipase A2 is activated after experimental closed head injury in the rat. J Neurochem 53: 1541-1546.

Song C, Chang XJ, Bean KM, Proia MS, Knopf JL, et al. 1999. Molecular characterization of cytosolic phospholipase A2-beta. J Biol Chem 274: 17063-17067.

Springer JE, Azbill RD, Mark RJ, Begley JG, Waeg G, et al. 1997. 4-hydroxynonenal, a lipid peroxidation product, rapidly accumulates following traumatic spinal cord injury and inhibits glutamate uptake. J Neurochem 68: 2469-2476.

Stephenson D, Rash K, Smalstig B, Roberts E, Johnstone E, et al. 1999. Cytosolic phospholipase A2 is induced in reactive glia following different forms of neurodegeneration. Glia 27: 110-128.

Stephenson DT, Manetta JV, White DL, Chiou XG, Cox L, et al. 1994. Calcium-sensitive cytosolic phospholipase A2 (cPLA2) is expressed in human brain astrocytes. Brain Res 637: 97-105.

Stewart A, Ghosh M, Spencer DM, Leslie CC. 2002. Enzymatic properties of human cytosolic phospholipase A(2) gamma. J Biol Chem 277: 29526-29536.

Streit WJ, Semple-Rowland SL, Hurley SD, Miller RC, Popovich PG, et al. 1998. Cytokine mRNA profiles in contused spinal cord and axotomized facial nucleus suggest a beneficial role for inflammation and gliosis. Exp Neurol 152: 74-87.

Sun GY, Xu J, Jensen MD, Simonyi A. 2004. Phospholipase A2 in the central nervous system: Implications for neurodegenerative diseases. J Lipid Res 45: 205-213.

Sundstrom E, Mo LL. 2002. Mechanisms of glutamate release in the rat spinal cord slices during metabolic inhibition. J Neurotrauma 19: 257-266.

Svensson CI, Lucas KK, Hua XY, Powell HC, Dennis EA, et al. 2005. Spinal phospholipase A2 in inflammatory hyperalgesia: Role of the small, secretory phospholipase A2. Neuroscience 133: 543-553.

Svensson CI, Yaksh TL. 2002. The spinal phospholipase-cyclooxygenase-prostanoid cascade in nociceptive processing. Annu Rev Pharmacol Toxicol 42: 553-583.

Tabuchi S, Uozumi N, Ishii S, Shimizu Y, Watanabe T, et al. 2003. Mice deficient in cytosolic phospholipase A2 are less susceptible to cerebral ischemia/reperfusion injury. Acta Neurochir Suppl 86: 169-172.

Taketo MM, Sonoshita M. 2002. Phospolipase A2 and apoptosis. Biochim Biophys Acta 1585: 72-76.

Tang J, Kriz RW, Wolfman N, Shaffer M, Seehra J, et al. 1997. A novel cytosolic calcium-independent phospholipase

A2 contains eight ankyrin motifs. J Biol Chem 272: 8567-8575.

Tator CH. 1991. Review of experimental spinal cord injury with emphasis on the local and systemic circulatory effects. Neurochirurgie 37: 291-302.

Tator CH, Fehlings MG. 1991. Review of the secondary injury theory of acute spinal cord trauma with emphasis on vascular mechanisms. J Neurosurg 75: 15-26.

Thwin MM, Ong WY, Fong CW, Sato K, Kodama K, et al. 2003. Secretory phospholipase A2 activity in the normal and kainate injected rat brain, and inhibition by a peptide derived from python serum. Exp Brain Res 150: 427-433.

Toborek M, Malecki A, Garrido R, Mattson MP, Hennig B, et al. 1999. Arachidonic acid-induced oxidative injury to cultured spinal cord neurons. J Neurochem 73: 684-692.

Tonai T, Taketani Y, Ueda N, Nishisho T, Ohmoto Y, et al. 1999. Possible involvement of interleukin-1 in cyclooxygenase-2 induction after spinal cord injury in rats. J Neurochem 72: 302-309.

Tucker DE, Stewart A, Nallan L, Bendale P, Ghomashchi F, et al. 2005. Group IVC cytosolic phospholipase A2gamma is farnesylated and palmitoylated in mammalian cells. J Lipid Res 46: 2122-2133.

Underwood KW, Song C, Kriz RW, Chang XJ, Knopf JL, et al. 1998. A novel calcium-independent phospholipase A2, cPLA2-gamma, that is prenylated and contains homology to cPLA2. J Biol Chem 273: 21926-21932.

van Tienhoven M, Atkins J, Li Y, Glynn P. 2002. Human neuropathy target esterase catalyzes hydrolysis of membrane lipids. J Biol Chem 277: 20942-20948.

Waldmeier PC. 2003. Prospects for antiapoptotic drug therapy of neurodegenerative diseases. Prog Neuropsychopharmacol Biol Psychiatry 27: 303-321.

Wang CX, Olschowka JA, Wrathall JR. 1997. Increase of interleukin-1beta mRNA and protein in the spinal cord following experimental traumatic injury in the rat. Brain Res 759: 190-196.

Wang H, Li J, Follett PL, Zhang Y, Cotanche DA, et al. 2004. 12-Lipoxygenase plays a key role in cell death caused by glutathione depletion and arachidonic acid in rat oligodendrocytes. Eur J Neurosci 20: 2049-2058.

Watanabe M, Aoki J, Manya H, Arai H, Inoue K. 1998. Molecular cloning of cDNAs encoding alpha1, alpha2, and beta subunits of rat brain platelet-activating factor acetylhydrolase. Biochim Biophys Acta 1401: 73-79.

Weerasinghe GR, Coon SL, Bhattacharjee AK, Harry GJ, Bosetti F. 2006. Regional protein levels of cytosolic phospholipase A2 and cyclooxygenase-2 in Rhesus monkey brain as a function of age. Brain Res Bull 69: 614-621.

Xiao J, Zhao D, Hou T, Wu K, Zeng H. 1998. Synergetic protective effects of combined blockade by two kinds of autolesion mediator receptor on neurological function after cervical cord injury. Chin Med J (Engl) 111: 443-446.

Xu GY, McAdoo DJ, Hughes MG, Robak G, de Castro R Jr. 1998. Considerations in the determination by microdialysis of resting extracellular amino acid concentrations and release upon spinal cord injury. Neuroscience 86: 1011-1021.

Xu Y, Tao YX. 2004. Involvement of the NMDA receptor/nitric oxide signal pathway in platelet-activating factor-induced neurotoxicity. Neuroreport 15: 263-266.

Yagami T, Ueda K, Asakura K, Hata S, Kuroda T, et al. 2002a. Human group IIA secretory phospholipase A2 induces neuronal cell death via apoptosis. Mol Pharmacol 61: 114-126.

Yagami T, Ueda K, Asakura K, Hayasaki-Kajiwara Y, Nakazato H, et al. 2002b. Group IB secretory phospholipase A2 induces neuronal cell death via apoptosis. J Neurochem 81: 449-461.

Yang HC, Mosior M, Ni B, Dennis EA. 1999. Regional distribution, ontogeny, purification, and characterization of the Ca^{2+}-independent phospholipase A2 from rat brain. J Neurochem 73: 1278-1287.

Yedgar S, Cohen Y, Shoseyov D. 2006. Control of phospholipase A2 activities for the treatment of inflammatory conditions. Biochim Biophys Acta 1761: 1373-1382.

Yoshida S, Inoh S, Asano T, Sano K, Shimasaki H, et al. 1983. Brain free fatty acids, edema, and mortality in gerbils subjected to transient, bilateral ischemia, and effect of barbiturate anesthesia. J Neurochem 40: 1278-1286.

Young W. 1993. Secondary injury mechanisms in acute spinal cord injury. J Emerg Med 11(Suppl 1): 13-22.

16 Axonal Damage due to Traumatic Brain Injury

K. E. Saatman · G. Serbest · M. F. Burkhardt

1	*Diffuse Axonal Injury: Diagnosis and Detection*	344
2	*Modeling Traumatic Axonal Injury*	345
2.1	In Vivo Models of Axonal Injury	345
2.1.1	Inertial Acceleration Injury in Large Animals	345
2.1.2	Closed Skull Impact Brain Injury	346
2.1.3	Cortical Impact and Fluid Percussion Brain Injury	346
2.1.4	Optic Nerve Stretch Injury	346
2.2	In Vitro Models of Traumatic Axonal Injury	347
3	*Posttraumatic Alterations in the Axolemma and Myelin Sheath*	347
3.1	Primary and Secondary Axotomy	347
3.2	Distortion of the Axolemma	347
3.3	Increased Nonspecific Permeability of the Axolemma	348
3.4	Membrane-Bound Ion Channels and Pumps	349
3.5	Myelination	350
4	*The Axonal Cytoskeleton*	350
4.1	Neurofilaments	350
4.2	Microtubules	351
4.3	Disruption of Axonal Transport	352
4.3.1	Ultrastructural Evidence	352
4.3.2	Use of Anterograde and Retrograde Tracers	352
4.3.3	Use of β-Amyloid Precursor Protein as a Marker	353
5	*Proteolysis*	353
5.1	Calpains	353
5.2	Caspase-3 and Other Proteases	354
6	*Treatment of Traumatic Axonal Injury*	354
6.1	Cyclosporin A and FK506	354
6.2	Hypothermia	355
6.3	Other Treatment Strategies	355
7	*The Need for Additional Assessment Tools*	355
7.1	Functional Assessment Tools	356
7.2	In Vivo Imaging	356
7.3	Biomarkers	356
8	*Conclusions*	356

© 2009 Springer Science+Business Media, LLC.

Abstract: Shearing motions during head injury often result in damage to axons scattered throughout the brain parenchyma. Although diffuse axonal injury (DAI) is thought to be a major contributor to morbidity after traumatic brain injury, no clinically approved treatment strategies are currently available to target dysfunctional or axotomized axons. Here, histological and imaging methods for detecting traumatic axonal injury (TAI) are reviewed, along with experimental models that have been developed to study the relationship between mechanical deformation and axonal injury. Data from clinical and experimental studies are summarized, which describe acute and progressive alterations in the myelin sheath and the axolemma, including modifications of axolemma permeability characteristics and membrane-bound proteins. The cytoskeleton is markedly affected by TAI, as evidenced by disruption of both neurofilament and microtubule networks and impairment of axonal transport. Several studies have demonstrated a role for calpains in posttraumatic axonal damage, while evidence for activity of other proteases is more preliminary. To date, most experimental studies exploring treatment strategies for TAI have focused on either the inhibition of mitochondrial permeability transition or the attenuation of axonal pathology through hypothermia. These studies and other novel approaches are reviewed. In closing, recent developments in methodologies for assessing axonal injury are outlined and opportunities for future investigations are suggested.

List of Abbreviations: APP, β-amyloid precursor protein; CCI, controlled cortical impact; CNS, central nervous system; CsA, cyclosporin A; CT, computerized tomography; DAI, diffuse axonal injury; FP, fluid percussion injury; GCS, Glasgow Coma Scale; GOS, Glasgow Outcome Scale; HRP, horseradish peroxidase; MRI, magnetic resonance imaging; MRS, magnetic resonance spectroscopy; NFH, neurofilament protein heavy subunit; NFL, neurofilament protein light subunit; NFM, neurofilament protein medium subunit; TAI, traumatic axonal injury; TBI, traumatic brain injury

1 Diffuse Axonal Injury: Diagnosis and Detection

Diffuse axonal injury (DAI) results from the shearing of axons during rapid acceleration/deceleration of the brain. A clinical diagnosis that is definitive only after postmortem examination, DAI is associated with high rates of mortality and morbidity. For example, one neuropathological study found that 50% of severely disabled patients and 80% of patients in a vegetative state exhibited moderate or severe DAI (Jennett et al., 2001). Importantly, DAI is also associated with mild head injuries (Blumbergs et al., 1994; Mittl et al., 1994), which represent as much as 80% of all traumatic brain injuries (TBI) (Kraus et al., 1996).

DAI was first appreciated clinically through postmortem evaluations of patients that suffered immediate coma following an acceleration/deceleration type of injury (Strich, 1956; Adams et al., 1977). Pathological features included hemorrhagic tears in the corpus callosum, an absence of mass lesions, diffuse degeneration of white matter, and swollen, disconnected axons or so-called "retraction balls" (Strich, 1961). In order to differentiate severity of injury, three grades of DAI were defined based on the regional distribution and extent of axonal damage, determined using histological silver staining for degenerating axons (Adams et al., 1977). Grade 1 comprises microscopic axonal damage in the cerebral hemispheres including the corpus callosum, the brain stem, and occasionally the cerebellum. In grade 2, a focal lesion in the corpus callosum is also present. Grade 3 DAI is characterized by diffuse axonal damage and focal lesions in both the corpus callosum and the brain stem. Examination of brains from patients that died within 24 h of sustaining a TBI revealed that axonal injury was most frequently detected in the brain stem, followed by the internal capsule, thalamus, and parasagittal white matter (McKenzie et al., 1996).

Initial studies of DAI relied on the use of silver stains to detect fragmented, degenerating, and swollen axons, but the technique was insensitive for short survival times (Povlishock et al., 1983). In the 1990s, a number of other immunohistochemical markers for axonal injury were evaluated in clinical studies, including β-amyloid precursor protein (APP), neurofilament proteins, ubiquitin, SNAP-25, chromogranin A, cathepsin D, and neuron-specific enolase (Grady et al., 1993; Gultekin and Smith, 1994; Sherriff et al., 1994a; Blumbergs et al., 1995; Ogata and Tsuganezawa, 1999). Amyloid precursor protein immunohistochemistry emerged as the most sensitive, reliable and easily interpreted approach, and is currently the most widely used method for postmortem detection of DAI. Axonal injury can be detected with APP antibodies in cases with survival times as short as 1–3 h (Sherriff et al., 1994b; McKenzie et al., 1996; Oehmichen et al., 1998; Ogata and Tsuganezawa, 1999; Wilkinson et al., 1999). However, studies in animal models now suggest that different immunohistochemical markers may detect distinct subsets of injured axons (vide infra), suggesting that multiple markers may be needed for a complete mapping of DAI.

The availability of sensitive axonal injury markers has facilitated qualitative and quantitative assessments of the degree, time course, and distribution of axonal injury. For example, using an axonal injury sector scoring method, Blumbergs et al. (1995) showed that patients with mild head injury, Glasgow Coma Scale (GCS) 13–15, exhibited less extensive axonal injury than those with severe head injury (GCS 3–8). Furthermore, the amount of axonal damage and the size of axonal swellings generally increase with increasing survival time (McKenzie et al., 1996; Wilkinson et al., 1999). Compared with head-injured patients with 3-h survival times, patients with longer survival times show a similar incidence of minor axonal injury but an increased incidence of axonal bulb formation, suggesting that axotomy and subsequent bulb formation are delayed after head injury (Oehmichen et al., 1998). Delayed secondary axotomy, which has been more completely described in animal models of TBI (vide infra), suggests a potential window for therapeutic intervention.

2 Modeling Traumatic Axonal Injury

A number of experimental models of axonal injury have been developed to better understand the progression of axonal pathology and to define the relationship between mechanical forces and the structural and functional responses of axons. These models present an opportunity to identify cellular mediators of injury and to test potential therapeutic agents in a controlled environment. Because DAI is a clinical diagnosis, the terminology of traumatic axonal injury (TAI) will be used within this review to denote axonal injury induced by an experimental traumatic insult. This review focuses exclusively on models that produce traumatic injury, such as that would occur during impact or acceleration/deceleration of the head. Studies of penetrating injury (gunshot or stab models) or primary axotomy (transection or crush models) are not included.

2.1 In Vivo Models of Axonal Injury

2.1.1 Inertial Acceleration Injury in Large Animals

The in vivo model that has most closely approximated DAI in humans is the inertial rotational head injury model. Developed by Gennarelli et al. (1982) for use in nonhuman primates, this model used a rapid, nonimpact head acceleration/deceleration to produce diffusely distributed axonal injury and focal lesions in the corpus callosum and cerebellar peduncle, accompanied by coma. The severity of axonal injury depended on the degree and direction of head rotation, and correlated with the length of unconsciousness, suggesting that axonal injury may be the pathological substrate for posttraumatic coma. A similar model was subsequently developed in the miniature swine (Smith et al., 1997). Using a coronal head rotation, widespread axonal injury was produced, most notably in the root of gyri and at the interfaces of gray and white matter. While conventional magnetic resonance imaging (MRI) revealed no abnormalities, magnetic resonance spectroscopy (MRS) detected a transient decrease in intracellular Mg^{2+} and a persistent decline in N-acetylaspartate (NAA) (Smith et al., 1998), as is observed after human head injury (Brooks et al., 2001). Using an axial, rather than coronal, plane of head rotation in the pig increased the proportion of axonal injury in the brain stem and induced prolonged coma, the severity of which correlated with the amount of brain stem axonal injury (Smith et al., 2000). Axial rotational head injury has also been shown to result in transient coma and axonal injury throughout the frontal and temporal lobes and the midbrain of neonatal piglets (Raghupathi and Margulies, 2002). Despite the clear clinical relevance of inertial acceleration models of TBI, their requirement for large animals such as nonhuman primates or miniature swine has limited their use.

2.1.2 Closed Skull Impact Brain Injury

In rodents, two general approaches to modeling TBI have been used: (1) dropping a weight onto or impacting the closed skull and (2) impacting the dural surface using a rigid impactor or a fluid bolus. These models produce varying degrees of axonal injury as well as neuronal cell death, vascular damage and gliosis. In the first type of model, the intact skull distributes the impact force across the brain, resulting in a more diffuse pattern of injury than that observed in the second type of model, which employs direct impact to the exposed brain (Kupina et al., 2003; Hall et al., 2005). However, when the head is fixed during closed skull impact injury, high impact velocities or forces can produce skull fracture, greatly increasing variability in the pattern of injury produced.

In the impact acceleration model in rats, translational movement of the head is permitted during weight drop onto the helmeted skull, thereby coupling distributed impact forces with acceleration (Foda and Marmarou, 1994; Marmarou et al., 1994). As a result, extensive TAI is produced with relatively minor neuronal and vascular damage in the brain parenchyma (Foda and Marmarou, 1994). The highest concentration of axonal injury is found in the long tracts of the brain stem; therefore, severe brain injury can be associated with high mortality rates unless animals are mechanically ventilated (Foda and Marmarou, 1994). This model of diffuse brain injury is now the most widely used for studying axonal pathology in rodents. Similar principles have been employed to create a diffuse brain injury model with brain stem axonal injury in the immature rat (Adelson et al., 1996, 2001).

2.1.3 Cortical Impact and Fluid Percussion Brain Injury

Commonly used brain injury models that employ a craniotomy and impact to the dural surface include controlled cortical impact (CCI) and fluid percussion (FP). Each simulates closed head injury, but creates injury by directly deforming the brain parenchyma using either a rigid rod (CCI) or fluid bolus (FP). CCI injury performed over the midline results predominantly in brain stem axonal injury (Dixon et al., 1991), although TAI has also been observed in the subcortical white matter, internal capsule, thalamus, and cerebellum (Lighthall et al., 1990). Lateral CCI injury produces TAI primarily in the hemisphere ipsilateral to the impact, in the cortex, corpus callosum, striatum, and thalamus (Dunn-Meynell and Levin, 1997). However, bilateral axonal injury can be achieved by using a bilateral craniotomy to direct tissue deformation across the midline (Meaney et al., 1994). Generally, FP injury results in a more diffuse pattern of tissue damage, and thus TAI, than observed with CCI injury. Midline FP is associated with TAI in the cerebellum, brain stem, internal and external capsules, and corpus callosum (Povlishock et al., 1983; Erb and Povlishock, 1988; Yaghmai and Povlishock, 1992), whereas lateral FP produces TAI in these same regions as well as in the striatum, cortex, hippocampus, and thalamus, albeit concentrated in the ipsilateral hemisphere (Pierce et al., 1996; Bramlett et al., 1997; Saatman et al., 1998). While CCI and FP brain injury produce TAI in multiple brain regions, conclusions about axonal pathology drawn from these models must be interpreted with consideration for superimposed effects of contusion and hemorrhage.

2.1.4 Optic Nerve Stretch Injury

Whole brain models of TAI, as described earlier, are necessary when addressing characteristics of injured axons in a specific brain region, the synergistic effects of TAI and contusion or hemorrhage, or physiological or behavioral outcomes. However, they present difficulties in sampling, quantification, and correlation of axonal injury, due to the complexity of axonal pathways in the brain and the scattered distribution of injury. To overcome some of these inherent challenges, a model of TAI was developed in guinea pigs (Gennarelli et al., 1989) and subsequently in mice (Saatman et al., 2003) in which the optic nerve is subjected to rapid, transient stretch injury. The large, easily accessible tract of myelinated central nervous system (CNS) axons in the optic nerve provides several advantages for evaluating and modulating TAI, including simplified geometry and well-defined connectivity. As opposed to more widely used models of nerve crush or transection, the stretch injury model can be used to study graded, progressive TAI with delayed axotomy (Maxwell et al., 1991; Jafari et al., 1997; Bain et al., 2001).

2.2 In Vitro Models of Traumatic Axonal Injury

Axonal responses to stretch injury have been evaluated in vitro using peripheral nerve fibers (Gray and Ritchie, 1954; Haftek, 1970) and large, single axons from invertebrates (Koike, 1987; Galbraith et al., 1993). However, in vitro studies of traumatic injury to vertebrate CNS axons are limited in number, due in part to challenges associated with studying isolated CNS axons, which are smaller in diameter than peripheral fibers, and applying clinically relevant deformations such as dynamic stretch. Smith et al. (1999) have developed a CNS axonal injury model using N-Tera2 cells, a human neuronal cell line, cultured on a flexible membrane. Following dynamic stretch applied selectively to clusters of axons traversing the membrane, axonal swellings evolved in a delayed fashion. One limitation of the model is that the cultured axons are unmyelinated, and therefore may not reproduce the full spectrum of axonal pathology observed in vivo. However, obvious advantages of this and other in vitro models include the capability to follow the progression of axonal change in real time and the relative ease of introducing probes or compounds with which to monitor or alter biochemical cascades. The continued use and development of in vitro models will complement the use of in vivo models, promoting greater understanding of underlying mechanisms that contribute to axonal dysfunction and degeneration.

3 Posttraumatic Alterations in the Axolemma and Myelin Sheath

3.1 Primary and Secondary Axotomy

The physical tolerance and biological responses of the axolemma to rapid deformation are not well understood. Data from neuronal cultures suggest that unmyelinated axons can withstand dynamic elongation of up to 65% of their original length without disconnection (Smith et al., 1999). Nevertheless, primary axotomy or tearing of the axolemma may result during severe traumatic injury. Tears in the subcortical white matter and fimbria of the rat after moderate to severe brain injury (Bramlett et al., 1997; Saatman et al., 1998) mimic white matter lesions observed in head-injured patients (Graham et al., 2000). However, diffusely distributed primary axotomy has been reported only in nonhuman primates after severe acceleration/deceleration brain injury (Maxwell et al., 1993). In this case, fragmentation of the axolemma occurred preferentially at the nodes of Ranvier or the paranodal regions in small caliber myelinated fibers, and fibers resealed within 60 min after injury. It is now generally accepted that primary axotomy is rare after TBI, and that the predominant axonal pathology is that of progressive damage culminating in secondary axotomy. Nonetheless, the true incidence of primary axotomy in DAI may be underappreciated due to technical and experimental difficulties in detecting primary axotomy, resulting from (1) the diffuse nature of the pathology, (2) the short time period that the axolemmal damage may be visible by electron microscopy, if membrane resealing occurs, and (3) the paucity of animal models capable of inducing a correlate to severe DAI.

3.2 Distortion of the Axolemma

In intact traumatically injured axons, damage to the axolemma may include distention, infolding, altered permeability, and/or changes in intramembrane protein content or distribution. Historically, one of the earliest hallmarks of axonal injury was localized distention of the axolemma associated with axonal swelling (❷ *Figure 16-1b–d*). Characteristic axonal swellings, or "axonal bulbs" if the axon has disconnected, have been described in human head injury and in numerous injury models involving different species. Despite the seemingly universal nature of axonal swellings after TAI, little is understood about the initiating events that cause distention of the axolemma, although physical stretching and osmotic imbalances have been suggested to play a role. While axonal swelling and concomitant axolemma distention often occur in the internodal region, the nodes of Ranvier appears to be especially vulnerable to axolemmal distortion, as illustrated by massive protrusions or blebbing of the nodal axolemma visualized in electron micrographs of traumatically injured myelinated axons (Gennarelli et al., 1989; Maxwell, 1996). Infolding of the axolemma may also occur in injured axons, either at sites of axonal swellings or in other damaged regions of the axon

Figure 16-1

Immunohistochemical detection of traumatic axonal injury. (a) In a low magnification image (40×), diffusely distributed axonal damage is evident in the subcortical white matter at one week after lateral fluid percussion (FP) brain injury in the rat. Arrows denote a subset of the many axonal swellings and bulbs immunolabelled with anti-NF200. (b) At higher magnification (100×), a characteristic morphology of axonal bulbs is illustrated. Neurofilament accumulation is noted in the region of localized axonal diameter enlargement in axotomized axons (arrows) at one week after lateral FP brain injury in the rat. (c) Neurofilament-immunopositive axonal bulbs (arrows) found in the mouse optic nerve one day after stretch injury show a morphology similar to those found in the rat after FP injury (see b). An axon with two swollen regions is clearly visible (double arrows). (d) Another commonly used marker of axonal injury, βAPP, reveals axonal swellings within intact axons (demarcated by * symbols) and axonal bulbs (arrow) in the rat after FP brain injury. Scale bars denote 50 μm in (a), and 20 μm in (b), (c), and (d)

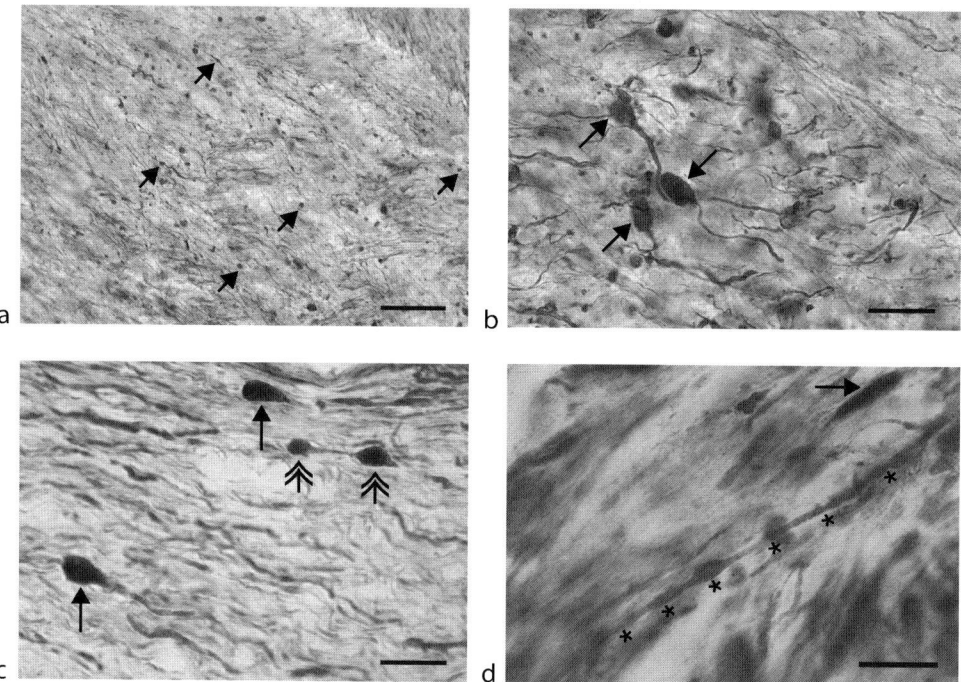

(Pettus and Povlishock, 1996; Maxwell et al., 1997; Povlishock et al., 1997). The extent to which the axolemma can recover from such dramatic distention or infolding is currently unclear. However, cultured unmyelinated axons have been shown to recover from elongation-induced undulations of the axon cylinder and overlying axolemma (Smith et al., 1999). In addition, blebbing of the nodal axolemma, evident acutely (within minutes to hours) after axonal injury, resolves over a time period of hours to days in the guinea pig optic nerve, suggestive of recovery processes (Gennarelli et al., 1993; Maxwell, 1996).

3.3 Increased Nonspecific Permeability of the Axolemma

The axolemma is functionally, as well as structurally, compromised after TAI. One of the primary functions of the axolemma is to create a selectively permeable barrier between the axoplasm and the extracellular space. Numerous experimental studies, pioneered by Povlishock and colleagues, have demonstrated a nonspecific axolemmal permeability increase in subpopulations of injured axons when the applied insult reaches a critical threshold of severity (Pettus et al., 1994; Pettus and Povlishock, 1996; Povlishock et al., 1997;

Stone et al., 2004). Using an intrathecal injection to distribute horseradish peroxidase (HRP) extracellularly before injury, Pettus et al. (1994) demonstrated posttraumatic influx of HRP to axons, specifically in regions showing mitochondrial damage and neurofilament compaction, within minutes after midline FP brain injury in cats. Uptake of HRP occurred before axonal disconnection, confirming that rapid axonal flooding was not due to diffusion at sites of axotomy. No influx of HRP was noted following mild brain injury, suggesting a threshold may exist for the initiation of the axolemmal permeability change. However, not all axons appear to be equally vulnerable to stretch-induced permeability changes. In axons that later exhibit neurofilament compaction, influx of normally excluded large molecules precedes overt damage to the axonal cytoskeleton (Povlishock et al., 1997; Stone et al., 2004). In contrast, in axons that swell and undergo secondary axotomy, induction of nonspecific axolemma permeability may occur in a much more delayed fashion (Stone et al., 2004). Furthermore, while large molecules on the order of 3–44 kDa have been shown to flood into the axoplasm of injured CNS axons in animal models of TBI (Pettus et al., 1994; Stone et al., 2004), axons of cultured human N-Tera2 neurons exhibited no permeability of the axolemma to a 570 Da dye at any level of stretch injury below that required for primary disconnection (Smith et al., 1999).

Many questions remain regarding the mechanism and duration of trauma-induced axolemma permeability changes, the relationship of increased permeability to acute and chronic axonal pathology, and the extent to which this phenomenon is a function of myelination, axonal caliber, species, injury severity, or other variables. A potential for membrane resealing or recovery following a diffuse, nonspecific permeability increase is suggested by an electroporation study in the squid giant axon which documented membrane resealing and functional recovery after moderate, but not severe, injury (Gallant and Galbraith, 1997). However, acute recovery after TAI may be limited, as Povlishock and Pettus (1996) have shown that the axolemma remains permeable to HRP for up to several hours after TBI in the cat.

3.4 Membrane-Bound Ion Channels and Pumps

As described earlier, nonspecific breaches of the axolemma may be an important pathological characteristic of TAI in subsets of axons after moderate to severe injury. Nevertheless, changes in the number, distribution or function of axonal membrane channels and pumps may be just as critical in initiating and prolonging axonal dysfunction after trauma, particularly for mild to moderate injuries. In myelinated CNS axons, specialized localization of membrane proteins is essential for efficient axonal electrical conductance and maintenance of ionic homeostasis. Within hours after TAI, the number of intramembranous particles, as assessed by freeze fracture, decreases in the nodes of Ranvier (Maxwell et al., 1988, 1999; Maxwell, 1996). More specifically, activity of the membrane pumps Ca-ATPase and Na,K-ATPase decreased in the nodal axolemma, and calcium accumulation was noted within nodal blebs (Maxwell, 1996). In contrast, increased numbers of intramembranous particles were observed in the internodal axolemma within minutes after injury, followed by the novel expression of Ca-ATPase (Maxwell et al., 1995, 1999). The most marked increases in Na,K-ATPase and Ca-ATPase activity occurred at sites of myelin disruption (Maxwell et al., 1999). These intriguing findings are suggestive of an acute redistribution of axolemmal membrane pumps from the nodes of Ranvier to the internodal region after TAI. Because Na,K-ATPase plays an important role in the restoration and maintenance of Na^+ and K^+ gradients in myelinated axons, a decrease in nodal Na,K-ATPase might compromise axonal conductance. A loss of Ca-ATPase, which along with the Na/Ca exchanger plays a role in maintaining low intracellular Ca^{2+}, might lead to Ca^{2+} accumulation at the node.

Under conditions of anoxia or ischemia, calcium-dependent pathology in myelinated axons has been linked to overactivation of voltage-dependent sodium channels resulting in the reversal of the Na/Ca exchangers, which are concentrated at the nodes of Ranvier, and activation of voltage-dependent calcium channels (Stys, 2005). A similar mechanism for aberrant calcium influx has been proposed for TAI. Increased intraaxonal calcium has been demonstrated in both in vivo and in vitro models of axonal trauma (Maxwell et al., 1995; Maxwell, 1996; Wolf et al., 2001). Posttraumatic elevations in intracellular Ca^{2+} were dependent on extracellular calcium and could be eliminated by blocking sodium channels with tetrodotoxin or reduced by blocking either voltage-dependent calcium channels or the Na/Ca exchanger (Wolf et al., 2001).

Recently, Iwata et al. (2004) showed that TAI results in proteolytic cleavage of axonal sodium channels in cultured neurons, potentially leading to noninactivation of the channel and exacerbation of calcium influx. These data demonstrating proteolytic modification of a membrane channel, along with experimental evidence that calpains and other proteases may be activated after axonal trauma (vide infra), provide a rationale for additional investigations into posttraumatic structural and functional alterations of axolemmal proteins.

3.5 Myelination

In addition to damage to the axolemma, traumatic injury leads to progressive disruption of the myelin sheath. Electron microscopy has been used to demonstrate separation of the myelin sheath from shrunken or collapsed axons (Maxwell et al., 1993; Pettus and Povlishock, 1996; Povlishock et al., 1997), separation of myelin lamellae and intrusion of myelin into axolemma invaginations (Maxwell et al., 1993, 1995, 2003; Stone et al., 2001), and thinning of the myelin sheath (Erb and Povlishock, 1988; Rodriguez-Paez et al., 2005). Dissociation of myelin lamellae has been correlated with loss of ecto-Ca-ATPase activity in the myelin in the acute posttraumatic period (Maxwell et al., 1995, 1999). In the weeks and months following brain injury, white matter atrophy, loss or disruption of myelin, and the presence of microglia/macrophages in white matter tracts have been described clinically and in experimental models of TBI (Levin et al., 1990; Bramlett and Dietrich, 2002; Wilson et al., 2004; Rodriguez-Paez et al., 2005). Investigation of chronic alterations in myelin in models of isolated TAI should help determine whether long-term white matter changes after TBI are a direct result of axonal trauma or a secondary consequence of neuronal death.

4 The Axonal Cytoskeleton

4.1 Neurofilaments

Under normal circumstances, neurofilament protein subunits light (NFL), medium (NFM), and heavy (NFH), named according to their relative molecular weights, coassemble to form neurofilaments. Both NFM and NFH have heavily phosphorylated carboxyl-terminal, or "sidearm," domains (Julien and Mushynski, 1983). Neurofilaments provide mechanical stability to the axon and influence axonal diameter through their number and interfilament spacing. Because TAI results from a rapid stretching of axons, structural elements such as neurofilaments are likely to play a role in the posttraumatic response.

Accumulation of neurofilament proteins in damaged axons has been well established using immunohistochemical approaches in postmortem tissue from humans with DAI (Grady et al., 1993; Christman et al., 1994) and in numerous animal models of axonal injury (Yaghmai and Povlishock, 1992; Meaney et al., 1994; Povlishock et al., 1997; Chen et al., 1999; Saatman et al., 2003) (❍ *Figure 16-1a–c*). Following DAI in humans, neurofilament accumulation can be detected as early as 6 h in damaged, swollen axons, and by 1 week in axonal bulbs (Grady et al., 1993; Christman et al., 1994). In animal models, axonal swellings and bulbs show robust neurofilament immunoreactivity within several hours after TBI (Yaghmai and Povlishock, 1992; Povlishock et al., 1997; Saatman et al., 1998; Chen et al., 1999; Smith et al., 1999). Accumulation of NFL precedes that of NFM and NFH in injured axons following inertial brain injury in the pig (Chen et al., 1999), and subunits are predominantly, although not exclusively, dephosphorylated (Chen et al., 1999; Saatman et al., 2003).

In complement to immunohistochemical studies demonstrating localized neurofilament accumulation in injured axons, electron microscopic studies illustrate alterations to neurofilament structure and organization in greater detail. Within hours following DAI in humans, the typical linear, parallel organization of axonal neurofilaments is disturbed in some axons (Christman et al., 1994). Data from animal studies suggest that such misalignment of neurofilaments may initially reflect mild TAI, as it is most often observed in axons that do not show increased nonspecific axolemma permeability, microtubule loss, or overt mitochondrial swelling (Erb and Povlishock, 1988; Yaghmai and Povlishock, 1992; Pettus et al., 1994; Pettus and Povlishock, 1996; Jafari et al., 1997; Povlishock et al., 1997; Maxwell et al., 2003). However, misalignment of neurofilaments has also been observed at sites of organelle accumulation, axonal swelling, and delayed axotomy (Christman et al., 1994; Pettus et al., 1994). Therefore, disruption of the neurofilament cytoskeleton may contribute to or be associated with progressive axonal damage that ultimately leads to permanent axonal dysfunction through secondary axotomy.

While some injured axons exhibit neurofilament misalignment, others maintain their linear alignment but show regions of reduced interfilament spacing, or neurofilament compaction. This is thought to represent a phenotype of TAI distinct from neurofilament misalignment (Stone et al., 2001). Because compaction often coincides with increased nonspecific axolemmal permeability, loss of microtubules, and mitochondrial swelling and has also been reported in regions of axolemma tears or axotomy, it is presumed to reflect more severe TAI (Pettus et al., 1994; Pettus and Povlishock, 1996; Jafari et al., 1997; Maxwell and Graham, 1997; Maxwell et al., 2003). Within the guinea pig optic nerve, compaction has been reported to be more prevalent in small to intermediate caliber myelinated axons, although it occurs in some large caliber axons exhibiting periaxonal spaces (Jafari et al., 1997, 1998). In contrast, in the brain stem of rats, neurofilament compaction has been reported in both small and large caliber axons (Buki et al., 1999b; Marmarou et al., 2005), although compaction was also associated with marked vacuolization in larger axons (Stone et al., 2001). Interestingly, a single axon may contain multiple sites of compacted neurofilaments separated by zones of apparently normal neurofilament spacing (Jafari et al., 1998; Maxwell et al., 2003), and the filaments can remain compacted for several hours after trauma in rats, after which axonal swelling and neurofilament misalignment may occur (Pettus and Povlishock, 1996; Stone et al., 2001). Although the initiating factors for neurofilament compaction are not understood at present, compacted neurofilaments have significantly shorter sidearm projections (Pettus and Povlishock, 1996; Povlishock et al., 1997; Okonkwo et al., 1998), which may be a result of posttraumatic dephosphorylation, because phosphorylation state is known to regulate sidearm positioning relative to the neurofilament core and interfilament spacing distances (Nixon et al., 1994). Alternatively, neurofilament sidearm modifications may be due to limited proteolysis.

The pathology of neurofilaments after TAI is complex and multifactorial. While some axons show neurofilament misalignment or compaction, others show increased spacing of neurofilaments in association with separation of the myelin sheath, a decrease in neurofilament number, or a loss of immunoreactivity (Maxwell et al., 2003; Saatman et al., 1998, 2003). In certain cases, especially those of more severe injury leading to primary or secondary axotomy, dissolution of both neurofilaments and microtubules takes place as a result of proteolysis, even if neurofilament compaction or misalignment precedes this end-stage process (Maxwell et al., 1993; Povlishock et al., 1997; Jafari et al., 1998).

4.2 Microtubules

The integrity of microtubules after TAI is of obvious interest due to the dependence of retrograde and anterograde fast axonal transport on an intact microtubule cytoskeleton. As with neurofilaments, the response of microtubules to traumatic injury is varied, and may depend on the severity of the initial insult, the caliber and myelination of the axon, the axonal region affected, species and/or brain region examined, and interactions with other secondary injury cascades. In addition, some differences in microtubule changes described in the literature may be a function of time points examined, as postmortem examinations and current experimental approaches have precluded continuous monitoring of the evolution of axonal pathology over time. Nonetheless, a great deal of information on microtubule damage has been gleaned from light and electron microscopic investigations.

The most commonly described alteration in the microtubule cytoskeleton after TAI is a loss of microtubules. Decreased microtubule density and numbers have been observed in nodal, paranodal, and internodal regions, although the extent of loss depends on axonal caliber and the time elapsed after injury (Povlishock et al., 1997; Maxwell et al., 2003). Loss of microtubules is pronounced in nodal blebs and at sites of axolemmal infolding (Maxwell, 1996; Maxwell and Graham, 1997). Regions of neurofilament compaction, separation of the myelin sheath from the axon cylinder, and increased nonspecific axolemma permeability are also frequent locations of acute microtubule loss, reported within minutes to hours after

TAI (Pettus and Povlishock, 1996; Jafari et al., 1997). However, microtubule loss can also occur in regions of increased neurofilament spacing (Maxwell, 1996) and neurofilament compaction can precede, or occur in the absence of, microtubule loss (Jafari et al., 1997; Maxwell et al., 2003). Furthermore, at mild levels of injury in which the axolemma remains impermeable to tracer molecules such as HRP, no loss of microtubules was noted despite a misalignment of neurofilaments (Pettus and Povlishock, 1996). Together these observations suggest that the mechanisms underlying microtubule loss and neurofilament disturbances are distinct.

While acute axonal microtubule loss persisted up to 6 h after moderate to severe FP brain injury in cats (Pettus and Povlishock, 1996), numbers of microtubules were restored in subsets of damaged guinea pig optic nerve axons within the first 24 h after stretch injury (Maxwell and Graham, 1997; Maxwell et al., 2003). In degenerating axons, both microtubule and neurofilament numbers decline (Maxwell et al., 2003), while in a subpopulation of severely injured or axotomized axons, dissolution of the cytoskeleton is observed (Maxwell et al., 1993, 2003). Less frequent observations of microtubule pathology after TAI include a reduction in axonal microtubule density in the absence of significant microtubule loss, and a misalignment of microtubules in damaged regions of axons (Maxwell et al., 2003).

4.3 Disruption of Axonal Transport

Because microtubules function as an efficient and highly specialized transport system within axons, trauma-induced damage to the microtubules would be expected to interrupt axonal transport. Vesicles containing membrane-bound proteins and lipids, neurotransmitters, enzymes, and cellular organelles are transported from the cell body to the synapse at rates of $\sim$200–400 mm/day (Lasek et al., 1984; Brown, 2003). This fast anterograde transport is dependent on the kinesin superfamily of microtubule-associated motor proteins (Cyr and Brady, 1992). In contrast, the motor protein dynein transports cargo such as activated growth factor receptors, degraded vesicular membrane components, and recycled proteins along microtubules in the retrograde direction (Pfister, 1999). Transport of cytoskeletal and other soluble cytosolic proteins is slower, proceeding at rates of $\sim$0.1–10 mm/day, but is also thought to be dependent on kinesins and dynein (Shea, 2000; Shah and Cleveland, 2002; Brown, 2003).

4.3.1 Ultrastructural Evidence

Within 30–60 h after head injury in humans, accumulation of organelles in damaged axons, suggestive of disrupted axonal transport, has been demonstrated in postmortem samples by electron microscopy (Christman et al., 1994). Similar axonal accumulations of organelles have been demonstrated in multiple species using several different models of TBI. Localized accumulation is noted as early as 2 h after experimental TAI, and becomes more pronounced with time (Maxwell et al., 1991; Yaghmai and Povlishock, 1992; Pettus et al., 1994). The aggregates of organelles colocalized with increased neurofilament immunoreactivity and disorganization of the filament network (Yaghmai and Povlishock, 1992) were not associated with regions of increased axolemmal permeability to HRP (Pettus et al., 1994).

4.3.2 Use of Anterograde and Retrograde Tracers

Ultrastructural evidence of interrupted axonal transport has been directly supported by studies using anterograde and retrograde tracer molecules. Povlishock and colleagues introduced the anterograde tracer HRP into target neuronal populations before traumatic injury and demonstrated pooling in their damaged axons within 1 h of postinjury which colocalized with accumulation of organelles under electron microscopy (Povlishock et al., 1983; Erb and Povlishock, 1988). Axons laden with HRP initially exhibited unilobular swellings that progressed to multilobular swellings and then to disconnection. In optic nerve stretch models of TAI, anterograde (Gennarelli et al., 1989) or retrograde (Saatman et al., 2003) tracers have been employed to demonstrate reduced delivery to the target neurons in the superior colliculus or the retina, respectively, confirming bidirectional functional impairment in axonal transport. Retrograde transport impairment preceded accumulation of neurofilaments in axonal swellings, and did not significantly worsen or abate during a period of 2 weeks after TAI (Saatman et al., 2003), suggesting an acute and persistent disruption in transport.

4.3.3 Use of β-Amyloid Precursor Protein as a Marker

Currently, the most widely used marker of impaired axonal transport in models of TAI is accumulation of APP, a membrane spanning glycoprotein which is transported in the anterograde direction by fast axonal transport (Koo et al., 1990). Because APP is transported more rapidly than are neurofilament proteins, APP has provided improved temporal resolution for detection of impaired transport. Accumulation of APP in damaged axons can be detected as early as 2–3 h after head injury in humans (Gentleman et al., 1993; Sherriff et al., 1994b; McKenzie et al., 1996), or 30 min to 1 h in animal models (Pierce et al., 1996; Stone et al., 2004). The size of APP-positive axonal swellings correlates with survival time after head injury in humans (Wilkinson et al., 1999), implying that although transport is disrupted locally, the microtubule machinery is operational upstream of the focal axonal damage, resulting in a progressive accumulation of APP. Although impaired axonal transport has been identified in the very early posttraumatic period, it may also contribute to ongoing, chronic neuronal damage following TBI, as APP-positive axonal swellings or bulbs have been identified for several months to a year following injury in humans (Blumbergs et al., 1994), rodents (Pierce et al., 1998), and pigs (Chen et al., 2004).

While APP is often observed in the axonal bulbs of axons having undergone secondary axotomy, APP may also accumulate in the node of Ranvier, paranodal region or internode of damaged, but continuous axons (Stone et al., 2000, 2001) (❯ *Figure 16-1d*). APP accumulation can occur in the absence of a notable increase in nonspecific axolemma permeability, although posttraumatic transport impairment may lead to delayed membrane damage and enhanced permeability (Stone et al., 2004). In addition, the phenotype of the APP-positive axon may depend on its caliber. Following impact acceleration brain injury in the rat, small caliber axons exhibited APP accumulation in the absence of neurofilament compaction, while these two pathologies were colocalized in some larger caliber axons (Stone et al., 2001; Marmarou et al., 2005).

Despite a great deal of clinical and experimental evidence of alterations in microtubules and disruption of axonal transport in traumatically injured axons, little is known about the initiating factors for these events. Changes in kinase or phosphatase activities could alter the phosphorylation state of proteins critical to microtubule structure and function, leading to depolymerization or altered transport dynamics. Alternatively, increased intracellular calcium and associated increases in proteolysis may contribute to degradation or modification of microtubules, motor proteins, or microtubule-stabilizing proteins such as tau.

5 Proteolysis

5.1 Calpains

Calpains are ubiquitous, nonlysosomal, cysteine proteases that, when activated by sufficient levels of intracellular free calcium, cleave numerous cellular proteins. The two major calpain isoforms in the CNS are μ-calpain and m-calpain, named for their calcium requirements for activation (near 1 μM and 1 mM, respectively). Many axonal proteins are potential substrates for calpains, including the neurofilament subunit proteins, tubulin, tau, spectrin, myelin basic protein, and certain ion channels, receptors and pumps, as well as signaling molecules.

Calpain activation in traumatically injured axons, as detected by calpain-specific cleavage of the cytoskeletal protein spectrin, was first reported after TBI in rats (Saatman et al., 1996). Axons that labeled for calpain-mediated spectrin fragments exhibited focal swellings and were disorganized within damaged white matter tracts at 90 min, but were clearly degenerating by 1 day after injury. Subsequently, calpain-mediated spectrin degradation was localized to the subaxolemmal region of damaged axons at 15 min following TBI in the rat (Buki et al., 1999c), consistent with acute calcium influx from the extracellular space. By 30–120 min after injury, evidence of calpain activation was also observed within the axoplasm, particularly near damaged mitochondria, suggestive of a delayed calcium efflux from the mitochondria. Furthermore, the damaged regions of axons that exhibited calpain activation also contained compacted neurofilaments. The clinical relevance of axonal calpain activation was established through postmortem studies confirming calpain-mediated spectrin proteolysis and neurofilament degradation in the corpus callosum, a primary site of axonal pathology in DAI, of humans that died following blunt head injury (McCracken et al., 1999). Interestingly, calpain activation following TAI may be biphasic, as suggested by a recent study in which an early, transient phase of spectrin proteolysis in normal appearing axons was followed by more pronounced calpain-mediated spectrin degradation in degenerating axons of injured mouse optic nerves (Saatman et al., 2003).

5.2 Caspase-3 and Other Proteases

In addition to calpains, other proteases are activated in traumatically injured axons. Caspase-3, a cysteine protease specifically linked to the execution of apoptosis, is activated in brain stem axons after impact acceleration brain injury in rats and colocalizes with the release of cytochrome *c* from mitochondria (Buki et al., 2000). In most cases, axonal calpain activation appeared to precede caspase-3 activation, suggesting that calpain activity may trigger or be coregulated with activity of caspases. Parallel activation of calpains and caspases and calpain-dependent caspase-3 activation have been demonstrated in several in vitro cell death models (Ferrand-Drake et al., 2003; Sharma and Rohrer, 2004; Gomez-Vicente et al., 2005). While the role of caspase-3 activation in axonal pathology is not yet clear, caspase-3 cleaves spectrin and APP in injured axons (Buki et al., 2000; Stone et al., 2002; Chen et al., 2004), and may accumulate with other proteases such as the β-site APP-cleaving enzyme and presenilin-1 in swollen axons for up to 6 months after TAI in pigs (Chen et al., 2004).

Future studies are likely to provide additional insight into the relative roles of μ-calpain, m-calpain, caspases, and other axonal proteases in the acute and chronic phases of TAI. A greater understanding is needed with regard to the in vivo substrates for these proteases, their initiating factors, their axonal localization and regulation, and interprotease interactions.

6 Treatment of Traumatic Axonal Injury

6.1 Cyclosporin A and FK506

Following axonal injury, one consequence of an imbalance in calcium homeostasis is mitochondrial damage. Within minutes to hours after TAI, swollen mitochondria are detected within damaged axons (Maxwell et al., 1995; Pettus and Povlishock, 1996). Evidence of mitochondrial calcium accumulation within this same time frame (Maxwell et al., 1995; Xiao-Sheng et al., 2004) supports the hypothesis that mitochondria act as calcium sinks in the acute posttraumatic period, potentially compromising mitochondrial function and triggering opening of the mitochondrial permeability transition pore (Okonkwo and Povlishock, 1999). Cyclosporin A (CsA) binds to cyclophilin D and prevents the opening of the permeability transition pore, and has been shown to attenuate axonal mitochondria damage when given before TBI in rats (Okonkwo and Povlishock, 1999). Furthermore, a single intrathecal dose administered 30 min following impact acceleration brain injury in rats reduced the numbers of axons exhibiting APP accumulation, calpain-mediated spectrin breakdown, and neurofilament compaction (Buki et al., 1999b). However, a dose–response study in rats demonstrated seizure activity and mortality at an intravenous dose only five times greater than the maximally efficacious dose (Okonkwo et al., 2003), raising concerns about the clinical usage of this treatment paradigm for TAI. In addition to its ability to block mitochondrial permeability transition, CsA also inhibits calcineurin, a calmodulin-dependent phosphatase. The immunophilin ligand FK506 inhibits calcineurin without inhibiting mitochondrial permeability transition and has been shown in a pretreatment paradigm to reduce the number of APP-positive swollen axons following TBI in the rat (Singleton et al., 2001). These data suggest that some of the protective effects of CsA in TAI may be due to its ability to inhibit calcineurin. Although calcineurin has been hypothesized to contribute to neurofilament dephosphorylation in injured axons, intravenous administration of FK506 during posttraumatic hypothermia and rapid rewarming was effective in limiting APP accumulation but not neurofilament compaction (Suehiro et al., 2001).

6.2 Hypothermia

The protective effects of hypothermia have been documented in numerous clinical and experimental studies of TBI (for reviews, see Marion, 2002; Clifton, 2004). While only a small number of these studies incorporated measures of axonal injury, collectively they indicate that temperature is a potent modulator of axonal pathology. For example, moderate hypothermia (32°C) for 4 h after CCI brain injury in rats reduced the number of neurofilament-positive injured axons in the internal capsule if temperature reduction was initiated earlier than 40 min postinjury (Marion and White, 1996). A slightly longer therapeutic window was demonstrated following impact acceleration brain injury in rats; lowering body temperature to 32°C for 1 h within 1 h after injury significantly attenuated the number of APP-positive injured axons in the pontomedullary junction (Koizumi and Povlishock, 1998). Acute hypothermia has also been shown to reduce the incidence of calpain-mediated spectrin proteolysis and neurofilament compaction in axons after TBI (Buki et al., 1999a). Conversely, posttraumatic hyperthermia (40°C for 4 h) exacerbates axonal injury, as evidenced by increases in APP-labeled axons in rats subjected to FP brain injury (Suzuki et al., 2004). Although acute posttraumatic hypothermia appears to slow or prevent axonal damage, rewarming too rapidly may cause secondary damage to the cytoskeleton (Suehiro et al., 2001; Maxwell et al., 2005).

6.3 Other Treatment Strategies

Several other strategies have been tested in a preliminary manner for efficacy in TAI. Pretreatment with a calpain inhibitor reduced numbers of damaged axons exhibiting neurofilament compaction or APP accumulation at 2 h after TBI in rats (Buki et al., 2003). Acute intravenous administration of the 21-aminosteriod U-74389G decreased the number of axons exhibiting neurofilament accumulation within the internal capsule in a rat model of TBI (Marion and White, 1996). Intracerebroventricular, but not intravenous, treatment with pituitary adenylate cyclase activating polypeptide immediately following impact acceleration TBI reduced the density of APP-positive axons in the corticospinal tract but not the medial lemniscus of rats (Farkas et al., 2004). In contrast, pretreatment of rats subjected to a mild weight drop brain injury with the NMDA antagonist, MK801, was not effective in reducing numbers of APP-positive injured axons (Lewen et al., 1996).

7 The Need for Additional Assessment Tools

The paucity of studies evaluating treatment strategies targeting traumatic axonal pathology reflects, in part, our incomplete understanding of the initiating factors and downstream mediators of TAI. For the most part, therapeutic strategies for ameliorating axonal injury have mimicked those tested for reduction of neuronal cell death in models of TBI. Although a number of pathological features of TAI, such as membrane and myelin damage, cytoskeletal disruption and mitochondrial damage, are well established, it appears that not all axons undergo the same combination of trauma-induced alterations at the same rate, perhaps due to axonal properties specific to neurons of certain brain regions, differences in size or extent of myelination, or other as yet undetermined factors. Therefore, any given marker of axonal injury, such as neurofilament or APP accumulation, mitochondrial damage, or calpain activation may represent only a subset of all injured axons. This presents challenges, not only for understanding the true prevalence of TAI in different models of brain trauma, but also in accurately assessing the efficacy of treatment paradigms, when all phenotypes of axonal injury may not respond equally well to a given therapy.

7.1 Functional Assessment Tools

In evaluating therapeutic strategies for TAI, functional assessments are greatly needed to augment the histological evaluation of axonal pathology. Currently only a handful of studies have directly measured physiological dysfunction in axons after traumatic injury. Using electrophysiological approaches, decreases in compound action potential amplitude (Baker et al., 2002; Reeves et al., 2005) or latency shifts in evoked potentials (Bain et al., 2001) have been demonstrated in damaged white matter. Interestingly, myelinated axons in the corpus callosum recovered more quickly than unmyelinated axons (Reeves et al., 2005). Increased use of electrophysiological approaches may provide important information about the relationship between impairment of axonal conductance and damage to the cytoskeleton, myelin sheath or the axolemma and its membrane channels.

7.2 In Vivo Imaging

In order to better understand the progression of TAI and its relationship to functional outcome, in vivo techniques for assessing axonal injury over multiple time points in patients or animals are needed. Great strides have been made in the development of imaging techniques with improved sensitivity and specificity for detecting white matter damage (for review see Hurley et al., 2004). Although computerized tomography (CT) and conventional MRI can detect axonal injury, correlations with clinical outcome have been poor (Mittl et al., 1994; Paterakis et al., 2000). In contrast, parameters of axonal injury measured using modified imaging techniques such as diffusion tensor imaging and T2*-weighted gradient-echo imaging have correlated significantly with acute GCS (Scheid et al., 2003; Huisman et al., 2004). Furthermore, the number of white matter lesions detected with T2*-weighted gradient-echo imaging has been correlated with the duration of unconsciousness and Glasgow Outcome Scale (GOS) (Yanagawa et al., 2000). The continued refinement of imaging modalities for clinical and experimental use will facilitate longitudinal studies to improve our understanding of the chronic effects of axonal injury and allow correlation of the extent and progression of TAI with behavioral or neuropsychological outcomes.

7.3 Biomarkers

Another potential approach to monitoring the progression of brain damage in vivo is the use of biomarkers. Several candidate biomarkers are currently under evaluation for TBI (Ingebrigtsen and Romner, 2003; Pineda et al., 2004). However, while proteins that are enriched in axons have been detected in the serum after traumatic injury (Gabbita et al., 2005; Shaw et al., 2005), no specific biomarkers for TAI have been reported.

8 Conclusions

Significant progress has made in diagnosing and detecting axonal injury in TBI, in the development of models of axonal trauma, and in understanding the progression and heterogeneity of the axonal membrane and cytoskeletal response to TAI. However, many questions remain unanswered. It is unclear to what extent the pathological events initiated by rapid stretching of CNS axons are reversible, leaving open the possibility of spontaneous recovery. And in contrast to the field of spinal cord injury, in which a great deal of effort has been directed toward understanding and promoting recovery of damaged axons, very little is known about the ability of axons in the brain to regenerate after traumatic injury, or what factors may regulate this response. Evidence of sprouting of swollen, disconnected axons in brain-injured cats (Christman et al., 1997), and of corticospinal tracts in response to TBI in rats (Lenzlinger et al., 2005) has been reported, but this area of investigation is conspicuously underdeveloped. Similarly, little is known about the plasticity of the injured brain in response to axonal damage. A recent study suggests that TAI may not result in acute neuronal cell death, even if axotomy occurs close to the cell body (Singleton et al., 2002). Therefore, strategies promoting plasticity or regeneration after TAI may be efficacious in improving functional outcome. Integrating investigations of chronic alterations in injured axons with the growing knowledge of acute pathological mechanisms should maximize the potential for development of therapies to reduce or reverse progressive axonal damage associated with TBI.

Acknowledgments

The authors thank Heather N. Foozer for assistance in manuscript preparation. This work was supported by NIH R01 NS045131, KSCHIRT 6-12, and T32 NS043126.

References

Adams JH, Mitchell DE, Graham DI, Doyle D. 1977. Diffuse brain damage of immediate impact type. Its relationship to 'primary brain-stem damage' in head injury. Brain 100: 489-502.

Adelson PD, Jenkins LW, Hamilton RL, Robichaud P, Tran MP, et al. 2001. Histopathologic response of the immature rat to diffuse traumatic brain injury. J Neurotrauma 18: 967-976.

Adelson PD, Robichaud P, Hamilton RL, Kochanek PM. 1996. A model of diffuse traumatic brain injury in the immature rat. J Neurosurg 85: 877-884.

Bain AC, Raghupathi R, Meaney DF. 2001. Dynamic stretch correlates to both morphological abnormalities and electrophysiological impairment in a model of traumatic axonal injury. J Neurotrauma 18: 499-511.

Baker AJ, Phan N, Moulton RJ, Fehlings MG, Yucel Y, et al. 2002. Attenuation of the electrophysiological function of the corpus callosum after fluid percussion injury in the rat. J Neurotrauma 19: 587-599.

Blumbergs PC, Scott G, Manavis J, Wainwright H, Simpson DA, et al. 1994. Staining of amyloid precursor protein to study axonal damage in mild head injury. Lancet 344: 1055-1056.

Blumbergs PC, Scott G, Manavis J, Wainwright H, Simpson DA, et al. 1995. Topography of axonal injury as defined by amyloid precursor protein and the sector scoring method in mild and severe closed head injury. J Neurotrauma 12: 565-572.

Bramlett HM, Dietrich WD. 2002. Quantitative structural changes in white and gray matter 1 year following traumatic brain injury in rats. Acta Neuropathol (Berl) 103: 607-614.

Bramlett HM, Kraydieh S, Green EJ, Dietrich WD. 1997. Temporal and regional patterns of axonal damage following traumatic brain injury: A β-amyloid precursor protein immunocytochemical study in rats. J Neuropathol Exp Neurol 56: 1132-1141.

Brooks WM, Friedman SD, Gasparovic C. 2001. Magnetic resonance spectroscopy in traumatic brain injury. J Head Trauma Rehabil 16: 149-164.

Brown A. 2003. Axonal transport of membranous and non-membranous cargoes: A unified perspective. J Cell Biol 160: 817-821.

Buki A, Farkas O, Doczi T, Povlishock JT. 2003. Preinjury administration of the calpain inhibitor MDL-28170 attenuates traumatically induced axonal injury. J Neurotrauma 20: 261-268.

Buki A, Koizumi H, Povlishock JT. 1999a. Moderate posttraumatic hypothermia decreases early calpain-mediated proteolysis and concomitant cytoskeletal compromise in traumatic axonal injury. Exp Neurol 159: 319-328.

Buki A, Okonkwo DO, Povlishock JT. 1999b. Postinjury cyclosporin A administration limits axonal damage and disconnection in traumatic brain injury. J Neurotrauma 16: 511-521.

Buki A, Okonkwo DO, Wang KK, Povlishock JT. 2000. Cytochrome c release and caspase activation in traumatic axonal injury. J Neurosci 20: 2825-2834.

Buki A, Siman R, Trojanowski JQ, Povlishock JT. 1999c. The role of calpain-mediated spectrin proteolysis in traumatically induced axonal injury. J Neuropathol Exp Neurol 58: 365-375.

Chen XH, Meaney DF, Xu BN, Nonaka M, McIntosh TK, et al. 1999. Evolution of neurofilament subtype accumulation in axons following diffuse brain injury in the pig. J Neuropathol Exp Neurol 58: 588-596.

Chen XH, Siman R, Iwata A, Meaney DF, Trojanowski JQ, et al. 2004. Long-term accumulation of amyloid-β, β-secretase, presenilin-1, and caspase-3 in damaged axons following brain trauma. Am J Pathol 165: 357-371.

Christman CW, Grady MS, Walker SA, Holloway KL, Povlishock JT. 1994. Ultrastructural studies of diffuse axonal injury in humans. J Neurotrauma 11: 173-186.

Christman CW, Salvant JB Jr, Walker SA, Povlishock JT. 1997. Characterization of a prolonged regenerative attempt by diffusely injured axons following traumatic brain injury in adult cat: A light and electron microscopic immunocytochemical study. Acta Neuropathol (Berl) 94: 329-337.

Clifton GL. 2004. Is keeping cool still hot? An update on hypothermia in brain injury. Curr Opin Crit Care 10: 116-119.

Cyr JL, Brady ST. 1992. Molecular motors in axonal transport. Cellular and molecular biology of kinesin. Mol Neurobiol 6: 137-155.

Dixon CE, Clifton GL, Lighthall JW, Yaghmai AA, Hayes RL. 1991. A controlled cortical impact model of traumatic brain injury in the rat. J Neurosci Meth 39: 253-262.

Dunn-Meynell AA, Levin BE. 1997. Histological markers of neuronal, axonal and astrocytic changes after lateral rigid impact traumatic brain injury. Brain Res 761: 25-41.

Erb DE, Povlishock JT. 1988. Axonal damage in severe traumatic brain injury: An experimental study in cat. Acta Neuropathol (Berl) 76: 347-358.

Farkas O, Tamas A, Zsombok A, Reglodi D, Pal J, et al. 2004. Effects of pituitary adenylate cyclase activating polypeptide in a rat model of traumatic brain injury. Regul Pept 123: 69-75.

Ferrand-Drake M, Zhu C, Gido G, Hansen AJ, Karlsson JO, et al. 2003. Cyclosporin A prevents calpain activation despite increased intracellular calcium concentrations, as well as translocation of apoptosis-inducing factor, cytochrome c and caspase-3 activation in neurons exposed to transient hypoglycemia. J Neurochem 85: 1431-1442.

Foda MA, Marmarou A. 1994. A new model of diffuse brain injury in rats. II. Morphological characterization. J Neurosurg 80: 301-313.

Gabbita SP, Scheff SW, Menard RM, Roberts K, Fugaccia I, et al. 2005. Cleaved-tau: A biomarker of neuronal damage after traumatic brain injury. J Neurotrauma 22: 83-94.

Galbraith JA, Thibault LE, Matteson DR. 1993. Mechanical and electrical responses of the squid giant axon to simple elongation. J Biomech Eng 115: 13-22.

Gallant PE, Galbraith JA. 1997. Axonal structure and function after axolemmal leakage in the squid giant axon. J Neurotrauma 14: 811-822.

Gennarelli TA, Thibault LE, Adams JH, Graham DI, Thompson CJ, et al. 1982. Diffuse axonal injury and traumatic coma in the primate. Ann Neurol 12: 564-574.

Gennarelli TA, Thibault LE, Tipperman R, Tomei G, Sergot R, et al. 1989. Axonal injury in the optic nerve: A model simulating diffuse axonal injury in the brain. J Neurosurg 71: 244-253.

Gennarelli TA, Tipperman R, Maxwell WL, Graham DI, Adams JH, et al. 1993. Traumatic damage to the nodal axolemma: An early, secondary injury. Acta Neurochir Suppl (Wien) 57: 49-52.

Gentleman SM, Nash MJ, Sweeting CJ, Graham DI, Roberts GW. 1993. β-amyloid precursor protein (β-APP) as a marker for axonal injury after head injury. Neurosci Lett 160: 139-144.

Gomez-Vicente V, Donovan M, Cotter TG. 2005. Multiple death pathways in retina-derived 661W cells following growth factor deprivation: Crosstalk between caspases and calpains. Cell Death Differ 12: 796-804.

Grady MS, McLaughlin MR, Christman CW, Valadka AB, Fligner CL, et al. 1993. The use of antibodies targeted against the neurofilament subunits for the detection of diffuse axonal injury in humans. J Neuropathol Exp Neurol 52: 143-152.

Graham DI, Raghupathi R, Saatman KE, Meaney D, McIntosh TK. 2000. Tissue tears in the white matter after lateral fluid percussion brain injury in the rat: Relevance to human brain injury. Acta Neuropathol (Berl) 99: 117-124.

Gray JA, Ritchie JM. 1954. Effects of stretch on single myelinated nerve fibres. J Physiol 124: 84-99.

Gultekin SH, Smith TW. 1994. Diffuse axonal injury in craniocerebral trauma. A comparative histologic and immunohistochemical study. Arch Pathol Lab Med 118: 168-171.

Haftek J. 1970. Stretch injury of peripheral nerve. Acute effects of stretching on rabbit nerve. J Bone Joint Surg Br 52: 354-365.

Hall ED, Sullivan PG, Gibson TR, Pavel KM, Thompson BM, et al. 2005. Spatial and temporal characteristics of neurodegeneration after controlled cortical impact in mice: More than a focal brain injury. J Neurotrauma 22: 252-265.

Huisman TA, Schwamm LH, Schaefer PW, Koroshetz WJ, Shetty-Alva N, et al. 2004. Diffusion tensor imaging as potential biomarker of white matter injury in diffuse axonal injury. AJNR Am J Neuroradiol 25: 370-376.

Hurley RA, McGowan JC, Arfanakis K, Taber KH. 2004. Traumatic axonal injury: Novel insights into evolution and identification. J Neuropsychiatry Clin Neurosci 16: 1-7.

Ingebrigtsen T, Romner B. 2003. Biochemical serum markers for brain damage: A short review with emphasis on clinical utility in mild head injury. Restor Neurol Neurosci 21: 171-176.

Iwata A, Stys PK, Wolf JA, Chen XH, Taylor AG, et al. 2004. Traumatic axonal injury induces proteolytic cleavage of the voltage-gated sodium channels modulated by tetrodotoxin and protease inhibitors. J Neurosci 24: 4605-4613.

Jafari SS, Maxwell WL, Neilson M, Graham DI. 1997. Axonal cytoskeletal changes after non-disruptive axonal injury. J Neurocytol 26: 207-221.

Jafari SS, Nielson M, Graham DI, Maxwell WL. 1998. Axonal cytoskeletal changes after nondisruptive axonal injury. II. Intermediate sized axons. J Neurotrauma 15: 955-966.

Jennett B, Adams JH, Murray LS, Graham DI. 2001. Neuropathology in vegetative and severely disabled patients after head injury. Neurology 56: 486-490.

Julien JP, Mushynski WE. 1983. The distribution of phosphorylation sites among identified proteolytic fragments of mammalian neurofilaments. J Biol Chem 258: 4019-4025.

Koike H. 1987. The extensibility of aplysia nerve and the determination of true axon length. J Physiol 390: 469-487.

Koizumi H, Povlishock JT. 1998. Posttraumatic hypothermia in the treatment of axonal damage in an animal model of traumatic axonal injury. J Neurosurg 89: 303-309.

Koo EH, Sisodia SS, Archer DR, Martin LJ, Weidemann A, et al. 1990. Precursor of amyloid protein in Alzheimer disease undergoes fast anterograde axonal transport. Proc Natl Acad Sci USA 87: 1561-1565.

Kraus JF, McArthur DL, Silverman TA, Jayaraman M. 1996. Epidemiology of brain injury. Neurotrama. Narayan RK, Wilberger JE, Povlishock JT, editors. New York: McGraw-Hill; pp. 13-30.

Kupina NC, Detloff MR, Bobrowski WF, Snyder BJ, Hall ED. 2003. Cytoskeletal protein degradation and neurodegeneration evolves differently in males and females following experimental head injury. Exp Neurol 180: 55-73.

Lasek RJ, Garner JA, Brady ST. 1984. Axonal transport of the cytoplasmic matrix. J Cell Biol 99: 212s-221s.

Lenzlinger PM, Shimizu S, Marklund N, Thompson HJ, Schwab ME, et al. 2005. Delayed inhibition of Nogo-A does not alter injury-induced axonal sprouting but enhances recovery of cognitive function following experimental traumatic brain injury in rats. Neuroscience 134: 1047-1056.

Levin HS, Williams DH, Valastro M, Eisenberg HM, Crofford MJ, et al. 1990. Corpus callosal atrophy following closed head injury: Detection with magnetic resonance imaging. J Neurosurg 73: 77-81.

Lewen A, Li GL, Olsson Y, Hillered L. 1996. Changes in microtubule-associated protein 2 and amyloid precursor protein immunoreactivity following traumatic brain injury in rat: Influence of MK-801 treatment. Brain Res 719: 161-171.

Lighthall JW, Goshgarian HG, Pinderski CR. 1990. Characterization of axonal injury produced by controlled cortical impact. J Neurotrauma 7: 65-76.

Marion DW. 2002. Moderate hypothermia in severe head injuries: The present and the future. Curr Opin Crit Care 8: 111-114.

Marion DW, White MJ. 1996. Treatment of experimental brain injury with moderate hypothermia and 21-aminosteroids. J Neurotrauma 13: 139-147.

Marmarou A, Foda MA, van den Brink W, Campbell J, Kita H, et al. 1994. A new model of diffuse brain injury in rats. I. Pathophysiology and biomechanics. J Neurosurg 80: 291-300.

Marmarou CR, Walker SA, Davis CL, Povlishock JT. 2005. Quantitative analysis of the relationship between intraaxonal neurofilament compaction and impaired axonal transport following diffuse traumatic brain injury. J Neurotrauma 22: 1066-1080.

Maxwell WL. 1996. Histopathological changes at central nodes of Ranvier after stretch-injury. Microsc Res Tech 34: 522-535.

Maxwell WL, Domleo A, McColl G, Jafari SS, Graham DI. 2003. Post-acute alterations in the axonal cytoskeleton after traumatic axonal injury. J Neurotrauma 20: 151-168.

Maxwell WL, Graham DI. 1997. Loss of axonal microtubules and neurofilaments after stretch-injury to guinea pig optic nerve fibers. J Neurotrauma 14: 603-614.

Maxwell WL, Irvine A, Graham, DI, Adams JH, Gennarelli TA, et al. 1991. Focal axonal injury: The early axonal response to stretch. J Neurocytol 20: 157-164.

Maxwell WL, Kansagra AM, Graham DI, Adams JH, Gennarelli TA. 1988. Freeze-fracture studies of reactive myelinated nerve fibres after diffuse axonal injury. Acta Neuropathol (Berl) 76: 395-406.

Maxwell WL, Kosanlavit R, McCreath BJ, Reid O, Graham DI. 1999. Freeze-fracture and cytochemical evidence for structural and functional alteration in the axolemma and myelin sheath of adult guinea pig optic nerve fibers after stretch injury. J Neurotrauma 16: 273-284.

Maxwell WL, McCreath BJ, Graham DI, Gennarelli TA. 1995. Cytochemical evidence for redistribution of membrane pump calcium-ATPase and ecto-Ca-ATPase activity, and calcium influx in myelinated nerve fibres of the optic nerve after stretch injury. J Neurocytol 24: 925-942.

Maxwell WL, Povlishock JT, Graham DL. 1997. A mechanistic analysis of nondisruptive axonal injury: A review. J Neurotrauma 14: 419-440.

Maxwell WL, Watson A, Queen R, Conway B, Russell D, et al. 2005. Slow, medium, or fast re-warming following post-traumatic hypothermia therapy? An ultrastructural perspective. J Neurotrauma 22: 873-884.

Maxwell WL, Watt C, Graham DI, Gennarelli TA. 1993. Ultrastructural evidence of axonal shearing as a result of lateral acceleration of the head in non-human primates. Acta Neuropathol (Berl) 86: 136-144.

McCracken E, Hunter AJ, Patel S, Graham DI, Dewar D. 1999. Calpain activation and cytoskeletal protein breakdown in the corpus callosum of head-injured patients. J Neurotrauma 16: 749-761.

McKenzie KJ, McLellan DR, Gentleman SM, Maxwell WL, Gennarelli TA, et al. 1996. Is β-APP a marker of axonal damage in short-surviving head injury?. Acta Neuropathol (Berl) 92: 608-613.

Meaney DF, Ross DT, Winkelstein BA, Brasko J, Goldstein D, et al. 1994. Modification of the cortical impact model to produce axonal injury in the rat cerebral cortex. J Neurotrauma 11: 599-612.

Mittl RL, Grossman RI, Hiehle JF, Hurst RW, Kauder DR, et al. 1994. Prevalence of MR evidence of diffuse axonal injury in patients with mild head injury and normal head CT findings. AJNR Am J Neuroradiol 15: 1583-1589.

Nixon RA, Paskevich PA, Sihag RK, Thayer CY. 1994. Phosphorylation on carboxyl terminus domains of neurofilament proteins in retinal ganglion cell neurons in vivo: Influences on regional neurofilament accumulation, interneurofilament spacing, and axon caliber. J Cell Biol 126: 1031-1046.

Oehmichen M, Meissner C, Schmidt V, Pedal I, Konig HG, et al. 1998. Axonal injury – a diagnostic tool in forensic neuropathology? A review. Forensic Sci Int 95: 67-83.

Ogata M, Tsuganezawa O. 1999. Neuron-specific enolase as an effective immunohistochemical marker for injured axons after fatal brain injury. Int J Legal Med 113: 19-25.

Okonkwo DO, Melon DE, Pellicane AJ, Mutlu LK, Rubin DG, et al. 2003. Dose–response of cyclosporin A in attenuating traumatic axonal injury in rat. Neuroreport 14: 463-466.

Okonkwo DO, Pettus EH, Moroi J, Povlishock JT. 1998. Alteration of the neurofilament sidearm and its relation to neurofilament compaction occurring with traumatic axonal injury. Brain Res 784: 1-6.

Okonkwo DO, Povlishock JT. 1999. An intrathecal bolus of cyclosporin A before injury preserves mitochondrial integrity and attenuates axonal disruption in traumatic brain injury. J Cereb Blood Flow Metab 19: 443-451.

Paterakis K, Karantanas AH, Komnos A, Volikas Z. 2000. Outcome of patients with diffuse axonal injury: The significance and prognostic value of MRI in the acute phase. J Trauma 49: 1071-1075.

Pettus EH, Christman CW, Giebel ML, Povlishock JT. 1994. Traumatically induced altered membrane permeability: Its relationship to traumatically induced reactive axonal change. J Neurotrauma 11: 507-522.

Pettus EH, Povlishock JT. 1996. Characterization of a distinct set of intra-axonal ultrastructural changes associated with traumatically induced alteration in axolemmal permeability. Brain Res 722: 1-11.

Pfister KK. 1999. Cytoplasmic dynein and microtubule transport in the axon: The action connection. Mol Neurobiol 20: 81-91.

Pierce JE, Smith DH, Trojanowski JQ, McIntosh TK. 1998. Enduring cognitive, neurobehavioral and histopathological changes persist for up to one year following severe experimental brain injury in rats. Neuroscience 87: 359-369.

Pierce JE, Trojanowski JQ, Graham DI, Smith DH, McIntosh TK. 1996. Immunohistochemical characterization of alterations in the distribution of amyloid precursor proteins and β-amyloid peptide after experimental brain injury in the rat. J Neurosci 16: 1083-1090.

Pineda JA, Wang KK, Hayes RL. 2004. Biomarkers of proteolytic damage following traumatic brain injury. Brain Pathol 14: 202-209.

Povlishock JT, Becker DP, Cheng CL, Vaughan GW. 1983. Axonal change in minor head injury. J Neuropathol Exp Neurol 42: 225-242.

Povlishock JT, Marmarou A, McIntosh T, Trojanowski JQ, Moroi J. 1997. Impact acceleration injury in the rat: Evidence for focal axolemmal change and related neurofilament sidearm alteration. J Neuropathol Exp Neurol 56: 347-359.

Povlishock JT, Pettus EH. 1996. Traumatically induced axonal damage: Evidence for enduring changes in axolemmal permeability with associated cytoskeletal change. Acta Neurochir Suppl 66: 81-86.

Raghupathi R, Margulies SS. 2002. Traumatic axonal injury after closed head injury in the neonatal pig. J Neurotrauma 19: 843-853.

Reeves TM, Phillips LL, Povlishock JT. 2005. Myelinated and unmyelinated axons of the corpus callosum differ in vulnerability and functional recovery following traumatic brain injury. Exp Neurol 196: 126-137.

Rodriguez-Paez AC, Brunschwig JP, Bramlett HM. 2005. Light and electron microscopic assessment of progressive atrophy following moderate traumatic brain injury in the rat. Acta Neuropathol (Berl) 109: 603-616.

Saatman KE, Abai B, Grosvenor A, Vorwerk CK, Smith DH, et al. 2003. Traumatic axonal injury results in biphasic calpain activation and retrograde transport impairment in mice. J Cereb Blood Flow Metab 23: 34-42.

Saatman KE, Bozyczko-Coyne D, Marcy V, Siman R, McIntosh TK. 1996. Prolonged calpain-mediated spectrin breakdown occurs regionally following experimental brain injury in the rat. J Neuropathol Exp Neurol 55: 850-860.

Saatman KE, Graham DI, McIntosh TK. 1998. The neuronal cytoskeleton is at risk after mild and moderate brain injury. J Neurotrauma 15: 1047-1058.

Scheid R, Preul C, Gruber O, Wiggins C, von Cramon DY. 2003. Diffuse axonal injury associated with chronic traumatic brain injury: Evidence from T2*-weighted gradient-echo imaging at 3 T. AJNR Am J Neuroradiol 24: 1049-1056.

Shah JV, Cleveland DW. 2002. Slow axonal transport: Fast motors in the slow lane. Curr Opin Cell Biol 14: 58-62.

Sharma AK, Rohrer B. 2004. Calcium-induced calpain mediates apoptosis via caspase-3 in a mouse photoreceptor cell line. J Biol Chem 279: 35564-35572.

Shaw G, Yang C, Ellis R, Anderson K, Parker MJ, et al. 2005. Hyperphosphorylated neurofilament NF-H is a serum biomarker of axonal injury. Biochem Biophys Res Commun 336: 1268-1277.

Shea TB. 2000. Microtubule motors, phosphorylation and axonal transport of neurofilaments. J Neurocytol 29: 873-887.

Sherriff FE, Bridges LR, Gentleman SM, Sivaloganathan S, Wilson S. 1994a. Markers of axonal injury in post mortem human brain. Acta Neuropathol (Berl) 88: 433-439.

Sherriff FE, Bridges LR, Sivaloganathan S. 1994b. Early detection of axonal injury after human head trauma using immunocytochemistry for β-amyloid precursor protein. Acta Neuropathol (Berl) 87: 55-62.

Singleton RH, Stone JR, Okonkwo DO, Pellicane AJ, Povlishock JT. 2001. The immunophilin ligand FK506 attenuates axonal injury in an impact-acceleration model of traumatic brain injury. J Neurotrauma 18: 607-614.

Singleton RH, Zhu J, Stone JR, Povlishock JT. 2002. Traumatically induced axotomy adjacent to the soma does not result in acute neuronal death. J Neurosci 22: 791-802.

Smith DH, Cecil KM, Meaney DF, Chen XH, McIntosh TK, et al. 1998. Magnetic resonance spectroscopy of diffuse brain trauma in the pig. J Neurotrauma 15: 665-674.

Smith DH, Chen XH, Xu BN, McIntosh TK, Gennarelli TA, et al. 1997. Characterization of diffuse axonal pathology and selective hippocampal damage following inertial brain trauma in the pig. J Neuropathol Exp Neurol 56: 822-834.

Smith DH, Nonaka M, Miller R, Leoni M, Chen XH, et al. 2000. Immediate coma following inertial brain injury dependent on axonal damage in the brainstem. J Neurosurg 93: 315-322.

Smith DH, Wolf JA, Lusardi TA, Lee VM, Meaney DF. 1999. High tolerance and delayed elastic response of cultured axons to dynamic stretch injury. J Neurosci 19: 4263-4269.

Stone JR, Okonkwo DO, Dialo AO, Rubin DG, Mutlu LK, et al. 2004. Impaired axonal transport and altered axolemmal permeability occur in distinct populations of damaged axons following traumatic brain injury. Exp Neurol 190: 59-69.

Stone JR, Okonkwo DO, Singleton RH, Mutlu LK, Helm GA, et al. 2002. Caspase-3-mediated cleavage of amyloid precursor protein and formation of amyloid β peptide in traumatic axonal injury. J Neurotrauma 19: 601-614.

Stone JR, Singleton RH, Povlishock JT. 2000. Antibodies to the C-terminus of the β-amyloid precursor protein (APP): A site specific marker for the detection of traumatic axonal injury. Brain Res 871: 288-302.

Stone JR, Singleton RH, Povlishock JT. 2001. Intra-axonal neurofilament compaction does not evoke local axonal swelling in all traumatically injured axons. Exp Neurol 172: 320-331.

Strich SJ. 1956. Diffuse degeneration of the cerebral white matter in severe dementia following head injury. J Neurol Neurosurg Psychiatry 19: 163-185.

Strich SJ. 1961. Shearing of nerve fibers as a cause of brain damage due to head injury: A pathological study of twenty cases. Lancet 2: 443-448.

Stys PK. 2005. General mechanisms of axonal damage and its prevention. J Neurol Sci 233: 3-13.

Suehiro E, Singleton RH, Stone JR, Povlishock JT. 2001. The immunophilin ligand FK506 attenuates the axonal damage associated with rapid rewarming following posttraumatic hypothermia. Exp Neurol 172: 199-210.

Suzuki T, Bramlett HM, Ruenes G, Dietrich WD. 2004. The effects of early post-traumatic hyperthermia in female and ovariectomized rats. J Neurotrauma 21: 842-853.

Wilkinson AE, Bridges LR, Sivaloganathan S. 1999. Correlation of survival time with size of axonal swellings in diffuse axonal injury. Acta Neuropathol (Berl) 98: 197-202.

Wilson S, Raghupathi R, Saatman KE, MacKinnon MA, McIntosh TK, et al. 2004. Continued in situ DNA fragmentation of microglia/macrophages in white matter weeks and months after traumatic brain injury. J Neurotrauma 21: 239-250.

Wolf JA, Stys PK, Lusardi T, Meaney D, Smith DH. 2001. Traumatic axonal injury induces calcium influx modulated by tetrodotoxin-sensitive sodium channels. J Neurosci 21: 1923-1930.

Xiao-Sheng H, Xiang Z, Zhou F, Luo-An F, Shuang W. 2004. Calcium overloading in traumatic axonal injury by lateral head rotation: A morphological evidence in rat model. J Clin Neurosci 11: 402-407.

Yaghmai A, Povlishock J. 1992. Traumatically induced reactive change as visualized through the use of monoclonal antibodies targeted to neurofilament subunits. J Neuropathol Exp Neurol 51: 158-176.

Yanagawa Y, Tsushima Y, Tokumaru A, Un-no Y, Sakamoto T, et al. 2000. A quantitative analysis of head injury using T2*-weighted gradient-echo imaging. J Trauma 49: 272-277.

17 Blood–Central Nervous System Barriers: The Gateway to Neurodegeneration, Neuroprotection and Neuroregeneration

H. S. Sharma

1	**Introduction**	**366**
1.1	"Neuroprotection" Misleading?	368
1.1.1	Endothelial, Glial, and Neuronal Interaction	370
2	**The Brain Extracellular Space**	**370**
2.1	CSF Influences Extracellular Microenvironment of the Brain	371
2.1.1	CSF is a Conduit between Brain and Peripheral Endocrine Function	372
2.2	BBB Restricts Protein Entry into the Brain and CSF Microenvironments	373
2.2.1	Size-Dependent Entry of Proteins across the BBB and BCSFB	374
2.2.2	Low Interstitial Protein Concentration is Neuroprotective	375
2.3	The BCSFB vs. BBB	375
2.4	The BSCB vs. BBB	376
2.4.1	Luminal vs. Abluminal Barrier Function	378
3	**CNS Edema Formation**	**381**
3.1	Basic Concepts of Edema Formation	382
3.1.1	Biomechanics of Brain Edema Formation	382
3.1.2	Biophysical Factors Influencing Brain Edema Formation	383
3.2	Types of Brain Edema Formation	383
3.2.1	Osmotic, Metabolic, and Traumatic Edema	383
3.3	Biochemical Mediators of Brain Edema Formation	384
3.4	Does BBB Disruption Always Accompany Brain Edema Formation?	385
4	**Alterations in BCSNB in Stressful Conditions**	**385**
4.1	Stress Enhances Penetration of Viruses into the Brain	386
4.2	Emotional Stress and the BBB	386
4.2.1	Stress Induced Hypoactivity: A Model of Depression	386
4.2.2	BBB Disruption in Immobilization Stress	386
4.2.3	Selective BBB Breakdown to Protein Tracers in Immobilization Stress	387
4.2.4	BSCB Permeability in Immobilization Stress	387
4.2.5	Alterations in Spontaneous EEG Activity	392
4.2.6	Structural Changes in the Brain in Immobilization Stress	393
4.3	Forced Swimming: Another Model of Depression	394
4.3.1	Continuous Forced Swimming Disrupts BBB to Proteins	394
4.3.2	BSCB Permeability in Forced Swimming	397
4.3.3	Morphological Changes in the Brain Following Forced Swimming	397

© 2009 Springer Science+Business Media, LLC.

4.4	Sleep Deprivation Stress	397
4.4.1	BBB and Brain Pathology in Sleep Deprivation Stress	398
4.5	Environmental Heat Stress	398
4.5.1	Hyperthermia: An Adjunct Therapy for Cancer Treatment	398
4.5.2	BBB Permeability in Heat Stress	399
4.5.3	BSCB Permeability in Heat Stress	399
4.5.4	Brain Edema Formation in Heat Stress	399
4.5.5	Spontaneous EEG Alters in Heat Stress	400
4.5.6	Structural Changes in the Brain Following Heat Stress	401
4.5.7	Ultratstructural Damage of Neuropil	401
4.5.8	Ultratstructural Changes in the Cerebral Endothelium	402
4.6	Functional Significance of BBB Disruption in Stress	402
5	***Psychostimulants Influence BCNSB Permeability***	**405**
5.1	Morphine Dependence and Withdrawal Influences BBB Function	405
5.1.1	Morphine Withdrawal Stress	405
5.1.2	BBB Permeability to Evans Blue and Radioiodine	406
5.1.3	Leakage of Albumin across the BBB during Morphine Withdrawal	406
5.1.4	Extravasation of Lanthanum at the Ultrastructural Level	409
5.2	Morphological Changes in Morphine Dependence or Withdrawal	409
5.2.1	Nerve Cell Reaction	409
5.2.2	Glial Cell Reaction	409
5.2.3	HSP Expression	409
5.2.4	Ultratstructural Changes in the Neuropil	410
5.3	Methamphetamine Induces Albumin Leakage across the BBB	412
5.3.1	Methamphetamine Induces Edema Formation and Structural Changes in the CNS	412
5.4	MDMA Alters BBB Permeability	412
5.4.1	MDMA Induces Brain Edema Formation	413
5.4.2	MDMA Induces Cell Injury and Upregulates HSP Expression	413
5.5	Psychostimulants Alter Cellular Stress Response	413
5.6	Psychostimulants Induced Decrease in BBB Function: Unanswered Questions	414
6	***Nanoparticles Influence BCNSB Function***	**414**
6.1	Nanoparticles and Neurotoxicity	415
6.2	Nanoparticles Alter BBB Permeability	415
6.2.1	EBA Staining	415
6.2.2	Radiotracer Extravasation	415
6.2.3	Endogenous Albumin Leakage	416
6.2.4	Lanthanum Extravasation	416
6.3	Nanoparticles Induce Edema Formation	420
6.4	Nanoparticles Induce Morphological Changes in the Brain	420
6.4.1	Nanoparticles and Stress Protein Reaction	420
6.4.2	Nanoparticles and Astrocytic Reaction	420
6.4.3	Nanoparticles and Myelin Changes	420
6.4.4	Ultrastructural Observations	421
6.5	Nanoparticles and Neurodegeneration	421
7	***Alterations in the BCNSB in Traumatic Injuries***	**421**
7.1	Spinal Cord Injury	421
7.1.1	Spinal Cord Microvessels Differ from Brain and Contain Collagen	421
7.1.2	Vascular Reaction in SCI	422
7.1.3	BSCB Permeability in SCI	422

7.1.4	Spinal Cord Edema Formation	422
7.1.5	New Model of SCI	423
7.1.6	BSCB Disturbances in Human Cases of SCI	428
7.2	Peripheral Nerve Lesion	428
7.2.1	Spinal Nerve Lesion Induces Leakage of Albumin across the BSCB	428
7.2.2	Spinal Nerve Lesion Influences Astrocytic Activation	429
7.2.3	Functional Significance of BSCB Disturbances in Nerve Lesion	431
7.3	Traumatic Brain Injuries	431
7.3.1	Open Brain Injury	432
7.3.2	Closed Head Injuries	435
8	***Pharmacological Manipulation of BCNSB and Neuroprotection***	**436**
8.1	Neurotrophins Influence BCSNB Function to Enhance Neuroregeneration	436
8.2	Antibodies against Injury Factors Attenuate BCNSB Dysfunction and CNS Pathology	439
8.3	Neurochemical Synthesis Inhibitors Reduce BCNSB Disruption and CNS Pathology	440
8.4	Neurochemical Receptor Modulation Influence BCNSB and CNS Pathology	443
8.5	Antioxidants Attenuate BCSNB Breakdown and CNS Pathology	443
8.6	Biogenic Amine Neurotoxins Exacerbate BCNSB Breakdown and CNS Pathology	444
9	***Mechanisms of BCSNB Breakdown in CNS Insults***	**444**
9.1	Passage of Tracers across the BCNSB	445
10	***General Conclusion***	**445**

Abstract: The microenvironment of the central nervous system (CNS) is precisely and meticulously maintained by a set of dynamic physiological barriers located within the cerebral microvessels of the brain (blood–brain barrier, BBB) and the spinal cord (blood–spinal cord barrier, BSCB), as well as within the epithelial cells of the choroid plexus separating the blood and cerebrospinal fluid (CSF) interface (blood–CSF barrier, BCSFB). The physicochemical properties of these cellular barriers are quite comparable to that of an extended plasma membrane. The BBB and the BSCB are quite tight to small molecules (12 Å, Lanthanum ion), whereas BCSFB is less restrictive in nature. On the other hand, the ependymal cell linings of the cerebral ventricles and spinal canal referred to as CSF–brain barrier do not normally restrict passage of several molecules of small sizes. However, protein transport across these blood–CNS barriers (BCNSB) is severely restricted. Entry of proteins into the CNS microenvironment induces vasogenic edema formation that is primarily responsible for cell and tissue injury. These BCNSB are often compromised under a wide variety of psychological, traumatic, metabolic, ischemic, environmental, or chemical insults leading to neuronal, glial, and axonal damage. Opening of the BCSNB to various endogenous or exogenous substances and proteins alters the molecular, cellular, biochemical, immunological, and metabolic environment of the CNS leading to abnormal neuronal function and/or brain pathology. This review is focused on current status of the BCSNB breakdown in experimental models of emotional stress, traumatic injuries, psychostimulants as well as key environmental health hazards, i.e., nanoparticles and heat exposure. Breakdown of the BCNSB in these conditions altered gene expression and induced brain pathology leading to neurodegeneration. Attenuation of the BCSNB disruption with drugs or antibodies affecting neurochemical metabolism and/or neurotrophic factors markedly reduced the development of brain pathology. Taken together, these novel observations strongly point out the role of BCNSB as a "gateway" to the neurodegeneration, neuroprotection, and/or neuroregeneration in neurological diseases.

List of Abbreviations: BBB, Blood–brain barrier; BCNSB, blood–central nervous system barrier; BCSFB, blood–CSF barrier; BDNF, brain derived neurotrophic factor; CBF, cerebral blood flow; CHI, closed head injury; CNS, central nervous system; CSF, cerebrospinal fluid; GFAP, glial fibrillary acidic protein; HS, heat stress; HSP, heat shock protein; HVSA, high voltage slow activity; IGF-1, insulin like growth factor-1; LHRF, lutenizing hormone-releasing factor; LVFA, low voltage fast activity; MBP, myelin basic protein; SCI, spinal cord injury; TBI, traumatic brain injury; TRH, thyrotropin-releasing hormone; WBH, whole body hyperthermia

1 Introduction

The "barriers" in the central nervous system (CNS) are the key regulators of neuronal transport in a highly specialized manner and are located at multiple interfaces (Sharma and Westman, 2004). These specialized barriers carry out specific functions that are essential for brain metabolism as well as extra- or intracellular fluid homeostasis (Spector and Johanson, 1989; Jones et al., 1992; Ghersi-Egea et al., 2001; Sharma and Westman, 2004; Sharma and Johanson, 2007). These transport interfaces is localized either within the endothelium or in the epithelium between the blood and various regions of the CNS (Rapoport, 1976; Sharma and Westman, 2004). The endothelial interfaces which regulate the "solute traffic" between the blood and the brain or the spinal cord regions (Sharma, 2005a) are true "barriers", and are referred to as the blood–brain barrier (BBB) and blood–spinal cord barrier (BSCB), respectively (Sharma, 2004a, b). Whereas, the epithelial interfaces located in the choroid plexuses and control the movement of substances between blood and cerebrospinal fluid (CSF) (Parandoosh and Johanson, 1982; Sharma and Johanson, 2007) is relatively leaky and known as the blood–CSF barrier (BCSFB). The permeability properties of BBB, BSCB, or BCSFB are largely regulated through the tight junctions for the intercellular transport; whereas, the transcellular passage of substances depends on the physicochemical properties of the cell membrane (Rapoport, 1976; Bradbury, 1979). Thus, the permeability properties of these CNS barriers are very similar to those of an extended plasma membrane (Sharma and Westman, 2004). Accordingly, the lipid-soluble solutes easily penetrate the BCNSB interface, and the lipid insoluble nonelectrolytes or proteins are strictly excluded at the barrier sites (Rapoport, 1976). On the other hand, the BCSFB appears to be less stringent, as intravascular substances can enter into the CSF at faster rates compared to the BBB (Davson, 1967; Rapoport, 1976; Sharma and Westman, 2004).

However, these BCNSB become leaky in a wide variety of noxious insults to the brain or spinal cord, and/or in several brain diseases of diverse origin (see ◉ *Table 17-1*; Rapoport, 1976; Bradbury, 1979;

◘ Table 17-1
Summary of various experimental and disease conditions in which the BBB is disrupted to various tracers

Disease/conditions	BBB breakdown	Possible mechanisms
A. Neurodegeneration		
Alzheimer's disease	Serum proteins	Vesicular transport
Brain tumors, neoplasms	HRP	Endothelial cell permeability
Schizophrenia	Microperoxidase	
Dementia	Lanthanum	
Ischemia, infarction	Radiotracers	
Peripheral nerve lesion		
Leukemia		
B. Trauma		
Mechanical, hypoxia	HRP, radiotracers	Vesicular transport
Hyperoxia, ischemia	Evans blue	Widening of tight junctions[a]
Metabolic insults	Trypan blue	Endothelial cell permeability
Incision[b]	Lanthanum	
Stab wounds, concussion		
Cryogenic lesions, thermocoagulations		
C. Influence of chemicals		
Serotonin[b], histamine	HRP, Evans blue	Vesicular transport
Protamine, norepinephrine	Lanthanum	Widening of tight junctions[a]
5-HTP[b], bradykinin	Microperoxidase	Endothelial cell permeability
Prostaglandins, leukotrienes	Radiotracers	
Glutamate, L-NAME, chemical induced		
Convulsions, cAMP, dibutyric cAMP		
Adrenaline, 6-OHDA, indomethacin,		
Bicuculline, angiotensin, amphetamine,		
Matrimonial, pentylenetetrazol		
D. Hyperosmotic solutions		
Infusion of various electrolytes	Radiotracers, HRP	Widening of tight junctions[a]
Nonelectrolytes	Evans blue, Lanthanum	Vesicular transport
	Microperoxidase	Endothelial cell permeability[c]
E. Irradiations		
X-ray irradiation	Radiotracers	Vesicular transport
a-, b-particle irradiation	Evans blue	Widening of tight junctions[a]
Microwave irradiation	Microperoxidase	Endothelial cell permeability
Ultrasonic irradiation		
F. Drugs and venoms		
Alcohol and other lipid solvents	Radiotracers	Vesicular transport
Bile salts, saponin, lysolecithin	Evans blue	Widening of tight junctions[a]
Cobra venoms, *Escherichia coli* endotoxin	Microperoxidase	Endothelial cell permeability
G. Toxicity to chemicals		
Diodrast, iodopyract, mercuric chloride	HRP, Evans blue	Vesicular transport
Nickel chloride, lead, manganese	Radiotracers	Endothelial cell permeability

◘ Table 17-1 (continued)

Disease/conditions	BBB breakdown	Possible mechanisms
H. Lesion/stimulation		
Locus coeruleus	Water	Not known
I. Vascular diseases		
Hypertension[a] (mechanical, Chemical, or metabolic)	Radiotracers	Widening of tight junctions[a]
	Evans blue, HRP	Endothelial cell permeability
Hypotension	Lanthanum	Vesicular transport
Carotid artery occlusion		
Air embolism, gas embolism		
Atherosclerosis, periarteritis nodosa,		
Thromboangitis obliterans, diabetic		
Vasculitis		
J. Loss of autoregulation		
Acute hypertension	HRP, Evans blue	Widening of tight junctions[a]
Hypertensive encephalopathy	Radiotracers	Vesicular transport
Intracranial hypertension		Endothelial cell permeability[c]
Hypovolaemic shock, hypervolaemia		
K. Autoimmune diseases		
Viral encephalitis, experimental allergic	HRP, Evans blue	Widening of tight junctions[c]
Encephalomyelitis, polyneuritis,	Radiotracers	Vesicular transport
Multiple sclerosis		
L. Stressful situations		
Immobilization[b], forced swimming[b]	HRP, Evans blue	Endothelial cell permeability
Heat exposure[b], seizures, training	Radiotracers	Vesicular transport
In water maze, adrenalectomy	Lanthanum	Widening of tight junctions[c]
Electroconvulsive shock		
Morphine withdrawal/dependence[c]		
M. Electromagnetic radiation		
Mobile telephony	Evans blue, HRP	Vesicular transport
Microwave radiation	Evans blue, HRP, sucrose	Endothelial cell permeability

Compiled from various sources: Rapoport (1976); Sharma (1982, 1999, 2004d); Sharma et al. (1998a)
[a]Known to occur
[b]Authors own investigations

Sharma and Westman, 2004). Breakdown of these BCNSB will allow proteins and other unwanted substances to enter into fluid microenvironment of the CNS (Sharma, 2004a). This would result in vasogenic brain edema formation and alterations in the molecular, cellular, biochemical, and immunological homeostasis causing cell and tissue injuries (Sharma, 2004b). Exposure of neural, glial, or axonal environments to edema fluid and other toxic substances, normally prevented by the active barrier, could be instrumental in inducing neurodegenerative changes (Sharma et al., 1998a). Thus, prevention of the BCSNS leakage in brain disease would be of paramount importance to thwart neurodegeneration and to enhance neuroprotection and/or neuroregeneration. This review is focused on the importance of BCNSB as one of the key factors in regulating neuroprotection in experimental models of traumatic, emotional, chemical, or environmental factors leading to brain pathology.

1.1 "Neuroprotection" Misleading?

The term "Neuroprotection" used to denote rescue of nerve cells (Sharma, 2004a, b, c). However in the CNS, the nonneural cells, i.e., glial cells and endothelial cells are equally important for brain function (See ❷ *Figure 17-1*, Nag et al., 2002; Moon and Bunge, 2005; Sofroniew, 2005; Taub and Uswatte, 2005).

◘ Figure 17-1
(a) Diagrammatic representation showing spatial relationships between cerebral capillary, neuron, glial cells and the extracellular space around them. The glial cells surround more than 85% of the cerebral capillary. The number of glial cells and capillary endothelial cells far exceeds that the number of neurons. Thus, the cellular integrity of the glial and cerebral endothelial cells (non-neural cells) is crucial to maintain healthy neuronal functions (Modified after Schmidt, 1978; Sharma, 1982, 1999, 2004b). (b) Schematic diagram showing structure and function of the blood–brain barrier (*BBB, left*) and brain–blood barrier (*right*). The endothelial cells (E) in the brain microvessels are connected with tight junction and are surrounded by a thick basement membrane. The glial cells (G) and nerve cells (N) around the cerebral endothelial cell are also seen. The glial cell covering of the brain microvessels is less intense in the spinal cord compared to the brain, particularly in large diameter microvessels (for details see text). Intravascular tracer is stopped at the tight junction and donot penetrate the luminal endothelial cell membrane to reach extracellular space indicating a very tight BBB. Similarly, intrathecal tracer does not pass the abluminal endothelial cell membrane and/or the tight junctions to reach the vascular compartment indicating that the brain–blood barrier effectively regulates the exchange of substances between CNS microenvironment and the vascular system. Whether disease processes equally affect the integrity of the BBB and brain–blood barrier is not known (Modified after Rapoport, 1976; Sharma, 1982, 1999; Sharma, 2007b)

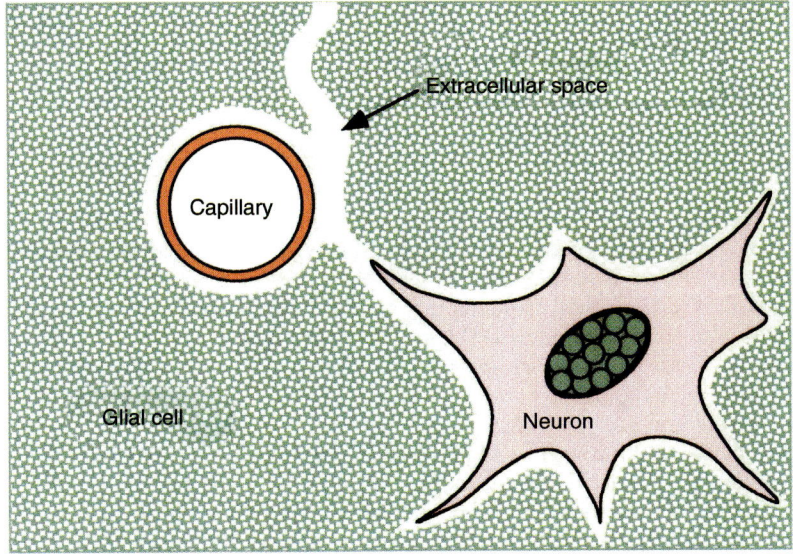

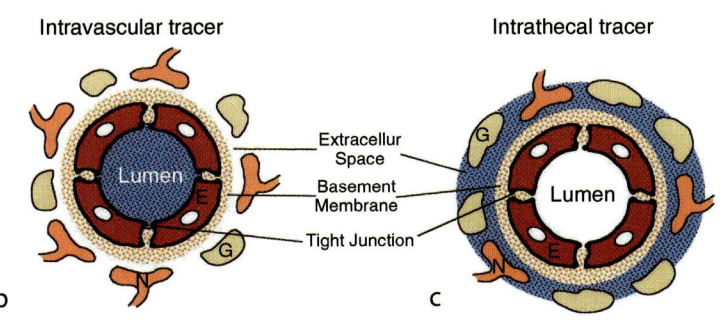

The number of nonneural cells far exceeds the number of neural cells in the CNS (Martin et al., 2001; Heins et al., 2002, ◐ *Figure 17-1*). However, most attention is still focused to rescue nerve cells in neurodegenerative diseases. Obviously to restore the normal function of the CNS, revival of neurons, glial cells, and endothelial cells are equally important (Sharma and Westman, 2004; Sharma, 2005a, 2007a). This is because of the fact that the nerve cell function is largely dependent on the normal endothelial and glial cells function (Sharma, 1999; Sharma and Westman, 2004). It is quite likely that reducing the damage of endothelial cells and/or glial cells by drugs or pharmacological agents in brain diseases will also attenuate nerve cell injury (Sharma, 1982, 1999, 2004a, b, c, d, e, 2005a, b, 2007a, b; Sharma and Westman, 2000). Thus, the term "*neuroprotection*" is misleading as neurons are in the minority in the CNS and their function depends on the survival of nonneural cells and *vice versa*. In this chapter the term "neuroprotection" denotes protection of all the "neural" and "nonneural" components in the CNS.

1.1.1 Endothelial, Glial, and Neuronal Interaction

The neurons are suspended in a pool of glial cells and endothelial cells (◐ *Figure 17-1a*). The number of glial cells are about 6–10 times higher that the number of neurons. The endothelial cells could further exceed in number than the neurons and the glial cells in the CNS of several mammalian species (◐ *Figure 17-1*). The endothelial cells in the CNS are connected with tight junctions (◐ *Figure 17-1b*) and regulate the transport of various substances across the BBB from the blood to the brain as well as from the brain to the blood compartments (see ◐ *Figure 17-1b, c*).

1.1.1.1 The Brain–Blood Barrier The endothelial cell membrane offers equal resistance for transport of nonessential substrates and protein from the blood to the brain and *vice versa* (Sharma, 2004a, 2007b). Thus, release of several neurochemicals within the brain under normal or physiological conditions, do not normally enter into the blood stream because of the abluminal endothelial cell membrane, referred to as the "brain–blood barrier" (Sharma, 1999, 2004a, 2007b). However, most studies on drug transport across the BBB are limited to the *luminal* cell membrane function and alteration in the *abluminal* properties of the cerebral endothelium is still largely ignored (Sharma, 2004a; ◐ *Figure 17-1c*).

Normal function of the brain–blood barrier appears to be very important for the maintenance of the CNS function (Sharma, 2005a). A leaky abluminal barrier will allow passage of several vasoactive or toxic substances into the blood stream that may have rebound consequences on the brain circulation leading to CNS pathology (Sharma, 2007b). Whether, luminal barrier disruption is always accompanied with identical damage to the abluminal barrier function is entirely a new subject and requires detailed investigations (Sharma, 2007a). Obviously, maintenance of normal barrier function across the luminal and abluminal membranes of the cerebral endothelial cells is necessary for healthy neuronal functions. However, breakdown of the *abluminal* barrier permeability is still a new concept and requires detailed investigations.

2 The Brain Extracellular Space

During brain pathology, extravasation of endogenous serum proteins occurs in the brain extracellular environment influencing neuronal, glial, and axonal functions (See ◐ *Figures 17-1* and ◐ *17-2*; Rapoport, 1976; Bradbury, 1979; Sharma and Westman, 2004). The brain extracellular space in the adult monkey is about 18% of the total brain weight in the gray matter and approximately 15% in the white matter (Fenstermacher 1970). The adult rabbits, cats, and dogs exhibit extracellular spaces in the brain between 17% and 23% in the gray matter based on inulin and sucrose profile analysis (Levin et al., 1970). Electron microscopic studies together with cortical conductance measurement suggest a value of 15–25% of brain extracellular space in the gray matter of several mammalian species (van Harreveld, 1966; Nevis and Collins, 1967; Rapoport, 1976). Leakage of BBB or BCSFB allows protein and other unwanted substances into the extracellular space (see ◐ *Figures 17-1* and ◐ *17-2*) leading to altered neuronal functions.

◘ Figure 17-2
Anatomical localization of the blood–brain barrier (BBB), the blood–CSF barrier (BCSFB), and other exchange interfaces in the CNS (After Sharma, 1982). The BBB resides in the endothelial cells of the cerebral capillaries (1) and is responsible for exchange (*arrows*) of essential solutes and nutrients from blood to the brain. The BCSFB is located within the choroidal epithelial cells that are connected with the tight junctions (2). BCSFB allows exchange of solutes (*arrows*) and nutrients between blood, CSF and the cerebral compartments (Rapoport, 1976). The ependymal linings of cerebral ventricles often referred to as CSF–brain interface permit exchange of essential solutes and drugs between CSF and cerebral compartments up to certain limit freely (*arrows*). Modified after Woodbury, 1974; Sharma, 1982

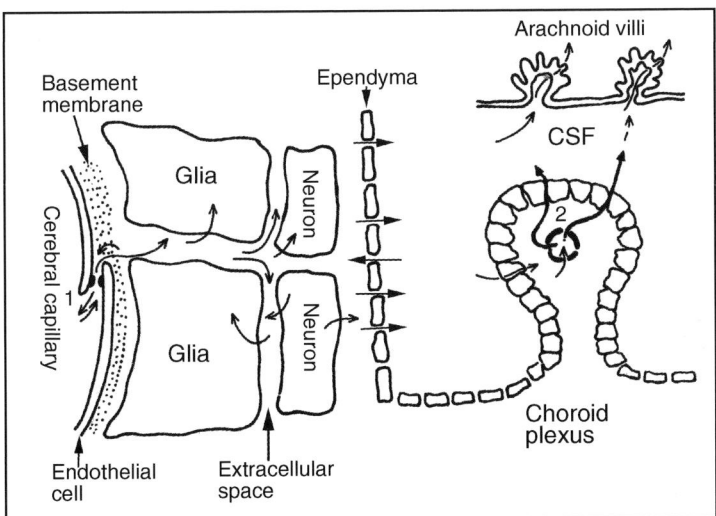

Desynchronization of neuronal function results in abnormalities in the electroencephalogram (EEG) activity reflecting brain pathology (see 4.2.5 for details, Sharma and Dey, 1988; Sharma, 2004b, d). In experimental animals studies, breakdown of the BBB is associated with desynchronization of EEG activity that is well correlated with the duration of altered BBB function (Sharma, 2004b, d). These observations indicate that leakage of tracers within the brain extracellular environment affecting brain function.

2.1   CSF Influences Extracellular Microenvironment of the Brain

The microenvironment of the neuronal and glial cells at the brain CSF interfaces, i.e., ependymal linings of the ventricles and at the pial–glial surfaces of the brain (see ❷ *Figure 17-2*) are rapidly influenced by the composition of the CSF (see ❷ *Figure 17-3*; Davson, 1967; Rapoport, 1976). Thus, solutes within the CSF can reach to the cerebral microenvironment by diffusion up to 15 mm in the adult human brain and much less (2–5 mm) in smaller animals (Davson, 1967; Cserr, 1971). The diffusion between CSF and brain largely occur through the extracellular space of the brain (see ❷ *Figure 17-3*) that roughly equals to 15–18% of the brain weight (see earlier, Fenstermacher and Rall, 1983). This suggests that alterations in the CSF composition will influence CNS functions.

The composition of CSF is largely regulated by several factors e.g., production, metabolism, and uptake of solutes at the choroid plexus; restriction of intercellular diffusion and/or specific transport at the choroid plexus or the cerebrovascular interfaces; and the rate of CSF production or excretion by bulk flow (Davson, 1967; Rapoport, 1976).

Figure 17-3
Schematic representation of exchange interfaces among plasma, CSF, extracellular fluid (ECF), and brain cells (Modified after Rapoport, 1976; Sharma, 1982). The blood–brain barrier (BBB) represents exchange interface (1) between plasma and the ECF and is very tight (*thick line, thin arrows*). Whereas, the Blood–CSF barrier provides exchange interfaces between plasma and the CSF (2) and is relatively less tight (*thin line, thick arrows*) compared to the BBB (for details, see text). On the other hand, the ECF exchange interfaces with the CSF (3) and the Brain cells (4) are not very restrictive in nature (*thin line and thick arrows*). Obviously, the exchange between plasma and brain cells (*thin arrows*) is restricted to only essential substances (see Rapoport, 1976; Sharma, 1982 for details)

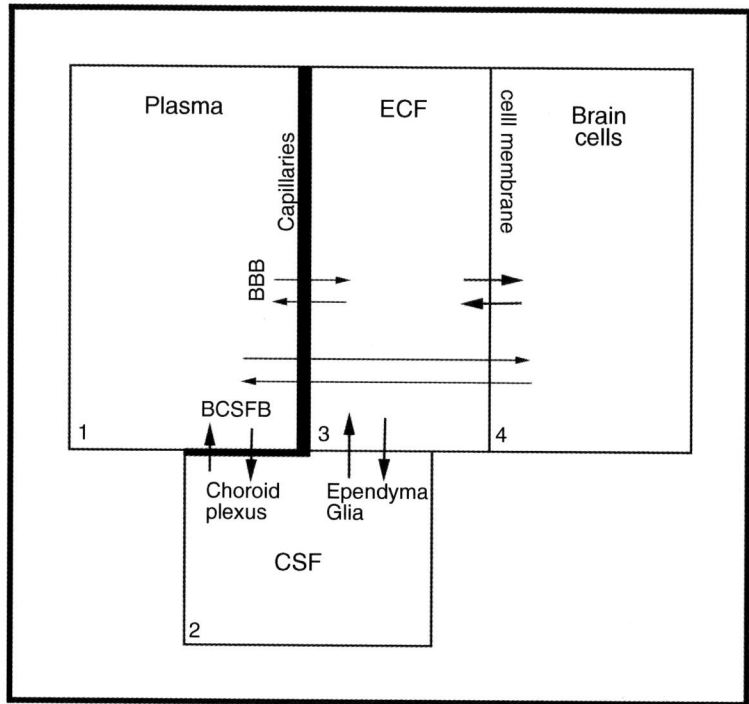

Thus, even slight changes in the CSF ion compositions, e.g., Ca^{++}, Mg^{++}, or K^+ affect the emotional states, heart rate, respiration, blood pressure, vasomotor response, muscle tone, and gastrointestinal motility (Leusen, 1972; Rapoport, 1976). Alterations in CSF pH influences brain metabolism, autoregulation of cerebral blood flow (CBF), and respiration (Cserr, 1971). This indicates that the CSF system regulates the chemical environment of the neurons and glia effectively under normal conditions. An altered CSF composition during CNS diseases will lead to disturbances in the neurochemical microenvironment resulting in serious brain dysfunction (see below).

2.1.1 CSF is a Conduit between Brain and Peripheral Endocrine Function

The CSF plays an important conduit for integration of endocrine functions between brain and the periphery (see ❯ *Figures 17-3* and ❯ *17-4*). Thus, thyrotropin-releasing hormone (TRH) and lutenizing hormone-releasing factor (LHRF) synthesized in the hypothalamus and released into the CSF are

Figure 17-4

Transport of various substances across the blood–CSF barrier (Modified after Rapoport, 1976; Adapted from Sharma, 1982). The BCSFB is less restrictive to small electrolytes and water (*thick arrows*). Ions and proteins in very small quantities can be transported in and out across the choroidal epithelium (*thin arrows*). The tight junctions are present between the apical cells of the choroidal epithelium. It appears that the tight junctions between the choroidal epithelium are less tight as compared to the endothelial tight junctions, a subject that requires additional investigation (for details see text; Sharma, 1982, 2004b)

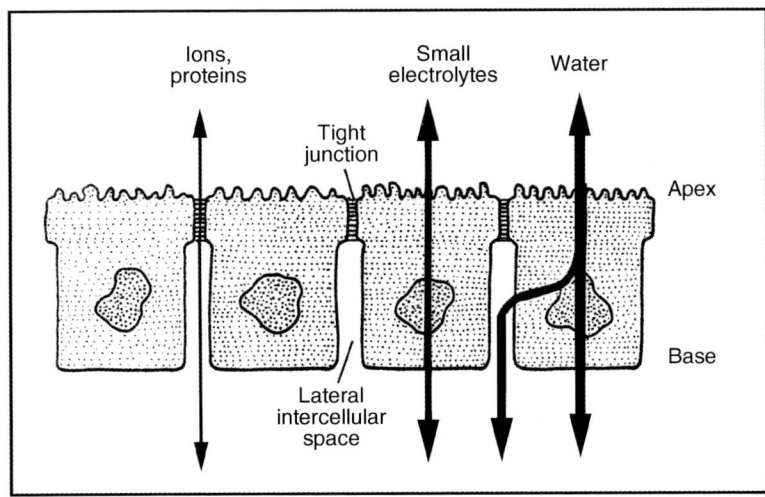

transported to the median eminence at the floor of the third ventricle through the CSF (Vigh-Teichmann and Vigh, 1970). From there, these hormones are transported into the portal blood via ependymal tanycytes and gain access to the anterior lobe of the pituitary (Rapoport, 1976). This ependymal transport of hormones to the pituitary occurs even when the neuronal connections between hypothalamus and median eminence is severed. However, increased concentration of thyroxin and catecholamines in the vascular system markedly alters the transport and accumulation of TRH by the ependymal tanycytes (Knigge and Joseph, 1974). This indicates that the CSF effectively regulates brain and peripheral endocrine functions.

Similarly, peripheral endocrine organs release hormones in response to pituitary hormone interactions via CSF (Rapoport, 1976) and exert feedback control over the pituitary, hypothalamus, and other higher CNS centers through CSF system (Knigge and Silverman, 1972; Porter, 1973; Rapoport, 1976). These observations indicate that CSF actively integrates endocrine functions between brain and the periphery. Whether, these hormonal transport and integration between brain and blood compartments by CSF is altered during brain diseases is still not known and require additional investigation.

2.2 BBB Restricts Protein Entry into the Brain and CSF Microenvironments

The BBB located within the cerebral endothelium severely restricts the protein transport into the brain fluid microenvironment that is size-related (Westergaard and Brightman, 1973). Thus, the normal BBB prevents the entry of immune γ-globulins G (IgG), IgA (Mol wt. ≈ 180 kD), and IgM (Mol wt. 900 kD). However, some arterioles in the brain can transfer proteins, i.e., Na^+ fluorescein (Mol wt. 376 kD) through pinocytosis (Cutler et al., 1970). In addition, non-barrier regions the brain and the choroid plexus allow some transport of proteins into the brain or CSF compartments (Hochwald and Wallenstein, 1967a, b). Transport of proteins across the choroid plexus depends on the rate of CSF secretion and involves intercellular diffusion (● *Figure 17-4*) as well as increased vesicular transport (Klatzo et al., 1965; Rapoport, 1976). These restrictions of proteins across the BBB and BCSFB is largely responsible for a very low protein content in the CSF (0.4%) and almost negligible content within the brain extracellular fluid compared to the plasma (see ● *Table 17-2*). Thus, normal levels of CSF proteins, e.g., IgG or IgA accounts for only 0.2–0.4% of their plasma levels and the level of IgM is present only in trace amounts.

◘ Table 17-2
Plasma and lumbar CSF concentration of solutes in human subjects

Solutes	CSF, mM	Plasma, mM	CSF/plasma
Protein, mg/100 cm^3	28	6800	0.004
Glucose, mg/100 cm^3	50–80	110	0.6
Osmolality	0.289	0.289	1
pH	7.307	7.397	–
PaCO$_2$ torr	50.5	41.1	–
Na$^+$	147	150	0.98
K$^+$	2.86	4.63	0.62
Cl$^-$	113	99	1.1
Mg^{++}	4.46	3.22	1.4
Ca^{++}	4.56	9.40	0.49
HCO$_3$	23.3	26.8	0.87
P (inorganic), mg/100 cm^3	3.40	4.70	0.73

Data after Davson (1967); Rapoport (1976); Sharma and Westman (2004)

On the other hand, in several diseases of the CNS the level of the IgG are considerably elevated in the CSF (Rapoport, 1976). Depending on the severity of the diseases, several immune γ-globulins are also found within the brain extracellular spaces (Davson, 1976). In some neurological diseases, β-globulins and prealbumin are seen in both the brain extracellular space and in the CSF (Frick and Scheid-Seydel, 1960; Rapoport, 1967). It is quite likely that these proteins enter into the brain extracellular space through CSF and then taken up by the brain cells. Once entered into the cells, these proteins are degraded intracellularly by lysosomes within the glial cells and/or in the pericytes (Rapoport, 1976). Alternatively, these proteins are either accumulated within the vesicles or transported to other cells, i.e., glia, neurons or arachnoids (Novikoff, 1960; Klatzo et al., 1965; Kristensson and Olsson, 1973; Sharma and Westman, 2004; see ● *Figure 17-2*, Woodbury, 1974). However, the detailed mechanism of protein transport or degradation in the CNS is still not well known.

2.2.1 Size-Dependent Entry of Proteins across the BBB and BCSFB

Entry of proteins into the CSF or in brain extracellular space is dependent on their sizes (see ● *Figures 17-3* and ● *17-4*; Lowenthal, 1972; Rapoport, 1976). Breakdown of the BBB and BCSFB thus facilitates entry of several proteins according to their sizes and molecular diameter (Rapoport, 1976; Bradbury, 1979; Sharma and Westman, 2004). After BBB disruption, accumulation of low concentrations of very large sized proteins and rapid entry of albumin compared to globulins occurs into the brain compartment (Rapoport, 1976). This protein transport from the periphery into the brain fluid microenvironment is largely due to facilitated diffusion or filtration and not by the vesicular transport. This is because of the fact that the vesicular transport of proteins does not discriminate between the sizes of the molecules (Rapoport, 1976).

However, several proteins in the sizes of 20 Å in diameter, such as microperoxidase do not cross the BCSFB in the rat (Brightman et al., 1970; Rapoport, 1976). Studies carried out in our laboratory suggest that lanthanum (Mol Diam 12 Å) is stopped at the BBB and BSCB interfaces in rats and mice (Sharma, 2005a, 2007b). However, some leakage of lanthanum can be seen occasionally at the BCSFB (Sharma and Johanson, 2007; see later). This suggests that the BBB and BSCB are very tight to several molecules compared to BCSFB and strictly regulate the size dependent entries of various substances under normal conditions.

2.2.2 Low Interstitial Protein Concentration is Neuroprotective

Maintenance of low protein concentrations in the CSF and in brain extracellular environment appears to be neuroprotective in nature (Sharma et al., 1998a; Sharma and Westman, 2004). Earlier studies demonstrated that elevation of interstitial proteins attenuates the ability of the brain to avoid edema formation (Rapoport, 1976). Furthermore, increased penetration of proteins into the brain fluid microenvironment results in edema formation (Klatzo et al., 1965; Brightman et al., 1970). In addition, rapid release of protein antigens from the CNS could induce peripheral autoimmune diseases (Cohen, 1968; Rapoport, 1976). This is further supported by the findings that high blood titers of antibodies are seen in brain diseases associated with brain pathology, i.e., encephalitis, brain injuries, Tay–Sachs diseases and multiple sclerosis (Cohen, 1968; Rapoport, 1976). It is believed that high blood level of antibodies to protein induced damage to the BBB and BCSFB (Rapoport, 1976; Sharma and Westman, 2004). Disruption of the BBB will further allow some lymphocytes to enter into the brain microfluid environment as evident in several neurological diseases (❷ Table 17-1; Connolly et al., 1968). These lymphocytes once enter into the brain induce destruction of the cellular immune reaction and generate antibodies that will further enhance cell or tissue injury and inhibit axonal regeneration (Feringa et al., 1974; Sharma, 2007a, b).

2.3 The BCSFB vs. BBB

The BBB and BCSFB perform crucial functions to regulate brain fluid microenvironment in selective and specific manners in health and/or disease (❷ Figures 17-2 and ❷ 17-4; Sharma and Johanson, 2007). Thus, the transport properties across the BBB or BCSFB under normal conditions are quite different in nature.

However, the 'leakiness' of the BBB and BCSFB across the tight junctions to large molecular tracers such as microperoxidase (20 kD; 2 nm diameter), horseradish peroxidase (HRP; 40 kD, 6 nm diameter) and Evans blue albumin (EBA; 68 kD) are very similar in nature (Rapoport, 1976; Sharma, 2004a, b). Thus, these tracers are stopped at the tight junctions within the endothelium across the BBB whereas, they easily diffuse into lateral intercellular spaces between the choroidal epithelial cells, but their passage is stopped at the apical tight junctions (near the CSF) by the BCSFB so that their transport into the ventricles is thwarted (see ❷ Figures 17-2 and ❷ 17-3).

However, the choroid plexus epithelial lining is relatively less tight compared to the BBB (see ❷ Figures 17-3 and ❷ 17-4). This is supported from the findings that the electrical resistance of choroidal epithelium is only 150–175 $\Omega \cdot cm^2$ (Strazielle and Ghersi-Egea, 1999; Shu et al., 2002) compared to 8,000 $\Omega \cdot cm^2$ for the cerebral endothelium (Smith and Rapoport, 1986; Butt and Jones, 1992). Based on these observations and calculation of the influx rate constants for large molecules into corresponding permeability-area products (ml/min) indicates a 100-fold greater permeability for inulin across the BCSFB compared to the BBB (see ❷ Figure 17-4; Rapoport, 1976; Sharma and Johanson, 2007).

Furthermore, perfusion of buffered lanthanum chloride (Mol Diam 12 Å) solutions into the internal carotid artery showed presence of the electron-dense tracer occasionally into the ventricular lumen, whereas the passage of this tracer is stopped at the tight junctions within the cortex (Castel et al., 1974; Sharma et al., 1998a). Conversely, when lanthanum was administered into the cerebral ventricles, the tracer passed paracellularly through the tight junctions between the epithelial cells but not within vesicular profiles or cell cytoplasm in the choroidal epithelium (Castel et al., 1974). This indicates the leakiness of the tight junctions of the choroid plexus epithelial cells (Castel et al., 1974; Rapoport, 1976). On the other hand, tight junctions between brain endothelial cells stopped the penetration of lanthanum (120 nm diameter) at the tight junctions in brain or spinal cord (Brightman and Reese, 1969; Rapoport, 1976; Sharma, 2004a, b, c, d, 2005a, b).

Other fundamental differences between BBB and BCSFB includes active transports of organic solutes and ions across the blood–brain and the blood–CSF interfaces (Sharma, 1999; Sharma and Westman, 2004; Sharma and Johanson, 2007). In general, the cerebral capillaries are responsible for providing neurons with energy substrate and amino acids for protein synthesis, whereas the choroid plexuses transport water, ions, vitamins, and growth factors to the CNS (Johanson et al., 2000).

Taken together, it appears that the normal function of BBB and BCSFB are needed to effectively regulate the water and ion homeostasis to strictly maintain the fluid microenvironment of the brain for healthy neuronal functions.

2.4 The BSCB vs. BBB

The characteristics of the BSCB appear to be quite similar to that of the BBB (Brightman et al., 1970; Rapoport, 1976; Bradbury, 1979; Cervós-Navarro and Ferstz, 1980; Sharma, 1982, 1999). However, the detailed structure and function of the BSCB with regard to the enzymes and neurochemical receptors located on the spinal versus cerebral microvessels in normal and in pathological conditions are still not well known (● *Table 17-3*; Sharma, 2004a, 2005a).

The spinal cord endothelial cells are connected with tight junctions and lack vesicular transport (● *Figure 17-5*). A thick basement membrane surrounds the endothelial cells of spinal cord, similar to the brain (● *Figure 17-5*; Sharma, 2005a). However, minor differences in astrocytes–microvessel interactions are seen in superficial spinal cord microvessels (see later; Griffiths, 1975, 1976, 1978). The large superficial vessels of the spinal cord contain enough deposits of glycogen, a feature not normally seen in the brain microvessels (Majno and Palade, 1961; Beggs and Waggener, 1975; Noble et al., 2002). The functional significance of such glycogen deposits in relation with the barrier properties of the cord is not known in all its details (see ● *Figure 17-5b*; Sharma, 2005a).

Several exogenous tracers such as EBA, radioiodine, HRP, and lanthanum when administered into the blood stream or added to the fixative (i.e., lanthanum) unable to cross the endothelial cells of the spinal cord in normal animals (Noble and Wrathall, 1987, 1988, 1989; Olsson et al., 1990, 1992; Sharma et al., 1993a, 1995a, 1998a, b; Olsson et al., 1995; Sharma, 1999, 2000a, b; Sharma and Hoopes, 2003). The electron dense tracer, lanthanum is confined within the lumen of the spinal cord microvessel (● *Figure 17-5*). The spinal cord endothelial cells do not show microvesicular profile containing lanthanum and the passage of tracer is stopped at the tight junction (● *Figure 17-5*).

However, the spinal microvessels react differently to ischemia as compared to the brain microvessels (see Sharma, 2004a for review). Thus, local impairment of spinal cord circulation results in less severe cellular damage compared to the brain ischemia (Nemecek, 1978; Osterholm, 1978; Windle, 1980; Osterholm et al., 1987). This appears to be due to a lack of regional differences in spinal cord microcirculation and metabolisms compared to the brain (Windle, 1980).

Although, leakage of proteins across the cerebral microvessels following brain injury leads to the development of vasogenic edema formation, (Klatzo et al., 1965; Brightman et al., 1970; Olsson et al., 1992, 1995; Sharma et al., 1998a, b, c, d; Sharma, 1999; 2004a, 2005a), it is still uncertain whether of BSCB disruption to proteins will initiate edema formation in the cord in identical manner (Wagner et al., 1971; Parker et al., 1973; Wagner and Stewart, 1981; Noble and Wrathall, 1989; Olsson et al., 1990; Sharma and Olsson, 1990). The vertebral canal provides space to accommodate edematous cord tissue up to some extent compared to the calvarium where edematous expansion of the brain compresses vital centers causing instant death (Sypert 1990a, b; Sharma, 2004a).

Furthermore, the spinal cord capillaries lack contractile elements (Windle, 1980). Thus, the blood flow is normally regulated by the arterioles that contain smooth muscle cells in their walls. The spinal cord venous system comprises venules with thin walls that lack valves (Windle, 1980). The dorsal longitudinal

Table 17-3
Available information on the structure and function of the BBB and BSCB at a glance

Components	BBB	BSCB
A. Endothelial cells		
Tight junction	+++	+++
Lack of vesicular transport	++++	++++
Basement membrane	Thick	Thick
Surrounded by	+++++	++++
Astrocytes	+++++	?
B. Enzymes present		
C. Permeability to tracers		
HRP	−	−
Iodine	−	−[a]
Albumin	−	−
Dextran	−	−
Fluorescein	−	−
Evans blue	−	−
D. Increased permeability		
Trauma	+++	+++[a]
Ischemia	+++	+++
Hypoxia	++++	+++
Irradiation	+++	+++
Stress	+++[a]	+/?[a]
E. Neurochemicals		
Serotonin	++++[a]	++++[a]
Histamine	+++[a]	++?[a]
cAMP	++++	Not known
NO	+++	Not known
F. Disease conditions		
Chemical induced		
Seizures	++++	Not known
Hypertension	++++	Not known
Tumours	++++	Not examined
Incision	++++[a]	++++[a]
Bacterial infection	++++	Not examined
Alzheimer's disease	++++	Not examined
EAE	Not known	++++
Transection	Not examined	++++
Compression	++/?	++++
Contusion	++/?	+++
Weight drop	+/?	+++
Systemic hyperthermia	++++[a]	+?[a]
Hyperthermia treatment	++++	++++
G. Experimental conditions		
Local		
Endotoxin infusion	++++	Not examined
Brain injury	+++++	Not examined
Spinal injury	Not examined	++++

Table 17-3 (continued)

Components	BBB	BSCB
Osmotic shrinkage	++++	Not examined
Nanoparticles[a]		
Al	++	++
Ag	++	+++
Cu	+++	++++

Compiled from various sources (for details see text): + = present; ++ = mild; +++ = moderate, ++++ = strong; +++++ = extensive; ? = more data needed

[a]Author's own investigation, modified after Sharma 2004d

venous trunk receives blood supply from the dorsal funiculi and the dorsal spinal gray matter. The ventral longitudinal venous trunk receives blood from most of the spinal cord through segmental sulcal veins. At few places, the dorsal and ventral venous trunks are connected through anastomotic venules (Batson, 1942; Windle, 1980). Due to these anatomical differences, the effect of trauma to the cord and its reaction to the injuries greatly vary in the spinal cord segments (❯ *Table 17-3*; Sharma, 2004a).

Ultrastructural studies in cats showed quite similarity between spinal and brain microvasculature (Dohrmann et al., 1971; Beggs and Waggener, 1975, 1976). However, many spinal cord microvessels (<60%) with the lumen size of capillaries (<8 mm; Rhodin, 1968) have a complete or partial perivascular space containing collagen (Rhodin, 1968; Beggs and Waggener, 1976; Griffiths et al., 1978a). The remaining spinal cord microvessels, the vascular and glial basement membranes are similar to that found in brain capillaries (Griffiths, 1976; Griffiths et al., 1978a, b). Taken together, these observations suggest that the BBB and BSCB properties are quite similar in nature and both offer severe restriction for transport across the blood and tissue interface under normal conditions (Sharma, 2004a).

2.4.1 Luminal vs. Abluminal Barrier Function

There are reasons to believe that both luminal and abluminal cell membranes of the brain or spinal cord endothelial cells constitute effective barrier functions in the CNS. This is further supported by our observations showing breakdown of the BBB or BSCB to Evans blue and [131]Iodine in model experiments using intravenous (luminal) infusion of serotonin or its topical (abluminal) application on the brain and spinal cord microvessels in rats or mice with advancing age (Sharma et al., 1990a, 1995b; Sharma, 2004c).

2.4.1.1 **Luminal Application of Serotonin** Intravenous infusion of 10 or 20 μg/kg/min serotonin for 10 min in the rats induces BBB and BSCB disruption to protein tracers (see ❯ *Figure 17-6*; Sharma et al., 1990a; Sharma, 2004c). This indicates that application of serotonin into the luminal side of the brain or spinal cord microvessels is capable to induce BBB and BSCB breakdown. However, when 10 or 20 μg/kg/min dose of serotonin was infused in adult rats (30–32 weeks), the BSCB permeability to protein tracers was much less evident compared to the young rats (12–18 weeks old) (see ❯ *Tables 17-4* and ❯ *17-5*). This observation supports the idea that receptor sensitivity to serotonin is considerably reduced with advancing age (Sharma et al., 1992a; Simonsen et al., 2004). Alternatively, serotonin-induced signal transduction mechanisms or stimulation of other neurochemical mediators, e.g., prostaglandins or NO production is less pronounced with aging (Sharma, 2004c, d; Sharma and Alm, 2004).

This effect of serotonin on BBB or BSCB function appears to be species dependent. Thus, intravenous infusion of serotonin (10 μg/kg/min for 10 min) was unable to induce breakdown of the BBB or BSCB permeability in the mouse. However, significant increase in tracer extravasation was observed in mice following high concentration of serotonin (20 μg or 30 μg/kg/min for 10 min) in a dose-dependent manner (❯ *Tables 17-4* and ❯ *17-5*). These observations suggest that almost two times higher dose of serotonin is needed to induce BBB or BSCB breakdown in mice compared to rats. Less sensitivity

◘ Figure 17-5

Anatomy of the blood–spinal cord barrier (BSCB) (Adapted after Sharma, 2005a). Ultrastructure of one spinal cord endothelial cell profile and its surroundings in the rat (a) The endothelial cells in the spinal cord are connected with tight junction (TJ, *thick arrow*). A thick basement membrane (BM, 10–12 nm, *thin arrow*), the glial cell (G) and the surrounding astrocytic processes (Asp) around the spinal cord endothelial cells are clearly visible. One nerve cell with prominent nucleus (N) in the close proximity of the microvessel is apparent. The glial cell covering of the spinal cord microvessels is less intense compared to the cerebral microvessels, particularly in large diameter microvessels (Windle, 1980; Sharma, 2004b). The passage of lanthanum (La) from lumen is stopped at the tight junction (TJ) and no infiltration of the tracer in seen in this normal rat either within the cell cytoplasm and/or in the basal lamina. Another important difference between cerebral and spinal microvessels is the presence of collagen in the large superficial microvessels of the spinal cord; (b, c) The collagen (Col) fibers are elastic fibers that are arranged in layers that are arranged either transversely (Col 1) or longitudinally and/or obliquely (Col 2) as seen after transverse sectioning (see Sharma, 2005a for details). In these collagen containing microvessels, the tight junction (b, arrow) and the endothelial cell membrane (c, arrow) is impermeable to the lanthanum tracer. Basal lamina (arrow head) clearly separates the endothelial layers and collagen fibers (b, c). Myelin sheath (MS) below the layer of collagen is also seen (c). Bar: a = 1 μm (modified after Sharma, 2004a); b, c = 200 nm. Reproduced with permission from Sharma, 2005a

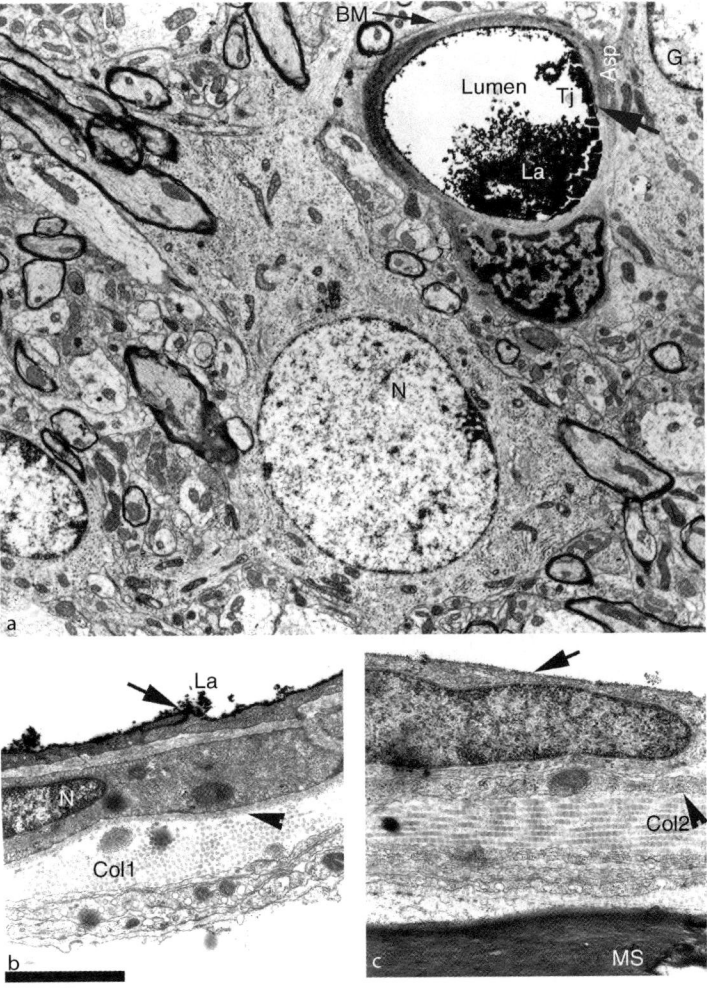

◻ **Figure 17-6**

Ultratstructural changes in the blood–brain barrier (BBB, A) and blood–spinal cord barrier (BSCB, B) to lanthanum tracer following intravenous infusion of serotonin in the rat and mice (for details see text; Sharma, 2004c). Lanthanum seen as dark black particles is infiltrated into the extra cellular space within the neuropil (A: *a–c, arrows*) after leaking into the basal lamina (A: *b, arrows*). Membrane damage, edema and vacuolation (*) are prominent in the regions exhibiting lanthanum exudation. However, the tight junction is not widened to allow passage of lanthanum across the BBB or BSCB (A, B). Extravasation of lanthanum across the spinal microvessel is also evident in mice after serotonin infusion (B). Spinal microvessels in mice show infiltration of lanthanum into the endothelial cell cytoplasm and in the basal lamina (B: *a–f, arrows*). Perivascular edema and myelin damage (*) are clearly seen. Extravasation of lanthanum into the basal lamina seen in spinal microvessel (B: a,b,e, *arrows*) without widening the tight junction (*arrow heads*). Bars: 0.5 μm. Only a selected part of the endothelial cell exhibits lanthanum infiltration within the cytoplasm or in the basal lamina in the rat or mice brain and spinal cord (A, B, *arrows*). Extravasation of lanthanum in spinal cord neuropil in mice is also seen (B: e). Presence of lanthanum in the endothelial cell cytoplasm as well as in the basal lamina in selects parts of the endothelial cell membrane are very similar in brain and spinal cord microvessels in rats or mice following serotonin infusion (A: b; B: e,f). Few microvesicular profiles containing lanthanum (B,d) are also seen following serotonin infusion. Perivascular edema, vacuolation and cell damage (*) are frequent around the microvessels showing lanthanum extravasation. Bars: A: a = 1.6 μm; b = 0.5 μm; c = 1.5 μm; B: a–d = 1.5 μm; e = 1.6 μm; f = 500 nm (Data modified after Sharma, 2004c)

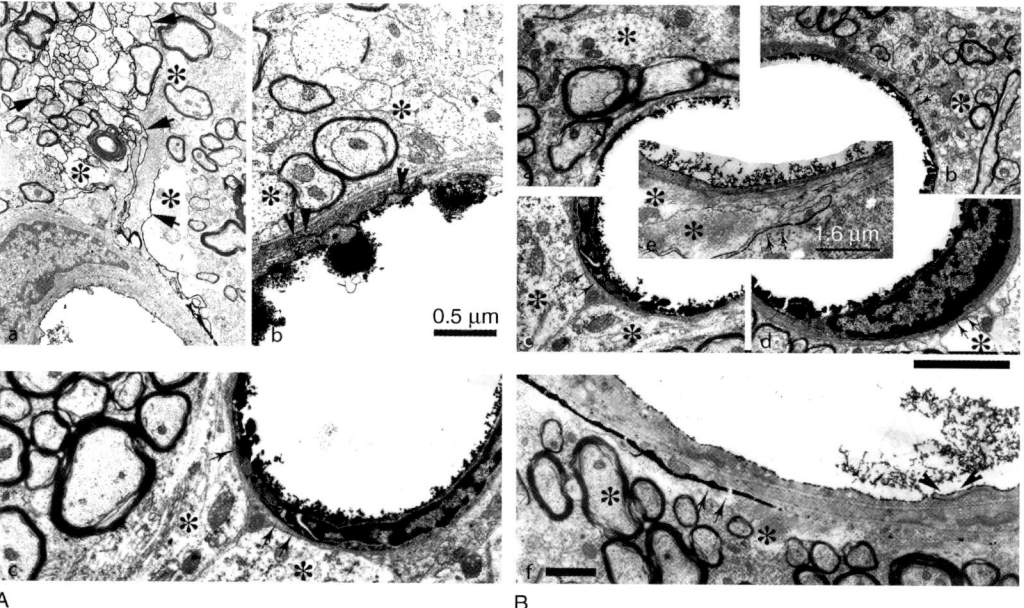

of serotonergic receptors and/or density on the spinal cord microvessels in mice compared to rats could play important roles (Sharma et al., 1995b).

2.4.1.2 Abluminal Application of Serotonin To study the influence of abluminal application of serotonin on the BSCB permeability, the amine was applied on either the exposed parietal cerebral cortex or on the exposed T10–11 segment of the spinal cord in rats or mice (Sharma et al., 1989, 1990a, 1992a, b, 1995b) (❯ *Tables 17-4* and ❯ *17-5*). Topical application of serotonin (0.1 μg/kg/min for 10 min) was able to induce breakdown of the BBB or BSCB to Evans blue and [131]Iodine tracers in rats within 15 min (❯ *Tables 17-4* and ❯ *17-5*). This effect of serotonin on BCNSB breakdown was reversible within 2 h.

Table 17-4
Effect of serotonin on blood–brain barrier permeability after different modes of application in the rat

Type of experiment	5-HT dose/min, 10 min	BBB permeability			
		Evans blue, mg %		[131I]-Sodium, %	
		Right half	Left half	Right half	Left half
A. Saline infusion					
Intravenous (6)	–	0.18 ± 0.02	0.20 ± 0.04	0.24 ± 0.04	0.26 ± 0.03
Internal carotid artery (6)	–	0.18 ± 0.04^a	0.19 ± 0.05	0.22 ± 0.06^a	0.25 ± 0.04
Topical application (5)	–	0.22 ± 0.06^a	0.18 ± 0.04	0.25 ± 0.04^a	0.28 ± 0.06
B. Serotonin infusion					
Intravenous (8)	10 μg/kg	1.56 ± 0.15**	1.48 ± 0.16**	1.88 ± 0.21**	1.92 ± 0.18**
Internal carotid artery (6)	2 μg/kg	1.44 ± 0.12**^{a}	0.32 ± 0.12	1.64 ± 0.23***^{a}	0.45 ± 0.22
Topical application (7)	0.05 μg	0.89 ± 0.14**^{a}	0.26 ± 0.08	0.93 ± 0.08***^{a}	0.24 ± 0.06

Source: Data from Sharma et al. (1995b)
**$P < 0.01$, Dunnet test for multiple group comparison (compared from corresponding saline controls)
Values are mean ± SD; Figures in parentheses indicate number of animals
^{a}Perfused hemisphere

This suggests that brain or spinal cord microvessels in one species are equally sensitive to serotonin induced BBB or BSCB breakdown (● Table 17-4). However, when serotonin (0.1 μg/kg/min) was applied on the spinal cord surface in older animals (30–35 weeks of age) it failed to induce any leakage of either tracer (● Table 17-5) suggesting that age of animals is an important factor in serotonin induced BSCB permeability.

On the other hand, in mice, high dose of serotonin on the abluminal side is needed to induce BSCB breakdown (● Tables 17-4 and ● 17-5). Thus, application of 0.2 μg and 0.3 μg/kg/min serotonin on brain or spinal cord microvessels in mice (age 12–20 weeks old) for 10 min resulted in a dose dependent extravasation of Evans blue and [131I]Iodine in the spinal cord that was reversible within 2 h (● Tables 17-4 and ● 17-5).

These observations support the hypothesis that the luminal or abluminal surfaces of the microvessels in brain and spinal cord exhibit similar barrier properties. However, the permeability properties of the BBB or BSCB depend on the age of animals, and/or the species used. Since identical concentration of serotonin induces breakdown of the BBB or BSCB permeability in one animal species, the permeability properties of the brain and spinal cord microvessels appears to be quite similar in nature.

3 CNS Edema Formation

Disruption of the BBB or BSCB leads to brain and spinal cord edema formation. Brain edema is one of the serious complications of several neurological diseases and following traumatic, ischemic, or hypoxic injuries to the CNS (Bakay and Lee, 1965; Klatzo and Seitelberger, 1967; Reulen and Kreysch, 1973; Cervós-Navarro and Ferszt, 1980; Sharma et al., 1998a). A progressive brain edema compresses the cerebral components causing an increase in the intracranial pressure and brain tissue softening (Davson, 1967; Aukland, 1973; Cervós-Navarro and Ferszt, 1980 Sharma et al., 1998a; Sharma, 2005a). Expansion of volume swelling of the brain in the closed cranium depresses the vital centers causing instant death (Klatzo et al., 1958, 1965; Rapoport, 1976). No effective therapy for brain edema has still been worked out.

◘ Table 17-5
Effect of luminal (i.v.) and abluminal (topical) application of serotonin on the Blood–Spinal Cord Barrier (BSCB) breakdown in rats and mice

Experiment type	n	Serotonin dose 10 min, μg/kg/min	Route	Species	BSCB permeability Evans blue, μg %	[131]Iodine, %
A. Rat experiments						
Control	8	Saline	i.v.	Rat	0.24 ± 0.06	0.36 ± 0.07
5-HT infusion	6	5 μg/kg/min	i.v.	Rat	0.32 ± 0.08	0.32 ± 0.09
	8	10 μg/kg/min	i.v.	Rat	1.56 ± 0.23**	1.86 ± 0.12**
	6	20 μg/kg/min	i.v.	Rat	2.06 ± 0.14**	2.89 ± 0.14**
	8	10 μg/kg/min	i.v.	Rat[a]	0.34 ± 0.14	0.56 ± 0.12
	6	10 μg/kg/min	i.v.	Rat[b]	0.65 ± 0.12*	0.89 ± 0.08*
Control	5	Saline	Topical	Rat	0.26 ± 0.04	0.34 ± 0.04
5-HT application	6	0.1 μg/kg/min	Topical	Rat	0.89 ± 0.03**	1.04 ± 0.02**
	8	0.1 μg/kg/min	Topical	Rat[a]	0.38 ± 0.04	0.46 ± 0.02
	6	0.1 μg/kg/min	Topical	Rat[b]	0.35 ± 0.08	0.49 ± 0.06
NaCl	6	0.06 mM	Topical	Rat	0.24 ± 0.04	0.38 ± 0.06
B. Mice experiments						
Control	6	Saline	i.v.	Mice	0.28 ± 0.08	0.38 ± 0.06
5-HT infusion	8	10 μg/kg/min	i.v.	Mice	0.35 ± 0.08	0.44 ± 0.12
	10	20 μg/kg/min	i.v.	Mice	1.67 ± 0.11**	2.06 ± 0.12**
	8	30 μg/kg/min	i.v.	Mice	2.14 ± 0.14**	2.48 ± 0.17**
	8	20 μg/kg/min	i.v.	Mice[b]	0.54 ± 0.12	0.67 ± 0.14
Control	5	Saline	Topical	Mice	0.32 ± 0.05	0.34 ± 0.08
5-HT application	10	0.2 μg/kg/min	Topical	Mice	0.87 ± 0.07**	0.98 ± 0.06**
	8	0.3 μg/kg/min	Topical	Mice	1.13 ± 0.10**	1.54 ± 0.12**
	8	0.2 μg/kg/min	Topical	Mice[b]	0.44 ± 0.02	0.57 ± 0.04

*$P < 0.01$, ANOVA followed by Dunnet's test from serotonin infused group, **$P < 0.01$, ANOVA followed by Dunnet's test from one control; Values are mean ± SD
[a]Old rats (26–34 weeks Old)
[b]tracers were given 2 h after the end of serotonin infusion

3.1 Basic Concepts of Edema Formation

Edema is defined as an increase in water content of the brain or spinal cord that may occur in any specific regions following a local or global insult to the CNS (Klatzo, 1972, 1967; Cervós-Navarro and Ferszt, 1980; Joó, 1987; Klatzo, 1987; Sharma et al., 1993a). Depending on the magnitude and severity of the primary insult, the edema fluid spreads within the extracellular space resulting in swelling of the whole brain within 24–72 h (Bakay and Lee, 1965; Klatzo et al., 1965; Hirano, 1971; Cervós-Navarro and Ferszt, 1980). Resolution of brain edema often takes weeks and months after the primary insult (Cervós-Navarro and Ferszt, 1980 Reulen et al., 1990; Sharma, 2004a, b, c, d; Sharma et al., 1998a).

The early symptoms of brain edema include headache, nausea, vomiting, disturbances of consciousness, and occasionally coma (Cervós-Navarro and Ferstz, 1980; Reulen et al., 1990). Increased intracranial pressure and papilledema further complicates the matter and leads to herniation of brain and vascular infarction leading to death (Davson, 1967; Rapoport, 1976; Bradbury, 1979; Cervós-Navarro and Ferstz, 1980).

3.1.1 Biomechanics of Brain Edema Formation

Brain edema formation can be explained on the basis of the basic principles of capillary filtration as suggested by Starling about 110 years ago (Starling, 1896). Thus, edema formation could occur in any tissue when the rate of capillary filtration exceeds the rate of fluid removal from the perivascular interstitium (Starling, 1896; Rapoport, 1976; Sharma et al., 1998a). Since brain is devoid of any lymphatic system, the accumulated fluid from cerebral capillaries percolate through the extracellular spaces of the brain fluid microenvironment in order to reach CSF spaces (Davson, 1967).

The rate of capillary filtration and the driving force of water across the capillary cell membrane are dependent on the differences of hydrostatic and osmotic pressures between the vascular and cerebral compartments (Wiederhielm, 1968). Due to presence of tight junctions, the cerebral capillaries are almost impermeable to proteins, electrolytes, and other water soluble nonelectrolytes (Aukland, 1973). Due to this reason, an effective osmotic pressure is created between plasma and brain compartments that are equal to the total sums of contributions from salts, impermeable nonelectrolytes and proteins (Lieb and Stein, 1971). Although, the venous beds are comparatively more permeable than the arterial beds (Aukland, 1973), the details about local differences in the osmotic and/or hydrostatic pressure gradients across the intracerebrovascular beds are still not well known (Staub, 1974; Rapoport, 1976; Sharma et al., 1998a).

Since the protein osmotic pressure in brain is zero and CSF protein content is negligible, the CSF osmolality due to nonprotein solutes is very similar to that of plasma osmolality (see ❯ *Table 17-2*; Davson, 1967; Klatzo and Seitelberger, 1967; Rapoport, 1976). Thus, the driving force for water permeability in brain is largely dependent on capillary hydrostatic pressure and protein osmotic pressure (Staub, 1974; Rapoport, 1976). Accordingly, when fluid accumulates in the brain, edema develops and increases the brain volume. The rate of fluid accumulation however, depends on brain compliance that is very low (Miller, 1976) and depends on local tissue pressures, which varies in different brain regions (Guyton, 1963; Sharma et al., 1998a).

3.1.2 Biophysical Factors Influencing Brain Edema Formation

Several biophysical factors, such as hydrostatic and osmotic pressure differences and brain compliance influence brain edema formation (as above Rapoport, 1976). Perturbation in the autoregulation of CBF due to various kinds of CNS insult aggravates the biomechanics of brain edema formation (❯ *Table 17-1*; Rapoport, 1976). A defective autoregulation of CBF allows increase in hydrostatic pressure and thus breakdown of the BBB permeability that induces rapid brain edema formation irrespective of the forces of hydrostatic pressure and brain compliances (Rapoport, 1976; Bradbury, 1979; Cervós-Navarro and Ferszt, 1980). However, the exact relationship between capillary filtration, driving force of water due to osmotic or hydrostatic gradients, hydraulic conductivity, and tissue compliance on brain edema formation is still largely unknown (Rapoport, 1976; Sharma, 1982, 1999; Reulen et al., 1990).

3.2 Types of Brain Edema Formation

Brain edema was classified 40 years ago by Klatzo to be either "vasogenic" or "cytotoxic" in origin (Klatzo, 1967, 1972). Leakage of plasma proteins and water into the cerebral extracellular space following traumatic or osmotic insults to the blood vessels leads to formation of vasogenic edema. Whereas, intracellular accumulation of water caused by alterations in cell metabolism induces development of cytotoxic edema (Klatzo et al., 1965; Klatzo, 1987). Cytotoxic edema represents swelling of glial cells or astrocytes, as well as the neurons, microglia, and vascular endothelial cells (Rapoport, 1976; Sharma et al., 1998a).

However, a considerable overlap exists between these two types of brain edema following severe injury, e.g., hyperthermic insults (Jóo, 1987; Klatzo, 1987; Sharma et al., 1998a). Thus, in vasogenic brain edema intracellular accumulation of water can also occur preceding the BBB breakdown (Rapoport, 1976; Sharma and Hoopes, 2003).

3.2.1 Osmotic, Metabolic, and Traumatic Edema

Disturbances in the osmotic balance between plasma and brain lead to development of osmotic edema that interferes with brain metabolism leading to metabolic edema (Rapoport, 1976). In osmotic edema, a decrease in plasma osmolality enhances transport of water into the cerebral compartment allowing brain swelling at the expense of CSF or blood volumes and raises the intracranial pressure (Rosomoff and Zugibe, 1963). However, the cell membranes of the cerebrovascular endothelium and brain cells are intact (Rymer and Fishman, 1973). The inflow of water into brain dilutes Na^+ and K^+ concentrations per gram wet weight without affecting their dry weight concentration (van Harreveld and Dubrovsky, 1967; Sharma et al., 1998a).

Inhibition of brain metabolism with pharmacological agents or ischemic brain injuries result in cell swelling and induce metabolic edema (Stern et al., 1949; Stern, 1959). Swelling of cells and subcellular organelles including mitochondria is due to inhibition of active transport mechanisms at the local membranes (Leaf, 1973). Intracellular accumulation of Na^+ and loss of K^+ occurs due to inhibition of active transport of Na^+ and drag Cl^- and water into the cells (Rapoport, 1976; Sharma et al., 1998a). Thus, inhibition of oxidative enzymes by cyanide causes gray and white matter edema in which the glial cells swell at the expense of extracellular space and reduces the cortical conductivity (Hirano et al., 1967; Hirano, 1971; Baethmann and Van Harreveld, 1973; Rapoport, 1976). Axonal swelling and vesiculation of myelin in which water filled vacuoles appear between myelin lamellae represents white matter edema (Hirano et al., 1967; Hirano, 1971). The brain water content is increased by more than 3% of the wet weight (Reulen and Kreysch, 1973). However, the cerebrovascular permeability to Trypan blue or ^{131}I-albumin is not altered (Pappius, 1970).

On the other hand, trauma to the brain or cerebral vessels leads to the development of traumatic edema (Pappius, 1970). The BBB largely remains intact in osmotic and metabolic edema, whereas leakage of plasma proteins correlates well with the development of traumatic brain edema (Bakay and Lee, 1965). Traumatic brain edema involves complex processes both at the cerebral capillaries and the brain cells resulting in a progressive and irreversible pathological and metabolic changes (Cervós-Navarro et al., 1988; Sharma, 2004a, b, c). In traumatic edema, vasogenic factors are predominating and the cytotoxic factors significantly contribute to edema formation by damaging the cell membranes and their metabolism (Klatzo, 1972).

3.3 Biochemical Mediators of Brain Edema Formation

Several endogenous neurotransmitters or factors (see ◐ Table 17-6) play important roles in brain edema formation as described elsewhere (for review, see Wahl et al., 1988; Olesen, 1989). Interestingly, these neurochemical mediators are known to alter the BBB permeability as well (Wahl et al., 1988; Olesen, 1989;

◘ Table 17-6
Neurochemical mediators of Brain edema and BBB dysfunction

Chemical mediators	Receptors mediated	BBB permeability tracers	Signal transduction mechanisms	Vasogenic edema
Bradykinin	B_2-kininergic	Na^+-Fluroscein	PI, PKC	Yes
Arachidonic Acid	–	Na^+-Fluorescein FITC-dextran EBA	–	Yes
Free radicals	–	EBA, Na-Fluorescein	–	Yes
Leukotrienes	–	EBA, Na-Fluorescein	–	?
Cytokines	–	Na-Fluorescien	–	?
Histamine	H_2	EBA, Na-Fluorescein FITC-dextran, HRP	cAMP	Yes
Serotonin[a]	$5-HT_2$	EBA, HRP, La^{++}	cAMP	Yes
Endothelin	ET_1	$^{51}Cr, ^{45}Ca$	Prostanoids cAMP, Ca^{++}	?
Nitric oxide[a]	–	MEAP, EBA	–	Yes
Dynorphin[a]	–	EBA, La^{++}	–	Yes

Compiled from various sources
[a]Authors own investigations: Modified after Sharma et al. (1998a, c); Sharma 2004d

Bradbury, 1992; Sharma et al., 1998a, c). Experiments carried out in our laboratory suggest that these biochemical mediators influence brain edema formation by an increase in the blood–brain transport of solutes, ions, and proteins causing alterations in osmotic gradient across the cerebrovascular compartments (Sharma et al., 1990b, 1995b). The altered osmotic gradient will further enhance water transport from the vascular to the cerebral fluid microenvironment (Sharma, 1982, 1999).

It is believed that these neurochemicals act on the cerebral microvessels leading to the breakdown of the BBB permeability (Sharma et al., 1998a). Brain edema develops due to leakage of proteins or alterations in the endothelial cell membrane permeability through neurochemical receptors-mediated changes in signal transduction mechanisms (Rapoport, 1976; Edvinsson and McKenzie, 1977; Bradbury, 1979, 1992; Sharma et al., 1990a, b). The neurochemicals may directly alter the autoregulation of CBF as well (Rapoport, 1976; Bradbury, 1979; Johansson et al., 1990) and affect the brain tissue compliance to induce brain edema formation (Rapoport, 1976; Johansson et al., 1990). However, further studies are needed to understand the molecular mechanism of brain edema formation by neurochemicals and their receptors.

3.4 Does BBB Disruption Always Accompany Brain Edema Formation?

There are reasons to believe that breakdown of the BBB permeability to large molecules results in brain edema formation that could be instrumental in precipitating further cell and tissue injuries (Sharma et al., 1998a, c; Sharma, 2004a, b, c, d, 2005a, b, c, 2006a, 2007c, d). Extravasation of serum proteins together with many unwanted substances when enters into the brain fluid microenvironment leads to several biochemical, immunological, cellular and molecular reactions resulting in adverse cell and tissue reaction and brain dysfunction. Alterations in the brain fluid microenvironment thus open the gates for early neurodegenerative changes that may have long-term consequences on brain function.

Almost all kinds of neurological diseases, e.g., brain tumors, infarction, ischemia, disturbances of cerebral autoregulation, metabolic disturbances, infections of the CNS by viruses and bacteria, autoimmune diseases, hypertensive encephalopathy, traumatic injuries, hypoxia, hyperoxia, hypercapnia, concussion, raised intracranial pressure, or hydrocephalus etc., exhibit breakdown of the BBB function and brain edema formation (see ❿ *Table 17-1*; Rapoport, 1976; Cervós-Navarro and Ferszt, 1980 Dey and Sharma, 1983, 1984; Mohanty et al., 1985, 1989; Reulen et al., 1990; Sharma and Olsson, 1990; Sharma et al., 1992a, b, 1993a, b, 1995a, b, c, 1996a, b). The spread of edema fluid and leakage of plasma proteins in these experimental or clinical situations progresses with time. The magnitude and severity of edematous swelling and concurrently brain damage depends on the intensity and duration of the primary insult (Rapoport, 1976; Bradbury, 1979) and is responsible for neurodegenerative changes. Thus, restoration of the BBB dysfunction in brain pathology is likely to attenuate brain edema development and induce neuroprotection and/or neuroregeneration (Sharma, 2003, 2005d, 2006b, 2007b).

4 Alterations in BCSNB in Stressful Conditions

Long-term mental disturbances are well known to precipitate neurodegeneration (Selye, 1976; see Ballok, 2007 for review) probably by inducing BBB disturbances (Sharma, 1982, 1999). However, investigations on the status of BBB function during emotional stress, anxiety, or depression are still rudimentary (Sharma, 2004b). Few experimental evidences suggest that the BBB is modified in specific brain regions in certain stressful conditions (❿ *Table 17-1*; Gilbert, 1965; Lorenzo et al., 1965; Angel, 1966, 1969; Cutler et al., 1968; Basch and Fazekas, 1970; Bondy and Purdy, 1974; Selye, 1976; Christensen et al., 1981). Thus, stimulation or lesion of central catecholaminergic neurons (Raichle et al., 1975), photic stimulation, and convulsive agents increase the uptake of radiotracers from the blood into specific brain areas (Cutler et al., 1968; Bondy and Purdy, 1974). A selective increase in the BBB permeability is also seen as early as 1 h after monocular eyelid suturing in 1-day-old chick's brain (Bondy and Purdy, 1974). These observations support the idea that stress could lead to increased penetration of blood-borne substances into the brain by modifying BBB permeability and to induce neuronal dysfunction.

4.1 Stress Enhances Penetration of Viruses into the Brain

Several forms of stress increase the susceptibility of animals to bacterial or viral infections (Dantzer and Kelley, 1989; Cohen and Williamson, 1991; Sheridan et al., 1994). Thus, increased morbidity and mortality (Friedman et al., 1970; Sheridan et al., 1994) by various infectious agents, e.g., herpes simplex virus (Rasmussen et al., 1957), influenza virus (Feng et al., 1991; Hermann et al., 1994), and encephalitic viruses (Ben-Nathan et al., 1991, 1998) is seen under a wide variety of stressful conditions compared to normal group (Sharma, 2004b). Mice when inoculated with attenuated variant of West-Nile virus (WN-25) or neuroadapted noninvasive Sindbis strain (SVN) subjected to either cold, isolation, or administration of corticosterone results in enhanced mortality by 50–80% compared to the nonstressed group (Ben-Nathan et al., 1998). The brain titters of viruses in the stressed mice are 3–4-fold higher compared to the unstressed group (Ben-Nathan et al., 1996, 1998). These stressors also resulted in fatal encephalitis by avirulent strain of Semliki Forest virus (SFV-A7) in mice compared to unstressed group where no mortality is seen (see Sharma, 2004b for review).

Stress-induced immunosupression that enhances proliferation of the viruses into the CNS is believed to be the leading cause of exacerbation of the infection-induced pathogenesis. However, an increased permeability of the BBB caused by stress could be another key factor in exacerbation of virus-induced mortality.

4.2 Emotional Stress and the BBB

Our laboratory was the first to examine the hypothesis of stress induced BBB disruption leading to brain damage about 27 years ago (Sharma and Dey 1980; see Sharma, 2004b). Thus, works done in our laboratory clearly showed that several kinds of emotional and environmental stressors induce a selective and specific leakage of proteins in discrete brain regions exhibiting cellular damage. This suggests that opening of the BBB is crucial for cell and tissue injury in stress.

4.2.1 Stress Induced Hypoactivity: A Model of Depression

Animal models of depression and/or simulation of posttraumatic stress disorders (PTSD) are based on stress induced hypoactivity that can be induced in animals precisely using immobilization stress or restraint as well as forced swimming in a limited pool of water (Porsolt et al., 1977, 1978; Kofman et al., 1995; Miyazato et al., 2000).

Restriction of movement in rats by placing them in a tube is often referred to as "restraint" while fastening of limbs on a wooden board is generally known as "immobilization" (Groves and Thompson, 1970; Kvetnansky and Mikulaj, 1970; Sharma and Dey, 1981; Sharma, 1982). Immobilization on a wooden board induces severe stress compared to restraint in a plastic tube (Marti et al., 2001). A decrease in motor activity during immobilization induces sleep disorders in animals thus mimicking depression like behavior (Zebrowska-Lupina et al., 1990).

4.2.2 BBB Disruption in Immobilization Stress

Several animal models of immobilization result in BBB disruption to various tracers at different time points (Sharma and Dey, 1981, 1986a; Belova and Jonsson, 1982; Dvorská et al., 1992; Esposito et al., 2001, 2002). Mild changes in the BBB to ^{14}C-iodoantipyrine in a few brain regions was observed in adult rats subjected to 5 or 15 min immobilization stress (Ohata et al., 1981, 1982). On the other hand, 2 and 6 h immobilization is needed to increase the tracer transport of peptides into different brain regions (Belova and Jonsson, 1982; Dvorská et al., 1992). Leakage of protein tracers, i.e., EBA or $^{[131]}$Iodine occurs in several brain regions after 7–9 h of immobilization in young rats (Sharma and Dey, 1981; Sharma and Dey 1981, 1987a,b). These observations indicate that the magnitude and intensity of the BBB breakdown in immobilization depend on the severity of the stress and the size of tracers used (Sharma, 2004b).

4.2.3 Selective BBB Breakdown to Protein Tracers in Immobilization Stress

Experiments carried out in our laboratory for the first time show that long-term immobilization stress induces a selective increase in the BBB permeability to several protein tracers, i.e., Evans blue, bromophenol blue, HRP, and radioiodine in rats (Sharma and Dey 1981, 1986a, 1987). This increase in BBB permeability depends on the duration and age of the animals (Sharma and Dey, 1981; Sharma, 1982; Sharma and Dey, 1986a).

Young rats (age 8–9 weeks) when subjected to immobilization stress for 2–4 h exhibited leakage of Evans blue or bromophenol blue in the brain in 1 out of 8 rats. When duration of stress is increased to 7–9 h, extravasation of Evans blue and bromophenol blue was seen in 20 out of 24 rat brains (Sharma and Dey, 1981). Continuation of stress further to 12–14 h markedly diminished the occurrence of BBB permeability. Thus, only 3 out of 10 young rats showed extravasation of Evans blue during this period (Sharma, 1982). On the other hand, subjection of adult rats (age 30–32 weeks) to 8–9 h of stress resulted in leakage of Evans blue in the brains of only 4 out of 12 rats (Sharma and Dey, 1981; Sharma, 1982).

Leakage of Evans blue and Bromophenol blue represents extravasation of serum albumin (Rawson, 1943; Rinder, 1968; Rapoport, 1976) as these tracers bind to endogenous serum proteins when administered into the circulation (Sharma, 2005e). Slight differences in the pattern of extravasation of these dyes into the brain of stressed animals represent minor differences in protein binding capacity of dyes when administered into the circulation (Rawson, 1943; Sharma and Dey, 1981; Sharma, 1982; Bonate, 1988).

4.2.3.1 Pattern of Evans Blue Extravasation Visual inspection of Evans blue after 8-h immobilization stress was confined to the cingulate, occipital, parietal, and frontal cortices. The cerebellum also took moderate staining (see ❯ *Figures 17-6* and ❯ *17-7*). The cerebroventricular walls of the lateral and fourth ventricles showed mild to moderate blue staining indicating a disruption of the blood–CSF barrier (BCSFB) in immobilization stress. Mild staining of the dorsal surface of the hippocampus and caudate nucleus further confirm this finding (see ❯ *Figure 17-7*). The massa intermedia, hypothalamus, and surrounding areas of third ventricle also took mild to moderate blue staining (❯ *Figures 17-7* and ❯ *17-8*). Coronal section of the brain passing through basal ganglia, hippocampus, brain stem, and cerebellum showed mild to moderate staining in the deeper tissues as well as across the dorsal, lateral, and ventral surfaces of the brain (❯ *Figures 17-7* and ❯ *17-8*). These observations suggest a selective disruption of the BBB in immobilization stress (Sharma, 2004d).

4.2.3.2 Radiotracer Extravasation On the other hand, extravasation of radiotracer was much more widespread in immobilization stress. (❯ *Figure 17-9*). Thus, studies on the regional BBB (rBBB) permeability showed radiotracer extravasation in 12 out of 14 brain regions examined after 8 h immobilization. These regions according to tracer accumulation are the occipital cortex (1018%) followed by parietal cortex (622%), hippocampus (563%), inferior colliculi (268%), cerebellum (255%), superior colliculus (225%), hypothalamus (102%), thalamus (94%), caudate nucleus (85%), cingulate cortex (61%), temporal cortex (55.5%), and frontal cortex (25%). The pons and medulla did not show extravasation of radiotracers (Sharma and Dey, 1986a). These observations suggest that the permeability of the BBB in immobilization stress is selective and specific. However, this selectivity depends on that nature of tracers used to evaluate the BBB leakage (Sharma, 2005a, 2007c).

4.2.4 BSCB Permeability in Immobilization Stress

Our laboratory was the first to demonstrate that BSCB permeability to protein tracers is also increased by immobilization stress (Sharma, 2004b). This indicates that the spinal cord is equally sensitive to stress induced BSCB disruption in a highly selective and specific manner (Sharma, 2004b; ❯ *Figures 17-7* and ❯ *17-8*, ❯ *Table 17-7*).

Thus, a mild to moderate degree of Evans blue extravasation in the dorsal surface of the spinal cord in the cervical and thoracic regions is apparent after 8 h immobilization stress (❯ *Figures 17-6* and ❯ *17-7*). The ventral surface of the cord exhibited light blue staining (❯ *Figures 17-7* and ❯ *17-8*). Cross section of

◘ Figure 17-7
Diagrammatic representation of Evans blue albumin (EBA) extravasation across the BBB and BSCB in rats following stress. (A) BSCB breakdown in rats subjected to 30-min forced swimming (a) 4-h heat stress (b), and 8-h immobilization (c) exhibited different pattern of Evans blue staining caused by various stressors. Disruption of the BBB to Evans blue on the dorsal (B: a–c) and on the ventral (B: d–f) surfaces of the brain following forced swimming (a,d), heat stress (b,e) and immobilization (c, f) also shows wide variation in the distribution of the tracer. (C) Mapping of Evans blue staining in the brains of rats following 30-min forced swimming (C: a); 4-h heat stress (C: b), 8-h immobilization (C: c) and 96-h sleep deprivation (C: d) showed breakdown of the BCSFB in these stressful situations. However the extent of blue staining in the cerebellum, areas of cerebral cortex, and in brain stem varied in these stressors. (D) Penetration of the Evans blue dye in the deeper areas of brain and is shown following forced swimming (a), heat stress (b), and immobilization stress (c). Spread of dye varies largely in the cortical and subcortical regions. (E) Extravasation of Evans blue dye also varied in different regions of the spinal cord following stressful situations. Marked variation in blue staining is seen in the gray matter of the thoracic spinal cord following forced swimming (a), heat stress (b), and immobilization (c). Data modified after Sharma, 2004b. The mapping of EBA extravasation in the brain and spinal cord is based on 6–8 individual experiments in each category. The pattern of EBA extravasation in different areas of the brain and spinal cord represents most common findings in each stress group. Bar = 5 mm

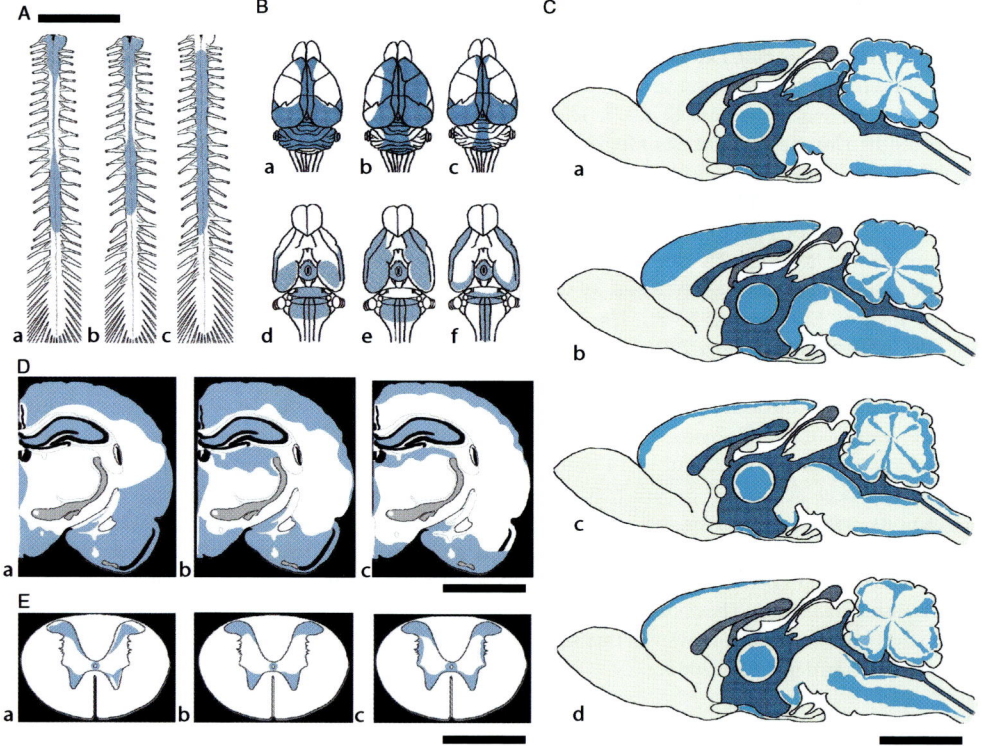

spinal cord showed a diffuse blue staining located mainly in the gray matter (❯ *Figure 17-7*). This increase in Evans blue extravasation is no longer observed in animals subjected to 11 h or 14 h immobilization, indicating that similar mechanisms are operating in disruption of the BBB and BSCB during immobilization stress.

Figure 17-8

Representative example of Evans blue albumin (EBA) extravasation on the dorsal and ventral surfaces of the spinal cord (A: a–f) and brain (B: a–f) of rats subjected to forced swimming, heat stress, and immobilization. Breakdown of the BSCB is evident on the dorsal (a,c,e) and the ventral (b,d,f) surfaces of the spinal cord that varied in different stressful situations. Likewise, the pattern and intensity of EBA extravasation also showed wide variations in the dorsal (B: a–c) and in the ventral (B: d–f) surfaces of the brain according to stressors. However, staining of somatosensory cortex, piriform cortex, cerebellum, hypothalamus, pons, and brain stem is common in all stress experiments. C. Representative examples of coronal sections of rat brain from 3 different levels showing Evans blue albumin (EBA) extravasation in deeper parts of the brain following forced swimming (C: a–c); heat stress (C:d–f); and immobilization (C: g–i). The pattern and intensity of EBA showed selective variations in different stressful situations. Modified after Sharma, 2004b. Coordinates for coronal sections: a = 5.25–6.65; b = −7.10–7.60; c = −10.60 to −11.90 from Bregma, that was almost identical for sets d–f and g–i. Bar = 5 mm

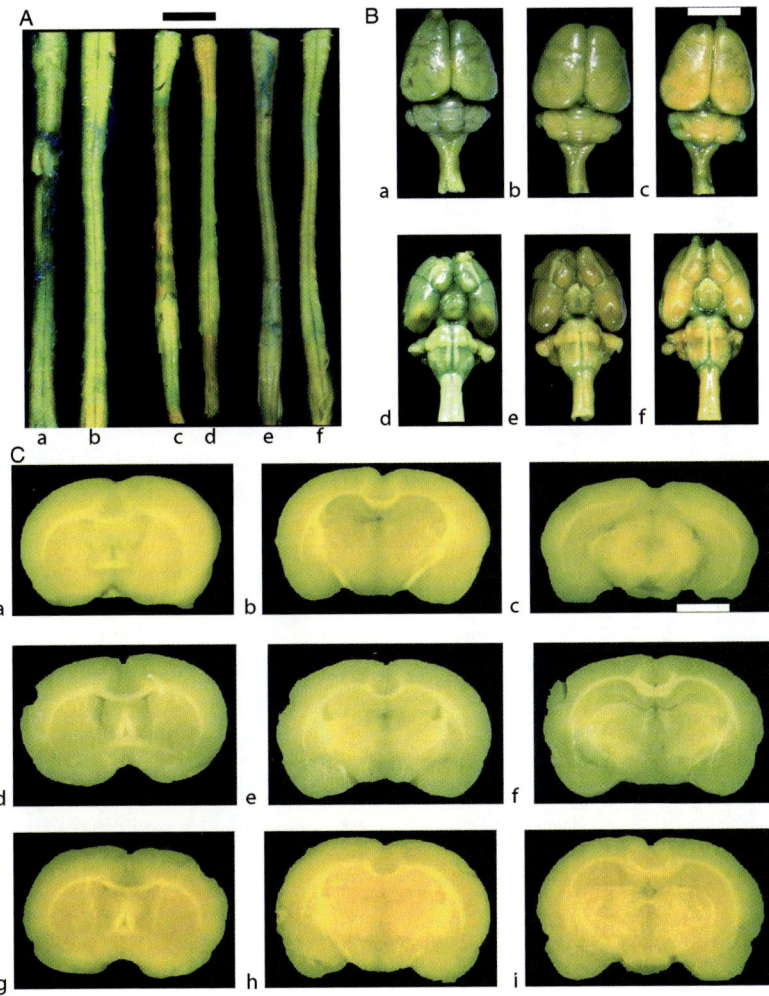

● Figure 17-9

Changes in BBB permeability in the whole brain (A) and in 14 different brain regions (B) following forced swimming, heat stress and immobilization stress. The whole brain BBB disruption to Evans blue albumin and [131]iodine is evident following 30 min after forced swimming (a), 4 h after heat stress (b) and 8 h after immobilization (c). No apparent increase in the BBB permeability was evident after 1-h rest following forced swimming (a), or 5-h rest following immobilization stress (c), whereas, 2-h rest after heat stress resulted in a reduction in BBB function (b). This indicates that the BBB permeability in stress appears to be reversible in nature. (B) The rBBB permeability showed significant increase in radiotracer extravasation in 9 out of 14 brain regions (b,c,e-i, k, l) following 30-min forced swimming (B: a), in all the brain regions after 4-h heat stress (B: b) and in 12 out of 14 brain regions (a-j) following 8-h immobilization stress (B: c). Brain regions (Forced Swimming): a = frontal cortex, b = parietal cortex, c = occipital cortex, d = anterior cingulate cortex, e = posterior cingulate cortex, f = cerebellar vermis, g = cerebellar cortex, h = caudate nucleus, i = hippocampus, j = colliculi, k = thalamus, l = hypothalamus, m = medulla, n = brain stem; (heat stress or immobilization): a = frontal cortex, b = parietal cortex, c = occipital cortex, d = temporal cortex, e = cingulate cortex, f = hippocampus, g = caudate nucleus, h = thalamus, i = hypothalamus, j = superior colliculus, k = inferior colliculus, l = cerebellum, m = pons, n = medulla. Values are mean ± SD of 10–12 rats. Data modified after Sharma and Dey, 1986a, 1987b, 1991a; Sharma, 2004b

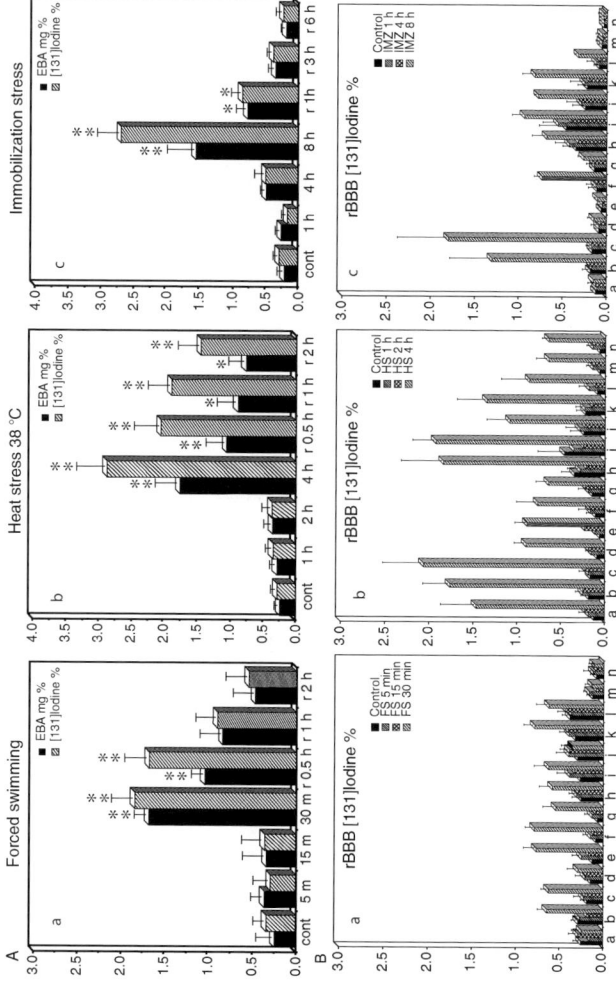

◘ Table 17-7
Blood–spinal cord barrier permeability in normal and stressed young rats

Spinal cord regions	Level	BSCB permeability [131]Iodine, %					
		Control	Swimming, 30 min	Heat stress, 4 h	Immobilization, 8 h	[a]Sleep deprivation, 48 h	[a]Morphine withdrawal, 48 h
		$n = 6$	$n = 8$	$n = 6$	$n = 8$		
Cervical	1–4	0.26 ± 0.04	0.66 ± 0.04**	0.85 ± 0.12**	0.60 ± 0.08**	0.76 ± 0.05**	0.680.08**
Thoracic	8–11	0.34 ± 0.06	78 ± 0.12**	0.96 ± 0.17**	0.84 ± 0.10**	0.89 ± 0.08**	0.84 ± 0.06**
Lumbar	1–4	0.39 ± 0.04	0.58 ± 0.08*	0.68 ± 0.14*	0.70 ± 0.10*	0.72 ± 0.06**	0.68 ± 0.08**

Source: Data modified after Sharma 2004e; [a]Sharma (unpublished observations)
Values are mean ± SD; *$P < 0.05$; **$P < 0.01$ ANOVA followed by Dunnet's test for multiple group comparison

4.2.5 Alterations in Spontaneous EEG Activity

A breakdown of the BBB function in stress could lead to mental abnormalities (Sharma, 2005b). Since spontaneous electrical activity of the brain is an important indicator of brain function, our laboratory examined the effect of stress induced BBB breakdown on the spontaneous EEG in rats (Sharma and Dey, 1988; Winkler et al., 1995).

The cortical EEG was recorded in male rats using transverse stainless steel screw electrodes (length 4 mm, tip o.d. 1 mm) chronically implanted on the parietal cortex and on the cerebellar cortex (● Figure 17-10; Sharma and Dey, 1988). The spontaneous EEG activity in normal rats between 8:00 AM

◘ Figure 17-10
Representative example of EEG changes in immobilization (A) and heat stress (B). The EEG activity recorded from the bipolar electrodes placed on the parietal cortex (PC) and on the cerebellar cortex (CC) showed high voltage slow activity (HVSA) at the onset of stress (A). Appearance of low voltage and fast activity (LVFA) is seen after 6-h immobilization stress. Flattening of EEG is apparent after 8-h immobilization. At this time, extravasation of Evans blue albumin (EBA) is seen in the parietal and cerebellar cortex (A: for details see text). Modified after Sharma and Dey, 1988. (B) Representative example of EEG changes in acute heat stress. The EEG is recorded from bipolar electrodes placed on the frontal cortex (FC) and parietal cortex (PC). The EEG activity shows high voltage and slow activity (HVSA) before heat exposure (−30 min). An increase in EEG voltage and frequency is apparent after 30-min heat exposure. Flattening of EEG appeared at the end of 4-h heat exposure. At this time profound extravasation of Evans blue dye is seen in the brain. Partial recovery in EEG activity is evident following 2-h rest after heat exposure. Data after Sharma, 2004b

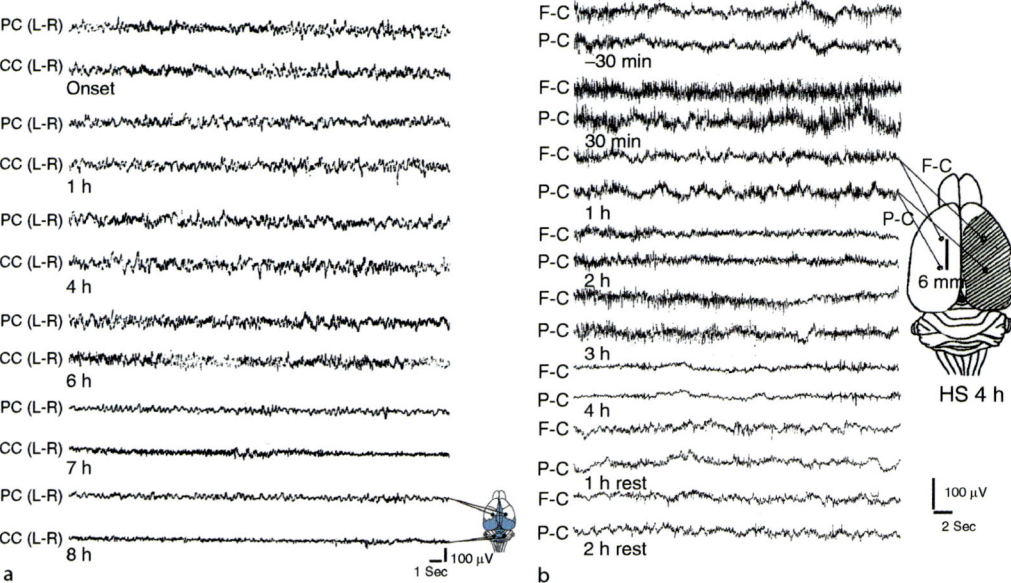

and 8:00 PM did not alter significantly (Sharma and Dey, 1988; Sharma, 2004b). However, subjection of rats to immobilization stress markedly affected the spontaneous EEG activity (Sharma and Dey, 1988). Thus, the bipolar EEG activity of the restrained rats from the onset to 1 h was predominantly high voltage slow activity (HVSA, 60–70 μV; 6–7 Hz) (● Figure 17-10). The EEG voltage increased slightly at the end of 4–5.5 h (70–90 μV; 5–6 Hz) (● Figure 17-10). From 6 h and onward, the EEG voltage declined markedly

and appearance of low voltage fast activity (LVFA) was seen at the end of 7-h immobilization. The most pronounced LVFA occurred at the end of 8-h immobilization (10–15 μV, 9–11 Hz) that persisted up to 8.5 h (● *Figure 17-10*; ● *Table 17-8*). Examination of BBB permeability at this time (8-h stress) revealed marked blue staining in the cerebral cortex and in the cerebellar vermis (● *Figure 17-10*). This indicates that

◘ Table 17-8
EEG changes in stress in relation to the BBB permeability

Type of experiment	n	EEG activity					BBB permeability Whole brain Evans blue, mg %
		Parietal cortex			Cerebellar cortex		
		Amplitude, μV	Frequency, Hz		Amplitude, μV	Frequency, Hz	
Immobilization							
Onset	6	68 ± 5	6 ± 1		75 ± 4	6 ± 2	0.23 ± 0.03
1 h	6	70 ± 4	6 ± 2		73 ± 3	7 ± 1	
4 h	6	86 ± 4**	5 ± 2		89 ± 4**	5 ± 2	
8 h	6	10 ± 5**	9 ± 2		15 ± 3**	9 ± 3	1.21 ± 0.22**
9 h	5	30 ± 5	8 ± 2		45 ± 3	8 ± 3	
11	5	54 ± 4	7 ± 1		64 ± 5	7 ± 2	0.34 ± 0.11
		Frontal cortex			Parietal cortex		
[a]Heat stress							
Onset	5	64 ± 3	6 ± 1		60 ± 2	6 ± 2	
30 min	5	90 ± 4**	7 ± 2		88 ± 4**	8 ± 2	
1 h	5	53 ± 2*	8 ± 3		55 ± 2*	8 ± 2	
2 h	5	50 ± 4**	9 ± 2		52 ± 4**	9 ± 3	
3 h	5	54 ± 3	8 ± 2		53 ± 3	8 ± 3	
4 h	5	12 ± 2**	8 ± 2		10 ± 2**	9 ± 3	1.67 ± 0.21**
Rest 1 h	4	30 ± 3**	7 ± 2		32 ± 4**	7 ± 3	
Rest 2 h	4	38 ± 4**	6 ± 3		35 ± 3**	6 ± 2	0.87 ± 0.12**

Source: Data modified after Sharma and Dey (1988); Sharma 2004e; Sharma ([a]unpublished observations)
Values are mean ± SD; *$P < 0.05$; **$P < 0.01$ ANOVA followed by Dunnet's test for multiple group comparison

induction of LVFA reflects the breakdown of the BBB caused by immobilization stress and/or alterations in the brain fluid microenvironment (Sharma, 2004b, 2005b, c, 2006a).

This idea is further supported by the fact that the LVFA gradually disappeared and HVSA ensued again in separate groups of animals at the end of 9-h stress (20–50 μV; 8–10 Hz). Almost 80% recovery in EEG activity was apparent at the end of 11-h immobilization (● *Figure 17-10*). No extravasation of Evans blue dye in any part of the brain is seen at the end of 11-h stress (Sharma and Dey, 1988; Winkler et al., 1995; for details Sharma, 2004b). These observations suggest that BBB breakdown could induce abnormal brain function leading to brain pathology.

4.2.6 Structural Changes in the Brain in Immobilization Stress

To examine whether leakage of BBB to proteins could result in brain pathology, morphological analysis of brains following 8-h immobilization stress was carried out in our laboratory (Sharma, 2004d). Our observations for the first time showed specific nerve cell damage in the cerebral cortex, hippocampus, cerebellum, and in brain stem following immobilization stress (❯ *Figure 17-11*). Several dark and distorted nerve cells are seen in the superficial layers of the cerebral cortex that were prominent in exhibiting Evans blue extravasation, viz., cingulate, parietal, temporal, and in occipital cortex (❯ *Figure 17-11*) indicting that BBB leakage in stress is associated with brain pathology.

Several damaged nerve cells were seen in the hippocampal dentate gyrus, CA4 and CA3 sectors as well as CA1 subfields (❯ *Figure 17-11*). Dark neurons are also frequent in the brain stem reticular formation. Cerebellar Purkinje cells and granule cells showed signs of neurodegeneration (❯ *Figure 17-11*).

The epithelial cells of the choroid plexuses and the ependyma around the lateral and third ventricles exhibited degenerative changes. In these regions, sponginess and edema is frequent (❯ *Figure 17-11*). These observations support the idea that immobilization stress induces ependymal damage and disruption of the blood–CSF barrier as well, as mentioned above (see earlier). At the ultratstructural level, areas showing BBB disruption exhibited membrane damage, vacuolation and distortion of the nerve cells (❯ *Figure 17-12*). These observations strongly indicate that the stress induced BBB disruption is associated with structural changes in the neuropil. Whether these structural changes will induce permanent brain deficit or initiate a cascade of events leading to widespread neurodegeneration in the CNS is not known and require further investigations.

4.3 Forced Swimming: Another Model of Depression

Forced swimming is frequently used as an animal model of severe stress leading to depression, as mentioned earlier (Dawson and Horvath, 1970; Porsolt et al., 1979; Gruner and Altman, 1980; Harri and Kuusela, 1986; Lalonde, 1986). Swimming involves movement of head and body positions for proper orientation and balance in water (Gruner and Altman, 1980). This activates the cerebellum and associated brain stem regions for coordinated limb movements (Porsolt et al., 1979; Gruner and Altman, 1980; Lalonde, 1986). Interestingly, clinical cases of cerebral circulatory defects, vertebral artery injury, cerebellar stroke, abnormal neurological function and behavior while swimming are well documented in the literature (Toole and Tucker 1960; Tramo et al., 1985; Sherman et al., 1981). Whether these clinical symptoms during swimming are related with the BBB dysfunction is still unclear.

When rodents are forced to swim in a restricted pool, they quickly acquire an immobility response (Lalonde, 1986). This hypoactivity caused by swim stress represents the state of depression and is associated with altered neurochemical metabolism in the brain (Sharma, 2004b). Our laboratory was the first to show that rats subjected to forced swimming exhibit selective disruption of the BBB breakdown (Sharma and Dey, 1980; Sharma et al., 1991a). However, it is unclear whether forced swimming is equally responsible for opening the BSCB to proteins and precipitate brain and spinal cord pathology.

4.3.1 Continuous Forced Swimming Disrupts BBB to Proteins

Using a rat model, we examined the relationship between BBB disruption and brain pathology in forced swimming (Sharma and Dey, 1980; Sharma et al., 1991a, 1995c). The animals were allowed to swim individually in a corning glass cylinder (150 cm height, 20 cm internal diameter) containing water (18 cm depth) maintained at $30 \pm 1°C$ (Sharma et al., 1991a). The water is manually stirred with a glass rod from time to time (every 4 or 5 min) so that the animals swim continuously in the pool during the entire stress session (Sharma et al., 1991a, 1995c).

Subjection of animals to 30 min forced swimming resulted in extravasation of EBA in 5 brain regions, viz., cingulate cortex, parietal cortex, occipital cortex, cerebellum, and the dorsal surface of the hippocampus (❯ *Figure 17-7*). The cerebellar vermis took moderate blue staining compared to the lateral cerebellar cortex (❯ *Figure 17-7*). The deep cerebellar nuclei were unstained (❯ *Figures 17-7* and ❯ *17-8*).

Forced swimming also disrupted the BCSFB. This is evident from the findings that the walls of lateral ventricle stained mild blue, whereas the fourth ventricle exhibited deep blue staining (❯ *Figure 17-7*). The areas around the third ventricle were stained blue moderately (❯ *Figure 17-8*).

◘ Figure 17-11

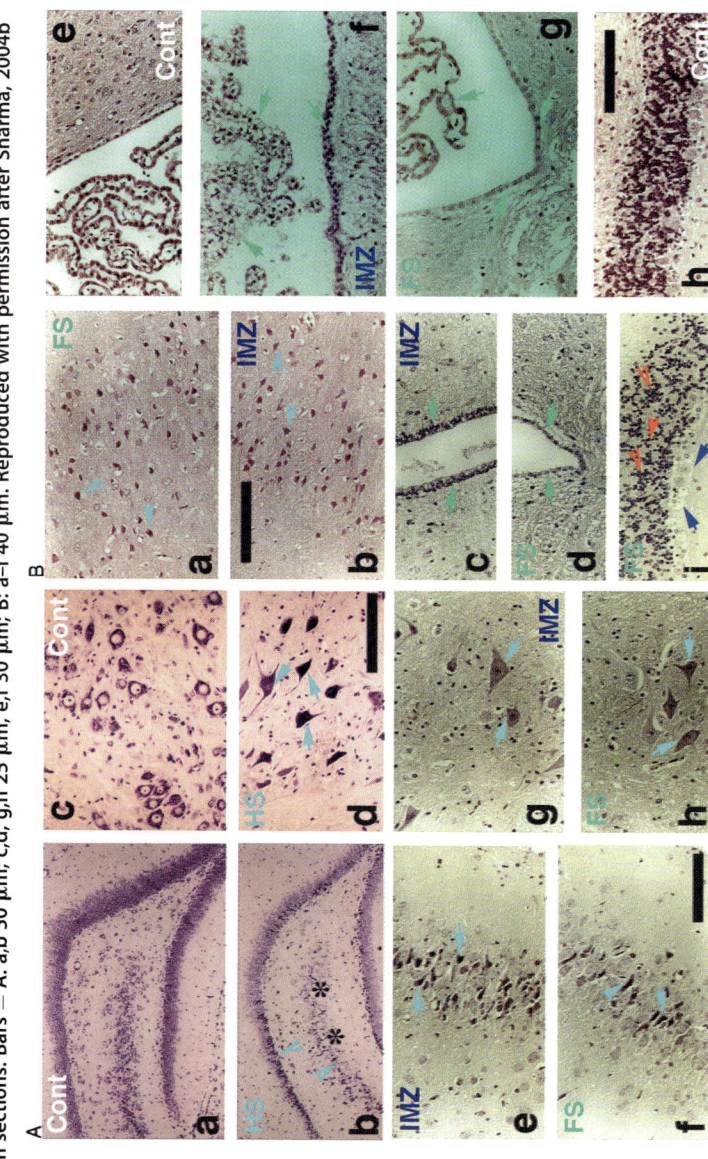

Representative examples of light microscopic changes in several brain regions following 4-h heat stress (HS), 8-h immobilization (IMZ), or 30-min forced swimming (FS). Marked degeneration (arrow heads) in dentate gyrus and CA4 region (*) of the hippocampus is seen following HS (A: b) compared to control (A: a). Many damaged and distorted nerve cells in the brain stem regions following HS is apparent (A: d, arrows) compared to control (A: c). Mild to moderate cell damage in the cortical region is seen following IMZ (A: e, arrows) or FS (A: f, arrow heads). IMZ (B: a) and FS (B: b) induce specific and selective cell damage (arrow heads) in the cerebral cortex. Cell damage in ependymal cells (arrows) in the median eminence following IMZ (B: c) and FS (B: d) is clearly evident. Degeneration (arrows) of choroid plexus epithelial cells and ependymal cells in the lateral cerebral ventricle by IMZ (B: f) and FS (B: g) is apparent compared to control (B: e). The cerebellar Purkinje cells and granule cells show marked degenerative changes following FS (B: i, arrows) compared to control (B: h). A: a–d Nissl stain; others Hematoxylin and Eosin stain on 3 mm thick paraffin sections. Bars = A: a,b 50 μm; c,d, g,h 25 μm; e,f 30 μm; B: a–i 40 μm. Reproduced with permission after Sharma, 2004b

■ Figure 17-12

Representative examples of ultrastructural changes in the rat brain following 4-h heat stress (HS), 30-min forced swimming (FS), or 8-h immobilization (IMZ) stress. In the cerebral cortex, one nerve cell with dark and electron dense cytoplasm (*arrows*) is seen following HS (A: a) Many degenerative changes can be seen in the neuropil (A: a). High power electron micrograph from the parietal cerebral cortex cellular layer III showing damaged synapses (*arrows*), vesiculation of myelin, and membrane damage following HS (A: b). A completely collapsed microvessel (arrow heads) and distorted granule cells in the cerebellum are seen following HS (A: c). Perivascular edema (*), membrane damage and leakage of lanthanum across the microvessels (A: d) are quite frequent in HS. Vacuolation, membrane damage (*arrows*) and perivascular edema (*) are frequent following FS (B: b, d). Myelin vesiculation (B: e) and nerve cell damage (B: f) is very common in hippocampus and in thalamus following HS. Bars: A: a = 0.6 μm; A: c,d = 1 μm; A: b = 0.4 μm nm; B: a–e = 1 μm; B: f = 0.5 μm. Data modified after Sharma and Cervós-Navarro, 1990 (A: d); Sharma et al., 1998a (B: f). Reproduced with permission from Sharma, 2004b

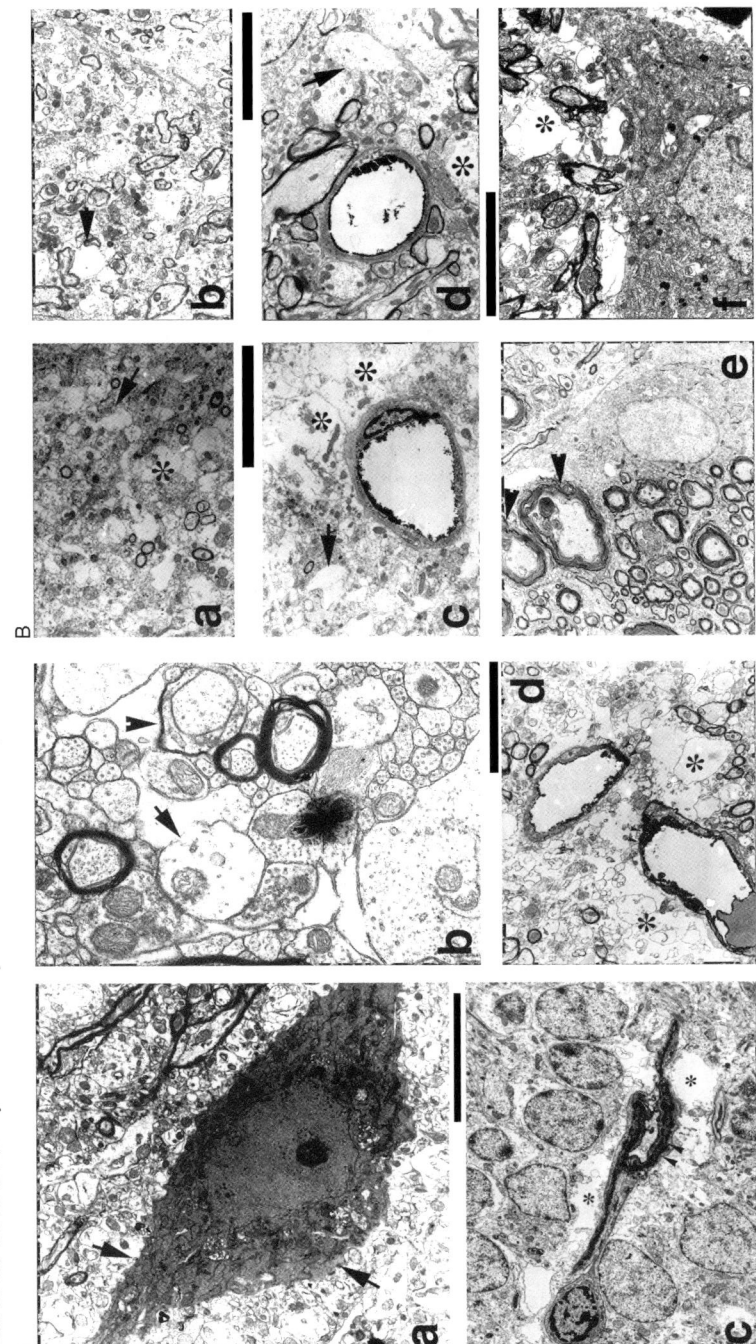

Extravasation of radioiodine tracer was more widespread and seen in eight brain regions including, caudate nucleus, thalamus, and hypothalamus besides the other five blue stained regions (❯ *Figure 17-9*). The BBB disruption following forced swimming in young rats is reversible in nature. Thus, when the BBB permeability is no longer observed in rats subjected to 2 h-rest after 30 min swimming (Sharma et al., 1991a, 1995c).

On the other hand, subjection of animals to 5 or 15 min forced swimming did not show extravasation of protein tracers in any brain regions (❯ *Figure 17-9*). Furthermore, adult rats when subjected to 30-min swim stress did not exhibit any significant leakage of the BBB (Sharma et al., 1995c). These observations suggest that forced swimming-induced BBB disruption is dependent on the duration and age of the animals (Sharma et al, 1995c).

4.3.2 BSCB Permeability in Forced Swimming

A significant increase in EBA in the spinal cord of rats subjected to 30-min forced swimming (see ❯ *Figures 17-7* and ❯ *17-8*) indicates opening of the BSCB at the same time of the BBB disruption. Interestingly, this increase in BSCB disruption was not restored after 2-h rest (Sharma, unpublished observations). This indicates that the opening of BSCB permeability during forced swimming is prolonged as compared to the BBB disruption. The basic mechanism behind this discrepancy between BBB and BSCB permeability in forced swimming are not known and requires additional investigation.

4.3.3 Morphological Changes in the Brain Following Forced Swimming

Light and electron microscopy in the brains of rats subjected to 30-min forced swimming revealed selective changes in the neuropil, e.g., damage of specific nerve cells in the cerebral cortex, hippocampus, and in brains stem in the regions showing BBB disturbances (❯ *Figures 17-11* and ❯ *17-12*). The selective nerve cell damage in the cortex is seen in layer III–V in the cingulate, occipital, and piriform cortices. Degenerative changes in few nerve cells are apparent in the brain stem reticular formation and granule cells of the cerebellum (❯ *Figure 17-10*). In the hippocampus, distortion of nerve cells is common in the dentate gyrus, CA3 and CA4 sectors. Damage to ependymal cells around the lateral and third ventricle and degenerative changes in the choroid plexuses is also seen in animals subjected to forced swimming for 30 min (❯ *Figure 17-10*) indicating breakdown of the blood–CSF barriers.

The cellular changes seen at the light microscopic level are further confirmed at the ultrastructural level (❯ *Figure 17-12*). Thus, membrane vacuolation and damage to neuropil is present in the cortex and in brain stem of rats after 30-min forced swimming (❯ *Figure 17-12*). These observations suggest that breakdown of the BBB following forced swim is associated with selective nerve cell damage. However, it is unclear whether these cell changes reflect acute nerve cell death or permanent neurodegenerative changes, a subject currently being investigated in our laboratory.

4.4 Sleep Deprivation Stress

Sleep deprivation for 96 h induces profound stress and alters expression of c-fos, Fos protein, as well as GABergic and serotonergic neurons in the brain stem reticular formation in rats (Maloney et al., 2000). However, the influence of sleep deprivation on BBB function is still unknown.

Our laboratory was the first to examine BBB function and brain pathology in a rat model of sleep deprivation ranging between 24 h and 96 h (Sharma, 2004b). In this model, each rat is placed on an inverted flower pot (6.5 cm in diameter) surrounded by water filled in a Plexiglas box up to 1 cm of the surface with free access to food and water (Mendelson et al., 1974; Maloney et al., 1999). The water temperature is maintained at $30 \pm 1°C$ (Sharma et al., 1991a). The animals will undergo slow wave sleep (SWS) but not the paradoxical sleep (PS, Maloney et al., 2000). The loss of muscle tonus with PS onset will cause the rats to fall into the water and thus awaken them (Maloney et al., 2000).

4.4.1 BBB and Brain Pathology in Sleep Deprivation Stress

Significant leakage of EBA in the brain and spinal cord is seen in rats subjected to 48–96 h sleep deprivation (Sharma, 2004d). Mild blue staining of the frontal cortex, temporal cortex, and cingulate cortex was noted at 48 h after sleep deprivation. The cerebellar cortex also took faint staining. This increase in Evans blue extravasation was further intensified at 96 h after sleep deprivation. Thus, moderate Evans blue staining in the cingulate, frontal, parietal, and temporal cortices is observed (Sharma, 2004b; Sharma, unpublished observations). The walls of lateral cerebral ventricles, dorsal surface of the hippocampus and massa intermedia showed mild blue staining (❷ Figure 17-7) indicating breakdown of the BCSFB as well (❷ Table 17-7; Sharma unpublished observation). Some areas in the brain stem reticular system took mild to moderate staining (❷ Figure 17-7).

Extravasation of endogenous albumin using immunohistochemistry showed a good relationship with the exogenous Evans blue extravasation and albumin leakage in the brain of sleep deprived rats (Sharma, unpublished observation). The albumin immunoreactivity was localized mainly around the microvessels in the cerebral cortex, hippocampus, brain stem, and thalamus. In 96-h sleep-deprived rats, the albumin immunoreactivity was also seen around few nerve cells in the cortex, hippocampus, brainstem, cerebellum, and thalamus (Sharma, unpublished observation). These novel observations suggest that the sleep deprivation stress depending on its duration is able to induce BBB disruption in specific regions. Further studies are in progress in our laboratory to see whether the BBB breakdown in sleep deprivation is associated with structural changes in the CNS.

4.5 Environmental Heat Stress

Environmental heat stress associated hyperthermia ($>40°C$) during summer seasons in many parts of the Globe is a life-threatening illness and is commonly associated with heat stroke (Sharma, 2007e). Severe CNS dysfunction such as delirium, convulsion coma, and even death ($<50\%$ of the victims) are quite common in environmental heat-related illnesses (Bouchama and Knochel, 2002; Sharma, 2007c, e). Those who survive heat stroke, often show permanent neurological deficits (Sharma, 2006c, 2007c, d). With recent increase in Global warming, the intensities and frequencies of heat waves and related heat illnesses have further affected large populations world-wide causing serious health problems (Sharma and Westman, 1998; Sharma and Hoopes, 2003; Sharma, 2007e).

Inspite of serious CNS consequences of hyperthermia, the state of BBB or BCSFB function is still largely ignored in victims of heat stress and heat stroke (Sharma and Hoopes, 2003). Our laboratory was the first to show that experimental or environmental heat stress without heat stroke induces BBB disruption (Sharma and Dey, 1978, 1984; Sharma, 1982) leading to brain pathology (Sharma and Cervós-Navarro, 1990). However, influence of heat stress on BSCB permeability is still not well known (Sharma, 2004b).

4.5.1 Hyperthermia: An Adjunct Therapy for Cancer Treatment

Whole body hyperthermia (WBH) is often used as an adjunct to cytotoxic therapy for cancer treatment (Katschinski et al., 1999; Sharma and Hoopes, 2003). However, several harmful side effects of WBH, e.g., increased DNA adduct formation, inhibition of DNA repair, increased drug permeability, and decreased resistance to DNA damaging agents, are a matter of great concern (Robins et al., 1995, 1997). The WBH has also been shown to enhance cytotoxic effects of ionizing radiation chemotherapy and induce brain dysfunction (Sharma and Hoopes, 2003 for review; Sminia et al., 1994; Sharma, 2007e).

Localized brain heating for cancer treatment alters BBB permeability to various tracers (for review see Blackwell and Saunders, 1986; Sminia et al., 1994; Sharma et al., 1998a, c). Low levels of microwave radiation causing rise in brain temperature up to $44-45°C$ in conscious or anaesthetized animals show deposits of sodium fluorescien (Frey et al., 1975; Merritt et al., 1978; Williams et al., 1984) or HRP in extracellular spaces around the cerebral blood vessels (Albert, 1979; Sutton and Carroll, 1979; Blackwell and

Saunders, 1986). Interestingly, these results were interpreted for long time as either perfusion artefacts or redistribution of local blood flow changes during hyperthermia but never considered to be due to the BBB disruption (Sharma et al., 1998a, c; Sharma, 2004b, 2006b, c).

4.5.2 BBB Permeability in Heat Stress

Our laboratory was the first to show that subjection of rats to WBH comprising a model to simulate environmental heat stress during summer season induces selective leakage of protein tracers in the brain and in the CSF (Sharma and Dey, 1978, 1984; Sharma, 1982). The model essentially consists exposure of conscious rats in a biological oxygen demand (BOD) incubator at 38°C (relative humidity 45–47%; wind velocity 20–26 cm/sec) for 4 h (Sharma, 1982, 1999; Sharma and Dey, 1986b, 1987b).

Leakage of EBA was apparent in eight brain regions, viz., cingulate cortex, occipital cortex, parietal cortex, cerebellum, temporal cortex, frontal cortex, hypothalamus, and thalamus (❯ *Figure 17-7*). Mild to moderate blue staining of the ventricular walls was also observed. The fourth ventricle showed deep blue staining and the structures around the third ventricle were moderately stained (❯ *Figure 17-5*). Occasionally the dorsal surface of the hippocampus took mild stain (❯ *Figures 17-7* and ❯ *17-8*) indicating disruption of the blood–CSF barrier in heat stress as well (Sharma, 2004b). Extravasation of radioiodine is present in all the 14 regions examined. Thus, besides the eight blue stained regions, other six regions viz., hippocampus, caudate nucleus, superior and inferior colliculi, pons, and medulla also showed an increase in radioactivity (❯ *Figure 17-9*; Sharma and Dey, 1987b).

On the other hand, subjection of rats to shorter periods of heat stress, i.e., 1 h or 2 h did not induce BBB disruption (❯ *Figure 17-9*). Adult animals when exposed to 4-h heat stress at 38°C exhibited only a mild increase in the BBB to Evans blue and radiotracer (Sharma, 2004b). These observations suggest that the like other stressors, the duration of heat stress and age of animals are important factors in BBB dysfunction.

The BBB permeability to Evans blue or radioiodine was not completely closed in heat stress even when the tracers were administered 2 h periods after a 4-h heat exposure session (Sharma, 2004b; Sharma, unpublished observation). These animals are still lethargic at room temperature, although their body temperature returned to near normal values (Sharma, unpublished observations). This indicates that the magnitude and intensity of stressors are important factors in reversibility of the BBB leakage.

4.5.3 BSCB Permeability in Heat Stress

Young rats subjected to 4-h heat stress exhibited marked increase in the BSCB permeability to Evans blue in specific regions of the cord (see ❯ *Figures 17-7* and ❯ *17-8*). This indicates that the BBB and BSCB permeability are equally sensitive to heat stress in young rats (*Table 17-7*). This increase in the BSCB permeability was not observed in adult animals subjected to 4-h heat stress (Sharma, unpublished observation). Also short duration of heat exposure to young rats did not result in BSCB breakdown (Sharma, unpublished observation). These observations suggest that the permeability properties of the BSCB and BBB in heat stress are quite similar in nature.

However, when animals are allowed to 3-h rest after 4-h heat stress, only a mild reduction in the BSCB permeability to Evans blue was observed (results not shown) indicating that the BSCB breakdown is long-lasting in heat stress than the BBB disruption. The molecular mechanisms behind such differences are unknown and require further investigation (Sharma, 2004b).

4.5.4 Brain Edema Formation in Heat Stress

Leakage of proteins into the fluid microenvironment could induce vasogenic edema formation, as mentioned earlier (Sharma et al., 1998a, c). Thus, a significant increase in brain edema formation occurs in rats following 4-h heat stress (❯ *Figures 17-13*). No increase in brain water is seen in animals subjected to 1-h or 2-h

Figure 17-13

Changes in BBB permeability and brain water content in acute and chronic heat stress. Marked increase in Evans blue albumin (a) and brain water (b) is seen following acute 4-h heat stress and following repeated heat exposure of 4-h heat stress on day 2. This indicates that leakage of Evans blue albumin closely correlate with the development of brain edema formation. Values are mean ± SD of 6–8 rats. ** $P < 0.01$, ANOVA followed by Dunnet's test for multiple group comparison from one control. Data modified after Sharma, 2004b

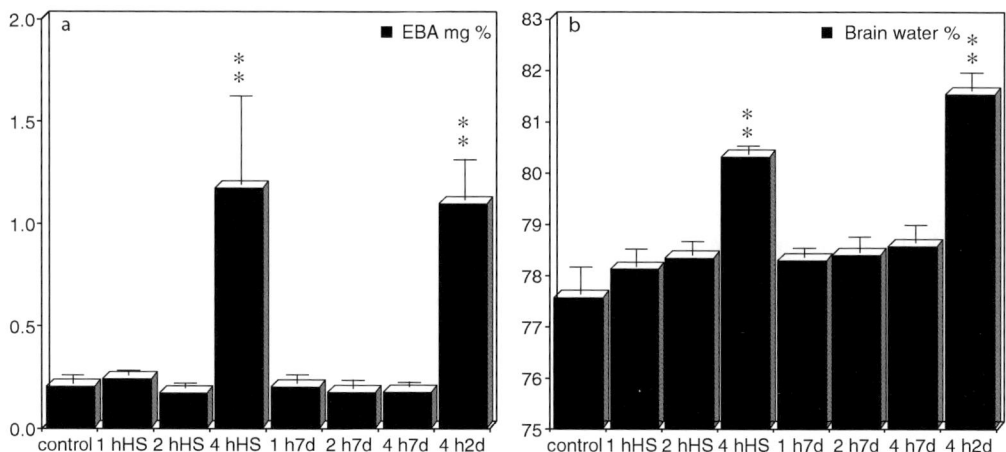

heat exposure. The magnitude and severity of edema formation are much less evident in adult animals following 4-h heat exposure (◉ Figure 17-13). These observations confirm the hypothesis that extravasation of protein tracers are primarily responsible for vasogenic edema formation in heat stress (Sharma et al., 1998a).

Regional increase in brain water showed a close correlation in the brain areas exhibiting Evans blue leakage. Thus, prominent edema formation was seen in the cerebral cortex, hippocampus, cerebellum, and brain stem and in spinal cord after 4-h heat exposure in young rats (◉ Figure 17-13). It is quite likely that profound volume swelling of the brain in closed cranium will compress the vital centers in the brain leading to high mortally in hyperthermia. Massive brain swelling in human victims who died following heat stroke is in line with this idea (for details see Sharma and Hoopes, 2003; Sharma, 2006b, c, 2007c, d).

4.5.5 Spontaneous EEG Alters in Heat Stress

Brain hyperthermia induces fatigue and alters the spontaneous EEG (Gonzalez-Alonso et al.,1999; Nielsen et al., 2001; Sharma, 2007e) in human subjects (Dubois et al., 1980, 1981; Febbraio et al., 1994). Breakdown of the BBB and edema formation in heat stress will further result in brain dysfunction and could be reflected in EEG abnormalities. We examined EEG in young rats following heat stress using bipolar screw electrodes placed over the right and left cingulate cortex and the parietal cortex (◉ Figure 17-10) (Winkler et al., 1995; Sharma, 2004b).

Normal EEG in conscious rats exhibited amplitude of 50–60 µV and a frequency of 6–7 Hz (◉ Figure 17-10). The EEG recordings from cingulate and parietal cortex were very similar in nature (Sharma, 2004b). Exposure of rats to heat stress at 38°C resulted in a significant increase in EEG amplitude (70–80 µV) and frequency (7–9 Hz, ◉ Figure 17-10) of EEG at 30 min. After 1 h of heat stress, the EEG voltage started to reduce (40–50 µV) while the frequency increases slightly (8–9 Hz) (◉ Table 17-8). At the end of 2-h heat exposure, the EEG voltage further reduced in the cingulate cortex (20–30 µV; 8–9 Hz)

(● *Figure 17-10*). A mild increase in EEG amplitude (30–40 μV) and frequency (10–11 Hz) is seen at 3 h after heat exposure (● *Table 17-8*). However, the EEG amplitude reduced considerably (10–12 μV) at the end of 4-h heat exposure without any apparent change in frequency (10–12 Hz) (● *Figure 17-10*). This effect was equally pronounced on both the cingulate and parietal cortex recording. At this time, extravasation of EBA is prominent in the cerebral cortex of heat-stressed rats indicating that increased BBB permeability and flattening of EEG are interrelated (Sharma, 2004b).

Interestingly, the EEG recovery is initiated following 30-min rest at room temperature after 4-h heat exposure. A complete recovery however was not evident in the heat-exposed animals even after 2-h rest at the room temperature (● *Table 17-8*; ● *Figures 17-10*). This observation suggests that hyperthermia can induce long-lasting changes in the brain electrical activity that could be well correlated with the BBB disruption in heat stress. These observations indicate that alteration in the BBB is associated with abnormal neuronal function.

4.5.6 Structural Changes in the Brain Following Heat Stress

Breakdown of the BBB in heat stress is associated with neuronal damage. Thus, profound neuronal, glial, and myelin changes are seen in the brain areas showing BBB disruption in young animals after 4-h heat exposure (see ● *Figure 17-11* and ● *17-12*). Nerve cell injury, edematous expansion, and sponginess of the neuropil are prominent in several brain areas, e.g., such as cerebral cortex, brain stem, cerebellum, thalamus, and hypothalamus (● *Figures 17-10* and ● *17-11*). A selective nerve cell damage in the hippocampus is most pronounced within the CA4 subfield compared to other regions (● *Figure 17-11*), although edematous swelling and general sponginess are present throughout this region. This indicates that BBB leakage is associated with brain damage in heat stress.

In addition, damage to ependymal cells around the lateral and third ventricle is quite distinct in heat stress (● *Figure 17-11*). The choroid plexus from the lateral ventricle, third ventricle, and fourth ventricles exhibited degenerative changes (● *Figure 17-10*) suggesting that heat stress induces profound alterations in the BCSFB.

Upregulation of glial fibrillary acidic protein (GFAP) in many parts of the brain in showing BBB permeability is seen at 4-h heat exposure (● *Figure 17-14*; Sharma et al., 1992c, d). However, the magnitude of heat induced glial cell reaction in different brain regions does not normally coincide with the severity of BBB breakdown or nerve cell injury. This indicates a selective difference in the vulnerability of neurons and glial cells in heat stress (Sharma et al., 1992d).

Profound axonal injuries in young rats following heat exposure are seen using myelin basic protein (MBP) immunostaining, (Sharma et al., 1992c, 1998a). A decrease in MBP immunostaining representing degradation of myelin is most pronounced in the brain stem reticular formation and the spinal cord (● *Figure 17-14*).

These structural changes in the brain were not seen following 1- and 2-h heat exposure in young rats. Furthermore, the magnitude and intensity of these morphological changes were much less apparent in adult animals after 4-h heat exposure. Taken together these observations indicate a role of BBB breakdown in brain damage in heat stress.

4.5.7 Ultratstructural Damage of Neuropil

Ultratstructural changes in heat stress show damaged nerve cells, degenerated nuclei often accompanied by eccentric nucleolus in the cerebral cortex, hippocampus, cerebellum, thalamus, hypothalamus, and brain stem (● *Figure 17-12*). The nerve cells are dark in appearance and contain vacuolated cytoplasm. The nuclear membrane contains many irregular foldings and the nucleolus showed signs of degeneration (● *Figure 17-12*). Interestingly, in many brain regions, one nerve cell is severely damaged whereas the adjacent neuron could be almost normal in appearance. This indicates a selective vulnerability of nerve cells in heat exposure.

Swollen synapses with damage in both pre- and postsynaptic membranes are frequent in thalamus, brain stem, hypothalamus, cerebellum, hippocampus, and in cerebral cortex (❯ *Figure 17-12*). In most of these regions damage to postsynaptic dendrites and disruption of synaptic membranes is common.

Axonal damage, demyelination, and vesiculation of myelin are the most prominent feature following heat stress, particularly in the brain stem reticular formation, pons, medulla, and the spinal cord (❯ *Figure 17-12*). Many unmyelinated axons are also swollen. Most of these ultrastructural changes are present in the brain of young rats after 4-h heat stress. The magnitude and intensity of cell injuries are considerably reduced in adult animals subjected to heat exposure (Sharma, 2004b). On the other hand, young animals exposed to short duration of heat stress did not show signs of ultrastructural damages in the CNS (Sharma, unpublished observations). These observations are in line with the idea that breakdown of the BBB is an important factor in brain damage.

4.5.8 Ultratstructural Changes in the Cerebral Endothelium

The breakdown of the BBB is crucial in brain damage following heat stress is further strengthened by the findings showing profound ultrastructural alterations in the cell membrane of the endothelial cells of brain and spinal cord capillaries (Sharma et al., 1998a; Sharma and Hoopes, 2003; Sharma, 2004b). Thus, several microvessels show leakage of lanthanum across the cerebral endothelium in brain areas exhibiting structural changes (❯ *Figure 17-12* and ❯ *17-15*). However, exudation of lanthanum could be evident often in one endothelial cell, whereas the rest of the vessel or the adjacent endothelial cells are entirely normal in appearance (see ❯ *Figure 17-15*). This indicates a highly specific nature of the endothelial cell membrane permeability to lanthanum in stress. Activation of specific endothelial cell transporters, permeability factors, neurochemical receptors, or ions channels located on the selected area of the endothelial cell membrane many be responsible for such a selective increase in the lanthanum permeability (Sharma, 2004b, 2006b, c, 2007c, d).

It appears that the tight junctions do not widen to allow lanthanum leakage into the basal lamina or within the microfluid environment of the brain in heat stress (Sharma et al., 1998a). Thus, in several vascular profiles, the lanthanum tracer is stopped at the luminal side of the tight junctions (❯ *Figure 17-15*). On the other hand, these microvessels often showed infiltration of lanthanum across the endothelial cell membranes covering the tight junctions without widening them (❯ *Figure 17-15*). These observations are the first to support the idea of specific receptor mediated increase in microvascular permeability across the cell membrane (Sharma et al., 1998a). Since receptors can also be present on the membranes apposing tight junctions, increased microvascular permeability around the junctions is possible via activation of such receptors (Sharma et al., 1998a; Sharma, 2004b, 2007c, d).

4.6 Functional Significance of BBB Disruption in Stress

Our observations suggest that several stressful conditions depending on their duration and age of the animals are able to disrupt BBB function to large molecules, e.g., proteins. Thus, 30-min forced swimming, 4-h heat stress, 8-h immobilization stress, and 48-h sleep deprivation cause BBB breakdown to protein tracers in selective and specific brain regions. The regional changes in the BBB permeability vary according to the stressors used. Thus, each stressor has some specific effects in particular brain region that is evident with a selective breakdown of the BBB in that area (Sharma et al., 1995a, b, c, 1998a, b, c; Sharma, 2004a, b, c, d, 2006a, b, c).

This stress induced BBB permeability appears to be reversible in nature. This suggests that a brief increase in BBB function could help the organisms to cope with the stress. Alternatively, it reflects a failure of brain homeostasis, leading to harmful effects on the brain structure and function. Profound nerve cell changes in stress suggest that opening of the BBB is harmful to the organism. Flattening of EEG activity at the time of BBB opening in stress further supports this idea.

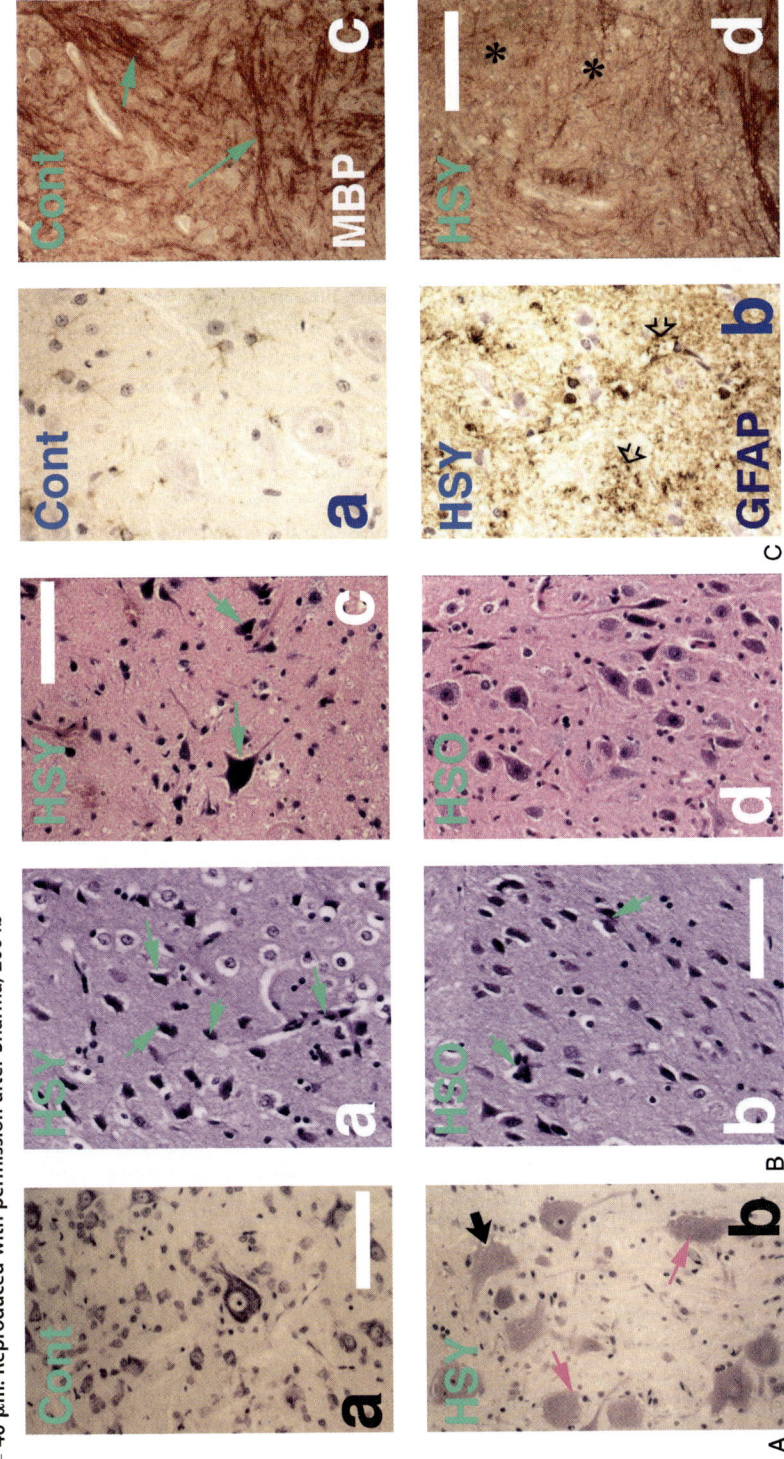

◨ Figure 17-14
Representative example of neuronal (A, B), glial, and myelin (C) changes following heat stress in young (HSY) and old (HSO) rats compared to controls (Cont). Heat stress for 4 h at 38 °C induces profound cell damage in young rats compared to old animals. Degeneration of nerve cells (*arrows*) in the brain stem reticular formation (A: b), in cerebral cortex (B: a) and in spinal cord (B: c) of young rats (HSY) is more pronounced compared to the same regions in old (HSO) animals (B: b; B: d). (C) Glial (C: a,b) and axonal (C: c,d) changes following heat stress in young rats (HSY) are also most pronounced compared to control (Cont). Glial fibrillary acidic protein (GFAP) immunoreactivity (*blank arrows*), a marker of astrocytes is upregulated following heat stress. Degradation of myelin basic protein (MBP) reflects damage or degeneration of myelin (*) is evident in heat stress compared to control (*arrows*) group. Bars: A = 25 μm (Nissl stain); B = 50 μm (Hematoxylin and Eosin); C = 40 μm. Reproduced with permission after Sharma, 2004b

◘ Figure 17-15

Ultratstructural changes in the cerebral endothelial cell membrane permeability to lanthanum and adjacent neuropil in various brain regions in heat stress. Data modified after Sharma et al., 1998a; Sharma, 2004b. (A) Lanthanum, an electron dense tracer (seen as dark black particles) is seen across the endothelial cell membrane containing tight junction (a, *blank arrows*). Infiltration of lanthanum across the endothelial cell membrane is evident (*solid arrows,* A: b,c). Interestingly, lanthanum diffusely infiltrated across the cell membranes constituting the tight junction complexes to enter into the endothelial cell cytoplasm (A: a–c; Sharma et al., 1998a). This new phenomena of endothelial cell membrane permeability across the tight junctions without widening them are not described before (for details see Sharma et al., 1998a; Sharma, 2004b). That the junctions are not widened is further evident with the fact that the lanthanum is clearly stopped within the intercellular cleft at the tight junction (*blank arrows,* A: a). In some cases, only one endothelial cell membrane covering tight junction exhibit diffuse infiltration of lanthanum across leaving its counterpart completely intact (*thick black arrows,* A: b). Occasionally, lanthanum is seen in the basal lamina and the adjacent neuropil across the endothelial cell membrane showing infiltration of lanthanum in a certain segment of the cell (A: d, *arrow*). (B) Almost all cerebral endothelial cell profiles examined (>400) showed blockade of lanthanum at the proximal tight junctions (*arrow heads,* B: a–d). However, basal lamina of these endothelial cells exhibited lanthanum exudation (see A: a–d). In addition to lanthanum infiltration within the endothelial cytoplasm, several cerebral endothelial cells show presence of lanthanum in the microvesicular (*) profiles within the cell cytoplasm (B: e,f). Normally, the tight junction in these microvessels is closed to lanthanum (B: a–d). Infiltration of lanthanum within the endothelial cell cytoplasm and the adjacent neuropil is also common in several brain areas (B: g,h). Bars: A=300 nm; B: a–e=100 nm; g,h=200 nm. Modified after Sharma et al., 1998a; Sharma, 1999, 2004b

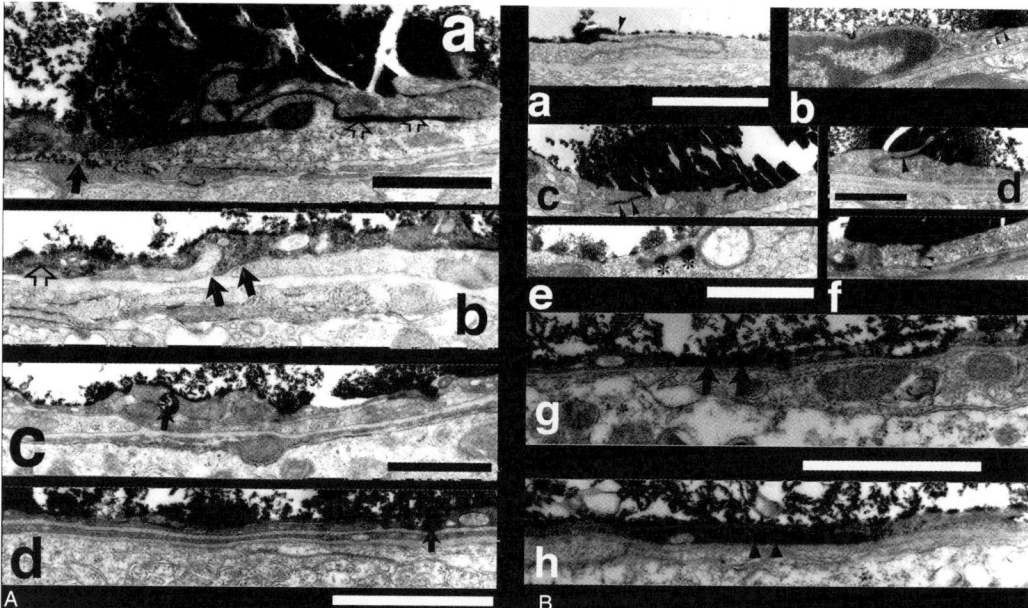

A significant higher permeability of radioiodine tracer across the CNS microvessels compared to Evans blue in stressed rats appears to be due to a difference in the molecular size of the tracers and/or due to a difference in the proteins to which they bind in vivo (Mayhan and Heistad, 1985; Sharma et al., 1990a, b). Thus, leakage of exogenous protein tracers in stress is likely to reflect endogenous protein extravasation and spread of edema fluid. Obviously, edema formation in a close cranium alone will induce adverse cell and tissue injury.

There are reasons to believe that opening of the BBB in stressful situation may induce premature aging and/or neurodegeneration. Mild to moderate nerve cell damage in the cerebral cortex or in hippocampus in stress are quite comparable to those observed during ageing and other neurodegenerative disorders, are in line with this idea.

It may be that children exposed to stressful environments are more vulnerable to mental abnormalities than adults (Essman, 1978; Sharma et al., 1995a, b, c). Once the BBB permeability is increased, these children are susceptible to more adverse neuronal dysfunction leading to various mental abnormalities. Furthermore, increased BSCB permeability will alter the spinal cord microenvironment resulting in damage to motor or sensory neurons causing long-term disability.

5 Psychostimulants Influence BCNSB Permeability

The psychostimulants, such as morphine, methamphetamine, and 3,4-methylenedioxymethamphetamine (MDMA, "ecstasy") administration and/or their withdrawal induce severe stress situations in humans and in animals (Sharma et al., 2004, 2006a; Sharma and Ali, 2006; Kiyatkin et al., 2007). These drugs also induce profound hyperthermia and behavioral alterations by modulating brain function (Sharma et al., 2004 for review). Although, use of these psychostimulants and its long-term effects are a huge burden on our society, our knowledge on the cellular and molecular mechanisms of their action in the CNS are still lacking. Thus, to develop suitable therapeutic strategies to treat this menace of the society, further investigation on the possible mechanisms of psychostimulants induced CNS dysfunction is urgently needed.

The drug dependence by regular intake of psychostimulants and the "withdrawal symptoms" caused by an abrupt cessation of the drugs was first described by Himmelsbach (1943) about 64 years ago. However, the consequences of drug dependence and their withdrawal response on alterations in the brain fluid microenvironment are still not well-known. Since, use of psychostimulants and their withdrawal induce severe stress, it is likely that the brain microvascular permeability is altered by these drugs of abuse (Sharma et al., 2004; Sharma and Ali, 2006). Keeping these views in consideration, our laboratory was the first to demonstrate breakdown of the BCSNB disturbances following morphine withdrawal or methamphetamine or MDMA administration in rodent models (Sharma et al., 2006a, 2007).

5.1 Morphine Dependence and Withdrawal Influences BBB Function

In naive rats, administration of morphine (7.5 mg/kg, i.p.) induces analgesia (Sharma et al., 2004). Thus, morphine tolerance and dependence could be developed using a daily 10 mg/kg, intraperitoneal dose on the fourth day and onward (● Figure 17-16). The full tolerance was developed by the 12th day of morphine treatment (● Figure 17-16). Acute administration of morphine (10 mg/kg, i.p.) induces profound hyperthermia (+1.8°C; from 37.6 ± 0.43 to 39.4 ± 0.23°C) at 60 min that continues until 90 min. The body temperature remained elevated (>1°C) even at 5 h after the morphine administration (● Figure 17-15). However, hyperthermia diminishes in chronic morphine-treated rats from the day 6 to 8 and onward (Sharma et al., 2004).

5.1.1 Morphine Withdrawal Stress

Abrupt cessation of morphine injection on day 12 in the morphine-dependent rats, result in appearance of spontaneous withdrawal symptoms within 12 h (● Table 17-9) e.g., loss of body weight, "wet shake" phenomena, piloerection, writhing, teeth chattering, diarrhea, and aggressive behavior and jumping (● Table 17-9). These symptoms continue to remain worsen over days. Thus, the withdrawal symptoms aggravated at 24 and 48 h after the cessation of morphine and apparent reduction in symptoms are observed at 72 h after cessation of morphine administration (● Table 17-9).

Figure 17-16

Morphine analgesia (*left*) and dependence (*right*) in rats. Intraperitoneal administration of morphine (7.5 mg/kg) results in analgesia as seen with tail flick latency using hot-plate (52°C) technique with 15 sec cutoff time (see Sharma et al., 2004 for details). Morphine analgesia developed in naive animals within 30 min after morphine injection that lasted up to 75 min. During chronic treatment with morphine (10 mg/kg, i.p.), development of analgesia is seen in animals up to day 3 (*right*). From the day 4 and onward, the animals started showing signs of morphine dependence as the tail-flick latency decreased from 15 (day 1–3) sec to 12 sec (day 4). This decrease in tail-flick latency continued until day 11. On the day 12, all animals showed complete tolerance to morphine analgesia (*right*). * = $P < 0.05$; ** = $P < 0.01$, ANOVA followed by Dunnet's test. Reproduced with permission from Sharma et al., 2004

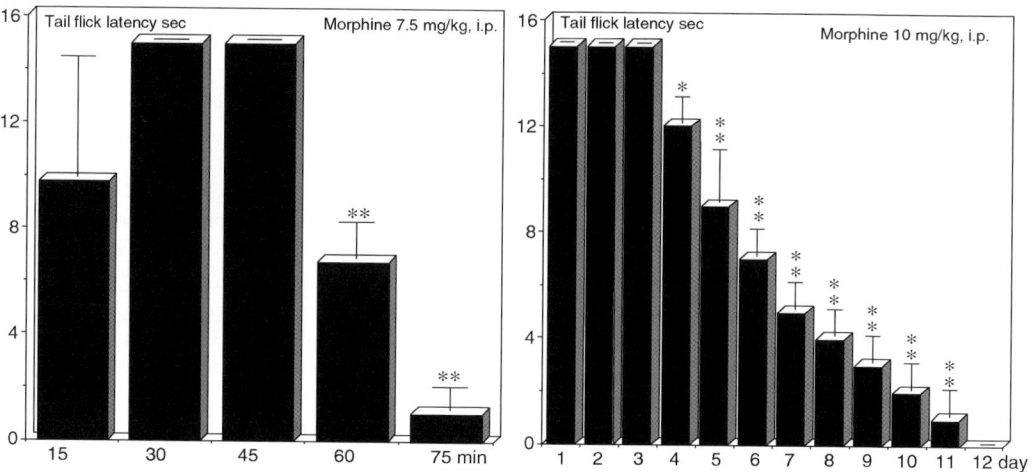

5.1.2 BBB Permeability to Evans Blue and Radioiodine

The BBB permeability to Evans blue and [131]Iodine in normal animals remained very low in several brain regions (◉ Table 17-10), although the magnitude and tracer extravasation varied in different CNS regions. The cerebellum, brain stem, and spinal exhibited least extravasation of the Evans blue and radioactive iodine tracers; whereas, thalamus and hypothalamus showed moderately high concentration compared to the cerebral cortex (◉ Table 17-10). Chronic morphine treatment did not result in any increase in the tracer extravasation in the CNS. However, in some brain regions, a significant decrease in the tracer extravasation was observed. This decrease was significant in the thalamus, hypothalamus, and in the cerebral cortex (◉ Table 17-10). On the other hand, morphine withdrawal resulted in a significant increase in the BBB permeability to Evans blue and radioactive iodine on day 1 in the cerebral cortex, hippocampus, and the cerebellum. The hypothalamus and thalamus showed only moderate increase in the permeability (◉ Table 17-10). On the second day, a significantly higher increase in the permeability to these tracers was in the cerebellum followed by cerebral cortex, thalamus, and hypothalamus at this time compared to the other regions (Sharma et al., 2004).

5.1.3 Leakage of Albumin across the BBB during Morphine Withdrawal

Profound leakage of albumin immunohistochemistry is seen in several brain areas (◉ Figure 17-17) in the neuropil. Few nerve cells exhibited albumin immunoreactivity; whereas, some glial cells, possibly astrocytes also show albumin immunostaining. The intensity of albumin immunoreactivity was much more pronounced on the second and third day of morphine withdrawal. The nerve cells showing albumin immunoreactivity were largely distorted and edematous in appearance and located in the edematous areas of the brain or spinal cord (Sharma, unpublished observations).

Table 17-9
Effect of Morphine dependence and withdrawal on stress symptoms

Parameters	Control	Morphine dependence			Morphine withdrawal			
		1st day	10th day	12th day	12 h	24 h	48 h	72 h
Stress symptoms	n = 6	n = 6	n = 8	n = 12	n = 8	n = 14	n = 16	n = 8
Body temperature °C	37.61 ± 0.42	39.42 ± 0.41**	38.64 ± 0.51	38.42 ± 0.31	39.54 ± 0.23**	40.23 ± 0.18**	39.28 ± 0.11**	38.67 ± 0.22
Heart rate, beats/min	280 ± 12	320 ± 18*	330 ± 12**	338 ± 10**	304 ± 8*	310 ± 8*	308 ± 7*	296 ± 14
Respiration, cycles/min	76 ± 6	80 ± 8	84 ± 6**	86 ± 7**	89 ± 5*	94 ± 4**	92 ± 5**	80 ± 8
Wet-shakes n	Nil	Nil	Nil	Nil	4 ± 2	8 ± 3[a]	6 ± 2	5 ± 3
Piloerection	Nil	Nil	Nil	Nil	+++	+++++	++++	+++
Writhing	Nil	Nil	Nil	Nil	+++	+++++	++++	+++
Teeth chattering	Nil	Nil	Nil	Nil	+++	+++++	+++++	+++
Diarrhoea	Nil	Nil	Nil	Nil	+++	+++++	+++	Nil
Aggressive behavior	Nil	Nil	?	?	+++	+++++	+++++	+++++
Microhemorrhages in stomach	Nil	6 ± 5	8 ± 5	12 ± 8	48 ± 12	68 ± 18[b]	85 ± 14[b]	23 ± 8[b]

Values are mean ± SD
[a]Significantly different ($P < 0.05$) from Morphine withdrawal 12 h
[b]Many microhemorrhages
+++ = mild, ++++ = moderate, +++++ = severe, ? = unclear, nil = absent
For details see text

Table 17-10
Effect of morphine dependence and withdrawal on blood–brain and the blood–spinal cord barrier permeability

CNS regions	Control		Morphine dependence		Morphine withdrawal			
			12th day		24 h		48 h	
	EBA, mg %	[131]I, %	EBA, mg %	[131]I, %	EBA, mg %	[131]I, %	EBA mg %	[131]I, %
Cingulate cortex	0.24 ± 0.06	0.38 ± 0.04	0.18 ± 0.06	0.20 ± 0.04*	0.54 ± 0.04**	0.68 ± 0.08**	0.89 ± 0.10**	0.98 ± 0.13**
Frontal cortex	0.28 ± 0.08	0.41 ± 0.05	0.08 ± 0.02**	0.10 ± 0.09**	0.68 ± 0.12**	0.75 ± 0.10**	1.06 ± 0.13**	1.32 ± 0.14**
Hippocampus	0.26 ± 0.06	0.36 ± 0.04	0.18 ± 0.08	0.24 ± 0.05	0.58 ± 0.11*	0.67 ± 0.14*	0.87 ± 0.12**	0.94 ± 0.08**
Amygdala	0.21 ± 0.04	0.30 ± 0.07	0.18 ± 0.04	0.22 ± 0.05	0.48 ± 0.11*	0.56 ± 0.08*	0.59 ± 0.11*	0.76 ± 0.14*
Thalamus	0.45 ± 0.08[a]	0.56 ± 0.06[a]	0.23 ± 0.08**	0.32 ± 0.09**	0.48 ± 0.08	0.64 ± 0.06	0.78 ± 0.10*	0.89 ± 0.12*
Striatum	0.23 ± 0.05	0.32 ± 0.06	0.28 ± 0.10	0.34 ± 0.08	0.56 ± 0.14*	0.65 ± 0.08*	0.89 ± 0.12**	0.96 ± 0.14*
Cerebellum	0.08 ± 0.02[a]	0.09 ± 0.04[a]	0.06 ± 0.04	0.08 ± 0.03	0.34 ± 0.12*	0.44 ± 0.12**	0.54 ± 0.18**	0.62 ± 0.16**
Hypothalamus	0.52 ± 0.08[a]	0.67 ± 0.05[a]	0.34 ± 0.11*	0.38 ± 0.10*	0.76 ± 0.11*	0.87 ± 0.15*	0.89 ± 0.21*	0.96 ± 0.19*
Brain stem	0.23 ± 0.06	0.28 ± 0.05	0.11 ± 0.04*	0.14 ± 0.03*	0.34 ± 0.05*	0.42 ± 0.06*	0.44 ± 0.08**	0.56 ± 0.06**
Spinal cord C-5	0.20 ± 0.04	0.28 ± 0.06	0.10 ± 0.02*	0.14 ± 0.05*	0.48 ± 0.08*	0.59 ± 0.08*	0.76 ± 0.10**	0.88 ± 0.09**

Source: Data from Sharma et al. (2004); Sharma (unpublished observations)

Data from 6 to 8 animals in each group; values are mean ± SD

*$P < 0.05$, **$P < 0.01$, significantly different from control group, ANOVA followed by Dunnet test from multiple group comparison

[a]Significantly different ($P < 0.05$) from cingulate cortex

5.1.4 Extravasation of Lanthanum at the Ultrastructural Level

Using ultrastructural investigation, lanthanum exudation across the endothelial cell membranes in the cerebral cortex, cerebellum, and the spinal cord is seen following second day of morphine withdrawal (see ❯ *Figure 17-17*). The lanthanum is present into the endothelial cell cytoplasm, whereas in some cases, the tracer was found in the basal lamina as well. However, the tight junctions were not altered to lanthanum (❯ *Figure 17-16*). Occasionally, the tracer is also seen in vesicular profiles located within the endothelial cell cytoplasm in some large superficial vessels (Sharma and Ali, 2006; Sharma et al., 2007).

This endothelial cell membrane permeability to lanthanum is very selective and specific in nature. Thus, microvessels, often comprising only one endothelial cell show only a part of the membrane permeable to lanthanum leaving other areas of the endothelial cell completely normal (see ❯ *Figure 17-17*). This indicates that selective alterations in membrane receptor functions or transporters are affected by morphine withdrawal stress.

5.2 Morphological Changes in Morphine Dependence or Withdrawal

Rats subjected to spontaneous morphine treatment or withdrawal showed mild to moderate degree of nerve cell and glial cell reaction at light and electron microscopy.

5.2.1 Nerve Cell Reaction

Few nerve cells showed degenerative changes in animals subjected to chronic morphine treatment in the regions showing altered BBB function (❯ *Figure 17-17*). During morphine withdrawal most pronounced nerve cell damage was seen in the brain stem reticular formation (❯ *Figure 17-17*). However, scattered damaged nerve cells were also present in the cerebral cortex, hippocampus, thalamus, and hypothalamus (Sharma and Ali, 2006; Sharma et al., 2007).

These changes in nerve cells were most prominent on the second day of morphine withdrawal (❯ *Figure 17-17*). The damaged nerve cells were largely seen in the brain areas showing leakage of albumin, sponginess, and edema. Perivascular regions around large vessels often show degeneration of the neuropil (❯ *Figure 17-17*). The epithelial cell layers of the choroid plexus also exhibited degenerative changes indicating breakdown of the BCSFB following morphine withdrawal (❯ *Figure 17-17*).

5.2.2 Glial Cell Reaction

Pronounced cell damage of the perivascular astrocytes and upregulation of GFAP is seen in several brain regions exhibiting changes in the BBB function during morphine dependence or withdrawal (Sharma et al., 2004). Chronic morphine treatment resulted in mild to moderate activation of astrocytes in specific brain or spinal cord regions (❯ *Figure 17-17*). However, most pronounced activation of astrocytes is seen during second day of spontaneous morphine withdrawal (❯ *Figure 17-17*). These observations suggest that alteration in the BBB function alters astrocytic activity (Sharma, 2006b, c; Sharma and Ali, 2006; Sharma et al., 2006a, 2007).

5.2.3 HSP Expression

We examined heat shock protein 72 (HSP 72kD) expression to determine the magnitude and severity of cellular stress during morphine treatment or withdrawal. Chronic treatment with morphine resulted in a mild upregulation of HSP in several CNS regions. However, the most pronounced upregulation of HSP was seen following morphine withdrawal on the second day (❯ *Figure 17-17*) that continued until 72 h after cessation of morphine administration (Sharma, unpublished observation). The magnitude and intensity of HSP expression is most pronounced in the regions exhibiting nerve cell damage and activation of astrocytes in the edematous areas showing extravasation of albumin (Sharma et al., 2004; Sharma and Ali, 2006).

5.2.4 Ultratstructural Changes in the Neuropil

In morphine-dependent animals, vacuolation, membrane disruption, and damage to cellular structures are seen in some regions of the brain (❯ *Figure 17-17*). These cellular changes are most prominent in the brain stem reticular information (Sharma et al., 2004) Morphine withdrawal further aggravated these changes

◘ Figure 17-17
Changes in nerve cell (A), heat shock protein (HSP 72) and glial fibrillary acidic protein (GFAP) expression (C), lanthanum (B), and albumin extravasation (D) in rodents following morphine or methamphetamine administration or withdrawal (Sharma and Ali, 2006). A. Nerve cell changes in the rat brain following morphine dependence (day 12; MD12; e) or after its spontaneous withdrawal following day 1 (MWD1; a,c) or day 2 (MWD2; b,d,f). Nissl stained paraffin sections (3 μm thick) from rat brain passing through piriform area (−3.70 mm from bregma) shows marked selective nerve cell damage following 1 day after morphine withdrawal (MWD1; *arrows*, A: a). The magnitude and intensity of nerve cell damage further increased on the second day of morphine withdrawal (MWD2; *arrows*, A: b). Hematoxylin and Eosin stained 3 μm thick paraffin sections passing through hippocampus (−3.25 mm from bregma) showing selective cell damage in the CA1 subfield from rats following morphine withdrawal (day 1, MWD1; A: c and day 2, MWD2; A: d). Sponginess and edema (*) are clearly visible. In morphine dependent rats on the day 12 (MD 12; A: e) damage to hippocampal dentate granule cells (−4.20 mm from bregma) are clearly visible (Nissl stain, *arrows*). Paraffin section passing through the mesencephalic reticular nucleus (−7.10 mm from bregma) show many damaged nerve cells (Hematoxylin and Eosin stain, *arrows*), sponginess and edema (*) on the second day of spontaneous morphine withdrawal in rats (MWD2; A: f). Bars = A: a,b,c,d = 25 μm; A: e,f = 30 μm. Data modified after Sharma et al., 2004. (C) Immunoreactivity of heat shock protein (HSP, *arrows*) 72 kD, a marker of cellular stress in the spinal cord (C-5) of morphine dependent rats exhibited profound upregulation in the ventral (C: a) and in lateral (C: e) horns in the morphine dependent rat on the day 12 (MD12) as well as in the ventral horn (C: c) following second day of spontaneous morphine withdrawal (MWD2). The HSP immunostaining is largely present in the cell cytoplasm; however, in some cases nerve cell nucleus also exhibited HSP expression. Bar = B: a,e = 40 μm, B: c = 30 μm. Data modified after Sharma et al., 2004. Increased expression of glial fibrillary acidic protein (GFAP, a marker of astrocytes activation) immunoreactivity in the spinal cord of in one 12 day morphine-dependent (MD 12, C.f) rat and following second day of morphine withdrawal MWD2, C: d) rat compared to control (C: b). Activation of astrocytes as evident with expression of GFAP is most intense in rats following second day of morphine withdrawal (MWD2). Bar = B: b = 25 μm, d = 30 μm, f = 40 μm. Data modified after Sharma et al., 2004. B, D. Lanthanum Infiltration is seen (*arrow heads*) across the endothelial cell of one microvessel from the cerebral cortex (B: b) and from the cervical spinal cord (B: a) following 48 h after spontaneous morphine withdrawal in dependent rat. Several microvesicular profiles within the endothelium are clearly seen (B: a). The tight junctions closed to lanthanum (*arrows*) in the cerebral (D: b) and spinal cord (D: a) microvessel. Edematous swelling of perivascular astrocytes is clearly evident (B: b; D: b). Bars = 500 nm (C: a); 600 nm (C: b), 300 nm (D: a); 400 nm (D: b). Data modified after Sharma et al., 2004. Extravasation of albumin as seen using immunohistochemistry is apparent in the cerebral cortex of the rat subjected to 48 h of spontaneous morphine withdrawal (B: c). Leakage of albumin seen around the nerve cell and within the nerve cell cytoplasm (*arrows*) located within the edematous region in morphine withdrawal rat on the second day (C: c). Interestingly, single injection of methamphetamine (MTH) also evoked leakage of albumin in the nerve cell cytoplasm and in the neuropil (*arrows*) in the mouse cerebral cortex (B: d). The magnitude and intensity of albumin leakage however is much less compared to the rats subjected top spontaneous morphine withdrawal (for details see text). Methamphetamine showed pronounced neurotoxicity in brain regions showing breakdown of the BBB function (D: c.,d). Thus, perineuronal edema, distorted and damaged nerve cells (*arrows*) are quite frequent in the edematous regions showing albumin leakage (D: c,d). Bars = C: c,d = 50 μm; D: c,d = 40 μm. Reproduced with permission from Sharma and Ali, 2006

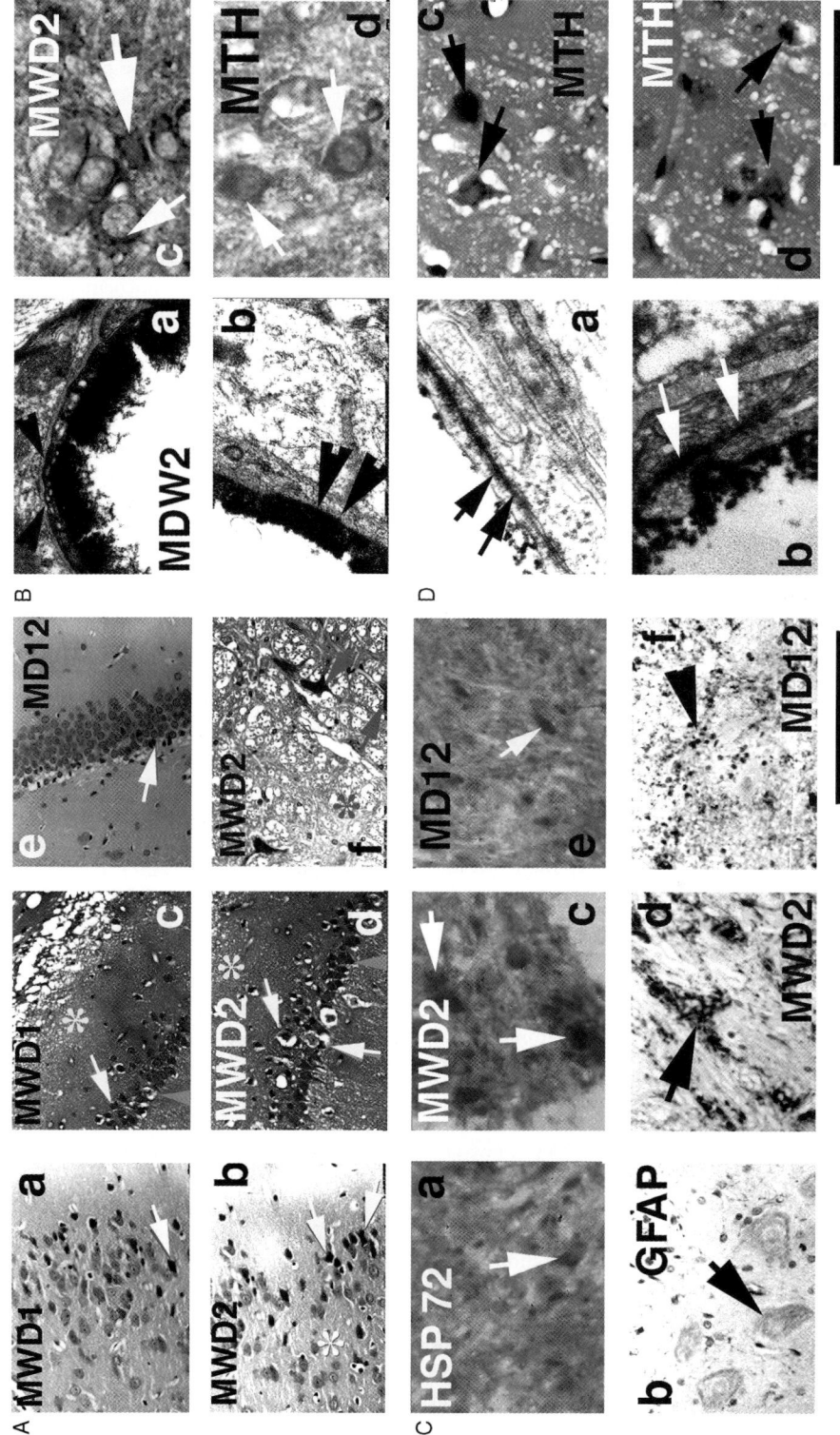

■ Figure 17-17 (continued)

and thus vacuolation, edema, cell membrane damage, and appearance of swollen cells are much more frequent in the neuropil in several brain regions on the day 2 of opioid withdrawal (Sharma et al., 2004, 2007). These observations suggest that alterations in the BBB functions, i.e., either increased or decrease in the permeability of the cell membrane is responsible for adverse cell reactions in the brain. However, whether these changes are reversible or permanent, is not known.

5.3 Methamphetamine Induces Albumin Leakage across the BBB

Methamphetamine (40 mg/kg) administered into inbred B57 mice (35–50 g body weight) induces marked hyperthermia (>40°C) and behavioral dysfunctions (Sharma and Ali, 2006; Sharma et al., 2007). The BBB permeability was examined at 4 h after drug administration using albumin immunohistochemistry (Sharma and Ali, 2006). Profound leakage of albumin is seen in the cerebral cortex, hippocampus, cerebellum, caudate nucleus, thalamus, and hypothalamus in methamphetamine treated mice. (❯ *Figure 17-17*). Few nerve cells took albumin staining as well. However, the magnitude and intensity of albumin immunostaining was much less severe compared to the rats subjected to morphine withdrawal (❯ *Figure 17-17*). Interestingly, the area showing BBB disruption also exhibited pronounced nerve cell damage, sponginess, and edema (❯ *Figure 17-17*).

In separate investigations, we administered methamphetamine in rats at normal room temperature and compared the results of albumin leakage in animals in which the drug was administered at high environmental temperature (29°C, Kiyatkin et al., 2007). Methamphetamine-induced leakage of albumin and Evans blue were much more pronounced when the psychostimulant was administered at 29°C compared to normal room temperature (Kiyatkin et al., 2007). In these rats, brain hyperthermia correlates well with the intensity of albumin leakage. These observations suggest that methamphetamine-induced hyperthermia is directly or indirectly contributing to BBB disruption (Kiyatkin et al., 2007).

5.3.1 Methamphetamine Induces Edema Formation and Structural Changes in the CNS

Measurement of brain edema formation in rats and mice after methamphetamine administration shows a pronounced increase from the control group (Sharma, unpublished observations). The magnitude of edema formation correlates well with the intensity of albumin leakage. Thus, rats showed greater accumulation of brain water when methamphetamine was given at higher ambient temperature compared to the rats received the drug at normal room temperature. Interstingly, an increase in Na^+ and K^+ concentration of the tissues correlated well with the increase in brain water content. The Cl^- content did not differ significantly (Kiyatkin et al., 2007).

Morphological examination revealed profound neuronal, glial, and myelin damage in rats and mice after methamphetamine treatment. The most marked neuronal and glial cell damage is seen in rats that received methamphetamine at high environmental temperature (Sharma, unpublished observations). These observations suggest that psychostimulants affects brain function that depends on the external environments, e.g., heat stress.

5.4 MDMA Alters BBB Permeability

The psychostimulant MDMA induces hyperthermia and brain dysfunction. Thus, we examined the acute effects of MDMA treatment on BBB dysfunction, brain edema, and cell injury in rats and mice. Male C57B mice or male Wistar rats kept at normal laboratory environment ($21 \pm 1°C$) were administered MDMA (40 mg/kg, i.p.) and allowed to survive 4 h after the drug administration. The BBB permeability, brain edema, ion content (Na, K, Cl), and cell injury were examined using standard protocol. The rats and mice showed profound behavioral disturbances (hyperactivity and hyperlocomotion) and hyperthermia (mice $36.8 \pm 0.08°C$ to $41.44 \pm 0.38°C$, $P < 0.001$; Rats $37.2 \pm 0.06°$ to $40.86 \pm 0.12°C$, $P<0.001$) at 4 h.

Massive extravasation of Evans blue in several brain regions, particularly in cerebellum, hippocampus, cortex, thalamus, and hypothalamus were observed at this time. The magnitude and intensity of Evans blue staining by MDMA in mice was much more pronounced than in rats (cortex: mice 0.23 ± 0.08–1.89 ± 0.21 mg%, $P < 0.001$; rat 0.28 ± 0.06–0.98 ± 0.12 mg %; $P < 0.001$) (Sharma and Ali, 2007). These observations are the first to show that like other psychostimulants, MDMA also disrupts the BBB function in rodents. It is unclear from this investigation whether MDMA directly alters the BBB function by inducing brain hyperthermia or indirectly via releasing neurochemical mediators, e.g., serotonin prostaglandins or histamine to disrupt the BBB function. This is a subject that is currently being investigated in our laboratory.

5.4.1 MDMA Induces Brain Edema Formation

MDMA induced a significant increase in brain water in the cerebral cortex, hippocampus, and cerebellum in the regions showing Evans blue leakage. Edema formation was most pronounced by MDMA in mice compared to rats (cortex: mice 76.34 ± 0.21–$78.43 \pm 0.12\%$; $P < 0.001$; rat 76.76 ± 0.23–$78.32 \pm 0.32\%$; $P < 0.001$). In edematous brain tissues a significant profound increase in Na^+ content and a moderate increase in K^+ level was also seen indicating vasogenic edema formation due to protein leakage (Rapoport, 1976; Sharma et al., 1998a, c; Sharma, 2006b). However, the Cl^- content did not differ significantly. These observations suggest that breakdown of the BBB to proteins is instrumental in edema formation.

5.4.2 MDMA Induces Cell Injury and Upregulates HSP Expression

Morphological examination revealed several distorted neuronal and glial cells in the brain areas showing BBB disruption. This is further evidenced by immunohistochemistry of albumin immunostaining in these edematous brain regions (Sharma and Ali, 2007). Activation of astrocytes is evident with upregulation of GFAP immunoreactvity. Expression of HSP immunoreactivity was also seen in the damaged and distorted neurons and glial cells located in the edematous neuropil. The HSP immunostaining was most pronounced in the nuclei of neurons compared to the cytoplasm (Sharma and Ali, 2007). These observations are the first to show that MDMA has the capacity to disrupt BBB permeability to proteins and induce brain edema formation. It appears that breakdown of the BBB is instrumental in inducing edema formation and cell injury. Upregulation of HSP in the edematous regions showed that cellular stress and hyperthermia following MDMA administration could primarily be responsible for BBB disruption and brain injury.

5.5 Psychostimulants Alter Cellular Stress Response

The detailed mechanisms of increased BBB permeability following morphine withdrawal or methamphetamine or MDMA administration are not fully understood. However, there are reasons to believe that drug induced hyperthermia, release of several neurochemical mediators, cytokines, immunomodulators, and/or signal transduction molecules are playing important roles in BBB breakdown (Sharma et al., 2004, 2007; Sharma and Ali, 2006).

Psychostimulants activate the hypothalamic–pituitary–adrenal (HPA) axis and thus are able to induce considerable stress response in both experimental and in clinical situations (Houshyar et al., 2001a, b; Sharma et al., 2004). There are evidences that acute psychostimulants increase corticotrophin releasing hormone (CRH) release, which in turn induces the secretions of adrenocorticotrophic hormone (ACTH) and corticosterone (Houshyar et al., 2001a, b). Similar activation of HPA as observed during other chronic stress conditions is also seen following repeated exposure to morphine (Sharma and Ali, 2006). Our studies provide new lines of evidences that support the idea that psychostimulants induce profound cellular stress as evident with the overexpression of HSP as well as massive alterations in the neuronal and glial cell function.

Likewise, withdrawal of morphine in chronically morphine-intoxicated rats also induces severe stress reactions (Sharma et al., 2004, 2006a, 2007). Thus, rats undergoing 12-h spontaneous morphine withdrawal exhibited a potentiated and prolonged corticosterone response following restraint despite of a high basal corticosterone levels in their plasma during chronic morphine treatment (Houshyar, 2001a). Elevated hormone level acts as negative feedback causing further release of corticosterone under stressful situations (Sharma et al., 2007).

An enhanced activity of paraventricular thalamic and hypothalamic nuclei, activation of CRH neurons in the hypothalamus, increased sensitivity of adrenal to ACTH, and reduced glucocorticoid receptor (GR) expression in the hippocampus following psychostimulants treatment or withdrawal is in line with this idea (see Sharma et al., 2007 for review). However, the anatomical neural network behind psychostimulants induced stress response is still unclear and requires further investigation.

5.6 Psychostimulants Induced Decrease in BBB Function: Unanswered Questions

Interestingly, chronic morphine treatment decreases the BBB function and induces neuronal damage. The mechanism behind a decrease in the BBB permeability and its possible functional significance is not known at all (Sharma et al., 2004). It seems quite likely that if the barrier is tightened enough to influence normal transport mechanisms, the CNS homeostasis are perturbed leading to adverse cell and tissue reactions (Sharma et al., 2007). Thus, a normal function of the BBB is needed for optimal function of the CNS.

Our novel findings using morphine thus further show that the opioid dependence is associated with tightening of the barrier and withdrawal is associated with the opposite effects, i.e., breakdown of the BBB. Under both the conditions, adverse neuronal changes and cell damage are evident. Interestingly, blood flow studies showed profound ischemia during morphine dependence (Sharma et al., 2004, 2007). Thus, the cell changes seen during morphine dependence may represent selective neuronal injury due to morphine toxicity and/or ischemia associated with opioid administration. Morphine withdrawal aggravates neuronal damage in these animals probably due to breakdown of the BBB function (Sharma and Ali, 2006). It may be that during morphine dependence, the cell changes seen could be due to local alterations in neuronal energy metabolism, toxicity, or alteration in cellular membrane function (Sharma and Ali, 2006, 2007). Similar cellular damage during early phase of ischemic injuries without clear disruption of the BBB to large molecular tracers is in line with this idea.

On the other hand, it is unclear whether the increased BBB permeability by psychostimulants is necessary for the altered physiological demand or it represents adverse pathological condition leading to brain damage. Structural alterations seen in the CNS during psychostimulants treatment or withdrawal suggest that opening of the BBB contributes to the adverse cell reaction in the CNS. Whether, these cell changes and the BBB dysfunction is reversible in nature still remains to be investigated.

To further clarify these points additional studies using pharmacological blockade of withdrawal symptoms and/or BBB at various time intervals after morphine withdrawal is needed, a feature that is currently being investigated in our laboratory.

6 Nanoparticles Influence BCNSB Function

Recently, nanoparticles and their influence on the human brain function have attracted wide attention among the scientific community. However, influence of nanoparticles on brain function in vivo conditions is still lacking. There are reasons to believe that nanoparticles from the environment can enter into the CNS through inhalation from the environment and will alter the brain function (Borm et al., 2006; Lam et al., 2006; Mills et al., 2006). However, the potential risks and the cellular toxicity imposed by these nanoparticles *per se* or through modifying the BBB functions are still unknown (Xia et al., 2006; Sharma and Sharma, 2007).

6.1 Nanoparticles and Neurotoxicity

The nanoparticles because of their small sizes (10–100 nm) when exposed to the cells and tissues have higher inflammatory potential compared to the larger particles of the same materials (for review see Sharma and Sharma, 2007). Thus, titanium dioxide (TiO2) particles in the range of 20 nm when applied intratracheally into rats or mice, induces intense inflammatory neutrophil response in the lung compared to the TiO2 in the range of 250 nm size at the identical doses (Oberdörster, 1996; Oberdörster et al., 2000, 2005a, b; Sharma and Sharma, 2007). The toxicological effects of nanoparticles are largely due to their particle chemistry, especially the surface chemistry in addition to their particle size (Oberdörster, 1996, 2004).

The engineered nanomaterials in different shapes induce toxicity depending on their dose, dimension, durability and/or distribution as well as the species examined (Zhu et al., 2006). Thus, carbon nanotubes (2–50 nm) administered into the mice trachea cause acute inflammatory effects in the lung (0.3–1.3 mg/kg; Shvedova et al., 2005) but not in rats (1–5 mg/kg; Warheit et al., 2004). Whether the acute nanotoxicity observed in mammalian species are due to direct effects of nanoparticles, or alterations in cellular and molecular functions by these particles are still unclear (Sharma and Sharma, 2007).

6.2 Nanoparticles Alter BBB Permeability

We have used engineered nanoparticles from metals, Al, Ag, and Cu in the range of 50–60 nm and examined their effects on the BBB disruption in rats and mice (Sharma et al., 2006b). The nanoparticles are either suspended in sterile water or mixed with Tween 80, and dose of each nanoparticle was adjusted so that each animal (rats or mice) received 30 mg/kg dose for intravenous route, 50 mg/kg for intraperitoneal route, or 20 μg in 10 μl on the cortical superfusion. The nanoparticle concentration in blood was approximately 0.5 mg/ml. Molarity of nanoparticles ranges from 0.01–0.05 M (Sharma and Dey, 1986a, b, 1987a, b). Thus, no hyperosmolarity factor is involved (Rapoport, 1976). For control group equimolar NaCl solution or carbonized microspheres suspended in Tween 80 or in sterile saline was used in this investigation (Sharma and Sharma, 2007).

6.2.1 EBA Staining

Intravenous (30 mg/kg), intraperitoneal (50 mg/kg), or intracerebral (20 μg in 10 μl) administration of either Ag, Cu, or Al nanoparticles (size 50–60 nm) influenced the BBB function to EBA in rats and mice in a highly selective and specific manner. The leakage of Evans blue dye was observed largely in the ventral surface of the brain and in the proximal frontal cortex. The dorsal surfaces of cerebellum and the thoracic spinal cord show mild to moderate Evans blue staining (❯ *Figure 17-18*). The ventral surface of the spinal cord exhibited mild staining, largely in the thoracic and lumbar segments of the spinal cord (Sharma and Sharma, unpublished observations). The effect of aluminum nanoparticles on the BBB function was much less intense compared to the silver and copper nanoparticles. Intraperitoneal administration of nanoparticles least influenced the visual disruption of the BBB to the Evans blue dye (see ❯ *Table 17-10*).

Cortical superfusion with nanoparticles resulted in mild to moderate opening of the BBB to Evans blue dye that was largely seen on the ipsilateral side. However, the cerebellum, dorsal parts of the brain stem, and cervical spinal cord showed leakage of EBA as well (Sharma, unpublished observations). This effect was most pronounced by Ag and Cu nanoparticles. The Al nanoparticles showed faint to mild blue staining (❯ *Table 17-11*).

6.2.2 Radiotracer Extravasation

Intravenous administration of Al, Cu, and Ag nanoparticles induces extravasation of radioiodine tracer in selected brain areas (❯ *Table 17-11*). There is no difference in radiotracer extravasation when the nanoparticles were administered as a suspension in water or mixed with Tween 80 (results not shown). Tween 80 or NaCl given in equimolar concentration did not induce radiotracer extravasation in any brain region compared with control group (● Table 17-11).

6.2.3 Endogenous Albumin Leakage

Using immunohistochemistry, pronounced leakage of endogenous serum albumin in the brain and spinal cord is seen in rats that received Al or Ag nanoparticles (● Figure 17-18). Our observations suggest that intravenous administration of nanoparticles caused extensive leakage of endogenous serum albumin in these animals in several parts of the brain (● Table 17-12). The albumin immunoreactivity was seen in the hippocampus, cortical cell layers III and V, as well as some regions in the brain stem (Sharma, unpublished observations).

6.2.4 Lanthanum Extravasation

Extravasation of lanthanum across the cerebral endothelial cells in Ag nanoparticle treated rats is clearly seen at the ultrastructural level (Sharma and Sharma, 2007). Thus, few endothelial cells showed infiltration of lanthanum within the cell cytoplasm, whereas other adjacent or neighboring microvessels were devoid of lanthanum extravasation. The tight junctions were normally intact (Sharma and Sharma, 2007). Thus, it

◘ Figure 17-18
(A) Shows extravasation of Evans blue on the dorsal (a) and ventral (b) surfaces of rat brain after silver (Ag) nanoparticles (≈50–60 nm) treatment. The Ag nanoparticles was administered intravenously (35 mg/kg) and the rat is allowed to survive 24 h after injection. Coronal sections of the brain passing through hippocampus (c) and caudate nucleus (d) are also shown. Leakage of Evans blue dye can be seen in various brain regions (*arrows*). The deeper parts of the brain, e.g., hippocampus, caudate nucleus, thalamus, hypothalamus, cortical layers including pyriform, cingulate, parietal, and temporal cortices showed moderate blue staining. This indicates widespread leakage of Evans blue albumin within the brains after Ag treatment. Bar = 3 mm (Sharma HS, unpublished observations). (B) Shows albumin immunoreactivity (a,b) and glial fibrillary acidic protein (GFAP) expression (c,d) in the cerebral cortex rats receiving copper (Cu; a,c) or Ag (b,d) treatment (≈50–60 nm). These nanoparticles were administered separately (35 mg/kg) intravenously in rats and the animals allowed to survive 24 h after the administration. Paraffin (3-μm thick) sections showed massive uptake of albumin (a,b, *arrows*) by the nerve cells and along with its extravasation in the neuropil. Damage to nerve cell and leakage of albumin in the neuropil is more conspicuous by Ag treatment (B: b) compared to Cu nanoparticles administration (B: a). Overexpression of astrocytes, as seen by GFAP immunoreactivity is present in the neuropil of both Ag (B: d) and Cu (B: c) treated animals. Star shaped astrocytes are more frequent in the neuropil of Ag treated rats compared to Cu treatment. Bar = 50 μm. Sharma and Sharma, unpublished observations. (C) Transmission electronmicrographs of myelin (a,b), nerve cell (c) and one microvessel (d) following Ag or Cu nanoparticles treatment. (D) Low power light micrograph showing immunostaining of heat shock protein (HSP, a,b) and glial fibrillary acidic protein (GFAP, c,d) in the spinal cord following Ag or Cu nanoparticles treatment. Ag (C,a) or Cu (C,b) nanoparticles selectively induce damage to myelinated fibers (arrowheads) in rats 24 h after their administration. However, normal myelinated fibers (*arrows*) are also seen in the neuropil (C: c). Bar a,b = 600 nm. Degeneration of one nerve cell and disintegration of nucleolus (*arrows*) is clearly seen in the cerebral cortex by Cu treatment (C: c). Damage to myelinated fibers and vacuolation of the neuropil are evident (C: c). A completely collapsed microvessel (*arrow*) with perivascular edema and myelin damage (*arrow*) is seen in Cu treated rat (C,d). Bars: c,d = 1 μm. Expression of stress protein seen using heat shock protein expression (HSP 72 kD) is seen in the spinal cord after Ag (D,a) or Cu (D,b) treatment. HSP expression within the neurons (*arrows*) as well as in glial cells is clearly visible. Bars: a,b = 40 μm. Expression of GFAP, marker of astrocytes is present in the longitudinal section of the spinal cord (*arrows*) of Ag (D; c) or Cu (D: d) treated rats. Bars: g,h = 50 μm. Data from Sharma et al., unpublished observations

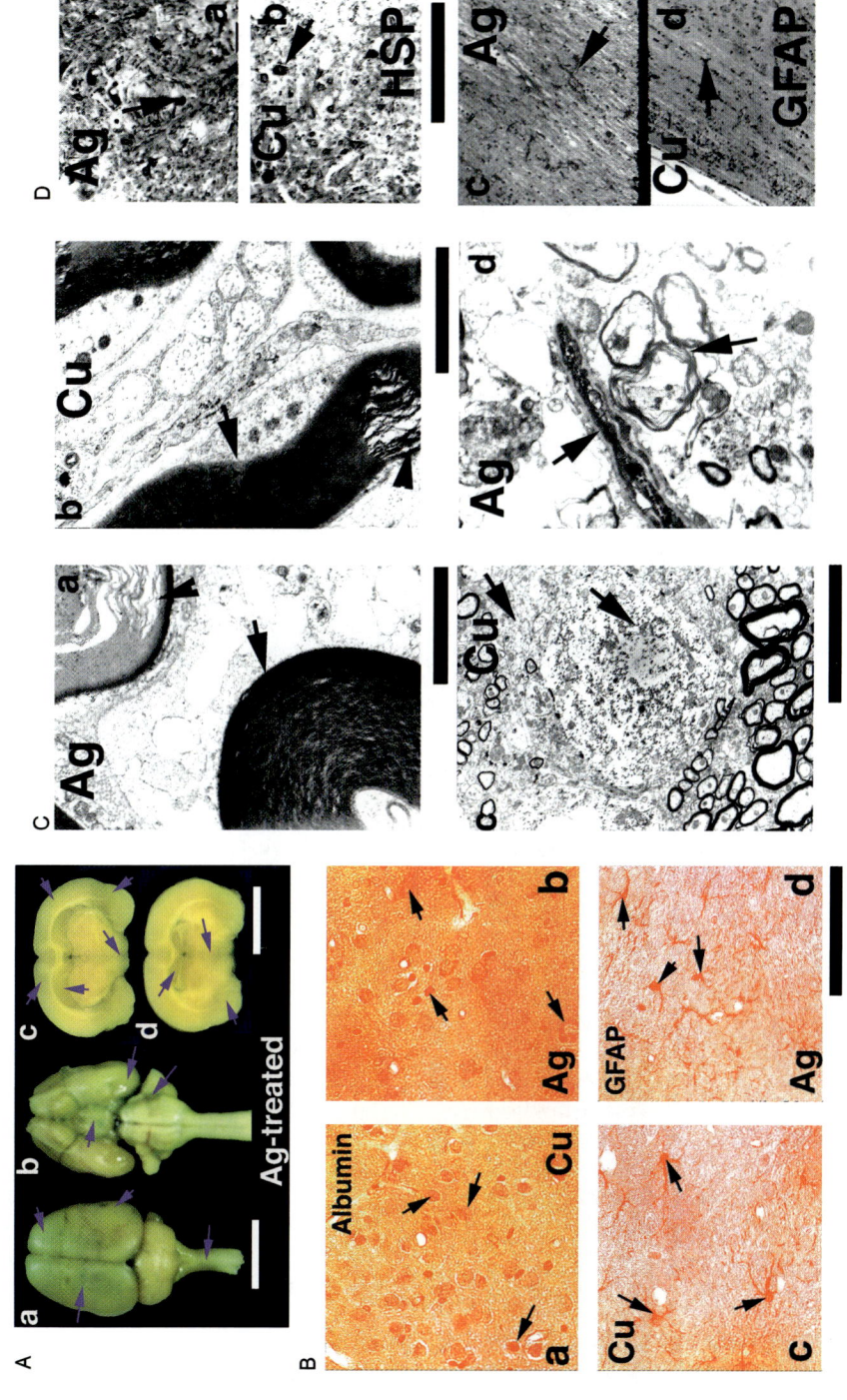

Figure 17-18 (continued)

◘ Table 17-11
Effect of nanoparticles on BBB permeability, cerebral blood flow and brain edema in normal rats

Experiment type	N	BBB permeability		BSCB Permeability		Blood flow		Edema formation		Volume swelling	
		Evans blue, mg %	[131]Iodine, %	Evans blue, mg %	[131]Iodine, %	Brain, ml/g/min	Spinal cord, ml/g/min	Brain water, %	Spinal cord water %	Brain, %f	Spinal cord, %f
Saline	6	0.28 ± 0.08	0.36 ± 0.06	0.22 ± 0.04	0.30 ± 0.05	1.16 ± 0.06	0.94 ± 0.04	74.23 ± 0.86	65.25 ± 0.21	Nil	Nil
Cu	8	0.56 ± 0.13**	0.68 ± 0.11**	0.67 ± 0.08**	0.76 ± 0.10**	0.98 ± 0.08*	0.84 ± 0.09*	75.14 ± 0.13*	66.16 ± 0.08*	3.53	2.61
Ag	8	0.64 ± 0.12**	0.74 ± 0.11**	0.64 ± 0.08**	0.75 ± 0.05**	0.96 ± 0.10*	0.88 ± 0.06*	75.04 ± 0.08*	66.08 ± 0.11*	3.14	2.38
Al	8	0.46 ± 0.08*	0.58 ± 0.06*	0.58 ± 0.06*	0.64 ± 0.08*	1.04 ± 0.06*	0.88 ± 0.08*	74.84 ± 0.13*	65.75 ± 0.31*	2.36	1.43

Source: Data modified after Sharma and Sharma (2007); Sharma (unpublished observations)
The nanoparticles were suspended in Tween 80 and administered separately intraperitoneally in rats daily once in a dose of 50 mg/kg (weight/volume)
Values are Mean ± SD from 5 to 9 animals at each data point. *$P < 0.05$; **$P < 0.01$ compared from saline in normal animals
ANOVA followed by Dunnett's test for multiple group comparison from one control group

Table 17-12

Effect of nanoparticles on morphological changes in the neurons, astrocytes, myelin, and endothelial cells in normal rats

Experiment type	Neuronal reaction					Glial reaction		Axonal reaction		Lanthanum extravasation	
	Distortion cell shape	Chromatolysis degeneration+	Nuclear damage	Eccentric nucleolus+	Sponginess	Reactive gliosis	perivascular gliosis+	Loss of myelin	Vesiculation of myelin	Endothelial cell cytoplasm	Basal lamina
Saline	Nil	Nil	Nil	Nil	Nil	±	±	Nil	±	Nil	Nil
Cu	++	±	+	±	+	++	++	++	+	++	++
Ag	+++	+	+	+	+	++	++	+++	++	++	++
Al	+	++	++	++	+	++	+++	+	++	+	+

Source: Data modified after Sharma and Sharma 2007; Sharma (unpublished observations)

Nil = absent, +++ = severe; ++ = moderate; + = mild; ± = present

The nanoparticles were suspended in Tween 80 and administered separately intraperitoneally in rats daily once in a dose of 50 mg/kg (wt/vol)

Light microscopy and immunohistochemistry was used to asses neuronal, glial, and myelin damage. Ultrastructural changes were used to determine eccentric nucleolus, perivascular gliosis, myelin vesiculation, and lanthanum extravasation. Lanthanum as dark electron dense particle can easily be seen within endothelial cell cytoplasm or in the basal lamina as dark particulate deposits under transmission electron microscope

appears that the Ag nanoparticle somehow influence the endothelial cell membrane permeability in few microvessels. Studies are in progress to evaluate changes in lanthanum extravasation following other nanoparticle administration.

6.3 Nanoparticles Induce Edema Formation

Small tissue pieces from the rat or mice brain showing Evans blue leakage was analyzed for water content to reveal edema formation. Intravenous administration of Cu nanoparticle resulted in a mild but significant edema formation in parts of the cortex compared to the control group (▶ *Tables 17-11* and ▶ *17-12*). Thus, about 0.5–1.2% increase in brain water content was noted in the cingulate, pyriform, and temporal cortices. Administration of Al nanoparticle induced only 0.4–0.6% increase in water content compared to the controls (see ▶ *Table 17-11*).

Our data further show that Cu treatment increased Na^+ content in the sample with a slight decrease in K^+ content confirming water movement from the vascular to cerebral compartments reflecting true edema formation (▶ *Table 17-11*). Ag treatment also altered ion content in the brain in similar way. Al treatment showed minimum changes in Na^+ and K^+ contents in the brain compared to the control groups (▶ *Table 17-11*). This indicates that nanoparticle-induced BBB disruption is associated with brain edema formation.

6.4 Nanoparticles Induce Morphological Changes in the Brain

Morphological studies done on paraffin embedded sections exhibit loss of myelin, dark and distorted nerve cell and glial cell damage in several brain regions showing albumin extravasation (▶ *Figure 17-18*). Ag and Cu nanoparticles showed pronounced effects on nerve cell, glial cell, and myelin changes compared with Al nanoparticle (Sharma, unpublished observations). These observations suggest that leakage of BBB and edema formation leads to cell and tissue injury (Sharma and Sharma, 2007).

6.4.1 Nanoparticles and Stress Protein Reaction

Immunohistochemical analysis of brain samples from Ag nanoparticle-treated rat and mice brains showed increased expression of HSP immunoreactivity in neurons in the regions showing leakage of the BBB (Sharma, unpublished observations). Cu and Al nanoparticles induced only a mild reaction of HSP expression (▶ *Figure 17-18*).

6.4.2 Nanoparticles and Astrocytic Reaction

Marked activation of astrocytes was seen using GFAP immunoreactivity in rats treated with Cu and Ag nanoparticles given intravenously (▶ *Figure 17-18*). The astrocytic reaction was mainly seen in the regions showing leakage of albumin. The Al nanoparticle shows weakest GFAP activation (▶ *Table 17-12*).

6.4.3 Nanoparticles and Myelin Changes

Luxol fast blue histochemistry showed loss of myelin in the cortex of superfused mice with Ag and Cu nanoparticles. Mild to moderate loss of myelin is also seen in animals that received intravenous Cu and Ag (Sharma, unpublished observations). These histochemical staining is partly confirmed using MBP immunostaining (see ▶ *Table 17-12*).

6.4.4 Ultrastructural Observations

Transmission electronmicroscopy showed degeneration of neural cells and damage to cell nucleus (❯ *Figure 17-18*) in Ag- or Cu-treated rats. Myelin vesiculation is often seen in animals treated with Ag and Cu nanoparticles in the neuropil (see ❯ *Figure 17-18*). Damage to cell membranes, perivascular edema, and astrocytic swelling is common in the neuropil of Ag- and Cu-treated rat brains (see ❯ *Figure 17-18*).

Normal animals or animals administered with equimolar NaCl did not show any change in the BBB function, immunomarkers, or cell and tissue reaction (see ❯ *Table 17-12*).

6.5 Nanoparticles and Neurodegeneration

These observations are the first to suggest that nanoparticles when administered either intravenously or superfused over the cortical surface are able to induce selective breakdown of the BBB. This BBB leakage allows large size proteins to enter into the brain causing vasogenic edema formation. Obviously, leakage of BBB and exposure of neurons, glial cells, and myelin to exogenous serum factors will induce cell reaction in the brain. It appears that cells are stressed following nanoparticle administration as seen using HSP over expression. This suggests that nanoparticles may induce mild to moderate cellular stress causing leakage of BBB leading to brain edema formation and abnormal reaction to cell and tissues in the brain.

7 Alterations in the BCNSB in Traumatic Injuries

Brain and spinal cord injury (SCI) is a complex event that includes physical destruction of microvessels, alterations in local and global microcirculation, as well as the permeability changes of the vessel walls leading to the leakage of plasma constituents into the brain microenvironment (Cervós-Navarro and Ferstz, 1980; Klatzo, 1987; Sharma, 2004a, 2005a, 2006a, 2007a, b). In addition, the brain or spinal cord trauma is influenced by a number of chemical compounds that are released or become activated in and around the primary lesion (Wahl et al., 1988; Sharma, 2004a, 2005a). These chemical mediators induce breakdown of the BBB or BSCB permeability and induce vasogenic edema formation (Wahl et al., 1988). Spread of edema fluid within the microfluid environment of the CNS is instrumental in inducing cell and tissue injury leading to neurodegeneration.

7.1 Spinal Cord Injury

Trauma to spinal cord-employing different experimental models result in breakdown of the BSCB leading to formation of vasogenic edema (Wagener et al., 1971; Osterholm, 1974; Griffiths, 1976; Yeo, 1976; Griffiths et al., 1978a, b). Immediately after an impact injury to the spinal cord, leakage of proteins occur across the BSCB (referred to as "BBB" in early investigations; Goodman et al., 1976). Griffiths et al. (1978b) using graded impact SCI studied extravasation of tracers within the cord in relation to vasogenic edema formation. These early observations suggest that profound vascular damage in the spinal cord following trauma occurs irrespective of the type of the initial nature of injury used (see Sharma, 2005a for review). Thus, the microvascular damage appears to be the most important factor in determining the pathological outcome of traumatic spinal injuries (Sharma, 2004a, 2007b).

7.1.1 Spinal Cord Microvessels Differ from Brain and Contain Collagen

The spinal cord and brain microvasculature, although very similar in nature differ slightly in structure as seen in ultrastructural investigations (Griffiths et al., 1976). Thus, the majority of the spinal cord microvessels (<60%) with the lumen size of capillaries (<8 μm) contain collagen in their perivascular spaces (Rhodin, 1968;

Beggs and Waggener 1976; Griffiths et al., 1976). This is a feature not seen in brain microvessels (see Sharma, 2005a). Whereas, the vascular and glial basement membranes in the remaining spinal cord capillaries are very similar to that found in the brain (Griffiths, 1976; Griffiths et al., 1978a, b). The functional significance of collagen deposits in spinal cord microvessels in relation to the barrier function in normal and in pathological conditions are still not known (Sharma, 2005a).

7.1.2 Vascular Reaction in SCI

Impact injury to the cord results in microhemorrhages within 3 min resulting in leakage of erythrocytes in the perivascular space of arterioles, veins, and venules as well as in the neuropil (Griffiths et al., 1978a, b). Depending on the magnitude of impact injury (>200 g/cm), neutrophils and basophills are also seen within the neuropil of the traumatized cord.

Within 6–30 min after injury, damage to cell membranes, disruption of neuropil, swollen astrocytes, and ruptured basement membrane occur (for details see Sharma, 2004a). The endothelial cells of capillaries and venules become much more electron dense and exhibit large number of vesicles (60–70 nm diameter) (Sharma, 2004a). Swollen endothelial cells surrounded by fibrin strands are often seen containing proteinaceous fluid in the perivascular space. However, endothelial tight junctions are not widened (Griffiths et al., 1978a). After 4–6 h trauma, the endothelial balloons are seen in some microvessels.

These microvascular changes after injury are confined to the capillaries and post capillary venules (vessels with an internal diameter of <15 μm). The arteries and arterioles are relatively unaffected although the perivascular space of these arteries and arterioles often contain erythrocytes, proteins, and neutrophils (Griffiths et al., 1978b). These observations suggest that microhemorrhages alone are not the principal cause of protein extravasation and edema formation (for details see Sharma, 2004a, 2005a).

7.1.3 BSCB Permeability in SCI

Noble and coworkers (Noble and Wrathall, 1987, 1988, 1989; Mautes et al., 2000; Mautes and Noble, 2000) using HRP as exogenous tracer, for the first time showed extravasation of protein in 0.5–2.0 cm proximal and distal segments of the cord after transection at the T8 site (Noble and Wrathall, 1987, 1988). This leakage of HRP was extended in 0.5 cm proximal and distal segments after 15 min transection and reached to the segment located 1 cm away from the lesion site between 30 min and 3 h. In these experiments, an increase in vesicular transport but not widening of the tight junctions was responsible for HRP leakage within the spinal cord neuropil (Noble and Wrathall, 1987, 1988). Thus, breakdown of the BSCB induces leakage of proteins in the cord that spread within the neuropil over time and contribute to vasogenic edema formation.

7.1.4 Spinal Cord Edema Formation

A close parallelism between BSCB leakage and spinal cord edema formation is seen in several injury models (for review see Sharma, 2005a, 2007b). Thus, edema formation in the cord occurs as early as 30 sec after injury and becomes prominent within 2–5 min. A significant increase in edema formation using spinal cord water content is seen as early as 5 min after a 300 g/cm impact injury in monkeys that persisted 15 days after the initial insult (Yashon et al., 1973). This suggests that the process of edema formation may last up to 15 days after the initial insult (Nolan, 1969; Demediuk et al., 1989).

Permanent paraplegia results due to an increase in tissue water content above and below the lesion site (Griffiths, 1980; Griffiths and McCulloh, 1983). The adjacent spinal cord segments are associated with tissue damage and exhibit profound leakage of albumin and dextran (Wagner and Stewart, 1981). Increased water content near the impact site after trauma, and in the adjacent segments not injured directly by the physical lesion represent true edema formation (Rapoport, 1976; Reulen, 1977a, b). Spread of edema in white matter is further influenced by tissue pressure gradients (Shapiro et al., 1977) and is mediated via bulk flow (Cserr and Ostrach, 1975). Fluorescein labelled dextran of different molecular weights migrated the same distance in the white matter over time is in line with this idea (Reulen, 1977a, b; Reulen et al., 1977).

7.1.5 New Model of SCI

We developed a new model of SCI in rats to study spread of edema fluid and BSCB permeability in the rostro-caudal direction (❯ *Figure 17-19*) in which a longitudinal incision to the right dorsal horn of the T10–11 segments was made. The deepest part of the lesion is limited to the Rexed's lamina VIII (Rexed, 1952, 1954; Sharma and Olsson, 1990). This model offers new possibilities in which the primary injury, i.e., the knife wound, is limited exclusively to the gray matter leaving white matter largely intact (Sharma and Olsson, 1990; Sharma et al., 1991b). In other models of impact, compression, hemisection, or transection induced injury, a direct damage to white matter results in immediate loss of spinal cord conduction in the ascending and descending tracts.

7.1.5.1 Leakage of EBA In this model, a focal injury to the spinal cord on the T10–11 segments resulted in widespread extravasation of EBA in various segments of the spinal cord following 1–12 h after trauma (❯ *Figure 17-19*). A progressive increase in EBA extravasation was most pronounced in the injured (T10–11) and the caudal T12 segments compared to the rostral T9 segment (❯ *Figure 17-19*). The reasons behind

◼ **Figure 17-19**
Blood–spinal cord barrier (BSCB) disruption and edema formation in a new model of spinal cord injury. (a) Schematic representation of spinal cord injury and sampling of tissues for biochemical and morphological analysis. The spinal cord injury was made by a longitudinal incision into the right dorsal horn of the T10–11 segments. The deepest part of the lesion (L) is located within the Rexed's lamina VIII (Sharma and Olsson, 1990). The BSCB and spinal cord edema was analyzed in several rostral and caudal segments from the lesion site. For morphological analyses, small tissue pieces (about 5 mm) from the T9 and T12 segments were dissected out and processed for light and electron microscopy (for details see Sharma, 2004a, 2005a). For high resolution light microscopy, tissue pieces were embedded in Epon and about 1 μm thick whole spinal cord sections were cut and stained with toluidine blue. These sections were examined at light microscopy and photographed (see b). For ultrastructural studies, small areas either from the contralateral dorsal horn (1) or ventral horn (2) was selected for ultrathin sectioning, counterstained with uranyl acetate and lead citrate and examined at the transmission electron microscope. The spinal cord edema formation was seen by measuring water content of the spinal cord in various segments above or below the lesion site (b). A focal trauma to the spinal cord showed a progressive increase in the BSCB permeability to Evans blue albumin (EBA, C) and $^{[131]}$Iodine (d) in different spinal cord segment between 5 min to 12 h-injury periods. The BSCB permeability to EBA and radioiodine is very low in several spinal cord segments in the normal rat. However, a progressive increase in BSCB to these tracers was observed during the 12-h observation period. Increased permeability of BSCB to radioiodine is apparent as soon as 5 min after spinal cord injury (SCI) in the traumatized and the adjacent segments. Whereas, a significant increase in EBA extravasation is seen after 1 h SCI. Several remote segments located rostral and caudal to the lesion site exhibited significant increase in tracer extravasation. The intensity and spread of radioiodine across the BSCB following SCI is much more extensive than the EBA tracer (for details see text). Significant increase in spinal cord water content (b) in the segments located near the lesion site can be seen as soon as 5 min after injury that continued progressively throughout the observation period of 12 h. Although, the remote caudal segment (L5) showed a profound increase in BSCB permeability, the spinal cord water content and edema formation is not seen in these remote rostral segments beyond the T9 level. Data are mean ± SD of 6–8 animals in each group. * = $P < 0.01$, ANOVA followed by Dunnet's test from the respective spinal cord segments in control group. Data modified after Sharma, 2004a

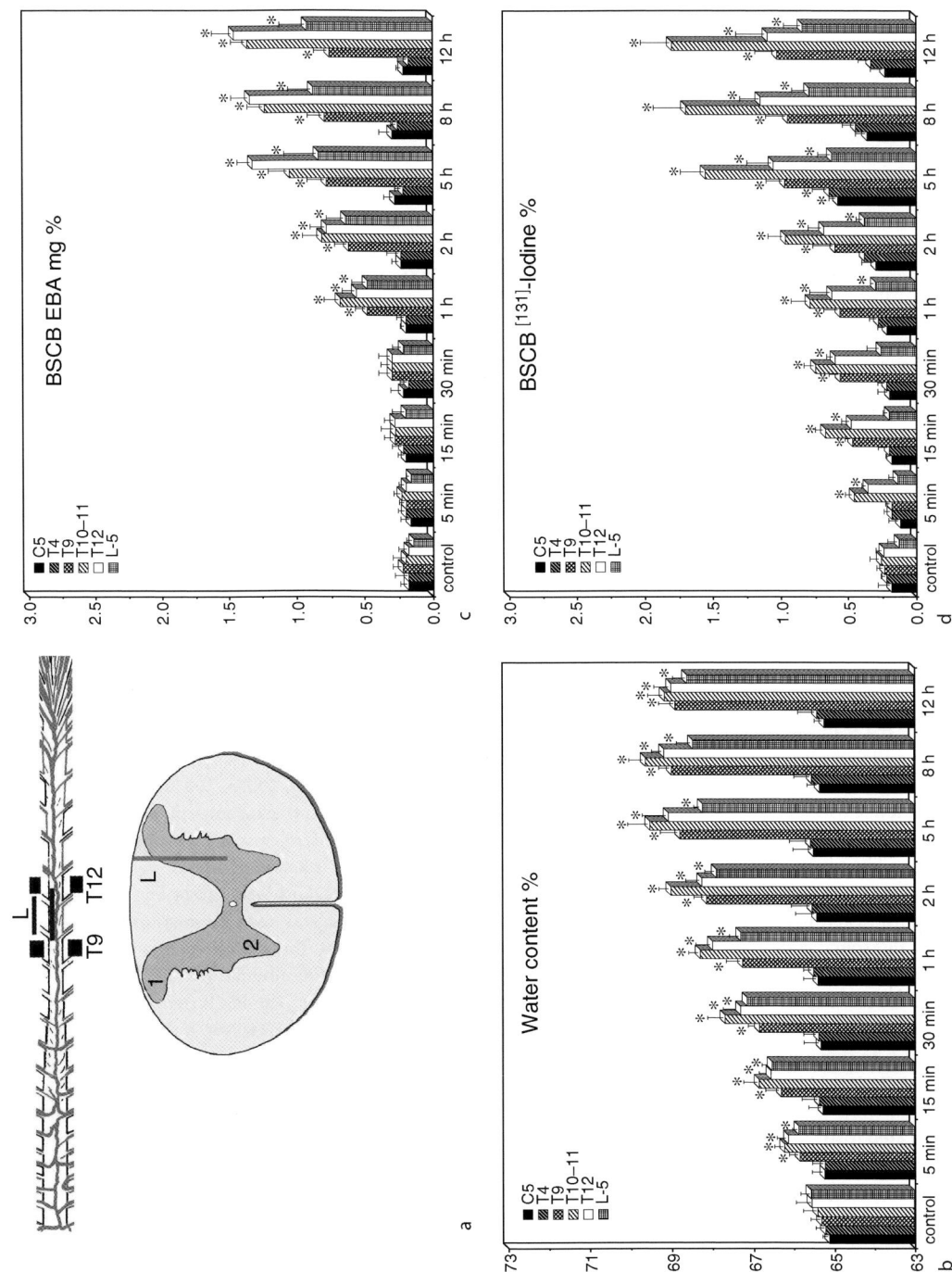

Figure 17-19 (continued)

such discrepancy between rostral and caudal segments of the cord are unclear. It may be that lesion of the dorsal horn results in accumulation of several neurochemicals below the lesion site compared to the rostral segments. Progressive increase in EBA extravasation in the remote L-5 segment is in line with this idea (Sharma, 2005a).

7.1.5.2 Leakage of Radioiodine Extravasation of radioiodine in this model of SCI is present as early as 5 min after trauma that was progressive in nature (❯ *Figure 17-19*). Leakage of radiotracer in distal segments, e.g., L-5 is seen after 1 h of trauma, whereas remote segments, i.e., C5 and T4 segments exhibited radiotracer extravasation not before 5 h of injury (❯ *Figure 17-19*). This indicates that the caudal segments exhibit an early and more pronounced BSCB disruption compared to the rostral segments (Sharma et al., 1995a; Noble and Wrathall, 1987).

The extravasation of tracers in the segments located far away from the traumatized region may represent local mechanisms of BSCB opening, as longitudinal spread of the tracer in the extracellular space from the injured cord is unlikely within such a short period (Sharma, 2004a).

Marked differences in the distribution of EBA and radioiodine in the cord after injury could be due to differences in their protein binding capacity in the circulation (Rawson, 1943; Rapoport, 1976; Sharma and Dey, 1986a, b). Obviously, the size of tracers and their binding capacity to proteins in circulation will determine the extent of tracer extravasation in the cord after injury (Sharma et al., 1995a, c; for details see Sharma, 2004a).

7.1.5.3 Leakage of Lanthanum across the BSCB Trauma to the spinal cord markedly increased the number of lanthanum filled vesicles within the endothelial cell cytoplasm (❯ *Figure 17-20*). Usually 5–8 microvesicles filled with lanthanum is seen in one vesicular profile in the segments adjacent to the lesion site. In addition, increase in the endothelial cell infiltration of lanthanum also occurred in a precise and specific manner following injury. Thus, one endothelial cell often exhibit lanthanum infiltration within the cell cytoplasm, whereas the neighboring endothelial cell could be normal (❯ *Figure 17-20*). Furthermore, several microvascular profiles showed lanthanum exudation in the basal lamina usually in its amorphous form. However, the tight junctional permeability to lanthanum remained intact (❯ *Figure 17-20*).

The diffused endothelial infiltration with lanthanum particles seen in some endothelial cell cytoplasm suggests that the cell membrane permeability is altered after injury (Olsson et al., 1990; Sharma et al., 1995a, 1998a, c). This suggests that increased endothelial cell membrane permeability could be one additional way of vascular leakage in vivo not well emphasized earlier (Olsson et al., 1990; Sharma et al., 1998a).

7.1.5.4 Edema Formation in SCI SCI showed a close parallelism between BSCB leakage and edema formation in several spinal cord segments in this model (See ❯ *Figure 17-19*). Thus, edema development was seen near the lesion site within 30 min after injury that was progressive with time (❯ *Figure 17-19*). Interestingly, the remote L-5 segment exhibited pronounced edema development compared to the C5 and T4 segments (❯ *Figure 17-19*). These observations suggest that trauma-induced edema formation closely corresponds to the breakdown of the BSCB. Alternatively, leakage of proteins across the BSCB contributes to spinal cord edema formation (Nemecek et al., 1977; Martinez et al., 1981; Stokes and Somerson, 1987; Sharma et al., 1998a, c).

7.1.5.5 Pathophysiology of Cell Injury in SCI High resolution light microscopy on toloudine stained Epon embedded sections (1 μm thick) showed profound expansion of the spinal cord and loss of a clear distinction between gray and white matter after 5-h injury (❯ *Figure 17-21*). Damage of nerve cells and axons as well as sponginess and edema are present throughout the cord (❯ *Figure 17-21*). These cell changes are most pronounced in both the rostral (T9) and the caudal (T12) segments of the spinal cord (Sharma and Olsson, 1990).

7.1.5.6 Nerve Cell Reactions Nissl or Haematoxylin and Eosin staining show several dark and distorted neurons in the spinal cord at 5 h (❯ *Figure 17-21*). Some nerve cells were swollen and few are shrunken. Most of them were devoid of a clear nucleus (❯ *Figure 17-21*). Degeneration of Nissl substance and

◘ Figure 17-20

Blood–spinal cord barrier (BSCB) to electron dense tracer lanthanum after 5 h spinal cord injury (SCI). High power electron micrograph showing presence of lanthanum in the basal lamina (*arrows*) of one spinal cord microvessel from the ventral horn of the T9 segment (a). Another spinal cord microvessel from the same region shows infiltration of lanthanum within the cell cytoplasm in one endothelial cell (*arrows*) while the adjacent endothelial cell is negative to lanthanum (b). Low power electron micrograph from the dorsal horn of the T9 segment showing lanthanum exudation in the basal lamina (*arrows, c*). Signs of perivascular edema (*) and myelin damage are prominent. High power electron micrograph from the T9 and T12 segments shows that the tight junctions are not altered to lanthanum after SCI (d,e, *arrows*). However, lanthanum is seen in several microvesicular profiles within the spinal cord endothelium (f,g) after SCI. Bar=a,b=600 nm; c=1 μm; d,f=800 nm; e,g=500 nm. Data modified from Olsson et al., 1990, 1992. Reproduced after permission from Sharma, 2004b

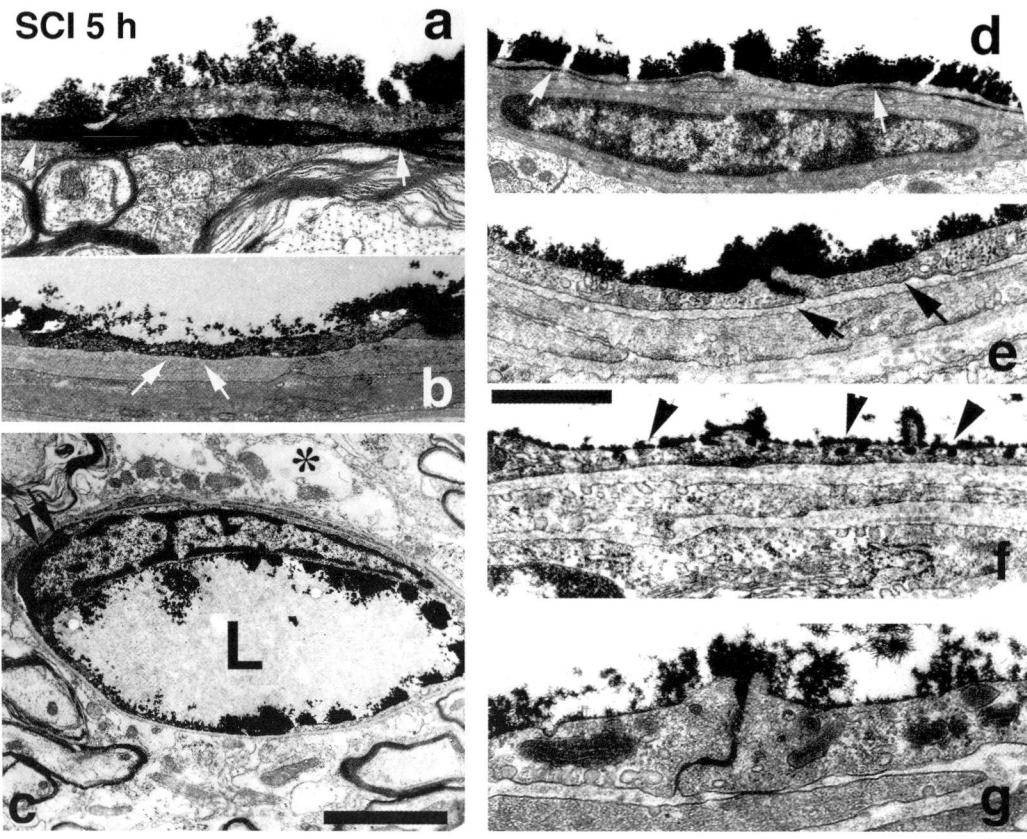

chromatolysis is frequent in many nerve cells around the lesion site (❯ *Figure 17-21*). The damaged nerve cells are located in the edematous regions of the spinal cord.

7.1.5.7 Glial Cell Reactions Upregulation of GFAP in the perifocal segments of the cord is seen at 5 h after injury (❯ *Figure 17-21*) indicating that astrocytes are sensitive indicators to trauma-induced alterations in the spinal cord microfluid environment (Sharma et al., 1993b; Sharma, 2005a). Alternatively, alterations in the fluid microenvironment due to disruption of the BSCB in trauma activate astrocytes to maintain homeostasis (Sharma et al., 1992b; Sharma, 2004a).

◻ **Figure 17-21**
Structural changes in the spinal cord after 5-h injury. (A) Epon embedded toludine stained sections of spinal cord from the T9 (a) and T12 (b) segment from untreated 5-h injured (about 2–4 mm away from the lesion site) rats. Marked expansion of the spinal cord segments and loss of distinction between gray and white matter is apparent. (B) Nissl stain showing dark and distorted nerve cells (arrow heads) in the T9 ventral horn of a 5-h spinal cord injured rats (b) compared to the control (a). Sponginess and edema (*) are clearly visible. (D) Degradation of myelin basic protein (MBP) immunoreactivity in one 5-h spinal cord injured rat (T12 segment, D: a) compared to control (D: b). Massive upregulation of GFAP in the ipsilateral (D: c) and in contralateral (D: d) ventral horn of the injured T9 segment after 5-h injury. Astrocytic damage (*arrows*) is clearly visible that is prominent in the ipsilateral cord compared to the contralateral side (Sharma et al., 1993c). (C) Ultrastructural changes in the spinal cord dorsal horn following 5-h spinal cord injury (SCI). Damage to neuropil is most pronounced in the ipsilateral (T9 R) dorsal horn compared to the contralateral side (T9 L). Activation of microglia (arrowhead; a), collapse of microvessels (*arrows*); perivascular edema (*; b) and microhameorrhages (c) are much more pronounced in the dorsal horn at 5 h. Myelin vesiculation (*arrows*; d–f) and membrane damage (*) in the ipsilateral dorsal horn is most pronounced. Bars: a–c=1 μm; d,e=600 nm; f=1.5 μm. Data modified after Sharma, 2004b

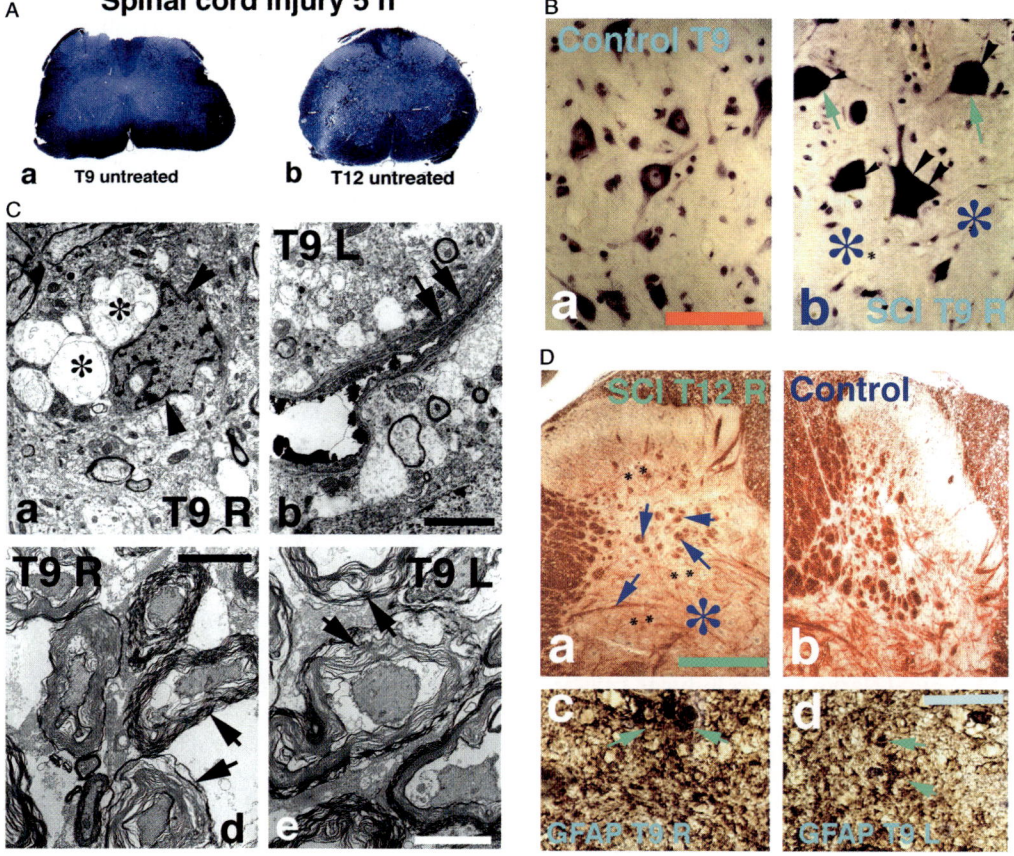

7.1.5.8 Axonal Damage SCI induced myelin vesiculation and axonal damage as early as 5 min and reaches its peak within 6–8 h after the primary injury (Balentine, 1988; Banik et al., 1985, 1987; Winkler et al., 1998; Sharma et al., 2003a, b, c). Degradation of MBP in the spinal cord segments located above and below the lesion site is most pronounced at 5 h after trauma in our model (❯ *Figure 17-21*). Loss of MBP is particularly prominent in the regions showing edematous expansion and sponginess in the cord (⬦ *Figure 17-21*). These observations suggest that injury induced spread of edema fluid contributes to myelin damage.

7.1.5.9 Ultratstructural Changes Selected tissue pieces obtained from the dorsal and ventral horns of the T9 or T12 segments show profound edema, vacuolation, and membrane damage in many parts of the neuropil at the ultrastructural level (⬦ *Figure 17-21*). Myelin vesiculation, axonal damage microhemorrhages, distortion of nerve cells, and glial cell damage are most pronounced in the spinal cord dorsal horn near the lesion site (⬦ *Figure 17-21*). These ultrastructural changes in the neuropil are most marked in the ipsilateral cord compared to the contralateral side indicating that trauma induced BSCB disruption and spread of edema fluid are important factors in inducing structural changes in the spinal cord (Balentine, 1988; Sharma and Olsson, 1990; Sharma et al., 1995a, b, 1996a, 1998d).

7.1.6 BSCB Disturbances in Human Cases of SCI

To further confirm the role of BSCB in cell damage, extravasation of endogenous albumin in human cases of SCI was examined using immunohistochemistry (Sharma and Kakulas, unpublished observations; Sharma, 2004a). The results for the first time show that victims of SCI that survived 5–8 h after the initial insult exhibited profound extravasation of albumin in the spinal cord gray matter (⬦ *Figure 17-22*). Most of the nerve cells and neuropil exhibited albumin extravasation in the region showing edema and sponginess within the cord (⬦ *Figure 17-22*). These observations strongly support the idea that disruption of BSCB plays important role in spinal cord cell and tissue damage following trauma.

7.2 Peripheral Nerve Lesion

Peripheral nerve lesions influence the spinal cord cell and tissue microenvironment probably by influencing receptors and/or neurotransmitters in the spinal cord (Govrin-Lippmann and Devor, 1978; Woolf, 1983; Woolf et al., 1992; Hökfelt et al., 1994). In addition, alterations in nonneural cells, e.g., astrocytes, endothelial cells, and microglia (Winkelstein and DeLeo, 2002; Inoue et al., 2004; Tsuda et al., 2004) also play important roles in structural and functional changes in the cord following nerve lesions. However, alterations in the BSCB permeability by peripheral nerve lesion are still not well known (Gordh and Sharma, 2006; Gordh et al., 2006). Our laboratory is the first to demonstrate that peripheral nerve lesion is capable to induce BSNB disruption in the spinal cord that could be crucial to influence glial cell function and morphological alteration in the cord.

7.2.1 Spinal Nerve Lesion Induces Leakage of Albumin across the BSCB

Spinal nerve lesion at L4 affects both sensory and motor fibers (Gordh et al., 1998, 2000) causing the "neuropathic pain" like syndrome, e.g., hypersensitivity to mechanical stimulation and cold stimuli (Gordh et al., 2006). Using this model, we examined the BSCB disruption to endogenous albumin extravasation (Sharma, 2004a; Sharma and Alm, 2004). Significant increase in albumin immunoreactivity in the spinal cord occurs after 1 week of nerve lesion (⬦ *Figure 17-23*). The albumin positive cells are more frequent in the ventral horn compared to the dorsal horn in both sides (Gordh et al., 2006; ⬦ *Figure 17-23*). Two weeks after the nerve lesion, albumin immunostaining is seen in the neuronal cytoplasm and in the surrounding neuropil (⬦ *Figure 17-23*), the intensity of which is most prominent in the ipsilateral ventral horn of the cord (⬦ *Figure 17-23*). The nucleus and karyoplasm of most of the nerve cells were devoid of albumin immunoreactivity (⬦ *Figure 17-23*).

◘ Figure 17-22
Blood–spinal cord barrier (BSCB) disruption in one representative Human case of 6-h spinal cord injury (SCI). Spinal cord tissue (T8–9) from spinal cord injured victims with 6-h survival period are obtained from Professor Byron Kakulas, Chief, Spinal Cord Tissue Bank, University of Western Australia, Perth, Australia. About 4-μm thick paraffin sections were processed for albumin immunohistochemistry to detect BSCB leakage. Massive extravasation of endogenous albumin (brown reaction product) within the spinal cord gray matter and around central canal (a,b) is seen in these specimens. Many damaged nerve cells took positive staining indicating that BSCB disruption is prominent in the spinal cord at 6 h in Human cases. Reproduced with permission from Sharma, 2004b

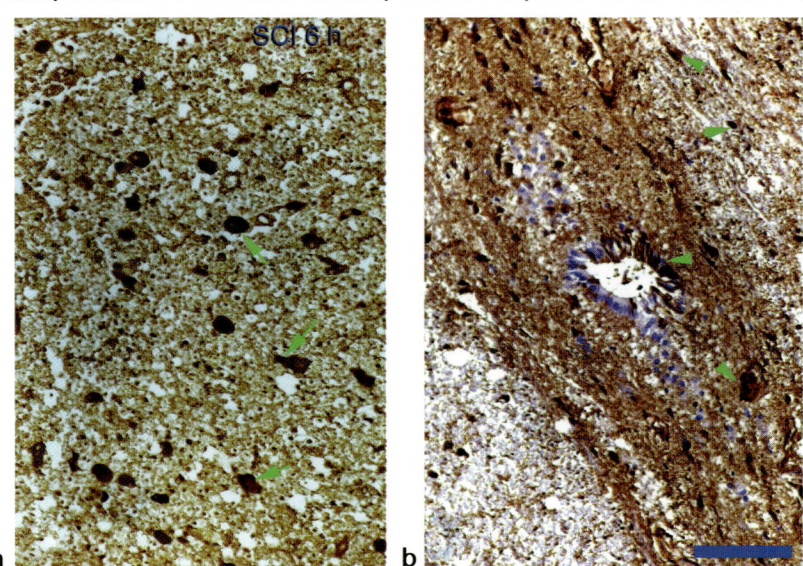

The nerve lesion induced BSCB permeability appears to be reversible in nature. Thus, After 4–10 weeks of nerve lesion the albumin immunostaining considerably decreased in the ventral horn compared to the dorsal horn. This suggests that chronic nerve lesion induces selective alterations in the BSCB permeability in the spinal cord in a time-related fashion (❯ *Figure 17-23*). Leakage of serum proteins could also be responsible for vasogenic edema formation and secondary cell and tissue injury in the spinal cord after nerve lesion (Barber et al., 2000; Li et al., 2001; Sharma, 2004a, b, c, d). This idea is in line with the fact that increased leukocyte adhesion (Lossinsky and Shivers, 2004; Scott et al., 2004) and apoptosis of both neural and vascular cells (Li et al., 2004; Sharma, 2005a) following BSCB breakdown are important elements that precede neurodegeneration (Gordh and Sharma, 2006; Gordh et al., 2006). Furthermore, nerve lesion will trigger alterations and/or release of several neurochemical mediators of the BBB permeability, e.g., serotonin, prostaglandins, opioid and nonopioid peptides, amino acids, and cytokines (Sharma, 2005a) that could contribute to the BSCB breakdown either directly or through specific receptor mediated mechanisms (Sharma, 2005a). A reduction in the magnitude of BSCB permeability over time could be due to slow recovery of microvascular function and/or loss of vasogenic effects of neurochemicals released over time.

7.2.2 Spinal Nerve Lesion Influences Astrocytic Activation

The glial cells and blood vessels in the CNS are in close apposition and thus breakdown of the BSCB is likely to affect glial cell functions (Krum, 1994; Whetstone et al., 2003; Sharma, 2004a). The astrocytes enhance

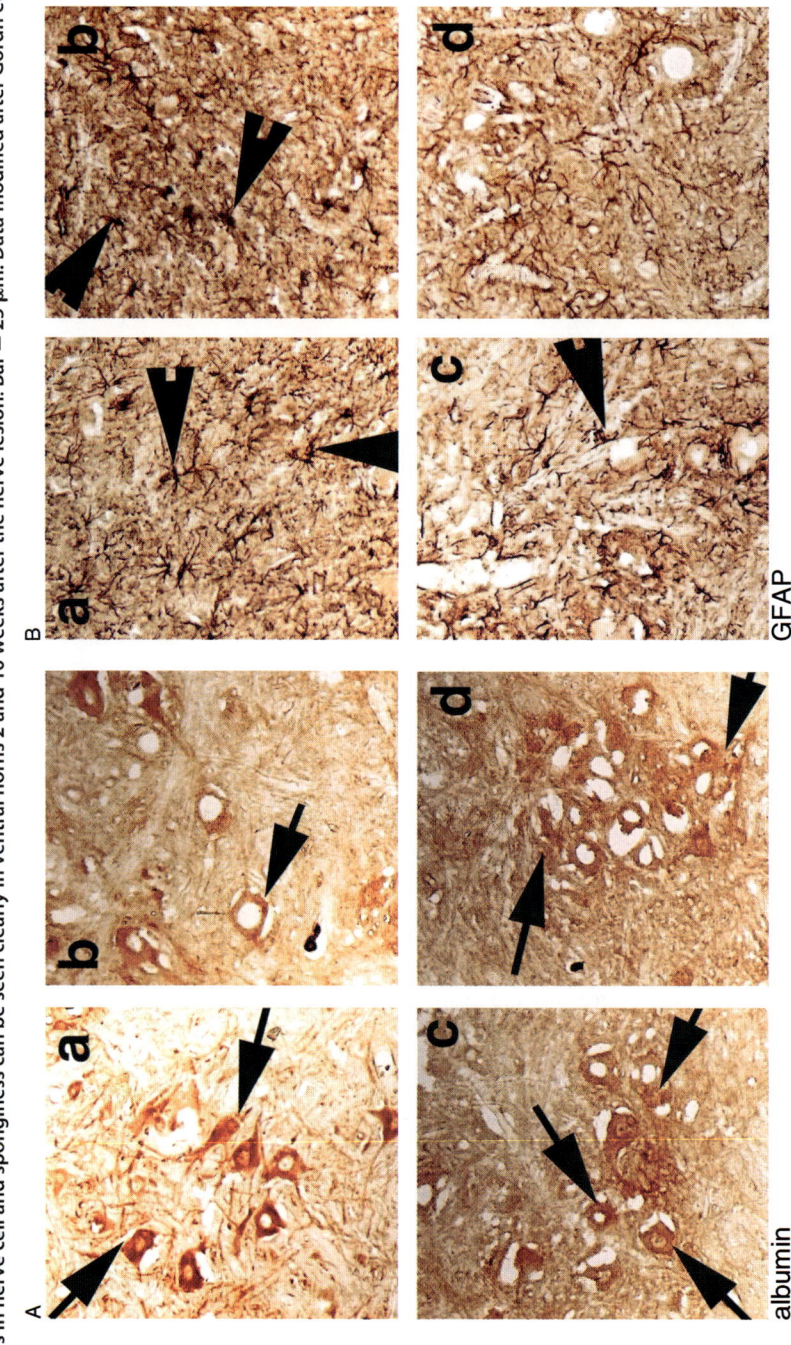

■ Figure 17-23
Representative examples of albumin and GFAP immunoreactivities into the ipsilateral and contralateral ventral horn of rats subjected to L-4 lesion for 2 and 10 weeks. The albumin and GFAP immunoreactivities are most prominent in the ipsilateral (*left panels*) side compared to the contralateral side (*right panels*). Structural changes in nerve cell and sponginess can be seen clearly in ventral horns 2 and 10 weeks after the nerve lesion. Bar = 25 μm. Data modified after Gordh et al., 2006

expression of tight junctional proteins that constitute one of the anatomical sites of the normal BBB (Gardner et al., 1997; Zenker et al., 2003). Thus, astrocytic reaction in the spinal cord following nerve lesion is also responsible for alteration in the spinal cord fluid microenvironment.

This idea is supported by a massive increase in GFAP-positive astrocytes that exhibited a close parallelism with the BSCB disruption to albumin in the cord (Gordh et al., 2006). Thus, GFAP immunoreactive cells are significantly higher in the ipsilateral ventral horn of spinal cord compared to the contralateral side (❯ *Figures 17-23*). The most pronounced GFAP immunoreaction occurs in nerve lesioned animals after 2 weeks (Gordh et al., 2006). Star shaped astrocytes are present around the nerve cells as well in the perivascular regions within the gray matter of both dorsal and ventral horns (❯ *Figure 17-23*). A significant decline in GFAP-positive astrocytes in the spinal cord is seen after 4–10 weeks of nerve lesion. The magnitude of reduction in GFAP-positive cells is much stronger in the contralateral cord compared to the ipsilateral side (❯ *Figure 17-23*). At this time, morphological changes include degeneration of nerve cells in the dorsal and ventral gray matter as well as, vacuolation in the neuropil and distortion of motor neurons (Gordh et al., 2006).

7.2.3 Functional Significance of BSCB Disturbances in Nerve Lesion

A breakdown of the BSCB following nerve lesion exposes the neural and nonneural cells in the spinal cord to numerous factors including neurotransmitters, ions, and many blood-borne substances normally prevented by the intact barrier. Access of various bioactive substances within the spinal cord cell and tissue microenvironment activate astrocytes and/or induce swelling of nerve cell and astrocytic foot processes and lead to nerve cell swelling and/damage (Sharma et al., 1992a, b, c, 1993a, b; Cervós-Navarro et al., 1998; Barbeito et al., 2004; Chisari et al., 2004; Sharma and Alm, 2004; Conrad et al., 2005). A decrease in GFAP immunostaining and reduction in albumin extravasation at the same time in the spinal cord suggests that BSCB disruption as well as the effect of chemicals and blood-borne substances on the astrocytes are neutralized over time (Sharma and Alm, 2004; Gordh and Sharma, 2006; Zai and Wrathall, 2005).

In conclusion, peripheral chronic nerve lesion like other insults to the CNS induces breakdown of the BSCB and activation of astrocytes in the spinal cord. Disruption of the BSCB permeability reflecting a widespread alteration in the fluid microenvironment plays an instrumental role in the spinal cord cell and tissue injury leading to neurodegeneration.

7.3 Traumatic Brain Injuries

Traumatic brain injuries (TBI) are either caused by trauma to the head and/or to the brain resulting in brain pathology. TBI is classified either as closed head injury (CHI), or open brain injury (OBI) depending on the impact delivered on the skull or on the brain respectively (Sharma et al., 2000a, 2006b; Vannemreddy et al., 2006). In CHI, an impact is delivered to the skull without fracture. In this situation, the kinetic energy of impact is transmitted to the underlying brain that induces concussive brain injury of much more severe in nature (Dey and Sharma, 1983, 1984; Sharma et al., 2007a). On the other hand, TBI is caused by penetrating head injury, e.g., bullets or knife wounds and can damage the brain structures, i.e., cerebral hemispheres, cerebellum, hippocampus etc., directly (Dey and Sharma, 1983; Sharma et al., 2000a, b). Symptoms of CHI and OBI include headache, mental confusion, dizziness, memory impairment, fatigue etc., that ranges from mild to severe depending on the magnitude and intensity of the primary insult and is often correlated well with extent of the brain damage (Sharma, 2004d). Permanent disability or death is the common outcome of TBI.

TBI is a major public health problem as its victims are often young male aged between 15 and 24 or elderly people of both sexes beyond 60–75 years of age (Vannemreddy et al., 2006). Male populations account for two thirds of childhood and adolescent head trauma victims whose recovery are largely partial and require lifetime support causing a huge burden on the society. CHI is a frequent cause of major long-term disability, particularly in individuals surviving head injuries sustained in war zones. The brain damage from CHI could be focal or diffuse, involving more than one area of the brain and is associated with diffuse axonal injury. On the other hand, localized injuries in OBI are often associated with neurobehavioral manifestations, hemiparesis, or other focal neurologic deficits.

TBI is often associated with contusion and intracranial hemorrhage or hematoma (Dey and Sharma, 1984). This is largely caused by microhemorrhages due to rupture of microvessels in the brain. All these factors either alone or in combination leads to brain pathology and major disability including death. Thus, understanding on the cellular and molecular mechanisms of brain damage in relation to the BBB disruption and brain edema formation is important in TBI for the development of suitable therapeutic strategies in the future.

7.3.1 Open Brain Injury

Our laboratory has developed a new model of OBI in rats in which a mild incision of the right parietal cortex is made after opening the parietal skull bone (Dey and Sharma, 1983; Sharma et al., 2000). Under urethane anaesthesia (1.5 g/kg, i.p.) a 4-mm^2 burr hole was made in the right parietal bone and the underlying cerebral cortex was exposed after careful removal of dura (● *Figure 17-24*). A stab wound 3-mm deep and 3-mm long was inflicted under stereotaxic guidance using a sharp sterile scalpel blade (Dey and Sharma, 1983). The lesion was limited to the cerebral cortex and/or the superficial parts of the subcortical white matter (Mohanty et al., 1989). Identical burr hole was made on the left parietal bone to expose the parietal cortex but no injury is made. The exposed brain areas in both hemispheres are covered with cotton soaked in saline throughout the period of study to prevent drying of the exposed brain tissues (Dey and Sharma, 1983; Mohanty et al., 1989).

7.3.1.1 OBI Induces Brain Edema Formation
Using brain water and electrolyte contents we determined brain edema following 1 h, 2 h, and 5 h after OBI, in the injured and in the contralateral hemispheres (Mohanty et al., 1989). A focal TBI is associated with profound edema formation in the traumatized half within 5 h. The untraumatized cerebral cortex also exhibited a significant increase in the brain water content at this time (● *Table 17-13*). About 3% increase in brain water content comparable to 12% increase in volume swelling in the traumatized hemisphere was observed (Sharma et al., 2000; Sharma, 2004c, d). On the other hand, the uninjured half exhibited a volume swelling of 5% (Mohanty et al., 1989). The brain water content or volume swelling was mild but significantly increased from the control group 1 h or 2 h after OBI that was largely seen in the ipsilateral half of the brain (Sharma et al., 1998a; Sharma et al., 2000; Sharma and Alm, 2004). The traumatized cortex exhibited a significant increase in Na$^+$ at 5 h; whereas, the K$^+$ did not show any change from the control value at any time point in either hemispheres (● *Table 17-13*).

7.3.1.2 BBB Disruption in OBI
Breakdown of the BBB following OBI was examined using extravasation of Evans blue and [131]Iodine as exogenous protein tracers (Sharma, 2004d). The passage of tracer across cerebral endothelium was investigated using lanthanum ion at the ultrastructural level (Olsson et al., 1990; Sharma et al., 1990a).

A progressive increase in the leakage of Evans blue and [131]Iodine tracers in cerebral cortex was observed in the injured as well as the uninjured hemispheres following 1 h, 2 h, and 5 h after trauma (● *Table 17-13*). The magnitude of tracer extravasation was significantly higher in the injured cortex compared to the uninjured side. These observations suggest a close correlation between extravasation of protein tracers and brain edema formation in OBI.

7.3.1.3 Lanthanum Extravasation in OBI At the ultratstructural level, infiltration of lanthanum into the endothelial cell cytoplasm reflecting an increase in the endothelial cell membrane permeability was seen in the traumatized cortex at 5 h in many microvessels in the vicinity of the lesion. The electron dense tracer was present in the basal lamina of several microvessels in the perifocal brain areas of the injured cortex (❯ *Figure 17-24*). However, in several microvessels around the lesion site, the tracer is infiltrated into the endothelial cytoplasm or present within the microvesicular profiles (❯ *Figure 17-24*). The tight junctions between the endothelial cells are intact to the lanthanum (❯ *Figure 17-24*). In few brain regions, exudation of lanthanum is seen within the neuropil (❯ *Table 17-13*). Perivascular edema, membrane damage and collapse of microvessels following OBI are commonly seen in the traumatized hemisphere exhibiting

◘ **Figure 17-24**
Traumatic brain injuries (TBI) induced BBB disruption and cell changes in the rat. (A) Shows model of closed head injury (CHI) in rats. Under Equithesin anaesthesia, silicon coated, iron bar weighing 114.6 g was dropped over the right parietal skull bone (o) from a 20 cm height through an aluminum guide tube (Dey and Sharma, 1984; Vannemreddy et al., 2006). This impact induces concussive injury in animals and do not normally induces fracture of the skull. After injury, the rats were allowed to survive under anaesthesia for 5 h (for details see text). (C) Shows Nissl stained nerve cells in the parietal cerebral cortex of one untreated injured (a,b) and one serotonin (5-hydroxytryptamine, 5-HT) antibodies (abs) treated (c,d) rat. The impact injury (0,224 N) on the right parietal skull bone was made under Equithesin anaesthesia. Five hours after closed head injury (CHI), profound edema (*), sponginess and neuronal damage (*arrows*) is seen in both the hemispheres (C: a,b). However, due to countercoup concussive insult, the injury severity in terms of edematous swelling and loss of nerve cells are most pronounced in the contralateral hemisphere (*left half, LH*) compared to the injured right half (RH). Intracerebroventricular administration of monoclonal 5-HT abs (1:20) into the left cerebral ventricle 30 min after CHI markedly attenuated neuronal loss (arrow heads) and reduced the development of edema in the cerebral cortex (C: c,d). Thus, more healthy and dense nerve cell population can be seen in the 5-HT abs treated injured rat (C: c,d). This effect of 5-HTabs was most pronounced on the left hemisphere. Probably local administration of 5-HT abs in left side could be partially responsible for this (for details see text). Bars: a–d = 50 μm. Data modified after Sharma et al., 2007a. E. Rat model of open brain injury (OBI) that was inflicted on the right parietal cerebral cortex after making a burr hole on the right and left parietal bones (4 mm2) (E:d). A longitudinal incision (3-mm deep and 5-mm long) was made on the right parietal cerebral cortex. The left parietal cortex was left intact. Animals were allowed to survive 5 h after the lesion (A). Extravasation of Evans blue albumin (EBA) and visual swelling following OBI is seen in rats (E: a) and its modification with the nitric oxide synthase (NOS) antiserum (E: b) Coronal section from the injured rat brain passing through hippocampus (−3.7 μm from Bregma, E: c) shows traumatic edema (*arrows*) and extravasation of Evans blue albumin (blue areas) and visual swelling around the lesion site (E: c). Administration of neuronal NOS antiserum into the traumatized region 5 min after the lesion markedly attenuated the leakage of EBA around the lesion site and the visual swelling (E: b). Data on HS modified after Alm and Sharma, 2002. Bar = 5 mm. Bar = 4 mm. D. High power electron micrograph from the perifocal parietal cortex in one rat following open brain injury (OBI). Collapse of microvessel, perivascular edema (*) and leakage of lanthanum across the endothelial cell (*arrows*) is prominent (D: a). The cerebral endothelium shows two microvesicular pits containing lanthanum (D: b, *arrow heads*). The tight junction remained intact (D: b, *arrows*). B: Ultrastructural changes in the blood-brain barrier (BBB) permeability to lanthanum in the cerebral cortex of one rat following open brain injury (OBI, B: a) and its modification with neuronal NOS antiserum treatment (B: b). Collapse of microvessels and leakage of lanthanum across the endothelial cell membrane (*arrows*), perivascular edema (*) and cell damage is apparent in the untreated traumatized rat (B: a). Intracerebral administration of neuronal NOS antiserum 5 min after TBI significantly reduced the perivascular edema and extravasation of lanthanum across the cerebral endothelium (arrow heads, B: b). The neuropil in the NOS antiserum treated rat is compact and signs of vacuolation and membrane damage (*) are much less apparent (B: b) compared to the untreated traumatized rat (B: a). Data modified after Sharma and Alm, 2004

□ Figure 17-24 (continued)

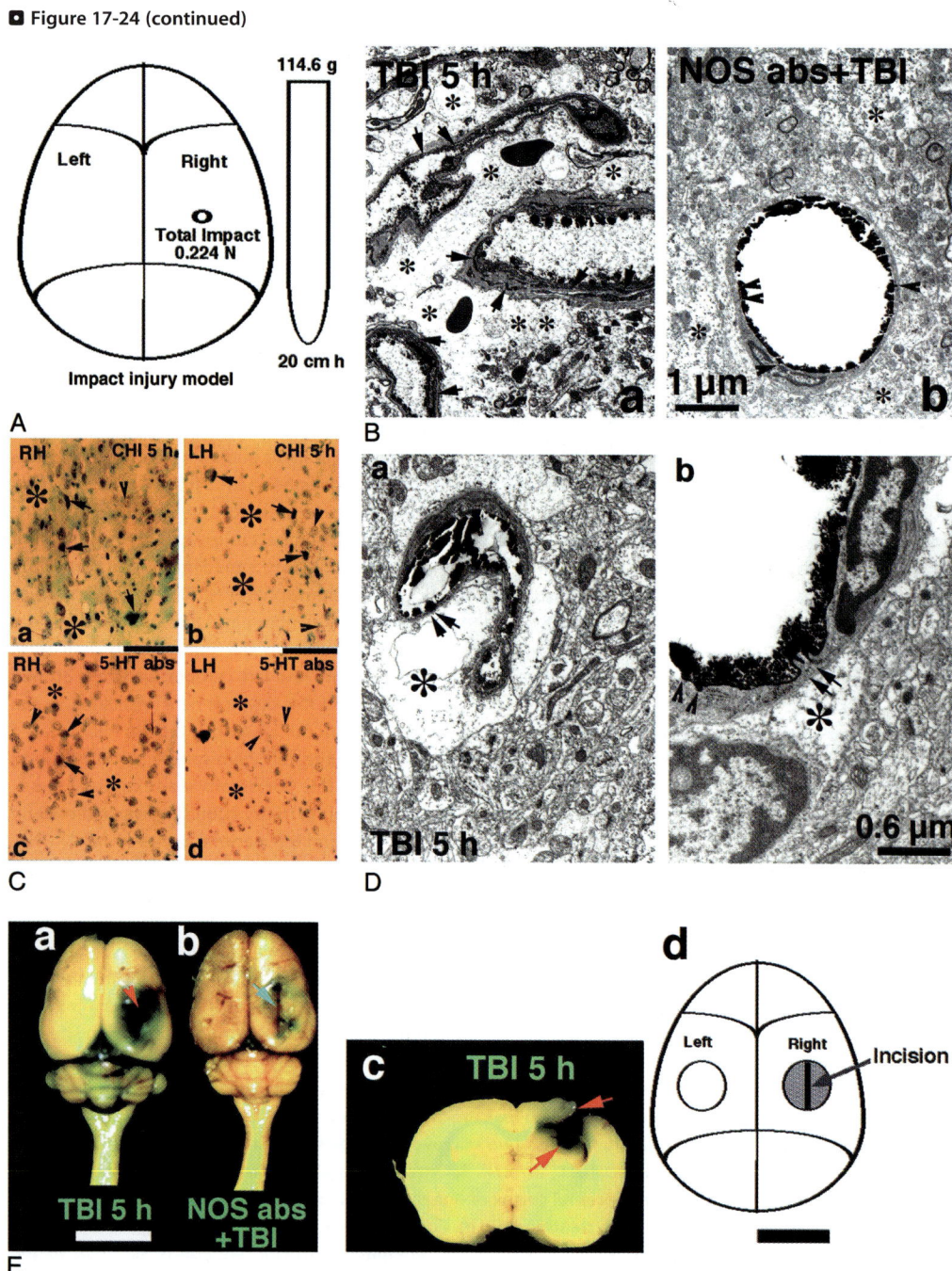

edematous cell or membrane swelling (● *Figure 17-24*). These observations suggest that trauma induced breakdown of the BBB induces endothelial cell membrane permeability without widening the tight junctions. Furthermore, breakdown of the BBB function appears to be instrumental in brain edema formation and cell injury (Sharma, 2004d).

◻ Table 17-13
Effect of 5-h OBI and CHI on blood–brain barrier permeability, brain edema, and ion content in the rat

Type of experiment	BBB permeability		Brain edema formation		Ion content, mM/kg	
	EBA, mg %	[131]Iodine, %	Water content, %	% f	Na^+	K^+
A. Control						
Right half	0.28 ± 0.04	0.38 ± 0.08	77.89 ± 0.23	Nil	310 ± 12	287 ± 12
Left half	0.26 ± 0.08	0.40 ± 0.06	77.83 ± 0.34	Nil		
B. 5-h CHI						
Right half[a]	1.34 ± 0.12**	1.86 ± 0.23**	81.67 ± 0.56**	+16	390 ± 18**	245 ± 8**
Left half	1.78 ± 0.23**[b]	2.34 ± 0.28**[b]	82.76 ± 0.44 [b,c]	+19	420 ± 16**	225 ± 13**[b]
C. 5-h OBI						
Right half[a]	2.34 ± 0.23**	2.89 ± 0.28**	80.34 ± 0.12**	+12	387 ± 17**	240 ± 21*
Left half	1.21 ± 0.18**	1.54 ± 0.21**	79.18 ± 0.23**	+8	324 ± 21	236 ± 17**

Source: Data modified after Sharma 2000, 2004, 2007; Sharma (unpublished observations)
Values are Mean ± SD of 5–8 rats *$P < 0.05$; **$P < 0.01$, compared from saline control
[a]Injured
[b]$P < 0.05$ compared from right injured half, compared from right injured half, Student's paired t-test
[c]$P < 0.05$, compared from CHI, Student's unpaired t-test

7.3.2 Closed Head Injuries

CHI results in brain swelling in a close cranial compartment causing compression of vital centers that could lead to instant deaths (Dey and Sharma, 1984; Vannemreddy et al., 2006). Microhemorrhage within the CNS, breakdown of the BBB permeability, and brain edema formation appears to be the key factors in inducing cell and tissue injury in CHI (Sharma et al., 2007).

We developed a new model of CHI in rats that allow rapid development of brain edema and cell injury (Dey and Sharma, 1984; Sharma et al., 2007a). The model is essentially based on delivery of an impact of 0.224 N Equithesin anesthesia on the right parietal skull bone (see ❯ Figure 17-24). This is achieved by dropping a weight of 114.6 g on the parietal bone of anesthetized rats from a height of 20 cm through a guide tube (❯ Figure 17-24; Dey and Sharma, 1984; Sharma et al., 2007a). The biomechanical force generated by this impact on the right parietal skull bone is diffusely distributing to the underlying brain tissues to induce a powerful countercoup concussive brain injury inside the closed cranium (Vannemreddy et al., 2006; Sharma et al., 2007a).

7.3.2.1 BBB Permeability Following CHI
Using Evans blue and [131]Iodine tracers, we measured BBB permeability in CHI (Sharma et al., 2007). Most pronounced increase in the permeability of Evans blue and radioiodine in both the cerebral hemisphere were observed at 5 h after CHI (❯ Table 17-13). This increase in BBB dysfunction to the protein tracers were most pronounced into the contralateral hemisphere compared to the ipsilateral side (see ❯ Table 17-13). Only mild to moderate extravasation of tracers were seen after 1 and 2 h of CHI. Interestingly, the tracer extravasation were prominent in the contralateral side, also during these early hours following CHI (❯ Table 17-13). This suggests that concussive brain injury induces BBB disruption (Rapoport, 1976).

7.3.2.2 Brain Edema Following CHI
Measurement of brain water content and the electrolytes (Na^+, K^+, Cl^-) in ipsilateral and the contralateral half showed a close correlation with BBB disruption to protein tracers. Thus, marked edema formation in the contralateral side following 1 and 2 h after CHI. However, most pronounced edema formation is seen at 5 h. At this time period, most of the brain was in semifluid

state indicating massive brain swelling and tissue softening (Dey and Sharma, 1984). A significant increase in Na$^+$ and a slight elevation in K$^+$ was noted after 5 h, whereas changes in Cl$^-$ content was not significant from the control value (◉ *Table 17-13*). These observations suggest that leakage of protein tracers across the BBB somehow contributes to brain edema formation.

We observed a greater accumulation of brain water into the contralateral hemisphere following CHI compared to the injured side (◉ *Table 17-13*). This suggests that the concussive countercoup insults are primarily responsible for the increased brain swelling in the contralateral side (Sharma, 2004a, b, c, d). Obviously, an impact of 0.224 N delivered on the right side skull will induce kinetic impact energy on the other side of the brain due to radiating concussive injury (Vannemreddy et al., 2006).

7.3.2.3 Brain Pathology Following CHI Using Nissl or Haematoxylin and Eosin staining on the paraffin sections, light microscopy revealed profound brain pathology, i.e., changes in nerve cells and edema formation following 5 h CHI (◉ *Figure 17-24*). The brain pathology is apparent in both the traumatized and the contralateral half of the cerebral cortex (see ◉ *Table 17-13*). Thus, neuronal distortion, glial cell damage, and myelin disruption were prominent in the cortical and sub cortical regions of the brain in both hemispheres (Sharma, unpublished observations). The intensities of nerve cellular injury were most prominent in the injured cortex. These nerve cell, glial cell, and myelin changes were much less apparent during early hours of CHI (Sharma, unpublished observations).

Taken together, these observations suggest that BBB disruption in TBI is instrumental in brain edema formation and cell injury.

8 Pharmacological Manipulation of BCNSB and Neuroprotection

Based on the above observations, it is clear that disruption of the BBB, BSCB, or BCSFB in diverse conditions, e.g., stress situations, psychostimulants abuse, nanoparticles treatment or traumatic injuries to the brain and spinal cord are associated with brain edema formation, gene expression, and cell or tissue injuries (see ◉ *Table 17-14*). This is evident from the fact that leakage of protein tracers or ionic lanthanum is seen in the areas of brain or spinal cord showing cellular injury and/or neurodegeneration. Furthermore, overexpression of HSP or GFAP and degradation of MBP is commonly seen in the areas of protein extravasation exhibiting cell injury. This suggests that BCNSB could play important roles in cell and tissue injuries in CNS insults leading to neurodegeneration.

To further confirm this idea, we examined the effects of several pharmacological agents, neurotrophic factors and antibodies directed against neurochemicals and enzymes on the BCNSB breakdown in relation to brain pathology in the above animal models of CNS insults.

8.1 Neurotrophins Influence BCSNB Function to Enhance Neuroregeneration

Neurotrophic factors are well known neuroprotective agents in CSN injuries (see Sharma, 2005a, 2007a, b for review). However, their role on maintenance of the BBB, BCSFB, or BSCB integrity is not well known (Sharma, 2004a). We examined the role of several neurotrophin factors either alone or in combination on their ability to influence BCNSB in several conditions of CNS insult. Our observations are the first to suggest that these neurotrophic factors are able to attenuate BBB dysfunction in CNS injury to achieve neuroprotection and/or enhance neuroregeneration (Sharma et al., 1998a, b, 2000a, b). We used several neurotrophins, e.g., brain derived neurotrophic factor (BDNF), glia derived neurotrophic factor (GDNF), and insulin like growth factor-1 (IGF-1) either alone or in combination in SCI (Sharma, 2003, 2005a, b, 2006a, b, 2007a, b) or in heat stress-induced brain pathology (Sharma, 2007e; Sharma et al., 2007). Topical application of these neurotrophic factors separately 5–30 min after spinal cord trauma markedly attenuated injury induced BSCB breakdown to EBA, radioiodine, and lanthanum tracers (Sharma, 2003, 2005a, b). However, no reduction in

◘ Table 17-14
Disruption of BCNSB is associated with brain edema and cell changes in the CNS

Stressors	BCNSB permeability			Endothelium	CNS Edema formation			Cell changes			
	BBB	BCSFB	BSCB	Lanthanum	Brain edema	Spinal edema	Nerve cell damage	Glial cell damage	Axon damage	HSP positive	GFAP positive	MBP loss
A. Immobilization												
1 h	No	No	No	No	No	No	No	No	No	No	No	No
4 h	No	No	No	No	No	No	No	No	No	No	No	No
8 h	Yes	Yes	Yes	Yes	Yes	Yes	Yes	Yes	Yes	Yes	Yes	Yes
B. Forced swimming												
15 min	No	No	No	No			No	No	No	No	No	No
30 min	Yes	Yes	Yes	Yes	Yes		Yes	Yes	Yes		Yes	
C. Heat stress												
2 h	No	No	No	No	No	No	No	No	No	No	No	No
4 h	Yes	Yes	Yes	Yes	Yes	Yes	Yes	Yes	Yes	Yes	Yes	Yes
D. Psychostimulants												
Morphine withdrawal	Yes	Yes	Yes	Yes	Yes		Yes	Yes	Yes	Yes	Yes	Yes
Methamphetamine	Yes	Yes	Yes	Yes	Yes		Yes	Yes	Yes	Yes	Yes	Yes
MDMA	Yes	Yes	Yes		Yes		Yes	Yes	Yes	Yes	Yes	Yes
E. CNS injury												
Spinal cord injury		Yes	Yes	Yes		Yes	Yes	Yes	Yes	Yes	Yes	Yes
Spinal nerve lesion		Yes	Yes				Yes	Yes	Yes	Yes	Yes	Yes
Traumatic brain injury	Yes	Yes		Yes	Yes		Yes	Yes	Yes	Yes	Yes	Yes
Closed head injury	Yes	Yes		Yes	Yes		Yes	Yes	Yes	Yes	Yes	Yes
F. Nanoparticles exposure												
Ag	Yes	Yes	Yes	Yes	Yes	Yes	Yes	Yes	Yes	Yes	Yes	Yes
Cu	Yes	Yes	Yes		Yes	Yes	Yes	Yes	Yes	Yes	Yes	Yes
Al	Yes	Yes	Yes		Yes	Yes	Yes	Yes	Yes	Yes	Yes	Yes

Blank column: No data available; Data from various sources (Sharma published and unpublished observations, for details see text)

the BSCB breakdown was observed when these neurotrophins were applied individually 60–90 min after injury (Sharma, 2006a, b, 2007a, b). Interestingly, when BDNF and GDNF in combination were applied topically over the injured spinal cord either 60 min or 90 min after trauma, these neurotrophins were able to significantly reduce the BSCB disruption to protein tracers (Sharma, 2006b, 2007b). A reduction in the BSCB breakdown by neurotrophins following SCI is always associated with a reduction in edema formation and neuronal, glial, or myelin damage (Sharma, 2007b). Whereas, no reduction in spinal cord pathology is seen when the neurotrophins failed to attenuate BSCB disturbances in SCI (Sharma, 2007). These observations point out that a reduction in BSCB by neurotrophins is essential in attenuating spinal cord pathology and/or enhancing spinal cord neuroregeneration (See ● *Table 17-14*; ● *Figure 17-25*).

Likewise, when BDNF, GDNF, or IGF-1 were administered intracerebroventriculalry in rats 30 min before a 4-h heat stress, significant reduction in the BBB and BCSFB dysfunction to protein tracers was observed (Sharma et al., 2007a). On the other hand, when these growth actors were administered either 30 min or 60 min after heat stress, no reduction in the BBB or BCSFB breakdown could be seen (Sharma, unpublished observation). Reduction in brain pathology, e.g., less neuronal glial or myelin reaction was seen in rats treated with neurotrophins 30 min before heat stress, but not in animals in which the growth factors were given 30 min or 60 min after the heat exposure.

Taken together these observations suggest that breakdown of the BBB, BCSFB, or BSCB permeability following CNS injuries is crucial in determining cell and tissue injury (see ● *Table 17-14*). A reduction in BCNSB is thus responsible for neuroprotection and/or neuroregeneration in stress or trauma.

◘ **Figure 17-25**
Morphological changes at light microscopy following spinal cord injury (*left upper panel*) and ultratstructural changes in stress (*right, upper panel*) and their modification with drugs. *Left Upper panel*: Neuronal NOS expression (a,c) after 5-h SCI and its modification with NOS antiserum (b) or brain derived neurotrophic factor (BDNF, d). Topical application of neuronal NOS antiserum prevented the NOS expression following SCI (b). Repeated topical application of BDNF (d) or IGF-1 (h) markedly attenuated SCI induced NOS upregulation in the spinal cord (d). In the BDNF treated rat, elongation of motor neuron axons is prominent (green arrow heads; d). Data modified after Sharma et al., 1996, 1997b, 1998c, d; Sharma, 2004a. (C) Representative examples of HO-2 immunoreactivity in SCI (a,e,f,g) and its modification with BDNF (b), IGF-1 (c,d) and *p*-CPA (h). Pronounced upregulation of HO-2 expression is seen in nerve cells in the spinal cord in both rostral (T9, a) and caudal (T12, e) segments following injury (T10–11) at 5 h. The HO-2 expression is most pronounced in the ipsilateral cord (a,e). However, the contralateral dorsal (f) and ventral (g) horns also exhibit marked HO-2 immunostaining after trauma. *Right upper panel*: (A) Pretreatment with p-CPA markedly attenuated cell damage following 8 h immobilization (IMZ) stress (a). However, treatment with 5,7-DHT (i.c.v.; b) or 6-OHDA (i.c.v.; c) did not prevent IMZ induced cell changes (A: b,c). Perivascular edema (arrowheads), membrane damage (*arrows*) and vacuolation (*) are quite common in these drug treated rats. B. Pretreatment with indomethacin (B.a), *p*-CPA (B: d), naltrexone (B: e) or nimodipine (B: f) significantly attenuated 4-h heat stress (HS) induced cell damage at the ultrastructural levels compared to the untreated group (B: c). Pretreatment with 6-OHDA (i.c.v.; B.b) did not reduce cell damage following HS. Bars = A: a–c = 1 μm; d,e = 0.6 μm; B: a–f = 1 μm; Data after Sharma et al., 1998a; Sharma, 1999. *Lower left and right panels*: Ultratstructural changes in the rat brain following hyperthermic brain injury (HBI) and their modification with histamine H1 and H2 receptor antagonists. One normal nerve cell and surrounding neuropil is apparent in control rat (A). Subjection of rats to 4-h heat stress at 38 °C results in HBI (B). Collapse of microvessel (*arrow*), perivascular edema (*), degeneration of nerve cell and synapses are quite prominent in the heat stressed rat (B). Pretreatment with mepyramine (1 mg/kg, i.p. 30 min before) did not reduce HBI induced brain damage (C). Myelin vesiculation (arrow heads), collapse of microvessel (arrow), perivascular edema (*) and degeneration of neuropil are frequent in mepyramine treated heat stressed rat (C). On the other hand, treatment with cimetidine (10 mg/kg/i.p., 30 min before) markedly reduced the ultrastructural damage caused by HBI (D). Partially collapsed microvessel with relatively compact neuropil is seen in cimetidine treated heat injured rat. Signs of perivascular edema (*) and myelin vesiculation following HBI are less apparent in cimetidine treated rat. Bar: 1 μm. Data after Sharma, 2004e

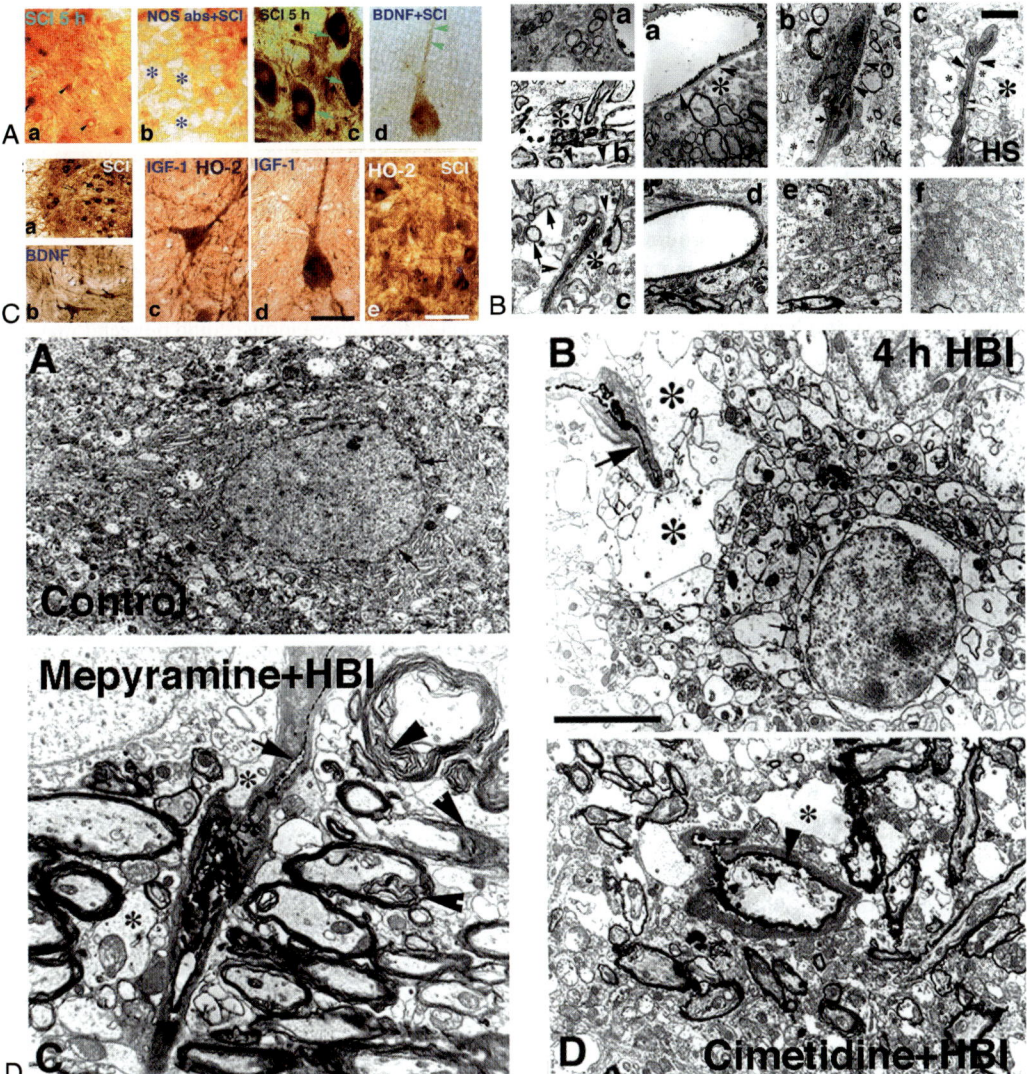

◘ Figure 17-25 (continued)

8.2 Antibodies against Injury Factors Attenuate BCNSB Dysfunction and CNS Pathology

Use of antibodies therapy to neutralize the endogenous effects of their antigens in various disease conditions has been emphasized recently (Sharma et al., 1995d, 1996a; Faden, 1990). The antibodies are considered much more effective in neutralizing the effects of the neurochemicals in vivo than pharmacological blockade of their receptors (Faden, 1990; Sharma et al., 1995a). Our laboratory took one of the earliest initiatives to study the in vivo effects of antibodies directed against several potential injury factors, e.g., neuropeptide dynorphin A 1–17 (Dyn A; Sharma et al., 1995d); neuronal nitric oxide synthase (nNOS, Sharma et al., 1996a); serotonin (5-HT, Sharma et al., 1997a); and the tumor necrosis factor-α (Sharma et al., 2000b) to induce neuroprotective effects in CNS injuries (Sharma, 2004a, b, c, d, 2007c).

The antibodies directed against Dyn A, nNOS, or 5-HT (1:20 dilution) when administered topically over the spinal cord either 5 min before or 10 min after injury induce a marked reduction in the BSCB disruption and cell damage at 5 h (Sharma et al., 1995d, 1996a, 1997a). On the other hand, when these antibodies were administered 30–60 min after injury, no reductions in BSCB breakdown or cell injury were noted (Sharma, unpublished observations). These observations suggest that antibodies to Dyn A, nNOS, and 5-HT if effectively neutralize the harmful effects of their endogenous antigens, i.e., dynorphin, nitric oxide, and serotonin within minutes after spinal cord trauma results in neuroprotection (see ❯ *Table 17-14*). Whereas, delayed neutralization of these injury factors with antibodies is ineffective. The findings support the idea that the antibodies could be potential neurotherapeutic agents to minimize BSCB dysfunction in SCI to achieve neuroprotection.

Using TNF-α antibodies, trauma-induced BSCB disruption and cord pathology were considerably reduced when the antiserum was applied 10–30 min after SCI only (Sharma et al., 2003). On the other hand, treatment with TNF-α antibodies before SCI exacerbated the trauma induced BSCB breakdown and cell or tissue injury (Sharma et al., 2003b). These observations are the first to suggest that early neutralization of TNF-α could be injurious to the cord. However, TNF-α neutralization with antibodies during 10–30 min after the trauma is neuroprotective. This suggests that TNF-α has a dual role in inducing spinal cord pathology (Sharma et al., 2003b) and this effect is largely dependent on its ability to influence BSCB permeability in spinal cord trauma (Sharma et al., 2003b).

The idea that antibodies are able to neutralize their antigens in vivo in brain trauma to achieve neuroprotective is further supported by the results obtained using 5-HT antibodies in our rat model of CHI (Sharma et al., 2007a). Thus, when concentrated 5-HT antibodies was administered intracerebroventricularly either 30 min before or 30 min after CHI, the brain edema formation, BBB disruption, and the cell injury were largely prevented (see ❯ *Table 17-14*; Sharma et al., 2007). On the other hand, when 5-HT antibodies were administered 60 min after CHI, no such neuroprotection or reduction in BBB disruption was noted (Sharma et al., 2007a). These observations suggest that antibodies against 5-HT could neutralize the endogenous effects of the amine in brain during early periods of brain trauma to induce neuroprotection. Taken together it appears that a reduction in the BBB or BSCB disruptions induced by antibodies appears to be largely responsible for neuroprotection in CNS injuries.

8.3 Neurochemical Synthesis Inhibitors Reduce BCNSB Disruption and CNS Pathology

The neurochemical mediators such as serotonin or prostaglandins are known to induce BBB disruption and brain edema formation (Sharma et al., 1990a, b, 1996; Sharma et al., 1994, 1997a, 1998a, b; Wahl et al., 1998; Sharma 2004a, b, c, d). Thus, it is likely that inhibition of prostaglandin or serotonin synthesis prior to CNS insults is neuroprotective (❯ *Figure 17-25*). We used p-chlorophenylalanine (p-CPA) and indomethacin to inhibit serotonin and prostaglandin synthesis, respectively in separate groups of rats (Sharma and Dey, 1986a, b, 1987a, b; Sharma and Olsson, 1990; Sharma et al., 1990a, b, c, 1993a, b, c) and then subjected them to either stress or trauma (see ❯ *Tables 17-14* and ❯ *17-15*). In these animals, the BBB, BCSFB, or BSCB permeability in relation to edema formation and cell injury was determined (Sharma, 2004a, b, c, d).

Pretreatment with p-CPA or indomethacin significantly attenuated BBB and BCSFB disruption in rats subjected to 8-h immobilization, 4-h heat stress, or 30-min forced swimming (Sharma and Dey, 1986a, b, 1987a, b; Sharma et al., 1991a, 1995c). Similarly, these drug treatments were also able to attenuate BSCB and BBB permeability disturbances in spinal cord or brain injuries, respectively (Dey and Sharma, 1983; 1984; Mohanty et al., 1985; Sharma and Olsson, 1990; Sharma et al., 1990a, 2000a, 2007a). Brain edema formation and cell injury were also considerably reduced in these drug treated stressed or traumatized animals (See ❯ *Tables 17-14* and ❯ *17-15*). These observations suggest that serotonin and prostaglandin participate in BBB disturbances and brain edema formation in stress and trauma. Furthermore, a reduction in BBB, BSCB, or BCSFB disturbances with p-CPA or indomethacin is largely responsible for reduction in brain pathology caused by stress or trauma.

Table 17-15
Pharmacological manipulation of the BCNSB and neuroprotection

Drug treatments	Stressors						Psychostimulants		Nanoparticles		CNS lesion					
	Restraint 8 h		Forced swimming		Heat stress		Morphine withdrawal				Spinal cord injury		Brain injury		Head injury	
	BBB	NP	BBB	NP	BBB	NP	BBB	NP	BBB	NP	BSCB	NP	BBB	NP	BBB	NP
A. Neurotrophic factors																
BDNF					Yes	Yes					Yes	Yes				
GDNF					Yes	Yes					Yes	Yes				
IGF-1					Yes	Yes					Yes	Yes				
B. Antibodies																
5-HT															Yes	Yes
nNOS											Yes	Yes	Yes	Yes		
Dyn A (1–17)											Yes	Yes				
TNF-α											Yes	Yes				
C. Synthesis inhibitors																
5-HT (pCPA)	Yes	Yes	Yes	Yes	Yes	Yes					Yes	Yes	Yes	Yes	Yes	Yes
PGs (Indomethacin)	Yes	Yes	Yes	Yes	Yes	Yes					Yes	Yes	Yes	Yes	Yes	Yes
D. Receptor blockers																
Ketanserin	Yes	Yes	Yes	Yes	Yes	Yes					Yes	Yes	Yes	Yes	Yes	Yes
Ritanserin			Yes	Yes	Yes	Yes					Yes	Yes			Yes	Yes
Cyproheptadine	No	No	No	No	No	No					No	No	No	No	No	No
Naloxone			Yes	Yes	Yes	Yes	Yes	Yes			Yes	Yes	Yes	Yes		
Cimetidine					Yes	Yes	Yes	Yes			Yes	Yes	Yes	Yes	Yes	Yes
Ranitidine					Yes	Yes					Yes	Yes	Yes	Yes	Yes	Yes

continued

■ Table 17-15 (continued)

Drug treatments	Stressors						Psychostimulants		Morphine withdrawal		Nanoparticles		CNS lesion					
	Restraint 8 h		Forced swimming		Heat stress								Spinal cord injury		Brain injury		Head injury	
	BBB	NP	BBB	NP	BBB	NP	BBB	NP	BBB	NP	BBB	NP	BSCB	NP	BBB	NP	BBB	NP
Mepyramine	No	No	No	No	No	No							No	No	No	No	No	No
HOE-140					Yes	Yes							Yes	Yes				
E. Antioxidants																		
H-290/51					Yes	Yes	Yes	Yes					Yes	Yes	Yes	Yes	Yes	Yes
EGb-761					Yes	Yes			Yes	Yes	Yes	Yes						
BN-52121					Yes	Yes												
F. Neurotoxins																		
5,7-DHT	No	No	No	No	No	No									No	No	No	No
6-OHDA	No	No	No	No	No	No									No	No	No	No
6-OHDA, i.v.	No	No	No	No	No	No												

Data compiled from various sources (Sharma published and unpublished observations)

The compounds able to reduce BBB (BCSF) or BSCB (Yes) are also able to induce neuroprotection (NP) following CNS insults. On the other hand those compounds failed to attenuate BBB (BCSFB) or BSCB (No) are unable induce neuroprotection (No) after stress or trauma. BBB, blood–brain and blood–CSF barriers; BSCB, blood–spinal cord barrier; NP, Neuroprotection; Blank column, no data available; YES, Induce neuroprotection; F, Failed

8.4 Neurochemical Receptor Modulation Influence BCNSB and CNS Pathology

There are reasons to believe that several neurochemicals released following stress or trauma will influence the BCNSB function and contribute to edema formation through specific receptor-mediated mechanisms (see Sharma, 2004a, b, c, d, 2005a, b, c, 2007a, b, c for review). Using pharmacological approaches we examined the effects of serotonin, histamine, bradykinin, and opioids receptor blockers on the BBB function and brain edema formation in stress or CNS injury (Dey and Sharma, 1983, 1984; Sharma and Dey, 1986a, b, 1987a, b, 1988; Sharma et al., 1990a, b, c, 1991a, 1992a, b, c, 2000a, 2006a).

Blockade of serotonin 2 (5-HT2) receptors with ketanserin or ritanserin (Sharma et al., 1995b); histamine 2 (H2) receptors with cimetidine or ranitidine (Sharma et al., 1992a; Sharma, 2004c, d); bradykinin 2 (BK2) receptors with HOE-140 (Sharma et al., 2000a); or multiple opioid receptor blocker with naloxone (Sharma et al., 1997b; Sharma, 2006c) significantly attenuated heat stress or SCI induced BCNSB disturbances and CNS pathology. However, blockade of histamine 1 (H1) receptor with mepyramine (Sharma et al., 1992a, b, c; Sharma et al., 2003) or a combination of H1 and 5-HT2 receptor blockers cyproheptadine (Dey and Sharma, 1983, 1984; Sharma et al., 1991a, 1995b) did not reduce BBB or BSCB breakdown and cell injury either following heat stress, brain injury, or spinal cord trauma (See *Table 17-14, 17-15;* ● *Figure 17-25*). Interestingly cyproheptadine and mepyramine treatment exacerbated BCSNB function and cell injury in stress and trauma (Sharma, 2004a, b, c, d). These observations indicate that blockade of 5-HT2, H2, BK2, and multiple opioid receptors before CNS insults are neuroprotective in nature. Whereas, antagonism of H1 or a combination of H1 and 5-HT2 receptor before injury or stress may have neurodestructive effects. These observations confirm the idea that BCSNB disruption in CNS insults play key roles in cell and tissue injuries.

8.5 Antioxidants Attenuate BCSNB Breakdown and CNS Pathology

Trauma, psychostimulant abuse, or other kinds of stressful situations induce oxidative stress and formation of free radicals and lipid peroxidation in the CNS (Sharma et al., 1997a, 2001; Sharma and Sjöquist, 2002). Generation of free radicals and lipid peroxidation causes neuronal, glial, and endothelial cell membrane damages (for review Sharma, 2004d, 2005a, 2006a). Disruption of endothelial cell membrane will induce BCNSB breakdown leading to brain edema formation and cell injury. Thus, antioxidants either capable to inhibit lipid peroxidation or are scavengers of free radicals would induce neuroprotection caused by CNS insults (Banik et al., 1985; Sharma et al., 2001; Sharma, 2004d). However the role of antioxidants in BCNSB disruption following CNS injuries is still not well known. Our laboratory was the first to examine the effects of antioxidants on BCSNB dysfunction in a wide variety of stress or traumatic injuries to the CNS (Sharma et al., 2001; Sharma, 2004d, 2005a, 2006a, 2007b).

Using potent inhibitor of lipid peroxidation (H-290/51; Sharma, 2004a) or scavengers of free radicals (EGB-761 and BN 22021; Sharma, 2004b) we found that antioxidants are capable to attenuate BCNSB disruption and brain pathology in various CNS injury models (see ● *Tables 17-14,* ● *17-15;* Sharma, 2004a, b, c, d, 2005a, 2007a, b). Thus, pretreatment with H-290/51 significantly attenuated BSCB permeability, edema formation, and cell damage after 5 h of spinal cord trauma (Sharma et al., 2001). Similarly, pretreatment with this compound reduced brain pathology and BBB disputation in CHI (Sharma et al., 2006; Sharma, unpublished observation) and following morphine withdrawal (Sharma et al., 2007). Brain and spinal cord pathology together with BBB or BSCB disturbances in heat stress was also reduced when H-290/51 was administered 30 min before heat exposure (Sharma et al., 1997; Sharma, 2004d). Interestingly, the compound was also effective in reducing BBB disruption and brain pathology in nanoparticles (Ag, Cu, or Al) treated rats (see ● *Tables 17-14* and ● *17-15;* Sharma, unpublished observations). These observations suggest that inhibition of lipid peroxidation in stress, trauma, or nanoparticle administration prevents BCNSB disruption and cell injury.

Similarly, daily treatment with EGB-761 or BN22123 separately in rats for 5 days reduced the BBB disruption and brain pathology in heat stress (Sharma, 2004b). However, the magnitude and intensity of

BBB disruption and brain pathology were much less pronounced than the H-290/51 treatment. This indicates that the compounds able to inhibit lipid peroxidation have superior neuroprotective ability than drugs able to scavenge the free radicals after their generation. These observations strongly indicate a close correlation between BCNSB disruption and pathological outcome in various kinds of CNS insults.

8.6 Biogenic Amine Neurotoxins Exacerbate BCNSB Breakdown and CNS Pathology

The microvessels of the brain and spinal cord are innervated with serotoninergic, histaminergic, and catecholaminergic nerve fibers (Rapoport, 1976; Edvinsson and McKenzie, 1977; Sharma and Westman, 2004). These nerve fibers control the vascular reactivity of neurochemicals and my influence the BBB function (Edvinsson and McKenzie, 1977). Thus, we examined the influence of serotoninergic and catecholaminergic nerve terminal degeneration on the BBB disruption in stress and trauma in relation to brain pathology.

Destruction of central 5-HT neurons with 5,7-dihydroxytruptamine (5,7-DHT) significantly enhanced the BBB disruption to EBA and radioiodine in the rat brains following immobilization, forced swimming, or heat stress (Sharma, 1982; Sharma and Dey, 1986; Sharma et al., 1995c; Sharma, 2004a, b, c, d). Similar enhancement of brain edema formation and BBB leakage was seen in 5,7-DHT treated rats when subjected to 5-h brain or head injuries (Dey and Sharma, 1983, 1984; Sharma, 2004d, see ❯ *Table 17-13*). These observations suggest that destruction of central serotoninergic nerve terminals enhances the BBB leakage and thus exacerbate brain pathology (Sharma, 1982, 1999).

Interestingly, exacerbation of BBB permeability, brain edema, and brain pathology was also observed in rats subjected to heat stress or TBI after destruction of either central or peripheral catecholaminergic nerve terminals with 6-hydroxydopamine (6-OHDA) (Sharma, 1982, 1989, 2004d; Sharma, unpublished observation).

These observations suggest that serotoninergic and catecholaminergic nerve terminals regulate BBB function in stress and trauma. Thus, chemical degeneration of serotoninergic and catecholaminergic nerve terminals aggravate BBB dysfunction leading to enhanced brain pathology. This indicates that the BBB disruption is instrumental in cell and tissue injury following CSN insults.

9 Mechanisms of BCSNB Breakdown in CNS Insults

It appears that alterations in neurochemicals following trauma, stress, psychostimulants abuse, or nanoparticle administration are responsible for the BCNSB breakdown. CNS insults releases several neurochemicals, i.e., serotonin, prostaglandin, histamine, opioids, amino acid neurotransmitters, and other injury factors in the CNS and in periphery (Sharma, 2004a, b, c, d), These neurochemicals then will act on the luminal and abluminal sides of the cerebrovascular endothelium in the CNS (Edvinsson and McKenzie, 1977). This would induce a cascade of effects leading to alteration in the membrane permeability (Sharma, 2004c, d). Release of neurochemicals in the CNS and in periphery will also induce oxidative stress, lipid peroxidation, and generation of free radicals that could directly or through indirect mechanisms affect the endothelial membrane permeability and induce cell damage. This idea is supported by the results obtained using pharmacological approaches. Thus, blockade of serotonin, prostaglandin, histamine, and opioid effects in various kinds of insults attenuated BCNSB dysfunction. The neurotrophic factors modulate neurotransmission and influence intracellular signaling pathways. Thus, it appears that cross talk between neurotrophins and neurochemical receptors lead to alteration in cell signaling pathways in neurotrophins treated animals following CNS injury. The neurotrophins may be responsible for cell membrane stabilization resulting in reduction in BCSNB permeability in CNS injuries.

The neurochemicals after binding to their receptors located on both the luminal and the abluminal surfaces of the CNS microvessels induce intracellular signaling, e.g., stimulation of prostaglandins, nitric oxide synthase (NOS), cAMP, or cGMP synthesis and release (Olesen, 1989; Sharma et al., 1990a, Sharma,

2004c, d). The CNS capillaries contain all necessary enzymes for synthesis and catabolism of prostaglandins, NOS as well as cAMP (Sharma and Alm, 2004). Local accumulation of prostaglandins, cGMP or cAMP in cerebral capillaries induces marked vasodilatation and an increases vesicular transport (Rapoport, 1976). Obviously, various pharmacological agents influence this mode of tracer transport. Alteration in the cell membrane transport caused by cAMP and cGMP could also account for increased tracer transfer across the BCSNB (Sharma and Westman, 2004). Whether the neurochemicals influence membrane transporters to enhance transport of lanthanum or proteins, e.g., albumin is unclear and requires further investigation.

Taken together, our observations suggest that the basic mechanisms of BCNSB disruption either caused by stress, trauma, psychostimulants abuse, or nanoparticles exposure are similar in nature. This is further supported by the facts that the pharmacological compounds that are capable to induce neuroprotection and BCSNB disturbances following CNS injuries, or stress are also able to prevent BCSNB disruption and brain pathology following nanoparticles or psychostimulants exposure. Thus, the BCSNB function strictly regulates the cell and tissue injury in various CNS insults or diseases.

9.1 Passage of Tracers across the BCNSB

Ultrastructural studies using lanthanum as electron dense tracer clearly show that the increase in endothelial cell membrane permeability, rather than widening of the tight junctions play important roles in BCSNB breakdown in stress, trauma, nanoparticles, or psychostimulants abuse. Thus, infiltration of lanthanum is seen in cerebral or spinal cord microvessels that can be extended to the basal lamina and even in the extracellular spaces in the adjacent neuropil. In severe trauma or serotonin infusion, intracellular accumulation of lanthanum was also observed (Sharma, 2004b). This suggests that altered membrane permeability of the CNS microvessels are playing major role in BCNSB breakdown in CNS diseases. In hyperthermia, this increase in membrane permeability is also seen in the regions connected with the tight junctions (Sharma et al., 1998a). However, the tight junctional morphology appears to be normal. Furthermore, widening of tight junctions is not seen in any microvessels following stress or trauma. Thus, the available evidences so far points out the role of endothelial ell membrane as important anatomical barrier between the vascular and cerebral compartments. Obviously, the cell membrane transport is increased under pathological conditions and this kind of tracer transfer could be modified using pharmacologic approaches.

10 General Conclusion

In conclusion, it appears that breakdown of the BCNSB either caused by stress, trauma, nerve lesion, nanoparticles, or psychostimulants abuse is instrumental in inducing edema formation leading to in cell and tissue injury in the CNS (see ❯ Figure 17-26). There are reasons to believe that breakdown of the BCNSB in during diverse CNS insults is achieved by a cascade of events involving several neurochemicals, viz., serotonin, PGs, amino acids, and opioids as well as various other endogenous factors, e.g., cellular and oxidative stress, generation of free radicals, and lipid peroxidation. These neurochemicals/factors act on microvessels either directly or indirectly through various signal transducing agents, e.g., cAMP or cGMP to enhance vesicular transport and/or to increase endothelial cell membrane permeability. Transport of macromolecules like proteins from the vascular compartment to the spinal cord microenvironment will induce vasogenic edema formation. Exposure of restricted chemicals, ions, molecules within the brain or spinal cord extra- and intracellular environment will initiate a series of events leading to cell and tissue injury. Spread of edema fluid within the uninjured portion of the CNS is likely to induce adverse cell reactions, apoptosis, gene expression and eventually cell death. Stress induced BCNSB dysfunction is primarily responsible for alterations in the bioelectrical activity.

However, our observations further suggest that no single chemical compound is solely responsible for all the pathophysiological changes seen in CNS injuries. Numerous endogenous substances that are released in following CNS insults may act as either endogenous "neuroprotective agents", or as endogenous "neurodestructive factors". A balance between these neuroprotective and neurodestructive agents/factors is crucial to determine the magnitude and extent of cell and tissue injury in the brain or spinal cord.

◻ **Figure 17-26**
Schematic diagram showing probable interrelationship between BSCB, edema formation, and cell death. Disruption of the BSCB following SCI appears to be crucial for edema formation, necrotic or apoptotic changes, gene expression, and cell death. Solid arrows represent possible interactive pathways supported by experimental observations. Broken arrow indicates a possibility based on assumption that requires further investigation

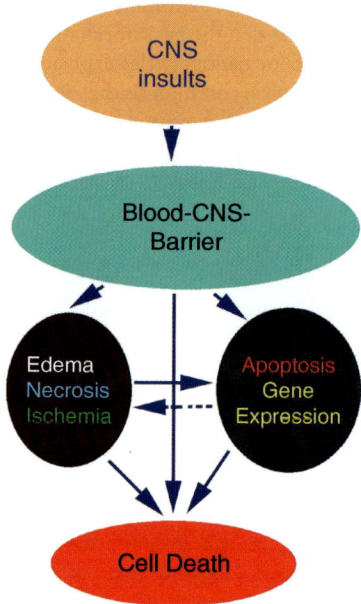

Taken together these observations strongly point out that the BCNSB appears to be a gateway for neurodegeneration, neuroregeneration, and neuroprotection. Thus, maintenance of near normal function of BCNSB is crucial for cell survival and breakdown of the barrier leads to cell death in the CNS.

Acknowledgement

Author's research is supported by Grants from Swedish Medical Research Council, Stockholm Sweden; Alexander von Humboldt Foundation, Germany; European Office of Aerospace Research and Development (EOARD), London Office, UK. Secretarial Assistance of Aruna Sharma and Graphics support of Suraj Sharma is highly appreciated.

References

Albert EN. 1979. Current status of microwave effects on the blood–brain barrier. J Microw Power 14(3): 281-285.

Angel C. 1966. Adrenalectomy, stress and the blood–brain barrier. Dis Nerv Syst 27: 389-393.

Angel C. 1969. Starvation, stress and the blood–brain barrier. Dis Nerv Syst 30: 94-97.

Aukland K. 1973. Autoregulation of interstitial fluid volume: Edema-preventing mechanisms. Scand J Clin Lab Invest 31: 247-254.

Baethmann A, Van Harreveld A. 1973. Water and electrolyte distribution in gray matter rendered edematous with a metabolic inhibitor. J Neuropathol Exp Neurol 32: 408-423.

Bakay L, Lee JC. 1965. Cerebral Edema. Springfield, IL: Charles C Thomas.

Balentine JD. 1988. Spinal cord trauma: In search of the meaning of granular axoplasm and vesicular myelin. J Neuropathol Exp Neurol 47: 77-92.

Ballok DA. 2007. Neuroimmunopathology in a murine model of neuropsychiatric lupus. Brain Res Rev 54(1): 67-79.

Banik NL, Hogan EL, Hsu CY. 1985. Molecular and anatomical correlates of spinal cord injury. Cent Nerv Syst Trauma 2(2): 99-107.

Banik NL, Hogan EL, Hsu CY. 1987. The multimolecular cascade of spinal cord injury. Studies on prostanoids, calcium, and proteinases. Neurochem Pathol 7(1): 57-77.

Barber AJ, Antonetti DA, Gardner TW, The Penn State Retina Research Group, 2000. Altered expression of retinal occludin and glial fibrillary acidic protein in experimental diabetes. Invest Ophthalmol Vis Sci 41: 3561-3568.

Barbeito LH, Pehar M, Cassina P, Vargas MR, Peluffo H, et al. 2004. A role for astrocytes in motor neuron loss in amyotrophic lateral sclerosis. Brain Res Brain Res Rev 47(1–3): 263-274 (Review).

Brightman MW, Hori M, Rapoport SI, Reese TS, Westergaard E. 1973. Osmotic opening of tight junctions in cerebral endothelium. J Comp Neurol 152(4): 317-325.

Basch A, Fazekas G. 1970. Increased permeability of the blood–brain barrier following experimental thermal injury of the skin. Angiologica 7: 357-364.

Batson OV. 1942. The vertebral vein system as a mechanism for the spread of metastases. Am J Roentgenol Rad Ther 48: 715-718.

Beggs JL, Waggener JD. 1975. Vasogenic edema in the injured spinal cord: A method of evaluating the extent of blood–brain barrier alteration to horseradish peroxidase. Exp Neurol 49(1 Pt 1): 86-96.

Beggs JL, Waggener JD. 1976. Transendothelial vesicular transport of protein following compression injury to the spinal cord. Lab Invest 34(4): 428-439.

Belova I, Jonsson G. 1982. Blood–brain barrier permeability and immobilization stress. Acta Physiol Scand 116: 21-29.

Ben-Nathan D, Kobiler D, Loria RM, Lustig S. 1998. Stress-induced central nervous system penetration by non-invasive attenuated encephalitis viruses. New Frontiers in Stress Research, Modulation of Brain Function. Levy A, Grauer E, Ben-Nathan D, de Kloet ER, editors. The Netherlands: Harwood Academic Publishers; pp. 277-2283.

Ben-Nathan D, Lustig S, Danenberg H. 1991. Stress-induced neuroinvasiveness of a neurovirulent non invasive Sindbis virus in cold or isolation subjected mice. Life Sci 48: 1493-1500.

Ben-Nathan D, Lustig S, Kobiler D. 1996. Cold stress-induced neuroinvasiveness of attenuated arboviruses is not solely mediated by corticosterone. Arch Virol 141: 459-469.

Blackwell RP, Saunders RD. 1986. The effects of low-level radiofrequency and microwave radiation on brain tissue and animal behaviour. Int J Radiat Biol Relat Stud Phys Chem Med 50(5): 761-787.

Bonate PL. 1988. Quantification of albumin in cerebrospinal fluid. Anal Biochem 175(1): 300-304.

Bondy SC, Purdy JL. 1974. Selective regulation of the blood–brain barrier by sensory input. Brain Res 76: 542-545.

Borm PJ, Robbins D, Haubold S, Kuhlbusch T, Fissan H, et al. 2006. The potential risks of nanomaterials: A review carried out for ECETOC. Part Fibre Toxicol 3: 11.

Bouchama A, Knochel JP. 2002. Heat stroke. N Engl J Med 346(25): 1978-1988.

Bradbury MWB. 1979. The Concept of a Blood–Brain Barrier. Chicester, London: John Wiley & sons.

Bradbury MWB. 1992. Physiology and Pharmacology of the Blood–Brain Barrier. Handbook of Experimental Pharmacology, Vol. 103. Heidelberg: Springer-Verlag; pp. 1-450.

Brightman MW, Klatzo I, Olsson Y, Reese TS. 1970. The blood–brain barrier to proteins under normal and pathological conditions. J Neurol Sci 10(3): 215-239.

Brightman MW, Reese TS. 1969. Junctions between intimately apposed cell membranes in the vertebrate brain. J Cell Biol 40: 648-677.

Butt AM, Jones HC. 1992. Effect of histamine and antagonists on electrical resistance across the blood–brain barrier in rat brain-surface microvessels. Brain Res 569: 100-105.

Castel M, Sahar A, Erlij D. 1974. The movement of lanthanum across diffusion barriers in the choroid plexus of the cat. Brain Res 67(1): 178-184.

Cervós-Navarro J, Ferszt R. 1980. Brain edema: Pathology, diagnosis, and therapy. Adv Neurol 28: 1-450.

Cervós-Navarro J, Kannuki S, Nakagawa Y. 1988. Blood–brain barrier (BBB): Review from morphological aspect. Histol Histopathol 3: 203-213.

Cervós-Navarro J, Sharma HS, Westman J, Bongcam-Rudloff E. 1998. Glial reactions in the central nervous system following heat stress. Prog Brain Res 115: 241-274.

Chisari M, Salomone S, Laureanti F, Copani A, Sortino MA. 2004. Modulation of cerebral vascular tone by activated glia: Involvement of nitric oxide. J Neurochem 91(5): 1171-1179.

Christensen TG, Diemer NH, Laursen H, Gjedde A. 1981. Starvation accelerates blood-brain glucose transfer. Acta Physiol Scand 112: 221-223.

Cohen S. 1968. The immune response in relation to the nervous system. Biochemical Aspects of Neurological Disorders, Third series. Cumings JN, Kremer M, editors. Oxford: Blackwell; pp. 10-22.

Cohen S, Williamson GM. 1991. Stress and infectious disease in human. Psychol Bull 109: 5-24.

Connolly JH, Allen IV, Hurwitz LJ, Millar JH. 1968. Subacute sclerosing panencephalitis. Clinical, pathological, epidemiological, and virological findings in three patients. Q J Med 37(148): 625-644.

Conrad S, Schluesener HJ, Adibzahdeh M, Schwab JM. 2005. Spinal cord injury induction of lesional expression of profibrotic and angiogenic connective tissue growth factor confined to reactive astrocytes, invading fibroblasts and endothelial cells. J Neurosurg Spine 2(3): 319-326.

Cserr HF. 1971. Physiology of the choroid plexus. Physiol Rev 51(2): 273-311.

Cserr HF, Ostrach LH. 1975. Bulk flow of interstitial fluid after intracranial injection of Blue Dextran 2000. Exp Neurol 45: 50-60.

Cutler RW, Murray JE, Cornick LR. 1970. Variations in protein permeability in different regions of the cerebrospinal fluid. Exp Neurol 28(2): 257-265.

Cutler RW, Lorenzo AV, Barlow CF. 1968. Changes in blood–brain barrier permeability during pharmacologically induced convulsions. Prog Brain Res 29: 367-384.

Dantzer R, Kelley KW. 1989. Stress and immunity: An integrated view of the relationships between the brain and immune system. Life Sci 44: 1995-2008.

Davson H. 1967. Physiology of the Cerebrospinal Fluid. London: Churchill.

Davson H. 1976. Physiology of the Cerebrospinal Fluid. London: Churchill.

Dawson CA, Horvath SM. 1970. Swimming in small laboratory animals. Med Sci Sports 2(2): 51-78.

Demediuk P, Faden AI, Vink R, Romhanyi R, McIntosh TK. 1989. Effects of traumatic brain injury on arachidonic acid metabolism and brain water content in the rat. Ann N Y Acad Sci 559: 431-432.

Dey PK, Sharma HS. 1983. Ambient temperature and development of traumatic brain oedema in anaesthetized animals. Indian J Med Res 77: 554-563.

Dey PK, Sharma HS. 1984. Influence of ambient temperature and drug treatments on brain oedema induced by impact injury on skull in rat. Indian J Physiol Pharmacol 28: 177-186.

Dohrmann GJ, Wagner FC Jr, Wick KM, Bucy PC. 1971. Fine structural alterations in transitory traumatic paraplegia. Proc Veterans Adm Spinal Cord Inj Conf 18: 6-8.

Dubois M, Coppola R, Buchsbaum MS, Lees DE. 1981. Somatosensory evoked potentials during whole body hyperthermia in humans. Electroencephalogr Clin Neurophysiol 52(2): 157-162.

Dubois M, Sato S, Lees DE, Bull JM, Smith R, et al. 1980. Electroencephalographic changes during whole body hyperthermia in humans. Electroencephalogr Clin Neurophysiol 50(5–6): 486-495.

Dvorská I, Brust P, Hrbas P, Rühle HJ, Barth T, et al. 1992. On the blood–brain barrier to peptides: Effects of immobilization stress on regional blood supply and accumulation of labelled peptides in the rat brain. Endocr Regul 26(2): 77-82.

Edvinsson E, McKenzie ET. 1977. Amine mechanisms in the cerebral circulation. Pharmacol Rev 28: 275-348.

Esposito P, Chandler N, Kandere K, Basu S, Jacobson S, et al. 2002. Corticotropin-releasing hormone and brain mast cells regulate blood–brain barrier permeability by acute stress. J Pharmacol Exp Ther 303: 1061-1066.

Esposito P, Gheorghe D, Kandere K, Pang X, Conally R, et al. 2001. Acute stress increase permeability of the blood–brain barrier through activation of mast cells. Brain Res 888: 117-127.

Essman W. 1978. Serotonin in Health and Disease. The Central Nervous System, Vol. 3. New York: Spectrum.

Faden AI. 1990. Opioid and non-opioid mechanisms may contribute to dynorphin's pathophysiological actions in the spinal cord injury. Ann Neurol 27: 64-74.

Febbraio MA, Snow RJ, Stathis CG, Hargreaves M, Carey MF. 1994. Effect of heat stress on muscle energy metabolism during exercise. J Appl Physiol 77(6): 2827-2831.

Feng N, Pagniano R, Tovar CA, Bonneaue RH, Glasser R, et al. 1991. The effect of restraint stress on the kinetics, magnitude, and isotype of the humoral immune response to influenza virus infection. Brain Behav Immun 5: 370-382.

Fenstermacher JD. 1970. Extracellular space of the cerebral cortex of normothermic and hypothermic cats. Exp Neurol 27(1): 101-114.

Fenstermacher JD, Rall DP. 1983. Physiology and pharmacology of cerebrospinal fluid. Pharmacology of Cerebral Circulation, Vol. 1. Oxford: Pergamon Press; pp. 35-79.

Feringa ER, Wendt JS, Johnson RD. 1974. Immunosuppressive treatment to enhance spinal cord regeneration in rats. Neurology 24(3): 287-293.

Frey AH, Feld SR, Frey B. 1975. Neural function and behavior: Defining the relationship. Ann N Y Acad Sci 247: 433-439.

Frick E, Scheid-Seydel L. 1960. [Research with I 131-labelled gamma-globupin on the problem of the origin of cerebrospinal fluid proteins.] Klin Wochenschr 38: 1240-1243.

Friedman SB, Glasgow LA, Ader R. 1970. Differential susceptibility to viral agent in mice housed alone or in group. Psychosom Med 32: 285-299.

Gardner TW, Lieth E, Khin SA, Barber AJ, Bonsall DJ, et al. 1997. Astrocytes increase barrier properties and ZO-1 expression in retinal vascular endothelial cells. Invest Ophthalmol Vis Sci 38(11): 2423-2427.

Ghersi-Egea JF, Strazielle N, Murat A, Edwards J, Belin MF. 2001. Are blood–brain interfaces efficient in protecting the brain from reactive molecules? Adv Exp Med Biol 500: 359-364.

Gilbert GJ. 1965. Focal breakdown of the blood–brain barrier by specific sensory stimulation. Trans Am Neurol Assoc 90: 246-248.

Griffiths HJ, Bushueff B, Zimmerman RE. 1976. Investigation of the loss of bone mineral in patients with spinal cord injury. Paraplegia 14(3): 207-212.

Gonzalez-Alonso J, Teller C, Andersen SL, Jensen FB, Hyldig T, et al. 1999. Influence of body temperature on the development of fatigue during prolonged exercise in the heat. J Appl Physiol 86(3): 1032-1039.

Goodman JH, Bingham WG Jr, Hunt WE. 1976. Ultrastructural blood–brain barrier alterations and edema formation in acute spinal cord trauma. J Neurosurg 44(4): 418-424.

Gordh T, Chu H, Sharma HS. 2006. Spinal nerve lesion alters blood–spinal cord barrier function and activates astrocytes in the rat. Pain 124(1–2): 211-221.

Gordh T, Sharma HS. 2006. Chronic spinal nerve ligation induces microvascular permeability disturbances, astrocytic reaction, and structural changes in the rat spinal cord. Acta Neurochir Suppl 96: 335-340.

Gordh T, Sharma HS, Alm P, Westman J. 1998. Spinal nerve lesion induces upregulation of neuronal nitric oxide synthase in the spinal cord. An immunohistochemical investigation in the rat. Amino Acids 14(1–3): 105-112.

Gordh T, Sharma HS, Azizi M, Alm P, Westman J. 2000. Spinal nerve lesion induces upregulation of constitutive isoform of heme oxygenase in the spinal cord. An immunohistochemical investigation in the rat. Amino Acids 19(1): 373-381.

Govrin-Lippmann R, Devor M. 1978: Ongoing activity in severed nerves: Source and variation with time. Brain Res 159: 406-410.

Griffiths IR. 1975. Vasogenic edema following acute and chronic spinal cord compression in the dog. J Neurosurg 42(2): 155-165.

Griffiths IR. 1976. Spinal cord blood flow after acute experimental cord injury in dogs. J Neurol Sci 27(2): 247-259.

Griffiths IR. 1978. Spinal cord injuries: A pathological study of naturally occurring lesions in the dog and cat. J Comp Pathol 88(2): 303-315.

Griffiths IR. 1980. Trauma of the spinal cord. Vet Clin North Am Small Anim Pract 10(1): 131-146.

Griffiths IR, Burns N, Crawford AR. 1978b. Early vascular changes in the spinal grey matter following impact injury. Acta Neuropathol (Berl) 41(1): 33-39.

Griffiths IR, McCulloch MC. 1983. Nerve fibres in spinal cord impact injuries, Part 1. Changes in the myelin sheath during the initial 5 weeks. J Neurol Sci 58(3): 335-349.

Griffiths IR, McCulloch M, Crawford RA. 1978a. Ultrastructural appearances of the spinal microvasculature between 12 hours and 5 days after impact injury. Acta Neuropathol (Berl) 43(3): 205-211.

Groves PM, Thompson RF. 1970. Habituation: A dual-process theory. Psychol Rev 77(5): 419-450.

Gruner JA, Altman J. 1980. Swimming in the rat: Analysis of locomotor performance in comparison to stepping. Exp Brain Res 40(4): 374-382.

Guyton AC. 1963. A concept of negative interstitial pressure based on pressures in implanted perforated capsules. Circ Res 12: 399-412.

Harri M, Kuusela P. 1986. Is swimming exercise or cold exposure for rats? Acta Physiol Scand 126(2): 189-197.

Heins N, Malatesta P, Cecconi F, Nakafuku M, Tucker KL, et al. 2002. Glial cells generate neurons: The role of the transcription factor Pax6. Nat Neurosci 5(4): 308-315.

Hermann G, Tovar CA, Beck FM, Sheridan JF. 1994. Kinetics of glucocorticoid response to restraint stress and/or experimental influenza viral infection in two inbred strains of mice. J Neuroimmunol 49: 25-33.

Himmelsbach CK. 1943. With reference to physical dependence. Fed Proc 2: 201-203.

Hirano A, Levine S, Zimmerman HM. 1967. Experimental cyanide encephalopathy: Electron microscopic observations of early lesions in white matter. J Neuropathol Exp Neurol 26(2): 200-213.

Hirano A. 1971. Edema damage. Neurosci Res Prog Bull 9: 493-496.

Hochwald GM, Wallenstein MC. 1967a. Exchange of gamma-globulin between blood, cerebrospinal fluid and brain in the cat. Exp Neurol 19(1): 115-126.

Hochwald GM, Wallenstein M. 1967b. Exchange of albumin between blood, cerebrospinal fluid, and brain in the cat. Am J Physiol 212(5): 1199-1204.

Hökfelt T, Ceccatelli S, Gustafsson L, Hulting AL, Verge V, et al. 1994. Plasticity of NO synthase expression in the nervous and endocrine systems. Neuropharmacology 33(11): 1221-1227.

Houshyar H, Cooper ZD, Woods JH. 2001b. Paradoxical effects of chronic morphine treatment on the temperature and pituitary–adrenal responses to acute restraint stress: A chronic stress paradigm. J Neuroendocrinol 13(10): 862-874.

Houshyar H, Galigniana MD, Pratt WB, Woods JH. 2001a. Differential responsivity of the hypothalamic–pituitary–adrenal axis to glucocorticoid negative-feedback and corticotropin releasing hormone in rats undergoing morphine withdrawal: Possible mechanisms involved in facilitated and attenuated stress responses. J Neuroendocrinol 13(10): 875-886.

Inoue K, Tsuda M, Koizumi S. 2004. Chronic pain and microglia: The role of ATP. Novartis Found Symp 261:55-64; discussion 64–67: 149–154.

Johansson BB, Owman Ch, Widner H. 1990. Pathophysiology of the blood–brain barrier. Fernström Foundation Series 14, Amsterdam: Elsevier; pp. 1-340.

Johanson CE, Palm DE, Primiano MJ, McMillan PN, Chan P, et al. 2000. Choroid plexus recovery after transient forebrain ischemia: Role of growth factors and other repair mechanisms. Cell Mol Neurobiol 20(2): 197-216.

Jones HC, Keep RF, Butt AM. 1992. The development of ion regulation at the blood–brain barrier. Prog Brain Res 91: 123-131.

Joó F. 1987. A unifying concept on the pathogenesis of brain oedemas. Neuropathol Appl Neurobiol 13: 161-167.

Katschinski DM, Wiedemann GJ, Longo W, d'Oleire FR, Spriggs D, et al. 1999. Whole body hyperthermia cytokine induction: A review, and unifying hypothesis for myeloprotection in the setting of cytotoxic therapy. Cytokine Growth Factor Rev 10(2): 93-97.

Kiyatkin E, Brown L, Sharma HS. 2007. Brain edema and breakdown of the blood–brain barrier during methamphetamine intoxication: Critical Role of brain hyperthermia. Eur J Neurosci 26(5): 1242-1253.

Klatzo I. 1967. Presidential address. Neuropathological aspects of brain edema. J Neuropathol Exp Neurol 26: 1-14.

Klatzo I. 1972. Pathophysiological aspects of brain edema. Steroids and Brain Edema. Reulen H, Schurmann K, editors. Berlin: Springer-Verlag; pp. 1-8.

Klatzo I. 1987. Pathophysiological aspects of brain edema. Acta Neuropathol (Berl) 72: 236-239.

Klatzo I, Piraux A, Laskowski EJ. 1958. The relationship between edema, blood–brain barrier and tissue elements in local brain injury. J Neuropathol Exp Neurol 17: 548-564.

Klatzo I, Seitelberger F. 1967. Brain Edema. Berlin: Springer-Verlag; pp. 1-350.

Klatzo I, Wisniewski H, Smith DE. 1965. Observations on penetration of serum proteins into the central nervous system. Prog Brain Res 15: 73-88.

Knigge KM, Joseph SA. 1974. Thyrotrophin releasing factor (TRF) in cerebrospinal fluid of the 3rd ventricle of rat. Acta Endocrinol (Copenh) 76(2): 209-213.

Knigge KM, Silverman AJ. 1972. Transport capacity of the median eminence. Brain Endocrine Interaction, Median Eminence: Structure and Function. Knigge KM, Scott DE, Weindl A, editors. Basel: Karger; pp. 350-363.

Kofman O, Levin U, Alpert C. 1995. Lithium attenuates hypokinesia induced by immobilization stress in rats. Prog Neuropsychopharmacol Biol Psychiatry 19(6): 1081-1090.

Kristensson K, Olsson Y. 1973. Uptake and retrograde axonal transport of protein tracers in hypoglossal neurons. Fate of the tracer and reaction of the nerve cell bodies. Acta Neuropathol (Berl) 23(1): 43-47.

Krum JM. 1994. Experimental gliopathy in the adult rat CNS: Effect on the blood–spinal cord barrier. Glia 11(4): 354-366.

Kvetnansky R, Mikulaj L. 1970. Adrenal and urinary catecholamines in rats during adaptation to repeated immobilization stress. Endocrinology 87(4): 738-743.

Lalonde R. 1986. Acquired immobility response in weaver mutant mice. Exp Neurol 94(3): 808-811.

Lam CW, James JT, McCluskey R, Arepalli S, Hunter RL. 2006. A review of carbon nanotube toxicity and assessment of potential occupational and environmental health risks. Crit Rev Toxicol 36(3): 189-217.

Leaf A. 1973. Cell swelling: A factor in ischemic tisuue injury. Circulation 48: 455-458.

Leusen I. 1972. Regulation of cerebrospinal fluid composition with reference to breathing. Physiol Rev 52(1): 1-56.

Levin VA, Fenstermacher JD, Patlak CS. 1970. Sucrose and inulin space measurements of cerebral cortex in four mammalian species. Am J Physiol 219(5): 1528-1533.

Li YQ, Ballinger JR, Nordal RA, Su Zi-Fen, Wong CS. 2001. Hypoxia in radiation-induced blood–spinal cord barrier breakdown. Cancer Res 61: 3348-3354.

Lieb WR, Stein WD. 1971. The molecular basis of simple diffusion within biological membranes. Curr Top Membr Transp 2: 1-39.

Lorenzo AV, Fernandez C, Roth LJ. 1965. Physiologically induced alterations of sulfate penetration into brain. Arch Neurol 12: 128-132.

Lossinsky AS, Shivers RR. 2004. Structural pathways for macromolecular and cellular transport across the blood–brain barrier during inflammatory conditions. Histol Histopathol 19(2): 535-564.

Lowenthal A. 1972. Chemical physiopathology of the cerebrospinal fluid. Handbook of Neurochemistry, Vol. 7. Lajtha A, editor. New York: Plenum Press; pp. 429-464.

Majno G, Palade GE. 1961. Studies on inflammation. I. The effect of histamine and serotonin on vascular permeability: An electron microscopic study. J Biophys Biochem Cytol 11: 571-605.

Maloney KJ, Mainville L, Jones BE. 1999. Differential c-Fos expression in cholinergic, monoaminergic, and GABAergic cell groups of the pontomesencephalic tegmentum after paradoxical sleep deprivation and recovery. J Neurosci 19(8): 3057-3072.

Maloney KJ, Mainville L, Jones BE. 2000. c-Fos expression in GABAergic, serotonergic, and other neurons of the pontomedullary reticular formation and raphe after paradoxical sleep deprivation and recovery. J Neurosci 20(12): 4669-4679.

Marti O, Garcia A, Velles A, Harbuz MS, Armario A. 2001. Evidence that a single exposure to aversive stimuli triggers long-lasting effects in the hypothalamus–pituitary–adrenal axis that consolidate with time. Eur J Neurosci 13(1): 129-136.

Martin R, Wallace BG, Fuch PA, Nicholls JG, editors. 2001. From Neuron to Brain: A Cellular and Molecular Approach to the Function of the Nervous System, Fourth edition (Hardcover). New York, USA: Sinauer Associates; pp. 1-617.

Martinez AJ, Alderman JL, Kagan RS, Osterholm JL. 1981. Spatial distribution of edema in the cat spinal cord after impact injury. Neurosurgery 8: 450-453.

Mautes AE, Bergeron M, Sharp FR, Panter SS, Weinzierl M, et al. 2000. Sustained induction of heme oxygenase-1 in the traumatized spinal cord. Exp Neurol 166(2): 254-265.

Mautes AE, Noble LJ. 2000. Co-induction of HSP70 and heme oxygenase-1 in macrophages and glia after spinal cord contusion in the rat. Brain Res 883(2): 233-237.

Mayhan WG, Heistad DD. 1985. Permeability of blood–brain barrier to various sized molecules. Am J Physiol 248: H712-H718.

Mendelson WB, Guthrie RD, Frederick G, Wyatt RJ. 1974. The flower pot technique of rapid eye movement (REM) sleep deprivation. Pharmacol Biochem Behav 2(4): 553-556.

Merritt JH, Chamness AF, Allen SJ. 1978. Studies on blood–brain barrier permeability after microwave-radiation. Radiat Environ Biophys 15(4): 367-377.

Miller JD. 1976. Pressure-volume response-clinical aspects. Chicago Conference on Neural Trauma. McLaurin R, editor. New York: Grunne & Stratton; pp. 35-46.

Mills NL, Amin N, Robinson SD, Anand A, Davies J, et al. 2006. Do inhaled carbon nanoparticles translocate directly into the circulation in humans? Am J Respir Crit Care Med 173(4): 426-431.

Miyazato H, Skinner RD, Garcia-Rill E. 2000. Locus coeruleus involvement in the effects of immobilization stress on the p13 midlatency auditory evoked potential in the rat. Prog Neuropsychopharmacol Biol Psychiatry 24(7): 1177-1201.

Mohanty S, Dey PK, Sharma HS, Ray AK. 1985. Experimental brain edema: Role of 5-HT. Brain Edema. Mohanty S, Dey PK, editors. India: Varanasi; Banaras Hindu University, Bhargava Bhushan Press; pp. 19-27.

Mohanty S, Dey PK, Sharma HS, Singh S, Chansouria JP, et al. 1989. Role of histamine in traumatic brain edema. An experimental study in the rat. J Neurol Sci 90: 87-97.

Moon L, Bunge MB. 2005. From animal models to humans: Strategies for promoting CNS axon regeneration and recovery of limb function after spinal cord injury. J Neurol Phys Ther 29(2): 55-69.

Nag S, Eskandarian MR, Davis J, Eubanks JH. 2002. Differential expression of vascular endothelial growth factor-A (VEGF-A) and VEGF-B after brain injury. J Neuropathol Exp Neurol 61(9): 778-788.

Nemecek S. 1978. Morphological evidence of microcirculatory disturbances in experimental spinal cord trauma. Adv Neurol 20: 395-405.

Nemecek S, Petr R, Sube P, Rozseval V, Melka O. 1977. Longitudinal extension of edema in experimental spinal cord injury: Evidence for two types of post-traumatic edema. Acta Neurochir (Wien) 37: 7-16.

Nevis AH, Collins GH. 1967. Electrical impedance and volume changes in brain during preparation for electon microscopy. Brain Res 5(1): 57-85.

Nielsen B, Hyldig T, Bidstrup F, Gonzalez-Alonso J, Christoffersen GR. 2001. Brain activity and fatigue during prolonged exercise in the heat. Pflugers Arch 442(1): 41-48.

Noble LJ, Donovan F, Igarashi T, Goussev S, Werb Z. 2002. Matrix metalloproteinases limit functional recovery after spinal cord injury by modulation of early vascular events. J Neurosci 22(17): 7526-7535.

Noble LJ, Wrathall JR. 1987. The blood–spinal cord barrier after injury: Pattern of vascular events proximal and distal to a transection in the rat. Brain Res 424(1): 177-188.

Noble LJ, Wrathall JR. 1988. Blood–spinal cord barrier disruption proximal to a spinal cord transection in the rat: Time course and pathways associated with protein leakage. Exp Neurol 99(3): 567-578.

Noble LJ, Wrathall JR. 1989. Distribution and time course of protein extravasation in the rat spinal cord after contusive injury. Brain Res 482(1): 57-66.

Nolan RT. 1969. Traumatic oedema of the spinal cord. Br Med J 1: 710

Novikoff AB. 1960. Biochemical and staining reactions of cytoplasmic constituents. Developing Cell System and Their Control. Rudnic D, editor. New York: Ronald Press; pp. 167-203.

Oberdörster E. 2004. Manufactured nanomaterials (fullerenes, C60) induce oxidative stress in brain of juvenile largemouth bass. Environ Health Perspect 112: 1058-1062.

Oberdörster G. 1996. Significance of particle parameters in the evaluation of exposure-dose-response relationships of inhaled particles. Inhal Toxicol 8 Suppl: 73-89.

Oberdörster G, Finkelstein JN, Johnston C, Gelein R, Cox C, et al. 2000. Acute pulmonary effects of ultrafine particles in rats and mice. Res Rep Health Eff Inst 96: 5-74; disc. 75–86

Oberdorster G, Maynard A, Donaldson K, Castranova V, Fitzpatrick J, et al. 2005a. ILSI Research Foundation/Risk Science Institute Nanomaterial Toxicity Screening Working Group. Principles for characterizing the potential human health effects from exposure to nanomaterials: Elements of a screening strategy. Part Fibre Toxicol 6; 2: 8.

Oberdorster G, Oberdorster E, Oberdorster J. 2005b. Nanotoxicology: An emerging discipline evolving from studies of ultrafine particles. Environ Health Perspect 113(7): 823-839.

Ohata M, Fredericks WR, Sundaram U, Rapoport SI. 1981. Effects of immobilization stress on regional cerebral blood flow in the conscious rat. J Cereb Blood Flow Metab 1(2): 187-194.

Ohata M, Takei H, Fredericks WR, Rapoport SI. 1982. Effects of immobilization stress on cerebral blood flow and cerebrovascular permeability in spontaneously hypertensive rats. J Cereb Blood Flow Metab 2(3): 373-379.

Olesen SP. 1989. An electrophysiological study of microvascular permeability and its modulation by chemical mediators. Acta Physiol Scand Suppl 579: 1-28.

Olsson Y, Sharma HS, Nyberg F, Westman J. 1995. The opioid receptor antagonist naloxone influences the pathophysiology of spinal cord injury. Progress in Brain Research Spinal Cord Monitoring: Basic Principles, Regeneration, Pathophysiology and Clinical Aspects, Vol. 104. Nyberg F, Sharma HS, Wissenfeld-Halin Z, editors. Amsterdam: Elsevier; pp. 381-399.

Olsson Y, Sharma HS, Pettersson Å, Cervós-Navarro J. 1992. Endogenous release of neurochemicals may increase vascular permeability, induce edema and influence on cell changes in trauma to the spinal cord. Circumventricular organs and brain fluid environment. Prog Brain Res 91: 197-203.

Olsson Y, Sharma HS, Pettersson CÅV. 1990. Effects of p-chlorophenylalanine on microvascular permeability changes in spinal cord trauma. An experimental study in the rat using ^{131}I-sodium and lanthanum tracers. Acta Neuropathol (Berl) 79: 595-603.

Osterholm JL. 1978. The Pathophysiology of Spinal Cord Trauma. Springfield, IL: Thomas.

Osterholm JL. 1974. The pathophysiological response to spinal cord injury. The current status of related research. J Neurosurg 40(1): 5-33.

Osterholm JL, Alderman JL, Northrup BE. 1987. Acute spinal cord injury. Spinal cord Injury Medical Engineering. Ghista DN, Frankel HL, Charles C, editors. Illinois: Charles C. Thomas Inc; pp. 5-46.

Pappius HM. 1970. The chemistry and fine structure in various types of cerebral edema. Riv Patol Nerv Ment 91: 311-322.

Parandoosh Z, Johanson CE. 1982. Ontogeny of blood–brain barrier permeability to, and cerebrospinal fluid sink action on, [14C]urea. Am J Physiol 243(3): R400-R407.

Parker AJ, Park RD. Stowater JL. 1973. Reduction of trauma induced edema of spinal cord in dogs given mannitol. Am J Vet Res 34: 1355-1357.

Porter JC. 1973. Neuroendocrine system. The need for precise identification and rigorous description of their operation. Prog Brain Res 92: 1-6.

Porsolt RD, Anton G, Blavet N, Jalfre M. 1978. Behavioural despair in rats: A new model sensitive to antidepressant treatments. Eur J Pharmacol 47(4): 379-391.

Porsolt RD, Bertin A, Blavet N, Deniel M, Jalfre M. 1979. Immobility induced by forced swimming in rats: Effects of agents which modify central catecholamine and serotonin activity. Eur J Pharmacol 57(2–3): 201-210.

Porsolt RD, Bertin A, Jalfre M. 1977. Behavioral despair in mice: A primary screening test for antidepressants. Arch Int Pharmacodyn Ther 229(2): 327-336.

Raichle ME, Hartmann BK, Eichling JO, Sharpe LG. 1975. Central noradrenergic regulation of cerebral blood flow and vascular permeability. Proc Natl Acad Sci USA 72: 3726-3730.

Rapoport SI. 1976. Blood–brain barrier in physiology and medicine. New York: Raven Press; pp. 1-272.

Rasmussen AF, March JT, Brill NQ. 1957. Increased susceptibility to herpes simplex in mice subjected to avoidance-learning stress or restrain. Proc Soc Exp Biol Med 96: 183-189.

Rawson RA. 1943. The binding of T-1824 and structurally related diazo dyes by plasma proteins. Am J Physiol 138: 708-717.

Reulen HJ. 1977a. Vasogenic brain oedema. New aspects in its formation, resolution and therapy. Br J Anaesth 48(8): 741-752.

Reulen HJ. 1977b. [Development and correction of water-electrolyte and acid base equilibrium disorders in brain edema (proceedings)] [Article in German] Klin Anasthesiol Intensivther (15): 109-122.

Reulen HJ, Baethmann A, Fenstermacher J, Marmarou A, Spatz M. 1990. Brain Edema VIII, Acta Neurochir (Wien) Suppl 51: 1-414, Wien: Springer-Verlag.

Reulen HJ, Graham R, Spatz M, Klatzo I. 1977. Role of pressure gradients and bulk flow in dynamics of vasogenic brain edema. J Neurosurg 46(1): 24-35.

Reulen HJ, Kreysch HG. 1973. Measurement of brain tissue pressure in cold induced cerebral oedema. Acta Neurochir (Wien) 29: 29-40.

Rexed BA. 1952. The cytoarchitectonic organization of the spinal cord in the cat. J Comp Neurol 96: 415-496.

Rexed BA. 1954. Cytoarchitectonic atlas of the spinal cord of the cat. J Comp Neurol 100: 297-379.

Rhodin JA. 1968. Ultrastructure of mammalian venous capillaries, venules, and small collecting veins. J Ultrastruct Res 25(5): 452-500.

Rinder L. 1968. Artefactitious extravasation of fluorescent indicators in the investigation of vascular permeability in brain and spinal cord. Acta Pathol Microbiol Scand 74(3): 333-339.

Robins HI, Kutz M, Wiedemann GJ, Katschinski DM, Paul D, et al. 1995. Cytokine induction by 41.8 degrees C whole body hyperthermia. Cancer Lett 97(2): 195-201.

Robins HI, Rushing D, Kutz M, Tutsch KD, Tiggelaar CL, et al. 1997. Phase I clinical trial of melphalan and 41.8 degrees C whole-body hyperthermia in cancer patients. J Clin Oncol 15(1): 158-164.

Rosomoff HL, Zugibe FT. 1963. Distribution of intracranial contents in experimental edema. Arch Neurol 9: 26-34.

Rymer MM, Fishman RA. 1973. Protective adaptation of brain to water intoxication Arch Neurol 28: 49-54.

Scott GS, Kean RB, Fabis MJ, Mikheeva T, Brimer CM, et al. 2004. ICAM-1 upregulation in the spinal cords of PLSJL mice with experimental allergic encephalomyelitis is dependent upon TNF-alpha production triggered by the loss of blood–brain barrier integrity. J Neuroimmunol 155(1–2): 32-42.

Selye H. 1976. Stress in Health and Disease. London: Butterworths.

Shapiro K, Shulman K, Marmarou A, Poll W. 1977. Tissue pressure gradients in spinal cord injury. Surg Neurol 7(5): 275-279.

Sharma HS. 1982. Blood–Brain Barrier in Stress, PhD Thesis. Varanasi, India: Banaras Hindu University; pp. 1-85.

Sharma HS, Westman J, Nyberg F. 1998a. Pathophysiology of brain edema and cell changes following hyperthermic brain injury. Prog Brain Res 115: 351-412 (Review).

Sharma HS. 1999. Pathophysiology of blood–brain barrier, brain edema and cell injury following hyperthermia: New role of heat shock protein, nitric oxide and carbon monoxide. An experimental study in the rat using light and electron microscopy. Acta Univ Ups 830: 1-94.

Sharma HS. 2000a. A bradykinin BK2 receptor antagonist HOE-140 attenuates blood–spinal cord barrier permeability following a focal trauma to the rat spinal cord. An experimental study using Evans blue, ^{131}I-sodium and lanthanum tracers. Acta Neurochir Suppl 76: 159-163.

Sharma HS. 2000b. Degeneration and regeneration in the CNS. New roles of heat shock proteins, nitric oxide and carbon monoxide. Amino Acids 19: 335-337.

Sharma HS. 2003. Neurotrophic factors attenuate microvascular permeability disturbances and axonal injury following trauma to the rat spinal cord. Acta Neurochir Suppl 86: 383-388.

Sharma HS. 2004a. Pathophysiology of the blood–spinal cord barrier in traumatic injury. The Blood–Spinal Cord and Brain Barriers in Health and Disease. Sharma HS, Westman J, editors. San Diego: Elsevier Academic Press; pp. 437-518.

Sharma HS. 2004b. Blood–brain and spinal cord barriers in stress. The Blood–Spinal Cord and Brain Barriers in Health and Disease. Sharma HS, Westman J, editors. San Diego: Elsevier Academic Press; pp. 231-298.

Sharma HS. 2004c. Int J Neuropotec Neuroregen 1(1): 8 (Editorial).

Sharma HS. 2004d. Influence of serotonin on the blood-brain and blood–spinal cord barriers. The Blood–Spinal Cord and Brain Barriers in Health and Disease. Sharma HS, Westman J, editors. San Diego: Elsevier Academic Press; pp. 117-158.

Sharma HS. 2004e. Histamine influences the blood–spinal cord and brain barriers following injuries to the central nervous system. The Blood–Spinal Cord and Brain Barriers in Health and Disease. Sharma HS, Westman J, editors. San Diego: Elsevier Academic Press; pp. 159-190.

Sharma HS. 2005a. Pathophysiology of blood–spinal cord barrier in traumatic injury and repair. Curr Pharm Des 11(11): 1353-1389.

Sharma HS. 2005b. Selective neuronal vulnerability, blood–brain barrier disruption and heat shock protein expression in stress induced neurodegeneration. *Invited Review*: Depression and Dementia: Progress in Brain Research, Clinical Applications and Future Trends. Sarbadhikari SN, editor. NY: Nova Science Publishers; pp. 97-152.

Sharma HS. 2005c. Heat-related deaths are largely due to brain damage. Indian J Med Res 121(5): 621-623.

Sharma HS. 2005d. Neuroprotective effects of neurotrophins and melanocortins in spinal cord injury: An experimental study in the rat using pharmacological and morphological approaches. Ann NY Acad Sci 1053: 407-421.

Sharma HS. 2005e. Methods to induce brain hyperthermia. Current Protocols in Toxicology, Suppl 23, Unit 11.14: 1-26.

Sharma HS. 2006a. Hyperthermia induced brain oedema: Current status and future perspectives. Indian J Med Res 123(5): 629-652.

Sharma HS. 2006b. Post-traumatic application of brain–derived neurotrophic factor and glia-derived neurotrophic factor on the rat spinal cord enhances neuroprotection and improves motor function. Acta Neurochir Suppl 96: 329-334.

Sharma HS. 2006c. Hyperthermia influences excitatory and inhibitory amino acid neurotransmitters in the central nervous system. An experimental study in the rat using behavioural, biochemical, pharmacological, and morphological approaches. J Neural Transm 113(4): 497-519.

Sharma HS, Sjöquist PO, Mohanty S, Wiklund L. 2006d. Post-injury treatment with a new antioxidant compound H-290/51 attenuates spinal cord trauma-induced c-fos expression, motor dysfunction, edema formation, and cell injury in the rat. Acta Neurochir Suppl 96: 322-328.

Sharma HS, Patnaik R, Patnaik S, Mohanty S, Sharma A, et al. 2007. Antibodies to serotonin attenuate closed head injury induced blood–brain barrier disruption and brain pathology. Ann N Y Acad Sci 1122: 295-312.

Sharma HS. 2007a. Neurodegeneration and neuroregeneration: Recent advancements and future perspectives. Curr Pharm Des 13(18): 1325-1327.

Sharma HS. 2007b. Neurotrophic factors in combination: A possible new therapeutic strategy to influence pathophysiology of spinal cord injury and repair mechanisms. Curr Pharm Des 13(18): 1841-1874.

Sharma HS. 2007c. Methods to produce hyperthermia-induced brain dysfunction. Prog Brain Res 162: 173-200.

Sharma HS. 2007d. Interaction between amino acid neurotransmitters and opiod receptors in hyperthermia induced brain pathology. Prog Brain Res 162: 295-320.

Sharma HS (editor). 2007e. Neurobiology of hyperthermia. Progress in Brain Research. Elsevier Amsterdam, North Hotland. Vol. 162. pp. 1–510.

Sharma HS, Ali SF. 2006. Alterations in blood–brain barrier function by morphine and methamphetamine. Ann NY Acad Sci 1074: 198-224.

Sharma HS, Ali SF. 2007. MDMA (ecstasy) induces blood–brain barrier disruption, brain edema formation and cell injury. An experimental study in rats and mice. Ann NY Acad Sci (in press).

Sharma HS, Ali SF, Schlager J, Hussain S. 2006b. Effect of nanoparticles on the blood–brain barrier. Int J Neuroprotec Neuroregen 2(3): 78.

Sharma HS, Alm P. 2004. Role of nitric oxide on the blood–brain and the spinal cord barriers. The Blood–Spinal Cord and Brain Barriers in Health and Disease. Sharma HS, Westman J, editors. San Diego: Elsevier Academic Press; pp. 191-230.

Sharma HS, Alm P, Westman. 1998d. Nitric oxide and carbon monoxide in the pathophysiology of brain functions in heat stress. Brain Functions in Hot Environment. Sharma HS, Westman J, editors. Prog Brain Res 115: 297–333.

Sharma HS, Cervós-Navarro J, Dey PK. 1991a. Increased blood–brain barrier permeability following acute short-term forced-swimming exercise in conscious normotensive young rats. Neurosci Res 10: 211-221.

Sharma HS, Cervós-Navarro J, Gosztonyi G, Dey PK. 1992b. Role of serotonin in traumatic brain injury. An experimental study in the rat. The Role of Neurotransmitters in Brain Injury. Globus M, Dietrich WD, editors. New York: Plenum Press; pp. 147-152.

Sharma HS, Dey PK. 1978. Influence of heat and immobilization stressors on the permeability of blood–brain and blood–CSF barriers. Indian J Physiol Pharmacol 22 (Suppl2): 59-60.

Sharma HS, Dey PK. 1980. Increased permeability of blood–brain barrier (BBB) in stress: Blockade by p-CPA pretreatment. Indian J Physiol Pharmacol 24 (Suppl 1): 423-424.

Sharma HS, Dey PK. 1981. Impairment of blood–brain barrier (BBB) in rat by immobilization stress: Role of serotonin (5-HT). Indian J Physiol Pharmacol 25(2): 111-122.

Sharma HS, Dey PK. 1984. Role of 5-HT on increased permeability of blood–brain barrier under heat stress. Indian J Physiol Pharmacol 28: 259-267.

Sharma HS, Dey PK. 1986a. Influence of long-term immobilization stress on regional blood–brain barrier permeability, cerebral blood flow and 5-HT level in conscious normotensive young rats. J Neurol Sci 72: 61-76.

Sharma HS, Dey PK. 1986b. Probable involvement of 5-hydroxytryptamine in increased permeability of blood–brain barrier under heat stress. Neuropharmacology 25: 161-167.

Sharma HS, Dey PK. 1987a. Increased blood–brain and blood-CSF barrier permeability following long-term immobilization stress in conscious rats. Wissenschaftliche Zeitschrift Karl-Marx Universität Leipzig, Mathematisch-Naturwissenschaftliche Reihe 36: 104-106.

Sharma HS, Dey PK. 1987b. Influence of long-term acute heat exposure on regional blood–brain barrier permeability, cerebral blood flow and 5-HT level in conscious normotensive young rats. Brain Res 424: 153-162.

Sharma HS, Dey PK. 1988. EEG changes following increased blood–brain barrier permeability under long-term immobilization stress in young rats. Neurosci Res 5(3): 224-239.

Sharma HS, Dey PK, Kumar A. 1986. Role of circulating 5-HT and lung MAO activity in physiological processes of heat adaptation in conscious young rats. Biomedicine 6: 31-40.

Sharma HS, Dey PK, Olsson Y. 1989. Brain edema, blood–brain barrier permeability and cerebral blood flow changes following intracarotid infusion of serotonin: Modification with cyproheptadine and indomethacin. Pharmacology of Cerebral Ischemia 1988. Krieglstein J, editor. Boca Raton, Florida: CRC Press; pp. 317-323.

Sharma HS, Hoopes PJ. 2003. Hyperthermia induced pathophysiology of the central nervous system. Int J Hypertherm 19: 325-354.

Sharma HS, Johanson CE. 2007. Blood–cerebrospinal fluid barrier in hyperthermia. Prog Brain Res 162: 459-480.

Sharma HS, Kretzschmar R, Cervós-Navarro J, Ermisch A, Rühle H-J, et al. 1992c. Age-related pathophysiology of the blood–brain barrier in heat stress. Prog Brain Res 91: 189-196.

Sharma HS, Lundstedt T, Boman A, Lek P, Seifert E, et al. 2006a. A potent serotonin-modulating compound AP-267 attenuates morphine withdrawal-induced blood–brain barrier dysfunction in rats. Ann NY Acad Sci 1074: 482-496.

Sharma HS, Lundstedt T, Flardh M, Westman J, Post C, et al. 2003c. Low molecular weight compounds with affinity to melanocortin receptors exert neuroprotection in spinal cord injury – an experimental study in the rat. Acta Neurochir Suppl 86: 399-405.

Sharma HS, Nyberg F, Cervós-Navarro J, Dey PK. 1992a. Histamine modulates heat stress induced changes in blood–brain barrier permeability, cerebral blood flow, brain oedema and serotonin levels: An experimental study in conscious young rats. Neuroscience 50: 445-454.

Sharma HS, Nyberg F, Gordh T, Alm P, Westman J. 1998c. Neurotrophic factors attenuate neuronal nitric oxide synthase upregulation, microvascular permeability disturbances, edema formation and cell injury in the spinal cord following trauma. Spinal Cord Monitoring. Basic Principles, Regeneration, Pathophysiology and Clinical Aspects. Stålberg E, Sharma HS, Olsson Y, editors. New York: Springer Wien; pp. 118-148.

Sharma HS, Nyberg F, Westman J, Alm P, Gordh T, et al. 1998b. Brain derived neurotrophic factor and insulin like growth factor-1 attenuate upregulation of nitric oxide synthase and cell injury following trauma to the spinal cord. Amino Acids 14: 121-130.

Sharma HS, Olsson Y. 1990. Edema formation and cellular alteration in spinal cord injury in the rat and their modification with p-chlorophenylalanine. Acta Neuropathol (Berl) 79: 604-610.

Sharma HS, Olsson Y, Cervós-Navarro J. 1993b. p-Chlorophenylalanine, a serotonin synthesis inhibitor, reduces the response of glial fibrillary acidic protein induced by trauma to the spinal cord. Acta Neuropathol (Berl) 86: 422-427.

Sharma HS, Olsson Y, Dey PK. 1990a. Blood–brain barrier permeability and cerebral blood flow following elevation of circulating serotonin level in the anaesthetized rats. Brain Res 517: 215-223.

Sharma HS, Olsson Y, Dey PK. 1990b. Early accumulation of serotonin in rat spinal cord subjected to traumatic injury. Relation to edema and blood flow changes. Neuroscience 36: 725-730.

Sharma HS, Olsson Y, Dey PK. 1995b. Serotonin as a mediator of increased microvascular permeability of the brain and spinal cord. Experimental observations in anaesthetised rats and mice. New Concepts of a Blood–Brain Barrier. Greenwood J, Begley D, Segal M, Lightman S, editors. New York: Plenum Press; pp. 75-80.

Sharma HS, Olsson Y, Nyberg F. 1995d. Influence of dynorphin-A antibodies on the formation of edema and cell changes in spinal cord trauma. Progress in Brain Research, Vol. 104. Nyberg F, Sharma HS, Wissenfeld-Halin Z, editors. Amsterdam: Elsevier; pp. 401-416.

Sharma HS, Olsson Y, Nyberg F, Dey PK. 1993a. Prostaglandins modulate alterations of microvascular permeability, blood flow, edema and serotonin levels following spinal cord injury. An experimental study in the rat. Neuroscience 57: 443-449.

Sharma HS, Olsson Y, Pearsson S, Nyberg F. 1995a. Trauma induced opening of the blood–spinal cord barrier is reduced by indomethacin, an inhibitor of prostaglandin synthesis. Experimental observations in the rat using ^{131}I-sodium, Evans blue and lanthanum as tracers. Restor Neurol Neurosci 7: 207-215.

Sharma HS, Patnaik R, Ray AK, Dey PK. 2004. Blood–central nervous system barriers in morphine dependence and withdrawal. The Blood-Spinal Cord and Brain Barriers in Health and Disease. Sharma HS, Westman J, editors. San Diego: Elsevier Academic Press; pp. 299-328.

Sharma HS, Sharma A. 2007. Nanoparticles aggravate heat stress induced cognitive deficits, blood–brain barrier disruption, edema formation and brain pathology. Prog Brain Res 162: 245-276.

Sharma HS, Sjöquist PO. 2002. A new antioxidant compound H-290/51 modulates glutamate and GABA immunoreactivity in the rat spinal cord following trauma. Amino Acids 23: 261-272.

Sharma HS, Sjöquist PO, Ali SF. 2007. Drugs of abuse-induced hyperthermia, blood–brain barrier dysfunction and neurotoxicity: Neuroprotective effects of a new antioxidant compound H-290/51. Curr Pharm Des 13(18): 1903-1923.

Sharma HS, Sjoquist PO, Alm P. 2003a. A new antioxidant compound H-290151 attenuates spinal cord injury induced expression of constitutive and inducible isoforms of nitric oxide synthase and edema formation in the rat. Acta Neurochir Suppl 86: 415-420.

Sharma HS, Sjöquist PO, Westman J. 2001. Pathophysiology of the blood–spinal cord barrier in spinal cord injury. Influence of a new antioxidant compound H-290/51. Blood–Brain Barrier: Drug Delivery and Brain Pathology. Kobiler D, Lustig S, Shapra S, editors. New York: Kluwer Academic/Plenum Publishers; pp. 401-416.

Sharma HS, Westman J. 1998. Brain function in hot environment. Prog Brain Res 115: 1-617.

Sharma HS, Westman J. 2000. Pathophysiology of hyperthermic brain injury. Current concepts, molecular mechanisms and pharmacological strategies. Research in Legal Medicine. Hyperthermia, Burning and Carbon Monoxide, Vol. 21. Oehmichen M, editor. Lübeck, Germany: Lübeck Medical University Publications, Schmidt-Römhild Verlag; pp. 79-120.

Sharma HS, Westman J. 2004. The Blood–Spinal Cord and Brain Barriers in Health and Disease. San Diego: Academic Press; pp. 1-617.

Sharma HS, Westman J, Cervós-Navarro J, Dey PK, Nyberg F. 1995c. Probable involvement of serotonin in the increased permeability of the blood–brain barrier by forced swimming. An experimental study using Evans blue and ^{131}I-sodium tracers in the rat. Behav Brain Res 72: 189-196.

Sharma HS, Westman J, Cervós-Navarro J, Nyberg F. 1996b. A 5-HT$_2$ receptor mediated breakdown of the blood–brain barrier permeability and brain pathology in heat stress. An experimental study using cyproheptadine and ketanserin in young rats. Biology and Physiology of the Blood–Brain Barrier. Couraud P, Scherman A, editors. New York: Plenum Press; pp. 117-124.

Sharma HS, Westman J, Gordh T, Alm P. 2000b. Topical application of brain derived neurotrophic factor influences upregulation of constitutive isoform of heme oxygenase in the spinal cord following trauma. An experimental study using immunohistochemistry in the rat. Acta Neurochir Suppl 76: 365-369.

Sharma HS, Westman J, Nyberg F. 1997a. Topical application of 5-HT antibodies reduces edema and cell changes following trauma to the rat spinal cord. Acta Neurochir Suppl (Wien) 70: 155-158.

Sharma HS, Westman J, Nyberg F. 1998a. Pathophysiology of brain edema and cell changes following hyperthermic brain injury. Brain functions in hot environment. Sharma HS, Westman J, editors. Progress in Brain Research. 115: 351-412.

Sharma HS, Westman J, Olsson Y, Alm P. 1996a. Involvement of nitric oxide in acute spinal cord injury: An immunohistochemical study using light and electron microscopy in the rat. Neurosci Res 24: 373-384.

Sharma HS, Wiklund L, Badgaiyan RD, Mohanty S, Alm P. 2006c. Intracerebral administration of neuronal nitric oxide synthase antiserum attenuates traumatic brain injury-induced blood–brain barrier permeability, brain edema formation, and sensory motor disturbances in the rat. Acta Neurochir Suppl 96: 288-294.

Sharma HS, Winkler T, Stalberg E, Gordh T, Alm P, et al. 2003b.Topical application of TNF-alpha antiserum attenuates spinal cord trauma induced edema formation, microvascular permeability disturbances and cell injury in the rat. Acta Neurochir Suppl 86: 407-413.

Sharma HS, Winkler T, Stålberg E, Mohanty S, Westman J. 2000a. p-Chlorophenylalanine, an inhibitor of serotonin synthesis reduces blood–brain barrier permeability, cerebral blood flow, edema formation and cell injury following trauma to the rat brain. Acta Neurochir Suppl 76: 91-95.

Sharma HS, Winkler T, Stålberg E, Olsson Y, Dey PK. 1991b. Evaluation of traumatic spinal cord edema using evoked potentials recorded from the spinal epidural space. An experimental study in the rat. J Neurol Sci 102: 150-162.

Sharma HS, Zimmer C, Westman J, Cervós-Navarro J. 1992d. Acute systemic heat stress increases glial fibrillary acidic protein immunoreactivity in brain. An experimental study in the conscious normotensive young rats. Neuroscience 48: 889-901.

Sheridan JF, Dobbs C, Brown D, Swilling B. 1994. Psychoneuroimmunology: Stress effects on pathogenesis and immunity during infection. Clin Microbiol Rev 7: 202-212.

Sherman DG, Hart RG, Easton JD. 1981. Abrupt change in head position and cerebral infarction. Stroke 12(1): 2-6.

Shu C, Shen H, Keep RF, Smith DE. 2002. Role of PEPT2 in peptide/mimetic trafficking at the blood-CSF barrier: Studies in rat choroid plexus epithelial cells in primary culture. J Pharmacol Exp Ther 301: 820-829.

Shvedova AA, Kisin ER, Mercer R, Murray AR, Johnson VJ, et al. 2005. Unusual inflammatory and fibrogenic pulmonary responses to single-walled carbon nanotubes in mice. Am J Physiol Lung Cell Mol Physiol 289(5): L698-L708.

Simonsen AH, Sheykhzade M, Berg Nyborg NC. 2004. Age- and endothelium-dependent changes in coronary artery reactivity to serotonin and calcium. Vascul Pharmacol 41(2): 43-49.

Sminia P, Zee van der J, Wondergem J, Haveman J. 1994. Effect of hyperthermia on the central nervous system: A review. Int J Hyperthermia 10(1): 1-30.

Smith QR, Rapoport SI. 1986. Cerebrovascular permeability coefficients to sodium, potassium, and chloride. J Neurochem 46: 1732-1742.

Sofroniew MV. 2005. Reactive astrocytes in neural repair and protection. Neuroscientist 11(5): 400-407.

Spector R, Johanson C. 1989. The mammalian choroid plexus. Sci Am 261(5): 68-74.

Starling EH. 1896. On the absorption of fluids from the connective tissue spaces. J Physiol 19: 312-326.

Staub NC. 1974. Pulmonary edema. Physiol Rev 54: 678-811.

Stern JR, Eggleston LV, Hems R, Krebs HA. 1949. Accumulation of glutamic acid in isolated brain tissue. Biochem J 44: 410-418.

Stern WE. 1959. Studies in experimental brain swelling and brain compression. J Neurosurg 16: 676-704.

Stokes BT, Somerson SK. 1987. Spinal cord extracellular microenvironment: Can the changes resulting from trauma be graded? Neurochem Pathol 7: 47-55.

Strazielle N, Ghersi-Egea JF. 1999. Demonstration of a coupled metabolism-efflux process at the choroid plexus as a mechanism of protection toward xenobiotics. J Neurosci 19: 6275-6289.

Sutton CH, Carroll FB. 1979. Effects of microwave-induced hyperthermia on the blood–brain barrier of the rat. Radio Sci 14: 329-334.

Sypert GW. 1990a. Thoracolumbar fusion techniques. Clin Neurosurg 36: 186-216.

Sypert GW. 1990b. Stabilization and management of cervical injuries. Pitt LH, Wagner FC, editors. Craniospinal Trauma. Thieme; New York: pp. 363-370.

Taub E, Uswatte G. 2005. Use of CI therapy for improving motor ability after chronic CNS damage: A development prefigured by paul Bach-y-Rita. J Integr Neurosci 4(4): 465-477.

Toole JF, Tucker SH. 1960. Influence of head position upon cerebral circulation. Arch Neurol 2: 42-49.

Tramo MJ, Hainline B, Petito F, Lee B, Caronna J. 1985. Vertebral artery injury and cerebellar stroke while swimming: Case report. Stroke 16(6): 1039-1042.

Tsuda M, Mizokoshi A, Shigemoto-Mogami Y, Koizumi S, Inoue K. 2004. Activation of p38 mitogen-activated protein kinase in spinal hyperactive microglia contributes to pain hypersensitivity following peripheral nerve injury. Glia 45(1): 89-95.

van Harreveld A. 1966. Extracellular space in the central nervous system. Proc K Ned Akad Wet C 69(1): 17-21.

van Harreveld A, Dubrovsky BO. 1967. Water and electrolytes in hydrated gray and white matter. Brain Res 4: 81-86.

Vannemreddy P, Ray AK, Patnaik R, Patnaik S, Mohanty S, et al. 2006. Zinc protoporphyrin IX attenuates closed head injury-induced edema formation, blood–brain barrier disruption, and serotonin levels in the rat. Acta Neurochir Suppl 96: 151-156.

Vigh-Teichmann I, Vigh B. 1970. Structure and function of the liquor contacting neurosecretory system. Aspects of Neuroendocrinology. Bargmann W, Scharrer B, editors. Berlin: Springer-Verlag; pp. 329-337.

Wagner FC, Green BA, Bucy PC. 1971. Spinal cord edema associated with paraplegia. Proc Veterans Adm Spinal Cord Inj Conf 18: 9-10.

Wagner FC Jr, Stewart WB. 1981. Effect of trauma dose on spinal cord edema. J Neurosurg 54: 802-806.

Wahl M, Unterberg A, Baethmann A, Schilling L. 1988. Mediators of blood–brain barrier dysfunction and formation of vasogenic brain edema. J Cereb Blood Flow Metab 8: 621-634.

Whetstone WD, Hsu JY, Eisenberg M, Werb Z, Noble-Haeusslein LJ. 2003. Blood–spinal cord barrier after spinal cord injury: Relation to revascularization and wound healing. J Neurosci Res 74(2): 227-239.

Warheit DB, Laurence BR, Reed KL, Roach DH, Reynolds GA, Webb TR. et al. 2004. Comparative pulmonary toxicity assessment of single-wall carbon nanotubes in rats. Toxicol Sci 77(1): 117-125.

Wiederhielm CA. 1968. Dynamics of transcapillary fluid exchange. J Gen Physiol 52: 29S-63S.

Williams WM, Hoss W, Formaniak M, Michaelson SM. 1984. Effect of 2450 MHz microwave energy on the blood–brain barrier to hydrophilic molecules. A. Effect on the permeability to sodium fluorescein. Brain Res 319(2): 165-170.

Windle WF. 1980. The spinal cord and its reaction to traumatic injury. Anatomy-physiology-pharmacology-therapeutics. Modern Pharmacology-Toxicology, Vol. 18., New York, Basel: Marcel Dekker.

Winkelstein BA, De Leo JA. 2002. Nerve root injury severity differentially modulates spinal glial activation in a rat lumbar radiculopathy model: Considerations for persistent pain. Brain Res 956(2): 294-301.

Winkler T, Sharma HS, Stålberg E, Olsson Y, Dey PK. 1995. Impairment of blood–brain barrier function by serotonin induces desynchronisation of spontaneous cerebral cortical activity. Experimental observations in the anaesthetised rat. Neuroscience 68: 1097-1104.

Winkler T, Sharma HS, Stålberg E, Westman. 1998. Spinal cord bioelectrical activity, edema and cell injury following a focal trauma to the spinal cord. An experimental study using pharmacological and morphological approach. Spinal Cord Monitoring. Basic Principles, Regeneration, Pathophysiology and Clinical Aspects. Stålberg E, Sharma HS, Olsson Y, editors. New York: Springer Wien; pp. 283-363.

Woodbury DM. 1974. Maturation of the blood–brain barrier and blood–CSF barrier. Drugs and Developing Brain. Vernadakis A, Weiner N, editors. New York: Plenum Press; pp. 259-280.

Woolf CJ. 1983. Evidence for a central component of post-injury pain hypersensitivity. Nature 306: 686-688.

Woolf CJ, Shortland P, Coggeshall RE. 1992. Peripheral nerve injury triggers central sprouting of myelinated afferents. Nature 355: 75-78.

Xia T, Kovochich M, Brant J, Hotze M, Sempf J, et al. 2006. Comparison of the abilities of ambient and manufactured nanoparticles to induce cellular toxicity according to an oxidative stress paradigm. Nano Lett 6(8): 1794-1807.

Yashon D, Bingham WG Jr, Faddoul EM, Hunt WE. 1973. Edema of the spinal cord following experimental impact trauma. J Neurosurg 38(6): 693-697.

Yeo JD. 1976. A review of experimental research in spinal cord injury. Paraplegia 14(1): 1-11.

Zai LJ, Wrathall JR. 2005. Cell proliferation and replacement following contusive spinal cord injury. Glia 50(3): 247-257.

Zebrowska-Lupina I, Stelmasiak M, Porowska A. 1990. Stress-induced depression of basal motility: Effects of antidepressant drugs. Pol J Pharmacol Pharm 42(2): 97-104.

Zenker D, Begley D, Bratzke H, Rubsamen-Waigmann H, Briesen von H. 2003. Human blood-derived macrophages enhance barrier function of cultured primary bovine and human brain capillary endothelial cells. J Physiol 551(Pt 3): 1023-1032.

Zhu S, Oberdorster E, Haasch ML. 2006. Toxicity of an engineered nanoparticle (fullerene, C60) in two aquatic species, Daphnia and fathead minnow. Mar Environ Res 62 (Suppl): S5-S9. Epub 2006 Apr 22.

18 Engineered Antibody Fragments as Potential Therapeutics against Misfolded Proteins in Neurodegenerative Diseases

E. Kvam · A. Messer

1	Introduction	460
2	Single-chain Fv antibodies	460
2.1	Utility and Structure	460
2.2	Methods for Generating and Improving scFvs as Therapies	462
3	scFvs that Inhibit the Aggregation of Misfolded Proteins in Neurodegenerative Disease Models	463
3.1	Amyloidogenic Diseases	463
3.2	Huntington's Disease	464
3.3	Parkinson's Disease	464
3.4	Alzheimer's Disease	465
3.5	Prions	465
3.6	Other Approaches	466
3.6.1	ER Intrabodies	466
3.6.2	scFvs that Influence Posttranslational Modifications	466
4	Applications for Conformation-Specific scFvs in Neurodegenerative Diseases	466
5	Perspective	467

© 2009 Springer Science+Business Media, LLC.

List of Abbreviations: AAV, adeno-associated virus; AD, Alzheimer's disease; APP, amyloid precursor protein; CDR, complementarity-determining region; ER, endoplasmic reticulum; Fab, antigen-binding fragment; FACS, fluorescence-activated cell sorting; Fc, constant (crystallizable) fragment; Fv, variable fragment; HD, Huntington's disease; PD, Parkinson's disease; PrP, prion protein; PrPSc, scrapie prion protein; scFv, single-chain variable fragment; VH, variable heavy chain; VL, variable light chain

1 Introduction

Immunoglobulins or antibodies are powerful tools for the research, diagnosis, and treatment of human diseases due to their unique ability to identify and neutralize specific targets with high affinity. Of late, monoclonal antibodies derived from the immune system have become the gold standard for targeted drug therapies because they afford a potent combination of selective specificity and minimal immunogenicity. Consequently, antibody-based therapeutics exist for a range of human disorders, including rheumatoid arthritis and cancer, and engineered antibodies (designed and produced using recombinant technologies) represent more than 30% of biopharmaceuticals currently in clinical trials.

Targeted antibody-based therapeutics are now being investigated for the treatment of diverse "protein misfolding" disorders of the human nervous system. These disorders share a common pathogenetic mechanism in which the aggregation of specific mutant proteins is linked to the neurodegeneration of specific regions of the brain. Engineered antibody fragments generated against neurodegenerative proteins have shown promise as potential therapeutics by altering the folding, localization, and binding dynamics of their neurotoxic protein targets. Here we review the therapeutic potential of two general types of single-chain variable fragment antibodies (scFvs) tested in neurodegenerative disease models, namely (1) anti-aggregation scFvs that block the formation and deposition of misfolded protein aggregates, and (2) conformation-specific scFvs that bind to or inhibit the production of specific disease-associated protein isoforms.

2 Single-chain Fv antibodies

2.1 Utility and Structure

Antibodies that are chief in immunity against invading pathogens consist of four polypeptide chains; two identical heavy chains (H) and two identical light chains (L) that are stabilized by intrachain covalent bonds (see ❷ *Figure 18-1a*). Each heavy and light chain is further comprised of a constant domain (C), which is identical among a specific immunoglobulin family, and a variable domain (V), which is unique to each antibody (❷ *Figure 18-1a*). Both heavy and light chain variable domains (VH and VL) confer specificity to an antibody by forming antigen-binding pockets called complementarity determining regions (CDRs) that form unique surfaces for recognizing and interacting with specific targets or antigens. Thus, the mutation of a single residue within a CDR amino acid sequence may drastically alter the efficacy of an antibody.

Recent advances in antibody engineering have fostered the development of recombinant antibody fragments that retain the binding specificity of much larger immunoglobulins within the framework of a single polypeptide. These small recombinant antibodies are derived from the heavy and light chain variable domains (VH and VL) of the antigen-binding site (Fab) in an immunoglobulin (see ❷ *Figure 18-1a*). The smallest antibody fragment capable of binding an antigen with modest affinity is a single VH or VL chain (❷ *Figure 18-1a*), but these fragments are typically unstable and prone to aggregation (Tanaka et al., 2003). Two exceptions are (1) highly stable camelid VH chains derived from naturally occurring camel immunoglobulins lacking VL chains (Hamers-Casterman et al., 1993), and (2) engineered VH or VL domains in which protein stability has been improved by targeted amino acid replacement (Davies and Riechmann, 1996). In general, Fv antibody fragments composed of both VH and VL domains afford greater stability and antigen recognition. A common method for assembling stable Fv fragments is to fuse a short, flexible

◼ **Figure 18-1**
(a) A monoclonal antibody (mAb), engineered single-chain variable fragment (scFv), and single-domain variable fragment (dAb). The variable (V) and constant (C) domains of the light (L, darker shading) and heavy (H, lighter shading) chains are indicated. dAbs may be either VH or VL single domains. Fab, antigen-binding fragment; Fc, constant or crystallizable fragment. (b) Domain structure of three different types of engineered scFvs. Antigen-binding VH and VL domains are connected by a flexible peptide linker (link). For secreted scFvs and ER intrabodies but not cytoplasmic scFvs, a leader sequence (LDR) or signal peptide designates synthesis in the ER. ER intrabodies also contain an ER-retention signal (KDEL)

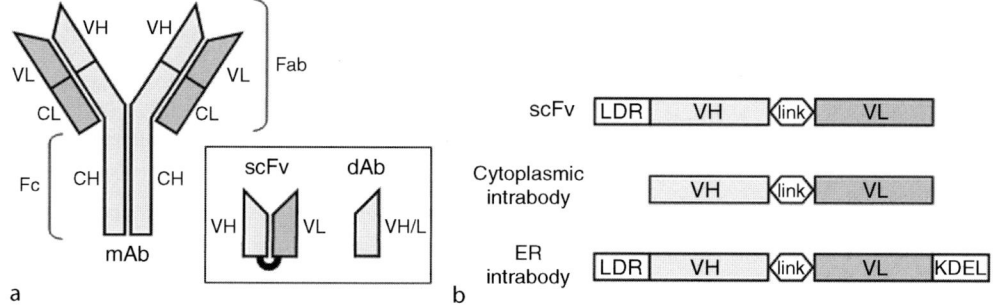

peptide linker between two variable domains via recombinant DNA techniques (❯ Figure 18-1). This linker serves to covalently bond VH and VL domains together while allowing them to fold independently. The end result is a scFv antibody fragment that comprises the antigen-binding site of a full immunoglobulin within the framework of a single nucleic acid coding sequence (Bird et al., 1988). Importantly, scFvs can be cloned and selected as single genes, which simplifies genetic manipulation.

Recombinant genes encoding scFvs can be expressed in many cell types and targeted to a variety of subcellular compartments. Typically, scFvs derived from VH and VL genes contain leader sequences or signal peptides (❯ Figure 18-1b) that direct the synthesis of nascent scFvs within the lumen of the endoplasmic reticulum (ER). The redox environment of the ER initiates the formation of covalent disulfide bonds that are critical to scFv stability. Matured scFvs exit the ER by vesicular trafficking and are usually secreted from the cell into the extracellular milieu. However, by removing the leader sequences of VH or VL genes, scFvs can be assembled and retained inside the cell cytoplasm (❯ Figure 18-1b). Cytoplasmic scFvs and single-domain fragments generated in this manner are called "intrabodies." Generally, intrabodies are prone to misfolding as a result of the high reducing potential of the cytoplasm, which impairs disulfide bond formation. High-throughput screening techniques have greatly improved intrabody stability by identifying naturally occurring VH or VL domains that fold correctly due to compensatory electrostatic and hydrophobic interactions (Visintin et al., 2002). These rigid scaffolds, termed intrabody consensus sequences, support a variety of antigen-binding CDRs, and have aided in the development of highly stable intrabodies against a variety of intracellular targets. Thus, scFvs can be easily modified to recognize antigens either inside or outside the cell.

Intracellular intrabodies and secreted scFvs are powerful tools for altering the biophysical dynamics of intracellular and extracellular protein antigens. The interaction of an scFv with a target protein may elicit any of several possible effects, including altering (1) protein folding and/or molecular interactions, (2) posttranslational modifications, (3) proteolysis, or (4) subcellular localization. As a result, scFvs are currently being investigated as therapeutics for a variety of human diseases, including neurodegenerative disorders. The genetic manipulability of scFvs also provides a platform to investigate the role of specific protein conformations and isoforms in disease pathogenesis. ❯ Table 18-1 summarizes how scFvs have been applied to target and neutralize pathogenic proteins associated with human neurodegenerative diseases, which will be discussed in greater detail later in this chapter.

◘ Table 18-1
Current literature on scFvs and single-domain antibody fragments specific to human neurodegenerative disease proteins

Neurodegenerative disorder	Protein target	Therapeutic potential	Publications
Huntington's disease	Huntingtin	Prevents pathogenic misfolding	Lecerf et al. (2001) Khoshnan et al. (2002) Colby et al. (2004b) Miller et al. (2005) Wolfgang et al. (2005)
Parkinson's disease (synucleinopathy)	α-synuclein	Prevents pathogenic misfolding; recognizes toxic conformations	Emadi et al. (2004) Zhou et al. (2004) Barkhordarian et al. (2006) Emadi et al. (2007)
Alzheimer's disease	β-amyloid	Prevents deposition; alters proteolysis; recognizes toxic conformations	Liu et al. (2004a) Liu et al. (2004b) Paganetti et al. (2005) Fukuchi et al. (2006a) Fukuchi et al. (2006b) Levites et al. (2006) Zameer et al. (2006)
Prion diseases	Prion (PrP)	Prevents pathogenic misfolding; alters trafficking; recognizes toxic conformations	Cardinale et al. (2005) Donofrio et al. (2005) Vetrugno et al. (2005) Filesi et al. (2007) Miyamoto et al. (2007)

2.2 Methods for Generating and Improving scFvs as Therapies

The simplest method for generating a scFv of known specificity is to clone the variable domains of an existing monoclonal antibody by PCR using mRNA from an antibody-producing hybridoma cell line (Orlandi et al., 1989). This approach is particularly applicable to neurodegenerative diseases because a variety of monoclonal antibodies have been cataloged against disease-associated proteins. Once cloned, VH and VL recombinant cDNAs are assembled around a spacer encoding a flexible peptide linker, typically a (Gly$_4$Ser)$_3$ decapentapeptide (Huston et al., 1993). The orientation of the variable domains plays a significant role in the expression and stability of an scFv, as single-chain VH–linker–VL polypeptides generally show greater functionality than VL–linker–VH polypeptides (Tsumoto et al., 1994; Merk et al., 1999). However, modifying a linker sequence such that accommodates specific structural or folding constraints between the VH and VL domains can improve stability and affinity (Robinson and Sauer, 1998). Alternatively, site-specific or random mutagenesis may be employed to "affinity mature" scFvs by introducing amino acid changes within the VH and VL domains that improve folding and functionality (Chowdhury et al., 1998).

A second method for generating scFvs, particularly those with novel epitopes for which a monoclonal antibody does not exist, is to screen a repertoire of recombinant VH and VL genes by in vitro selection. This high-throughput technology, which mimics the natural process of antibody selection in the immune system, relies on the construction of diverse scFv libraries from the mRNA of "naïve" or nonimmunized antibody-producing cells (e.g., peripheral blood lymphocytes, bone marrow, or spleen cells) (Marks et al., 1991). Alternatively, semisynthetic libraries can be constructed by inserting randomized sequences into the antigen-binding regions (CDRs) of one or more scFv frameworks (Barbas et al., 1992; Knappik et al., 2000). Naïve or semisynthetic scFv libraries are then screened against a desired antigen using a variety of in vitro display methods that select for specific binding. During this process, antibody fragments encoded by combinatorial libraries are individually displayed on the surface of biological particles and selected either by adsorption onto an immobilized antigen (through a process known as "panning") or by binding to a fluorescent-labeled soluble antigen, which can be detected by fluorescence-activated cell sorting (FACS) (reviewed in Maynard and Georgiou, 2000). High-affinity antibody fragments are then selected and enriched by successive rounds of panning or FACS analysis. Thus far, recombinant antibody libraries displayed on the surfaces of ribosomes, bacteriophages, or single cell organisms such as bacteria and yeast have shown great success in identifying antibody fragments with tailor-made binding to a specific antigen (Boder and Wittrup, 1997; Daugherty et al., 1998; Hoogenboom et al., 1998; He and Taussig, 2005). However, scFvs derived from high-throughput screens may be subject to the same expression and folding challenges as scFvs derived from monoclonal antibodies, and therefore might require further engineering for biological applications.

Protein folding is a critical issue for scFvs that are to be expressed and retained within cells as intrabodies. Many scFvs are intrinsically unstable when expressed within the complex reducing environment of the cytoplasm, largely due to the redox inhibition of disulfide bond formation. A simple way to select for stable intrabodies in a high-throughput manner is by intracellular antibody capture, which uses an in vivo two-hybrid screen to select for antigen–scFv interactions within the cytoplasm of yeast (Visintin et al., 1999; der Maur et al., 2002). Alternatively, the antigen-binding CDRs of an unstable scFv can be recombinantly grafted into several well-characterized intrabody consensus scaffolds that exhibit improved intracellular folding and stability (Ewert et al., 2004). Unstable intrabodies can also be engineered for improved folding using random mutagenesis or amino acid replacement of cysteine residues in order to generate electrostatic and hydrophobic interactions that compensate for a loss of disulfide bonds (Colby et al., 2004a). To date, intrabodies specific to huntingtin, α-synuclein, and other intracellular human neurodegenerative disease proteins have been developed using some of these antibody-engineering techniques.

3 scFvs that Inhibit the Aggregation of Misfolded Proteins in Neurodegenerative Disease Models

3.1 Amyloidogenic Diseases

More than 20 human diseases are linked to the aggregation of proteins into highly ordered and insoluble amyloid fibrils, including neurodegenerative disorders like Alzheimer's disease (AD), Parkinson's disease (PD), Huntington's disease (HD), and prion encephalopathies. Despite a lack of sequence homology, all amyloid fibrils share a common structure called a "cross-β" pattern, which is formed through the polymerization of β-strand-rich polypeptides into rigid, zipper-like perpendicular arrays (reviewed in Fandrich, 2007). The end result is a highly structured protein polymer that can self-associate with other amyloid fibrils to form even larger protein aggregates. The kinetics of protein fibrillogenesis is thought to involve soluble aggregate intermediates called oligomers, protofibrils, or filaments that nucleate or "seed" the polymerization of higher-order amyloid fibrils (reviewed in Bellotti et al., 2007). The oligomeric and fibrillar species of several pathogenic proteins are directly implicated in the etiology of many neurodegenerative diseases, although the mechanism by which these protein aggregates mediate toxicity remains fiercely debated. On one hand, studies have shown that amyloid fibrillogenesis can be neuroprotective, perhaps due to the sequestration of toxic protein oligomers (see Ferreira et al., 2007). However, numerous studies have demonstrated that blocking fibrillogenesis through the use of small molecules is also neuroprotective (for example, see Desai et al., 2006). Given these conflicting models, scFvs are promising tools for research and treatment of neurodegenerative diseases because they can be selected against a full range of protein epitopes and structural conformations that are associated with pathogenesis.

3.2 Huntington's Disease

HD is an autosomal dominant trinucleotide repeat disease that is caused by an expansion in the polyglutamine tract of the huntingtin protein. Expansions of greater than 37 consecutive glutamines confer a toxic gain-of-function mutation in huntingtin that promotes its intracellular aggregation and results in the neurodegeneration of GABA-ergic neurons in the striatum (reviewed in Li and Li, 2006). Currently, there is no effective treatment for HD, and existing therapies target only the symptoms of the disease. Based on the hypothesis that intrabody technology in clinical trials for cancer can be applied to neurodegenerative diseases, a collaboration between the Messer and Huston labs used a large naïve human spleen scFv phage display library to generate and test the effects of anti-huntingtin intrabodies in a culture model of HD. Intrabodies selected against the mutant polyglutamine tract showed both a lack of specificity for huntingtin and generalized toxicity. However, an intrabody selected against the N-terminal 17 amino acid residues that precede the polyglutamine tract of huntingtin was successful in counteracting the aggregation of an N-terminal huntingtin fragment in several different cell lines, including striatal progenitor cells isolated from rat striatum (Lecerf et al., 2001; Miller et al., 2005). This intrabody, denoted C4, also inhibited aggregation in an organotypic slice culture model of HD and provided functional protection against mutant huntingtin-specific malonate toxicity (Murphy and Messer, 2004). When expressed in the nervous system of a *Drosophila* HD model, the C4 intrabody completely rescued an eclosion defect specific to mutant huntingtin larvae and partially rescued adult life span while significantly delaying neuronal degeneration (Wolfgang et al., 2005). Further studies revealed that C4 binds preferentially to the monomeric form of an N-terminal huntingtin fragment and inhibits its aggregation while eliciting a faster turnover rate (Miller et al., 2005). Ongoing studies with C4 continue to show promise with little apparent toxicity.

The therapeutic potential of anti-huntingtin intrabodies has been confirmed by independent studies. ScFvs generated from monoclonal antibodies against the polyproline region of huntingtin, which flanks the polyglutamine tract, also reduced aggregation and apoptosis when expressed as intrabodies in cells, while those against the expanded polyglutamine tract are toxic (Khoshnan et al., 2002). Additional anti-huntingtin intrabodies have been isolated and affinity-matured using yeast surface display techniques. For example, a single variable light chain intrabody (VL), derived from a nonfunctional scFv to the N-terminal 20 residues of huntingtin, significantly inhibited aggregation and increased cell survival in a culture model of HD (Colby et al., 2004b). Reengineering this single VL domain by targeted replacement of cysteine residues has allowed for the creation of a disulfide bond-free intrabody optimized for intracellular folding. Consequently, this intrabody shows greater efficacy for blocking aggregation in situ (Colby et al., 2004a). These and other studies demonstrate small molecules like intrabodies that inhibit the aggregation of mutant huntingtin offer significant therapeutic potential.

3.3 Parkinson's Disease

PD and related neurodegenerative synucleinopathies are characterized by the presence of Lewy bodies, which are intracellular protein inclusions composed primarily of α-synuclein. α-Synuclein duplication or mutation events are both causative to PD and Lewy body formation, and lead to the neurodegeneration of dopaminergic neurons in the substantia nigra (reviewed in Takeda et al., 2006). Although surgical and pharmacological treatments exist for PD, none address the underlying process of neurodegeneration, which eventually renders these treatments ineffective. Thus, small molecules that block the intracellular aggregation and toxicity of α-synuclein may possess greater therapeutic potential. The Sierks lab first showed that an scFv selected against monomeric α-synuclein from a naïve phage display library inhibited the aggregation of α-synuclein in vitro. This scFv interacted with the C-terminus of α-synuclein with nanomolar affinity and prevented the formation of toxic oligomeric and protofibrillar species in vitro by stabilizing monomeric α-synuclein (Emadi et al., 2004). A related study by the Messer lab demonstrated that a human scFv raised against monomeric α-synuclein and expressed intracellularly as an intrabody was effective in blocking aggregation in situ. This latter intrabody, named D10, also rescued a cell adhesion defect that is specific to α-synuclein overexpression in cultured cells, thereby indicating functional correction of a cellular defect (Zhou et al., 2004). Additional human scFvs have been isolated against monomeric and dopamine-adduct forms of synuclein but await characterization (Maguire-Zeiss et al., 2006).

3.4 Alzheimer's Disease

AD is the most prominent neurodegenerative disorder associated with aging. It is characterized by the extracellular deposition of β-amyloid into "senile" protein plaques as well as the intracellular aggregation of tau protein into neurofibrillary tangles. Senile plaques and neurofibrillary tangles are arguably causative to the neurodegeneration of cerebral cortex and subcortical regions during the progression of AD, though other models exist (Castellani et al., 2007). Therefore, one strategy for combating AD has focused on developing molecules that block the misfolding and aggregation of β-amyloid and tau. Anti-tau intrabodies have been selected in yeast by intracellular capture technology but their efficacy in AD models is unknown (Visintin et al., 2002). On the other hand, human scFvs selected against β-amyloid epitopes by phage display are largely effective in blocking aggregation in vitro, and reduce the toxic effects of β-amyloid deposition in neuronal cell lines (Liu et al., 2004a; Zameer et al., 2006). These results support prior studies demonstrating that full-length antibodies specific to β-amyloid can disaggregate amyloid fibrils and alter toxicity (Solomon et al., 1997).

In vivo therapies using monoclonal antibodies against β-amyloid are effective in reducing plaques and slowing cognitive defects in animal models of AD but can elicit adverse inflammatory side effects through the complement response (Schenk, 2002). This reaction is known to be mediated by the constant region (Fc) of an antibody. Recombinant scFvs are attractive therapeutic alternatives because they lack the Fc effector domains of natural antibodies (see ❍ *Figure 18-1a*). Recently, the safety and efficacy of scFvs that block the aggregation of β-amyloid have been evaluated in transgenic mouse models of AD using gene therapy viruses approved for human clinical use. Work by the Golde lab demonstrated that injection of adeno-associated virus (AAV) encoding anti-β-amyloid scFvs into the ventricles of neonatal transgenic mice significantly attenuated widespread β-amyloid deposition in the brain with no apparent neuroinflammation (Levites et al., 2006). A similar study employed AAV technology to deliver an anti-β-amyloid scFv into the corticohippocampal region of the murine brain and observed significantly fewer amyloid deposits on hippocampal neurons with no added neurotoxicity (Fukuchi et al., 2006a). These early results hold promise for the development of an effective gene therapy for AD using scFvs.

3.5 Prions

Cell-surface prion proteins (PrP) are susceptible to misfolding and aggregation, leading to the formation of extracellular amyloid plaques that are linked to encephalitis and neurodegeneration. It is generally thought that the misfolding of PrP is elicited by infection of a mutant "scrapie" prion protein (denoted PrP^{Sc}), which results in the precipitous buildup of toxic PrP^{Sc} isoforms by subversion of wild-type PrP (reviewed in Westergard et al., 2007). In models of passive immunization, full-length antibodies against PrP are able to protect mice against PrP^{Sc} infection by antagonizing the toxic conversion of wild-type PrP on the neuronal surface (Heppner et al., 2001; White et al., 2003). However, cross-linking of PrP on the cell surface by bivalent anti-PrP monoclonal antibodies can induce neurotoxicity through a mechanism that perhaps mimics the aggregation of PrP during prion infection (Solforosi et al., 2004). scFvs offer the significant immunotherapeutic advantage of being monovalent. High-affinity scFvs with picomolar avidity for PrP have been isolated by several rounds of genetic selection using ribosome display and affinity maturation (Luginbuhl et al., 2006). Additional anti-PrP scFvs have been generated as diagnostic tools for distinguishing between normal and scrapie prion protein (Miyamoto et al., 2007). In a model of paracrine immunotherapy, a secreted anti-PrP scFv was recently shown to block the propagation of PrP^{Sc} among chronically infected neuronal cells, presumably by shielding wild-type PrP from conversion (Donofrio et al., 2005). Similar results are observed when using an scFv against a cell surface scrapie protein receptor (laminin receptor precursor or LRP/LR) to antagonize the binding and propagation of PrP^{Sc} (Zuber et al., 2007).

3.6 Other Approaches

An alternative strategy for inhibiting protein aggregation is to engineer novel scFvs that neutralize early events in the maturation and trafficking of neurotoxic protein prior to aggregation. When expressed in the ER lumen at the site of protein synthesis and secretion, these scFvs can prevent the generation or cellular transport of toxic protein species de novo.

3.6.1 ER Intrabodies

The trafficking and secretion of newly synthesized protein can be altered by novel scFvs called "ER intrabodies" that are expressed and retained in the ER lumen (reviewed in Boldicke, 2007). Compartmentalization of these scFvs is conferred via small polypeptide signals (e.g., signal peptides, KDEL) that interact with resident ER receptors (see ❯ *Figure 18-1b*). As a consequence, ER intrabodies are generally more stable than cytoplasmic intrabodies due to the favorable redox environment of the ER. To date, ER intrabodies have been engineered and tested in two neurodegenerative disease models. Work by the Biocca lab demonstrated that an ER-restricted anti-PrP intrabody efficiently blocked the transport of prion protein to the cell surface, thereby trapping nascent prions inside the cell (Cardinale et al., 2005). In neurons subsequently infected with PrP^{Sc}, this ER intrabody interfered with the spread of infectious PrP^{Sc} in both cell culture and mouse models of prion disease (Cardinale et al., 2005; Vetrugno et al., 2005). Similarly, a secreted version of this anti-PrP intrabody was found to increase the proteasomal degradation of PrP and reroute newly synthesized PrP into secretory vesicles, thus antagonizing PrP^{Sc} infectivity (Filesi et al., 2007). In a cellular model of AD, a KDEL-tagged ER intrabody selected against β-amyloid precursor protein (APP) elicited the disposal of newly synthesized APP from the ER (Paganetti et al., 2005). Because the processing of APP leads to the generation of β-amyloid, use of this anti-APP ER intrabody was found to abolish the formation of toxic β-amyloid species in a cell culture model of AD (Paganetti et al., 2005).

3.6.2 scFvs that Influence Posttranslational Modifications

In addition to altering protein trafficking, scFvs can be used to regulate protein posttranslational modifications by influencing substrate–enzyme interactions. For example, the Molinari lab completely inhibited the generation of β-amyloid in situ using an scFv that blocked the pathologic endoproteolysis of APP. This anti-APP scFv, which was selected against an epitope near the β-secretase cleavage site of APP, remained associated with newly synthesized APP throughout the secretion pathway. Consequently, expression of this anti-APP intrabody shielded the β-secretase recognition site of nascent APP, thereby blocking the formation and aggregation of β-amyloid in a cell culture model of AD (Paganetti et al., 2005). Alternatively, scFvs can be generated from proteolytic antibodies to catalyze enzymatic functions when bound to substrate targets (see review by Paul et al., 2006). The Sierks lab identified a single variable light chain that exhibited carboxypeptidase-like activity when incubated with β-amyloid in vitro (Rangan et al., 2003). This proteolytic light chain was found to reduce the aggregation and cytotoxicity of β-amyloid by sequentially cleaving the amyloidogenic carboxyl terminus of β-amyloid (Rangan et al., 2003; Liu et al., 2004b). Strategies like these that exploit enzyme–substrate interactions are important precedents for other neurotoxic proteins like mutant huntingtin, α-synuclein, and infectious prions.

4 Applications for Conformation-Specific scFvs in Neurodegenerative Diseases

The development of scFvs into an effective immunotherapy demands the creation of novel scFvs with disease-specific epitopes; otherwise, an scFv might target normal cellular protein. Several of the anti-aggregation scFvs described in this chapter were selected against peptide sequences common to both pathogenic and wild-type protein, and thus carry significant risk of binding nonpathogenic proteins with unknown physiological consequences. However, some anti-aggregation scFvs have demonstrated little or no specificity for normal cellular protein, likely due to differences in the accessibility of protein epitopes between pathogenic and nonpathogenic states (Miller et al., 2005). A more efficient strategy for generating disease-specific scFvs is to target the conformational superstructure of aggregation-prone proteins rather than particular peptide sequences. Such conformation-specific scFvs have been generated by combining antibody display techniques with atomic force microscopy or electron microscopy to select against oligomeric and fibrillar misfolded protein intermediates in vitro (Barkhordarian et al., 2006). Using this method, scFvs against oligomeric and fibrillar forms of α-synuclein, which are associated with Lewy body formation and neurotoxicity in PD, have been isolated (Barkhordarian et al., 2006; Emadi et al., 2007). Likewise, a conformation-specific scFv with affinity for the oligomeric and fibrillar forms of β-amyloid involved in plaque deposition and toxicity in AD has been generated (Fukuchi et al., 2006b). Importantly, these conformation-specific scFvs were found to bind exclusively to disease-associated oligomeric or fibrillar species rather than monomeric protein and inhibit protein deposition and toxicity in vitro. Moreover, an antioligomeric scFv has shown significant therapeutic potential in a mouse model of AD by reducing the number of β-amyloid plaques in vivo (Fukuchi et al., 2006b). Because amyloidogenic diseases of the nervous system are thought to share similar folding intermediates, conformation-specific scFvs could be engineered to recognize a common amyloid intermediate from divergent neurodegenerative diseases. In support of this notion, conformation-specific monoclonal antibodies against β-amyloid are known to cross-react with oligomeric and fibrillar species of nonhomologous neurotoxic proteins like α-synuclein, prion peptide, and polyglutamine (O'Nuallain and Wetzel, 2002; Kayed et al., 2003).

5   Perspective

Engineered antibody fragments are new classes of antibody reagents that show significant potential both as direct therapeutics and drug discovery tools for a wide range of neurological disorders, many of which currently lack viable long-term treatment options. Being derived from full-length antibodies, scFvs are distinguished from other small molecule and peptide therapeutics by their targeted specificity, enhanced stability, and low immunogenicity. Preliminary studies of scFv efficacy demonstrate that these engineered antibody fragments are proficient in altering the folding, localization, and binding dynamics of toxic amyloidogenic proteins in several neurological disease models, including Huntington's, Parkinson's, Alzheimer's, and prion-mediated diseases. However, challenges associated with the delivery of therapeutics, including monoclonal antibodies, to the brain currently hinder the clinical application of engineered antibody fragments as treatments for human neurological disorders. Rapidly advancing techniques in viral- and nonviral-mediated gene therapy hold great promise for the development of scFv-based immunotherapies in the near future.

Acknowledgments

We thank members of the Messer lab for helpful discussions of the manuscript. Work in the Messer lab was supported in part by grants from NIH/NINDS, High Q Foundation, National Parkinson Foundation, Hereditary Disease Foundation/Cure HD Initiative, and Huntington's Disease Society of America.

References

Barbas CF III, Bain JD, Hoekstra DM, Lerner RA. 1992. Semisynthetic combinatorial antibody libraries: A chemical solution to the diversity problem. Proc Natl Acad Sci USA 89: 4457-4461.

Barkhordarian H, Emadi S, Schulz P, Sierks MR. 2006. Isolating recombinant antibodies against specific protein morphologies using atomic force microscopy and phage display technologies. Protein Eng Des Sel 19: 497-502.

Bellotti V, Nuvolone M, Giorgetti S, Obici L, Palladini G, et al. 2007. The workings of the amyloid diseases. Ann Med 39: 200-207.

Bird RE, Hardman KD, Jacobson JW, Johnson S, Kaufman BM, et al. 1988. Single-chain antigen-binding proteins. Science 242: 423-426.

Boder ET, Wittrup KD. 1997. Yeast surface display for screening combinatorial polypeptide libraries. Nat Biotechnol 15: 553-557.

Boldicke T. 2007. Blocking translocation of cell surface molecules from the ER to the cell surface by intracellular antibodies targeted to the ER. J Cell Mol Med 11: 54-70.

Cardinale A, Filesi I, Vetrugno V, Pocchiari M, Sy MS, et al. 2005. Trapping prion protein in the endoplasmic reticulum impairs PrPC maturation and prevents PrPSc accumulation. J Biol Chem 280: 685-694.

Castellani RJ, Zhu X, Lee HG, Moreira PI, Perry G, et al. 2007. Neuropathology and treatment of Alzheimer disease: Did we lose the forest for the trees? Expert Rev Neurother 7: 473-485.

Chowdhury PS, Vasmatzis G, Beers R, Lee B, Pastan I. 1998. Improved stability and yield of a Fv-toxin fusion protein by computer design and protein engineering of the Fv. J Mol Biol 281: 917-928.

Colby DW, Chu Y, Cassady JP, Duennwald M, Zazulak H, et al. 2004a. Potent inhibition of huntingtin aggregation and cytotoxicity by a disulfide bond-free single-domain intracellular antibody. Proc Natl Acad Sci USA 101: 17616-17621.

Colby DW, Garg P, Holden T, Chao G, Webster JM, et al. 2004b. Development of a human light chain variable domain (VL) intracellular antibody specific for the amino terminus of huntingtin via yeast surface display. J Mol Biol 342: 901-912.

Daugherty PS, Chen G, Olsen MJ, Iverson BL, Georgiou G. 1998. Antibody affinity maturation using bacterial surface display. Protein Eng 11: 825-832.

Davies J, Riechmann L. 1996. Single antibody domains as small recognition units: Design and in vitro antigen selection of camelized, human VH domains with improved protein stability. Protein Eng 9: 531-537.

der Maur AA, Zahnd C, Fischer F, Spinelli S, Honegger A, et al. 2002. Direct in vivo screening of intrabody libraries constructed on a highly stable single-chain framework. J Biol Chem 277: 45075-45085.

Desai UA, Pallos J, Ma AA, Stockwell BR, Thompson LM, et al. 2006. Biologically active molecules that reduce polyglutamine aggregation and toxicity. Hum Mol Genet 15: 2114-2124.

Donofrio G, Heppner FL, Polymenidou M, Musahl C, Aguzzi A. 2005. Paracrine inhibition of prion propagation by anti-PrP single-chain Fv miniantibodies. J Virol 79: 8330-8338.

Emadi S, Barkhordarian H, Wang MS, Schulz P, Sierks MR. 2007. Isolation of a human single chain antibody fragment against oligomeric α-synuclein that inhibits aggregation and prevents α-synuclein-induced toxicity. J Mol Biol 368: 1132-1144.

Emadi S, Liu R, Yuan B, Schulz P, McAllister C, et al. 2004. Inhibiting aggregation of α-synuclein with human single chain antibody fragments. Biochemistry 43: 2871-2878.

Ewert S, Honegger A, Pluckthun A. 2004. Stability improvement of antibodies for extracellular and intracellular applications: CDR grafting to stable frameworks and structure-based framework engineering. Methods 34: 184-199.

Fandrich M. 2007. On the structural definition of amyloid fibrils and other polypeptide aggregates. Cell Mol Life Sci 64: 2066-2078.

Ferreira ST, Vieira MNN, De Felice FG. 2007. Soluble protein oligomers as emerging toxins in alzheimer's and other amyloid diseases. IUBMB Life 59: 332-345.

Filesi I, Cardinale A, Mattei S, Biocca S. 2007. Selective re-routing of prion protein to proteasomes and alteration of its vesicular secretion prevent PrPSc formation. J Neurochem 101: 1516-1526.

Fukuchi K, Accavitti-Loper MA, Kim HD, Tahara K, Cao Y, et al. 2006b. Amelioration of amyloid load by anti-Aβ single-chain antibody in Alzheimer mouse model. Biochem Biophys Res Commun 344: 79-86.

Fukuchi K, Tahara K, Kim HD, Maxwell JA, Lewis TL, et al. 2006a. Anti-Aβ single-chain antibody delivery via adeno-associated virus for treatment of Alzheimer's disease. Neurobiol Dis 23: 502-511.

Hamers-Casterman C, Atarhouch T, Muyldermans S, Robinson G, Hamers C, et al. 1993. Naturally-occurring antibodies devoid of light chains. Nature 363: 446-448.

He M, Taussig MJ. 2005. Ribosome display of antibodies: Expression, specificity and recovery in a eukaryotic system. J Immunol Methods 297: 73-82.

Heppner FL, Musahl C, Arrighi I, Klein MA, Rulicke T, et al. 2001. Prevention of scrapie pathogenesis by transgenic expression of anti-prion protein antibodies. Science 294: 178-182.

Hoogenboom HR, de Bruine AP, Hufton SE, Hoet RM, Arends JW, et al. 1998. Antibody phage display technology and its applications. Immunotechnology 4: 1-20.

Huston JS, Tai MS, McCartney J, Keck P, Oppermann H. 1993. Antigen recognition and targeted delivery by the single-chain Fv. Cell Biophys 22: 189-224.

Kayed R, Head E, Thompson JL, McIntire TM, Milton SC, et al. 2003. Common structure of soluble amyloid oligomers implies common mechanism of pathogenesis. Science 300: 486-489.

Khoshnan A, Ko J, Patterson PH. 2002. Effects of intracellular expression of anti-huntingtin antibodies of various specificities on mutant huntingtin aggregation and toxicity. Proc Natl Acad Sci USA 99: 1002-1007.

Knappik A, Ge L, Honegger A, Pack P, Fischer M, et al. 2000. Fully synthetic human combinatorial antibody libraries (HuCAL) based on modular consensus frameworks and CDRs randomized with trinucleotides. J Mol Biol 296: 57-86.

Lecerf JM, Shirley TL, Zhu Q, Kazantsev A, Amersdorfer P, et al. 2001. Human single-chain Fv intrabodies counteract in situ huntingtin aggregation in cellular models of Huntington's disease. Proc Natl Acad Sci USA 98: 4764-4769.

Levites Y, Jansen K, Smithson LA, Dakin R, Holloway VM, et al. 2006. Intracranial adeno-associated virus-mediated delivery of anti-pan amyloid beta, amyloid beta40, and amyloid beta42 single-chain variable fragments attenuates plaque pathology in amyloid precursor protein mice. J Neurosci 26: 11923-11928.

Li S, Li XJ. 2006. Multiple pathways contribute to the pathogenesis of Huntington disease. Mol Neurodegener 1: 19.

Liu R, McAllister C, Lyubchenko Y, Sierks MR. 2004b. Proteolytic antibody light chains alter beta-amyloid aggregation and prevent cytotoxicity. Biochemistry 43: 9999-10007.

Liu R, Yuan B, Emadi S, Zameer A, Schulz P, et al. 2004a. Single chain variable fragments against beta-amyloid (Abeta) can inhibit Abeta aggregation and prevent Abeta-induced neurotoxicity. Biochemistry 43: 6959-6967.

Luginbuhl B, Kanyo Z, Jones RM, Fletterick RJ, Prusiner SB, et al. 2006. Directed evolution of an anti-prion protein scFv fragment to an affinity of 1 pM and its structural interpretation. J Mol Biol 363: 75-97.

Maguire-Zeiss KA, Wang CI, Yehling E, Sullivan MA, Short DW, et al. 2006. Identification of human α-synuclein specific single chain antibodies. Biochem Biophys Res Commun 349: 1198-1205.

Marks JD, Hoogenboom HR, Bonnert TP, McCafferty J, Griffiths AD, et al. 1991. By-passing immunization: Human antibodies from V-gene libraries displayed on phage. J Mol Biol 222: 581-597.

Maynard J, Georgiou G. 2000. Antibody engineering. Annu Rev Biomed Eng 2: 339-376.

Merk H, Stiege W, Tsumoto K, Kumagai I, Erdmann VA. 1999. Cell-free expression of two single-chain monoclonal antibodies against lysozyme: Effect of domain arrangement on the expression. J Biochem 125: 328-333.

Miller TW, Zhou C, Gines S, Mac Donald ME, Mazarakis ND, et al. 2005. A human single-chain Fv intrabody preferentially targets amino-terminal huntingtin fragments in striatal models of Huntington's disease. Neurobiol Dis 19: 47-56.

Miyamoto K, Kimura S, Nakamura N, Yokoyama T, Horiuchi H, et al. 2007. Chicken antibody against a restrictive epitope of prion protein distinguishes normal and abnormal prion proteins. Biologicals 35: 303-308.

Murphy RC, Messer A. 2004. A single-chain Fv intrabody provides functional protection against the effects of mutant protein in an organotypic slice culture model of Huntington's disease. Brain Res Mol Brain Res 121: 141-145.

O'Nuallain B, Wetzel R. 2002. Conformational Abs recognizing a generic amyloid fibril epitope. Proc Natl Acad Sci USA 99: 1485-1490.

Orlandi R, Gussow DH, Jones PT, Winter G. 1989. Cloning immunoglobulin variable domains for expression by the polymerase chain reaction. Proc Natl Acad Sci USA 86: 3833-3837.

Paganetti P, Calanca V, Galli C, Stefani M, Molinari M. 2005. beta-site specific intrabodies to decrease and prevent generation of Alzheimer's Abeta peptide. J Cell Biol 168: 863-868.

Paul S, Nishiyama Y, Planque S, Taguchi H. 2006. Theory of proteolytic antibody occurrence. Immunol Lett 103: 8-16.

Rangan SK, Liu R, Brune D, Planque S, Paul S, et al. 2003. Degradation of beta-amyloid by proteolytic antibody light chains. Biochemistry 42: 14328-14334.

Robinson CR, Sauer RT. 1998. Optimizing the stability of single-chain proteins by linker length and composition mutagenesis. Proc Natl Acad Sci USA 95: 5929-5934.

Schenk D. 2002. Amyloid-beta immunotherapy for Alzheimer's disease: The end of the beginning. Nat Rev Neurosci 3: 824-828.

Solforosi L, Criado JR, McGavern DB, Wirz S, Sanchez-Alavez M, et al. 2004. Cross-linking cellular prion protein triggers neuronal apoptosis in vivo. Science 303: 1514-1516.

Solomon B, Koppel R, Frankel D, Hanan-Aharon E. 1997. Disaggregation of Alzheimer beta-amyloid by site-directed mAb. Proc Natl Acad Sci USA 94: 4109-4112.

Takeda A, Hasegawa T, Matsuzaki-Kobayashi M, Sugeno N, Kikuchi A, et al. 2006. Mechanisms of neuronal death in synucleinopathy. J Biomed Biotechnol 2006: 19365.

Tanaka T, Lobato MN, Rabbitts TH. 2003. Single domain intracellular antibodies: A minimal fragment for direct in vivo selection of antigen-specific intrabodies. J Mol Biol 331: 1109-1120.

Tsumoto K, Nakaoki Y, Ueda Y, Ogasahara K, Yutani K, et al. 1994. Effect of the order of antibody variable regions on the expression of the single-chain HyHEL10 Fv fragment in Escherichia coli and the thermodynamic analysis of its antigen-binding properties. Biochem Biophys Res Commun 201: 546-551.

Vetrugno V, Cardinale A, Filesi I, Mattei S, Sy MS, et al. 2005. KDEL-tagged anti-prion intrabodies impair PrP lysosomal degradation and inhibit scrapie infectivity. Biochem Biophys Res Commun 338: 1791-1797.

Visintin M, Settanni G, Maritan A, Graziosi S, Marks JD, et al. 2002. The intracellular antibody capture technology (IACT): Towards a consensus sequence for intracellular antibodies. J Mol Biol 317: 73-83.

Visintin M, Tse E, Axelson H, Rabbitts TH, Cattaneo A. 1999. Selection of antibodies for intracellular function using a two-hybrid in vivo system. Proc Natl Acad Sci USA 96(21): 11723-11728.

Westergard L, Christensen HM, Harris DA. 2007. The cellular prion protein (PrPC)): Its physiological function and role in disease. Biochim Biophys Acta 1772: 629-644.

White AR, Enever P, Tayebi M, Mushens R, Linehan J, et al. 2003. Monoclonal antibodies inhibit prion replication and delay the development of prion disease. Nature 422: 80-83.

Wolfgang WJ, Miller TW, Webster JM, Huston JS, Thompson LM, et al. 2005. Suppression of Huntington's disease pathology in Drosophila by human single-chain Fv antibodies. Proc Natl Acad Sci USA 102: 11563-11568.

Zameer A, Schulz P, Wang MS, Sierks MR. 2006. Single chain Fv antibodies against the 25–35 Abeta fragment inhibit aggregation and toxicity of Abeta42. Biochemistry 45: 11532-11539.

Zhou C, Emadi S, Sierks MR, Messer A. 2004. A human single-chain Fv intrabody blocks aberrant cellular effects of overexpressed alpha-synuclein. Mol Ther 10: 1023-1031.

Zuber C, Knackmuss S, Rey C, Reusch U, Rottgen P, et al. 2008. Single chain Fv antibodies directed against the 37 kDa/67 kDa laminin receptor as therapeutic tools in prion diseases. Mol Immunol 45: 144–151.

19 Pathogenesis and Treatment of HIV-associated Dementia: Recent Studies in a SCID Mouse Model

W. R. Tyor

1	***Overview of HIV-Associated Dementia***	***472***
1.1	Introduction	472
1.2	Clinical Features	472
1.3	Pathological Features	473
1.4	Pathogenesis of HAD	474
2	***Current Treatment of HIV-Associated Dementia***	***477***
2.1	Highly Active Antiretroviral Therapy	477
2.2	Other Potential Treatments for HAD	478
2.3	Summary of Treatments	479
3	***Animal Models of Lentivirus Encephalitis***	***479***
3.1	The SCID Mouse Hive Model	480

List of Abbreviations: (AIDS), acquired immunodeficiency syndrome; (BBB), blood brain barrier; (CNS), central nervous system; (FIV), feline immunodeficiency virus; (HAART), highly active antiretroviral therapy; (HIVE), HIV encephalitis; (HAD), HIV-associated dementia; (MCMD), mild cognitive-motor disorder; (NNRTIs), non-nucleoside reverse transcriptase inhibitors; (NRTIs), Nucleoside reverse transcriptase inhibitors; (RT-PCR), Reverse transcriptase polymerase chain reaction; (SCID), severe combined immunodeficient; (SIV), simian immunodeficiency virus

1 Overview of HIV-Associated Dementia

1.1 Introduction

HIV is a retrovirus of the lentivirus family that causes acquired immunodeficiency syndrome (AIDS) (Darr, 1998). Over 40 million people worldwide live with HIV infection; 1.5 million Americans are infected with the virus. As of 2005, HIV has claimed the lives of over 22 million people (WHO, 2005). Despite intensive research over the last 20–25 years a vaccine and/or treatment(s) that will prevent or cure HIV have not been developed. It is, therefore, likely that in the foreseeable future additional, specific treatments have to be developed, which address the complications of HIV infection.

HIV-associated dementia (HAD) is a common complication of HIV infection with cognitive impairment developing in approximately 30% of people with AIDS and frank dementia in about 15% (McArthur, 2003; Wong, 2007). Fortunately, AIDS death rates have declined by 40% since 1996, apparently due to highly active antiretroviral therapy (HAART) (Dore, 1999). There has also been a decreased incidence of HAD since HAART was introduced (Sacktor, 2002). Despite this recent decline in the incidence of HAD, it is still the leading cause of dementia in adults under 60 in the United States (McArthur, 1993). A great concern is that this decline in death rate could ultimately result in an aging population of HIV-infected individuals who are more susceptible to HAD (Cook, 2005a). In fact, recent studies in a Hawaiian cohort showing that HAD risk for adults older than 50 is more than 3 times that for younger HIV-positive patients suggests that these concerns are well founded (Valcour, 2004). Although this Hawaiian cohort was small and may not be applicable to all populations of HIV-positive individuals, this data in addition to evidence that the prevalence of HAD is actually increasing (McArthur, 2003) has prompted increasing concern among neuro-AIDS specialists.

To further complicate matters, HAART failures are common and frequent side effects of antiretroviral agents often lead to poor medication compliance (Maher, 1999). In addition, evidence that HAART does not penetrate the brain sufficiently to eradicate HIV (Cysique, 2004; Cook, 2005b) suggests that drug resistance may increase (Ellis, 2000). It seems likely that HAD will become a larger problem in the future. Clearly new and innovative treatments for HAD are needed. In addition, a better appreciation of the ability of HAART to suppress CNS (central nervous system) infection and HAD is warranted, given that this is currently the only treatment for HAD.

1.2 Clinical Features

Dementia is an AIDS-defining illness most commonly observed in patients who are significantly immunocompromised (CD4 count less than 200/mm^3) (McArthur, 1993). However, there are exceptions to this and there is evidence that HAD occurs with higher CD4 counts in the HAART era. Early, bradyphrenia, or slowness of mental functions, is commonly seen and patients may appear apathetic or depressed (Atkinson, 1997). Memory impairment, impairment of executive functions and mood abnormalities typically develop (Navia, 1986a). Neuropsychological testing suggests frontal/subcortical deficits (Schmitt, 1988). In addition to subcortical and frontal lobe dysfunction, prominent impairment of memory functions alludes to other areas of brain dysfunction including the hippocampus. Recent fMRI studies have confirmed the clinical impression that the hippocampus is involved (Castelo, 2006) and this concept is further validated by histopathological correlations (Petito, 2003). Memory impairment becomes more apparent as the disease progresses, eventually culminating in severe neurological sequelae – a bedridden, mute state and death over several months (Navia, 1986a).

Prior to HAART the average life expectancy of HAD patients was approximately 6 months (Harrison, 1995). However, since the advent of HAART, in general, life expectancy for HAD patients has increased significantly. Severe HAD is usually seen in HAART-naïve patients and those that have been taken off of HAART due to viral resistance. A milder form of cognitive impairment that can be chronic has also been described in patients on HAART (Paul, 2002). This has been given the term "mild cognitive-motor disorder" (MCMD). MCMD impacts the patient's activities of daily living but does not result in complete debilitation like HAD (Wachtel, 1992; Meehan, 2001). Taken together, these clinical features suggest that frontal lobes, including cortex, basal ganglia, and hippocampus are affected.

Imaging has confirmed both a cortical and subcortical influence of HIV infection in the brain, indicating that cortical and caudate atrophy is associated with HAD (Dal Pan, 1992). MRI is significantly more sensitive than CT scans of the brain for detection and aids in diagnosis of HIV related brain disease. On routine MRI of HAD patients, diffuse atrophy is the most common manifestation although cerebral white matter lesions seen on T2 weighted images are also common (Olsen, 1988). These areas tend to be fairly confluent and symmetric in the deep white matter, especially later in the course of disease. In addition, there is no evidence of edema or enhancement after contrast infusion. They likely represent increased myelin water content (see Pathology below), which is of unclear significance from a pathogenetic standpoint.

1.3 Pathological Features

For the purposes of this discussion, the designation HAD will apply to the clinical manifestations of HIV brain infection, and HIV encephalitis (HIVE) is the term that will be used for the pathological findings. Associated pathology of HIVE includes HIV-infected mononuclear phagocytes (i.e., macrophages and microglia), multinucleated giant cells, microglial nodules, diffuse myelin pallor, gliosis, and neuronal abnormalities (Navia, 1986a; Wiley, 1991; Masliah, 1992; Petito, 1995). Gliosis includes increased numbers and activity of astrocytes and microglia and is a prominent feature of HIVE (Sharer, 1985; Budka, 1991). Gliosis is, however, not specific to viral infection since it is commonly seen as a response to a number of CNS diseases (e.g., stroke). Nevertheless, it may have specific implications with respect to HIVE (see Pathogenesis below).

Although diffuse myelin pallor and multinucleated giant cells are often attributed to HIVE, their presence is actually not always associated with dementia, casting some doubt on the significance of the role of these findings in disease pathogenesis (Glass, 1993). In addition, diffuse myelin pallor is not associated with decreases in myelin proteins suggesting that it does not represent destruction of myelin but perhaps reflects increased water content, a finding which is of unclear significance (Power, 1993). The Power et al. study also emphasizes that the blood brain barrier (BBB) is disrupted in HIV infection of the CNS and reminds us that the virus is able to gain entry early during HIV infection, in part because the BBB is relatively "leaky" (Gray, 1996).

Mononuclear phagocytes are the primary population of cells infected in the CNS (Sharer, 1985; Shaw, 1985; Gabuzda, 1986; Koenig, 1986; Navia, 1986b; Wiley, 1986; Balestra, 2001). In addition to HIV spreading infection of mononuclear phagocytes within the CNS, HIV-infected astrocytes are also found (Tornatore, 1994). Although astrocyte infection is not productive in the sense that virions are not produced, the infection of astrocytes could result in dysfunction and death of these cells (Thompson, 2001). Astrocytes could therefore be an additional source of potential neurotoxins (see below) and their important function of serving to clear potentially toxic substances (e.g., glutamate) within the CNS could be compromised by HIV infection (Lipton, 1995). Studies also suggest that other CNS cell types such as endothelial cells may be infected (Moses, 1994). In addition, some studies have suggested that small populations of neurons are nonproductively infected by HIV (Nuovo, 1994; Trillo-Pazos, 2003; Kolson, 2004), while other studies have failed to demonstrate neuronal infection (Sharer, 1996; Takahashi, 1996). The bottom line is that the overwhelming majority of cells infected in the brain with HIV are mononuclear phagocytes. These cells are also probably the only productively infected cells and major source of HIV-related neurotoxins (e.g., tat, see Pathogenesis below). It also seems likely that mononuclear phagocytes are the major source of other, non-HIV protein, neurotoxins (e.g., TNF-α). The dominant effect on neuronal function is probably related to these putative neurotoxins and not infection of neurons with HIV.

The neuroanatomy of HIV brain infection suggested by clinical and imaging studies (see above) is corroborated by histopathology. HIVE has been consistently found in deep white matter and correlated with both ventricular enlargement and cortical atrophy (Sharer, 1985; Budka, 1991; Gelman, 1993; Patsouris, 1993; Archibald, 2004). Clearly, increased mononuclear phagocytes are found in the basal ganglia with concomitant increases in HIV protein detection in these cells (Kure, 1990; Neuen-Jacob, 1993) perhaps attesting to the subcortical nature of the dementia. Nigral degeneration has been specifically described (Reyes, 1991) and clinical evidence also suggests a prominent role of the basal ganglia (Aylward, 1993; Berger, 2000; Wang, 2004). The frontal lobes, prefrontal cortex, and hippocampus are implicated clinically by the prominent memory loss exhibited by these patients and other studies indicating these areas are affected (Frederick, 1988; Selnes, 1995; Stern, 1995; Hall, 1996; Chang, 2000; Stankoff, 2001; Castello, 2006). Petito et al. have also shown involvement of the hippocampus (Petito, 2003). However, the severity of dementia may be better determined by the presence of HIVE in neocortex (Bell, 1998).

Several studies have reported decreased neuronal cell counts in cerebral cortex during HIV infection (Everall, 1991; Wiley, 1991; Masliah, 1992; Asare, 1996). Neuronal abnormalities include decreased neuronal cell counts, abnormal dendritic arborizations and dendritic injury, and neuronal apoptosis (Wiley, 1991; Masliah, 1992; Petito, 1995; Shi, 1996; Masliah, 1997). However, whether neuronal death is prerequisite for dementia is unclear (Subbiah, 1996). Indeed, Seilhean et al. were unable to find an association with neuronal loss and dementia (Seilhean, 1993). Studies that have demonstrated improvement of cognitive function after initiation of antiretroviral therapy support the concept that neuronal death is not involved in early HAD (Schmitt, 1988; Tozzi, 1999; Sacktor, 2006). Our own studies in a severe combined immunodeficient (SCID) mouse model of HIVE (see below) would also suggest that neuronal death is not a prominent feature of early HAD (Cook, 2007). It seems more likely that neuronal death mechanisms, including apoptosis, are relatively late features of HAD/HIVE. While it is important to consider treatments that reverse these neuronal death mechanisms, it is equally important to consider other pathogenetic mechanisms of HAD pathogenesis that result in neuronal dysfunction but not frank death.

1.4 Pathogenesis of HAD

This section highlights only some of the potential pathogenetic mechanisms that have been implicated in HAD (❯ *Figure 19-1*). To begin a discussion of HAD pathogenesis it is appropriate to briefly review HIV entry into the CNS. The most prominent hypothesis about HIV entry into the CNS is the so called Trojan Horse theory. Infected mononuclear phagocytes enter the CNS through a leaky BBB and virus then spreads to other cells. The primary infiltrating cell carrying HIV in this scenario is the mononuclear phagocyte; but it is also possible that HIV invades the CNS by crossing the BBB as free virus and infecting endothelial cells and/or other perivascular cells such as pericytes, microglia, or astrocytes. Furthermore there is increasing evidence that T cells play a role in HIVE (Weidenheim, 1993; Katsetos, 1999; Petito, 2003; Petito, 2006) and so infected CD4+ T cells could also carry HIV into the CNS. What is clear is that CNS infection occurs early after systemic infection and the virus likely resides in the CNS throughout the remaining lifespan of the infected individual (Gray, 1996). The events that eventually lead to full blown HIVE and HAD are only partially appreciated. Determining the trigger(s) for the development of HAD is an active area of research.

To date, studies strongly suggest that neurons are not significantly infected by HIV (see above). Therefore, hypotheses for HAD pathogenesis have centered around the production of neurotoxins by mononuclear phagocytes and/or astrocytes (Tyor, 1995; Power, 2001; Cook, 2005a). These putative toxins include quinolinic acid, neopterin, arachidonic acid metabolites, Ntox, nitric oxide, HIV proteins, cytokines, chemokines, and others (Heyes, 1991; Genis, 1992; Tyor, 1992; Griffin, 1994; Adamson, 1996; Gelman, 1997; Kaul, 1999; Zhang, 2003; Lee, 2004). It also seems likely that host factors, such as genetics, play a role in the development and severity of HAD (Corder, 1998). Our own studies in SCID mice with

◻ **Figure 19-1**
Diagram of HAD pathogenesis. BBB = blood brain barrier

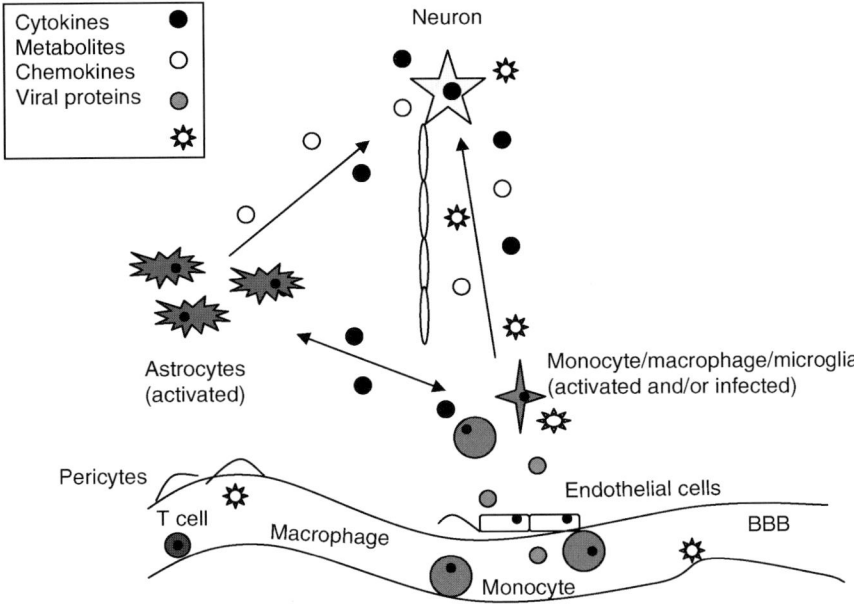

HIVE would also suggest that this is the case since cognitive dysfunction is more evident in some strains of mice compared with others.

HIV and its proteins represent a potential major source of neurotoxins in HAD (Werner, 1991; Tardieu, 1992; Mordelet, 2004; Singh, 2004). The most prominent of these HIV proteins are tat and gp120 (Brenneman, 1988; Sabatier, 1991). It is beyond the scope of this article to provide a complete review of these well-studied viral proteins; but briefly, gp120 is associated with neuronal injury that appears to be mediated through the NMDA receptor (Dreyer, 1990; Kaiser, 1990). These actions have been shown both in vitro and in vivo (Sundar, 1991; Toggas, 1994) and can result in neuronal apoptosis (Lannuzel, 1997). Recently tat has emerged as the leading candidate HIV protein involved in HAD pathogenesis. The potentially deleterious actions of lentiviral transactivating proteins are well known (Hayman, 1993; Kolson 1993). Although tat may render its neurotoxic affects through non-NMDA receptors (Magnuson, 1995), tat and gp120 appear to have synergistic affects on NMDA receptor mediated neurotoxicity (Nath, 2000). Both gp120 and tat may also exert their neurotoxic effects through indirect mechanisms (Dawson, 1993; Giulian, 1993; Ranga, 2004). Importantly, tat can be detected in brains of HIV-infected individuals (Nath, 2000) and when infused into rodents, causes cognitive dysfunction (Tian, 2004). However, there remains doubt as to whether HIV directly causes HAD (Bernton, 1992) or indirectly results in the development of HAD by stimulating the production of other substances that are neurotoxic.

One might think that if HIV protein(s) is the predominant source of neurotoxicity in HAD then the development of HAD would be dependent on viral load. However, demonstrating a correlation between CNS viral load and HAD has been difficult. Reports showing the reversibility of cognitive dysfunction after introduction of antiretroviral therapy suggest viral load is important (Schmitt, 1988; Sacktor, 1999; Sacktor, 2006), but the reductions in viral load are primarily systemic. It is unclear how systemic reduction in HIV impacts CNS HIV. Certainly CSF viral load can be reduced during antiretroviral therapy and can be associated with cognitive improvements (Marra, 2003). However, not all antiretrovirals are equal in their CNS penetration. In addition, CSF viral load probably does not accurately reflect brain parenchymal viral load and the exact mechanisms involved in improvement of cognitive dysfunction after reduction of

CSF viral load may or may not be due to direct HIV affects. Some studies have reported a direct relationship between viral load in the CNS and severity of dementia (Wiley, 1994; Ellis 1997; McArthur, 1997), whereas others have not (Portegies, 1989; Johnson, 1996). Given these disparate results regarding the importance of viral load in HAD, it has been suggested that while the presence of HIV is necessary to induce neurological disease, the amount of virus present may not be important (Tyor, 1995; Persidsky, 2001). A recent report treating SCID mice with HIVE with HAART (see below) also suggests that viral load is not important to the development and maintenance of HAD (Cook, 2007). The bottom line is the jury is still out on whether viral load and HIV proteins directly or indirectly play a prominent role in HAD pathogenesis and whether the amount of HIV present in the brain determines the severity of dementia. Other viral factors, such as cell tropism or specific sequence and/or strain differences may also play a role in HAD pathogenesis (Nath, 2002; Ranga, 2004; van Marle, 2005).

Mononuclear phagocytes (i.e., macrophages and microglia) are the primary population of CNS cells infected with HIV (Navia, 1986b; Takahashi, 1996), and the number of activated mononuclear phagocytes is increased in HIVE (Kure, 1991; Tyor, 1992; Glass, 1995). Astrogliosis is also prominent in HIVE and both astrocytes and mononuclear phagocytes in the brain can produce proinflammatory cytokines such as TNF-α and IL-1, especially in the setting of a chronic viral infection such as HIV (Merrill 1992; Tyor, 1992). TNF-α is perhaps the leading cytokine proposed to be involved in HAD pathogenesis. TNF-α protein is increased in the brain during HIVE (Tyor, 1992) and TNF-α mRNA is elevated in the brains of individuals with HAD compared with HIV-infected non-demented patients (Wesselingh, 1993). TNF-α is toxic to oligodendrocytes, alters neuronal function, and stimulates mononuclear phagocytes (Selmaj, 1988; Benveniste, 1992; Soliven, 1992). However, a recent trial of CPI-1189, a TNF-α blocker, in patients with mild to moderate HIV-associated cognitive impairment failed to show significant beneficial effects (Clifford, 2002). In addition, studies in the SCID mouse model of HIVE suggest that TNF-α may not be a prominent contributor to early cognitive dysfunction in HAD (Cook, 2007). Nevertheless, it is likely that proinflammatory cytokines like TNF-α do play a role in HAD over the course of the disease because they contribute to overall neurotoxicity directly or indirectly by stimulating a substantial number of other putative neurotoxins (Tyor, 1995).

IFN-α is another cytokine that has also been proposed to be involved in HAD pathogenesis (Rho, 1995). Interferons were originally characterized by their ability to interfere with virus replication, slow viral proliferation, and profoundly alter the immune response to a viral infection. Interferons are a group of soluble hormone-like proteins synthesized and secreted by a wide variety of cells including macrophages, monocytes, T lymphocytes, glial cells, and neurons (Dafny, 1998). IFN-α is a pleomorphic cytokine that acts as an innate immune response to viral infection. IFN-α produces an antiviral state in infected and surrounding uninfected cells through binding to the IFN-α receptor and activating the JAK/STAT pathway. This pathway results in production of viral replication interfering proteins, 2′–5′ oligoadenylate synthetase, dsRNA-dependent protein kinase, and more IFN-α (Ransohoff, 1998).

IFN-α has been used as a therapy in a variety of other viral infections, such as Herpes virus and hepatitis B and C infections, and is also used as a treatment against chronic myelogenous leukemia, among other cancers. During prolonged treatment with IFN-α, CNS side effects associated with the therapy include depression, anxiety, cognitive slowing, amnesia, and impaired executive function (Rho, 1995; Valentine, 1998). These are key symptoms consistent with subcortical dementia, which is the hallmark of HAD. These symptoms can be reversed when IFN-α treatment is discontinued.

IFN-α was measured by ELISA in the CSF of patients with HIV infection, with and without HAD (Rho, 1995). Individuals with HAD had significantly elevated levels of IFN-α compared with controls. This has been corroborated by Perrella et al. (Perrella, 2001). Immunohistochemistry for IFN-α in the brains of HIV infected individuals at autopsy revealed staining present on macrophages and astrocytes (Rho, 1995), consistent with previously reported staining in multiple sclerosis brains (Traugott, 1988). Reverse transcriptase polymerase chain reaction (RT-PCR) for IFN-α mRNA also indicated that IFN-α is actively produced within the brains of individuals with HAD.

Other data also support the theory that IFN-α is involved in HAD pathogenesis. Transgenic mice that overexpress IFN-α develop severe encephalitis and cognitive abnormalities (Campbell, 1999). Repeated injections of IFN-α into mice results in cognitive dysfunction (Andrea, 1993). Preliminary data using the SCID mouse HIVE model indicates that increased IFN-α associated with HIV in the brains of these mice correlates with their cognitive dysfunction (see below) (Sas, 2006). Furthermore, exposure of neurons in vitro to IFN-α impairs long-term potentiation (Mendoza-Fernandez, 2000). Finally, IFN-α production is increased in the presence of HIV (Capobianchi, 1993; Perrella, 2001), although there is contradictive data (Jiang, 2005). Nevertheless the overwhelming preponderance of data regarding IFN-α expression during HIV infection provides compelling information to support the hypothesis that HIV results in increased local production of IFN-α, which results in worsening of HIVE and neuronal dysfunction.

As discussed, there are numerous potential neurotoxins that have been implicated in HAD pathogenesis. Only a few of these have been discussed above in order to focus on aspects that are highlighted later in the article. This is not to suggest that many of these other proposed substances are not in part involved in the initiation, development, and/or end stages of HAD. It seems likely that given HIV's ability to stimulate cytokines and other inflammatory molecules, a vicious cycle of neurotoxin production is incited (Vitkovic, 1994). Thus, the pool of cells capable of producing neurotoxin(s) and resultant neuronal dysfunction or even death is exponentially increased. The production of proinflammatory cytokines may be further exacerbated by a relative paucity of IL-10 production, which would normally serve to suppress mononuclear phagocyte production of inflammatory cytokines and neurotoxic metabolites (Tyor, 1995). What is most important at this point in investigating new ideas for HAD therapy, is that each specific putative neurotoxin, one by one, be thoroughly examined to determine if it is a major contributor. This can only be done by careful pre-clinical and clinical therapeutic studies.

2 Current Treatment of HIV-Associated Dementia

2.1 Highly Active Antiretroviral Therapy

HAART is typically composed of at least three antiretroviral agents and has been shown to delay the onset of HAD and improve cognitive function in those with relatively early HAD (Dore, 1999; Cook, 2005a). Among the currently available antiretroviral drugs there are three main classes of agents and they are categorized based on their mechanism of action. Nucleoside reverse transcriptase inhibitors (NRTIs) reduce viral replication by incorporation into the elongating strand of viral DNA, causing chain termination. Protease inhibitors prevent viral maturation. The third class of antiretroviral agents, termed non-nucleoside reverse transcriptase inhibitors (NNRTIs), bind noncompetitively to HIV reverse transcriptase to inhibit viral replication. There are several other classes in use and/or under investigation, including those that prevent the virus from fusing to the cell membrane and those that prevent the integration of viral DNA into the host genome (Portegies, 2002). The ability of these agents to penetrate the BBB varies widely, although the methods used to determine BBB penetration in human studies are not optimal since only CSF can be routinely accessed for measurement (Cook, 2005b). Therefore, there is a great need to delineate the ability of antiretroviral agents to penetrate the BBB (by brain tissue accumulation) and suppress CNS virus. This is particularly important when one considers that the CNS could be a reservoir for HIV in an aging population of HIV-positive individuals who are probably at progressively greater risk for HAD (McArthur, 2003; Antinori, 2004, Valcour, 2004). In addition to the CNS potentially providing a sanctuary for HIV, it may enable viral [resistance] mutations to develop when selective antiretroviral agents have poor CNS penetration; although some data suggest this may not be important (van Marle, 2005).

Despite significant limitations of estimating the penetrability of individual HAART agents, studies which have determined CSF to plasma ratios have attempted to give clinicians some indication of antiretrovirals that may be active within the CNS (❯ *Table 19-1*) (Kandanearatchi, 2003). Brain autopsy evaluation of patients treated with HAART suggests HIVE has also decreased in the HAART era (Gray, 2003). However, there is evidence which suggests that HAART agents that are presumably active within the CNS have no significant benefit over agents with less CNS penetrability in stabilizing neuropsychological performance (Cysique, 2004). Recent studies in SCID mice with HIVE would also suggest that suppressing CNS viral load with HAART is not a critical component in treating cognitive dysfunction, although this is by no means clear (see above) (Cook, 2007). Other studies have shown that HAART has reduced

◘ Table 19-1
Various drug concentrations found within the CSF and their CSF:Plasma ratio

Drug	CSF Concentration	CSF:Plasma (%)	Author
Zidovudine	0.12–0.17 µM	60	Foudraine (1998), Rolinski (1997)
Stavudine	0.3 µM	40	Brady (1998), Haworth (1998)
Abacavir	2–5 µM	30	McDowell (1999)
Didanosine	0.17 mM	20	Clifford (1998)
Lamivudine	0.29–0.35 µM	10	Foudraine (1998), Clifford (1998)
Nevirapine		40	Enzensberger (1999)
Efavirenz	6.6–58.9 nm	0.69	Tashima (1999), Enzenzberger (1999)
Indinavir	0.09–0.66 µM	7	Stahle (1997), Shafer (1997)
Saquinavir	Negligible	1–2	Shafer (1999)

NRTI (zidovudine, stavudine, didanosine, lamivudine and abacavir), NNRTI (nevirapine and efavirenz), and PI (idinavir and saquinavir) drug concentrations found in the cerebrospinal fluid and their CSF:Plasma ratios

the incidence of HAD and does indeed have an effect on cognitive outcomes during HIV infection (Ellis, 2007). Indeed, clinical data suggests that HAART may have shifted HAD to milder conditions which may presage HAD, such as asymptomatic neurocognitive impairment and mild neurocognitive impairment.

It may be that *high viral loads* do influence the onset of relatively *severe dementia* (i.e., HAD), associated with *neuronal death*. Lesser [CNS] viral loads, which still occur during HAART, could result in mild cognitive impairment, as recent clinical data would suggest (Ellis, 2007). It is possible that these comparatively smaller CNS viral loads do not result in frank neuronal death, but rather a condition short of death (i.e., reversible neurophysiological dysfunction). Lower CNS viral loads may therefore serve to stimulate other neurotoxins (besides HIV proteins) that result in mild cognitive dysfunction. So perhaps in this way, on the low end of the viral load spectrum there is no correlation with amount of virus and severity of cognitive dysfunction because another substance (e.g., IFN-α) is more closely correlated with cognitive dysfunction. Then as viral load increases a threshold is reached whereby the impact of CNS viral load has reached a different phase and reducing CNS HIV becomes more effective. Is it at this point where HAART shows its effects in reversing dementia? The answer really is unknown at this time. Other factors (e.g., TNF-α) including HIV proteins (e.g., tat) may then come into play at this later stage. Eventually prolonged or high level exposure of neurons to one or more neurotoxins results in neuronal death and severe dementia. At that point nothing short of neural regeneration will improve outcome.

2.2 Other Potential Treatments for HAD

Several agents, which address potential pathogenetic mechanisms of HAD have been investigated in small clinical trials. CPI-1189 blocks TNF-α in vitro but had no significant beneficial effects in humans with HAD (Clifford, 2002). A platelet activating factor antagonist, lexipafant, also underwent Phase II testing but failed to show statistical effect (Shafitto, 1999). A phase II trial of deprenyl, which has a trophic effect on injured neurons, and thioctic acid (hydroxyl radical scavenger), suggested, at best, modest improvement (The Dana Consortium, 1998). Another phase II trial involved OPC-14117, an antioxidant, and again nonsignificant trends in improvement were noted (The Dana Consortium, 1997). Among other actions, valproic acid inhibits apoptosis; but a phase I/II trial in HAD showed nonsignificant improvements in cognitive function (Shafitto, 2006). D-Ala1-peptide T-amide, a purported neuroprotective agent, was shown to reduce peripheral HIV load but not CSF HIV (Goodkin, 2006). A pilot trial of transdermal selegiline, which may prevent O_2 radical formation, showed possible improvements (Sacktor, 2000).

The results of these trials, although perhaps somewhat promising in showing trends, are also somewhat disappointing. They suggest that mechanisms related to TNF-α, platelet activating factor, reactive oxygen species, and apoptosis may not be prominent factors in HAD pathogenesis, or that possibly they are not prominently involved at the time point examined (i.e., relatively early cognitive dysfunction). Alternatively, these mechanisms may be involved in these patient populations but the agents used did not penetrate the CNS well enough or are not potent enough in their actions.

2.3 Summary of Treatments

HAART has had positive effects on HAD, such as the ability to reverse the disorder, at least to a degree, especially in less severe cases (Schmitt, 1988). In addition, the incidence of HAD has fallen considerably since HAART was introduced (Dore, 1999; Sacktor, 2002). However, there are considerable concerns about HIV infection of the brain even in this era of HAART. Antiretrovirals are expensive, have troublesome side effects, resistance frequently develops to specific agents, and patients are often noncompliant (Cook, 2005a). Most of these antiretrovirals are not even available in the third world where HAD incidence may be greater than the US (Wong, 2007). As mentioned above, there is thought to be great variability in the ability of specific antiretroviral agents to penetrate the BBB and suppress or eradicate HIV. Little is actually known about most of these agents in this regard. It is conceivable that, especially for agents with low CNS penetration, mutated CNS HIV strains could evolve and cause more severe systemic or CNS disease. Although current data would suggest that these types of mutations are not significant (van Marle, 2005), there is concern that the chances of such a scenario are becoming greater. This is because although the incidence of HAD has decreased in the United States, the prevalence is actually increasing, probably due to an aging population of HIV+ patients who are more susceptible to developing HAD (Valcour, 2004). These individuals represent a growing population of HIV+ patients, in addition to those in the third world, who will have HAD or milder forms of it (see above), with or without HAART. In addition and importantly, it seems unlikely that a highly effective vaccine will be developed in the near future. So HIV and HAD [or milder forms of it] are going to be around for a long time. Therefore, there is a tremendous need for treatments to be developed that address the pathogenetic mechanisms of HAD. These treatments could be additive to HAART regimens or, if inexpensive, may be the sole source of therapy for HAD in some third world countries where HAART remains unavailable.

3 Animal Models of Lentivirus Encephalitis

It is beyond the scope of this article to thoroughly review all of the currently available lentivirus animal models or other in vitro systems for the study of retroviral neuropathogenesis. Several animal models of CNS retroviral disease have been developed but like all animal models, they have their advantages and disadvantages. Nevertheless, despite the drawbacks of all of these models they have contributed to our understanding of the pathogenesis of HAD and their continued use should be encouraged.

The simian immunodeficiency virus (SIV) macaque model and the feline immunodeficiency virus (FIV) model enable researchers to study the effects of these viruses both systemically and within the CNS in their natural hosts (Lackner, 1991; Power, 1998). The viral genomes of SIV and FIV are similar to HIV and these models have similarities to HAD in humans. However, there are some aspects of SIV and FIV CNS infection that are dissimilar to HAD. Initial attempts to recapitulate the CNS effects of HIV using wild type SIV strains resulted in little neuropathology. The development of SIV encephalitis was facilitated by deriving neurovirulent strains. The study of SIV and FIV encephalitis has been particularly instructive in terms of determining the roles of chemokines, viral invasion of the CNS, and mechanisms of lentiviral neurotoxicity (Lackner, 1991; Sasseville, 1996; Power, 1998).

Less expensive and more easily manipulated murine models have been developed, some of which involve xenografts or the use of other retroviruses such murine leukemia virus (MuLV) (Bazler, 1991; Morse, 1992; Tyor, 1993; Potash, 2005). Like the SIV and FIV models, MuLV infection in mice utilizes a retrovirus in which the mouse is the natural host. MuLV encephalitis also has histopathological similarities to HIVE. However, as with SIV and FIV the efficacy of antiretrovirals and other therapeutic agents may vary when comparing viruses which only have partially analogous genomes to HIV.

HIV transgenic mouse models have also been developed expressing single HIV proteins such as gp120, tat, and HIV-LTR (Corboy, 1992; Toggas, 1994; Maragos, 2003). These transgenic models enable the study of the effects of single viral proteins, which are neurotoxic. This is important when dissecting the effects of single potential neurotoxins from the myriad of neurotoxins which are likely to be in effect in HAD. However, these models may be relatively inconsequential assuming the pathogenesis of HAD is related to the combination of viral genes/proteins and other factors. In addition, there may be effects specific to the entire HIV genome, which would then render transgenic and other animal retrovirus models relatively irrelevant when testing therapies.

Nevertheless, these models have increased our knowledge of the pathogenesis of retroviral encephalitis. Therefore, the use of these models will continue to aid in the understanding of CNS viral, inflammatory, and neurodegenerative processes. And as mentioned, these models offer approaches to the study of lentiviral encephalitis that the HIVE SCID mouse model cannot.

3.1 The SCID Mouse Hive Model

The SCID mouse model of HIVE was developed initially at Johns Hopkins in the late 1980s (Tyor, 1993). Its advantages include that infectious HIV is used in a relatively inexpensive and easily manipulated mouse model. In addition, it enables one to study the effects of HAART and other potential therapeutic agents after the onset of HIVE on both histopathological and behavioral parameters (Cook, 2005b; Cook, 2007). This is important because simply monitoring pathological parameters may not predict a clinical (i.e., cognitive improvement) response to therapy since many of these pathological parameters (see Pathology above) have not been strictly correlated with cognitive impairment.

However, there are drawbacks to the SCID mouse HIVE model. Mice are obviously not the natural hosts for HIV and presently with the way the model is configured one cannot study the factors that determine entry of the virus into the brain. So like all models, it is artificial. In addition, it has been argued that introduction of human cells directly into the mouse brain results in tissue destruction and that the human cells themselves have detrimental effects. This is true to a small degree, which is why it is important to control for the effects of intracerebral (IC) inoculation of human cells (see below).

The pathology of HIVE in SCID mice is similar to that seen in humans (Navia, 1986a, b; Tyor, 1993; Persidsky, 1996; Avgeropoulos, 1998a; Griffin, 2004; Cook, 2005a, b; Cook, 2006; Griffin, 2006). HIV-infected human macrophages, which are injected IC into the brains of weanling SCID mice, are HIV-positive by immunocytochemistry in brain sections. Furthermore, virus can be recovered from brain as long as 2 months after infection and spliced mRNA for HIV indicates ongoing viral nucleic acid synthesis (Tyor 1993; Avgeropoulos, 1998b). Together, these results reveal that viral proteins and virions are actively produced for weeks after inoculation of HIV-infected human monocytes into SCID mouse brain. Other pathological parameters of encephalitis include HIV-positive human multinucleated giant cells, striking diffuse astrogliosis and microgliosis. There is evidence of increased human and mouse proinflammatory cytokine production such as TNF-α, IL-6, and IFN-α. These pathological features are prominent in humans with HIVE and indicate that the model portrays many important features of HIVE. Importantly, behavioral abnormalities have also demonstrated that are reminiscent of those seen in humans with HAD (Avgeropoulos, 1998a; Griffin, 2004; Sas, 2006; Griffin, 2007). These studies have suggested that, similar to HAD in humans, the neuroanatomical basis of HIVE cognitive dysfunction in mice lies in frontal, subcortical, and hippocampal circuits (Cook, 2005a).

Control mice that are injected with uninfected human macrophages also develop mild pathology. However, it is imperative to note that mice injected with HIV-infected human macrophages have significantly greater pathology than these control mice and *this can only be attributed to the presence of HIV.* In addition, control mice are able to successfully perform cognitive tests such as the Morris Water maze and the water radial arm maze (Avgeropoulos, 1998a; Griffin, 2004; Sas, 2006; Cook, 2007; Griffin, 2007), whereas mice injected with HIV-infected human macrophages perform poorly on these tasks. These findings are consistently reproducible and irrefutably show that HIVE in SCID mice is related to the direct or indirect effects of HIV infection in the brains of these mice.

More recent studies have verified these behavioral abnormalities and focused on the effects of HAART in SCID mice with HIVE (Cook, 2005a, b; Cook, 2007). A HAART regimen of AZT, lamivudine and indinavir was tested using HPLC to determine individual amounts of these agents in brain parenchyma and serum (Cook, 2005b). All three agents were detected in brain parenchyma suggesting that individually they are present in sufficient amounts to suppress HIV. In addition to reducing viral load, this regimen resulted in decreased astrogliosis in SCID mice with HIVE compared with HIVE mice given saline (❯ *Figure 19-2*).

◘ Figure 19-2
Severity of astrogliosis in infected and uninfected (data represents one and/or 2 weeks since the scores were similar) mice treated with HAART or with vehicle (*$p = 0.01$, †$p = 0.05$). Data represent mean ± SE

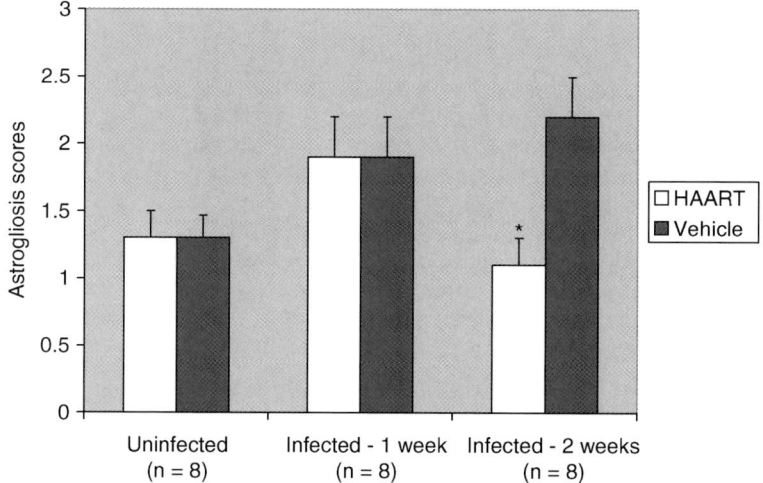

(Cook, 2005b). There was a trend for reduction of microgliosis. Importantly, HAART does not eradicate HIV in the brains of these mice suggesting as others recently have (Clements, 2005) that the brain is a sanctuary for HIV. This may have important implications, as noted above, for the development of resistant or more virulent HIV strains in an aging population of HIV-infected, HAART exposed patients.

These studies, which primarily examined histopathological parameters and viral load, were extended to include analyses of TNF-α expression, neuronal dendritic integrity (MAP2 expression) and behavior (Morris maze testing) (Cook, 2007). TNF-α mRNA was found to be elevated in HIVE mice and HAART reduced TNF-α mRNA levels (❯ *Figure 19-3*). This is consistent with the ability of HAART to decrease viral load and the two are likely to be related since HIV can increase TNF-α production (Merrill, 1992).

HIVE mice developed cognitive problems on Morris maze testing (Cook, 2007) and, importantly, these behavioral abnormalities were not ameliorated by HAART, even though HAART significantly reduced HIV brain parenchymal load. As discussed above, this suggests that HIV brain viral load may not be a predominant, determinant of early cognitive dysfunction. Furthermore, these behavioral abnormalities occurred in the presence of significant decreases in MAP2 immunostaining in the area where human cells were injected (❯ *Figure 19-4*). The decrease in MAP2 staining was significantly greater in HIVE mice compared with controls indicating that the MAP2 decrease in HIVE mice is a consequence of the presence of HIV. Just as HAART did not improve behavioral abnormalities in HIVE mice, the MAP2 decrease was also not affected by HAART (❯ *Figure 19-4d*). It is also interesting to note that neuronal apoptosis was not found at the time the mice underwent the retention phase (memory component) of cognitive testing

Figure 19-3

HIV induces production of mouse TNF-α mRNA and HAART decreases TNF-α mRNA. Ratios of mouse TNF-α to GAPDH for brains of UV – uninfected vehicle ($n = 7$), UH – uninfected HAART ($n = 5$), IV – infected vehicle ($n = 7$), and IH – infected HAART ($n = 6$) are summarized as medians. A nonparametric Median test confirmed the group differences noted in the graph: $X^2 = 10.463$ (df,3), $p > 0.015$. Follow-up comparison of the IV and IH groups with a Kolmogorov–Smirnov test confirmed the reduced levels for the HAART – treated group: $X^2 = 8.972$(df,1), $*p > 0.025$

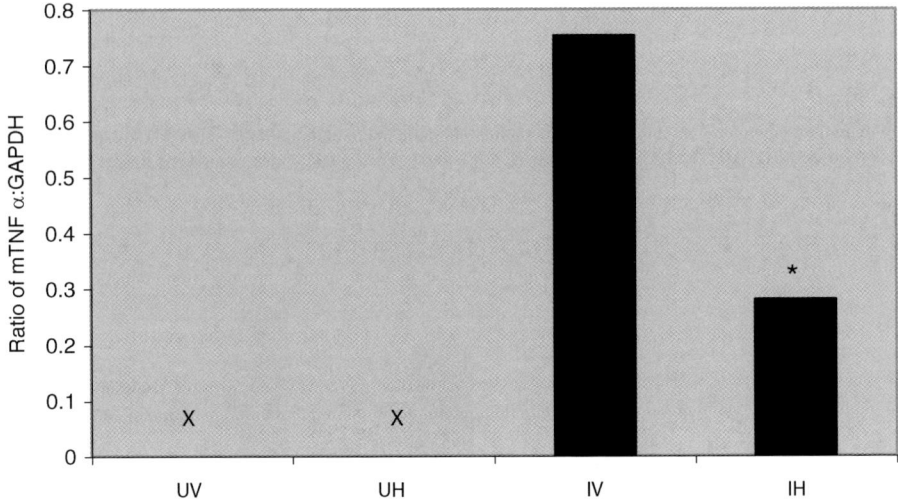

and pathological evaluation for MAP2 staining. These findings may have important implications regarding the fundamental basis of early cognitive dysfunction in HAD.

The decrease in MAP2 staining likely signifies disruption of dendritic arborization of these frontal neurons and probably reflects damage that relates to poor performance on Morris Maze and water radial arm maze testing. This is consistent with reports in humans of dendritic abnormalities in frontal cortex in HIVE (Wiley, 1991; Masliah, 1992; Masliah, 1997). Although neuronal apoptosis has been described in humans with HIVE it has not been correlated with HAD. Our findings in SCID mice with HIVE suggest that neuronal apoptosis is not a significant component in *early* cognitive dysfunction. These findings do not rule out other mechanisms of neuronal death relating to cognitive dysfunction or the possibility that neuronal apoptosis is an important event late in HAD. Nevertheless, it is plausible that early cognitive dysfunction in HAD is not related to neuronal death, but rather reversible neurophysiological dysfunction. It seems likely that frank neuronal death is a relatively late occurrence in HAD. This potential scenario is important because it suggests a significant window of opportunity for reversing cognitive dysfunction in HAD and clinical data support this assumption as well. The above data in SCID mice also suggest that rather than primarily focusing on viral factors, TNF-α, and neuronal apoptosis mechanisms when attempting to treat *early* cognitive dysfunction in HAD, other pathogenetic mechanisms should also be thoroughly addressed.

In terms of investigating other potential neurotoxins involved in HAD, we have recently demonstrated that IFN-α is significantly elevated in HIVE SCID mouse brain compared with control mice (Sas, 2006). This increase in IFN-α in HIVE mice correlates with the severity of cognitive dysfunction on water radial arm maze testing. As detailed above there is substantial evidence both clinically and in animal and in vitro experimental systems that IFN-α causes cognitive dysfunction and is intimately involved in the pathogenesis of HAD (Rho, 1995; Valentine, 1998; Campbell, 1999; Mendoza-Fernandez, 2000). Future studies in SCID mice with HIVE will determine whether blocking IFN-α results in amelioration of their cognitive deficits.

◘ Figure 19-4
HIV infected mice exhibit cognitive deficits regardless of treatment. Percent change in time required to reach the goal in a spatial learning task after a 6-day hiatus for UV – uninfected vehicle treated ($n = 7$), IV – infected vehicle treated ($n = 8$), UH – uninfected HAART treated ($n = 7$), and IH – infected HAART treated ($n = 8$) mice (Mean ± SE). Performance was enhanced after the break in training for uninfected mice (UV and UH), but not infected mice (Infection: $F(1,25) = 6.35$, *$p < 0.019$). The HAART factor and its interaction with Infection were not significant

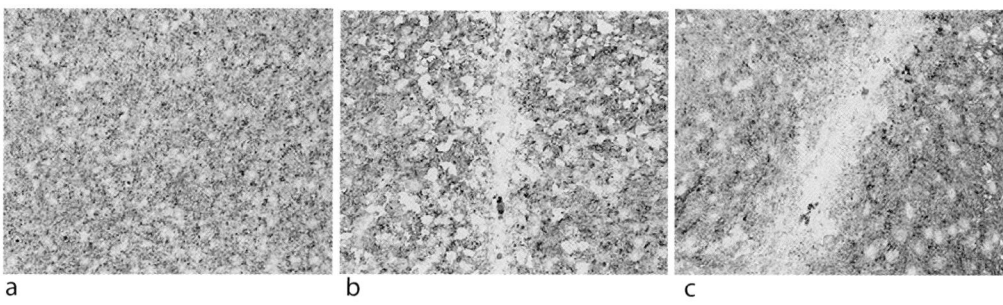

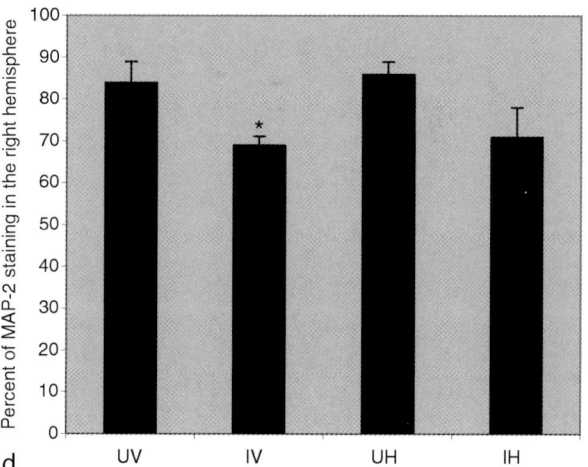

References

Adamson DC, Wildemann B, Sasaki M, et al. 1996. Immunologic NO synthase: Elevation in severe AIDS dementia and induction by HIV-1 gp41. Science 274: 1917-1921.

Andrea L, Conic LS. 1993. Repeated Injections of Interferon α A/D in Balb/c Mice: Behavioral Effects. Brain Behav Immun 7: 104-111.

Antinori A, Larussa D, Lorenzini P, Quaranta L, Uccella I, et al. 2004. Prevalence and risk factors of HIV-related cognitive disorders as a function of age in the era of HAART. J Neurovirol 10(Suppl3): H.6.

Atkinson JH, Grant II. 1997. Mood disorder due to human immunodeficiency virus: Yes, no, or maybe? Semin Clin Neuropsychiatry 2: 276-284.

Avgeropoulos N, Kelley B, Middaugh L, et al. 1998a. SCID mice with HIV encephalitis develop behavioral abnormalities. J Acq Imm Def Syn Hum Retrovirol 18: 13-20.

Avgeropoulos N, Burris G, Ohlandt G, et al. 1998b. Potential relationships between the presence of HIV, macrophages and astrogliosis in SCID mice with HIV encephalitis. J NeuroAIDS 2: 1-20.

Archibald SL, Masliah E, Fennema-Notestine C, et al. 2004. Correlation of in vivo neuroimaging abnormalities with postmortem human immunodeficiency virus encephalitis and dendritic loss. Arch Neurol 61: 369-376.

Asare E, Dunn G, Glass J, et al. 1996. Neuronal pattern correlates with severity of human immunodeficiency virus-associated dementia complex. Am J Pathol 148: 31-38.

Aylward EH, Henderer JD, McArthur JC, et al. 1993. Reduced basal ganglia volume in HIV-1-associated dementia: Results from quantitative neuroimaging. Neurology 43: 2099-2104.

Balestra E, Perno CF, Aquaro S, et al. 2001. Macrophages: A crucial reservoir for human immunodeficiency virus in the body. J Biol Regul Homeost Agents 15: 272-276.

Baszler TV, Zachary JF. 1991. Murine retroviral neurovirulence correlates with an enhanced ability of virus to infect selectively, replicate in, and activate resident microglial cells. Am J Pathol 138: 655-671.

Bell JE, Brettle RP, Chiswick A, Simmonds P. 1998. HIV encephalitis, proviral load and dementia in drug users and homosexuals with AIDS: Effect of neocortical involvement. Brain 121: 2043-2052.

Benveniste EN. 1992. Inflammatory cytokines within the central nervous system: Sources, function and mechanisms of action. Am J Physiol 263: C1-C16.

Bernton EW, Bryant HU, Decoster MA, et al. 1992. No direct neuronotoxicity by HIV-1 virions or culture fluids from HIV-1-infected T cells or monocytes. AIDS Res Hum Retroviruses 8: 495-503.

Brenneman DE, Westbrook G, Fitzgerald SF, et al. 1988. Human immunodeficiency virus envelope protein-induced neuronal cell death and its prevention by vasoactive intestinal peptide. Nature 335: 639-642.

Budka H. 1991. Neuropathology of human immunodeficiency virus infection. Brain Pathol 1: 163-175.

Campbell IL, Krucker T, Stefensen S, Akwa Y, Powell HC, et al. 1999. structural and functional neuropathology in transgenic mice with CNS expression of IFN-α. Brain Res 835: 46-61.

Capobianchi MR, Ameglio F, Fei PC, Castiletti C, Mercuri F, et al. 1993. Coordinate induction of interferon α and γ by recombinant HIV-1 and glycoprotein 120. AIDS Res Hum Retrovir 9: 10: 957-961.

Castello JMB, Sherman SJ, Courtney MA, Melrose RJ, Stern CE. 2006. Altered hippocampal-prefrontal activation in HIV patients during episodic memory encoding. Neurology 66: 1688-1695.

Chang L, Ernst T, Leonido-Yee M, Speck O. 2000. Perfusion MRI detects rCBF abnormalities in early stages of HIV-cognitive motor complex. Neurology 54: 389-396.

Clements JE, Li M, Gama L, et al. 2005. The central nervous system is a viral reservoir in simian immunodeficiency virus-infected macaques on combined antiretroviral therapy: A model for human immunodeficiency virus patients on highly active antiretroviral therapy. J Neurovirol 11: 180-189.

Clifford DB, McArthur JC, Schiffito G, et al. 2002. A randomized trial of CPI-1189 for HIV-associated cognitive-motor impairment. Neurology 59: 1568-1573.

Cook J, Tyor WR. 2005a. The pathogenesis of HIV-associated dementia: Recent advances using a SCID mouse model of HIV-encephalitis. Einstein J Biol Med 22: 32-40.

Cook J, Dasgupta S, Terry E, Gorry P, Wesselingh S, et al. 2005b. Highly active antiretroviral therapy and HIV encephalitis. Ann Neurol 57: 795-803.

Cook J, Middaugh LD, Griffin WC, Khan I, Tyor WR. 2007. Highly active antiretroviral therapy (HAART) and its effects in HIV-associated cognitive dysfunction in a SCID mouse model. Exp Neurol. 205: 506-512.

Corboy JR, Buzy JM, Zink MC, Clements JE. 1992. Expression directed from HIV long terminal repeats in the central nervous system of transgenic mice. Science, 258: 1804-1808.

Corder EH, Robertson K, Lannfelt L, et al. 1998. HIV-infected subjects with the E4 allele for APOE have excess dementia and peripheral neuropathy. Nat Med 4: 1182-1184.

Cysique LAJ, Maruff P, Brew BJ. 2004. Antiretroviral therapy in HIV infection: Are neurologically active drugs important ? Arch Neurol 61: 1699-1704.

Dafny N. 1998. Is Inteferon-α a Neuromodulator? Brain Res Rev 26: 1-15.

Dawson VL, Dawson TM, Uhl GR, Snyder SH. 1993. Human immunodeficiency virus type 1 coat protein neurotoxicity mediated by nitric oxide in primary cortical cultures. Proc Natl Acad Sci, USA 90: 3256-3259.

Dore GJ, Correll PK, Li Y, et al. 1999. Changes to AIDS dementia complex in the era of highly active antiretroviral therapy. AIDS 13(10): 1249-1253.

Dreyer EB, Kaiser PK, Offermann JT, Lipton SA. 1990. HIV-1 coat protein neurotoxicity prevented by calcium channel antagonists. Science 248: 364-367.

Ellis RJ, Hsia K, Spector SA, et al. 1997. Cerebrospinal fluid human immunodeficiency virus type 1 RNA levels are elevated in neurocognitively impaired individuals with acquired immunodeficiency syndrome. HIV Neurobehavioral Research Center Group. Ann Neurol 42: 679-688.

Ellis RJ, Gamst AC, Capparelli E, et al. 2000. Cerebrospinal fluid HIV RNA originates from both local CNS and systemic sources. Neurology 54: 927-936.

Ellis R, Langford D, Masliah E. 2007. HIV and antiretroviral therapy in the brain: Neuronal injury and repair. Nature Reviews Neurosci 8: 33-44.

Everall IP, Luthert PJ, Lantos PL. 1991. Neuronal loss in the frontal cortex in HIV infection. The Lancet 337: 1119-1121.

Gabuzda DH, Ho DD, dela Monte SM, et al. 1986. Immunohistochemical identification of HTLV-III antigen in brains of patients with AIDS. Ann Neurol 20: 289-295.

Gelman BB. 1993. Diffuse microgliosis associated with cerebral atrophy in the acquired immune deficiency syndrome. Ann Neurol 34: 65-70.

Gelman BB, Wolf DA, Rodriguez-Wolfe M, west AB, Haque AK, et al. 1997. Mononuclear phagocyte hydrolytic enzyme activity associated with cerebral HIV-1 infection. Am J Pathol. 151: 1437-1446.

Genis P, Jett M, Bernton EW, et al. 1992. Cytokines and arachidonic acid metabolites produced during human immunodeficiency virus (HIV)-infected macrophage-astroglia interactions: Implications for the neuropathogenesis if HIV disease. J Exp Med 176: 1703-1718.

Giulian D, Wendt E, Vaca K, Noonan CA. 1993. The envelope protein of human immunodeficiency virus type 1 stimulates release of neurotoxins from monocytes. Pro Natl Acad Sci USA 90: 2769-1273.

Glass JD, Fedor H, Wesselingh SL, McArthur JC. 1995. Immunocytochemical quantification of human immunodeficiency virus in the brain: Correlations with dementia. Ann Neurol 38: 755-762.

Glass JD, Wesselingh SL, Selnes OA, McArthur JC. 1993. Clinical-neuropathologic correlation in HIV-associated dementia. Neurology 43: 2230-2237.

Goodkin K, Vitiello B, Lyman WD, et al. 2006. Cerebrospinal and peripheral human immunodeficiency virus type 1 load in a multisite, randomized, double-blind, placebo-controlled trial of D-Ala1-peptide T-amide for HIV-1-associated cognitive-motor impairment. J NeuroVirol 12: 178-189.

Griffin DE, Wesselingh SL, McArthur JC. 1994. Elevated central nervous system prostaglandins in human immunodeficiency virus-associated dementia. Ann Neurol 35: 592-597.

Griffin WC, Cook JE, Middaugh L, Tyor WR. 2004. The SCID mouse model of HIV encephalitis: Deficits in cognitive function. J NeuroVirol 10: 109-115.

Griffin WC, Middaugh LD, Tyor WR. 2007. Chronic cocaine exposure in a SCID mouse model of HIV encephalitis. 2007. Brain Res 1134: 214-219.

Gray F, Keohane C. 2003. The neuropathology of HIV infection in the era of highly active antiretroviral therapy (HAART). Brain Pathol 13: 79-83.

Gray F, Scaravelli F, Everall I, et al. 1996. Neuropathology of early HIV-1 infection. Brain Pathol 6: 1-15.

Guilian D, Vaca K, Noonan CA. 1990. Secretion of neurotoxins by mononuclear phagocytes infected with HIV-1. Science 250: 1593-1596.

Hall M, Whaley R, Robertson K, Hamby S, Wilkins J, et al. 1996. The correlation between neuropsychological and neuroanatomic changes over time in asymptomatic and symptomatic HIV-1-infected individuals. Neurology 46: 1697-1702.

Harrison MJG, McArthur JC. 1995. HIV-associated dementia complex. AIDS and Neurology. New York: Churchill Livingstone; pp 31-64.

Hayman M, Arcbuthnott G, Harkiss G, et al. 1993. Neurotoxcity of peptide analogues of the transactivating protein tat from Maedi-Visna virus and human immunodeficiency virus. Neuroscience 53: 1-6.

Heyes MP, Brew BJ, Martin A, et al. 1991. Quinolinic acid in cerebrospinal fluid and serum in HIV-1 infection: Relationship to clinical and neurological status. Ann Neurol 29: 202-209.

Jiang W, Lederman MM, Salkowitz JR, et al. 2005. Impaired monocyte maturation in response to CpG oligonucleotide is related to viral RNA levels in human immunodeficiency virus disease and is at least partially mediated by deficiencies in alpha/beta interferon responsiveness and production. J Virol 79: 4109-4119.

Johnson RT, Glass JD, McArthur JC, Chesebro BW. 1996. Quantitation of human immunodeficiency virus in brains of demented and nondemented patients with acquired immunodeficiency syndrome. Ann Neurol 39: 392-395.

Kaiser PK, Offermann JT, Lipton SA. 1990. Neuronal injury due to HIV-1 envelope protein is blocked by anti-gp120 antibodies but not by anti-CD4 antibodies. Neurology 40: 1757-1761.

Kandanearatchi A, Williams B, Everall IP. 2003. Assessing the efficacy of highly active antiretroviral therapy in the brain. Brain Pathol 13: 104-110.

Katsetos CD, Fincke JE, Legido A, et al. 1999. Angiocentric CD3 + T-cell infiltrates in human immunodeficiency virus type 1-associated central nervous system disease in children. Clin Diag Lab Immunol 6: 105-114.

Kaul M, Lipton SA. 1999. Chemokines and activated macrophages in HIV gp120-induced neuronal apoptosis. Proc Natl Acad Sci USA 96: 8212-8216.

Kaul M, Garden GA, Lipton SA. 2001. Pathways to neuronal injury and apoptosis in HIV-associated dementia. Nature 410: 988-994.

Koenig S, Gendelman HE, Orenstein JM, et al. 1986. Detection of AIDS virus in macrophages in brain tissue from AIDS patients with encephalopathy. Science 233: 1089-1093.

Kolson DL, Buchwalter J, Collman R, et al. 1993. HIV-1 tat alters normal organization of neurons and astrocytes in primary rodent brain cell cultures: RGD sequence dependence. AIDS Res Hum Retroviruses 9: 677-685.

Kolson DL, Sabnekar P, Baybis M, Crino PB. 2004. Gene expression in TUNEL-positive neurons in human immunodeficiency virus-infected brain. J Neurovirol 10: 102-107.

Kure K, Llena JF, Lyman WD, et al. 1991. Human immunodeficiency virus-1 infection of the nervous system: An autopsy study of 268 adult, pediatric and fetal brains. Hum Pathol 22: 700-710.

Lackner AA, Dandekar S, Gardner MB. 1991. Neurobiology of simian and feline immunodeficiency virus infections. Brain Pathol 1: 201-212.

Lannuzel A, Barnier JV, Hery C, et al. Human immunodeficiency virus type 1 and its coat protein gp 120 induce apoptosis and activate JNK and ERK mitogen-activated protein kinases in human neurons. Ann Neurol 42: 847–856.

Lee ES, Kalantari P, Tsutsui S, et al. 2004. RON receptor tyrosine kinase, a negative regulator of inflammation, inhibits HIV-1 transcription in monocytes/macrophages and is decreased in brain tissue from patients with AIDS. J Immunol 173: 6864-6872.

Limoges J, Persidsky Y, Poluektova L, et al. 2000. Evaluation of antiretroviral drug efficacy for HIV-1 encephalitis in SCID mice. Neurology 54: 379-389.

Lipton SA. 1992. Memantine prevents HIV coat protein-induced neuronal injury in vitro. Neurology 42: 1403-1405.

Lipton SA, Gendelman HE. 1995. Dementia associated with the acquired immunodeficiency syndrome. N Eng J Med 332: 934-940.

Magnuson DSK, Knudsen BE, Geiger JD, Brownstone RM, Nath A. 1995. Human immunodeficiency type 1 tat activates non-N-methyl-D-aspartate excitatory amino acid receptors and causes neurotoxicity. Ann Neurol 37: 373-380.

Maher K, Klimas N, Fletcher MA. 1999. Disease progression, adherence and response to protease inhibitor therapy for HIV infection in an urban Veterans Affairs Medical Center. J Acquir Immune Defic Syndr 22: 358-363.

Major EO, Rausch D, Marra C, Clifford D. 2000. HIV-associated dementia. Science 288: 440-442.

Maragos WF, Tillman P, Jones M, et al. 2003. Neuronal injury in hippocampus with human immunodeficiency virus transactivating protein, Tat. Neuroscience 117(1): 43-53.

Marra CM, Lockhart D, Zunt JR, et al. 2003. Changes in CSF and plasma HIV-1 RNA and cognition after starting potent antiretroviral therapy. Neurology 60: 1388-1390.

Masliah E, Heaton RK, Marcotte TD, et al. 1997. Dendritic injury is a pathological substrate for human immunodeficiency virus-related cognitive disorders. Ann Neurol 42: 963-972.

Masliah E, Morey M, DeTeresa R. 1992. Cortical dendritic pathology in human immunodeficiency virus encephalitis. Lab Invest 66: 285-291.

McArthur JC, Haughey N, Gartner S, et al. 2003. Human immunodeficiency virus-associated dementia: An evolving disease. J Neurovirol 9(2): 205-221.

McArthur JC, Hoover DR, Bacellar H. 1993. Dementia in AIDS patients: Incidence and risk factors. Neurology 43: 2245-2252.

McArthur JC, McClernon DR, Cronin MF, et al. 1997. Relationship between human immunodeficiency virus-associated dementia and viral load in cerebrospinal fluid and brain. Ann Neurol 42: 689-698.

Meehan RA, Brush JA. 2001. An overview of AIDS dementia complex. Am J Alzheimers Dis Other Demen 16: 225-229.

Mendoza-Fernandez V, Andrew R, Barajas-Lopez C. 2000. IFNα inhibits long term potentiation and unmasks a long term depression in the rat hippocampus. Brain Res 885: 14-24.

Merrill JE, Koyanagi Y, Zack J, Thomas L, Martin F, et al. 1992. Induction of IL-1 and TNF in brain cultures by human immunodeficiency virus type 1. J Virol 66: 2217-2225.

Mordelet E, Kissa K, Cressant A, et al. 2004. Histopathological and cognitive defects induced by Nef in the brain. FASEB J 18: 1851-1861.

Morse HC, Chadpadhyay SK, Makino M, et al. 1992. Retrovirus-induced immunodeficiency in the mouse: MAIDS as a model for AIDS. AIDS 6: 607-621.

Moses AV, Nelson JA. 1994. HIV infection of human brain capillary endothelial cells – implications for AIDS dementia. Adv Neuroimmunol 239–247.

Nath A, Haughey NJ, Jones M, et al. 2000. Synergistic neurotoxicity by human immunodeficiency virus proteins tat and gp120: Protection by memantine. Ann Neurol 47: 186-194.

Nath A. 2002. Human immunodeficiency virus (HIV) proteins in neuropathogenesis of HIV dementia. J Infect Dis 186: S193-S198.

Navia BA, Cho E, Petito CK, Price RW. 1986b. The AIDS dementia complex: II Neuropathology. Ann Neurol 19: 525-535.

Navia BA, Jordan BD, Price RW. 1986a. The AIDS dementia complex: I Clinical features. Ann Neurol 19: 517-524.

Nuovo GJ, Gallery F, MacConnell P, Braun A. 1994. In situ detection of polymerase chain reaction - amplified HIV-1 nucleic acids and tumor necrosis factor-RNA in the central nervous system. Am J Pathol 144: 659-666.

Olsen WL, Longo FM, Mills CM, Norman D. 1988. White matter disease in AIDS: Findings at MR imaging. Radiology 169: 445-448.

Patsouris E, Kretzschmar HA, Stavrou D, Mehraein P. 1993. Cellular composition and distribution of gliomesenchymal nodules in the CNS of AIDS patients. Clin Neuropathol 12: 130-137.

Paul RH, Cohen RA, Stern RA. 2002. Neurocognitive manifestations of human immunodeficiency virus. CNS Spectr 7: 860-866.

Perrella O, Carrieri PB, Perrella A, et al. 2001. Transforming growth factor-beta1 and interferon-alpha in the AIDS dementia complex (ADC): Possible relationship with cerebral viral load? Eur Cytokine Netw 12: 51-55.

Persidsky Y, Limoges J, McComb R, et al. 1996. Human immunodeficiency virus encephalitis in SCID mice. Am J Pathol 149: 1027-1053.

Petito CK, Roberts B. 1995. Evidence of apoptotic cell death in HIV encephalitis. Am J Pathol 146: 1121-1130.

Petito CK, Adkins B, McCarthy M, Khamis I. 2003. CD4 + and CD8 + cells accumulate in the brains of acquired immunodeficiency syndrome patients with human immunodeficiency virus encephalitis. J NeuroVirol 9: 36-44.

Petito CK, Torres-Munoz JE, Ziegler F, McCarthy M. 2006. Brain CD8 + and cytolytic T lymphocytes are associated with, and may be specific for, human immunodeficiency virus type 1 encephalitis in patients with acquired immune deficiency syndrome. J NeuroVirol 12: 272-283.

Portegies P. 2002. Antiretroviral Therapuetics. J Neurovirol 8: 148-150.

Portegies P, Epstein LG, Hung ST, de Gans J, Goudsmit J. 1989. Human immunodeficiency virus type 1 antigen in cerebrospinal fluid. Correlation with clinical neurologic status. Arch Neurol 46: 261-264.

Potash MJ, Chao W, Bentsman G, Paris N, Saini M, et al. 2005. A mouse model for study of systemic HIV-1 infection, antiviral immune responses, and neuroinvasiveness. Proc Natl Acad Sci USA 102: 3760-3765.

Power C, Kong P-A, Crawford TO, et al. 1993. Cerebral white matter changes in acquired immunodeficiency syndrome dementia: Alterations of the blood brain barrier. Ann Neurol 34: 339-350.

Power C, Buist R, Johnston JB, Del Bigio MR, Ni W, et al. 1998. Neurovirulence in feline immunodeficiency virus-infected neonatal cats is viral strain specific and dependent on systemic immune suppression. J Virol 72: 9109-9115.

Power C, Johnson RT. 2001. Neuroimmune and neurovirological aspects of human immunodeficiency virus infection. Adv Virus Res 56: 389-433.

Ranga U, Shankarappa R, Siddappa NB, et al. 2004. Tat protein of human immunodeficiency virus type 1 subtype C strains is a defective chemokine. J Virol 78: 2586-2590.

Ransohoff RM. 1998. Cellular responses to interferons and other cytokines: The jak-stat paradigm. N Eng J Med 338: 616-618.

Reyes MG, Faraldi F, Senseng CS, Flowers C, Fariello R. 1991. Nigral degeneration in acquired immune deficiency syndrome (AIDS). Acta Neuropathol 82: 39-44.

Rho M, Wesselingh S, Glass JD, McArthur JC, Griffin JW, et al. 1995. The potential role of interferon-alpha in the pathogenesis of HIV-1 associated dementia. Brain Behav Immun 9: 366-377.

Sabatier J, Vives E, Mabrouk K, et al. 1991. Evidence of neurotoxic activity of tat from human immunodeficiency virus type 1. J Virol 65: 961-967.

Sacktor N, Lyles RH, Skolasky RL, et al. 1999. Combination antiretroviral therapy improves psychomotor speed performance in HIV-seropositive homosexual men. Neurology 52: 1640-1647.

Sacktor N, Shafitto G, Mc Dermott MP, et al. 2000. Transdermal selegiline in HIV-associated cognitive impairment: Pilot, placebo-controlled study. Neurology 54: 233-235.

Sacktor N, McDermott MP, Marder K, et al. 2002. HIV – associated cognitive impairment before and after the advent of combination therapy. J Neurovirol 8: 136-142.

Sacktor N, Nakasujja N, Skolasky MA, . et al. 2006. Antiretroviral therapy improves cognitive impairment in HIV + individuals in sub-Saharan Africa. Neurology 67: 311-314.

Samuel CE. 2001. Antiviral actions of interferons. Clin Microbiol Rev 14: 778.

Sas A, Bimonte-Nelson H, Tyor WR. 2006. Interferon-alpha in the pathogenesis of HIV encephalitis. Boston, MA: American Association of Immunologists meeting.

Sasseville VG, Smith MM, Mackay CR, et al. 1996. Chemokine expression in simian immunodeficiency virus-induced AIDS encephalitis. Am J Pathol 149: 1459-1467.

Schmitt FA, Bigley JW, McKinnis R. 1988. Neuropsychological outcome of Zidovudine (AZT) treatment of patients with AIDS and AIDS-related complex. N Eng J Med 319: 1573-1578.

Seilhean D, Duyckaerts C, Vazeux R, et al. 1993. HIV-1-associated cognitive motor complex: Absence of neuronal loss in the cerebral neocortex. Neurology 43: 1492-1499.

Selmaj KW, Raine CS. 1988. Tumor necrosis factor mediates myelin and oligodendrocyte damage in vitro. Ann Neurol 23: 339-346.

Selnes O, Galai N, Bacellar H, et al. Cognitive performance after progression to AIDS: A longitudinal study from the Multicenter AIDS Cohort Study. Neurology 45: 267–275.

Shafitto G, Saktor N, Marder K, et al. 1999. Randomized trial of the platelet activating factor antagonist lexipafant in HIV-associated cognitive impairment. Neurology 53: 391-396.

Shafitto G, Peterson DR, Zhong J, et al. 2006. Valproic acid adjunctive therapy for HIV-associated cognitive impairment: A first report. Neurology 66: 919-921.

Sharer LR, Cho E-S, Epstein LG. 1985. Multinucleated giant cells and HTLV-III in AIDS encephalopathy. Hum Pathol 16: 780.

Sharer LR, Saito Y, Da Cunha A, et al. 1996. In situ amplification and detection of HIV-1 DNA in fixed pediatric AIDS brain tissue. Hum Pathol 27: 614-617.

Shaw GM, Harper ME, Hahn BH, et al. 1985. HTLV-III infection in brains of children and adults with AIDS encephalopathy. Science 227: 177-182.

Shi B, De Girolami U, He J, et al. 1996. Apoptosis induced by HIV-1 infection of the central nervous system. J Clin Invest 98: 1979-1990.

Singh I, Goody RJ, Dean C, et al. 2004. Apoptotic death of striatal neurons induced by human immunodeficiency virus-1 Tat and gp120: Differential involvement of caspase-3 and endonuclease G. J NeuroVirol 10: 141-151.

Soliven B, Albert J. 1992. Tumor necrosis factor modulates Ca2 + currents in cultured sympathetic neurons. J Neurosci 12: 2665-2671.

Staeheli P. 1990. Interferon-induced proteins and the antiviral state. Adv Virus Res 38: 147-200.

Stankoff B, Tourbah A, Suarez S, et al. 2001. Clinical and spectroscopic improvement in HIV-associated cognitive impairment. Neurology 56: 112-115.

Stern y, Liu X, Marder K, et al. 1995. Neuropsychological changes in a prospectively followed cohort of homosexual and bisexual men with and without HIV infection. Neurology 45: 467-472.

Subbiah P, Mouton P, Fedor H, McArthur JC, Glass JD. 1996. Stereological analysis of cerebral atrophy in human immunodeficiency virus-associated dementia. J Neuropathol Exp Neurol 55: 1032-1037.

Sundar SK, Cierpial MA, Kamaraju LS, et al. 1991. Human immunodeficiency virus glycoprotein (gp120) infused into rat brain induces interleukin 1 to elevate pituitary-adrenal activity and decrease peripheral cellular immune responses. Proc Natl Acad Sci, USA 88: 11246-11250.

Takahashi K, Wesselingh SL, Griffin DE, et al. 1996. Localization of HIV-1 in human brain using polymerase chain reaction/in situ hybridization and immunocytochemistry. Ann Neurol 39: 705-717.

Tardieu M, Hery C, Peudenier S, Boespflug O, Montagnier L. 1992. Human immunodeficiency virus type 1-infected monocytic cells can destroy human neural cells after cell to cell adhesion. Ann Neurol 32: 11-17.

The Dana Consortium on the Therapy if HIV Dementia and Related Cognitive Disorders. 1997. safety and tolerability of the antioxidant OPC-14117 in HIV-associated cognitive impairment. Neurology 49: 142-146.

The Dana Consortium on the Therapy if HIV Dementia and Related Cognitive Disorders. 1998. A randomized, double blind, placebo-controlled trial of deprenyl and thioctic acid in human immunodeficiency virus-associated cognitive impairment. Neurology 50: 645-651.

Thompson KA, McArthur JC, Wesselingh SL. 2001. Correlation between neurological progression and astrocyte apoptosis in HIV-associated dementia. Ann Neurol 49: 745-752.

Tian S, Matsushita M, Moriwaki A, et al. 2004. HIV-1 inhibits long-term potentiation and attenuates spatial learning. Ann Neurol 55: 362-371.

Toggas SM, Masliah E, Rockenstein EM, et al. 1994. Central nervous system damage produced by expression of the HIV-1 coat protein gp120 in transgenic mice. Nature 367: 188-193.

Tornatore C, Chandra R, Berger JR, Major EO. 1994. HIV-1 infection of subcortical astrocytes in the pediatric central nervous system. Neurology 44: 481-487.

Tozzi V, Balestra P, Galgani S, et al. 1999. Positive and sustained effects of highly active antiretroviral therapy on HIV-1-associated neurocognitive impairment. AIDS 13: 1889-1897.

Traugott U, Lebon P. 1988. Multiple sclerosis: Involvement of interferons in lesion pathogenesis. Ann Neurol 24: 243-251.

Trillo-Pazos G, Diamanturos A, Rislove L, et al. 2003. Detection of HIV-1 DNA in microglia/macrophages, astrocytes and neurons isolated from brain tissue with HIV-1 encephalitis by laser capture microdissection. Brain Pahtol 13: 144-154.

Tyor WR, Glass JD, Griffin JW, et al. 1992. Cytokine expression in the brain during AIDS. Ann Neurol 31: 349-360.

Tyor WR, Power C, Gendelman HE, Markham RB. 1993. A model of human immunodeficiency virus encephalitis in SCID mice. Proc Natl Acad Sci USA 90: 8658-8662.

Tyor WR, Wesselingh SL, Griffin JW, et al. 1995. Unifying hypothesis for the pathogenesis of HIV-associated dementia complex, vacuolar myelopathy and sensory neuropathy. J Acq Immune Def Syn Hum Retroviral 9: 379-388.

Valcour V, Shikuma C, Shiramizu B, Watters M, Poff P, et al. 2004. Higher frequency of dementia in older HIV-1 individuals: The Hawaii Aging with HIV-1 Cohort. Neurology 63: 822-827.

Valentine AD, Meyers CA, Kling MA, Richelson E, Hauser P. 1998. Mood and cognitive side effects of interferon-α therapy. Semin Oncol 25: 39-47.

Van Marle G, Power C. 2005. HIV type 1 genetic diversity in the nervous system: Evolutionary epiphenomenon or disease determinant? J Neurovirol 11: 107-128.

Vitkovic L, da Cunha A, Tyor WR. 1994. Cytokine expression and pathogenesis in AIDS brain. HIV, AIDS and the Brain. Price RW, Berry SW, editors. New York: Raven Press, pp. 203-222.

Wachtel T, Piette J, Mor V, Stein M, Fleishman J, et al. 1992. Quality of life in persons with human immunodeficiency virus infection: Measurement by the Medical Outcomes Study instrument. Ann Intern Med 116: 129-137.

Wang GJ, Chang L, Volkow ND, et al. 2004. Decreased brain dopaminergic transporters in HIV-associated dementia patients. Brain 127: 2452-2458.

Weidenheim KM, Epshteyn I, Lyman WD. 1993. Immunocytochemical identification of T-cells in HIV-1 encephalitis: Implications for pathogenesis of CNS disease. Modern Pathol 6: 167-174.

Werner T, Ferroni S, Saermark T, et al. 1991. HIV-1 Nef protein exhibits structural and functional similarity to scorpion peptides interacting with K + channels. AIDS 5: 1301-1308.

Wesselingh SL, Power C, Glass J, et al. 1993. Intracerebral cytokine mRNA expression in AIDS dementia. Ann Neurol 33: 576-582.

Wiley CA, Achim C. 1994. Human immunodeficiency virus encephalitis is the pathological correlate of dementia in acquired immunodeficiency syndrome. Ann Neurol 36: 673-676.

Wiley CA, Masliah E, Morey M, et al. 1991. Neocortical damage during HIV infection. Ann Neurol 29: 651-657.

Wiley CA, Schrier RD, Nelson JA, Lampert PW, Oldstone MBA. 1986. Cellular localization of human immunodeficiency virus infection within the brains of acquired immune deficiency syndrome parients. Proc Natl Acad Sci USA 83: 7089-7093.

Wong MH, Robertson K, Nakasujja N, et al. 2007. Frequency of and risk factors for HIV dementia in an HIV clinic in sub-Saharan Africa. Neurology 68: 350-355.

Zhang K, McQuibban GA, Silva C, et al. 2003. HIV-induced metalloproteinase processing of the chemokine stromal cell derived factor-1 causes neurodegeneration. Nature Neurosci 6: 1064-1071.

Zink W, Anderson E, Boyle J. 2002. Impaired spatial cognition and synaptive potentiation in a murine model of HIV type I encephalitis. J Neurosci 22(6): 2096-2105.

20 Stem Cell Transplantation Therapy for Neurological Diseases

X.-Y. Hu · J.-A. Wang · K. Francis · M. E. Ogle · L. Wei · S. P. Yu

1	Introduction	492
2	*Stem Cell Characterization*	*492*
2.1	Embryonic Stem Cells	492
2.2	Fetal Stem Cells	493
2.3	Adult Stem Cells	493
2.3.1	Mesenchymal Stem Cells	493
2.3.2	Neural Stem Cells	493
3	*Potential Mechanisms of Tissue Repair and Functional Recovery Induced by Stem Cell Transplantation Therapy*	*494*
3.1	Cell Replacement and Integration with Host Circuitry	494
3.2	Increased Trophic Support	494
3.3	Enhanced Neurogenesis and Angiogenesis	495
3.4	Induction of Host Tissue Plasticity	495
4	*Stem Cell Transplantation Therapy for Neurological Diseases in Animal Models*	*496*
4.1	Stem Cell Therapy in Stroke	496
4.1.1	ES Cell Transplantation	496
4.1.2	MSC Transplantation in Stroke	497
4.1.3	Neural Stem Cell Therapy for Stroke	498
4.2	Stem Cell Therapy in Parkinson's Disease	500
4.2.1	ES Cell Transplantation in Parkinson's Disease	500
4.2.2	MSC Transplantation in Parkinson's Disease	500
4.3	Stem Cell Therapy in SCI	501
4.3.1	ES Cells Transplantation in Spinal Cord Injury	501
4.3.2	MSC Therapy in Spinal Cord Injury	501
4.4	Stem Cell Therapy in Peripheral Nerve Disease	502
4.4.1	ES Cell Transplantation	502
4.4.2	MSC Transplantation in Peripheral Injuries	502
5	*Human Embryonic Stem Cell Transplantation in Neurodegenerative Diseases*	*504*
6	*Clinical Application*	*505*
7	*Problems and Perspectives*	*506*

© 2009 Springer Science+Business Media, LLC.

Abstract: One great aspiration in modern medical research is the use of stem cell transplantation therapy for the treatment of neurological diseases. Due to their remarkable pluripotent/multipotent differentiation potential, their inherent plasticity and susceptibility to microenvironmental cues, stem cells provide an ideal resource for treatment by cell replacement, trophic support, and tissue repair in degenerative, ischemic, and traumatic injury disorders. The successful application of stem cell transplantation therapy, however, still requires a great deal of effort in basic and clinical research in order to understand the mechanisms associated with cell differentiation, cell survival, cell regeneration, and/or cellular integration. This technology also faces a number of safety and ethical issues. Stem cells derived from various species and a range of developmental origins have been tested for their potential in transplantation therapy. This review provides up-to-date information on some important areas in basic and preclinical investigations. It is hoped that with extraordinary efforts and further understanding of the mechanisms involved, stem cell transplantation, in combination with other strategies such as gene therapy and neuroprotective/regenerative treatments, will eventually be applied clinically for neurological diseases.

List of Abbreviations: ASC, adipose-derived stem cells; BBB, Basso–Beattie–Bresnehan; BDNF, brain-derived neurotrophic factor; bFGF, basic fibroblast growth factor; BNP, brain natriuretic peptide; CNS, central nervous system; CSF, cerebrospinal fluid; EPO, erythropoietin; ES, embryonic stem; GDNF, glial cell-derived neurotrophic factor; GFAP, glial fibrillary acidic protein; hESC, human embryonic stem cell; LIF, leukemia inhibitory factor; MAP2, microtubulin associated protein 2; MSCs, Mesenchymal stem cells; NeuN, neuron specific nuclear protein; NPCs, neural progenitor cells; NSCs, Neural stem cells; PD, Parkinson's disease; PET, Positron emission tomography; PNS, peripheral nervous system; RA, retinoic acid; SCI, spinal cord injury; 6-OHDA, 6-hydroxydopamine; VEGF, vascular endothelial growth factor

1 Introduction

Neurological diseases such as stroke, spinal cord injury (SCI), peripheral nerve injury, and Parkinson's disease (PD) lead to an array of sensory–motor dysfunctions and are a major cause of human disability. In the last two decades, many studies have focused on neuroprotective strategies to rescue injured cells in neurological diseases. Clinical trials, however, have demonstrated that preventing neurovascular damage during the acute phase of central nervous system (CNS) injury is a difficult task in the clinical setting. Thus, attention has increasingly shifted to more flexible post hoc therapy, including facilitating regeneration and repair of the damaged nervous system, which may be more clinically relevant.

Stem cells are capable of proliferation and differentiation into progenitor cells and ultimately mature cells. Stem cell transplantation has been tested experimentally in neurological disorders and provides new hope for the treatment of neurological diseases. The aim of stem cell transplantation includes enhancing trophic support for host tissue, replacement of injured cells, and tissue repair in disorders affecting the brain, spinal cord, and the peripheral nervous system (PNS).

2 Stem Cell Characterization

Stem cells differ from other precursor or progenitor cells in two essential ways. First, stem cells are self-renewing, theoretically possessing the ability to proliferate and reproduce themselves limitlessly. Second, stem cells are multi- or pluripotent, meaning they can not only divide to replenish themselves, but also have the potential to differentiate into multiple cell types. The plasticity and differentiation of stem cells are regulated by the microenvironment in which they reside. Developmental origin dictates the existence of three main categories of stem cells: embryonic stem (ES) cells, fetal stem cells, and adult stem cells.

2.1 Embryonic Stem Cells

Embryonic stem cells are those cells derived from the inner cell mass of a blastocyst. They are pluripotent in nature, as defined by their ability to differentiate into cells of the endoderm, mesoderm, or ectoderm including neurogenic, cardiogenic, and myogenic lineages. ES cells have been neurally induced in vitro by a number of protocols (Fraichard et al., 1995; Dinsmore et al., 1996; Brustle et al., 1997; Brustle et al., 1999). For example, treatment with retinoic acid (RA) induces ES cells to differentiate into neural progenitors that can give rise to three types of nerve cells: neurons, astrocytes, and oligodendrocytes (Bain et al., 1995). The pluripotency of ES cells makes them a good source for cellular replacement therapies in various diseases, such as ischemic stroke, SCI, Parkinson's disease, and heart failure.

2.2 Fetal Stem Cells

Fetal stem cells are isolated from fetal organs and represent more restricted progenitor cells compared to ES cells. Currently, the use of human fetal or embryonic stem cells faces ethical and legal issues in the clinic, limiting the widespread application and study of these cells.

2.3 Adult Stem Cells

Adult stem cells can be obtained from multiple adult organs, such as bone marrow, blood, muscle, skin, and adipose tissue. The adult stem cells that reside in each of these organs have distinct properties and characteristics. One advantage of the use of stem cells derived from adult tissue is they do not carry the same ethical stigma as fetal and embryonic-derived tissues. These type of cells also hold promise in circumventing the issue of immunologic rejection in transplantation therapy since adult stem cells can be harvested from a patient's own supply. Mesenchymal stem cells (MSCs), for example, can be obtained from a patient's bone marrow under local anesthesia and have become a very attractive source for autologous cell transplantation.

2.3.1 Mesenchymal Stem Cells

Bone marrow is a major source of two types of adult stem cells: hematopoietic stem cells and the less differentiated marrow MSCs. In addition to the property of self-renewal, marrow-derived MSCs have a number of defining characteristics that are useful in their isolation. They are plastic adherent when they are cultured under standard in vitro conditions. They have a characteristic surface marker profile, which is the expression of CD105, CD73, and CD90, but not CD45, CD34, CD14 or CD11b, CD79α or CD19 and HLA class II. These stem cells have been shown to differentiate into osteoblasts, adipocytes, and chondroblasts under in vitro differentiating conditions (Dominici et al., 2006). A number of reports have also shown the ability of these marrow MSCs, under the appropriate conditions, to differentiate into cells with neuronal phenotypes in vivo (Kopen et al., 1999; Brazelton et al., 2000) and in vitro (Wislet-Gendebien et al., 2005; Lei et al., 2007).

Another multipotent population of MSCs with cell surface markers similar to bone marrow MSCs has been isolated from subcutaneous fat deposits (Tholpady et al., 2003; Katz et al., 2005). These adipose-derived stem cells (ASC) were shown to differentiate into adipocytes, osteoblasts, chondrocyes, myoblasts, and neural-like cells in vitro (Zuk et al., 2001; Tholpady et al., 2003).

2.3.2 Neural Stem Cells

Neural stem cells (NSCs) have been identified in the developing and mature CNS, as well as the PNS. In the CNS, NSCs can be isolated from various regions, including the subventricular zone (SVZ), dentate gyrus, cortex, striatum, and spinal cord (Palmer et al., 1995; Palmer et al., 1999; Yamamoto et al., 2001). NSCs derived from the SVZ and dentate gyrus exist throughout the life span of mammals (Kuhn et al., 1996; Eriksson et al., 1998). NSCs have the ability to self-renew and differentiate, although they are more restricted than ES cells, giving rise to neurons, astrocytes, and oligodendrocytes (Gage, 2000). Because they are already partially lineage defined and require little manipulation to achieve neural differentiation, NSCs have an advantage over ES cells in regenerative therapy of the nervous system.

3 Potential Mechanisms of Tissue Repair and Functional Recovery Induced by Stem Cell Transplantation Therapy

In recent years, a growing number of studies have shown that stem cell transplantation therapy can restore lost function in various neurological diseases. Several mechanisms may be responsible for the therapeutic benefits observed (❯ *Figure 20-1*).

◘ Figure 20-1
Potential mechanisms of stem cell transplantation therapy. Several mechanisms may contribute to the therapeutic effects of stem cell transplantation, depending on cell types, stages of differentiation, gene expression, administration routes, and the host microenvironment

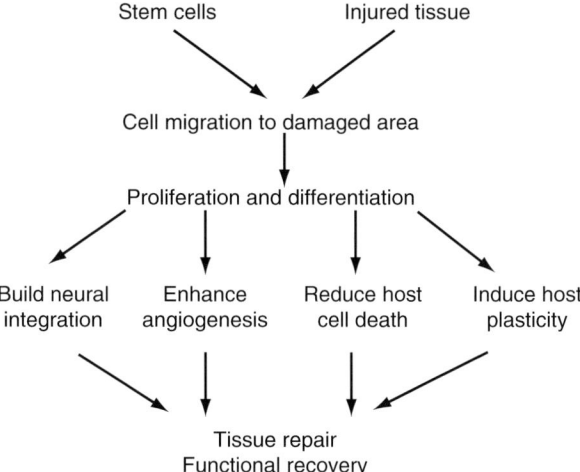

3.1 Cell Replacement and Integration with Host Circuitry

The principle attraction of stem cell transplantation is the promise to replace lost tissue and thus benefit long-term functional recovery. It has been demonstrated that transplanted stem cells can differentiate into neurons, astrocytes, and oligodendrocytes in vivo, which may replace and regenerate the injured tissue (Snyder et al., 1997). Engrafted neurons may establish functionally appropriate connections in the adult brain (Sotelo and Alvarado-Mallart, 1991; Hernit-Grant and Macklis, 1996). The capacity of transplanted stem cells to make such connections was found to be increased when the endogenous networks are damaged. This suggests that CNS injury or degenerative changes may trigger mechanisms regulating neuronal differentiation and connectivity similar to those that occur during development (Sotelo and Alvarado-Mallart, 1991; Hernit-Grant and Macklis, 1996; Snyder et al., 1997). Electron microscopy studies showed that human neural progenitor cells (NPCs) formed synapses with host circuits when transplanted into the ischemic brain (Ishibashi et al., 2004). Early functional recovery, however, can't be explained by these newly formed synapses (Englund et al., 2002; Song et al., 2002), indicating that other mechanism besides neural integration may be involved in recovery.

3.2 Increased Trophic Support

In addition to rebuilding damaged structures through cell replacement, the stem cells may play an important role in creating an enhanced local trophic environment. Both ES cells and MSCs have been shown to secrete various cytokines and growth factors, such as vascular endothelial growth factor (VEGF), basic fibroblast growth factor (bFGF), angiopoietin-1 (Ang1), erythropoietin (EPO), glial cell-derived neurotrophic factor (GDNF), and brain-derived neurotrophic factor (BDNF) (Hamano et al., 2000; Kinnaird et al., 2004). These factors provide the host tissue with trophic/regenerative support and prevent cell death of host tissue. Several studies have shown that cells transplanted during the acute phase of ischemic injury result in a neuroprotective phenomenon, including a reduction of endogenous cell death and diminished lesion size (Johnston et al., 2001; Llado et al., 2004; Kurozumi et al., 2005). Trophic factor release by transplanted stem cells may be enhanced by gene modification and stem cell preconditioning before transplant. For example, our recent investigations show that in vitro hypoxic preconditioning of ES cells or MSCs significantly increases the expression of hypoxia-inducible factor 1, Ang-1, VEGF, and its receptor fetal liver kinase-1 (Flk-1), EPO, and the antiapoptotic genes *Bcl-2* and *Bcl-xL* in these cells. The preconditioned cells demonstrate increased survival and promote angiogenesis both in vitro and after transplantation (Hedrick et al., 2004; Keogh et al., 2004).

Stem cells can also synthesize and release brain natriuretic peptide (BNP) (Song et al., 2004), which, like its close homolog atrial natriuretic peptide (ANP), exerts powerful natriuretic, diuretic, and vasodilatory effects. Transplanted MSCs may facilitate recovery from brain and spinal cord lesions by releasing BNP and other vasoactive factors that reduce edema, decrease intracranial pressure, and improve cerebral perfusion (Song et al., 2004).

3.3 Enhanced Neurogenesis and Angiogenesis

It has been established that stem cells residing in the SVZ and dentate gyrus of the adult mammalian brain have the capacity to divide, differentiate into neural cell types, and migrate to sites of injury (Palmer et al., 1995; Palmer et al., 1999; Yamamoto et al., 2001). The precise mechanisms of this endogenous response to injury are still being elucidated. One recent study showed that injection of BDNF, a factor expressed by certain stem cell populations (Crigler et al., 2006), enhanced the proliferation and migration of NSCs in the brain after stroke (Schabitz et al., 2007). Many groups have shown that transplantation of stem cells increases endogenous neurogenesis putatively through the expression and release of certain growth factors such as BDNF (Li et al., 2000).

An angiogenic environment is also essential for tissue repair and functional recovery after an ischemic insult. It is well known that ischemic insults can activate angiogenesis (Krupinski et al., 1994; Senior, 2001; Wei et al., 2001). However, injury-induced angiogenesis is not sufficient to support transplanted cells and susceptible host cells for long-term survival. It is also not sufficient for the tissue plasticity required for functional recovery (Chen et al., 2003). In this regard, angiogenesis induced by transplanted stem cells may be important for therapeutic benefits (Hamano et al., 2000; Shen et al., 2006). The effect of stem cell therapy on neovascularization can be explained by multiple nonexclusive pathways. First, transplanted stem cells differentiate into endothelial cells and smooth muscle cells to support and/or form new vessel components, as shown in our recent investigations (❯ *Figure 20-2*). Second, transplanted cells may release angiogenic and/or trophic factors, which facilitate angiogenesis (see later). It has been reported that human bone marrow stromal cells may promote angiogenesis by increasing endogenous levels of VEGF and one of its receptors (Chen et al., 2003).

Endothelial progenitor cells (EPCs) have been engineered ex vivo to overexpress angiogenic growth factors. The gene-modified EPCs were shown to rescue impaired neovascularization in an animal model of limb ischemia (Iwaguro et al., 2002). Understanding the regulation of angiogenesis and the contribution of transplanted cells to this process are necessary to promote a controlled angiogenic response, resulting in the appropriate formation of a functional vascular network in the desired location and time.

3.4 Induction of Host Tissue Plasticity

Recent studies suggest that cell transplantation may produce beneficial increases in host plasticity (Li et al., 2002; Howells, 2003). Such events include increasing neural connections between the injury site and both adjacent and contralateral regions, restoration of local synaptic activity, and enhancement of existing

◘ Figure 20-2
ES cell-derived Glut-1-positive cells in the ischemic region. (a) Seven days after transplantation, the ischemic core was filled with Hoechst prelabeled ES cells (Hoechst-positive, blue) and contained Glut-1-positive vascular-like structures of endothelial cells (green; *arrow*). Scale bar = 50 μm. (b) Enlarged view of a vascular structure in a, positively stained with Glut-1 (green). Scale bar = 10 μm. (c) Glut-1-positive endothelial cells (green) were also colabeled with Hoechst (blue), verifying their origination from transplanted ES cells. The pink color is from neighboring NeuN-positive cells. Images were taken using an inverted fluorescence microscope (Wei et al., 2005)

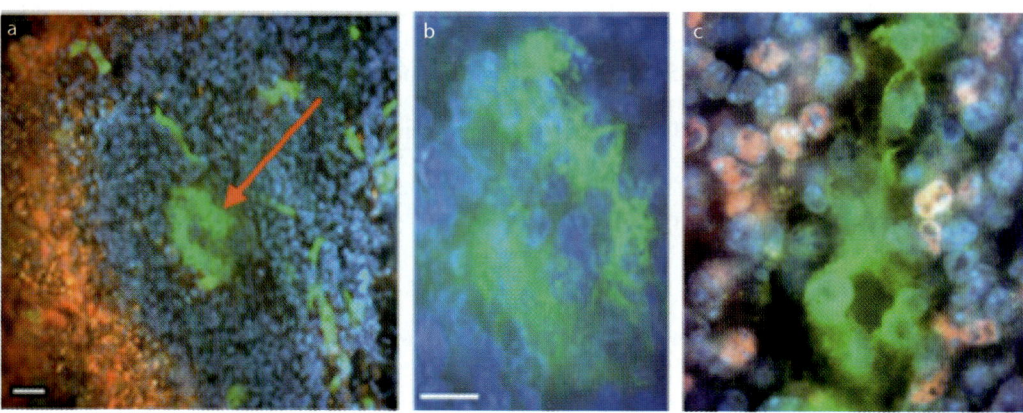

synapses, as well as activation of silent synapses (Carmichael, 2003; Dancause et al., 2005; Carmichael, 2006). In rat SCI models, neural stem cell transplantation after insult promotes host axonal growth, at least in part due to growth factor secretion (Lu, et al., 2003). Human cord blood cell transplantation to the ischemic cortex increased the growth of nerve fibers from the contralateral to the ischemic hemisphere (Xiao et al., 2005). Shen et al (2006) reported that intracarotid administration of human bone marrow stromal cells promoted synaptophysin expression in the penumbra region surrounding the ischemic core.

4 Stem Cell Transplantation Therapy for Neurological Diseases in Animal Models

4.1 Stem Cell Therapy in Stroke

Stem cell transplantation may aid in the restoration of lost brain function by integrating into functional synaptic networks within host tissues or contributing trophic support for neurogenesis and angiogenesis in and around damaged areas (Kordower et al., 1995; Borlongan et al., 1998).

4.1.1 ES Cell Transplantation

ES cells exposed to RA in vitro have been shown to differentiate into NPCs. Studies from our group and others have shown that neurally differentiated ES cells transplanted in animal models of focal ischemia could survive, migrate, and differentiate into various types of neuronal and glial cells, re-establish connections with target areas, and contribute to functional recovery (Liu et al., 2000; Ikeda et al., 2005; Wei et al., 2005). Hayashi et al transplanted monkey ES cells into the ischemic mouse brain and found that transplanted cells migrated widely throughout the ischemic brain. They also demonstrated that 28 days after transplantation, there were ten times more cells in the graft labeled with Fluoro-Gold (FG) following stereotactic focal injection into the anterior thalamus and substantia nigra, when compared with 14 days post transplant. These data suggested that ES cell transplantation results in network formation in the ischemic brain (Hayashi et al., 2006). Transplanted cells displayed characteristics of electrophysiological activities of voltage-gated sodium currents. Additionally, spontaneous excitatory postsynaptic currents were observed in graft-derived cells during the first 7 weeks after transplantation, indicating synaptic input (Buhnemann et al., 2006).

The therapeutic benefits achieved by transplantation can be improved through modification of stem cells to increase the survival of transplanted cells. Our group showed that transplantation of ES cells overexpressing the antiapoptotic gene *Bcl-2* enhanced the survival of ES cells transplanted into the ischemic core. With this increased survival, we also observed increased neuronal differentiation of transplanted ES cells and an overall improved functional recovery of the animals. Collectively, these studies indicate a benefit for ES cell transplantation in stroke, as well as a need to further study mechanisms to enhance cell survival upon transplantation. (Wei et al., 2005) (❯ *Figure 20-3* and ❯ *20-4*).

4.1.2 MSC Transplantation in Stroke

Adult MSCs have been transplanted into the lateral ventricles of the brain of E15.5 mouse embryos in utero. Transplanted cells showed survival, migration, engraftment, and differentiation into neurons (Munoz-Elias et al., 2004). MSCs transplanted into the lateral ventricles of postnatal day 3 mice also showed neuronal and astroglial differentiation (Kopen et al., 1999). These studies indicate that transplanted MSCs are able to integrate and differentiate into neurons and glia within the developing CNS.

The effect of MSC transplantation has been tested following cerebral ischemia in animal models. Previous investigations have shown that intrastriatally transplanted MSCs 4 days after middle cerebral artery occlusion can survive, migrate, and differentiate into neuronal or glial cells in a rodent model. Although there was no difference in infarct volume between animals which did or did not receive cell transplantation, mice receiving MSC transplant showed significant improvement of behavior 28 days later compared with focal ischemia alone (Li et al., 2000). Both intravenous and intracarotid administration of MSCs improve functional recovery in stroke animal models (Chen et al., 2001; Li et al., 2001). MSCs exposed to neurotrophic factors before transplantation produced greater neurological functional recovery compared with control MSCs. Our studies have also shown that hypoxic preconditioning of MSCs prior to transplantation enhanced cell survival compared to cells cultured under normoxic (21% O_2) conditions.

Immunohistochemical staining of brain section revealed that a small percentage of transplanted MSCs expressed the neuron specific nuclear protein (NeuN), microtubulin associated protein 2 (MAP2), and the astrocytic marker glial fibrillary acidic protein (GFAP). Chopp and colleagues reported that only 1% of labeled MSCs expressed neuronal markers after transplant via intrastriatal administration. In a different study, only 0.02% of two million MSCs transplanted through the carotid artery stained positively for neural markers.

Despite the low rate of expression of neuronal markers, subsequent studies have shown that MSC transplantation reduced host cell death and increased the expression levels of BDNF, nerve growth factor, and bFGF in the ischemic brain hemisphere, as well as increased the number of mitotic cells within the ipsilateral SVZ (Li et al., 2002; Chen et al., 2003). Therefore, the functional recovery achieved from MSC transplantation therapy may be largely related to the effects of trophic support. This concept is supported by the evidence that GDNF or BDNF overexpression in transplanted MSCs protects against injury and result in greater functional recovery in cerebral ischemia models (Kurozumi et al., 2004; Horita et al., 2006). The observed benefit likely lies in the reduction of parenchymal cell death and stimulation of endogenous cell proliferation, promoting tissue repair and regeneration.

In addition, transplantation of MSCs enhanced angiogenesis in postischemic tissues (Hu et al., 2007). As discussed in Section 3.3, increased angiogenesis may enhance delivery of oxygen and neurotrophic factors to the local environment, thereby benefiting functional recovery.

Figure 20-3

Transplanted ES cells in the postischemic brain. ES cells in the ischemic cavity and immunostaining 7 days after transplantation. (a) Distribution of transplanted ES cells prelabeled with Hoechst in the postischemic basal ganglia revealed by the blue fluorescence. Note the ES cell narrow aggregations along injection paths from two transplantation sites (site 1 and site 2; *arrows*). (b) Distribution of transplanted ES cells in the postischemic cortex. Hoechst-positive ES cells (blue) were present next to undamaged tissue (*arrow* and the area to the *left*). NeuN staining (red) shows native neurons in nonischemic cortical region (*left* of the *arrow*) and differentiated neuron-like cells in the ischemic region (*right* of the *arrow*). The NeuN staining on the *right* appears pink in color due to overlap with the blue Hoechst staining; the mouse ES cells are also visibly smaller than the NeuN-positive neurons in the adjacent normal rat brain. (c) Mouse cell-specific antibodies M2 and M6 (both green) and Hoechst staining (blue) verified that the cells surviving 7 days in the rat host ischemic cortex originated from transplantation. NeuN staining (pink) identified differentiated neuronal cells. (d) Higher magnification of a confocal image shows doublelabeling of Hoechst (blue) and mouse antibody M2/M6 (green), confirming the murine origin of these cells (white *arrow*). Triple-labeling with the neuronal marker NeuN (pink) identifies the ES cell-derived neurons (red *arrow*). Scale bar equals 800, 200, 20, and 10 μm, in a, b, c, and d, respectively (Wei et al., 2005)

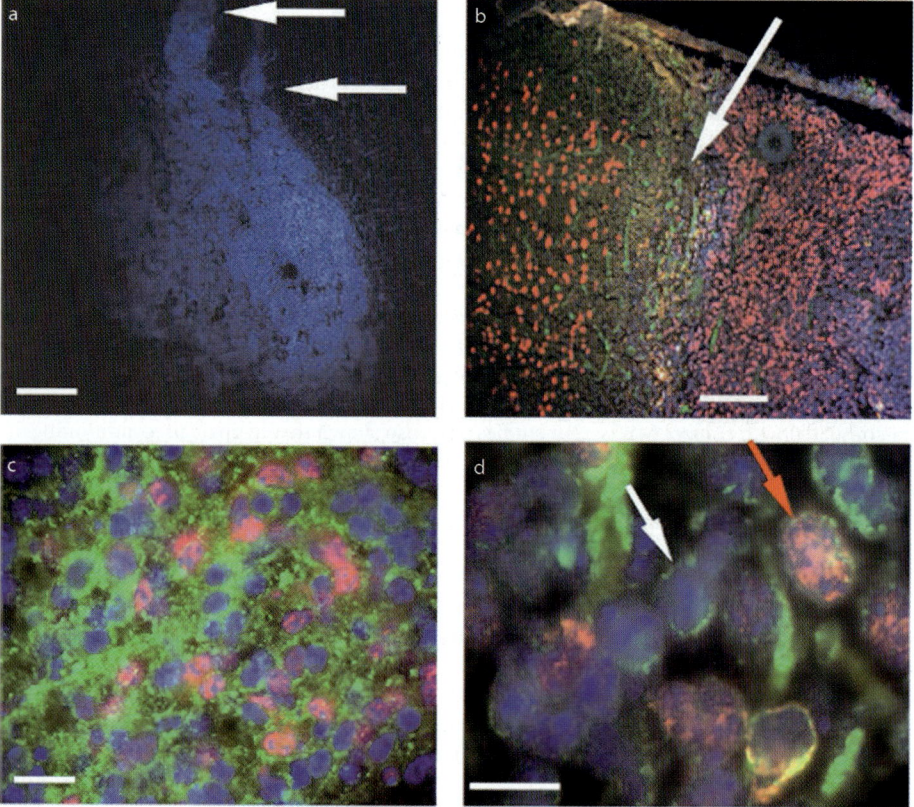

4.1.3 Neural Stem Cell Therapy for Stroke

4.1.3.1 Endogenous NSCs The mobilization of endogenous NSCs is one potential solution for achieving neural repair. Various diseases can stimulate the proliferation and migration of endogenous NSCs. Stroke itself is able to elicit the migration of SVZ cells (Yagita et al., 2001; Li et al., 2002; Takasawa et al., 2002).

◘ Figure 20-4
Dendrite growth after ES cell transplantation in the ischemic cortex. NeuN-positive cells and apical dendrite distribution in the contralateral cortex (a, b, and c) and ipsilateral cortex (d, e, and f). (a and b) In the contralateral nonischemic cortex, normal neurons were stained with NeuN, appearing yellow or orange in color. Neuronal processes and dendrites were labeled with neurofilament (NF) antibody (green). (c) Confocal image of a normal neuron in an 8-μm thick slice, showing the NeuN-positive cell body (red) and its extended apical dendrites stained by NF (green). Several more neurons can be seen in the background. (d and e) Seven weeks after ES cell transplantation, NeuN and NF double immunostaining in the ischemic core region of the ipsilateral cortex under an inverted fluorescence microscope. Many NeuN-positive cells (yellow or orange) were surrounded by NF labeled processes (green); the distribution of these processes, however, was less organized compared with that in image A. (f) Confocal image of NeuN/NF-positive ES cell in the 8-μm thick slice of ischemic cortex. The cell body was positive for NeuN staining (red), cell processes were positive to NF staining (green). The Hoechst labeling (blue) showed several transplanted ES cells and indicated the ES cell origin of the NeuN/NF-positive cell. Scale bar=60 μm in a and d, 20 μm in b, c and e. 10 μm in f (Wei et al., 2005)

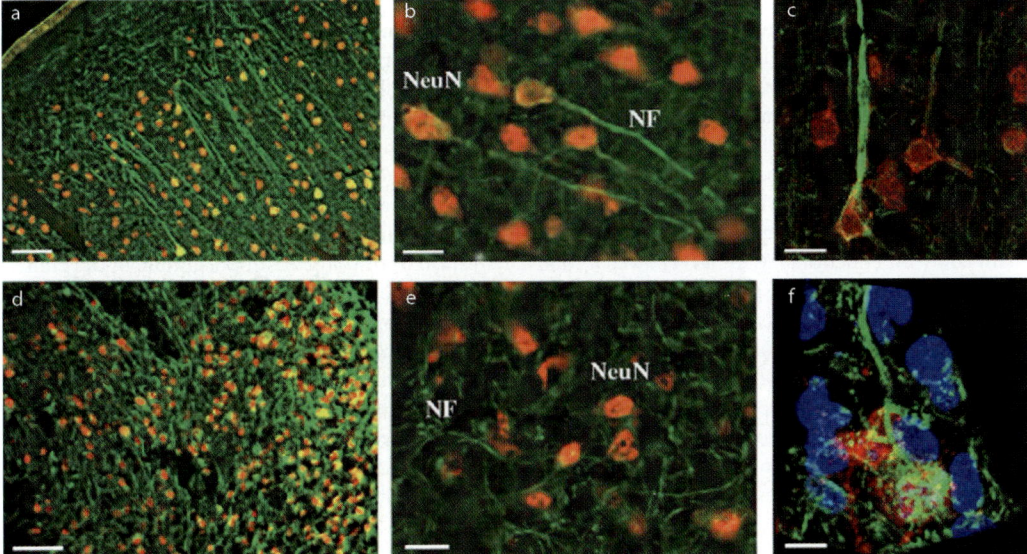

The mechanisms guiding NSC migration to the lesion site are not very clear. The migrating endogenous NSCs can replace lost neurons and enhance neurogenesis after stroke, but the fraction of cell replenishment is low. Experimental data suggested that only 0.2% of lost cells are replaced by newly generated neurons (Arvidsson et al., 2002). Strategies that can promote neurogenesis and cell migration will greatly enhance endogenous repair mechanisms.

4.1.3.2 Neural Stem Cell Transplantation Previous studies showed transplantation of exogenous NSCs into the damaged brain could potentially restore neuronal networks. Human embryonic or fetal-derived NSCs were shown to differentiate into neurons and astroglia after transplantation into the ischemic brain, thereby resulting in functional recovery (Mattsson et al., 1997; Kelly et al., 2004). Adult NSCs have been transplanted into multiple animal models to assess the therapeutic effect. These studies showed transplanted cells survived and migrated toward the ischemic area in adult rats, enhancing functional recovery (Taupin and Gage, 2002; Zhang et al., 2003) and indicating their potential for stroke therapy.

4.2 Stem Cell Therapy in Parkinson's Disease

Parkinson's disease is a chronic neurodegenerative disorder characterized by bradykinesia, tremors, rigidity, and impaired postural reflexes. In addition to motor symptoms, mental disorders like depression or psychosis, as well as autonomic and gastrointestinal dysfunction, may occur (Schrag et al., 2000). The main pathological phenomena in PD is a rather selective degeneration of nigrostriatal neurons leading to the severe loss of dopaminergic (DA) neurons in the striatum, combined with the presence of intraneuronal inclusions, or lewy bodies (Braak et al., 2003; Lindvall and Kokaia, 2004). Cell transplantation strategy in PD has been based on restoration of DA neurotransmission in the striatum by cellular grafts. Implantation of fetal dopamine neurons has shown a reduction of Parkinsonian symptoms in patients (Freed et al., 1992; Kordower et al., 1995). The lack of this cell source limits its utility and therefore alternate stem cell transplantation therapies provide another option for cell replacement in PD.

4.2.1 ES Cell Transplantation in Parkinson's Disease

ES cells can differentiate into DA neurons after in vitro neural induction. Several studies have provided evidence that DA neurons derived from mouse and nonhuman primate ES cells can integrate and benefit behavioral recovery after transplant in animal PD models (Bjorklund et al., 2002; Kim et al., 2002; Barberi et al., 2003; Sanchez-Pernaute et al., 2005; Takagi et al., 2005; Kim et al., 2006). When undifferentiated mouse ES cells were transplanted into a PD rat model, a portion of these implanted cells could differentiate into DA neurons spontaneously. Positron emission tomography (PET) and carbon-11-labeled 2β-carbomethoxy-3β(4-fluorophenyl) tropane ($[^{11}C]CFT$) were used to assess the functional imaging of ES-derived DA neurons. It showed a significant increase of $[^{11}C]CFT$ binding in the grafted striatum as compared with sham controls. Functional MRI also showed a robust activation in response to amphetamine in the grafted striatum and ipsilateral sensorimotor of ES cell grafted animals. The functional effects possibly produced by ES cell-derived DA neurons resulted in behavioral recovery of amphetamine-induced motor asymmetry 9 weeks after transplantation (Bjorklund et al., 2002). It was reported that neurons derived from mouse ES cells survived over 32 weeks after transplantation and improved behavioral effects. Microdialysis in the grafted animals showed depolarization and pharmacological stimulation-induced dopamine release. PET showed the normalization of upregulated postsynaptic dopaminergic D2 receptors and the expression of presynaptic dopamine transporters in the graft. These data indicate that DA neurons derived from transplanted ES cells can restore both pre- and postsynaptic functions (Rodriguez-Gomez et al., 2007).

4.2.2 MSC Transplantation in Parkinson's Disease

MSCs are able to differentiate into DA neurons via induction or genetic modification. MSCs modified by an adeno-associated viral vector with tyrosine hydroxylase (TH) and GTP cyclohydrolase I genes synthesized 3,4-dihydroxyphenylalanine (L-DOPA) both in vitro and in vivo. The expression of transgenes, however, ceased about 9 days after transplantation and the beneficial functional effect was only seen within 1 week of transplantation (Schwarz et al., 1999). Notch intracellular domain (NICD) gene transfection and subsequent treatment of MSCs with bFGF, forskolin, and ciliary neurotrophic factors showed MSCs could be efficiently and specifically induced to become cells with neuronal characteristics. Some of these neuron-like cells exhibited voltage-gated fast sodium and delayed rectifier potassium currents and action potentials comparable to currents of functional neurons. Further treatment of induced neuronal cells with GDNF increased the proportion of TH-positive and dopamine-producing cells. Transplantation of these GDNF-treated cells increased functional recovery following intrastriatal implantation in a 6-hydroxy dopamine rat model of PD (Dezawa et al., 2004). Animals with MSCs transplanted into the body of the striatum significantly reduced the number of turns induced by amphetamine (Pavon-Fuentes et al., 2004). Together these data indicate that induced MSCs exhibiting neuronal-like phenotype and physiological behavior can ameliorate the symptoms of PD in animal models.

4.3 Stem Cell Therapy in SCI

SCI occurs with an annual incidence of 15–40 cases per million individuals throughout the world, resulting from traffic accidents, community violence, sports, recreational activities, and workplace-related injuries (Sekhon and Fehlings, 2001). The potential therapeutic objectives for SCI fall into several categories, including cell protection, axonal regeneration, rehabilitation, and the repair of damaged tissues by transplantation therapy (Ramer et al., 2005).

4.3.1 ES Cells Transplantation in Spinal Cord Injury

ES cells neurally induced by the 4−/4+ RA protocol (4 days of leukemia inhibitory factor (LIF) withdrawal followed by 4 days in RA) can reliably produce oligodendrocytes. Using immnohistochemical markers and scanning and transmission electron microscopy, it was found that these oligodendrocytes could rapidly myelinate axons in vitro (Liu et al., 2000). After transplantation into the injured rat spinal cord, ES cells survived, differentiated into oligodendrocytes, and migrated away from the lesion edge. Animals receiving ES cell transplantation also showed partial functional recovery, showing higher Basso–Beattie–Bresnehan (BBB) locomotor rating score 6 weeks after transplantation (McDonald et al., 1999). When ES cells were transplanted into chemical demyelinated spinal cord or myelin-deficient shiverer mutant mice, a number of transplanted cells survived and differentiated into mature oligodendrocytes which produced myelin and myelinated host axons (Brustle et al., 1999; Liu et al., 2000; Nistor et al., 2005). A previous study by McDonald et al. showed around 60% of transplanted ES cells developed into oligodendrocytes 2 weeks after transplantation, approximately 10% turned into neurons, and about 30% became astrocytes (McDonald et al., 2004). These data indicate ES cell transplantation is able to replace lost myelin in the injured spinal cord.

4.3.2 MSC Therapy in Spinal Cord Injury

Stem cells are able to give rise to neurons and glia after transplantation into the spinal cord, although the percentage of differentiation toward neurons versus glia differs among studies (Lee et al., 2003; Kinnaird et al., 2004; Zhao et al., 2004). About 10% of transplanted MSCs may be NeuN- or MAP2-positive, while 15% may be GFAP-positive. Some studies have reported that most of the cells transplanted into the spinal cord differentiated into glial cells. In general, MSC transplanted animals with spinal cord injury show increases in the degree of hindlimb locomotor recovery, although the exact mechanisms by which MSCs may be beneficial are not well-defined. Neurotrophic and neurite elongation-facilitating actions may play important roles. It has also been shown that a tissue matrix formed by MSC grafts supported greater numbers of axons and better longitudinally directed axonal growth than that found in controls (Ankeny et al., 2004). Direct microinjection of these cells into the demyelinated spinal cord of immunosuppressed rats resulted in remyelination and improved conduction velocity. The remyelinated axons showed characteristics of both central and peripheral myelination. MSCs colocalized with myelin basic protein and myelin protein zero (P0) cellular elements, indicating that these cells can form myelin in the demyelinated spinal cord (Akiyama et al., 2002). Transplantation of MSCs alone may not be sufficient to stimulate axonal growth beyond lesion sites. Gene-modified MSCs may provide additional advantages over native MSCs, as transduction of MSCs overexpressing BDNF resulted in a significant increase in the extent and diversity of host axonal growth, enhancing the growth of host serotonergic, cerulospinal, and dorsal column sensory axons (Lu et al., 2005).

The route of cell administration is another key issue when considering therapeutic MSC transplantation for SCI. On one hand, the anatomy of focal injury would suggest that direct intralesional cell transplantation might facilitate regeneration or protection of demyelinated axons within a specific segment of the spinal cord. After SCI, intralesionally transplanted MSCs integrate with the host tissue, differentiate into neurons and glial cells, release neurotrophic growth factors, promote neurogenesis, and inhibit reactive astrogliosis, thereby favoring motor recovery (Wu et al., 2003; Ankeny et al., 2004; Lu et al., 2004;

Zurita and Vaquero, 2004). On the other hand, there are always concerns about the risk of direct local injection of cells into the lesion causing additional damage to the spinal cord. Thus, other transplantation routes focusing on minimizing damage to the host tissue have been developed, such as intravenous injection or intrathecal infusion via lumbar puncture (LP). Intravenously delivered MSCs survived and populated the lesion 4 weeks after cell implantation and improved scoring in both the BBB locomotor rating and the plantar reflex test (Sykova and Jendelova, 2005). After infusion into the cerebrospinal fluid (CSF), MSCs were conveyed through the CSF to the spinal cord, where most MSCs attached to the spinal surface but few invaded into the lesion (Ohta et al., 2004). Rats with cell transplantation had a smaller cavity volume and higher BBB score than control rats. This suggested that MSCs can exert effects to reduce lesion cavitation and improve behavioral function (Ohta et al., 2004) by secreting trophic factors into the CSF or by contact with host spinal tissues. A study by Bakshi et al compared the transplantation routes of intravenous, intraventricular, and LP injection, finding the use of LP and intraventricular routes allowed more efficient targeting of cells to the injured tissue compared with IV injection (Bakshi et al., 2004).

4.4 Stem Cell Therapy in Peripheral Nerve Disease

Peripheral nerve injuries are common and often result in incomplete or no functional recovery (Kline and Hudson, 1995). The reason for incomplete recovery lies in the progressive decline in the ability of neurons to sustain axon growth, as chronically denervated Schwann cells atrophy and fail to provide a supportive axonal growth environment (Fu and Gordon, 1997). Increasing evidence supports the concept that Schwann cells provide a highly preferred substrate for axonal migration, releasing neurotrophic factors that further enhance nerve migration. Schwann cell transplantation can promote axonal regeneration, but the inadequate supply of cells for transplantation has limited clinical applications. Recently, stem cell transplantation, including ES cells, NPCs, and MSCs were investigated as potential cellular components for a bioartificial nerve grafts (Mosahebi et al., 2002).

4.4.1 ES Cell Transplantation

Our group has shown that by using the 4−/4+ RA neural induction protocol, ES cells can differentiate into neural lineage cells (Wei et al., 2005). Three days after RA induction, many of these cells express the oligodendrocyte marker O4 and Schwann cell marker S100. To assess ES cell potential in vivo, cells underwent 4−/4+ induction and were then transplanted into rats with sciatic nerve transection. One to three months later, immunostaining revealed that the implanted cells colabeled with the Schwann cell marker S100, suggesting that transplanted ES cells differentiated into myelinating cells. Regenerated axons were myelinated and showed a uniform connection between proximal and distal stumps. Three months after transplantation, gross examination showed marked differences between animals that received vehicle injection and those that received ES cell transplantation. Nerve stumps receiving ES cell implantation had near normal diameter and appeared healthy with longitudinally oriented, densely packed Schwann cell-like phenotype (❷ *Figure 20-5* and ❷ *20-6*). Neurons retrogradely labeled with Fluoro-Gold were found in the spinal cord and dorsal root ganglia after ES cell transplantation, suggesting reconnection of axons across the transection. Electrophysiological recordings showed functional activity was recovered across the injury gap. These data suggest that transplanted neurally induced ES cells differentiate into myelin-forming cells and provide a potential therapy for severely injured peripheral nerves (Cui et al., 2007).

4.4.2 MSC Transplantation in Peripheral Injuries

Recent investigations have demonstrated the transplantation of MSCs in animal models of peripheral nerve injury. With RA treatment and culture in the presence of forskolin, bFGF, PDGF, and heregulin, MSCs differentiated to Schwann-like cells and expressed p75, S100B, GFAP, and the oligodendrocyte marker O4.

◘ Figure 20-5

Sciatic nerve damage, regeneration, remyelination, and effects of ES-NPC transplantation. (a–c) Gross anatomy of the sciatic nerve after axotomy and removal of a 1 cm nerve segment. (a) Three months after axotomy, a significant decrease in the diameter of the sciatic nerve was observed in rats receiving medium. (b) Repaired, normal looking sciatic nerve in rat 3 months after axotomy and ES cell transplantation. (c) Normal (nontransected) sciatic nerve (*left*) in comparison with sciatic nerve 3 months after axotomy plus ES cell transplantation (*right*). (d) Quantification of sciatic nerve diameter at proximal stump site 1 and 3 months after axotomy and transplantation. In rats that received ES cell transplantation, nerve diameter recovered to normal levels while medium injection alone showed no such effect; *$P<0.05$ compared with sham controls. (e–g) Sciatic nerve regeneration and remyelination. Three months after sciatic nerve injury, nerve regeneration and remyelination were revealed by immunohistochemical double-staining with the neurofilament marker SMI-31 (green) and myelin (red, appeared as orange). (e) Normal nerve after sham operation. (f) Lack of regeneration of transected nerve, showing scattered SMI-31 and myelin staining within deficient gap. (g) Repaired sciatic nerve after nerve axotomy and ES cell transplantation. Continuous SMI-31 and myelin staining demonstrates vigorous axonal regeneration and remyelination across the transection gap. (h–j) H&E histology of sciatic nerve damage and the effect of ES cell transplantation. Three months after sciatic nerve transection, H&E staining showed visible differences in diameter and axonal density of sciatic nerve in rats. Bar=150 μm. (h) Sciatic nerve from sham-operated animal. (i) Sciatic nerve from animal that had transection and injection with culture medium. (j) Sciatic nerve after axotomy plus ES cell transplantation, showing regenerating nerve fibers (*) extending form proximal stump (*arrow*) (Cui et al., 2007)

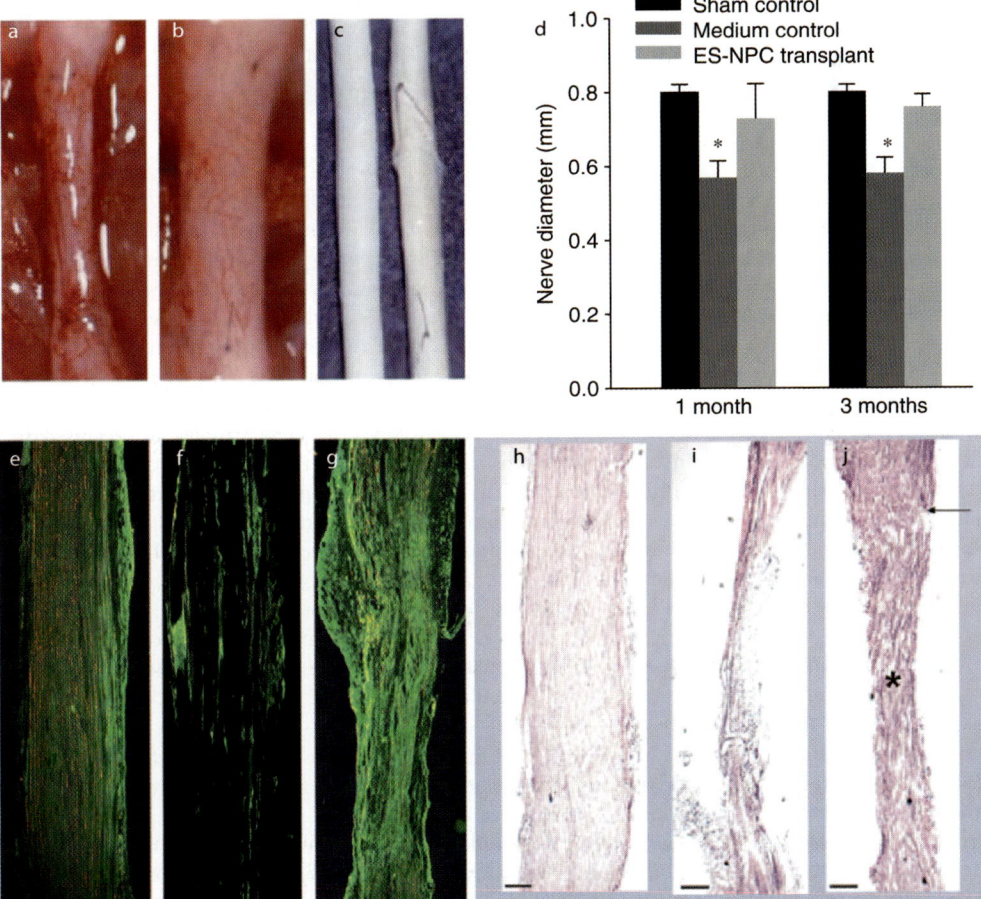

Figure 20-6

Transplanted ES cells and their differentiation into Schwann-like cells. Immunostaining with a mouse cell specific antibody (M2/M6) showed mouse ES cells 1 month after their transplantation into rats of sciatic nerve axotomy. (a) M2/M6-positive cells demonstrate transplanted ES cells; (b) Staining of glia/Schwann cell marker S-100 in the same section of the sciatic nerve. (c) Merged image of a and b. (d) Higher magnification image from a different section of the sciatic nerve. (e–g) M2/M6 and S-100 staining in the transplantation region of the sciatic nerve 3 months after transplantation. The majority of M2/M6-positive cells were also S-100-positive, indicating that transplanted ES cells survived and differentiated into Schwann-like cells. (h) Quantification of S-100 and M2/M6 double-stained cells. ES cell transplanted rats showed increasing number of double staining during 3 months recovery (Cui et al., 2007)

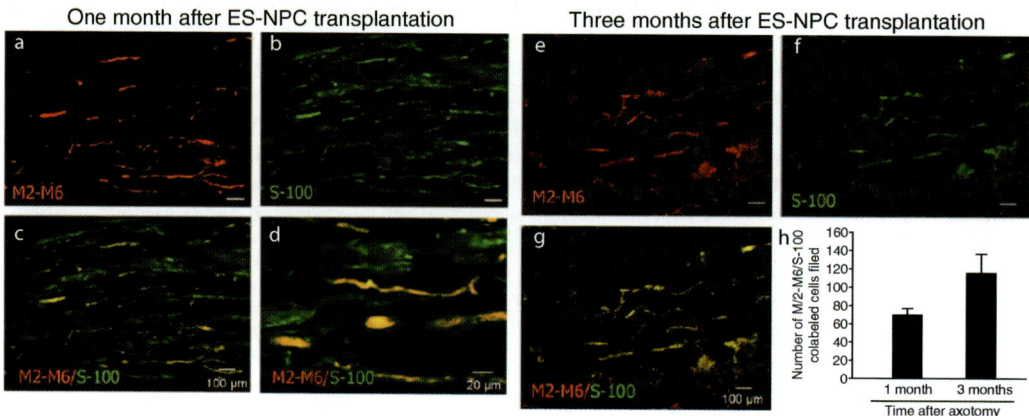

These MSCs were able to differentiate into myelinating cells capable of supporting nerve fiber regrowth in the injured sciatic nerve (Dezawa et al., 2001). It was reported that MSCs survived in the injected area of the transected sciatic nerve at least 33 days after implantation and almost 5% of cells expressed a Schwann cell-like phenotype (Cuevas et al., 2002; Cuevas et al., 2004). The migration and differentiation of MSCs following injection into the site of a sciatic nerve axotomy resulted in significant improvement on a walking track test at 18 and 33 days following transplantation compared with controls. Another study observed the long-term effect of MSC transplantation in peripheral nerve injury. MSC-derived Schwann cells (M-Sch) were suspended in Matrigel and transferred into permeable tubes to make an artificial graft. This artificial graft was interposed into the transected sciatic gap. Six months later, the mean sciatic nerve functional index (SFI) and mean motor nerve conduction velocity (MNCV) were significantly improved in transplantation animals compared to control groups. Immunohistochemical analysis revealed that transplanted MSCs were positive for P0 and myelin associate glycoprotein (MAG), reconstructed nodes of Ranvier, and remyelinated regenerating nerve axons (Mimura et al., 2004).

5 Human Embryonic Stem Cell Transplantation in Neurodegenerative Diseases

In contrast to the promising results of mouse ES cell transplantation studies, human embryonic stem cell (hESC) transplantation to both injury and noninjury models has been fraught with difficulties. Since isolation of the first hESC line by Thomson et al. (1998), research has been undertaken by many groups to better understand these cells (Thomson et al., 1998). A multitude of challenges have arisen concerning the expansion and differentiation of these cells, including epigenetic alterations affecting genomic stability, feeder cell contamination, and karyotypic abnormalities (Draper et al., 2004; Klimanskaya et al., 2005; Maitra et al., 2005).

To compound matters further, parallel studies comparing multiple cell lines demonstrate differences between NIH approved hESC lines in cell growth, differentiation, maintenance of pluripotency, transfection efficiency, and genomic stability (Ware et al., 2006). hESC lines have also been shown less responsive to known mESC regulatory pathways such as the LIF–signal transducer and activator of transcription-3 (STAT-3) and express vastly different markers of pluripotency (Daheron et al., 2004; Koestenbauer et al., 2006).

In spite of these difficulties, a number of groups have been able to direct neural differentiation to achieve a high yield of specific neuronal phenotypes. Studies have used various combinations of growth factors and media supplements, including retinoic acid, fibroblast growth factors, sonic hedgehog and N2, to achieve directed neuronal differentiation (Reubinoff et al., 2001; Schulz et al., 2004; Itsykson et al., 2005; Benzing et al., 2006). For example, a published protocol by Benjamin Reubinoff's group utilizes a high concentration of the bone morphogenetic protein (BMP) antagonist Noggin to direct floating cultures of isolated hESCs to a neural fate (Itsykson et al., 2005).

To date, transplantation of hESC-derived neural precursors into rodent models has achieved limited success. Early studies found significant hESC death, limited differentiation, and formation of teratomas in transplanted animals, though more recent work suggests significant promise. Fred Gage's group was able to demonstrate by immunohistochemistry and measurement of action potential that hESC-derived neural precursors transplanted in utero could develop into functional neurons (Muotri et al., 2005). Green fluorescent protein-tagged hESC-derived neural precursors have also been shown to survive transplantation and migrate to the SVZ and olfactory bulb following striatal administration (Benzing et al., 2006).

The transplantation of hESCs to Parkinson's disease models has attracted significant research with some early success. One group was able to induce a large portion of dopaminergic-like neurons from the BGO1 and BGO3 hESC lines using combinations of serum-free media, neural growth factors, and bFGF, producing tyrosine hydroxylase expressing neurons in vivo (Schulz et al., 2004). hESC-derived dopaminergic precursors have recently been shown to differentiate to a TH-positive dopaminergic phenotype in vivo and potentially ameliorate functional deficit in a 6-hydroxydopamine (6-OHDA)-lesioned Parkinsonian model (Roy et al., 2006). However, some of these cells appeared to remain immature and retain the potential for cell division, suggesting the possibility of tumor formation. A second group using an alternate model for dopaminergic differentiation observed TH-positive neurons after cell engraftment, but no significant behavioral recovery was observed after 6-OHDA-induced lesion (Sonntag et al., 2007).

hESC transplantation to other CNS injury models has met similar difficulties. To date, one published paper has demonstrated hESCs transplanted after cerebral ischemia can survive and possibly differentiate into mature neurons (Kim et al., 2007). However, these hESCs did not express mature neuronal markers in vitro and no functional neuronal activity was presented. The amelioration of motor deficit following spinal cord injury (SCI) through hESC transplantation has also been researched heavily. hESC-derived motoneurons have been generated through cellular aggregation and growth factor cocktail supplementation (Li et al., 2005). hESC-derived oligodendrocytes have also been shown to promote remyelination and enhance motor function following SCI in a rat model (Keirstead et al., 2005).

A major hurdle limiting the clinical usefulness of hESC transplantation has been immune rejection of transplanted hESCs. Several recent publications have begun to circumvent this issue through nuclear reprogramming of an individual's differentiated cells to a pluripotent, stem cell-like state (Keirstead et al., 2005; Egli et al., 2007; Maherali et al., 2007; Wernig et al., 2007). While this research has only been demonstrated in mice, the potential clinical implications for cell transplantation and neurodegenerative disorders are extremely exciting.

6 Clinical Application

Nonembryonic stem cells have been applied in clinical studies, including ischemic heart disease, Parkinson's disease, and spinal cord injury. Phase I clinical trials in ischemic heart disease indicate stem cell transplantation is relatively safe and feasible; Phase II/III trials are ongoing. Human fetal dopaminergic cell transplantation in Parkinson's disease patients showed a transient benefit (Freed et al., 2001). ES cells have not been used clinically to this point, though the first clinical trial will likely start this year.

The remarkable repair capability of transplanted stem cells signifies an exciting possibility for clinical application. However, many questions associated with clinical application of stem cell transplantation need to be answered. For example: Which patients are the optimal candidates for stem cell therapy? When is the optimal time point for stem cell delivery? What is the best cell type for transplantation? What is the best route of delivery? Further animal studies are required to address these questions. Caution must be used in the translation of these animal studies into clinical trials as the results obtained in animal studies can not be directly extended to human diseases due to the variation in neuronal plasticity across different species. Eventually, the effect of stem cell therapies in clinical applications will be assessed by the ability to provide patients with safe, long-lasting improvements to their quality of life.

7 Problems and Perspectives

A growing number of animal studies utilizing stem cell transplantation have brought new hope for neurological disease treatment. However, several issues remain to be investigated. First, a major dilemma in stem cell therapy is the low survival rate of transplanted cells. Up to 90% of transplanted cells died following intracerebral grafts (Sortwell et al., 2000; Emgard et al., 2003), which could be a key factor resulting in insufficient functional improvement. Gene modification can enhance the survival of grafted cells, but whether there is long-term risk of tumorigenesis is unclear (Meuillet et al., 2003). In this respect, the priming method of preconditioning stem cells before transplantation may provide needed protection of transplanted cells during the initial days after transplantation (Hedrick et al., 2004). Second, the exact fate of transplanted stem cells and the molecular/environmental mechanisms that control stem cell differentiation in vivo is poorly defined. The proliferation and differentiation of stem cells into specific phenotypes must be able to be controlled in vivo. Third, the long-term survival and integration of transplanted stem cells with host neuronal and vascular networks should be fully elucidated at cellular and ultrastructural levels. The efficacy, efficiency, and underlying mechanisms of stem cell transplantation need to be further investigated in animal models. Finally, it is necessary to exclude even the rare possibility of graft associated side effects, such as teratoma formation or formation of inappropriate neural or nonneural tissues.

References

Akiyama Y, Radtke C, Kocsis JD. 2002. Remyelination of the rat spinal cord by transplantation of identified bone marrow stromal cells. J Neurosci 22: 6623-6630.

Ankeny DP, McTigue DM, Jakeman LB. 2004. Bone marrow transplants provide tissue protection and directional guidance for axons after contusive spinal cord injury in rats. Exp Neurol 190: 17-31.

Arvidsson A, Collin T, Kirik D, Kokaia Z, Lindvall O. 2002. Neuronal replacement from endogenous precursors in the adult brain after stroke. Nat Med 8: 963-970.

Bain G, Kitchens D, Yao M, Huettner JE, Gottlieb DI. 1995. Embryonic stem cells express neuronal properties in vitro. Dev Biol 168: 342-357.

Bakshi A, Hunter C, Swanger S, Lepore A, Fischer I. 2004. Minimally invasive delivery of stem cells for spinal cord injury: Advantages of the lumbar puncture technique. J Neurosurg Spine 1: 330-337.

Barberi T, Klivenyi P, Calingasan NY, Lee H, Kawamata H, et al. 2003. Neural subtype specification of fertilization and nuclear transfer embryonic stem cells and application in parkinsonian mice. Nat Biotechnol 21: 1200-1207.

Benzing C, Segschneider M, Leinhaas A, Itskovitz-Eldor J, Brustle O. 2006. Neural conversion of human embryonic stem cell colonies in the presence of fibroblast growth factor-2. Neuroreport 17: 1675-1681.

Bjorklund LM, Sanchez-Pernaute R, Chung S, Andersson T, Chen IY, et al. 2002. Embryonic stem cells develop into functional dopaminergic neurons after transplantation in a Parkinson rat model. Proc Natl Acad Sci USA 99: 2344-2349.

Borlongan CV, Tajima Y, Trojanowski JQ, Lee VM, Sanberg PR. 1998. Transplantation of cryopreserved human embryonal carcinoma-derived neurons (NT2N cells) promotes functional recovery in ischemic rats. Exp Neurol 149: 310-321.

Braak H, Del Tredici K, Rub U, de Vos RA, Jansen Steur EN, et al. 2003. Staging of brain pathology related to sporadic Parkinson's disease. Neurobiol Aging 24: 197-211.

Brazelton TR, Rossi FM, Keshet GI, Blau HM. 2000. From marrow to brain: Expression of neuronal phenotypes in adult mice. Science 290: 1775-1779.

Brustle O, Jones KN, Learish RD, Karram K, Choudhary K, et al. 1999. Embryonic stem cell-derived glial precursors: A source of myelinating transplants. Science 285: 754-756.

Brustle O, Spiro AC, Karram K, Choudhary K, Okabe S, et al. 1997. In vitro-generated neural precursors participate in mammalian brain development. Proc Natl Acad Sci USA 94: 14809-14814.

Buhnemann C, Scholz A, Bernreuther C, Malik CY, Braun H, et al. 2006. Neuronal differentiation of transplanted embryonic stem cell-derived precursors in stroke lesions of adult rats. Brain 129: 3238-3248.

Carmichael ST. 2003. Plasticity of cortical projections after stroke. Neuroscientist 9: 64-75.

Carmichael ST. 2006. Cellular and molecular mechanisms of neural repair after stroke: Making waves. Ann Neurol 59: 735-742.

Chen J, Li Y, Katakowski M, Chen X, Wang L, et al. 2003. Intravenous bone marrow stromal cell therapy reduces apoptosis and promotes endogenous cell proliferation after stroke in female rat. J Neurosci Res 73: 778-786.

Chen J, Li Y, Wang L, Zhang Z, Lu D, et al. 2001. Therapeutic benefit of intravenous administration of bone marrow stromal cells after cerebral ischemia in rats. Stroke 32: 1005-1011.

Chen J, Zhang ZG, Li Y, Wang L, Xu YX, et al. 2003. Intravenous administration of human bone marrow stromal cells induces angiogenesis in the ischemic boundary zone after stroke in rats. Circ Res 92: 692-699.

Crigler L, Robey RC, Asawachaicharn A, Gaupp D, Phinney DG. 2006. Human mesenchymal stem cell subpopulations express a variety of neuro-regulatory molecules and promote neuronal cell survival and neuritogenesis. Exp Neurol 198: 54-64.

Cuevas P, Carceller F, Dujovny M, Garcia-Gomez I, Cuevas B, et al. 2002. Peripheral nerve regeneration by bone marrow stromal cells. Neurol Res 24: 634-638.

Cuevas P, Carceller F, Garcia-Gomez I, Yan M, Dujovny M. 2004. Bone marrow stromal cell implantation for peripheral nerve repair. Neurol Res 26: 230-232.

Cui L, Jiang J, Wei L, Zhou X, Fraser JL, et al. 2007. Transplantation of embryonic stem cells improves nerve repair and functional recovery after severe sciatic nerve axotomy in rats. Stem Cells in revision.

Daheron L, Opitz SL, Zaehres H, Lensch WM, Andrews PW, et al. 2004. LIF/STAT3 signaling fails to maintain self-renewal of human embryonic stem cells. Stem Cells 22: 770-778.

Dancause N, Barbay S, Frost SB, Plautz EJ, Chen D, et al. 2005. Extensive cortical rewiring after brain injury. J Neurosci 25: 10167-10179.

Dezawa M, Kanno H, Hoshino M, Cho H, Matsumoto N, et al. 2004. Specific induction of neuronal cells from bone marrow stromal cells and application for autologous transplantation. J Clin Invest 113: 1701-1710.

Dezawa M, Takahashi I, Esaki M, Takano M, Sawada H. 2001. Sciatic nerve regeneration in rats induced by transplantation of in vitro differentiated bone-marrow stromal cells. Eur J Neurosci 14: 1771-1776.

Dinsmore J, Ratliff J, Deacon T, Pakzaban P, Jacoby D, et al. 1996. Embryonic stem cells differentiated in vitro as a novel source of cells for transplantation. Cell Transplant 5: 131-143.

Dominici M, Le Blanc K, Mueller I, Slaper-Cortenbach I, Marini F, et al. 2006. Minimal criteria for defining multipotent mesenchymal stromal cells. The International Society for Cellular Therapy position statement. Cytotherapy 8: 315-317.

Draper JS, Smith K, Gokhale P, Moore HD, Maltby E, et al. 2004. Recurrent gain of chromosomes 17q and 12 in cultured human embryonic stem cells. Nat Biotechnol 22: 53-54.

Egli D, Rosains J, Birkhoff G, Eggan K. 2007. Developmental reprogramming after chromosome transfer into mitotic mouse zygotes. Nature 447: 679-685.

Emgard M, Hallin U, Karlsson J, Bahr BA, Brundin P, et al. 2003. Both apoptosis and necrosis occur early after intracerebral grafting of ventral mesencephalic tissue: A role for protease activation. J Neurochem 86: 1223-1232.

Englund U, Bjorklund A, Wictorin K, Lindvall O, Kokaia M. 2002. Grafted neural stem cells develop into functional pyramidal neurons and integrate into host cortical circuitry. Proc Natl Acad Sci USA 99: 17089-17094.

Eriksson PS, Perfilieva E, Bjork-Eriksson T, Alborn AM, Nordborg C, et al. 1998. Neurogenesis in the adult human hippocampus. Nat Med 4: 1313-1317.

Fraichard A, Chassande O, Bilbaut G, Dehay C, Savatier P, et al. 1995. In vitro differentiaton of embryonic stem cells into glial cells and functional neurons. J Cell Sci 108 (Pt 10): 3181-3188.

Freed CR, Breeze RE, Rosenberg NL, Schneck SA, Kriek E, et al. 1992. Survival of implanted fetal dopamine cells and neurologic improvement 12 to 46 months after transplantation for Parkinson's disease. N Engl J Med 327: 1549-1555.

Freed CR, Greene PE, Breeze RE, Tsai WY, Du Mouchel W, et al. 2001. Transplantation of embryonic dopamine neurons for severe Parkinson's disease. N Engl J Med 344: 710-719.

Fu SY, Gordon T. 1997. The cellular and molecular basis of peripheral nerve regeneration. Mol Neurobiol 14: 67-116.

Gage FH. 2000. Mammalian neural stem cells. Science 287: 1433-1438.

Hamano K, Li TS, Kobayashi T, Kobayashi S, Matsuzaki M, et al. 2000. Angiogenesis induced by the implantation of self-bone marrow cells: A new material for therapeutic angiogenesis. Cell Transplant 9: 439-443.

Hayashi J, Takagi Y, Fukuda H, Imazato T, Nishimura M, et al. 2006. Primate embryonic stem cell-derived neuronal progenitors transplanted into ischemic brain. J Cereb Blood Flow Metab 26: 906-914.

Hedrick ML, Yu SP, Wei L. 2004. Hypoxia preconditioning enhances survival of neural derived embryonic stem cells. Society for Neurosci Abstr 905.4.

Hernit-Grant CS, Macklis JD. 1996. Embryonic neurons transplanted to regions of targeted photolytic cell death in adult mouse somatosensory cortex re-form specific callosal projections. Exp Neurol 139: 131-142.

Horita Y, Honmou O, Harada K, Houkin K, Hamada H, et al. 2006. Intravenous administration of glial cell line-derived neurotrophic factor gene-modified human mesenchymal stem cells protects against injury in a cerebral ischemia model in the adult rat. J Neurosci Res 84: 1495-1504.

Howells D. 2003. Stem cells: Do they replace or stimulate? Stroke 34: 2082-2083.

Hu X, Yu SP, Wang J, Fraser JL, Wei L. 2007. Hypoxic preconditioning promotes bone marrow mesenchymal stem cells survival by enhancing stem cell autocrine action. J Am Coll Cardiol 49: 234A.

Ikeda R, Kurokawa MS, Chiba S, Yoshikawa H, Ide M, et al. 2005. Transplantation of neural cells derived from retinoic acid-treated cynomolgus monkey embryonic stem cells successfully improved motor function of hemiplegic mice with experimental brain injury. Neurobiol Dis 20: 38-48.

Ishibashi S, Sakaguchi M, Kuroiwa T, Yamasaki M, Kanemura Y, et al. 2004. Human neural stem/progenitor cells, expanded in long-term neurosphere culture, promote functional recovery after focal ischemia in Mongolian gerbils. J Neurosci Res 78: 215-223.

Itsykson P, Ilouz N, Turetsky T, Goldstein RS, Pera MF, et al. 2005. Derivation of neural precursors from human embryonic stem cells in the presence of noggin. Mol Cell Neurosci 30: 24-36.

Iwaguro H, Yamaguchi J, Kalka C, Murasawa S, Masuda H, et al. 2002. Endothelial progenitor cell vascular endothelial growth factor gene transfer for vascular regeneration. Circulation 105: 732-738.

Johnston RE, Dillon-Carter O, Freed WJ, Borlongan CV. 2001. Trophic factor secreting kidney cell lines: In vitro characterization and functional effects following transplantation in ischemic rats. Brain Res 900: 268-276.

Katz AJ, Tholpady A, Tholpady SS, Shang H, Ogle RC. 2005. Cell surface and transcriptional characterization of human adipose-derived adherent stromal (hADAS) cells. Stem Cells 23: 412-423.

Keirstead HS, Nistor G, Bernal G, Totoiu M, Cloutier F, et al. 2005. Human embryonic stem cell-derived oligodendrocyte progenitor cell transplants remyelinate and restore locomotion after spinal cord injury. J Neurosci 25: 4694-4705.

Kelly S, Bliss TM, Shah AK, Sun GH, Ma M, et al. 2004. Transplanted human fetal neural stem cells survive, migrate, and differentiate in ischemic rat cerebral cortex. Proc Natl Acad Sci USA 101: 11839-11844.

Keogh CL, Whitaker VR, Yu SP, Wei L. 2004. Hypoxia preconditioning enhances angiogenesis in the neuronatal ischemic rat brain. Society for Neurosci Abstr 457.2.

Kim JH, Auerbach JM, Rodriguez-Gomez JA, Velasco I, Gavin D, et al. 2002. Dopamine neurons derived from embryonic stem cells function in an animal model of Parkinson's disease. Nature 418: 50-56.

Kim DW, Chung S, Hwang M, Ferree A, Tsai HC, et al. 2006. Stromal cell-derived inducing activity, Nurr1, and signaling molecules synergistically induce dopaminergic neurons from mouse embryonic stem cells. Stem Cells 24: 557-567.

Kim DY, Park SH, Lee SU, Choi DH, Park HW, et al. 2007. Effect of human embryonic stem cell-derived neuronal precursor cell transplantation into the cerebral infarct model of rat with exercise. Neurosci Res 58: 164-175.

Kinnaird T, Stabile E, Burnett MS, Shou M, Lee CW, et al. 2004. Local delivery of marrow-derived stromal cells augments collateral perfusion through paracrine mechanisms. Circulation 109: 1543-1549.

Klimanskaya I, Chung Y, Meisner L, Johnson J, West MD, et al. 2005. Human embryonic stem cells derived without feeder cells. Lancet 365: 1636-1641.

Kline DG, Hudson AR. 1995. Vertebral artery compression. J Neurosurg 83: 759.

Koestenbauer S, Zech NH, Juch H, Vanderzwalmen P, Schoonjans L, et al. 2006. Embryonic stem cells: Similarities and differences between human and murine embryonic stem cells. Am J Reprod Immunol 55: 169-180.

Kopen GC, Prockop DJ, Phinney DG. 1999. Marrow stromal cells migrate throughout forebrain and cerebellum, and they differentiate into astrocytes after injection into neonatal mouse brains. Proc Natl Acad Sci USA 96: 10711-10716.

Kordower JH, Freeman TB, Snow BJ, Vingerhoets FJ, Mufson EJ, et al. 1995. Neuropathological evidence of graft survival and striatal reinnervation after the transplantation of fetal mesencephalic tissue in a patient with Parkinson's disease. N Engl J Med 332: 1118-1124.

Krupinski J, Kaluza J, Kumar P, Kumar S, Wang JM. 1994. Role of angiogenesis in patients with cerebral ischemic stroke. Stroke 25: 1794-1798.

Kuhn HG, Dickinson-Anson H, Gage FH. 1996. Neurogenesis in the dentate gyrus of the adult rat: Age-related decrease of neuronal progenitor proliferation. J Neurosci 16: 2027-2033.

Kurozumi K, Nakamura K, Tamiya T, Kawano Y, Ishii K, et al. 2005. Mesenchymal stem cells that produce neurotrophic factors reduce ischemic damage in the rat middle cerebral artery occlusion model. Mol Ther 11: 96-104.

Kurozumi K, Nakamura K, Tamiya T, Kawano Y, Kobune M, et al. 2004. BDNF gene-modified mesenchymal stem cells promote functional recovery and reduce infarct size in the rat middle cerebral artery occlusion model. Mol Ther 9: 189-197.

Lee J, Kuroda S, Shichinohe H, Ikeda J, Seki T, et al. 2003. Migration and differentiation of nuclear fluorescence-labeled bone marrow stromal cells after transplantation into cerebral infarct and spinal cord injury in mice. Neuropathology 23: 169-180.

Lei Z, Yongda L, Jun M, Yingyu S, Shaoju Z, et al. 2007. Culture and neural differentiation of rat bone marrow mesenchymal stem cells in vitro. Cell Biol Int 31: 916-923.

Li Y, Chen J, Chen XG, Wang L, Gautam SC, et al. 2002. Human marrow stromal cell therapy for stroke in rat: Neurotrophins and functional recovery. Neurology 59: 514-523.

Li Y, Chen J, Chopp M. 2002. Cell proliferation and differentiation from ependymal, subependymal and choroid plexus cells in response to stroke in rats. J Neurol Sci 193: 137-146.

Li Y, Chen J, Wang L, Lu M, Chopp M. 2001. Treatment of stroke in rat with intracarotid administration of marrow stromal cells. Neurology 56: 1666-1672.

Li Y, Chopp M, Chen J, Wang L, Gautam SC, et al. 2000. Intrastriatal transplantation of bone marrow nonhematopoietic cells improves functional recovery after stroke in adult mice. J Cereb Blood Flow Metab 20: 1311-1319.

Li XJ, Du ZW, Zarnowska ED, Pankratz M, Hansen LO, et al. 2005. Specification of motoneurons from human embryonic stem cells. Nat Biotechnol 23: 215-221.

Lindvall O, Kokaia Z. 2004. Recovery and rehabilitation in stroke: Stem cells. Stroke 35: 2691-2694.

Liu S, Qu Y, Stewart TJ, Howard MJ, Chakrabortty S, et al. 2000. Embryonic stem cells differentiate into oligodendrocytes and myelinate in culture and after spinal cord transplantation. Proc Natl Acad Sci USA 97: 6126-6131.

Llado J, Haenggeli C, Maragakis NJ, Snyder EY, Rothstein JD. 2004. Neural stem cells protect against glutamate-induced excitotoxicity and promote survival of injured motor neurons through the secretion of neurotrophic factors. Mol Cell Neurosci 27: 322-331.

Lu P, Jones LL, Snyder EY, Tuszynski MH. 2003. Neural stem cells constitutively secrete neurotrophic factors and promote extensive host axonal growth after spinal cord injury. Exp Neurol 181: 115-129.

Lu P, Jones LL, Tuszynski MH. 2005. BDNF-expressing marrow stromal cells support extensive axonal growth at sites of spinal cord injury. Exp Neurol 191: 344-360.

Lu P, Yang H, Jones LL, Filbin MT, Tuszynski MH. 2004. Combinatorial therapy with neurotrophins and cAMP promotes axonal regeneration beyond sites of spinal cord injury. J Neurosci 24: 6402-6409.

Maherali N SR, Xie W, Utikal J, Eminli S, Arnold K, et al. 2007. Directly reprogrammed fibroblasts show global epigenetic remodeling and widespread tissue contribution. Cell Stem Cell 1: 55-70.

Maitra A, Arking DE, Shivapurkar N, Ikeda M, Stastny V, et al. 2005. Genomic alterations in cultured human embryonic stem cells. Nat Genet 37: 1099-1103.

Mattsson B, Sorensen JC, Zimmer J, Johansson BB. 1997. Neural grafting to experimental neocortical infarcts improves behavioral outcome and reduces thalamic atrophy in rats housed in enriched but not in standard environments. Stroke 28: 1225-1231; discussion 1231–1232.

Meuillet EJ, Mahadevan D, Vankayalapati H, Berggren M, Williams R, et al. 2003. Specific inhibition of the Akt1 pleckstrin homology domain by D-3-deoxy-phosphatidyl-myo-inositol analogues. Mol Cancer Ther 2: 389-399.

McDonald JW, Liu XZ, Qu Y, Liu S, Mickey SK, et al. 1999. Transplanted embryonic stem cells survive, differentiate and promote recovery in injured rat spinal cord. Nat Med 5: 1410-1412.

McDonald JW, Becker D, Holekamp TF, Howard M, Liu S, et al. 2004. Repair of the injured spinal cord and the potential of embryonic stem cell transplantation. J Neurotrauma 21: 383-393.

Mimura T, Dezawa M, Kanno H, Sawada H, Yamamoto I. 2004. Peripheral nerve regeneration by transplantation of bone marrow stromal cell-derived Schwann cells in adult rats. J Neurosurg 101: 806-812.

Mosahebi A, Fuller P, Wiberg M, Terenghi G. 2002. Effect of allogeneic Schwann cell transplantation on peripheral nerve regeneration. Exp Neurol 173: 213-223.

Munoz-Elias G, Marcus AJ, Coyne TM, Woodbury D, Black IB. 2004. Adult bone marrow stromal cells in the embryonic brain: Engraftment, migration, differentiation, and long-term survival. J Neurosci 24: 4585-4595.

Muotri AR, Nakashima K, Toni N, Sandler VM, Gage FH. 2005. Development of functional human embryonic stem cell-derived neurons in mouse brain. Proc Natl Acad Sci USA 102: 18644-18648.

Nistor GI, Totoiu MO, Haque N, Carpenter MK, Keirstead HS. 2005. Human embryonic stem cells differentiate into oligodendrocytes in high purity and myelinate after spinal cord transplantation. Glia 49: 385-396.

Ohta M, Suzuki Y, Noda T, Ejiri Y, Dezawa M, et al. 2004. Bone marrow stromal cells infused into the cerebrospinal fluid promote functional recovery of the injured rat spinal cord with reduced cavity formation. Exp Neurol 187: 266-278.

Palmer TD, Markakis EA, Willhoite AR, Safar F, Gage FH. 1999. Fibroblast growth factor-2 activates a latent neurogenic program in neural stem cells from diverse regions of the adult CNS. J Neurosci 19: 8487-8497.

Palmer TD, Ray J, Gage FH. 1995. FGF-2-responsive neuronal progenitors reside in proliferative and quiescent regions of the adult rodent brain. Mol Cell Neurosci 6: 474-486.

Pavon-Fuentes N, Blanco-Lezcano L, Martinez-Martin L, Castillo-Diaz L, de la Cuetara-Bernal K, et al. 2004. [Stromal cell transplant in the 6-OHDA lesion model]. Rev Neurol 39: 326-334.

Ramer LM, Ramer MS, Steeves JD. 2005. Setting the stage for functional repair of spinal cord injuries: A cast of thousands. Spinal Cord 43: 134-161.

Reubinoff BE, Itsykson P, Turetsky T, Pera MF, Reinhartz E, et al. 2001. Neural progenitors from human embryonic stem cells. Nat Biotechnol 19: 1134-1140.

Rodriguez-Gomez JA, Lu JQ, Velasco I, Rivera S, Zoghbi SS, et al. 2007. Persistent dopamine functions of neurons derived from embryonic stem cells in a rodent model of Parkinson disease. Stem Cells 25: 918-928.

Roy NS, Cleren C, Singh SK, Yang L, Beal MF, et al. 2006. Functional engraftment of human ES cell-derived dopaminergic neurons enriched by coculture with telomerase-immortalized midbrain astrocytes. Nat Med 12: 1259-1268.

Sanchez-Pernaute R, Studer L, Ferrari D, Perrier A, Lee H, et al. 2005. Long-term survival of dopamine neurons derived from parthenogenetic primate embryonic stem cells (cyno-1) after transplantation. Stem Cells 23: 914-922.

Schabitz WR Steigleder T, Cooper-Kuhn CM, Schwab S, Sommer C, Schneider A, et al. 2007. Intravenous brain-derived neurotrophic factor enhances poststroke sensorimotor recovery and stimulates neurogenesis. Stroke 38: 2165-72.

Schrag A, Jahanshahi M, Quinn N. 2000. How does Parkinson's disease affect quality of life? A comparison with quality of life in the general population. Mov Disord 15: 1112-1118.

Schulz TC, Noggle SA, Palmarini GM, Weiler DA, Lyons IG, et al. 2004. Differentiation of human embryonic stem cells to dopaminergic neurons in serum-free suspension culture. Stem Cells 22: 1218-1238.

Schwarz EJ, Alexander GM, Prockop DJ, Azizi SA. 1999. Multipotential marrow stromal cells transduced to produce L-DOPA: Engraftment in a rat model of Parkinson disease. Hum Gene Ther 10: 2539-2549.

Sekhon LH, Fehlings MG. 2001. Epidemiology, demographics, and pathophysiology of acute spinal cord injury. Spine 26: S2-S12.

Senior K. 2001. Angiogenesis and functional recovery demonstrated after minor stroke. Lancet 358: 817.

Shen LH, Li Y, Chen J, Zhang J, Vanguri P, et al. 2006. Intracarotid transplantation of bone marrow stromal cells increases axon-myelin remodeling after stroke. Neuroscience 137: 393-399.

Snyder EY, Yoon C, Flax JD, Macklis JD. 1997. Multipotent neural precursors can differentiate toward replacement of neurons undergoing targeted apoptotic degeneration in adult mouse neocortex. Proc Natl Acad Sci USA 94: 11663-11668.

Song S, Kamath S, Mosquera D, Zigova T, Sanberg P, et al. 2004. Expression of brain natriuretic peptide by human bone marrow stromal cells. Exp Neurol 185: 191-197.

Song HJ, Stevens CF, Gage FH. 2002. Neural stem cells from adult hippocampus develop essential properties of functional CNS neurons. Nat Neurosci 5: 438-445.

Sonntag KC, Pruszak J, Yoshizaki T, Arensbergen van J, Sanchez-Pernaute R, et al. 2007. Enhanced yield of neuroepithelial precursors and midbrain-like dopaminergic neurons from human embryonic stem cells using the bone morphogenic protein antagonist noggin. Stem Cells 25: 411-418.

Sortwell CE, Pitzer MR, Collier TJ. 2000. Time course of apoptotic cell death within mesencephalic cell suspension grafts: Implications for improving grafted dopamine neuron survival. Exp Neurol 165: 268-277.

Sotelo C, Alvarado-Mallart RM. 1991. The reconstruction of cerebellar circuits. Trends Neurosci 14: 350-355.

Sykova E, Jendelova P. 2005. Magnetic resonance tracking of implanted adult and embryonic stem cells in injured brain and spinal cord. Ann N Y Acad Sci 1049: 146-160.

Takagi Y, Takahashi J, Saiki H, Morizane A, Hayashi T, et al. 2005. Dopaminergic neurons generated from monkey embryonic stem cells function in a Parkinson primate model. J Clin Invest 115: 102-109.

Takasawa K, Kitagawa K, Yagita Y, Sasaki T, Tanaka S, et al. 2002. Increased proliferation of neural progenitor cells but reduced survival of newborn cells in the contralateral hippocampus after focal cerebral ischemia in rats. J Cereb Blood Flow Metab 22: 299-307.

Taupin P, Gage FH. 2002. Adult neurogenesis and neural stem cells of the central nervous system in mammals. J Neurosci Res 69: 745-749.

Tholpady SS, Katz AJ, Ogle RC. 2003. Mesenchymal stem cells from rat visceral fat exhibit multipotential differentiation in vitro. Anat Rec A Discov Mol Cell Evol Biol 272: 398-402.

Thomson JA, Itskovitz-Eldor J, Shapiro SS, Waknitz MA, Swiergiel JJ, et al. 1998. Embryonic stem cell lines derived from human blastocysts. Science 282: 1145-1147.

Ware CB, Nelson AM, Blau CA. 2006. A comparison of NIH-approved human ESC lines. Stem Cells 24: 2677-2684.

Wei L, Cui L, Snider BJ, Rivkin M, Yu SS, et al. 2005. Transplantation of embryonic stem cells overexpressing Bcl-2 promotes functional recovery after transient cerebral ischemia. Neurobiol Dis 19: 183-193.

Wei L, Erinjeri JP, Rovainen CM, Woolsey TA. 2001. Collateral growth and angiogenesis around cortical stroke. Stroke 32: 2179-2184.

Wernig M, Meissner A, Foreman R, Brambrink T, Ku M, et al. 2007. In vitro reprogramming of fibroblasts into a pluripotent ES-cell-like state. Nature 448: 318-324.

Wislet-Gendebien S, Hans G, Leprince P, Rigo JM, Moonen G, et al. 2005. Plasticity of cultured mesenchymal stem cells: Switch from nestin-positive to excitable neuron-like phenotype. Stem Cells 23: 392-402.

Wu S, Suzuki Y, Ejiri Y, Noda T, Bai H, et al. 2003. Bone marrow stromal cells enhance differentiation of cocultured neurosphere cells and promote regeneration of injured spinal cord. J Neurosci Res 72: 343-351.

Xiao J, Nan Z, Motooka Y, Low WC. 2005. Transplantation of a novel cell line population of umbilical cord blood stem cells ameliorates neurological deficits associated with ischemic brain injury. Stem Cells Dev 14: 722-733.

Yagita Y, Kitagawa K, Ohtsuki T, Takasawa K, Miyata T, et al. 2001. Neurogenesis by progenitor cells in the ischemic adult rat hippocampus. Stroke 32: 1890-1896.

Zhang ZG, Jiang Q, Zhang R, Zhang L, Wang L, et al. 2003. Magnetic resonance imaging and neurosphere therapy of stroke in rat. Ann Neurol 53: 259-263.

Zhao ZM, Li HJ, Liu HY, Lu SH, Yang RC, et al. 2004. Intraspinal transplantation of CD34+human umbilical cord blood cells after spinal cord hemisection injury improves functional recovery in adult rats. Cell Transplant 13: 113-122.

Zurita M, Vaquero J. 2004. Functional recovery in chronic paraplegia after bone marrow stromal cells transplantation. Neuroreport 15: 1105-1108.

Zuk PA, Zhu M, Mizuno H, Huang J, Futrell JW, et al. 2001. Multilineage cells from human adipose tissue: Implications for cell-based therapies. Tissue Eng 7: 211-228.

Yamamoto S, Nagao M, Sugimori M, Kosako H, Nakatomi H, et al. 2001. Transcription factor expression and Notch-dependent regulation of neural progenitors in the adult rat spinal cord. J Neurosci 21: 9814-9823.

21 Ubiquitin/Proteasome and Autophagy/Lysosome Pathways: Comparison and Role in Neurodegeneration

N. Myeku · M. E. Figueiredo-Pereira

1	Introduction	514
2	Comparison Between the UPP and ALP	515
2.1	Degradation Sites	515
2.1.1	UPP	515
2.1.2	ALP	516
2.2	Degradation Mechanisms	517
2.2.1	UPP	517
2.2.2	ALP	517
2.3	Substrate Targeting	517
2.3.1	UPP	517
2.3.2	ALP	518
2.4	Substrate Identification	519
2.4.1	UPP	519
2.4.2	ALP	519
2.5	Substrate Delivery (UPP and ALP)	519
3	Role of the UPP and APL in Neurodegeneration	520
4	Overall Conclusions	522

© 2009 Springer Science+Business Media, LLC.

Abstract: Neurodegenerative disorders, such as Alzheimer's, Parkinson's, and Huntington's diseases as well as amyotrophic lateral sclerosis, are a heterogeneous group of clinical diseases that are characterized by the selective loss of neurons in specific regions of the CNS. Despite their variability they have similar features including the accumulation of misfolded proteins that eventually develop into inclusion bodies. Whether these protein deposits are pathogenic or represent a coping mechanism to prolong survival of the affected neurons is a hotly debated issue. One important point to consider is that these protein deposits are indicative of a disease-state as they are not prevalent in healthy cells. Ubiquitinated proteins are major components of these proteinaceous cytoplasmic or nuclear inclusions suggesting that impaired proteasome activity and the ubiquitination machinery may be main players in this process. Emerging data revealed that autophagosomes are also components of inclusion bodies, implicating the autophagy/lysosome pathway in neurodegenerative disorders as well. Herein, we compare some of the most important characteristics of these two pathways for intracellular protein degradation and discuss their potential role in neurodegeneration. When the proteasome is impaired, it is possible that autophagy may be the alternate pathway for clearing out aggregated ubiquitinated proteins. The question emerges if this potential "survival" mechanism can be explored as a strategy to overcome the most common feature shared by various neurodegenerative disorders, i.e., protein aggregation manifested as inclusion bodies. One potential drawback is that degradation through autophagy seems to be a "bulky," nonspecific process. A thorough knowledge of the mechanisms involved in the targeting of substrates to autophagy will provide clues to the putative specificity of this pathway so that its ectopic manipulation will target only abnormal protein aggregates and not critical intracellular components, the removal of which may cause cell death.

List of Abbreviations: AD, Alzheimer's disease; ALS, amyotrophic lateral sclerosis; ALP, autophagy/lysosome pathway; Atg, autophagy related genes; CMA, chaperone-mediated autophagy; E1, ubiquitin activating enzyme; E2, ubiquitin conjugating enzyme; E3, ubiquitin ligase; E4, ubiquitin-chain elongating factor; HD, Huntington's disease; HDAC6, histone deacetylase 6; LC3, microtubule-associated protein-light chain 3; LIR, LC3-interacting region; PB1, protein binding domain1; PD, Parkinson's disease; PE, phosphatidylethanolamine; p62/SQSTM1, p62/sequestosome1; Ub, ubiquitin; UBA, ubiquitin-associating domain; UBL, ubiquitin-like; UPP, ubiquitin/proteasome pathway

1 Introduction

Many neurodegenerative disorders are associated with formation of protein aggregates, resulting ultimately in proteinaceous inclusions, such as Lewy bodies in Parkinson's disease (PD) and neurofibrillary tangles in Alzheimer's disease (AD) (Lowe et al., 1988). While the composition of these abnormal inclusion bodies varies with the disorder, a general feature is that these aggregates contain ubiquitinated proteins. Thus, although selective sets of neurons are affected in different neurodegenerative disorders, they are associated with an accumulation and aggregation of ubiquitinated proteins. In general, high levels of ubiquitinated proteins do not accumulate in healthy cells as they are rapidly degraded. The formation of these inclusion bodies is thus attributed to disturbed protein degradation (Alves-Rodrigues et al., 1998).

Eukaryotic cells contain two major intracellular pathways for protein degradation: the ubiquitin/proteasome pathway (UPP) and the lysosome pathway. The UPP mainly degrades short-lived proteins such as cell cycle regulators and transcription factors, as well as misfolded proteins from the cytosol, nucleus, and endoplasmic reticulum, while the lysosome degrades long-lived proteins and cellular organelles. In mammalian cultured cells, 80–90% of protein degradation is carried out by the UPP (Lee and Goldberg, 1998), while only 10–20% is attributable to lysosomes (Gronostajski et al., 1985). Lysosomes are responsible for the turnover of extracellular proteins that enter cells by endocytosis and pinocytosis, as well as degradation of strictly intracellular proteins and organelles, by autophagy. Three distinct types of autophagy have been described so far: microautophagy, chaperone-mediated autophagy (CMA), and macroautophagy (Larsen and Sulzer, 2002). Microautophagy is a constitutive form of autophagy better characterized in yeast (Abeliovich and Klionsky, 2001). Its molecular details and functional importance in mammalian cells are largely unknown. CMA differs from the other two forms of autophagy because it does not require vesicular trafficking. Instead cytoplasmic proteins are delivered to the lysosome for degradation by a chain of molecules. In addition, CMA is a selective form of lysosomal degradation, since it is restricted to the elimination of proteins that possess the penta peptide KFERQ in their sequence (Massey et al., 2006). Macroautophagy is inducible and is the best studied form of autophagy, so far (Mizushima and Klionsky, 2007). Our discussion herein will focus exclusively on this latter form of autophagy known as the autophagy/lysosome pathway (ALP).

The UPP and the ALP have been viewed as two distinct proteolytic pathways with no molecular links between them. However, this view was challenged by the finding of prominent ubiquitin-positive pathology evident in autophagy-deficient mice despite a functional UPP (Komatsu et al., 2006; Hara et al., 2006a). These studies strongly suggest that the ALP also participates in the degradation of ubiquitinated proteins. The challenge is to determine the impact of each of the two proteolytic pathways, i.e., the UPP and the ALP, in the accumulation/aggregation of ubiquitinated proteins as well as in their removal under conditions that lead to neurodegeneration.

In this review, we will compare some of the mechanisms involved in protein degradation by both pathways and discuss evidence supporting a crosstalk between them and their role in neurodegeneration. A better understanding of the relationship between these two major proteolytic systems may reveal new strategies to prevent or ameliorate the devastating effects of a decline in protein turnover and therefore contribute to treatment of neurodegenerative diseases associated with the accumulation and aggregation of ubiquitinated proteins.

2 Comparison Between the UPP and ALP

2.1 Degradation Sites

2.1.1 UPP

Through the UPP, intracellular proteins are degraded by a large multiprotein complex known as the 20S proteasome, which has a native molecular mass of ∼700 kDa (Chen and Hochstrasser, 1996; Coux et al., 1996). In eukaryotic cells, proteasomes are found in the cytoplasm both as free and ER-attached particles as well as in the nucleus (Wojcik and DeMartino, 2003). Translocation into the nucleus is mediated by nuclear localization signals found on proteasome subunits (α subunits) (Bader et al., 2007). The 20S proteasome is composed of 28 subunits arranged in four heptameric-stacked rings forming a cylindrical structure with a hollow center in which proteolysis takes place (DeMartino and Slaughter, 1999). The barrel-shaped 20S particle contains three internal chambers, but peptide hydrolysis occurs only in the middle one. The function of the two outer chambers remains to be elucidated, although they can store one or two substrate molecules, depending on substrate size (Sharon et al., 2006). It is not clear how the substrates move from the outer chambers to the middle chamber which contains the proteolytic active sites. This "enclosed" architecture, where the active sites are sheltered and face the inside of the proteolytic chamber, excludes degradation of proteins that come into contact with the outside of the 20S proteasome. By itself, the 20S proteasome is autocatalytically inhibited, as the openings (one at each end) to the inside chambers are occluded (Groll et al., 2000). The pores open up when the 20S proteasome associates with other particles such as the 19S and 11S regulatory particles (Brooks et al., 2000). The former binds to polyubiquitinated proteins promoting their degradation, while the latter facilitates degradation of nonubiquitinated short substrates. It has been suggested by in vitro studies, that unfolded hydrophobic substrates may also induce pore opening if they come into contact with the 20S proteasome (Liu et al., 2003). Whether or not this is an in vivo event remains to be established. Overall, degradation of polyubiquitinated proteins through the UPP is carried out by the 26S proteasome, which includes the 20S core particle capped at one end or both by 19S regulatory particles or by a 19S particle at one end and an 11S particle at the other (❶ *Figure 21-1*).

Figure 21-1

(a) Hybrid 26S proteasome: The 20S proteasome, which is the catalytic core, can be singly capped or capped at both ends by two 19S particles (PA700) (*not shown*) or by one 19S and one 11S (PA28) particle (*shown on the left*). The latter configuration corresponds to a hybrid 26S proteasome. The 19S and 11S are regulatory particles. Assembly of the 26S proteasome requires ATP hydrolysis. Three of the 20S proteasome subunits, namely β1, β2, and β5 exhibit the N-terminal threonine active sites functional at a neutral pH. (b) Lysosome: This organelle is bounded by a membrane consisting of a phospholipid bilayer and contains a variety of hydrolytic enzymes, among them more than 12 different types of cathepsins. The latter are cysteine, serine, aspartyl, or metalloproteases optimally active at an acidic pH. Lysosomes maintain an internal acidic environment through hydrogen ion pumps at the membrane that drive the ions from the cytoplasm into the lumenal space in an ATP-dependent manner

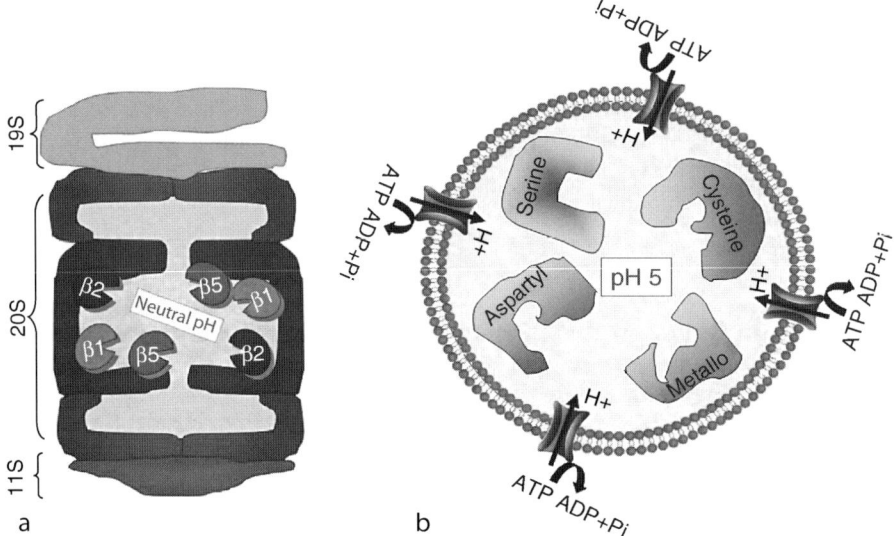

It is well-established that proteasome assembly to a functional 26S proteasome as well as the proteolytic operation of the proteasome, require substantial amounts of ATP (300–400 molecules per molecule of globular or unfolded substrate) (Benaroudj et al., 2003), suggesting that ATP depletion in cells may affect proteasome function.

2.1.2 ALP

Through the ALP, intracellular proteins are degraded by cathepsins that are proteases enclosed in the lysosome, an organelle surrounded by a phospholipid bilayer (● *Figure 21-1*). The interior of the lysosome is maintained at an acidic pH (pH 5) by proton pumps embedded in the lysosomal membrane (McGrath, 1999). Unlike the proteasome that lies unprotected within the cytoplasm and nucleus, cathepsins are sheltered by the lysosomal membrane and therefore there is no need for a complex architecture, such as the one exhibited by the proteasome. Accordingly, most cathepsins are monomeric (as an exception, cathepsin C is a tetramer), their molecular weight ranges between 20 and 30 kDa, and the active sites on cathepsins reside in a cleft that is easily accessible to the substrates within the lysosome (McGrath, 1999). Furthermore, cathepsins are optimally active at an acidic pH, a property that prevents them from degrading cytoplasmic proteins in the event that the lysosomal membrane falls apart and cathepsins are spilled into the neutral physiological pH of the cytoplasm. Moreover, cathepsins are highly unstable at a neutral pH. Although peptide bond hydrolysis by cathepsins does not require ATP, their activity is indirectly dependent on ATP, because the proton pumps required to maintain an intralysosomal acidic pH, are ATP-dependent.

2.2 Degradation Mechanisms

2.2.1 UPP

The 20S proteasome hydrolyzes most peptide bonds at a neutral pH (Orlowski and Wilk, 2000). The proteolytic activity of the 20S proteasome resides in three different β subunits, each with distinct specificity exhibiting caspase-like, trypsin-like, and chymotrypsin-like activities, respectively (❯ *Figure 21-1*). The three proteolytically active β subunits (β1, β2, and β5) bear the active sites consisting of N-terminal threonines (Groll et al., 1997). These three subunits are first synthesized as inactive precursors with a propeptide that, upon full incorporation of the subunits into 20S proteasomes, is autocatalytically removed to expose the active site Thr + 1 (Schmidtke et al., 1996; Sharon et al., 2007). Substrate proteolysis by the proteasome involves processive (Kisselev et al., 1999), nonprocessive (Wang et al., 1999) as well as endoproteolytic (Liu et al., 2003) activities but does not always lead to the degradation of the substrate to small 8–9 amino acid fragments. In some cases, when certain domains of the protein substrate remain folded, the proteasome degrades only the unfolded segments. This process is essential for regulating the activity of certain transcription factors (Rape and Jentsch, 2004).

2.2.2 ALP

More than 12 different kinds of cathepsins within lysosomes can, as a group, hydrolyze most peptide bonds on proteins (Johnson, 2000). Cathepsins are distributed throughout the four protease classes. The majority of cathepsins are cysteine proteases (cathepsins B, C, H, K, L, S, and T). From the remaining cathepsins some are aspartyl proteases, such as cathepsins D and E, metalloproteases, such as cathepsin III, and serine proteases, such as cathepsin A, G, and R (Johnson, 2000). Cathepsins are synthesized as inactive zymogens and activation within the Golgi involves their proteolytic processing. The hierarchy for protein degradation by the diverse groups of cathepsins is not always clear and their redundancy makes the function and necessity for each of these enzymes quite ambiguous (McGrath, 1999).

2.3 Substrate Targeting

2.3.1 UPP

The targeting of protein substrates for degradation by the UPP is a highly regulated process and, in most cases, requires their ubiquitination (❯ *Figure 21-2*).

Protein ubiquitination involves: (1) the formation of a high energy thioester bond between ubiquitin and a ubiquitin-activating enzyme (E1) in an ATP-dependent reaction; (2) a thioester bond between the activated ubiquitin and ubiquitin-conjugating enzymes (E2) is created; (3) covalent attachment of the C-terminal of ubiquitin, usually to the ε-amino group of a lysine on a protein substrate via an isopeptide bond mediated by ubiquitin ligases (E3); and (4) assembly of multiubiquitin chains by a family of ubiquitination factors (E4) which promote the elongation of the ubiquitin-chains (Hershko and Ciechanover, 1998). Ubiquitin may also be transferred directly to proteins by ubiquitin-conjugating enzymes.

Degradation of polyubiquitinated proteins is enhanced when more than one ubiquitin is attached to the target protein. The minimal signal for efficient degradation is a tetraubiquitin chain (Thrower et al., 2000). Removal of two ubiquitins from a tetraubiquitinated substrate by deubiquitinating enzymes can decrease substrate/26S proteasome affinity by approximately 100-fold, allowing the substrate to escape degradation. Longer chains do not increase substrate/26S proteasome affinity, but optimize their interaction time

◘ **Figure 21-2**
(a) Ubiquitination: Protein ubiquitination is a complex ATP-dependent process in which ubiquitin (Ub) is sequentially activated by ubiquitin-activating enzymes (E1), transferred to ubiquitin-conjugating enzymes (E2), and ligated to protein substrates by ubiquitin ligases (E3). Ubiquitin chains can be elongated by E4 enzymes. (b and c) Atg conjugating system: Two ubiquitin-like (UBL) conjugation pathways, Atg12 (*shown in the middle*) and Atg8 (*shown on the right*), regulate autophagosome formation. Both Atg 12 and Atg 8 are activated by an E1-like enzyme (Atg7), are then transferred to E2-like enzymes, Atg10 and Atg3, respectively, and are finally attached to the specific substrates, Atg5 or phosphatidylethanolamine (PE). See text for details

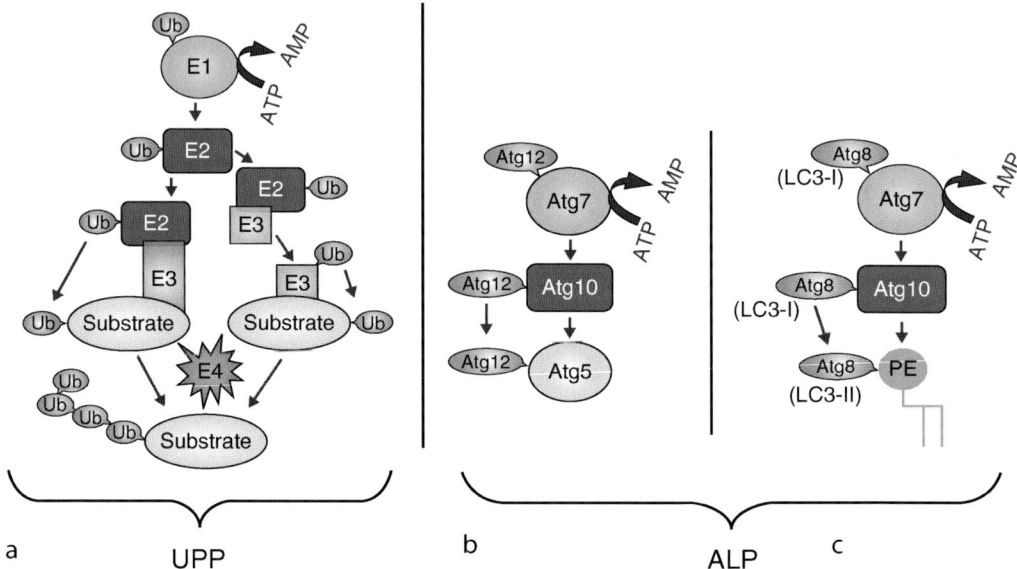

(Thrower et al., 2000). The interaction of the polyubiquitin chain with the 26S proteasome involves hydrophobic patches on the surface of the tetraubiquitin chain, generated by Leu8, Ile44, and Val70 in each ubiquitin moiety, and two hydrophobic sequences with the motif LeuAlaLeuAlaLeu in the PA700 subunit S5a (Young et al., 1998). Additional ubiquitin-binding subunits on the 26S proteasome exist since S5a is not an essential protein in yeast (Young et al., 1998). The rate at which protein substrates of this pathway are degraded depends on the interplay between their deubiquitination and their unfolding (Thrower et al., 2000).

2.3.2 ALP

Based on the current knowledge, targeting of protein substrates to degradation by the ALP seems to be a random process. However, ALP suppression or activation in animal cells is highly regulated. Autophagy appears to be constitutively active and is subject to suppression or further induction in response to extracellular stimuli, such as nutrient or growth factor deprivation, stress or pathogenic invasion, specific hormones, and other factors (Wang and Klionsky, 2003). One target of these stimuli is mTOR (mammalian target of rapamycin), a large molecular weight kinase that is phosphorylated via the PI3 kinase/AKT-signaling pathway, and acts as a negative regulator of autophagy. Dephosphorylation of mTOR by nutrient deprivation or rapamycin leads to the induction of autophagy (Kirkegaard et al., 2004).

Two ubiquitin-like (UBL) conjugation pathways, Atg12 and Atg8, directly regulate the formation of autophagosomes (❷ *Figure 21-2*). In the *Atg12 system*, the C-terminal Gly of Atg12 is activated by Atg7, an E1-like enzyme leading to the formation of a thioester intermediate (Ohsumi, 2001). Atg12 is then transferred to Atg10, an E2-like enzyme, forming a second thioester intermediate. Finally the C-terminal

Gly of Atg12 is covalently attached to Lys130 of Atg5, an acceptor molecule, via an isopeptide bond (Mizushima et al., 2002). The Atg12–Atg5 conjugate is present on the isolation membrane/phagophore and helps autophagosome expansion. Subsequently, the Atg12–Atg5 complex binds Atg16 noncovalently and the self-interaction of Atg16 promotes complex multimerization. This high molecular weight complex is distributed along the external lipid bilayer, thus helping the elongation process of the autophagosome (Klionsky, 2005; Reggiori and Klionsky, 2005).

The *Atg8 system* is the second UBL conjugation pathway. The mammalian orthologue of the yeast Atg8 is MAP-LC3 (microtubule-associated protein light chain), and is the only Atg protein to remain associated with fully formed autophagosomes. MAP-LC3, thus serves as a specific marker for autophagy in mammalian cells (Ohsumi, 2001; Mizushima and Klionsky, 2007). The C-terminal region of MAP-LC3 is cleaved by the mammalian Atg4 cysteine protease. This processed form of nascent MAP-LC3 is known as LC3-I and has a Gly residue exposed at the C-terminus necessary for further activation of this protein by Atg7 (Ohsumi, 2001; Boland and Nixon, 2006). An active LC3-I is then transiently transferred to a particular E2-like enzyme, Atg3, and finally LC3-I is transferred to the double membrane specifically to its acceptor molecule, phosphatidylethanolamine (PE). This lipidated form of LC3 is known as LC3-II (Mizushima et al., 2002; Kirkegaard et al., 2004). LC3-II is located on the autophagosome membrane on both sides of the lipid bilayer. Once the autophagosome is complete the Atg5–Atg12–Atg16 complex dissociates from the membrane, whereas Atg4 releases LC3-II from the external lipid bilayer into the cytoplasm by cleaving it off from the lipid molecule PE. Released LC3 can be reused for the biogenesis of new autophagosomes. Mature autophagosomes then swiftly fuse with lysosomes where intralysosomal hydrolases are available for degradation.

2.4 Substrate Identification

2.4.1 UPP

Three key characteristics are identified on proteins to be degraded the UPP: (1) misfolding due to mutations or damaging events; (2) constitutively active ubiquitination signals; and (3) posttranslational modifications such as phosphorylation/dephosphorylation events or cofactor binding (Wilkinson, 1999). The unfolding of normal substrates precedes their degradation. This step is required to allow entry into the proteolytic chamber of the 20S proteasome through its narrow openings (Thrower et al., 2000). Unfolding activities may be provided by ATPase subunits in the base of the 19S particle or by extraproteasomal chaperones.

2.4.2 ALP

As far as we know, this proteolytic pathway is considered to be a bulky, nonspecific degradation process (❯ *Figure 21-3*).

ALP involves the sequestration of cytoplasmic regions containing proteins and organelles into double membrane vacuoles known as autophagosomes (Hara et al., 2006). Autophagosomes then rapidly fuse with lysosomes, which provide all of the hydrolases required for degradation. Formation of the autophagosome is a de novo process and is one of the least understood steps of autophagosome biogenesis. In this process, a membrane of unknown origin, called phagophore, expands leading to autophagosome formation. As discussed earlier, (see "Substrate targeting") Atg proteins play a role in phagophore formation as well as in causing it to expand into a double membrane sphere that will become an autophagosome (Reggiori and Klionsky, 2005).

2.5 Substrate Delivery (UPP and ALP)

Initially, noncanonical chaperones were identified as molecules that deliver ubiquitinated proteins to the 26S proteasome. These potential shuttles for polyubiquitinated proteins contain an UBL domain at

Figure 21-3
Autophagosome formation. The *induction* of autophagy leads to the *formation* of the autophagosome resulting in the sequestration of bulk cytoplasm, including organelles. Upon completion, the autophagosome *docks on* and *fuses* with the lysosome, releasing the inner autophagic body, which is then *broken down*, allowing for the degradation of the cytoplasmic material by the lysosomal hydrolases, such as cathepsins, with recycling of the products

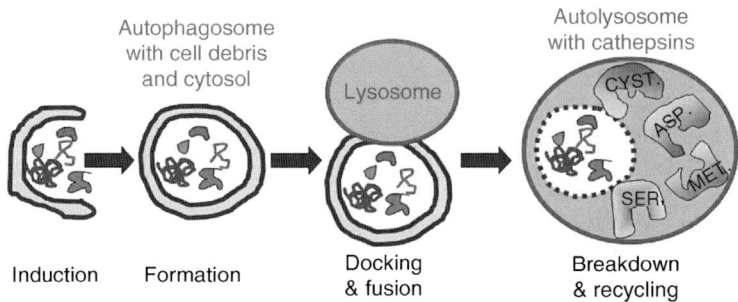

the N-terminus and at least one ubiquitin-associated (UBA) domain at the C-terminus (Hartmann-Petersen et al., 2003; Madura, 2004). The UBL domain is known to interact with the 19S particle of the proteasome, in particular with the subunit S5a/Rpn10 (Seibenhener et al., 2004). The UBA domain noncovalently binds polyubiquitin chains up to 300-times more tightly than mono-ubiquitin (Wilkinson et al., 2001; Raasi et al., 2004).

In the context of neurodegeneration, one of the UBL/UBA proteins that is best characterized as a delivery molecule for polyubiquitinated proteins is the sequestosome1, also known as p62 (p62/SQSTM1). This protein was first identified in human tissues by Shin and colleagues (Park et al., 1995) and was found to be a stress-inducible protein that contributes to the sequestration of polyubiquitinated proteins into aggregates (Seibenhener et al., 2004). At its C-terminus, p62/SQSTM1 has a UBA domain that binds noncovalently to polyubiquitin chains. At its N-terminus, p62/SQSTM1 has a PB1 domain, which is a protein–protein interaction domain (● *Figure 21-4*). The PB1 domain assumes UBL folding and can directly bind to the proteasome and also polymerize with other PB1-containing proteins including homopolymerization (Wooten et al., 2006). In this regard, p62/SQSTM1 may play an important role as a scaffold and/or shuttle molecule sorting polyubiquitinated proteins and delivering them to the proteasome for degradation (Wang and Figueiredo-Pereira, 2005).

Recent data reported that p62/SQSTM1 binds directly to LC3 on the autophagosome membrane, via the LC3-interacting region (LIR), a 22-amino acid sequence. Thus, p62/SQSTM1 seems to be specifically recruited to the autophagosome for degradation (Pankiv et al., 2007). By binding directly polyubiquitinated proteins via its C-terminal UBA domain, proteasomes via its N-terminal PB1 domain, and LC3 via its LIR domain, p62/SQSTM1 may be shuttling polyubiquitinated proteins to an alternate degradation pathway, i.e., the ALP, when proteasomes are impaired or overwhelmed. Due to its binding versatility, the p62/SQSTM1 may play an important role as a scaffold and/or shuttle molecule storing polyubiquitinated proteins and delivering them to the UPP or ALP in a regulated manner.

3 Role of the UPP and APL in Neurodegeneration

Neurodegenerative disorders such as Alzheimer's disease, Parkinson's disease, Huntington's disease (HD), and amyotrophic lateral sclerosis (ALS) are characterized by selective loss of neurons in specific regions of the brain and usually manifest themselves in the later stages of life. The most common feature shared by these disorders is aberrant protein aggregation. Many of the proteins that cause these proteinopathies are

Figure 21-4

Sequestosome (p62/SQSTM1): A molecular LINK between UPP and ALP? These "noncanonical" chaperones exhibit a UBL domain at the N-terminus and at least one UBA (ubiquitin-associated) domain at the C-terminus. These proteins are thought to shuttle polyubiquitinated proteins to the proteasome and to the autophagosome. While the UBA domain is a "receptor" for polyubiquitin chains, the UBL domain is a "receptor" for particular subunits of the 19S particle of the 26S proteasome. Other domains, such as LIR (LC3-interacting region) may directly bind to molecules on the autophagosome membrane

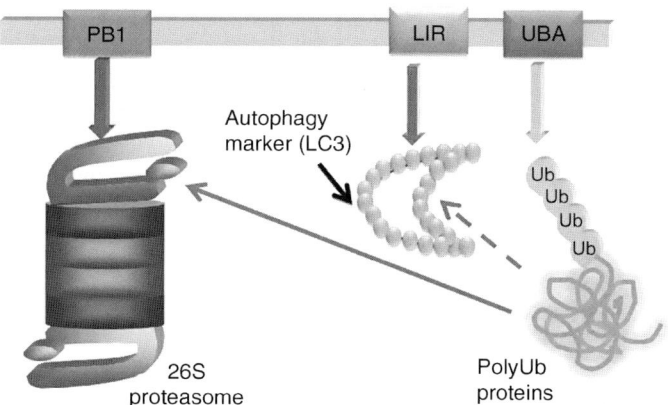

dependent on the UPP for their degradation (Rubinsztein, 2006). However, proteasome function is impaired in the affected brain areas in patients with AD (Keller et al., 2000), PD (McNaught et al., 2003), HD (Seo et al., 2004) and declines with age (Keller et al., 2002) as well as with oxidative stress (Jenner, 2003). Collectively, these findings support the notion that UPP impairment is a risk factor in a variety of neurodegenerative disorders (Moore et al., 2003).

In the past, the ALP was not studied in the brain, since brain appears to be a protected tissue where nutrients can be delivered to it from other organs even under starvation conditions (Boland and Nixon, 2006). Hence, induction of the ALP was not thought to play a role in brain homeostasis. However, recent histological data from neuronal-specific Atg5 and Atg7 knockout mice revealed the presence of neurodegenerative changes such as loss of cerebellar Purkinje cells and cerebral cortex neurons as well as the presence of ubiquitinated inclusions in many regions of the brain (Hara et al., 2006; Komatsu et al., 2006). These data support a role for basal ALP in brain homeostasis. Furthermore, loss of ALP in the brain first leads to the accumulation of diffused abnormal proteins followed by the generation of inclusion bodies, which are hallmarks of numerous neurodegenerative diseases (Hara et al., 2006). In conclusion, both the UPP and the ALP seem to play important roles in the development of protein aggregates detected in neurodegenerative disorders, although the relationship between these two proteolytic pathways remains poorly defined.

Recent studies propose that when the UPP is not functioning properly, ALP seems to be a compensatory mechanism for intracellular protein degradation (Pandey et al., 2007b). In these studies overexpression of a microtubule-associated protein known to bind polyubiquitinated proteins, i.e., histone deacetylase 6 (HDAC6), was sufficient to rescue degeneration in an autophagy-dependent manner, in a *Drosophila* model of neurodegeneration. In this *Drosophila* model the UPP was impaired by a temperature sensitive, dominant negative mutant of the β2 subunit of the 20S proteasome, which is responsible for the trypsin-like activity of the proteasome, or by expression of a toxic polyQ-expanded androgen receptor. HDAC6 overexpression seemed to accelerate the turnover of the mutant androgen receptor as well as of the high molecular weight aggregates that formed upon proteasome inhibition. Further studies by the same group (Pandey et al., 2007a) suggest that HDAC6 promotes the degradation of toxic proteins and mediates the potentially neuroprotective role of autophagy.

4 Overall Conclusions

We addressed the relationship between two proteolytic systems, the UPP and the ALP, which are relevant to the development of protein aggregates detected in a variety of neurodegenerative disorders. The postulated crosstalk between the two proteolytic pathways may inevitably be mediated by molecules, such as p62/SQSTM1 and HDAC6, which may directly interact with these two systems. The following model is proposed (● Figure 21-5).

◘ Figure 21-5
Scheme depicting a putative crosstalk between the UPP and ALP in cell "death" or "recovery" pathways. Environmental and genetic insults in the CNS may initiate a cascade of events that leads to increased levels of ubiquitinated proteins and UPP impairment in neuronal cells, culminating in the development of protein aggregates. These events may lead to neuronal cell death if compensatory mechanisms fail to remove the protein aggregates. Some of the compensatory or "rescue" mechanisms may include the induction of autophagy and upregulation of proteins that deliver the polyubiquitinated protein aggregates to autophagosomes for removal, thus promoting cell recovery (see text for a more detailed explanation)

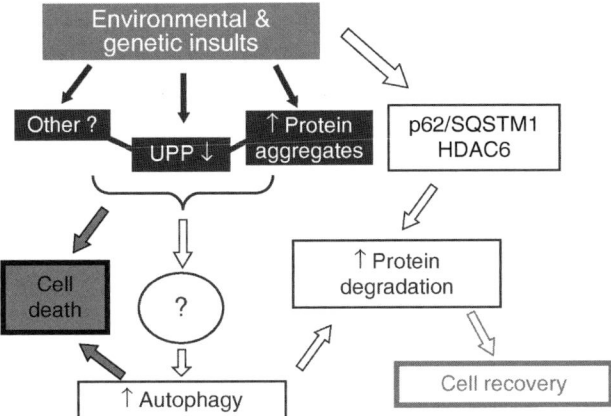

Environmental and genetic insults may affect important cellular pathways involved in neuronal homeostasis, such as the UPP, leading to the formation of proteins aggregates. Under mild (nonlethal) conditions the cell will initiate a "prosurvival/repair" response, which may include, among others, increased expression of p62/SQSTM1 and HDAC6. In addition, an alternate protein degradation pathway, i.e., autophagy, may be induced. These events indicate a cellular attempt to rescue and/or remove protein aggregates that, due to their bulky nature, are precluded from entering and being degraded by the proteasome. If the protein aggregates cannot be removed by these repair mechanisms and the proteasome is impaired as well, prodeath pathways, including apoptosis, may be activated most likely to remove the damaged cells. The resulting neuronal cell death may have devastating effects as, in the vast majority of cases, neurons lost to disease processes cannot be replaced. The ultimate result will be development and exacerbation of neurodegenerative disorders.

Overall, a better understanding of the relationship between the UPP and the ALP will be critical to the development of novel and more effective therapies that prevent and potentially rescue the pathological phenotypes of neurodegenerative disorders, such as AD, PD, HD, and ALS, which are characterized by the accumulation and aggregation of ubiquitinated proteins in a variety of inclusion bodies.

Acknowledgments

Please note that this review is not intended to be comprehensive and we apologize to the authors whose work is not mentioned. This work was supported by NIH [NS41073 (SNRP) to M.E.F.P. (head of sub-project) from NINDS, and RR03037 to Hunter College from NIGMS/RCMI].

References

Abeliovich H, Klionsky DJ. 2001. Autophagy in yeast: Mechanistic insights and physiological function. Microbiol Mol Biol Rev 65: 463-479.

Alves-Rodrigues A, Gregori L, Figueiredo-Pereira ME. 1998. Ubiquitin, cellular inclusions and their role in neurodegeneration. Trends Neurosci 21: 516-520.

Bader N, Jung T, Grune T. 2007. The proteasome and its role in nuclear protein maintenance. Exp Gerontol 42: 864-870.

Benaroudj N, Zwickl P, Seemuller E, Baumeister W, Goldberg AL. 2003. ATP hydrolysis by the proteasome regulatory complex PAN serves multiple functions in protein degradation. Mol Cell 11: 69-78.

Boland B, Nixon RA. 2006. Neuronal macroautophagy: From development to degeneration. Mol Aspects Med 27: 503-519.

Brooks P, Fuertes G, Murray RZ, Bose S, Knecht E, et al. 2000. Subcellular localization of proteasomes and their regulatory complexes in mammalian cells. Biochem J 346 (Pt 1): 155-161.

Chen P, Hochstrasser M. 1996. Autocatalytic subunit processing couples active site formation in the 20S proteasome to completion of assembly. Cell 86: 961-972.

Coux O, Tanaka K, Goldberg AL. 1996. Structure and functions of the 20S and 26S proteasomes. Annu Rev Biochem 65: 801-847.

DeMartino GN, Slaughter CA. 1999. The proteasome, a novel protease regulated by multiple mechanisms. J Biol Chem 274: 22123-22126.

Groll M, Bajorek M, Kohler A, Moroder L, Rubin DM, et al. 2000. A gated channel into the proteasome core particle. Nat Struct Biol 7: 1062-1067.

Groll M, Ditzel L, Lowe J, Stock D, Bochtler M, et al. 1997. Structure of 20S proteasome from yeast at 2.4 A resolution. Nature 386: 463-471.

Gronostajski RM, Pardee AB, Goldberg AL. 1985. The ATP dependence of the degradation of short- and long-lived proteins in growing fibroblasts. J Biol Chem 260: 3344-3349.

Hara T, Nakamura K, Matsui M, Yamamoto A, Nakahara Y, et al. 2006. Suppression of basal autophagy in neural cells causes neurodegenerative disease in mice. Nature 441: 885-889.

Hartmann-Petersen R, Semple CA, Ponting CP, Hendil KB, Gordon C. 2003. UBA domain containing proteins in fission yeast. Int J Biochem Cell Biol 35: 629-636.

Hershko A, Ciechanover A. 1998. The ubiquitin system. Annu Rev Biochem 67: 425-479.

Jenner P. 2003. Oxidative stress in Parkinson's disease. Ann Neurol 53 (Suppl 3): S26–S36.

Johnson DE. 2000. Noncaspase proteases in apoptosis. Leukemia 14: 1695-1703.

Keller JN, Gee J, Ding Q. 2002. The proteasome in brain aging. Ageing Res Rev 1: 279-293.

Keller JN, Hanni KB, Markesbery WR. 2000. Impaired proteasome function in Alzheimer's disease. J Neurochem 75: 436-439.

Kirkegaard K, Taylor MP, Jackson WT. 2004. Cellular autophagy: Surrender, avoidance and subversion by microorganisms. Nat Rev Microbiol 2: 301-314.

Kisselev AF, Akopian TN, Castillo V, Goldberg AL. 1999. Proteasome active sites allosterically regulate each other, suggesting a cyclical bite-chew mechanism for protein breakdown. Mol Cell 4: 395-402.

Klionsky DJ. 2005. The molecular machinery of autophagy: Unanswered questions. J Cell Sci 118: 7-18.

Komatsu M, Waguri S, Chiba T, Murata S, Iwata J, et al. 2006. Loss of autophagy in the central nervous system causes neurodegeneration in mice. Nature 441: 880-884.

Larsen KE, Sulzer D. 2002. Autophagy in neurons: A review. Histol Histopathol 17: 897-908.

Lee DH, Goldberg AL. 1998. Proteasome inhibitors: Valuable new tools for cell biologists. Trends Cell Biol 8: 397-403.

Liu CW, Corboy MJ, DeMartino GN, Thomas PJ. 2003. Endoproteolytic activity of the proteasome. Science 299: 408-411.

Lowe J, Blanchard A, Morrell K, Lennox G, Reynolds L, et al. 1988. Ubiquitin is a common factor in intermediate filament inclusion bodies of diverse type in man, including those of Parkinson's disease, Pick's disease, and Alzheimer's disease, as well as Rosenthal fibres in cerebellar astrocytomas, cytoplasmic bodies in muscle, and Mallory bodies in alcoholic liver disease. J Pathol 155: 9-15.

Madura K. 2004. Rad23 and Rpn10: Perennial wallflowers join the melee. Trends Biochem Sci 29: 637-640.

Massey AC, Zhang C, Cuervo AM. 2006. Chaperone-mediated autophagy in aging and disease. Curr Top Dev Biol 73: 205-235.

McGrath ME. 1999. The lysosomal cysteine proteases. Annu Rev Biophys Biomol Struct 28: 181-204.

McNaught KS, Belizaire R, Isacson O, Jenner P, Olanow CW. 2003. Altered proteasomal function in sporadic Parkinson's disease. Exp Neurol 179: 38-46.

Mizushima N, Klionsky DJ. 2007. Protein turnover via autophagy: Implications for metabolism (*). Annu Rev Nutr 27: 19-40.

Mizushima N, Ohsumi Y, Yoshimori T. 2002. Autophagosome formation in mammalian cells. Cell Struct Funct 27: 421-429.

Moore DJ, Dawson VL, Dawson TM. 2003. Role for the ubiquitin-proteasome system in Parkinson's disease and other neurodegenerative brain amyloidoses. Neuromol Med 4: 95-108.

Ohsumi Y. 2001. Molecular dissection of autophagy: Two ubiquitin-like systems. Nat Rev Mol Cell Biol 2: 211-216.

Orlowski M, Wilk S. 2000. Catalytic activities of the 20S proteasome, a multicatalytic proteinase complex. Arch Biochem Biophys 383: 1-16.

Pandey UB, Batlevi Y, Baehrecke EH, Taylor JP. 2007a. HDAC6 at the intersection of autophagy, the ubiquitin-proteasome system and neurodegeneration. Autophagy 3: 643-645.

Pandey UB, Nie Z, Batlevi Y, McCray BA, Ritson GP, et al. 2007b. HDAC6 rescues neurodegeneration and provides an essential link between autophagy and the UPS. Nature 447: 859-863.

Pankiv S, Clausen TH, Lamark T, Brech A, Bruun JA, et al. 2007. p62/SQSTM1 binds directly to Atg8/LC3 to facilitate degradation of ubiquitinated protein aggregates by autophagy. J Biol Chem 282: 24131-24145.

Park I, Chung J, Walsh CT, Yun Y, Strominger JL, et al. 1995. Phosphotyrosine-independent binding of a 62-kDa protein to the src homology 2 (SH2) domain of p56lck and its regulation by phosphorylation of Ser-59 in the lck unique N-terminal region. Proc Natl Acad Sci USA 92: 12338-12342.

Raasi S, Orlov I, Fleming KG, Pickart CM. 2004. Binding of polyubiquitin chains to ubiquitin-associated (UBA) domains of HHR23A. J Mol Biol 341: 1367-1379.

Rape M, Jentsch S. 2004. Productive RUPture: Activation of transcription factors by proteasomal processing. Biochim Biophys Acta 1695: 209-213.

Reggiori F, Klionsky DJ. 2005. Autophagosomes: Biogenesis from scratch? Curr Opin Cell Biol 17: 415-422.

Rubinsztein DC. 2006. The roles of intracellular protein-degradation pathways in neurodegeneration. Nature 443: 780-786.

Schmidtke G, Kraft R, Kostka S, Henklein P, Frommel C, et al. 1996. Analysis of mammalian 20S proteasome biogenesis: The maturation of beta-subunits is an ordered two-step mechanism involving autocatalysis. EMBO J 15: 6887-6898.

Seibenhener ML, Babu JR, Geetha T, Wong HC, Krishna NR, et al. 2004. Sequestosome 1/p62 is a polyubiquitin chain binding protein involved in ubiquitin proteasome degradation. Mol Cell Biol 24: 8055-8068.

Seo H, Sonntag KC, Isacson O. 2004. Generalized brain and skin proteasome inhibition in Huntington's disease. Ann Neurol 56: 319-328.

Sharon M, Witt S, Felderer K, Rockel B, Baumeister W, et al. 2006. 20S proteasomes have the potential to keep substrates in store for continual degradation. J Biol Chem 281: 9569-9575.

Sharon M, Witt S, Glasmacher E, Baumeister W, Robinson CV. 2007. Mass spectrometry reveals the missing links in the assembly pathway of the bacterial 20S proteasome. J Biol Chem 282: 18448-18457.

Thrower JS, Hoffman L, Rechsteiner M, Pickart CM. 2000. Recognition of the polyubiquitin proteolytic signal. EMBO J 19: 94-102.

Wang CW, Klionsky DJ. 2003. The molecular mechanism of autophagy. Mol Med 9: 65-76.

Wang R, Chait BT, Wolf I, Kohanski RA, Cardozo C. 1999. Lysozyme degradation by the bovine multicatalytic proteinase complex (proteasome): Evidence for a non-processive mode of degradation. Biochemistry 38: 14573-14581.

Wang Z, Figueiredo-Pereira ME. 2005. Inhibition of sequestosome 1/p62 up-regulation prevents aggregation of ubiquitinated proteins induced by prostaglandin J2 without reducing its neurotoxicity. Mol Cell Neurosci 29: 222-231.

Wilkinson CR, Seeger M, Hartmann-Petersen R, Stone M, Wallace M, et al. 2001. Proteins containing the UBA domain are able to bind to multi-ubiquitin chains. Nat Cell Biol 3: 939-943.

Wilkinson KD. 1999. Ubiquitin-dependent signaling: The role of ubiquitination in the response of cells to their environment. J Nutr 129: 1933-1936.

Wojcik C, DeMartino GN. 2003. Intracellular localization of proteasomes. Int J Biochem Cell Biol 35: 579-589.

Wooten MW, Hu X, Babu JR, Seibenhener ML, Geetha T, et al. 2006. Signaling, polyubiquitination, trafficking, and inclusions: Sequestosome 1/p62's role in neurodegenerative disease. J Biomed Biotechnol 2006: 62079.

Young P, Deveraux Q, Beal RE, Pickart CM, Rechsteiner M. 1998. Characterization of two polyubiquitin binding sites in the 26S protease subunit 5a. J Biol Chem 273: 5461-5467.

22 Experimental Autoimmune Encephalomyelitis in the Pathogenesis of Optic Neuritis: Is Calpain Involved?

M. K. Guyton · A. W. Smith · S. K. Ray · N. L. Banik

1	Introduction	526
2	Overview of Calpains	526
3	The Inflammatory Arm of Calpain Pathophysiology	527
3.1	T Cell Activation	527
3.2	Immune Cell Migration	528
4	The Neurodegenerative Arm of Calpain Pathophysiology	528
4.1	Pathophysiology of EAE	528
4.2	Demyelination and Axonal Damage	529
4.3	Roles of Calpain in Apoptosis	529
5	Evidence for Calpain in ON	531
5.1	Evidence for Calpain in MS and EAE	531
5.2	Cell Death in Experimental Models of ON	531
5.3	Ca^{2+}-Mediated Damage to the Optic Nerve	532
5.4	Role of Calpain Following Ca^{2+}-Mediated Damage to the Optic Nerve	532
5.4.1	Effects of Calpain on Optic Nerve and Retina in EAE-ON	533
6	Potential for Calpain Inhibitors as Therapeutic Agents	533
7	Efficacy of Calpain Inhibitors in Neurodegeneration and Inflammation	534
8	Conclusion	535

© 2009 Springer Science+Business Media, LLC.

Abstract: Multiple sclerosis (MS) is a debilitating, autoimmune disease of the central nervous system (CNS) that attacks 1 in every 1,000 Americans each year. Patients suffer clinical symptoms including fatigue, paralysis, and visual dysfunction as a result of demyelination and degeneration of axons and neurons. Since visual dysfunction results from pathophysiological changes in the optic nerve, the animal model of MS, experimental autoimmune encephalomyelitis (EAE), is also used as a tool to study optic neuritis (ON). Previous reports have indicated that the calcium (Ca^{2+})-activated neutral protease calpain may be involved in multiple pathways leading to the development of EAE and EAE-induced ON. These mechanisms include T cell activation and migration, demyelination, and apoptosis of neurons and glial cells. Upregulation of calpain in MS patients further supports this hypothesis. The purpose of this chapter is to discuss and summarize literature involving calpain in EAE and MS, with special emphasis on how calpain may be involved in the development of ON in EAE animals and MS patients.

List of Abbreviations: AIF, apoptosis-inducing factor; Ca^{2+}, calcium; CFA, complete Freund's adjuvant; CNS, central nervous system; CSF, cerebrospinal fluid; de-NFP, dephosphorylation of NFP; EAE, experimental autoimmune encephalomyelitis; ER, endoplasmic reticulum; IFN-γ, interferon-gamma; IL, interleukin; IV, intravenous; LFA, leukocyte function-associated antigen; MAG, myelin-associated glycoprotein; MBP, myelin basic protein; MCP-1, monocyte chemotactic protein-1; MHC II, major histocompatibility complex II; MOG, myelin oligodendrocyte glycoprotein; MS, multiple sclerosis; NFκB, nuclear factor kappa-B; NFP, neurofilament protein; ON, optic neuritis; PBMCs, peripheral blood mononuclear cells; PKC, protein kinase C; PLP, proteolipid protein; R/R, relapsing/remitting; RGCs, retinal ganglion cells; RRMS, relapsing remitting MS; SPMS, secondary progressive MS; STAT, signal transducer and activator of signaling; TBI, traumatic brain injury; TCR, T cell receptor; TNF-α, tumor necrosis factor-alpha

1 Introduction

Multiple sclerosis (MS) is an autoimmune demyelinating disease of the central nervous system (CNS) with clinical symptoms that include fatigue, paralysis, and visual dysfunction (Williams et al., 1994). MS affects 1 in every 1,000 Americans and is the leading cause of neurologic disability in early-to-middle adulthood (Kang, 2006). Optic neuritis (ON), an inflammation of the optic nerve, is strongly associated with MS, and is the initial presentation of the disease in 15–20% of patients. Moreover, 38–50% of MS patients eventually develop ON during subsequent relapses (Arnold, 2005). MS and ON are thought to result from inflammatory attacks on the myelin sheath by auto-reactive T cells and other immune cells, resulting in demyelination, axonal damage, and neuronal death (Compston and Sadovnick, 1992). Recent studies also suggest that later in the development of disease, neurodegenerative events occur even in the absence of an immune attack (Bitsch et al., 2000; Bruck, 2007). Although the mechanisms of immune dysregulation and neurodegeneration in MS and ON are still not fully characterized, studies utilizing the animal model of MS and ON, known as experimental autoimmune encephalomyelitis (EAE), have led to an increased understanding of the pathophysiology associated with these devastating neurodegenerative diseases. Experiments with EAE animals (Shields and Banik, 1998b, 1999; Guyton et al., 2005a), as well as subsequent examinations of brain lesions from MS patients (Shields et al., 1999b; Diaz-Sanchez et al., 2006), have focused on the calcium (Ca^{2+})-dependent neutral protease calpain as a potential candidate for the damaging events in MS and ON. Further support of a role for calpain in MS is derived from evidence that the activities of calpain and other neutral and acid proteases are increased in cerebrospinal fluid (CSF) from MS patients (Hogan et al., 1987). The purpose of this chapter is to offer a general overview of inflammatory and neurodegenerative events mediated by calpain, with specific emphasis on the multiple roles of calpain in EAE-induced ON.

2 Overview of Calpains

Calpains belong to a family of at least 15 cysteine proteases that are activated at neutral conditions in the cell by Ca^{2+} (Wu et al., 2007). Both ubiquitous and tissue-specific calpains have been identified, which play many roles in cell proliferation and differentiation, signal transduction, platelet activation, membrane fusion, necrosis, and apoptosis (Saido et al., 1994; Ray and Banik, 2003). Calpain dysregulation has been linked to multiple disorders including retinal degeneration, ON, MS, lupus, spinal cord injury, and ischemia (Huang and Wang, 2001; Ray and Banik, 2003; Paquet-Durand et al., 2007). In addition, genetic mutations in the tissue-specific calpains, calpain-3 and calpain-10, are associated with limb-girdle muscular dystrophy type 2A (Richard et al., 1995) and type 2 diabetes (Horikawa et al., 2000), respectively. Thus, calpains have been implicated in a variety of immunologic and neurodegenerative diseases and injuries, requiring researchers to dissect the multiple immunologic and neurodegenerative mechanisms associated with calpain-mediated pathophysiology.

The most common isoforms of calpain are μ-calpain and m-calpain, requiring micromolar and millimolar Ca^{2+} concentrations for activation, respectively (Murachi, 1984; Suzuki et al., 2004). Each isoform of calpain contains two subunits: an 80-kD catalytic subunit and a 30-kD regulatory subunit. Their activities, like those of tissue-specific calpains (e.g., muscle, lens), are regulated by Ca^{2+}, calpastatin (an endogenous inhibitor), lipids, and an activator protein (Murachi, 1984; Chakrabarti et al., 1990; Suzuki et al., 2004). Increased calpain activity correlates with increased calpain protein expression, but m-calpain and μ-calpain mRNA levels are not affected (Shields and Banik, 1998a; Shields et al., 1999b). In contrast, neither transcriptional nor translational levels of calpastatin isoforms were changed in MS lesions, which indicates that the regulation of calpain activity in MS is altered to support a pathophysiological change. Calpains have been linked to many physiological events, including T cell activation (Schaecher et al., 2001), cell migration (Glading et al., 2002), cell cycle regulation (Santella et al., 1998), and apoptosis (Ray, 2006; Wu et al., 2007). Knocking out the small calpain subunit is embryonically lethal, which suggests that calpains are necessary to sustain life (Zimmerman et al., 2000). Both μ-calpain and m-calpain are the focus of interest in MS and EAE studies since these are the calpain isoforms that are upregulated in these diseases. Thus, all subsequent references to calpains will pertain to μ-calpain and m-calpain, unless otherwise noted.

3 The Inflammatory Arm of Calpain Pathophysiology

3.1 T Cell Activation

The activation of myelin-specific T cells is thought to be a major event in MS disease development and progression. T cell activation occurs through induction of a complex series of signaling cascades that begin on the surface of the T cell. These cascades are activated when the T cell receptor (TCR) on T cells and the major histocompatibility complex II (MHC II) on the surface of antigen presenting cells interact (Gardner, 1989; Thome and Acuto, 1995). Depending on the microenvironment, CD4+ T cells produce either Th1 or Th2 cytokines following activation. Production of Th1 cytokines, which are pro-inflammatory and include interleukin (IL)-2, IL-12, interferon-gamma (IFN-γ), and tumor necrosis factor-alpha (TNF-α), are increased in MS patients. In contrast, anti-inflammatory cytokines such as IL-4, IL-10, and IL-13 are downregulated. Alterations in calpain expression and activity also correlate with relapse states of MS patients, with increased Th1 cytokine and decreased Th2 cytokine levels (Imam et al., 2007). The potential roles of calpain in T cell activation are multifaceted and include direct modulation of signaling proteins that lead to cytokine production (Hendry and John, 2004; Schaecher et al., 2004) as well as indirect modulation, via cleavage of myelin protein into antigenic peptides that could potentially activate myelin basic protein (MBP)-specific T cells (Deshpande et al., 1995a).

One important consequence of T cell activation is the induction of transcription factor nuclear factor kappa B (NFκB), which is regulated by its inhibitors, the IκBα and IκBβ (Baldwin, 1996). IκB binds and inhibits NFκB in resting cells, but upon activation of the TCR, IκB is degraded, allowing NFκB to translocate to the nucleus and initiate transcription of proteins important for T cell activity. NFκB plays an important role in the transcription of two cytokines that are crucial to T lymphocyte activation, IL-2 and CD25 (Algarte et al., 1995; Gerondakis et al., 1998; Zuckerman et al., 1998). It has been previously reported that calpain inhibition decreases IL-2 and CD25 mRNA expression dose dependently, both at early time points following the initial activation, and over extended periods of time in activated human peripheral blood mononuclear cells (PBMCs) (Schaecher et al., 2001). Furthermore, studies involving a cell-permeable calpain inhibitor, calpeptin, inhibited IκBα degradation in a time-dependent and dose-dependent manner, leading to the identification of calpain-cleavage sites on the IκBα protein (Schaecher et al., 2004). Subsequent studies demonstrated that constitutive degradation of IκBα in human T cells is mediated by calpain, independent of involvement of the 26S proteosome (Ponnappan et al., 2005). Thus, pathophysiological levels of calpain may indirectly induce T cell activation by decreasing available levels of IκBα, thereby permitting unchecked transcription of NFκB, with subsequent synthesis of IL-2 and CD25.

Another consequence of T cell activation that leads to the transcription of Th1 and Th2 cytokines and other mediators important for immunomodulation in both normal and autoimmune states is the induction of members of the signal transducer and activator of signaling (STAT) family of proteins (Chitnis and Khoury, 2003; Land et al., 2004). Calpain activates STAT3 and STAT5, which are involved in both pro-inflammatory and anti-inflammatory cytokine production (Oda et al., 2002). In contrast, STAT6 is inactivated and degraded by calpain (Hendry and John, 2004). STAT6 is specific for inducing transcription of Th2 cytokines such as IL-4, which can in turn lead to inhibition of NFκB. These studies demonstrate the interplay between signal transduction factors that regulate Th1 and Th2 production and highlight the potential for increased calpain activity to shift cytokine production toward a Th1 profile in pro-inflammatory diseases such as MS, ON, and EAE. Since pro- and anti-inflammatory cytokines differentially regulate chemokines and their receptors, Th1 and Th2 profiles help to control another important event in MS and ON pathophysiology—the migration of T cells and other immune cells into the CNS.

3.2 Immune Cell Migration

Migration of immune cells into the CNS is a hallmark of MS and ON pathology (Noseworthy et al., 2000). The levels of chemokine and chemokine receptors, which regulate cell migration, are altered in response to the cytokines present in the cell environment (Castellino et al., 2006; Mikhak et al., 2006). In addition to T cell activation, calpain has also been linked to the migration of immune cells during normal and pathophysiological conditions (Glading et al., 2002). Calpain targets many proteins involved in cell adhesion and migration, altering their activation states (Glading et al., 2002). Calpain inhibitor blocked migration of a T cell hybridoma through modulation of a signaling pathway involving leukocyte function-associated antigen (LFA) (Stewart et al., 1998; Soede et al., 1999). Calpain correlated with T cell and macrophage migration into the CNS following EAE induction in an acute rat model (Shields et al., 1999a). This protease was also associated with the migration of neutrophils (Lokuta et al., 2003) and fibroblasts (Dourdin et al., 2001). Calpain inhibitor reduced the production of IL-8 and monocyte chemotactic protein-1 (MCP-1), chemokines that induce migration, by reducing NFκB translocation and subsequent transcription of these chemokines in epithelial cells in response to viral and parasitic infection (Kim et al., 2000; Kim et al., 2001). These studies highlight the various mechanisms by which calpain modulates the activation and migration of the signaling proteins responsible for pathogenesis of demyelinating diseases.

4 The Neurodegenerative Arm of Calpain Pathophysiology

4.1 Pathophysiology of EAE

EAE is an animal model commonly used for studying the pathophysiology of MS and EAE related ON (EAE-ON). EAE can be induced in rodents by administering spinal cord homogenate and myelin proteins, including MBP, myelin oligodendrocyte glycoprotein (MOG), and proteolipid protein (PLP), or by adoptive transfer of MBP or MOG-specific T cells to generate a relapsing/remitting model of MS (Bullington and Waksman, 1958; Rao, 1981; de Rosbo and Ben-Nun, 1998; Shields and Banik, 1998b; Dasgupta et al., 2003). We studied an acute EAE-ON model in which male Lewis rats (200–230 g) were injected with an emulsion of Complete Freund's Adjuvant (CFA) containing *Mycobacterium tuberculosis* (10 mg/ml) plus guinea pig spinal cord homogenate (10 mg/rat) and MBP (10 μg/rat) (Shields and Banik, 1998b).

Animals lose weight and develop clinical symptoms of paralysis at days 9–12 post-EAE induction, then recover with no further episodes of the disease. As observed in MS patients, EAE-ON animals also demonstrate decreased visual evoked potential and electroretinogram recordings, implicating visual dysfunction due to pathophysiological changes in the optic nerve and retina (Bilbool et al., 1983; Meyer et al., 2001; Hobom et al., 2004). Perivascular cuffing of immune cells, splitting of the myelin lamellae, and cleavage of myelin and cytoskeletal proteins, such as α-spectrin and neurofilament protein (NFP), have been shown in optic nerve from EAE-ON animals (von Sallmann et al., 1967; Banik and Shields, 1999). Unfortunately, the molecular events leading to axonal damage and apoptotic cell death in MS and EAE-ON are not fully understood, but studies have demonstrated that Ca^{2+}-associated events, including calpain activation, may be at least partially responsible.

4.2 Demyelination and Axonal Damage

Demyelination of axons leads to loss of axonal conductivity and is a major component of EAE. Demyelination and the subsequent impairment of conduction are thought to be responsible for disabilities associated with MS and EAE (McFarlin and McFarland, 1982; Raine, 1982), although magnetic resonance imaging studies fail to consistently correlate demyelinating lesions with significant clinical deficits in MS (Antel, 1999; Hickey, 1999). These studies do indicate possible damage to axons as well as abnormalities in gray matter contributing to dysfunction. In addition, recent magnetic resonance spectroscopy studies demonstrated axonal damage with functional deficit in MS at the early phase as well as at the progressive phase of the disease (De Stefano et al., 1999; Bozzali et al., 2002; Kuhlmann et al., 2002). As mentioned earlier, calpain has been shown to degrade MBP as well as other axonal and cytoskeletal proteins, including NFP. Loss of NFP has long been associated with damage to axons and dysfunction in MS (Newcombe et al., 1982). In fact, increases in dephosphorylation of NFP (de-NFP) can be used as an indicator of axonal damage since NFPs are dephosphorylated before degradation (Gehrmann et al., 1995; Kornek et al., 2000). Recently, we demonstrated that increased Ca^{2+} influx and calpain expression in spinal cord from Lewis rats with acute EAE correlated with increased axonal damage, as assessed by increases in de-NFP expression (Guyton et al., 2005b).

In MS and EAE, there is not only axonal degeneration, but also loss of neuron cell bodies and glia as well, indicating that the disease has a neurodegenerative component and is more complex than originally presumed. The mechanisms by which cells and axons are damaged in MS or EAE are not clearly understood. Since there are many destructive pathways, it is likely that different mechanisms are involved, including apoptosis. In EAE, glutamate cytotoxicity was found to cause neuron and oligodendrocyte death and axonal damage (Peterson et al., 2001; Smith and Hall, 2001). It is postulated that Ca^{2+} influx due to excessive glutamate release from activated macrophages and microglia leads to activation of intracellular stores of calpain, which may then greatly affect the disease process. In addition, calpain released directly from immune cells and resident glia may participate in myelin protein degradation for antigen presentation (contributing to epitope spreading-immune arm) and also directly or indirectly cause damage to axons, neurons, and oligodendrocytes (Shields and Banik, 1999). Further, axonal damage at the early phase of the disease may be secondary to neuronal death. Examination of control and acute EAE spinal cord indicated that Ca^{2+} influx and calpain expression correlated with axonal damage, as indicated by increases in de-NFP and EM analysis of granular degeneration of axons, mitochondrial disruption, and loss of structural integrity of microtubules and filaments (Guyton et al., 2005b). It would be important to expand the examination of axonal damage in acute EAE after animals have recovered from paralysis and during remission and the second relapse (chronic phase) in relapsing/remitting (R/R) EAE animals. Such studies in acute EAE have shown the persistence of axonal degeneration in spinal cord several months after the animals recovered from clinical manifestations of the disease (Guyton et al., 2006).

4.3 Roles of Calpain in Apoptosis

Both necrosis and apoptosis have been demonstrated in MS and EAE, and elucidating the mechanisms of cell death will aid in developing therapeutic agents for these neurodegenerative diseases. Specific cell types, including T cells and macrophages, are known to undergo programmed cell death in EAE (Nguyen et al., 1994; Sarin et al., 1994; Smith et al., 1996). In apoptosis, there is no inflammatory reaction, but other mediators participate, including Ca^{2+}, free radicals, cytokines, and proteases including calpain. Increased calpain activity and expression in EAE spinal cord suggest that calpain is one of the cysteine proteases responsible for cell death in EAE. This is supported by many in vitro studies employing various cell types, including lymphocytes, cardiomyocytes, neurons, and glial cells (Saido et al., 1994; Sarin et al., 1994; Squier et al., 1994; Ray et al., 1999a, 2000) in which calpain inhibitors blocked cell death. Calpains can cleave talin, filamin, fodrin, protein kinase C (PKC), c-Mos, c-Jun, c-Fos, and p53, all of which are related to apoptosis (Watanabe et al., 1989; Kishimoto, 1990; Croall and DeMartino, 1991; Hirai et al., 1991; Watt and Molloy, 1993; Kubbutat and Vousden, 1997). Although the role of activated calpain is potentially important in apoptosis, its role has not yet been thoroughly investigated in demyelinating diseases. The apoptosis of glial cells has been demonstrated in MS plaques (Dowling et al., 1997), injured spinal cord (Li et al., 1996; Crowe et al., 1997), and EAE spinal cord (Pitt et al., 2000; Meyer et al., 2001), and death of oligodendrocytes and neurons in EAE spinal cord leads to impaired function and disability. The involvement of calpain in apoptosis may be attenuated by agents that inhibit its activity and block other pathways responsible for cell death and tissue degeneration.

Several apoptotic pathways have been implicated in apoptosis in neurons and other cells in neurodegenerative diseases. ◯ *Figure 22-1* gives an overview of these potential pathways, which involve Ca^{2+} influx, calpain activation, mitochondrial dysfunction, and caspase activation, and how these pathways may

◻ Figure 22-1
Calpain-induced apoptotic signaling pathways

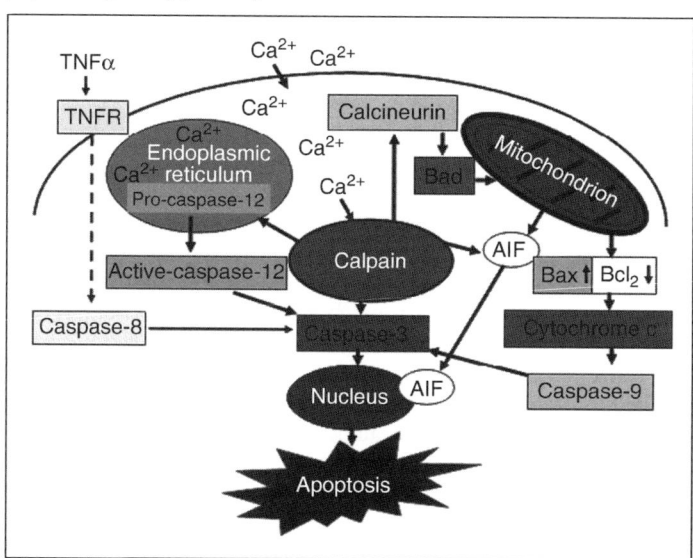

intersect. Calpain has been shown to activate calcineurin indirectly through cleavage of cain/caibin1 in the Jurkat T cell line (Kim et al., 2002) and more recently through direct cleavage of calcineurin in Alzheimer's brain to induce cell death (Liu et al., 2005). Calpain also induces increases in the ratio of the pro-apoptotic protein Bax to the anti-apoptotic protein Bcl-2 by directly cleaving Bax through a mitochondrial pathway (Wood and Newcomb, 2000) or by inactivating Bcl-2, which results in the release of cytochrome c and induction of apoptosis (Gao and Dou, 2000). Release of the apoptosis-inducing factor (AIF) from the mitochondrion is also dependent on calpain (Polster et al., 2005). Further, there is cross-talk between calpain and caspase-3, the executioner in the final step of cell death, and calpain has been shown to cleave caspase-3 into active forms (Blomgren et al., 2001). Caspase-3 degrades calpastatin, the endogenous inhibitor of calpain, through proteolysis, which further increases calpain activity by keeping it unregulated (Wang et al., 1998). Calpain also cleaves and activates caspase-12, which can in turn activate caspase-3-induced apoptosis (Nakagawa and Yuan, 2000). Caspase-12, an endoplasmic reticulum (ER) stress-associated caspase, has not been identified in humans; however, the identification of caspase-12 homologues in human cancer cell lines (Bitko and Barik, 2001; Kitamura et al., 2003) may suggest that ER caspases are activated in apoptosis in human cells and could potentially be attenuated by calpain inhibition.

5 Evidence for Calpain in ON

5.1 Evidence for Calpain in MS and EAE

Proteases have been implicated in myelin breakdown, axonal degeneration, and cell death in demyelinating ON, MS, and EAE (Shields et al., 1998, 1999b; Shields and Banik, 1999). More specifically, calpain has been implicated in the degradation of myelin proteins, neurofilament, cytoskeletal proteins (fodrin, talin actin), and nuclear laminins (Saito et al., 1993; Posmantur et al., 1994; Ray and Banik, 2003). It also degrades substrates of eye tissue and fodrin in retinal ganglion cells (RGCs), neurons, and optic glia (Nixon, 1986; Shearer and David, 1990). Increased m-calpain activity and immunoreactivity in MS plaques (Sato et al., 1984), CSF (Hogan et al., 1987; Inuzuka et al., 1987), and EAE optic nerve (Banik and Shields, 1999) has also been demonstrated. Regarding lymphoid cell involvement, calpain secreted from activated T cells and mononuclear phagocytes participates in proteolytic cleavage of myelin proteins (e.g., MBP) for antigen presentation (promoting epitope spreading), which either directly or indirectly damages axons, neurons, and oligodendrocytes (Shields and Banik, 1999; Imam et al., 2007). Moreover, increased calpain expression has been found in activated human lymphoid cell lines (Deshpande et al., 1995a, b). Analysis of MS patient PBMCs collected during periods of relapse and remission also revealed increased calpain expression and activity compared with controls (Imam et al., 2007). Furthermore, it was observed that MBP-specific T cells are a source of activated calpain, and degradation of MBP is diminished upon pretreatment with a calpain inhibitor.

Many cell types undergo programmed cell death in EAE, including T cells and macrophages (Nguyen et al., 1994; Smith et al., 1996), RGCs, and glial cells (Pitt et al., 2000; Meyer et al., 2001). It is clear that mediators participate in this process, such as Ca^{2+}, free radicals, cytokines, and proteases. Increased calpain expression has been measured in ON optic nerve, suggesting that it is a mediator responsible for apoptosis in ON, and the pro-apoptotic capacity of calpain is supported in many in vitro studies using various cell types, including lymphocytes, cardiomyocytes, neurons, and glial cells (Watanabe et al., 1989; Croall and DeMartino, 1991; Saido et al., 1994; Sarin et al., 1994; Squier et al., 1994; Ray et al., 1999b, 2000). Although in vitro studies support the idea that calpain proteolysis is important in apoptosis, this has only been investigated to a limited extent in demyelinating diseases.

5.2 Cell Death in Experimental Models of ON

As mentioned earlier, different clinical courses of EAE can be induced depending on the myelin protein used for immunization and the animals chosen, including monophasic/acute, chronic progressive, and relapsing/remitting disease (Mendel et al., 1995; Storch et al., 1998; Wujek et al., 2002). The incidence of ON has been observed in both chronic and R/R EAE (Raine et al., 1980; O'Neill et al., 1998; Sakuma et al., 2004; Shao et al., 2004). In addition, a proportion of MOG-specific TCR transgenic mice were found to develop spontaneous isolated ON in the absence of clinical or histological manifestations of EAE (Bettelli et al., 2003; Guan et al., 2006). It is clear that apoptotic death of RGCs accompanies inflammation in both isolated and EAE-ON, though the sequence of events is unclear. For instance, in transgenic MOG and R/R mouse models, neurodegeneration apparently occurs secondary to the inflammatory process (Guan et al., 2006; Shindler et al., 2006). In contrast, a rat model of chronic MOG-induced EAE demonstrated loss of RGCs for an entire week prior to histopathological evidence of optic nerve inflammatory cell infiltration, although loss of RGCs increased greatly following inflammation and onset of clinical symptoms (Meyer et al., 2001; Hobom et al., 2004). Understanding the cellular and molecular mechanisms preceding optic nerve degeneration may provide potential therapeutic benefit by ushering the development of improved agents to protect RGCs, preserve retinal nerve fiber layer (RNFL) thickness and macular volume, and ultimately restore visual function.

5.3 Ca^{2+}-Mediated Damage to the Optic Nerve

Increased levels of intracellular free Ca^{2+} have been implicated in many vision disorders, including cataract formation (Shearer and David, 1982), photoreceptor degeneration (Fox et al., 1999; Sharma and Rohrer, 2004), retinal ischemia (Nihard, 1982), and EAE-ON (Shields and Banik, 1998b; Diem et al., 2003). Selenite cataract formation in rats was shown to be dependent on increased intracellular Ca^{2+} levels (Shearer and David, 1982). Increased levels of intracellular free Ca^{2+} led to neuronal apoptosis through mitochondrial depolarization and caspase-3 activation in the rd mouse, a strain that contains a recessive mutation in the rd gene that encodes for a cGMP phosphodiesterase. In this scenario of Ca^{2+} toxicity, activity of the Ca^{2+}-dependent protease calpain has been found to be increased up to six-fold in the rd mouse retina, and the elevated activity followed the known time course of photoreceptor degeneration (Sharma and Rohrer, 2004).

Injury to the optic nerve due to increased Ca^{2+} causes irreversible damage to white matter (Ransom et al., 1993). Treatment with Ca^{2+} channel blockers only partially ameliorated the damage associated with Ca^{2+} toxicity in patients with retinal ischemia, with some improvement in visual function (Nihard, 1982).

The mechanisms of Ca^{2+} influx in MS and EAE-ON are unclear; however, membrane damage, glutamate neurotoxicity, and alterations in Ca^{2+} channels may be involved. For instance, treatment with a voltage-gated Ca^{2+} channel blocker inhibited apoptosis of RGCs in EAE-ON (Diem et al., 2003). Once Ca^{2+} influx occurs, signaling pathways are activated that lead to axonal damage and cell death in the optic nerve and retina. One Ca^{2+}-induced event, which is upregulated in EAE-ON, is increased expression and activation of the Ca^{2+}-dependent protease calpain.

5.4 Role of Calpain Following Ca^{2+}-Mediated Damage to the Optic Nerve

Calpain cleaves many proteins in the optic nerve, including myelin proteins [MBP, myelin-associated glycoprotein (MAG)], cytoskeletal proteins (α-spectrin, actin), NFPs, and cell signaling proteins (Ray and Banik, 2003). In the retina, calpain has been found to degrade endogenous substrates of eye tissue (Shearer and David, 1990), α-crystalline, vimentin, actin (Yoshida et al., 1984), and other intrinsic membrane proteins. Calpain degrades fodrin in RGCs, neurons, and optic glia (Nixon, 1986), and its activity is six times higher in neurons than that in glial cells, suggesting greater vulnerability of neurons in neuropathological states. Proteases including calpain are involved in cataract formation and in retinal and corneal degeneration (Hightower et al., 1987; Twining et al., 1993; Wagner and Margolis, 1993; Azarian and Williams, 1995).

A decrease in MAG has been found in EAE-ON (Shields and Banik, 1998b). However, no significant changes in calpastatin transcriptional or translational expression were found, suggesting that increases in calpain expression and activity in EAE-ON were not due to decreased calpastatin expression. Calpain-dependent degradation of NFP, a marker for axonal damage, in optic nerve in vitro was also demonstrated (Shields et al., 1997). Immunohistolabeling of calpain expression in EAE-ON indicated that both inflammatory (T cell, macrophages) and glial (astrocytes) cells expressed increased levels of calpain, as compared with controls (Banik and Shields, 1999). Since calpain has been shown to be involved in apoptosis in many CNS diseases (Ray et al., 1999b; Sharma and Rohrer, 2004), the data suggest that calpain upregulation may also play a role in apoptosis of cell types involved in EAE-ON, including glial cells and RGCs.

5.4.1 Effects of Calpain on Optic Nerve and Retina in EAE-ON

MS has long been considered a disease in which early inflammatory events damage myelin, but spare axons. There is compelling recent evidence that in MS patients and EAE animals, significant axonal damage and loss of neurons occur during early stages of the disease and is the primary reason for chronic disability (Trapp et al., 1998; Bjartmar et al., 2000; Wujek et al., 2002; De Stefano et al., 2003). However, little is known about the mechanisms leading to neuronal loss. Since ON (impairment of vision) is the first sign in the manifestation of MS and since the majority of ON patients develop MS (Kang, 2006), it is likely that RGCs and glial cells are damaged quite early in the disease process. Increased calpain expression has been detected in peripheral T cells at 4–5 days after EAE challenge (Shields et al., 1999a), and activated T cells may then migrate into the optic nerve at 7–8 days after disease induction. In fact, T cell migration appears to depend on calpain activation, since calpain-specific inhibitors block integrin-mediated cell detachment (Huttenlocher et al., 1997; Shields et al., 1999a). This suggests that increased calpain may facilitate damage to RGCs, oligodendrocytes, and axons in the optic nerve, leading to visual dysfunction. Calpain-mediated damage to RGCs has been demonstrated in vitro, and functionality of these cells was restored when they were pretreated with calpain inhibitor (Das et al., 2006). If calpain is involved in the visual impairment of early phase MS, changes in several parameters would be expected in the ON optic nerve and RGCs, including an influx of Ca^{2+}, increased calpain activity and expression, and upregulated expression of pro-apoptotic proteins.

The cumulative evidence suggests that the optic nerve is affected early, and treatment of EAE-ON animals with calpain inhibitors prior to the onset of clinical symptoms may prevent cell death and axonal damage and salvage visual function.

6 Potential for Calpain Inhibitors as Therapeutic Agents

The chief goal of MS and ON therapy is to prevent permanent neurological disability. Currently, the treatment of choice for acute ON is intravenous (IV) methylprednisolone (Balcer, 2006). Although this medication promotes rapid recovery of vision, it does not affect long-term visual outcome (Beck et al., 1992, 2004; Foroozan et al., 2002; Arnold, 2005; Frohman et al., 2005). Oral corticosteroid therapy (i.e., prednisone) has been considered as an alternative intervention for exacerbations of ON. However, several recent randomized clinical trials found that oral prednisone was associated with a significantly increased risk of recurrent ON, suggesting that this therapy may in fact be detrimental (Beck et al., 1992, 2004; Cole et al., 2000). Furthermore, many patients receiving corticosteroid therapy report serious side effects, including sleep disturbances, mood changes, stomach upset, weight gain, hyperglycemia, and worsening hypertension (Balcer, 2006).

Though the limitations of corticosteroid therapy are obvious, it is the only well-characterized option for short-term treatment. Regarding long-term options, early administration of the MS disease-modifying agents interferon beta-1a and interferon beta-1b has been shown to reduce the development of MS in patients presenting with both acute ON and demyelinating lesions in MRI (Jacobs et al., 2000, CHAMPS Study Group 2001; Comi et al., 2001). This strategy does not address the immediate visual disability, but it holds promise as a way to delay disease onset and the appearance of new lesions.

The FDA-approved disease-modifying agents for MS include beta-interferon-1a (Avonex), or beta-interferon-1a (Rebif), beta-interferon-1b (Betaseron), glatiramer acetate (Copaxone), and mitoxantrone (Novatrone) (Johnson et al., 1995; Jacobs et al., 1996; PRISMS Study Group 1998). The beta-interferon class and glatiramer acetate effectively reduce the number of attacks in relapsing remitting MS (RRMS), but they provide no benefit in primary or secondary progressive forms (SPMS) of the disease (Inglese, 2006). Only mitoxantrone treatment has shown promise in retarding disease progression and the onset of disability (Calabresi, 2002). Currently, there are roughly 150 new immunomodulatory compounds being used or in development for MS therapy, some of which have dramatic effects as measured clinically using MRI. To date, all of these pharmaceuticals carry significant side effects, so their widespread use is limited. Thus, for MS patients who have crossed the threshold from RRMS to SPMS, there are few options; namely treatment with immunosuppressants such as mitoxantrone if tolerated (Hartung et al., 2002; Goodin et al., 2003; Cohen and Mikol, 2004) and management of symptoms during acute attacks. Additionally, 10–15% of MS patients experience a primary progressive course, which is characterized by a "slow burn" of low-level inflammation and myelopathy, and is typically refractory to immunomodulatory agents (Montalban, 2005).

Despite the limitations of standard treatment, the current status of research in MS is auspicious. Efforts over the past 15 years have led to an abundance of compounds in development, ongoing randomized clinical trials, and improved sophistication of diagnostic imaging. Since MS is a chronic disease with a progressive component in most cases, questions of treatment efficacy and cost effectiveness must be answered after assessing long-term outcomes. The newest agents offer hope in the short term, but they have not yet withstood the test of time.

As we understand more about the cellular and molecular pathology of MS, this will allow a more specific, individualized approach to therapy. Recent studies are investigating the process of axonal degeneration and plaque formation in the CNS, as well as the potential for neuroprotection, repair, and recovery of damaged tissue. We propose that a combination of anti-inflammatory compounds with agents that reduce neurodegeneration may significantly augment the current treatment modality and possibly delay the progression of MS and ON. Among the pharmaceutics that block neurodegeneration are specific inhibitors of calpain. Calpain inhibitors such as calpeptin and CYLA [cyzteic-leucyl-arginal that reading crosses blood brain barrier (BBB)] have been found to prevent axonal degeneration and cell death in EAE. Moreover, they attenuated disease severity, as evidenced by a reduction of clinical symptoms (Hassen et al., 2006; Imam et al., 2007). These recent findings indicate that calpain inhibitors may have therapeutic potential in the treatment of MS and ON.

7 Efficacy of Calpain Inhibitors in Neurodegeneration and Inflammation

Studies using calpain-specific inhibitors, particularly SJA6017, in ischemia, cataracts, traumatic brain injury (TBI), photoreceptor degeneration, and glutamate excitotoxicity have shown neuroprotective effects, indicating that the inhibitors cross the BBB and are cell permeable (Kupina et al., 2001; Das et al., 2005, 2006; Sharma and Rohrer, 2007). In the rd1 mouse, a model of retinal photoreceptor degeneration, combination therapy with calpain inhibitors drastically blocked apoptosis by interfering with AIF and caspase-12 activation (Sanges et al., 2006). Furthermore, calpain inhibition can reduce the development of acute and chronic inflammation in vivo (Cuzzocrea et al., 2000). As mentioned earlier, calpain is essential for the degradation of IκBα in human T cells (Ponnappan et al., 2005). When availability of IκBα is decreased, NFκB translocation to the nucleus of PBMCs continues unchecked, which allows expression of NFκB-dependent genes, including those for iNOS (Griscavage et al., 1996; Kengatharan et al., 1996; Milligan et al., 1996), COX-2 (Yamamoto et al., 1995; Crofford et al., 1997), and cytokines crucial for T cell activation (Algarte et al., 1995; Gerondakis et al., 1998; Zuckerman et al., 1998). The generation of such pro-inflammatory factors, particularly the hydroxyl radicals, leads to tissue injury (Moncada et al., 1991; Youn et al., 1991; Nathan, 1992; McCord, 1993; Zingarelli et al., 1996; Cuzzocrea et al., 2000), DNA damage, and ultimately cell death. The cell-permeable calpain inhibitor, calpeptin (Schaecher et al., 2004), and calpain inhibitor I can prevent NFκB transcription. Because inflammation is central to tissue injury and cell death in MS and EAE, calpain inhibition holds promise as a novel therapeutic approach. The effects of calpain inhibitors in ON are not known, though a recent in vitro study demonstrated that calpeptin provides functional neuroprotection to RGCs (Das et al., 2005). Thus, EAE-ON studies are needed, which investigate the efficacy of cell-permeable calpain inhibitors to attenuate apoptosis of RGCs, reduce inflammation by mitigating NFκB transcription by PBMCs, inhibit major basic protein (MBP)-antigen presentation, and prevent the migration of activated T cells.

In addition to the calpain-specific inhibitors, which have been investigated over the past few years, more novel derivatives are being developed (Shirasaki et al., 2006). The most novel inhibitor is improved in measures of potency, aqueous solubility, oral bioavailability, and extended plasma half-life in rats. Combination therapy with a novel calpain inhibitor and an approved immunomodulatory agent may be a rational strategy for future studies in EAE and MS. First, combination therapy will be dose sparing, which may help to limit side effects as long as the toxic effects of each drug are non-overlapping. Second, since calpain produces tissue damage via multiple mechanisms, a supplementary calpain inhibitor will have at least one mode of action, which differs from the standard medications. Finally, by instituting a multimodal regimen, there is an enhanced ability to inhibit or otherwise influence several major steps in the injury cascade of MS. Interference at multiple points may interrupt the coordinated pathogenesis so severely that it prevents or retards the occurrence of inflammation, demyelination, axonal degeneration, and permanent neurological disability.

8   Conclusion

This chapter intended to provide an overview of the inflammatory and neurodegenerative events mediated by calpain in EAE and EAE-ON. The manifold roles of calpain in processes as diverse as immune cell migration, T cell activation, demyelination, axonal damage, and apoptosis make it a compelling target, which may be considered for clinical trials of calpain inhibitors as a novel approach to MS therapy. Indeed, calpain-specific inhibition has already shown promise in several preclinical models of EAE; it will be important to determine if calpain inhibitors, in combination with approved disease-modifying agents, can prevent the devastating neurological impairment of MS and ON in a synergistic fashion.

Acknowledgments

This work was supported in part by the R01 grants (CA-91460, NS-31622, NS-38146, NS-41088, NS-45967, NS-56176, and NS-57811) from the National Institutes of Health and Spinal Cord Injury Research Foundation grants (SCIRF-0803, SCIRF-1205, and SCIRF-607) from the State of South Carolina.

References

Algarte M, Lecine P, Costello R, Plet A, Olive D, et al. 1995. In vivo regulation of interleukin-2 receptor alpha gene transcription by the coordinated binding of constitutive and inducible factors in human primary T cells. EMBO J 14: 5060-5072.

Antel J. 1999. Multiple sclerosis—emerging concepts of disease pathogenesis. J Neuroimmunol 98: 45-48.

Arnold AC. 2005. Evolving management of optic neuritis and multiple sclerosis. Am J Ophthalmol 139: 1101-1108.

Azarian SM, Williams DS. 1995. Calpain activity in the retinas of normal and RCS rats. Curr Eye Res 14: 731-735.

Balcer LJ. 2006. Clinical practice. Optic neuritis. N Engl J Med 354: 1273-1280.

Baldwin AS Jr. 1996. The NF-kappa B and I kappa B proteins: New discoveries and insights. Annu Rev Immunol 14: 649-683.

Banik NL, Shields DC. 1999. A putative role for calpain in demyelination associated with optic neuritis. Histol Histopathol 14: 649-656.

Beck RW, Cleary PA, Anderson MM Jr., Keltner JL, Shults WT, et al. 1992. A randomized, controlled trial of corticosteroids in the treatment of acute optic neuritis. The Optic Neuritis Study Group. N Engl J Med 326: 581-588.

Beck RW, Gal RL, Bhatti MT, Brodsky MC, Buckley EG, et al. 2004. Visual function more than 10 years after optic neuritis: Experience of the optic neuritis treatment trial. Am J Ophthalmol 137: 77-83.

Bettelli E, Pagany M, Weiner HL, Linington C, Sobel RA, et al. 2003. Myelin oligodendrocyte glycoprotein-specific T cell receptor transgenic mice develop spontaneous autoimmune optic neuritis. J Exp Med 197: 1073-1081.

Bilbool N, Kaitz M, Feinsod M, Soffer D, Abramsky O. 1983. Visual evoked potentials in experimental allergic encephalomyelitis. J Neurol Sci 60: 105-115.

Bitko V, Barik S. 2001. An endoplasmic reticulum-specific stress-activated caspase (caspase-12) is implicated in the apoptosis of A549 epithelial cells by respiratory syncytial virus. J Cell Biochem 80: 441-454.

Bitsch A, Schuchardt J, Bunkowski S, Kuhlmann T, Bruck W. 2000. Acute axonal injury in multiple sclerosis. Correlation with demyelination and inflammation. Brain 123 (Pt 6): 1174-1183.

Bjartmar C, Kidd G, Mork S, Rudick R, Trapp BD. 2000. Neurological disability correlates with spinal cord axonal loss and reduced N-acetyl aspartate in chronic multiple sclerosis patients. Ann Neurol 48: 893-901.

Blomgren K, Zhu C, Wang X, Karlsson JO, Leverin AL, et al. 2001. Synergistic activation of caspase-3 by m-calpain after neonatal hypoxia-ischemia: A mechanism of "pathological apoptosis"? J Biol Chem 276: 10191-10198.

Bozzali M, Cercignani M, Sormani MP, Comi G, Filippi M. 2002. Quantification of brain gray matter damage in different MS phenotypes by use of diffusion tensor MR imaging. AJNR Am J Neuroradiol 23: 985-988.

Bruck W. 2007. New insights into the pathology of multiple sclerosis: Towards a unified concept? J Neurol 254 (Suppl. 1): I3-I9.

Bullington SJ, Waksman BH. 1958. Uveitis in rabbits with experimental allergic encephalomyelitis; results produced by injection of nervous tissue and adjuvants. AMA Arch Ophthalmol 59: 435-445.

Calabresi PA. 2002. Considerations in the treatment of relapsing-remitting multiple sclerosis. Neurology 58: S10-S22.

Castellino F, Huang AY, Altan-Bonnet G, Stoll S, Scheinecker C, et al. 2006. Chemokines enhance immunity by guiding naive CD8+ T cells to sites of CD4+ T cell-dendritic cell interaction. Nature 440: 890-895.

Chakrabarti AK, Dasgupta S, Banik NL, Hogan EL. 1990. Regulation of the calcium-activated neutral proteinase (CANP) of bovine brain by myelin lipids. Biochim Biophys Acta 1038: 195-198.

CHAMPS Study Group. 2001. Interferon beta-1a for optic neuritis patients at high risk for multiple sclerosis. Am J Ophthalmol 132: 463–471.

Chitnis T, Khoury SJ. 2003. Cytokine shifts and tolerance in experimental autoimmune encephalomyelitis. Immunol Res 28: 223-239.

Cohen BA, Mikol DD. 2004. Mitoxantrone treatment of multiple sclerosis: Safety considerations. Neurology 63: S28-S32.

Cole SR, Beck RW, Moke PS, Gal RL, Long DT. 2000. The National Eye Institute Visual Function Questionnaire: Experience of the ONTT. Optic Neuritis Treatment Trial. Invest Ophthalmol Vis Sci 41: 1017-1021.

Comi G, Filippi M, Barkhof F, Durelli L, Edan G, et al. 2001. Effect of early interferon treatment on conversion to definite multiple sclerosis: A randomised study. Lancet 357: 1576-1582.

Compston A, Sadovnick AD. 1992. Epidemiology and genetics of multiple sclerosis. Curr Opin Neurol Neurosurg 5: 175-181.

Croall DE, DeMartino GN. 1991. Calcium-activated neutral protease (calpain) system: Structure, function, and regulation. Physiol Rev 71: 813-847.

Crofford LJ, Tan B, McCarthy CJ, Hla T. 1997. Involvement of nuclear factor kappa B in the regulation of cyclo-oxygenase-2 expression by interleukin-1 in rheumatoid synoviocytes. Arthritis Rheum 40: 226-236.

Crowe MJ, Bresnahan JC, Shuman SL, Masters JN, Beattie MS. 1997. Apoptosis and delayed degeneration after spinal cord injury in rats and monkeys. Nat Med 3: 73-76.

Cuzzocrea S, McDonald MC, Mazzon E, Siriwardena D, Serraino I, et al. 2000. Calpain inhibitor I reduces the development of acute and chronic inflammation. Am J Pathol 157: 2065-2079.

Das A, Garner DP, Del Re AM, Woodward JJ, Kumar DM, et al. 2006. Calpeptin provides functional neuroprotection to rat retinal ganglion cells following Ca2+ influx. Brain Res 1084: 146-157.

Das A, Sribnick EA, Wingrave JM, Del Re AM, Woodward JJ, et al. 2005. Calpain activation in apoptosis of ventral spinal cord 4.1 (VSC4.1) motoneurons exposed to glutamate: Calpain inhibition provides functional neuroprotection. J Neurosci Res 81: 551-562.

Dasgupta S, Zhou Y, Jana M, Banik NL, Pahan K. 2003. Sodium phenylacetate inhibits adoptive transfer of experimental allergic encephalomyelitis in SJL/J mice at multiple steps. J Immunol 170: 3874-3882.

de Rosbo NK, Ben-Nun A. 1998. T-cell responses to myelin antigens in multiple sclerosis; relevance of the predominant autoimmune reactivity to myelin oligodendrocyte glycoprotein. J Autoimmun 11: 287-299.

De Stefano N, Matthews PM, Filippi M, Agosta F, De Luca M, et al. 2003. Evidence of early cortical atrophy in MS: Relevance to white matter changes and disability. Neurology 60: 1157-1162.

De Stefano N, Narayanan S, Matthews PM, Francis GS, Antel JP, et al. 1999. In vivo evidence for axonal dysfunction remote from focal cerebral demyelination of the type seen in multiple sclerosis. Brain 122 (Pt 10): 1933-1939.

Deshpande RV, Goust JM, Chakrabarti AK, Barbosa E, Hogan EL, et al. 1995b. Calpain expression in lymphoid cells. Increased mRNA and protein levels after cell activation. J Biol Chem 270: 2497-2505.

Deshpande RV, Goust JM, Hogan EL, Banik NL. 1995a. Calpain secreted by activated human lymphoid cells degrades myelin. J Neurosci Res 42: 259-265.

Diaz-Sanchez M, Williams K, DeLuca GC, Esiri MM. 2006. Protein co-expression with axonal injury in multiple sclerosis plaques. Acta Neuropathol 111: 289-299.

Diem R, Hobom M, Maier K, Weissert R, Storch MK, et al. 2003. Methylprednisolone increases neuronal apoptosis during autoimmune CNS inflammation by inhibition of an endogenous neuroprotective pathway. J Neurosci 23: 6993-7000.

Dourdin N, Bhatt AK, Dutt P, Greer PA, Arthur JS, et al. 2001. Reduced cell migration and disruption of the actin cytoskeleton in calpain-deficient embryonic fibroblasts. J Biol Chem 276: 48382-48388.

Dowling P, Husar W, Menonna J, Donnenfeld H, Cook S, et al. 1997. Cell death and birth in multiple sclerosis brain. J Neurol Sci 149: 1-11.

Foroozan R, Buono LM, Savino PJ, Sergott RC. 2002. Acute demyelinating optic neuritis. Curr Opin Ophthalmol 13: 375-380.

Fox DA, Poblenz AT, He L. 1999. Calcium overload triggers rod photoreceptor apoptotic cell death in chemical-induced and inherited retinal degenerations. Ann N Y Acad Sci 893: 282-285.

Frohman EM, Frohman TC, Zee DS, McColl R, Galetta S. 2005. The neuro-ophthalmology of multiple sclerosis. Lancet Neurol 4: 111-121.

Gao G, Dou QP. 2000. N-terminal cleavage of bax by calpain generates a potent proapoptotic 18-kDa fragment that promotes bcl-2-independent cytochrome C release and apoptotic cell death. J Cell Biochemi 80: 53-72.

Gardner P. 1989. Calcium and T lymphocyte activation. Cell 59: 15-20.

Gehrmann J, Banati RB, Cuzner ML, Kreutzberg GW, Newcombe J. 1995. Amyloid precursor protein (APP) expression in multiple sclerosis lesions. Glia 15: 141-151.

Gerondakis S, Grumont R, Rourke I, Grossmann M. 1998. The regulation and roles of Rel/NF-kappa B transcription factors during lymphocyte activation. Curr Opin Immunol 10: 353-359.

Glading A, Lauffenburger DA, Wells A. 2002. Cutting to the chase: Calpain proteases in cell motility. Trends Cell Biol 12: 46-54.

Goodin DS, Arnason BG, Coyle PK, Frohman EM, Paty DW. 2003. The use of mitoxantrone (Novantrone) for the treatment of multiple sclerosis: Report of the Therapeutics and Technology Assessment Subcommittee of the American Academy of Neurology. Neurology 61: 1332-1338.

Griscavage JM, Wilk S, Ignarro LJ. 1996. Inhibitors of the proteasome pathway interfere with induction of nitric oxide synthase in macrophages by blocking activation of transcription factor NF-kappa B. Proc Natl Acad Sci USA 93: 3308-3312.

Guan Y, Shindler KS, Tabuena P, Rostami AM. 2006. Retinal ganglion cell damage induced by spontaneous autoimmune optic neuritis in MOG-specific TCR transgenic mice. J Neuroimmunol 178: 40-48.

Guyton MK, Das A, Matzelle DD, Samantaray S, Azuma M, et al. 2006. SJA6017 attenuates immune cell infiltration and neurodegeneration in EAE. Eighth International Congress of Neuroimmunology, Medimond, Nagoya, Japan.

Guyton MK, Sribnick EA, Ray SK, Banik NL. 2005a. A role for calpain in optic neuritis. Ann N Y Acad Sci 1053: 48-54.

Guyton MK, Wingrave JM, Yallapragada AV, Wilford GG, Sribnick EA, et al. 2005b. Upregulation of calpain correlates with increased neurodegeneration in acute experimental auto-immune encephalomyelitis. J Neurosci Res 81: 53-61.

Hartung HP, Gonsette R, Konig N, Kwiecinski H, Guseo A, et al. 2002. Mitoxantrone in progressive multiple sclerosis: A placebo-controlled, double-blind, randomised, multi-centre trial. Lancet 360: 2018-2025.

Hassen GW, Feliberti J, Kesner L, Stracher A, Mokhtarian F. 2006. A novel calpain inhibitor for the treatment of acute experimental autoimmune encephalomyelitis. J Neuroimmunol 180: 135-146.

Hendry L, John S. 2004. Regulation of STAT signalling by proteolytic processing. Eur J Biochem 271: 4613-4620.

Hickey WF. 1999. The pathology of multiple sclerosis: A historical perspective. J Neuroimmunol 98: 37-44.

Hightower KR, David LL, Shearer TR. 1987. Regional distribution of free calcium in selenite cataract: Relation to calpain II. Invest Ophthalmol Vis Sci 28: 1702-1706.

Hirai S, Kawasaki H, Yaniv M, Suzuki K. 1991. Degradation of transcription factors, c-Jun and c-Fos, by calpain. FEBS Lett 287: 57-61.

Hobom M, Storch MK, Weissert R, Maier K, Radhakrishnan A, et al. 2004. Mechanisms and time course of neuronal degeneration in experimental autoimmune encephalomyelitis. Brain Pathol 14: 148-157.

Hogan EL, Banik NL, Goust JM, Lobo D. 1987. Enzymes in cerebrospinal fluid: Evidence for a calcium activated neutral proteinase in CSF. Cellular and Humoral Components of Cerebrospinal Fluid in Multiple Sclerosis, Vol. 129. Lowenthal AN, Raus J, editors. New York: Plenum Press; pp. 479-487.

Horikawa Y, Oda N, Cox NJ, Li X, Orho-Melander M, et al. 2000. Genetic variation in the gene encoding calpain-10 is associated with type 2 diabetes mellitus. Nat Genet 26: 163-175.

Huang Y, Wang KK. 2001. The calpain family and human disease. Trends Mol Med 7: 355-362.

Huttenlocher A, Palecek SP, Lu Q, Zhang W, Mellgren RL, et al. 1997. Regulation of cell migration by the calcium-dependent protease calpain. J Biol Chem 272: 32719-32722.

Imam SA, Guyton MK, Haque A, Vandenbark A, Tyor WR, et al. 2007. Increased calpain correlates with Th1 cytokine profile in PBMCs from MS patients. J Neuroimmunol. 190: 139-145

Inglese M. 2006. Multiple sclerosis: New insights and trends. AJNR Am J Neuroradiol 27: 954-957.

Inuzuka TSS, Baba H, Miyatake T. 1987. Degradation of myelin basic protein in myelin by cerebrospinal fluid and effect of protease inhibitors. Cellular and Humoral Components of Cerebrospinal Fluid in Multiple Sclerosis, Vol. 129. Lowenthal AN, Raus. J, editors. New York: Plenum Press; pp. 489-523.

Jacobs LD, Beck RW, Simon JH, Kinkel RP, Brownscheidle CM, et al. (CHAMPS Study Group.) 2000. Intramuscular interferon beta-1a therapy initiated during a first demyelinating event in multiple sclerosis. N Engl J Med 343: 898-904.

Jacobs LD, Cookfair DL, Rudick RA, Herndon RM, Richert JR, et al. (The Multiple Sclerosis Collaborative Research Group (MSCRG)). 1996. Intramuscular interferon beta-1a for disease progression in relapsing multiple sclerosis. Ann Neurol 39: 285-294.

Johnson KP, Brooks BR, Cohen JA, Ford CC, Goldstein J, et al. (The Copolymer 1 Multiple Sclerosis Study Group.) 1995. Copolymer 1 reduces relapse rate and improves disability in relapsing-remitting multiple sclerosis: Results of a phase III multicenter, double-blind placebo-controlled trial. Neurology 45: 1268-1276.

Kang P. 2006. Optic neuritis. http://www.emedicine.com/radio/topic488.htm.

Kengatharan M, De Kimpe SJ, Thiemermann C. 1996. Analysis of the signal transduction in the induction of nitric oxide synthase by lipoteichoic acid in macrophages. Br J Pharmacol 117: 1163-1170.

Kim J, Sanders SP, Siekierski ES, Casolaro V, Proud D. 2000. Role of NF-kappa B in cytokine production induced from human airway epithelial cells by rhinovirus infection. J Immunol 165: 3384-3392.

Kim JM, Oh YK, Kim YJ, Cho SJ, Ahn MH, et al. 2001. Nuclear factor-kappa B plays a major role in the regulation of chemokine expression of HeLa cells in response to Toxoplasma gondii infection. Parasitol Res 87: 758-763.

Kim MJ, Jo DG, Hong GS, Kim BJ, Lai M, et al. 2002. Calpain-dependent cleavage of cain/cabin1 activates calcineurin to mediate calcium-triggered cell death. Proc Natl Acad Sci USA 99: 9870-9875.

Kishimoto A. 1990. Limited proteolysis of protein kinase C by calpain, its possible implication. Adv Second Messenger Phosphoprotein Res 24: 472-477.

Kitamura Y, Miyamura A, Takata K, Inden M, Tsuchiya D, et al. 2003. Possible involvement of both endoplasmic reticulum-and mitochondria-dependent pathways in thapsigargin-induced apoptosis in human neuroblastoma SH-SY5Y cells. J Pharmacol Sci 92: 228-236.

Kornek B, Storch MK, Weissert R, Wallstroem E, Stefferl A, et al. 2000. Multiple sclerosis and chronic autoimmune encephalomyelitis: A comparative quantitative study of axonal injury in active, inactive, and remyelinated lesions. Am J Pathol 157: 267-276.

Kubbutat MH, Vousden KH. 1997. Proteolytic cleavage of human p53 by calpain: A potential regulator of protein stability. Mol Cell Biol 17: 460-468.

Kuhlmann T, Lingfeld G, Bitsch A, Schuchardt J, Bruck W. 2002. Acute axonal damage in multiple sclerosis is most extensive in early disease stages and decreases over time. Brain 125: 2202-2212.

Kupina NC, Nath R, Bernath EE, Inoue J, Mitsuyoshi A, et al. 2001. The novel calpain inhibitor SJA6017 improves functional outcome after delayed administration in a mouse model of diffuse brain injury. J Neurotrauma 18: 1229-1240.

Land KJ, Moll JS, Kaplan MH, Seetharamaiah GS. 2004. Signal transducer and activator of transcription (Stat)-6-dependent, but not Stat4-dependent, immunity is required for the development of autoimmunity in Graves' hyperthyroidism. Endocrinology 145: 3724-3730.

Li GL, Brodin G, Farooque M, Funa K, Holtz A, et al. 1996. Apoptosis and expression of Bcl-2 after compression trauma to rat spinal cord. J Neuropathol Exp Neurol 55: 280-289.

Liu F, Grundke-Iqbal I, Iqbal K, Oda Y, Tomizawa K, et al. 2005. Truncation and activation of calcineurin A by calpain I in Alzheimer disease brain. J Biol Chem 280: 37755-37762.

Lokuta MA, Nuzzi PA, Huttenlocher A. 2003. Calpain regulates neutrophil chemotaxis. Proc Natl Acad Sci USA 100: 4006-4011.

McCord JM. 1993. Oxygen-derived free radicals. New Horiz 1: 70-76.

McFarlin DE, McFarland HF. 1982. Multiple sclerosis (second of two parts). N Engl J Med 307: 1246-1251.

Mendel I, Kerlero de Rosbo N, Ben-Nun A. 1995. A myelin oligodendrocyte glycoprotein peptide induces typical chronic experimental autoimmune encephalomyelitis in H-2b mice: Fine specificity and T cell receptor V beta expression of encephalitogenic T cells. Eur J Immunol 25: 1951-1959.

Meyer R, Weissert R, Diem R, Storch MK, de Graaf KL, et al. 2001. Acute neuronal apoptosis in a rat model of multiple sclerosis. J Neurosci 21: 6214-6220.

Mikhak Z, Fleming CM, Medoff BD, Thomas SY, Tager AM, et al. 2006. STAT1 in peripheral tissue differentially regulates homing of antigen-specific Th1 and Th2 cells. J Immunol 176: 4959-4967.

Milligan SA, Owens MW, Grisham MB. 1996. Inhibition of IkappaB-alpha and IkappaB-beta proteolysis by calpain inhibitor I blocks nitric oxide synthesis. Arch Biochem Biophys 335: 388-395.

Moncada S, Palmer RM, Higgs EA. 1991. Nitric oxide: Physiology, pathophysiology, and pharmacology. Pharmacol Rev 43: 109-142.

Montalban X. 2005. Primary progressive multiple sclerosis. Curr Opin Neurol 18: 261-266.

Murachi T. 1984. Calcium-dependent proteinases and specific inhibitors: Calpain and calpastatin. Biochem Soc Symp 49: 149-167.

Nakagawa T, Yuan J. 2000. Cross-talk between two cysteine protease families. Activation of caspase-12 by calpain in apoptosis. J Cell Biol 150: 887-894.

Nathan C. 1992. Nitric oxide as a secretory product of mammalian cells. FASEB J 6: 3051-3064.

Newcombe J, Glynn P, Cuzner ML. 1982. The immunological identification of brain proteins on cellulose nitrate in human demyelinating disease. J Neurochem 38: 267-274.

Nguyen KB, McCombe PA, Pender MP. 1994. Macrophage apoptosis in the central nervous system in experimental autoimmune encephalomyelitis. J Autoimmun 7: 145-152.

Nihard P. 1982. Effect of calcium-entry-blockers on arterioles, capillaries, and venules of the retina. Angiology 33: 37-45.

Nixon RA. 1986. Fodrin degradation by calcium-activated neutral proteinase (CANP) in retinal ganglion cell neurons and optic glia: Preferential localization of CANP activities in neurons. J Neurosci 6: 1264-1271.

Noseworthy JH, Lucchinetti C, Rodriguez M, Weinshenker BG. 2000. Multiple sclerosis. N Engl J Med 343: 938-952.

O'Neill JK, Baker D, Morris MM, Gschmeissner SE, Jenkins HG, et al. 1998. Optic neuritis in chronic relapsing experimental allergic encephalomyelitis in Biozzi ABH mice: Demyelination and fast axonal transport changes in disease. J Neuroimmunol 82: 210-218.

Oda A, Wakao H, Fujita H. 2002. Calpain is a signal transducer and activator of transcription (STAT) 3 and STAT5 protease. Blood 99: 1850-1852.

Paquet-Durand F, Johnson L, Ekstrom P. 2007. Calpain activity in retinal degeneration. J Neurosci Res 85: 693-702.

Peterson JW, Bo L, Mork S, Chang A, Trapp BD. 2001. Transected neurites, apoptotic neurons, and reduced inflammation in cortical multiple sclerosis lesions. Ann Neurol 50: 389-400.

Pitt D, Werner P, Raine CS. 2000. Glutamate excitotoxicity in a model of multiple sclerosis. Nat Med 6: 67-70.

Polster BM, Basanez G, Etxebarria A, Hardwick JM, Nicholls DG. 2005. Calpain I induces cleavage and release of apoptosis-inducing factor from isolated mitochondria. J Biol Chem 280: 6447-6454.

Ponnappan S, Cullen SJ, Ponnappan U. 2005. Constitutive degradation of IkappaBalpha in human T lymphocytes is mediated by calpain. Immun Ageing 2: 15.

Posmantur R, Hayes RL, Dixon CE, Taft WC. 1994. Neurofilament 68 and neurofilament 200 protein levels decrease after traumatic brain injury. J Neurotrauma 11: 533-545.

PRISMS (Prevention of Relapses and Disability by Interferon beta-1a Subcutaneously in Multiple Sclerosis) Study Group. 1998. Randomised double-blind placebo-controlled study of interferon beta-1a in relapsing/remitting multiple sclerosis. Lancet 352: 1498–1504.

Raine CS. 1982. Multiple sclerosis and chronic relapsing EAE: Comparative ultrastrucural neuropathology. Multiple Sclerosis: The Patient, the Disease, and the Treatment. Hallpike J, Adams C, Tourtellotte W, editors. London: Chapman and Hall; pp. 411-458.

Raine CS, Barnett LB, Brown A, Behar T, McFarlin DE. 1980. Neuropathology of experimental allergic encephalomyelitis in inbred strains of mice. Lab Invest 43: 150-157.

Ransom BR, Waxman SG, Stys PK. 1993. Anoxic injury of central myelinated axons: Some mechanisms and pharmacology. Molecular and Cellular Approaches to the Treatment of Neurological Disease. Waxman SG, editor. New York: Raven Press; pp. 121-151.

Rao NA. 1981. Chronic experimental allergic optic neuritis. Invest Ophthalmol Vis Sci 20: 159-172.

Ray SK. 2006. Currently evaluated calpain and caspase inhibitors for neuroprotection in experimental brain ischemia. Curr Med Chem 13: 3425-3440.

Ray SK, Banik NL. 2003. Calpain and its involvement in the pathophysiology of CNS injuries and diseases: Therapeutic potential of calpain inhibitors for prevention of neurodegeneration. Curr Drug Targets CNS Neurol Disord 2: 173-189.

Ray SK, Fidan M, Nowak MW, Wilford GG, Hogan EL, et al. 2000. Oxidative stress and Ca2+ influx upregulate calpain and induce apoptosis in PC12 cells. Brain Res 852: 326-334.

Ray SK, Wilford GG, Crosby CV, Hogan EL, Banik NL. 1999a. Diverse stimuli induce calpain overexpression and apoptosis in C6 glioma cells. Brain Res 829: 18-27.

Ray SK, Wilford GG, Matzelle DC, Hogan EL, Banik NL. 1999b. Calpeptin and methylprednisolone inhibit apoptosis in rat spinal cord injury. Ann N Y Acad Sci 890: 261-269.

Richard I, Broux O, Allamand V, Fougerousse F, Chiannilkulchai N, et al. 1995. Mutations in the proteolytic enzyme calpain 3 cause limb-girdle muscular dystrophy type 2A. Cell 81: 27-40.

Saido TC, Sorimachi H, Suzuki K. 1994. Calpain: New perspectives in molecular diversity and physiological-pathological involvement. FASEB J 8: 814-822.

Saito K, Elce JS, Hamos JE, Nixon RA. 1993. Widespread activation of calcium-activated neutral proteinase (calpain) in the brain in Alzheimer disease: A potential molecular basis for neuronal degeneration. Proc Natl Acad Sci USA 90: 2628-2632.

Sakuma H, Kohyama K, Park IK, Miyakoshi A, Tanuma N, et al. 2004. Clinicopathological study of a myelin oligodendrocyte glycoprotein-induced demyelinating disease in LEW.1AV1 rats. Brain 127: 2201-2213.

Sanges D, Comitato A, Tammaro R, Marigo V. 2006. Apoptosis in retinal degeneration involves cross-talk between apoptosis-inducing factor (AIF) and caspase-12 and is blocked by calpain inhibitors. Proc Natl Acad Sci USA 103: 17366-17371.

Santella L, Kyozuka K, De Riso L, Carafoli E. 1998. Calcium, protease action, and the regulation of the cell cycle. Cell Calcium 23: 123-130.

Sarin A, Clerici M, Blatt SP, Hendrix CW, Shearer GM, et al. 1994. Inhibition of activation-induced programmed cell death and restoration of defective immune responses of HIV+ donors by cysteine protease inhibitors. J Immunol 153: 862-872.

Sato S, Quarles RH, Brady RO, Tourtellotte WW. 1984. Elevated neutral protease activity in myelin from brains of patients with multiple sclerosis. Ann Neurol 15: 264-267.

Schaecher K, Goust JM, Banik NL. 2004. The effects of calpain inhibition on IkB alpha degradation after activation of PBMCs: Identification of the calpain cleavage sites. Neurochem Res 29: 1443-1451.

Schaecher KE, Goust JM, Banik NL. 2001. The effects of calpain inhibition upon IL-2 and CD25 expression in human peripheral blood mononuclear cells. J Neuroimmunol 119: 333-342.

Shao H, Huang Z, Sun SL, Kaplan HJ, Sun D. 2004. Myelin/oligodendrocyte glycoprotein-specific T-cells induce severe optic neuritis in the C57BL/6 mouse. Invest Ophthalmol Vis Sci 45: 4060-4065.

Sharma AK, Rohrer B. 2004. Calcium-induced calpain mediates apoptosis via caspase-3 in a mouse photoreceptor cell line. J Biol Chem 279: 35564-35572.

Sharma AK, Rohrer B. 2007. Sustained elevation of intracellular cGMP causes oxidative stress triggering calpain-mediated apoptosis in photoreceptor degeneration. Curr Eye Res 32: 259-269.

Shearer TR, David LL. 1982. Role of calcium in selenium cataract. Curr Eye Res 2: 777-784.

Shearer TR, David LL. 1990. Calpain in lens and cataract. Intracellular Calcium-Dependent Proteolysis. Mellgren RL, editor. CRC Press; pp. 265–274.

Shields DC, Banik NL. 1998a. Upregulation of calpain activity and expression in experimental allergic encephalomyelitis: A putative role for calpain in demyelination. Brain Res 794: 68-74.

Shields DC, Banik NL. 1998b. Putative role of calpain in the pathophysiology of experimental optic neuritis. Exp Eye Res 67: 403-410.

Shields DC, Banik NL. 1999. Pathophysiological role of calpain in experimental demyelination. J Neurosci Res 55: 533-541.

Shields DC, Leblanc C, Banik NL. 1997. Calcium-mediated neurofilament protein degradation in rat optic nerve in vitro: Activity and autolysis of calpain proenzyme. Exp Eye Res 65: 15-21.

Shields DC, Schaecher KE, Goust JM, Banik NL. 1999a. Calpain activity and expression are increased in splenic inflammatory cells associated with experimental allergic encephalomyelitis. J Neuroimmunol 99: 1-12.

Shields DC, Schaecher KE, Saido TC, Banik NL. 1999b. A putative mechanism of demyelination in multiple sclerosis by a proteolytic enzyme, calpain. Proc Natl Acad Sci USA 96: 11486-11491.

Shields DC, Tyor WR, Deibler GE, Banik NL. 1998. Increased calpain expression in experimental demyelinating optic neuritis: An immunocytochemical study. Brain Res 784: 299-304.

Shindler KS, Guan Y, Ventura E, Bennett J, Rostami A. 2006. Retinal ganglion cell loss induced by acute optic neuritis in a relapsing model of multiple sclerosis. Mult Scler 12: 526-532.

Shirasaki Y, Nakamura M, Yamaguchi M, Miyashita H, Sakai O, et al. 2006. Exploration of orally available calpain inhibitors 2: Peptidyl hemiacetal derivatives. J Med Chem 49: 3926-3932.

Smith KJ, Hall SM. 2001. Factors directly affecting impulse transmission in inflammatory demyelinating disease: Recent advances in our understanding. Curr Opin Neurol 14: 289-298.

Smith T, Schmied M, Hewson AK, Lassmann H, Cuzner ML. 1996. Apoptosis of T cells and macrophages in the central nervous system of intact and adrenalectomized Lewis rats during experimental allergic encephalomyelitis. J Autoimmun 9: 167-174.

Soede RD, Driessens MH, Ruuls-Van Stalle L, Van Hulten PE, Brink A, et al. 1999. LFA-1 to LFA-1 signals involve zeta-associated protein-70 (ZAP-70) tyrosine kinase: Relevance for invasion and migration of a T cell hybridoma. J Immunol 163: 4253-4261.

Squier MK, Miller AC, Malkinson AM, Cohen JJ. 1994. Calpain activation in apoptosis. J Cell Physiol 159: 229-237.

Stewart MP, McDowall A, Hogg N. 1998. LFA-1-mediated adhesion is regulated by cytoskeletal restraint and by a Ca2+-dependent protease, calpain. J Cell Biol 140: 699-707.

Storch MK, Stefferl A, Brehm U, Weissert R, Wallstrom E, et al. 1998. Autoimmunity to myelin oligodendrocyte glycoprotein in rats mimics the spectrum of multiple sclerosis pathology. Brain Pathol 8: 681-694.

Suzuki K, Hata S, Kawabata Y, Sorimachi H. 2004. Structure, activation, and biology of calpain. Diabetes 53 (Suppl 1): S12-S18.

Thome M, Acuto O. 1995. Molecular mechanism of T-cell activation: Role of protein tyrosine kinases in antigen receptor-mediated signal transduction. Res Immunol 146: 291-307.

Trapp BD, Peterson J, Ransohoff RM, Rudick R, Mork S, et al. 1998. Axonal transection in the lesions of multiple sclerosis. N Engl J Med 338: 278-285.

Twining SS, Kirschner SE, Mahnke LA, Frank DW. 1993. Effect of Pseudomonas aeruginosa elastase, alkaline protease, and exotoxin A on corneal proteinases and proteins. Invest Ophthalmol Vis Sci 34: 2699-2712.

von Sallmann L, Myers RE, Lerner EM 2nd, Stone SH. 1967. Vasculo-occlusive retinopathy in experimental allergic encephalomyelitis. Arch Ophthalmol 78: 112-120.

Wagner BJ, Margolis JW. 1993. Thermal stability and activation of bovine lens multicatalytic proteinase complex (proteasome). Arch Biochem Biophys 307: 146-152.

Wang KK, Posmantur R, Nadimpalli R, Nath R, Mohan P, et al. 1998. Caspase-mediated fragmentation of calpain inhibitor protein calpastatin during apoptosis. Arch Biochem Biophys 356: 187-196.

Watanabe N, Vande Woude GF, Ikawa Y, Sagata N. 1989. Specific proteolysis of the c-mos proto-oncogene product by calpain on fertilization of Xenopus eggs. Nature 342: 505-511.

Watt F, Molloy PL. 1993. Specific cleavage of transcription factors by the thiol protease, m-calpain. Nucleic Acids Res 21: 5092-5100.

Williams KC, Ulvestad E, Hickey WF. 1994. Immunology of multiple sclerosis. Clin Neurosci 2: 229-245.

Wood DE, Newcomb EW. 2000. Cleavage of Bax enhances its cell death function. Exp Cell Res 256: 375-382.

Wu HY, Tomizawa K, Matsui H. 2007. Calpain-calcineurin signaling in the pathogenesis of calcium-dependent disorder. Acta Med Okayama 61: 123-137.

Wujek JR, Bjartmar C, Richer E, Ransohoff RM, Yu M, et al. 2002. Axon loss in the spinal cord determines permanent neurological disability in an animal model of multiple sclerosis. J Neuropathol Exp Neurol 61: 23-32.

Yamamoto K, Arakawa T, Ueda N, Yamamoto S. 1995. Transcriptional roles of nuclear factor kappa B and nuclear factor-interleukin-6 in the tumor necrosis factor alpha-dependent induction of cyclooxygenase-2 in MC3T3-E1 cells. J Biol Chem 270: 31315-31320.

Yoshida H, Murachi T, Tsukahara I. 1984. Degradation of actin and vimentin by calpain II, a Ca2+-dependent cysteine proteinase, in bovine lens. FEBS Lett 170: 259-262.

Youn YK, LaLonde C, Demling R. 1991. Use of antioxidant therapy in shock and trauma. Circ Shock 35: 245-249.

Zimmerman UJ, Boring L, Pak JH, Mukerjee N, Wang KK. 2000. The calpain small subunit gene is essential: Its inactivation results in embryonic lethality. IUBMB Life 50: 63-68.

Zingarelli B, O'Connor M, Wong H, Salzman AL, Szabo C. 1996. Peroxynitrite-mediated DNA strand breakage activates poly-adenosine diphosphate ribosyl synthetase and causes cellular energy depletion in macrophages stimulated with bacterial lipopolysaccharide. J Immunol 156: 350-358.

Zuckerman LA, Pullen L, Miller J. 1998. Functional consequences of costimulation by ICAM-1 on IL-2 gene expression and T cell activation. J Immunol 160: 3259-3268.

23 Protein Carbonylation in Neurodegenerative and Demyelinating CNS Diseases

O. A. Bizzozero

1	Introduction	544
2	Chemical and Biochemical Aspects of Protein Carbonylation	545
2.1	Chemistry of Protein Carbonylation	545
2.2	Measurement of Protein Carbonyls	546
2.3	Experimental Models for Induction of Protein Carbonyls	548
2.4	Removal of Protein Carbonyls from Oxidized Tissues	549
3	Functional Consequences of Protein Carbonylation	550
4	CNS Disorders Associated with Carbonylated Proteins	551
4.1	Alzheimer's Disease	551
4.2	Parkinson's Disease (PD)	552
4.3	Amyotrophic Lateral Sclerosis	553
4.4	Huntington's Disease	553
4.5	Multiple Sclerosis	553
4.6	Other CNS Conditions Associated with Carbonyl Accumulation	555
4.7	Protein Carbonylation Profiles in Different CNS Diseases	555
5	Pharmacological Approaches to Prevent Protein Carbonyl Formation	555
6	Concluding Remarks and Perspectives	556

© 2009 Springer Science+Business Media, LLC.

Abstract: Carbonylation is the most common protein modification that takes place as consequence of severe oxidative stress. Carbonyl (C = O) groups are introduced into proteins by direct metal-catalyzed oxidation of certain amino acids or indirectly by reaction with reactive carbonyl species derived from the oxidation of lipids (acrolein, 4-hydroxynonenal, malondialdehyde) and sugars (glyoxal, methylglyoxal). Carbonylation can alter protein function or lead to deleterious intermolecular cross-links and aggregates that preclude their degradation by intracellular proteases. Accumulation of carbonylated proteins has been implicated in the etiology and/or progression of several chronic central nervous system (CNS) disorders including Alzheimer's disease, Parkinson's disease, amyotrophic lateral sclerosis, and multiple sclerosis. In this chapter, the basic aspects of carbonyl chemistry and metabolism and their role in the pathophysiology of CNS neurodegenerative and demyelinating diseases are reviewed.

List of Abbreviations: ACR, acrolein; AD, Alzheimer's disease; ALS, amyotrophic lateral sclerosis; ANT, adenine nucleotide translocator; BCNU, 1,2-bis(2-chloroethyl)-1-nitrosourea; BHT, butylated hydroxytoluene; CAPE, caffeic acid phenethyl ester; CEL, N^ε-(1-carboxyethyl)lysine; CK, creatine kinase; CML, N^ε-(1-carboxymethyl)lysine; CSF, cerebrospinal fluid; DARP-2, dihydropyrimidinase-related-protein-2; DEM, diethyl maleate; DNP, dinitrophenyl; EAE, experimental autoimmune encephalomyelitis; ETC, electron transport chain; GAPDH, glyceraldehyde-3-phosphate dehydrogenase; GFAP, glial acidic fibrilary protein; GO, glyoxal; GS, glutamine synthetase; GSH, glutathione; HD, Huntington's disease; 4-HNE, 4-hydroxy-2-nonenal; Hsp, heat shock protein; MDA, malondialdehyde; MGO, methylglyoxal; MS, multiple sclerosis; NFH, neurofilament heavy (200 kDa) protein; NFL, neurofilament light (69 kDa) protein; NFM, neurofilament medium (150 kDa) protein; NFT, neurofibrilary tangles; PD, Parkinson's disease; PDI, protein disulfide isomerase; RCS, reactive carbonyl species; ROS, reactive oxygen species; SOD, superoxide dismutase; TPI, triose phosphate isomerase; UCH-L, ubiquitin C-terminal hydrolase L; VDAC, voltage-dependent anion channel

1 Introduction

The principal outcome from oxidative stress is the chemical transformation of lipids, proteins, and nucleic acids by reactive oxygen species (ROS). In recent years particular attention has been given to the oxidation of enzymes and structural proteins because they play a significant role in the pathogenesis of various human disorders (Stadtman and Berlett, 1997; Shacter, 2000). Although the protein backbone and the side-chains of most amino acids are susceptible to oxidation, the nonenzymatic introduction of aldehyde or ketone functional groups to specific amino acid residues (i.e., carbonylation) constitutes the major and most common oxidative alteration (Berlett and Stadtman, 1997; Dalle-Donne et al., 2003). However, because carbonyls are relatively difficult to induce compared to other oxidative modifications of proteins (e.g., oxidation of thiols), they are normally detected in conditions of considerable oxidative stress such as in neurodegenerative and chronic inflammatory diseases (Levine, 2002). Accumulation of protein carbonyls has been implicated in the etiology and/or progression of several neurodegenerative disorders such as Alzheimer's disease (Hensley et al., 1995; Montine et al., 1997; Smith et al., 1998), Parkinson's disease (Alam et al., 1997; Keeney et al., 2006), and amyotrophic lateral sclerosis (ALS) (Bowling et al., 1993, Shaw et al., 1995). Our recent discovery that protein carbonylation is augmented in multiple sclerosis (MS) and its animal model experimental autoimmune encephalomyelitis (Bizzozero et al., 2005; Smerjac and Bizzozero, 2007) suggests that this type of protein modification may play a critical pathophysiological role in inflammatory demyelinating diseases as well. In this chapter, the most relevant chemical and biochemical aspects of protein carbonylation and the functional consequences that this modification may have in neurodegenerative and demyelinating central nervous system (CNS) disorders are summarized.

2 Chemical and Biochemical Aspects of Protein Carbonylation

2.1 Chemistry of Protein Carbonylation

Carbonyl (C = O) groups are introduced into proteins by two different mechanisms (Adams et al., 2001) (❯ *Figure 23-1*). The first one involves the direct metal-catalyzed oxidation of amino acid side-chains through Fenton chemistry. By this process, threonine side-chains are oxidized to α-amino-β-ketobutyric acid while oxidative deamination of lysine, arginine, and proline generates α-aminoadipic semialdehyde and glutamic semialdehyde (❯ *Figure 23-2*). Protein carbonyl derivatives are also produced through oxidative cleavage of proteins via the α-amidation pathway, although the occurrence of this process in CNS disorders is uncertain (Stadtman, 1990a). A small portion of the protein-based carbonyls may also arise from decomposition of side-chain hydroperoxides of valine and leucine (Fu and Dean, 1997). The second mechanism involves the reaction of the nucleophilic centers in cysteine, histidine, or lysine residues with reactive carbonyl species (RCS) derived from the oxidation of lipids (e.g., 4-hydroxynonenal (4-HNE), malondialdehyde (MDA), acrolein (ACR)), and carbohydrates (e.g., glyoxal (GO), methylglyoxal (MGO)) (Adams et al., 2001) (❯ *Figure 23-1*). These nonoxidative processes of carbonyl formation are often termed lipoxidation and glycoxidation, respectively.

RCS are bifunctional aldehydes that fall into two groups: α,β-unsaturated aldehydes or alkenals (e.g., 4-HNE and ACR) and dialdehydes (e.g., MDA, GO, and MGO) (❯ *Figure 23-3a*). 4-HNE, an oxidation product of ω-6 polyunsaturated (linoleoyl and arachidonyl) chains, is an abundant RCS produced in neuroinflammatory and neurodegenerative disorders (Uchida, 2003). Both 4-HNE and ACR have the propensity to form stable Michael adducts with nucleophilic amino acids thereby introducing an aldehyde function in the protein (Esterbauer et al., 1991; Uchida and Stadtman, 1992) (❯ *Figure 23-3b*). The bifunctional aspect of 4-HNE and ACR also allows to cross-link proteins by Michael addition of cysteine, histidine, or lysine at C3 and Schiff base condensation with lysine at the C1 carbonyl (Uchida and Stadtman, 1993). The mechanism of carbonyl addition into proteins by the dialdehydes is rather similar: one of carbonyls in MDA, GO, and MGO reacts with amino groups forming a Schiff base and leaves the other functional group free (❯ *Figure 23-3c*). However, in contrast to Michael adducts formed with the C3

◻ Figure 23-1
Mechanisms of formation of protein-bound carbonyls

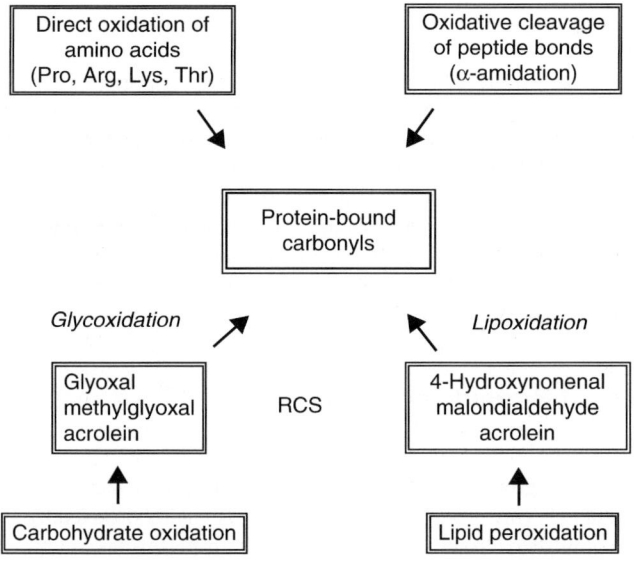

◘ **Figure 23-2**
Introduction of carbonyl groups into proteins by direct metal ion-catalyzed oxidation of threonyl, prolyl, arginyl, and lysyl residues

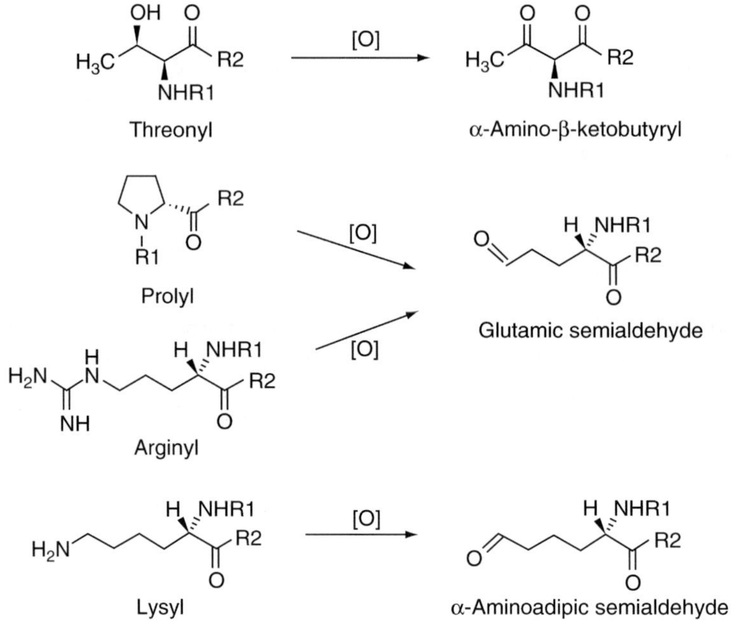

of 4-HNE and ACR, Schiff bases are sometimes reversible and may not lead to protein cross-links (Slatter et al., 1998). Moreover, in the case of GO and MGO, the Schiff bases undergo a Cannizzaro rearrangement reaction to produce N^ε-(1-carboxymethyl)lysine (CML) (Fu et al., 1996) and N^ε-(1-carboxyethyl)lysine (CEL) (Ahmed et al., 1997), respectively.

Data on the role of lipoxidation and glycoxidation in inflammatory and neurodegenerative diseases (Sayre et al., 2005) together with the fact that GO and MGO can also result from lipid peroxidation (Fu et al., 1996), strongly suggest that lipoxidative damage to proteins is much more prominent than glycoxidative damage. However, there is controversy as to the relative contribution of oxidative (direct) and nonoxidative (indirect) pathways to the total amount of carbonyls present in a protein. On one hand, glutamic and α-aminoadipic semialdehydes, products of metal ion-catalyzed oxidation of various amino acids, have been identified as the major carbonyl-containing amino acids in rat liver (Requena et al., 2001) and human brain proteins (Pamplona et al., 2005). On the other hand, inhibition of lipid peroxidation reduces significantly protein carbonyl levels in diseased tissues (Yan et al., 1997; Bizzozero et al., 2007), suggesting that modification of protein side-chains by bifunctional carbonyl-containing products of lipoxidation is a viable mechanism. Moreover, proteins containing 2-pentylpyrrole (a stable product derived from the HNE-lysine adduct) accumulate in human neurodegenerative disorders (Yoritaka et al., 1996; Sayre et al., 1997; Pedersen et al., 1998).

2.2 Measurement of Protein Carbonyls

The most commonly utilized procedure for assessing protein carbonyls in biological samples involves derivatization of the carbonyl group with 2,4-dinitrophenylhydrazine, which forms a stable dinitrophenyl (DNP) hydrazone product (Levine et al., 1990). This moiety is readily detected by absorbance at 370 nm, and the assay can be coupled to protein fractionation by high-performance liquid chromatography for

◻ **Figure 23-3**
Introduction of carbonyl groups into proteins via an indirect mechanism involving the reaction of RCS with nucleophilic residues (e.g., lysine)

a – Chemical structures of most common RCS

b – Reaction between an amino groups (lysine) and an α,β-unsaturated alkenal (ACR)

c – Reaction between an amino groups (lysine) and dialdehyde (MDA)

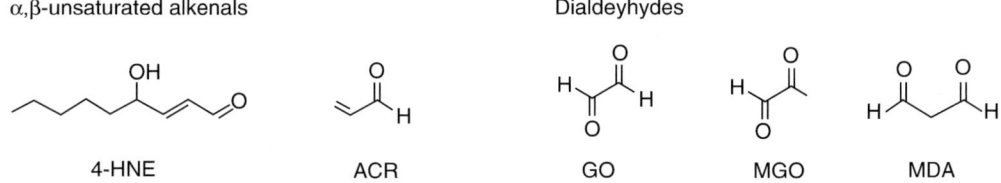

greater sensitivity and specificity (Levine et al., 1994). DNP-labeled proteins are also detected with specific antibodies and reactivity measured by enzyme-linked immunosorbent assay, slot blotting, or immunohistochemistry (Shacter, 2000). One-dimensional/two-dimensional gel electrophoresis followed by immunoblotting with anti-DNP antibodies (a.k.a. oxyblot) is normally used to detect the presence of carbonyls in specific proteins. The extent of carbonylation in total or individual proteins is also determined by reduction with sodium borotritide (NaB^3H_4), which converts carbonyl groups into tritiated alcohols. Proteins can be radiolabeled in solution (Lenz et al., 1989) or reduced just prior to gel electrophoresis (Yan and Sohal, 1998a). While the NaB^3H_4 method allows one to determine the amount of carbonylated protein, some noncarbonyl moieties present in proteins may also incorporate tritium when reduced with this reagent. For example, the noncarbonyl methylenelysine, formed by reaction of lysine residues with formaldehyde, is reduced by NaB^3H_4 producing tritiated methyllysine (Floor et al., 2006). Thus, the identity of radiolabeled amino acids must be determined when using this method to evaluate the levels of protein carbonylation in a particular sample.

The above two carbonyl assays (DNPH and NaB^3H_4) generally measure any carbonyl group present in a protein, without distinguishing between direct or indirect amino acid modifications. Specific techniques have been developed to identify the structure of the carbonyl-containing amino acids. For example, glutamic and α-aminoadipic semialdehydes are measured with a very sensitive gas-liquid chromatography/mass spectrometry technique (Requena et al., 2001). Lipid peroxidation adducts are detected using specific antibodies against MDA (Palinski et al., 1989), 4-HNE (Waeg et al., 1996), and ACR (Uchida et al., 1998). These antibodies are commercially available and can be used in ELISAs, western blots, or immunochemistry. Similarly, antibodies are available for detection of the glycoxidative products CEL and CML (Koito et al., 2004).

Identification of carbonylated proteins constitutes a critical step of studies addressing the functional consequences of protein oxidation. The major obstacle for identifying carbonylated proteins is their relatively low abundance. To circumvent this problem, Butterfield (2004) devised a DNP-based redox proteomic approach. In this technique, proteins from a tissue homogenate are first separated into single detectable species by two-dimensional gel electrophoresis. Carbonylated proteins are localized on the stained gel by comparison with the image corresponding to a two-dimensional oxyblot, and then identified by peptide mass fingerprinting. This scheme is relatively straightforward but is based on the assumption that neither carbonylation nor the introduction of the DNP moiety affects the protein's mobility on two-dimensional gels, which may not always be the case. A preferable approach is to separate the carbonylated species from the total proteins. One commonly employed method is to convert the protein carbonyl groups into biotinylated residues by reaction with the carbonyl reagent biotin hydrazide (Soreghan et al., 2003). Biotinylated proteins are then isolated by avidin-affinity chromatography and separated by one-dimensional/two-dimensional gel electrophoresis, followed by protein identification using specific antibodies (Bizzozero et al., 2007) or conventional proteomic techniques (Soreghan et al., 2003). Biotinylated proteins can also be digested with trypsin and the oxidation sites determined by reversed-phase chromatography and tandem mass spectrometry (Mirzaei and Regnier, 2005).

2.3 Experimental Models for Induction of Protein Carbonyls

Molecular mechanisms of protein carbonyl formation in cells or tissues are studied experimentally by (1) addition of oxidants (e.g., hydrogen peroxide, hypochlorite, peroxynitrite, organic peroxides) or RCS (e.g., ACR, 4-HNE, MGO), (2) depletion of cellular antioxidants (e.g., GSH), (3) inhibition of antioxidant enzymes (e.g., catalase, superoxide dismutase (SOD)), and (4) stimulation of endogenous production of ROS (e.g., mitochondrial toxins). Induction of protein carbonyls by GSH depletion is an experimental model, particularly relevant to neurodenegative and demyelinating CNS disorders characterized by reduced levels of this antioxidant (see later). Acute (partial or total) depletion of brain GSH is attained with dimethyl maleate (DEM), an α,β-unsaturated dicarboxylic acid that conjugates to GSH via a reaction catalyzed by glutathione-*S*-transferase (Buchmuller-Rouiller et al., 1995) or with 1,2-bis(2-chloroethyl)-1-nitrosourea (BCNU) which inhibits glutathione reductase (Karplus et al., 1988). Incubation of brain sections with these agents augments the levels of superoxide and hydrogen peroxide by increased mitochondrial production rather than by decreased detoxification of these species (Bizzozero et al., 2006). These ROS increase both lipid peroxidation and protein carbonylation by a metal ion-catalyzed process likely involving the formation of hydroxyl radical. A diagram shown in ❯ *Figure 23-4* depicts the most likely mechanism of protein carbonyl formation in this experimental paradigm. While several proteins are oxidized during GSH depletion, β-actin, α/β-tubulin, and glial fibrillary acidic protein (GFAP) are the major targets of carbonylation (Bizzozero et al., 2007). Interestingly the lipid peroxidation scavengers caffeic acid phenethyl ester (CAPE) and butylated hydroxytoluene (BHT) inhibit the carbonylation of these cytoskeletal proteins but not that of soluble proteins suggesting the existence of different mechanisms of carbonyl formation (Bizzozero et al., 2007). Carbonylation of cytoskeletal proteins in this system does not proceed by an indirect lipooxidative mechanism, but rather by the direct oxidation of amino acid side-chain with lipid hydroperoxides. Lipid hydroperoxide-induced protein carbonylation was initially proposed by Refsgaard et al. (2000), who discovered that metal-catalyzed oxidation of proteins is greatly enhanced by addition of polyunsaturated fatty acids. These investigators speculated that alkoxyl radicals, derived from metal-catalyzed heterolytic cleavage of lipid hydroperoxides (Davies and Slater, 1987), are responsible for the introduction of carbonyls into proteins by a mechanism that might involve site-specific interaction with lysine residues. Direct protein carbonylation induced by lipid hydroperoxides may explain why glutamic and α-aminoadipic semialdehydes are the major carbonyl-containing amino acids in many oxidative stress paradigms and yet protein carbonylation is inhibited by lipid peroxidation scavengers (❯ *Section 2.1*).

◘ Figure 23-4

A model for induction of protein carbonyls in neural cells by GSH depletion. Acute depletion of brain GSH with BCNU or DEM increases mitochondrial production of superoxide and hydrogen peroxide. In the presence of transition metal ions (Me), hydrogen peroxide produces highly reactive hydroxyl radicals (HO*), which can draw a hydrogen atom from polyunsaturated lipids (LH) and proteins (Pr) forming lipid-based (L*) and protein-based carbon-centered radicals (Pr*), respectively. Once formed, and in the presence of oxygen and Me, Pr* generate protein carbonyls (PrCO) by a series of reactions depicted on the *right* side of the figure. Carbonylation of water-soluble proteins in this experimental paradigm takes place mostly by this process. In contrast, most cytoskeletal proteins are carbonylated by a mechanism that involves lipid peroxidation since it is inhibited by the lipid peroxidation scavengers BHT or CAPE. Although RCS are also formed during oxidation of lipids, they react with proteins very slowly as compared to lipid hydroperoxides (LOOH). Thus, the most likely mechanism for carbonylation of cytoskeletal proteins in this system is that depicted by thick arrows. In this process Pr* is formed by reaction of proteins with lipid alkoxyl radicals (LO*) which derive from metal-catalyzed decomposition of LOOH. Experimental data supporting this model is presented in Bizzozero et al. (2007). Other abbreviations are LOO*, lipid hydroperoxyl radicals; LOH, hydroxylated lipid; PrOO*, protein peroxyl radicals; PrOOH, protein hydroperoxides, PrO*, protein alkoxyl radicals

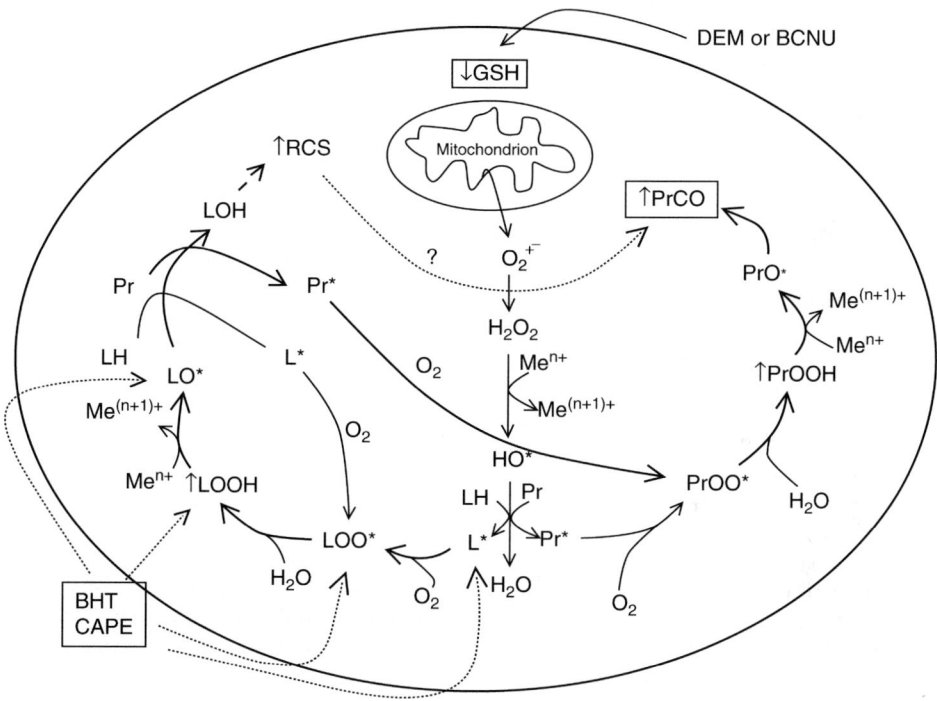

2.4 Removal of Protein Carbonyls from Oxidized Tissues

Due to the toxic effects of carbonyl groups, they need to be efficiently removed from tissues where they are formed. Enzymatic reduction of low molecular weight carbonyls (RCS) to alcohols is carried out by NADPH-dependent aldo-keto reductase (O'Connor et al., 1999) and NADH-dependent alcohol dehydrogenase (Sellin et al., 1991), while the oxidation of aldehydes to carboxylic acid is catalyzed by NAD^+-dependent aldehyde dehydrogenases (Mitchell and Petersen, 1987). These enzymes can reduce/oxidize not only free RCS but also glutathione-conjugated RCS. However, there is no evidence for enzymatic reduction of protein-bound carbonyls. Consequently, proteolytic degradation appears to be the only physiological mechanism for elimination of carbonylated proteins (❷ *Figure 23-5*) Degradation of

Figure 23-5
Diagram depicting the possible outcomes of increased protein carbonylation. Note that while of RCS can be enzymatically reduced to alcohols, oxidized to carboxylic acid or conjugated to GSH, protein carbonyls cannot be repaired and need to be removed by proteolysis

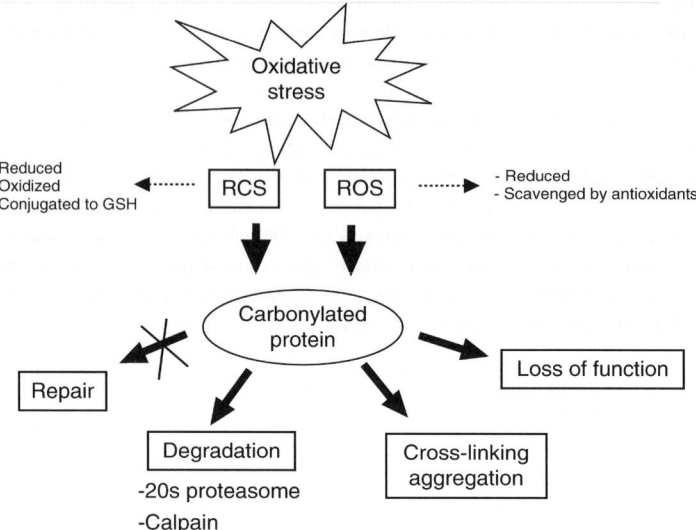

carbonylated proteins is carried out in an ATP- and ubiquitin-independent manner by the 20S proteasome, which selectively recognizes and digests partially unfolded (denatured) oxidized proteins (Rivett, 1985; Pacifici et al., 1993; Grune et al., 1995, 1997). Indeed, inhibition of the 20S proteosome results in accumulation of carbonylated proteins (Lee et al., 2001; Divald and Powell, 2006). Efficient and preferential removal of denatured proteins by proteolysis ensures low steady-state levels of carbonylated proteins in cells. However, conditions where proteosomal activity is reduced (e.g., aging, severe oxidative stress) lead to a buildup of these oxidized species. While proteins are more susceptible to degradation by cellular proteases upon carbonylation of one or more residues (Stadtman, 1990b), heavily oxidized proteins and cross-linked protein aggregates are not only more resistant to proteolysis (Friguet et al., 1994) but they also inhibit the activity of proteases that degrade them (Powell et al., 2005). For instance, HNE-containing proteins have been shown to form conjugates with the proteosome and cause significant loss of activity (Friguet and Szweda, 1997; Shringarpure et al., 2000). In addition, components of the proteosome can themselves become carbonylated during oxidative stress leading to further buildup of oxidized protein. These cycles of protein carbonylation/aggregation, proteosome inhibition, and accumulation of oxidized protein are one of the most important factors leading to the demise of affected cells.

3 Functional Consequences of Protein Carbonylation

Historically, carbonylation has been used just as a marker of oxidative stress. Thus, little efforts have been devoted to investigate the effect of this abundant modification on protein function or to determine if carbonylation has major physiological consequences. The relatively few studies conducted on isolated proteins have shown that introduction of carbonyl groups has a significant impact on protein folding and protein function. However, in many cases the type of moiety introduced and the modified amino acid residue are not the same as those found in vivo. Moreover, the great chemical diversity of protein carbonyls (e.g., glutamic acid semialdehyde, 4-HNE-adducts, MDA-adducts) is likely to affect protein function in different ways.

Cytoskeletal proteins, due to their abundance and relatively slow turnover (Wataya et al., 2002), are readily susceptible to carbonylation in various experimental conditions and CNS disorders characterized by severe oxidative stress (see ❯ Section 4). However, the significant aspect of this phenomenon is that addition of carbonyls groups to major cytoskeletal proteins (tubulin, actin, and neurofilament proteins) alters significantly the function of these macromolecules. For example, NFkB activation of intestinal epithelium cells causes carbonylation of tubulin, which in turn, leads to microtubule disassembly and instability (Banan et al., 2004). The same phenomenon takes place upon incorporation 4-HNE into neuronal microtubules (Neely et al., 2005). Actin filaments, the major cellular target of HClO-, metal-, and 4-HNE-mediated oxidation (Dalle-Done et al., 2001; Ozeki et al., 2005) are also easily depolymerized upon carbonylation (Banan et al., 2001). Neurofilament proteins, particularly NFM and NFH, are physiological substrates of HNE and other RCOs (Wataya et al., 2002), and carbonylation causes major changes in their secondary structure (Gelinas et al., 2000) leading to aggregation (Smith et al., 1995) and increased susceptibility to calpain proteolysis (Troncoso et al., 1995).

Less studied but equally important are the biological consequences of carbonylation of mitochondrial proteins. Due to their proximity to a major source of ROS, proteins of the mitochondrial electron transport system are particularly vulnerable to oxidation and carbonylation. Indeed, a number of electron transport chain (ETC) proteins are oxidatively modified by carbonylation and HNE-adduction (Choksi et al., 2004). In murine heart infected with *Trypanosme cruzi*, the carbonylation of certain protein subunits from complex I and III correlate with the loss of their catalytic activity, which in turn may lead to further generation of ROS by mitochondria. Also, heart cytochrome c oxidase is inactivated by lipoxidation during ischemia/reperfusion (Chen et al., 2001). Two additional mitochondrial proteins, the adenine nucleotide translocator (ANT) and the voltage-dependent anion channel (VDAC), which are not components of the ETC are also important carbonylation substrates. These proteins are the major components of the mitochondrial permeability transition pore, a nonspecific channel whose opening results in Ca^{2+}-dependent decrease in mitochondrial inner membrane potential. In flies, ANT carbonylation increases with age and under hyperoxic conditions leading to decrease in activity of the transporter (Yan and Sohal, 1998b). Also, carbonylation of ANT with 4-HNE in vitro enhances permeability transition pore opening in proteoliposomes, suggesting that this modification may have some regulatory function in vivo (Vieira et al., 2001). Aconitase and ATP synthetase are also modified by MDA during aging with a concomitant decrease in their enzymatic activities (Yarian et al., 2005). Several enzymes of glycolytic pathway are carbonylated under conditions of cellular oxidative stress, during etopside-induced apoptosis of HL60 cells, which correlates with decreased glycolytic rate and ATP synthesis (England et al., 2004).

Endoplasmic reticulum (ER) proteins are also carbonylated during oxidative stress. In HL-60 cells, a number of physiologically relevant chaperones including Hsp70, Hsp90, Grp-78, protein disulfide isomerase (PDI), and calreticulin are carbonylated in response to oxidative stress induced by hydrogen peroxide (England and Cotter, 2004). Grp-78, PDI, and calreticulin are also carbonylated in aged mouse liver (Rabek et al., 2003). While these studies suggest the possibility that carbonylation of chaperone proteins may induce the unfolded protein response, there is still no experimental evidence that links directly carbonylation with loss of chaperone activity.

4 CNS Disorders Associated with Carbonylated Proteins

A large number of human conditions are associated with accumulation of protein carbonyls including atherosclerosis, rheumatoid arthritis, ischemia/reperfusion injury, muscular dystrophy, chronic ethanol ingestion, emphysema, adult respiratory distress syndrome, and aging. Increased levels of protein carbonyls are also found in a number of chronic neurological disorders and these elevations correlate well with the progression or severity of the disease.

4.1 Alzheimer's Disease

Alzheimer's disease (AD) is a common neurodegenerative disorder characterized clinically by loss of memory, progressive intellectual deterioration, and dementia. Pathologically there is neuronal and synaptic loss in the cerebral cortex and hippocampus, together with the formation of neurofibrillary tangles (NFT) of the hyperphosphorylated microtubule-associated protein tau, and of extracellular deposits of amyloid β-peptides (senile plaques) that are derived from amyloid precursor proteins. Oxidative stress appears to have major role in the pathogenesis of AD, manifested by protein oxidation (both carbonylation and nitration), lipid peroxidation, and ROS formation (Markesbery, 1997; Butterfield and Lauderback, 2002). However, like in other neurodegenerative disorders, there is still discussion as to whether or not oxidative stress is a primary event that causes neuronal loss in AD. Increased amounts of carbonylated proteins are found in both the hippocampus and the inferior parietal lobule of AD patients relative to the AD cerebellum, a brain region that has little degenerative change, and the corresponding brain areas from controls (Hensley et al., 1995). Moreover, carbonylated proteins are associated with paired helical filaments and NFT, hallmarks of AD (Montine et al., 1997; Smith et al., 1998; Calingasan et al., 1999). The buildup of proteins carbonyls in regions with severe histopathological alterations suggests that carbonylation may play a pathogenic role in the development of lesions in AD. The accumulation of carbonylated proteins may arise not only from increased oxidative stress but also from oxidative proteasomal inactivation that takes place during aging (Farout and Friguet, 2006). Proteomic studies initially identified cytosolic creatine kinase (CK) and β-actin as brain proteins that are specifically carbonylated in AD (Aksenov et al., 2001). Subsequent work identified other carbonylation targets in AD including glutamine synthase (GS), ubiquitin C-terminal hydrolase L (UCH-L), dihydropyrimidinase-related-protein-2 (DARP-2), α-enolase, phosphoglycerate mutase-1, triose phosphate isomerase (TPI), glyceraldehyde-3-phosphate dehydrogenase (GAPDH), VDAC, carbonic anhydrase, and ATP synthetase (Castegna et al., 2002a, b; Butterfield et al., 2006). Most of these proteins are also carbonylated in various in vitro and in vivo model of AD (Butterfield et al., 2006). These data suggest that possible mechanisms for neurodegeneration in AD brain include: energy depletion (CK, TPI, VDAC, GAPDH, and α-enolase), protein misfolding and aggregation (UCH-L), decreased neuronal communication (DARP-2), and excitotoxic damage (GS). However, future studies should investigate the effect of carbonylation on the function of these proteins. A recent study showed that important proteins involved in both glial and neuronal homeostasis such as neurofilament light chain (NFL), GFAP, α-tubulin, ubiquinol-cytochrome c reductase complex protein I, and ATP synthetase are targeted by MDA in the cortex of AD patients (Pamplona et al., 2005). However, the concentration of MDA-modified amino acids is 50-times lower than that of glutamic semialdehyde, the major product derived by metal catalyzed oxidation of arginine and proline (❥ Section 2.1).

4.2 Parkinson's Disease (PD)

Parkinson's disease is a common neurodegenerative disorder affecting 1% of individuals over the age of 65. Most of the cases are sporadic but there is a small group of patients with the familial form caused by mutations in α-synuclein. PD is characterized by loss of neurons of the brain's substantia nigra and formation of protein deposits called Lewy bodies, which are made mostly of aggregated α-synuclein. Mitochondrial dysfunction and oxidative stress are prominent features of PD. In fact, mitochondrial neurotoxins like rotenone and 1-methyl-4-phenyl-tetrahydropyridine are used to create animal models of PD by inhibiting the ETC at complex I. Protein carbonyls accumulate primarily in areas associated with PD, such as substantia nigra, caudate nucleus, and putamen (Alam et al., 1997). However, increased carbonyl levels were also found in areas of the brain thought to be unaffected in PD. While the entire carbonyl proteome in PD has not been characterized, a recent study showed that five proteins from the mitochondrial complex I are carbonylated to a higher extent in the frontal cortex of PD patients relative to unaffected controls (Keeney et al., 2006). Furthermore, the oxidation of catalytic proteins of complex I in PD mitochondria correlate with reduced function of the complex and subunit misassembly. These findings provide some support to the idea that oxidative damage to proteins involved in energy production may underlie the pathophysiology of sporadic PD. This notion may also apply to familial PD since the metabolic related enzymes α-enolase, lactate dehydrogenase, and carbonic anhydrase are selective carbonylated in the brain of mice overexpressing the mutant human α-synuclein (Poon et al., 2005). α-Enolase along with β-actin are also the major carbonylated proteins in the substantia nigra of animals injected with 6-OH dopamine (De Iuliis et al., 2005), another animal model of PD.

4.3 Amyotrophic Lateral Sclerosis

ALS is a progressive neurodegenerative disorder that affects primarily spinal cord motor neurons. There are sporadic and familial forms of ALS, with sporadic ALS (sALS) being the most common type (>90% of all cases). Familal ALS (fALS) is caused by a number of mutations in the Cu,Zn-SOD1 gene, being the substitution gycine 93→alanine (G93A) the most common. Protein carbonyls are present at an increased level in the spinal cord of both sALS (Shaw et al., 1995; Lyras et al., 1996, Ferrante et al., 1997, Pedersen et al., 1998) and fALS (Bowling et al., 1993, Ferrante et al., 1997) compared to normal controls. A more recent study revealed that there are four major carbonylated species in sALS, one of them identified as the 68-kDa NFL (Niebroj-Dobosz et al; 2004). Protein carbonyls also accumulate in the spinal cord of fALS mice overexpressing the SOD1 mutation G93A (Andrus et al., 1998). This study showed that the most heavily carbonylated species is the SOD1 protein itself. Other proteins also susceptible to carbonylation in these animals were later identified as the translationally controlled tumor protein, HCH-L, and α-B-crystallin (Poon et al., 2005). In a related proteomic study, several 4-HNE-labeled proteins were also identified in the spinal cord of the mutant animals, including DRP-2, HSP70, and α-enolase (Perluigi et al., 2005a). Despite these findings, little is known regarding the role that these modifications may play in the neurodegeneration of ALS.

4.4 Huntington's Disease

Huntington's disease (HD) is a rare autosomal dominant disorder characterized by selective basal ganglia degeneration. The disease is caused by a CAG triplet expansion in exon 4 of the huntingtin gene. The encoded protein has an expanded polyglutamine repeat that may result in a toxic gain of function. The precise mechanisms underlying neuronal damage in HD are still elusive but there is evidence of decreased activity of mitochondrial complexes II, III, and IV in the caudate of HD patients, suggesting that oxidative stress plays a role in disease pathogenesis. This notion is supported by the accumulation of nuclear 8-hydroxydeoxyguanosine in the caudate nucleus of HD patients and the presence of lipofuscin deposits in striatal and cortical HD neurons (Browne et al., 1999). Despite this early evidence of oxidative stress in the human disease, no alterations in the levels of lipid peroxidation and protein carbonylation are found in the caudate nucleus, putamen, or frontal cortex of HD patients compared with normal controls (Alam et al., 2000). Opposite results are obtained in animal models of HD. For example, the succinate dehydrogenase (complex II) inhibitor 3-nitropropionic acid induces striatal degeneration similar to that observed in HD and thus has been used to model the disease (Palfi et al., 1996). The systemic administration of 3-nitropropionic acid into rats led to severe oxidative stress and carbonylation of not only striatal but also cortical synapstosomal proteins (LaFontaine et al., 2000). Increased protein oxidation was also detected in transgenic mice overexpressing human huntingtin compared with nontransgenic mice (Perluigi et al., 2005b). Proteins whose carbonylation levels are augmented in this model of HD were identified as α-enolase, γ-enolase, aconitase, CK, VDAC, and heat shock protein 90 (Hsp-90). These investigators proposed that the loss of activity, caused by oxidative modification, of the enzymes involved in glucose metabolism (α/γ-enolase), TCA cycle (aconitase), energy transfer (CK), and mitochondrial homeostasis (VDAC) might lead to a reduced ATP production, mitochondrial dysfunction, and further oxidative stress.

4.5 Multiple Sclerosis

MS is an inflammatory disease of the CNS characterized by demyelination, death of oligodendrocytes, and axonal damage (Hellings et al., 2002; Keegan and Noseworthy, 2002). The disease is heterogeneous both in its clinical presentation (Hafler, 2004) and pathological features (Lucchinetti et al., 1996). Perivenular infiltration of lymphocytes and macrophages in CNS white matter with the ensuing development of well-defined areas of demyelination and astrocytic proliferation (plaques) is the hallmark of this disorder (Noseworthy et al., 2000). Experimental allergic (autoimmune) encephalomyelitis (EAE) shares a number of clinical and pathological features with MS, and is routinely employed to study the mechanistic bases of disease and to test therapeutic approaches (Gold et al., 2000).

Like in several neurodegenerative disorders oxidative stress is also a player in the pathogenesis of inflammatory demyelination (Gilgun-Sherki et al., 2004). Generation of ROS in vivo has been inferred from (1) the presence of the lipid peroxidation products MDA (Hunter et al., 1985; Naidoo and Knapp, 1992; Calabrese et al., 1994) and isoprostanes (Greco et al., 1999) in the cerebrospinal fluid (CSF) and plasma of MS patients, (2) reduced plasma levels of antioxidants (e.g., ubiquinone, vitamin E) (Skukla et al., 1997) and antioxidant enzymes (e.g., catalase and glutathione peroxidase) (LeVine, 1992) in MS, (3) low levels of antioxidants like glutathione, α-tocopherol, and uric acid in MS plaques (Langemann et al., 1992), and (4) the damage to mitochondrial DNA in active chronic plaques (Vatassery et al., 1998). In addition to accumulation of oxidized macromolecules and reduction of antioxidants, MS and EAE are characterized by dysregulation of transition metals like iron (LeVine, 1997) and copper (Melo et al., 2003), which promote oxidative damage via the Fenton and Haber–Weiss reactions (Lewen et al., 2000). Additional evidence for a pathogenic role of ROS in demyelinating disease comes from the effectiveness of various antioxidant treatments at ameliorating EAE. For example, oral administration of the oxidant-scavenger N-acetylcysteine, which raises intracellular glutathione levels, inhibits the induction of acute EAE (Lehmann et al., 1994). Therapeutic success in the treatment of EAE has also been reported with other antioxidants such as α-lipoic acid (Marracci et al., 2002), bilirubin (Liu et al., 2003), uric acid (Hooper et al., 1998), and flavonoids (Hendriks et al., 2004). Administration of the antioxidant enzymes like metallothionein (Penkowa and Hidalgo, 2003), catalase (Ruuls et al., 1995), and synthetic catalytic scavengers with combined SOD and catalase activity (Malfroy et al., 1997) are effective as well. Likewise, chelation of the pro-oxidant iron with deferoxamine (Pedchenko and LeVine, 1998) and copper with N-acetylcysteine amide (Offen et al., 2004) reduces tissue damage and improve the clinical course of EAE.

While the aforementioned studies clearly show that oxidative stress is a major pathophysiological phenomenon in inflammatory demyelination, it was not until recently that oxidation of proteins in MS was demonstrated (Bizzozero et al., 2005). This study showed that protein carbonyls accumulate in the brain of normal-appearing white matter and gray matter of patients with chronic MS as compared with unaffected controls. The levels of carbonylation are particularly prominent in areas with high GFAP levels, suggesting that oxidative damage is related to the presence of small lesions. The increase in protein carbonyls in MS is likely a consequence of reduced levels of antioxidants since GSH concentration in the diseased white matter and gray matter is reduced by 30%. Identification of carbonylated proteins in the brain white matter of five MS patients (40.6 ± 3.6 years) and three age-matched controls (43.0 ± 6.3 years) revealed that GFAP, β-actin, and α/β-tubulin are the major targets of carbonylation (Bizzozero, 2007). Carbonylation of NFH and NFM are also increased in the brain white matter of MS, whereas that of NFL is not detected. $2',3'$-Cyclic nucleotide-$3'$-phospho-diesterase, an oligodendrocyte-specific protein associated to the myelin cytoskeleton, is also carbonylated in MS white matter, while the abundant myelin basic protein is not oxidized in either MS or control white matter. The finding that a number of cytoskeletal proteins, including the neuronal cytoskeleton, are carbonylated in MS tissues may be of particular significance since in recent years it has become clear that in this disorder there is not only loss of myelin, but also neuronal/oligodendroglial apoptosis and axonal damage (Ferguson et al., 1997; Trapp et al., 1998; Peterson and Trapp, 2005). In fact, it is the extent of axonal damage, rather that the lack of myelin, which correlates with functional deficits in MS both at the early and progressive phases of the disease (Kuhlmann et al., 2002; Sarchieli et al., 2002; De Stefano et al., 2003). The finding that carbonylated cytoskeletal proteins, including axon-specific species like NFM and NFH, are present in MS white matter and that oxidative modifications alter the function of these polypeptides, suggests that this posttranslational chemical modification of proteins may be involved in the axonal pathology characteristic of this disorder.

Protein carbonylation is also detected in the spinal cord of Lewis rats with acute EAE (Smerjac and Bizzozero, 2007). In this model the accumulation of protein carbonyls occurs not only around inflammatory lesions but also throughout the gray and white matter. The concentration of protein carbonyls parallels the disease course and is inversely proportional to that of GSH. As the inflammatory process subsides and animals recover for neurological deficits, protein carbonyl levels return to normal values. This finding suggests that the mechanism for removal of the oxidized protein is still operative in this acute animal model of MS. Interestingly, the concentration of RCS in the spinal cord remains elevated throughout disease course, even after animals have recovered. The lack of protein carbonylation in the presence of sustained RCS suggests that carbonylation of proteins in EAE takes place via a direct mechanism and not indirectly by reaction with dialdehydes (MDA, GO, MGO). A similar mechanism of protein carbonylation seems to operate in GSH depleted brain sections (❯ *Section 2.3*). Like in the human disease, GFAP, β-actin, and β-tubulin are the major targets of carbonylation in EAE, suggesting the possibility of profound cytoskeletal abnormalities in this inflammatory disorder.

4.6 Other CNS Conditions Associated with Carbonyl Accumulation

Accumulation of protein carbonyls as a marker of oxidative stress is described in other neurological conditions. For example, increased protein carbonylation is observed in spinal cord injury both at the center (Aksenova et al., 2002) and in the rostral and caudal segments of the damaged tissue (Kamencic et al., 2001; Leski et al., 2001). Augmented carbonyl content is also found in the cerebral cortex of patients with aceruloplasminemia, an autosomal recessive disorder characterized by ceruloplasmin deficiency and iron overload, being GFAP the major carbonylated species (Kaneko et al., 2002). GFAP is also modified by both lipoxidation and glycoxidation in a subset of patients with fronto-temporal dementia (Pick's disease) (Muntané et al., 2006).

4.7 Protein Carbonylation Profiles in Different CNS Diseases

While many disorders with considerable oxidative stress show accumulation of protein carbonyls, there is specificity to the process. For example, β-actin, CK, and GS are the major carbonylated brain proteins in Alzheimer's patients, while GFAP and β-tubulin are moderately oxidized (Aksenov et al., 2001). In contrast, we have found that α/β-tubulin, GFAP, and β-actin are among the major carbonylation targets in MS brains (Bizzozero, 2007). Also, NFL is not carbonylated in MS but is the main carbonylated species in the spinal cord of patients with sALS (Niebroj-Dobosz et al., 2004). GFAP, on the other hand, has been identified as the major target of lipoxidative damage in Pick's disease (Muntané et al., 2006), while a number of mitochondrial proteins are carbonylated in Parkinson's disease. In summary, although carbonylation can be used as a general marker of oxidative stress, different groups of proteins are modified in each condition. This relative specificity likely depends on various factors including the cellular and subcellular origin of the ROS, chemical nature of the oxidant, type and quantity of the RCS, the region of the CNS affected, duration of the disease as well as the ability of proteolytic enzymes to remove the modified proteins.

5 Pharmacological Approaches to Prevent Protein Carbonyl Formation

There are only three potential therapeutic strategies to prevent the occurrence of these deleterious modifications: (1) the inhibition of protein carbonyl synthesis either by chelation of transition metals or by scavenging ROS with antioxidants (Picklo et al., 2002), (2) induction of endogenous detoxifying enzymes such as NAD(P)H:quinone oxidoreductase 1, glutathione S-transferase, UDP-glucuronosyltransferase, and heme oxygenase-1 by phenylethyl isothiocyanate or sulforophane (Hu et al., 2004), and (3) the trapping of free (RCS) and protein-bound carbonyls with nucleophilic drugs (Shapiro, 1998; Dukic-Stefanovic et al., 2001). In recent years carbonyl trapping has received great attention as a specific treatment for conditions with severe carbonyl stress (e.g., diabetes, Alzheimer's, aging) (Onorato et al., 2000; Dukic-Stefanovic et al., 2001; Miyata et al., 2005). Carbonyl trapping agents react with RCS at a faster rate than do cell macromolecules, thereby ensuring the safe excretion of drug–carbonyl conjugates. They also react with protein-based carbonyls, thereby preventing deleterious cross-linking reactions. This latter process, however, may have damaging consequences since the introduction of a foreign moiety on a protein could elicit undesired immune responses. Chemically, carbonyl traps are compounds containing amino groups of relative low pKa so that at physiological pH they are mostly in the unprotonated, reactive form. Therapeutic amino-containing agents include hydralazine (Burcham et al., 2004), aminoguanidine (Abdel-Rahman and Bolton, 2002), metformin (Ruggiero-Lopez et al., 1999), OPB-9195 (Miyata et al., 2000), methoxylamine (Burcham et al., 2004), D-penicillamine (Wondrak et al., 2002), carnosine (Hipkiss and Brownson, 2000), pyridoxamine (Voziyan et al., 2002), and 2,3-diaminophenazine (Soulis et al., 1999) (❯ *Figure 23-6*) The mechanism of action of these carbonyl scavengers are discussed in detail in a recent review (Aldini et al., 2007). While most of these drugs display significant protection against carbonylation in cell-free systems, the hydrazines (aminoguanidine, hydralazine, and OPB-9195) are generally more effective in vivo. The reason for this phenomenon is that (1) the reaction between carbonyls and amines is rather slow and (2) the resulting Schiff bases (R–N = CHR') are readily reversible. In contrast, carbonyls and hydrazines (R–NH–NH$_2$) react very rapidly with each other and the resulting hydrazones (R–NH–N = CH–R') are chemically more stable. Moreover, in the case of hydralazine and aminoguanidine, the hydrazones undergo further reaction to yield stable condensation products. However, it should be noted that the carbonyl scavengers currently available are not specific. For example, hydralazine is a NADPH oxidase inhibitor and antioxidant, while aminoguanidine is a potent inhibitor of inducible nitric oxide synthetase. It is expected that in the future more specific carbonyl-trapping agents will be developed and that these will be administered in combination with antioxidants, anti-inflammatory or neuroactive substances for an improved clinical management of chronic neurological disorders.

6 Concluding Remarks and Perspectives

Nonenzymatic protein modification by acquisition of carbonyl functional groups has been known for a long time. Since protein carbonyls are formed by a number of oxidative pathways and are chemically stable and easy to measure, their accumulation is normally used as a reliable nonspecific marker of severe

◼ Figure 23-6
Chemical structures of classical RCS scavengers

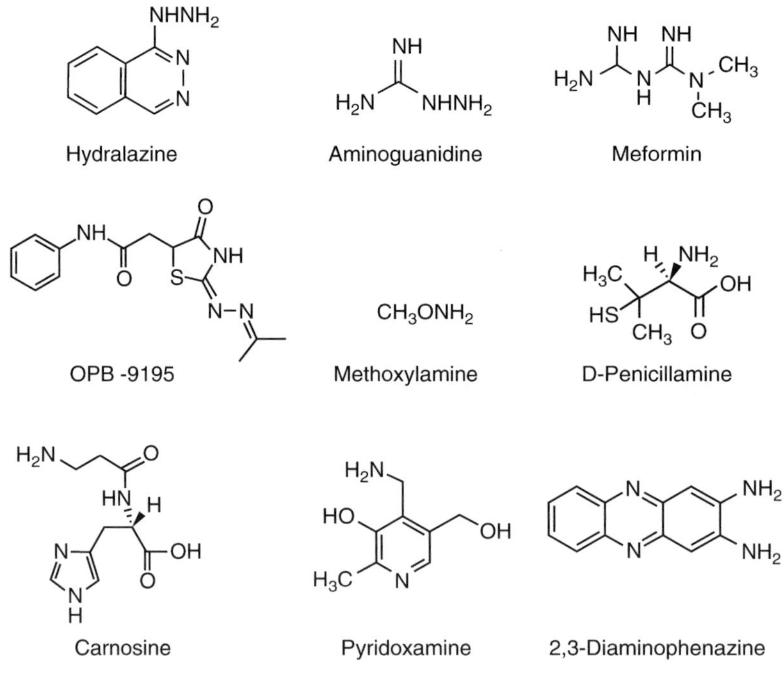

oxidative stress. In the last decade a number of proteomic and mass spectrometric techniques have made possible to identify the modified proteins. This, along with the number of studies demonstrating significant effect of carbonylation on protein function, has led to a renewed interest in the field. However, the most important issue still remains whether carbonylation plays an important role in the development of tissue damage or is merely the result of severe oxidative stress without deleterious functional consequences. While in a number of neurological disorders there is a clear temporal/spatial correlation between carbonyl accumulation and tissue damage, a direct pathogenic role for this modification will have to be established by specifically preventing protein carbonylation. Unfortunately the RCS scavengers currently available are not specific. Furthermore, they do not prevent protein carbonylation unless the modification occurs via lipoxidation or glycoxidation, which seems to be a minor modification when compared to direct carbonylation. Thus, a better understanding of the pathways of carbonyl formation is necessary to identify processes that can be selectively blocked. To this end, it is essential that future studies identify not only the modified polypeptides but also the structure of the carbonylated residues and the functional consequences of these modifications.

Acknowledgements

The author's work is supported by PHHS grant, NS047448 and NS057755 from NIH.

References

Abdel-Rahman E, Bolton WK. 2002. Pimagedine: A novel therapy for diabetic nephropathy. Exp Opin Invest Drugs 11: 565-574.

Adams S, Green P, Claxton R, Simcox S, Williams MV, et al. 2001. Reactive carbonyl formation by oxidative and non-oxidative pathways. Front Biosci 6: 17-24.

Ahmed MU, Brinkmann E, Degenhardt TP, Thorpe SR, Baynes JW. 1997. N-ε-(carboxyethyl)lysine, a product of the chemical modification of proteins by methylglyoxal, increases with age in human lens proteins. Biochem J 324: 565-570.

Aksenov MY, Aksenova MV, Butterfield DA, Geddes JW, Markesbery WR. 2001. Protein oxidation in the brain in Alzheimer's disease. Neuroscience 103: 373-383.

Aksenova M, Butterfield DA, Zhang SX, Underwood M, Geddes JW. 2002. Increased protein oxidation and decreased creatine kinase BB expression and activity after spinal cord contusion injury. J Neurotrauma 19: 491-502.

Alam ZI, Daniel SE, Lees AJ, Marsden DC, Jenner P, et al. 1997. A generalised increase in protein carbonyls in the brain in Parkinson's but not incidental Lewy body disease. J Neurochem 69: 1326-1329.

Alam ZI, Halliwell B, Jenner P. 2000. No evidence for increased oxidative damage to lipids, proteins or DNA in Huntington's disease. J Neurochem 75: 840-846.

Aldini G, Dalle-Donne I, Facino RM, Milzani A, Carini M. 2007. Intervention strategies to inhibit protein carbonylation by lipoxidation-derived reactive carbonyls. Med Res Rev 27: 817-868.

Andrus PK, Fleck TJ, Gurney ME, Hall ED. 1998. Protein oxidative damage in a transgenic mouse model of familial amyotrophic lateral sclerosis. J Neurochem 71: 2041-2048.

Banan A, Fitzpatrick L, Zhang LJ, Keshavarzian A. 2001. OPC-compounds prevent oxidant-induced carbonylation and depolymerization of the F-actin cytoskeleton and intestinal barrier hyerpermeability. Free Radic Biol Med 30: 287-298.

Banan A, Zhang LJ, Shaikh M, Fields JZ, Farhadi A, et al. 2004. Novel effect of NF-kB activation: Carbonylation and nitration injury to cytoskeleton and disruption of monolayer barrier in intestinal epithelium. Am J Physiol Cell Physiol 287: C139–C151.

Berlett BS, Stadtman ER. 1997. Protein oxidation in aging, disease and oxidative stress. J Biol Chem 272: 20313-20316.

Bizzozero OA. 2007. Major cytoskeletal proteins are carbonylated in multiple sclerosis, in ISN-Satellite Meeting on "Myelin development and function". Chichen Itza, Mexico; pp. 33.

Bizzozero OA, DeJesus G, Callahan K, Pastuszyn A. 2005. Elevated protein carbonylation in the brain white matter and gray matter of patients with multiple sclerosis. J Neurosci Res 81: 687-695.

Bizzozero OA, Reyes S, Ziegler J, Smerjac S. 2007. Lipid peroxidation scavengers prevent the carbonylation of cytoskeletal brain proteins induced by glutathione depletion. Neurochem Res 32: 2114-2122.

Bizzozero OA, Ziegler JL, De Jesus G, Bolognani F. 2006. Acute depletion of reduced glutathione causes extensive carbonylation of rat brain proteins. J Neurosci Res 83: 656-667.

Bowling AC, Schulz JB, Brown RH, Beal MF. 1993. Superoxide dismutase activity, oxidative damage and mitochondrial energy metabolism in familial and sporadic amyotrophic lateral sclerosis. J Neurochem 61: 2322-2325.

Browne SE, Ferrante RJ, Beal MF. 1999. Oxidative stress in Huntington's disease. Brain Pathol 9: 147-163.

Buchmuller-Rouiller Y, Corrandin SB, Smith J, Schneider P, Ransijn A, et al. 1995. Role of glutathione in macrophage activation: Effect of cellular glutathione depletion on nitrite production and leishmanicidal activity. Cell Immunol 164: 73-80.

Burcham PC, Fontaine FR, Kaminskas LM, Petersen DR, Pyke SM. 2004. Protein adduct-trapping by hydrazinophthalazine drugs: Mechanisms of cytoprotection against acrolein-mediated toxicity. Mol Pharmacol 65: 655-664.

Butterfield DA. 2004. Proteomics: A new approach to investigate oxidative stress in Alzheimer's disease brain. Brain Res 1000: 1-7.

Butterfield DA, Abdul HM, Newman S, Reed T. 2006. Redox proteomics in some age-related neurodegenerative disorders and models thereof. NeuroRx 3: 344-357.

Butterfield DA, Lauderback CM. 2002. Lipid peroxidation and protein oxidation in Alzheimer's disease brain: Potential causes and consequences involving amyloid beta-peptide-associated free radical oxidative stress. Free Radic Biol Med 32: 1050-1060.

Calabrese V, Raffaele R, Cosentino E, Rizza V. 1994. Changes in cerebrospinal fluid levels of malondialdehyde and glutathione reductase activity in multiple sclerosis. Int J Clin Pharmacol Res 14: 119-123.

Calingasan N, Uchida K, Gibson G. 1999. Protein-bound acrolein: A novel marker of oxidative stress in Alzheimer's disease. J Neurochem 72: 751-756.

Castegna A, Aksenov M, Aksenova M, Thongboonkerd V, Klein J, et al. 2002a. Proteomic identification of oxidatively modified proteins in Alzheimer's disease brain. I. creatine kinase BB, glutamine synthase, and ubiquitin carboxy-terminal hydrolase L-1. Free Radic Biol Med 33: 562-571.

Castegna A, Aksenov M, Thongboonkerd V, Klein JB, Pierce WM, et al. 2002b. Proteomic identification of oxidatively modified proteins in Alzheimer's disease brain. II. Dihydropyrimidinase-related protein 2, alpha-enolase and heat shock cognate 71. J Neurochem 82: 1524-1532.

Chen J, Henderson GI, Freeman GL. 2001. Role of 4-hydroxynonenal in modification of cytochrome c oxidase in ischemia/reperfused rat heart. J Mol Cell Cardiol 33: 1919-1927.

Choksi KB, Boyston WH, Rabek JP, Widger WR, Papaconstantinou J. 2004. Oxidatively damaged proteins of heart mitochondrial electron transport complexes. Biochim Biophys Acta 1688: 95-101.

Dalle-Donne I, Giustarini D, Colombo R, Rossi R, Milzani A. 2003. Protein carbonylation in human diseases. Trends Mol Med 9: 169-176.

Dalle-Done I, Rossi R, Giustarini D, Gagliano N, Lusini L, et al. 2001. Actin carbonylation: From a simple marker of protein oxidation to relevant signs of severe functional impairment. Free Radic Biol Med 31: 1075-1083.

Davies MJ, Slater TF. 1987. Studies on the metal-ion and lipoxygenase-catalysed breakdown of hydroperoxides using electron-spin-resonance spectroscopy. Biochem J 245: 167-173.

De Iuliis A, Grigoletto J, Recchia A, Giusti P, Arslan P. 2005. A proteomic approach in the study of an animal model of Parkinson's disease. Clin Chim Acta 357: 202-209.

De Stefano N, Matthews PM, Filippi M, Agosta F, De Luca M, et al. 2003. Evidence of early cortical atrophy in MS: Relevance to white matter changes and disability. Neurology 60: 1157-1162.

Divald A, Powell SR. 2006. Proteosome mediates removal of proteins oxidized during myocardial ischemia. Free Radic Biol Med 40: 156-164.

Dukic-Stefanovic S, Schinzel R, Riederer P, Münch G. 2001. AGES in brain aging: AGE-inhibitors as neuroprotective and anti-dementia drugs? Biogerontology 2: 19-34.

England K, Cotter TG. 2004. Identification of carbonylated proteins by MALDI-TOF mass spectroscopy reveals susceptibility of ER. Biochem Biophys Res Commun 320: 123-130.

England K, O'Driscoll C, Cotter TG. 2004. Carbonylation of glycolytic proteins is a key response to drug-induced oxidative stress and apoptosis. Cell Death Differ 11: 252-260.

Esterbauer H, Schaur RJ, Zollner H. 1991. Chemistry and biochemistry of 4-hydroxynonenal, malondialdehyde and related aldehydes. Free Radic Biol Med 11: 81-128.

Farout L, Friguet B. 2006. Proteasome function in aging and oxidative stress: Implications in protein maintenance failure. Antioxid Redox Signal 8: 205-216.

Ferguson B, Matyszak MK, Esiri MM, Perry VH. 1997. Axonal damage in acute multiple sclerosis lesions. Brain 120: 393-399.

Ferrante RJ, Browne SE, Shinobu LA, Bowling AC, Baik MJ, et al. 1997. Evidence of increased oxidative damage in both sporadic and familial amyotrophic lateral sclerosis. J Neurochem 69: 2064-2074.

Floor E, Maples AM, Rankin CA, Yaganti VM, Shank SS, et al. 2006. A one-carbon modification of protein lysine associated with elevated oxidative stress in human substantia nigra. J Neurochem 97: 504-514.

Friguet B, Szweda LI. 1997. Inhibition of multicatalytic proteinase (proteosome) by 4-hydroxynonenal cross-linked protein. FEBS Lett 405: 21-25.

Friguet B, Szweda LI, Stadtman ER. 1994. Susceptibility of glucose-6-phosphate dehydrogenase modified by 4-hydroxy-2-nonenal and metal-catalyzed oxidation to proteolysis by the multicatalytic protease. Arch Biochem Biophys 311: 168-173.

Fu SL, Dean RT. 1997. Structural characterization of the products of hydroxyl-radical damage to leucine and their detection on proteins. Biochem J 324: 41-48.

Fu MX, Requena JR, Jenkins AJ, Lyons TJ, Baynes JW, et al. 1996. The advanced glycation end product, N-ε-(carboxymethyl)lysine is a product of both lipid peroxidation and glycoxidation reactions. J Biol Chem 271: 9982-9986.

Gelinas S, Chapados C, Beauregard M, Gosselin I, Martinoli MG. 2000. Effect of oxidative stress on stability and structure of neurofilament proteins. Biochem Cell Biol 78: 667-674.

Gilgun-Sherki Y, Melamed E, Offen D. 2004. The role of oxidative stress in the pathogenesis of multiple sclerosis: The need for effective antioxidant therapy. J Neurol 251: 261-268.

Gold R, Hartung HP, Toyka KV. 2000. Animal models for autoimmune demyelinating disorders of the nervous system. Mol Med Today 6: 88-91.

Greco A, Minghetti L, Sette G, Fieschi C, Levi G. 1999. Cerebrospinal fluid isoprostane shows oxidative stress in patients with multiple sclerosis. Neurology 53: 1876-1879.

Grune T, Reinheckel T, Davies KJ. 1997. Degradation of oxidized proteins in mammalian cells. FASEB J 11: 526-534.

Grune T, Reinheckel T, Joshi M, Davies KJ. 1995. Proteolysis in cultured liver epithelial cells during oxidative stress. Role of the multicatalytic proteinase complex, proteasome. J Biol Chem 270: 2344-2351.

Hafler DA. 2004. Multiple sclerosis. J Clin Invest 113: 788-794.

Hellings N, Raus J, Stinissen P. 2002. Insights into the immunopathogenesis of multiple sclerosis. Immunol Res 25: 27-51.

Hendriks JJ, Alblas J, van der Pol SM, van Tol EA, Dijkstra CD, et al. 2004. Flavonoids influence monocytic GTPase activity and are protective in experimental allergic encephalitis. J Exp Med 200: 1667-1672.

Hensley K, Hall N, Subramaniam R, Cole P, Harris M, et al. 1995. Brain regional correspondence between Alzheimer's disease histopathology and biomarkers of protein oxidation. J Neurochem 65: 2146-2156.

Hipkiss AR, Brownson C. 2000. A possible new role for the anti-aging peptide carnosine. Cell Mol Life Sci 57: 747-753.

Hooper DC, Spitsin S, Kean RB, Champion JM, Dickson GM, et al. 1998. Uric acid, a natural scavenger of peroxynitrite, in experimental allergic encephalomyelitis and multiple sclerosis. Proc Natl Acad Sci USA 95: 675-680.

Hu R, Hebbar V, Kim BR, Chen C, Winnik B, et al. 2004. In vivo pharmacokinetics and regulation of gene expression profiles by isothiocyanate sulforaphane in the rat. J Pharmacol Exp Ther 310: 263-271.

Hunter MS, Neldmadin S, Davidson DLW. 1985. Lipid peroxidation products and antioxidant proteins in plasma and cerebrospinal fluid from multiple sclerosis patients. Neurochem Res 10: 1645-1652.

Kamencic H, Griebel RW, Lyon AW, Paterson PG, Juurlink BH. 2001. Promoting glutathione synthesis after spinal cord trauma decreases secondary damage and promotes retention of function. FASEB J 15: 243-250.

Kaneko K, Nakamura A, Yoshida K, Kametani F, Higuchi K, et al. 2002. Glial fibrillary acidic protein is greatly modified by oxidative stress in aceruloplasminemia brain. Free Radic Res 36: 303-306.

Karplus PA, Krauth-Siegel RL, Schirmer RH, Schulz GE. 1988. Inhibition of human glutathione reductase by the nitrosourea drugs 1,3-bis(2-chloroethyl)-1-nitrosourea and 1-(2-chloroethyl)-3-(2-hydroxyethyl)-1-nitrosourea. Eur J Biochem 171: 193-198.

Keegan BM, Noseworthy JH. 2002. Multiple sclerosis. Annu Rev Med 53: 285-302.

Keeney PM, Xie J, Capaldi RA, Bennett JP. 2006. Parkinson's disease brain mitochondrial complex I has oxidatively damaged subunits and is functionally impaired and misassembled. J Neurosci 26: 5256-5264.

Koito W, Araki T, Horiuchi S, Nagai R. 2004. Conventional antibody against $N\varepsilon$-(carboxymethyl)lysine (CML) shows cross-reaction to $N\varepsilon$-(carboxyethyl)lysine (CEL): Immunochemical quantification of CML with a specific antibody. J Biochem 136: 831-837.

Kuhlmann T, Lingfeld G, Bitsch A, Schuchardt J, Bruck W. 2002. Acute axonal damage in multiple sclerosis is most extensive in early disease stages and decreases over time. Brain 125: 2202-2012.

LaFontaine MA, Geddes JW, Banks A, Butterfield DA. 2000. 3-Nitropropionic acid induced in vivo protein oxidation in striatal and cortical synaptosomes: Insights into Huntington's disease. Brain Res 858: 356-362.

Langemann H, Kabiersch A, Newcombe J. 1992. Measurement of low molecular weight antioxidants, uric acid, tyrosine and tryptophan in plaques from patients with multiple sclerosis. Eur Neurol 32: 248-252.

Lee MH, Hyun DH, Jenner P, Halliwell B. 2001. Effect of proteasome inhibition on cellular oxidative damage, antioxidant defences and nitric oxide production. J Neurochem 78: 32-41.

Lehmann D, Karussis D, Misrachi-Koll R, Shezen E, Ovadia H, et al. 1994. Oral administration of the oxidant-scavenger N-acetyl-L-cysteine inhibits acute experimental autoimmune encephalomyelitis. J Neuroimmunol 50: 35-42.

Lenz AG, Costabel U, Shaltiel S, Levine RL. 1989. Determination of carbonyl groups in oxidatively modified proteins by reduction with tritiated sodium borohydride. Anal Biochem 177: 419-425.

Leski ML, Bao F, Wu L, Qian H, Sun D, et al. 2001. Protein and DNA oxidation in spinal injury: Neurofilaments – an oxidation target. Free Radic Biol Med 30: 613-624.

LeVine SM. 1992. The role of reactive oxygen species in the pathogenesis of multiple sclerosis. Med Hypotheses 39: 271-274.

LeVine SM. 1997. Iron deposits in multiple sclerosis and Alzheimer's disease brains. Brain Res 760: 298-303.

Levine RL. 2002. Carbonyl modified proteins in cellular regulation, aging, and disease. Free Radic Biol Med 32: 790-796.

Levine RL, Garland D, Oliver CN, Amici A, Climent I, et al. 1990. Determination of carbonyl content in oxidatively modified proteins. Methods Enzymol 186: 464-478.

Levine RL, Williams J, Stadtman ER, Shacter E. 1994. Carbonyl assays for determination of oxidatively modified proteins. Methods Enzymol 233: 346-357.

Lewen A, Matz P, Chan PH. 2000. Free radical pathways in CNS injury. J Neurotrauma 17: 871-890.

Liu Y, Zhu B, Wang X, Luo L, Li P, et al. 2003. Bilirubin as a potent antioxidant suppresses experimental autoimmune encephalomyelitis: Implications for the role of oxidative stress in the development of multiple sclerosis. J Neuroimmunol 139: 27-35.

Lucchinetti CF, Bruck W, Rodriguez M, Lassman H. 1996. Distinct patterns of multiple sclerosis pathology indicate heterogeneity in pathogenesis. Brain Pathol 6: 259-274.

Lyras L, Evans PJ, Shaw PJ, Nice PG, Halliwell B. 1996. Oxidative damage and motor neuron disease: Difficulties in the measurement of protein carbonyl in human tissues. Free Radic Rex 24: 397-406.

Malfroy B, Doctrow SR, Orr PL, Tocco G, Fedoseyeva EV, et al. 1997. Prevention and suppression of autoimmune encephalomyelitis by EUK-8, a synthetic catalytic scavenger of oxygen-reactive metabolites. Cell Immunol 177: 62-68.

Markesbery WR. 1997. Oxidative stress hypothesis in Alzheimer's disease. Free Radic Biol Med 23: 134-147.

Marracci GH, Jones RE, McKeon GP, Bourdette DN. 2002. Alpha-lipoic acid inhibits T cell migration into the spinal cord and suppresses and treats experimental autoimmune encephalomyelitis. J Neuroimmunol 131: 104-114.

Melo TM, Larsen C, White LR, Aasly J, Sjobakk TE, et al. 2003. Manganese, copper, and zinc in cerebrospinal fluid from patients with multiple sclerosis. Biol Trace Elem Res 93: 1-8.

Mitchell D, Petersen D. 1987. The oxidation of unsaturated aldehydic products in lipid peroxidation by rat liver aldehyde dehydrogenases. Toxicol Appl Pharmacol 87: 403-410.

Miyata T, Ueda Y, Asahi K, Izuhara Y, Inagi R, et al. 2000. Mechanism of the inhibitory effect of OPB-9195 [(+/−)-2-isopropylidenehydrazono-4-oxo-thiazolidin-5-ylacetanilide] on advanced glycation end product and advanced lipoxidation end product formation. J Am Soc Nephrol 11: 1719-1725.

Miyata T, Yamamoto M, Izuhara Y. 2005. From molecular footprints of disease to new therapeutic interventions in diabetic nephropathy. Ann N Y Acad Sci 1043: 740-749.

Mirzaei H, Regnier F. 2005. Affinity chromatographic selection of carbonylated proteins followed by identification of oxidation sites using tandem mass spectrometry. Anal Chem 77: 2386-2392.

Montine KS, Olson SJ, Amarnath V, Whetsell WO, Graham DJ, et al. 1997. Immunohistochemical detection of 4-hydroxy-2-nonenal adducts in Alzheimer's disease is associated with inheritance of APOE4. Am J Pathol 150: 437-443.

Muntané G, Dalfó E, Martínez A, Rey MJ, Avila J, et al. 2006. Glial fibrillary acidic protein is a major target of glycoxidative and lipoxidative damage in Pick's disease. J Neurochem 99: 177-185.

Naidoo R, Knapp ML. 1992. Studies of lipid peroxidation products in cerebrospinal fluid and serum in multiple sclerosis and other conditions. Clin Chem 38: 2449-2454.

Neely MD, Boutte A, Milatovic D, Montine TJ. 2005. Mechanisms of 4-hydroxynonenal-induced neuronal microtubule dysfunction. Brain Res 1037: 90-98.

Niebroj-Dobosz I, Dziewulska D, Kwiecinki H. 2004. Oxidative damage to proteins in the spinal cord in amyotrophic lateral sclerosis (ALS). Folia Neuropathol 42: 151-156.

Noseworthy JH, Lucchinetti C, Rodriguez M, Weinshenker BG. 2000. Multiple sclerosis. N Engl J Med 343: 938-952.

O'Connor T, Ireland LS, Harrison DJ, Hayes JD. 1999. Major differences exist in the function and tissue-specific expression of human aflatoxin B1 aldehyde reductase and the principal human aldo-keto reductase AKR1 family members. Biochem J 343: 487-504.

Offen D, Gilgun-Sherki Y, Barhum Y, Benhar M, Grinberg L, et al. 2004. A low molecular weight copper chelator crosses the blood–brain barrier and attenuates experimental autoimmune encephalomyelitis. J Neurochem 89: 1241-1251.

Onorato JM, Jenkins AJ, Thorpe SR, Baynes JW. 2000. Pyridoxamine, an inhibitor of advanced glycation reactions, also inhibits advanced lipoxidation reactions. Mechanism of action of pyridoxamine. J Biol Chem 275: 21177-21184.

Ozeki M, Miyagawa-Hayashino A, Akatsuka S, Shirase T, Lee WH, et al. 2005. Susceptibility of actin to modification by 4-hydroxy-2-nonenal. J Chromatogr 827: 119-126.

Pacifici RE, Kono Y, Davies KJ. 1993. Hydrophobicity as the signal for selective degradation of hydroxyl radical-modified hemoglobin by the multicatalytic proteinase complex, proteasome. J Biol Chem 268: 15405-15411.

Palfi S, Ferrante RJ, Broulliet E, Beal MF, Dolan R, et al. 1996. Chronic 3-nitropropionic acid treatment in babooms replicates the cognitive and motor deficits of Huntington's disease. J Neurosci 16: 3019-3025.

Palinski W, Rosenfeld ME, Yla-Herttuala S, Gurtner GC, Socher SS, et al. 1989. Low density lipoprotein undergoes oxidative modification in vivo. Proc Natl Acad Sci USA 86: 1372-1376.

Pamplona R, Dalfo E, Ayala V, Bellmunt MJ, Prat J, et al. 2005. Proteins in human brain cortex are modified by oxidation, glycoxidation, and lipoxidation. J Biol Chem 280: 21522-21530.

Pedchenko TV, LeVine SM. 1998. Desferrioxamine suppresses experimental allergic encephalomyelitis induced by MBP in SJL mice. J Neuroimmunol 84: 188-197.

Pedersen WA, Fu W, Keller JN, Markesbery WR, Appel S, et al. 1998. Protein modification by the lipid peroxidation product 4-hydroxynonenal in the spinal cords of amyotrophic lateral sclerosis patients. Ann Neurol 44: 819-824.

Penkowa M, Hidalgo J. 2003. Treatment with metallothionein prevents demyelination and axonal damage and increases oligodendrocyte precursors and tissue repair during experimental autoimmune encephalomyelitis. J Neurosci Res 72: 574-586.

Perluigi M, Fai Poon H, Hensley K, Pierce WM, Klein JB, et al. 2005a. Proteomic analysis of 4-hydroxy-2-nonenal-modified proteins in G93A-SOD1 transgenic mice – a model of familial amyotrophic lateral sclerosis. Free Radic Biol Med 38: 960-968.

Perluigi M, Poon HF, Maragos W, Pierce WM, Klein JB, et al. 2005b. Proteomic Analysis of protein expression and oxidative modification in R6/2 transgenic mice – a model of Huntington's disease. Mol Cell Proteomics 4: 1849-1861.

Peterson JW, Trapp BD. 2005. Neuropathobiology of multiple sclerosis. Neurol Clin 23: 107-129.

Picklo MJ, Montine TJ, Amarnath V, Neely MD. 2002. Carbonyl toxicology and Alzheimer's disease. Toxicol Appl Pharmacol 184: 187-197.

Poon HF, Frasier M, Shreve N, Calabrese V, Wolozin B, et al. 2005. Mitochondrial associated metabolic proteins are selectively oxidized in A30P α-synuclein transgenic mice: A model of familial Parkinson's disease. Neurobiol Dis 18: 492-498.

Powell SR, Wang P, Divald A, Teichberg S, Haridas V, et al. 2005. Aggregates of oxidized proteins (lipofuscin) induce apoptosis through proteosome inhibition and dysregulation of proapoptotic proteins. Free Radic Biol Med 38: 1093-1101.

Rabek JP, Boylston WH, Papaconstantinou J. 2003. Carbonylation of ER chaperone proteins in aged mouse liver. Biochem Biophys Res Commun 305: 566-572.

Refsgaard HF, Tsai L, Stadman ER. 2000. Modification of proteins by polyunsaturated fatty acid peroxidation products. Proc Natl Acad Sci USA 97: 611-691.

Requena JS, Chao CC, Levine R, Stadtman ER. 2001. Glutamic and aminoadipic semialdehydes are the main carbonyl products of metal-catalyzed oxidation of proteins. Proc Natl Acad Sci USA 98: 69-74.

Rivett AJ. 1985. Preferential degradation of the oxidatively modified form of glutamine synthetase by intracellular mammalian proteases. J Biol Chem 260: 300-305.

Ruggiero-Lopez D, Lecomte M, Moinet G, Patereau G, Lagarde M, et al. 1999. Reaction of metformin with dicarbonyl compounds. Possible implication in the inhibition of advanced glycation end product formation. Biochem Pharmacol 58: 1765-1773.

Ruuls SR, Bauer J, Sontrop K, Huitinga I, Hart BA, et al. 1995. Reactive oxygen species are involved in the pathogenesis of experimental allergic encephalomyelitis in Lewis rats. J Neuroimmunol 56: 207-217.

Sarchieli P, Presciutti O, Tarducci R, Gobbi G, Alberti A, et al. 2002. Localized ^{1}H magnetic resonance spectroscopy in mainly cortical gray matter of patients with multiple sclerosis. J Neurol 249: 902-910.

Sayre LM, Moreira PI, Smith MA, Perry G. 2005. Metal ions and oxidative protein modification in neurological disease. Ann Ist Super Sanita 41: 143-164.

Sayre LM, Zelasko DA, Harris PL, Perry G, Salomon RG, et al. 1997. 4-Hydroxynonenal-derived advanced lipid peroxidation end products are increased in Alzheimer's disease. J Neurochem 68: 2092-2097.

Sellin S, Holmquist B, Mannervik B, Vallee B. 1991. Oxidation and reduction of 4-hydroxyalkenals catalyzed by isozymes of human alcohol dehydrogenase. Biochemistry 30: 2514-2518.

Shacter E. 2000. Quantification and significance of protein oxidation in biological samples. Drug Metab Rev 32: 307-326.

Shapiro HK. 1998. Carbonyl-trapping therapeutic strategies. Am J Ther 5: 323-353.

Shaw PJ, Ince PG, Falkous G, Mantle D. 1995. Oxidative damage to protein in sporadic motor neuron disease spinal cord. Ann Neurol 38: 691-695.

Shringarpure R, Grune T, Sitte N, Davies KJ. 2000. 4-Hydroxynonenal-modified amyloid-beta peptide inhibits the proteasome: Possible importance in Alzheimer's disease. Cell Mol Life Sci 57: 1802-1808.

Skukla UK, Jensen GE, Clausen J. 1997. Erythrocyte glutathione peroxidase deficiency in multiple sclerosis. Acta Neurol Scand 56: 542-550.

Slatter DA, Murray M, Bailey AJ. 1998. Formation of a dihydropyridine derivative as a potential cross-link derived from malondialdehyde in physiological systems. FEBS Lett 421: 180-184.

Smerjac SM, Bizzozero OA. 2008. Cytoskeletal protein carbonylation and degradation in experimental autoimmune encephalomyelitis. J Neurochem 105: 763-773.

Smith MA, Sayre LM, Anderson V, Harris PL, Beal MF, et al. 1998. Cytochemical demonstration of oxidative damage in Alzheimer disease by immunochemical enhancement of the carbonyl reaction with 2,4-dinitrophenylhydrazine. J Histochem Cytochem 46: 731-735.

Smith MA, Rudnicka-Nawrot M, Richey PL, Praprotnik D, Mulvihill P, et al. 1995. Carbonyl-related posttranslational modification of neurofilament protein in the neurofibrillary pathology of Alzheimer's disease. J Neurochem 64: 2660-2666.

Soreghan BA, Yang F, Thomas SN, Hsu J, Yang AJ. 2003. High-throughput proteomic-based identification of oxidatively induced protein carbonylation in mouse brain. Pharm Res 20: 1713-1720.

Soulis T, Sastra S, Thallas V, Mortensen SB, Wilken M, et al. 1999. A novel inhibitor of advanced glycation end-product formation inhibits mesenteric vascular hypertrophy in experimental diabetes. Diabetologia 42: 472-479.

Stadtman ER. 1990a. Metal ion-catalyzed oxidation of proteins: Biochemical mechanisms and biological consequences. Free Radic Biol Med 9: 315-325.

Stadtman ER. 1990b. Covalent modification reactions are making steps for protein turnover. Biochem J 29: 6323-6331.

Stadtman ER, Berlett BS. 1997. Reactive oxygen-mediated protein oxidation in aging and disease. Chem Res Toxicol 10: 485-494.

Trapp BD, Peterson J, Ransohoff RM, Rudick R, Mork S, et al. 1998. Axonal transection in the lesions of multiple sclerosis. N Engl J Med 338: 278-285.

Troncoso JC, Costello AC, Kim JH, Johnson GV. 1995. Metal-catalyzed oxidation of bovine neurofilaments in vitro. Free Radic Biol Med 18: 891-899.

Uchida K. 2003. 4-Hydroxy-2-nonenal: A product and mediator of oxidative stress. Prog Lipid Res 42: 318-343.

Uchida K, Kanematsu M, Sakai K, Matsuda T, Hattori N, et al. 1998. Protein-bound acrolein: Potential markers for oxidative stress. Proc Natl Acad Sci USA 95: 4882-4887.

Uchida K, Stadtman ER. 1992. Modification of histidine residues in proteins by reaction with 4-hydroxynonenal. Proc Natl Acad Sci USA 89: 4544-4548.

Uchida K, Stadtman ER. 1993. Covalent attachment of 4-hydroxynonenal to glyceraldehyde-3-phosphate dehydrogenase. A possible involvement of intra- and intermolecular cross-linking reaction. J Biol Chem 268: 6388-6393.

Vatassery GT, Younoszai R, Vladimirova O, O'Connor J, Cahill A, et al. 1998. Oxidative damage to DNA in plaques of MS brains. Mult Scler 4: 413-418.

Vieira HLA, Belzacq AS, Haouzi D, Bernassola F, Cohen I, et al. 2001. The adenine nucleotide translocator: A target of nitric oxide, peroxynitrite, and 4-hydroxynonenal. Oncogene 20: 4305-4316.

Voziyan PA, Metz TO, Baynes JW, Hudson BG. 2002. A post-amadori inhibitor pyridoxamine also inhibits chemical modification of proteins by scavenging carbonyl intermediates of carbohydrate and lipid degradation. J Biol Chem 277: 3397-3403.

Waeg G, Dimsity G, Esterbauer H. 1996. Monoclonal antibodies for detection of 4-hydroxynonenal modified proteins. Free Radical Res 25: 149-159.

Wataya T, Nunomura A, Smith MA, Siedlak SL, Harris PL, et al. 2002. High molecular weight neurofilament proteins are physiological substrates of adduction by the lipid peroxidation product hydroxynonenal. J Biol Chem 277: 4644-4648.

Wondrak GT, Cervantes-Laurean D, Roberts MJ, Qasem JG, Kim M, et al. 2002. Identification of alpha-dicarbonyl scavengers for cellular protection against carbonyl stress. Biochem Pharmacol 63: 361-373.

Yan LJ, Lodge JK, Traber MG, Packer L. 1997. Apolipoprotein B carbonyl formation is enhanced by lipid peroxidation during copper-mediated oxidation of human low-density lipoproteins. Arch Biochem Biophys 339: 165-171.

Yan LJ, Sohal RS. 1998a. Gel electrophoretic quantitation of protein carbonyls derivatized with tritiated sodium borohydride. Anal Biochem 265: 176-182.

Yan LJ, Sohal RS. 1998b. Mitochondrial adenine nucleotide translocase is modified oxidatively during aging. Proc Natl Acad Sci USA 95: 12896-12901.

Yarian CS, Rebrin I, Sohal RS. 2005. Aconitase and ATP synthase are targets of malondialdehyde nodification and undergo an age-related decrease in activity in mouse heart mitochondria. Biochim Biophys Res Commun 330: 151-156.

Yoritaka A, Hattori N, Uchida K, Tanaka M, Stadtman ER, et al. 1996. Immunohistochemical detection of 4-hydroxynonenal protein adducts in Parkinson's disease. Proc Natl Acad Sci USA 93: 2696-2701.

24 Clinical Considerations in Translational Research with Chronic Spinal Cord Injury: Intervention Readiness and Intervention Impact

J. S. Krause · S. D. Newman · S. S. Brotherton

1	Introduction	564
2	Translation of Breakthroughs into Practice	564
3	Ambulation and SCI	565
3.1	Gait Training Interventions	565
3.2	Benefits and Limitations	566
3.3	Ambulation and Community Outcomes	567
4	Intervention Readiness and Intervention Impact	568
4.1	Intervention Readiness	568
4.2	Intervention Impact	568
5	Time Since SCI Onset, SCI Severity, and Intervention Readiness	569
6	Functional Decline and Intervention Readiness	570
6.1	Musculo-Skeletal Changes After SCI	570
6.2	Cardiovascular Function	572
7	Allostatic Load and Telomeres	573
7.1	Allostatic Load	573
7.2	Telomeres	575
8	Conclusion	575
8.1	Future Directions	575

© 2009 Springer Science+Business Media, LLC.

Abstract: Despite the extensive efforts to identify interventions that can be successfully translated to promote functional recovery after chronic spinal cord injury (SCI), investigators have neglected the viability of interventions given the cumulative physiologic decline resulting from the SCI. Two concepts are introduced including intervention readiness and intervention impact. The former concept refers to the accumulation of secondary conditions and health decline over time after SCI and the extent to which these factors may be evaluated to determine whether a given intervention is feasible. The latter concept refers to the actual impact of an intervention on all outcomes, including overall health, participation, and quality of life. We have highlighted an existing body of literature on physiologic decline after SCI (e.g., loss in bone density and muscle mass), as well as research on individuals with chronic, ambulatory, incomplete SCI that identifies heightened risk of secondary conditions associated with incomplete ambulation. Individuals with incomplete ambulatory SCI represent a type of natural experiment that can be used to glimpse the future of SCI and help us prepare for the increasing portion of individuals with incomplete injuries. We must balance our zeal for identification of interventions to restore function with research that helps us to realistically appraise the extent to which those with chronic SCI are likely to be ready for and benefit from such interventions.

List of Abbreviations: BMD, bone mineral density; BWSTT, body weight supported treadmill training; CAD, coronary artery disease; CSAs, cross sectional areas; CVD, cardiovascular disease; DEXA, dual energy X-ray absorptiometry; FES, functional electrical stimulation; IMF, intramuscular fat; RGO-II, reciprocal gait orthosis II; SCI, spinal cord injury; SCI-FAI, spinal cord injury functional ambulation inventory; WISCI, walking index for spinal cord injury

1 Introduction

The number of studies designed to identify interventions to either preserve or promote recovery after neurologic injury has increased dramatically during recent years. There has been a parallel emphasis in clinical research to rehabilitate individuals who have neurologic injuries in order to promote optimal health, well-being, and longevity. The clear emphasis of the roadmap from the National Institutes of Health (NIH) is for an increase in translational research. This requires that scientists and clinicians find common ground and work together toward solutions to promote better outcomes. Although frequently described as from bench to bedside, translational studies may also go from bedside to community. In fact, promoting outcomes at the community level is the ultimate goal and the endpoint of translational work.

The purpose of this chapter is to discuss potential barriers in the translation of discoveries to promote functional recovery among individuals with chronic spinal cord injury (SCI) due to the accumulation of physiologic changes and secondary conditions resulting from SCI. In the context of this chapter, we use the term *restorative intervention* to define any intervention that is designed to promote recovery or function after SCI (rather than preserve function). Our emphasis is on the potential application of these interventions among those with chronic SCI. We challenge the assumption that there will be universal readiness among those with chronic SCI to receive restorative interventions and that the application of such interventions will necessarily result in enhanced health, participation, and quality of life. This is a significant word of caution to both basic and applied scientists who hold the premise that any increase in function will ultimately lead to better outcomes, regardless of circumstance.

2 Translation of Breakthroughs into Practice

Popular beliefs reinforced through the media lead to the perception that a single medical breakthrough will someday lead to a "cure" for SCI. This perception has been fueled by injuries to high visibility individuals, some of which have resulted in the development of foundations to fund research directed at such a "cure." Few scientists would adhere to this view. On the other hand, few scientists actively dispute such a contention in public forums. Promoting greater functional recovery after SCI will almost certainly result from multiple discoveries using diverse techniques to preserve, promote, and reinforce recovery. Also contrary to popular belief, successful translation of new discoveries is likely to lead to a greater need for SCI rehabilitation and will present a number of new challenges.

Perhaps the most visible benefit of years of research and clinical efforts to improve recovery after SCI is the decrease in the number of neurologically complete SCI. There has been a slight increase of incomplete injuries over the past three decades (NSCISC, 2006). We cannot pinpoint whether this has resulted from steroid protocols, enhanced emergency medical treatment, superior treatment at the trauma center, or a combination of factors. As new discoveries are translated into clinical practice, we would expect that the portion of neurologically complete injuries will continue to decline.

There will almost certainly be a corresponding increase in the portion of individuals who are able to ambulate after SCI. However, given the complex nature of SCI, quality of ambulation is likely to vary substantially according to level of recovery and the nature of the initial injury. Therefore, clinicians will be challenged to better describe and measure function, particularly among those with neurologically incomplete injuries that allow for ambulation. We have already seen an increase in measurement tools to evaluate ambulation after SCI.

The Spinal Cord Injury Functional Ambulation Inventory (SCI-FAI) (Field-Fote et al., 2001) is a tool developed specifically to inventory walking parameters in SCI. It contains three sections related to gait parameters (e.g., observation of weight shift, step length), use of assistive devices and orthoses, and temporal/distance measures. Because of the observational methods, the SCI-FAI is appropriate for studies of intensive interventions of small numbers of participants but is too labor-intensive for studies that use larger participant samples and require only more general information on gait, such as use of specific devices and orthoses. The Walking Index for Spinal Cord Injury (WISCI; Ditunno et al., 2000) provides a 21-level measure based on assistive devices and orthoses required for walking, and it appears to have gained acceptance in the rehabilitation community (Wirz et al., 2005). The initial ASIA score is predictive of mobility potential as measured by the WISCI (Morganti et al., 2005). Van Hedel et al. (2005) compared the WISCI to the other existing tests (Timed Up and Go test, the 6-min walk test, and the 10-m walk test) and found the three timed tests highly correlated with each other (/r/>0.88), and moderately correlated with the WISCI (/r/>0.60). The 6-min walk test and the 10-m walk test have been found to be more responsive than the WISCI to demonstrating walking improvement after incomplete SCI. These are only initial efforts to measure ambulation and challenges in assessing SCI outcomes will certainly increase in the years to come.

3 Ambulation and SCI

There have been a substantial number of intervention studies to promote ambulation through intensive physical therapy. These studies are generally targeted toward the period immediately after injury, and outcomes are measured for short term effects. In contrast, the number of investigations of the relationship between ambulation and outcomes in the community is much more limited. These studies, although based on secondary analysis of existing data, differ from the intensive gait training in that the outcomes are not tied to specific interventions and are measured years or even decades later.

3.1 Gait Training Interventions

In recent years, considerable effort has been devoted to finding effective treatments to promote walking recovery. Advances in the knowledge of neuronal control of walking led to the development of body weight supported treadmill training (BWSTT) to facilitate ambulation. Another intervention used to improve walking is functional electrical stimulation (FES) of the nerves and muscles used to perform stepping movements. Lower extremity orthotics and assistive devices for ambulation are also used to restore walking. These interventions may be used alone or in combination for more effective treatment outcomes.

The current principles of locomotor training after human SCI are based on studies of locomotion in mammals with a thoracic spinal cord transaction (Lovely et al., 1986). In these studies, animals relearned how to step with their hind legs on a treadmill when trained with complex temporal patterns of sensory input related to stepping. If humans have similar neural networks, they should be able to integrate and interpret sensory information associated with walking and produce functional efferent output with repetitive training.

The notion that locomotion can be trained using lower extremity and trunk movements to generate sensory input consistent with ambulation to facilitate a stepping response led to the development of BWSTT (Harkema et al., 1997). During BWSTT, a portion of the body weight is supported with a harness system while the individual walks on a motorized treadmill (Finch et al., 1991). Manual facilitation is required to assist the patient to achieve upright posture and perform stepping movements associated with walking (Finch et al., 1991). Benefits of a 12-week program of BWSTT include increased gait speed and walking endurance and decreased oxygen costs (Protas et al., 2001). Even after a single session of BWSTT, self-selected and maximum overground walking speeds increased 26 and 25%, respectively (Trimble et al., 2001).

Despite the potential benefits of BWSTT, the procedure may require up to three persons to manually assist the patient and is physically demanding and uncomfortable for those providing help (Behrman and Harkema, 2000). Motorized, computer-controlled devices that provide assistance with stepping movements have been developed to improve the usefulness of BWSTT in the clinical setting. Recent studies of locomotor training using robotic-assisted BWSTT for those with incomplete SCI revealed improvements in gait velocity, endurance (Hornby et al., 2005), and performance of functional tasks (Wirz et al., 2005).

Another strategy for improving walking in people with incomplete SCI is the use of FES (Fouad and Pearson, 2004). In a study of the use of FES for 1 year to strengthen muscles and augment walking, adolescents with incomplete SCI had increased voluntary strength, decreased energy cost, increased maximum walking distance and speed, increased step length, and improved joint kinematics during gait (Johnston et al., 2003). Long-term FES assisted walking was also found to decrease reflex and intrinsic ankle stiffness in persons with incomplete SCI when compared with a non-FES SCI control group (Mirbagheri et al., 2002). When FES was combined with BWSTT, improvement was noted in intralimb coordination (Field-Fote and Tepavac, 2002) and overground and treadmill walking speeds, treadmill walking distance, and lower extremity motor scores (Field-Fote, 2001). FES combined with use of a hinged ankle-foot orthosis resulted in improved walking in persons with incomplete SCI as evidenced by increased gait speed and endurance, while application of FES alone led to improved foot clearance (Kim et al., 2004).

A number of different orthotic devices are available for use with individuals with SCI. In a study of the effectiveness of walking exercise with a weight bearing control orthosis, energy consumption and cost were measured and compared to values obtained with conventional orthosis (Kawashima et al., 2003). When compared to conventional orthosis, investigators found the weight bearing control orthosis allowed thoracic level paraplegic patients to walk at a higher speed with similar energy expenditure (Kawashima et al., 2003). The Walkabout, a hip-knee-ankle-foot-orthotic, proved convenient for standing and walking and compatible with wheelchair use (Saitoh et al., 1996). After a relatively short training period with this orthotic, individuals with paraplegia were able to ambulate independently with loftstrand crutches with energy consumption comparable to normal walking (Saitoh et al., 1996). The Reciprocal Gait Orthosis II (RGO-II) was found to be useful for home ambulation with individuals with complete paraplegia (Thoumie et al., 1995), but both the Walkabout and RGO-II were used more often for standing and were less useful for other tasks (Thoumie et al., 1995; Harvey et al., 1997).

3.2 Benefits and Limitations

Ambulation training has been associated with improvement in physical function, physiological adaptations, and subjective well-being. Over the ground, walking speed improved for people with incomplete SCI following a combination of BWSTT and FES (Field-Fote, 2001). Lower extremity strength increased with FES (Johnston et al., 2003), and lower limb coordination improved for individuals who used BWSTT and FES (Field-Fote and Tepavac, 2002). Gains in speed and efficiency of walking, slower heart rates, and decreased load on the upper extremities were reported 1 year after discharge from rehabilitation (Yakura et al., 1990). In addition to enhanced ambulatory capacity, BWSTT resulted in increased muscle fiber size and oxidative capacity and reduction in total plasma and low density lipoprotein cholesterol (Stewart et al., 2004). Standing with FES for persons with paraplegia provided sufficient cardiorespiratory stress for aerobic conditioning (Jacobs et al., 2003; Kawashima et al., 2003). Additional beneficial adaptations included improved resting blood flow and an augmented hyperemic response to an ischemic stimulus (Nash et al., 1997). Increased satisfaction with life and with physical function was associated with improved treadmill walking ability (Hicks et al., 2005).

It must be noted that these are short-term effects, measured during or immediately after treatment on a limited number of participants. Furthermore, there were generally no other treatment groups that could be used to determine if alternative interventions, such as rigorous nonambulatory exercise programs, may produce many of the same effects (e.g., conditioning, satisfaction, and quality of life). The long-term benefits for the months, years, and even decades after intervention have not been demonstrated but are often assumed based on relatively time-limited therapeutic effects.

Although walking ability improved with the various interventions, carry-over to functional ambulation was limited. Less than half of those who increased their walking speed and distance and required less support during BWSTT were able to improve their capacity for over the ground walking (Hicks et al., 2005). Even with an increase in gait speed with ambulation training, individuals with incomplete SCI had difficulty attaining the required velocity to safely cross an intersection (Lapointe et al., 2001). At their preferred walking speed, these subjects also had greater oxygen uptake and increased lactate concentration, indicating use of anaerobic pathways and less efficiency with walking. Post-injury osteopenia, a common concern after SCI, did not improve with weight bearing activity (Scremin et al., 1999).

In sum, studies of intensive ambulation training have demonstrated some beneficial effects on short-term outcomes, but they do not address the more important question of how ambulation relates to secondary health conditions and quality of life over a lifetime. It is generally accepted that, as individuals age with severe mobility impairments, functional decline may be accelerated compared with those who do not have mobility impairments. This was first observed among long-term polio survivors and has been described as post-polio syndrome (Dalakas et al., 1986). Similar observations have been made among people with SCI but have not received the same attention, as it is only within the past couple of decades that the life expectancy of individuals with SCI has increased to the point, where substantial numbers of survivors are reaching aging milestones (Strauss et al., 2006). Clearly, it would be ill-advised to generalize from ambulation trials that are conducted over short periods of time with carefully selected participants to long-term survivors of varying ages, physical characteristics, and degree of preserved function.

3.3 Ambulation and Community Outcomes

Unlike the everincreasing number of studies focusing on short-term gains associated with gait training interventions, naturalistic studies of outcomes among ambulators in community settings have been rare. This type of study holds the key to understanding the potential long-term complications associated with successful interventions to increase function resulting in ambulation.

Krause (2004) investigated risk of subsequent injuries among participants with SCI. Injuries were described as "broken bones, burns, or cuts that happened as a result of some type of mishap or event such as a fall, collision, motor vehicle wreck, or act of violence" (Krause, 2004, p. 1504). Of the 1,328 participants, 19% reported one or more subsequent injury in the previous year severe enough to require medical attention in a clinic, emergency room, or hospital (Krause, 2004). An unanticipated finding was that injury severity was related to the risk of subsequent injuries. Participants who reported neurologically incomplete injuries sufficient to allow ambulation had 2.1 greater odds of a subsequent injury after controlling for behavioral risk factors. In a follow-up to this study, 75% of ambulatory participants reported at least one fall in the previous 12 months, and 18% of those who fell sustained a fracture as a result of a fall (Brotherton et al., 2007).

In another study using secondary analysis of existing data (Krause et al., 2007a), lower pain interference scores and six life areas were noted among 261 participants who were independent in assistance from others compared with 102 participants who reported themselves to be partially dependent on others for assistance.

The results were clinically relevant, as nearly all participants scoring on the low end of the total pain score distribution were independent in ambulation, whereas over half of those at the opposite extreme were dependent. The analyses were extended to include wheelchair-bound participants in order to test the hypothesis that pain interference mediated the relationship between ambulatory status and two depressive disorders (Krause et al., 2007b). After controlling for biographic characteristics, the partially dependent ambulation group had 2.3 greater odds of clinically significant symptomatology than the independent ambulation group and 2.0 greater odds of probable major depression (the wheelchair dependent group was not significantly different from the independent ambulation group). However, after including pain interference as a moderating variable, the differences between the two ambulation groups were no longer statistically significant (OR = 1.52, 1.23, respectively).

Taken together, these findings suggest that individuals who are ambulatory are at elevated risk for subsequent injuries, particularly falls, and those who are not independent in their ambulation are at risk for other secondary conditions including pain interference and depression. Clearly, more research needs to be done to evaluate a broader number of parameters of ambulation in relation to a wider array of secondary conditions after SCI. Nevertheless, the preliminary findings suggest that caution should be taken when projecting dramatic increases in quality of life with enhanced ability to ambulate. We simply do not know whether basic science research will be translated into successful interventions to enhance function and how those enhancements will ultimately affect the lives of people of SCI.

4 Intervention Readiness and Intervention Impact

Intervention readiness and intervention impact are two new constructs intricately related to translational research and SCI. They reflect the extent to which the individual is physiologically and psychologically ready for a restorative intervention (intervention readiness) and the secondary or unintended outcomes of such interventions (intervention impact). As will become apparent, the two concepts are interrelated and necessary to consider if translation of new discoveries is to lead to interventions that, not only increase function, but enhance overall health, participation, and quality of life.

4.1 Intervention Readiness

Intervention readiness refers to individual differences in ability to benefit from a restorative intervention as the result of the accumulation of changes in physiologic and psychological function resulting from SCI. Several things should be immediately apparent from this definition. First, not everyone will be equally prepared to receive a restorative intervention. Individual differences are pervasive in all areas of life, and this will be no exception. Second, since there is accumulation of changes which will affect readiness, then readiness will be affected by the length of time between SCI onset and implementation of the intervention. This implicitly acknowledges that as time passes after SCI onset, the readiness for an intervention will decline, and there may be a point where, regardless of what the intervention may do to restore neurologic function, it is no longer practical for the individual. Third, physiologic factors are not the sole consideration. Psychological changes in response to SCI are also important. It should not be assumed that all individuals would choose to have a neurologic intervention, particularly those who have lived a substantial amount of time after SCI.

4.2 Intervention Impact

Intervention impact refers to the changes in outcomes, both favorable and unfavorable, that directly and indirectly result from a restorative intervention. While intervention readiness refers to the net sum of factors that mediate the extent to which any given individual is prepared to receive a restorative intervention, intervention impact refers to the outcomes that will result from the interaction between the restorative intervention and the individual's readiness to receive an intervention. The term interaction specifies the outcome as depending on the combination of the specific intervention and readiness status of the individual. Therefore, any given intervention may have differential effects depending on the individual. Any given individual may respond differently to different types of interventions. The ultimate benefit of any restorative intervention must be evaluated, not only in terms of its ability to restore function, but its ability to improve associated outcomes including health, participation, and quality of life. We have focused our discussion on intervention readiness, as there is a body of research to guide our discussion.

5 Time Since SCI Onset, SCI Severity, and Intervention Readiness

How do we measure intervention readiness – what are the primary factors that relate to physiologic changes? What factors mediate decline? Are there critical cutoff points after which time interventions might be contraindicated?

Even the most general review of the physiologic consequences of SCI would lead to several key indicators of physiologic decline. These include, but are not limited to, changes in muscle mass, bone loss, orthopedic integrity, and cardiovascular function. We provide an overview of the evidence for each of these. There are also more general concepts, which are indicators of physiologic decline that are linked to multiple indicators of decline or accelerated aging. Two such prominent constructs, allostatic load (AL) and telomeres, are discussed later in this chapter.

Injury severity and time since injury are two key factors that may be mediators of decline. Injury severity certainly would be associated with differential outcomes in many of these areas. It might also be argued that the more severe the injury, the lower the probability of successfully intervening to promote functional recovery (i.e., may have further to go to reach recovery). On the other hand, it could be argued that those with more severe injuries are in greater need of the intervention and that the risk/benefit ratio is better with more severe injuries because of the potential for larger gains in function. Regardless of which perspective is taken, it is also important to remember that injury severity is difficult to quantify in some respects, and qualitative differences as a function of type of injury also exist. For instance, individuals with higher level injuries often retain spasticity which is associated with retention of muscle tone and this could actually be an advantage in terms of readiness.

We turn our attention to the time variable as it is more clearly related to physiologic decline and, therefore, intervention readiness. The permanence or potential reversibility of the physiologic changes are also an important consideration, yet clearly more difficult to project in the absence of current interventions to promote recovery.

❯ *Figure 24-1* demonstrates several potential relationships between physiologic decline and time since injury. Consider the example of an individual who is injured at the age of 20 and has now reached 50 years of age or 30 years post-injury. Four hypothetical decline functions are projected on a graph, each of which varies by: (1) the time post-injury at which the decline first occurs, (2) the type of decline function (i.e., linear, curvilinear), (3) if curvilinear, whether the decline is accelerated or decelerated, and (4) the slope of the decline. Each hypothetical function could represent a single aspect of physiologic decline, such as a change in bone density or muscle mass, or could represent a cumulative line that is the average of all important parameters (i.e., overall decline).

Line A summarizes a function where physiologic decline is delayed and minimal until it begins to decline gradually about 25 years post-injury. Under this circumstance, intervention readiness would be high and there would be no need for early intervention or reason that years lived post-injury would preclude an intervention.

Line B represents a linear function where the physiologic factor declines steadily with years lived post-injury. Under this scenario, the importance of years lived post-injury would depend on the slope of the line (i.e., the rate of decline). However, interventions could occur at any point with proportional loss of function over time prior to initiation of the intervention.

Line C represents the type of function where intervention readiness is compromised relatively quickly after SCI onset. Decline occurs rather rapidly after onset of SCI, then plateaus for a number of years where

Figure 24-1
Potential relationships between physiologic decline and time since injury

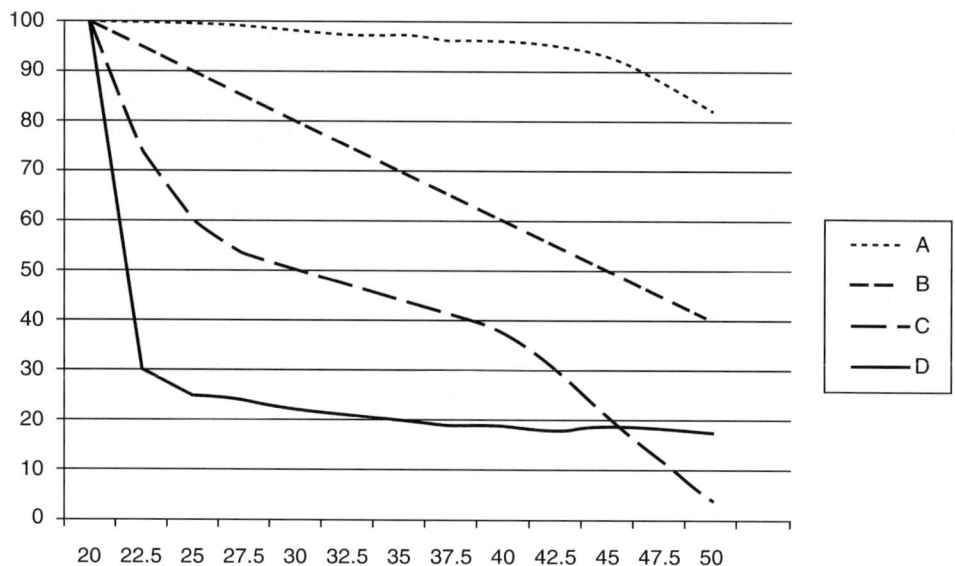

decline is minimal, followed by additional declines thereafter. This type of relationship between preserved function and years lived post-injury is complex, but possible.

Lastly, line D represents the worst-case scenario for intervention readiness, yet is entirely consistent with some types of loss of function after SCI (e.g., skeletal muscle atrophy; Gorgey and Dudley, 2006). It represents the case where the decline is immediate and profound. Under this scenario, with the exception of immediate intervention after SCI onset, there would be significant concern with any intervention and the possibility of complications related to that parameter after initiation of the intervention (i.e., there would be adverse intervention impact). Of course, if the condition were reversible, then intervention readiness may not be compromised.

6 Functional Decline and Intervention Readiness

In the following paragraphs we have summarized literature regarding several physiologic parameters related to intervention readiness. This should not be taken as an exhaustive list of potential complications, nor should the literature review be taken as comprehensive or definitive. Our goal is to outline some key research and how it relates to the concepts of intervention readiness and intervention impact.

6.1 Musculo-Skeletal Changes After SCI

SCI leads to significant alteration in the proportion of lean muscle and adipose (fat) tissue in the overall body composition of individuals with SCI compared with able-bodied (AB) individuals. Skeletal muscle atrophy occurs rapidly after SCI due to the loss of central activation and the ensuing unloading of mechanical forces below the level of injury (Gorgey and Dudley, 2006). In individuals who were only 6 weeks post-SCI, average muscle cross sectional areas (CSAs) were 18–46% lower than in controls. Prospective evaluation of these individuals, up to 24 weeks post-injury, revealed further declines in average gastrocnemius and soleus muscle CSAs of 24 and 12%, respectively. Likewise, from 6 to 24 weeks post-injury, the average decreases in quadriceps, hamstrings, and adductor muscle CSAs were 16, 14, and 16%, respectively (Castro et al., 1999). A prospective study utilizing dual energy X-ray absorptiometry (DEXA) revealed a 15% lower extremity lean tissue mass decrease by 1 year after injury onset. (Wilmet et al., 1995). Independent of age, total body and regional lean mass was lower, and fat mass was found to be higher in individuals with SCI compared with AB controls. Advancing age was found to be strongly related to less lean mass and more fat mass in individuals with SCI and minimally related to the same measures in AB individuals (Spungen et al., 2003). Modlesky et al. (2004) noted approximately 15% less muscle mass in the fat free soft tissue of men with long-term, complete SCI compared with matched able bodied controls. The accumulation of intramuscular fat (IMF) has been demonstrated to occur at a much higher rate after SCI as compared with AB controls. IMF has been found to be strongly correlated with glucose intolerance in the SCI population. Body composition changes after SCI may be a contributing factor associated with adverse consequences such as the development of metabolic syndrome and Type 2 diabetes (Gorgey and Dudley, 2006).

Loss of bone density is a major complication after SCI, placing the individual at increased risk for fractures. Bone loss after SCI is site-specific, with the largest deficiency apparent in the lower limbs. Significant differences have been noted in upper extremity bone status when comparisons were made between individuals with cervical and noncervical injuries. Upper extremity loss is only noted with cervical injuries (Demirel et al., 1998). Imaging technology most often used to measure the impact of osteoporosis in persons with SCI is DEXA. The "gold standard" for defining degrees of osteoporosis is DEXA whose standards were set by the World Health Organization (WHO). These definitions are based on being below the population mean for bone density as measured by standard deviations or t-scores. Osteoporosis is defined as bone with a t-score <-2.5. (Kirk et al., 2002). Measurement of bone structure in individuals with SCI requires consideration of potential confounding variables, such as heterotopic ossification (HO), that may falsely elevate bone mineral density (BMD) as measured by DEXA (Jaovisidha et al., 1998; Liu et al., 2000). Heterotopic ossification, formation of bony tissue in abnormal locations, is a common orthopedic complication after SCI causing variable degrees of functional limitation in affected joints (Banovac and Gonzalez, 1997; Banovac et al., 2004). Incidence of HO after SCI is estimated at 40–50%, and in 10–20% of these patients, HO develops in a more severe form, which usually needs surgical treatment. (Banovac et al., 2004).

Bone loss begins within the first few days after the onset of SCI. Rapid bone loss continues in the first 4–6 months after injury with approximately 4% loss of the trabecular bone per month and 2% loss of cortical bone per month (Wilmet et al., 1995). A prospective study of 30 subjects with SCI, followed for 6 months from onset of injury, demonstrated alteration in biochemical markers of bone resorption in the first week post-injury, gradually peaking around 10–16 weeks at levels that were as high as 10 times the upper limits of normal for these markers (Jiang et al., 2006). Presently, there is insufficient evidence to determine whether a "steady-state" of bone formation and resorption occurs after SCI (Demirel et al., 1998; Bauman et al., 1999c). Earlier studies indicated that the rate of bone loss appears to level off during the first 12–18 months post-injury (Garland et al., 1992; Garland et al., 1993). However, significant bone loss has been reported many years after SCI indicating that bone loss may not plateau as previously suggested (Sabo et al., 2001; de Bruin et al., 2005).

Sublesional osteoporosis places individuals with SCI at a greater risk of low-trauma fracture (Giangregorio and McCartney, 2006). Low-energy fractures during events that would not normally cause fracture, such as a wheelchair transfer or being turned in bed, have been reported in individuals with SCI (Vestergaard et al., 1998). Fracture prevalence has been reported to increase with time post-SCI, from 1% in the first 12 months to 4.6% in individuals 20 years post-injury (Zehnder et al., 2004). Specific criteria for assessing fracture risk in individuals with SCI have yet to be defined via prospective studies (Giangregorio and McCartney, 2006). The risk of fracture for individuals with SCI who participate in activities such as FES, standing frames, and treadmill walking has not been studied extensively (Giangregorio and McCartney, 2006). A case report documented a femoral fracture that occurred during measurement of maximal isometric quadriceps torque using electrical stimulation (Hartkopp et al., 1998). Incidents such as this emphasize the need for a comprehensive evaluation of risk for fracture, performed by a qualified practitioner, before initiation of activities that involve the application of mechanical forces of any type to the lower limbs of individuals with SCI (Giangregorio and McCartney, 2006).

The exact etiology of bone density loss after SCI is not known and the changes seen are most likely the result of many combined factors (Giangregorio and McCartney, 2006). The extent of bone loss appears to be associated with the degree of immobility and the time post-injury (Kannisto et al., 1998; Dauty et al., 2000). Individuals with neurologically complete SCI lose bone more than individuals with incomplete injuries (Garland et al., 1994; Demirel et al., 1998). A cross-sectional study demonstrated that BMD in individuals with SCI was positively correlated with mobility assessed with a mobility index ranging from complete paralysis to unlimited ambulation (Saltzstein et al., 1992).

A number of interventions, including standing, electrically stimulated cycling or resistance training, and walking exercises have been explored with the aim of reducing bone loss and/or increasing bone mass and muscle mass in individuals with SCI (Giangregorio and McCartney, 2006). FES is a common therapeutic modality employed with individuals with SCI to produce isometric contractions, assist gait, or produce contractions against resistance during cycling or leg extensions. Despite inconsistency in the intensity, duration, and frequency of the exercise interventions, the positive effects of FES exercise on muscle are fairly well-recognized (Hjeltnes et al., 1997; Baldi et al., 1998; Dudley et al., 1999; Scremin et al., 1999). In contrast to a reported positive influence on muscle, the effects of FES exercise on bone in acute and chronic SCI are inconclusive (Giangregorio and McCartney, 2006). Several studies have demonstrated no effect of FES-induced strengthening or cycle ergometry on measures of bone health (Leeds et al., 1990; BeDell et al., 1996; Eser et al., 2003), while others have demonstrated increases in bone mass after FES-induced muscle strengthening (Belanger et al., 2000) and FES cycle ergometry (Mohr et al., 1997; Chen et al., 2005). The studies demonstrating improvement in bone mass as a result of FES were longer in duration than the other studies and employed a higher frequency of exercise.

There are few published studies that report the effect of standing or walking interventions on muscle mass after SCI (Giangregorio and McCartney, 2006). A recent case series reported increases in lean mass and muscle area in individuals with acute SCI (defined as less than 1 year post-injury) as a result of early weight bearing via body weight-supported treadmill training (Giangregorio et al., 2005). Stewart and colleagues (2004) reported that body weight-supported treadmill training 3 times per week for 6 months produced increases in muscle fiber size and a shift toward a less fatigable fiber type.

Studies of changes in bone mass associated with weight-bearing activities after SCI are also limited. Individuals with SCI who participated in 12–20 weeks of training with an ambulation device that combined FES and a modified walker (Needham-Shropshire et al., 1997), or participated in regular standing (with a standing frame) (Kunkel et al., 1993) did not experience changes in BMD. No exercise intervention has demonstrated restoration of bone micro-architecture after it has been lost. Currently, there are no interventions that have consistently demonstrated efficacy for preventing or reversing the dramatic bone loss occurring after SCI that can easily be implemented in the clinical setting. Studies evaluating the effectiveness of pharmacologic interventions, including calcium, calcitonin, or biphosphonates, to prevent osteoporosis after SCI have been inconsistent as well (Jiang et al., 2006).

6.2 Cardiovascular Function

Abnormalities of the cardiovascular system have become a major concern for individuals with chronic SCI (Myers et al., 2007). Morbidity from cardiovascular causes, chiefly coronary artery disease (CAD), is high relative to able-bodied individuals. CAD also tends to occur earlier in individuals with SCI than among nondisabled populations (DeVivo et al., 1992; Garshick et al., 2005). Whiteneck and colleagues (1992) reported that cardiovascular disease (CVD) was the leading cause of mortality in persons with SCI of more than 30-year duration. An evaluation of data from the Model Systems indicated that CVD accounted for 20.6% of all deaths, falling just behind respiratory complications (22%) as the leading cause of death (DeVivo et al., 1999). A more recent prospective study of mortality after chronic SCI indicated the most common underlying and contributing causes of death were diseases of the circulatory system (Garshick et al., 2005). A study evaluating the relationship between neurological level of injury and symptomatic CVD risk in a sample of 545 subjects with SCI, at least 25 years post-injury, revealed that the risk of developing CVD was associated with both the level and extent of injury(Groah et al., 2001). Tetraplegia was related to a

16% higher risk of all CVD (CAD, hypertension, cerebrovascular disease, valvular disease, and dysrhythmias). Complete injury imparted a 44% greater risk of overall CVD (Groah et al., 2001).

A key contributor to the heightened risk of CVD in the presence of SCI is the fact that risk factors, including abnormal lipid profiles, obesity, and diabetes, have been shown to be comparatively high among individuals with SCI (Bauman et al., 1999a; Demirel et al., 2001; Lee et al., 2005; LaVela et al., 2006). SCI is also characterized by a disruption of the normal autonomic cardiovascular control mechanisms. Loss of autonomic control can result in a variety of physiologic changes related to cardiovascular control that are observed in SCI, including loss of normal regulation of the peripheral vasculature, autonomic dysreflexia (AD), and a higher prevalence of cardiac rhythm disturbances (Mallory,1994; Bauman et al., 1999b). The autonomic dysfunction that occurs with SCI also underlies several cardiovascular irregularities, including an accelerated decline in cardiovascular function with aging, that contribute to CVD (Myers et al., 2007).

The reduced physical function associated with SCI, an additional contributing factor to high cardiovascular morbidity and mortality, leads to a greater sedentary lifestyle and lower energy expenditure (Jacobs and Nash, 2004; Myers et al., 2007). The level of the physiologic response to exercise is reduced in SCI; thus, altered cardiovascular response to activity most likely reduces the well-known gains achieved by able-bodied individuals engaged in regular physical activity (Washburn and Figoni, 1998). Abnormal cardiovascular responses to arm cycling exercise and transient postexercise hypotension were seen in cervical, but not thoracic, SCI and may be partly related to loss of descending sympathetic nervous control of the heart and vasculature following cervical level injuries (Claydon et al., 2006). Loss of sympathetic tone also causes reduced myocardial preload and myocardial contractility, resulting in reduced stroke volume and cardiac output. Chronically, these changes can lead to cardiac atrophy. Left ventricular myocardial atrophy and diminished cardiac function have been shown to develop in the presence of chronic tetraplegia (Nash et al., 1991). These factors contribute significantly to a reduced capacity to adapt effectively to exercise, resulting in early fatigue, general avoidance of physical exertion, and overall cardiac deconditioning (Washburn and Figoni, 1998; Jacobs and Nash, 2004). The cardiac deconditioning associated with chronic SCI may seriously compromise intervention readiness of the individual and the impact of interventions requiring even minimal physical exertion in the effort to facilitate functional recovery.

7 Allostatic Load and Telomeres

7.1 Allostatic Load

Investigators have historically focused on biomarkers of health after SCI using either a single measure or multiple parameters targeted to a single area of loss, such as bone density, body composition, or cardiovascular risk (Tsuzuku et al., 1999; Bauman et al., 1999b; Spungen et al., 2003). Given the general absence of studies utilizing multiple biomarkers and the absence of theories of physiologic decline, it is not surprising that there is also an absence of integration of behavioral, psychological, or coping theories with physiologic decline after SCI. Allostatic load is a summary of biomarkers, generally 10, that has the promise of providing a comprehensive means to assess risk of morbidity and mortality after SCI (McEwen and Stellar, 1993; Seeman et al., 2001; McEwen and Wingfield, 2003). It may also represent the single most important index in determining intervention readiness.

The concept of AL was developed to provide a framework for clarifying the concept of stress by defining the cumulative, lifelong, physiologic decline associated with prolonged or repeated exposures to stress over time. Two concepts – allostasis and allostatic load – relate to adaptation and the cost of adaptation for the body and brain (McEwen, 1998). The construct of "allostasis" was first developed in an attempt to understand the *physiological* basis for incongruent patterns of morbidity and mortality that could not be expained by socioeconomic status, accessibility of health care, or lifestyle choices. Allostasis, a process of healthy functioning, requires constant adjustments of the internal physiologic environment, with physiologic systems demonstrating varying levels of activity with response and adaptation to environmental demands (Sterling and Eyer, 1988). From a perspective of the health and well-being of the individual, the most important aspect of mediators associated with allostasis is that they have protective effects for limited periods of time. Conversely, these mediators can have damaging effects over extended periods of activation, as in a continued allostatic state that leads to allostatic overload (McEwen and Wingfield, 2003).

McEwen and Stellar (1993) proposed the construct of "allostatic load" to express the cumulative impact of physiologic "wear and tear" over time that could predispose organisms to disease. A fundamental assumption is that the allostatic load of any organism will increase over time. This implies that a resilient organism, with adaptive flexibility, will be able to minimize physiologic deterioration (Carlson and Chamberlain, 2005). However, as with theories of stress-coping, there are important individual differences in ability to deal with stressful events, such that we would anticipate that those who successfully or more quickly adapt to stressors will accumulate AL at a slower rate.

Measurement of AL was initially operationalized to reflect levels of physiologic activity across the hypothalamic–pituitary–adrenal axis (HPA), the sympathetic nervous system (SNS), cardiovascular system, and metabolic processes through a set of ten biomarkers, each having been previously linked to increased risk for pathology. The summary index of AL predicted mortality as well as risk for CVD and declines in cognitive and physical functioning over time (Seeman et al., 2004). The ten traditional biomarkers of AL include: (1) waist to hip circumference ratio (WHR), (2) systolic blood pressure (SBP), (3) diastolic blood pressure (DBP), (4) urinary cortisol, (5) urinary norepinephrine (NE), (6) urinary epinephrine (EPI), (7) serum dehydroepiandosterone (DHEA), (8) glycosylated hemoglobin (Hb_{A1c}), (9) HDL cholesterol, and (10) total serum cholesterol. The AL score was initially calculated as a summary measure (0–10) of the ten measurements for which the individual is in the highest risk quartile. None of the ten components of AL proved to have significant associations with health outcomes on their own; however, the summary measure was found to be significantly correlated with new cardiovascular events, decline in cognitive functioning, decline in physical functioning, and mortality over both 2.5 and 7-year follow-ups (Seeman et al., 1997, Seeman et al., 2001). Additional research has demonstrated that an increase in AL score over time is related to increased all-cause mortality risk compared to individuals whose AL score decreased over the same time period. Each unit increment in the AL change score was associated with mortality odds ratio of 3.3 (95% CI, 1.1–9.8), adjusted for age and baseline AL (Karlamangla et al., 2006).

SCI has consistently been associated with increased morbidity and decreased life expectancy. Recent research has indicated that CVD is quickly emerging as the leading cause of mortality in chronic SCI (Garshick et al., 2005), surpassing renal and pulmonary conditions, the primary causes of mortality in previous decades (Myers et al., 2007). A comprehensive review of studies evaluating various risk factors for CVD (Myers et al., 2007) revealed that risk factors, including diabetes, obesity, and hyperlipidemia, have been shown to have higher incidence among individuals with SCI when compared with able-bodied controls. Dyslipidemias and glucose intolerance are common among individuals with SCI and contribute to cardiovascular risk (Tharion et al., 1998). Elevated LDL cholesterol and total cholesterol, as well as lower HDL cholesterol, occur more frequently among individuals with SCI as compared with able-bodied persons, with the most unfavorable lipid profiles correlating with the severity of neurological deficit. Persons with tetraplegia were found to have a greater number of lipid abnormalities than persons with paraplegia (Bauman et al., 1999a; Bauman et al., 1999b; Demirel et al., 2001). Metabolic syndrome (MS) and insulin resistance (IR) are also disproportionately high in the SCI population. Lee and colleagues (2005) noted a prevalence of MS and IR in nearly 25% of their sample of individuals with SCI. Biomarkers of inflammation, such as CRP and IL-6, are commonly present at elevated levels in individuals with SCI and appear to be related to the use of indwelling urinary catheters, the presence of pressure ulcers, or heterotopic ossification (Segal et al., 1997; Estores et al., 2004; Scivoletto et al., 2004; Frost et al., 2005; Lee et al., 2005). Manns and colleagues (2005) noted that lower physical activity levels in paraplegic men were associated with higher fasting glucose, lower HDL-C levels, and larger abdominal sagittal diameter (ASD), which in turn was associated with higher triglycerides and higher CRP.

The concepts of allostasis and AL are highly significant to SCI, as SCI is associated with significant lifelong stressors, both physiological and psychological, and prolonged exposure to situations where there is substantial health risk. Events that may be relatively minor to an able-bodied person may become substantial health risks, even life-threatening, for individuals with SCI, within a short period of time. The stressors on the average person tend to be transitory, where as those on a person with SCI are constant. From the perspective of allostasis and AL, people with SCI experience repeated, almost constant, demands to maintain allostasis with repeated cycles that increase AL over time. This is a fertile area for future investigation.

7.2 Telomeres

The biomarkers that are used to define insulin resistance, oxidative stress, and inflammation in a clinical setting are most indicative of the metabolic and inflammatory status of the individual at the time of the sample collection. This prompts consideration of whether there is a biomarker that better portrays the cumulative lifelong burden of oxidative stress and inflammation, which are considered to be major determinants of lifespan as key elements of age related diseases. Leukocyte telomere length appears to be such a biomarker and may provide a better appraisal of the cumulative burden of oxidative stress. (Blasco, 2005; Demissie et al., 2006).

Telomeres are DNA–protein complexes that cap chromosomal ends, promoting chromosomal stability. When cells divide, the telomere is not fully replicated because of limitations of the DNA polymerases in finishing the replication of the ends of the linear molecules, leading to telomere shortening with every replication (Chan and Blackburn, 2004). Telomeres shorten with age in all replicating somatic cells that have been examined, including fibroblasts and leukocytes (Frenck et al., 1998). Consequently, telomere length can serve as a biomarker of a cell's biological (vs. chronological) "age" (Epel et al., 2004).

Age-related diseases are characterized by short telomeres, which can compromise cell viability. This loss of cell viability associated with telomere shortening is thought to contribute to the onset of diseases of aging through mechanisms involving oxidative stress, inflammation, and progression to CVD (Blasco, 2005; Fitzpatrick et al., 2007). There is evidence that psychological stress – both perceived stress and chronicity of stress – is significantly associated with higher oxidative stress, lower telomerase activity, and shorter telomere length (Epel et al., 2004). Chronic stress associated with mood disorders may contribute to increased susceptibility for diseases of aging such as CVD (Simon et al., 2006). Individuals in lower SES experience more exposure to risk factors that accelerate telomere shortening which is compatible with the concept that SES influences longevity and health (World Health Organization, 2002; Adams and White, 2004; Cherakas et al., 2006).

Presently, there are no published investigations of telomere length in a cohort with SCI. In light of the established existence of elevated levels of stress, decreased employment and lower SES, and accelerated morbidity and mortality in individuals with SCI, investigations of telomere length is certainly warranted.

8 Conclusion

8.1 Future Directions

In this chapter, we have focused on critical issues of the translation of new discoveries into viable interventions at the human level by identifying potential limitations of the applicability of these interventions to chronic SCI. In doing so, we have highlighted issues related to intervention readiness and presented a summation of studies that have indicated clear barriers to implementation of new interventions to enhance function.

> Should we be discouraged? No.
> Should we be prepared? Yes.
> Then, how do we proceed?

Studies of intervention readiness now will help us to predict the future impact of translational research to enhance functional recovery. The body of research on complications associated with chronic SCI represents a giant step in this direction. However, an even greater opportunity for research lies in the population of people living with motor functional SCI, particularly ambulatory SCI. Understanding the long-term problems faced by this population is in essence a "natural experiment" that will help us to prepare for the parallel problems that will inevitably follow translation of successful interventions to restore function. This is a unique opportunity to glimpse the future of SCI. Unfortunately, we have not taken advantage of this opportunity, as minimal research has been initiated to help us understand the problems of this population, particularly in contrast to the ever growing body of research on short-term interventions to promote ambulation. It would be shortsighted indeed if we were to overlook this opportunity. Failure to perform the research required to help us fully understand the complications that inevitably will follow may result in a rehabilitation system unprepared to deal with SCI, as a vital link in the circle of translational research will be broken.

When do we begin to proceed? Now.

References

Adams J, White M. 2004. Biological ageing. A fundamental, biological link between socioeconomic status and health? Eur J Pub Health 14: 331-334.

Baldi JC, Jackson RD, Moraille R, Mysiw WJ. 1998. Muscle atrophy is prevented in patients with acute spinal cord injury using functional electrical stimulation. Spinal Cord 36(Suppl 7): 463-469.

Banovac K, Gonzalez F. 1997. Evaluation and management of heterotopic ossification in patients with spinal cord injury. Spinal Cord 35: 158-162.

Banovac K, Williams J, Patrick L, Levi A. 2004. Prevention of heterotopic ossification after spinal cord injury with COX-2 selective inhibitor (rofecoxib). Spinal Cord 42: 707-710.

Bauman WA, Adkins RH, Spungen AM, Waters R. 1999a. The effect of residual neurological deficit on oral glucose tolerance in persons with chronic spinal cord injury. Spinal Cord 37: 765-771.

Bauman WA, Kahn NN, Grimm DR, Spungen AM. 1999b. Risk factors for atherogenesis and cardiovascular autonomic function in persons with spinal cord injury. Spinal Cord 37: 601-616.

Bauman W, Spungen A, Wang J, Pierson R, Schwartz E. 1999c. Continuous loss of bone during chronic immobilization: A monozygotic twin study. Osteoporosis International 10: 123-127.

BeDell KK, Scremin AM, Perell KL, Kunkel CF. 1996. Effects of functional electrical stimulation-induced lower extremity cycling on bone density of spinal cord-injured patients. Am J Phys Med Rehabil 75(Suppl 1): 29-34.

Behrman AL, Harkema SJ. 2000. Locomotor training after human spinal cord injury. Phys Ther 80: 688-700.

Belanger M, Stein RB, Wheeler GD, Gordon T, Leduc B. 2000. Electrical stimulation: Can it increase muscle strength and reverse osteopenia in spinal cord injured individuals? Arch Phys Med Rehabil 81(Suppl 8): 1090-1098.

Blasco M. 2005. Telomeres and human disease: Ageing, cancer and beyond. Nat Rev Genet 6: 611-622.

Brotherton SS, Krause JS, Nietert PJ. 2007. Falls in individuals with incomplete spinal cord injury. Spinal Cord 45: 37-40.

Carlson E, Chamberlain R. 2005. Allostatic load and health disparities: A theoretical orientation. Res Nurs Health 28: 306-315.

Castro MJ, Apple DF Jr, Hillegass EA, Dudley GA. 1999. Influence of complete spinal cord injury on skeletal muscle cross sectional area within the first 6 months of injury. Eur J Appl Physiol Occup Physiol 80(Suppl 4): 373-378.

Chan S, Blackburn E. 2004. Telomeres and telomerase. Philos Trans R Soc Lond B 359: 109-121.

Chen SC, Lai CH, Chan WP, Huang MH, Tsai HW, et al. 2005. Increases in bone mineral density after functional electrical stimulation cycling exercises in spinal cord injured patients. Disabil Rehabil 27(Suppl 22): 1337-1341.

Cherakas L, Aviv A, Valdes A, Hunkin J, Gardner J, et al. 2006. The effect of social status on biological aging as measured by white-blood-cell telomere length. Aging Cell 5: 361-365.

Claydon VE, Hol AT, Eng JJ, Krassioukov AV. 2006. Cardiovascular responses and postexercise hypotension after arm cycling exercise in subjects with spinal cord injury. Arch Phys Med Rehabil 87: 1106-1114.

Dalakas M, Elder G, Hallett M, et al. 1986. A long-term follow-up study of patients with post-poliomyelitis neuromuscular symptoms. N Engl J Med 314: 959-963.

Dauty M, Perrouin VB, Maugars Y, Dubois C, Mathe JF. 2000. Supralesional and sublesional bone mineral density in spinal cord-injured patients. Bone 27(Suppl 2): 305-309.

de Bruin ED, Vanwanseele B, Dambacher MA, Dietz V, Stussi E. 2005. Long-term changes in the tibia and radius bone mineral density following spinal cord injury. Spinal Cord 43(Suppl 2): 96-101.

Demirel G, Yilmaz H, Parker N, Onel S. 1998. Osteoporosis after spinal cord injury. Spinal Cord 36: 822-825.

Demirel S, Demirel G, Tukek T, Erk O, Yilmaz H. 2001. Risk factors for coronary heart disease in patients with spinal cord injury in Turkey. Spinal Cord 39: 134-138.

Demissie S, Levy D, Benjamin E, Cupples L, Gardner J, et al. 2006. Insulin resistance oxidative stress, hypertension, and leukocyte telomere length in men from the Framingham Heart Study. Aging Cell 5: 325-330.

DeVivo MJ, Shewchuk RM, Stover SL, Black K, Go B. 1992. A cross sectional study of the relationship between age and current health status for persons with spinal cord injuries. Paraplegia 30: 820-827.

DeVivo M, Krause J, Lammertse D. 1999. Recent trends in mortality and causes of death in people with spinal cord injury. Arch Phys Med Rehab 80: 1411-1419.

Ditunno JF Jr, Bituanno PL, Graziani V, Scieoletto G, Bernardi M, et al. 2000. Walking index for spinal cord injury (WISCI): An international multicenter validity and reliability study. Spinal Cord 38: 234-243.

Dudley GA, Castro MJ, Rogers S, Apple DF Jr. 1999. A simple means of increasing muscle size after spinal cord injury: A pilot study. Eur J Appl Physiol Occup Physiol 80(Suppl 4): 394-396.

Epel E, Blackburn E, Lin J, Dhabhar F, Adler N, et al. 2004. Accelerated telomere shortening in response to life stress. Proc Natl Acad Sci USA 101: 17312-17315.

Eser P, de Bruin ED, Telley I, Lechner HE, Knecht H, et al. 2003. Effect of electrical stimulation-induced cycling on bone mineral density in spinal cord-injured patients. Eur J Clin Invest 33(Suppl 5): 412-419.

Estores I, Harrington A, Banovac K. 2004. C-Reactive protein and erythrocyte sedimentation rate in patients with heterotopic ossification after spinal cord injury. J Spinal Cord Med 27: 434-437.

Field-Fote EC. 2001. Combined use of body weight support, functional electric stimulation, and treadmill training to improve walking ability in individuals with chronic incomplete spinal cord injury. Arch Phys Med Rehabil 82: 818-824.

Field-Fote EC, Fluet GG, Schafer SD, Schneider EM, Smith R, et al. 2001. The spinal cord injury functional ambulation inventory (SCI-FAI). J Rehabil Med 33: 177-181.

Field-Fote EC, Tepavac D. 2002. Improved intralimb coordination in people with incomplete spinal cord injury following training with body weight support and electrical stimulation. Phys Ther 82: 707-715.

Finch L, Barbeau H, Arsenault B. 1991. Influence of body weight support on normal human gait: Development of a gait retraining strategy. Phys Ther 71: 842-855.

Fitzpatrick A, Kronmal R, Gardner J, Psaty B, Jenny N, et al. 2007. Leukocyte telomere length and cardiovascular disease in the cardiovascular health study. Am J Epidemiol 165: 14-21.

Fouad K, Pearson K. 2004. Restoring walking after spinal cord injury. Prog Neurobiol 73: 107-126.

Frenck R, Blackburn E, Shannon K. 1998. The rate of telomere sequence loss in human leukocytes varies with age. Proc Natl Acad Sci USA 95: 5607-5610.

Frost F, Roach M, Kushner I, Screiber P. 2005. Inflammatory C-reactive protein and cytokine levels in asymptomatic people with chronic spinal cord injury. Arch Phys Med Rehabil 86: 312-317.

Garland DE, Foulkes GD, Adkins RH, Stewart CA, Yakura JS. 1994. Regional osteoporosis following incomplete spinal cord injury. Contemp Orthop 28(Suppl 2): 134-139.

Garland DE, Maric Z, Adkins RH, Stewart CA. 1993. Bone mineral density about the knee in spinal cord injured patients with pathologic fractures. Contemp Orthop 26(Suppl 4): 375-379.

Garland DE, Stewart CA, Adkins RH, et al. 1992. Osteoporosis after spinal cord injury. J Orthop Res 10: 371-378.

Garshick E, Kelley A, Cohen S, Garrison A, Tun C, et al. 2005. A prospective assessment of mortality in chronic spinal cord injury. Spinal Cord 43: 408-416.

Giangregorio LM, Hicks AL, Webber CE, et al. 2005. Body weight supported treadmill training in acute spinal cord injury: Impact on muscle and bone. Spinal Cord 43(Suppl 11): 649-657.

Giangregorio L, McCartney N. 2006. Bone loss and muscle atrophy in spinal cord injury: Epidemiology, fracture prediction, and rehabilitation strategies. J Spinal Cord Med 29: 489-500.

Gorgey A, Dudley G. 2006. Skeletal muscle atrophy and increased intramuscular fat after incomplete spinal cord injury. Spinal Cord 45: 304-309.

Groah SL, Weitzenkamp D, Sett P, Soni B, Savic G. 2001. The relationship between neurological level of injury and symptomatic cardiovascular disease risk in the aging spinal injured. Spinal Cord 39: 310-317.

Harkema S, Hurley SL, Patel UK, Requejo PS, Dobkin BH, et al. 1997. Human lumbosacral spinal cord interprets loading during stepping. J Neurophysiol 77: 797-811.

Hartkopp A, Murphy R, Mohr T, Kjoer M, Biering-Sorensen F. 1998. Bone fracture during electrical stimulation of the quadriceps in a spinal cord injured subject. Arch Phys Med Rehabil. 79: 1133-1136.

Harvey LA, Newton-John T, Davis GM, Smith MB, Engel S. 1997. A comparison of the attitude of paraplegic individuals to the walkabout orthosis and the isocentric reciprocal gait orthosis. Spinal Cord 35: 580-584.

Hicks AL, Adams MM, Martin Ginis K, Giangregorio L, Latimer A, et al. 2005. Long-term body-weight-supported treadmill training and subsequent follow-up in persons with chronic SCI: Effects on functional walking ability and measures of subjective well-being. Spinal Cord 43: 291-298.

Hjeltnes N, Aksnes AK, Birkeland KI, Johansen J, Lannem A, et al. 1997. Improved body composition after 8 wk of electrically stimulated leg cycling in tetraplegic patients. Am J Physiol. 273(Suppl 3, pt 2): R1072-R1079.

Hornby TG, Zemon DH, Campbell D. 2005. Robotic-assisted, body-weight-supported treadmill training in individuals following motor incomplete spinal cord injury. Phys Ther 85: 52-66.

Jacobs PL, Johnson B, Mahoney ET. 2003. Physiologic responses to electrically assisted and frame-supported standing in persons with paraplegia. J Spinal Cord Med 26: 384-389.

Jacobs PL, Nash MS. 2004. Exercise recommendations for individuals with spinal cord injury. Sports Med 34: 727-751.

Jaovisidha S, Sartoris D, Martin E, Foldes K, Szollar S, et al. 1998. Influence of heterotopic ossification of the hip on bone densiometry: A study of spinal cord injured patients. Spinal Cord 36: 647-653.

Jiang S, Dai L, Jiang L. 2006. Osteoporosis after spinal cord injury. Osteoporos Int 17: 180-192.

Johnston TE, Betz RR, Smith BT, Mulcahey MJ. 2003. Implanted functional electrical stimulation: An alternative for standing and walking in pediatric spinal cord injury. Spinal Cord 41: 144-152.

Kannisto M, Alaranta H, Merikanto J, Kroger H, Karkkainen J. 1998. Bone mineral status after pediatric spinal cord injury. Spinal Cord 36(Suppl 9): 641-646.

Karlamangla A, Singer B, Seeman T. 2006. Reduction in allostatic load in older adults is associated with lower all-cause mortality risk: MacArthur Studies of Successful Aging. Psychosomatic Medicine 68: 500-507.

Kawashima N, Sone Y, Nakazawa K, Akai M, Yano H. 2003. Energy expenditure during walking with weight-bearing control (WBC) orthosis in thoracic level of paraplegic patients. Spinal Cord 41: 506-510.

Kim CM, Eng JJ, Whittaker MW. 2004. Effects of a simple functional electric system and/or a hinged ankle-foot orthosis on walking in persons with incomplete spinal cord injury. Arch Phys Med Rehabil 85: 1718-1723.

Kirk JK, Nichols M, Spanger JG. 2002. Use of a peripheral DEXA measurement for osteoporosis screening. Fam Med 34: 201.

Krause J. 2004. Factors associated with risk for subsequent injuries after the onset of traumatic spinal cord injury. Arch Phys Med Rehab 85: 1503-1508.

Krause JS, Morrisette D, Brotherton S, Karakostas T, Apple D. 2007a. Pain interference in ambulatory spinal cord injury. Top Spinal Cord Inj Rehabil 12: 91-96.

Krause JS, Brotherton S, Morrisette D, Newman S, Karakostas T. 2007b. Does pain interference mediate the relationship of independence in ambulation with depressive symptoms after spinal cord injury? Rehabil Psychol 52: 162-169.

Kunkel CF, Scremin AM, Eisenberg B, Garcia JF, Roberts S, et al. 1993. Effect of "standing" on spasticity, contracture, and osteoporosis in paralyzed males. Arch Phys Med Rehabil 74(Suppl 1): 73-78.

Lapointe R, Lajoie Y, Serresse O, Barbeau H. 2001. Functional community ambulation requirements in incomplete spinal cord injured subjects. Spinal Cord 39: 327-335.

LaVela S, Weaver F, Goldstein B, Chen K, Miskevics S, et al. 2006. Diabetes in individuals with spinal cord injury or disorder. J Spinal Cord Med 29: 387-395.

Lee M, Myers J, Hayes A, Madan S, Froelicher V, et al. 2005. C-Reactive protein, metabolic syndrome, and insulin resistance in individuals with spinal cord injury. J Spinal Cord Med 28: 20-25.

Leeds EM, Klose KJ, Ganz W, Serafini A, Green BA. 1990. Bone mineral density after bicycle ergometry training. Arch Phys Med Rehabil. 71(Suppl 3): 207-209.

Liu CC, Theodorou DJ, Theodorou SJ, et al. 2000. Quantitativecomputed tomography in the evaluation of spinalosteoporosis following spinal cord injury. Osteoporos Int 11(Suppl 10): 889-896.

Lovely RG, Gregor RJ, Roy RR, Edgerton VR. 1986. Effects of training on the recovery of full-weight-bearing stepping in the adult spinal cat. Exp Neurol 92: 421-435.

Mallory B. 1994. Autonomic function in the isolated spinal cord. The Physiological Basis of Rehabilitation Medicine. Downey JA, Myers SJ, Gonzalez EG, Lieberman JS, editors. Boston, Butterworth-Heinemann; pp. 519-542.

Manns P, McCubbin J, Williams D. 2005. Fitness, inflammation, and the metabolic syndrome in men with paraplegia. Arch Phys Med Rehabil 86: 1176-1181.

McEwen BS. 1998. Stress, adaptation, and disease. Allostasis and allostatic load. Ann N Y Acad Sci 840: 33-44.

McEwen BS, Stellar E. 1993. Stress and the individual: Mechanisms leading to disease. Arch Int Med 153: 2093-2101.

McEwen BS, Wingfield J. 2003. The concept of allostasis in biology and biomedicine. Horm Behav 43: 2-15.

Mirbagheri MM, Ladouceur M, Barbeau H, Kearney RE. 2002. The effects of long-term fes-assisted walking on intrinsic and reflex dynamic stiffness in spastic spinal-cord-injured subjects. IEEE Trans Neural Syst Rehabil Eng 10: 280-289.

Modlesky CM, Bickel CS, Slade JM, Meyer RA, Cureton KJ, et al. 2004. Assessment of skeletal muscle mass in men with spinal cord injury using dual-energy X-ray absorptiometry and magnetic resonance imaging. J Appl Physiol 96: 561-565.

Mohr T, Podenphant J, Biering-Sorensen F, Galbo H, Thamsborg G, et al. 1997. Increased bone mineral density after prolonged electrically induced cycle training of paralyzed limbs in spinal cord injured man. Calcif Tissue Int 61(Suppl 1): 22-25.

Morganti B, Scivoletto G, Ditunno P, Ditunno JF, Molinari M. 2005. Walking index for spinal cord injury (WISCI): Criterion validation. Spinal Cord 43: 27-33.

Myers J, Lee M, Kiratli J. 2007. Cardiovascular disease in spinal cord injury: An overview of prevalence, risk, evaluation, and management. Am J Phys Med Rehabil 86: 142-152.

Nash MS, Bilsker S, Marcillo AE, Isaac SM, Botelho LA, et al. 1991. Reversal of adaptive left ventricular atrophy following electrically-stimulated exercise training in human tetraplegics. Paraplegia 29: 590-599.

Nash MS, Jacobs PL, Montalvo BM, Klose KJ, Guest RS, et al. 1997. Evaluation of a training program for persons with SCI paraplegia using the Parastep 1 ambulation system. V. Lower extremity blood flow and hyperemic responses to occlusion are augmented by ambulation training. Arch Phys Med Rehabil 78: 808-814.

National Spinal Cord Injury Statistical Center. 2006. Spinal Cord Injury: Facts and Figures at a Glance. Birmingham: University of Alabama.

Needham-Shropshire BM, Broton JG, Klose J, Lebwohl N, Guest RS, et al. 1997. Evaluation of a training program forpersons with SCI paraplegia using the Parastep 1Ambulation System. III. Lack of effect on bone mineral density. Arch Phys Med Rehabil 78: 799-803.

Protas EJ, Holmes SA, Qureshy H, Johnson A, Lee D, et al. 2001. Supported treadmill ambulation training after spinal cord injury: A pilot study. Arch Phys Med Rehabil 82: 825-831.

Sabo D, Blaich S, Wenz W, Hohmann M, Loew M, et al. 2001. Osteoporosis in patients with paralysis after spinal cord injury: A cross-sectional study in 46 male patients with dual-energy X-ray absorptiometry. Arch Orthop Trauma Surg 121(Suppl 1–2): 75-78.

Saitoh E, Suzuki T, Sonoda S, Fujitani J, Tomita Y, et al. 1996. Clinical experience with a new hip-knee-ankle-foot orthotic system using a medial single hip joint for paraplegic standing and walking. Am J Phys Med Rehabil 75: 198-203.

Saltzstein RJ, Hardin S, Hastings J. 1992. Osteoporosis in spinal cord injury: Using an index of mobility and its relationship to bone density. J Am Paraplegia Soc 15(Suppl 4): 232-234.

Scivoletto G, Fuoco U, Morganti B, Cosentino E, Molinari M. 2004. Pressure sores and blood and serum dysmetabolism in spinal cord injury patients. Spinal Cord 42: 473-476.

Scremin AM, Kurta L, Gentili A, Wiseman B, Perell K, et al. 1999. Increasing muscle mass in spinal cord injured persons with a functional electrical stimulation exercise program. Arch Phys Med Rehabil 80(Suppl 12): 1531-1536.

Seeman T, Crimmins E, Huang M, Singer B, Bucur A, et al. 2004. Cumulative biological risk and socio-economic differences in mortality: MacArthur studies of successful aging. Soc Sci Med 58: 1985-1997.

Seeman T, McEwen B, Rowe J, Singer B. 2001. Allostatic load as a marker of cumulative biological risk: MacArthur studies of successful aging. Proc Natl Acad Sci USA 98: 4770-4775.

Seeman T, Singer B, Rowe J, Horwitz R, McEwen B. 1997. Price of adaptation – allostatic load and its health consequences. Arch Int Med 157: 2259-2268.

Segal J, Gonzales E, Yousefi S, Jamishidipour L, Brunneman S. 1997. Circulating levels of IL-2R, ICAM-1, and IL-6 in spinal cord injuries. Arch Phys Med Rehabil 78: 44-47.

Simon N, Smoller J, McNamara K, Maser R, Zalta A, et al. 2006. Telomere shortening and mood disorders: Preliminary support for a chronic stress model of accelerated aging. Biol Psychiatry 60: 432-435.

Spungen A, Adkins R, Stewart C, Wang J, Pierson R, et al. 2003. Factors influencing body composition in persons with spinal cord injury: A cross-sectional study. J Appl Physiol 95: 2398-2407.

Sterling P, Eyer J. 1988. Allostasis: A new paradigm to explain arousal pathology. Handbook of Life Stress, Cognition, and Health. Fisher S, Reason J, editors. New York: John Wiley; pp. 629-651.

Stewart BG, Tarnopolsky MA, Hicks AL, McCartney N, Mahoney DJ, et al. 2004. Treadmill training-induced adaptations in muscle phenotype in persons with incomplete spinal cord injury. Muscle Nerve 30(Suppl 1): 61-68.

Strauss D, DeVivo M, Paculdo D, Shavelle R. 2006.Trends in life expectancy after spinal cord injury. Arch Phys Med Rehabil 87: 1079-1085.

Tharion G, Prasad K, Gopalan L, Bhattacharji S. 1998. Glucose intolerance and dyslipidemias in persons with paraplegia and tetraplegia in South India. Spinal Cord 36: 228-230.

Thoumie P, Perrouin-Verbe B, Le Claire G, Bedoiseau M, Busnel M, et al. 1995. Restoration of functional gait in paraplegic patients with the rgo-ii hybrid orthosis. A multicentre controlled study. I. Clinical evaluation. Paraplegia 33: 647-653.

Trimble MH, Behrman AL, Flynn SM, Thigpen MT, Thompson FJ. 2001. Acute effects of locomotor training on overground walking speed and h-reflex modulation in individuals with incomplete spinal cord injury. J Spinal Cord Med 24: 74-80.

Tsuzuku S, Ikegami Y, Yabe K. 1999. Bone mineral density differences between paraplegic and quadriplegic patients: A cross sectional study. Spinal Cord 37: 358-361.

Van Hedel HJ, Wirz MM, Dietz V. 2005. Assessing walking ability in subjects with spinal cord injury: Validity and reliability of 3 walking tests. Arch Phys Med Rehabil 86: 190-196.

Vestergaard P, Krogh K, Rejnmark L, Mosekilde L. 1998. Fracture rates and risk factors for fractures in patients with spinal cord injury. Spinal Cord 36: 790-796.

Washburn RA, Figoni SF. 1998. Physical activity and chronic cardiovascular disease prevention in spinal cord injury: A comprehensive literature review. Top Spinal Cord Inj Rehabil 3: 16-32.

Whiteneck GG, Charlifue SW, Frankel HL, Fraser MH, Gardner BP, et al. 1992. Mortality, morbidity, and psycho-social outcomes of persons spinal cord injured more than 20 years ago. Paraplegia 30: 617-630.

Wilmet E, Ismail A, Heilporn A, Welraeds D, Bergmann P. 1995. Longitudinal study of the bone mineral content and soft tissue composition after spinal cord section. Paraplegia 33: 674-677.

Wirz M, Zemon DH, Rupp R, Scheel A, Colombo G, et al. 2005. Effectiveness of automated locomotor training in patients with chronic incomplete spinal cord injury: A multicenter trial. Arch Phys Med Rehabil 86: 672-680.

World Health Organization. 2002. Reducing Risks, Promoting Healthy Life: World Health Report 2002. Geneva: World Health Organization.

Yakura JS, Waters RL, Adkins RH. 1990. Changes in ambulation parameters in spinal cord injury individuals following rehabilitation. Paraplegia 28: 364-370.

Zehnder Y, Luthi M, Michel D, Knecht H, Perrelet R, et al. 2004. Long-term changes in bone metabolism, bone mineral density, quantitative ultrasound parameters, and fracture incidence after spinal cord injury: A cross-sectional observational study in 100 paraplegic men. Osteoporos Int 15(Suppl 3): 180-189.

25 Estrogen as a Promising Multi-Active Agent for the Treatment of Spinal Cord Injury

E. A. Sribnick · D. D. Matzelle · S. K. Ray · N. L. Banik

1	Introduction	582
2	**Tissue Destruction in SCI**	**582**
2.1	Primary and Secondary Damage	582
2.2	Calcium in SCI	583
2.3	Calpain Activation in SCI	584
3	**Current Approach of Pharmacological Intervention for SCI**	**585**
3.1	Efficacy of Various Therapeutic Agents in SCI	585
3.2	Estrogen as a Neuroprotective Agent	585
3.3	Mechanisms of Estrogen Action	586
3.4	Estrogen as an Anti-Inflammatory and Anti-apoptotic Agent	586
3.5	Estrogen as a Ca^{2+} Modulator	586
3.6	Estrogen as an Angiogenic Agent	587
4	**Estrogen Efficacy**	**587**
4.1	Effects of Estrogen In Vitro in Neurons and Glial Cells	587
4.2	In Vivo Animal Studies in Ischemia, Stroke, TBI, and Clinical Trials	587
5	**Estrogen in Neuroprotection**	**588**
5.1	Estrogen as a Neuroprotectant in SCI	588
5.2	Estrogen as Anti-inflammatory Agent	589
5.3	Estrogen Protects Axon and Myelin	590
6	*Conclusion*	591

© 2009 Springer Science+Business Media, LLC.

Abstract: Spinal cord injury (SCI) causes neurological deficits leading to devastating functional disabilities. The only clinical treatment available is high dose methylprednisolone, which has limited efficacy. Recent studies have recognized estrogen, a steroid hormone, with multi-active properties in neurodevelopment, cell function, and the prevention of pathophysiology of central nervous system injury and diseases. Estrogen efficacy, at high doses or at a physiologic dose, has been investigated in animal models of injury, ischemia and stroke, as well as cell culture. Many of these investigations have provided convincing evidence that estrogen has significant beneficial effects. This short review describes the results obtained from our laboratory on the effect of estrogen in the treatment of experimental SCI in rats. The data indicated that estrogen reduced edema and inflammation, acted as an antioxidant and antiapoptotic agent, modulated intracellular Ca^{2+}, inhibited calpain and caspase activities, promoted angiogenesis, preserved axons and myelin, and improved motor function following SCI in rats. These observations on the multi-active properties of estrogen in CNS trauma and diseases strongly suggest that estrogen will have a significant role in the treatment of patients with SCI.

List of Abbreviations: AD, Alzheimer's disease; AMPA, α-amino-3-hydroxy-5-methyl-4-isoxazolepropionic acid; BDGF, brain-derived growth factor; CNS, central nervous system; EAE, experimental autoimmune encephalomyelitis; ER, estrogen receptor; MAG, myelin-associated glycoprotein; MBP, myelin basic protein; MP, methylprednisolone; MS, multiple sclerosis; NGF, nerve growth factor; NFP, neurofilament proteins; NMDA, N-methyl-D-aspartate; NT-3, neurotrophin-3; PD, Parkinson's disease; PLP, proteolipid protein; PMN, polymorphonuclear cell; SCI, spinal cord injury; TBI, traumatic brain injury; TNF, tumor necrosis factor

1 Introduction

Spinal cord injury (SCI) is a devastating and debilitating problem with over 250,000 sci patients in the United States, and the average age at injury is 29 years (National SCI Statistical Center, 2008). The debilitation and loss of function depends on the extent of severity of injury and often leads to lifelong disability at the time of life's highest productivity. In addition to neurological deficits, physiological deficits include sphincter dysfunction, infection, ulcers, etc. Although our understanding of SCI has increased over the years, effective therapies for functional recovery remained the primary goal for researchers as well as clinicians. The only pharmacological treatment available is high dose methylprednisolone (MP) (Bracken et al., 1997), which has limited clinical efficacy and become controversial (Hurlbert, 2000). Although many clinical and investigational agents have been used for the treatment, no single pharmacological compound has been found that effectively reverses SCI dysfunction. This may be due to the complex pathophysiology of tissue destruction following SCI as there are many destructive pathways that participate in the process. SCI occurs in two stages: the primary impact injury disrupts blood vessels, destroys tissue and kills cells and is followed by a devastating secondary injury that causes tissue damage and apoptosis of cells by releasing excess glutamate, free radicals, Ca^{2+}, and other harmful substances (Ray and Banik, 2003; Ray et al., 2003a). Because of this multi-destructive nature, it is difficult to develop an effective therapy, and blocking one pathway may not effectively prevent spinal cord damage. Therefore, it is of utmost importance to develop a multi-active agent or to employ a mixture of agents. One multi-active agent that meets these criteria is estrogen, an endogenous steroid hormone, that has been found to improve neurological function following treatment of experimental SCI in rats.

2 Tissue Destruction in SCI

2.1 Primary and Secondary Damage

Damage to spinal cord following injury causes neurological deficits, sphincter dysfunction, and may lead to paralysis. Destruction of tissue following injury is a two-stage process with an initial primary impact (mechanical) injury and a delayed secondary injury that develops hours, days, and weeks later. The mechanical injury elicits several destructive events at the site of the injury (lesion), including disruption of cell membranes causing necrotic cell death (neurons, astrocytes, microglia, and oligodendroglia), damage/degeneration of the axon-myelin fibers, disruption of blood vessels (will loss of blood supply leading to ischemia), edema, and ultimately neuron death and paralysis. The damage to the tissue does not end with the primary impact injury – the lesion size continues to increase over time following injury. The increase of the lesion size is caused by secondary damage events initiated from the primary injury. Thus, release of detrimental factors from necrotic cells and disruption of blood vessels lead to secondary inflammation and reactive gliosis in the adjacent areas of lesion (penumbra) help promote expansion of the size of damage.

The secondary inflammation then causes an influx of neutrophils (polymorphonuclear cells, PMNs) into the lesion site and activation of inflammatory cells, astrocytes, and microglia. These cells are responsible for producing a number of secondary injury factors, including proteases, lipases, free radicals, and inflammatory cytokines that kill neurons and oligodendrocytes and destroy myelinated axons. Biochemical and morphological changes have been thoroughly studied in the injured cord. Extensive morphological changes include granular degeneration of axons, vesiculation of myelin, mitochondrial calcification, accumulation of hydroxyapatite crystals of calcium, and phagocytosis by infiltrating macrophages (Balentine, 1978; Bresnahan, 1978; Banik et al., 1980). Morphologic changes with axonal degeneration and loss of myelin sheath were also evident in the penumbra of the injured cord (Wingrave et al., 2003). Histological studies demonstrated necrotic cells not only in the gray matter, but also in the central white matter. Hemorrhage, edema, damage to surrounding areas of white matter, and infiltration of PMNs were seen at 8 h following injury. At longer times following injury, neutrophils were found to be stripping myelin from degenerated axons, and glial cells, macrophages, and neurons were found to express increased amounts of Ca^{2+}-activated protease (calpain) in the lesion and penumbra (Li et al., 1996; Springer et al., 1997; Ray et al., 2003a). Damage due to increased inflammatory response and glutamate excitotoxicity led to an increased influx of Ca^{2+} into neuron and axon, activating many detrimental pathways that are involved in cell death and mitochondrial damage (Azbill et al., 1997; Wingrave et al., 2003; Ray et al., 2003a). An increased level of Ca^{2+} was found in the lesion cord when compared with control (Happel et al., 1981; Young and Flamm, 1982; Stokes et al., 1983). Elevated Ca^{2+} was also demonstrated in the penumbra of the injured spinal cord (Wingrave et al., 2003). The loss of ionic homeostasis of Na^+, K^+, Mg^{2+}, and Ca^{2+} in SCI has been suggested as a major cause for cell demise (Agrawal and Fehlings, 1996; Li et al., 2000; Ray and Banik, 2003).

The paramount biochemical changes found in the SCI tissue, including lipids, prostaglandins, glutamic acid, proteins, and proteases and lipases, have been extensively reviewed previously (Ray and Banik, 2003). Readers are advised to consult respective reviews for additional information. Changes in the loss of axonal (neurofilament protein, microtubular protein), myelin (myelin basic protein, proteolipid protein), and cytoskeletal (fodrin) proteins were correlated with ultrastructural alterations seen in axons and myelin in SCI (Banik et al., 1980; Springer et al., 1997; Yu and Geddes, 2007). These and other cellular proteins have been found to be excellent substrates of calpain. Thus, their protection is of immense importance to maintain the structural integrity of cells and membrane for functional recovery after SCI. This chapter, however, will deal mainly with the Ca^{2+}-dependent protease, calpain, and will examine the efficacy of estrogen in SCI by regulating Ca^{2+} influx and calpain-mediated proteolysis.

2.2 Calcium in SCI

The findings of hydroxyapatite crystals of Ca^{2+} in the lesion of spinal cord prompted investigators to determine the level of Ca^{2+} in SCI (Banik et al., 1980). An eightfold increase in the total Ca^{2+} level in SCI compared with control suggested a crucial role for this divalent ion in cell and tissue damage in SCI (Happel et al., 1981; Stokes et al., 1983), and increased intracellular free Ca^{2+} is known to cause cell death. While the free intracellular Ca^{2+} cannot be controlled in the lesion because of primary impact injury, influx of free Ca^{2+} may be modulated in the penumbra. Subsequently, influx of intracellular Ca^{2+} was demonstrated in the penumbra of SCI, suggesting its important role in secondary damage (Wingrave et al., 2003). Although increased total Ca^{2+} level in the lesion is likely due to damage/disruption of cell membranes, the mechanism of the rise in intracellular Ca^{2+} concentration is SCI is not clear. However, it has been suggested that the influx of Ca^{2+} may be due to voltage-gated Ca^{2+} channels, a reversed action of the Na^+/Ca^{2+} exchanger, L-type Ca^{2+} channels, glutamate-mediated excitotoxicity with increased activation of N-methyl-D-aspartate (NMDA) receptor and α-amino-3-hydroxy-5-methyl-4-isoxazolepropionic acid (AMPA) receptor (Bennett et al., 1996; Li et al., 2000; Ahmed et al., 2002). Ca^{2+} plays many roles in the physiology of cell function, and its one of the important roles is the activation of enzymes, lipases, and proteases. Thus, increased Ca^{2+} found in SCI plays such a role, i.e., activation of calpain, and controlling Ca^{2+} may prevent activation of Ca^{2+}-dependent events that destroy cells and tissues in SCI.

2.3 Calpain Activation in SCI

Calpain is a cysteine protease and requires Ca^{2+} for activation. It exists as ubiquitous forms, microcalpain and millicalpain, requiring μM and mM Ca^{2+} concentrations for activation, respectively. Both isoforms have two subunits, a regulatory and catalytic subunit (Ray and Banik, 2003). Increased calpain activity has been implicated in the pathogenesis of many diseases, including Alzheimer's disease (AD), Parkinson's disease (PD), multiple sclerosis (MS), ischemia and stroke, traumatic brain injury (TBI), and SCI (Saito et al., 1993; Crowe et al., 1997; Liu et al., 1997; Ray and Banik, 2003; Guyton et al., 2005). In SCI, calpain activity and expression is progressively increased with loss of cytoskeletal proteins and myelin proteins. Calpain has been shown to degrade myelin proteins (myelin basic protein, MBP; myelin-associated glycoprotein, MAG; proteolipid protein, PLP); neurofilament proteins (hNFP, mNFP, and lNFP), cytoskeletal proteins (fodrin/spectrin, talin, and actin), and nuclear laminin (Khachaturian, 1989; Saito et al., 1993; Bartus et al., 1994; Springer et al., 1997). Progressive increases in calpain activity were also found in penumbra, indicating that the partially damaged cells and fibers in the penumbra may be protected if treated with proper agents at appropriate time after injury (Ray et al., 2000a; Sribnick et al., 2007). Calpain activity in the cells is tightly regulated by calpastatin, an endogenous inhibitor of calpain. A trigger for an increase in calpastatin activity can inhibit the pathogenic calpain activity. In juvenile SCI, there was more calpastatin activity than calpain activity (Wingrave et al., 2004), suggesting that calpain activity is controlled by calpastatin and blocked the calpain-mediated damage in young spinal cord after injury. In contrast, there is more calpain activity than calpastatin activity in the lesion and penumbra of adult injured spinal cord. With a greater ratio of calpain:calpastatin, the inhibitor becomes a suicide substrate of calpain. The end result is that calpain becomes unregulated and destroys cells and tissue following SCI (Wingrave et al., 2003; Ray et al., 2003a). Therefore, calpain inhibitor is of paramount importance for protection of cells and their processes. Calpain inhibitor treatment of SCI as well as TBI animals prevented degradation of proteins and protected cells, further confirming a crucial role for calpain in CNS injuries (Bartus et al., 1994; Saatman et al., 1996; Springer et al., 1997; Pike et al., 1998; Ray et al., 2000a).

Calpain also has been found to mediate both apoptotic and necrotic cell death (Kohli et al., 1999; Johnson, 2000). In penumbra following SCI, cells may die of delayed apoptotic death, while in the lesion, cells die by necrosis, an irreversible death (Crowe et al., 1997; Liu et al., 1997; Springer et al., 1999; Ray et al., 1999a; Ray et al., 2000a). Such apoptotic death of neurons and other cells due to increased Ca^{2+} concentration and calpain activation has been shown and treatment with various calpain inhibitors provided protection (Ray et al., 2000a; Das et al., 2005; Das et al., 2006; Sribnick et al., 2007; Guyton et al., 2008). The notion that calpain is involved in cell death in SCI and pathophysiology of diseases has been supported by numerous in vitro cell culture studies as well, including cardiomyocytes, lymphocytes, neurons, and glial cells (Saido et al., 1994; Sarin et al., 1994; Squier et al., 1994; Ray et al., 1999b; Ray et al., 2000b). Neurons, whether cortical neurons, motoneurons, or retinal ganglion cells damaged due to different stress stimuli were protected by calpain inhibitors with recovery of function, as demonstrated by electrophysiological techniques (Ray et al., 2003b; Sribnick et al., 2004; Das et al., 2005).

Calpain is substantially involved in the mechanisms of apoptotic cell death. Several players in this cell death pathway are substrates of calpain. For example, while calpain degrades the Ca^{2+}-calmodulin-dependent protein kinase IV and Bax leading to neuronal and glial apoptosis, the breakdown of the anti-apoptotic Bcl-X_L by calpain converts it into a pro-apoptotic mediator (McGinnis et al., 1998; Nakagawa and Yuan, 2000; Chen et al., 2001; Choi et al., 2001). Calpain also cleaves Bid into tBid (Nicotera, 2000). Calpain also activates caspase-12 and promotes apoptosis involving endoplasmic reticulum. It activates caspase-3, the final common executioner of cell death (Nakagawa and Yuan, 2000; Chen et al., 2001). Thus, agents that can control intracellular Ca^{2+} influx and Ca^{2+}-mediated events are necessary for protection in SCI, and estrogen is one such agent that has been found to be beneficial (Sribnick et al., 2004; Sribnick et al., 2005).

3 Current Approach of Pharmacological Intervention for SCI

3.1 Efficacy of Various Therapeutic Agents in SCI

Many different agents have been used for the treatment of SCI in animals as well as in patients, including thyrotropin-releasing hormone by clinical investigators (Pitts et al., 1995; Dumont et al., 2001). Only administration of high-dose methylprednisolone, which has limited efficacy and is controversial, is currently used for the treatment of acute SCI patients (Bracken et al., 1997; Hurlbert, 2000). Similar results were found with administration of 21-aminosteroid tirilizad mesylate (Dumont et al., 2001). In addition, many secondary cell damage inhibitors or antagonists have been used as therapeutic agents in animal SCI studies. NMDA receptor antagonist MK801 has been used to block excitotoxic cell death in SCI and was found to reduce the number of cell death and improve function (Yanase et al., 1995; Wada et al., 1999). While AMPA receptor antagonist as well as tetrodotoxin have been shown to protect neurons and axons in SCI (Rosenberg and Wrathall, 2001), the NMDA receptor antagonist and tetrodotoxin are toxic and not suitable for clinical use. Overexpression of the anti-apoptotic *bcl-2* gene has been induced in mice and found to protect neurons following axotomy (Saavedra et al., 2000).

Calpain has also been used as a therapeutic target since its increased activity is associated with cytoskeletal protein degradation and neuron and oligodendrocyte death in SCI (Ray et al., 2003a). Thus calpain inhibitors have been extensively used to treat animals following SCI and TBI. These studies have provided neuronal protection and some recovery of function (Saatman et al., 1996; Posmantur et al., 1997; Springer et al., 1997; Ray et al., 2000a; Kupina et al., 2001; Yu and Geddes, 2007). Melatonin, a pineal hormone with anti-oxidant properties, has been found to reduce tissue damage and cell death and to improve function following treatment of SCI animals (Kaptanoglu et al., 2000; Reiter et al., 2001). Other studies carried out with Nogo or stem cell transplantation have not produced any substantial improvement of function in SCI (Dasari et al., 2007; Dasari et al., 2008). The therapeutic efficacy of melatonin, a pineal hormone with its anti-oxidant and other properties, has been examined in experimental SCI by a number of investigators (Fujimoto et al., 2000; Kaptanoglu et al., 2000; Cayli et al., 2004). Results from our current studies indicated melatonin to be neuroprotective as it protected neurons and fibers and improved motor function following SCI in rats (Samantaray et al., 2008).

3.2 Estrogen as a Neuroprotective Agent

Estrogen has recently been suggested to be neuroprotective in AD, ischemia, stroke, PD, and CNS injuries (Emerson et al., 1993; Chao et al., 1994; Kondo et al., 1997; Xu et al., 1998; Stein, 2001). Both progesterone and estrogen are anti-inflammatory and act as antioxidants; they reduce excitotoxic damage and Ca^{2+} flux and act as neurotrophic agents. The mechanisms of their neuroprotective actions are not clearly known (Stein, 2001). Estrogen can penetrate cells, affecting both neurons and glial cells, and bind to nuclear receptors such as estrogen receptor α (ER-α) and estrogen receptor β (ER-β). Both receptors have two distinct domains, an N-terminal activation domain and a C-terminal DNA binding domain that binds to DNA via two zinc finger motifs. Both receptors bind to DNA as dimers.

3.3 Mechanisms of Estrogen Action

While no single mechanism of action of estrogen has been found, many pathways have been discovered for the neuroprotective effects of estrogen, suggesting it to be a multi-active neuroprotectant. One of the nonreceptor-mediated effects of estrogen is that it functions as a potent antioxidant. In CNS injury and diseases, the increases in damaging reactive oxygen species (ROS) have been found to be attenuated by the antioxidant effects of estrogen (Barut et al., 1993; Moosmann and Behl, 1999; Xu et al., 2004). Interestingly, the antioxidant effect is due to the hydroxyl group of the phenolic ring present in the estrogen structure (Behl et al., 1997). It also intercalates in the membrane and prevents lipid peroxidation (Golden et al., 1998). One of the important mechanisms in estrogen mediated neuroprotection is binding to its receptor, translocating to nucleus, and upregulating transcription (Alkayed et al., 2001). The involvement of signal transduction pathways (MAP-kinase) are thought to be major participants in the estrogen-mediated neuroprotection following glutamate toxicity (Singer et al., 1999; Nilsen et al., 2002).

3.4 Estrogen as an Anti-Inflammatory and Anti-apoptotic Agent

Estrogen has been found to protect neurons from damage by exerting its antiapoptotic actions both in vitro and in vivo (Singer et al., 1999; Honda et al., 2001; Jover et al., 2002; Kaja et al., 2003; Sribnick et al., 2004; Sribnick et al., 2006b). This protection has been linked to increased ER-α expression, Bcl-2 expression, activation of MAP-kinase, phosphatidylinositol-3-kinase (PI3 kinase), inhibition of calpain and caspase activities (Singer et al., 1999; Honda et al., 2001; Sribnick et al., 2006a; Sribnick et al., 2007), and increased antiapoptotic Bcl-xL expression following cytokine-induced cell death (Koski et al., 2004). Estrogen also has been shown to protect neurons exposed to amyloid β by activating protein kinase C (PKC), a pro-survival protein (Cordey et al., 2003).

ER-α, and not ER-β, has been suggested to provide neuroprotection in TBI, and recent studies indicated that neuroprotection mediated by ER-α in experimental autoimmune encephalomyelitis (EAE) is due to its anti-inflammatory action (Dubal et al., 2001; Tiwari-Woodruff et al., 2007). In contrast, ER-β has been thought to be involved in neurodevelopment and may have a role in protection and repair (Wang et al., 2003; Sribnick et al., 2007). ER-β, but not ER-α, also has been indicated to play a major neuroprotective role in the neurodegenerative phase of EAE (Tiwari-Woodruff et al., 2007). Nevertheless, both receptors are important for protecting neurons and fibers from damage caused by inflammation and degeneration.

Estrogen has been found to inhibit macrophage infiltration and adhesion molecule expression, as well as to down regulate tumor necrosis factor alpha (TNF-α) that is involved in the mechanisms of inflammatory response (Miyamoto et al., 1999; Ito et al., 2001). At very low concentrations, estrogen is known to inhibit microglial activation and subsequent inflammation by blocking inducible nitric oxide synthase (iNOS), lipid peroxidation, and other inflammatory factors (Bruce-Keller et al., 2000; Vegeto et al., 2000). Although NF-κB regulates many genes involved in inflammation, estrogen has been found to inhibit translocation of NF-κB to the nucleus and block activation of inflammatory cells (Bruce-Keller et al., 2000). Since calpain and proteosome are involved in the cleavage of the IκB/NF-κB complex to facilitate NF-κB translocation, the inhibition of these proteases by estrogen may be responsible for preventing NF-κB translocation to nuclear membrane (Palombella et al., 1994; Chen et al., 1997). Another anti-inflammatory effect of estrogen may involve estrogen receptor-dependent mitogen-activated protein (MAP) kinase activation (Singer et al., 1999; Bruce-Keller et al., 2000; Dubal et al., 2001).

3.5 Estrogen as a Ca^{2+} Modulator

Maintaining the Ca^{2+} homeostasis is vital for normal cell function, and one of the many roles of Ca^{2+} is activation of enzymes, e.g., lipases and proteases. Alterations or increases in intracellular concentration of Ca^{2+} have been implicated in the pathophysiology of several disease states and injury models. Such increases in Ca^{2+} have been found in SCI, TBI, EAE, and other neurodegenerative disorders (Wingrave et al., 2003; Guyton et al., 2005), causing axonal degeneration and cell death. Therefore, it is imperative to maintain or restore the Ca^{2+} homeostasis in injuries and diseases. One of the protective roles of estrogen is suppression of intracellular Ca^{2+} levels through several mechanisms following oxidative or excitotoxic damage to neurons (Goodman et al., 1996; Mermelstein et al., 1996; Nilsen et al., 2002; Sribnick et al., 2004), possibly by modulating L-type Ca^{2+} channel (Goodman et al., 1996; Mermelstein et al., 1996), voltage-gated Ca^{2+} channels (Kurata et al., 2001), and Ca^{2+} release from intracellular Ca^{2+} stores (Morales et al., 2003). Attenuation of Ca^{2+}-influx in glutamate toxicity has been found to protect cells and restore function (Sur et al., 2003; Sribnick et al., 2004).

3.6 Estrogen as an Angiogenic Agent

Since disruption of blood vessels following SCI leads to ischemia in the injured tissue, promoting angiogenesis, e.g., microvessel growth and restoration of blood supply, is critical to cell survival and functional recovery in SCI (Balentine, 1978; Holtz et al., 1990; Tator, 1991). While angiogenesis is slow or nonexistent in adults, estrogen is thought to play a vital role in angiogenesis in the female reproductive system. Recurrent neovascularization is a normal process in females, and estrogen mediates this process by upregulating expression of vascular endothelial growth factor (VEGF), basic fibroblast growth factor (bFGF), nitric oxide, and endothelial nitric oxide synthase (Duckles and Krause, 2003).

4 Estrogen Efficacy

4.1 Effects of Estrogen In Vitro in Neurons and Glial Cells

In order to understand the mechanisms of the neuroprotective actions of estrogen, the presence of both receptors, ER-α and ER-β, were found in neurons and glial cells, including astrocytes, microglia, and oligodendrocytes (Santagati et al., 1994; Mor et al., 1999). Since cytotoxic factors, including H_2O_2, nitric oxide, and superoxide free radicals are released by activated macrophages and microglia in CNS injury and diseases (e.g., AD), neuroprotective efficacy of estrogen has been examined in CNS cells (e.g., neurons and glial cells) undergoing stress stimuli (Behl et al., 1995; Mermelstein et al., 1996; Singer et al., 1999; Singh et al., 1999; Sur et al., 2003). Estrogen treatment reduced the production of nitric oxide synthase in microglial cells stimulated with cytokines, e.g., IL-1, TNF-α, IFN-γ, and lipopolysaccharide (LPS) (Vegeto et al., 2000). At nanomolar concentrations, estrogen (17β-estradiol) has been shown to prevent LPS-induced superoxide release, phagocytic activity, and iNOS expression (Bruce-Keller et al., 2000). Estrogen has been found to protect cortical neurons and glia from damage due to excitotocicity and oxidative damage, reducing Ca^{2+} concentration, and activating growth factor signaling pathways. Protection of motoneurons from these types of damage by estrogen restored electrophysiological function (e.g., membrane potential) (Sribnick et al., 2004). It also acts as a growth factor and promotes cell survival and neuronal outgrowth (Honjo et al., 1992; Weaver et al., 1997). Estrogen upregulates insulin-like growth factor (IGF-1) in glial cells, and its receptors are known to be colocalized with several growth factor receptors, including brain-derived growth factor (BDGF), neurotrophin-3 (NT-3), and nerve growth factor (NGF), which promote cell survival and growth (Toran-Allerand et al., 1992; Toran-Allerand, 1999).

4.2 In Vivo Animal Studies in Ischemia, Stroke, TBI, and Clinical Trials

While the estrogen efficacy in studies involving animal models of different CNS injuries (including SCI, TBI, stroke, and ischemia) have provided evidence of beneficial effect, estrogen has not received much attention in clinical situations. This may have been due to estrogen's detrimental effect in some clinical situations such as aggravating breast cancer or causing thromboembolic injuries. Nonetheless, the clinical efficacy of estrogen in treatment of AD and estrogen replacement therapy indicated that estrogen reduced risk and played a neuroprotective role in neurodegenerative diseases (Kawas et al., 1997; Cholerton et al., 2002). These studies remain controversial. However, interesting and important epidemiological studies on SCI and TBI indicated female patients with either of these injuries have significantly better recovery of function than males (Hillier et al., 1997; Groswasser et al., 1998). The estrogen-related gender differences, such as better functional recovery and survival rate, have been found in rats following head injury (Roof and Hall, 2000), suggesting that better functional outcome seen in female patients with SCI and TBI may be due to higher levels of estrogen. These gender differences have been subsequently confirmed in female mice with TBI, demonstrating increased survival rate and delayed calpain-mediated formation of spectrin degradation product (Kupina et al., 2003). The early neuroprotection seen in female rats following TBI has been suggested to be due to the presence of endogenous circulation of estrogen (Wagner et al., 2004). A faster functional recovery of function from sciatic nerve injury has been seen in ovariectomized mice following treatment with estrogen (Islamov et al., 2002). Estrogen treatment has been found to improve working memory in ovariectomized rats and estrogen replacement therapy has also been suggested to improve memory in postmenopausal women with Alzheimer's disease (Asthana et al., 2001; Robertson et al., 2001). These studies strongly suggest an important role for estrogen conferring early neuroprotection in animals and humans.

Estrogen has been shown to reduce inflammation and neurodegeneration in ischemia, stroke, and TBI. Acute and chronic administration of estrogen has been shown to significantly reduce infarct volume, edema, and neuronal loss in rats following ischemia (middle cerebral artery occlusion), and brain contusion (Mor et al., 1999; Wen et al., 2004). These effects of estrogen have been attributed to many factors, including anti-apoptotic, anti-oxidant, and anti-inflammatory properties. Estrogen has been found to inhibit macrophage infiltration in endotoxin-induced uveitis (Miyamoto et al., 1999). Also, estrogen treatment has been found to down regulate TNF-α production and cytotoxic T cell activation and to reduce the severity of the experimental demyelinating disease EAE, an animal model of MS (Ito et al., 2001). This protective action of estrogen, both at the inflammatory phase as well as at the neurodegenerative phase, has been found to be mediated via ER-α and ER-β, respectively (Dubal et al., 2001; Polanczyk et al., 2003; Tiwari-Woodruff et al., 2007). In support of this finding, patients with MS during pregnancy have been found to have diminished clinical symptoms, possibly due to increased levels of hormones, especially estrogen (Birk and Rudick, 1986; Manson et al., 2007).

5 Estrogen in Neuroprotection

5.1 Estrogen as a Neuroprotectant in SCI

While the advances made over the years in understanding the pathophysiology of SCI are enormous, ameliorating dysfunction from such injuries by pharmacological agents or cure is lacking. The only treatment available for SCI patients is high-dose methylprednisolone (Bracken et al., 1997) with limited efficacy, and the treatment is controversial (Hurlbert, 2000). The difficulties in developing therapies to restore or improve functional outcome may be due to several unavoidable situations following injury. For example, cells and the processes at the impact injury site are necrotic and suffer irreversible damage, becoming nonfunctional. In contrast, the cells in the area surrounding the injury (lesion) site, or the penumbra, may be partially damaged or apoptotic and reversible. They may be protected for functional recovery if treated early. Another major problem in SCI is that damage to neurons and the axon-myelin structural units is caused by many destructive pathways. The many devastating secondary injury factors involved in this mechanism include release of increased excitotoxic glutamic acid, ROS, inflammation, intracellular Ca^{2+} influx, activation of Ca^{2+}-dependent events, e.g., lipases and proteinases (including calpain), and many others (Ray et al., 2003a; Sribnick et al., 2004).

Because many destructive pathways are activated and involved in cell death and tissue damage following SCI, treatment with a drug that attenuate one pathway may not provide any improvement in function. Therefore, a drug with multiple actions or a combination of several drugs may be needed for effective treatment of SCI. Estrogen is one such multi-active agent and is noted to be neuroprotective, anti-apoptotic, antioxidant, anti-inflammatory, and a Ca^{2+}-channel modulator. Estrogen has been shown to be neuroprotective in TBI, ischemia, stroke, and pre- and post-injury treatment of SCI (Roof and Hall, 2000; Dumont et al., 2001; Yune et al., 2004; Sribnick et al., 2005; Sribnick et al., 2006b) as well as in cell culture (Sribnick et al., 2004).

Since cysteine proteases are involved in apoptotic cell death, our focus has been on calpain and caspase-3 which are the downstream effectors of apoptosis (Ray et al., 2003a). To determine the role estrogen may play in inhibiting protease activity and attenuating cell death, we have examined the efficacy of high-dose (4 mg/kg) estrogen treatment in both acute and chronic SCI in rats and determined calpain activation and activity (Sribnick et al., 2006b; Sribnick et al., 2007). Our studies indicated prevention of NFP (68 kD) and spectrin degradation production formation by estrogen treatment of SCI for 48 h in lesion and penumbra, indicating inhibition of increased calpain activity and activation, a common finding after SCI (Springer et al., 1997; Ray et al., 2003b; Wingrave et al., 2003; Yu and Geddes, 2007). Since NFP is degraded by other proteases, including cathepsin D (Nixon and Marotta, 1984), its breakdown can not be taken solely due to calpain activity. Thus, caspase activity was also determined in SCI treated with estrogen. In addition to calpain, estrogen treatment also inhibited caspase-3 activity, which is increased in spinal cord after injury (Springer et al., 1999). It was shown previously that estrogen treatment inhibited calpain-like protease activity in lower-limb muscle following exercise and estrogen receptor (ER-α) blocked degradation of calpain substrates (Tiidus et al., 2001; Morelli et al., 2003). We have found estrogen treatment restored the normal level of ER-β in the penumbra, which was substantially reduced following injury. Since ER activation has been found to increase the anti-apoptotic Akt and Bcl-2 expression, restoring the normal level of estrogen receptor in SCI following estrogen treatment may have protected cells in the penumbra (Nakamizo et al., 2000; Pugazhenthi et al., 2000; Sribnick et al., 2006b). Such upregulation of anti-apoptotic Bcl-2 by estrogen has been demonstrated in vitro in neurons and in tissue (Alkayed et al., 2001; Honda et al., 2001). The anti-apoptotic effect of estrogen may also have been due to inhibition of calcineurin, which promotes apoptosis in SCI (Springer et al., 2000). We have also found estrogen treatment prevented cytochrome c release from the mitochondria in injured spinal cord (Banik et al., 1980; Wingrave et al., 2003; Sribnick et al., 2006b) and perhaps preserved mitochondrial function. This estrogen-mediated protection may have been due to inhibition of ROS generated from mitochondria following injury since estrogen is a powerful anti-oxidant and may also work via receptor-mediated pathways as estrogen receptors since ER-α and ER-β are present in mitochondria (Moosmann and Behl, 1999; Honda et al., 2001; Chen et al., 2004). As the damage to the mitochondria, whether by increased protease activation or release of ROS after injury, is the "kiss of death" to cells, protecting mitochondrial function by estrogen treatment is essential for cell survival. In addition to assessing the effect of estrogen treatment on different apoptotic parameters, our laboratory has also determined the extent of apoptotic neurons in SCI animals after estrogen administration (Sribnick et al., 2006b). Estrogen treatment has been found to significantly protect neurons in SCI compared with untreated animals. The apoptotic cell death, as assayed by DNA fragmentation (laddering), has also been found to be markedly attenuated following treatment with calpain inhibitors (Sribnick et al., 2007).

5.2 Estrogen as Anti-inflammatory Agent

Studies in our laboratory on estrogen efficacy in both acute and chronic SCI have been focused on three major areas: (1) estrogen's anti-inflammatory properties and preservation of axon and myelin, (2) its mechanisms of protection of cells and fibers, and (3) its effects on functional outcome. In SCI, there is edema, infiltration of inflammatory cells, activated microglia and macrophages and phagocytic cells (Banik et al., 1980; Hsu et al., 1985; Kwo et al., 1989; Ray et al., 2003b). We demonstrated that at 48 h after injury, there was a significantly increased level of edema (water) not only in the lesion, but also in the caudal penumbra section of the cord compared with control (❯ *Figure 25-1*; Sribnick et al., 2005). There was an increased number of inflammatory macrophages and microglia in both lesion and penumbra, as identified by OX-42 and ED2+ antibodies, respectively. High-dose estrogen (4 mg/kg) treatment of rat sci significantly reduced edema

◘ Figure 25-1

Induction of SCI increased edema (water) in the lesion and treatment of SCI rats with estrogen (4 mg/kg) decreased edema. The extent of tissue edema was determined in terms of percent water content. This is a modified figure from Sribnick et al., 2005

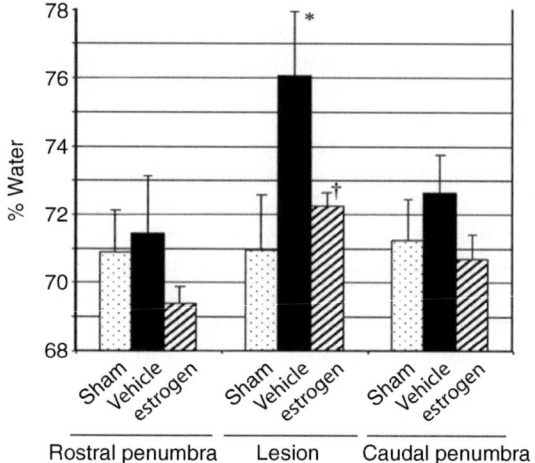

and the number of inflammatory cells (Sribnick et al., 2005). Progesterone at a high dose has been found to reduce edema in TBI and estrogen treatment increased the survival rate and decreased lesion volume in ischemia (Simpkins et al., 1997; Roof and Hall, 2000; Grossman et al., 2004). While the mechanisms by which estrogen reduces edema are not clear, the anti-inflammatory effect of high-dose estrogen via modulation of the NF-κB pathway may be one of them. Nonetheless, there is activation of NF-κB in SCI as previously shown (Bethea et al., 1998), and it can have both detrimental and protective effects (Pizzi et al., 2002; Kucharczak et al., 2003). We have shown there are decreases in NF-κB and IκB levels in SCI with an increased in nuclear NF-κB. Estrogen treatment reversed the trend – there is an increase in the cytosolic NF-κB and IκB and significantly decreased nuclear NF-κB (Sribnick et al., 2005). In agreement with this finding, prevention of NF-κB nuclear translocation has been found to protect neurons and estrogen treatment also prevented NF-κB nuclear translocation in ischemia in rats (Schneider et al., 1999; Wen et al., 2004) in vivo as well as in vitro (Dodel et al., 1999). Thus, estrogen may render protection by inhibiting inflammatory response following injury, since estrogen receptors ER-α and ER-β have been found to be both anti-inflammatory and neuroprotective (Dubal et al., 2001; Vegeto et al., 2003; Tiwari-Woodruff et al., 2007).

5.3 Estrogen Protects Axon and Myelin

The degeneration of axon and myelin is a common feature in SCI – destruction of the axon-myelin structural unit impairs conduction of nerve impulses and may lead to paralysis (Balentine, 1978; Bresnahan, 1978; Banik et al., 1980; Samantaray et al., 2007a). Therefore, preservation of cells and the axon-myelin unit is of utmost importance for recovery of function. Previous studies have used anti-apoptotic agents in SCI and have shown protection of neurons and oligodendrocytes, particularly in the penumbra, for functional recovery (Ray et al., 2000a; Dumont et al., 2001; Sribnick et al., 2007; Yu and Geddes, 2007). Estrogen treatment at early times following SCI for 48 h significantly reduced the loss of myelin in the lesion as well as in both the rostral and caudal penumbra compared to untreated animals (❯ *Figure 25-2*; Sribnick et al., 2005). The disruption of the architecture of spinal cord seen in the injured animals was better preserved by estrogen treatment. These findings on high-dose estrogen efficacy in SCI as neuroprotective in attenuating

Figure 25-2
Induction of SCI caused loss of myelin (demyelination) in the lesion as well as in both the rostral and caudal penumbra but treatment of SCI rats with estrogen (4 mg/kg) decreased demyelination. The extent of demyelination was examined by Luxol-fast-blue staining. This is a modified figure from Sribnick et al., 2005

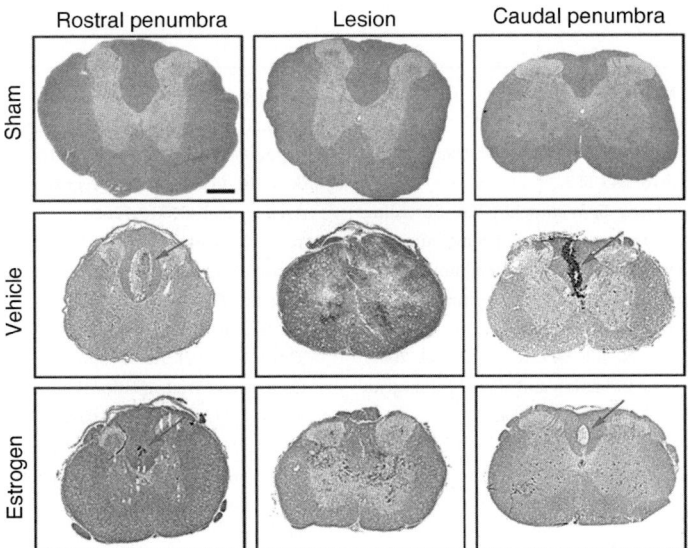

inflammation, cell death, and myelin loss are in agreement with other in vivo and in vitro injury models. Similar studies on the neuroprotective role of high-dose estrogen treatment have been carried out in chronic SCI animals. These studies indicated protection of cells, axons, and myelin preservation and restoration of motor function, as assessed by monitoring BBB parameters (Samantaray et al., 2007a). Although supraphysiologic doses of estrogen were used in these studies, beneficial effects were demonstrated. Recent studies with low-dose estrogen (10 μg/kg) protected cells and axons and restored motor function (Samantaray et al., 2007b). These various studies suggested that estrogen is a potential therapeutic agent for the treatment of SCI.

6 Conclusion

Current therapy with high dose methylprednisolone for the treatment of SCI has been found to have limited efficacy. It is therefore essential to develop a novel and more effective therapy to ameliorate dysfunction following injury. Developing a single strategic therapy is difficult because of the activation of many complex pathways following injury that lead to inflammation, oxidative and excitotoxic damage to cells, ischemia, apoptotic and necrotic cell death. Thus, in order to abrogate the complex and devastating pathophysiological cascades in SCI, use of a combination drug therapy or identification of a single agent with multi-action properties is important. Estrogen is such an agent, a multi-active steroid hormone for which substantial evidence has been found to support its neuroprotective capabilities in TBI, SCI, stroke, and ischemia in vivo as well as protecting cells from damage in vitro. Its effect on the devastating consequences of SCI by reducing excitotoxic damage, promoting angiogenesis, enhancing anti-oxidant and anti-inflammatory activities, decreasing Ca^{2+} influx and calpain activity, protecting cells and preserving axons and myelin, and finally, leading to recovery of function has been very encouraging. These observations, as well as regeneration potential in CNS injuries and diseases, suggest that estrogen with its multi-action properties will have a significant impact in the treatment of patients with SCI.

Acknowledgments

This work was supported in parts by the grants from the NIH and the state of SC.

References

Agrawal SK, Fehlings MG. 1996. Mechanisms of secondary injury to spinal cord axons in vitro: Role of Na+, Na(+)-K(+)-ATPase, the Na(+)-H+ exchanger, and the Na(+)-Ca^{2+} exchanger. J Neurosci 16: 545-552.

Ahmed SM, Weber JT, Liang S, Willoughby KA, Sitterding HA, et al. 2002. NMDA receptor activation contributes to a portion of the decreased mitochondrial membrane potential and elevated intracellular free calcium in strain-injured neurons. J Neurotrauma 19: 1619-1629.

Alkayed NJ, Goto S, Sugo N, Joh HD, Klaus J, et al. 2001. Estrogen and *Bcl-2*: Gene induction and effect of transgene in experimental stroke. J Neurosci 21: 7543-7550.

Asthana S, Baker LD, Craft S, Stanczyk FZ, Veith RC, et al. 2001. High-dose estradiol improves cognition for women with AD: Results of a randomized study. Neurology 57: 605-612.

Azbill RD, Mu X, Bruce-Keller AJ, Mattson MP, Springer JE. 1997. Impaired mitochondrial function, oxidative stress and altered antioxidant enzyme activities following traumatic spinal cord injury. Brain Res 765: 283-290.

Balentine JD. 1978. Pathology of experimental spinal cord trauma. II. Ultrastructure of axons and myelin. Lab Invest 39: 254-266.

Banik NL, Powers JM, Hogan EL. 1980. The effects of spinal cord trauma on myelin. J Neuropathol Exp Neurol 39: 232-244.

Bartus RT, Hayward NJ, Elliott PJ, Sawyer SD, Baker KL, et al. 1994. Calpain inhibitor AK295 protects neurons from focal brain ischemia. Effects of postocclusion intra-arterial administration. Stroke 25: 2265-2270.

Barut S, Canbolat A, Bilge T, Aydin Y, Cokneseli B, et al. 1993. Lipid peroxidation in experimental spinal cord injury: Time-level relationship. Neurosurg Rev 16: 53-59.

Behl C, Widmann M, Trapp T, Holsboer F. 1995. 17-Beta estradiol protects neurons from oxidative stress-induced cell death in vitro. Biochem Biophys Res Commun 216: 473-482.

Behl C, Skutella T, Lezoualc'h F, Post A, Widmann M, et al. 1997. Neuroprotection against oxidative stress by estrogens: Structure-activity relationship. Mol Pharmacol 51: 535-541.

Bennett MV, Pellegrini-Giampietro DE, Gorter JA, Aronica E, Connor JA, et al. 1996. The GluR2 hypothesis: Ca(++)-permeable AMPA receptors in delayed neurodegeneration. Cold Spring Harb Symp Quant Biol 61: 373-384.

Bethea JR, Castro M, Keane RW, Lee TT, Dietrich WD, et al. 1998. Traumatic spinal cord injury induces nuclear factor-kappaB activation. J Neurosci 18: 3251-3260.

Birk K, Rudick R. 1986. Pregnancy and multiple sclerosis. Arch Neurol 43: 719-726.

Bracken MB, Shepard MJ, Holford TR, Leo-Summers L, Aldrich EF, et al. 1997. Administration of methylprednisolone for 24 or 48 hours or tirilazad mesylate for 48 hours in the treatment of acute spinal cord injury. Results of the Third National Acute Spinal Cord Injury Randomized Controlled Trial. National Acute Spinal Cord Injury Study. JAMA 277: 1597-1604.

Bresnahan JC. 1978. An electron-microscopic analysis of axonal alterations following blunt contusion of the spinal cord of the rhesus monkey (*Macaca mulatta*). J Neurol Sci 37: 59-82.

Bruce-Keller AJ, Keeling JL, Keller JN, Huang FF, Camondola S, et al. 2000. Antiinflammatory effects of estrogen on microglial activation. Endocrinology 141: 3646-3656.

Cayli SR, Kocak A, Yilmaz U, Tekiner A, Erbil M, et al. 2004. Effect of combined treatment with melatonin and methylprednisolone on neurological recovery after experimental spinal cord injury. Eur Spine J 13: 724-732.

Chao HM, Spencer RL, Frankfurt M, McEwen BS. 1994. The effects of aging and hormonal manipulation on amyloid precursor protein APP695 mRNA expression in the rat hippocampus. J Neuroendocrinol 6: 517-521.

Chen F, Lu Y, Kuhn DC, Maki M, Shi X, et al. 1997. Calpain contributes to silica-induced I kappa B-alpha degradation and nuclear factor-kappa B activation. Arch Biochem Biophys 342: 383-388.

Chen JQ, Delannoy M, Cooke C, Yager JD. 2004. Mitochondrial localization of ERα and ERβ in human MCF7 cells. Am J Physiol Endocrinol Metab 286: E1011-E1022.

Chen M, He H, Zhan S, Krajewski S, Reed JC, et al. 2001. Bid is cleaved by calpain to an active fragment in vitro and during myocardial ischemia/reperfusion. J Biol Chem 276: 30724-30728.

Choi WS, Lee EH, Chung CW, Jung YK, Jin BK, et al. 2001. Cleavage of Bax is mediated by caspase-dependent or -independent calpain activation in dopaminergic neuronal cells: Protective role of *Bcl-2*. J Neurochem 77: 1531-1541.

Cholerton B, Gleason CE, Baker LD, Asthana S. 2002. Estrogen and Alzheimer's disease: The story so far. Drugs Aging 19: 405-427.

Cordey M, Gundimeda U, Gopalakrishna R, Pike CJ. 2003. Estrogen activates protein kinase C in neurons: Role in neuroprotection. J Neurochem 84: 1340-1348.

Crowe MJ, Bresnahan JC, Shuman SL, Masters JN, Beattie MS. 1997. Apoptosis and delayed degeneration after spinal cord injury in rats and monkeys. Nat Med 3: 73-76.

Das A, Sribnick EA, Wingrave JM, Del Re AM, Woodward JJ, et al. 2005. Calpain activation in apoptosis of ventral spinal cord 4.1 (VSC4.1) motoneurons exposed to glutamate: Calpain inhibition provides functional neuroprotection. J Neurosci Res 81: 551-562.

Das A, Garner DP, Del Re AM, Woodward JJ, Kumar DM, et al. 2006. Calpeptin provides functional neuroprotection to rat retinal ganglion cells following Ca^{2+} influx. Brain Res 1084: 146-157.

Dasari VR, Spomar DG, Li L, Gujrati M, Rao JS, et al. 2008. Umbilical cord blood stem cell mediated downregulation of fas improves functional recovery of rats after spinal cord injury. Neurochem Res 33: 134-149.

Dasari VR, Spomar DG, Gondi CS, Sloffer CA, Saving KL, et al. 2007. Axonal remyelination by cord blood stem cells after spinal cord injury. J Neurotrauma 24: 391-410.

Dodel RC, Du Y, Bales KR, Gao F, Paul SM. 1999. Sodium salicylate and 17beta-estradiol attenuate nuclear transcription factor NF-kappaB translocation in cultured rat astroglial cultures following exposure to amyloid A beta(1–40) and lipopolysaccharides. J Neurochem 73: 1453-1460.

Dubal DB, Zhu H, Yu J, Rau SW, Shughrue PJ, et al. 2001. Estrogen receptor alpha, not beta, is a critical link in estradiol-mediated protection against brain injury. Proc Natl Acad Sci USA 98: 1952-1957.

Duckles S, Krause S. 2003. Vascular and endothelial function: Role of gonadal steroids in neuronal and vascular plasticity. Neuronal and Vascular Plasticity: Elucidating Basic Cellular Mechanisms for Future Therapeutic Discovery. Maiese K, editor. New York: Kluwer Press; pp. 95-115.

Dumont RJ, Okonkwo DO, Verma S, Hurlbert RJ, Boulos PT, et al. 2001. Acute spinal cord injury. I. Pathophysiologic mechanisms. Clin Neuropharmacol 24: 254-264.

Emerson CS, Headrick JP, Vink R. 1993. Estrogen improves biochemical and neurologic outcome following traumatic brain injury in male rats, but not in females. Brain Res 608: 95-100.

Fujimoto T, Nakamura T, Ikeda T, Takagi K. 2000. Potent protective effects of melatonin on experimental spinal cord injury. Spine 25: 769-775.

Golden GA, Mason PE, Rubin RT, Mason RP. 1998. Biophysical membrane interactions of steroid hormones: A potential complementary mechanism of steroid action. Clin Neuropharmacol 21: 181-189.

Goodman Y, Bruce AJ, Cheng B, Mattson MP. 1996. Estrogens attenuate and corticosterone exacerbates excitotoxicity, oxidative injury, and amyloid beta-peptide toxicity in hippocampal neurons. J Neurochem 66: 1836-1844.

Grossman KJ, Goss CW, Stein DG. 2004. Effects of progesterone on the inflammatory response to brain injury in the rat. Brain Res 1008: 29-39.

Groswasser Z, Cohen M, Keren O. 1998. Female TBI patients recover better than males. Brain Inj 12: 805-808.

Guyton MK, Sribnick EA, Wingrave M, Ray SK, Banik NL. 2005. Axonal damage and neuron death in MS and EAE: A pivotal role for calpain in demyelination. Multiple Sclerosis as a Neuronal Disease. Waxman S, editor. New York: Academi Press; pp. 293-303.

Guyton MK, Das A, Pahan K, Ray SK, Banik NL. 2008. Calpain inhibition attenuates immunological and neurodegenerative pathologies of EAE. J Neurochem 104 (suppl 1):78 (Abstrat OP02-07).

Happel RD, Smith KP, Banik NL, Powers JM, Hogan EL, et al. 1981. Ca^{2+}-accumulation in experimental spinal cord trauma. Brain Res 211: 476-479.

Hillier SL, Hiller JE, Metzer J. 1997. Epidemiology of traumatic brain injury in South Australia. Brain Inj 11: 649-659.

Holtz A, Nystrom B, Gerdin B. 1990. Relation between spinal cord blood flow and functional recovery after blocking weight-induced spinal cord injury in rats. Neurosurgery 26: 952-957.

Honda K, Shimohama S, Sawada H, Kihara T, Nakamizo T, et al. 2001. Nongenomic antiapoptotic signal transduction by estrogen in cultured cortical neurons. J Neurosci Res 64: 466-475.

Honjo H, Tamura T, Matsumoto Y, Kawata M, Ogino Y, et al. 1992. Estrogen as a growth factor to central nervous cells. Estrogen treatment promotes development of acetylcholinesterase-positive basal forebrain neurons transplanted in the anterior eye chamber. J Steroid Biochem Mol Biol 41: 633-635.

Hsu CY, Hogan EL, Gadsden RH Sr, Spicer KM, Shi MP, et al. 1985. Vascular permeability in experimental spinal cord injury. J Neurol Sci 70: 275-282.

Hurlbert RJ. 2000. Methylprednisolone for acute spinal cord injury: An inappropriate standard of care. J Neurosurg 93: 1-7.

Islamov RR, Hendricks WA, Jones RJ, Lyall GJ, Spanier NS, et al. 2002. 17Beta-estradiol stimulates regeneration of sciatic nerve in female mice. Brain Res 943: 283-286.

Ito A, Bebo BF Jr, Matejuk A, Zamora A, Silverman M, et al. 2001. Estrogen treatment down-regulates TNF-alpha production and reduces the severity of experimental autoimmune encephalomyelitis in cytokine knockout mice. J Immunol 167: 542-552.

Johnson DE. 2000. Noncaspase proteases in apoptosis. Leukemia 14: 1695-1703.

Jover T, Tanaka H, Calderone A, Oguro K, Bennett MV, et al. 2002. Estrogen protects against global ischemia-induced neuronal death and prevents activation of apoptotic signaling cascades in the hippocampal CA1. J Neurosci 22: 2115-2124.

Kaja S, Yang SH, Wei J, Fujitani K, Liu R, et al. 2003. Estrogen protects the inner retina from apoptosis and ischemia-induced loss of Vesl-1L/Homer 1c immunoreactive synaptic connections. Invest Ophthalmol Vis Sci 44: 3155-3162.

Kaptanoglu E, Tuncel M, Palaoglu S, Konan A, Demirpence E, et al. 2000. Comparison of the effects of melatonin and methylprednisolone in experimental spinal cord injury. J Neurosurg 93: 77-84.

Kawas C, Resnick S, Morrison A, Brookmeyer R, Corrada M, et al. 1997. A prospective study of estrogen replacement therapy and the risk of developing Alzheimer's disease: The Baltimore Longitudinal Study of Aging. Neurology 48: 1517-1521.

Khachaturian ZS. 1989. The role of calcium regulation in brain aging: Reexamination of a hypothesis. Aging (Milano) 1: 17-34.

Kohli V, Madden JF, Bentley RC, Clavien PA. 1999. Calpain mediates ischemic injury of the liver through modulation of apoptosis and necrosis. Gastroenterology 116: 168-178.

Kondo Y, Suzuki K, Sakuma Y. 1997. Estrogen alleviates cognitive dysfunction following transient brain ischemia in ovariectomized gerbils. Neurosci Lett 238: 45-48.

Koski CL, Hila S, Hoffman GE. 2004. Regulation of cytokine-induced neuron death by ovarian hormones: Involvement of antiapoptotic protein expression and c-JUN N-terminal kinase-mediated proapoptotic signaling. Endocrinology 145: 95-103.

Kucharczak J, Simmons MJ, Fan Y, Gelinas C. 2003. To be, or not to be: NF-kappaB is the answer – role of Rel/NF-kappaB in the regulation of apoptosis. Oncogene 22: 8961-8982.

Kupina NC, Detloff MR, Bobrowski WF, Snyder BJ, Hall ED. 2003. Cytoskeletal protein degradation and neurodegeneration evolves differently in males and females following experimental head injury. Exp Neurol 180: 55-73.

Kupina NC, Nath R, Bernath EE, Inoue J, Mitsuyoshi A, et al. 2001. The novel calpain inhibitor SJA6017 improves functional outcome after delayed administration in a mouse model of diffuse brain injury. J Neurotrauma 18: 1229-1240.

Kurata K, Takebayashi M, Kagaya A, Morinobu S, Yamawaki S. 2001. Effect of beta-estradiol on voltage-gated $Ca(2+)$ channels in rat hippocampal neurons: A comparison with dehydroepiandrosterone. Eur J Pharmacol 416: 203-212.

Kwo S, Young W, Decrescito V. 1989. Spinal cord sodium, potassium, calcium, and water concentration changes in rats after graded contusion injury. J Neurotrauma 6: 13-24.

Li S, Jiang Q, Stys PK. 2000. Important role of reverse $Na(+)$-$Ca(2+)$ exchange in spinal cord white matter injury at physiological temperature. J Neurophysiol 84: 1116-1119.

Li Z, Hogan EL, Banik NL. 1996. Role of calpain in spinal cord injury: Increased calpain immunoreactivity in rat spinal cord after impact trauma. Neurochem Res 21: 441-448.

Liu XZ, Xu XM, Hu R, Du C, Zhang SX, et al. 1997. Neuronal and glial apoptosis after traumatic spinal cord injury. J Neurosci 17: 5395-5406.

Manson JE, Allison MA, Rossouw JE, Carr JJ, Langer RD, et al. 2007. Estrogen therapy and coronary-artery calcification. N Engl J Med 356: 2591-2602.

McGinnis KM, Whitton MM, Gnegy ME, Wang KK. 1998. Calcium/calmodulin-dependent protein kinase IV is cleaved by caspase-3 and calpain in SH-SY5Y human neuroblastoma cells undergoing apoptosis. J Biol Chem 273: 19993-20000.

Mermelstein PG, Becker JB, Surmeier DJ. 1996. Estradiol reduces calcium currents in rat neostriatal neurons via a membrane receptor. J Neurosci 16: 595-604.

Miyamoto N, Mandai M, Suzuma I, Suzuma K, Kobayashi K, et al. 1999. Estrogen protects against cellular infiltration by reducing the expressions of E-selectin and IL-6 in endotoxin-induced uveitis. J Immunol 163: 374-379.

Moosmann B, Behl C. 1999. The antioxidant neuroprotective effects of estrogens and phenolic compounds are independent from their estrogenic properties. Proc Natl Acad Sci USA 96: 8867-8872.

Mor G, Nilsen J, Horvath T, Bechmann I, Brown S, et al. 1999. Estrogen and microglia: A regulatory system that affects the brain. J Neurobiol 40: 484-496.

Morales A, Diaz M, Ropero AB, Nadal A, Alonso R. 2003. Estradiol modulates acetylcholine-induced Ca^{2+} signals in LHRH-releasing GT1-7 cells through a membrane binding site. Eur J Neurosci 18: 2505-2514.

Morelli C, Garofalo C, Bartucci M, Surmacz E. 2003. Estrogen receptor-alpha regulates the degradation of insulin receptor substrates 1 and 2 in breast cancer cells. Oncogene 22: 4007-4016.

Nakagawa T, Yuan J. 2000. Cross-talk between two cysteine protease families. Activation of caspase-12 by calpain in apoptosis. J Cell Biol 150: 887-894.

Nakamizo T, Urushitani M, Inoue R, Shinohara A, Sawada H, et al. 2000. Protection of cultured spinal motor neurons by estradiol. Neuroreport 11: 3493-3497.

National SCI Statistical Center. 2008. Spinal Cord Injury: Facts and Figures at a Glance. National SCI Statistical Center. University of Alabama at Birmingham, AL.

Nicotera P. 2000. Caspase requirement for neuronal apoptosis and neurodegeneration. IUBMB Life 49: 421-425.

Nilsen J, Chen S, Brinton RD. 2002. Dual action of estrogen on glutamate-induced calcium signaling: Mechanisms requiring interaction between estrogen receptors and src/mitogen activated protein kinase pathway. Brain Res 930: 216-234.

Nixon RA, Marotta CA. 1984. Degradation of neurofilament proteins by purified human brain cathepsin D. J Neurochem 43: 507-516.

Palombella VJ, Rando OJ, Goldberg AL, Maniatis T. 1994. The ubiquitin-proteasome pathway is required for processing the NF-kappa B1 precursor protein and the activation of NF-kappa B. Cell 78: 773-785.

Pike BR, Zhao X, Newcomb JK, Wang KK, Posmantur RM, et al. 1998. Temporal relationships between de novo protein synthesis, calpain and caspase 3-like protease activation, and DNA fragmentation during apoptosis in septo-hippocampal cultures. J Neurosci Res 52: 505-520.

Pitts LH, Ross A, Chase GA, Faden AI. 1995. Treatment with thyrotropin-releasing hormone (TRH) in patients with traumatic spinal cord injuries. J Neurotrauma 12: 235-243.

Pizzi M, Goffi F, Boroni F, Benarese M, Perkins SE, et al. 2002. Opposing roles for NF-kappa B/Rel factors p65 and c-Rel in the modulation of neuron survival elicited by glutamate and interleukin-1beta. J Biol Chem 277: 20717-20723.

Polanczyk M, Zamora A, Subramanian S, Matejuk A, Hess DL, et al. 2003. The protective effect of 17beta-estradiol on experimental autoimmune encephalomyelitis is mediated through estrogen receptor-alpha. Am J Pathol 163: 1599-1605.

Posmantur R, Kampfl A, Siman R, Liu J, Zhao X, et al. 1997. A calpain inhibitor attenuates cortical cytoskeletal protein loss after experimental traumatic brain injury in the rat. Neuroscience 77: 875-888.

Pugazhenthi S, Nesterova A, Sable C, Heidenreich KA, Boxer LM, et al. 2000. Akt/protein kinase B up-regulates Bcl-2 expression through cAMP-response element-binding protein. J Biol Chem 275: 10761-10766.

Ray SK, Banik NL. 2003. Calpain and its involvement in the pathophysiology of CNS injuries and diseases: Therapeutic potential of calpain inhibitors for prevention of neurodegeneration. Curr Drug Targets CNS Neurol Disord 2: 173-189.

Ray SK, Hogan EL, Banik NL. 2003a. Calpain in the pathophysiology of spinal cord injury: Neuroprotection with calpain inhibitors. Brain Res Brain Res Rev 42: 169-185.

Ray SK, Wilford GG, Matzelle D, Hogan EL, Banik NL. 1999a. Calpeptin and methylprednisolone inhibit apoptosis in rat spinal cord injury. Ann N Y Acad Sci 890: 261-269.

Ray SK, Wilford GG, Crosby CV, Hogan EL, Banik NL. 1999b. Diverse stimuli induce calpain overexpression and apoptosis in C6 glioma cells. Brain Res 829: 18-27.

Ray SK, Matzelle DC, Wilford GG, Hogan EL, Banik NL. 2000a. E-64-d prevents both calpain upregulation and apoptosis in the lesion and penumbra following spinal cord injury in rats. Brain Res 867: 80-89.

Ray SK, Fidan M, Nowak MW, Wilford GG, Hogan EL, et al. 2000b. Oxidative stress and Ca^{2+} influx upregulate calpain and induce apoptosis in PC12 cells. Brain Res 852: 326-334.

Ray SK, Matzelle DD, Sribnick EA, Guyton MK, Wingrave JM, et al. 2003b. Calpain inhibitor prevented apoptosis and maintained transcription of proteolipid protein and myelin basic protein genes in rat spinal cord injury. J Chem Neuroanat 26: 119-124.

Reiter RJ, Acuna-Castroviejo D, Tan DX, Burkhardt S. 2001. Free radical-mediated molecular damage. Mechanisms for the protective actions of melatonin in the central nervous system. Ann N Y Acad Sci 939: 200-215.

Robertson DM, van Amelsvoort T, Daly E, Simmons A, Whitehead M, et al. 2001. Effects of estrogen replacement therapy on human brain aging: An in vivo 1H MRS study. Neurology 57: 2114-2117.

Roof RL, Hall ED. 2000. Estrogen-related gender difference in survival rate and cortical blood flow after impact-acceleration head injury in rats. J Neurotrauma 17: 1155-1169.

Rosenberg LJ, Wrathall JR. 2001. Time course studies on the effectiveness of tetrodotoxin in reducing consequences of spinal cord contusion. J Neurosci Res 66: 191-202.

Saatman KE, Murai H, Bartus RT, Smith DH, Hayward NJ, et al. 1996. Calpain inhibitor AK295 attenuates motor and cognitive deficits following experimental brain injury in the rat. Proc Natl Acad Sci USA 93: 3428-3433.

Saavedra RA, Murray M, de Lacalle S, Tessler A. 2000. In vivo neuroprotection of injured CNS neurons by a single injection of a DNA plasmid encoding the *Bcl-2* gene. Prog Brain Res 128: 365-372.

Saido TC, Sorimachi H, Suzuki K. 1994. Calpain: New perspectives in molecular diversity and physiological-pathological involvement. FASEB J 8: 814-822.

Saito K, Elce JS, Hamos JE, Nixon RA. 1993. Widespread activation of calcium-activated neutral proteinase (calpain) in the brain in Alzheimer disease: A potential molecular basis for neuronal degeneration. Proc Natl Acad Sci USA 90: 2628-2632.

Samantaray S, Das A, Sribnick EA, Janardhanan R, Matzelle DD, et al. 2007a. Neuroprotective efficacy of estrogen in acute and chronic spinal cord injury in rats. Joint ASN-ISN Meeting.

Samantaray S, Das A, Sribnick EA, Janardhanan R, Matzelle DD, et al. 2007b. Low dose of estrogen provides effective neuroprotection in chronic spinal cord injury in rats. Program No. 902.11, 2007 Neuroscience Meeting Planer, San Diego, CA: Society for Neuroscience, 2007. Online.

Samantaray S, Sribnick EA, Das A, Knaryan VH, Matzelle DD, et al. 2008. Melatonin attenuates calpain upregulation, axonal damage and neuronal death in spinal cord injury in rats. J Pineal Res 44: 348-357.

Santagati S, Melcangi RC, Celotti F, Martini L, Maggi A. 1994. Estrogen receptor is expressed in different types of glial cells in culture. J Neurochem 63: 2058-2064.

Sarin A, Clerici M, Blatt SP, Hendrix CW, Shearer GM, et al. 1994. Inhibition of activation-induced programmed cell death and restoration of defective immune responses of HIV+ donors by cysteine protease inhibitors. J Immunol 153: 862-872.

Schneider A, Martin-Villalba A, Weih F, Vogel J, Wirth T, et al. 1999. NF-kappaB is activated and promotes cell death in focal cerebral ischemia. Nat Med 5: 554-559.

Simpkins JW, Rajakumar G, Zhang YQ, Simpkins CE, Greenwald D, et al. 1997. Estrogens may reduce mortality and ischemic damage caused by middle cerebral artery occlusion in the female rat. J Neurosurg 87: 724-730.

Singer CA, Figueroa-Masot XA, Batchelor RH, Dorsa DM. 1999. The mitogen-activated protein kinase pathway mediates estrogen neuroprotection after glutamate toxicity in primary cortical neurons. J Neurosci 19: 2455-2463.

Singh M, Sotalo G Jr, Guan X, Warren M, Toran-Allerand CD. 1999. Estrogen-induced activation of mitogen-activated protein kinase in cerebral cortical explants: Convergence of estrogen and neurotrophin signaling pathways. J Neurosci 19: 1179-1188.

Springer JE, Azbill RD, Knapp PE. 1999. Activation of the caspase-3 apoptotic cascade in traumatic spinal cord injury. Nat Med 5: 943-946.

Springer JE, Azbill RD, Nottingham SA, Kennedy SE. 2000. Calcineurin-mediated BAD dephosphorylation activates the caspase-3 apoptotic cascade in traumatic spinal cord injury. J Neurosci 20: 7246-7251.

Springer JE, Azbill RD, Kennedy SE, George J, Geddes JW. 1997. Rapid calpain I activation and cytoskeletal protein degradation following traumatic spinal cord injury: Attenuation with riluzole pretreatment. J Neurochem 69: 1592-1600.

Squier MK, Miller AC, Malkinson AM, Cohen JJ. 1994. Calpain activation in apoptosis. J Cell Physiol 159: 229-237.

Sribnick EA, Ray SK, Banik NL. 2006a. Estrogen prevents glutamate-induced apoptosis in C6 glioma cells by a receptor-mediated mechanism. Neuroscience 137: 197-209.

Sribnick EA, Matzelle DD, Ray SK, Banik NL. 2006b. Estrogen treatment of spinal cord injury attenuates calpain activation and apoptosis. J Neurosci Res 84: 1064-1075.

Sribnick EA, Matzelle DD, Banik NL, Ray SK. 2007. Direct evidence for calpain involvement in apoptotic death of neurons in spinal cord injury in rats and neuroprotection with calpain inhibitor. Neurochem Res 32: 2210-2216.

Sribnick EA, Ray SK, Nowak MW, Li L, Banik NL. 2004. 17beta-estradiol attenuates glutamate-induced apoptosis and preserves electrophysiologic function in primary cortical neurons. J Neurosci Res 76: 688-696.

Sribnick EA, Wingrave JM, Matzelle DD, Wilford GG, Ray SK, et al. 2005. Estrogen attenuated markers of inflammation and decreased lesion volume in acute spinal cord injury in rats. J Neurosci Res 82: 283-293.

Stein DG. 2001. Brain damage, sex hormones and recovery: A new role for progesterone and estrogen? Trends Neurosci 24: 386-391.

Stokes BT, Fox P, Hollinden G. 1983. Extracellular calcium activity in the injured spinal cord. Exp Neurol 80: 561-572.

Sur P, Sribnick EA, Wingrave JM, Nowak MW, Ray SK, et al. 2003. Estrogen attenuates oxidative stress-induced apoptosis in C6 glial cells. Brain Res 971: 178-188.

Tator CH. 1991. Review of experimental spinal cord injury with emphasis on the local and systemic circulatory effects. Neurochirurgie 37: 291-302.

Tiidus PM, Holden D, Bombardier E, Zajchowski S, Enns D, et al. 2001. Estrogen effect on post-exercise skeletal muscle neutrophil infiltration and calpain activity. Can J Physiol Pharmacol 79: 400-406.

Tiwari-Woodruff S, Morales LB, Lee R, Voskuhl RR. 2007. Differential neuroprotective and antiinflammatory effects of estrogen receptor (ER)alpha and ERbeta ligand treatment. Proc Natl Acad Sci USA 104: 14813-14818.

Toran-Allerand CD. 1999. Novel mechanisms of estrogen action in the developing brain: Role of steroid-neruotrophin interactions. Neurosteroids: A New Regulatory Function in the Nervous System. Baulieu EE, Schumacher M, Robel P, editors. Humana Press, Totowa, NJ, pp. 293–316.

Toran-Allerand CD, Miranda RC, Bentham WD, Sohrabji F, Brown TJ, et al. 1992. Estrogen receptors colocalize with low-affinity nerve growth factor receptors in cholinergic neurons of the basal forebrain. Proc Natl Acad Sci USA 89: 4668-4672.

Vegeto E, Pollio G, Ciana P, Maggi A. 2000. Estrogen blocks inducible nitric oxide synthase accumulation in LPS-activated microglia cells. Exp Gerontol 35: 1309-1316.

Vegeto E, Belcredito S, Etteri S, Ghisletti S, Brusadelli A, et al. 2003. Estrogen receptor-alpha mediates the brain antiinflammatory activity of estradiol. Proc Natl Acad Sci USA 100: 9614-9619.

Wada S, Yone K, Ishidou Y, Nagamine T, Nakahara S, et al. 1999. Apoptosis following spinal cord injury in rats and preventative effect of N-methyl-D-aspartate receptor antagonist. J Neurosurg 91: 98-104.

Wagner AK, Willard LA, Kline AE, Wenger MK, Bolinger BD, et al. 2004. Evaluation of estrous cycle stage and gender on behavioral outcome after experimental traumatic brain injury. Brain Res 998: 113-121.

Wang L, Andersson S, Warner M, Gustafsson JA. 2003. Estrogen receptor (ER)beta knockout mice reveal a role for ERbeta in migration of cortical neurons in the developing brain. Proc Natl Acad Sci USA 100: 703-708.

Weaver CE Jr, Park-Chung M, Gibbs TT, Farb DH. 1997. 17beta-Estradiol protects against NMDA-induced excitotoxicity by direct inhibition of NMDA receptors. Brain Res 761: 338-341.

Wen Y, Yang S, Liu R, Perez E, Yi KD, et al. 2004. Estrogen attenuates nuclear factor-kappa B activation induced by transient cerebral ischemia. Brain Res 1008: 147-154.

Wingrave JM, Schaecher KE, Sribnick EA, Wilford GG, Ray SK, et al. 2003. Early induction of secondary injury factors causing activation of calpain and mitochondria-mediated neuronal apoptosis following spinal cord injury in rats. J Neurosci Res 73: 95-104.

Wingrave JM, Sribnick EA, Wilford GG, Matzelle DD, Mou JA, et al. 2004. Higher calpastatin levels correlate with resistance to calpain-mediated proteolysis and neuronal apoptosis in juvenile rats after spinal cord injury. J Neurotrauma 21: 1240-1254.

Xu GY, Hughes MG, Ye Z, Hulsebosch CE, McAdoo DJ. 2004. Concentrations of glutamate released following spinal cord injury kill oligodendrocytes in the spinal cord. Exp Neurol 187: 329-336.

Xu H, Gouras GK, Greenfield JP, Vincent B, Naslund J, et al. 1998. Estrogen reduces neuronal generation of Alzheimer beta-amyloid peptides. Nat Med 4: 447-451.

Yanase M, Sakou T, Fukuda T. 1995. Role of N-methyl-D-aspartate receptor in acute spinal cord injury. J Neurosurg 83: 884-888.

Young W, Flamm ES. 1982. Effect of high-dose corticosteroid therapy on blood flow, evoked potentials, and extracellular calcium in experimental spinal injury. J Neurosurg 57: 667-673.

Yu CG, Geddes JW. 2007. Sustained calpain inhibition improves locomotor function and tissue sparing following contusive spinal cord injury. Neurochem Res 32: 2046-2053.

Yune TY, Kim SJ, Lee SM, Lee YK, Oh YJ, et al. 2004. Systemic administration of 17beta-estradiol reduces apoptotic cell death and improves functional recovery following traumatic spinal cord injury in rats. J Neurotrauma 21: 293-306.

26 Immunotherapy Strategies for Lewy Body and Parkinson's Diseases

L. Crews · B. Spencer · E. Masliah

1	Introduction	600
2	*Role of α-Syn in the Pathogenesis and Treatment of LBD*	*602*
2.1	α-Syn Oligomerization in LBD/PD	602
2.2	Novel Strategies for the Treatment of LBD Based on the Lysosomal Clearance of α-Syn Aggregates: A Role for Immunotherapy	603
3	*In vivo Evaluation of Immunotherapies in Synucleopathies*	*606*
3.1	The use of Transgenic Models Expressing α-Syn to Investigate Immunotherapeutic Approaches in LBD/PD	606
4	*Conclusions*	*609*

© 2009 Springer Science+Business Media, LLC.

Abstract: Accumulation of α-synuclein (α-syn) resulting in the formation of oligomers and protofibrils has been linked to the neurodegenerative process in Parkinson's disease and Lewy body dementia. Genetic and environmental factors affecting the rate of α-syn production, aggregation, and clearance might play an important role. Therefore, development of new therapies will require either reducing α-syn synthesis and rate of aggregation, or enhancing clearance. Clearance of α-syn aggregates depends on lysosomal and nonlysosomal pathways. Among the various approaches to promote clearance of α-syn aggregates, immunotherapy is of special interest because antibodies can specifically target abnormal α-syn aggregates. Experimental studies have shown that active as well as passive immunization might reduce α-syn accumulation and associated deficits. Similarly, strategies employing anti-α-syn intrabodies, where cells are genetically modified to produce anti-α-syn antibodies, rescue the phenotype associated with α-syn aggregates. In addition, cellular immunotherapy with copolymer has been shown to reduce neurodegeneration in acute models such as 1-methyl 4-phenyl 1,2,3,6-tetrahydropyridine (MPTP) toxicity. The mechanisms through which antibodies against α-syn aggregates might ameliorate the deficits in mice and cellular models are under investigation. Antibodies might act by recognizing α-syn oligomers accumulating in the neuronal membranes and promote degradation of α-syn aggregates via endosomal–lysosomal pathways and autophagy. Therapeutic approaches focusing on the combination of antibodies and regulation of cellular immune responses might prove in the future to be effective.

List of Abbreviations: CMA, chaperone-mediated autophagy; DLB, dementia with LBs; LBD, Lewy body disease; LRRK2, leucine-rich repeat kinase-2; MPTP, 1-methyl 4-phenyl 1,2,3,6-tetrahydropyridine; MSA, multiple systems atrophy; PDD, PD dementia; PDGF-β, platelet-derived growth factor-β; PINK1, PTEN-induced putative kinase-1; PrP, prion protein; UCH-L1, ubiquitin carboxyl-terminal esterase-L1

1 Introduction

In recent years new hope for understanding the pathogenesis of Parkinson's disease (PD) and Lewy body disease (LBD) has emerged with the discovery of mutations and duplications in the α-synuclein (α-syn) gene that are associated with rare familial forms of parkinsonism (Polymeropoulos et al., 1997; Kruger et al., 1998; Singleton et al., 2003). Moreover, it has been shown that α-syn is centrally involved in the pathogenesis of both sporadic and inherited forms of PD and LBD because this molecule accumulates in Lewy bodies (LBs) (Spillantini et al., 1997; Wakabayashi et al., 1997; Takeda et al., 1998), synapses, and axons. Overexpression of α-syn in transgenic (tg) mice (Masliah et al., 2000; Lee et al., 2002; Lee et al., 2004) and Drosophila (Feany and Bender, 2000) mimics several aspects of PD. α-Syn is the major component of the LBs and Lewy neurites, which also contain ubiquitin and neurofilament, among several other components (Trojanowski and Lee, 1998).

LBD is a heterogeneous group of disorders that includes PD and dementia with LBs (DLB) (Kosaka et al., 1984; Hansen and Galasko, 1992; McKeith, 2000) and is characterized by the degeneration of dopaminergic systems (Shastry, 2001), nondopaminergic systems (Halliday et al., 2005; Burn, 2006), motor alterations (Braak et al., 2002), cognitive impairments (Salmon et al., 1989), and formation of LBs in cortical and subcortical regions (Trojanowski and Lee, 1998). The new consortium criterion for the classification of LBD recognizes two clinical entities, the first denominated DLB and the second PD dementia (PDD) (McKeith et al., 1996; Aarsland et al., 2004; Burn, 2006; Lippa et al., 2007). While in patients with DLB the clinical presentation is of dementia followed by parkinsonism, in patients with PDD the initial signs are of parkinsonism followed by dementia (McKeith et al., 1996; Litvan et al., 1998; Janvin et al., 2006).

The mechanisms through which α-syn leads to neurodegeneration and the characteristic symptoms of LBD are unclear. However, recent evidence indicates that abnormal accumulation of misfolded α-syn in the synaptic terminals and axons play an important role (Iwatsubo et al., 1996; Trojanowski et al., 1998; Hashimoto and Masliah, 1999; Lansbury, 1999) (❯ *Figure 26-1*). These studies suggest that α-syn oligomers and protofibrils rather than fibrils might be the neurotoxic species (Conway et al., 2000).

◘ Figure 26-1
Pathogenesis of Parkinson's disease

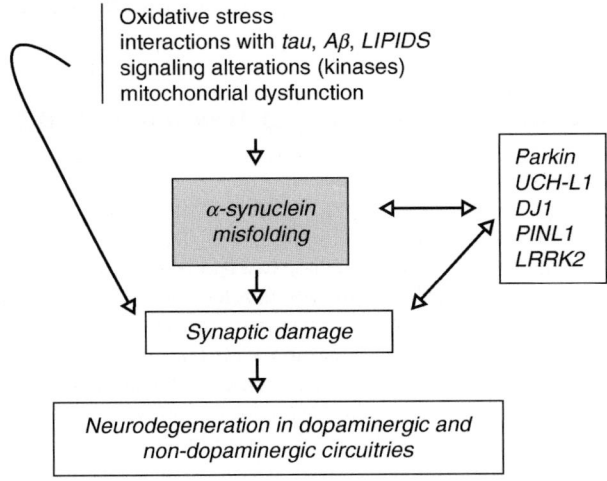

In addition to α-syn, recent studies in families with inherited forms of parkinsonism have uncovered the involvement of mutations in several genes involved in regulating mitochondrial and proteasomal function (Gasser, 2007; Thomas and Beal, 2007). The most common mutations are on Parkin, followed by leucine-rich repeat kinase-2 (LRRK2), PTEN-induced putative kinase-1 (PINK1), DJ1, ubiquitin carboxyl-terminal esterase-L1 (UCH-L1), and others (Kitada et al., 1998; Hattori et al., 2003; Golbe and Mouradian, 2004) (❯ *Figure 26-1*). Although these studies suggest a polygenetic origin for LBD/PD, it is worth noting that in most cases of sporadic and familial parkinsonism, regardless of the genetic causes, accumulation of the α-syn protein is a central event. Moreover, supporting a primary role of accumulation of α-syn in disorders with LBs and parkinsonism, other studies have shown that in acute models, the presence of α-syn is necessary for toxicity. For example, while α-syn-deficient mice are resistant to 1-methyl 4-phenyl 1,2,3,6-tetrahydropyridine (MPTP) and 6-hydroxy dopamine (6-OHDA) (Klivenyi et al., 2006), α-syn tg mice are more sensitive to the toxic insult of MPTP (Song et al., 2004; Nieto et al., 2006; Yu et al., 2008).

Thus, decreasing the accumulation of α-syn oligomers and protofibrils is a major target for the development of novel therapies for LBD/PD. Accumulation of α-syn is governed by the rate of synthesis, aggregation, and clearance of this molecule (❯ *Figure 26-2*). Therapies directed at reducing accumulation of α-syn oligomers could target any of these three factors. Immunotherapy might be effective by promoting

◘ Figure 26-2
Factors influencing α-synuclein (α-syn) aggregation

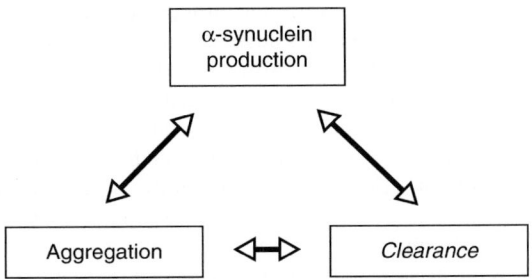

the clearance of α-syn oligomers and protofibrils. Therefore, although it is important to consider in the future therapies targeting other pathways and genes involved (Parkin, LRRK2, PINK1) in PD, the main focus of the present handbook will be on immunotherapy strategies directed at reducing the accumulation of α-syn.

2 Role of α-Syn in the Pathogenesis and Treatment of LBD

2.1 α-Syn Oligomerization in LBD/PD

α-Syn is an abundant presynaptic molecule (Iwai et al., 1994) that plays a role in modulating vesicular synaptic release (Murphy et al., 2000). Synucleins belong to a family of related proteins including α-, β-, and γ-syn. α-Syn belongs to a class of so-called "naturally unfolded proteins" (Lansbury, 1999; Wright and Dyson, 1999). Such proteins do not have a stable tertiary structure and during their existence change their conformations. Human α-syn is a protein of 140 amino acids (aa) in length, β-syn contains 134 aa, and γ-syn has 127 aa. Each of the synucleins is composed of an N-terminal lipid-binding domain containing 11 residue repeats and a C-terminal acidic domain that has been proposed to be involved in protein–protein interactions. It has been shown (Davidson et al., 1998; Chandra et al., 2003; Ulmer et al., 2005; Kamp and Beyer, 2006) that at the lipid–protein interface, α-syn has a conformation characterized by two helical domains interrupted by a short nonhelical turn. α-Syn contains a highly amyloidogenic hydrophobic domain in the N-terminus region (aa 60–95), which is partially absent in β-syn and might explain why β-syn has a reduced ability to self-aggregate and form oligomers and fibrils (Hashimoto et al., 2001; Uversky et al., 2002). Moreover, previous studies have shown that β-syn interacts with α-syn and is capable of preventing α-syn aggregation and related deficits both in vitro and in vivo (Hashimoto et al., 2001).

Various lines of evidence support the contention that abnormal aggregates arise from a partially folded intermediate precursor that contains hydrophobic patches. It has been proposed that the intermediate α-syn oligomers form annular protofibrils and pore-like structures (Ding et al., 2002; Volles and Lansbury, 2002; Lashuel et al., 2003; Rochet et al., 2004). The mechanisms through which monomeric α-syn converts into a toxic oligomer and later into fibrils is currently under intense investigation. Recent studies suggest that α-syn oligomerization might occur on the membrane and involves interactions between hydrophobic residues of the amphipathic α-helices of α-syn (Zhu et al., 2003). These studies indicate that the hydrophobic lipid binding domains in the N-terminal region might be important in modulating α-syn aggregation (Conway et al., 1998; Lansbury, 1999; Uversky et al., 2001; Jao et al., 2004).

The conformational state of α-syn at the initial and end stages of fibrillation has been characterized in some detail and recent studies have shown that early stage oligomers are globular structures with variable height (2–6 nm), and that prolonged incubation results in the formation of elongated protofibrils, which disappear upon fibril formation (Apetri et al., 2006). Molecular modeling and molecular dynamics simulations showed that α-syn homodimers could adopt nonpropagating (head-to-tail) and propagating (head-to-head) conformations (Tsigelny et al., 2007) (❷ *Figure 26-3*). Propagating α-syn dimers on the membrane incorporate additional α-syn molecules, leading to the formation of pentamers and hexamers, which form rings suggestive of pore-like structures (Tsigelny et al., 2007) (❷ *Figure 26-3*). Cells expressing α-syn displayed increased ion current activity consistent with the formation of Zn^{2+}-sensitive nonselective cation channels. More recent studies showed that human neuronal cells expressing mutant α-syn have high plasma membrane ion permeability that was sensitive to calcium chelators (Furukawa et al., 2006). Oligomers form complexes in the membranes of neurons that facilitate abnormal calcium currents that might disturb synaptic and neuronal function leading to neurodegeneration (Danzer et al., 2007).

In conclusion, it is likely that α-syn oligomers rather than fibrils might be responsible for the neurodegenerative process in LBD/PD. The LBs, which primarily contain α-syn fibrils, might represent a cellular mechanism to isolate more toxic oligomers. The α-syn oligomers most likely associate with the neuronal membranes and synapses, interfering with neurotransmission and plasticity. Thus, better understanding the steps involved in the process of α-syn aggregation is important in order to develop intervention strategies that might prevent or reverse α-syn oligomerization and toxic conversion.

◘ Figure 26-3
Steps involved in α-syn oligomerization

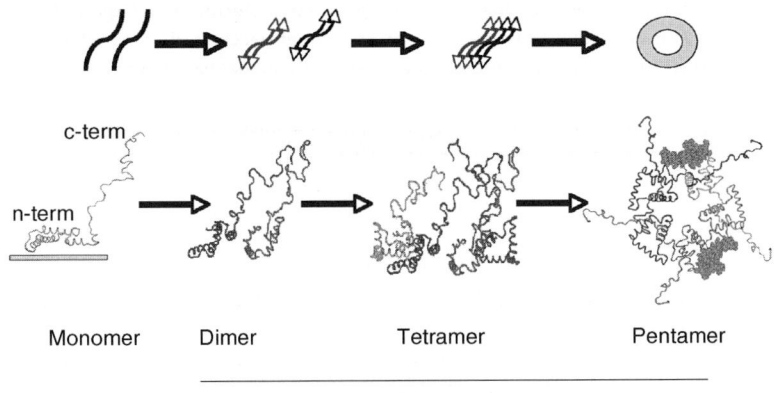

2.2 Novel Strategies for the Treatment of LBD Based on the Lysosomal Clearance of α-Syn Aggregates: A Role for Immunotherapy

Treatments for Alzheimer's disease (AD), PD, and LBD have been centered on improving neurotransmission by utilizing agonists of the cholinergic and dopaminergic systems, and more recently by delivering neurotrophic factors that will protect selective neuronal populations from the toxic effects of the AD-related amyloid-β protein (Aβ). In view of the growing body of evidence indicating that accumulation of abnormal aggregates, such as oligomers and protofibrils, plays a central role in the pathogenesis of these disorders (Lashuel et al., 2002; Lashuel et al., 2003), considerable efforts have been devoted toward developing new treatment strategies designed around either blocking the toxicity of aggregates, preventing aggregation or promoting their degradation and clearance. In the case of AD, because Aβ is secreted into the extracellular space, significant progress has been made in promoting Aβ clearance with the development of several vaccination approaches, where antibodies directed against Aβ target this molecule to the microglia/macrophages for lysosomal degradation (Schenk et al., 1999; Games et al., 2000; Morgan et al., 2000; Schenk et al., 2000). In addition to this mechanism, it has been proposed that immunotherapy might work by altering the balance between Aβ in the CNS and plasma (Holtzman et al., 1999).

In the case of LBD, the challenge is greater because misfolded α-syn accumulates intracellularly. To begin tackling this problem with an anti-aggregation tactic, we developed a gene therapy approach, where a lentiviral vector construct was used to deliver β-syn into the neocortex and hippocampus of α-syn tg animals (Hashimoto et al., 2004). Although this method demonstrated effectiveness in protecting the neurons and in promoting the clearance of toxic α-syn aggregates, the effects are limited to the site of the lentivirus injection since β-syn is not a secreted protein. Since the neurodegenerative process and the patterns of neurodegeneration in LBD are more widespread than originally suspected, there is the need for developing alternative approaches that will target the toxic α-syn in multiple neuronal populations simultaneously. For this reason, we began to explore the possibility of developing an immunotherapy approach for LBD.

In support of the feasibility of such a therapy, previous studies have shown in vitro that intracellular antibodies (intrabodies) can inhibit α-syn aggregation (Emadi et al., 2004; Zhou et al., 2004) and in vivo immunotherapy with copolymer-1 reduces neurodegeneration in the MPTP model of PD (Benner et al., 2004; Reynolds et al., 2007). More recently, we found that in vaccinated α-syn tg mice that produced high-affinity antibodies against α-syn, there was decreased accumulation of aggregated human α-syn in neuronal cell bodies and synapses that was associated with reduced neurodegeneration (Masliah et al., 2005) (❶ *Figure 26-4*). Remarkably, double-labeling immunocytochemistry and passive immunization

Figure 26-4

Epitope mapping of anti-α-syn antibodies generated by the vaccinated mice. The most effective antibodies were those against the C-terminus. (a) Diagram of α-synuclear molecule and regions that the antibodies recognized. (b) α-syn immunoactivity in the brain of an animal vaccinated using CFA (Complete Found's Adjuvant). (c) Reduced α-syn immunoreactivity in the brain of an animal vaccinated with α-syn

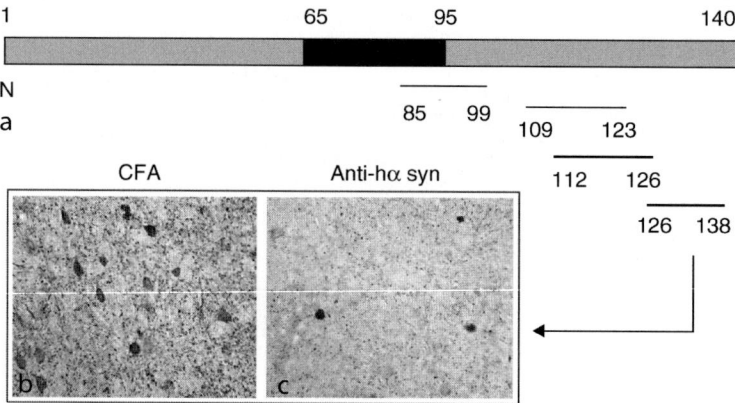

experiments showed that the antibodies recognized α-syn associated with the neuronal membrane and promoted the degradation of α-syn aggregates via lysosomal pathways (Masliah et al., 2005). The most effective antibodies appear to be those recognizing the C-terminus of α-syn (❷ *Figure 26-4*), which is of interest because this domain of α-syn appears to play an important role in the process α-syn aggregation and oligomerization (Masliah et al., 2005) (❷ *Figure 26-4*).

Motor deficits are predominant in patients with LBD/PD; however, studies have shown that nonmotor alterations precede the motor symptoms. Moreover, in addition to the degeneration of the dopaminergic system, patients with LBD/PD display extensive degeneration of cholinergic (nucleus basalis), serotoninergic, (raphe nucleus), adrenergic (locus ceruleus), and glutaminergic cells (hippocampus). For these reasons, although most of the emphasis when evaluating therapeutic approaches in LBD/PD models is on assessment of motor behavior, dopaminergic cells, and inclusion formation, other endpoints need to be considered. For example, it is important to evaluate: (1) the effects on memory and learning tests, (2) the titers of the antibodies, (3) their affinity for aggregated versus monomeric forms of α-syn, (4) the levels of aggregated insoluble and soluble α-syn, and (5) the effects on selected nondopaminergic neuronal populations, and synaptic and dendritic density.

The mechanisms as to how the antibodies against α-syn specifically identify affected neurons are not completely clear. However, antibodies might be able to recognize abnormal α-syn accumulating in the neuronal plasma membrane (Masliah et al., 2005). Previous studies have shown that α-syn is secreted in vitro (Lee et al., 2005), and that α-syn is present in the CSF of α-syn tg mice and in patients with LBD (Borghi et al., 2000). Given that circulating antibodies might be able to recognize membrane-bound or secreted α-syn, there are several possibilities as to how they might promote the clearance of intracellular aggregates. One such possibility involves antibodies in close opposition with the neuronal surface that might enter the neurons alone or in association with membrane-bound α-syn, and promote lysosomal degradation. The intracellular mechanisms that mediate the clearance of α-syn aggregates by antibodies are not well-understood either; however, preliminary data supports the possibility that activation of the autophagy pathway might be involved (❷ *Figure 26-5*).

The autophagy pathway is a regulated process of turnover of cellular components that occurs in long-lived cells, during development and under stress conditions such as starvation (Meijer and Codogno, 2004). There are three distinct autophagic pathways (Larsen and Sulzer, 2002; Cuervo, 2004): (1) macroautophagy, (2) microautophagy, and (3) chaperone-mediated autophagy (CMA) (❷ *Figure 26-5*). In macroautophagy, organelles and macromolecular components are first surrounded by a double membrane, designated

◘ Figure 26-5
Pathways that might promote the clearance of α-syn aggregates

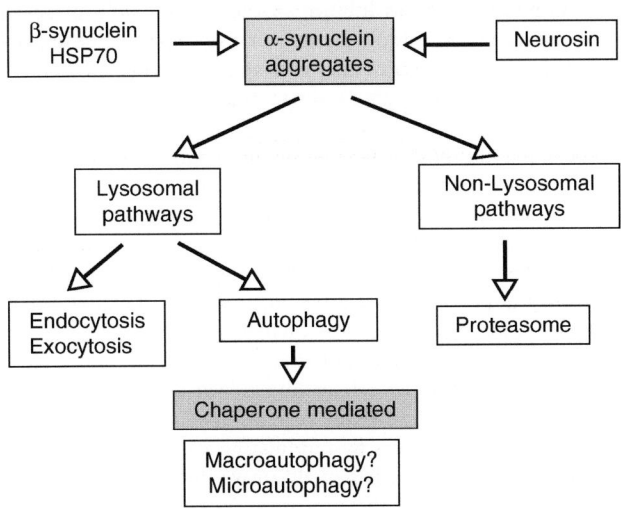

the autophagosome or autophagic vacuole (AV), which then fuses with lysosomes to form autophagolysosomes. In microautophagy, the lysosome invaginates its own membrane, resulting in the uptake of segments of the cytoplasm (Shintani and Klionsky, 2004). Finally, in CMA, individual proteins are targeted to lysosomes for degradation (Majeski and Dice, 2004). This process is tightly regulated, but under extreme conditions can result in cell death and carcinogenesis (Gozuacik and Kimchi, 2004). Autophagy has been linked to neuronal cell death (Edinger and Thompson, 2004; Komatsu et al., 2006) and is activated in mouse models of neurodegeneration and in neurodegenerative disorders such as AD, PD, and Huntington's disease (HD) (Bahr and Bendiske, 2002; Nixon et al., 2005). Recent studies have suggested that alterations in lysosomal functioning and autophagy might participate in the mechanisms of α-syn-mediated neurodegeneration (Stefanis et al., 2001; Meredith et al., 2002; Cuervo et al., 2004; Rideout et al., 2004; Nakajima et al., 2005; Rockenstein et al., 2005) (❯ *Figure 26-5*). Further supporting a role for lysosomal dysfunction in LBD, recent studies have shown that in lysosomal storage disorders, such as Gaucher disease (Tayebi et al., 2001; Varkonyi et al., 2003) and Niemann-Pick disease (Saito et al., 2004), there is increased susceptibility to develop parkinsonism and α-syn accumulation.

Autophagy participates both in protecting the CNS by mediating the lysosomal degradation of α-syn aggregates, and in injuring the CNS by playing a role in the mechanisms of neurodegeneration. This apparent paradox might be resolved by considering that under basal conditions, when low levels of α-syn oligomers are formed, the autophagic pathways, in combination with other clearance mechanisms [proteasome, α-syn-degrading enzymes, β-syn, heat shock protein-70 (HSP70)] might be sufficient to eliminate toxic α-syn aggregates (❯ *Figure 26-5*). While all forms of α-syn may be cleared by autophagy, perturbations in the autophagy pathways have the greatest impact on the clearance of A53T mutant α-syn compared with A30P and wild-type (wt) forms (Stefanis et al., 2001; Webb et al., 2003; Cuervo et al., 2004). Recent studies suggest that α-syn aggregates might be degraded primarily via CMA and that mutant-type α-syn binds to the receptor for this pathway, blocking lysosomal degradation (Cuervo et al., 2004), suggesting that unlike wt α-syn, mutant forms are not cleared by CMA. However, it is possible that other autophagic mechanisms might be involved, since cross-talk among autophagic pathways occurs.

Recent studies have shown that the clearance of aggregate-prone proteins such as α-syn is dependent on macroautophagy (Webb et al., 2003; Williams et al., 2006). Furthermore, clearance of mutant forms of α-syn is retarded when autophagy is inhibited (Webb et al., 2003). Therefore, induction of the autophagic process through disinhibition of mTOR, a signaling regulator of autophagy, is a primary target of therapeutic approaches aimed at stimulating the autophagic clearance of α-syn (Ravikumar et al., 2002;

Webb et al., 2003; Sarkar et al., 2005; Berger et al., 2006; Williams et al., 2006). In some studies, rapamycin, an activator of macroautophagy, enhanced the clearance of all forms of α-syn (Webb et al., 2003). Other studies show that other compounds such as lithium (Sarkar et al., 2005) or the disaccharide trehalose (Sarkar et al., 2007) might also stimulate autophagic clearance of α-syn. While lithium, like rapamycin, acts on the mTOR pathway, trehalose-induced autophagy of α-syn may be mTOR-independent, or alternatively may act downstream of mTOR (Sarkar et al., 2007), suggesting the potential involvement of other pathways in the autophagic clearance of α-syn.

Immunotherapy has been previously shown to induce elimination of viral particles and other antigens via autophagy (Paludan et al., 2005; Komatsu et al., 2007). Similarly, antibodies against α-syn might trigger clearance of oligomers via lysosomal pathways. Supporting this possibility, our previous studies have shown that immunotherapy increases cathepsin-D immunoreactivity and ultrastructural analysis demonstrated the presence of autophagosomes containing α-syn. This suggests that combining immunotherapy with small compounds promoting autophagy (e.g., rapamycin) might be of significant interest. In conclusion, immunotherapeutic approaches for LBD/PD can be divided into those targeting the α-syn aggregates in the membranes of affected neurons by using anti-α-syn antibodies transferred via passive or active immunization, or by engineering vectors to deliver humanized antibodies directly to the affected cells (intrabodies) or from external sources. Alternatively, regulation of T-cell and macrophage responses might also ameliorate the neurodegenerative process by modulating neuroinflammatory and cytotoxic cellular mediators. The combination of these approaches might prove in the future to be more effective than the application of an individual immunotherapeutic approach.

3 In vivo Evaluation of Immunotherapies in Synucleopathies

3.1 The use of Transgenic Models Expressing α-Syn to Investigate Immunotherapeutic Approaches in LBD/PD

Animal models have been proven to be of critical importance to the development of novel therapies for neurodegenerative disorders. Careful consideration of the pros and cons of each of the models in terms of testing immunotherapies is critical in order to consider the best models for a given approach. While for AD and fronto-temporal dementia (FTD), accumulation of Aβ protein and phosphorylated Tau has become a major target for drug development and tg animal modeling, in PD and LBD the central target has been α-syn in tg models. In pharmacological models the target has been dopaminergic neurons using MPTP as a neurotoxin. In addition to α-syn, mutations in Parkin, UCH-L1, LRRK2, DJ1, and PINK1 have been also linked to familial forms of parkinsonism and are potential targets for treatment development and modeling in rodents.

Since progressive intraneuronal aggregation of α-syn has been proposed to play a central role in the pathogenesis of PD and related disorders (Hashimoto and Masliah, 1999; Trojanowski and Lee, 2000; Volles and Lansbury, 2002), most tg models have been focused on investigating the in vivo effects of α-syn accumulation utilizing neuron-specific promoters. Several recent reviews have been published dealing with this subject (Hashimoto et al., 2003; Fernagut and Chesselet, 2004). Among these models, overexpression of wt α-syn under the regulatory control of the platelet-derived growth factor-β (PDGF-β) promoter results in motor deficits, dopaminergic loss, and formation of inclusion bodies (Masliah et al., 2000) (❏ Figures 26-6a, b). Mice with the highest levels of expression (line D) showed intraneuronal accumulation of α-syn that started at 3 months of age and was accompanied by the loss of tyrosine hydroxylase (TH) fibers in the caudo-putamen region and synapses in the temporal cortex. Although no apparent neuronal loss was detected in the substantia nigra (SN), measurements of dopamine levels in the caudo-putamen region showed a 25–50% reduction at 12 months of age. Analysis of locomotor activity in the open field showed that these tg mice had increased thigmotaxis, a behavior that has been associated with dopaminergic activity. Supporting a role for alterations in the dopaminergic system, this abnormal behavior was ameliorated upon treatment with apomorphine, a compound known to promote dopamine

Figure 26-6

α-Syn models of Lewy body disease/Parkinson's disease. (a) Diagram of constructs used to generate mThy1 and PDGF-αsyn trasgain mice. (b) Damage to dendritic structure, and α-syn inclusions in the brains of α-syn tg mice. (c) Behavioral alteration in α-syn tg mice

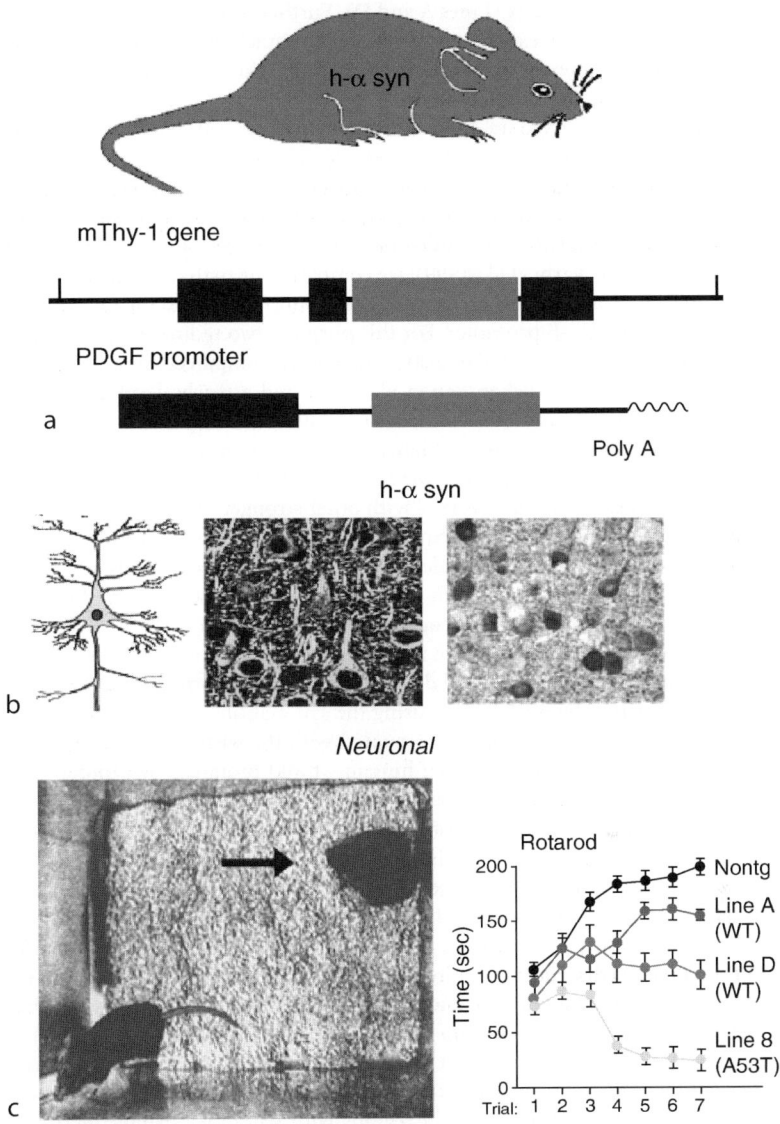

release. Furthermore, tg mice showed mild to moderate motor deficits in the rotarod (● Figure 26-6b), particularly in mice with the greatest loss of dopamine, indicating that more substantial deficits (>75%) of this transmitter might be necessary for more overt deficits to appear.

To determine whether different levels of human α-syn expression from distinct promoters might result in neuropathology mimicking other synucleopathies, we also compared patterns of human α-syn accumulation in the brains of tg mice expressing this molecule from the murine (m)Thy-1 and PDGF-β promoters

(Rockenstein et al., 2002) (◉ *Figure 26-6a*). In mThy-1-α-syn tg mice, α-syn accumulation occurs primarily in synapses and neurons throughout the brain, including the thalamus, basal ganglia, SN, and brainstem. Expression of α-syn from the PDGF-β promoter resulted in accumulation in synapses of the neocortex, limbic system, and olfactory regions as well as formation of inclusion bodies in neurons in deeper layers of the neocortex (Lines A and D). Furthermore, one of the intermediate expressor lines (line M) displayed human α-syn expression in glial cells, mimicking some features of multiple systems atrophy (MSA). These results show a more widespread accumulation of α-syn in tg mouse brains. Taken together, these studies support the contention that α-syn expression in tg mice might mimic some neuropathological alterations observed in LBD and other synucleopathies such as MSA (Rockenstein et al., 2002).

However, from these studies, the potential relationship among synaptic damage, α-syn oligomerization and fibrillation, inclusion formation, and neurological deficits is still unclear. Because previous studies have shown that mutations associated with familial parkinsonism accelerate α-syn aggregation and oligomerization (Conway et al., 1998; Narhi et al., 1999), we compared the patterns of neurodegeneration, α-syn aggregation, and neurological alterations in tg mice expressing wt or mutant (A53T) human α-syn at comparable levels under the PDGF-β promoter. For this purpose, two tg lines that express moderate levels of α-syn in a similar anatomical and cellular distribution were compared: Line A mice that express wt human α-syn (Masliah et al., 2000; Rockenstein et al., 2002) and a newly developed Line 8 that express mutant human α-syn (A53T) (Hashimoto et al., 2003). Additional comparisons were performed against our well-described Line D mice that express higher levels of wt human α-syn (Masliah et al., 2000; Rockenstein et al., 2002). Lower expresser lines were selected for these experiments to avoid unintended effects of very high levels of expression as observed with other stronger promoters such as the mThy-1 and the prion protein (PrP) promoters. These mice express about one third of the levels of α-syn compared with our higher expresser Line D, which is equivalent to a ratio of approximately 1.5–1 with respect to endogenous levels. Remarkably, we found that mice expressing mutant α-syn developed progressive motor deficits (◉ *Figure 26-6b*) and neurodegeneration associated with α-syn accumulation in synapses and neurons, but very few or no inclusions were found. In contrast, mice from Line A (wt human α-syn) did not show neurodegenerative or neurological deficits, but displayed formation of inclusions. Analysis of the patterns of α-syn aggregation by western blot using the syn-1 antibody showed that in the mutant line there was greater formation of α-syn oligomers compared with the wt line. This is consistent with recent studies comparing the neuropathogenic effects of human wt and mutant α-syn under the control of the PrP promoter. For example, mice expressing mutant, but not wt, α-syn developed a severe and complex motor impairment leading to paralysis and death (Giasson et al., 2002). These animals developed age-dependent intracytoplasmic neuronal α-syn inclusions paralleling disease onset, and the α-syn inclusions recapitulated features of human disorders. Moreover, immunoelectron microscopy revealed that the α-syn inclusions contained 10–16 nm wide fibrils similar to human pathological inclusions. These mice demonstrate that A53T α-syn leads to the formation of toxic filamentous α-syn neuronal inclusions that cause neurodegeneration (Giasson et al., 2002).

Similarly, other studies have shown that under the PrP promoter, mice expressing the A53T human α-syn, but not the wt or A30P variants, develop adult-onset neurodegenerative disease with a progressive locomotor dysfunction leading to death (Lee et al., 2002). Affected mice exhibit neuronal abnormalities (in perikarya and neurites) including pathological accumulations of α-syn and ubiquitin. Consistent with abnormal neuronal accumulation of α-syn, brain regions with pathology exhibit increases in detergent-insoluble α-syn and the presence of α-syn aggregates (Lee et al., 2002).

Under the control of the mThy-1 promoter, expression of either wt or mutant α-syn (van der Putten et al., 2000) results in extensive α-syn accumulation throughout the CNS including, in some cases, in the SN or motor neurons (Rockenstein et al., 2002). In the model developed by van der Putten et al. (2000) mice develop early-onset motor decline, axonal degeneration, and α-syn accumulation in the spinal cord but not in the SN. In contrast, in our models (Line 61) we observed less severe motor deficits, but considerable accumulation of α-syn in cortical and subcortical regions equivalent to approximately 10 times human expression levels (Rockenstein et al., 2002). In both models there is accumulation of detergent-insoluble α-syn; however, its effects in the nigral system in terms of cell death are not completely clear.

To address the question of selective neuronal vulnerability, tg mice expressing either wt or mutated (A53T or A30P) forms of human α-syn under control of the 9-kb rat TH promoter were developed (Richfield et al., 2002). Initial studies in these mice showed accumulation of α-syn in neurons of the SN but no neurodegeneration or motor deficits. More recent studies in tg mice expressing either wt or doubly mutated forms of human α-syn under the control of the rat TH promoter have shown that the expression of α-syn in nigrostriatal terminals resulted in increased density of the dopamine transporter and enhanced susceptibility to the neurotoxin MPTP. Expression of a double mutant form of α-syn reduced locomotor responses to repeated doses of amphetamine and blocked the development of sensitization compared with adult wt α-syn tg mice (Richfield et al., 2002). Expression of double mutant human α-syn adversely affects the integrity of dopaminergic terminals and leads to age-related declines in motor coordination and dopaminergic markers (Richfield et al., 2002).

Further studies to investigate the role of α-syn mutations and selective neuronal vulnerability in the SN have been performed in rats using lentiviral and adeno-associated viral vectors (Kirik et al., 2002; Klein et al., 2002; Lo Bianco et al., 2002). In contrast to tg mice models, a selective loss of nigral dopaminergic neurons associated with a dopaminergic denervation of the striatum was observed in animals expressing either wt or mutant forms of α-syn. This neuronal degeneration correlates with the appearance of abundant α-syn-positive inclusions and extensive neuritic pathology detected with both α-syn and silver staining. Similarly, rat α-syn leads to protein aggregation but without cell loss, suggesting that inclusions are not the primary cause of cell degeneration in PD (Lo Bianco et al., 2002). This suggests that although the mouse tg models may accurately represent the accumulation of α-syn and the loss of cortical and caudo-putamen neurons, the rat models may better represent the loss of dopaminergic neurons in the nigro-striatal pathway.

In summary, under the PDGF-β promoter, both wt and mutant α-syn tg mice develop behavioral deficits and synaptic alterations following a limbic and cortical pattern; however, mutant α-syn is more toxic despite the fact that very few inclusions are formed relative to wt α-syn tg mice. Under the PrP promoter, mutant α-syn is more toxic than wt, inclusion-like structures develop and more widespread involvement of the lower motor system is observed. With the Thy-1 construct both wt and mutant α-syn are toxic and inclusion-like structures develop that can affect cortical or subcortical regions or spinal cord. Furthermore, these in vivo models support the contention that α-syn-dependent neurodegeneration is associated with abnormal accumulation of detergent-insoluble α-syn (probably representing oligomeric forms) rather than with inclusion formation representing fibrillar polymeric α-syn. The specific accumulation of detergent-insoluble α-syn in these tg mice recapitulates a pivotal feature of LBD (Kahle et al., 2001) and it is of significant importance in the future development and evaluation of novel treatments.

4 Conclusions

Immunotherapeutic approaches for AD have been under development for the past several years, and more recently various lines of investigation support the possibility of developing similar treatment approaches for HD, prion disease, tauopathies, and LBD/PD. The advantages of this approach for PD are that: (1) antibodies delivered via active or passive immunization can specifically target α-syn accumulation in various brain regions with peripheral delivery, (2) antibody-mediated clearance may be capable of targeting intracellular proteins such as α-syn via interactions with membrane-associated forms, and (3) there is potential for the development of combination therapies utilizing immunization with pharmacological treatment with regulators of autophagy such as rapamycin, which might enhance the effects of immunotherapy in LBD/PD. Taken together, immunotherapy approaches targeting α-syn aggregates represent an important step forward in the search for viable treatments for devastating neurodegenerative disorders such as LBD and PD.

Acknowledgments

This work was supported by NIH Grants AG18440, AG022074, and AG10435.

References

Aarsland D, Ballard CG, Halliday G. 2004. Are Parkinson's disease with dementia and dementia with Lewy bodies the same entity? J Geriatr Psychiatry Neurol 17: 137-145.

Apetri MM, Maiti NC, Zagorski MG, Carey PR, Anderson VE. 2006. Secondary structure of alpha-synuclein oligomers: Characterization by raman and atomic force microscopy. J Mol Biol 355: 63-71.

Bahr BA, Bendiske J. 2002. The neuropathogenic contributions of lysosomal dysfunction. J Neurochem 83: 481-489.

Benner EJ, Mosley RL, Destache CJ, Lewis TB, Jackson-Lewis V, et al. 2004. Therapeutic immunization protects dopaminergic neurons in a mouse model of Parkinson's disease. Proc Natl Acad Sci USA 101: 9435-9440.

Berger Z, Ravikumar B, Menzies FM, Oroz LG, Underwood BR, et al. 2006. Rapamycin alleviates toxicity of different aggregate-prone proteins. Hum Mol Genet 15: 433-442.

Borghi R, Marchese R, Negro A, Marinelli L, Forloni G, et al. 2000. Full length alpha-synuclein is present in cerebrospinal fluid from Parkinson's disease and normal subjects. Neurosci Lett 287: 65-67.

Braak H, Del Tredici K, Bratzke H, Hamm-Clement J, Sandmann-Keil D, et al. 2002. Staging of the intracerebral inclusion body pathology associated with idiopathic Parkinson's disease (preclinical and clinical stages). J Neurol 249 (Suppl 3): III/1-5.

Burn DJ. 2006. Cortical Lewy body disease and Parkinson's disease dementia. Curr Opin Neurol 19: 572-579.

Chandra S, Chen X, Rizo J, Jahn R, Sudhof TC. 2003. A broken alpha -helix in folded alpha -Synuclein. J Biol Chem 278: 15313-15318.

Conway K, Harper J, Lansbury P. 1998. Accelerated in vitro fibril formation by a mutant alpha-synuclein linked to early-onset Parkinson disease. Nat Med 4: 1318-1320.

Conway KA, Lee SJ, Rochet JC, Ding TT, Williamson RE, et al. 2000. Acceleration of oligomerization, not fibrillization, is a shared property of both alpha-synuclein mutations linked to early-onset Parkinson's disease: Implications for pathogenesis and therapy. Proc Natl Acad Sci USA 97: 571-576.

Cuervo AM. 2004. Autophagy: In sickness and in health. Trends Cell Biol 14: 70-77.

Cuervo AM, Stefanis L, Fredenburg R, Lansbury PT, Sulzer D. 2004. Impaired degradation of mutant alpha-synuclein by chaperone-mediated autophagy. Science 305: 1292-1295.

Danzer KM, Haasen D, Karow AR, Moussaud S, Habeck M, et al. 2007. Different species of alpha-synuclein oligomers induce calcium influx and seeding. J Neurosci 27: 9220-9232.

Davidson W, Jonas A, Clayton D, George J. 1998. Stabilization of alpha-synuclein secondary structure upon binding to synthetic membranes. J Biol Chem 273: 9443-9449.

Ding TT, Lee SJ, Rochet JC, Lansbury PT Jr. 2002. Annular alpha-synuclein protofibrils are produced when spherical protofibrils are incubated in solution or bound to brain-derived membranes. Biochemistry 41: 10209-10217.

Edinger AL, Thompson CB. 2004. Death by design: Apoptosis, necrosis and autophagy. Curr Opin Cell Biol 16: 663-669.

Emadi S, Liu R, Yuan B, Schulz P, McAllister C, et al. 2004. Inhibiting aggregation of alpha-synuclein with human single chain antibody fragments. Biochemistry 43: 2871-2878.

Feany M, Bender W. 2000. A *Drosophila* model of Parkinson's disease. Nature 404: 394-398.

Fernagut PO, Chesselet MF. 2004. Alpha-synuclein and transgenic mouse models. Neurobiol Dis 17: 123-130.

Furukawa K, Matsuzaki-Kobayashi M, Hasegawa T, Kikuchi A, Sugeno N, et al. 2006. Plasma membrane ion permeability induced by mutant alpha-synuclein contributes to the degeneration of neural cells. J Neurochem 97: 1071-1077.

Games D, Bard F, Grajeda H, Guido T, Khan K, et al. 2000. Prevention and reduction of AD-type pathology in PDAPP mice immunized with A beta 1–42. Ann NY Acad Sci 920: 274-284.

Gasser T. 2007. Update on the genetics of Parkinson's disease. Mov Disord 22: S343-S350.

Giasson BI, Duda JE, Quinn SM, Zhang B, Trojanowski JQ, et al. 2002. Neuronal alpha-synucleinopathy with severe movement disorder in mice expressing A53T human alpha-synuclein. Neuron 34: 521-533.

Golbe LI, Mouradian MM. 2004. Alpha-synuclein in Parkinson's disease: Light from two new angles. Ann Neurol 55: 153-156.

Gozuacik D, Kimchi A. 2004. Autophagy as a cell death and tumor suppressor mechanism. Oncogene 23: 2891-2906.

Halliday GM, Macdonald V, Henderson JM. 2005. A comparison of degeneration in motor thalamus and cortex between progressive supranuclear palsy and Parkinson's disease. Brain 128: 2272-2280.

Hansen LA, Galasko D. 1992. Lewy body disease. Curr Opin Neurol Neurosurg 5: 889-894.

Hashimoto M, Masliah E. 1999. Alpha-synuclein in Lewy body disease and Alzheimer's disease. Brain Pathol 9: 707-720.

Hashimoto M, Rockenstein E, Mante M, Crews L, Bar-On P, et al. 2004. An antiaggregation gene therapy strategy for Lewy body disease utilizing beta-synuclein lentivirus in a transgenic model. Gene Ther 11: 1713-1723.

Hashimoto M, Rockenstein E, Mante M, Mallory M, Masliah E. 2001. β-Synuclein inhibits α-synuclein aggregation: A possible role as an anti-parkinsonian factor. Neuron 32: 213-223.

Hashimoto M, Rockenstein E, Masliah E. 2003. Transgenic models of alpha-synuclein pathology: Past, present, and future. Ann NY Acad Sci 991: 171-188.

Hattori N, Kobayashi H, Sasaki-Hatano Y, Sato K, Mizuno Y. 2003. Familial Parkinson's disease: A hint to elucidate the mechanisms of nigral degeneration. J Neurol 250 (Suppl 3): III2-III10.

Holtzman DM, Bales KR, Wu S, Bhat P, Parsadanian M, et al. 1999. Expression of human apolipoprotein E reduces amyloid-β deposition in a mouse model of Alzheimer's disease. J Clin Invest 103: R15-R21.

Iwai A, Masliah E, Yoshimoto M, De Silva R, Ge N, et al. 1994. The precursor protein of non-Ab component of Alzheimer's disease amyloid (NACP) is a presynaptic protein of the central nervous system. Neuron 14: 467-475.

Iwatsubo T, Yamaguchi H, Fujimuro M, Yokosawa H, Ihara Y, et al. 1996. Purification and characterization of Lewy bodies from brains of patients with diffuse Lewy body disease. Am J Pathol 148: 1517-1529.

Janvin CC, Larsen JP, Salmon DP, Galasko D, Hugdahl K, et al. 2006. Cognitive profiles of individual patients with Parkinson's disease and dementia: Comparison with dementia with lewy bodies and Alzheimer's disease. Mov Disord 21: 337-342.

Jao CC, Der-Sarkissian A, Chen J, Langen R. 2004. Structure of membrane-bound alpha-synuclein studied by site-directed spin labeling. Proc Natl Acad Sci USA 101: 8331-8336.

Kahle PJ, Neumann M, Ozmen L, Muller V, Odoy S, et al. 2001. Selective insolubility of alpha-synuclein in human Lewy body diseases is recapitulated in a transgenic mouse model. Am J Pathol 159: 2215-2225.

Kamp F, Beyer K. 2006. Binding of alpha-synuclein affects the lipid packing in bilayers of small vesicles. J Biol Chem 281: 9251-9259.

Kirik D, Rosenblad C, Burger C, Lundberg C, Johansen TE, et al. 2002. Parkinson-like neurodegeneration induced by targeted overexpression of alpha-synuclein in the nigrostriatal system. J Neurosci 22: 2780-2791.

Kitada T, Asakawa S, Hattori N, Matsumine H, Yamamura Y, et al. 1998. Mutations in the parkin gene cause autosomal recessive juvenile parkinsonism. Nature 392: 605-608.

Klein RL, King MA, Hamby ME, Meyer EM. 2002. Dopaminergic cell loss induced by human A30P alpha-synuclein gene transfer to the rat substantia nigra. Hum Gene Ther 13: 605-612.

Klivenyi P, Siwek D, Gardian G, Yang L, Starkov A, et al. 2006. Mice lacking alpha-synuclein are resistant to mitochondrial toxins. Neurobiol Dis 21: 541-548.

Komatsu M, Ueno T, Waguri S, Uchiyama Y, Kominami E, et al. 2007. Constitutive autophagy: Vital role in clearance of unfavorable proteins in neurons. Cell Death Differ 14: 887-894.

Komatsu M, Waguri S, Chiba T, Murata S, Iwata J, et al. 2006. Loss of autophagy in the central nervous system causes neurodegeneration in mice. Nature 441: 880-884.

Kosaka K, Yoshimura M, Ikeda K, Budka H. 1984. Diffuse type of Lewy body disease. Progressive dementia with abundant cortical Lewy bodies and senile changes of varying degree – A new disease? Clin Neuropathol 3: 183-192.

Kruger R, Kuhn W, Muller T, Woitalla D, Graeber M, et al. 1998. Ala30Pro mutation in the gene encoding a-synuclein in Parkinsons's disease. Nat Genet 18: 106-108.

Lansbury PTJ. 1999. Evolution of amyloid: What normal protein folding may tell us about fibrillogenesis and disease. Proc Natl Acad Sci USA 96: 3342-3344.

Larsen KE, Sulzer D. 2002. Autophagy in neurons: A review. Histol Histopathol 17: 897-908.

Lashuel HA, Hartley DM, Petre BM, Wall JS, Simon MN, et al. 2003. Mixtures of wild-type and a pathogenic (E22G) form of Abeta40 in vitro accumulate protofibrils, including amyloid pores. J Mol Biol 332: 795-808.

Lashuel HA, Petre BM, Wall J, Simon M, Nowak RJ, et al. 2002. Alpha-synuclein, especially the Parkinson's disease-associated mutants, forms pore-like annular and tubular protofibrils. J Mol Biol 322: 1089-1102.

Lee HJ, Patel S, Lee SJ. 2005. Intravesicular localization and exocytosis of alpha-synuclein and its aggregates. J Neurosci 25: 6016-6024.

Lee MK, Stirling W, Xu Y, Xu X, Qui D, et al. 2002. Human alpha-synuclein-harboring familial Parkinson's disease-linked Ala-53→Thr mutation causes neurodegenerative disease with alpha-synuclein aggregation in transgenic mice. Proc Natl Acad Sci USA 99: 8968-8973.

Lee VM, Giasson BI, Trojanowski JQ. 2004. More than just two peas in a pod: Common amyloidogenic properties of tau and alpha-synuclein in neurodegenerative diseases. Trends Neurosci 27: 129-134.

Lippa CF, Duda JE, Grossman M, Hurtig HI, Aarsland D, et al. 2007. DLB and PDD boundary issues: Diagnosis, treatment, molecular pathology, and biomarkers. Neurology 68: 812-819.

Litvan I, MacIntyre A, Goetz CG, Wenning GK, Jellinger K, et al. 1998. Accuracy of the clinical diagnoses of Lewy body disease, Parkinson disease, and dementia with Lewy bodies: A clinicopathologic study. Arch Neurol 55: 969-978.

Lo Bianco C, Ridet JL, Schneider BL, Deglon N, Aebischer P. 2002. α-Synucleinopathy and selective dopaminergic neuron loss in a rat lentiviral-based model of Parkinson's disease. Proc Natl Acad Sci USA 99: 10813-10818.

Majeski AE, Dice JF. 2004. Mechanisms of chaperone-mediated autophagy. Int J Biochem Cell Biol 36: 2435-2444.

Masliah E, Rockenstein E, Adame A, Alford M, Crews L, et al. 2005. Effects of alpha-Synuclein immunization in a mouse model of Parkinson's disease. Neuron 46: 857-868.

Masliah E, Rockenstein E, Veinbergs I, Mallory M, Hashimoto M, et al. 2000. Dopaminergic loss and inclusion body formation in alpha-synuclein mice: Implications for neurodegenerative disorders. Science 287: 1265-1269.

McKeith I, Galasko D, Kosaka K, Perry E, Dickson D, et al. 1996. Clinical and pathological diagnosis of dementia with Lewy bodies (DLB): Report of the CDLB International Workshop. Neurology 47: 1113-1124.

McKeith IG. 2000. Spectrum of Parkinson's disease, Parkinson's dementia, and Lewy body dementia. Neurol Clin 18: 865-902.

Meijer AJ, Codogno P. 2004. Regulation and role of autophagy in mammalian cells. Int J Biochem Cell Biol 36: 2445-2462.

Meredith GE, Totterdell S, Petroske E, Santa Cruz K, Callison RC Jr, et al. 2002. Lysosomal malfunction accompanies alpha-synuclein aggregation in a progressive mouse model of Parkinson's disease. Brain Res 956: 156-165.

Morgan D, Diamond DM, Gottschall PE, Ugen KE, Dickey C, et al. 2000. A beta peptide vaccination prevents memory loss in an animal model of Alzheimer's disease. Nature 408: 982-985.

Murphy D, Reuter S, Trojanowski J, Lee V-Y. 2000. Synucleins are developmentally expressed, and a-synuclein regulates the size of the presynaptic vesicular pool in primary hippocampal neurons. J Neurosci 20: 3214-3220.

Nakajima T, Takauchi S, Ohara K, Kokai M, Nishii R, et al. 2005. Alpha-synuclein-positive structures induced in leupeptin-infused rats. Brain Res 1040: 73-80.

Narhi L, Wood SJ, Steavenson S, Jiang Y, Wu GM, et al. 1999. Both familial Parkinson's disease mutations accelerate alpha-synuclein aggregation. J Biol Chem 274: 9843-9846.

Nieto M, Gil-Bea FJ, Dalfo E, Cuadrado M, Cabodevilla F, et al. 2006. Increased sensitivity to MPTP in human alpha-synuclein A30P transgenic mice. Neurobiol Aging 27: 848-856.

Nixon RA, Wegiel J, Kumar A, Yu WH, Peterhoff C, et al. 2005. Extensive involvement of autophagy in Alzheimer disease: An immuno-electron microscopy study. J Neuropathol Exp Neurol 64: 113-122.

Paludan C, Schmid D, Landthaler M, Vockerodt M, Kube D, et al. 2005. Endogenous MHC class II processing of a viral nuclear antigen after autophagy. Science 307: 593-596.

Polymeropoulos M, Lavedan C, Leroy E, Ide S, Dehejia A, et al. 1997. Mutation in the a-synuclein gene identified in families with Parkinson's disease. Science 276: 2045-2047.

Ravikumar B, Duden R, Rubinsztein DC. 2002. Aggregate-prone proteins with polyglutamine and polyalanine expansions are degraded by autophagy. Hum Mol Genet 11: 1107-1117.

Reynolds AD, Banerjee R, Liu J, Gendelman HE, Mosley RL. 2007. Neuroprotective activities of CD4+CD25+regulatory T cells in an animal model of Parkinson's disease. J Leukoc Biol 82: 1083-1094.

Richfield EK, Thiruchelvam MJ, Cory-Slechta DA, Wuertzer C, Gainetdinov RR, et al. 2002. Behavioral and neurochemical effects of wild-type and mutated human alpha-synuclein in transgenic mice. Exp Neurol 175: 35-48.

Rideout HJ, Lang-Rollin I, Stefanis L. 2004. Involvement of macroautophagy in the dissolution of neuronal inclusions. Int J Biochem Cell Biol 36: 2551-2562.

Rochet JC, Outeiro TF, Conway KA, Ding TT, Volles MJ, et al. 2004. Interactions among alpha-synuclein, dopamine, and biomembranes: Some clues for understanding neurodegeneration in Parkinson's disease. J Mol Neurosci 23: 23-34.

Rockenstein E, Mallory M, Hashimoto M, Song D, Shults CW, et al. 2002. Differential neuropathological alterations in transgenic mice expressing alpha-synuclein from the platelet-derived growth factor and Thy-1 promoters. J Neurosci Res 68: 568-578.

Rockenstein E, Schwach G, Ingolic E, Adame A, Crews L, et al. 2005. Lysosomal pathology associated with alpha-synuclein accumulation in transgenic models using an eGFP fusion protein. J Neurosci Res 80: 247-259.

Saito Y, Suzuki K, Hulette C, Murayama S. 2004. Aberrant phosphorylation of alpha-synuclein in human Niemann-Pick type C1 disease. J Neuropathol Exp Neurol 63: 323-328.

Salmon D, Hansen L, Masliah E, Galasko D, Butters N, et al. 1989. Neurophsychological characteristics of a Lewy body variant of Alzheimer's disease. Soc Neurosci Abstr 15: 863.

Sarkar S, Davies JE, Huang Z, Tunnacliffe A, Rubinsztein DC. 2007. Trehalose, a novel mTOR-independent autophagy enhancer, accelerates the clearance of mutant huntingtin and alpha-synuclein. J Biol Chem 282: 5641-5652.

Sarkar S, Floto RA, Berger Z, Imarisio S, Cordenier A, et al. 2005. Lithium induces autophagy by inhibiting inositol monophosphatase. J Cell Biol 170: 1101-1111.

Schenk D, Barbour R, Dunn W, Gordon G, Grajeda H, et al. 1999. Immunization with amyloid-beta attenuates Alzheimer-disease-like pathology in the PDAPP mouse. Nature 400: 173-177.

Schenk DB, Seubert P, Lieberburg I, Wallace J. 2000. β-Peptide immunization: A possible new treatment for Alzheimer disease. Arch Neurol 57: 934-936.

Shastry BS. 2001. Parkinson disease: Etiology, pathogenesis and future of gene therapy. Neurosci Res 41: 5-12.

Shintani T, Klionsky DJ. 2004. Autophagy in health and disease: A double-edged sword. Science 306: 990-995.

Singleton AB, Farrer M, Johnson J, Singleton A, Hague S, et al. 2003. alpha-Synuclein locus triplication causes Parkinson's disease. Science 302: 841.

Song DD, Shults CW, Sisk A, Rockenstein E, Masliah E. 2004. Enhanced substantia nigra mitochondrial pathology in human alpha-synuclein transgenic mice after treatment with MPTP. Exp Neurol 186: 158-172.

Spillantini M, Schmidt M, Lee V-Y, Trojanowski J, Jakes R, et al. 1997. α-Synuclein in Lewy bodies. Nature 388: 839-840.

Stefanis L, Larsen KE, Rideout HJ, Sulzer D, Greene LA. 2001. Expression of A53T mutant but not wild-type alpha-synuclein in PC12 cells induces alterations of the ubiquitin-dependent degradation system, loss of dopamine release, and autophagic cell death. J Neurosci 21: 9549-9560.

Takeda A, Mallory M, Sundsmo M, Honer W, Hansen L, et al. 1998. Abnormal accumulation of NACP/a-synuclein in neurodegenerative disorders. Am J Pathol 152: 367-372.

Tayebi N, Callahan M, Madike V, Stubblefield BK, Orvisky E, et al. 2001. Gaucher disease and parkinsonism: A phenotypic and genotypic characterization. Mol Genet Metab 73: 313-321.

Thomas B, Beal MF. 2007. Parkinson's disease. Hum Mol Genet 16 (Spec No. 2): R183-R194.

Trojanowski J, Goedert M, Iwatsubo T, Lee V. 1998. Fatal attractions: Abnormal protein aggregation and neuron death in Parkinson's disease and lewy body dementia. Cell Death Differ 5: 832-837.

Trojanowski J, Lee V. 1998. Aggregation of neurofilament and alpha-synuclein proteins in Lewy bodies: Implications for pathogenesis of Parkinson disease and Lewy body dementia. Arch Neurol 55: 151-152.

Trojanowski JQ, Lee VM. 2000. "Fatal attractions" of proteins. A comprehensive hypothetical mechanism underlying Alzheimer's disease and other neurodegenerative disorders. Ann NY Acad Sci 924: 62-67.

Tsigelny IF, Bar-On P, Sharikov Y, Crews L, Hashimoto M, et al. 2007. Dynamics of alpha-synuclein aggregation and inhibition of pore-like oligomer development by beta-synuclein. FEBS J 274: 1862-1877.

Ulmer TS, Bax A, Cole NB, Nussbaum RL. 2005. Structure and dynamics of micelle-bound human alpha-synuclein. J Biol Chem 280: 9595-9603.

Uversky VN, Lee HJ, Li J, Fink AL, Lee SJ. 2001. Stabilization of partially folded conformation during alpha-synuclein oligomerization in both purified and cytosolic preparations. J Biol Chem 276: 43495-43498.

Uversky VN, Li J, Souillac P, Millett IS, Doniach S, et al. 2002. Biophysical properties of the synucleins and their propensities to fibrillate: Inhibition of alpha-synuclein assembly by beta- and gamma-synucleins. J Biol Chem 277: 11970-11978.

van der Putten H, Wiederhold KH, Probst A, Barbieri S, Mistl C, et al. 2000. Neuropathology in mice expressing human alpha-synuclein. J Neurosci 20: 6021-6029.

Varkonyi J, Rosenbaum H, Baumann N, MacKenzie JJ, Simon Z, et al. 2003. Gaucher disease associated with parkinsonism: Four further case reports. Am J Med Genet A 116: 348-351.

Volles MJ, Lansbury PT Jr. 2002. Vesicle permeabilization by protofibrillar alpha-synuclein is sensitive to Parkinson's disease-linked mutations and occurs by a pore-like mechanism. Biochemistry 41: 4595-4602.

Wakabayashi K, Matsumoto K, Takayama K, Yoshimoto M, Takahashi H. 1997. NACP, a presynaptic protein, immunoreactivity in Lewy bodies in Parkinson's disease. Neurosci Lett 239: 45-48.

Webb JL, Ravikumar B, Atkins J, Skepper JN, Rubinsztein DC. 2003. Alpha-Synuclein is degraded by both autophagy and the proteasome. J Biol Chem 278: 25009-25013.

Williams A, Jahreiss L, Sarkar S, Saiki S, Menzies FM, et al. 2006. Aggregate-prone proteins are cleared from the cytosol by autophagy: Therapeutic implications. Curr Top Dev Biol 76: 89-101.

Wright PE, Dyson HJ. 1999. Intrinsically unstructured proteins: Re-assessing the protein structure-function paradigm. J Mol Biol 293: 321-331.

Yu WH, Matsuoka Y, Sziraki I, Hashim A, Lafrancois J, et al. 2008. Increased dopaminergic neuron sensitivity to 1-methyl-4-phenyl-1,2,3,6-tetrahydropyridine (MPTP) in transgenic mice expressing mutant A53T alpha-synuclein. Neurochem Res 33: 902-911.

Zhou C, Emadi S, Sierks MR, Messer A. 2004. A human single-chain Fv intrabody blocks aberrant cellular effects of overexpressed alpha-synuclein. Mol Ther 10: 1023-1031.

Zhu M, Li J, Fink AL. 2003. The association of alpha-synuclein with membranes affects bilayer structure, stability, and fibril formation. J Biol Chem 278: 40186-40197.

27 Clinical Outcomes After Spinal Cord Injury

J. S. Krause · S. D. Newman

1	Introduction	616
2	Epidemiology of SCI	616
3	Nature of Clinical Outcomes	617
4	Biographic and Injury Factors	618
5	Psychological and Environmental Factors	620
5.1	Personality	620
5.2	Environment	621
5.3	Stress-Coping Models	622
6	Protective and Risk Behaviors	622
7	Health Outcomes and Secondary Conditions	624
8	Mortality	626
9	Summary and Conclusions	627

© 2009 Springer Science+Business Media, LLC.

Abstract: Spinal cord injury may result in prolonged hospitalization and affects all areas of life. After the initial rehabilitation hospitalization, a new set of concerns arise as the recently injured individual must work to maintain his/her health and avoid secondary health conditions. Early mortality generally occurs, particularly to those with the most severe injuries, and is mediated by psychological, environmental, and behavioral factors. The focus of this chapter is on outlining sets of factors that predict long-term outcomes and survival among people with spinal cord injury. An enhanced version of a previously published general empirical model is outlined that relates risk and protective factors to morbidity and mortality. This model provides a general scheme for both reviewing current research and for outlining areas in need of future research. In this chapter, we describe the model, outline each subsequent component of the model, and end with a discussion of longevity.

List of Abbreviations: BAC, blood alcohol concentration; CDC, Centers for Disease Control and Prevention; CHART, Craig Handicap Assessment Reporting Technique; CHIEF, Craig Hospital Inventory of Environmental Factors; MDD, major depressive disorder; MSCIS, Model Spinal Cord Injury System; MVC, motor vehicle crash; NSCISC, National Spinal Cord Injury Statistical Center; PTSD, posttraumatic stress disorder; QOL, quality of life; SCI, spinal cord injury; UTI, urinary tract infection

1 Introduction

Spinal cord injury (SCI) has immediate and profound consequences and generally results in permanent functional loss. The primary symptoms of SCI include loss of sensory and motor function, the extent of which is related both to the neurologic level and neurologic completeness of injury. Bowel and bladder function are typically impaired with neurologically complete injuries. Sexual function may also be impaired. In addition to the immediate complications, there are also a number of secondary complications that may occur in the months, years, and decades after injury. Whereas the primary complications are a direct function of the SCI, secondary complications occur only indirectly as a result of SCI and are frequently related to the extent to which the individual performs appropriate health-maintenance behaviors and avoids risk behaviors. Even in cases where injury type and severity are highly similar, there is a wide array of individual differences in the extent to which SCI will result in long-term secondary health complications. Life expectancy is also diminished to the extent which relates to the nature and severity of injury, as well as a diversity of individual factors related to behaviors, access to care, and level of adaptation.

In this chapter we present an overview of clinical research after SCI using the framework of a general empirical risk model that links biographic and injury factors at onset to health and longevity. The model identifies various components or layers of predictive factors that form a sequence of events that may either accelerate or protect the individual from development of secondary health conditions and early mortality. We begin with a discussion of the epidemiology of SCI.

2 Epidemiology of SCI

The National Spinal Cord Injury Statistical Center (NSCISC) estimates that there are approximately 12,000 new incidences of SCI annually. The number of people in the United States who are alive in 2007 with SCI has been estimated to be approximately 255,702 persons, with a range of 227,080–300,938 persons (NSCISC, 2008). Since 2005, motor vehicle crashes (MVC) have accounted for 42% of new cases of SCI. Other etiologies include falls (27.1%), acts of violence, principally gunshot wounds (15.3%), sports (7.4%), and other reasons (7%). Since 1973, injuries due to MVC and sporting activities have declined steadily, while injuries related to falls have increased. Acts of violence were the cause of 13.3% of SCI before 1980, peaking at 24.8% between 1990 and 1999, before declining to the present rate (15.3%) since 2000 (NSCISC, 2008).

Severity of SCI is defined by neurologic level of injury and neurologic completeness of injury. An incomplete injury results when there is partial preservation of motor and/or sensory function below the level of injury including the lowest sacral segments (Jackson et al., 2004a). Based on a 2 × 2 combination of injury level, broken-down into cervical (tetraplegia) and noncervical (paraplegia), and neurologic completeness of injury (complete versus incomplete), the majority of individuals have incomplete tetraplegia (34.1%), followed by complete paraplegia (23%), complete tetraplegia (18.3%), and incomplete paraplegia (18.5%). There has been a slight increase of incomplete cervical injuries since the beginning of data collection by the Statistical Center, while complete paraplegia and tetraplegia have decreased slightly (NSCISC, 2008).

The greater portion of cases are male (77.8%), although there has been a slight drift toward a decreasing incidence of SCI in males since the inception of the database (before 1980, 81.8% of new SCIs were male). Caucasians account for 63% of the cases, 22.7% African-American, 11.8% Hispanic, and 2.4% from other racial and ethnic groups. From 1973 to 1979, the average age at injury was 28.7 years, and most injuries occurred between the ages of 16 and 30. However, as the median age of the general population of the United States has increased since the mid-1970s, the average age at injury has also increased. Since 2005, the average age at time of injury is 39.5 years. The percentage of persons older than 60 years of age at injury has increased from 4.7% prior to 1980 to 11.5% among injuries occurring since 2000 (NSCISC, 2008).

At present, 87.9% of all discharges are to private noninstitutional residences in the community. Lifetime costs of SCI, which account for the average yearly health care and living expenses, vary greatly depending on level and severity of the injury. Estimated lifetime costs for a person injured at age 25, based on level of injury, are $3,059,184 for high tetraplegia (C1–C4), $1,729,754 for low tetraplegia (C5–C8), and $1,022,138 for paraplegia. Initial average yearly expenses alone for a person with high tetraplegia are $775,567. These numbers do not reflect any indirect costs, such as loss in wages, benefits, and productivity, which average $62,270 annually, though greatly differ based on severity of injury, preinjury employment history, and education (NSCISC, 2008). In addition to the great expense to the SCI survivor, SCI treatment is also very costly to the nation. The financial burden of SCI to the United States is estimated to be $9.7 billion dollars annually. According to the Centers for Disease Control and Prevention (CDC), the cost of treating pressure sores, a common secondary complication after SCI, is an estimated $1.2 billion annually (CDC, 2002).

3 Nature of Clinical Outcomes

In contrast to basic science research, where function or recovery is the primary outcome, there is a greater diversity of outcomes that are important in clinical settings. Functional independence is a primary goal in rehabilitation and, historically, may be the most prominent goal. However, function is primarily a means of promoting other types of outcomes that include health, participation, and quality of life (QOL). Although conceptually distinct, these outcomes generally co-vary among individuals, in that, those who are able to achieve a favorable outcome in one area generally achieve favorable outcomes in other areas as well. For instance, participation in community activities is highly correlated with subjective well-being or QOL, much more so than is either impairment (i.e., severity of injury) or disability (functional status; Dijkers, 1997). These parameters are also associated with longevity (Krause et al., 2004).

In order to gain a better understanding of clinical outcomes, it is necessary to consider several types of factors that are associated with or predictive of these outcomes. An understanding of epidemiology of SCI, not simply from the perspective of the incidence and prevalence of SCI, but also the nature of risk and protective factors for desired outcomes, is central to SCI rehabilitation. Risk factors are those that are associated with a greater likelihood of an adverse outcome; whereas protective factors are those that are associated with a diminished likelihood of an adverse outcome. Several types of risk and protective factors must be considered with SCI, and theoretical and empirical models are needed to guide research relating predictive factors to key outcomes.

Krause (1996) proposed an empirical model for guiding investigations on clinical outcomes after SCI. ❶ *Figure 27-1* forwards an enhanced model that includes additional parameters, not included in the original model. There are five types or layers of factors, the last of which is mortality. The most basic level consists of biographic characteristics, such as race, gender, age, and injury characteristics. These include severity, etiology, and timing of SCI (i.e., age at injury onset and years lived since onset).

◘ Figure 27-1
"An enhanced general risk model for morbidity and mortality"

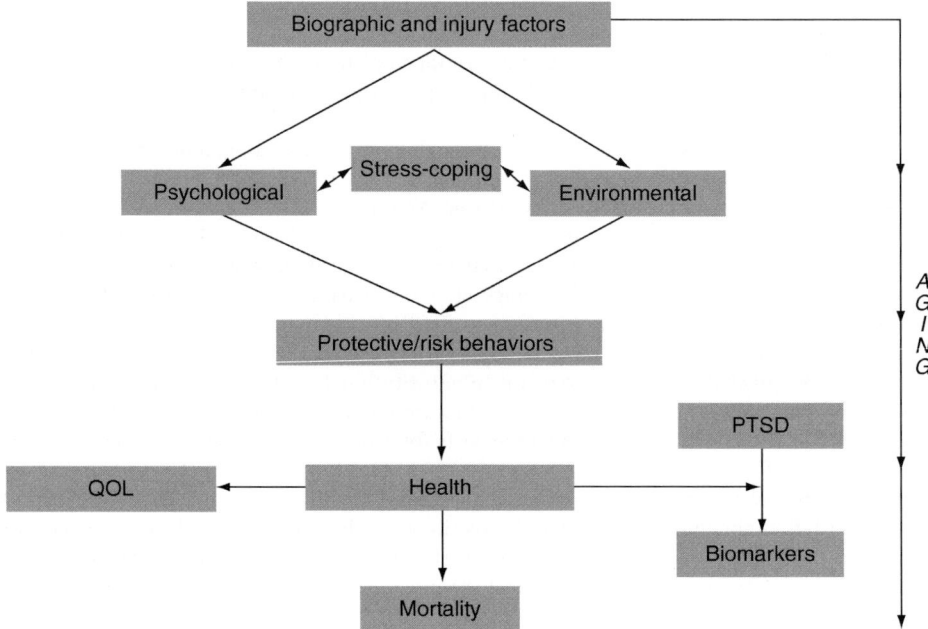

The next three levels of predictor variables include psychological and environmental variables (level 2), risk and protective behaviors (level 3), and secondary conditions and general health indicators (level 4). The ultimate outcome or endpoint is mortality (level 5). The enhanced model also includes participation and QOL and refinements that are encompassed within the area of health outcomes. Although the model was developed for SCI, it is clearly applicable to other neurologic disorders including traumatic brain injury and stroke, with modification of some of the basic parameters related to the nature and severity of the condition.

The general model proposes a testable sequence of associations where, for a given type of injury and a given biographic profile, psychological and environmental factors are associated with differential patterns of risk and protective behaviors. These behavioral patterns may directly impact health and the development of secondary conditions which, in turn, will impact longevity. Participation and subjective well-being are viewed as associated areas of the model, adjacent to health and less directly linked to mortality. Each stage of the model is discussed in the following paragraphs along with relevant research.

4 Biographic and Injury Factors

There has been a substantial amount of attention paid to biographic and injury factors and their relationship with outcomes after SCI. The primary biographic factors are gender, race/ethnicity, and age. Socioeconomic status, particularly education, has also received attention in the literature (Krause et al., 2000a; Krause et al., 2006). Studies of gender and race/ethnicity are, in essence, a focus on disparities in outcomes related to SCI, whereas aging studies relate to decline in health status (Krause and Crewe, 1991; Davis et al., 1995; Krause, 2000; Krause and Broderick, 2004). Education and socioeconomic status are related more to opportunities and access to health care.

Despite the importance of investigations of disparities, there have been relatively few studies of the association of gender and race/ethnicity with SCI outcomes. The aforementioned epidemiology of SCI has resulted in an availability of participants who are predominantly young Caucasian males. Simply due to convenience sampling, there have not been substantial numbers of women and minorities in research, although there has been an increase in recent years, particularly with women.

For instance, a recent study evaluating the relationship of minority status and SCI revealed a demographic profile of young, single, unemployed males with less than a high school education residing in an urban area among the minority group (>90% African-American) (Burnett et al., 2002). Violence was the leading cause of SCI in this population. The specific effect of minority status on various outcomes after SCI has also received recent attention. Miller et al. (2006) examined differences in life satisfaction and health-related QOL between minority and Caucasian individuals with SCI finding both to be lower for minorities. Rintala et al. (1998) investigated the association of racial and ethnic differences with community integration after SCI. Others have evaluated the effect of minority status on pain experience (Cardenas et al., 2004), perceived quality of care (Johnston et al., 2004), and depressive symptoms (Krause et al., 2000b) after SCI. Findings of these studies generally revealed less favorable outcomes in minority populations when compared with their nonminority counterparts. However, results of one study that compared the subjective well-being of Caucasians and African-American participants suggested that the differences were limited to areas where there are traditionally disparities in the general population, such as employment and economic status (Krause, 1998).

Although the majority of studies have examined differences only between Caucasians and African-Americans, some attention has also been focused on Hispanic and American-Indian cohorts with SCI. Findings from the only study to compare all four groups on a diversity of outcomes (Krause and Broderick, 2004) suggested that while outcomes of African-Americans generally did not reach the level of Caucasians, they were superior in general to the outcomes of both Hispanics and American-Indians with SCI. However, geographic region was confounded with race/ethnicity in this study, and it is difficult to envision a study that could be performed without such confounding. An interesting component of this overall study was the focus on outcomes of American-Indians with SCI (Krause et al., 1999a; Krause et al., 1999b; Krause et al., 2000a). Of the more interesting findings was a high annual incidence of subsequent injuries, with 59% of these injuries requiring medical attention (Krause et al., 2000a). Another interesting finding was that the overall level of alcohol consumption was lower for American-Indians than the general population, but the incidence of binge drinking was higher (Krause et al., 1999b). Understanding these patterns of risk factors is central to the provision of service for any special population.

Estores and Sipski (2004) completed a comprehensive review of studies that investigated issues affecting women with SCI. The primary finding of their review is that most current medical practice is based upon nonrandomized trials or evidence from controlled studies done either in mixed samples of men and women or in the non-SCI population. Specific studies have examined depression (Hughes et al., 2001; Kalpakjian and Albright, 2006), sexuality and reproductive health (Broderick and Krause, 2003; Jackson, et al., 2004b; Forsythe and Horsewell, 2006; Leibowitz and Stanton, 2007), preventive healthcare (Lavela et al., 2006), stress and life satisfaction (Rahman et al., 2005), community integration (Forchheimer et al., 2004), pain (Norrbrink et al., 2003), osteoporosis (Weeks, 2001), and aging (Pentland et al., 2002) in women with SCI. The findings consistently support the argument that women with SCI have medical and psychosocial needs that differ from men. Further, well designed, gender-specific investigations are warranted to provide the evidence to support best SCI rehabilitation practices and ultimately outcomes.

In order to fully appreciate the importance of aging and SCI, it is first necessary to define the aging parameters accounting for SCI. Chronologic age is a linear combination of age at injury onset and the number of years lived postinjury. Outcomes may differ substantially for individuals with similar chronologic age, given differences in the other two parameters. For instance, Krause (2000) identified several health conditions (renal stones and orthopedic complications) that rarely occur among individuals who were injured later in life simply because these individuals do not live long enough to develop the condition, even if there is a higher risk due to the SCI. In contrast, a similar set of conditions, particularly orthopedic complications which included contractures and curvature of the spine, increased with years lived postinjury and only occurred among those who reach substantial milestones. Therefore, the percentage or portion of the time the individual has lived with the SCI is important to understanding outcomes. These types of considerations are very important when considering translational research, as the success of any new intervention to promote functional recovery will be mediated by the readiness of the individual for the intervention. The accumulation of secondary conditions that have occurred over the time the individual has lived with SCI may act as a substantial barrier to utilization of these interventions (discussed in detail in a separate chapter).

Further complicating any research on aging is the role of environmental change over time. This is particularly important in healthcare, since policies and reimbursement practices continually change and this requires more sophisticated research designs in order to account for this effect (Schaie, 1965). For instance, Crewe and Krause (1990) identified longitudinal changes in outcomes at two times of measurement over 11 years among 154 participants with SCI, noting several favorable changes. However, when the responses of the participant cohort from the first time of measurement were compared with that of a new matched cohort at the second time of measurement, the changes observed over the 11 years *between* different participant cohorts were similar to those observed within the *same* cohort (Krause and Crewe, 1991). The use of the multiple cohorts identified a time lag effect related to environmental change and suggested that the observed longitudinal changes within the same participant cohort were a reflection of broader environmental changes that produced more favorable outcomes overall. This is consistent with the rehabilitation climate at the time of the study. Interestingly, when similar investigations were conducted over the subsequent 9 years, again using a combination of old and new cohorts and longitudinal and timeline data, the trend toward improving outcomes had reversed (Krause and Sternberg, 1997). This also is consistent with the changing rehabilitative environment.

Although preserved neurologic function, measured both by neurologic level and neurologic completeness of injury, clearly relates to greater longevity (DeVivo et al., 1999; Strauss et al., 2006), it is not as strongly related to other parameters. For instance, neurologic level of injury has been shown to have no significant effect associated with risk of pressure ulcers (Chen et al., 2005). In a comprehensive review of the literature, Dijkers (1997) found that, although persons with SCI tend to report lower subjective well-being than nondisabled people, the relationship between impairment (i.e., diagnosis) and QOL was weak at best, with an average correlation of only -0.05. The association between disability (i.e., functional limitations) and QOL was stronger but still rather weak (average correlation of -0.21). The relationship between QOL and handicap (i.e., participation, activities) was strongest (-0.34).

5 Psychological and Environmental Factors

Psychological factors are important considerations in both clinical and translational research. From the standpoint of the general risk model of clinical outcomes, psychological factors serve as proxy variables for behaviors. For instance, measures such as the CAGE (Rumpf et al., 1997) are proxy variables for alcohol misuse. The availability of proxy variables allows clinicians to evaluate risk of particular behaviors, such as substance misuse or sensation seeking high-risk activities, without asking about specific behaviors themselves. This allows clinicians to mask the intention of the measurement and obtain more valid responses.

5.1 Personality

Personality inventories are designed to measure traits, which may be defined as predispositions or propensities to perform certain types of behaviors across situations. One such scale particularly relevant to SCI is the sensation seeking scale from the Zuckerman–Kuhlman personality questionnaire (Zuckerman et al., 1993) which measures propensity for engaging in high-risk activities, which has been related to the onset of SCI as well as to subsequent injuries from new events in the years and decades after the onset of SCI (Krause, 2004).

Personality types are even more general categories related to consistency of preferences for certain types of activities. For instance, the Holland typology (Holland, 1985) of six personality types is based on vocational interests and preferences for activities. One personality type (realistic theme) described more than half of the study sample in one of the only investigations of vocational interest and SCI (Rohe and Athelstan, 1982). This personality type persisted over an 11-year interval (Rohe and Krause, 1998), although the population was primarily Caucasian male and the findings might not hold for other groups. The latter finding is particularly interesting, since the personality type (realistic theme) that is most prominent among people with SCI reflects a preference for working with tools and hands-on types of occupations, such as auto mechanics or other skilled blue-collar occupations. These are the very activities that are most difficult to perform after SCI, yet the interest patterns have shown little change to date.

5.2 Environment

Environmental factors are important in clinical settings. In fact, there has been a major emphasis on the development of measures of environment that account for parameters that relate to them for clinical outcomes. Primary measures include the Craig Handicap Assessment Reporting Technique (CHART; Whiteneck et al., 1992a; Whiteneck et al., 1992c; Hall et al., 1998) and the Craig Hospital Inventory of Environmental Factors (CHIEF; Craig Hospital, 2001; Whiteneck et al., 2004).

The CHART is a well-established tool that provides a quantitative measure of societal participation after SCI and serves as a means to evaluate six dimensions of social participation, including physical independence, cognitive independence, mobility, occupation, social integration, and economic self-sufficiency. The CHART produces a total score as well as one for each of the individual subscales. Each subscale is scored on a 100-point scale with higher scores indicating higher levels of social participation/community integration (Hall et al., 1998; Anderson et al., 2003). The CHART demonstrates high test-retest reliability with a score of 0.93 for the total score and scores of 0.80–0.95 for the subscales (Hall et al., 1998). It has also demonstrated validity by successfully discriminating between groups of individuals evaluated by rehabilitation professionals as having high or low levels of impairment (Whiteneck et al., 1992c). The CHART has been utilized to investigate the relationship between community integration and coping strategies/life satisfaction (Hansen, Forchheimer, Tate and Luera, 1998), demographics and type of injury (Whiteneck et al., 1999), fitness and physical activity (Manns and Chadd, 1999), environmental factors (Dijkers et al., 2002), duration of injury (Anderson et al., 2003; Charlifue and Gerhart, 2004), gender (Forchheimer et al., 2004), life expectancy (Krause et al., 2004), and coping and QOL (Kennedy et al., 2006) in individuals with SCI.

The CHIEF, developed by experts on disability, is designed to assess the experience of environmental barriers in terms of their frequency of occurrence and the intensity of the problem. Five subscales of environmental barriers were derived through factor analysis: attitudes and support, service and assistance, physical and structural, policy, and work/school. Scale and total CHIEF scores range from 0 to 8 and indicate the interaction of barrier frequency and magnitude. Higher scores indicate more significant obstacles i.e., a score of 0 indicates no barriers and a score of 8 indicates major problems (Whiteneck et al., 2004). The CHIEF has high test-retest reliability (intraclass correlation coefficient = 0.93), high internal consistency (Cronbach α = 0.93), and has been demonstrated to detect differences among groups of individuals with known differences in levels of disability and disability categories (Whiteneck et al., 2004). The CHIEF has been utilized to assess the relationship between environmental factors and community participation in two countries, The United States and Turkey (Dijkers et al., 2002), gender (Forchheimer et al., 2004), and participation and life satisfaction (Whiteneck et al., 2004) in individuals with SCI.

Environmental factors may be powerful predictors of outcomes and nearly all theoretical schemes include some type of interaction between the person and environment. For instance, the aforementioned studies on vocational interests defined personality types in terms of preferences for working in particular environments (Holland, 1985). The stress-coping models described later also include the interaction of multiple environmental (e.g., events) and psychological (appraisals of events) parameters. Clearly, the environment provides the context or backdrop upon which psychological or other personal factors may be investigated.

5.3 Stress-Coping Models

Stress is a concept that is widely used in research on adaptation to trauma and aging, yet, due to a diversity of definitions, has lost some of its unique meaning. Stress-coping models are particularly attractive since they simultaneously address psychological and environmental components after neurologic injury. There is an extensive body of research that supports the use of the stress-coping model of adaptation with SCI (Galvin and Godfrey, 2001).

Lazarus and Folkman (1984) defined stress as the relationship between a person and their environment that has been appraised by the person as making substantial or excessive demands on their ability to cope or by endangering their well-being. Lazarus (1999) suggested that psychological stress may result from attempts to cope with three general types of event/conditions in the environment including harm/loss, threat, and challenge. Because the definition describes stress as the product of an interaction between an individual and their environment, stress occurs in response to events/conditions in the environment. Second, the construct of coping is linked with the experience of stress. Third, it emphasizes the subjective component of stress and individual differences in response to stress. For example, with regards to SCI, even though many people experience similar levels of physiologic loss, many of them report different levels of stress. Lastly, the definition provides a description of general classes of events that cause stress (e.g., harm/loss, threat, and challenge).

Several models of stress-coping have been advanced (e.g., Lazarus and Folkman, 1984; Hobfoll, 1989), each with their own strengths and weaknesses. An integrated model, consistent with Lazarus and Folkman (1984), asserts that there are five important factors determining the degree of stress a person will experience in relation to life experiences/events (i.e., their environment). These include: (1) life events, (2) appraisal, (3) social support, (4) coping strategies, and (5) personality. From the standpoint of the general risk model, life events and social support represent environmental or situational factors, whereas appraisals, coping, and personality are clearly psychological. There are varying degrees of research that have been focused on the various stress-coping parameters, and it is beyond the scope of this chapter to provide an extensive review beyond that already done (Galvin and Godfrey, 2001). Nevertheless, there has been a great deal of support for several parameters. For instance, social support has been linked to several outcomes after SCI, including being positively correlated with life satisfaction and physical well-being and negatively associated with depression (Wallston et al., 1983; Schulz and Decker, 1985; Elliott et al., 1991; Rintala et al., 1992). Promoting effective coping strategies has been found to reduce depression and anxiety after SCI (Kennedy et al., 2003). Sensation seeking, a personality trait and proxy variable for participation in risk behaviors, has been found to be associated with increased risk of subsequent injuries after SCI (Krause, 2004).

6 Protective and Risk Behaviors

Health behaviors are a particularly important component of the model because they are potent predictors of outcomes, yet they are changeable and may be the focus of interventions. Health behaviors are generally broken down into two categories – risk and protective. Risk behaviors are those that increase the likelihood of an unfavorable condition or outcome, whereas protective behaviors are those that are associated with a decreased likelihood of an unfavorable condition or outcome. For instance, smoking has been identified as a risk behavior for pressure ulcers. Salzberg et al. (1996) found that subjects with pressure ulcers were twice as likely to be current cigarette smokers and had a longer smoking history. In 88 patients who had never smoked, 73.9% had a pressure ulcer compared with 90.1% of the 71 current smokers ($P = 0.011$). Smoking also affected ulcer recurrence rates of the first pressure ulcer with much higher rates being seen in the current smoker group (42.2%) compared with the never-smoked group (26.9%).

Alcohol use is another prominent risk behavior. This is not surprising, since alcohol intoxication rates are substantial at the time of the onset of SCI. Estimates of alcohol and substance use at the time of injury are reported to be anywhere from 17 to 62% (O'Donnell et al., 1981; Heinemann et al., 1990; McKinley et al., 1999). Analysis of the relationship between blood alcohol concentration (BAC) status and the severity of neurological impairment indicated that individuals with a (+) BAC did have more severe impairments than persons with (−) BAC on admission to acute care (Forchheimer et al., 2005). Worse medical outcomes after SCI, such as pain and pressure ulcers, have been found to be linked to patterns of alcohol consumption, abuse, and substance use (Tate et al., 2004). Alcohol and prescription medication use for pain, spasticity, sleep, or depression were found to be associated with an elevated risk for subsequent injuries, which were defined as those that occur in the years and decades after the initial SCI (Krause, 2004). This suggests that there is continuity from the time period precipitating SCI onset into the years after SCI onset.

Conversely, multiple types of preventative behaviors are taught in an effort to prevent secondary conditions such as pressure ulcers. They include performing routine weight shifts in and out of the chair and heightened vigilance in routinely checking skin for redness and irritation. Other activities focus on preventing urinary tract infections (UTIs) by keeping supplies, such as urine bags, clean and checking urine for discoloration. Other types of protective behaviors such as maintaining proper nutrition and exercising are much more general in nature yet are relatively unexplored areas in populations with SCI.

Newly injured individuals tend to lose weight due to hypermetabolism and hypercatabolism following major trauma (Rodriguez et al., 1997). After the acute injury phase, resting energy expenditure decreases as a result of the loss of muscle mass (Buchholz et al., 2003a). The resulting relatively sedentary lifestyle associated with SCI also results in decreased energy needs for physical activity (Buchholz et al., 2003b). This decreased energy need makes proper adjustment in dietary intake essential after injury, as the caloric intake may easily exceed daily requirements, thus placing individuals with SCI at increased risk of weight gain (Chen et al., 2006). A survey of 348 persons with chronic SCI showed that approximately 40% were overweight or obese (Anson and Shepherd, 1996). Obesity-related health risks are a great concern in the SCI population. Cardiovascular disease, hypertension, diabetes, impaired glucose tolerance, and an abnormal lipid profile were reported to be more common in persons with SCI than in able-bodied control subjects (Bauman and Spungen, 2001; Myers et al., 2007). Considerable empirical evidence exists which demonstrates improvement in health outcomes after weight loss in the general population; however, there is a general lack of information regarding the health outcomes of weight loss in individuals with SCI (Chen et al., 2006). A recent intervention study evaluating the effectiveness of a 12-week weight loss program for individuals with SCI demonstrated that weight loss could be achieved even among persons with tetraplegia and long duration of injury (Chen et al., 2006). Various measures of physical health (abdominal fat, blood cholesterol level, blood pressure) as well as psychosocial and physical functioning (sense of attractiveness, transferring, dressing) showed improvement after program intervention. The investigators of this study expressed their concern over the difficulty of maintaining weight loss strategies without the assistance of a formal program supporting the argument for extended intervention programs in clinical or community settings to assist individuals with SCI with the complexity of long-term weight management (Chen et al., 2006). Much more research is needed to create nutritional guidelines that complement the unique metabolic needs of persons with SCI.

Physical activity programs and information on how physical activity can promote health after SCI are two of the services most desired but least available to people with SCI (Ginis et al., 2007). A number of physical activity interventions, including standing, electrically stimulated cycling or resistance training, and walking exercises have been explored with the aim of reducing bone loss and/or increasing bone mass and muscle mass in individuals with SCI (Giangregorio and McCartney, 2006). These studies, however, are generally of short duration and do not evaluate the effect of long-term physical activity programs on general health after SCI. The absence of large-scale, prospective cohort studies of the relationship between health and activity in persons with SCI impairs the development of specific physical activity guidelines and activity-enhancing interventions that meet the unique needs of persons with SCI (Ginis et al., 2007). A recent study evaluating the effect of *habitual* high-level physical activity in 14 college-aged athletes with SCI demonstrated maintenance of metabolic outcomes to the same level as sedentary able-bodied individuals despite the greatly reduced muscle mass activation in the SCI group (Mojtahedi et al., 2007). This is an encouraging finding as metabolic disease (i.e., Type II Diabetes) appears decades earlier in SCI individuals due to greatly reduced activity levels and increased risk for obesity (Bauman et al., 1999). Increasing physical activity among individuals with varying degrees of paralysis is not a straightforward issue; however, the increasing longevity of individuals with SCI necessitates further research into physical activity programs, especially accessible community-based programs that will help decrease morbidity and mortality and improve overall QOL after SCI.

7 Health Outcomes and Secondary Conditions

The Institute of Medicine defines secondary condition as a preventable disease, impairment, injury, or disability occurring at an increased frequency as a result of a primary disabling condition (Brandt and Pope, 1997). Examples of secondary conditions include depression, pressure ulcers, and fatigue. Because secondary conditions are common in persons with disabilities (Campbell et al., 1999; Rimmer, 1999; Coyle et al., 2000; Gajdosik and Cicirello, 2001; Wilber et al., 2002), persons with SCI have heightened vulnerability of developing them (DeLisa and Kirshblum, 1997; McKinley et al., 1999). Aging and neurological level and completeness of injury are reported to increase the risk of secondary conditions and further disability (Gerhart et al., 1993; Noreau and Fougeyrollas, 2000). Service utilization and health care costs associated with medical complications are reported to increase with age and time postinjury. (Menter et al., 1991). DeVivo et al. (1990) revealed that patients who were at least 61 years of age at the time of rehabilitation discharge after initial injury were more likely to develop pneumonia or pulmonary emboli and have renal stones or gastrointestinal hemorrhage than patients aged 16–30 years.

Anson and Shepherd (1996) reported patterns of secondary complications among 348 outpatients across all neurological levels of SCI. A clinical team evaluated the patients and documented that approximately 95% had at least one secondary condition and 58% had three or more. After controlling for age and level of injury, duration of injury was predictive of number and severity of complications. Patients who lived in rural areas and small towns were more likely to have more medical problems. The most prevalent medical conditions observed were pain with or without medication (45%), overweight or obese (40%), spasticity in 290 patients with cervical or thoracic injuries (74%), UTIs (27.2%), and pressure ulcers (22.4%).

Pressure ulcers are the primary secondary condition of concern after SCI because of their cost and detriment to overall health. Serious skin break down significantly increases the risk of death from infections. (Chan et al., 2003; Wall and Colley, 2003). Surgeries to repair pressure ulcers were identified as one of five primary risk factors for mortality (Krause et al., in press). A population-based study in Colorado that documented across all degrees of SCI severity determined that 23% of 368 persons injured for 5 years reported having at least one pressure ulcer (Johnson et al., 1998). A study of 633 nonambulatory participants 5 years after SCI onset reported a 13% recurrence of one or more pressure ulcers (Krause and Broderick, 2004). Of persons injured an average of 10 years, Fuhrer et al. (1993) noted that 32% reported pressure ulcers, 37% reported UTIs, and 50% reported joint degeneration. Whiteneck et al. (1992b), studying 834 persons with SCI in Great Britain who were 20–45 years postinjury, revealed a 23% average incidence of pressure ulcers and a 20% incidence of UTIs.

Completeness and level of lesion predict the type of secondary condition. A multi-center survey of 1,668 persons with tetraplegia revealed that persons with complete lesions more commonly reported pressure ulcers and UTIs. In contrast, persons with incomplete lesions reported contractures and pain (Klotz et al., 2002). Completeness and level of the injury are also significantly associated with the likelihood of hospitalization (Samsa et al., 1996; Middleton et al., 2004), which, although not a secondary condition per se, is a proxy variable for either an acute or chronic health problem requiring medical attention. Rates of hospitalization appear to be highest during the first year after injury (Davidoff et al., 1990), decline during the next 5–10 years (Ivie and DeVivo, 1994; Samsa et al., 1996), and increase during later life (Whiteneck et al., 1992b; Savic et al., 2000). In a population-based sample of persons with chronic SCI in the United Kingdom, Savic et al. (2000) revealed the most frequent reasons for readmissions were urinary complications (40.5%), followed by skin problems (17%), and digestive (10%), musculoskeletal (8.7%), and nervous system complications (6.9%). Davidoff et al. (1990) examined the incidence, cause, and monetary cost of hospitalizations during the first year after discharge from initial rehabilitative care at a Model SCI System (MSCIS) facility. They found 39% of patients were readmitted at least once in a year with an average length of stay per admission of 11.9 ± 2.1 days. The leading causes of hospitalization were additional rehabilitation (19.1%), symptomatic UTI (17%), deep venous thrombosis (12.8%), and removal of spinal instrumentation (12.8%). Thirty-four percent of all readmissions were considered preventable.

Middleton et al. (2004) found that, over an 11-year study period (1989–2000), 58.6% of 432 persons aged 16 years and older discharged with new SCI were readmitted for SCI-related causes. Readmissions accounted for 977 hospitalizations and an average length of stay of 15.5 days. Nine persons under age 30 years accounted for 50% of all readmissions. The most frequent causes of rehospitalizations were genitourinary and gastrointestinal complications, further rehabilitation, and skin conditions. Age, level, and completeness of neurological impairment most influenced rates of readmission. Skin-related conditions accounted for approximately 30% of all bed days. In the first 12 months after discharge, an overall 0.64 hospitalization rate was noted that averaged 12.6 bed-days per person and an average length of stay of 19.7 (median 6) days. A downward trend was noted for time since injury. Length of stay was significantly longer for persons with ASIA A, B, or C impairments (22.2–17.0 days) compared to persons with ASIA D impairments (11.3 days).

Ivie and DeVivo (1994) identified risk factors for unplanned hospitalizations during the previous year in a cross-sectional study of 2,305 persons who had been injured 1–8 years. Overall, 26% of the participants had been hospitalized during the most recent follow-up year. Cardenas et al. (2004) examined the frequency and reasons for hospitalizations at 1-, 5-, 10-, 15-, and 20-year follow-ups of 8,668 persons in the MSCIS centers. Overall, the leading cause of hospitalization was diseases of the genitourinary system such as UTIs. Patients with tetraplegia were more likely to be hospitalized due to diseases of the respiratory system such as pneumonia, while patients with paraplegia were more likely to be hospitalized due to disease of the skin, most likely pressure ulcers. Between 1995 and 2002, 28–37% of patients were hospitalized, with an average hospital stay of 14 days. A significantly greater number of hospitalizations occurred in the first year after injury than all other years. Factors that most influenced hospitalization were: being discharged from acute rehabilitation to a skilled nursing facility, having a lower motor score, having a state or federal program as health care provider.

Adverse psychological outcomes may also be considered a secondary condition. In particular, depressive disorders are perhaps the most common form of psychological distress in SCI and range in severity from minor depression and adjustment disorders to major depressive episodes. Estimating the prevalence of depression after SCI has indeed been challenging due to variations in the definitions, measures, and strategies utilized in the various studies. There is a consensus that depressive disorders occur at an elevated rate among people with SCI (Boekamp et al., 1996; Elliot and Frank, 1996). For instance, Kemp et al. (1999) found that 42% of respondents in their community sample of individuals with SCI reported clinically significant depressive symptomatology.

Longitudinal research (Craig et al., 1994) suggests that approximately 30% of persons with SCI have heightened levels of anxiety and depressive mood for up to 2 years after injury. Other longitudinal research (Kennedy and Rogers, 2000) examined the prevalence of anxiety and depression longitudinally in a sample of 104 participants using the Beck Depression Inventory (BDI). Their findings suggested a gradual increase between weeks 24 (BDI = 11) and 48 (BDI = 21), with a significant decrease thereafter. State Anxiety Inventory scores followed a similar pattern.

Another line of research used the Older Adult Health and Mood Questionnaire to measure depressive symptoms, as it was specifically designed to be used with populations that are older and those with physical impairments (Kemp and Adams, 1995). Because of the study populations utilized, these results have significant implications for minority participants. For instance, Kemp et al. (1999) found ethnic differences in depressive symptoms, with Latino participants reporting more symptoms than Caucasian and African-American participants. Krause et al. (2000b) investigated the relation among aging, gender, ethnicity, socioeconomic indicators, and depressive symptoms after SCI and found that 48% of the participants reported clinically significant symptoms. Minority participants, particularly women, were at a substantially higher risk for depressive symptoms. However, the risk diminished but did not disappear after controlling for years of education and income.

Weingardt et al. (2001) identified the rates of positive screens for psychological and substance abuse disorders among 117 veterans admitted to the inpatient SCI unit. They found 45% of respondents screened positive for any psychiatric disorder, 16% for depression, 40% for any anxiety disorder, and 6% for alcohol problems. Positive screens for depression and anxiety correlated positively with the number of recent hospitalizations. Bombardier et al. (2004) used an abbreviated version of a Patient Health Questionnaire (the PHQ-9) to identify the incidence of depressive symptoms and probable major depressive disorder (MDD) at 1 year postinjury using data from the MSCIS. They found an 11.4% rate of MDD, and 15.4% of participants reported suicidal ideation. As expected, depressive symptoms were negatively correlated with life satisfaction and health.

Another potentially important psychological outcome is posttraumatic stress disorder (PTSD). Most traumatic injuries are associated with some type of psychological distress (Mohta et al., 2003). PTSD was first described in the Diagnostic and Statistical Manual of Mental Disorders by the American Psychiatric Association (APA) in 1980. A diagnosis of PTSD may be applied to individuals that have been exposed to a traumatic event that involved actual or threatened death or serious injury, with an associated response of horror, fear, or helplessness, and who have subsequent symptoms of intrusive reexperiencing of the trauma, avoidant behaviors, and increased physiological arousal (APA, 1994).

The etiology of the majority of SCI (>60%) is related to potentially highly traumatic events – motor vehicle crash or violence (i.e., gunshot injury; NSCISC, 2008). The true prevalence of PTSD after SCI is not known, and there are several inherent methodological difficulties with these types of studies. First, when present, it is not apparent whether PTSD is a consequence of the trauma experience which resulted in the SCI, the SCI itself, or both (Mona et al., 2000; Kennedy and Duff, 2001). Second, the majority of investigations examining PTSD after SCI are cross-sectional, focusing primarily on establishing the occurrence of PTSD after SCI and the relationship to demographic characteristics of the study population, such as level of injury (Binks et al., 1997), veteran status, including combat and noncombat exposures (Radnitz et al., 1998; Radnitz et al., 2000), gender (Mona et al., 2000), time since injury (Kennedy and Evans, 2001), and culture (Lude et al., 2005). More recent investigations of post-SCI PTSD explore the relationship between psychological characteristics of the study population, such as death anxiety (Martz, 2004a), adaptation (Martz, 2004b), cognitive appraisal (Agar et al., 2006), and locus of control (Chung et al., 2006) and the development of PTSD after SCI. Lastly, using existing instruments, it is often difficult to differentiate whether the participant is responding to the trauma of SCI itself (i.e., the event), or simply the focus on the day-to-day consequences resulting from SCI (e.g., thinking about the need to schedule someone to provide personal assistance).

8 Mortality

Early mortality is a reality after SCI in the end point of the predictive model. It represents the culmination of the collective efforts of rehabilitation professionals, healthcare providers, and the individual himself/ herself to successfully adapt to SCI and to promote the best possible outcomes. It is also influenced by each type of variable within the model, at least conceptually, and therefore reflects multiple opportunities to intervene to enhance longevity.

Historically, individuals did not live long after SCI, particularly those with more severe injuries. UTIs and other urologic complications were primary contributors to early mortality. This changed over the first few decades after World War II, as the life expectancy of individuals with SCI has improved dramatically (Burke et al., 1960; Freed et al., 1966; Hardy, 1976; Hackler, 1977; Mesard et al., 1978; Ravichandron and Silver, 1982; DeVivo et al., 1987; Whiteneck et al., 1992a; DeVivo and Stover, 1995). There was also a shift in the primary cause of death from urologic complications to respiratory complications (Carter, 1979; Kiwerski et al., 1981; DeVivo et al., 1989; DeVivo et al., 1992; DeVivo et al., 1993). DeVivo and Stover (1995) found that several specific causes of death are more common among people with SCI. Pneumonia and influenza were the primary cause of mortality (17.7%), followed by nonischemic heart disease (16.5%), septicemia (12%), and diseases of the urinary tract (8.7%). They also calculated standardized mortality ratios (SMRs; the ratio of SCI deaths to expected deaths), noting the highest SMRs were for septicemia (64.2 times the general population), disease of the pulmonary circulation (47.1), pneumonia and influenza

(35.6), symptoms and ill-defined conditions (13.8), and diseases of the urinary system (10.9). It is noteworthy that suicides were 4.8 times that of the general population.

Whereas mortality rates have declined dramatically in the first year after SCI, life expectancies for those who survive the first year are essentially unchanged over more than two decades, except for those who are ventilator-dependent patients (DeVivo and Ivie, 1995; DeVivo et al., 1999; Strauss et al., 2006). Biographic and injury-related characteristics have by far received the greatest attention in relation to early mortality. Current age is the most critical risk factor for mortality, as would be expected, but calendar time has had only minimal effects (Strauss et al., 2000; Krause et al., 2004; Shavelle et al., 2006; Strauss et al., 2006; Shavelle et al., 2007). Neurologic level, completeness of injury, and ventilator-dependency are the significant predictors of mortality, with gender, race, and cause of injury being weaker but still significant risk factors (DeVivo et al., 1987; DeVivo et al., 1992; DeVivo and Ivie, 1995; DeVivo and Stover, 1995; DeVivo et al., 1999; Strauss et al., 2000; Krause et al., 2004; Shavelle et al., 2006; Shavelle et al., 2007).

When life expectancy was shorter after SCI and urologic complications led to early mortality in the majority of cases, there may have been less opportunity for psychological, environmental, and behavioral factors to mediate life expectancy. Conversely, since life expectancy was enhanced, the importance of maintaining appropriate health care maintenance behaviors and avoiding risk behaviors would likely be enhanced. According to the model, these factors would mediate the health status of the individual which would in turn impact longevity. Unfortunately, there is little data to test this assumption.

There have been essentially three manuscripts that have addressed the empirical risk model established at the outset of this chapter in relation to mortality. DeVivo et al. (1999) investigated traditional risk variables for early mortality, including biographic and injury related variables, as well as the role of violent etiology and type of health insurance coverage. Participants injured as the result of an act of violence had a greater odds of mortality than those with nonviolent etiologies (1.53 odds ratio for all 1 day admits; 1.15 odds ratio for all 1 year survivors), and participants with Medicare and Medicaid insurance had a greater odds of mortality than those with other insurance (excluding HMOs and unknown sponsors). In a more recent report that also utilized the study model proposed in this study, but post hoc with existing data, Krause et al. (2004) investigated a broader set of risk factors in relation to mortality. Consistent with the study model, health factors, participation, and socioeconomic factors all contributed to prediction of mortality, with the most substantial increase in prediction attributable to health factors.

Krause et al. (in press) identified the predictive efficacy of secondary conditions and other health factors (this set of risk factors most proximal to mortality in the general risk model) using data specifically collected to test the empirical model. They found the best set of health predictors included: (1) probable major depression, (2) surgeries to repair pressure ulcers, (3) fractures and/or amputations, (4) symptoms of infections, and (5) days hospitalized. When compared with the traditional model that included only injury severity and biographic characteristics as predictors, the comprehensive model that included secondary conditions and general health factors alone was superior as indicated by the pseudo-R^2 (health factors = 0.178; injury severity = 0.121) and the concordance R^2 (health factors = 0.776; injury severity = 0.730). Although additional analyses are in preparation related to other parameters in the model, none have been completed to date.

9 Summary and Conclusions

In this chapter, we have reviewed clinical outcomes after SCI from the perspective of a general risk model which outlines several sets of variables that may sequentially lead to the development of secondary health conditions and mortality. Despite an explosion in SCI research over the past decade, there continues to be several clinical areas that receive inadequate attention, and there is a lack of studies and models to integrate the various parameters. The model put forth in this chapter is an attempt to provide a structure for summarizing the current state of research, as well as a model for guiding future research. Only through persistent ongoing efforts we will more fully understand the problems faced by people with SCI and gain the knowledge necessary to develop evidence-based interventions to promote health, reduce the risk of secondary conditions, and enhance longevity after SCI.

References

Agar E, Kennedy P, King N. 2006. The role of negative cognitive appraisals in PTSD symptoms following spinal cord injuries. Behav Cogn Psychother 34: 437-452.

American Psychiatric Association (APA). 1994. Diagnostic and statistical manual of mental disorders, 4th edition. Washington, DC: APA.

Anderson CJ, Krajci KA, Vogel LC. 2003. Community integration among adults with spinal cord injuries sustained as children or adolescents. Dev Med Child Neurol 45: 129-134.

Anson CA, Shepherd C. 1996. Incidence of secondary complications in spinal cord injury. Int J Rehabil Res 19: 55-66.

Bauman W, Kahn N, Grimm D, Spungen A. 1999. Risk factors for atherogenesis and cardiovascular autonomic function in persons with spinal cord injury. Spinal Cord 37: 601-616.

Bauman WA, Spungen AM. 2001. Carbohydrate and lipid metabolism in chronic spinal cord injury. J Spinal Cord Med 24: 266-277.

Binks M, Radnitz C, Moran A, Vinciguerra V. 1997. Relationship between spinal cord injury level and post traumatic stress disorder symptoms. Ann N Y Acad Sci 821: 430-432.

Boekamp JR, Overholser JC, Schubert DS. 1996. Depression following a spinal cord injury. Int J Psychiatry Med 26: 329-349.

Bombardier CH, Richards JS, Krause JS, Tulsky D, Tate DG. 2004. Symptoms of major depression in people with spinal cord injury: Implications for screening. Arch Phys Med Rehabil 85: 1749-1756.

Broderick L, Krause J. 2003. Breast and gynecologic health screening behaviors among 191 women with spinal cord injuries. J Spinal Cord Med 26: 145-149.

Brandt E, Pope A. 1997. Enabling America. Washington, DC: North Academy Press.

Buchholz AC, McGillivray CF, Pencharz PB. 2003a. Differences in resting metabolic rate between paraplegic and able bodied subjects are explained by differences in body composition. Am J Clin Nutr 77: 371-378.

Buchholz AC, McGillivray CF, Pencharz PB. 2003b. Physical activity levels are low in free-living adults with chronic paraplegia. Obes Res 11: 563-570.

Burke MH, Hicks AF, Robbins M, Kessley H. 1960. Survival of patients with injuries to the spinal cord. JAMA 172: 121-124.

Burnett D, Kolakowsky S, White J, Cifu D. 2002. Impact of minority status following spinal cord injury. Neurorehabilitation 17: 187-194.

Campbell M, Sheets D, Strong P. 1999. Secondary health conditions among middle-aged individuals with chronic physical disabilities. Arch Phys Med Rehabil 11: 105-122.

Cardenas D, Bryce T, Shem K, Richards J, Elhefni H. 2004. Gender and minority differences in the pain experience of people with spinal cord injury. Arch Phys Med Rehab 85: 1774-1781.

Carter RE. 1979. Experiences with high tetraplegics. Paraplegia 17: 140-146.

Centers for Disease Control and Prevention (CDC). 2002. CDC spinal cord injury fact book 2001–2002. http://www.cdc.gov/ncipc/fact_book/25_Spinal_Cord_Injury.htm.

Chan JW, Virgo KS, Johnson FE. 2003. Hemipelvectomy for severe decubitus ulcers in patients with previous spinal cord injury. Am J Surg 185: 69-73.

Charlifue S, Gerhart K. 2004. Community integration in spinal cord injury of long duration. NeuroRehabilitation 19: 91-101.

Chen Y, DeVivo MJ, Jackson AB. 2005. Pressure ulcer prevalence in people with spinal cord injury: Age-period-duration effects. Arch Phys Med Rehabil 86: 1208-1213.

Chen Y, Henson S, Jackson A, Richards J. 2006. Obesity interventions in persons with spinal cord injury. Spinal Cord 44: 82-91.

Chung MC, Preveza E, Papandreou K, Prevezas N. 2006. The relationship between posttraumatic stress disorder following spinal cord injury and locus of control. J Affect Disord 93: 229-232.

Coyle CP, Santiago MC, Shank JW, Ma GX, Boyd R. 2000. Secondary conditions and women with physical disabilities: A descriptive study. Arch Phys Med Rehabil 81: 1380-1387.

Craig AR, Hancock KM, Dickson HG. 1994. A longitudinal investigation into anxiety and depression in the first 2 years following a spinal cord injury. Paraplegia 32: 675-679.

Craig Hospital. 2001. Craig hospital inventory of environmental factors, version 3.0. Available at http://www.craighospital.org/research/disability/CHIEF%20Manual.pdf. Cited 10 Jan 2006.

Crewe NM, Krause JS. 1990. An eleven-year follow-up of adjustment to spinal cord injury. Rehabil Psychol 35: 205-210.

Davidoff G, Schultz S, Lieb T, Andrews K, Wardner J, et al. 1990. Rehospitalization after initial rehabilitation for acute spinal cord injury: Incidence and risk factors. Arch Phys Med Rehabil 71: 121-124.

Davis M, Matthews B, Jackson WT, Fraser E, Richards JS. 1995. Self-concept as an outcome of spinal cord injury: The relation to race, hardiness, and locus of control. SCI Psychosoc Process 8: 90-101.

DeLisa J, Kirshblum S. 1997. A review: Frustrations and needs in clinical care of spinal cord injury patients. J Spinal Cord Med 4: 384-390.

DeVivo MJ, Black KJ, Stover SL. 1993. Causes of death during the first 12 years after spinal cord injury. Arch Phys Med Rehabil 74: 248-254.

DeVivo MJ, Ivie CS. 1995. Life expectancy of ventilator-dependent persons with spinal cord injuries. Chest 108: 226-232.

DeVivo MJ, Kartus PL, Rutt RD, Stover SL, Fine PR. 1990. The influence of age at time of spinal cord injury on rehabilitation outcome. Arch Neurol 47: 687-691.

DeVivo MJ, Kartus PL, Stover SL, Rutt RD, Fine PR. 1987. Seven year survival following spinal cord injury. Arch Neurol 44: 872-875.

DeVivo MJ, Kartus PL, Stover SL, Rutt RD, Fine PR. 1989. Cause of death for patients with spinal cord injuries. Arch Intern Med 149: 1761-1766.

DeVivo MJ, Krause JS, Lammertse DP. 1999. Recent trends in mortality and causes of death among persons with spinal cord injury. Arch Phys Med Rehabil 80: 1411-1419.

DeVivo MJ, Stover SL. 1995. Long-term survival and causes of death. Spinal cord injury: Clinical outcomes from the model systems. Stover SL, DeLisa JA, Whiteneck GG, editors. Gaithersburg, MD: Aspen Publishers; pp. 289-316.

DeVivo MJ, Stover SL, Black KJ. 1992. Prognostic factors for twelve-year survival following spinal cord injury. Arch Phys Med Rehabil 73: 156-162.

Dijkers M. 1997. Quality of life after spinal cord injury: A meta-analysis of the effects of disablement components. Spinal Cord 35: 829-840.

Dijkers MP, Yavuzer G, Ergin S, Weitzenkamp D, Whiteneck GG. 2002. A tale of two countries: Environmental impacts on social participation after spinal cord injury. Spinal Cord 40: 351-362.

Elliott T, Herrick S, Patti A, Witty T, Godshall F, et al. 1991. Assertiveness, social support, and psychological adjustment following spinal cord injury. Behav Res Ther 29: 485-493.

Elliott TR, Frank RG. 1996. Depression following spinal cord injury. Arch Phys Med Rehabil 77: 816-823.

Estores I, Sipski M. 2004. Women's issues after SCI. Top Spinal Cord Inj Rehabil 10: 107-125.

Forchheimer M, Cunningham R, Gater D, Maio R. 2005. The relationship of blood alcohol concentration to impairment severity in spinal cord injury. J Spinal Cord Med 28: 303-307.

Forchheimer M, Kalpakjian C, Tate D. 2004. Gender differences in community integration after spinal cord injury. Top Spinal Cord Inj Rehabil 10: 163-174.

Forsythe E, Horsewell J. 2006. Sexual rehabilitation of women with spinal cord injury. Spinal Cord 44: 234-241.

Freed MM, Bakst HJ, Barrie DL. 1966. Life expectancy, survival rates, and causes of death in civilian patients with spinal cord trauma. Arch Phys Med Rehabil 47: 457-463.

Fuhrer MJ, Garber SL, Rintala DH, Clearman R, Hart KA. 1993. Pressure ulcers in community resident persons with spinal cord injury. Arch Phys Med Rehabil 74: 1172-1177.

Gajdosik CG, Cicirello N. 2001. Secondary conditions of the musculoskeletal system in adolescents and adults with cerebral palsy. Phys Occup Ther Pediatr 21: 49-68.

Galvin L, Godfrey H. 2001. The impact of coping on emotional adjustment to spinal cord injury: Review of the literature and application of a stress appraisal and coping mode. Spinal Cord 29: 615-627.

Gerhart KA, Bergstrom E, Charlifue SW, Menter RR, Whiteneck GG. 1993. Long-term spinal cord injury: Functional changes over time. Arch Phys Med Rehabil 74: 1030-1034.

Giangregorio L, McCartney N. 2006. Bone loss and muscle atrophy in spinal cord injury: Epidemiology, fracture prediction, and rehabilitation strategies. J Spinal Cord Med 29: 489-500

Ginis K, Latimer A, Buchholz A, Bray S, Craven B, et al. 2007. Establishing evidence-based physical activity guidelines: Methods for the study of health and activity in people with spinal cord injury (SHAPE SCI). Spinal cord. Available at www.nature.com/sc. Cited 24 July 2007.

Hackler RH. 1977. A 25-year prospective mortality study in the spinal cord injured patient: Comparison with the long-term living paraplegic. J Urol 117: 486-488.

Hall K, Dijkers M, Whiteneck G, Brooks C, Krause J. 1998. The Craig handicap assessment and reporting technique (CHART): Metric properties and scoring. Top Spinal Cord Inj Rehabil 4: 16-30.

Hansen N, Forchheimer M, Tate D, Luera, G. 1998. Relationships among community reintegration, coping strategies, and life satisfaction in a sample of persons with spinal cord injury. Top Spinal Cord Inj Rehabil 4: 56-72.

Hardy AG. 1976. Survival periods in traumatic tetraplegia. Paraplegia 14: 41-46.

Heinemann A, Mamott B, Schnoll S. 1990. Substance use by persons with recent spinal cord injuries. Rehabil Psychol 35: 217-238.

Hobfoll SE. 1989. Conservation of resources. A new attempt at conceptualizing stress. Am Psychol 44: 513-524.

Holland J. 1985. Making vocational choices, 2nd edition. Odessa, FL.: Psychological Assessment Resources.

Hughes R, Swedlund N, Petersen N, Nosek M. 2001. Depression and women with spinal cord injury. Top Spinal Cord Inj Rehabil 7: 16-24.

Ivie CS, DeVivo MJ. 1994. Predicting unplanned hospitalizations in persons with spinal cord injury. Arch Phys Med Rehabil 75: 1182-1188.

Jackson A, Dijkers M, DeVivo M, Poczatek R. 2004a. A demographic profile of spinal cord injuries: Change and stability over 30 years. Arch Phys Med Rehabil 85: 1740-1748.

Jackson A, Lindsey L, Klebine P, Poczatek R. 2004b. Reproductive health for women with spinal cord injury: Preganancy and delivery. SCI Nurs 21: 88-91.

Johnson RL, Gerhart KA, McCray J, Menconi JC, Whiteneck GG. 1998. Secondary conditions following spinal cord injury in a population-based sample. Spinal Cord 36: 45-50.

Johnston M, Wood K, Millis S, Page S, Chen D. 2004. Perceived quality of care and outcomes following spinal cord injury: Minority status in the context of multiple predicators. J Spinal Cord Med 27: 241-251.

Kalpakjian C, Albright K. 2006. An examination of depression through the lens of spinal cord injury: Comparative prevalence rates and severity in women and men. Women's Health Issues 16: 380-388.

Kemp BJ, Adams BM. 1995. The older adult health and mood questionnaire: A measure of geriatric depressive disorder. J Geriatr Psychiatry Neurol 83: 162-167.

Kemp B, Krause J, Adkins R. 1999. Depression among African Americans, Latinos, and Caucasians with spinal cord injury: A exploratory study. Rehabil Psychol 44: 235-247.

Kennedy P, Duff J. 2001. Post traumatic stress disorder and spinal cord injuries. Spinal Cord 39: 1-10.

Kennedy P, Duff J, Evans M, Beedie A. 2003. Coping effectiveness training reduces depression and anxiety following traumatic spinal cord injuries. Br J Clin Psychol 42: 41-52.

Kennedy P, Evans M. 2001. Evaluation of post traumatic distress in the first 6 months following SCI. Spinal Cord 39: 381-386.

Kennedy P, Lude P, Taylor N. 2006. Quality of life, social participation, appraisals and coping post spinal cord injury: A review of four community samples. Spinal Cord 44: 95-105.

Kennedy P, Rogers, BA. 2000. Anxiety and depression after spinal cord injury: A longitudinal analysis. Arch Phys Med Rehabil 81: 932-937.

Kiwerski J, Weiss M, Chrostowska T. 1981. Analysis of mortality of patients after cervical spine trauma. Paraplegia 19: 347-351.

Klotz R, Joseph PA, Ravaud JF, Wiart L, Barat M, et al. 2002. The tetrafigap survey on the long-term outcome of tetraplegic spinal cord injured persons. III. Medical complications and associated factors. Spinal Cord 40: 457-467.

Krause JS. 1996. Secondary conditions and spinal cord injury: A model for prediction and prevention. Top Spinal Cord Inj Rehabil 2: 217-227.

Krause JS. 1998. Subjective well being after spinal cord injury: Relationship to gender, race/ethnicity, and chronologic age. Rehabil Psychol 43: 282-296.

Krause JS. 2000. Aging after spinal cord injury: An exploratory study. Spinal Cord 38: 77-83.

Krause JS. 2004. Factors associated with risk for subsequent injuries after the onset of traumatic spinal cord injury. Arch Phys Med Rehabil 85: 1503-1508.

Krause JS, Broderick LE. 2004. Outcomes after spinal cord injury: Comparisons as a function of gender and race ethnicity. Arch Phys Med Rehabil 85: 355-362.

Krause JS, Broderick LE, Saladin LK, Broyles J. 2006. Racial disparities in health outcomes after spinal cord injury: Mediating effects of education and income. J Spinal Cord Med 29: 17-25.

Krause JS, Carter R, Pickelsimer E. A prospective study of health and risk of mortality after spinal cord injury. Arch Phys Med Rehabil. In press.

Krause JS, Crewe NM. 1991. Chronologic age, time since injury, and time of measurement: Effect on adjustment after spinal cord injury. Arch Phys Med Rehabil 72: 91-100.

Krause JS, Coker JL, Charlifue S, Whiteneck GG. 1999a. Depression and subjective well being among 97 American Indians with spinal cord injury. Rehabil Psychol 44: 354-372.

Krause JS, Coker JL, Charlifue S, Whiteneck GG. 1999b. Selected health behaviors among American Indians with spinal cord injury: Comparison to 1996 data from the Behavioral Risk Factor Surveillance System. Arch Phys Med Rehabil 80: 1435-1440.

Krause JS, Coker JL, Charlifue S, Whiteneck GG. 2000a. Health outcomes among American Indians with spinal cord injury. Arch Phys Med Rehabil 81: 924-931.

Krause JS, DeVivo MJ, Jackson AB. 2004. Health status, community integration and economic risk factors for mortality after spinal cord injury. Arch Phys Med Rehabil 85: 1764-1773.

Krause J, Kemp B, Coker J. 2000b. Depression after spinal cord injury: Relation to gender, ethnicity, aging, and socioeconomic indicators. Arch Phys Med Rehabil 81: 1099-1109.

Krause JS, Sternberg M. 1997. Aging and adjustment after spinal cord injury: The roles of chronologic age, time since injury, and environmental change. Rehabil Psychol 42: 287-302.

Lavela S, Weaver F, Smith B, Chen K. 2006. Disease prevalence and use of preventive services: Female veterans in general and those with spinal cord injuries and disorders. J Women's Health 15: 301-311.

Lazarus R. 1999. Stress and emotion, 1st edition. London: Springer.

Lazarus R, Folkman S. 1984. Stress, appraisal, and coping. New York: Springer.

Leibowitz R, Stanton A. 2007. Sexuality after spinal cord injury: A conceptual model based on women's narratives. Rehabil Psychol 52: 44-55.

Lude P, Kennedy P, Evans M, Lude Y, Beedie A. 2005. Post traumatic distress symptoms following spinal cord injury: A comparative review of European samples. Spinal Cord 4: 102-108.

Manns P, Chadd K. 1999. Determining the relation between quality of life, handicap, fitness and physical activity for persons with spinal cord injury. Arch Phys Med Rehabil 80: 1566-1571.

Martz E. 2004a. Death anxiety as a predictor of posttraumatic stress levels among individuals with spinal cord injuries. Death Stud 28: 1-17.

Martz E. 2004b. Do post-traumatic stress symptoms predict reactions of adaptation to disability after a sudden-onset spinal cord injury? Int J Rehabil Res 27: 185-194.

McKinley WO, Jackson AB, Cardenas DD, DeVivo MJ. 1999. Long term medical complication after traumatic spinal cord injury: A regional model systems analysis. Arch Phys Med Rehabil 80: 1402-1410.

Menter RR, Whiteneck GG, Charlifue SW, et al. 1991. Impairment, disability, handicap and medical expenses of persons aging with spinal cord injury. Paraplegia 29: 613-619.

Mesard L, Carmody A, Mannarino E, Ruge D. 1978. Survival after spinal cord trauma. Arch Neurol 35: 78-83.

Middleton JW, Lim K, Taylor L, Soden R, Rutkowski S. 2004. Patterns of morbidity and rehospitalisation following spinal cord injury. Spinal Cord 42: 359-367.

Miller S, Rahman R, Dixon P, Forchheimer M, Tate D, et al. 2006. Differences in satisfaction with life and health-related quality of life between minority and Caucasian individuals with spinal cord injury. SCI Psychosoc Process 19(2): 9.

Mohta M, Sethi A, Tyagi A, Mohta A. 2003. Psychological care in trauma patients. Injury 34: 17-25.

Mojtahedi M, Valentine R, Arngrimsson S, Wilund K, Evans E. 2007. The association between regional body composition and metabolic outcomes in athletes with spinal cord injury. Spinal Cord. Available at www.nature.com/sc. 15 May 2007.

Mona L, Cameron R, Lesondak L, Norris F. 2000. Post-traumatic stress disorder symptomatology in men and women with spinal cord injury. Top Spinal Cord Inj Rehabil 6: 76-86.

Myers J, Lee M, Kiratli J. 2007. Cardiovascular disease in spinal cord injury: An overview of prevalence, risk, evaluation, and management. Am J Phys Med Rehabil 86: 142-152.

National Spinal Cord Injury Statistical Center (NSCISC). 2008. Spinal cord injury: Facts and figures at a glance. Birmingham: University of Alabama.

Noreau L, Fougeyrollas P. 2000. Long-term consequences of spinal cord injury on social participation: The occurrence of handicap situations. Disabil Rehabil 22: 170-180.

Norrbrink Budh C, Lund I, Hulting C, Levi R, Werhagen L, et al. 2003. Gender differences in pain in spinal cord injured individuals. Spinal Cord 41: 122-128.

O'Donnell J, Cooper J, Gessner J, Shehan I, Ashley J. 1981. Alcohol, drugs, and spinal cord injury. Alcohol Health Res World 82: 27-29.

Pentland W, Walker J, Minnes P, Tremblay M, Brouwer B, et al. 2002. Women with spinal cord injury and the impact of aging. Spinal Cord 40: 374-387.

Radnitz C, Hsu L, Willard J, Perez-Strumolo L, Festa J, et al. 1998. Post traumatic stress disorder in veterans with spinal cord injury: Trauma related risk factors. J Trauma Stress 11: 505-520.

Radnitz C, Schlein I, Hsu L. 2000. The effect of prior trauma exposure on the development of PTSD after spinal cord injury. J Anxiety Disord 14: 313-324.

Rahman R, Albright K, Yaroslavsky I. 2005. Perceived stress and life satisfaction in women with a spinal cord injury: An exploratory look at racial differences. SCI Psychosoc Process 18: 1, 6–8.

Ravichandron G, Silver JR. 1982. Survival following traumatic tetraplegia. Paraplegia 20: 264-269.

Rimmer JH. 1999. Health promotion for people with disabilities: The emerging paradigm shift from disability prevention to prevention of secondary conditions. Phys Ther 79: 495-502.

Rintala D, Hart K, Priebe M, Ballinger D. 1998. Racial and ethnic differences in community reintegration in a community-based sample of adults with spinal cord injury. Top Spinal Cord Rehabil 4: 1-17.

Rintala D, Young M, Hart K, Clearman R, Fuhrer M. 1992. Social support and the well-being of persons with spinal cord injury. Rehabil Psychol 37: 155-163.

Rodriguez DJ, Benzel EC, Clevenger FW. 1997. The metabolic response to spinal cord injury. Spinal Cord 35: 599-604.

Rohe D, Athelstan G. 1982. Vocational interests of persons with spinal cord injury. J Couns Psychol 29: 283-229.

Rohe D, Krause JS. 1998. Stability of interests after severe physical disability: An 11-year longitudinal study. J Vocat Behav 52: 45-58.

Rumpf H, Hapke U, Hill A, John U. 1997. Development of a screening questionnaire for the general hospital and general practices. Alcohol Clin Exp Res 21: 894-898.

Salzberg CA, Byrne DW, Cayten G, van Niewerburgh P, Murphy JG, et al. 1996. A new pressure ulcer risk assessment scale for individuals with spinal cord injury. Am J Phys Med Rehabil 75: 96-104.

Samsa G, Landsman P, Hamilton B. 1996. Inpatient hospital utilization among veterans with traumatic spinal cord injury. Arch Phys Med Rehabil 77: 1037-1043.

Savic G, Short DJ, Weitzenkamp D, Charlifue SW, Gardner BP. 2000. Hospital readmissions in people with chronic spinal cord injury. Spinal Cord 38: 371-377.

Schaie KW. 1965. A general model for the study of developmental problems. Psychol Bull 64: 92-107.

Schulz R, Decker S. 1985. Long-term adjustment to physical disability: The role of social support, perceived control, and self-blame. J Pers Soc Psychol 48: 1162-1172.

Shavelle RM, DeVivo MJ, Strauss DJ, Paculdo DR, Lammertse DP, et al. 2006. Long-term survival of persons ventilator dependent after spinal cord injury. J Spinal Cord Med 29: 511-519.

Shavelle RM, DeVivo MJ, Paculdo DR, Vogel LC, Strauss DJ. 2007. Long-term survival after childhood spinal cord injury. J Spinal Cord Med 30(Suppl): S48-S54.

Strauss D, DeVivo M, Paculdo D, Shavelle R. 2006. Trends in life expectancy after spinal cord injury. Arch Phys Med Rehabil 87: 1079-1085.

Strauss D, DeVivo MJ, Shavelle R. 2000. Long-term mortality risk after spinal cord injury. J Insur Med 32: 11-16.

Tate D, Forchheimer M, Krause J, Meade M, Bombardier C. 2004. Patterns of alcohol use and abuse in persons with spinal cord injury: Risk factors and correlates. Arch Phys Med Rehabil 85: 1837-1847.

Wallston B, Alagna S, DeVellis B, DeVellis, R. 1983. Social support and physical health. Health Psychol 2: 367-391.

Wall J, Colley T. 2003. Preventing pressure ulcers among wheelchair users: Preliminary comments on the development of a self-administered risk assessment tool. J Tissue Viability 13: 48-50, 52-54, 56.

Weeks C. 2001. Women, spinal cord injury, and osteoporosis. Top Spinal Cord Inj Rehabil 7: 53-63.

Weingardt KR, Hsu J, Dunn ME. 2001. Brief screening for psychological and substance abuse disorders in veterans with long-term spinal cord injury. Rehabil Psychol 46: 271-278.

Whiteneck G, Brooks C, Charlifue S, Gerhart K, Mellick D, et al. 1992a. Guide for use of the CHART: Craig handicap assessment and reporting technique. Englewood, CO: Craig Hospital.

Whiteneck GG, Charlifue SW, Frankel HL, et al. 1992b. Mortality, morbidity, and psychosocial outcomes of persons spinal cord injured more than 20 years ago. Paraplegia 30: 617-630.

Whiteneck G, Charlifue S, Gerhart K, Overhosler J, Richardson G. 1992c. Quantifying handicap: A new measure of long term rehabilitation outcomes. Arch Phys Med Rehabil 73: 519-526.

Whiteneck G, Harrison-Felix C, Mellick D, Brooks C, Charlifue S, et al. 2004. Quantifying environmental factors: A measure of physical, attitudinal, service, productivity, and policy barriers. Arch Phys Med Rehabil 85: 1324-1335.

Whiteneck G, Tate D, Charlifue S. 1999. Predicting community integration after spinal cord injury from demographic and injury characteristics. Arch Phys Med Rehabil 80: 1485-1491.

Wilber N, Mitra M, Walker DK, Allen D, Meyers AR, et al. 2002. Disability as a public health issue: Findings and reflections from the Massachusetts survey of secondary conditions. Milbank Q 80: 393-421.

Zuckerman M, Kuhlman D, Joireman J, Teta P, Kraft M. 1993. A comparison of three structural models for personality, the big three, the big five and the alternate five. J Pers Soc Psychol 65: 757-768.

28 Spinal Cord Injury – A Clinical Perspective

A. Varma

1	Introduction	635
2	Epidemiology	635
3	*Clinical Features*	635
3.1	Central Cord Syndrome	636
3.2	Brown-Sequard Syndrome	636
3.3	Anterior Cord Syndrome	637
3.4	Conus Medullaris Syndrome	637
3.5	Cauda Equina Syndrome	637
3.6	Spinal Shock	637
4	*Management*	638
4.1	Prehospital Management	638
4.2	Traction	638
4.3	Pharmacotherapy	638
4.3.1	Methylprednisolone	638
4.3.2	GM1 Ganglioside	639
4.3.3	Thyrotropin Releasing Hormone	639
4.3.4	Nimodipine	639
4.3.5	Gacyclidine	639
4.3.6	Minocycline	639
4.3.7	Estrogen	639
4.4	Imaging	640
4.4.1	SCI Without Radiographic Abnormality	641
4.5	Role of Surgery	641
4.6	Medical Management	642
4.6.1	Respiratory System	642
4.6.2	Cardiovascular System	642
4.6.3	Gastrointestinal System	642
4.6.4	Urinary System	643
4.6.5	Nutritional Support	643
4.6.6	Deep Venous Thrombosis and Pulmonary Embolism	643
4.6.7	Integumentary System	643
5	*Neuroregeneration*	643
5.1	Autologous Activated Macrophages	644
5.2	Peripheral Nerve Grafts and Implants	644

© 2009 Springer Science+Business Media, LLC.

5.3	Electrical Stimulation	644
5.4	Bone Marrow Stem Cells	644
5.5	Olfactory Ensheathing Cells	644
6	***Conclusion***	***645***

Abstract: SCI afflicts about 11,000 individuals in the United States every year. It imposes an enormous psychological and economic burden on the patient and the society. The severity of neurological deficit following SCI can vary from subtle deficit to complete loss of function. Discreet syndromes have been described for incomplete SCI. Different grading systems have been described for objective assessment of severity of SCI. ASIA system is one of the more popular systems. Management of SCI includes radiological evaluation (to establish injury to spinal cord, the vertebral column and surrounding supporting structures), surgery when indicated (for cord compression, spine instability and spinal deformity), management of medical complications and rehabilitation. Pharmacotherapy with MP has not proven to be of significant clinical benefit. Still no definitive treatment, for the neurological deficit resulting from SCI, is available. To improve the chances of neurological recovery newer treatment modalities are required to minimize secondary damage (neuroprotection), and enhance recovery of the damaged neural tissue (neuroregeneration).

List of Abbreviations: ASIA, American Spinal Injury Association; CNS, central nervous system; CT, computerized tomography; DPR, delayed plantar response; DTR, deep tendon reflex; DVT, deep venous thrombosis; GM-1, monosialotetrahexosylganglioside; IMSOP, International Medical Society of Paraplegia; MRI, magnetic resonance imaging; MP, methylprednisolone; OEC, olfactory ensheathing cells; PE, pulmonary embolism; PNS, peripheral nervous system; SCI, spinal cord injury; SCIWORA, spinal cord injury without radiographic abnormality; STIR, short T1 inversion recovery; T1WI, T1 weighted images; T2WI, T2 weighted images; TRH, thyrotropin releasing hormone

1 Introduction

Spinal cord injury (SCI) spans a clinical spectrum from minimal disability to severe incapacitation with lifelong consequences for not only the victim, but also for the family and society at large. Management of this condition starts at the moment of contact of the paramedics with the trauma victim and continues lifelong for those unfortunate patients with severe disability. This chapter focuses on the epidemiology, clinical features, and acute management of SCI including imaging, medical and surgical therapy. A brief summary of current status of research strategy and treatment options available for neuroprotection and neuroregeneration is also included. The pathophysiology and molecular biology of spinal injury is discussed in separate sections. Rehabilitation, a very critical part of long-term care of these patients, is also discussed in a separate chapter.

2 Epidemiology

The annual incidence of SCI in the United States is approximately 11,000 and prevalence of living subjects with SCI is 253,000 (2005). SCI primarily affects young adults. The average age of SCI has shown a steady rise, a reflection of rise in the median age of general population. The proportion of elderly subjects (older than 60 years) has also shown a steady rise. Since 2000, the average age at injury has been 37.6 years with 79.6% of subjects being male. The most common cause of SCI in the United States is motor vehicle accident (47.5%) followed by falls, violence (mainly gun shot injuries), and recreational sporting activities. Cervical spine is the most common site of injury, and incomplete tetraplegia is the most common (34.5%) discharge diagnosis (2005).

SCI injury imposes a huge economic burden on the society from not only the care of the injured patients but also indirect costs such as losses in wages and productivity. The cost of caring for the patient is influenced by the level of injury and age, with the highest estimated life time cost attributable to the care of elderly patients (more than 50 years) with high cervical (C1–C4 injury) (2005).

3 Clinical Features

The severity of neurological deficit following SCI depends upon the level of injury and the extent of damage to the cord or roots at the level of injury. Patients with total loss of motor, sensory, and autonomic function distal to the level of SCI are considered to have functionally complete SCI. Patients with incomplete injury can have a wide spectrum of neurological dysfunction. Various clinical models have been developed for assessment and documentation of neurological status following SCI (Frankel et al., 1969; Lucas and Ducker, 1979; Marino et al., 2003). The objective assessment offered by these grading systems allows follow up and comparison of patients with SCI. The ASIA (American Spinal Injury Association) scale (Marino et al., 2003) was endorsed by the International Medical Society of Paraplegia (IMSOP) in 1992 and is currently the most widely used gradation system for SCI. Components of assessment, for ASIA scale, include definitions of the neurologic level (motor and sensory), skeletal level, completeness of injury, the zone of partial preservation, determination of total motor and sensory scores, and the ASIA impairment scale (> *Table 28-1*). The neurologic level refers to the most caudal segment of the spinal cord with normal sensory and motor function on both sides of the body. Skeletal level refers to the level of vertebral damage by radiographic examination. If partial preservation of sensory and/or motor function is found below the neurological level and includes the lowest sacral segment, the injury is defined as incomplete. When there is no motor or sensory function in the lowest sacral segment, the injury is defined as complete. The term "zone of partial preservation" is used only with complete injuries. It refers to those dermatomes and myotomes caudal to the neurological level that remain partially innervated. Sensory and motor scores reflect the degree of neurological impairment, and are obtained by examining 28 paired dermatomes and 10 paired myotomes, as defined by ASIA.

Assessment of sacral motor and sensory function is therefore of paramount importance, as this may be the only evidence of neurological function distal to injury. Complete injury has important prognostic implications as chances of neurologically useful recovery after complete SCI are remote (Maynard et al., 1979).

Functionally incomplete injuries can have a wide spectrum of neurological findings depending on the anatomical parts of the cord involved. Discreet syndromes have been described for incomplete SCI.

3.1 Central Cord Syndrome

Seen almost exclusively following cervical spine SCI, the central cord syndrome is characterized by sacral sensory sparing and greater weakness in the upper limbs than the lower limbs. This results from a lesion within the cord that afflicts the central gray matter and the surrounding fiber tracts. The peripherally located corticospinal lumbosacral fibers are spared leading to relative preservation of lower extremity motor strength. Similarly, the laterally placed sacral spinothalamic tracts are preserved with preservation of sacral sensations.

3.2 Brown-Sequard Syndrome

This is characterized by physiologic hemisection of the cord producing ipsilateral motor weakness and impaired propioception with contralateral impairment of pain and temperature sensation. This syndrome

◘ Table 28-1

ASIA impairment scale

A	Complete: No sensory or motor function is preserved in the sacral segments S4-S5
B	Incomplete: Sensory but not motor function is preserved below the neurological level and includes the sacral segments S4-S5
C	Incomplete: Motor function is preserved below the neurological level, and more than half of key muscles below the neurological level have a muscle grade less than 3 (Grades 0–2)
D	Incomplete: Motor function is preserved below the neurological level, and *at least half* of key muscles below the neurological level have a muscle grade greater than or equal to 3
E	Normal: Sensory and motor function are normal

usually occurs with other SCI or nerve root injury (or both) or brachial plexus injury. It is most often seen in cervical spine and less frequently in thoracic spine. This syndrome carries a favorable prognosis except when resulting from a penetrating injury.

3.3 Anterior Cord Syndrome

This symptom complex results from physiological ventral cord section that spares the posterior columns. The patient has variable impairment of motor function and pain and temperature sensation with relative sparing of proprioception. It occurs with hyperflexion injuries, acute central disc herniation, and burst fractures with ventral cord compression.

3.4 Conus Medullaris Syndrome

This syndrome results from injury to the sacral spine segments and lumbar roots at the T11 and T12 vertebral body levels. This part of the vertebral column is prone to injury because it lies in the transitional zone between the rigid thoracic spine and mobile lumbar spine. The patients can have a combination of upper and lower motor neuron paralysis. The deficit is usually symmetrical. In the acute phase there is areflexic paralysis of lower limbs, flaccid rectal tone, and urinary retention. Sacral segments may occasionally show preserved reflexes, like bulbocavernosus and micturition reflexes, with lesions limited to more cephalad part of the conus. In the chronic phase combination of atrophy and hyperreflexia is seen. There is sensory impairment in lumbar and sacral dermatomes. Prognosis for recovery of sphincter function is poor.

3.5 Cauda Equina Syndrome

Injury to the lumbosacral nerve roots of cauda equina in the canal from L2 vertebral body level downwards, leads to asymmetrical areflexic paralysis of lower extremities, sensory impairment in lumbar and sacral dermatomes, and lower motor neuron type bowel and bladder incontinence. Cauda equina syndrome is usually associated with burst fractures of lumbar spine and/or acute disc herniation.

3.6 Spinal Shock

Immediately following complete SCI, all spinal reflexes distal to the injury level may be abolished temporarily. This stage of evolution of symptoms following SCI is known as spinal shock. Clinically it is manifested by areflexia and flaccidity. This phase is eventually replaced by spasticity and hyperreflexia. It is important to differentiate spinal shock from neurogenic shock. The latter results from disruption of sympathetic outflow from the spinal cord leading to vasomotor collapse. It is characterized by hypotension and bradycardia and is most often seen after injuries above T6 vertebral level.

The duration of spinal shock is variable and depends on what the accepted criterion for recovery is. If the duration of spinal shock is defined by the initial recovery of any reflex, then it probably lasts up to 1 h by which time the delayed plantar response (DPR) can be observed (Ditunno et al., 2004). According to Ditunno et al. (2004), "The DPR requires an unusually strong stimulus, in contrast to the Babinski sign or normal plantar response, and is elicited by stroking a blunt instrument upward from the heel toward the toes along the lateral sole of the foot and then continuing medially across the volar aspect of the metatarsal heads. In response to the stimulus, the toes flex and relax in a delayed sequence." If, however, spinal shock is defined as an absence of deep tendon reflexes (DTRs), the currently accepted standard, then its duration is several weeks (Hiersemenzel et al., 2000). Ko et al. (1999) prospectively studied reflex recovery in 50 subjects admitted within 24 h following SCI. They concluded that the DPR is generally the first reflex to appear and other reflexes tend to return in the following sequence: bulbocavernosus, cremasteric, ankle jerk, Babinski sign, and knee jerk. In recent literature, some researchers have suggested the term "spinal shock" be used to describe the entire phase of evolution from areflexia and flaccidity to spasticity and hyperreflexia following SCI (Ditunno et al., 2004).

4 Management

4.1 Prehospital Management

Advances in understanding of the mechanisms and pathophysiology of SCI have led to establishment of protocols for care of patients with SCI in the field prior to their transport to a specialized trauma care unit.

Any trauma victim with multiple injuries could potentially have a SCI, and measures to prevent further injury should be instituted at the earliest. The cervical spine should be immobilized in a hard collar prior to extrication. Movement of spine should be avoided as much as possible during extrication and transport.

4.2 Traction

Patients with unstable cervical spine injury, e.g., fracture dislocation, should be placed in skeletal traction to align and stabilize the spine (Hadley et al., 1992). Skeletal traction is an invasive procedure as it involves attaching the traction device to the calvarium. Contraindications to traction include occipitoatlantal dislocation and open skull fracture. Two commonly used traction devices are the Gardner Wells Tongs and halo traction ring. These are easy to use and can be applied in the emergency room. An advantage of the halo traction ring is that, once the deformity is corrected, it can be connected to a vest. This maintains the spine alignment while the patient is waiting for further imaging or surgery. There is some controversy as to whether it is safe to apply traction prior to magnetic resonance imaging (MRI) evaluation of incomplete cervical SCI. Risk of neurological deterioration following skeletal traction has been reported in patients with coexistent facet dislocation and disc herniation. It has been suggested that in such patients skeletal traction for closed reduction be avoided; hence, MRI should be performed prior to placing the patient in traction (Doran et al., 1993). Detailed review of literature seems to suggest it is safe to apply traction prior to MRI examination, in a patient with unstable incomplete cervical spine injury, as long as the patient can be neurologically monitored (2002a).

4.3 Pharmacotherapy

Pharmacotherapy early after SCI is aimed at neuroprotection to minimize the secondary injury. Several trials with different neuroprotective pharmacological agents have been reported in literature. These pharmacological agents block one or more of the mechanisms of secondary injury following trauma to the cord. Unfortunately, till date, no neuroprotective pharmacological agent has been conclusively shown to be clinically effective in human subjects with SCI (Tator, 2006).

4.3.1 Methylprednisolone

Animal studies have demonstrated that methylprednisolone (MP) reduces the secondary cord injury following trauma by inhibiting lipid peroxidation, improving spinal blood flow, enhancing the postinjury activity of Na^+/K^+-ATPase, and facilitating the recovery of extracellular calcium ions (Young, 1991; Hall et al., 1992). Three MP trials have been reported from North America (Bracken et al., 1984; Bracken et al., 1990; Bracken et al., 1997). The NASCIS 2 trial (Bracken et al., 1990) suggested clinical efficacy that did not translate into effectiveness. Subsequent study from France (Pointillart et al., 2000) and evidence-based analysis (Hurlbert, 2000; Short et al., 2000; Hurlbert, 2001; Hurlbert, 2006) have not supported the routine use of MP in acute SCI. Despite equivocal effectiveness, MP is widely used in many North American centers, for want of a better alternative. The drug is given within 8 h of injury in a bolus dose of 30 mg/kg followed by a maintenance dose of 5.4 mg/kg/h over next 23 h.

4.3.2 GM1 Ganglioside

Animal studies have shown that monosialotetrahexosylganglioside (GM-1) ganglioside enhances the functional recovery of damaged neurons (Ferrari and Greene, 1998). The first randomized control trial, with 34 patients, suggested beneficial effects of GM-1 (Geisler et al., 1991). However, a later adequately powered multicenter study failed to establish the effectiveness of this agent in SCI (Geisler et al., 2001).

4.3.3 Thyrotropin Releasing Hormone

Thyrotropin releasing hormone (TRH) is a partial endorphin antagonist. Endorphins are released after SCI and postulated to exacerbate posttraumatic ischemia by reducing spinal blood flow secondary to systemic hypotension (Faden et al., 1981). A smaller study (Pitts et al., 1995) suggested beneficial effects of TRH in SCI. A larger study to evaluate the efficacy has not been carried out.

4.3.4 Nimodipine

Nimodipine is a Ca^{2+} channel blocker. It can potentially minimize the injury following SCI by counteracting vasospam, ischemia, and infarction that contribute to secondary damage (Guha et al., 1987; Pointillart et al., 1993). A single center randomized trial that compared four treatment arms viz. nimodipine, MP, MP and nimodipine, and placebo failed to demonstrate any significant benefits of nimodipine alone or in combination with MP (Pointillart et al., 2000).

4.3.5 Gacyclidine

Gacyclidine is an *N*-methyl-D-aspartate antagonist. It blocks the toxic effects of glutamate which is released following SCI. A multicenter RCT failed to establish efficacy of Gacyclidine at 1 year following SCI (Tator, 2006).

4.3.6 Minocycline

Minocycline is a tetracycline derivative (a bacteriostatic antibiotic) that is currently in common clinical use for the treatment of acne and chronic periodontitis. Animal experiments have shown minocycline has a neuroprotective effect following SCI. It reduces axonal loss at the site of injury, decreases oligodendrocyte apoptosis, and prevents activation of microglia and macrophages (Lee et al., 2003; Wells et al., 2003; Stirling et al., 2004). A pilot study to evaluate minocycline in patients with acute spinal cord injuries has been initiated in Calgary, Alberta, Canada (Kwon et al., 2005).

4.3.7 Estrogen

Early treatment with the estrogen 17β-estradiol has been shown to have a neuroprotective effect following experimental SCI in rats (Yune et al., 2004; Sribnick et al., 2006). 17β-Estradiol reduces apoptosis in the penumbra around the zone of necrosis, thereby minimizing the secondary damage.

4.4 Imaging

Radiographs (plain films), computerized tomography (CT), and MRI are the cornerstones of radiological evaluation of spine injury. The radiographs and CT scan are most useful in evaluating the bony anatomy of the vertebral column. Vertebral body fractures are better visualized on radiographs than dorsal column fractures. Overall, the sensitivity of plain films in diagnosing spinal injuries, especially injuries in the cervical spine, varies, ranging from 39 to 94%, with variable specificity. Several studies have proved that if only plain radiographs are used, 23–57% of all fractures of the cervical spine are missed (Sundgren et al., 2007). Plain films are also used for checking stability of the spine in certain injuries without neurological deficit, where integrity of the soft tissues is in question.

CT (❯ *Figure 28-1*) detects fractures with higher sensitivity than plain films. CT scan software allows high quality coronal and sagittal reconstruction, from the axial images, allowing precise evaluation for any fracture or dislocation. CT scan is indispensable for evaluation of difficult to visualize parts of the vertebral column like the cervicothoracic junction and upper thoracic spine. With the introduction of spiral CT, the spine can be imaged very rapidly, making this modality the initial method of choice, in most trauma centers when evaluating the cervical spine for bone injuries after blunt trauma. With advances in imaging technology, the quality of CT images allows diagnosis of soft tissue abnormalities like disc herniation or extradural hematoma (Sundgren et al., 2007).

The MRI (❯ *Figure 28-2*) is the best imaging modality to evaluate soft tissues like the ligaments, intervertebral discs, muscles, spinal cord, and nerve roots (Sundgren et al., 2007). MRI also allows diagnosis of intra or extramedullary hematoma, helps differentiate spinal cord hemorrhage from edema, and can clearly establish the degree of spinal cord or nerve root compression. In most situations T1 weighted images (T1WI), T2 weighted images (T2WI), short T1 inversion recovery (STIR), and T2 gradient echo sequences are sufficient for adequate evaluation of SCI. Acute bone injuries show bright signal on T2WI and STIR sequences upon MRI.

The extent of radiological evaluation should be guided by the patient's clinical status. The whole spine should be evaluated in an obtunded or uncooperative patient with CT. Most trauma patients get CT of chest, abdomen, and pelvis. The CT software allows reconstruction of thoracic and lumbar spine from chest and abdominal series, thus avoiding additional radiation exposure for evaluation of thoracic and lumbar

◘ Figure 28-1
(a) axial CT image of C5 vertebra, in a patient with cervical spine injury, shows fracture of both laminae (*black arrows*) and pedicles (*white arrows*). (b) sagittal reconstruction shows anterior subluxation of C5 vertebra over C6 (*white diamond arrow*)

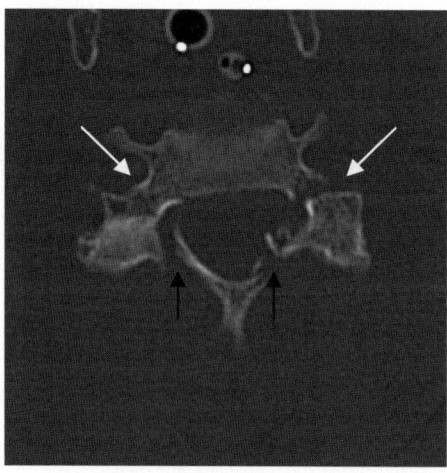
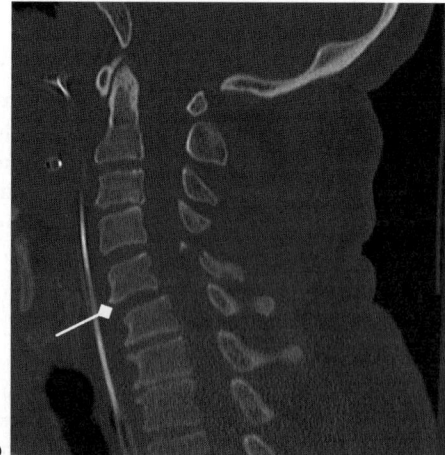

Figure 28-2
(a) T1WI (b) T2WI, and (c) STIR sequence, from MRI of the same patient, show anterior subluxation of C5 vertebra over C6, fluid anterior to the cord at C4 and C5 levels, cord edema at C5 and C6 levels, disruption of the anterior longitudinal ligament at C4-5 and C5-6 disc levels, and edema of the interspinous ligament and adjacent neck muscles in the upper cervical spine

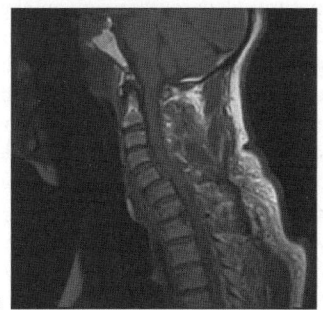
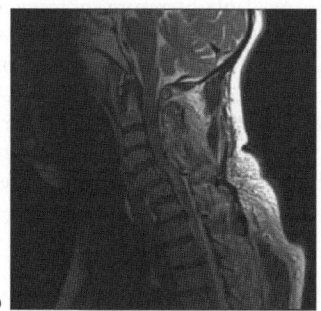
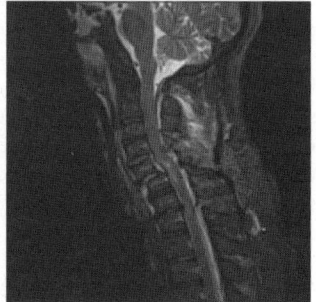

spine. Patients who demonstrate a significant bony injury with spinal canal compromise or those who have any neurological deficit should undergo a MRI.

Trauma protocols require separate "clearing" of cervical spine. In a fully conscious and cooperative asymptomatic patient, plain radiographs (anteroposterior, lateral, swimmer's, and open mouth views) are sufficient to rule out serious cervical spine injury. The combined negative predictive value of cervical spine X-ray assessment of asymptomatic patients for a significant cervical spine injury is virtually 100% (2002c). In fact, the American Association of Neurological Surgeons and Congress of Neurological Surgeons guidelines do not recommend radiographic assessment of the cervical spine in trauma patients who are awake, alert, and not intoxicated, who are without neck pain or tenderness, and who do not have significant associated injuries that detract from their general evaluation (2002c). In conscious but symptomatic patients, incidence of cervical spine injuries ranges from 1.9 to 6.2% (2002c), and the onus is really on the treating physician to ensure that a serious injury is not missed. A CT of cervical spine down to T1 level should be performed. If CT does not demonstrate a bone injury, dynamic plain films and, if required, MRI of cervical spine should be performed.

4.4.1 SCI Without Radiographic Abnormality

The term SCIWORA (SCI without radiographic abnormality) was first defined in 1982 to describe children who had objective signs of myelopathy as a result of trauma with no evidence of fracture or ligamentous instability on plain radiographs and tomography (Pang and Wilberger, 1982). The finding of fracture, subluxation, or abnormal intersegmental motion at the level of neurological injury excludes SCIWORA as a diagnosis. With subsequent availability of MRI, it became apparent that most of these patients have injury to the cord or the supporting soft tissues (Pang, 2004).

4.5 Role of Surgery

Broad indications for surgical intervention following SCI include decompression of neural elements, stabilization of spine, and deformity correction. Stabilization is indicated for an "unstable" injury. Injury is considered unstable when there is progressive neurological decline, pain, or deformity of the spine under physiological loading (Panjabi et al., 1988). Surgery can range from simple decompression to instrumented fusion. Under most circumstances, some kind of stabilization would be required as an isolated decompression can lead to persistent pain and late deformity (Malcolm et al., 1981).

The role of surgical intervention in complete SCI is mainly to stabilize the spine to control local pain, prevent or reduce deformity, and expeditiously mobilize the patient. Decompression very rarely leads to useful neurological recovery (Benzel and Larson, 1986a, b) except for recovery of some root(s) function in cervical spine (Benzel and Larson, 1986b; Anderson and Bohlman, 1992) or cauda equina (Tator, 2006).

The role of surgical decompression in incomplete SCI is also not clear. No phase III study has been carried out to answer this question. Available literature indicates that surgical decompression in these patients, even if performed late can lead to neurological improvement (Benzel and Larson, 1986a; Benzel and Larson, 1987; Bohlman and Anderson, 1992). There is no consensus on the timing of surgery following incomplete injury. A meta-analysis of SCI treatment series published between 1996 and 2000 suggested that surgical decompression performed within 24 h of incomplete SCI produced greater neurological improvement compared with surgery performed after 24 h (La Rosa et al., 2004). But, the authors also concluded that the evidence was not strong and robust and well-designed prospective, randomized controlled trials are still needed to answer the question of timing of decompression.

4.6 Medical Management

SCI affects multiple body systems. The degree of systemic complications is related to level and severity of SCI. Thorough knowledge of systemic affects of SCI is required to prevent, identify, and treat complications arising from multi-system involvement even later in the chronic phase. Pneumonia, septicemia, and pulmonary embolism (PE) are the three most common causes of mortality in this group (2005).

4.6.1 Respiratory System

SCI profoundly affects respiratory system, particularly cervical spine injury. High cervical spine injury (C1–C4) can lead to severe respiratory compromise as it compromises both the diaphragm (phrenic nerve) and the intercostal muscles. Cervical spine injury below C4 and thoracic spine injury can still compromise respiration through involvement of intercostal muscles, ileus, and loss of abdominal muscle tone. Additional chest and head injuries can compound the problem. Respiratory complications can develop in 51–67% of patients following cervical or thoracic injuries (Jackson and Groomes, 1994; Cotton et al., 2005). Common respiratory complications are atelectasis, pneumonia, pulmonary edema, and adult respiratory distress syndrome. Prevention includes respiratory support (including intubation and ventilation), pulmonary toilet, and chest physiotherapy. Use of rotating bed has also helped to reduce the incidence of respiratory complication (Reines and Harris, 1987). Vital signs, blood counts, chest radiographs, and respiratory cultures should be followed closely to diagnose infection as early as possible. Once the patients have stabilized, they can be weaned off the ventilator. Prolonged ventilator dependence mandates tracheostomy.

4.6.2 Cardiovascular System

SCI above T6 leads to neurogenic shock secondary to vasomotor paralysis from disruption of sympathetic outflow. It manifests as hypotension with bradycardia. Aggressive management with hemodynamic monitoring, volume expansion, and pressor therapy should be used to maintain mean arterial pressure above 85 mm Hg. Prolonged systemic hypotension can worsen the cord ischemia, further augmenting the cord damage (Tator and Fehlings, 1991).

4.6.3 Gastrointestinal System

Because of disruption of sympathetic outflow, patients with SCI are prone to developing ileus and gastric dilatation. Enteral feeding should be delayed till bowel movement returns. Nasogastric suctioning may be required to prevent respiratory compromise. There is also a higher incidence of stress ulcer after SCI (Leramo et al., 1982; Walters and Silver, 1986), and prophylaxis with H2-blocker is recommended, particularly if the patient is receiving steroids.

4.6.4 Urinary System

Following cervical or thoracic SCI, patients develop atonic and flaccid bladder during spinal shock. Slowly it evolves into upper motor neuron type bladder with small capacity. Initially an indwelling catheter is placed. Eventually it should be removed and intermittent catheterization started to reduce the risk of infection (Siroky, 2002). SCI patients also have a very high incidence of urinary infections, and urine microbiology should be part of every fever work up.

4.6.5 Nutritional Support

Hypermetabolism, an accelerated catabolic rate, and massive nitrogen losses are consistently seen following SCI. Consequently SCI patients develop muscle atrophy and marked losses in lean body mass and are susceptible to infection, impaired wound healing, and difficulty weaning from mechanical ventilation. Energy expenditure is best determined by indirect calorimetry in these patients. Nutritional support of the SCI patient to meet caloric and nitrogen needs, not to achieve nitrogen balance, is recommended and may reduce the deleterious effects of the catabolic, nitrogen-wasting process that occurs after acute SCI (2002b).

4.6.6 Deep Venous Thrombosis and Pulmonary Embolism

Incidence of deep venous thrombosis (DVT), following SCI, approaches 50–100% among patients not given any thromboprophylaxis. (Jones et al., 2005). Early institution of pharmacological and mechanical DVT prophylaxis helps to reduce the risk of DVT (Aito et al., 2002). The cumulative incidence of PE within first 3 months after SCI has remained around 5.4%, despite widespread use of thromboprophylaxis (Jones et al., 2005). Newer strategies are required to further reduce the incidence of this potentially fatal complication (Green, 2003).

4.6.7 Integumentary System

Combination of nitrogen loss and hypotension with decreased skin perfusion predispose SCI patients to pressure sores. An unattended patient left too long on a hard spine board can develop breakdown of skin. Frequent checks, cleanliness, change in position, padding of pressure points, and good nursing care are a prerequisite for prevention of decubiti. Use of adjuncts like air cushions and rotating beds can also help. Decubiti increase the morbidity due to increased protein loss and risk of infection, delay rehabilitation, and significantly add to the cost of treatment.

5 Neuroregeneration

The search for an answer to this devastating medical condition involves looking at multiple treatment options. Regeneration of the destroyed neural tissue is one of the approaches that has generated considerable interest. Translational studies on promoting regeneration in human subjects with SCI have also been carried out. These are mainly Phase I trials, and need further evaluation. Following is a brief summary of the work done in this field.

5.1 Autologous Activated Macrophages

Several studies have suggested that the central nervous system (CNS) may recover following a boosting and/or modulation of the immune response. While inflammation in the early postinjury period appears to be harmful and may potentiate cell death, there is evidence that in the subacute phase (1–2 weeks) after SCI, inflammatory cells including activated macrophages and T cells may help to promote repair (Rapalino et al., 1998). Exposure of the macrophages to the injured regenerative tissue (e.g., excised skin) is thought to "educate" the macrophages toward a unique wound-healing phenotype that is different from when they are activated by exposure to bacterial antigens (Bomstein et al., 2003). The skin-coincubated macrophages have been reported to demonstrate a distinctive profile of cytokine secretion and cell-surface markers indicative of antigen-presenting activity; in other words, these cells apparently have the potential for profound reparative influence on nerve cells, immune cells, and glial cells that are present in the injured spinal cord. In a rat model of spinal cord contusion, these skin-activated macrophages promoted neurological recovery and their application led to reduced cavity formation (Bomstein et al., 2003). In a Phase I trial, treatment with autologous activated macrophages, prepared from the peripheral blood of patients with SCI and incubated in tissue culture conditions in the presence of autologous skin, was found to be safe (Knoller et al., 2005).

5.2 Peripheral Nerve Grafts and Implants

Axons in the peripheral nervous system (PNS) and CNS form sprouts after injury. Richardson et al. (1980) demonstrated the capacity of transected axons originating in the CNS to regrow into nerve grafts containing Schwann cells. Several isolated reports of peripheral nerve grafts have been published subsequently (Tadie et al., 2002; von Wild and Brunelli, 2003; Cheng et al., 2004).

5.3 Electrical Stimulation

Electrical stimulation has been shown to induce axonal growth in experimental SCI (Borgens et al., 1987). A Phase I trial of pulsed oscillating current (within 18 days of injury through surgically implanted electrodes) was effective in some cases and free of major complications (Shapiro et al., 2005).

5.4 Bone Marrow Stem Cells

Animal studies have demonstrated that populations of bone marrow derived stem cells are capable of neuronal and glial differentiation, and can promote neurological recovery after transplantation into the injury site after SCI (Chopp et al., 2000; Hofstetter et al., 2002). These studies have generated Phase I clinical trials (Park et al., 2005; Tator, 2006). The major attractions of this strategy are the ease of obtaining autologous tissue for transplantation and the possibility that bone marrow cells have a homing instinct so that they may be effective after administration remote from the injury site, through either intravascular or intrathecal routes (Tator, 2006).

5.5 Olfactory Ensheathing Cells

In the olfactory system, the sensory neurons are replaced throughout adult life, and the newly formed axons continually reenter the CNS. The entry point of the olfactory axons into the olfactory bulb is associated with special glial cells known as olfactory ensheathing cells (OECs). When myelinated fiber tracts of the spinal cord are damaged, the cut axons produce local sprouts at the site of injury, but the sprouts do not reenter the distal part of the tract. Li et al. (1998) reported that in a rat model OECs induced rapid, aligned growth of cut corticospinal tract (CST) axon sprouts across the lesion and into the caudal CST. The OECs formed a bridge conveying them across the lesion and mediating their reentry. The technique is attractive because of the ease with which autologous tissue can be obtained for transplantation in patients with SCI. Transplantation of OEC has been tried on a larger scale in China (Huang et al., 2003; Dobkin et al., 2006), Portugal (Lima et al., 2006), Russia (Tator, 2006), and Australia (Feron et al., 2005).

Other treatment modalities to promote regeneration after SCI that are being investigated include Schwann cell transplantation, fetal porcine stem cell xenotransplantation, polyethylene glycol application, Rho antagonist – Cethrin, Anti-Nogo-A inhibitor and other inhibitors, and human embryonic stem cells (Tator, 2006).

6 Conclusion

Significant advances have been made in the supportive care of the severely injured SCI patients. More patients are now surviving because of better management at the site of accident and later in the acute care facility and rehabilitation centers. But definitive treatment of the neurological damage continues to be elusive. As outlined earlier, research is continuing on multiple fronts to minimize secondary damage (neuroprotection), and enhance recovery of the damaged neural tissue (neuroregeneration). It is a momentous task that requires persistence by the scientific community and unflinching support by the administrators and funding agencies. The solution is not easy, but appears more promising now than ever before.

References

2002a. Initial closed reduction of cervical spine fracture-dislocation injuries. Neurosurgery 50: S44-S50.

2002b. Nutritional support after spinal cord injury. Neurosurgery 50: S81-S84.

2002c. Radiographic assessment of the cervical spine in asymptomatic trauma patients. Neurosurgery 50: S30–S35.

2005. Spinal cord injury. Facts and figures at a glance. J Spinal Cord Med 28: 379-380.

Aito S, Pieri A, D'Andrea M, Marcelli F, Cominelli E. 2002. Primary prevention of deep venous thrombosis and pulmonary embolism in acute spinal cord injured patients. Spinal Cord 40: 300-303.

Anderson PA, Bohlman HH. 1992. Anterior decompression and arthrodesis of the cervical spine: Long-term motor improvement. II. Improvement in complete traumatic quadriplegia. J Bone Joint Surg Am 74: 683-692.

Benzel EC, Larson SJ. 1986a. Functional recovery after decompressive operation for thoracic and lumbar spine fractures. Neurosurgery 19: 772-778.

Benzel EC, Larson SJ. 1986b. Recovery of nerve root function after complete quadriplegia from cervical spine fractures. Neurosurgery 19: 809-812.

Benzel EC, Larson SJ. 1987. Functional recovery after decompressive spine operation for cervical spine fractures. Neurosurgery 20: 742-746.

Bohlman HH, Anderson PA. 1992. Anterior decompression and arthrodesis of the cervical spine: Long-term motor improvement. I. Improvement in incomplete traumatic quadriparesis. J Bone Joint Surg Am 74: 671-682.

Bomstein Y, Marder JB, Vitner K, Smirnov I, Lisaey G, et al. 2003. Features of skin-coincubated macrophages that promote recovery from spinal cord injury. J Neuroimmunol 142: 10-16.

Borgens RB, Blight AR, McGinnis ME. 1987. Behavioral recovery induced by applied electric fields after spinal cord hemisection in guinea pig. Science 238: 366-369.

Bracken MB, Collins WF, Freeman DF, Shepard MJ, Wagner FW, et al. 1984. Efficacy of methylprednisolone in acute spinal cord injury. JAMA 251: 45-52.

Bracken MB, Shepard MJ, Collins WF, Holford TR, Young W, et al. 1990. A randomized, controlled trial of methylprednisolone or naloxone in the treatment of acute spinal-cord injury. Results of the Second National Acute Spinal Cord Injury Study. N Engl J Med 322: 1405-1411.

Bracken MB, Shepard MJ, Holford TR, Leo-Summers L, Aldrich EF, et al. 1997. Administration of methylprednisolone for 24 or 48 hours or tirilazad mesylate for 48 hours in the treatment of acute spinal cord injury. Results of the third national acute spinal cord injury randomized controlled trial. National Acute Spinal Cord Injury Study. JAMA 277: 1597-1604.

Cheng H, Liao KK, Liao SF, Chuang TY, Shih YH. 2004. Spinal cord repair with acidic fibroblast growth factor as a treatment for a patient with chronic paraplegia. Spine 29: E284-E288.

Chopp M, Zhang XH, Li Y, Wang L, Chen J, et al. 2000. Spinal cord injury in rat: Treatment with bone marrow stromal cell transplantation. Neuroreport 11: 3001-3005.

Cotton BA, Pryor JP, Chinwalla I, Wiebe DJ, Reilly PM, et al. 2005. Respiratory complications and mortality risk associated with thoracic spine injury. J Trauma 59: 1400-1407; discussion 1407-1409.

Ditunno JF, Little JW, Tessler A, Burns AS. 2004. Spinal shock revisited: A four-phase model. Spinal Cord 42: 383-395.

Dobkin BH, Curt A, Guest J. 2006. Cellular transplants in China: Observational study from the largest human experiment in chronic spinal cord injury. Neurorehabil Neural Repair 20: 5-13.

Doran SE, Papadopoulos SM, Ducker TB, Lillehei KO. 1993. Magnetic resonance imaging documentation of coexistent traumatic locked facets of the cervical spine and disc herniation. J Neurosurg 79: 341-345.

Faden AI, Jacobs TP, Holaday JW. 1981. Thyrotropin-releasing hormone improves neurologic recovery after spinal trauma in cats. N Engl J Med 305: 1063-1067.

Feron F, Perry C, Cochrane J, Licina P, Nowitzke A, et al. 2005. Autologous olfactory ensheathing cell transplantation in human spinal cord injury. Brain 128: 2951-2960.

Ferrari G, Greene LA. 1998. Promotion of neuronal survival by GM1 ganglioside. Phenomenology and mechanism of action. Ann N Y Acad Sci 845: 263-273.

Frankel HL, Hancock DO, Hyslop G, Melzak J, Michaelis LS, et al. 1969. The value of postural reduction in the initial management of closed injuries of the spine with paraplegia and tetraplegia. I. Paraplegia 7: 179-192.

Geisler FH, Coleman WP, Grieco G, Poonian D. 2001. The Sygen multicenter acute spinal cord injury study. Spine 26: S87-S98.

Geisler FH, Dorsey FC, Coleman WP. 1991. Recovery of motor function after spinal-cord injury – a randomized, placebo-controlled trial with GM-1 ganglioside. N Engl J Med 324: 1829-1838.

Green D. 2003. Diagnosis, prevalence, and management of thromboembolism in patients with spinal cord injury. J Spinal Cord Med 26: 329-334.

Guha A, Tator CH, Piper I. 1987. Effect of a calcium channel blocker on posttraumatic spinal cord blood flow. J Neurosurg 66: 423-430.

Hadley MN, Fitzpatrick BC, Sonntag VK, Browner CM. 1992. Facet fracture-dislocation injuries of the cervical spine. Neurosurgery 30: 661-666.

Hall ED, Yonkers PA, Andrus PK, Cox JW, Anderson DK. 1992. Biochemistry and pharmacology of lipid antioxidants in acute brain and spinal cord injury. J Neurotrauma 9 (Suppl 2): S425-S442.

Hiersemenzel LP, Curt A, Dietz V. 2000. From spinal shock to spasticity: Neuronal adaptations to a spinal cord injury. Neurology 54: 1574-1582.

Hofstetter CP, Schwarz EJ, Hess D, Widenfalk J, El Manira A, et al. 2002. Marrow stromal cells form guiding strands in the injured spinal cord and promote recovery. Proc Natl Acad Sci USA 99: 2199-2204.

Huang H, Chen L, Wang H, Xiu B, Li B, et al. 2003. Influence of patients' age on functional recovery after transplantation of olfactory ensheathing cells into injured spinal cord injury. Chin Med J (Engl) 116: 1488-1491.

Hurlbert RJ. 2000. Methylprednisolone for acute spinal cord injury: An inappropriate standard of care. J Neurosurg 93: 1-7.

Hurlbert RJ. 2001. The role of steroids in acute spinal cord injury: An evidence-based analysis. Spine 26: S39-S46.

Hurlbert RJ. 2006. Strategies of medical intervention in the management of acute spinal cord injury. Spine 31: S16-S21; discussion S36.

Jackson AB, Groomes TE. 1994. Incidence of respiratory complications following spinal cord injury. Arch Phys Med Rehabil 75: 270-275.

Jones T, Ugalde V, Franks P, Zhou H, White RH. 2005. Venous thromboembolism after spinal cord injury: Incidence, time course, and associated risk factors in 16,240 adults and children. Arch Phys Med Rehabil 86: 2240-2247.

Knoller N, Auerbach G, Fulga V, Zelig G, Attias J, et al. 2005. Clinical experience using incubated autologous macrophages as a treatment for complete spinal cord injury: Phase I study results. J Neurosurg Spine 3: 173-181.

Ko HY, Ditunno JF Jr, Graziani V, Little JW. 1999. The pattern of reflex recovery during spinal shock. Spinal Cord 37: 402-409.

Kwon BK, Fisher CG, Dvorak MF, Tetzlaff W. 2005. Strategies to promote neural repair and regeneration after spinal cord injury. Spine 30: S3-S13.

La Rosa G, Conti A, Cardali S, Cacciola F, Tomasello F. 2004. Does early decompression improve neurological outcome of spinal cord injured patients? Appraisal of the literature using a meta-analytical approach. Spinal Cord 42: 503-512.

Lee SM, Yune TY, Kim SJ, Park DW, Lee YK, et al. 2003. Minocycline reduces cell death and improves functional recovery after traumatic spinal cord injury in the rat. J Neurotrauma 20: 1017-1027.

Leramo OB, Tator CH, Hudson AR. 1982. Massive gastroduodenal hemorrhage and perforation in acute spinal cord injury. Surg Neurol 17: 186-190.

Li Y, Field PM, Raisman G. 1998. Regeneration of adult rat corticospinal axons induced by transplanted olfactory ensheathing cells. J Neurosci 18: 10514-10524.

Lima C, Pratas-Vital J, Escada P, Hasse-Ferreira A, Capucho C, et al. 2006. Olfactory mucosa autografts in human spinal cord injury: A pilot clinical study. J Spinal Cord Med 29: 191-203; discussion 204-206.

Lucas JT, Ducker TB. 1979. Motor classification of spinal cord injuries with mobility, morbidity and recovery indices. Am Surg 45: 151-158.

Malcolm BW, Bradford DS, Winter RB, Chou SN. 1981. Post-traumatic kyphosis. A review of forty-eight surgically treated patients. J Bone Joint Surg Am 63: 891-899.

Marino RJ, Barros T, Biering-Sorensen F, Burns SP, Donovan WH, et al. 2003. International standards for neurological classification of spinal cord injury. J Spinal Cord Med 26 (Suppl 1): S50-S56.

Maynard FM, Reynolds GG, Fountain S, Wilmot C, Hamilton R. 1979. Neurological prognosis after traumatic quadriplegia. Three-year experience of California Regional Spinal Cord Injury Care System. J Neurosurg 50: 611-616.

Pang D. 2004. Spinal cord injury without radiographic abnormality in children, 2 decades later. Neurosurgery 55: 1325-1342; discussion 1342-1343.

Pang D, Wilberger JE. Jr. 1982. Spinal cord injury without radiographic abnormalities in children. J Neurosurg 57: 114-129.

Panjabi MM, Thibodeau LL, Crisco JJ III, White AA III. 1988. What constitutes spinal instability? Clin Neurosurg 34: 313-339.

Park HC, Shim YS, Ha Y, Yoon SH, Park SR, et al. 2005. Treatment of complete spinal cord injury patients by autologous bone marrow cell transplantation and administration of granulocyte-macrophage colony stimulating factor. Tissue Eng 11: 913-922.

Pitts LH, Ross A, Chase GA, Faden AI. 1995. Treatment with thyrotropin-releasing hormone (TRH) in patients with traumatic spinal cord injuries. J Neurotrauma 12: 235-243.

Pointillart V, Gense D, Gross C, Bidabe AM, Gin AM, et al. 1993. Effects of nimodipine on posttraumatic spinal cord ischemia in baboons. J Neurotrauma 10: 201-213.

Pointillart V, Petitjean ME, Wiart L, Vital JM, Lassie P, et al. 2000. Pharmacological therapy of spinal cord injury during the acute phase. Spinal Cord 38: 71-76.

Rapalino O, Lazarov-Spiegler O, Agranov E, Velan GJ, Yoles E, et al. 1998. Implantation of stimulated homologous macrophages results in partial recovery of paraplegic rats. Nat Med 4: 814-821.

Reines HD, Harris RC. 1987. Pulmonary complications of acute spinal cord injuries. Neurosurgery 21: 193-196.

Richardson PM, McGuinness UM, Aguayo AJ. 1980. Axons from CNS neurons regenerate into PNS grafts. Nature 284: 264-265.

Shapiro S, Borgens R, Pascuzzi R, Roos K, Groff M, et al. 2005. Oscillating field stimulation for complete spinal cord injury in humans: A phase 1 trial. J Neurosurg Spine 2: 3-10.

Short DJ, El Masry WS, Jones PW. 2000. High dose methylprednisolone in the management of acute spinal cord injury – a systematic review from a clinical perspective. Spinal Cord 38: 273-286.

Siroky MB. 2002. Pathogenesis of bacteriuria and infection in the spinal cord injured patient. Am J Med 113 (Suppl 1A): 67S-79S.

Sribnick EA, Matzelle DD, Ray SK, Banik NL. 2006. Estrogen treatment of spinal cord injury attenuates calpain activation and apoptosis. J Neurosci Res 84: 1064-1075.

Stirling DP, Khodarahmi K, Liu J, McPhail LT, McBride CB, et al. 2004. Minocycline treatment reduces delayed oligodendrocyte death, attenuates axonal dieback, and improves functional outcome after spinal cord injury. J Neurosci 24: 2182-2190.

Sundgren PC, Philipp M, Maly PV. 2007. Spinal trauma. Neuroimaging Clin N Am 17: 73-85.

Tadie M, Liu S, Robert R, Guiheneuc P, Pereon Y, et al. 2002. Partial return of motor function in paralyzed legs after surgical bypass of the lesion site by nerve autografts three years after spinal cord injury. J Neurotrauma 19: 909-916.

Tator CH. 2006. Review of treatment trials in human spinal cord injury: Issues, difficulties, and recommendations. Neurosurgery 59: 957-982; discussion 982-987.

Tator CH, Fehlings MG. 1991. Review of the secondary injury theory of acute spinal cord trauma with emphasis on vascular mechanisms. J Neurosurg 75: 15-26.

von Wild KR, Brunelli GA. 2003. Restoration of locomotion in paraplegics with aid of autologous bypass grafts for direct neurotisation of muscles by upper motor neurons – the future: Surgery of the spinal cord? Acta Neurochir Suppl 87: 107-112.

Walters K, Silver JR. 1986. Gastrointestinal bleeding in patients with acute spinal injuries. Int Rehabil Med 8: 44-47.

Wells JE, Hurlbert RJ, Fehlings MG, Yong VW. 2003. Neuroprotection by minocycline facilitates significant recovery from spinal cord injury in mice. Brain 126: 1628-1637.

Young W. 1991. Methylprednisolone treatment of acute spinal cord injury: An introduction. J Neurotrauma 8 (Suppl 1): S43-S46.

Yune TY, Kim SJ, Lee SM, Lee YK, Oh YJ, et al. 2004. Systemic administration of 17beta-estradiol reduces apoptotic cell death and improves functional recovery following traumatic spinal cord injury in rats. J Neurotrauma 21: 293-306.

29 Cubing the Brain: Mapping Expression Patterns Genome-Wide

M. H. Chin · D. J. Smith

1	Introduction	650
2	**Voxelation of the Mouse Brain**	**651**
2.1	Voxelation is Used to Analyze a Mouse Model of Disease	651
2.2	Use of Singular Value Decomposition to Analyze Large Amounts of Data	651
2.3	High-Resolution Rodent Brain Voxelation	652
2.4	Independent Confirmation of Voxelation Replicability	652
3	**Voxelation of the Human Brain**	**654**
3.1	Voxelation on the Brain of Alzheimer's Patient and Unaffected Individual	654
3.2	High-Resolution Voxelation of the Human Brain	655
4	*Conclusion*	*655*

© 2009 Springer Science+Business Media, LLC.

Abstract: Technologies for 3D genome-wide mapping of expression patterns will be increasingly important to understand the brain in health and disease. Here, we describe the use of voxelation to reach these goals. The brain is divided into spatially registered cubes which are subjected to high-throughput expression analysis using methods such as microarrays or real-time PCR. The data can then be used to reconstruct expression images reminiscent of those obtained from biomedical imaging systems. We discuss the insights obtained from voxelation of the mouse brain at a volumetric resolution of 11 µl and 1 µl and the human brain at 1 cm^3 and 87 µl. The human and mouse studies also incorporated Alzheimer's and Parkinson's disease specimens, giving a better understanding of the molecular pathology of these disorders. Furthermore, we describe useful analytic approaches to understanding the large datasets resulting from voxelation.

List of Abbreviations: Drd2, dopamine D2 receptor; Nfl, neurofilament light chain; qRT-PCR, quantitative real time-PCR; SVD, singular value decomposition

1 Introduction

Voxelation permits high-throughput acquisition of gene expression patterns in the mammalian brain. The method uses analysis of spatially registered voxels (cubes) to create multiple volumetric images similar to those obtained from biomedical imaging systems. Voxelation has been used to investigate normal and diseased brains in both mice and humans and the results suggest that the method will be a useful approach to understanding how the genome directs the molecular architecture of the brain.

The mammalian brain is the most complex organ known and the mechanisms of its development are poorly understood (Owen et al., 2000). The completion of the human and mouse genome projects have provided researchers with the sequence of all of the genes in both species. Now the challenging task of profiling all brain gene expression patterns is underway so that the role of the genome constructing the brain can be defined. Multiple approaches are currently being employed to create these atlases of gene expression in the mouse brain. The Allen Brain Atlas (Boguski et al., 2004) and Brain Gene Expression Map (BGEM) (www.stjudebgem.org) utilize in situ hybridization to investigate gene expression at the cellular level using probes to mRNA. In this method, genes are analyzed singly but throughput has been improved using automation. Cells are identified as expressing the gene or not, but there is little information on relative levels of gene expression. Another resource is the Gene Expression Nervous System Atlas. This atlas uses GFP (green fluorescent protein)-transgenic mice, in which an autoflourescent protein is expressed under the promoter of the gene of interest (Heintz, 2001; Gong et al., 2003). This technique requires the creation of transgenic mice, but has the advantage of autoflourescence; hence, binding of antibodies or probes is unnecessary. Advances in imaging can even allow scanning of live animals for regions of autoflouresence indicating gene expression. All of the techniques discussed earlier have the distinct advantage of up to cellular resolution but are relatively low throughput, examining one gene at a time. Microarrays are high-throughput (tens of thousands of genes analyzed in one experiment) and have been used to analyze both dissected regions of the brain as well as single cells obtained by laser capture microdissection combined with linear amplification of RNA (Bunney et al., 2003), (Bonner et al., 1997). However, there exists a need for an approach that is intermediate in resolution between the single cell level and regional dissection. Voxelation meets these criteria by combining the high-throughput of microarrays with the reconstruction methods of biomedical imaging to create 3D maps of gene expression in the brain.

In voxelation, the brain is separated into spatially registered voxels (cubes), which can then be assayed for gene expression. The methods currently in use to analyze the voxels are microarray technology or quantitative real time-PCR (qRT-PCR). qRT-PCR is relatively sensitive but can be conducted for only tens of genes at a time. It is commonly used as a validation for microarray gene expression levels. The profile of expression for each gene obtained from voxelation can be converted into images similar to those obtained from CT (X-ray computerized tomography), PET (positron emission tomography), and MRI (magnetic resonance imaging) (Gambhir et al., 1999; Herschman et al., 2000; Louie et al., 2000; Zacharias et al., 2000). Volumetric gene expression profiles in the brain will be of great interest in helping to understand how the genome directs the development of the brain. Furthermore, this information may also suggest insight into the genetic mechanisms of neurological diseases and provide potential therapeutic targets.

2 Voxelation of the Mouse Brain

2.1 Voxelation is Used to Analyze a Mouse Model of Disease

Voxelation has been performed at both high and low spatial resolution in the mouse brain. At low resolution, brains from control mice and those from a mouse model of Parkinson's disease were investigated. The model of Parkinson's disease was created using toxic doses of methamphetamine. The control and disease brains were each sliced into 10 coronal sections and divided into 4 voxels, for a total of 40 voxels with a resolution of ~11 µl (Brown et al., 2002). Gene expression in each of the voxels was analyzed using a 9,000-gene microarray. The large-scale data obtained from microarrays offered the opportunity to perform cluster analysis based on correlations of gene expression. Two distinct and mutually exclusive clusters of genes were identified as being expressed in either the anterior or the posterior half of the brain, suggesting that these clusters may be involved in rostral/caudal patterning (❷ *Figure 29-1a*).

❏ Figure 29-1
Replicability of datasets using voxelation. (a) High resolution analysis (1 mm³) of Thy-1 gene expression in a coronal section of mouse brain at the level of the striatum. Scatter plot shows voxel by voxel comparisons of Thy-1 expression in two replicates. (b) Low resolution analysis (40 voxels, ~11µl) of mouse brain using microarrays. Within microarray reliability was confirmed using the Nfl gene, which was represented with two separate spots. Reproducibility of expression levels was excellent for both the normal brain ($r = 0.96$, $F_{[1,38]} = 443.64$, $p < 0.0001$) and the MA (methamphetamine) brain ($r = 0.90$, $F_{[1,38]} = 153.80$, $p < 0.0001$)

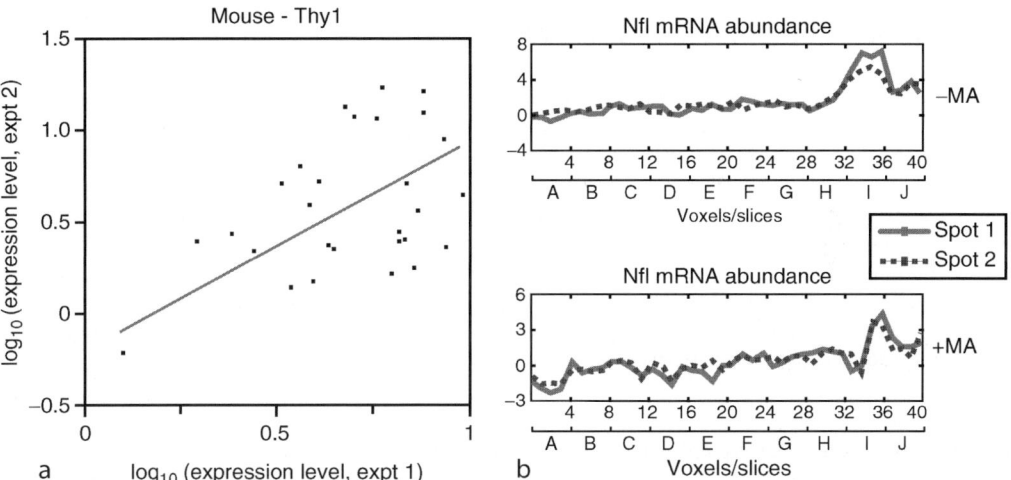

2.2 Use of Singular Value Decomposition to Analyze Large Amounts of Data

A less hypothesis driven approach to analyzing the data is a mathematical method called singular value decomposition (SVD) (Hendler et al., 1994). SVD reduces the dimensionality of the data by constructing linear combinations of the variables called vectors, while explaining the maximum amount of variance. Each of these gene vectors, or principal components, represents a portion of the overall data set. Applying this method to the normal and Parkinson's disease voxelation data showed good consistency for three of the four principal components between each set. However, the third principal component exhibited a striking shift in expression pattern in the Parkinson's disease model compared to control (❯ *Figure 29-2b*). The normal brain showed higher levels of expression in the striatum and cerebellum, whereas the brain from

◘ Figure 29-2
Mouse brain gene expression patterns obtained using qRT-PCR and high-resolution voxelation. (a) Thy-1 gene, (b) dopamine receptor D2 (Drd2) gene, and (c) tyrosine hydroxylase (TH) gene. IA-interaural coordinates (mm); Br-bregma. Mouse atlas sections from the Mouse Brain Library (Rosen et al., 2000; Williams, 2000)

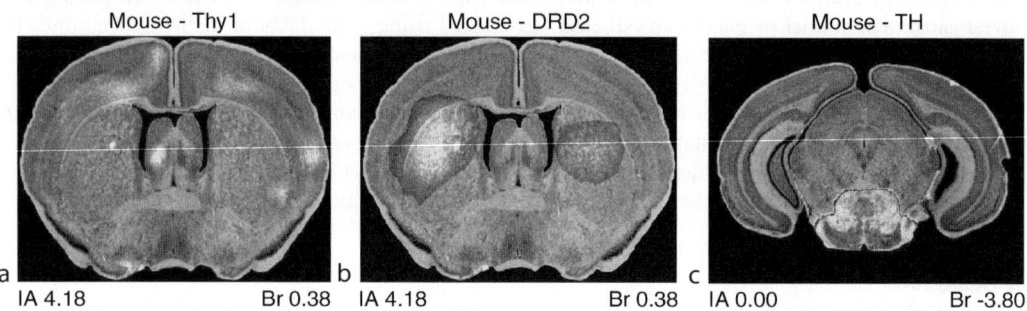

the disease model shifted toward a region containing the hippocampus. These data suggested that the physiological alterations that occur in the striatum of the Parkinson's disease mouse model are mirrored by a shift in expression of the cluster of genes in the third principal component.

2.3 High-Resolution Rodent Brain Voxelation

In the high resolution studies, a rodent voxelation device was used to dissect a coronal slice of the mouse brain at the level of the striatum and cortex (interaural or IA. 3.82 mm, bregma or Br. 0.02 mm) into 1μl cubes (1 mm^3) resulting in approximately 70 voxels (Singh et al., 2003). qRT-PCR was performed on each voxel to analyze mRNA transcript levels of three known genes; Thy-1, dopamine D2 receptor (Drd2) and tyrosine hydroxylase (TH). The results were properly normalized and an image reconstruction program was used to display the gene expression patterns (❯ *Figure 29-2*). Thy-1 was predominantly expressed in the cortex (Barlow et al., 2000), Drd2 in the striatum (Missale et al., 1998), and TH in the substantia nigra, ventral tegmental area, and other parasagittal nuclei (Min et al., 1994; Stork et al., 1994). The images of expression images for each of these three well-characterized genes were obtained using voxelation recapitulated the known expression patterns from in situ hybridization.

2.4 Independent Confirmation of Voxelation Replicability

Concerns about the reproducibility of voxelation techniques have been addressed in each of the studies. The high-resolution voxelation coupled with qRT-PCR was repeated in its entirety and showed good correlation in gene expression between each data set for the genes examined (❯ *Figure 29-3a*). In the experiment using the Parkinson's disease mouse model, excellent reproducibility of gene expression was found at the global level between the left and right sides of the brain. At the single gene level, a number of observations further confirmed the good replicability of the data. An example is provided by the neurofilament A (Nfl) gene (Yaworsky et al., 1997). Two probes for this gene were spotted on the microarray providing an internal control to measure within array reproducibility (❯ *Figure 29-2b*). The correlation coefficients of Nfl gene expression within the normal and Parkinson's disease brain were highly statistically significant.

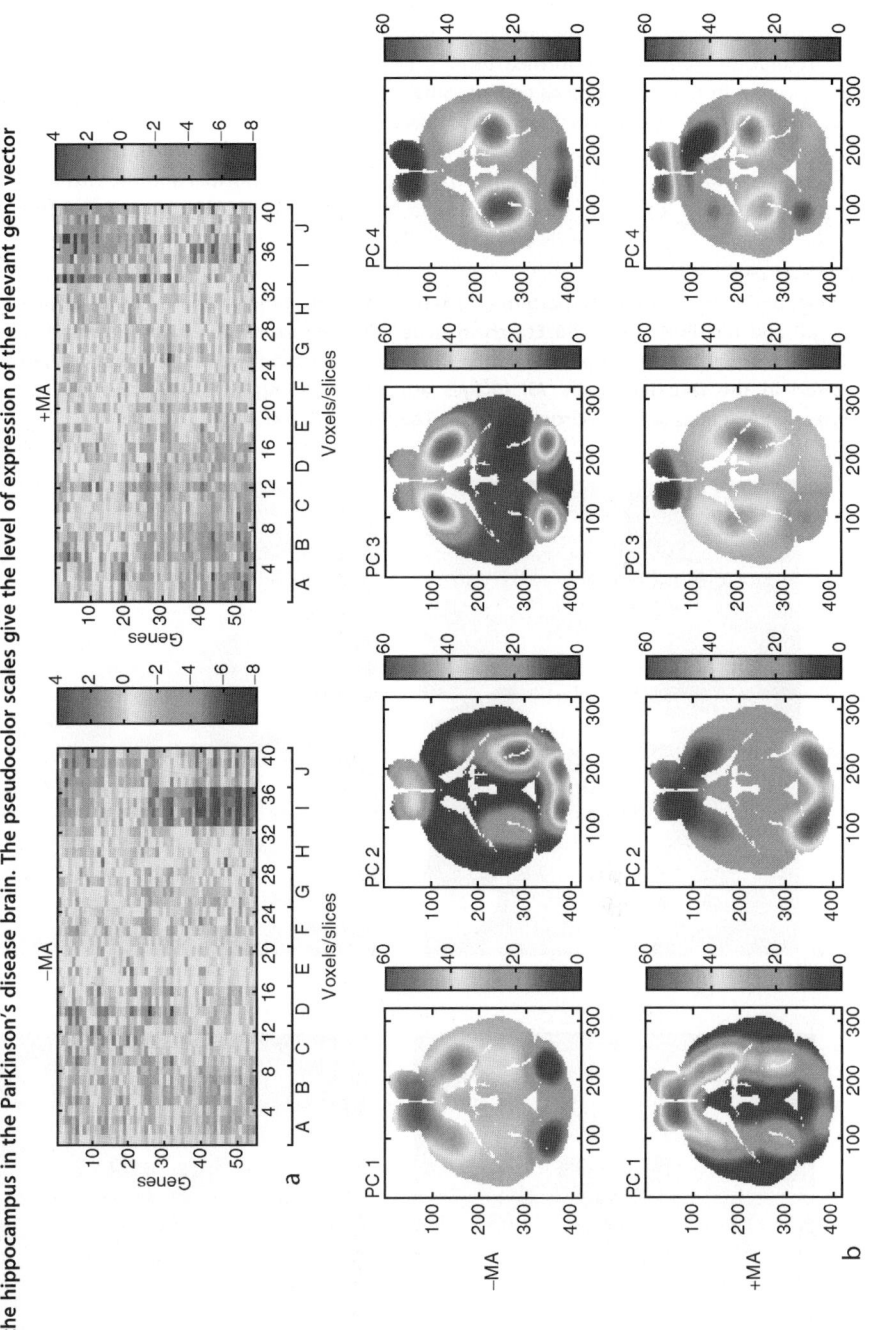

◘ Figure 29-3

(a) Spatial gene expression patterns for two clusters of genes in control (−MA) and methamphetamine-induced Parkinson's disease model (+MA) mouse brains. One cluster of genes is expressed anteriorly and the other posteriorly, in voxels corresponding to the cerebellum. The two clusters are comparable between the normal and Parkinson's disease brains with only minor differences. (b) SVD analysis identifies principal components (PC) defined by anatomical regions of the brain. These spatial patterns are generally conserved in the first, second, and fourth PCs. However, the third PC shows a shift from the striatum and cerebellum in the normal brain toward the hippocampus in the Parkinson's disease brain. The pseudocolor scales give the level of expression of the relevant gene vector

3 Voxelation of the Human Brain

3.1 Voxelation on the Brain of Alzheimer's Patient and Unaffected Individual

Voxelation experiments at low and high resolution were also performed on human specimens. The low resolution study was performed using a hemi-section from a normal human brain and a matching hemi-section from an Alzheimer's patient (Singh et al., 2003) at the level of the hippocampus. Each specimen was divided into $\sim$24 voxels of 1 cm^3 and subjected to microarray analysis. SVD identified gene expression vectors restricted to anatomical regions of the brain such as the cortex and striatum. In addition, a number of genes were found to be differentially expressed between the normal and the Alzheimer's specimens. The expression pattern for one of these, 14-3-3 Eta chain gene (YWHAH), is shown in ❷ *Figure 29-4b*.

◘ Figure 29-4
Human brain gene expression patterns. (a) THY-1 and Drd2 genes obtained using high-resolution voxelation (87 µl) and qRT-PCR. (b) 14-3-3 Eta chain gene (YWHAH) expression in the normal and Alzheimer's brain obtained using low-resolution voxelation (1cm^3) and microarrays. The expression pattern is dramatically altered in the diseased brain. All images are created using the voxelation data and warped onto a corresponding atlas section (Virtual Hospital: The Human Brain)

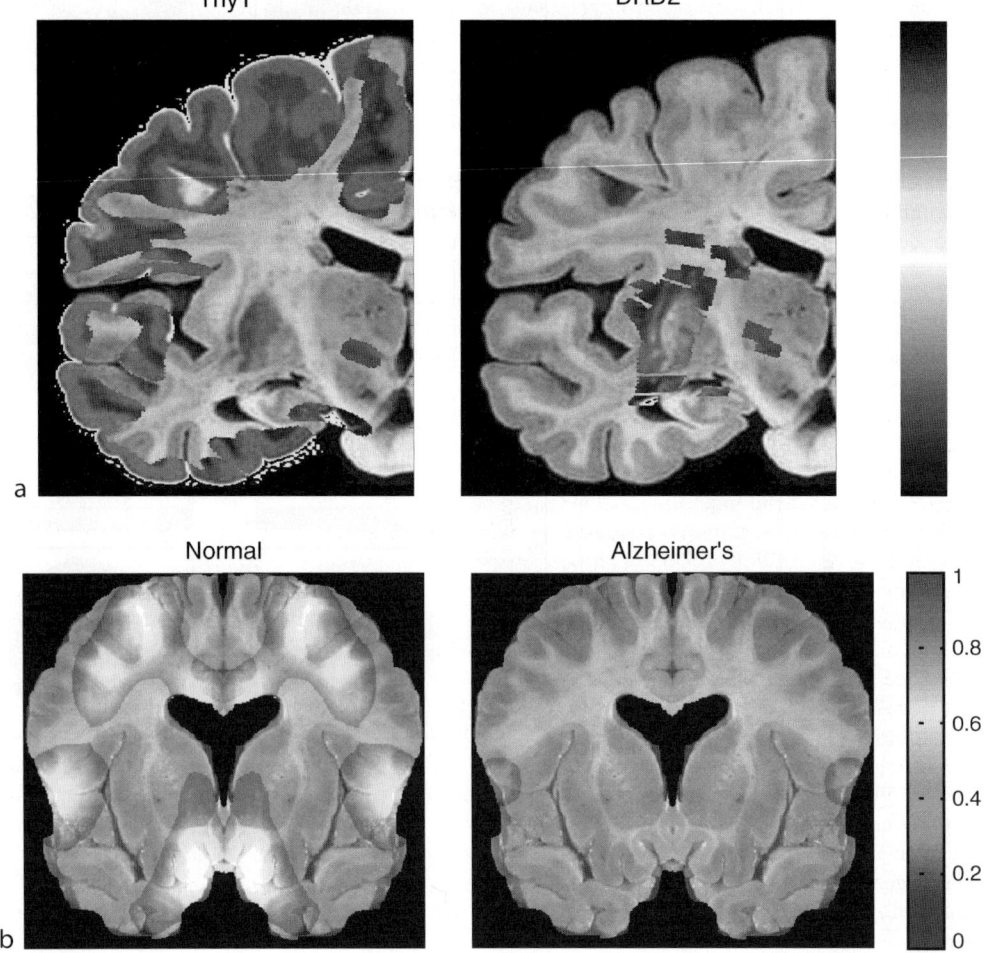

3.2 High-Resolution Voxelation of the Human Brain

Voxelation of postmortem human brain at high resolution involved the use of a cutting device producing voxels of size 3.3 × 3.3 × 8 mm^3 (depth), a volumetric resolution of 87 μl (Singh et al., 2003). This device was used to analyze a coronal hemi-section from a normal brain at the level of the striatum resulting in 425 voxels. Total RNA was isolated and qRT-PCR was performed for two genes with well-known expression patterns. As performed in the high-resolution mouse brain study, THY-1 (cortex) and Drd2 (striatum) were assayed for gene expression. Following image reconstruction the genes showed expression patterns matching those expected from the literature (❯ *Figure 29-4a*). These observations suggest that voxelation will help researchers to understand gene expression patterns occurring in normal and diseased human brains. Combining mouse and human data will further sharpen these insights to provide a better understanding of the genomic changes that occur in disease and reveal potential targets for drug therapy.

4 Conclusion

Voxelation can provide views of the molecular architecture of the brain not easily obtained using other methods. In the past, the structures of the brain have been determined by anatomical and functional boundaries. These boundaries may further be defined by groups of cells that share gene expression profiles. The global overview produced by voxelation can elucidate these domains. At its most basic level, voxelation provides a three-dimensional map of individual gene expression patterns. At a higher level, more advanced analyses such as clustering and SVD algorithms provide deeper insights into the complex datasets produced by voxelation coupled with microarrays.

Presently, voxelation is not capable of attaining the cellular resolution obtained by in situ hybridization or immunocytochemsitry. The finite size of each voxel creates heterogeneity in each sample (Raichle, 1998), where gene expression by a small subset of cells within a voxel leads to a partial volume effect in which its signal gets "diluted" by adjacent cells. However, further improvements in voxelation resolution and microarray technology will allow more efficient sample processing and better quality maps. The mouse voxelation device (1 μl^3) is near the strictly mechanical limit that can be attained. Alternative approaches, such as laser grids or microelectromechanical systems (MEMS) technologies, may be utilized in the future. Microarray technology is also continuing to improve by querying more splice variants, alternate start codons, microRNAs, and other exotic RNA species (Loring, 2005). The increasing ability to pack spots closer together on the chip also increases the array sensitivity and decreases the starting material requirements. This permits the amount of total RNA required from each voxel to be reduced, ultimately decreasing the size of each voxel and improving spatial resolution.

The gene expression patterns obtained from voxelation coupled with either microarray or qRT-PCR show good correlation with previously known patterns. Since microarray technology is still relatively new, more robust methods of analyzing the complex data structures obtained are continually being applied. SVD is one method that has shown very interesting results and may be used to explain changes that occur between control and disease samples. Furthermore, clustering programs not only decipher genes that are coregulated, but also begin to shed light on putative networks of gene regulation. Control regions shared by genes with similar expression profiles can be identified to allow for better understanding of the relevant transcriptional circuitry.

Future goals for voxelation include the creation of whole brain atlases of gene expression for both mice and humans. Furthermore, genes involved in many disorders affecting the human brain can be identified using voxelation including schizophrenia, Down syndrome, autism, and Parkinson's disease. Voxelation uses a simple dicing apparatus coupled with gene expression analysis technology to create maps of gene expression patterns in a high-throughput manner. This straightforward approach should give us insights into the molecular architecture of the brain as well as the genetic mechanisms of neurological disease.

References

Barlow JZ, Huntley GW. 2000. Developmentally regulated expression of thy-1 in structures of the mouse sensory-motor system. J Comp Neurol 421(2): 215-233.

Boguski MS, Jones AR. 2004. Neurogenomics: At the intersection of neurobiology and genome sciences. Nat Neurosci 7(5): 429-433. Website available at: http://www.brainatlas.org/.

Bonner RF, Emmert-Buck M, Cole K, Pohida T, Chuaqui R, et al. 1997. Laser capture microdissection: Molecular analysis of tissue. Science 278(5342): 1481, 1483.

Brown VM, Ossadtchi A, Khan AH, Yee S, Lacan G, et al. 2002. Multiplex three-dimensional brain gene expression mapping in a mouse model of parkinson's disease. Genome Res 12(6): 868-884.

Bunney WE, Bunney BG, Vawter MP, Tomita H, Li J, et al. 2003. Microarray technology: A review of new strategies to discover candidate vulnerability genes in psychiatric disorders. Am J Psychiatry 160(4): 657-666.

Gambhir SS, Barrio JR, Herschman HR, Phelps ME. 1999. Assays for noninvasive imaging of reporter gene expression. Nucl Med Biol 26(5): 481-490.

GongS, Zheng C, Doughty ML, Losos K, Didkovsky N, et al. 2003. A gene expression atlas of the central nervous system based on bacterial artificial chromosomes. Nature 425(6961): 917-925.

Heintz N. 2001. Bac to the future: The use of bac transgenic mice for neuroscience research. Nat Rev Neurosci 2(12): 861-870.

Hendler RW, Shrager RI. 1994. Deconvolutions based on singular value decomposition and the pseudoinverse: A guide for beginners. J Biochem Biophys Methods 28(1): 1-33.

Herschman HR, MacLaren DC, Iyer M, Namavari M, Bobinski K, et al. 2000. Seeing is believing: Non-invasive, quantitative and repetitive imaging of reporter gene expression in living animals, using positron emission tomography. J Neurosci Res 59(6): 699-705.

Loring JF. 2005. Evolution of microarray analysis. Neurobiol Aging 27(8):1084-1086.

Louie AY, Huber MM, Ahrens ET, Rothbacher U, Moats R, et al. 2000. In vivo visualization of gene expression using magnetic resonance imaging. Nat Biotechnol 18(3): 321-325.

Min N, Joh TH, Kim KS, Peng C, Son JH. 1994. 5′ upstream DNA sequence of the rat tyrosine hydroxylase gene directs high-level and tissue-specific expression to catecholaminergic neurons in the central nervous system of transgenic mice. Brain Res Mol Brain Res 27(2): 281-289.

Missale C, Nash SR, Robinson SW, Jaber M, Caron MG. 1998. Dopamine receptors: From structure to function. Physiol Rev 78(1): 189-225.

Owen MJ, Cardno AG, O'Donovan MC. 2000. Psychiatric genetics: Back to the future. Mol Psychiatry 5(1): 22-31.

Raichle ME. 1998. Behind the scenes of functional brain imaging: A historical and physiological perspective. Proc Natl Acad Sci USA 95(3): 765-772.

Singh RP, Brown VM, Chaudhari A, Khan AH, et al. 2003. High-resolution voxelation mapping of human and rodent brain gene expression. J Neurosci Methods 125(1–2): 93-101.

Stork O, Hashimoto T, Obata K. 1994. Haloperidol activates tyrosine hydroxylase gene-expression in the rat substantia nigra, pars reticulata. Brain Res 633(1–2): 213-222.

Yaworsky PJ, Gardner DP, Kappen C. 1997. Transgenic analyses reveal developmentally regulated neuron- and muscle-specific elements in the murine neurofilament light chain gene promoter. J Biol Chem 272(40): 25112-25120.

Zacharias DA, Baird GS, Tsien RY. 2000. Recent advances in technology for measuring and manipulating cell signals. Curr Opin Neurobiol 10(3): 416-421.

Index

AACOCF₃. *See*
  Arachidonyltrifluoromethyl ketone
ABC, half transporters, 19
ABC, transporters in, 19
Abluminal application, 378, 380
Acetylcholine
  – acute therapies, 180–181
  – choline acetyltransferase (ChAT), 181
  – chronic therapies, 181
  – injury responses, 180–181
  – muscarinic, 181
  – in traumatic brain injury, 181–182
  – vesicular ACh transporter, 181
Acute ammonia toxicity
  – AMPA receptors, 57
  – astrocyte swelling, 50, 58
  – ATP depletion in, 50, 52–54, 58
  – ATP synthesis in, 53, 54
  – brain energy metabolism in, 52
  – calcineurin in, 53
  – calpain in, 55
  – *vs.* chronic hyperammonemia, 46, 59
  – cyclic GMP in, 51
  – free radicals in, 53, 54, 58
  – GLAST, 57–58
  – GLT-1, 57–58
  – glutamate transporters, 57–58
  – glutamine synthetase in, 53, 56
  – glutathione, 54
  – kainate receptors, 57
  – MAP-2 degradation, 55
  – microtubules and, 55
  – mitochondrial function in, 54
  – molecular mechanisms, 59
  – Na$^+$/K$^+$-ATPase in, 52, 53, 62
  – neuronal depolarization, 51
  – nitric oxide in, 53, 56
  – nitric oxide synthase, 50, 53, 56
  – prevention by NMDA receptor antagonists, 50, 53, 54

  – role of NMDA receptors in brain, 50, 55
  – superoxide, 54, 55
  – swelling of astrocytes, 58
  – tyrosine nitration, 55, 56
Acyl-CoA: cholesterol *O*-acyltransferase (ACAT), 145
Adenosine
  – A2AR/D2R interactions, 190–191
  – in traumatic brain injury
    – adenosine responses to, 190
    – adenosine therapies for, 190
Adjunct therapy for cancer treatment, 398, 399
Adrenoleukodystrophy (ALD)
  – effect of proinflammatory cytokines on, 21
  – gene, 14
  – inflammatory mediators in, 21, 30, 32, 36
  – metabolism in, 14–19, 21, 31, 32
  – mutations in, 15, 16, 19
  – neuroinflammation in, 14, 20, 21, 30–32, 36
  – possible functions of, 20
  – protein (ALDP), 15, 17–20, 31, 36
  – related protein (ALDRP), 19, 31
Aerobic respiration, 139
Age-related macular degeneration (AMD)
  – all-*trans*-retinal, 139
  – amyloid deposits, 135, 143, 146
  – amyloid in drusen, 144
  – ApoE, 140, 143–147, 153
  – cholesterol in, 143–146
  – chromophore, 138
  – 11-*cis* retinal, 138–139
  – dry AMD, 140, 141
  – frequent retinal disease, 140
  – genetic risk factors in, 141, 149
  – HTRA1, 149
  – inflammation in, 146, 152
  – intermediate AMD, 141

  – NSAID, 146
  – opsin, 138
  – phagocytosis, 138
  – photoreceptors, 140
  – prevalence of drusen, 140, 141, 146
  – retinal pigmented epithelium (RPE), 133, 140, 141, 143, 146
  – rhodopsin, 146
  – shared features, of AD and AMD, 152–154
  – wet AMD, 141–142
Albinism, 151
Alkenylacyl glycerophospholipids, 72
Alkylacyl glycerophospholipids, 72
Alpha-1-antichymotrypsin, 146
Alpha-2-macroglobulin, 146
Alteration of peroxisomes biogenesis in, 15, 16, 26
Alterations in spontaneous EEG activity, 392
Alzheimer treatments
  – Alzhemed, 134
  – 5-HT agonists, 134
  – immunotherapy, 134
  – M1 agonists, 134
  – α-secretase, 134
  – statins, 134
Alzheimer's disease (AD), 465, 514, 520, 544, 551, 603, 654–552
  – Aβ, 133–135, 140, 144–147, 152–154
  – ApoE in, 134, 135, 140, 144, 147
  – cholesterol in, 133, 134, 143–145
  – diabetes, 134, 135
  – estrogen, 134, 135, 143
  – familial AD (FAD), 134, 153
  – faulty protein turnover, 146
  – glucocorticoids, 134
  – homocysteine, 134
  – hypertension, 134, 143
  – impaired processing in, 146
  – inflammation in, 146, 152, 153

- key triggers, 134
- leading cause of dementia, 134
- neurofibrillary tangles, 134
- presenilins 1 (PS1) and PS 2, 134
- prevalence rate, 134
- retina as a model, 135
- risk factors, 134, 135, 143, 152, 153
- β-secretase, 134, 135
- γ-secretase, 134, 145
- shared features of AD and AMD, 152–154
- SORL1, 135
- stroke, 134

Aminoguanidine, 171
α-Amino-3-hydroxy-5-methylisoxazole-4-propionate, 330
2-Amino-5-phosphonovalerate, 75
21-Aminosteroids, 204
Ammonia, 45–63. See also Hyperammonemia
- acute toxicity (see Acute ammonia toxicity)
- detoxification, 45, 48, 56
- and glutamate-glutamine cycle, 47, 48
- membrane permeability, 45, 46
- and pH, 45, 46, 48
- similarity to potassium ion, 46
- transporters, 46, 47

AMPA receptors, 57
AMPA. See α-amino-3-hydroxy-5-methylisoxazole-4-propionate
β-Amyloid, 462, 465, 603
Amyloidogenic diseases
- Alzheimer's disease, 465
- Huntington's disease, 464
- Parkinson's disease, 464, 465
- Prion diseases, 465
- synucleinopathies, 464

Amyloid P, 146, 154
Amyotropic lateral sclerosis, 514, 520, 544, 553
Angiogenesis, 2–8, 74
- in peripheral nerve sheath tumor formation, 117, 118

Animal models, 20, 21, 28, 32
- associated with anti-MAG/SGPG antibody, 239–240
- associated with anti-sulfatide antibodies, 241
- closed skull impact injury, 346
- controlled cortical impact injury, 346
- experimental neuropathies
  - induced by gangliosides, 240–241
- fluid percussion injury, 346

- hepatic encephalopathy, 48, 49, 63, 64
- hyperammonemia, 49, 50, 59, 60, 63, 64
- inertial acceleration injury, 345
- Lentivirus encephalitis of
  - FIV, 479
  - HIVE in SCID mice, 480–483
  - MuLV, 479
  - SIV, 479, 480
- optic nerve stretch injury, 346
- transgenic models, 480

Antibodies, against α-synuclein, 600, 603
Antibody fragments
- intrabody, 466
- potential therapeutics, for treatment neurodegenerative disease, 460
- single-chain Fv (scFv), 461
- single-domain antibody fragment, 461

Anti-inflammatory effects of, 31
Anti-Nogo-A inhibitor, 645
Antisense oligomer, 82
Apoptosis, 76, 81, 82
- calpain roles in, 529–531
- CNS neurological disorders, 334–335

Arachidonic acid, 75, 76, 78–83, 207
Arachidonoyl trifluoromethyl ketone, 79, 330
Ascorbate, 143
ASIA, 635, 636
Astrocytes, 73, 74, 76–78, 81
- gliosis, 473
- HIV cell tropism, 476
- neurotoxins, 473–474
- swelling, 50, 58

Astrocytic activation, 429–431
Atelectasis, 642
ATP, 72, 74, 76, 77
- binding of, 15, 19
- ammonia toxicity
  - consumption in, 52, 53, 56
  - depletion in, 52, 53
  - synthesis in, 53, 54

ATPases, 519
Autocrine mechanisms, 74
Autoimmune diseases, 368, 375, 385
Autologous activated macrophages, 644
Autophagosome, 514, 518–522
Autophagy, 151, 152, 514, 515, 518–522
- chaperone-mediated, 604
- macroautophagy, 604
- microautophagy, 604, 605
- mTOR, 605, 606

- rapamycin, 606
- related genes, 518
- role in clearance of α-synuclein, 604–606

Autoregulation, 212–214
Axolemma
- distortion, 347, 348
- injury effect on proteins localized to, 347
- permeability, 348, 349

Axonal transport
- β-amyloid precursor protein, 353
- anterograde/retrograde tracing as measure of, 352

Axotomy
- primary, 347
- secondary, 347

Babinski sign, 637, 638
Barriergenesis
- angiogenesis, 306
- differentiation, 305, 306
- shift of localization, 306
- tightening, 306
- transporter expression, 306, 307

BBB. See Blood-brain barrier
BCNSB. See Blood-CNS-barriers
BCSFB, 366, 370, 371, 373–376, 387, 388, 394, 398, 401, 409, 436–438, 440, 442
BCSNB breakdown, 366, 443–445
Best disease, 147, 148
Bestrophin, 148
Biomarkers, 356
Blind spot (scotoma), 136, 141
Blood-brain barrier (BBB), 73, 74, 78, 82, 83, 145
- cerebral endothelium, 370, 373, 432
- development of, 305–307
- disruption, 374, 385, 386, 390, 394, 397–399, 401, 402, 412, 413, 415, 420, 432, 433, 435, 436, 440, 443, 444
- entry of immune γ-globulins G (IgG), 373
- function of, 304–305
  - breakdown, 371, 385, 392, 410, 414, 434
  - during emotional stress, 385
- permeability in heat stress, 399
- regulation of, 313
- vascularity, 306

Blood-CNS-barriers (BCNSB), 366–368, 380, 405, 414, 421, 436–441, 443–446
- changes in CSF ion compositions, 372

- choroid plexus, 371, 376, 394, 397, 401, 409
- CSF production, 371
- passage of tracers across, 445

Blood-retina barrier, 170

Blood-spinal cord barrier (BSCB), 366, 375–382, 387–389, 391, 394, 397–399, 405, 408, 418, 421–423, 425, 426, 428, 429, 431, 436–438, 440–443, 446
- permeability, 378, 380–382, 387, 391, 397–399, 405, 418, 421–423, 428, 429, 431, 438, 440, 443

Blood-spinal cord barrier permeability in spinal cord injury, 391, 408

B lymphocytes, 230, 241

BN-80933, 222

Brain edema formation, 368, 382–385, 399, 400, 412, 413, 420, 421, 432, 434-436, 440, 443, 444

Brain fluid microenvironment, 373–375, 383, 385, 393, 405

Brain hyperthermia, induced fatigue, 400

Brain imaging
- anatomy of HAD, 474
- atrophy, 473
- MRI, 473

Brain injuries, 375, 384, 431, 433, 440

Brain mapping, 3-dimensional gene expression profiles, 650

Brain microdialysis, 50, 59, 60

Brain, 14–16, 19–26, 30, 32, 36

Brain-derived neurotrophic factor, 76

Brain pathology, 366, 368, 370, 371, 375, 385, 393, 394, 397, 398, 431, 432, 436, 438, 440, 443–445

Brain tumors, 2, 4, 5, 367, 385

Bromoenol lactone, 76

Bruch's membrane, 133, 139, 141, 142, 144–146, 148, 151

BSCB. See Blood-spinal cord barrier

Bulk flow, 371, 423

Ca^{++} ATPase, 210

Ca^{2+}-binding domain, 325

Ca^{2+}-independent PLA_2, 322, 326
- in mammalian CNS, 327, 328

Ca^{2+}-mediated damage of optic nerve, 532

CaLB. See Ca^{2+}-binding domain

Calcineurin, 52, 53, 62

Calcium, 74–77, 79, 81–83

Calorimetry, 643

Calpains, 55, 526, 527
- calcium activated, 251
- family members, 251–253
- functions, 253, 254

- inflammatory arm of pathophysiology of
  - immune cells migration, 528
  - inflammatory arm of, 527, 528
- inhibitors
  - in neurodegeneration and inflammation, 534, 535
  - as therapeutic agents, 533, 534
- neurodegenerative arm of pathophysiology of, 528–531

Capillary filtration, 382, 383

Carbonate radical ($\cdot CO_3$), 209

Carbon dioxide (CO_2), 209

Caspases, 72, 76, 77, 79, 83
- family members (initiators, effectors, inflammatory), 251, 252
- functions
  - inflammation and programmed cell death, 254, 255
  - targets, 255
- inhibitors
  - endogenous (inhibitors of apoptosis proteins (IAPs)), 255, 256
  - synthetic, 256, 257

Catalase, 143, 208

Catecholamines
- dopamine, 182, 183
  - transporter, 183, 184
- therapies for TBI
  - amantadine, 184, 185
  - amphetamines, 184
  - bromocriptine, 184–186
  - methylphenidate, 184–186
- in traumatic brain injury, 182–184
- tyrosine hydroxylase, 183

Cathepsin D, 146

Cathepsins, 516, 517, 520
- family members (aspartate, cysteine, serine), 257–259
- functions, 259, 260
  - antigen presentation, 258
  - bone remodeling, 258
  - hormone processing, 258
  - programmed cell death, 258
  - protein degradation, 258
- targets, inhibitors
  - endogenous (cystatins, thyropin), 259
  - synthetic and natural, 259, 260

CDP-choline (citicoline), 82

Cell death, 72, 74, 76, 77, 79–81, 83
- pathways of
  - endoplasmic (endoplasmic reticulum), 255

- extrinsic (cellular membrane), 254
- intrinsic (mitochondria), 255

Cell migration, 252

Central catecholaminergic neurons, stimulation/lesion of, 385

Central fovea, 139

Central nervous system (CNS), 366, 368–376, 378, 381–383, 385, 386, 394, 398, 402, 404–406, 408, 409, 412, 414, 421, 429, 431, 435–445
- dysfunction, 398, 405
- by viruses and bacteria, infections of, 385

Central nervous system (CNS), neurological disorders, PLA2 in
- future perspectives of, 335
- mechanisms of, 332–335
- multiple sclerosis
  - calpain in, 531
  - calpain inhibitors as therapeutic agents, 533, 534
  - demyelination of axons, 529
  - immune cells migration in CNS, 528
  - myelin-specific T cells activation, 527–528
  - necrosis and apoptosis, 529–531
  - pathophysiology of EAE, 528, 529
- in neurotrauma
  - ischemia, 332
  - traumatic brain injury, 330–332
  - traumatic spinal cord injury, 328–330

Central vision, 140, 141, 148, 149

Ceramide-1-phosphate, 79

Ceramides, 72, 79

Cerebral autoregulation, disturbances of, 385

Cerebral ischemia
- bone marrow stromal cell transplantation, 314
- matrix-metalloproteinases, 314
- permeability, 307

Cerebral microvessels, 366, 376, 379, 385

Cerebrospinal fluid (CSF), 330, 526

CFA. See Complete Freund's Adjuvant

Chaperone-mediated, 514

Chemical factors, 366–368, 372, 377, 384, 421, 444, 445

4-Chlorokynurenine, 98

7-Chlorokynurenic acid, 98, 99

Cholinergic, 93

Choriocapillaris, 139, 149
Chronic hyperammonemia
 – vs. acute hyperammonemia, 49
 – calcineurin, 62
 – cGMP, 59, 60, 64
 – cGMP-degrading phosphodiesterase, 60
 – cGMP-dependent protein kinase, 60
 – cognitive function, 59, 60, 64
 – GABA-ergic neurotransmission, 61
 – glutamate-nitric oxide-cGMP pathway, 59
 – glutamatergic neurotransmission, 59
 – learning ability, 49, 64, 65
 – long-term potentiation, 60
 – MAP-2 phosphorylation, 62
 – metabotropic glutamate receptors, 60
 – motor function, 63, 64
 – nitric oxide, 59
 – NMDA receptors in, 59, 60, 64
 – phosphorylation of neuronal proteins, 62
 – serotoninergic neurotransmission, 61, 62
 – soluble guanylate cyclase, 50, 51
Closed head injuries, 435, 436
CNQX. See 6-Cyano-7-nitroquinoxaline–2,3-dione
CNS. See Central nervous system
Cognitive dysfunction
 – anatomical basis, 480
 – behavioral abnormalities in HIVE SCID mice, 480, 481
 – clinical features in humans, 472, 473
 – IFN-α, 476
 – neurophysiologic dysfunction, 472
 – pathology of HAD, 473, 474
Cognitive function
 – cGMP in, 64, 65
 – restoration by cGMP, 59, 60, 64
 – restoration by phosphodiesterase inhibitors, 60, 64, 65
Colony stimulating factors, 74
Coma, 382, 398
Complement cascade, 146
Complement factor H (CFH), 146, 147, 149
Complement proteins, 146
Complete Freund's adjuvant, 528
Concussion, 367, 385
Cone-rod dystrophy (CRD), 148

Convulsion, 397, 398
Convulsive agents, 385
Core particle, 515
Cortical EEG, 392
cPLA2. See Cytosolic PLA2
C-reactive protein, 146, 149
Cryoglobulinemia, 239
CT scan, 640
Cu/Zn SOD, 208, 218
6-Cyano-7-nitroquinoxaline-2,3-dione (CNQX), 76, 330
Cyclic GMP
 – in acute ammonia toxicity, 50
 – in chronic hyperammonemia, 59, 60, 64, 65
 – and cognitive function, 60
 – -degrading phosphodiesterase, 60
 – -dependent protein kinase, 60
 – and learning ability, 64, 65
 – restoration of learning ability, 64, 65
Cyclooxygenase, 73, 78, 79
Cyclosporin A, 212
Cystine, 74
Cytochrome c, 79
Cytoglobin (Cygb), 139
Cytoskeletal remodeling, 252
Cytokine(s), 72–74, 77–83
 – effects on, 30
 – IFN-α, 476
 – TNF-α, 476
Cytosolic PLA2, 322, 325, 326
 – in mammalian CNS, 327
 – in spinal cord ischemic injury, 332
 – in traumatic spinal cord injury, 330–332
Cytotoxic edema, 383
Cytotoxicity, 74, 79
Death, 376, 381, 382, 397, 398, 431, 432, 435, 445, 446
Deep venous thrombosis, 643
Delayed plantar response, 637
Delirium, 398
Dementia
 – clinical, 472, 473
 – with Lewy bodies, 600
 – with parkinsonism, 600
 – subcortical, 473
Demyelination, 72, 73, 553, 554
Dephosphorylation, of NFP (de-NFP), 529
Deubiquitinating enzymes, 517
Dexanabinol, 218, 223
Dextrorphan, 76
Diabetic retinopathy, 135, 140
Diacyl glycerophospholipids, 72

Diacylglycerol lipase, 75, 76
Diffuse axonal injury
 – detection of, 344, 345
 – grades of, 344
 – pathological features of, 344
3,4-Dimethoxy-N-[4-(3-nitrophenyl) thiazol-2-yl]benzenesulfonamide, 100
Dimethylthiourea (DMTU), 220
Disrupts BBB to proteins, 394
Docosahexaenoic acid, 73, 75–77, 80, 207
Docosanoids, 77
Docosatrienes, 77
Down's syndrome, 97, 100, 140, 153
E1, ubiquitin activating enzyme, 517, 518
E2, ubiquitin conjugating enzyme, 517, 518
E3, ubiquitin ligase, 517, 518
E4, ubiquitin-chain elongation factor, 517, 518
EAA. See Excitatory amino acid
EAE. See Experimental autoimmune encephalomyelitis
EcSOD, 208
Ecstasy (MDMA), 405
Edema formation, 366, 368, 375, 376, 381–385, 399, 400, 404, 412, 413, 418, 420–423, 425, 429, 432, 434–438, 440, 443–446
Edematous expansion, 376, 401, 428
EEG abnormalities
 – ependymal cells around lateral/third ventricle, damage to, 401
 – myelin, degradation of, 401
 – post-synaptic dendrites, damage to, 402
EGF. See Epidermal growth factor
Eicosanoids, 72, 74, 77, 80, 81, 83
Electrical stimulation, 644
Electroencephalogram (EEG) activity, 371
Electron microscopic studies, 370
Electron transport, 74
Electrophysiology, 356
 – in TBI research
  – evoked potentials in TBI, 192, 193
  – long-term potentiation in TBI models, 194
  – TBI-induced alterations in spontaneous brain electrical activity, 192
ELISA, 232, 233
Emotional factors, 366, 368, 372, 385, 386
Emotional stress, and BBB, 385, 386

Encephalitis, 368, 375, 386
Endocrine mechanisms, 74
Endogenous albumin leakage, 416
Endogenous neurotransmitters, 384
Endoplasmic reticulum, 514
Endothelial cells, 2–4, 6, 7, 74, 169–171, 173, 368–371, 375–379, 383, 402, 404, 416, 419, 422, 428, 433
Endothelial, glial and neuronal interaction, 370
Environmental factors, 368
Environmental heat stress, 398–402
Ependyma around lateral and third ventricles, 394
Ependymal cell, 366, 395, 397, 401
Ependymal transport, of hormones, 373
Epidermal growth factor, 325
Epithelial cells of choroid plexuses, damage of, 394
ER intrabody, use in regulating post-translational protein modifications, 466
17β-Estradiol, 639
Estrogen, 134, 135, 143, 639
Evans blue albumin
 – leakage of, 400, 416
 – staining, 388, 416, 433
Excitatory amino acid, 74, 92, 93, 97–99, 328
Excitatory neurotoxicity, in CNS neurological disorders, 334
Excitotoxicity, 211
Excitotoxins, 93
Excretion, 371
Experimental autoimmune encephalomyelitis, 21, 526, 544
 – calpain in, 531
 – demyelination of axons, 529
 – pathophysiology of, 528, 529
Experimental neuropathies
 – associated with anti-MAG/SGPG antibody, 239, 240
 – associated with anti-sulfatide antibodies, 240, 241
 – induced by gangliosides, 239, 240
Expression profiling
 – dopamine D2 receptor (Drd2), 652, 654
 – 14-3-3 Eta chain Gene (YWHAH), 654
 – neurofilament light chain (Nfl), 651, 652
 – Thy-1 cell surface antigen (Thy-1), 651, 652, 654
 – tyrosine hydroxylase (TH), 652
Extravasation of endogenous serum proteins, 370

Fatty acids, 14–21, 25, 26, 28–32, 36
Fenton reaction, 73, 209
Ferritin, 206
Fibrils, 600, 602, 608
Fibroblast growth factor, 309, 310
FK506 treatment, 354, 355
Fluctuations, of ionic concentrations, 310
Forced swimming, 368, 386, 388–390, 394–397, 402, 440–442, 444
Formamidase, 94
Fos protein, 397
Free radicals, 53, 54, 58, 73, 74, 76, 81
GABA-ergic neurotransmission, 61
Gacyclidine, 639
Gamma-aminobutyric acid (GABA), in traumatic brain injury, 189, 190
Ganglioside
 – GM1, 82, 231
 – GM2, 236
 – GM3, 82
 – GD1a, 231, 236, 240
 – GD1b, 231, 240
Gastric dilatation, 642
Gateway to neurodegeneration, 366
Gavestinel, 98
Gene expression, 75, 76, 79, 82
Genes associated with Parkinson disease
 – DJ1, 601, 606
 – LRRK2, 601, 602, 606
 – Parkin, 601, 602, 606
 – PINK1, 601, 602, 606
 – UCH-L1, 601, 606
Genomics functional genomics, 655
Geranylgeranyl transferase, 146
GLAST, 57, 58
Glaucoma
 – anterior chamber, 136
 – intraocular pressure, 136, 140, 144
 – retinal ganglion cells, 133, 137
 – Schlemm's canal, 136, 140
 – vitreous chamber, 135
Glial, 168, 170
Gliosis
 – astrocytes, 473
 – inflammation, 477
 – microglia, 473
γ-Globulins, 373, 374
GLT-1, 57, 58
Glucocorticoid, 215
Glucosamine-kynurenic acid, 99
Glucose transporter-1
 – K^+, 311
 – Na^+, 311–312
 – P-glycoprotein, 313, 314
 – water transport, 313

Glucose-6-phosphate dehydrogenase, 76
Glutamate, 72–78, 81–83
 – excitotoxicity, 188
 – glutamate treatment targets in TBI
  – glutamate release inhibitors, 188
  – NMDA receptor antagonists, 188, 189
 – in traumatic brain injury, 188
Glutamate-glutamine cycle, 47, 48
Glutamate-nitric oxide-cGMP pathway
 – in cognitive function, 60, 65
 – in learning ability, 64, 65
Glutamatergic, 93, 97
Glutamatergic neurotransmission
 – in acute hyperammonemia, 50
 – in chronic hyperammonemia, 59–61
Glutamate transporters, 47, 51, 52, 57, 58
Glutamine synthetase, 45, 47, 53, 56
Glutathione, 54, 72, 74, 143, 208
Glutathione peroxidase (GSHPx), 208
Glutathione reductase (GSHRx), 208
Glutathione-S-transferases, 143
Glutathione synthetase, 208
Glycoproteins
 – J1, 232, 234
 – MAG, 232
 – NCAM, 232
 – PMP–22, 231, 232
 – Po, 232
Glycosphingolipids
 – gangliosides, 232
 – SGPG, 233, 234
 – sulfatide, 236, 237
GM1 Ganglioside, 639
Gp120, 475
Growth factor receptors, in peripheral nerve sheath tumor formation, 117
Growth factors, 6
 – hypersensitivity to, 115
GTPase activating protein, neurofibromin, 111
Head injury, 72–74, 77, 78, 80–83
Hematoma
 – extradural, 640
 – extramedullary, 640
 – intramedullary, 640
Hemoglobin, 139, 206
Hemorrhage, 206
Hepatic encephalopathy
 – animal models, 49, 63, 64
 – cognitive function, 59
 – learning ability, 49, 64

- motor function, 64
- neurological alterations, 45, 48, 49, 59, 61
- role of hyperammonemia in, 48, 61, 63, 64

High blood titers of antibodies, 375
High-density lipoprotein (HDL), 145
Highly active antiretroviral therapy
- antiretroviral classification, 477
- BBB penetration, 477

High voltage slow activity (HVSA), 392, 393
Hippocampus, long-term potentiation, 60
Histone deacetylase 6
HIV
- AIDS, 472
- cell tropism, 476
- HAART, 472
- HAD, 472
- viral proteins, 475

HIV-associated dementia
- clinical features, 472, 473
- pathogenesis, 474–477
- pathology, 473, 474
- treatment, 477, 478

HIVE, in SCID mice, 480–483
- animal model, 479, 480
- clinical features, 472, 473
- pathology, 473–474

4-HNE. See 4-hydroxynonenal
HNK-1, 232, 234
Human brain, Alzheimer's disease patient, 654
Human cases of spinal cord injury, 428
Huntingtin, 462, 464
Huntington's disease, 94, 96, 97, 464, 514, 520, 553
Hydrolases, 519, 520
Hydroperoxyl radical, 205
Hydrostatic and osmotic pressure differences, 383
3-Hydroxyanthranilate-oxygenase, 92, 96
3-Hydroxyanthranilic acid-oxygenase inhibitor, 98–100
8-Hydroxyguanine, 208
4-Hydroxyhexenal, 77, 78
3-Hydroxykynurenine, 92, 94
4-Hydroxynonenal (4-HNE), 72, 74, 76, 78, 207, 330
Hydroxyl radical (·OH), 205, 206
Hyperammonemia, 45–65. See also Ammonia
- acute (see Acute ammonia toxicity)
- acute vs. chronic, 49

- animal models, 49, 50, 59, 60, 63, 64
- chronic (see Chronic hyperammonemia)
- cognitive function, 59, 60, 64
- learning ability, 49, 64, 65
- in liver disease, 46, 48–50, 60–64
- motor function, 63, 64
- neurological alterations, 48, 49, 59, 61, 63–64
- nitric oxide in, 55, 59
- in pathological situations, 46, 48
- perinatal, 63
- role in hepatic encephalopathy, 48, 49, 59, 61, 63, 64
- in urea cycle deficits, 48, 62, 63

Hypercapnia, 385
Hypercholesterolemia, 145
Hyperglycemia, 140
Hyperoxia, 150, 367, 385
Hypertensive encephalopathy, 368, 385
Hyperthermia, 355, 398–401, 405, 412, 413, 445
- (> 40° C) during summer seasons, 398

Hypoxia, 367, 377, 385
Hypoxia-inducible factor 1 (HIF-1), 171, 172
IFN-α
- cytokine, 476
- HAD, 472
- HIVE SCID mice, 476
- neurophysiologic dysfunction, 472
- neurotoxin, 473, 474
- ognitive dysfunction, 476

IgA, 373, 374
IgA gammopathy, 237, 238
IgG gammopathy, 237, 238
IgM gammopathy, 236
IL–1. See Interleukin-1
Ileus, 642
Imaging (CT/MRI), 356
Immune cells migration, in CNS, 528
Immunoglobulins, 230, 241
Immunomodulatory proteins, 146
Immunotherapy
- in Alzheimer disease, 603
- in Parkinson disease, 602

In vitro models, 347
Increased penetration of proteins, 375
Increased vesicular transport
- neuronal transport, 366
- tight junctions, 367, 368, 376, 422

Indoleamine 2,3-dioxygenase, 92, 94
Induced by gangliosides, 240, 241
Infarction, 367, 382, 385

Inflammation, 14, 20, 78, 82, 146, 149, 151, 152
- calpain inhibitors in, 534, 535
- in PLA2 mediated CNS injuries, 333

Inflammatory cytokines, 146
Inhibitor of HMG-CoA reductase, 31
Injury
- head, 72–74, 77, 78, 80–83
- ischemic, 82
- mechanical, 73
- primary, 73
- secondary, 73, 74
- spinal cord, 72, 73, 76–79, 81, 82

Innate immunity, 146
Integrins, 6–8
Interferon-γ, 325
Interferons, 74
Interleukin–1 (IL-1), 146, 325
Interleukin–6 (IL-6), 146
Interleukins, 74, 80–82
Interplay, 81–82
Intrabody
- ER-retention of, 461
- structure of, 461
- as therapeutics for treatment of neurodegenerative disease, 462
- in vivo and in vitro selection, 462, 463

Intracranial pressure, 213, 381, 382, 384, 385
Intravenous immunoglobulin (IvIg), 238, 241
Ion channels, 72, 76
Ionic transporters
- amino acid regulation, 312, 313
- analyzing, 311
- Ca^{2+}, 312
- Cl^-, 312

iPLA2. See Ca^{2+}-independent PLA2
Iron, 205, 206
Ischemia, 93, 96–98, 100, 367, 376, 377, 385, 414
Ischemic brain injuries, 332
Isoprostanes, 72, 74, 76, 222
KA. See Kainic acid
Kainate receptors, 57
Kainic acid, 330
Kinases, protein, 74, 76
Kynurenic acid, 92–94, 96–100
Kynureninase, 94, 95
Kynureninase inhibitor, 99, 100
Kynurenine 3-hydroxylase, 94
Kynurenine 3-hydroxylase inhibitor, 99, 100
Kynurenine-aminotransferase, 92–94
Kynurenine aminotransferase I, 96

Kynurenine-aminotransferase knockout, 92, 93
Kynurenine aminotransferase II, 92, 93, 96, 100
Kynurenine aminotransferase inhibitor, 100
Kynurenine pathway, 92–96
Lactic acid, 205, 219
Lanthanum, 366–368, 375, 376, 379, 380, 396, 402, 404, 409, 410, 416, 419, 420, 425, 426, 432, 433, 436, 437, 445
Lanthanum across BSCB, leakage of, 425
Lanthanum extravasation, 380, 416, 419, 420, 433
Lanthanum ion, 366, 432
Lazaroids, 216
LC3-interacting region, 520, 521
Leakage of albumin across the BSCB, 428, 429
Leaky in noxious insults to the brain or spinal cord, 367
Learning ability. See Cognitive function
Lentivirus
 – FIV, 479
 – HIV, 472
 – MuLV, 479
 – SIV, 479, 480
Leukocyte function associated antigen, 528
Leukotrienes, 76, 80
Levels, of CSF proteins, 374
Lewy bodies, 514, 600–602, 607
Lewy body dementia, 600
LFA. See Leukocyte function associated antigen
Linoleic acid (18:2), 207
Lipid hydroperoxides, 76, 206, 207
Lipid mediator production, PLA2 effect on metabolic pathway of, 323
Lipid peroxidation, 206, 207
 – products of, 554
Lipid peroxyl radical (LOO•), 206, 207
Lipofuscin, 150–152
Lipopolysaccharide, 325
Lipoxygenase, 73, 76, 77
Long-term depression, 75
Long-term mental disturbances, 385
Long-term potentiation, 75
 – cGMP-degrading phosphodiesterase, 60
 – cGMP-dependent protein kinase, 60
 – in chronic hyperammonemia, 60
Lovastatin, 31–35

Low voltage fast activity (LVFA), 392, 393
LPS. See Lipopolysaccharide
Luminal application, 378
Lutenizing hormone-releasing factor (LHRF), 372
Lysine, 517
Lyso-PC. See Lysophosphatidylcholine
Lysophosphatidylcholine, 329
Lysosomal clearance, 603
Lysosome, 514–517, 519, 520
Macrophage-colony stimulating factor, 325
Macrophages, 75, 78–81
MAG. See Myelin-associated glycoprotein
Major histocompatibility complex II, 527
Malignant gammopathy
 – cryoglobulinemia, 239
 – multiple myeloma, 239
 – POEMS syndrome, 239
 – Waldenstrom's macroglobulinemia, 239
Malondialdehyde (MDA), 207
Mammalian PLA2 enzymes, 324
Mammalian target of rapamycin, 518
MAP-2
 – degradation, 55
 – phosphorylation, 61, 62
MAPK. See Mitogen-activated protein kinase
Mast cell, in peripheral nerve sheath tumor formation, 117, 118
Matrix metalloproteinases (MMPs or matrixins)
 – classes
  – collagenases, 260
  – gelatinases, 260
  – membrane-type, 260
  – other, 260
  – stromelysins, 260
 – family members, 260–261
  – subgroups (metallocarboxypeptidases, metalloendopeptidases), 260
 – functions
  – programmed cell death, 261
  – remodeling, 261
m-calpain, characteristics of, 527
MCP-1. See Monocyte chemotactic protein-1
M-CSF. See Macrophage-colony stimulating factor
MDL 101, 202, 220
MDMA
 – alters BBB permeability, 412

 – induces brain edema formation, 413
 – induces cell injury, 413
Melanin, 151
Melatonin, 220
Membrane
 – breakdown, in PLA2 mediated CNS injuries, 333
 – composition, 72, 73
 – permeability, 72
 – proteins, 72
 – pumps
  – Ca-ATPase, 349
  – Na, K-ATPase, 349
 – topology, 19
Metabolic disturbances, 385
Metabolic edema, 383, 384
Metabolism
 – of long chain fatty acids, 15, 18
 – of very long chain, 14, 15
Metabotropic glutamate receptors, 100
 – and motor function, 64
Metalloproteinases, 2, 7
Meta-nitrobenzoylalanine, 99, 100
Methamphetamine, 405, 410, 412, 413, 437
Methyl arachidonoyl fluorophosphates, 82
Methylprednisolone (MP), 216, 638, 639
MGUS, 230, 231
MHC II. See Major histocompatibility complex II
Microaneurysms, 166, 169
Microarray, 650–652, 654
Microglia, 73, 78, 79, 81
Microperoxidase, 367, 374, 375
Microtubules, 55, 351, 352
Microvascular pathophysiology, 212
Mild cognitive impairment (MCI), 145
Minocycline, 79, 639
Misleading, 368–370
Mitochondria
 – cyclosporin-A treatment targeting, 354
 – damage to, 354
Mitochondrial dysfunction, 211, 212
Mitochondrial function, 54
Mitochondrial NOS isoform (mtNOS), 212
Mitogen-activated protein kinase (MAPK), 76, 79, 325
MK–801, 76
MnSOD, 212
Modifier gene in, 14, 16, 36
MOG. See Myelin oligodendrocyte glycoprotein

Molecular mechanism, of acute ammonia toxicity, 52, 54, 59
Molecular species, 72, 73
Monoacylglycerol lipase, 76
Monoclonal gammopathy
- IgA gammopathy, 237–238
- IgG gammopathy, 238
- IgM gammopathy, 237
- malignant gammopathy, 238–239
- nonmaligIant gammopathy, 237–238
Monocular eyelid suturing, 385
Monocyte chemotactic protein-1, 528
Mononuclear phagocytes
- cytokines, 476
- HIV, 472
- macrophages, 476
- microglia, 473
- neurotoxins, 473, 474
Morphine, 368, 391, 405–410, 412–414, 437, 441–443
- dependence and withdrawal, 405, 407, 408
Morphological changes, 397, 401, 409, 419, 420, 431, 438
Motor function, metabotropic glutamate receptors and, 61
Mouse brain
- cortex, 652
- striatum, 651–653
Mouse models, of Parkinson disease
- MPTP, 606
- 6-OHDA, 601
- transgenic α-synuclein-expressing, 606
Movement of head and body positions, 394
MRI scan, 638, 640
MS. See Multiple sclerosis
Multiple myeloma, 239
Multiple sclerosis (MS), 21, 92, 97, 368, 375, 526, 544, 553–555
- calpain in, 531
- calpain inhibitors as therapeutic agents, 533, 534
- demyelination of axons, 529
- immune cells migration in CNS, 528
- myelin-specific T cells activation, 527, 528
- necrosis and apoptosis, 529–531
- pathophysiology of EAE, 528, 529
Mycobacterium tuberculosis, 528
Myelin, 72, 73, 78, 79, 347
Myelin-associated glycoprotein (MAG), 232, 532

Myelin oligodendrocyte glycoprotein, 528
Myelin-specific T cells activation, 527, 528
Myoglobin, 139
Na^+/Ca^{++} exchanger, 210
Na^+/K^+-ATPase, 76, 210
- phosphorylation in acute ammonia toxicity, 53
Na^+-fluorescein, 373, 384
N-Acetylneuraminic acid, 82
Nanoparticles
- alter BBB permeability, 415–420
- and astrocytic reaction, 420
- induce edema formation, 420
- induce morphological changes in the brain, 420, 421
- from metals
  - Ag, 415, 416
  - Al, 415
  - Cu, 415
- and myelin changes, 420
- and neurodegeneration, 421
- and neurotoxicity, 415
- and stress protein reaction, 420
- titanium dioxide (TiO2), 415
NASCIS I, 216
NASCIS II, 216
NASCIS III, 217, 218
Necroptosis, 267
Necrosis, 74, 76, 80, 81
- oncosis, 250
Neprilysin, 153
Nerve cells, damaged, 394, 395, 397, 401, 409, 410, 425, 426, 429
Neural antigens
- glycoproteins, 232
- glycosphingolipids, 232
Neurodegeneration, 72, 78, 80–83, 366–368, 385, 394, 405, 421, 429, 431, 436, 446, 514, 515, 520, 521
- calpain inhibitors in, 534, 535
Neurodegenerative disease(s)
- Alzheimer disease, 603, 650, 654
- dementia with Lewy bodies, 600
- Lewy body dementia, 600
- Parkinson disease, 600, 650–653, 655
- Parkinson disease dementia, 600
Neurofibromatosis type 1
- complications associated with, 110
- diagnostic criteria for, 108, 109
- malignancies associated with, 110
Neurofibromin
- deficiency in, 115
- GAP activity, 116

- isoforms, 111–112
- other possible functions of, 116, 117
- tissue-specific expression of isoforms, 112
Neurofibromin-deficient cells
- hypersensitivity to growth factors in, 115
- increased ras activation in, 114, 115
Neurofilament protein, 529
Neurofilaments
- accumulation of, 350
- compaction of, 351
- misalignment of, 350
- proteolysis of, 351, 353, 354
Neurogenic shock, 637, 642
Neuroglobin (Ngb), 139
Neuroinflammatory diseases, 94
Neurological alterations
- cognitive function, 64
- in hepatic encephalopathy, 48, 61, 63, 64
- in hyperammonemia, 48, 63–65
- learning ability, 64, 65
- motor function, 63, 64
- in perinatal hyperammonemia, 63
Neurological diseases, 366, 374, 375, 381, 385
Neuronal depolarization, 51
Neuronal function, desynchronization of, 371
Neurons, 73–76, 80, 82
- apoptosis, 482
Neuropathy associated
- with anti-ganglioside IgM paraproteins, 236
- with anti-MAG/SGPG IgM paraproteins, 234–236
- with anti-sulfatide IgM paraproteins, 236, 237
Neuroprostanes, 222
Neuroprotection, 77, 366, 368, 370, 385, 436, 438, 440, 442, 443, 445, 446, 635, 638
- glial cells, 368–370
Neuroprotective agents, 82
Neuroregeneration, 366, 368, 385, 436, 438, 446, 635, 643–645
Neurotoxins
- cytokines, 476
- viral proteins, 475
Neutrophils, 78
New model of spinal cord injury, 423

NF1 gene
- discovery of, 110
- genotype-phenotype correlations, 114
- loss of heterozygosity, 112, 113
- mutations, 113, 114
- structure of, 110, 111

NFkB. See Nuclear factor kappa B
NFP. See Neurofilament protein
α–7-Nicotonic receptor, 93
Nimodipine, 639
Nitric oxide (NO), clinical importance of, 296
Nitric oxide (·NO) radical, 74, 76, 77, 79, 212
- in acute ammonia toxicity, 56
- biochemical reactions, 291
- in chronic hyperammonemia, 59, 60, 64, 65
- modulation of glutamine synthetase, 56
- peroxynitrite formation, 284
- protein nitrosylation, 284
- protein tyrosination, 284

Nitric oxide synthase (NOS), 50, 51, 53, 55, 56, 74, 77, 79, 209
- eNOS, 283
- iNOS, 283
- nNOS, 283

Nitrogen dioxide (·NO_2), 209
Nitrosoperoxocarbonate (ONOOCO$_2^-$), 209
3-Nitrotyrosine (3-NT), 210
Nitroxides, 220
NMDA receptors
- and ammonia-induced death, 51, 59
- antagonists prevent ammonia-induced death, 51
- and ATP depletion in ammonia toxicity, 52, 53
- binding of ligands, 59
- in chronic hyperammonemia, 59, 60, 64
- and glutamine synthetase, 53, 56
- and long-term potentiation, 60
- and MAP-2 degradation, 55
- and Na^+/K^+-ATPase, 52–54
- and nitric oxide, 50, 55, 56
- role in acute ammonia toxicity, 52–54, 56, 59
- and tyrosine nitration, 55, 56, 58

NMDA. See N-methy-D-aspartate
N-methy-D-aspartate, 330
N-methyl-D-aspartate, 92–94, 96–100
NO and demyelination
- activation of caspase, 292
- cysteine modifications, 292, 293
- dysregulation of antioxidant enzymes, 292
- glutamate excitotoxicity, 293
- lipid peroxidation, 292
- oligodendrocyte damage, 291, 292

Nonmalignant gammopathy, 237, 238
- MGUS, 238

Nonsteroidal anti-inflammatory drugs (NSAIDs), 146

NO targets
- guanylate cyclase activation, 283
- MAP kinase activation, 283, 284
- oxidative stress, 284, 285

Nuclear factor kappa B (NFκB), 80, 527
Nucleus, 73, 78
NXY–059, 220
1-O-Octadecyl-2-methyl-glycero-3-phosphocholine, 79
Oculocutaneous albinism, 151
OGD. See Oxygen and glucose deprivation
Olfactory ensheathing cells, 644, 645
Oligodendrocytes, 73, 74
Oligomers
- composed of α-synuclein, 602
- pore formation, 602

Omega 3 fatty acids, 143
ON. See Optic neuritis
Open brain injury, 431–435
Optic neuritis, 526
- calpain in, 531–533
- calpain inhibitors as therapeutic agents, 533–534

Ortho-methoxybenzoylalanine, 92, 100
Osmotic edema, 383
Osmotic gradient, altered, 385
β-oxidation, of long chain fatty acids in, 15, 18
Oxidative injury, in CNS neurological disorders, 334
Oxidative stress, 72–74, 76, 77, 79, 82, 83, 139, 143, 150, 521
- in demyelinating disorders, 548, 554
- in neurodegenerative disorders, 544, 552, 554

Oxygen and glucose deprivation, 332
PAF. See Platelet-activating factor
PAF-acetylhydrolases, 322, 326
- in mammalian CNS, 328

PAF-AH. See PAF-acetylhydrolases
Paracrine mechanisms, 74
Paraprotein
- IgA, 238
- IgG, 237, 238
- IgM, 233–237

Paraproteinemia
- monoclonal protein, 231
- M-protein, 230
- paraprotein, 230, 231

Parkinson disease dementia, 600
Parkinson's disease (PD), 92, 97, 464, 465, 514, 520, 544, 552, 555, 600, 601, 607, 650–653, 655
Parkinsonism, with dementia, 600
Pathogenic role, of NO in EAE
- adhesion molecules, 287
- BBB breakdown, 286, 287
- gliosis, 287–291
- iNOS
  - in astrocytes, 290
  - in microglia, 288, 289
- NO
  - in astrogliosis, 291
  - in microgliosis, 289, 290
- Th switching, 285

Pathways of cell death
- endoplasmic (endoplasmic reticulum), 255
- extrinsic (cellular membrane), 254
- intrinsic (mitochondria), 255

PBMCs. See Peripheral blood mononuclear cells
PDGF. See Platelet-derived growth factor
Pegaptanib sodium (Macugen), 174
Pentraxins, 146
Pericytes, 169, 170
Perinatal hyperammonemia, 63
Peripheral blood mononuclear cells, 527, 528
Peripheral nerve grafts, 644
Peripheral nerve lesion
- BSCB disturbances, 431
- spinal, 428–431

Peripheral nerve sheath tumors
- aberrant angiogenesis, 119, 120
- aberrant receptor expression, 117–119
- paradigm for tumor formation, 117, 118
- role of mast cells, 120
- secondary genetic mutations, 120, 121
- therapeutic approaches, 121, 122
- types of, 109

Peripheral neuropathy, 239
Permeability
- distribution of material, 307
- measurements of, 307
- trafficking across BBB, 308

Peroxisomal disorder, 14–17

Peroxisomal membrane protein of 70-kDa (PMP-70), 15, 19, 30
Peroxisomal membrane protein of 70-kDa related (P70R), 19
Peroxisomes, 14, 15, 17–21, 25, 26, 28, 31, 32
Peroxynitrite (PN), 74, 209, 213
Peroxynitrous acid (ONOOH), 209
PGE2. See Prostaglandin E2
Pharmacological treatments, 19, 31, 33, 36
pH, effects of ammonia on, 46–48
α-phenyl-N-tert-butylnitrone (PBN), 220
Phosphatidic acids, 72
Phosphatidylcholines, 72, 80
Phosphatidylethanolamine, 72, 518
Phosphatidylinositols, 72
Phosphatidylserines, 72
Phosphodiesterase, 51, 60, 64, 65
Phosphodiesterase inhibitors, restoration of learning ability, 64
Phospholipases, 207
 – C, 75, 76, 80, 81
 – calcium-independent PLA2, 79
 – cytosolic PLA2, 78, 80
 – inhibitors, 72, 75, 82
 – plasmalogen-selective PLA2, 80
 – secretory PLA2, 75
 – sphingomyelinase, 78, 79
Phospholipases A2
 – classification, structure and properties of, 323–326
 – in mammalian CNS, 326–328
 – mediated CNS injuries
  – future perspectives of, 335
  – mechanisms of, 332–335
 – in neurotrauma
  – ischemia, 332
  – traumatic brain injury, 330–332
  – traumatic spinal cord injury, 328–330
Phospholipid glutathione peroxidase (PHGPx), 208
Phospholipids, 72, 73, 75, 76, 79–81, 83
Phosphorylation, 79, 82
 – of neuronal proteins, of MAP-2, 62
Photic stimulation, 385
Photosensitizers, 143
Photostasis, 151
Pigment epithelial derived factor (PEDF), 170, 172, 173
Pituitary-hormone interactions, 373
PKC. See Protein kinase C

PLA2s. See Phospholipases A2
Plasma cells, 230
Plasma exchange, 238, 239, 241
Plasmalogens, 72, 80
Plasmenylethanolamines, 72
Platelet-activating factor (PAF), 72, 74, 83, 322
Platelet-derived growth factor, 325
PLP. See Proteolipid protein
PMN. See Polymorphonuclear neutrophils
Pneumonia, 642
PNU 156561, 99
POEMS syndrome, 239
Polyethylene glycol, 645
Polyethylene glycol-conjugated superoxide dismutase (PEG-SOD), 218
Polymorphonuclear neutrophils, 328
Polyunsaturated fatty acid, 143, 206
Potential mechanisms of tissue repair and functional recovery induced by stem cell transplantation therapy
 – cell replacement and integration with host circuitry, 494
 – enhanced neurogenesis and angiogenesis, 495
 – increased trophic support, 494, 495
 – induction of host tissue plasticity, 495, 496
Prealbumin, 374
Prion
 – disease, 462
 – scrapie form of, 465
Programmed cell death (PCD), 252
 – apoptosis, 268
 – autophagy, 268
Prostaglandin E2, 329
Prostaglandin generating cyclooxygenases—COX-1 and COX-2, 146
Prostaglandins, 76, 78
Protease inhibitors, 146
Proteasome, 266, 267, 514–522
20S Proteasome, 515–519, 521
Protection, in multiple sclerosis patients, 21
Protection of "neural" and "non-neural" components, 370
Protective role, of NO, in EAE
 – immunosuppression, 283
 – inhibition of leukocyte infiltration, 287
 – inhibition of NF-kB, 295
Protein aggregates, prevention of using antibody fragments, 460

Protein aggregation, 514, 520
Protein binding domain 1, 520
Protein carbonyl, 208
Protein carbonylation
 – in experimental autoimmune encephalomyelitis, 544
 – functional consequences of, 544, 548, 550, 551
 – mechanism of, 545–549, 555
 – in multiple sclerosis, 544, 553–555
 – in neurodegenerative diseases, 546
 – profiles in different CNS disorders, 544, 545, 551–555
 – by reactive carbonyl species, 544, 545
Protein carbonyls
 – chemistry of, 545
 – in cytoskeleton, 554
 – in endoplasmic reticulum, 551
 – experimental models of, 548–550
 – measurement of, 546–548
 – in mitochondria, 548, 549, 555
 – pharmacological prevention of, 555, 556
 – physiological removal of, 549, 550
Protein kinase C, 325
Protein kinase, dependent on cGMP, 60
Protein oxidation, 208
Proteins, size-dependent entry of, 374, 375
Proteolipid protein, 528
Proteolysis
 – calpains, 353, 354
 – caspase-3, 354
 – inhibition of, 354, 355
Protofibrils, 600–603
Psychostimulants, 366, 405, 412–414, 436, 437, 441, 442, 444, 445
Pulmonary edema, 642
Pulmonary embolism, 642, 643
Pyknosis, 151
Pyrrolpyrimidine, 214
Quinacrine, 75, 79, 81
Quinolinic acid, 92–94, 96–100
Quinolinic acid phosphoribosyltransferase, 96
Rab proteins, 146
Radiograph, 636, 640–642
Radioiodine, leakage of, 425
Radiotracer extravasation, 387, 390, 415, 416, 425
Ranibizumab (Lucentis), 174
Ras, increased activation in NF1, 114, 115
Reactive nitrogen species (RNS), 204, 210

Reactive oxygen species (ROS), 72, 80, 82, 83, 138, 139, 143, 205–208, 544, 548, 551, 552, 554, 555
Reacylation, 76
Receptors
 – AMPA, 74, 81
 – cytokine, 72, 82
 – glutamate, 74, 81
 – kainate, 75, 80
 – nerve growth factor, 82
 – NMDA, 76
Reflex
 – ankle, 638
 – bulbocavernosus, 637, 638
 – cremasteric, 638
 – knee, 638
Regulatory particle, 515, 516
Relapsing remitting MS, 533
Resolvins, 77
Respiratory distress syndrome, 642
Restoration of learning ability, 64, 65
Restriction
 – filtration rate, 304
 – reflection coefficients, 304
Retinal ganglion cells, 531
Retinal nerve fiber layer, 532
Retinal pigment epithelium (RPE), 168, 170
Retinitis pigmentosa (RP), 139, 140, 147
RGCs. See Retinal ganglion cells
RHC80267, 76
Rho antagonist, 645
Rhodopsin, 138, 139, 146
Rituximab, 241
RNFL. See Retinal nerve fiber layer
Ro-61–8048, 100
Rotterdam study, 142, 143
RRMS. See Relapsing remitting MS
(R,S)-3,4-dichlorobenzoylalanine, 99
Ruboxistaurin mesylate, (LY333531), 173
Schizophrenia, 97, 100
Schwann cell, 644, 645
 – in peripheral nerve sheath tumor formation, 117, 118
SCI. See Spinal cord injury
SCIWORA, 641
Scotopic, 140
Secondary progressive forms, 533
Secretory PLA2, 322, 323, 325
 – in mammalian CNS, 326–327
 – in spinal cord ischemic injury, 332
 – in traumatic spinal cord injury, 330
Sequestosome, 520, 521
Serine proteases
 – family members, 263–265
 – functions, 265–266

 – Htra2/Omi, 263
 – neuroprotective, 265
 – programmed cell death, 263
 – thrombin, 263, 264
 – tissue-type plasminogen activator, 265
 – targets, inhibitors (serpins)
 – endogenous (PN-1, PI-12, PAI-1, AT), 265, 266
 – synthetic and natural (ucf–101, hirudin), 266
Serotonin, 367, 377, 378, 380–382, 384, 413, 429, 433, 439, 440, 443–445
 – 5-HT$_{1A}$ receptors, 186–188
 – serotonergic therapies for TBI
 – 8-OH-DPAT, 186, 187
 – repinotan, 186, 187
 – in traumatic brain injury, 186
Serotoninergic neurotransmission, 61, 62
SGLPG, 232, 234
SGPG, 233, 234
Signaling molecules, 75
Signaling pathway molecules, 4, 7
Signal transducer and activator of signaling, 528
Signal transduction cascade, 138, 152
Sindbis strain (SVN), 386
Single-chain Fv (scFv)
 – as potential therapeutics for the treatment of neurodegenerative disease, 462, 463
 – in vitro selection, 462
 – structure of, 461
Single-domain antibody fragment
 – as potential therapeutics for the treatment of neurodegenerative disease, 462
 – structure of, 461
Singular value decomposition
 – analysis of large data sets, 651, 652
 – principal components, 651–653
Skeletal traction, 638
Sleep deprivation stress, 397, 398
Sodium-calcium exchanger, 349
Sorsby's fundus dystrophy, 146, 150, 151
Sorsby's macular dystrophy, 148
Sphingomyelins, 72, 73, 78, 79
Sphingosine-1-phosphate, 79
Spinal cord
 – edema formation, 381, 422, 423, 425
 – membranes, 73

 – microvessels differs from brain and contain collagen, 421, 422
 – pathology, 394, 438, 440, 443
 – trauma, 72–74, 77, 78, 81–83
Spinal cord injury (SCI), 215–216, 322, 421–423, 425–429, 436, 438, 440–443, 446
 – angiogenesis and growth factor activation
  – upregulation of anti-apoptotic Bcl2, 589
  – upregulation of IGF, BDGF, NT3, 587
  – upregulation of VEGF, 587
 – behaviors, health
  – nutrition, 623
  – physical activity, 573, 623
 – behaviors, risk
  – alcohol use, 622
  – smoking, 622
  – cardiovascular system, 642
 – clinical features, 635–638
 – epidemiology, 616–618, 635, (In ref. only Chapter–24)
 – estrogen modulation of calcium
  – L-type calcium channels, 587
  – reduced calcium in SCI, 583, 587
  – voltage-gated calcium channel, 587
 – estrogen restores cell function
  – improve motor function in SCI, 585
  – oxidative excitotoxic damage, 585, 587
  – potential therapeutic agent, 591
  – preserved axon and myelin in SCI, 582, 589
  – prevention of cytochrome c release, 589
  – protection of cells in SCI, 589
  – protection of neurons, 590
  – reduces calpain/caspase activation in SCI, 589
 – gastrointestinal system, 642, 643
 – integumentary system, 643
 – management, 635, 638–643
 – mortality, secondary conditions
  – depression, 568, 624, 625
  – pressure ulcer, 574, 623, 624
  – rehospitalization, 625
 – nutritional support, 643
 – outcomes, general risk model
  – biographic factors, 618

- environmental factors, 621
- psychological factors (PTSD/ Stress-Coping), 620, 621, 626
- pharmacotherapy, 638, 639
- and PLA$_2$, 329, 330
- primary damage
  - disruption of blood vessels, 582
  - membrane disruption, 584
  - necrotic cell death, 583
- respiratory system, 642
- role of surgery, 641, 642
- secondary damage
  - apoptosis, 582
  - calcium influx, 583
  - calpain increase in SCI, 583, 584
  - caspase(s) increase in SCI, 589
  - glutamate excitoxicity, 583
  - protease activation, 589
- therapeutic agents for intervention
  - anti-inflammatory via ER-α action, 586
  - anti-oxidant, 585
  - ER-β involved in neurodevelopment, 586
  - estrogen as neuroprotectant (17β-estradiol), 587
  - inhibit microglial activation, 586
  - melatonin, 585
  - methylprednisolone, 585
  - neuroprotection in TBI, SCI, ischemia, stroke, 585, 586
  - NMDA and AMPA receptor antagonists, 585
  - prevent NF-kB translocation, 586, 590
- urinary system, 643
- vascular reaction in, 422

Spinal shock, 637, 638, 643
sPLA2. *See* Secretory PLA2
SPMS. *See* Secondary progressive forms
Sponginess, of the neuropil, 401
Spontaneous electroencephalogram alters in heat stress, axonal injuries, 401
Stargardt's disease, 147–149
STAT. *See* Signal transducer and activator of signaling
Stem cell
- bone marrow, 644
- human embryonic, 645

Stem cell, characterization
- adult stem cells
- mesenchymal stem cells, 493
- neural stem cells, 493
- embryonic stem cells
  - fetal stem cells, 492

Stem cell, transplantation therapy for neurological diseases in animal model
- clinical application, 505, 506
- human embryonic stem cell transplantation in neurodegenerative diseases, 504, 505
- problems and perspectives, 506
- stem cell therapy, in Parkinson's disease
  - ES cell transplantation in Parkinson's disease, 500
  - MSC transplantation in Parkinson's disease, 500
- stem cell therapy, in peripheral nerve disease
  - ES cell transplantation in peripheral nerve disease, 502
  - MSC therapy in peripheral nerve disease, 502–504
- stem cell therapy, in spinal cord injury
  - ES cell transplantation in spinal cord injury, 501
  - MSC therapy in spinal cord injury, 501, 502
- stem cell therapy, in stroke
  - ES cell transplantation in stroke, 496, 497
  - MSC transplantation in stroke, 497, 498
  - NSC therapy in stroke, 498, 499

Stilbazulenyl nitrone (STAZN), 220
STIR, 640, 641
Stress-induced immunosuppression, 386
Structure (zymogens)
- basic organization
  - catalytic domain, 260
  - pro-peptide, 257
  - signal peptide, 260
- heterotetramers, 254
- preferred cleavage sites, 255

Subcortical
- dementia, 473
- pathology, 473, 474

Sulfated glycopshingolipids
- SGLPG, 232, 234
- SGPG, 233, 234
- sulfatide, 236, 237

Sulfatide, 236, 237
Superoxide, 54, 55
Superoxide dismutase (SOD), 143, 205
Superoxide radical, 205
Swelling of astrocytes, 58
Swollen synapses with damage to both pre-and post-synaptic membranes, 402
Synaptosomal plasma membrane, 73
Syndrome
- anterior cord, 637
- Brown-Sequard, 636, 637
- cauda equine, 637
- conus medullaris, 637

α-Synuclein, 464
- accumulation in neurodegenerative disease, 602, 603, 608
- antibodies against, 600, 603, 604, 606
- clearance via autophagy pathways, 605
- insoluble forms, 604, 608, 609
- intrabodies against, 600, 603
- in Lewy body dementia, 600
- membrane-bound forms, 604
- mutated forms, 609
- oligomers, 600–602, 605, 606, 608
- in Parkinson disease, 600, 601, 605, 607, 608

Synucleinopathy, 462
Targets
- inhibitors
  - endogenous (calpastatins), 253
  - synthetic, 256
- matrix and non-matrix proteins, inhibitors
  - endogenous (TIMPs, RECK), 261, 262
  - synthetic and natural, 262

Tat, 475
Tay-Sachs diseases, 375
TBI. *See* Traumatic brain injury
T cell receptor, 527
TCR. *See* T cell receptor
Tempol, 220
Tetracycline, 79
TGF-β. *See* Transforming growth factor
Therapeutic agent, 82, 83
Therapy
- cytotoxic agents, 241
- intravenous immunoglobulin, 241
- plasma exchange, 241
- rituximab, 241
- steroids, 241

Thin-layer chromatogram (TLC)-overlay, 232, 233

Thromboxane A2, 329
Thromboxanes, 76
Thyrotropin releasing hormone (TRH), 372, 639
Tight and adherens junctions
 – components of, 305
 – junctional complex, 305
Tirilazad, 216, 217
TNF-α. *See* Tumor necrosis factor-α
Topical application, 380–382, 436, 438
Tracheostomy, 642
Transcription factors, 78, 79
Transferrin, 205, 206
Transforming growth factor (TGF-β), 325
Transgenic mice, models for α-synuclein aggregation, 600, 601, 603, 607, 609
Transmembrane domain of, 19
Transport, 15, 17–19, 36
Traumatic factors, 366, 368, 381, 383–386, 421, 431, 433, 436, 437, 443
Traumatic brain injury (TBI), 93, 97, 218–220, 322, 330–332
Traumatic edema, 383, 384, 433
Traumatic injuries, 366, 385, 421, 436, 443
Traumatic spinal cord injury, 328–330
Traumatic subarachnoid hemorrhage (SAH), 219

Triamcinolone acetonide, 173
Trisomy 21, 140
Tryptophan, 92, 94
Tryptophan dioxygenase, 92, 94
Tumor growth factors, 74
Tumor necrosis factor-α (TNFα), 146, 310, 325, 476
 – cytokine, 476
 – neurotoxin, 476
Tumor necrosis factors, 74
Tumor suppressor genes, 6
Tunnel vision, 141
T1 weighted images, 640
T2 gradient echo sequences, 640
T2 weighted images, 640
TXA2. *See* Thromboxane A2
Tyrosine nitration, 55–58
U101033, 214
U72099, 214, 216
U78517, 214, 219, 222
Ubiquitin associated domain, 521
Ubiquitin-like domain, 519–521
Ubiquitin/proteasome pathway (UPP), 514, 515, 517–522
Ultraviolet (UV) light, 142
Upregulation of GFAP, 409, 413, 426, 427
Vascular endothelial growth factor, 308, 309
Vasogenic brain edema, 213

Vasogenic edema, 366, 376, 383, 384, 399, 400, 413, 421, 422, 429, 445
Very long chain acyl-CoA synthetase, 14, 17
Viral proteins
 – gp120, 475
 – tat, 475
Viral vectors
 – adeno-associated viral vectors, 609
 – lentiviral vectors, 603
Vitamin C, 209
Vitamin E, 143, 209, 214
Vitelliform macular dystrophy, 148
VLDL, 143
Voxelation
 – human brain, 654, 655
 – mouse brain, 651, 653
Waldenstroms macroglobulinemia, 239
Western blotting, 232, 233
West-Nile virus (WN-25), 386
Whole body hyperthermia (WBH), 398
Withdrawal symptoms, 405, 414
Xenotransplantation, 645
X-linked adrenoleukodystrophy, 16, 19
Zellweger syndrome, 16
Zone of partial preservation, 636